Online Optimization
of Large Scale Systems

Springer
Berlin
Heidelberg
New York
Barcelona
Hong Kong
London
Milan
Paris
Tokyo

Martin Grötschel
Sven O. Krumke
Jörg Rambau
Editors

Online Optimization of Large Scale Systems

Springer

Martin Grötschel
Sven O. Krumke
Jörg Rambau
Konrad-Zuse-Zentrum für Informationstechnik Berlin (ZIB)
Takustraße 7
14195 Berlin-Dahlem
Germany
e-mail: groetschel@zib.de
krumke@zib.de
rambau@zib.de

Mathematics Subject Classification (2000): 90-02, 90C90, 90Bxx

Cataloging-in-Publication Data applied for
Die Deutsche Bibliothek – CIP-Einheitsaufnahme
Online optimization of large scale systems / Martin Grötschel ... ed.. –
Berlin; Heidelberg; New York; Barcelona; Hong Kong; London; Milan; Paris; Tokyo: Springer, 2001

http://www.springer.de

ISBN 978-3-642-07633-6 ISBN 978-3-662-04331-8 (eBook)
DOI 10.1007/978-3-662-04331-8

Originally published by Springer-Verlag Berlin Heidelberg New York in 2001.
Softcover reprint of the hardcover 1st edition 2001

Typeset by the authors using a Springer TEX macro package
Cover design: Design and Production GmbH, Heidelberg, title graphics by the authors
Printed on acid-free paper SPIN 10848963 46/3142/db – 5 4 3 2 1 0

PREFACE

In its thousands of years of history, mathematics has made an extraordinary career. It started from rules for bookkeeping and computation of areas to become the language of science. Its potential for decision support was fully recognized in the twentieth century only, vitally aided by the evolution of computing and communication technology. Mathematical optimization, in particular, has developed into a powerful machinery to help planners.

Whether costs are to be reduced, profits to be maximized, or scarce resources to be used wisely, optimization methods are available to guide decision making. Optimization is particularly strong if precise models of real phenomena and data of high quality are at hand – often yielding reliable automated control and decision procedures. But what, if the models are soft and not all data are around? Can mathematics help as well?

This book addresses such issues, e.g., problems of the following type:

- An elevator cannot know all transportation requests in advance. In which order should it serve the passengers?
- Wing profiles of aircrafts influence the fuel consumption. Is it possible to continuously adapt the shape of a wing during the flight under rapidly changing conditions?
- Robots are designed to accomplish specific tasks as efficiently as possible. But what if a robot navigates in an unknown environment?
- Energy demand changes quickly and is not easily predictable over time. Some types of power plants can only react slowly. When do you switch on which type of power plant to produce the right amount of electricity at every point in time?
- A complicated interplay between pressure, temperature, and feed determines the behavior of a chemical reaction over time. How do you adjust these parameters to minimize undesired by-products and energy consumption, and to keep the process safe?

The last two decades have witnessed a growing scientific interest in processes where – due to the nature of these processes – data are incomplete and uncertain and where decisions have to be made under tight time restrictions. This development was fueled, in particular, by crucial problems in computer science, engineering, and economics and has led to the emergence of online and real-time optimization. What do these terms mean?

In optimization the typical situation is as follows. We assume that an optimization problem is given in general terms (such as optimizing the moves of an elevator system) and that we want to solve a particular instance of this problem algorithmically. In "ordinary optimization" we require knowledge of all data of the instance before we start solving it (e.g., we need to see the list of all transportation tasks prior to scheduling the elevator moves). In "online optimization" the data, in contrast, arrives sequentially "in pieces". And each time a piece arrives, we have to make a decision that is irrevocable. (These are the usual side constraints under which elevator control systems operate.)

In *online optimization* we do allow (in theory) the use of unlimited computing power. The main issue is: incomplete data; and the scientific challenge: How well can an online algorithm perform? Can one guarantee solution quality, even without knowing all data in advance?

In *real-time optimization* there is an additional requirement, decisions have to be computed very fast, fast in relation to the time frame of the instance we consider. If the elevator control algorithm is slow people may have to wait a little longer, a nuisance, but if there is imminent danger of an explosion of a chemical reactor, our control algorithm ought to be quick.

Online and real-time optimization problems occur in all branches of optimization: linear, nonlinear, integer, stochastic. These areas have developed their own techniques but they are addressing the same issues: quality, stability, and robustness of the solutions.

To fertilize this emerging topic of optimization theory and to foster cooperation between the different branches of optimization, the *Deutsche Forschungsgemeinschaft (DFG)* has supported a Priority Programme (*Schwerpunktprogramm*) "Online Optimization of Large Systems". This program with a total budget of about DM 15,000,000 over six years has financed 25 projects in all areas of online optimization at 28 universities and research institutes all over Germany. It supported conference participation, mutual visits, and the organization of workshops for about 1,000 mathematicians that were involved in these projects for all or part of these six years.

Instead of writing a final report for the files of DFG, the participants decided to publish a book of a novel kind. It is neither a proceedings volume, nor a monograph or a textbook, it is neither a report of latest research nor a collection of survey articles. This book is all of this, to some extent, at the same time.

The aim of this book is to show the current state of online and real-time optimization to a broad audience. The book comprises 12 introductory articles that are written on the level of an advanced student. These surveys, marked with an "*" in the table of contents, cover all areas of online optimization. They introduce the basic techniques of the analysis of online problems in the individual areas, provide illustrative examples, and survey important results. The introductory papers provide, at the same time, the grounds for the research surveys contained in this book. The research-oriented articles summarize the results of the various topics of the *Schwerpunkt* and, in this way, demonstrate the progress achieved in the years 1996–2001.

The editors and authors hope that this book will find readers inside mathematics and, as well, in the application areas that are covered. The research results obtained in the *Schwerpunkt* have moved the area a big step forward. However, online and real-time optimization are still in their infancy. There are more questions than answers and a lot of research ground is ahead of us to be ploughed by concepts and techniques yet to be discovered.

We do hope that the book achieves its aim to popularize the area of online optimization, to get more mathematicians and computer scientists acquainted with this topic and to show to the application areas that there are already quite a number of topics where mathematics can make a difference.

The book is subdivided into various sections. Sections are defined by mathematical approach and not by application area. We suggest to those not too familiar with online optimization to start reading the articles with an "$*$".

We are indebted to numerous referees who provided valuable comments, evaluations, and suggestions on the material in this book. Without them this book would not have been possible.

June 2001 Martin Grötschel, Sven O. Krumke, Jörg Rambau
Editors

The research in this book was supported by a Priority Programme (*Schwerpunktprogramm SPP-469*) of the *Deutsche Forschungsgemeinschaft (DFG)*.

CONTENTS

Articles with an "∗" are introductory texts and aim at the non-specialist reader

I Optimal Control for Ordinary Differential Equations

II Optimal Control for Partial Differential Equations

III Moving Horizon Methods in Chemical Engineering

IV Delay Differential Equations in Medical Decision Support Systems

V Stochastic Optimization in Chemical Engineering

VI Stochastic Trajectory Planning in Robot Control

VII Integer Stochastic Programming

VIII Combinatorial Online Planning in Transportation

IX Real-Time Annealing in Image Segmentation

I

Optimal Control for Ordinary Differential Equations

*Sensitivity Analysis and Real-Time Optimization of Parametric Nonlinear Programming Problems

Christof Büskens[1] and Helmut Maurer[2]

[1] Lehrstuhl für Ingenieurmathematik, Universität Bayreuth, Germany
[2] Institut für Numerische Mathematik, Universität Münster, Germany

Abstract Basic results for sensitivity analysis of parametric nonlinear programming problems [11] are revisited. Emphasis is placed on those conditions that ensure the differentiability of the optimal solution vector with respect to the parameters involved in the problem. We study the explicit formulae for the sensitivity derivatives of the solution vector and the associated Lagrange multipliers. Conceptually, these formulae are tailored to solution algorithm calculations. However, we indicate numerical obstacles that prevent these expressions from being a direct byproduct of current solution algorithms. We investigate post-optimal evaluations of sensitivity differentials and discuss their numerical implementation. The main purpose of this paper is to describe an important application of sensitivity analysis: the development of real-time approximations of the perturbed solutions using Taylor expansions. Two elementary examples illustrate the basic ideas.

1 Introduction

The literature on optimization problems uses two different notions of sensitivity analysis. One conception of sensitivity analysis appears in the calculation of the objective function and the constraint partial derivatives for determining search directions and optimality conditions. The second form is related to *parameter sensitivity analysis* where one studies the impact of a change in the design parameters on the optimal solution vector and the objective function. In this paper, we review those results and conditions which ensure that the optimal solution is differentiable with respect to the design parameters. The fundamental results in this area go back to Fiacco [10, 11] and Robinson [14] who independently used the classical implicit function theorem for proving solution differentiability.

This approach also provided explicit formulae for the *parameter sensitivity derivatives* of the optimal solution vector, the associated Lagrange multiplier and the optimal value function. Fiacco [11] has spent a great effort in popularizing these ideas by pointing out that the sensitivity information is a natural and highly useful byproduct of solution algorithms. Armacost, Fiacco, Mylander [1, 2] have implemented sensitivity analysis by interfacing it to the optimization code SUMT which is based on penalty techniques. In the last two decades, considerable progress has been made towards the development of efficient optimization codes that are able to handle large-scale optimization problems. One prominent class of algorithms is that of SQP-methods (Sequential Quadratic Programming). Beltracchi and Gabriele [4, 5] have presented a more detailed study of sensitivity derivatives computations via SQP-methods. They found that the explicit formulae for the sensitivity derivatives can not be tied directly to iterative calculations in SQP-methods since the employed low rank updates of the Hessians usually do not converge to the exact Hessian. As a consequence, they propose a *post-optimal* analysis of the sensitivity derivatives.

In the present paper, we take up the idea of post-optimal sensitivity analysis and endeavour to advertise its practical importance by relating it to the current implementations of SQP-methods. We discuss the numerical issues of (a) checking second order sufficient conditions which constitute the theoretical basis of sensitivity analysis and (b) computing the Hessian with high accuracy. Beyond revitalizing the methods of sensitivity analysis, our main purpose with this article is to use them for the *real-time optimization* of perturbed optimal solutions. The principal idea behind real-time optimization is to approximate perturbed solutions by their first Taylor expansions with respect to the parameters. This approach will be further extended in [7–9] to develop *real-time control* approximations of perturbed optimal control solutions.

2 Parametric Nonlinear Optimization Problems

2.1 *Necessary Optimality Conditions for Nonlinear Optimization Problems*

We consider parametric nonlinear programming problems (NLP-problems) involving a parameter $p \in P \subset \mathbb{R}^{N_p}$. The optimization variable is denoted by $z \in \mathbb{R}^{N_z}$. The unusual notation z for the optimization variable and the dimensional notation N used here and hereafter in the constraints of the problem are adapted to the fact that the optimization problem often results from a suitable *discretisation* of a dynamic optimization problem; cf. [8]. The *parametric NLP-problem* with equality and inequality constraints is given by

$$\mathbf{NLP}(p) \quad \begin{cases} \min\limits_{z} & F(z,p), \\ \text{subject to} & G_i(z,p) = 0, \quad i = 1,\ldots,N_e, \\ & G_i(z,p) \leq 0, \quad i = N_e+1,\ldots,N_c. \end{cases} \tag{1}$$

For simplicity of exposition, the functions $F : \mathbb{R}^{N_z} \times P \to \mathbb{R}$ and $G_i : \mathbb{R}^{N_z} \times P \to \mathbb{R}$, $i = 1,\ldots,N_c$, are assumed throughout to be of class C^2 on $\mathbb{R}^{N_z} \times P$. The set

$$S(p) := \{z \in \mathbb{R}^n \mid G_i(z,p) = 0, \ i = 1,\ldots,N_e, \\ G_i(z,p) \leq 0, \ i = N_e+1,\ldots,N_c,\} \tag{2}$$

is called the *set of admissible vectors* or *points* or simply the *admissible set*. An admissible $\bar{z} \in S(p)$ is called a *local minimum* of the NLP-problem (1), if a neighbourhood $V \subset \mathbb{R}^{N_z}$ of $\bar{z}$ exists, such that

$$F(\bar{z},p) \leq F(z,p), \qquad \forall\, z \in S(p) \cap V. \tag{3}$$

A point $\bar{z} \in S(p)$ is a *strong local minimum* of the problem (1), if a neighbourhood $V \subset \mathbb{R}^{N_z}$ of $\bar{z}$ exists such that

$$F(\bar{z},p) < F(z,p), \qquad \forall\, z \in S(p) \cap V, z \neq \bar{z}. \tag{4}$$

Any (strong) local minimum $\bar{z}$ of **NLP**(p) is called an *optimal solution*. With $\bar{z}$ we associate the following *active sets* of indices:

$$J(\bar{z},p) := \{1,\dots,N_e\} \cup \{i \in \{N_e+1,\dots,N_c\} \mid G_i(\bar{z},p) = 0\}. \tag{5}$$

The Lagrangian function $L : \mathbb{R}^{N_z} \times \mathbb{R}^{N_c} \times P \longrightarrow \mathbb{R}$ for the constrained nonlinear optimization problem **NLP**(p) is defined as

$$L(z,\eta,p) := F(z,p) + \eta^T G(z,p), \tag{6}$$

where we have used the function $G(z,p) := (G_1(z,p),\dots,G_{N_c}(z,p))^T$ and the *Lagrange multiplier* $\eta = (\eta_1,\dots,\eta_{N_c})^T \in \mathbb{R}^{N_c}$. Henceforth, the symbol $(\cdot)^T$ denotes the transpose. We introduce the active constraints and the multiplier corresponding to the active constraints by

$$G^a := (G_i)_{i \in J(z,p)}, \quad \eta^a \in \mathbb{R}^{N_a}, \quad N_a := \#J(z,p). \tag{7}$$

In the sequel, partial derivatives of first or second order are denoted by subscripts referring to the specific variable. First order necessary optimality conditions for an optimal solution $\bar{z}$ of **NLP**(p) may be found in any textbook on nonlinear optimization; cf., e.g., Fletcher [12]. We shall recall these so-called KKT-conditions under the *strong regularity condition* (constraint qualification) that $\mathrm{rank}(G_z^a(\bar{z},p)) = N_a$ holds, i.e., the Jacobian of $G_z^a(\bar{z},p)$ has full rank. This strong regularity condition will be needed again for second order sufficient conditions (SSC) and the sensitivity analysis based thereupon.

Theorem 1 (Strong Necessary Optimality Conditions for NLP(p)**).** *Let $\bar{z}$ be an optimal solution of* ***NLP***(p) *for which the Jacobian $G_z^a(\bar{z},p)$ has full rank N_a. Then there exist a uniquely determined multiplier $\eta \in \mathbb{R}^{N_c}$ satisfying*

$$L_z(\bar{z},\eta,p) = F_z(\bar{z},p) + \eta^T G_z(\bar{z},p) = 0, \tag{8}$$

$$\eta_i \geq 0 \quad \forall\, i \in \{1,\dots,N_c\}, \qquad \eta_i = 0 \quad \forall\, i \notin J(\bar{z},p). \tag{9}$$

2.2 *Post Optimal Calculation of Lagrange Multipliers*

In general, numerical algorithms for computing an optimal solution $\bar{z}$ also yield a satisfactory approximation for the multipliers η. Otherwise, an accurate value of η can be calculated *post-optimally* once an optimal solution $\bar{z}$ has been determined. The procedure uses an appropriate QR-factorisation of a matrix as described in Gill, Murray and Saunders [13]. Since $\mathrm{rank}(G_z^a(\bar{z},p)) = N_a$, there exist a $N_a \times N_z$ matrix R and an orthogonal $N_z \times N_z$ matrix Q with

$$G_z^a(\bar{z},p) = RQ. \tag{10}$$

The matrices R and Q can be partitioned into

$$R = [R_1 : 0], \quad Q = \begin{bmatrix} Q_1 \\ Q_2 \end{bmatrix}, \tag{11}$$

with an upper triangular $N_a \times N_a$ matrix R_1 having positive diagonal elements and a $N_a \times N_z$ matrix Q_1, resp., a $(N_z - N_a) \times N_z$ matrix Q_2. In view of the fact that $\eta_i = 0$ for all $i \notin J(z,p)$, equation (8) gives:

$$-F_z(\bar{z},p) = (\eta^a)^T R_1 Q_1. \tag{12}$$

Since the matrix R_1 is regular, we get the explicit expression

$$\eta^a = \left(-F_z(\bar{z},p) Q_1^T R_1^{-1}\right)^T. \tag{13}$$

2.3 *Second Order Sufficient Optimality Conditions*

Second order sufficient conditions (SSC) are needed to ensure that any point $\bar{z}$ which satisfies the KKT-conditions (8) and (9) is indeed an optimal solution of problem (1). Another important aspect of SSC appears in the *sensitivity analysis* of the problem **NLP**(p) where SSC are indispensable for showing that the optimal solutions are differentiable functions of the parameter p; cf. Theorem 3 below. The following strong SSC are well known, cf. Fiacco [11].

Theorem 2 (Strong Second Order Sufficient Conditions for NLP(p)).
*For a given parameter p, let $\bar{z}$ be an admissible point for problem **NLP**(p) which satisfies the KKT-conditions (8) and (9).* Assume that
(a) the gradients in G_z^a are linearly independent, i.e., $\operatorname{rank}(G_z^a(\bar{z},p)) = N_a$,
(b) strict complementarity $\eta^a > 0$ of the Lagrange multipliers holds,
(c) the Hessian of the Lagrangian is positive definite on $\ker(G_z^a(\bar{z},p))$,

$$v^T L_{zz}(\bar{z},\eta,p) v > 0, \quad \forall\, v \neq 0 \ \text{with} \ G_z^a(\bar{z},p)v = 0. \tag{14}$$

Then there exist $\varepsilon > 0$ and a constant $c > 0$ such that

$$F(z,p) \geq F(\bar{z},p) + c\|z-\bar{z}\|^2, \quad \forall\, z \in S(p),\ \|z-\bar{z}\|^2 \leq \varepsilon. \tag{15}$$

*In particular $\bar{z}$ is a strong local minimum of **NLP**(p).*

It is not apparent how one may check the SSC in (14) since positive definiteness of the Hessian is restricted to the null space of the Jacobian of the active constraints. A numerical check of the SSC may be performed as follows. Consider the $N_z \times (N_z - N_a)$ matrix H with full column rank whose columns span the kernel $\ker(G_z^a(\bar{z},p))$. Any vector $v \in \ker(G_z^a(\bar{z},p))$ can be written as $v = Hw$ for some vector $w \in \mathbb{R}^{N_z - N_a}$. Then condition (14) may be restated as

$$w^T H^T L_{zz}(\bar{z},\eta,p) H w > 0, \quad \forall\, w \in \mathbb{R}^{N_z - N_a}, \quad w \neq 0. \tag{16}$$

The matrix $H^T L_{zz}(\bar{z},\eta,p)H$ is called the *projected Hessian*. It follows from (16) that the positive definiteness of the projected Hessian on the whole space $\mathbb{R}^{N_z - N_a}$ is equivalent to the positive definiteness of the Hessian $L_{zz}(\bar{z},\eta,p)$ on $\ker(G_z^a(\bar{z},p))$. Thus the test for SSC proceeds by showing that the projected Hessian has only *positive* eigenvalues.

One efficient method to compute the matrix H is the RQ-factorization as described in (10) and (11). Suppose, that a RQ-factorization (10) and (11) is given. Then $H := Q_2^T$ forms an orthogonal basis for $\ker(G_z^a(\bar{z},p))$ which follows from

$$G_z^a H = RQQ_2^T = R\begin{bmatrix} Q_1 \\ Q_2 \end{bmatrix} Q_2^T = [R_1 : 0]\begin{bmatrix} 0 \\ I_{N_z - N_a} \end{bmatrix} = 0. \qquad (17)$$

Here and in the sequel, $I_{N_z - N_a}$ denotes the identity matrix of dimension $N_z - N_a$. The matrices Q_1 and R_1 in (11) are unique while Q_2 is not. If $\tilde{Q}_2^T$ is another orthogonal basis for the null space of $G_z^a(\bar{z},p)$, then the corresponding projected Hessian is a matrix which is similar to $H^T L_{zz}(\bar{z},\eta,p)H$. Thus, the eigenvalues of the projected Hessian are independent of the orthogonal basis for $\ker(G_z^a(\bar{z},p))$.

3 Sensitivity Analysis for Parametric Nonlinear Optimization Problems

3.1 *First Order Sensitivity Analysis of the Optimal Solution*

For a fixed reference or *nominal* parameter $p_0 \in P$, the problem **NLP**(p_0) is called the *unperturbed* or *nominal* problem. We shall study the differential properties of optimal solutions to the perturbed problems **NLP**(p) and the related optimal values of the objective function with respect to parameters p in a neighbourhood of the nominal parameter p_0. The strong SSC presented in Theorem 2 have the important consequence that the optimal solution and the Lagrange multipliers become differentiable functions of the parameter. Moreover, explicit formulae for the parameter derivatives, the so-called *sensitivity differentials,* can be given in terms of quantities that depend alone on the nominal solution. The following principal result may be found in Fiacco [11].

Theorem 3 (Differentiability of optimal solutions). *Let the pair* (z_0,η_0) *be an admissible point and multiplier which satisfy the strong SSC of Theorem 2 for the nominal problem* **NLP**(p_0). *Then there exists a neighbourhood* $P_0 \subset P$ *of* p_0 *and continuously differentiable functions* $z : P_0 \longrightarrow \mathbb{R}^{N_z}$ *and* $\eta : P_0 \longrightarrow \mathbb{R}^{N_c}$ *with the following properties:*

(i) $z(p_0) = z_0, \quad \eta(p_0) = \eta_0,$
(ii) the active sets are constant in P_0, *i.e.* $J(z(p),p) \equiv J(z_0,p_0)\ \ \forall\, p \in P_0,$
(iii) the gradients in $G_z^a(z(p),p)$ *are linearly independent, i.e.* $\operatorname{rank}(G_z^a(z(p),p)) = N_a\ \ \forall\, p \in P_0,$
(iv) For all $p \in P_0$, *the pair* $(z(p),\eta(p))$ *satisfies the strong SSC in Theorem 2 for the perturbed problem* **NLP**(p). *In particular* $(z(p),\eta(p))$ *is a strong local minimum of* **NLP**(p).

It is instructive to sketch the main steps of the proof since the analysis additionally leads to explicit formulae for the sensitivity derivatives of the optimal solutions and multipliers. Due to the strict complementarity condition $\eta_0^a > 0$, the KKT-conditions (8) for the unknown pair $(z,\eta^a) = (z(p),\eta^a(p))$ can be written in the form

$$K(z,\eta^a,p) := \begin{pmatrix} L_z(z,\eta^a,p) \\ G^a(z,p) \end{pmatrix} = \begin{pmatrix} F_z(z,p) + (\eta^a)^T G_z^a(z,p) \\ G^a(z,p) \end{pmatrix} = 0. \quad (18)$$

At the nominal solution, the Jacobian of the mapping $K(z,\eta^a,p)$ with respect to the variable (z,η^a) is given by

$$\frac{\partial K}{\partial(z,\eta^a)}(z_0,\eta_0^a,p_0) = \begin{pmatrix} L_{zz}(z_0,\eta_0^a,p_0) & G_z^a(z_0,p_0)^T \\ G_z^a(z_0,p_0) & 0 \end{pmatrix}. \quad (19)$$

One easily shows that this so-called *Kuhn-Tucker matrix* is regular since the SSC in (14) are assumed to hold. Hence, we are in the position to apply the classical implicit function theorem to equation (18). This yields the existence of differentiable functions $z = z(p)$ and $\eta^a = \eta^a(p)$ satisfying the equation $K(z(p),\eta^a(p),p) = 0$ identically for all parameters p in a neighbourhood of p_0. The differentiation of the identity $K(z(p),\eta^a(p),p) \equiv 0$ at the nominal parameter p_0 then yields the following system of linear equations for the *sensitivity differentials* of the optimal solutions and multipliers:

$$\begin{pmatrix} L_{zz}(z_0,\eta_0^a,p_0) & G_z^a(z_0,p_0)^T \\ G_z^a(z_0,p_0) & 0 \end{pmatrix} \begin{pmatrix} \frac{dz}{dp}(p_0) \\ \frac{d\eta^a}{dp}(p_0) \end{pmatrix} + \begin{pmatrix} L_{zp}(z_0,\eta_0^a,p_0) \\ G_p^a(z_0,p_0) \end{pmatrix} = 0. \quad (20)$$

Hence, we obtain the *explicit formulae for the sensitivity differentials*

$$\begin{pmatrix} \frac{dz}{dp}(p_0) \\ \frac{d\eta^a}{dp}(p_0) \end{pmatrix} = - \begin{pmatrix} L_{zz}(z_0,\eta_0^a,p_0) & G_z^a(z_0,p_0)^T \\ G_z^a(z_0,p_0) & 0 \end{pmatrix}^{-1} \begin{pmatrix} L_{zp}(z_0,\eta_0^a,p_0) \\ G_p^a(z_0,p_0) \end{pmatrix}. \quad (21)$$

The structure of this formula brought Fiacco [11] to argue that "sensitivity analysis can be invariably be tailored to solution algorithm calculations" and that "it appears natural and efficient to calculate sensitivity as a byproduct" of solution point algorithms. This statement follows from the fact that, at least conceptually, the Kuhn-Tucker matrix appears in all iterative solution algorithms that apply Newton's method to the KKT-system (18), e.g., in SQP-methods. However, the exact computation of the Jacobian (19) in every iteration is by far too expensive. This has lead to the development of Quasi-Newton methods where one uses low rank approximations of the Jacobian; cf., e.g., the BFGS update matrices described in Gill et al. [13]. In general, the Quasi-Newton approximations do not converge to the Kuhn-Tucker matrix in (19). This means that the information gained in the solution process can not be used directly for an accurate computation of the sensitivity differentials (21); cf. the examples in Beltracchi and Gabriele [4,5]. This fact may have impaired a computational sensitivity analysis as a direct byproduct of modern solution algorithms.

Accurate values for the sensitivity differentials can be obtained from a *post-optimal* analysis. First, one has to perform an exact calculation of the Kuhn-Tucker matrix in (19). This needs a rather accurate calculation of the first and second order derivatives appearing in (19) which can be achieved by appropriate finite difference schemes; details may be found in Büskens [6]. Then sensitivity differentials can be computed through a LR-factorisation of the Kuhn-Tucker matrix. Another method is to use the techniques of RQ-factorisation for solving the linear equation (18). These methods are similar to the ones developed in section 2.3 and are described in greater detail in [6]; cf. also Fiacco [11].

3.2 First and Second Order Sensitivity Analysis of the Optimal Value Function

A formula for the first order sensitivity derivative of the optimal value function is found as follows. Since $L_z(z_0, \eta_0^a, p_0) = 0$ and $L_{\eta^a}(z_0, \eta_0^a, p_0) = G^a(z_0, p_0) = 0$, we obtain

$$\frac{dL}{dp}(z(p), \eta^a(p), p)|_{p=p_0} = L_p(z_0, \eta_0^a, p_0). \tag{22}$$

The sensitivity derivative of the objective function is given by

$$\frac{dF}{dp}(z(p), p)|_{p=p_0} = F_z(z_0, p_0)\frac{dz}{dp}(p_0) + F_p(z_0, p_0). \tag{23}$$

Upon evaluating the second component in relation (20), we get

$$G_z^a(z_0, p_0)\frac{dz}{dp}(p_0) = -G_p^a(z_0, p_0). \tag{24}$$

Hence, together with equations (18) and (22) relation (23) reduces to

$$\frac{dF}{dp}(z(p), p)\,|_{p=p_0} = L_p(z_0, \eta_0^a, p_0). \tag{25}$$

Note that in contrast to (23), where second order information is required to calculate $\frac{dz}{dp}(p_0)$, only first order information is needed in equation (25) which allows for a more efficient calculation of the optimal value sensitivities.

The differentiablity of optimal solutions is the basis for the second order sensitivity analysis of the optimal value function. Observe that the formula (25) holds identically in p, i.e., we have $\frac{dF}{dp}(z(p), p) = L_p(z(p), \eta^a(p), p)$ for all $p \in P_0$. A further differentiation of this identity yields the second order derivative

$$\frac{d^2F}{dp^2}[p_0] = \left(\frac{dz}{dp}(p_0)\right)^T L_{zp}[p_0] + \left(\frac{d\eta^a}{dp}(p_0)\right)^T G_p^a[p_0] + L_{pp}[p_0], \tag{26}$$

where the notation $[p_0]$ stands for all nominal arguments. Multiplying the upper part of equation (18) from the left by $\left(\frac{dz}{dp}(p_0)\right)^T$ and inserting the equation (24) into (26), we find the alternative formulae

$$\frac{d^2F}{dp^2}[p_0] = \left(\frac{dz}{dp}(p_0)\right)^T L_{zz}[p_0]\frac{dz}{dp}(p_0) + 2\left(L_{pz}[p_0]\frac{dz}{dp}(p_0)\right)^T + L_{pp}[p_0]. \tag{27}$$

The numerical advantage of this second representation is its independence from the Jacobian of the constraints and the derivative of the multipliers. Among others, equations (26) and (27) can be used to estimate the error of the first order real-time approximation (31) given below.

3.3 Linear Perturbations

The formula (21) for the sensitivity differentials of optimal solutions simplifies considerably if the constraints (1) involve *linear* perturbations in the form $G^a(z,p) = G^a(z) - p$ and if the objective is independent of p, i.e., $F(z,p) = F(z)$. Then we have $L_{zp}(z_0,\eta_0^a,p_0) = 0$ and $G_p^a(z_0,p_0) = -I_{N_a}$. Let

$$(l_{i,j})_{i,j=1,\dots,N_z+N_a} = \begin{pmatrix} L_{zz} & (G_z^a)^T \\ G_z^a & 0 \end{pmatrix}^{-1}$$

denote the inverse of the Kuhn-Tucker matrix in (18).

Then with $p = (p_1,\dots,p_{N_a})$ it follows that

$$\begin{aligned} \frac{dz_i}{dp_j}(p_0) &= l_{i,j+N_z}, \qquad i = 1,\dots,N_z,\; j = 1,\dots,N_a, \\ \frac{d\eta_i^a}{dp_j}(p_0) &= l_{i+N_z,j+N_z},\; i = 1,\dots,N_a,\; j = 1,\dots,N_a. \end{aligned} \tag{28}$$

Moreover, the sensitivity differential (25) of the optimal value function reduces to

$$\frac{dF}{dp_i}[p_0] = \begin{cases} -(\eta_0^a)_i, & \text{for } i \in J(z_0,p_0), \\ 0, & \text{for } i \notin J(z_0,p_0). \end{cases} \tag{29}$$

Noting that this relations also holds identically in p, we may differentiate again and obtain the second derivative which also follows directly from (26):

$$\frac{d^2F}{dp_i^2}[p_0] = \begin{cases} -\dfrac{d\eta_i^a}{dp}(p_0), & \text{for } i \in J(z_0,p_0), \\ 0, & \text{for } i \notin J(z_0,p_0). \end{cases} \tag{30}$$

Linear perturbations in the objective are not of interest as they represent only an additive factor.

4 Real-Time Optimization of NLP Problems

4.1 Real-Time Approximations by Taylor Expansions

The differentiability properties of optimal solutions to the parametric NLP problems which we discussed in the previous section are of fundamental importance for *real-time approximations* of perturbed solutions. The knowledge of the nominal solution $z_0 = z(p_0)$ and the sensitivity differentials $\frac{dz}{dp}(p_0)$ allow to approximate the perturbed solution $z(p)$ by its first order Taylor expansion:

$$z(p) \approx z_0 + \frac{dz}{dp}(p_0)\,(p - p_0)\,, \tag{31}$$

Both the nominal functions $z(p_0)$ and the sensitivity differentials $\frac{dz}{dp}(p_0)$ can be computed *off-line*. Hence, if the parameter p deviates from the nominal parameter p_0, equation (31) provides an *on-line* approximation whose computation is very fast since it requires only matrix-multiplications. Should the first order approximation not be accurate enough, a further improvement of the real-time approximations in view of optimality and admissibility can be achieved by applying the ideas in [7]. Otherwise it is advantageous to use the approximation as a starting point for calculating the exact solution.

One way to estimate the quality of the real-time approximation (31) is to evaluate the Taylor-expansion of the objective function

$$F(z(p), p) \approx F(z_0, p_0) + \frac{dF}{dp}[p_0](p - p_0). \tag{32}$$

The error in the first order approximation can further be estimated from the second order Taylor expansion

$$F(z(p), p) \approx F(z, p_0) + \frac{dF}{dp}[p_0](p - p_0) + \frac{1}{2}(p - p_0)^T \frac{d^2F}{dp^2}[p_0](p - p_0), \tag{33}$$

where $\frac{dF}{dp}[p_0]$ and $\frac{d^2F}{dp^2}[p_0]$ are calculated from equations (25) and (27).

4.2 A Numerical Example

Consider the optimization problem in two variables $z = (z_1, z_2)$:

$$\begin{array}{ll} \text{Maximize} & \tilde{F}(z,p) = (0.5 + p)\sqrt{z_1} + (0.5 - p) z_2 \\ \text{subject to} & z_1 + z_2 \le 1, \quad z_1 \ge 0.1. \end{array}$$

The nominal parameter is chosen as $p_0 = 0$. Before applying the KKT-conditions (8) and (9) we have to observe that we are minimizing the function $F(z, p) := -\tilde{F}(z, p)$. It is easy to see that the optimal solution for parameters p in a small neighbourhood of $p_0 = 0$ is given by

$$z_1(p) = \left(\frac{0.5 + p}{1 - 2p}\right)^2, \quad z_2(p) = 1 - z_1(p), \quad \eta^a(p) = \eta_1^a(p) = 0.5 - p.$$

The active set is $J(z(p), p) = \{1\}$ and hence the Lagrangian is

$$L(z_1, z_2, \eta^a) = -(0.5 + p)\sqrt{z_1} - (0.5 - p) z_2 + \eta^a (z_1 + z_2 - 1).$$

The nominal solution is $z_0 = (0.25, 0.75)$ and $\eta_0^a = 0.5$. Though the Hessian $L_{zz}(z_0, \eta_0^a, p_0) = \begin{pmatrix} 1 & 0 \\ 0 & 0 \end{pmatrix}$ is *not* positive on the whole space $\mathbb{R}^2$, the SSC in (14) are satisfied in view of $G_z^a(z_0, p_0) = (1,1)$ and $\ker(G_z^a(z_0,p_0)) = \mathbb{R} \cdot (1,-1)$. The matrices in (21) for the nominal solution are

$$\begin{pmatrix} L_{zz} & (G_z^a)^T \\ G_z^a & 0 \end{pmatrix} = \begin{pmatrix} 1 & 0 & 1 \\ 0 & 0 & 1 \\ 1 & 1 & 0 \end{pmatrix}, \quad \begin{pmatrix} L_{zp} \\ G_p^a \end{pmatrix} = \begin{pmatrix} -1 \\ 1 \\ 0 \end{pmatrix}.$$

Then the formula (21) for the sensitivity differentials yields $d(z_1, z_2, \eta^a)/dp = (2, -2, -1)^T$ from which we get the following first order approximation (31):

$$\begin{pmatrix} z_1(p) \\ z_2(p) \\ \eta^a(p) \end{pmatrix} \approx \begin{pmatrix} 0.25 \\ 0.75 \\ 0.50 \end{pmatrix} + \begin{pmatrix} 2 \\ -2 \\ -1 \end{pmatrix} p.$$

To illustrate the quality of this approximations, let us consider the perturbation $p = 0.05$. The exact solution is computed from the above formula as

$$(z(0.05), z_2(0.05), \eta^a(0.05)) = (0.373457, 0.626543, 0.45).$$

The first order approximation yields the value $(0.35, 0.65, 0.45)$ which is in an acceptable agreement with the exact value. The sensitivity analysis of the optimal value function is as follows. The exact functional value is $F[0.05] = -0.618056$. The first order formula (25) yields $dF/dp = L_p = -\sqrt{z_1} + z_2 = 0.25$ whereas the second derivative is computed from (30) as $d^2F/dp^2 = d\eta^a/dp = -1$. Then the first order expansion (32) gives the value -0.6125 which can be improved by the second order expansion (33) where we obtain the value -0.615.

4.3 Prediction of the Sensitivity Domain

When dealing with real-time approximations of the form (31) one has to ensure that a change of the parameter p does not change the set of the active constraints $J(z_0, p_0)$. In general one has to distinguish two cases: a constraint enters the active set or a constraint leaves the active set. This section is concerned with the prediction of the sensitivity domain to determine when a perturbation p is too large to apply formula (31). The following analysis is based on [4, 5]. A first order Taylor-expansion of the active Lagrange multipliers is given by

$$\eta^a(p) \approx \eta^a(p_0) + \frac{d\eta^a}{dp}(p_0)\,(p - p_0), \tag{34}$$

where the derivative $\frac{d\eta^a}{dp}(p_0)$ is obtained from (21). A constraint will leave the active set when the corresponding Lagrange multiplier will go from some nonzero value to zero. Hence, if one of the multipliers in (34) approximates zero,

$$0 = \eta_i^a(p) \approx \eta_i^a(p_0) + \frac{d\eta_i^a}{dp}(p_0)\,(p^i - p_0), \quad i \in J(z_0, p_0), \tag{35}$$

it follows from (35), that an approximation of the perturbation $p^i = (p_1^i, \ldots, p_{N_p}^i)^T$ causing a constraint G_i to leave the active set is given by

$$p_j^i \approx (p_0)_j - \frac{\eta_i^a(p_0)}{\frac{d\eta_i^a}{dp_j}(p_0)}, \quad i \in J(z_0, p_0), \ j \in \{1, \ldots, N_p\}, \tag{36}$$

provided that $\frac{d\eta_i^a}{dp_j}(p_0) \neq 0$. The situation where a constraint enters the active set is rather similar. In this case a constraint G_i, $i \notin J(z_0, p_0)$, is zero:

$$0 = G_i(z(p), p) \approx G_i(z_0, p_0) + \frac{dG_i}{dp}(z_0, p_0)\left(p^i - p_0\right), \quad i \notin J(z_0, p_0), \tag{37}$$

where

$$\frac{dG_i}{dp}(z_0, p_0) = \frac{\partial G_i}{\partial z}(z_0, p_0)\frac{dz}{dp}(p_0) + \frac{\partial G_i}{\partial p}(z_0, p_0), \quad i \notin J(z_0, p_0). \tag{38}$$

Hence an approximation of the perturbation $p^i = (p_1^i, \ldots, p_{N_p}^i)^T$ causing the constraint G_i to enter the active set is given by

$$p_j^i \approx (p_0)_j - \frac{G_i(z_0, p_0)}{\frac{dG_i}{dp_j}(z_0, p_0)}, \quad i \notin J(z_0, p_0), \ j \in \{1, \ldots, N_p\}, \tag{39}$$

provided that $\frac{dG_i}{dp_j}(z_0, p_0) \neq 0$. We summarize our findings as follows. The sensitivity domain P_0 is determined by those values p_j^i in (36) and (39) which are closest to the nominal perturbation $(p_0)_j$:

$$\begin{aligned}
&P_0 \approx P_0^1 \times P_0^2 \times \cdots \times P_0^{N_c}, \\
&P_0^j := \left[\max_{\bar{p}_j < (p_0)_j} \{\bar{p}_j \in \bar{P}_j\}, \min_{\bar{p}_j > (p_0)_j} \{\bar{p}_j \in \bar{P}_j\}\right], \quad j = 1, \ldots, N_p, \\
&\bar{P}_j := \left\{p_j^i \mid i = 1, \ldots, N_c\right\} \cup \{-\infty, +\infty\}.
\end{aligned} \tag{40}$$

4.4 Handling of Larger Perturbations

This section will be based on ideas in Beltracchi and Gabriele [4]. In general, sensitivity derivatives do not exist at points where the active set changes. However, one can show that at least *directional* sensitivity derivatives exist. Hence, one method for dealing with changes in the active set is based on calculating directional derivatives. Based on the arguments of the preceeding section, we propose the following strategy for dealing with constraints entering or leaving the active set:

1. Calculate the optimal solution $(z(p_0^1), \eta^a(p_0^1))$ and the sensitivity differentials $(\frac{dz}{dp}(p_0^1), \frac{d\eta^a}{dp}(p_0^1))$ at the nominal value $p_0^1 = p_0$.
2. Calculate the sensitivity domain as described in (40). Let p_0^2 denote the perturbation that causes a constraint to enter or to leave the active set.

3. Calculate the sensitivity differentials $(\frac{dz}{dp}(p_0^2), \frac{d\eta^a}{dp}(p_0^2))$ at the value p_0^2 with the active set updated to reflect the change indicated by step 2. Let η^u denote the updated Lagrange multiplier and remember $\frac{d\eta_i}{dp}(p_0) = 0, i \notin J(z_0, p_0)$.
4. Calculate the first order changes

$$\begin{aligned} \Delta z &:= \tfrac{dz}{dp}(p_0^1)(p_0^2 - p_0^1) + \tfrac{dz}{dp}(p_0^2)(p - p_0^2), \\ \Delta\eta^u &:= \tfrac{d\eta^u}{dp}(p_0^1)(p_0^2 - p_0^1) + \tfrac{d\eta^u}{dp}(p_0^2)(p - p_0^2). \end{aligned} \tag{41}$$

5. Calculate the new first order approximations by

$$z(p) \approx z_0 + \Delta z, \quad \eta^u(p) \approx \eta_0^u + \Delta\eta^u, \quad \text{if } p - p_0^2 \geq 0, \tag{42}$$

otherwise by (21).

Note that the approximations (42) also lead to first and second order Taylor approximations of the optimal value function.

4.5 A Numerical Example for Predicting the Sensitivity Domain

The purpose of this section is to illustrate the theoretical results presented in Sections 2–4 by a numerical example which is taken from [4,5]. Consider the following linearly perturbed NLP-problem with optimization variable $z = (z_1, z_2)^T \in \mathbb{R}^2$ and perturbation parameter $p \in \mathbb{R}$:

$$\begin{array}{ll} \min\limits_{z} & F(z,p) = (z_1 + 1)^2 + (z_2 - 2)^2, \\ \text{subject to} & G_1(z) = -z_1 + p \qquad \leq 0, \\ & G_2(z) = 2z_1 + z_2 - 6 \leq 0 \end{array} \tag{43}$$

The nominal perturbation is assumed to be $p_0 = 1$. The discussion of the necessary optimality conditions in Theorem 1 yields the optimal candidate $z_0 = (1,2)$ and $\eta_0^a = (4,0)$ with active constraint G_1. The optimal nominal value is $F(z_0, p_0) = 4$. The SSC in (14) obviously hold in view of $L_{zz} = I_2$. It is straightforward to see that the optimal perturbed solution is given by $(z_1(p), z_2(p)) = (p, 2)$, $\eta^a(p) = 2(p+1)$ for $-1 \leq p \leq 2$ and has the optimal value $F(z(p), p) = (p+1)^2$.

Let us compare this exact solution with the approximate solutions in (31)–(33). Applying formula (21), the sensitivities of the optimization variables and Lagrange multipliers are calculated as $\frac{dz}{dp}(1) = (1,0)^T$ and $\frac{d\eta^a}{dp}(1) = 2$. Since the perturbation appears linearly in the constraints we can employ equations (29) and (30) to get the first and second order sensitivity differentials of the objective: $\frac{dF}{dp}(1) = 4$ and $\frac{d^2F}{dp^2}(1) = 2$. The sensitivity domain is calculated from (36) and (39) as $P_0 = [-1, 2]$.

Now we apply the real-time approximation ideas of Section 4 for the perturbation $p = 2$. Since the perturbation is inside the sensitivity domain P_0, we can use formula (31) for real-time optimization. We get $z(2) \approx (2,2)^T$ which agrees with the exact solution. This is due to the fact that both the constraints and the perturbation are linear. The first order Taylor expansion (32) of the objective function yields

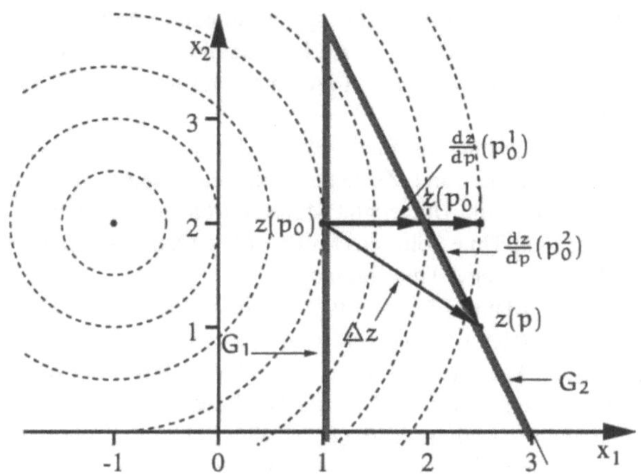

Figure 1. Real-time optimization of perturbed NLP problems

$F(z(2), 2) \approx 8$. The second order Taylor expansion (33) of the objective function gives $F(z(2), 2) \approx 9$ which agrees with the exact optimal value.

Next we investigate a perturbation $p = 2.5$ outside the sensitivity domain. Without using equation (42) we obtain from (31) the non-admissible point $z(2.5) = (2.5, 2) \notin S(2.5)$. Step 2 of the strategy for handling larger perturbations tells us that the constraint G_2 enters the active set when $p = p_0^2 = 2$. For values of p greater then p_0^2 the position of the optimum is along the intersection of constraint G_1 and G_2. The new search direction along constraints G_1 and G_2 is determined in step 3 as $\frac{dz}{dp}(2) = (1, -2)$. Then formula (42) gives an estimate for the new position of the optimum, namely $z(2.5) = (2.5, 1)$, which agrees with the optimal solution; cf. Figure 1).

We should probably not expect that good results in more general optimization applications. However, at least we can expect better predictions of the new optimum for small changes in those parameter which cause a change of the active set.

References

1. R. L. Armacost, A. V. Fiacco: Computational Experience in Sensitivity Analysis for Nonlinear Programming. Mathematical Programming **6** (1974) 301–326
2. R. L. Armacost, W. C. Mylander: A Guide to the SUMT-Version 4 Computer Subroutine for Implementing Sensitivity Analysis in Nonlinear Programming. Technical report **T-287**, Institute for Management Sciences and Engineering, George Washington University, Washington, D.C., (1973)

3. T. J. Beltracchi, G. A. Gabriele: A RQP Based Method for Estimating Parameter Sensitivity Derivatives. Advances in Design Automation, ASME DE **14** (1988) 155–164
4. T. J. Beltracchi, G. A. Gabriele: Observations on Extrapolations Using Parameter Sensitivity Derivatives. Advances in Design Automation, ASME DE **14** (1988) 165–174
5. T. J. Beltracchi, G. A. Gabriele: An Investigation of New Methods for Estimating Parameter Sensitivity Derivatives. NASA Contractor Report **4245**, Langley Research Center, USA (1989) 1–130
6. C. Büskens: Optimierungsmethoden und Sensitivitätsanalyse für optimale Steuerprozesse mit Steuer- und Zustands- Beschränkungen, Dissertation, Institut für Numerische Mathematik, Universität Münster, Münster, Germany, (1998)
7. C. Büskens: Real-Time Solutions for Perturbed Optimal Problems by a Mixed Open- and Closed- Loop Strategy. This volume.
8. C. Büskens, H. Maurer: Sensitivity Analysis and Real-Time Control of Parametric Optimal Control Problems Using Nonlinear Programming Methods. This volume.
9. C. Büskens, H. Maurer: Real-Time Control of an Industrial Robot. This volume.
10. A. V. Fiacco: Sensitivity Analysis for Nonlinear Programming Using Penalty Methods. Mathematical Programming **10**, (1976) 287–311
11. A. V. Fiacco: Introduction to Sensitivity and Stability Analysis in Nonlinear Programming. Mathematics in Science and Engineering **165**, Academic Press, New York, (1983)
12. R. Fletcher: Practical Methods of Optimization. 2nd ed., J. Wiley & Sons, Chichester/New York, (1997)
13. P. E. Gill, W. Murray, M. A. Saunders and M. H. Wright: Model Building and Practical Aspects of Nonlinear Programming. In Computational Mathematical Programming, Schittkowski, K., ed., Springer-Verlag, Berlin, Heidelberg (1985) 209–247
14. S. M. Robinson: Perturbed Kuhn-Tucker Points and Rate of Convergence for a Class of Nonlinear Programming Algorithms. Mathematical Programming **7** (1974) 1–16

*Sensitivity Analysis and Real-Time Control of Parametric Optimal Control Problems Using Boundary Value Methods

Helmut Maurer and Dirk Augustin

Institut für Numerische Mathematik, Westfälische Wilhelms-Universität Münster, Germany

Abstract Parametric nonlinear control problems subject to mixed control-state constraints and pure state constraints are investigated. Parameters are introduced to model perturbations of the control system and may appear in all system data. We review conditions under which the optimal solutions are differentiable functions of the parameter. In the theoretical part, these conditions are related to regularity conditions and to second order sufficient conditions. On the numerical side, the conditions are connected to shooting methods for solving the boundary value problems that characterize the optimal solution. We discuss methods for computing the sensitivity differentials of the optimal solutions with respect to parameters. The calculated sensitivity differentials can be used to construct real-time approximations of the perturbed solutions via first order Taylor expansions. Two numerical case studies are discussed in detail to illustrate the numerical methods for mixed control-state constraints and for pure state constraints.

1 Introduction

We study parametric optimal control problems that are subject to mixed control-state constraints and pure state constraints. There exist numerous papers on applications of optimal control problems with control-state constraints, e.g., [9, 30–33, 45, 48, 49, 57–60] or pure state constraints, e.g., [7, 9, 11, 12, 15, 16, 19, 20, 30, 32, 33, 42, 46, 47, 53, 55, 59, 60]. Parameters in the control system play the role of modelling perturbations of system data in a deterministic way. We adopt the deterministic point of view since the stochastic approach to perturbations is by far too difficult for the general control problem under consideration.

Stability analysis of parametric control problems is concerned with Lipschitz continuity of optimal solutions with respect to the parameters [17, 18, 38]. *Sensitivity analysis* aims at a stronger property of optimal solutions. Namely, conditions are derived that establish differentiability of optimal solutions [39–41, 49]. Moreover, the theory is tied to numerical methods which allow to compute the sensitivity differentials. The interest in conditions for solution differentiability originates in the real-time computation of perturbed solutions under parameter changes. The basic ideas of using first order approximations of perturbed solutions for real-time control go back to the sixties (see [9] and other work cited in this book) and have been refined later to treat control problems with control and state constraints [8, 31–33, 59, 60]. The approach in these papers uses boundary value methods for computing both the nominal unperturbed solution and the real-time approximations of perturbed solutions. The shooting method [10,54,65] turned out to be an efficient method for both tasks yielding highly accurate solutions.

Recently, the theory underlying solution differentiability was developed in [39–41] in such a way that all needed conditions were related to shooting methods. The essential prerequisite for solution differentiability is the property that the nominal solution satisfies second order sufficient conditions (SSC), cf., [48,50,51,56,68]. It was shown that SSC can be checked via a bounded solution of Riccati equations. This test could also be related to shooting methods.

The present article intends to give a survey on sensitivity analysis and the numerical tools to compute sensitivity differentials. The reader is assumed to have a basic knowledge on optimal control problems as conveyed in the textbooks [9], [27], [63]. The material is presented as to make the article self-contained both with respect to the theory and the numerical methods. In section 2, we consider a general optimal control problem subject to control and state constraints and sketch a basic idea for real-time computations for perturbed solutions which is based on Taylor expansions. Section 3 is devoted to sensitivity analysis of control problems with *mixed control-state constraints* and summarizes the work in [39,49]. Section 4 treats the Rayleigh problem (the optimal control of an electric oscillator) with a pure control constraint and with a mixed control-state constraint. We have strived for a complete account of both the nominal solution and the sensitivity differentials. State constraints are considered in section 5 where for simplicity of exposition only a scalar state constraint is studied. The theory is illustrated by a case study in section 6: the optimal control of a Van-der-Pol oscillator under a state constraint.

To readers, who are not familiar with optimal control theory, we recommend starting with the numerical examples in sections 4 and 6 which may provide a better insight into the technical assumptions needed for the theory in sections 3 and 5.

2 Parametric Optimal Control Problems and the Principle of Real-Time Control

We shall study optimal control processes on a time interval $[0, t_f]$ with a fixed final time $t_f > 0$. The *state of a system* at time $t \in [0, t_f]$ is denoted by the vector $x(t) \in \mathbb{R}^n$. The system is steered by a *control vector* $u(t) \in \mathbb{R}^m$. The control function $u : [0, t_f] \to \mathbb{R}^m$ is assumed to be an essentially bounded function, i.e., an element of the Banach space $L^\infty(0, t_f; \mathbb{R}^m)$ which is endowed with the norm $\|u\|_\infty := \operatorname{ess\,sup}\{\,|u(t)|,\ 0 \le t \le t_f\,\}$. Here $|\cdot|$ stands for the Euclidean norm in $\mathbb{R}^m$. The state function $x : [0, t_f] \to \mathbb{R}^n$ is an element of the space $W^{1,\infty}(0, t_f; \mathbb{R}^n)$ of uniformly Lipschitz-continuous functions which is a Banach space under the norm $\|x\|_{1,\infty} := |x(0)| + \|\dot{x}\|_\infty$.

All data of the control problem may depend on a parameter $p \in P$ where $P \subset \mathbb{R}^q$ is an open set of finite-dimensional parameters. The following *parametric optimal control problem* will be denoted by OC(p):

OC(p) Determine a pair $(x_p, u_p) \in W^{1,\infty}(0,t_f;\mathbb{R}^n) \times L^\infty(0,t_f;\mathbb{R}^m)$ that minimizes the cost functional

$$F(x,u,p) = g(x(0),x(t_f),p) + \int_0^{t_f} f_0(x(t),u(t),p)\,dt \tag{2.1}$$

subject to the dynamics, boundary conditions and constraints

$$\dot{x}(t) = f(x(t),u(t),p) \quad \text{for a.e.} \quad t \in [0,t_f], \tag{2.2}$$

$$\psi(x(0),x(t_f),p) = 0, \tag{2.3}$$

$$C(x(t),u(t),p) \le 0 \quad \text{for a.e.} \quad t \in [0,t_f], \tag{2.4}$$

$$S(x(t),p) \le 0 \quad \text{for all} \quad t \in [0,t_f], \tag{2.5}$$

where $g : \mathbb{R}^n \times \mathbb{R}^n \times P \to \mathbb{R}$, $f_0 : \mathbb{R}^n \times \mathbb{R}^m \times P \to \mathbb{R}$, $f : \mathbb{R}^n \times \mathbb{R}^m \times P \to \mathbb{R}^n$, $\psi : \mathbb{R}^n \times \mathbb{R}^n \times P \to \mathbb{R}^r$, $1 \le r \le 2n$, $C = (C^1, \ldots, C^k) : \mathbb{R}^n \times \mathbb{R}^m \times P \to \mathbb{R}^k$ and $S : \mathbb{R}^n \times P \to \mathbb{R}$. The inequality constraints (2.4) are called *mixed* control-state constraints whereas (2.5) constitutes a *pure* state constraint. At first sight, the distinction between mixed control-state (2.4) constraints and the *pure* state constraint (2.5) seems to be artificial since the constraint (2.5) is formally included in the constraint (2.4). However, for the mixed control-state constraint (2.4) we tacitly assume that any function $C^i(x,u,p)$, $i = 1, \ldots, k$, contains at least one control component, i.e., $D_u C^i \not\equiv 0$. The precise assumption will be formulated in the regularity condition (AC-1) in section 3. This distinction between mixed control-state and pure state constraints will lead to a significant difference in theoretical and numerical solution techniques for both types of constraints; cf. the examples in sections 4 and 6.

In (2.5) we have considered only a *scalar* state constraint since the general case of *vector-valued* state constraints would complicate the technical exposition considerably. Non-autonomous control problems can be reduced to the above autonomous problem by treating the time t as a new state variable. Similarly, control problems with a free final time t_f can be reduced to an augmented control problem with a fixed final time. This can be achieved by the time transformation $t = s \cdot t_f$ where the new time variable $s \in [0,1]$ is introduced; cf. [27,48].

The *general* standing assumptions (AG-1)–(AG-3) made in this section are supposed to be satisfied throughout the paper. Hereafter, assumptions referring to mixed constraints will be denoted by (AC-1)–(AC-5) in section 3, whereas those for the pure state constraint will be called (AS-1)–(AS-6) in section 5.

(AG-1) There exist open sets $\mathcal{R}^n \subset \mathbb{R}^n$ and $\mathcal{R}^m \subset \mathbb{R}^m$ such that the functions g, f_0, f, ψ, C, S are twice differentiable in (x,u) and once in p on $\mathcal{R}^n \times \mathcal{R}^n \times P$ and on $\mathcal{R}^n \times \mathcal{R}^m \times P$, respectively. All derivatives are assumed to be uniformly Lipschitz-continuous in (x,u,p). In case of a state constraint of order $l \ge 1$ (cf. section 5), the functions are assumed to be of class C^{2l+1}.

For a fixed reference parameter $p_0 \in P$, the problem $OC(p_0)$ is considered as the *unperturbed* or *nominal* problem.

(AG-2) There exists a local solution (x_0, u_0) of the reference problem $OC(p_0)$ such that the control u_0 is *continuous*.

The continuity assumption for the optimal control is restrictive but holds in many practical examples. The continuity property holds, e.g., for control problems with a *regular* Hamiltonian function. We note that the continuity assumption can be dropped, see [38], but this approach requires a refined technical exposition of sensitivity analysis which is beyond the scope of this paper.

One basic prerequisite for sensitivity analysis is an assumption on the *structure* of the optimal solution, i.e., on the number of active intervals for the inequality constraints (2.4) and (2.5). An interval $[t_1, t_2]$ is called a *boundary arc* for the i-th component C_i of the mixed constraint (2.4) if it is a maximal interval on which $C^i(x(t), u(t), p) = 0$ for all $t \in [t_1, t_2]$. The points t_1 and t_2 are called *entry* and *exit* point for the i-th constraint or, more simply, *junction points*. A *boundary arc* for the *pure state* constraint (2.5) is defined in a similar way by $S(x(t), p) = 0$ for all $t \in [t_1, t_2]$. A point $\tau \in [0, t_f]$ is called a *contact* point for the state constraint if it is an isolated point at which the trajectory touches the boundary. Taken together, entry, exit and contact points are called *junction* points. We denote by TC_i^0 the set of junction points for the i-th mixed inequality constraint (2.4) and put $TC^0 := TC_1^0 \cup \cdots \cup TC_k^0$. Similarly, TS^0 denotes the set of junction points for the pure state constraint (2.5). We assume:

(AG-3) The set $T^0 := TC^0 \cup TS^0 = \{t_1^0, \ldots, t_N^0\}$ of all junction points is finite with $0 \le t_1^0 < \cdots < t_N^0 \le t_f$. The intersections of these sets are disjoint and satisfy $TC_i^0 \cap TC_j^0 = \emptyset, \ \forall\, i \neq j$, and $TC^0 \cap TS^0 = \emptyset$.

The second part of this assumption means that every point t_i^0 is a junction point for *exactly one* constraint in either (2.4) or (2.5).

We briefly sketch now the sensitivity result that leads to a *real-time control approach of perturbed optimal solutions*. The next section summarizes conditions that ensure the property of *solution differentiability* for optimal solutions. Namely, it will be shown that there exists a neighborhood $P_0 \subset P$ of the nominal parameter p_0 such that the unperturbed solution (x_0, u_0) can be embedded into a family of perturbed solutions to the perturbed problems OC(p),

$$(x_0(t), u_0(t)) \Rightarrow (x(t,p), u(t,p)) \quad \forall\, p \in P_0,$$

with the property that the partial derivatives (sensitivity differentials)

$$y(t) := \frac{\partial x}{\partial p}(t, p_0) \quad v(t) := \frac{\partial u}{\partial p}(t, p_0)$$

exist for all t except at the junction points. Moreover, these partial derivatives satisfy a *linear* inhomogeneous boundary value problem which is easily solvable. This property allows to approximate the perturbed solutions by a first order Taylor expansion

$$x(t,p) \approx x_0(t) + \frac{\partial x}{\partial p}(t, p_0)(p - p_0), \quad u(t,p) \approx u_0(t) + \frac{\partial u}{\partial p}(t, p_0)(p - p_0), \tag{2.6}$$

for all t which are not junction points. The right sides represent quantities which allows for a very fast *on-line* computation since they require only matrix multiplications after the unperturbed solution and the sensitivity differentials have been computed *off-line*. This approximation of the perturbed solution has been used in a more or less general situation by various authors, cf., e.g., [8, 9, 31–33, 49, 57–60]. The practical benefit of this approach has strongly stimulated the systematic study of sensitivity analysis. The approximations in (2.6) represent the counterpart to the first order sensitivity analysis for *finite-dimensional* optimization problems [25]. This approach has been exploited in [13, 14] by applying nonlinear programming techniques to a discretized version of the optimal control problem.

We point out that some care has to be taken before using these approximations on the whole time interval. Pesch [57]– [60] gives a computationally efficient modification of the Taylor expansion in a neighborhood of the junction points.

3 Sensitivity Analysis for Optimal Control Problems with Mixed Control-State Constraints

3.1 Constraints Qualifications and Second Order Sufficient Conditions

To simplify notations, we shall treat the mixed control-state constraint (2.4) and the pure state constraint (2.5) separately. In this section, we concentrate on control problems with control-state constraints (2.4) only and give a survey of the sensitivity results in [38, 39]. The first order necessary conditions for an optimal pair (x_0, u_0) are well known in the literature [27, 52]. The *unconstrained* Hamiltonian function H^0, resp., the *augmented* Hamiltonian H are defined as

$$H^0(x,u,\lambda,p) = f_0(x,u,p) + \lambda^* f(x,u,p), \tag{3.1}$$

$$H(x,u,\lambda,\mu,p) = H^0(x,u,\lambda,p) + \mu^* C(x,u,p), \tag{3.2}$$

where $\lambda \in \mathbb{R}^n$ denotes the adjoint variable and $\mu \in \mathbb{R}^k$ is the multiplier associated with the control-state constraint (2.4); the asterisk denotes the transpose. Henceforth, partial derivatives will either be denoted by the symbols D_x, D_u or by subscripts. Likewise, second derivatives are denoted by D^2_{xx}, D^2_{xu} or by double subscripts. The nominal solution corresponding to the nominal parameter p_0 is marked by the subscript zero.

Now we need two *regularity assumptions* in order to formulate *first order necessary conditions* in a *normal form* with a non-zero cost multiplier.

Let $C = (C^1, \ldots, C^k)$ be the vector defining the constraint (2.4). Consider the set of active indices resp. the vector of active components:

$$I_0(t) = \{i \in \{1, \ldots, k\} \mid C^i(x_0(t), u_0(t), p_0) = 0\}, \quad C^a := (C^i)_{i \in I_0(t)}.$$

For an empty index set $I_0(t) = \emptyset$ the vector C^a is taken as the zero vector. The following regularity assumption concerns *linear independence* of gradients for active constraints; cf. [34–36, 39, 50, 68].

(AC-1) The gradients $D_u C^i(x_0(t), u_0(t), p_0)$ are uniformly linear independent for all $i \in I_0(t)$ and all $t \in [0, t_f]$.
(AC-2) (*Controllability condition*) For each $b \in \mathbb{R}^r$ there exist functions $v \in L^\infty(0, t_f; \mathbb{R}^m)$ and $y \in W^{1,\infty}(0, t_f; \mathbb{R}^n)$ which satisfy the following equations for a.e. $t \in [0, t_f]$:

$$\begin{aligned}
&\dot{y}(t) = D_x f(x_0(t), u_0(t), p_0)\, y(t) + D_u f(x_0(t), u_0(t), p_0)\, v(t), \\
&D_{x(0)}\psi(x_0(0), x_0(t_f), p_0)\, y(0) + D_{x(t_f)}\psi(x_0(0), x_0(t_f), p_0)\, y(t_f) = b, \\
&D_x C^a(x_0(t), u_0(t), p_0) y(t) + D_u C^a(x_0(t), u_0(t), p_0) v(t) = 0.
\end{aligned}$$

Usually, it is difficult to verify the controllability condition. A numerical check will be provided by the shooting method in section 3.2. The last two assumptions guarantee that the following first order necessary optimality conditions are satisfied. There exist Lagrange-multipliers

$$(\lambda_0, \mu_0, \rho_0) \in W^{1,\infty}(0, t_f; \mathbb{R}^n) \times L^\infty(0, t_f; \mathbb{R}^k) \times \mathbb{R}^r$$

such that the following conditions hold for a.e. $t \in [0, t_f]$,

$$\dot{\lambda}_0(t) = -H_x(x_0(t), u_0(t), \lambda_0(t), \mu_0(t), p_0)^*, \tag{3.3}$$

$$(-\lambda_0(0), \lambda_0(t_f)) = D_{(x(0),x(t_f))}\,(g + \rho_0^*\psi)^*(x_0(0), x_0(t_f), p_0), \tag{3.4}$$

$$H_u(x_0(t), u_0(t), \lambda_0(t), \mu_0(t), p_0) = 0, \tag{3.5}$$

$$\mu_0(t) \geq 0 \quad \text{and} \quad \mu_0(t)^* C(x_0(t), u_0(t), p_0) = 0, \tag{3.6}$$

$$H(x_0(t), u_0(t), \lambda_0(t), \mu_0(t), p_0) \equiv \text{const.} \tag{3.7}$$

The stationarity condition (3.5) can be sharpened to the following *minimum condition* for the optimal control

$$\begin{aligned}
&H^0(x_0(t), u_0(t), \lambda_0(t), p_0) \\
&\quad = \min_{u \in \mathbb{R}^m} \{ H^0(x_0(t), u, \lambda_0(t), p_0) \mid C(x_0(t), u, p_0) \leq 0 \}.
\end{aligned}$$

Let μ_0^i denote the i-th component of the multiplier μ_0 in (3.6) and consider the set of indices

$$J_0(t) := \{ i \in \{1, \ldots, k\} \mid \mu_0^i(t) > 0 \}.$$

To facilitate the analysis and to prepare sensitivity analysis, we introduce already at this point the following assumption on *strict complementarity*.
(AC-3) (*Strict complementarity condition*) The index sets $J_0(t)$ and $I_0(t)$ coincide for every $t \in [0, t_f]$ which is *not* a junction point with the boundary.

It follows from this hypothesis that the only points where the index sets $J_0(t)$ and $I_0(t)$ do *not* coincide are the junction points since the respective component of the multiplier $\mu(t)$ vanishes at a junction point; cf. Lemma 2 in section 3.1.

To simplify notations in the sequel, the argument of all functions evaluated at the reference solution (x_0, u_0, p_0) will be denoted by [t], e.g.,

$$f[t] := f(x_0(t), u_0(t), p_0) \quad \text{and} \quad H[t] := H(x_0(t), u_0(t), \lambda_0(t), \mu_0(t), p_0).$$

Under the hypothesis of the strict complementarity condition, the *modified strict Legendre-Clebsch condition* from [35,36,39,50,68] takes the following form where $|.|$ denotes the euclidean norm:

(AC-4) (*Strict Legendre-Clebsch condition*) There exists $c > 0$ such that for all $t \in [0, t_f]$ the estimate holds:

$$v^* H_{uu}[t]v \geq c|v|^2 \quad \forall\ v \in \mathbb{R}^m,\ C_u^a[t]v = 0.$$

We note that the strict Legendre-Clebsch condition is already a consequence of the stronger coercivity condition (AC-5) below. Nevertheless, it is convenient to introduce the strict Legendre-Clebsch condition explicitly since this condition will be needed in the formulation of the Riccati equations (3.12) and (3.16). The following coercivity condition constitutes the essential part of second order sufficient conditions (SSC) and can be retrieved from a study of the properties of the second variation with respect to the variational system associated with equations (2.2)–(2.4). We do not formulate this condition in full generality (cf. [23,24,48,50]) but present a form which is appropriate for practical verification. Let us introduce the function

$$G(x(0), x(t_f), p) := g(x(0), x(t_f), p) + \rho^*\, \psi(x(0), x(t_f), p), \qquad \rho \in \mathbb{R}^r.$$

In the sequel, we shall use the notations $\psi[0, t_f] = \psi(x_0(0), x_0(t_f), p_0)$ and $G[0, t_f] = G(x_0(0), x_0(t_f), p_0)$. Second order derivatives of the function G are denoted by

$$G_{00}[0, t_f] = D^2_{x(0)x(0)} G[0, t_f], \quad G_{0T}[0, t_f] = D^2_{x(0)x(t_f)} G[0, t_f], \quad \text{etc.}$$

Whenever it is convenient, we shall denote partial derivatives with respect to x and u by subscripts.

(AC-5) (*Coercivity conditions*)

(a) There exists a symmetric matrix $Q \in W^{1,\infty}(0, t_f; M_{n\times n})$ and $c > 0$ such that

$$(y^*, v^*) \begin{pmatrix} \dot{Q}(t) + Q(t)f_x[t] + f_x[t]^*Q(t) + H_{xx}[t] & | \ H_{xu}[t] + Q(t)f_u[t] \\ H_{ux}[t] + f_u[t]^*Q(t) & | \ H_{uu}[t] \end{pmatrix} \begin{pmatrix} y \\ v \end{pmatrix} \geq c|(y, v)|^2 \tag{3.8}$$

holds *uniformly* for all $t \in [0, t_f]$ and all vectors $(y, v) \in \mathbb{R}^n \times \mathbb{R}^m$ with

$$C_x^a[t]y + C_u^a[t]v = 0. \tag{3.9}$$

(b) The boundary condition

$$(\xi_0^*, \xi_1^*) \begin{pmatrix} G_{00}[0, t_f] + Q(0) & | \ G_{0T}[0, t_f] \\ G_{T0}[0, t_f] & | \ G_{TT}[0, t_f] - Q(t_f) \end{pmatrix} \begin{pmatrix} \xi_0 \\ \xi_1 \end{pmatrix} > 0 \tag{3.10}$$

is valid for all $(\xi_0, \xi_1) \in \mathbb{R}^n \times \mathbb{R}^n \setminus \{(0, 0)\}$ satisfying

$$D_{x(0)}\psi[0, t_f]\xi_0 + D_{x(t_f)}\psi[0, t_f]\xi_1 = 0. \tag{3.11}$$

The next theorem summarizes the SSC for a weak local minimum which are to be found in [39,50,61,68].

Theorem 1. (SSC for control problems with mixed control-state constraints.) *Let* (x_0, u_0) *be admissible for problem OC(*p_0*). Suppose that there exist multipliers* $(\lambda_0, \mu_0, \rho_0) \in W^{1,\infty}(0, t_f; \mathbb{R}^n) \times L^\infty(0, t_f; \mathbb{R}^k) \times \mathbb{R}^r$ *such that the necessary conditions (3.3)–(3.7) and assumptions (AC-1)–(AC-5) are satisfied. Then there exist* $c_0 > 0$ *and* $\alpha > 0$ *such that*

$$F(x,u) \geq F(x_0,u_0) + c_0\{\|x - x_0\|_{1,2}^2 + \|u - u_0\|_2^2\}$$

holds for all admissible (x,u) *with* $\|x - x_0\|_{1,\infty} + \|u - u_0\|_\infty \leq \alpha$*. In particular,* (x_0, u_0) *provides a strict weak local minimum for problem OC(*p_0*).*

The remarkable fact in the last theorem is that the estimate of the functional values $F(x,u)$ from below involves the L^2-norms in a L^∞-neighborhood of the nominal solution. This phenomenon is a consequence of the *two-norm discrepancy*; cf. Maurer [44], Malanowski [35–38].

Extensions of SSC to control problems with *free* final time or to multiprocess control systems may be found in [5,48]. The SSC in the previous theorem are usually not suitable for a *direct numerical verification* in practical control problems. In order to obtain *verifiable sufficient conditions* we shall strengthen the SSC in Theorem 1 in the following way. We observe that the definiteness condition (3.8) is valid if the matrix in (3.8) is positive definite on the *whole* space $\mathbb{R}^n \times \mathbb{R}^m$ and if conditions (3.10) and (3.11) are satisfied. Firstly, this leads to the requirement that the strict Legendre-Clebsch condition

$$H_{uu}[t] \geq c \cdot I_m \quad \forall\, t \in [0, t_f], \quad c > 0,$$

is valid on the *whole* interval $[0, t_f]$. Secondly, by evaluating the Schur complement of the matrix in (3.8) and using the continuous dependence of ODEs on system data, the estimate (3.8) follows from the following assumption: there exists a solution of the Riccati equation

$$\begin{aligned}\dot{Q} = &-Qf_x[t] - f_x[t]^*Q - H_{xx}[t] \\ &+ (H_{xu}[t] + Qf_u[t])H_{uu}[t]^{-1}(H_{xu}[t] + Qf_u[t])^*,\end{aligned} \tag{3.12}$$

which is *bounded* on $[0, t_f]$ and satisfies the boundary conditions (3.10) and (3.11); cf. [50], Theorem 5.2.

However, in some applications these conditions still may be too strong since the Riccati equation (3.12) may fail to have a *bounded* solution; cf. the Rayleigh problem in [45] which will be discussed in the next section. A weaker condition is obtained by introducing the following *modified* or *reduced* Riccati equation. Recall the definition of the vector $C^a = (C^i)_{i \in I_0(t)}$ of active components and let $\iota_0(t) = \#(I_0(t))$ denote the number of active components. Then the matrix $C_u^a[t]$ of partial derivatives has dimension $\iota_0(t) \times m$. For simplicity, the time argument will be omitted in the sequel. The *pseudo-inverse* of the matrix C_u^a is given by the $(m \times \iota_0)$-matrix

$$(C_u^a)^+ := (C_u^a(C_u^a)^*)^{-1}C_u^a.$$

Let $(C_u^a)^\perp$ denote the $(m \times (m - \iota_0))$-matrix whose column vectors form an orthogonal basis of $\mathrm{Ker}(C_u^a)$. Consider the matrices; cf. [23,24,50,68]:

$$D := -(C_u^a)^+ C_x^a, \quad A := f_x + f_u D, \quad P := (C_u^a)^\perp, \tag{3.13}$$
$$\mathcal{H}_{xx} := H_{xx} + H_{xu} D + D^* H_{ux} + D^* H_{uu} D, \quad \mathcal{H}_{xu} := H_{xu} + D^* H_{uu}, \tag{3.14}$$
$$(\mathcal{H}_{uu})^{(-1)} := P(P^* H_{uu} P)^{-1} P^*. \tag{3.15}$$

Note that the $m \times m$-matrix $(\mathcal{H}_{uu})^{(-1)}$ in (3.15) is well defined due to assumptions (AC-1) and (AC-4). Studying again the Schur complement of the matrix in (3.8), it follows that the estimate (3.8) holds if the matrix Riccati equation

$$\dot{Q} = -QA - A^* Q - \mathcal{H}_{xx} + (\mathcal{H}_{xu} + Q f_u)(\mathcal{H}_{uu})^{(-1)}(\mathcal{H}_{xu} + Q f_u)^* \tag{3.16}$$

has a *bounded* solution on $[0, t_f]$ which satisfies the boundary conditions (3.10) and (3.11).

In general it is rather tedious to elaborate this Riccati equation explicitly. To facilitate the numerical treatment in practical applications we discuss some special cases in more detail. On *interior arcs* with $C^i[t] < 0$, $i = 1, \ldots, k$, we have $\iota_0(t) = 0$ and thus the Riccati equation (3.16) reduces to the one introduced in (3.12). Consider now a *boundary arc* with $\iota_0(t) = m$ where we have as many control components as active constraints. Due to assumption (AC-1) the pseudo-inverse is given by $(C_u^a)^+ = (C_u^a)^{-1}$ and, hence, the matrix $P = (C_u^a)^\perp = 0$ in (3.13) vanishes. Then the matrices in (3.13)–(3.15) become

$$D = -(C_u^a)^{-1} C_x^a \quad A = f_x - f_u (C_u^a)^{-1} C_x^a, \quad (\mathcal{H}_{uu})^{(-1)} = 0, \tag{3.17}$$
$$\mathcal{H}_{xx} = H_{xx} - H_{xu}(C_u^a)^{-1} C_x^a + [(C_u^a)^{-1} C_x^a]^* [H_{uu}(C_u^a)^{-1} C_x^a - H_{ux}]. \tag{3.18}$$

It follows that the Riccati equation (3.16) reduces to the *linear* equation

$$\dot{Q} = -QA - A^* Q - \mathcal{H}_{xx}. \tag{3.19}$$

For *pure control* constraints with $C_x \equiv 0$, this equation further simplifies to the equation

$$\dot{Q} = -Q f_x - f_x^* Q - H_{xx}. \tag{3.20}$$

The Rayleigh problem treated in the next section will provide an illustrative application. Let us also evaluate the boundary conditions (3.10) and (3.11) in a special case of practical interest. Suppose that the boundary conditions are separated and that some components for the initial and final state are fixed according to

$$\begin{aligned} x_k(0) &= a_k, & k &\in K_0 \subset \{1, \ldots, n\}; \\ x_k(t_f) &= b_k, & k &\in K_f \subset \{1, \ldots, n\}. \end{aligned} \tag{3.21}$$

Denote the complements of the index sets by

$$K_0^c := \{1, \ldots, n\} \setminus K_0, \quad K_f^c := \{1, \ldots, n\} \setminus K_f.$$

Then it is easy to see that the boundary conditions (3.10), (3.11) are satisfied if the following submatrices are positive definite:

$$[Q(0)]_{(i,j)\in K_0^c\times K_0^c} > 0, \quad [-Q(t_f)]_{(i,j)\in K_f^c\times K_f^c} > 0. \tag{3.22}$$

By virtue of the continuous dependence of solutions to ODEs on systems data, one of these definiteness conditions can be relaxed. E.g., it suffices to require only positive semi-definiteness,

$$[Q(0)]_{(i,j)\in K_0^c\times K_0^c} \geq 0, \quad \text{resp.,} \quad [-Q(t_f)]_{(i,j)\in K_f^c\times K_f^c} \geq 0.$$

The case study in the next section will benefit from this relaxation.

3.2 Parametric Boundary Value Problem and Solution Differentiability

We formulate an appropriate parametric boundary value problem BVP(p) which characterizes optimal solutions to OC(p) for parameters p in a neighborhood of p_0. We begin with a parametric form of the minimum condition (3.5) and consider the following *parametric mathematical program* depending on the parameter $z = (x, \lambda, p) \in \mathbb{R}^n \times \mathbb{R}^n \times P$:

$$\text{MP}(z) \quad \text{minimize}_{u\in\mathbb{R}^m} H(x,u,\lambda,p) \text{ subject to } C(x,u,p) \leq 0.$$

By virtue of assumptions (AC-1) and (AC-4), the control $u_0(t)$ and the multiplier $\mu_0(t)$ are a solution, respectively a multiplier, for the problem MP(z) evaluated at $z = z_0(t) = (x_0(t), \lambda_0(t), p_0)$. Using in addition the strict complementarity assumption (AC-3), the following result has been shown in [39] using an appropriate implicit function theorem: for every $t \in [0, t_f]$ there exists a locally unique solution $u(z)$ of MP(z) and a unique associated Lagrange multiplier $\mu(z) \in \mathbb{R}^{\iota_0(t)}$, $\iota_0(t) = \#(I_0(t))$, which satisfy

$$u_0(t) = u(z_0(t)), \quad \mu_0(t) = \mu(z_0(t)).$$

The functions $u(x,\lambda,p)$ and $\mu(x,\lambda,p)$ are *Fréchet differentiable* functions of $z = (x,\lambda,p)$ at $z_0 = z_0(t)$ for $t \notin TC^0$, i.e., if t is *not* a junction point. Then, for any direction $d \in \mathbb{R}^n \times \mathbb{R}^n \times P$ and for $t \notin TC^0$ the differentials

$$D_z u(z_0(t))d \in \mathbb{R}^m \quad \text{and} \quad D_z\mu(z_0(t))d \in \mathbb{R}^{\iota_0(t)}$$

are given by the formulae

$$\begin{pmatrix} D_{(x,\lambda,p)}u(z_0(t))d \\ D_{(x,\lambda,p)}\mu(z_0(t))d \end{pmatrix} = K(t)^{-1} \begin{pmatrix} -H_{ux}[t]d \mid -f_u^*[t]d \mid -H_{up}[t]d \\ -C_x^a(t)d \mid \quad \mathbf{0} \mid -C_p^a[t]d \end{pmatrix}, \tag{3.23}$$

where the $(m + \iota_0(t)) \times (m + \iota_0(t))$-matrix

$$K_0[t] = \begin{pmatrix} H_{uu}[t] & C_u^a[t]^* \\ C_u^a[t] & 0 \end{pmatrix} \tag{3.24}$$

is *nonsingular* due to (AC-1) and (AC-4). The preceding formulae are derived again on the basis of an implicit function theorem. It is noteworthy to evaluate the last two formulae in the case of a *scalar* control and constraint, i.e., for $m = k = 1$. On an *interior arc* with non-active constraints we get

$$D_z u(z_0(t)) = -(H_{uu}[t])^{-1} H_{uz}[t] d, \quad \mu(t) = 0.$$

On a *boundary arc* we find a *boundary control* $u^b(x,p)$ which does *not* depend on the adjoint variable λ since it is the locally unique solution of the equation $C(x,u,p) = 0$. Then the minimum condition $H_u[t] = 0$ in (3.5) yields the multiplier explicitly,

$$\mu(x,\lambda,p) = -H_u^0(x,u^b(x,p),\lambda,p) \,/\, C_u(x,u^b(x,p),p).$$

From (3.23) and (3.24) we obtain the differentials

$$\begin{pmatrix} D_z u^b(z_0(t))d \\ D_z \mu(z_0(t))d \end{pmatrix} = \begin{pmatrix} -C_z[t]d \,/\, C_u[t] \\ H_{uu}[t]C_z[t]d/(C_u[t])^2 - H_{uz}[t]d \,/\, C_u[t] \end{pmatrix}.$$

Now we are able to specify a *parametric boundary value problem* for a solution to the perturbed problem OC(p). Recall the assumption (AG-3) where it was required that the set $TC^0 := \{t_1^0, t_2^0, \ldots, t_N^0\}$ of junction points for the nominal solution is finite with $0 := t_0^0 \le t_1^0 < t_2^0 < \cdots < t_N^0 \le t_{N+1}^0 := t_f$. Moreover, every junction t_i^0 is the junction for exactly one constraint $C^{j(i)}[t] \le 0$ with index $j(i)$. For simplicity we shall assume that $0 < t_1^0$ and $t_N^0 < t_f$ hold, i.e., the constraints are not active at the initial and final time. Hence, by construction we have

$$I_0(t) = \text{const.} \quad \text{for} \quad t_i^0 < t < t_{i+1}^0, \quad i = 0, 1, \ldots, N.$$

In the interval (t_i^0, t_{i+1}^0) the solution $u(z)$ and the multiplier $\mu(z)$ to the program MP(z) agree with the solution $u^{[i]}(z)$ and the multiplier $\mu^{[i]}(z)$ of the following parametric program with *equality* constraints:

$$MP^{[i]}(x,\lambda,p) \quad \text{minimize}_{u\in\mathbb{R}^m} \{H(x,u,\lambda,p) \mid C^j(x,u,p) = 0, \; j \in I_0(t)\}.$$

The notation $[i]$ in brackets is used to distinguish these functions from the i-th component. The conditions underlying sensitivity analysis have the consequence that the above structure of the optimal solution is preserved for parameters p in a neighborhood of p_0. Let t_i be a point in a neighborhood of t_i^0, $i = 1, \ldots, N$, such that $0 = t_0 < t_1 < \cdots < t_N < t_{N+1} = t_f$ holds. Then the reference boundary value problem (2.2), (2.3) and (3.3), (3.4) can be embedded into the following *parametric multipoint boundary value problem* BVP(p):
Differential equations for $t_i < t < t_{i+1}$, $i = 0, \ldots, N$:

$$\dot{x} = f(x, u^{[i]}(x,\lambda,p), p), \tag{3.25}$$

$$\dot{\lambda} = -H_x(x, u^{[i]}(x,\lambda,p), \lambda, \mu^{[i]}(x,\lambda,p), p)^*. \tag{3.26}$$

Boundary conditions and junction conditions:

$$\psi(x(0), x(t_f), p) = 0, \tag{3.27}$$

$$(-\lambda(0), \lambda(t_f)) = D_{(x(0),x(t_f))}(g + \rho^*\psi)^*(x(0), x(t_f), p), \tag{3.28}$$

$$\tilde{C}^i(x(t_i), \lambda(t_i), p) = 0, \quad i = 1, \ldots, N. \tag{3.29}$$

The functions $\tilde{C}^i$ in the last equations are defined in the following way: if t_i is an *entry-point* for the constraint $C^{j(i)}$ then

$$\tilde{C}^i(x, \lambda, p) = C^{j(i)}(x, u^{[i-1]}(x, \lambda, p), p), \tag{3.30}$$

and if t_i is an *exit-point* for $C^{j(i)}$ then

$$\tilde{C}^i(x, \lambda, p) = C^{j(i)}(x, u^{[i+1]}(x, \lambda, p), p). \tag{3.31}$$

An efficient method for solving BVP(p) is the *shooting method* developed in [10,65] which has been implemented in the code BNDSCO of [54]. The *shooting procedure* treats the initial value $x(0), \lambda(0)$, the multiplier ρ and the junction points t_i, $i = 1, \ldots, N$, as an *unknown vector variable*

$$s = (s_x, s_\lambda, \rho, t_1, \ldots, t_N) \in \mathbb{R}^n \times \mathbb{R}^n \times \mathbb{R}^r \times \mathbb{R}^N.$$

Let $x(t, s, p)$ and $\lambda(t, s, p)$ denote the solution of the ODEs (3.25) and (3.26) with initial conditions

$$x(0, s, p) = s_x, \quad \lambda(0, s, p) = s_\lambda.$$

Then the BVP(p) has a solution for p in a neighborhood of p_0 if the $2n + r + N$ nonlinear equations

$$F(s,p) := \begin{pmatrix} \psi(s_x, x(T, s, p), p) \\ s_\lambda + D_{x(0)}(g + \rho^*\psi)^*(s_x, x(T, s, p), p) \\ \lambda(T, s, p) - D_{x(t_f)}(g + \rho^*\psi)^*(s_x, x(T, s, p), p) \\ \left(\tilde{C}^i(x(t_i, s, p), \lambda(t_i, s, , p), p)\right)_{i=1,\ldots,N} \end{pmatrix} = 0 \tag{3.32}$$

can be solved for the shooting parameter s as a function of p.

The unperturbed solution (x_0, λ_0, ρ_0) corresponds to the shooting variable $s_0 = (x_0(0), \lambda_0(0), \rho_0, t_1^0, \ldots, t_N^0)$ for which $F(s_0, p_0) = 0$ holds by definition. The classical implicit function theorem is applicable to the parametric equation (3.32) if the Jacobian of $\partial F(s_0, p_0)/\partial s$ with respect to s is *regular*. One easily finds [39] that a *necessary condition* for the non-singularity of the Jacobian is the following condition on *non-tangential junctions* with the boundary:

(AC-6) (*Non-tangential junction conditions*) $\frac{d}{dt}\tilde{C}^i[t_i^0] \neq 0, \ i = 1, \ldots, N.$

The non-tangential condition and the regularity of the Jacobian usually can only be checked *numerically*. A convenient regularity test is provided, e.g., by the multiple shooting code BNDSCO in [54]. Nevertheless, Theorem 3 provide a *theoretical* characterization for the regularity of the Jacobian from which the property of solution differentiability ensues.

The following auxiliary result is useful for the numerical analysis; cf. [39], Lemma 4.

Lemma 2. *If assumptions (AC-1)–(AC-6) hold then* $\dot{u}_0$ *is discontinuous at* t_i^0, *i.e., it holds*

$$\dot{u}_0((t_i^0)+) \neq \dot{u}_0((t_i^0)-), \tag{3.33}$$

and the multiplier satisfies the relations

$$\mu_0^{j(i)}(t_i^0) = 0, \tag{3.34}$$
$$\dot{\mu}_0^{j(i)}((t_i^0)+) > 0, \quad \text{if } t_i^0 \text{ is an entry-point for } C^{j(i)}, \tag{3.35}$$
$$\dot{\mu}_0^{j(i)}((t_i^0)-) < 0, \quad \text{if } t_i^0 \text{ is an exit-point for } C^{j(i)}. \tag{3.36}$$

Now we are in the position to state the main sensitivity result for control problems with mixed control-state constraints; cf. [39]:

Theorem 3. (Solution differentiability for mixed control-state constraints)
Let (x_0, u_0) *be admissible for OC*(p_0) *such that assumptions (AC-1)–(AC-6) hold. Then there exists a neighborhood* $P_0 \subset P$ *of* p_0 *and a* C^1*-function* $s : P_0 \to \mathbb{R}^{2n+r+N}$ *such that the shooting equations* $F(s(p), p) = 0$ *in (3.32) are satisfied for all* $p \in P_0$. *The functions*

$$x(t,p) := x(t, s(p), p), \quad \lambda(t,p) := \lambda(t, s(p), p)$$

are C^1*-functions for all* $(t,p) \in [0, t_f] \times P_0$. *Together with the* C^1*-functions of multipliers* $\rho : P_0 \to \mathbb{R}^r$ *and junction points* $t_i : P_0 \to \mathbb{R}$, $i = 1, \ldots, N$, *they solve the parametric problem BVP*(p). *The continuous functions*

$$u : [0, t_f] \times P_0 \to \mathbb{R}^m, \quad \mu : [0, t_f] \times P_0 \to \mathbb{R}^k,$$

defined piecewisely by

$$u(t,p) := u^{[i]}(x(t,p), \lambda(t,p), p), \quad \mu(t,p) := \mu^{[i]}(x(t,p), \lambda(t,p), p)$$
$$\text{for all} \quad t_i(p)+ \leq t \leq t_{i+1}(p)- \quad (i = 0, 1, \ldots, N),$$

are of class C^1 *for* $(t,p) \neq (t_i(p), p)$, $i = 1, \ldots, N$. *For every* $p \in P_0$, *the triple* $x(\cdot, p)$, $\lambda(\cdot, p)$, $u(\cdot, p)$ *and the multipliers* $\mu(\cdot, p)$, $\rho(p)$ *satisfy the second-order sufficient conditions in Theorem 1 evaluated for the parameter* p *and, hence, the pair* $(x(\cdot, p), u(\cdot, p))$ *provides a local minimum for OC*(p).

This theorem provides a justification for the first order Taylor expansion (2.6) of the perturbed solution and thus will serve as a theoretical basis for the real-time control approximations of perturbed solutions.

3.3 Computational Sensitivity Analysis and Real-Time Control

Our aim is to perform a computational sensitivity analysis for the perturbed solutions to BVP(p) for a parameter $p = p_0 + d \in P$ and a direction d. Theorem 3 leads to the following first order Taylor expansions:

$$x(t,p) = x_0(t) + \frac{\partial x}{\partial p}(t,p_0)d + o(|d|),\ \lambda(t,p) = \lambda_0(t) + \frac{\partial \lambda}{\partial p}(t,p_0)d + o(|d|),$$

$$\rho(p) = \rho_0 + \frac{\partial \rho}{\partial p}(p_0)d + o(|d|),\ t_i(p) = t_i^0 + \frac{dt_i}{dp}(p_0)d + o(|d|),\ i = 1,\dots,N,$$

where $o(|d|)/|d| \to 0$ for $|d| \to 0$ uniformly with respect to $t \in [0,t_f]$. The differentials

$$y_d := \frac{\partial x}{\partial p}(p_0)d,\ \gamma_d := \frac{\partial \lambda}{\partial p}(p_0)d : [0,t_f] \to \mathbb{R}^n, \quad \rho_d := \frac{d\rho}{dp}(p_0)d \in \mathbb{R}^r$$

are C^1-functions which are found as solutions to a *linear* boundary value problem. Namely, Theorem 3 allows us to differentiate the boundary value problem BVP(p) in (3.25)–(3.29) formally with respect to p. This procedure yields the following linear inhomogeneous BVP:

$$\dot{y}_d = \mathcal{A}^0(t)y_d + \mathcal{B}^0(t)\gamma_d + \mathcal{R}_y^0(t)\ , \tag{3.37}$$

$$\dot{\gamma}_d = \mathcal{W}^0(t)y_d - \mathcal{A}^0(t)^*\gamma_d + \mathcal{R}_\gamma^0(t)\ , \tag{3.38}$$

$$0 = D_{x(0)}\psi[0,t_f]\,y_d(0) + D_{x(t_f)}\psi[0,t_f]\,y_d(t_f) + D_p\psi[0,t_f]d, \tag{3.39}$$

$$\begin{aligned}\gamma_d(0) = &-D^2_{x(0)x(0)}(g+\rho^*\psi)[0,t_f]\,y_d(0)\\ &-D^2_{x(0)x(t_f)}(g+\rho^*\psi)[0,t_f]\,y_d(t_f)\\ &-D^2_{x(0)p}(g+\rho_0^*\psi)[0,t_f]d - D_{x(0)}\,\psi[0,t_f]^*\rho_d,\end{aligned} \tag{3.40}$$

$$\begin{aligned}\gamma_d(t_f) = &\ D^2_{x(t_f)x(0)}(g+\rho^*\psi)[0,t_f]\,y_d(0)\\ &+D^2_{x(t_f)x(t_f)}(g+\rho^*\psi)[0,t_f]\,y_d(t_f)\\ &+D^2_{x(t_f)p}(g+\rho_0^*\psi)[0,t_f]d + D_{x(t_f)}\,\psi[0,t_f]^*\rho_d.\end{aligned} \tag{3.41}$$

The $n \times n$-matrices $\mathcal{A}^0(t), \mathcal{B}^0(t)$ and $\mathcal{W}^0(t)$ are given by

$$\mathcal{A}^0(t) = f_x[t] - (f_u[t],\ \mathbf{0})K_0[t]^{-1}\begin{pmatrix} H_{ux}[t] \\ C_x^a[t] \end{pmatrix},$$

$$\mathcal{B}^0(t) = -(f_u[t],\ \mathbf{0})K_0[t]^{-1}\begin{pmatrix} f_u^*[t] \\ \mathbf{0}^* \end{pmatrix},$$

$$\mathcal{W}^0(t) = -H_{xx}[t] + (H_{xu}[t], C_x^a[t]^*)K_0[t]^{-1}\begin{pmatrix} H_{ux}[t] \\ C_x^a[t] \end{pmatrix}.$$

The computation of the inhomogeneous terms $\mathcal{R}_z^0(t)$ and $\mathcal{R}_\gamma^0$ makes use of (3.23) and yields

$$\mathcal{R}_y^0(t) = -(f_u[t],\mathbf{0})K_0[t]^{-1}\begin{pmatrix} H_{up}[t]d \\ C_p^a[t]d \end{pmatrix} + f_p[t]d,$$

$$\mathcal{R}_\gamma^0(t) = (H_{ux}[t], C_x^a[t]^*)K_0[t]^{-1}\begin{pmatrix} H_{up}[t]d \\ C_p^a[t]d \end{pmatrix} - H_{xp}[t]d.$$

Formulae for the differentials $\frac{dt_i}{dp}(p_0)d$, $i = 1, \dots, N$, can be derived from the fact that the junction condition (3.29) holds identically in p,

$$\tilde{C}^i(x(t_i(p),p), \lambda(t_i(p),p), p) \equiv 0 \quad \text{for} \quad p \in P_0, \; i = 1, \dots, N.$$

The differentiation yields the relation

$$\frac{d}{dt}\tilde{C}^i[t_i^0]\,\frac{dt_i}{dp}(p_0)d + \tilde{C}^i_x[t_i^0]\,y_d(t_i^0) + \tilde{C}^i_\lambda[t_i^0]\,\gamma_d(t_i^0) + \tilde{C}^i_p[t_i^0]d = 0,$$

which can be solved for $\frac{dt_i}{dp}(p_0)d$ due to assumption (AC-6) (compare [59], formula (65)):

$$\frac{dt_i}{dp}(p_0)d = -\left(\tilde{C}^i_x[t_i^0]\,y_d(t_i^0) + \tilde{C}^i_\lambda[t_i^0]\,\gamma_d(t_i^0) + \tilde{C}_p[t_i^0]d\right) \big/ \frac{d}{dt}\tilde{C}^i[t_i^0]. \quad (3.42)$$

4 NUMERICAL CASE STUDY: THE RAYLEIGH PROBLEM WITH CONTROL CONSTRAINTS

Consider the electric circuit (tunnel-diode oscillator) shown in Figure 1 where L denotes inductivity, C capacity, R resistance, I electric current and where D is the diode. The state variable $x(t)$ represents the electric current I at time t. The voltage $v_0(t)$ at the generator is regarded as a control function. After a suitable transformation of $v_0(t)$ we arrive at the following specific Rayleigh equation with a scalar control $u(t)$, cf. [30], [66],

$$\ddot{x}(t) = -x(t) + \dot{x}(t)\,(\,1.4 - p\,\dot{x}(t)^2\,) + 4\,u(t).$$

The scalar $p > 0$ in this equation is considered as a perturbation parameter for which we choose the *nominal* value $p_0 = 0.14$. A numerical analysis reveals that the Rayleigh equation with nominal parameter $p_0 = 0.14$ and zero control $u(t) \equiv 0$ has a *limit cycle* in the $(x, \dot{x})$ -plane. Define the state variables as $x_1 = x$

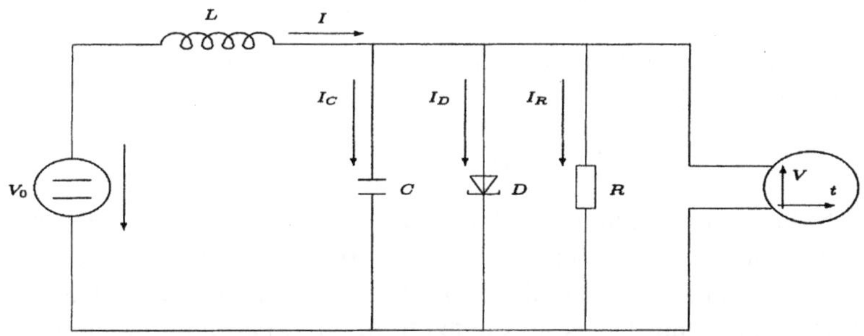

Figure 1. Tunnel-diode oscillator

and $x_2 = \dot{x}$. In [2, 45] we have performed a sensitivity analysis of the following parametric control problem with fixed final time $t_f > 0$ and a perturbation $p > 0$: minimize the functional

$$F(x,u,p) = \int_0^{t_f} (u(t)^2 + x_1(t)^2)\,dt \tag{4.1}$$

subject to

$$\dot{x}_1(t) = x_2(t),\; \dot{x}_2(t) = -x_1(t) + x_2(t)\,(1.4 - p\,x_2(t)^2) + 4u(t), \tag{4.2}$$

$$x_1(0) = x_2(0) = -5, \quad x_1(t_f) = x_2(t_f) = 0, \tag{4.3}$$

$$|u(t)| \le 1 \quad \text{for } t \in [0, t_f]. \tag{4.4}$$

The control constraint (4.4) can be written in the form (2.4) with two control constraints

$$C^1(x,u,p) := u - 1 \le 0, \quad C^2(x,u,p) := -u - 1 \le 0. \tag{4.5}$$

We choose the final time $t_f = 4.5$ instead of the final time $t_f = 2.5$ considered in [30], [66]. The larger final time $t_f = 4.5$ produces a richer structure of the optimal control subject to the constraint (4.5). We also note that the final conditions $x_1(t_f) = x_2(t_f) = 0$ are not incorporated in [30], [66].

4.1 *Boundary Value Problem for the Optimal Control*

First, we study the *unconstrained* control problem (4.1)–(4.3) and omit the control constraint (4.4). The Hamiltonian (3.1) becomes

$$H^0(x_1,x_2,u,\lambda_1,\lambda_2,p) = u^2 + x_1^2 + \lambda_1 x_2 + \lambda_2\,(-x_1 + x_2\,(1.4 - p x_2^2) + 4u), \tag{4.6}$$

where λ_1, λ_2 are the adjoint variables associated with x_1, x_2. The optimal control is determined via the minimum condition (3.5):

$$H_u = 2u + 4\lambda_2 = 0, \quad \text{i.e.} \quad u(t) = -2\lambda_2(t). \tag{4.7}$$

The strict Legendre-Clebsch condition (AC-4) holds in view of $H^0_{uu}(t) \equiv 2 > 0$. The adjoint equations (3.3) are

$$\dot{\lambda}_1 = \lambda_2 - 2x_1, \quad \dot{\lambda}_2 = 3p\,\lambda_2\,x_2^2 - 1.4\,\lambda_2 - \lambda_1. \tag{4.8}$$

Note that the boundary conditions (3.4) or (3.28) for the adjoint variables are redundant since the final state is specified. The *unconstrained solution* is then obtained by solving the boundary value problem (BVP) consisting of equations (4.2), (4.3) and (4.8) where the control u is substituted from (4.7). The multiple shooting code BNDSCO in [54] yields the following initial and final values for the adjoint variables when $p = p_0 = 0.14$ and $t_f = 4.5$:

$$\begin{aligned} \lambda_1(0) &= -9.00247067, & \lambda_2(0) &= -2.67303084, \\ \lambda_1(t_f) &= -0.04456054, & \lambda_2(t_f) &= -0.00010636. \end{aligned} \tag{4.9}$$

The nominal cost is

$$F(x_0, u_0, p_0) = 29.75107515.$$

The corresponding unconstrained control is depicted in the upper part of Figure 2.

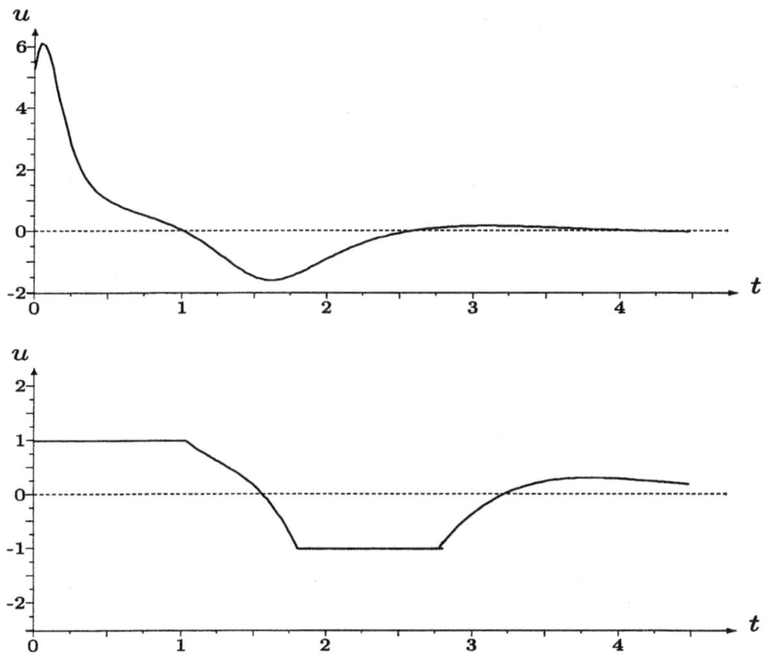

Figure 2. Unconstrained optimal control and constrained optimal control $|u(t)| \leq 1$

The unconstrained control in Figure 2 obviously violates the control constraint (4.4). Hence, when computing the *constrained* solution, we have to take into account boundary arcs for the constraints and consider the augmented Hamiltonian (3.2)

$$H(x, u, \lambda, \mu, p) = H^0(x, u, \lambda, p) + \mu_1(u - 1) + \mu_2(-u - 1), \qquad (4.10)$$

where μ_1, μ_2 are the multipliers associated with the inequality constraints (4.5). The multipliers satisfy $\mu_i(t) \geq 0$ for $0 \leq t \leq t_f$, $i = 1, 2$, and $\mu_1(t) = 0$ if $u(t) < 1$, resp., $\mu_2(t) = 0$ if $-1 < u(t)$. The shape of the unconstrained control suggests that the constraint (4.5) is active with $u(t) \equiv 1$ for $0 \leq t \leq t_1$ and $u(t) \equiv -1$ for $t_2 \leq t \leq t_3$ where the *junction points* t_i satisfy $0 < t_1 < t_2 < t_3 < t_f$. Hence we assume the following control structure; cf. the lower graph in

Figure 2:

$$u(t)=\left\{\begin{array}{ll}1 & , \quad 0\le t\le t_1\\ -2\lambda_2(t) & , \quad t_1\le t\le t_2\\ -1 & , \quad t_2\le t\le t_3\\ -2\lambda_2(t) & , \quad t_3\le t\le 4.5\end{array}\right\}. \tag{4.11}$$

This assumed structure leads us to formulate the parametric BVP(p) (3.25)–(3.29) with $N = 3$. Moreover, we see that the solutions and multipliers of the parametric programming problem $MP^{[i]}(x,\lambda,p)$ in Section 3.2 are given by

$$\begin{array}{lll} u^{[0]}(x,\lambda,p)=1, & \mu^{[0]}(x,\lambda,p)=(-2-4\lambda_2,0), & \\ u^{[2]}(x,\lambda,p)=-1, & \mu^{[2]}(x,\lambda,p)=(0,-2+4\lambda_2), & \\ u^{[i]}(x,\lambda,p)=-2\lambda_2, & \mu^{[i]}(x,\lambda,p)=(0,0), & \text{for } i=1 \text{ and } i=3. \end{array}$$

The unconstrained Hamiltonian (4.6) is *regular* in the sense that it admits a *unique* minimum with respect to u. Furthermore, the augmented Hamiltonian (4.10) is continuous along the optimal solution. From these properties it follows that the control is *continuous* at the junction points t_i, $i = 1,2,3$. This fact yields the following three *junction conditions* in view of (3.29) and (3.30), respectively (4.11):

$$-2\lambda_2(t_1)=1,\quad -2\lambda_2(t_2)=-1,\quad -2\lambda_2(t_3)=-1. \tag{4.12}$$

The adjoint equations $\dot{\lambda} = -H_x$ agree with those in (4.8) for the unconstrained solution. In summary, the constrained case requires the solving of a BVP which consists (1) of the differential equations (4.2) and (4.8) using the control law (4.11), (2) the boundary conditions (4.3) and (3) three junction conditions (4.12) for three unknowns t_1, t_2, t_3. We use the multiple shooting code BNDSCO in [54] to find the following solution:

$$\begin{array}{rlrl} \lambda_1(0) = & -12.70997710, & \lambda_2(0) = & -4.59631683,\\ \lambda_1(4.5) = & 0.05196024, & \lambda_2(4.5) = & -0.13990408,\\ t_1 = & 1.04198283, & t_2 = & 1.70663109,\\ t_3 = & 2.88494411, & F(x_0,u_0,p_0) = & 44.72093908. \end{array} \tag{4.13}$$

The *constrained control* is shown in the lower graph in Figure 2 whereas the state and adjoint variables are displayed in Figure 3. The constrained solution satisfies the strict complementarity condition (AC-3). Namely, inserting the values for the adjoint variable $\lambda_2(t)$, it follows that strict complementarity holds with

$$\begin{array}{ll} \mu_1(t)=-2-4\lambda_2(t)>0 & \text{for } 0\le t<t_1,\\ \mu_2(t)=-2+4\lambda_2(t)>0 & \text{for } t_2<t<t_3. \end{array} \tag{4.14}$$

In addition, the numerical results show that the *non-tangential conditions* in assumption (AC-6) are satisfied,

$$\dot{u}(t_i)=-2\dot{\lambda}_2(t_i)\neq 0,\quad i=1,2,3, \tag{4.15}$$

where the derivative is taken from the right for $i = 1$ and $i = 3$ and from the left for $i = 2$. Moreover, it is easily verified that the multiplier relations (3.34)–(3.36) hold at the junction points.

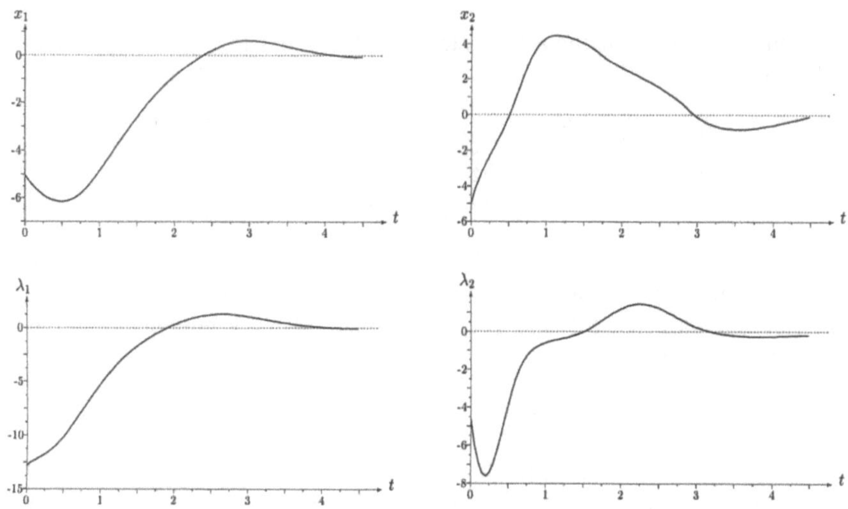

Figure 3. State variables x_1, x_2 and adjoint variables λ_1, λ_2 for $|u(t)| \leq 1$

4.2 Second Order Sufficient Conditions

The numerical evaluation of SSC requires a bounded solution of the Riccati equations (3.12) and (3.20) in view of the relation $C_x \equiv 0$. Since the Rayleigh problem has dimension $n = 2$, we consider a symmetric 2×2 matrix $Q(t)$ of the form

$$Q(t) = \begin{pmatrix} q_1(t) & q_2(t) \\ q_2(t) & q_4(t) \end{pmatrix}.$$

First, an optimality test is performed for the *unconstrained control* which corresponds to the initial conditions (4.9). The Riccati equations (3.12) are explicitly

$$\begin{aligned} \dot{q}_1 &= 2q_2 + 8q_2^2 - 2, \\ \dot{q}_2 &= -q_1 + (3p_0 x_2^2 - 1.4) q_2 + q_4 + 8 q_2 q_4, \\ \dot{q}_4 &= -2 q_2 + 2(3p_0 x_2^2 - 1.4) q_4 + 8 q_4^2 + 6p_0 \lambda_2 x_2. \end{aligned} \tag{4.16}$$

Since the final state is fixed, the boundary condition (3.22) for $Q(t_f)$ is automatically satisfied. Searching for a bounded solution that satisfies the boundary condition $Q(t_f) = 0$, we find the following initial conditions:

$$q_1(0) = 2.00632377, \quad q_2(0) = 0.47051349, \quad q_4(0) = -0.35166544.$$

Thus the unconstrained solution shown in Figure 2 is a local minimum.

Next we proceed to the *constrained control* characterized by (4.11) and (4.13). It is worth-while noting that we could not find a *bounded* solution of the Riccati equations (4.16) on the *whole* interval $[0, t_f]$. Fortunately, we may resort to the modified

Riccati equations (3.20) on the active intervals $[0, t_1]$ and $[t_2, t_3]$ which reduce to the following linear ODEs:

$$\begin{aligned} \dot{q}_1 &= 2q_2 - 2, \\ \dot{q}_2 &= -q_1 + (3p_0 x_2^2 - 1.4)\, q_2 + q_4, \\ \dot{q}_4 &= -2q_2 + 2(3p_0 x_2^2 - 1.4)\, q_4 + 6p_0 \lambda_2 x_2. \end{aligned} \tag{4.17}$$

Prescribing the boundary condition $Q(t_f) = 0$, we are able to show that these equations on the boundary intervals combined with the Riccati equations (4.16) on the interior intervals $[t_1, t_2]$ and $[t_3, t_f]$ possess indeed a bounded solution on $[0, t_f]$ with initial values

$$q_1(0) = 2.39858190, \quad q_2(0) = 0.89023233, \quad q_4(0) = -1.26620913.$$

The solutions are bounded by $|q_i(t)| \le 2.5$ for $t \in [0, t_f]$; cf. Figure 4 depicting the function $q_4(t)$. Thus we arrive at the conclusion that the constrained control shown in the lower graph of Figure 2 is a local minimum.

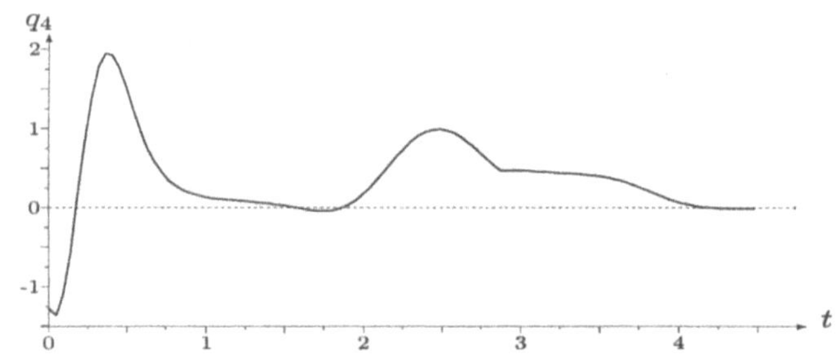

Figure 4. Solution $q_4(t)$ of the Riccati equations (4.16) and (4.17)

4.3 *Sensitivity Analysis and Real-Time Control*

In the two preceding sections, we have verified numerically that the *unperturbed (nominal)* solution for $p_0 = 0.14$ fulfills all assumptions for *solution differentiability* listed in Theorem 3. Thus the unperturbed solution can be embedded into a C^1-family of perturbed solutions

$$x_i(t, p), \quad \lambda_i(t, p) \ (i = 1, 2), \quad t_i(p) \ (i = 1, 2, 3),$$

to the perturbed problem OC(p). Our aim now is to compute the *sensitivity differentials*

$$y_i(t) = \frac{\partial x_i}{\partial p}(t, p_0), \ \ \gamma_i(t) = \frac{\partial \lambda_i}{\partial p}(t, p_0) \ (i = 1, 2), \ \ \frac{dt_i}{dp}(p_0) \ (i = 1, 2, 3),$$

where formally the direction $d = 1 \in \mathbb{R}$ is taken in section 3.3. The variational differential equations for $y_i(t)$ and $\gamma_i(t)$ are obtained by formally differentiating the BVP(p) (4.2), (4.3) and (4.8) with respect to the perturbation p. In this way we get the following ODE corresponding to (3.37) and (3.38):
Unconstrained arcs $[t_1, t_2]$ and $[t_3, t_f]$:

$$\begin{aligned} \dot{y}_1 &= y_2, & \dot{y}_2 &= -y_1 + 1.4y_2 - 3p_0x_2^2y_2 - x_2^3 - 8\gamma_2, \\ \dot{\gamma}_1 &= \gamma_2 - 2y_1, & \dot{\gamma}_2 &= 3x_2^2\lambda_2 + 3p_0\,(2x_2y_2\lambda_2 + x_2^2\gamma_2) - 1.4\gamma_2 - \gamma_1. \end{aligned} \tag{4.18}$$

Constrained arcs $[0, t_1]$ and $[t_2, t_3]$:

$$\begin{aligned} \dot{y}_1 &= y_2, & \dot{y}_2 &= -y_1 + 1.4y_2 - 3p_0x_2^2y_2 - x_2^3, \\ \dot{\gamma}_1 &= \gamma_2 - 2y_1, & \dot{\gamma}_2 &= 3x_2^2\lambda_2 + 3p_0(2x_2y_2\lambda_2 + x_2^2\gamma_2) - 1.4\gamma_2 - \gamma_1. \end{aligned} \tag{4.19}$$

Since the initial and final state is fixed, we get the boundary conditions from (3.39),

$$y_i(0) = y_i(t_f) = 0 \quad \text{for} \quad i = 1, 2. \tag{4.20}$$

Note that the boundary conditions (3.40) and (3.41) for γ_1, γ_2 are redundant because the initial and final state is fixed. The state and adjoint variables x_i and λ_i in these equations represent the nominal solution characterized by the data (4.13). The multiple shooting code BNDSCO in [54] provides the initial and final values

$$\begin{aligned} \gamma_1(0) &= -7.73953349, & \gamma_2(0) &= 27.59301812. \\ \gamma_1(t_f) &= -0.68941701, & \gamma_2(t_f) &= 1.98521500. \end{aligned}$$

Figure 5 displays the sensitivity differentials for state and adjoint variables. Formulae for the sensitivity differentials $\frac{dt_i}{dp}(p_0)$ of the junction points are obtained upon differentiating the junction conditions (4.12) and applying the general formulae (3.42). Hence, by differentiating the identities $-2\lambda_2(t_i(p), p) \equiv \pm 1$ we get

$$\dot{\lambda}_2(t_i(p), p) \cdot \frac{dt_i}{dp}(p) + \gamma_2(t_i(p), p) = 0,$$

which yields

$$\frac{dt_1}{dp}(p_0) = -0.38936098, \quad \frac{dt_2}{dp}(p_0) = 4.81518854, \quad \frac{dt_3}{dp}(p_0) = -4.07372212.$$

The *off-line* computation of the sensitivity differentials $y_i(t)$, $i = 1, 2$, leads to a *real-time* computation of the perturbed solution $x_i(t, p)$, $i = 1, 2$, as outlined in (2.6). The first order Taylor expansion

$$x_i(t, p) \approx x_i(t, p_0) + y_i(t)(p - p_0), \quad y_i(t) = \frac{\partial x_i}{\partial p}(t, p_0), \tag{4.21}$$

is used as an easily computable *on-line* approximation of the perturbed solution for parameters p in a suitable neighborhood of the nominal parameter p_0. In

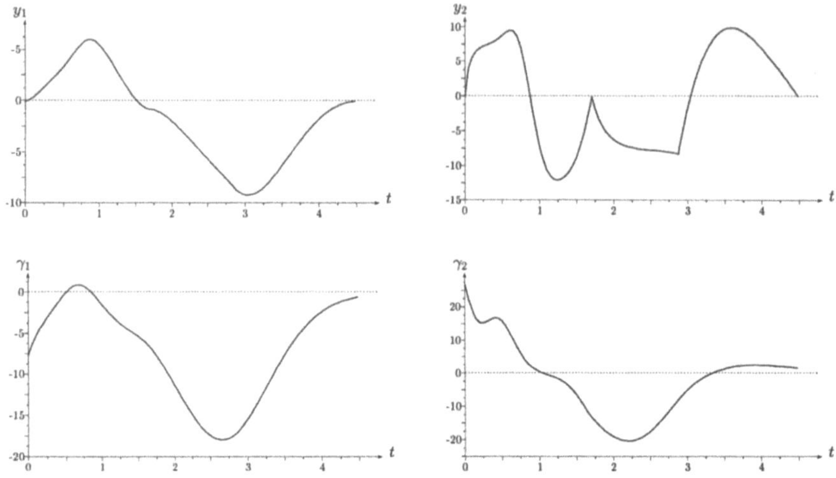

Figure 5. Sensitivity differentials of state and adjoint variables

the same way, Taylor expansions (4.21) exist for the perturbed adjoint variable $\lambda(t,p)$ and perturbed control $u(t,p)$. As noted at the end of section 2, these expansions need a numerical modification in a neighborhood of the junction points t_i; cf. the discussion in [59, 60]. Taking into account these modifications, Figure 6 demonstrates the quality of the control approximation for the perturbed parameter $p = 0.16$. The error in the difference of the perturbed and the unperturbed solution is given by $|x(t,p) - x(t,p_0)| \leq 0.2$. This error can be substantially reduced by the real-time approximation (4.21) which leads to the estimate $|x(t,p) - (x(t,p_0) + y(t)(p - p_0))| \leq 0.017$.

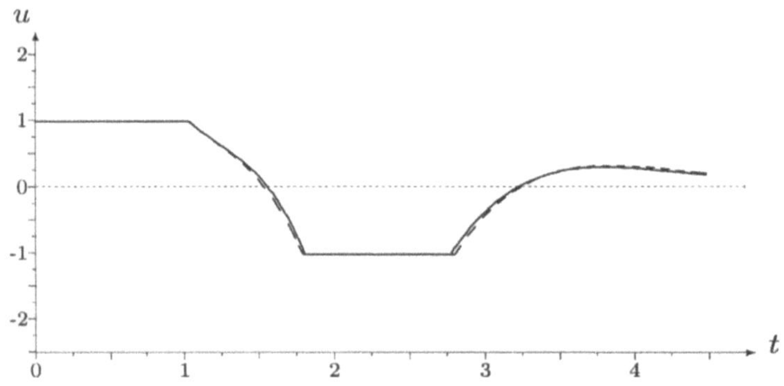

Figure 6. Approximate constrained control (solid line) and exact constrained control (dashed line) for perturbed parameter $p = 0.16$

4.4 The Rayleigh Problem with a Mixed Control-State Constraint

We consider again the Rayleigh problem (4.1)–(4.2) with given initial conditions

$$x_1(0) = -5, \quad x_2(0) = -5 \tag{4.22}$$

but with free final state $x(t_f)$. We replace the *pure* control constraint (4.4) by the following mixed control-state constraint, cf. [30, 66]:

$$C(x(t), u(t)) := u(t) + \frac{x_1(t)}{6} \leq 0 \quad \forall\, t \in [0, t_f]. \tag{4.23}$$

Again, the final time is $t_f = 4.5$. The augmented Hamiltonian (3.2) becomes

$$H(x, u, \lambda, \mu, p) = H^0(x, u, \lambda, p) + \mu(u + x_1/6), \quad \mu \in \mathbb{R}, \tag{4.24}$$

where H^0 is the unconstrained Hamiltonian (4.6). The adjoint equations (3.3) are

$$\dot{\lambda}_1 = \lambda_2 - 2x_1 - \mu/6, \quad \dot{\lambda}_2 = 3p\,\lambda_2\, x_2^2 - 1.4\,\lambda_2 - \lambda_1. \tag{4.25}$$

It is easy to see that the optimal solution has one boundary arc $[t_1, t_2]$, $0 < t_1 < t_2 < t_f$, with respect to the constraint (4.24). Then the optimal control has the structure

$$u(t) = \left\{ \begin{array}{ll} -2\lambda_2(t) \;, & t \notin [t_1, t_2] \\ -x_1(t)/6 \;, & t \in [t_1, t_2] \end{array} \right\}. \tag{4.26}$$

This structure leads us to a parametric BVP(p) (3.25)–(3.29) with $N = 2$. The multiplier μ is determined on the boundary by the condition $H_u = 0$ which gives

$$\mu = -2u - 4\lambda_2 = \frac{1}{3}x_1 - 4\lambda_2. \tag{4.27}$$

The junction conditions (3.29)–(3.31) are

$$u(t_i) + x_1(t_i)/6 = -2\lambda_2(t_i) + x_1(t_i)/6 = 0, \quad i = 1, 2. \tag{4.28}$$

The code BNDSCO from [54] yields the following numerical results for the nominal parameter $p_0 = 0.14$:

$$\begin{aligned} \lambda_1(0) &= -10.67759388, & \lambda_2(0) &= -4.19056427, \\ t_1 &= 1.27006745, & t_2 &= 2.97712730, \\ x_1(4.5) &= -0.39470571, & x_2(4.5) &= -1.18802201, \\ F(x_0, u_0, p_0) &= 44.80479861. \end{aligned}$$

The optimal control and the state variable x_1 are shown in Figure 7. We leave it as a numerical exercise to the reader to verify that this solution satisfies all conditions for solution differentiability in Theorem 3. In the same way as in section 4.3 one could compute the sensitivity differentials by formally differentiating the BVP (4.1), (4.2), (4.22) and (4.25)–(4.28). Also this procedure is left as an exercise to the reader.

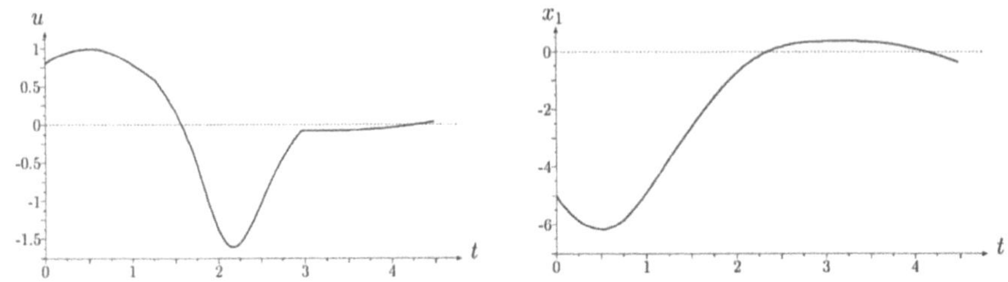

Figure 7. Optimal control u and state variable x_1 for the mixed constraint (4.23).

5 Sensitivity Analysis for Control Problems with Pure State Constraints

5.1 Constraint Qualifications and Second Order Sufficient Conditions

As noted in section 2, we shall treat only a scalar *pure* state constraint (2.5),

$$S(x(t), p) \leq 0 \quad \text{for all} \quad 0 \leq t \leq t_f. \tag{5.1}$$

A salient feature of state constraints is the so-called *order* l (see below) of a state constraint which is responsible for a variety of different phenomena. The exposition of *vector-valued* state constraints creates many technical difficulties due to the fact that the order may be different for different components of state constraints; see, e.g., the examples in [15, 16]. For *vector-valued* state constraints, a sensitivity analysis has been carried out in [40] for state constraints of order $l = 1$ whereas higher order state constraints are treated in [41] only in the scalar case. For notational convenience, we will assume in the sequel that there is only one boundary interval $[t_1, t_2]$ and one contact point $\tau \in (t_2, t_f)$. The analysis is exactly the same if there is any other finite number of junction points. Moreover, to avoid the degeneracy phenomenon reported in [1, 64] we assume that the initial and final point is not an active point for the state constraint, i.e., we have $0 < t_1 < t_2 < \tau < t_f$. Nevertheless, the case of an active initial or final point, $t_1 = 0$ or $t_2 = T$, is tractable provided that an additional regularity condition holds; cf. the example in [6].

Let us define recursively the following functions [26, 29, 43]:

$$\begin{aligned} &S^j : \mathbb{R}^n \times \mathbb{R}^m \times P \mapsto \mathbb{R}, \\ &S^0(x, u, p) := S(x, p), \quad S^{j+1}(x, u, p) := D_x S^j(x, u, p) f(x, u, p). \end{aligned} \tag{5.2}$$

The constraint $S(x, p) \leq 0$ is called to be of *order* $l \geq 1$ with respect to the dynamics $\dot{x} = f(x, u, p)$, if

$$D_u S^j(x, u, p) \equiv 0 \quad (j = 0, 1, \ldots, l-1) \quad \text{and} \quad D_u S^l(x, u, p) \not\equiv 0. \tag{5.3}$$

The preceding relations mean that the functions $S^j(x, u, p) \equiv S^j(x, p)$ are independent of u for $j = 0, 1, \ldots, l-1$. It can be easily seen that if $S(x, p)$ is of order

l and (x, u) is the solution of $\dot{x} = f(x, u, p)$, then

$$\frac{d^j}{dt^j} S(x(t), p) = \begin{cases} S^j(x(t), p), & j = 0, 1, \ldots, l-1, \\ S^l(x(t), u(t), p), & j = l. \end{cases} \tag{5.4}$$

In the sequel, we have to assume that *for any* p the constraint $S(x, p) \leq 0$ has a fixed order $l \geq 1$. For the further analysis, the order l dictates the differentiability requirement that all functions are of class C^{2l} with respect to their relevant arguments; cf. the general assumption (AG-1).

There exists a simple formal procedure whereby one obtains necessary and sufficient conditions for a state constrained problem in a way which is quite similar to that for mixed control-state constraints. Formally, one replaces the function $C(x, u, p)$ for mixed control-state constraints by the function $S^l(x, u, p)$ defined in (5.2) and adds junction conditions related to the functions S^j, $j = 0, 1, \ldots, l-1$. This procedure has been justified by the results in [26, 40, 41].

The following two assumptions are analogous to (AC-1) and (AC-2). Due to its very technical nature, assumption (AC-2) will not be stated explicitly.

(AS-1) (*Linear Independence Condition*) On the boundary arc we have

$$D_u S^l(x_0(t), u_0(t), p_0) \neq 0 \quad \text{for all } t \in [t_1, t_2]. \tag{5.5}$$

(AS-2) (*Controllability Condition*) See [40], assumption (I.5), and [41], assumption (A.7).

There are several ways of defining an *augmented* Hamiltonian depending on which function S^j, $j = 0, 1, \ldots, l$, is adjoined to the unconstrained Hamiltonian H^0 in (3.1); cf. [26, 43]. Here we choose the *augmented* Hamiltonian which is formed by adjoining the function $S^l(x, u, p)$ of highest order:

$$\begin{aligned} H(x, u, \lambda, \nu^l, p) &:= f_0(x, u, p) + \lambda^* f(x, u, p) + \nu^l S^l(x, u, p) \\ &= H^0(x, u, \lambda, p) + \nu^l S^l(x, u, p). \end{aligned} \tag{5.6}$$

Then by virtue of (AS-1) and (AS-2), there exist Lagrange-multipliers

$$(\lambda_0, \nu_0^l, \rho_0, \sigma_0, \phi_0) \in W^{1,\infty}(0, t_f; \mathbb{R}^n) \times L^\infty(0, t_f; \mathbb{R}) \times \mathbb{R}^r \times \mathbb{R}^l \times \mathbb{R}$$

such that the adjoint equations, transversality conditions and minimum condition hold for a.e. $t \in [0, t_f]$,

$$\dot{\lambda}_0(t) = -H_x(x_0(t), u_0(t), \lambda_0(t), \nu_0^l(t), p_0)^*, \tag{5.7}$$

$$(-\lambda_0(0), \lambda_0(t_f)) = D_{(x(0), x(t_f))}(g + \rho_0^* \psi)^*(x_0(0), x_0(t_f), p_0), \tag{5.8}$$

$$H_u(x_0(t), u_0(t), \lambda_0(t), \nu_0^l(t), p_0) = 0, \tag{5.9}$$

$$H(x_0(t), u_0(t), \lambda_0(t), \nu_0^l(t), p_0) = \text{const.} \tag{5.10}$$

We shall use again the notation that the symbol $[t]$ stands for all nominal arguments $(x_0(t), u_0(t), \lambda_0(t), \nu_0^l(t), p_0)$. At the entry point t_1, resp. at the contact point τ, the following *jump conditions* are satisfied with the jump variable $\sigma_0 = (\sigma_0^0, \ldots, \sigma_0^{l-1})$ resp. with ϕ_0:

$$\lambda_0(t_1+) = \lambda_0(t_1-) - \sum_{j=0}^{l-1} \sigma_0^j \, D_x S^j[t_1]^*, \quad \sigma_0^j \geq 0, \tag{5.11}$$

$$\lambda_0(\tau+) = \lambda_0(\tau-) - \phi_0 \, D_x S[\tau]^*, \quad \phi_0 \geq 0. \tag{5.12}$$

The multiplier $\nu_0^l(t)$ satisfies the relations

$$\nu_0^l(t) = 0 \;\; \forall\, t \notin [t_1, t_2], \;\; (-1)^j \frac{d^j}{dt^j} \nu_0^l(t) \geq 0 \;\; \forall\, t \in [t_1, t_2], \; j = 0, \ldots, l. \tag{5.13}$$

When adjoining the functions S^j, $j = 0, 1, \ldots, l-1$, to the unconstrained Hamiltonian instead of the function S^l, one obtains different multipliers and jump conditions. Explicit relations between these multipliers may be found in [26,41,43]. It is now appropriate to add some remarks on the occurence of boundary arcs and contact points with respect to the order l. It was shown in [29] that *boundary arcs* can not occur for *odd* orders $l \geq 3$ if the Hamiltonian is *regular*, i.e., admits a unique minimum, and if the control is piecewise analytic. However, for a problem with a *non-regular* Hamiltonian, [11, 12] present an example with a boundary arc of order $l = 3$. To our knowledge, boundary arcs for an *even* order $l \geq 4$ have not been reported in the literature. We omit a detailed discussion of this phenomenon. As a consequence, the following analysis treats boundary arcs only for the orders $l = 1$ and $l = 2$. The situation is different for *contact points* which usually do not occur for $l = 1$ but which may be present for all orders $l \geq 2$.

The analogon to the strict complementarity assumption (AC-3) is obtained by sharpening some inequalities in (5.13); cf. [41]:

(AS-3) (*Strict complementarity condition*) There exists $\alpha > 0$ such that the following relations hold;

$$\left.\begin{array}{rll} (-1)^l \frac{d^l}{dt^l} \nu_0^l(t) \geq \alpha > 0 & \text{for} & t \in [t_1, t_2], \\ \sigma_0^0 - (-1)^{l-1} \frac{d^{l-1}}{dt^{l-1}} \nu_0^l(t_1+) > 0 & \text{for} & l \geq 2, \\ (-1)^{l-1} \frac{d^{l-1}}{dt^{l-1}} \nu_0^l(t_2-) > 0 & \text{for} & l \geq 2, \\ \phi_0 > 0 & \text{for} & l \geq 2. \end{array}\right\} \tag{5.14}$$

(AS-4) (*Strict Legendre-Clebsch condition*) There exists $c > 0$ such that the following estimate holds for all $t \in [0, t_f]$,

$$v^* H_{uu}[t] v \geq c |v|^2 \quad \forall \; v \in \mathbb{R}^m, \;\; S_u^l[t] v = 0.$$

The *second order sufficient conditions* (SSC) for pure state constraints are similar to those for mixed control-state constraints except that we have to admit jumps for the matrix $Q(t)$ appearing in assumption (AC-5). We can shorten the presentation of SSC by appealing to the fact that one replaces the function $C(x, u, p)$ in (AC-5) formally by the function $S^l(x, u, p)$.

(AS-5) (*Coercivity condition*) There exists a symmetric $n \times n$ matrix $Q(t)$ which is of class C^1 except at the entry point t_1 or the contact point τ such that the definiteness condition (3.8)–(3.11) in (AC-5) are satisfied upon replacing the function $C(x,u,p)$ by $S^l(x,u,p)$. In addition, the following jump conditions hold with the jump multipliers σ_0^j from (5.11), resp. ϕ_0 from (5.12),

$$\begin{aligned} Q(t_1+) &= Q(t_1-) - \sum_{j=0}^{l-1} \sigma_0^j D_{xx}^2 S^j[t_1], \\ Q(\tau+) &= Q(\tau-) - \phi_0 D_{xx}^2 S[\tau]. \end{aligned} \tag{5.15}$$

Under these conditions, the SSC given in Theorem 1 carry over to control problems with pure state constraints. However, we refrain from giving an exact statement. In the same way, the sufficient conditions in (3.12)–(3.22) which are based on Riccati equations extend to pure state constraints. So far, no formal proof of this type of SSC can be found in the literature. We are going to show in a future paper that a proof evolves from a generalization of the techniques in [50, 61]. The work in [37, 44] gives a proof for strong SSC which are based on the Legendre-Clebsch condition and the Riccati equation (3.12) on the entire interval $[0, t_f]$.

5.2 *Parametric Boundary Value Problem and Solution Differentiability*

Recall that the nominal solution was supposed to contain only one boundary arc $[t_1, t_2]$ and one contact point τ in case $l \geq 2$. Under appropriate conditions to be specified below this property persists for all parameters p near p_0. On the *interior arcs* $[0, t_1]$ and $[t_2, t_f]$, the *free control* $u^f(z)$ is a function of the variable $z := (x, \lambda, p)$ which is determined by

$$u^f(x,\lambda,p) = \arg\min_{u \in \mathbb{R}^m} H(x,u,\lambda,p). \tag{5.16}$$

In particular, we have $u_0(t) = u^f(x_0(t), \lambda_0(t), p_0)$. The *boundary control* $u^b(z)$ and the multiplier $\nu^l(z)$ are the solution, resp., the multiplier of the mathematical programming problem depending on $z = (x, \lambda, p)$:

$$\text{MP}(z) \quad \text{minimize}_{u \in \mathbb{R}^m} H(x,u,\lambda,p) \quad \text{subject to } S^l(x,u,p) = 0.$$

Formulae for the differentials $D_z u^f(z)$, $D_z u^b(z)$ and $D_z \nu^l(z)$ are obtained in a way completely analogous to (3.23) and (3.24) upon substituting the active components C^a by S^l. Then the reference boundary value problem (2.2), (2.3) and (5.7), (5.8) can be embedded into the following *parametric multipoint boundary value problem* BVP(p):

Differential equations:

$$\dot{x} = \begin{cases} f(x, u^f(x,\lambda,p), p), & t \notin [t_1, t_2] \\ f(x, u^b(x,\lambda,p), p), & t \in [t_1, t_2] \end{cases} \tag{5.17}$$

$$\dot{\lambda} = \begin{cases} -H_x^0(x, u^f(x,\lambda,p), \lambda, p)^*, & t \notin [t_1, t_2] \\ -H_x(x, u^b(x,\lambda,p), \lambda, \nu^l(x,\lambda,p), p)^*, & t \in [t_1, t_2] \end{cases} \tag{5.18}$$

Boundary conditions:

$$\psi(x(0), x(t_f), p) = 0, \tag{5.19}$$
$$(-\lambda(0), \lambda(t_f)) = D_{(x(0),x(t_f))}(g + \rho^* \psi)^*(x(0), x(t_f), p). \tag{5.20}$$

Junction conditions:

$$S^j(x(t_1), p) = 0, \qquad j = 0, 1, \ldots, l-1, \tag{5.21}$$
$$\tilde{S}^l(x(t_1), \lambda(t_1-), p) = 0, \tag{5.22}$$
$$\tilde{S}^l(x(t_2), \lambda(t_2), p) = 0, \tag{5.23}$$
$$S(x(\tau), p) = 0, \tag{5.24}$$
$$S^1(x(\tau), p) = 0. \tag{5.25}$$

Jump conditions:

$$\lambda(t_1+) = \lambda(t_1-) - \sum_{j=0}^{l-1} \sigma^j D_x S^j(x(t_1), p)^*, \tag{5.26}$$
$$\lambda(\tau+) = \lambda(\tau-) - \phi D_x S(x(\tau), p)^*. \tag{5.27}$$

In (5.22) and (5.23) we have used the function

$$\tilde{S}^l(x, \lambda, p) := S^l(x, u^f(x, \lambda, p), p), \tag{5.28}$$

where the *free control* $u^f(x, \lambda, p)$ from (5.16) is inserted. Note that the additional condition (5.25) follows from the property that the function $S(x(t)$ attains a local maximum at $t = \tau$.

To apply the *shooting method* to BVP(p), we recall our earlier statement that a *boundary arc* usually occurs only for orders $l = 1$ and $l = 2$ whereas a *contact point* may be present for any order $l \geq 2$. The *shooting variable* is defined as the vector

$$s = (s_x, s_\lambda, \rho, \sigma, \phi, t_1, t_2, \tau) \in \mathbb{R}^n \times \mathbb{R}^n \times \mathbb{R}^r \times \mathbb{R}^l \times \mathbb{R} \times \mathbb{R}^3.$$

We obtain a shooting equation $F(s, p) = 0$ which is similar to (3.32) and takes into account the additional equations (5.21) and (5.25). The unperturbed solution corresponds to the shooting variable s_0 which satisfies the equation $F(s_0, p_0) = 0$. Again, we can apply the classical implicit function theorem to the shooting equation $F(s_0, p_0) = 0$ provided that the Jacobian $\partial F(s_0, p_0)/\partial s$ is *regular*. A *necessary condition* for the regularity of this Jacobian is the following *non-tangential junction* condition:

(AS-6) (*Non-tangential junctions*)

$$\left.\begin{array}{lll} \frac{d}{dt}\tilde{S}^l[t_i^0] \neq 0, \quad i = 1, 2, & \text{for a boundary arc,} \\ S^2[\tau^0] \neq 0 & \text{for a contact point.} \end{array}\right\} \tag{5.29}$$

For orders $l \geq 2$, the further assumption (A.11) has been imposed in [41]. Since this assumption is rather technical, we shall omit it and indicate only that it can be verified by checking the regularity of the shooting matrix. Now we can state the main sensitivity result for pure state constraints; cf. [40,41]:

Theorem 4. (Solution differentiability for pure state constraints)
Let (x_0, u_0) be admissible for OC(p_0). Suppose that (AS-1)–(AS-6) hold and assumption (A.11) in [41] is satisfied for $l \geq 2$. Then there exists a neighborhood $P_0 \subset P$ of p_0 and a C^1-function $s : P_0 \to \mathbb{R}^{2n+r+l+4}$ such that the shooting equations $F(s(p), p) = 0$ holds with

$$s(p) = (s_x(p), s_\lambda(p), \rho(p), \sigma(p), \phi(p), t_1(p), t_2(p), \tau(p))$$

for all $p \in P_0$. Let

$$D := [0, t_f] \times P_0 \quad \textit{and} \quad \tilde{D} := D \setminus \{(t_1(p), p), (t_2(p), p), (\tau(p), p)\}.$$

Then the state functions $x(t,p) := x(t, s(p), p)$ are C^1-functions on D whereas the adjoint functions $\lambda(t,p) := \lambda(t, s(p), p)$ and the multiplier

$$\nu^l(t,p) := \begin{cases} 0, & \textit{for} \quad t \notin [t_1(p), t_2(p)] \\ \nu^l(x(t,p), \lambda(t,p), p), & \textit{for} \quad t \in [t_1(p), t_2(p)] \end{cases}$$

are C^1-functions on $\tilde{D}$. The control function

$$u(t,p) := \begin{cases} u^f(x(t,p), \lambda(t,p), p), & \textit{for} \quad t \notin [t_1(p), t_2(p)] \\ u^b(x(t,p), \lambda(t,p), p), & \textit{for} \quad t \in [t_1(p), t_2(p)] \end{cases}$$

is continuous and is of class C^1 on $\tilde{D}$. For every $p \in P_0$, the triple $x(\cdot, p)$, $\lambda(\cdot, p)$, $u(\cdot, p)$ and the multipliers $\nu^l(\cdot, p)$, $\rho(p)$, $\sigma(p)$, $\phi(p)$ solve the parametric BVP(p) and satisfy the SSC analogous to those in Theorem 1 evaluated for the parameter p. Hence, the pair $(x(\cdot, p), u(\cdot, p))$ provides a local minimum for OC(p).

5.3 Computational Sensitivity Analysis and Real-Time Control

Let us perturb p_0 in a direction d with $p_0 + d \in P$. Theorem 4 assures us that the following sensitivity differentials exist:

$$y_d(t) := \tfrac{\partial x}{\partial p}(t, p_0)d, \quad \gamma_d(t) := \tfrac{\partial \lambda}{\partial p}(t, p_0)d$$
$$\rho_d := \tfrac{d\rho}{dp}(p_0)d, \qquad \sigma_d^j := \tfrac{d\sigma^j}{dp}(p_0)d \; (j = 0, \ldots, l-1), \quad \Phi_d := \tfrac{d\phi}{dp}(p_0)d,$$

These differentials satisfy a *linear* boundary value problem which is obtained by differentiating the boundary value problem BVP(p) in (5.17)–(5.27) formally with respect to p. We refrain from writing down explicitly this BVP since it is rather similar to the one for mixed control-state constraint given in (3.37)–(3.41). The case study in the next section will exemplify this procedure. However, we expand on the differentiation of the jump conditions (5.26) and (5.27). At the entry point t_1^0 we find

$$\gamma_d(t_1^0+) = \gamma_d(t_1^0-) - [\dot\lambda_0(t_1^0+) - \dot\lambda_0(t_1^0-)]\, \tfrac{dt_1}{dp}(p_0) - \sum_{j=0}^{l-1} \sigma_d^j\, D_x S^j[t_1^0]^* \\ - \sum_{j=0}^{l-1} \sigma_0^j \left\{ D^2_{xx} S^j[t_1^0]\big(f[t_1^0]\, \tfrac{dt_1}{dp}(p_0) + y_d(t_1^0)\big) + D^2_{xp} S^j[t_1^0] d \right\}. \tag{5.30}$$

At the contact point τ_1^0 we get

$$\gamma_d(\tau^0+) = \gamma_d(\tau^0-) - [\dot\lambda_0(\tau^0+) - \dot\lambda_0(\tau^0-)]\, \tfrac{d\tau}{dp}(p_0) - \Phi_d D_x S[\tau^0]^* \\ -\psi_0 \left\{ D^2_{xx} S[\tau^0]\big(f[\tau^0]\, \tfrac{d\tau}{dp}(p_0) + y_d(\tau^0)\big) + D^2_{xp} S[\tau^0] d \right\}. \tag{5.31}$$

Formulas for the differentials of junction and contact points can derived from the following identities which hold for all $p \in P_0$:

$$\tilde S^l(x(t_i(p),p), \lambda(t_i(p),p), p) \equiv 0 \quad (i = 1,2), \qquad S^1(x(\tau_1(p),p),p) \equiv 0.$$

Recall that the function $\tilde S^l$ is defined in (5.28). Upon differentiating these equations and using the non-tangential junction condition (AS-6), we obtain the directional derivatives

$$\frac{dt_i}{dp}(p_0)d = -(D_x \tilde S^l y_d + D_\lambda \tilde S^l \gamma_d + D_p \tilde S^l d)[t_i^0] \,/\, \frac{d}{dt}\tilde S^l[t_i^0], \tag{5.32}$$

$$\frac{d\tau_i}{dp}(p_0)d = -(D_x S^1 y_d + D_p S^1 d)[\tau^0] \,/\, S^2[\tau^0]. \tag{5.33}$$

The work [3] discusses numerical examples for state constraint of orders $l = 1$, $l = 2$ and $l = 4$ which illustrate the computational sensitivity analysis presented in the preceding sections.

6 Numerical Case Study: The Van der Pol Oscillator with a State Constraint

The following optimal control model for the Van-der-Pol oscillator with state constraints is taken from [67]. A verification of SSC and a sensitivity analysis has recently been carried out in [3,4] on which the presentation below will be based. The Van-der-Pol oscillator also refers to the electric circuit depicted in Figure 1 including a capacity, resistance, inductivity and a diode. However, in contrast to section 4, the state variable $x_1(t)$ now represents the voltage and we set $x_2(t) := \dot x_1(t)$. We consider the following control problem where the parameter p appears in the state constraint:

$$\text{Minimize} \quad F(x,u,p) = \int_0^5 \left(u^2(t) + x_1^2(t) + x_2^2(t)\right) dt \tag{6.1}$$

$$\text{subject to} \quad \dot{x}_1 = x_2, \; \dot{x}_2 = (1 - x_1^2)x_2 - x_1 + u, \quad \text{for } t \in [0,5], \tag{6.2}$$

$$x_1(0) = 1, \; x_2(0) = 0, \tag{6.3}$$

$$p \le x_2(t), \quad \text{for} \quad t \in [0,5]. \tag{6.4}$$

The state inequality constraint (6.4) is of the form (2.5) upon defining the function $S(x,p) := -x_2 + p \le 0$ with the scalar parameter p. The *nominal parameter* is chosen as $p_0 = -0.4$.

6.1 Boundary Value Problem

In a first step we study the *unconstrained* problem (6.1)–(6.3) and omit the state constraint (6.4). The shape of the unconstrained solution reveals that the *constrained* solution has one boundary arc with $x_2(t) \equiv p$ for $t \in [t_1, t_2]$ and $0 < t_1 < t_2 < 5$. The function S^1 is computed from (5.2) as $S^1(x,u,p) = -\dot{x}_2 = -(1-x_1^2)x_2 + x_1 - u$. Since $D_u S^1 \equiv -1$, the order of the state constraint is $l = 1$. From (5.7)–(5.9) we find the following adjoint equations, the control u and the multiplier ν^1 associated with the state constraint:

$$\dot{\lambda}_1 = -2x_1 + 2x_1x_2\lambda_2 + \lambda_2 - \nu^1(2x_1x_2 + 1), \quad \lambda_1(5) = 0, \tag{6.5}$$

$$\dot{\lambda}_2 = -2x_2 - \lambda_1 + \lambda_2(x_1^2 - 1) + \nu^1(1 - x_1^2), \quad \lambda_2(5) = 0, \tag{6.6}$$

where

$$u = \begin{cases} -\frac{\lambda_2}{2} & , \quad t \in [0,t_1] \cup [t_2,5], \\ (x_1^2 - 1)x_2 + x_1 & , \quad t \in [t_1,t_2], \end{cases} \tag{6.7}$$

$$\nu^1 = \begin{cases} 0 & , \quad t \in [0,t_1] \cup [t_2,5], \\ 2(x_1^2 - 1)x_2 + 2x_1 + \lambda_2 & , \quad t \in [t_1,t_2]. \end{cases} \tag{6.8}$$

The junction conditions (5.21)–(5.23) at the entry and exit point t_1 and t_2 are:

$$x_2(t_1) = p, \tag{6.9}$$

$$\left(x_1^2(t) - 1\right) x_2(t) + x_1(t) + 0.5\lambda_2(t) = 0, \quad \text{at } t = (t_1)-, \; t = t_2. \tag{6.10}$$

In the last equation, we have used both the control law $u = -\lambda_2/2$ on the interior arcs $[0,t_1] \cup [t_2,5]$ and the continuity of the control at the junction points. Due to relation (5.11), the adjoint variable λ_2 has a jump at the entry point t_1,

$$\lambda_2(t_1^+) = \lambda_2(t_1^-) + \sigma, \quad \sigma \ge 0, \tag{6.11}$$

whereas λ_1 is continuous at t_1. The multipoint boundary value problem (6.2), (6.3) and (6.5)–(6.11) can be solved using the code BNDSCO of [54]. For the nominal

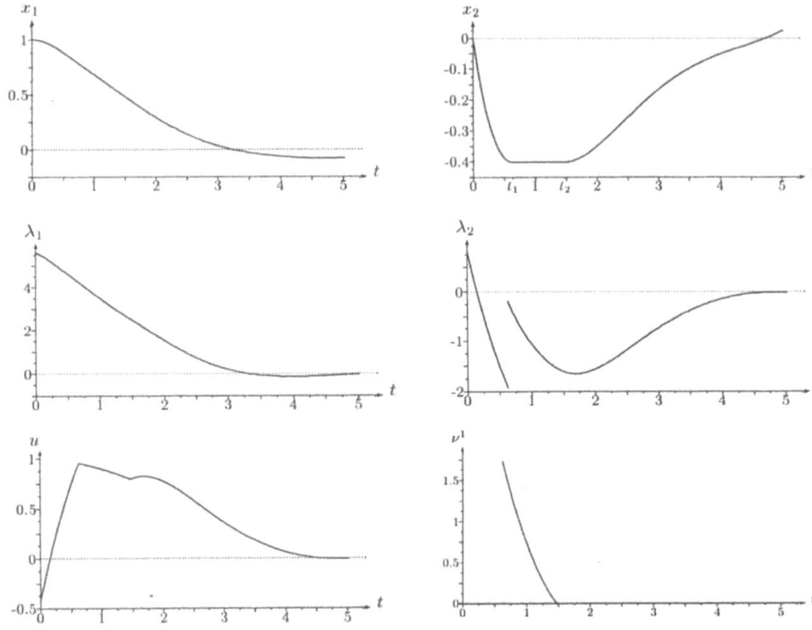

Figure 8. Nominal states $x_i(t)$ and adjoint variables $\lambda_i(t)$, $i = 1, 2$, optimal control $u(t)$ and multiplier $\nu^1(t)$

parameter $p_0 = -0.4$ we obtain the following results for the initial values of the adjoint variables and the junction points:

$$\begin{aligned} \lambda_1(0) &= 5.66576692 \;, & \lambda_2(0) &= 0.82392266 \;, \\ t_1 &= 0.62752939 \;, & t_2 &= 1.46908215 \;, \\ x_1(5) &= -0.07420849 \;, & x_2(5) &= 0.02929680 \;, \\ \sigma_0 &= 1.73011588 \;, & F(x_0, u_0, p_0) &= 2.95370134 \;. \end{aligned}$$

The nominal optimal solution with state and adjoint variables, control and multiplier are shown in Figure 8.

Next, we shall verify that the computed nominal solution satisfies all regularity conditions and the SSC in section 5. The regularity condition (5.5) in (AS-1) and the strict Legendre-Clebsch condition (AS-4) obviously hold. Moreover, the *non-tangential* junction condition (5.29) in (AS-6) is valid since

$$\begin{aligned} \frac{d}{dt} S^1(x(t_1), u(t_1^-), p) &= -1.62175826 \neq 0, \\ \frac{d}{dt} S^1(x(t_2), u(t_2^+), p) &= -0.52365061 \neq 0. \end{aligned}$$

The strict complementarity condition (5.14) in (AS-3) is satisfied since $-d\nu^1/dt > 0$ for $t \in [t_1^+, t_2^-]$, cf. Figure 8. In order to check SSC it suffices to find a bounded solution of the Riccati equation (3.12). With $n = 2$ we consider a symmetric 2×2 matrix

$$Q(t) = \begin{pmatrix} q_1(t) & q_2(t) \\ q_2(t) & q_4(t) \end{pmatrix}$$

and obtain the following explicit Riccati equations from (3.12):

$$\begin{aligned} \dot{q}_1 &= (4x_1x_2 + 2)q_2 - 2 + 2x_2(\lambda_2 - \nu^1) + \frac{1}{2}q_2^2, \\ \dot{q}_2 &= -q_1 - (1 - x_1^2)q_2 + (2x_1x_2 + 1)q_4 + 2x_1(\lambda_2 - \nu^1) + \frac{1}{2}q_2q_4, \\ \dot{q}_4 &= -2q_2 - 2(1 - x_1^2)q_4 - 2 + \frac{1}{2}q_4^2. \end{aligned}$$

The functions q_i, $i = 1,2,4$, are continuous at the entry point t_1 due to the jump relations (5.15) and $D^2_{xx}S \equiv 0$. We are allowed to use a relaxed form of the boundary conditions (3.22) and impose the boundary conditions $q_1(5) = q_2(5) = q_4(5) = 0$. Numerical integration along the nominal solution with $p_0 = -0.4$ shows indeed that there exist a bounded solution with $|q_i(t)| \le 6 \ \forall\, t \in [0, t_f]$, cf. Figure 9. Thus we arrive at the conclusion that the solution shown in Figure 8 is a local minimum.

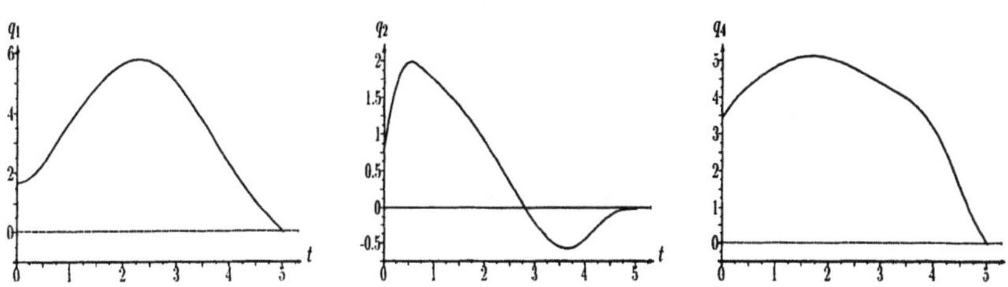

Figure 9. Solutions $q_i(t)$ $(i = 1,2,4)$ of the Riccati equation

6.2 *Computational Sensitivity Analysis and Real-Time Control*

In the preceding section we have verified that the *unperturbed (nominal)* solution with $p_0 = -0.4$ meets all assumptions for *solution differentiability* in Theorem 4. Thus the unperturbed solution can be embedded into a family of perturbed solutions

$$x_i(t,p), \quad \lambda_i(t,p) \ (i = 1,2), \quad t_j(p) \ (j = 1,2), \quad \nu^1(t,p), \quad \sigma(p),$$

to the perturbed problem $OC(p)$. The C^1 properties of these functions are specified in Theorem 4. The perturbed solution satisfies the boundary value problem (6.2),

(6.3) and (6.5)–(6.11) identically for all parameter p near p_0. The *sensitivity differentials*

$$y_i(t) = \frac{\partial x_i}{\partial p}(t, p_0), \quad \gamma_i(t) = \frac{\partial \lambda_i}{\partial p}(t, p_0), \ (i = 1, 2), \ \ p_0 = -0.4.$$

satisfy a linear BVP which we are going to derive now. The ODEs for $y_i(t)$ and $\gamma_i(t)$ are obtained by formally differentiating the nonlinear equations (6.2), (6.3) and (6.5)–(6.11) with respect to the perturbation p. This procedure yields the linear ODEs

$$\begin{aligned}
\dot{y}_1 &= y_2, \\
\dot{y}_2 &= -2x_1x_2y_1 + (1 - x_1^2)y_2 - y_1 + u_p, \\
\dot{\gamma}_1 &= -2y_1 + 2(x_2\lambda_2 y_1 + x_1x_2\gamma_2 + x_1\lambda_2 y_2) + \gamma_2 - 2\nu^1(x_2y_1 + x_1y_2) \\
&\quad -\nu^1_p(2x_1x_2 + 1), \\
\dot{\gamma}_2 &= -2y_2 - \gamma_1 + \gamma_2(x_1^2 - 1) + 2x_1\lambda_2 y_1 - 2\nu^1 x_1 y_1 + \nu^1_p(1 - x_1^2),
\end{aligned}$$

where

$$u_p = \frac{\partial u}{\partial p} = \begin{cases} -\frac{\gamma_2}{2} & , \ t \in [0, t_1] \cup [t_2, 5], \\ 2x_1x_2y_1 + (x_1^2 - 1)y_2 + y_1 , & t \in [t_1, t_2], \end{cases}$$

$$\nu^1_p = \frac{\partial \nu^1}{\partial p} = \begin{cases} 0 & , \ t \in [0, t_1] \cup [t_2, 5], \\ 4x_1x_2y_1 + 2(x_1^2 - 1)y_2 + 2y_1 + \gamma_2 , & t \in [t_1, t_2]. \end{cases}$$

Since the initial state is fixed and the final state is free, we find the following boundary conditions from (3.39) and (3.41),

$$y_i(0) = 0, \quad \gamma_i(5) = 0, \quad \text{for } i = 1, 2.$$

The differentiation of the junction condition $x_2(t_1(p), p) \equiv p$ yields the junction condition

$$y_2(t_1) = 1.$$

The jump relations for the sensitivity differentials γ_1, γ_2 at the entry point t_1 can be obtained from relation (5.30). When evaluating this formula the reader may check that the nominal derivatives $\dot{\lambda}_i, i = 1, 2,$ are *continuous* at t_1 due to equations (6.5)–(6.8). Then we get the jump conditions

$$\gamma_1(t_1^+) = \gamma_1(t_1^-), \quad \gamma_2(t_1^+) = \gamma_2(t_1^-) + \sigma, \quad \sigma = \frac{d\sigma}{dp}(p_0).$$

Finally, the differentials of the junction points $t_i(p)$, $i = 1, 2,$ are obtained by differentiating the junction conditions $S^1(x(t_j(p), p), p) = [-(1 - x_1^2)x_2 + x_1 +$

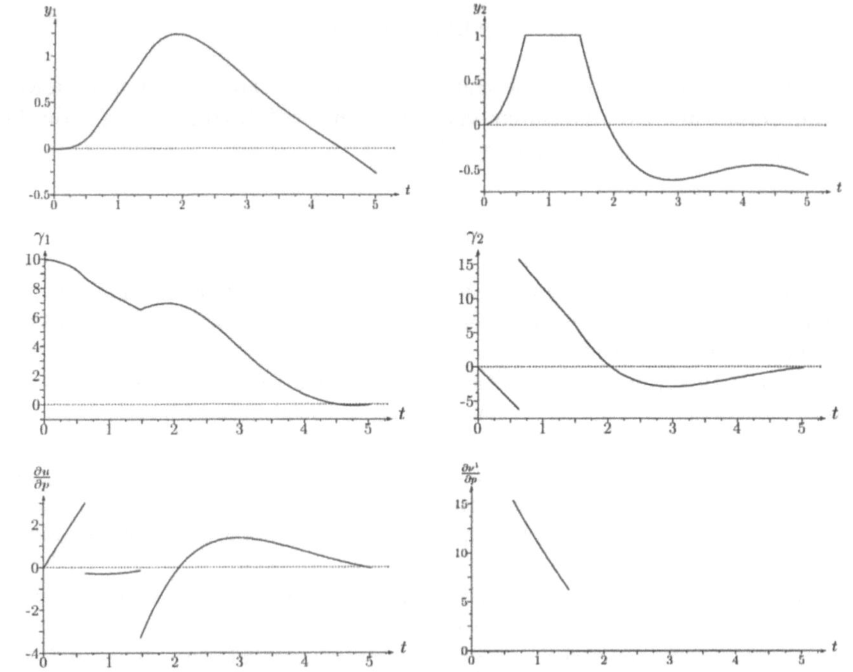

Figure 10. Sensitivity differentials $y_i = \partial x_i/\partial p$, $\gamma_i = \partial \lambda_i/\partial p$, $\partial u/\partial p$, $\partial \nu^1/\partial p$

$0.5\lambda_2](t_i(p), p) \equiv 0$. The general formulae (5.32) then read explicitly:

$$\frac{dt_i}{dp} = -\left. \frac{(4x_1x_2 + 2)y_1 + 2(x_1^2 - 1)y_2 + \gamma_2}{4x_1x_2^2 - 2(x_1^2 - 1)^2 x_2 + 2(1 - x_1^2)x_1 - \lambda_1} \right|_{t=t_i} \quad \text{at} \quad t_1^- \text{ and } t_2^+.$$

Again, the code BNDSCO in [54] is used to resolve the BVP for the sensitivity differentials:

$$\begin{aligned} \gamma_1(0) &= 9.88324529 \quad, & \gamma_2(0) &= -0.00482702 \,, \\ dt_1/dp &= -2.03284608 \,, & dt_2/dp &= 6.03498889 \quad, \\ y_1(5) &= -0.26590753 \,, & y_2(5) &= -0.55722168 \,, \\ \sigma = d\sigma/dp &= 22.0100099 \quad, & dF/dp &= 1.73011588 \quad. \end{aligned}$$

The sensitivity differentials are shown in Figure 10. On the basis of the computed sensitivity differentials we may perform a real-time approximation of a perturbed solution according to the Taylor expansion (2.6). The numerical results and the quality of the approximation are very similar to those for the Rayleigh problem in Section 4.3 so that we refrain from discussing them explicitly.

ACKNOWLEDGEMENT

The authors would like to thank Prof. Hans Josef Pesch and an anonymous reviewer for helpful comments. The first author wishes to thank Prof. Kazimierz Malanowski for a fruitful collaboration over many years.

REFERENCES

1. A. V. Arutyunov, S. M. Aseev: Investigation of the degeneracy phenomenon of the maximum principle for optimal control problems with state constraints. SIAM J. Control and Optimization **35** (1997) 930–952
2. D. Augustin: Hinreichende Optimalitätsbedingungen und Sensitivitätsanalyse bei optimalen Steuerproblemen mit Steuer- und Zustandsbeschränkungen. Diploma thesis, Institut für Numerische Mathematik, Universität Münster, Germany, 1996.
3. D. Augustin, H. Maurer: An example for computational sensitivity analysis for state constrained control problems, in Proceedings of Parametric Optimization and Related Topics V, Tokyo, 1998, Guddat, J., et al., eds, Peter Lang Verlag, Frankfurt am Main, 2000, 25–35
4. D. Augustin, H. Maurer: Computational sensitivity analysis for state constrained optimal control problems. Annals of Operations Research (2000).
5. D. Augustin, H. Maurer: Second order sufficient conditions and sensitivity analysis for optimal multiprocess control problems. Control and Cybernetics **29** (2000) 11–31
6. D. Augustin, H. Maurer: Sensitivity analysis and real-time control of a container crane under state constraints. This volume.
7. P. Berkmann, H. J. Pesch: Abort landing in windshear: Optimal control problem with third-order state constraint and varied switching structure. J. of Optimization Theory and Applications **85** (1995) 21–57
8. H. G. Bock, P. Krämer-Eis: An Efficient Algorithm for Approximate Computation of Feedback Control Laws in Nonlinear Processes. ZAMM, **61** (1981) T 330–T 332
9. A. E. Bryson, Y. C. Ho: Applied Optimal Control. Revised Printing, Hemisphere Publishing Corporation New York, New York, 1975.
10. R. Bulirsch: Die Mehrzielmethode zur numerischen Lösung von nichtlinearen Randwertproblemen und Aufgaben der optimalen Steuerung. Report of the Carl-Cranz Gesellschaft, Oberpfaffenhofen, Germany, 1971. Reprinted as report R1.06 of the Sonderforschungsbereich "Transatmosphärische Flugsysteme", TU München, Germany, 1993
11. R. Bulirsch, F. Montrone, and H. J. Pesch: Abort landing in the presence of a windshear as a minimax optimal control problem, Part 1: Necessary conditions. J. of Optimization Theory and Applications **70** (1991) 1–23
12. R. Bulirsch, F. Montrone, and H. J. Pesch: Abort landing in the presence of a windshear as a minimax optimal control problem, Part 2: Multiple shooting and homotopy, J. of Optimization Theory and Applications **70** (1991) 223–254
13. Ch. Büskens: Optimierungsmethoden und Sensitivitätsanalyse für optimale Steuerprozesse mit Steuer- und Zustands-Beschränkungen. Dissertation, Institut für Numerische Mathematik, Universität Münster, Germany, 1998.
14. Ch. Büskens, H. Maurer: Sensitivity analysis and real-time control of parametric optimal control problems using nonlinear programming methods. This volume

15. K. Chudej: Realistic modelled optimal control problems in aerospace engineering – a challenge to the necessary optimality conditions. Mathematical Modelling of Systems **2** (1996) 252–261
16. K. Chudej: Effiziente Lösungen zustandsbeschränkter Optimalsteuerungsaufgaben. Habilitationsschrift, Universität Bayreuth, Germany, 2000
17. A. L. Dontchev, W. W. Hager, P. A. Poore, and B. Yang: Optimality, stability and convergence in nonlinear control. Applied Math. and Optim. **31** (1995) 297–326
18. A. L. Dontchev, W. W. Hager: Lipschitzian stability for state constrained nonlinear optimal control. SIAM J. on Control and Optimization **36** (1998) 698–718
19. A. L. Dontchev, I. Kolmanovsky: State constraints in the linear regulator problem: a case study, J. of Optimization Theory and Applications **87** (1995) 327–347
20. A. L. Dontchev, I. Kolmanovsky: Best interpolation in a strip II: Reduction to unconstrained convex optimization, Computational Optimization and Applications **5** (1996) 233–251
21. J. C. Dunn: Second order optimality conditions in sets of L^∞ functions with range in a polyhedron. SIAM J. Control Optimization **33** (1995) 1603–1635
22. J. C. Dunn: On L^2 sufficient conditions and the gradient projection method for optimal control problems. SIAM J. Control Optimization **34** (1996) 1270–1290
23. U. Felgenhauer: Diskretisierung von Steuerungsproblemen unter stabilen Optimalitätsbedingungen. Habilitationsschrift, Dept. of Mathematics, Technische Universität Cottbus, Germany, 1999
24. U. Felgenhauer: On smoothness properties and approximability of optimal control functions. To appear in Annals of Operations Research
25. A. V. Fiacco: Introduction to Sensitivity and Stability Analysis in Nonlinear Programming, Mathematics in Science and Engineering **165**, Academic Press, New York, 1983
26. R. F. Hartl, S. P. Sethi, and R. G. Vickson: A survey of the maximum principle for optimal control problems with state constraints. SIAM Review **37** (1995) 181–218
27. M. Hestenes: Calculus of Variations and Optimal Control Theory. John Wiley, New York, 1966
28. K. Ito, K. Kunisch: Sensitivity analysis of solutions to optimization problems in Hilbert spaces with applications to optimal control and estimation. Journal of Differential Equations, **99** (1992) 1–40
29. D. H. Jacobson, M. M. Lele, and J. L. Speyer: New necessary conditions of optimality for control problems with state-variable inequality constraints, J. of Mathematical Analysis and Applications **35** (1971) 255–284
30. D. H. Jacobson, D. Q. Mayne: Differential Dynamic Programming. American Elsevier Publishing Company Inc., New York, 1970
31. P. Krämer-Eis: Ein Mehrzielverfahren zur numerischen Berechnung optimaler Feedback-Steuerungen bei beschränkten nichtlinearen Steuerungsproblemen. Bonner Mathematische Schriften **164**, 1985
32. B. Kugelmann, H. J. Pesch: A new general guidance method in constrained optimal control, Part 1: Numerical method. J. Optim. Theory and Appl. **67** (1990) 421–435
33. B. Kugelmann, H. J. Pesch: A new general guidance method in constrained optimal control, Part 2: Application to space shuttle guidance. J. Optim. Theory and Appl. **67** (1990) 437–446
34. K. Malanowski: Second order conditions and constraint qualifications in stability and sensitivity analysis of solutions to optimization problems in Hilbert spaces. Applied Math. Optimization **25** (1992) 51–79

35. K. Malanowski: Two-norm approach in stability and sensitivity analysis of optimization and optimal control problems. Advances in Math. Sciences and Applications **2** (1993) 397–443
36. K. Malanowski: Stability and sensitivity of solutions to nonlinear optimal control problems. Applied Math. Optim. **32** (1995) 111–141
37. K. Malanowski: Sufficient optimality conditions for optimal control problems subject to state constraints. SIAM J. on Control and Optimization **35** (1997), 205–227
38. K. Malanowski: Stability and sensitivity analysis for optimal control problems with control-state constraints. To appear
39. K. Malanowski, H. Maurer: Sensitivity analysis for parametric control problems with control-state constraints. Comput. Optim. and Applications **5** (1996) 253–283
40. K. Malanowski, H. Maurer: Sensitivity analysis for state constrained optimal control problems. Discrete and Continuous Dynamical Systems **4** (1998) 241–272
41. K. Malanowski, H. Maurer: Sensitivity analysis for optimal control problems subject to higher order state constraints. To appear in Annals of Operations Research, 2000.
42. H. Maurer: On optimal control problems with bounded state variables and control appearing linearly. SIAM J. Control and Optimization **15** (1977) 345–362
43. H. Maurer: On the minimum principle for optimal control problems with state constraints. Rechenzentrum Universität Münster, Schriftenreihe Nr. 41, Münster, Germany, 1979
44. H. Maurer: First- and second-order sufficient optimality conditions in mathematical programming and optimal control. Math. Programming Study **14** (1981) 43–62
45. H. Maurer, D. Augustin: Second order sufficient conditions and sensitivity analysis for the controlled Rayleigh problem. In Parametric Optimization and Related Topics IV, J. Guddat, H. Th. Jongen, F. Nozicka, G. Still, F. Twilt, eds., Peter Lang Verlag, 1996, 245–259
46. H. Maurer, W. Gillessen: Application of multiple shooting to the numerical solution of optimal control problems with bounded state variables. Computing **15** (1975) 105–126
47. H. Maurer, H. D. Mittelmann: The nonlinear beam via optimal control with bounded state variables, Optimal Control Applications & Methods **12** (1991) 19–31
48. H. Maurer, H. J. Oberle: Second order sufficient conditions for optimal control problems with free final time: the Riccati approach. Submitted to SIAM J. Control and Optimization
49. H. Maurer, H. J. Pesch: Solution differentiability for parametric nonlinear control problems with control-state constraints. J. Optimization Theory and Applications **86** (1995) 285–309
50. H. Maurer, S. Pickenhain: Second order sufficient conditions for optimal control problems with mixed control-state constraints. J. Optim. Theory and Applications **86** (1995) 649–667
51. A. A. Milyutin, N. P. Osmolovskii: Calculus of Variations and Optimal Control. Translations of Mathematical Monographs, Vol. 180, American Mathematical Society, Providence, 1998
52. L. W. Neustadt: Optimization: A Theory of Necessary Conditions. Princeton University Press, Princeton, 1976
53. H. J. Oberle: Numerical solution of minimax optimal control problems by multiple shooting technique. J. of Optimization Theory and Applications **50** (1986) 331–357
54. H. J. Oberle, W. Grimm: BNDSCO – A program for the numerical solution of optimal control problems. Institute for Flight Systems Dynamics, DLR, Oberpfaffenhofen, Germany, Internal Report **515–89/22**, 1989

55. G. Opfer, H. J. Oberle: The derivation of cubic splines with obstacles by methods of optimization and optimal control. Numerische Mathematik **52** (1988) 17–31.
56. N. P. Osmolovskii: Quadratic conditions for nonsingular extremals in optimal control (A theoretical treatment). Russian J. of Mathematical Physics **2** (1995) 487–516
57. H. J. Pesch: Numerical Computation of Neighboring Optimum Feedback Control Schemes in Real-Time. Applied Mathematics and Optimization **5** (1979), 231–252
58. H. J. Pesch: Neighboring Optimum Guidance of a Space-Shuttle-Orbiter-Type Vehicle. J. of Guidance and Control **3** (1980), 386–391
59. H. J. Pesch: Real-time computation of feedback controls for constrained optimal control problems, Part 1: Neighbouring extremals. Optimal Control Applications & Methods **10** (1989) 129–145
60. H. J. Pesch: Real-time computation of feedback controls for constrained optimal control problems, Part 2: A correction method based on multiple shooting. Optimal Control Applications & Methods **10** (1989) 147–171
61. S. Pickenhain: Sufficiency conditions for weak local minima in multidimensional optimal control problems with mixed control-state restrictions. Zeitschrift für Analysis und ihre Anwendungen **11** (1992) 559–568
62. S. Pickenhain, K. Tammer: Sufficient conditions for local optimality in multidimensional control problems with state restrictions. Zeitschrift für Analysis and ihre Anwendungen **10** (1991) 397–405
63. L. S. Pontrjagin, V. G. Boltjanskij, R. V. Gamkrelidze, E. F. Miscenko: Mathematische Theorie optimaler Prozesse. R.Oldenbourg, München Wien, 1967
64. F. Rampazzo, R. B. Vinter: Degenerate optimal control problems with state constraints. SIAM J. on Control and Optimization **39** (2000) 989–1007
65. J. Stoer, R. Bulirsch: Introduction to Numerical Analysis. Springer-Verlag, New York, 1980
66. T. Tun, T. S. Dillon: Extensions of the differential dynamic programming method to include systems with state dependent control constraints and state variable inequality constraints. J. of Applied Science and Engineering A, **3** (1978) 171–192
67. V. S. Vassiliadis, R. W. H. Sargent, and C. C. Pantelides: Solution of a Class of Multistage Dynamic Optimization Problems. Part 2: Problems with Path Constraints, Ind. Eng. Chem. Res. **33, No.9** (1994) 2123–2133
68. V. Zeidan: The Riccati equation for optimal control problems with mixed state-control constraints: necessity and sufficiency. SIAM J. Control and Optimization **32** (1994) 1297–1321

*Sensitivity Analysis and Real-Time Control of Parametric Optimal Control Problems Using Nonlinear Programming Methods

Christof Büskens[1] and Helmut Maurer[2]

[1] Lehrstuhl für Ingenieurmathematik, Universität Bayreuth, Germany
[2] Institut für numerische Mathematik, Universität Münster, Germany

Abstract We discuss nonlinear programming (NLP) methods for solving optimal control problems with control and state inequality constraints. Suitable discretizations of control and state variables are used to transform the optimal control into a finite dimensional NLP problem. In [8] we have proposed numerical methods for the post-optimal calculations of parameter sensitivity derivatives of optimal solutions to NLP problems. The purpose of this paper is to extend the methods of post-optimal sensitivity analysis and real-time optimization to discretized control problems. The dimension of the discretized control problem should be kept small to obtain accurate sensitivity results. This can be achieved by taking only the discretized control variables as optimization variables whereas the state variables are computed recursively through an appropriate integration routine. We discuss the implications of this approach for the calculations of parameter sensitivity derivatives with respect to optimal control, state and adjoint functions. The efficiency of the proposed methods are illustrated by two numerical examples.

1 Introduction

The sensitivity analysis of parametric optimal control problems has been discussed in Augustin and Maurer [1, 19] on the basis of *boundary value methods*. This so-called *indirect approach* is characterized by explicitly solving the necessary conditions in terms of state and adjoint equations to which one has to add an explicit set of equations for the sensitivity derivatives. Though these methods yield highly accurate results, the derivation of the boundary value problem may be cumbersome or may not be provided explictly. A further drawback of this approach is the difficulty of finding reasonable estimates for the adjoint variables.

Various *direct optimization methods* have been proposed to avoid the drawbacks of the indirect approach; cf., e.g., Barclay et al. [2], Betts [3], Bock and Plitt [4], Büskens [5,7], Enright and Conway [12], Evtushenko [13], von Stryk [22]. All direct methods proceed by a suitable discretization of the optimal control problem treating the discretized control and state variables as optimization variables in a nonlinear programming (NLP) problem. One advantage of direct approaches is that it does not need any estimates of the values for the adjoint variables. Section 3 discusses two discretization methods, the full discretization approach and the recursive approach, whose merits in connection with sensitivity analysis we are going to evaluate in this paper. In [8], we have proposed a *post-optimal* sensitivity analysis on the basis of well known formulae for the parameter sensitivity derivatives. It seems natural and appropriate to extend this sensitivity analysis to discretized control problems and to test its effiency in comparison with boundary value methods.

The evaluation of sensitivity formulae requires a computation of the exact Hessian of the Lagrangian. This fact obliges us to to keep the dimension of the resulting NLP problem as small as possible. For this reason, we concentrate on the *recursive approach* in Section 3.2 where only the discretized control variables are considered as optimization variables whilst the state variables are computed recursively through an appropriate integration routine as functions of the control variables. Adjoint variables can be obtained by a postoptimal calculation using the Lagrange multipliers of the resulting nonlinear optimization problem. The sensitivity analysis in [8] then enables us to compute the sensitivity derivatives of the control variables. In a second step, the parameter sensitivity derivatives for state and adjoint variables may be recovered recursively from the control sensitivities. Thus we obtain a complete picture of sensitivity analysis that is comparable to that gained from the indirect approach [19].

The articles [5,7,8,19] have documented the usefulness of a computational sensitivity analysis for the design of *real-time approximations* of perturbed optimal solutions. Thus the computational methods in this article can be applied directly to the computation of real-time control approximations in the original optimal control problem.

We point out that the sensitivity analysis in [19] is restricted to the special class of optimal control problems for which the strict Legendre-Clebsch condition is satisfied. Thus control problems with control appearing linearly in the system (bang-bang and singular controls) have to be excluded from the analysis so far. However, progress is being made towards developing second order conditions and sensitivity analysis also for this class of control problems. It is interesting to note that the following numerical sensitivity analysis using *direct methods* can be carried out for *any* class of optimal control problems though this approach is not yet backed up by a complete theory in all cases.

2 Parametric Optimal Control Problems with Constraints

2.1 *Problem Formulation*

We consider parametric control problems subject to control and state constraints. All data may be subject to perturbations that are modeled by a parameter $p \in P := \mathbb{R}^{N_p}$. The following parametric control problem will be referred to as problem **OCP**(p).

$$
\begin{aligned}
&\text{Minimize} \quad && F(x,u,p) = g(x(t_0),x(t_f),p) + \int_{t_0}^{t_f} f_0(x(t),u(t),p)\,dt \\
&\text{subject to} \quad && \dot{x}(t) = f(x(t),u(t),p) \quad \text{for all } t \in [0,t_f], \\
& && \psi(x(t_0),x(t_f),p) = 0, \\
& && C(x(t),u(t),p) \le 0 \quad \text{for all } t \in [0,t_f].
\end{aligned} \tag{1}
$$

Herein, $x(t) \in \mathbb{R}^n$ denotes the state of a system and $u(t) \in \mathbb{R}^m$ the control function in a time interval $[t_0, t_f]$. The functions $g : \mathbb{R}^{2n} \times P \to \mathbb{R}$, $f_0 : \mathbb{R}^{n+m} \times P \to \mathbb{R}$, $f : \mathbb{R}^{n+m} \times P \to \mathbb{R}^n$, $\psi : \mathbb{R}^{2n} \times P \to \mathbb{R}^r$, $0 \le r \le 2n$, and $C : \mathbb{R}^{n+m} \times P \to \mathbb{R}^k$ are assumed to be sufficiently smooth on appropriate open sets. The admissible class of control functions is that of piecewise continuous controls. The final time t_f is either fixed or free. A control problem with a free final time t_f can be reduced to an augmented control problem with a fixed final time by the time transformation $t = s \cdot t_f$ introducing the new time variable $s \in [0, 1]$. A *non-autonomous* control problem can be reformulated as an *autonomous* problem **OCP**(p) by considering the time variable t as an additional state variable. In the sequel, we assume for simplicity that problem (1) is given in Mayer form with $f_0(x, u, p) \equiv 0$. This form can be achieved by introducing an additional state variable x_{n+1} defined by the differential equation $\dot{x}_{n+1} = f_0(x, u, p)$ with initial value $x_{n+1}(t_0) = 0$.

A sensitivity analysis of parametric control problem **OCP**(p) in the neighbourhood of a fixed *nominal parameter* $p_0 \in P$ has been carried out in Maurer and Augustin [19] to which we shall refer hereafter for assumptions and results. In [19], mixed state-control constraints and *pure control* constraints are treated separately. To shorten the presentation in the present article, we suggest that the formulation of state-control constraints $C(x(t), u(t), p) \le 0$ in (1) also includes *pure state* constraints $C(x(t), p) \le 0$ if the function C does not depend on the control u. As in [19] we assume that there are only finitely many *boundary arcs* or *contact points* where the inequality constraints $C(x(t), u(t), p) \le 0$ becomes active. The finitely many *junction points* with the boundary arcs are denoted by t_j.

2.2 *First Order Necessary Optimality Conditions*

The theory of first order necessary conditions (minimum principle) for the control problem (1) is well developed, cf. Pontryagin [21], Neustadt [20], Maurer [18] and Hartl, Sethi and Vickson [16]. For the sake of simplicity, we suppose that the inequality constraints are not active at the initial and final state, i.e., we have $C(x(t_0), u(t_0), p) < 0$ and $C(x(t_f), u(t_f), p) < 0$. Let us define the *augmented Hamiltonian function* for problem (1) by

$$H(x, \lambda, \mu, u, p) := \lambda^T f(x, u, p) + \mu^T C(x, u, p), \tag{2}$$

where T denotes the transpose, $\lambda \in \mathbb{R}^n$ is the *adjoint variable* and $\mu \in \mathbb{R}^k$ is a multiplier associated with the inequality constraints. Henceforth, partial derivatives of first and second order are denoted by subscripts referring to the variable, e.g., by f_x and H_{xx}, or by the symbol $D_x f$ and $D^2_{xx} H$. The necessary conditions [16,20] for an optimal solution (x^*, u^*) to problem **OCP**(p) imply that there exists a piecewise continuous and piecewise continuously differentiable *adjoint function* $\lambda : [t_0, t_f] \to \mathbb{R}^n$, a piecewise continuous *multiplier function* $\mu : [t_0, t_f] \to \mathbb{R}^k$ with $\mu(t) \ge 0$ for $t \in [t_0, t_f]$, a multiplier $\rho \in \mathbb{R}^r$, and multipliers $\sigma(t_j) \in \mathbb{R}^k$, $\sigma(t_j) \ge 0$, for each *junction* or *contact point* t_j such that the following conditions hold: the *adjoint* *equations*

$$\dot{\lambda}(t)^T = -H_x(x^*(t), \lambda(t), \mu(t), u^*(t), p)$$
$$= -\lambda(t)^T f_x(x^*(t), u^*(t), p) - \mu(t)^T C_x(x^*(t), u^*(t), p), \tag{3}$$

the *transversality conditions*

$$\begin{aligned} \lambda(t_0)^T &= -D_{x(t_0)}(g + \rho^T \psi)(x^*(t_0), x^*(t_f), p), \\ \lambda(t_f)^T &= D_{x(t_f)}(g + \rho^T \psi)(x^*(t_0), x^*(t_f), p), \end{aligned} \tag{4}$$

the *optimality conditions*

$$0 = H_u(x^*(t), \lambda(t), \mu(t), u^*(t), p)$$
$$= \lambda(t)^T f_u(x^*(t), u^*(t), p) + \mu(t)^T C_u(x^*(t), u^*(t), p), \tag{5}$$

and the *junction conditions* for a pure state constraint

$$\lambda(t_j^+)^T = \lambda(t_j^-)^T - \sigma(t_j)^T C_x(x(t_j), p). \tag{6}$$

The notation + and − in (6) indicates the limit from the left, resp. from the right. Note that for pure state constraints $C(x(t), p) \leq 0$ the function $C(x, p)$ is *directly adjoined* to the Hamiltonian; cf. Hartl et al. [16], Section 4. Alternatively, one may adjoin a higher order time derivative of the function $C(x, p)$ to the Hamiltonian as in [19] and [16], Sections 5 and 6. Relations between the adjoint variables and multipliers corresponding to different Hamiltonians may also be found in [16].

3 Numerical Solution of Optimal Control Problems via NLP Methods

The numerical solution of optimal control problems with constraints by nonlinear programming (NLP) techniques is well developed and there exist a number of excellent methods; cf., e.g., Betts [3], Büskens [5,7], Barclay et al. [2]. These methods use a suitable *discretization* of the the control problem **OCP**(p) by which it is transcribed into a NLP problem. In principal these methods can be divided into two classes. The first type of methods is characterized by the fact that both the discretized state *and* control variables are taken as optimization variables. This approach leads to a high dimensional NLP problem which has a sparse structure in the Jacobian of the constraints and the Hessian of the Lagrangian. In the second class of NLP methods, only the discretized control variables are considered as optimization variables whereas the state variables are calculated as functions of the control variables using appropriate numerical integration methods. One obtains a small but dense NLP problem where the dimension of the NLP problem does not depend on the dimension of the differential equation.

To capture the main features of discretization methods, we restrict the discussion to Euler's method (dating back to 1744) applied to the control problem (1). Implementations of higher order approximations of control and state functions may be found in Büskens [5,7].

3.1 Numerical Solution by Full Discretization

Let $N_t > 0$ be a positive integer representing the meshsize. For notational simplicity we choose equidistant mesh points τ_i, $i = 1, \dots, N_t$, with

$$\tau_i = t_0 + (i-1)h, \quad i = 1, \dots, N_t, \quad h := \frac{t_f - t_0}{N_t - 1}. \tag{7}$$

Upon denoting approximations of the values $x(\tau_i)$ and $u(\tau_i)$ by $x^i \in \mathbb{R}^n$ and $u^i \in \mathbb{R}^m$, the control problem (1) in Mayer form with $f_0 \equiv 0$ is replaced by the problem:

$$\begin{array}{lll} \text{Minimize} & g(x^1, x^{N_t}, p) & \\ \text{subject to} & x^{i+1} = x^i + hf(x^i, u^i, p), & i = 1, \dots, N_t - 1, \\ & \psi(x^1, x^{N_t}, p) = 0, & \\ & C(x^i, u^i, p) \leq 0, & i = 1, \dots, N_t. \end{array} \tag{8}$$

Problem (8) defines a perturbed NLP problem NLP(p) of the form (1) in [8], if we choose the optimization variables, the objective function and the constraints as follows:

$$\begin{aligned} z &:= (x^1, \dots, x^{N_t}, u^1, \dots, u^{N_t}) \in \mathbb{R}^{N_z}, \\ F(z,p) &:= g(x^1, x^{N_t}, p), \\ G(z,p) &:= \begin{pmatrix} [-x^{i+1} + x^i + hf(x^i, u^i, p)]_{i=1,\dots,N_t-1} \\ \psi(x^1, x^{N_t}, p) \\ C(x^1, u^1, p) \\ \vdots \\ C(x^{N_t}, u^{N_t}, p) \end{pmatrix} \in \mathbb{R}^{N_c}. \end{aligned} \tag{9}$$

The dimensions are given by $N_z := (n+m)N_t$, $N_e := n(N_t - 1) + r$ and $N_c := (n+k)N_t - n + r$. In may control problems, the boundary conditions are separated in the form

$$\varphi(x(t_0), p) = 0, \quad \hat{\psi}(x(t_f), p) = 0,$$

with suitable functions φ and $\hat{\psi}$. It should be clear that one may remove from the optimization variable z all those components of the initial vector x^1 which are fixed by the initial condition $\varphi(x^1, p) = 0$. In particular, if the initial state is given by the condition $x(t_0) = \varphi(p)$ then the initial vector x^1 can be eliminated completely from the optimization variable z. In this way, a free final time t_f can be handled as an additional state variable in (9) for which the initial condition is not fixed, whereas other components of the initial state eventually are specified.

3.2 Numerical Solution by a Recursive Approach

In practice, optimal control problems often are of high dimension. In view of our aim to establish a computational sensitivity analysis it is mandatory to keep the dimension of the NLP problem as small as possible. Thus it is useful to reduce the dimension of the NLP problem (8). This can be done by considering only the *discretized control variables as optimization variables* and adding eventually the unknown initial state:

$$z := (x^1, u^1, \ldots, u^{N_t}) \in \mathbb{R}^{N_z}, \quad N_z = n + mN_t. \tag{10}$$

The state variables are computed recursively from the Euler approximation in (8) as functions of the control variables:

$$\begin{aligned} x^1(z,p) &:= x^1, \\ x^{i+1}(z,p) &:= x^i(z,p) + hf(x^i(z,p), u^i, p), \quad i = 1, \ldots, N_t - 1, \end{aligned} \tag{11}$$

This leads to the following approximation of (1):

$$\begin{aligned} \text{Minimize} \quad & g(x^1, x^{N_t}(z,p), p), \\ \text{subject to} \quad & \psi(x^1, x^{N_t}(z,p), p) = 0, \\ & C(x^i(z,p), u^i, p) \le 0, \quad i = 1, \ldots, N_t. \end{aligned} \tag{12}$$

Setting

$$\begin{aligned} F(z,p) &:= g(x^1, x^{N_t}(z,p), p), \\ G(z,p) &:= \begin{pmatrix} \psi(x^1, x^{N_t}(z,p), p) \\ C(x^1(z,p), u^1, p) \\ \vdots \\ C(x^{N_t}(z,p), u^{N_t}, p) \end{pmatrix} \in \mathbb{R}^{N_c}, \quad N_e = r,\ N_c = kN_t + r, \end{aligned} \tag{13}$$

we arrive again at a perturbed NLP problem NLP(p) of the form (1) in [7]. This recursive approximation of the control problem results in a dense structure of the Hessian of the Lagrangian, whereas about 50% of the elements in the Jacobian of the constraints are zero due to the relation $x^i(z,p) = x^i(x^1, u^1, \ldots, u^{i-1}, p)$ in (11).

All calculations described in this and the following sections were performed by the code NUDOCCCS of Büskens [5,7] which has implemented also various higher order approximations for state and control variables. The treatment of stiff ODEs or grid refinement techniques can also be found in [7]. The convergence of solutions discretized via Euler's method to solutions of the continuous control problem has been proven in Malanowski, Büskens and Maurer [17]. This article treats only mixed control-state constraints. Convergence properties for control problems with pure state constraints have recently been discussed in Dontchev and Malanowski [11]. Convergence of higher order Ritz discretization schemes has been established in Felgenhauer [14, 15].

4 Post-Optimal Estimation of Adjoint Variables

The general idea behind the post-optimal estimation of the adjoint variables in (2)–(6) is to employ the necessary conditions (8) and (9) in [8] and to use the Lagrange multipliers (13) in [8] related to the equality and inequality constraints of the discretized control problems. In this section two numerical methods are described which refer to the two discretization approaches in the previous section.

4.1 Estimating Adjoint Variables by the Full Discretization Approach

The Lagrange function for problem (8) is defined by $L : \mathbb{R}^{N_z} \times \mathbb{R}^{n(N_t-1)} \times \mathbb{R}^r \times \mathbb{R}^{kN_t} \times P \longrightarrow \mathbb{R}$,

$$L(z,\lambda,\rho,\mu,p) := g(x^1,x^{N_t},p) + \sum_{i=1}^{N_t-1} (\lambda^{i+1})^T \left(x^i + hf(x^i,u^i,p) - x^{i+1}\right) \\ +\rho^T\psi(x^1,x^{N_t},p) + \sum_{i=1}^{N_t} (\mu^i)^T C(x^i,u^i,p), \tag{14}$$

with multipliers $\lambda^i \in \mathbb{R}^n$, $\lambda = (\lambda^2,\dots,\lambda^{N_t})^T \in \mathbb{R}^{n(N_t-1)}$, $\rho \in \mathbb{R}^r$, $\mu^i \in \mathbb{R}^k$, $\mu = (\mu^1,\dots,\mu^{N_t})^T \in \mathbb{R}_+^{kN_t}$. Recall that $\mu_0 = \mu_N = 0$ since the inequality constraint was assumed to be non-active at t_0 and t_f. The evaluation of the KKT conditions in [8], Equations (8) and (9) yields estimates for the *continuous* adjoint variables $\lambda(t)$ in (2)–(6) as follows. The optimality conditions with respect to the variable x^i are for $i = 2,\dots,N_t-1$:

$$\begin{aligned} L_{x^i}(z,\lambda,\rho,\mu,p) &= (\lambda^{i+1})^T + h(\lambda^{i+1})^T f_x(x^i,u^i,p) - (\lambda^i)^T \\ &\quad + (\mu^i)^T C_x(x^i,u^i,p) \\ &= 0. \end{aligned} \tag{15}$$

These relations represent the discretized version of the adjoint equation $\dot{\lambda} = -\lambda f_x$ in (3) if we identify the multiplier $\mu(t)$ in the Hamiltonian (2) by $\mu(\tau_i) \approx \mu^i/h$. An approximation of the first transversality condition in (4) is obtained from the optimality condition

$$\begin{aligned} L_{x^1}(z,\lambda,\rho,\mu,p) &= D_{x^1}(g+\rho^T\psi)(x^1,x^{N_t},p) + (\lambda^2)^T + h(\lambda^2)^T f_x(x^1,u^1,p) \\ &= 0. \end{aligned} \tag{16}$$

Defining the multiplier

$$\lambda^1 := (\lambda^2)^T + h(\lambda^2)^T f_x(x^1,u^1,p), \tag{17}$$

we see that this multiplier λ^1 satisfies both the adjoint equation and the first transversality condition in (4). The second condition in (4) follows from

$$0 = L_{x^{N_t}}(z,\lambda,\rho,\mu,p) = D_{x^{N_t}}(g+\rho^T\psi)(x^1,x^{N_t},p) - (\lambda^{N_t})^T. \tag{18}$$

The optimality conditions with respect to the control variables yield for $i = 1,\dots,N_t-1$

$$L_{u^i}(z,\lambda,\rho,\mu,p) = h(\lambda^{i+1})^T f_u(x^i,u^i,p) + (\mu^i)^T C_u(x^i,u^i,p) = 0 \tag{19}$$

which agree with (5) if we use again the identification $\mu(\tau_i) \approx \mu^i/h$. In the case of a *pure state constraint* $C(x(t),p) \le 0$ we have to modify the identification as follows. Let τ_l be an approximation for the junction or contact point t_j and let μ^l be the corresponding multiplier in (15). Then there exists a splitting $\mu^l = \sigma_j + \tilde{\mu}^l \approx \sigma_j + h\mu(t_j)$ and herewith an approximation of the junction conditions

$$\begin{aligned} 0 = L_{x^l}(z,\lambda,\rho,\mu,p) \approx (\lambda^{l+1})^T - (\lambda^l)^T + h(\lambda^{l+1})^T f_x(x^l,u^l,p) \\ +(\tilde{\mu}^l)^T C_x(x^l,u^l,p) + \sigma_j C_x(x^l,u^l,p) \end{aligned} \tag{20}$$

where $\sigma_j \approx \sigma(t_j)$ is the jump in (6).

4.2 Estimating Adjoint Variables by the Recursive Approach

The Lagrangian function $L : \mathbb{R}^{N_z} \times \mathbb{R}^r \times \mathbb{R}^{kN_t} \times P \longrightarrow \mathbb{R}$ for the NLP problem (12) is

$$L(z,\rho,\mu,p) := (g + \rho^T\psi)(x^1, x^N(z,p),p) + \sum_{i=1}^{N_t} (\mu^i)^T C(x^i(z,p),u^i,p). \tag{21}$$

Büskens [7] has shown that an approximation of the adjoint variables $\lambda(\tau_i)$ for the optimal control problem (1) can be calculated *a posteriori* from the Lagrange function (21). Because of the lengthy proof we omit the details and sketch only the main ideas. Observe that $N_z = n + mN_t$ now represents the reduced number of optimization variables in (10). The recursive definition in (11) allows us to calculate approximations of the adjoint variables without using Lagrange multipliers for the discretized dynamics. One possibility is to use the expressions

$$\lambda(\tau^i) \approx \lambda^i := L_{x^i}(z,\rho,\mu,p), \quad i = 1,\ldots,N_t. \tag{22}$$

Again, a careful study of the KKT conditions for the NLP problem (12) shows that the vectors λ^i as defined by (22) satisfy the discrete adjoint equations (15)–(18), the minimum condition (19) and the junction conditions (20); cf. Büskens [7].

A second more intuitive approach for estimating the multipliers $\lambda^i \approx \lambda(\tau_i)$ is to calculate the vector $\lambda^{N_t} = D_{x^{N_t}}(g + \rho^T\psi)(x^1, x^{N_t}, p)$ where the multiplier $\rho \in \mathbb{R}^r$ is provided by the SQP-method; cf. Formula (13) in [8]. Then relations (15) and (16) are used for the recursive calculation of the vectors λ^i, $i = N_t,\ldots 1$. A detailed analysis of the optimality conditions $L_z(z,\rho,\mu,p) = 0$ then reveals that the necessary conditions (19) hold.

Especially the first method yields accurate approximations even for complicated and highly nonlinear problems. Readers who are interested in technical difficulties that occur in all direct optimization methods for control problems with mixed control-state or pure state constraints are referred to Büskens [7].

5 Sensitivity Analysis and Real-Time Control

For a fixed *reference* or *nominal* parameter p_0 we consider problem **OCP**(p_0) as the *unperturbed* or *nominal* problem. The sensitivity analysis of (1) in Maurer and

Augustin [19] provides conditions such that the unperturbed solution $x_0(t)$, $\lambda_0(t)$, $u_0(t)$ can be embedded into a family of optimal solutions $x(t,p)$, $\lambda(t,p)$, $u(t,p)$ to the perturbed problem $\mathbf{OCP}(p)$. The state functions $x(t,p)$ are of class C^1 with respect to both variables (t,p) whereas the control and adjoint functions $u(t,p)$, $\lambda(t,p)$ are at least piecewise of class C^1. This differentiability property allows us to construct an approximation of the perturbed solution $x(t,p), \lambda(t,p), u(t,p)$ by considering the following first order Taylor expansion where the variable y represents any one of the variables x, u, λ:

$$y(t,p) \approx y_0(t) + \frac{\partial y}{\partial p}(t,p_0)(p-p_0). \tag{23}$$

The optimal solution $y_0(t)$ and the sensitivity differentials $\frac{\partial y}{\partial p}(t,p_0)$ are computed *off-line*. In the case that an actual deviation p from the nominal parameter p_0 is detected, the expression (23) gives an approximation for the perturbed solution which is quickly computable since it requires only matrix-multiplications. There exist two methods to calculate the sensitivity differentials $\frac{\partial y}{\partial p}(t,p_0)$ in (23). One method is the boundary value method whose details may be found in [19]. To avoid the well-known drawbacks of boundary value methods, we propose to apply NLP techniques. The NLP approach is based on the Formulae (21)–(27) in our companion paper [8] which are applicable to the discretized control problem (12) and which allow to compute the sensitivity differentials for state, control and adjoint variables.

In a first step, we evaluate the expression $\frac{dz}{dp}(p_0)$ given in the sensitivity derivative Formula (21) in [8]. This formula yields approximations for the sensitivity of the perturbed optimal control at the mesh points:

$$\frac{\partial u}{\partial p}(\tau_i,p_0) \approx \frac{du^i}{dp}(p_0), \quad i=1,\ldots,N_t. \tag{24}$$

Then the state sensitivities are obtained by differentiating the recursive relation (11) with respect to the parameter.

$$\frac{\partial x}{\partial p}(\tau_i,p_0) \approx \frac{dx^i}{dp}(z(p_0),p_0) = \frac{\partial x^i}{\partial z}(z_0,p_0)\frac{dz}{dp}(p_0) + \frac{\partial x^i}{\partial p}(z_0,p_0). \tag{25}$$

In a final step we compute the sensitivity derivatives of the adjoint variables. Note that the multipliers λ^i given in (22) depend also on the Lagrange multipliers $\eta = (\rho,\mu)$, i.e., we have $\lambda^i = \lambda^i(z,\eta,p)$. The sensitivity derivatives $d\eta/dp$ can be obtained again from formula (21) in [8]. This procedure yields the following approximations for the sensitivity derivatives:

$$\frac{\partial \lambda}{\partial p}(\tau_i, p_0) \approx \frac{d\lambda^i}{dp}(z(p_0), \eta(p_0), p_0) \tag{26}$$

$$= \left(\frac{\partial \lambda^i}{\partial z}(z_0, \eta_0, p_0), \frac{\partial \lambda^i}{\partial \eta^a}(z_0, \eta_0, p_0) \right) \begin{pmatrix} \frac{dz}{dp}(p_0) \\ \frac{d\eta^a}{dp}(p_0) \end{pmatrix}$$

$$+ \frac{\partial \lambda^i}{\partial p}(z_0, \eta_0, p_0).$$

First and second order sensitivities for the objective function are provided by Formulae (23)–(27) in [8] which will not be further discussed here. In the special case that the perturbation p represents a deviation from the nominal trajectory at τ_i, i.e., $x(\tau_i) = x_0(\tau_i) + p$ (this includes also perturbations in the initial values $x(t_0) = x_0(t_0) + p$), Formula (25) in [8] for the objective functional reduces to

$$\frac{dF}{dp}(z(p_0), p_0) = \lambda^i(z_0, \eta_0, p_0) \tag{27}$$

with $\lambda^i(z_0, \eta_0, p_0)$ defined in (22).

This formula constitutes the well-known marginal interpretation of the adjoint variable; cf. Büskens and Maurer [9]. Moreover, differentiating (27) yields the second order sensitivity for the objective

$$\frac{d^2F}{dp^2}(z(p_0), p_0) = \frac{d\lambda^i}{dp}(z_0, \eta_0, p_0), \tag{28}$$

with $\frac{d\lambda^i}{dp}(z_0, \eta_0, p_0)$ from (26). Linear perturbations in the terminal conditions of the form $\psi(x(t_f), p) = \psi(x(t_f)) - p$ lead to simplifications similar to (28) in [8]. In particular, the first and second order sensitivities are given by

$$\frac{dF}{dp}(z(p_0), p_0) = \rho(p_0), \qquad \frac{d^2F}{dp^2}(z(p_0), p_0) = \frac{d\rho}{dp}(p_0). \tag{29}$$

6 Numerical Examples

The sensitivity analysis [19] of optimal control problems is based on boundary value (BVP) methods. The case studies in [19] and the more complex problem in [1] show that the implementation of sensitivity analysis via BVP methods can become a rather difficult task. However, these implementation efforts are compensated by the fact that the BVP approach yields highly accurate optimal solutions and parameter sensitivity derivatives.

In contrast to BVP methods, the implementation of NLP approach poses fewer problems to any user since NLP methods dispense with adjoint variables which can be recovered from a post-optimal analysis. To compare the efficiency of both approaches to sensitivity analysis we choose three numerical examples from [1, 19] as reference problems: the Rayleigh problem with a control constraint, the Van der Pol oscillator under a state constraint and the container crane with a state constraint. We do not want to dwell on these problems again and report only on some typical features of the numerical results.

Instead of the simple Euler discretization in (8)–(20) a fourth-order Runge-Kutta approximation for the state and a linear interpolation of the control variables was used. All calculations were performed by the code NUDOCCCS [5,7] which works with an automatic grid-refinement to have the local discretization error equally distributed and to localize and work out the unknown junction or contact points. The computations for the nominal solutions and the sensitivity differentials were done on a one processor PC and took only a few seconds of CPU time.

The grid-refinement was terminated between $N_t = 81$ and $N_t = 126$ meshpoints when the calculated optimal values of the objective function were exact within 8 digits as compared to the solutions in [1, 19]. The adapted grids were used also for the post-optimal calculation of the adjoints functions by means of Formulae (22) which gave a precision of about 6 digits. The evaluation of the sensitivity differentials (26) for the adjoint variables produced an accuracy of about 4 digits. Note that this accuracy is by far higher than that required for the first order real-time approximation in (23). More precise solutions can be obtained by a further refinement of the meshpoints and/or a scaling of Formulae (20) in [8].

For a more complex practical example, e.g., the real-time control of an industrial robot, the reader is referred to [9, 10]. In [6] a demanding problem from flight dynamics is discussed and, in addition, a method is proposed for reducing the error in the constraints which is caused by the first order real-time approximation (23).

References

1. D. Augustin, H. Maurer: Sensitivity Analysis and Real-Time Control of a Container Crane under State Constraints. This volume
2. A. Barclay, P. E. Gill, J. B. Rosen: SQP Methods and their Application to Numerical Optimal Control. In Variational Calculus, Optimal Control and Applications, W. H. Schmidt, Heier, K., Bittner, L., Bulirsch, R., eds., Birkhäuser Basel, Boston, Berlin (1998) 207–222
3. J. T. Betts: Survey of Numerical Methods for Trajectory Optimization. Journal of Guidance, Control, and Dynamics, **21** (1998) 193–207
4. H. G. Bock, K. J. Plitt: A Multiple Shooting Algorithm for Direct Solution of Optimal Control Problems. IFAC 9th World Congress, Budapest, Hungary (1984)
5. C. Büskens: Direkte Optimierungsmethoden zur numerischen Berechnung optimaler Steuerungen. Diploma thesis, Institut für Numerische Mathematik, Universität Münster, Münster, Germany (1993)
6. C. Büskens: Real-Time Solutions for Perturbed Optimal Control Problems by a Mixed Open- and Closed-Loop Strategy. This volume.
7. C. Büskens: Optimierungsmethoden und Sensitivitätsanalyse für optimale Steuerprozesse mit Steuer- und Zustands-Beschränkungen. Dissertation, Institut für Numerische Mathematik, Universität Münster, Münster, Germany (1998)

8. C. Büskens, H. Maurer: Sensitivity Analysis and Real-Time Optimization of Parametric Nonlinear Programming Methods. This volume.
9. C. Büskens, H. Maurer: Real-Time Control of Robots with Initial Value Perturbations via Nonlinear Programming Methods. Optimization **47** (2000) 383–405
10. C. Büskens, H. Maurer: Real-Time Control of an Industrial Robot. This volume.
11. A. L. Dontchev, W. W. Hager, K. Malanowski: Error Bounds for Euler Approximation of a State and Control Constrained Optimal Control Problem. Functional Analysis and Optimization **21** (2000) 653–682
12. P. J. Enright, B. A. Conway: Discrete Approximations to Optimal Trajectories Using Direct Transcription and Nonlinear Programming. AIAA Paper 90-2963-CP (1990)
13. Yu. G. Evtushenko: Numerical Optimization Techniques. Translation Series in Mathematics and Engineering, Optimisation Software Inc., Publications Division, New York (1985)
14. U. Felgenhauer: Diskretisierung von Steuerungsproblemen unter stabilen Optimalitätsbedingungen. Institut für Mathematik, Habilitation, Technische Universität Cottbus, Cottbus, Germany (1998)
15. U. Felgenhauer: On Higher Order Methods for Control Problems with Mixed Inequality Constraints. Institut für Mathematik, Preprint **M-01/1998**, Technische Universität Cottbus, Cottbus, Germany (1998)
16. R. F. Hartl, S. P. Sethi, R. G. Vickson: A Survey of the Maximum Principles for Optimal Control Problems with State Constraints. SIAM Review **37** (1995) 181–218
17. K. Malanowski, C. Büskens, H. Maurer: Convergence of Approximations to Nonlinear Optimal Control Problems. In: Mathematical Programming with Data Perturbations, A. V. Fiacco, ed., Lecture notes in pure and applied mathematics, Vol. **195**, Marcel Dekker, Inc. (1998) 253–284
18. H. Maurer: Optimale Steuerprozesse mit Zustandsbeschränkungen. Mathematisches Institut, Habilitation, Universität Würzburg, Würzburg, Germany (1976).
19. H. Maurer, D. Augustin: Sensitivity Analysis and Real-Time Control of Parametric Optimal Control Problems Using Boundary Value Methods. This volume.
20. L. W. Neustadt: Optimization: A Theory of Necessary Conditions. Princeton University Press, Princeton, New Jersey (1976)
21. L. S. Pontrjagin, V. G. Boltjanskij, R. V. Gamkrelidze, E. F. Miscenko: Mathematische Theorie optimaler Prozesse. R. Oldenbourg, München, Wien (1967)
22. O. von Stryk, Numerische Lösung optimaler Steuerungsprobleme: Diskretisierung, Parameteroptimierung und Berechnung der adjungierten Variablen. Fortschritt-Berichte VDI, Reihe 8, Nr. 441 VDI Verlag, Germany (1995)

Sensitivity Analysis and Real-Time Control of a Container Crane under State Constraints

Dirk Augustin and Helmut Maurer

Institut für Numerische Mathematik, Westfälische Wilhelms-Universität Münster, Germany

Abstract The sensitivity analysis for state constrained optimal control problems [7, 8] is illustrated by a practically relevant problem: the optimal control of a container crane based on a model developed in [12]. The container crane is subject to a state constraint on the vertical velocity. The multiple shooting method is used to determine a highly precise nominal solution which is still lacking in the literature. Second order sufficient conditions are checked by showing that an associated Riccati equation has a bounded solution. Sensitivity differentials of optimal solutions are computed with respect to perturbations in the swing angle and velocity. This allows for a fast computation of real-time approximations of perturbed optimal solutions.

1 Introduction

Cargo handling in ship or railroad terminals is mostly operated by container cranes. Sakawa and Shindo [12] have developed a dynamical control model to improve the efficiency of crane operations. The critical part of the crane motion, the diagonal motion, is optimally controlled such as to avoid a large swing of the container load. In addition, several control and state constraints are imposed. This model has been chosen by other authors [10, 13] as a test example for optimal control algorithms. However, the numerical solutions presented in [10, 12, 13] are not very accurate and are not complete in the sense that one might verify necessary optimality conditions of first order. This fact has motivated us to reconsider the problem and to determine a numerical solution that satisfies optimality conditions with high accuracy.

The numerical solution presented in Sakawa and Shindo [12] is not congruent with the cost functional (18) or (48) considered by the authors. The optimal controls seem to correspond to a cost functional including a quadratic penalty term for the controls. For this reason, we will consider in section 2 an optimal control problem that augments the cost functional in [12] by a control quadratic penalty term which also enforces the strict Legendre-Clebsch condition. With this modification the model serves as a rather complex example by which we can illustrate the *sensitivity analysis* for optimal control problems subject to *pure* state constraints [7], [8].

We present a numerical solution including the adjoint variables and the junction points with the state boundary. In section 3, the boundary value problem (BVP) is derived which characterizes the optimal solution. The formulation of this BVP incorporates the assumed structure of the optimal control with respect to the active state constraint. The numerical solution of this BVP via multiple shooting techniques [9] allows us to check first order conditions with high accuracy. This approach clearly shows the advantage of the explicit formulation of the BVP to other optimization approaches.

We will be able to verify all assumptions for solution differentiability of the optimal solutions when the system is subject to perturbations; cf. [7] and [8]. The crucial assumption for differentiability of solutions is that second order sufficient conditions (SSC) are satisfied for the nominal solution. In section 4, SSC are tested by showing that an associated matrix Riccati equation has a bounded solution. The rather complicated Riccati equations are given in more detail. Section 5 discusses the numerical methods for computing the sensitivity differentials of optimal solutions with respect to parameters. As a parameter of practical importance we choose a perturbation in the swing angle or velocity. In section 6, we use the sensitivity differentials to perform real-time control approximations of the perturbed solutions on the basis of first order Taylor expansions.

2 Optimal Control of a Container Crane

The dynamical model in [12] describes a container crane which is equipped with a trolly drive motor (acceleration: control variable u_1) and a hoist motor (acceleration: control variable u_2); see Figure 1. The aim of the control process is to keep the swing angle as small as possible since a large swing of the container load during the transfer may become dangerous. The critical part of the motion is the *diagonal motion* from point **B** to point **C** as shown in Figure 2 where the vertical motion is connected with the horizontal motion. The model comprises six state variables and two control variables:

x_1 : horizontal motion,	x_4 : horizontal velocity,
x_2 : vertical motion,	x_5 : vertical velocity,
x_3 : swing angle,	x_6 : swing velocity,
u_1 : control via trolly drive motor,	u_2 : control via hoist motor.

The dynamical model, boundary conditions and control resp. state constraints are considered in the time interval $[0, t_f]$ with fixed final time $t_f > 0$:

$$\begin{aligned} \dot{x}_1 &= x_4, \quad & \dot{x}_2 &= x_5, \\ \dot{x}_3 &= x_6, \quad & \dot{x}_4 &= u_1 + 17.2656 x_3, \\ \dot{x}_5 &= u_2, \quad & \dot{x}_6 &= -\frac{1}{x_2}(u_1 + 27.0756 x_3 + 2 x_5 x_6), \end{aligned} \tag{1}$$

$$x(0) = (0, 22, 0, 0, -1, 0)^*, \quad x(t_f) = (10, 14, 0, 2.5, 0, 0)^*, \tag{2}$$

$$|x_4(t)| \le 2.5, \quad |x_5(t)| \le 1, \tag{3}$$

$$|u_1(t)| \le 2.83374, \quad -0.80865 \le u_2(t) \le 0.71265. \tag{4}$$

The boundary conditions (2) correspond to point **B** resp. **C** in Figure 2. Note that the load arrives at point **B** with maximal vertical velocity which gives the initial condition $x_5(0) = -1$ at point **B**. Thus the state constraint $|x_5(t)| \le 1$ becomes active at $t_1 = 0$. The goal of the control process is to keep the swing angle and velocity small. As in [12], the final time is chosen as $t_f = 9$ sec.

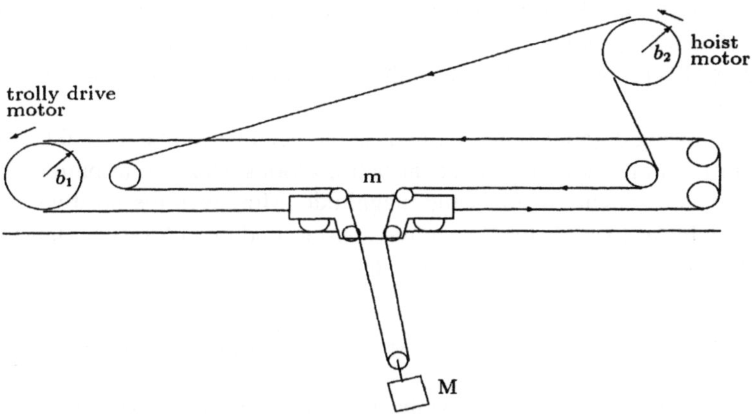

Figure 1. Control u_1 by trolly drive motor and control u_2 by hoist motor

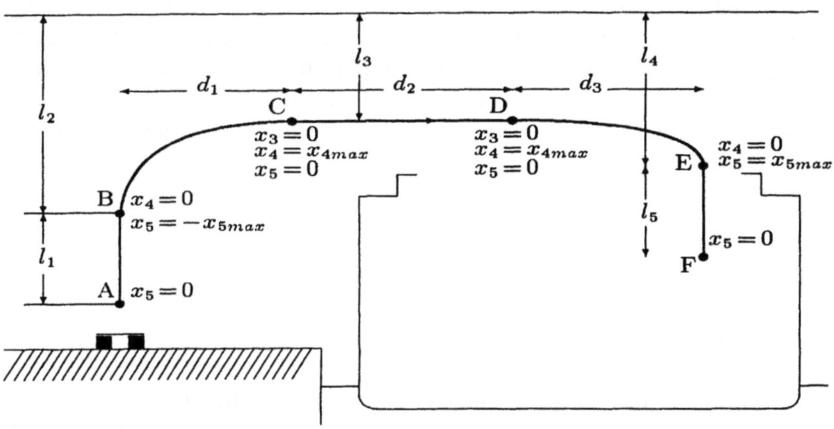

Figure 2. The diagonal motion **B** ⇒ **C** of the container crane

We consider the following cost criterion with a penalty parameter $c \geq 0$:

$$F(x, u) = \frac{1}{2} \int_0^9 \left[x_3^2(t) + x_6^2(t) + c(u_1^2(t) + u_2^2(t)) \right] dt. \tag{5}$$

The optimal control problem is to determine control functions $u_i \in L^\infty(0, t_f; \mathbb{R})$, $i = 1, 2$, which minimize the cost criterion (5) subject to the constraints (1)–(4). Sakawa and Shindo [12], formula (18) resp. (48) claim that they determine optimal controls for the cost criterion (5) with $c = 0$ by adding a quadratic penalty term for the terminal constraints. However, the computed optimal controls in Figure 4 of [12] are likely obtained by assuming a positive weight $c > 0$ in (5). The article of Teo and Jennings [13] presents a solution for the cost criterion (5) with $c = 0$ to which we refer later.

In the following computations, we choose the positive weight $c = 0.01$. This penalty parameter produces state and control trajectories whose contributions to the functional (5) are roughly of the same magnitude. Moreover, the positive weight $c > 0$ will enable us to apply the Legendre-Clebsch condition (AS-4) in Maurer and Augustin [8] which will be crucial for sensitivity analysis.

3 Boundary Value Problem for the Nominal Solution

The notations and equations refer to the theory of state constrained control problems which is summarized in section 5 of [8], cf. also [5]. The state constraint $-1 \le x_5(t)$ can be written as the constraint $S(x, p) := -1 - x_5 \le 0$. Then the function S^1 defined in [8], equation (5.2), is given by $S^1(x, u, p) = \frac{d}{dt} S(x, p) = -\dot{x}_5 = -u_2$. Hence, the state constraint has order $l = 1$. When determining the structure of the optimal solution we observe the following property: the boundary condition $x_5(0) = -1$ suggests that there exists a boundary arc $x_5(t) = -1,\ 0 \le t \le t_2$, with entry-time $t_1 = 0$ and exit-time $t_2 > 0$. Due to our choice of the weight $c = 0.01$ in the cost functional (5) it turns out that the interval $[0, t_2]$ is the *only interval* where the state and control constraints (3) and (4) become active.

Then the Hamiltonian function, resp., augmented Hamiltonian are given by, cf. [8], equation (3.1), resp. (5.6):

$$H^0 = \frac{1}{2}[x_3^2 + x_6^2 + 0.01(u_1^2 + u_2^2)] + \lambda_1 x_4 + \lambda_2 x_5 + \lambda_3 x_6 + \lambda_5 u_2$$
$$+\lambda_4(u_1 + 17.2656 x_3) - \frac{\lambda_6}{x_2}(u_1 + 27.0756 x_3 + 2 x_5 x_6), \tag{6}$$

$$H = H^0 + \nu^1 S^1 = H^0 - \nu^1 u_2, \quad \nu^1 \in \mathbb{R}. \tag{7}$$

Since the initial point lies on the boundary of the state constraint $|x_5(t)| \le 1$, the existence of non-trivial multipliers requires additional conditions to be satisfied [1, 11]. It can be easily checked that these conditions, e.g., relation (2.2) in [11], hold for the control problem considered here. Then from (5.7) in [8] we obtain the adjoint equations

$$\begin{aligned}
\dot{\lambda}_1 &= 0, & \dot{\lambda}_4 &= -\lambda_1, \\
\dot{\lambda}_2 &= -\frac{\lambda_6}{x_2^2}(u_1 + 27.0756 x_3 + 2 x_5 x_6), & \dot{\lambda}_5 &= -\lambda_2 + \frac{2 x_6 \lambda_6}{x_2}, \\
\dot{\lambda}_3 &= -x_3 - 17.2656 \lambda_4 + \frac{27.0756 \lambda_6}{x_2}, & \dot{\lambda}_6 &= -\lambda_3 - x_6 + \frac{2 x_5 \lambda_6}{x_2}.
\end{aligned} \tag{8}$$

No boundary conditions are imposed for the adjoint variables since the final state is fixed for the endtime $t_f = 9$ sec. On the *interior arc* $[t_2, 9]$, the control variables are determined from the relations $H_{u_i} = 0,\ i = 1, 2$. On the *boundary arc* $[0, t_2]$, the boundary controls satisfy the relations $S^1(x, u, p) = -u_2 = 0$ and $H_{u_1} = 0$. This leads to the control law

$$u_1 = 100\left(\frac{\lambda_6}{x_2} - \lambda_4\right) \text{ on } [0,9], \quad u_2 = \left\{\begin{array}{ccc} 0 & , & \text{on } [0, t_2] \\ -100\lambda_5 & , & \text{on } [t_2, 9] \end{array}\right\} \tag{9}$$

Because the entry point of the boundary arc is $t_1 = 0$ we can ignore the jump condition (5.11) in [8] for the adjoint variable. We recall that the adjoint variables are *continuous* in the *exit* point t_2. The continuity of the control u_2 at the exit-time t_2 yields the additional junction condition

$$\lambda_5(t_2) = 0. \tag{10}$$

The multiplier ν^1 in (7) for the state constraint is determined by the condition $0 = H_{u_2} = 0.01u_2 + \lambda_5 - \nu^1$. Since $u_2 = 0$ on $[0, t_2]$ we get explicitly

$$\nu^1(t) = \left\{\begin{array}{ll} \lambda_5(t), & t \in [0, t_2] \\ 0 \quad , & t \in [t_2, 9] \end{array}\right\}. \tag{11}$$

Hence, the optimal solution satisfies the BVP consisting of the ODEs and boundary conditions (1), (2), (8)–(10) and the unknown exit-time t_2. The multiple shooting code BNDSCO of Oberle and Grimm [9] yields the following solution with $c = 0.01$ in (5):

$$\begin{array}{rlrl}
\lambda_1(0) &= 2.431557994 \times 10^{-3}, & \lambda_2(0) &= 1.830176462 \times 10^{-3}, \\
\lambda_3(0) &= 1.340766745 \times 10^{-1}, & \lambda_4(0) &= -1.607058155 \times 10^{-2}, \\
\lambda_5(0) &= 9.369280372 \times 10^{-3}, & \lambda_6(0) &= -2.139967667 \times 10^{-1}, \\
\lambda_1(9) &= \lambda_1(0), & \lambda_2(9) &= 1.713132770 \times 10^{-3}, \\
\lambda_3(9) &= -1.136061218 \times 10^{-1}, & \lambda_4(9) &= -3.795460349 \times 10^{-2}, \\
\lambda_5(9) &= -6.899740864 \times 10^{-3}, & \lambda_6(9) &= -3.594353079 \times 10^{-1}, \\
t_2 &= 5.672903236, & F(x_0, u_0) &= 3.751946138 \times 10^{-2}.
\end{array}$$

The corresponding state, control and adjoint variables are displayed in Figures 3 and 4. Notice that the control and state variables are rather similar to those given in Figure 4 of Sakawa and Shindo [12].

Next, we test if this solution satisfies the assumptions in section 5 of [8]. The regularity condition (AS-1) trivially holds in view of $D_uS^1 = (0, -1)$. The controllability condition (AS-2) in [8] is a consequence of the fact that the Jacobian matrix of the shooting method is regular. The strict Legendre-Clebsch condition (AS-4) trivially holds due to $H_{uu} = 0.01 \cdot I_2 > 0$. The strict complementarity (AS-3) amounts to the requirement that the multiplier ν^1 associated with the state constraint $-1 \le x_5(t)$ satisfies the following inequality with some $\alpha > 0$:

$$-\frac{d}{dt}\nu^1(t) \ge \alpha > 0 \quad \text{for all } t \in [0, t_2].$$

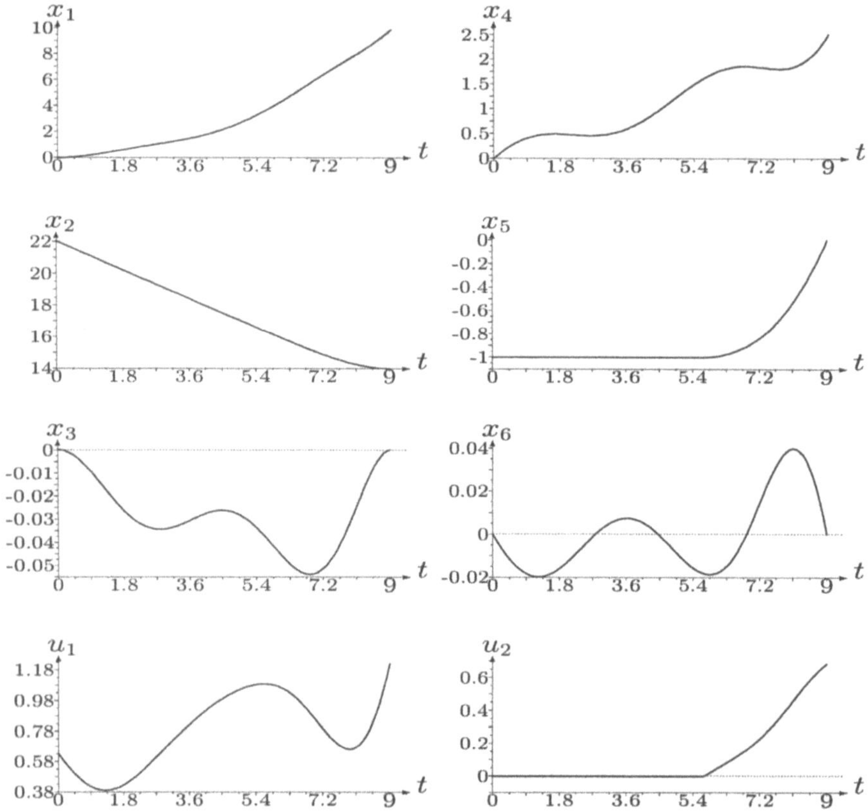

Figure 3. State variables $x_1(t),\dots,x_6(t)$ and control variables $u_1(t), u_2(t)$

In view of (11), the multiplier is given by $\nu^1(t) = \lambda_5(t),\ t \in [0, t_2]$. Then a look at Figure 4 may immediately convince the reader that the preceding inequality holds. The *non-tangential junction condition* (AS-6) requires that the derivative $\dot{u}_2(t)$ be *discontinuous* at the junction point t_2. The optimal control u_2 shown in Figure 3 confirms that this property holds. The test of the coercivity condition (AS-5) is deferred to the next section.

We point out that the dynamical system (1) constitutes a simplified model which is derived from a fully nonlinear model [12] by replacing $\sin(x) \approx x$ for small values of x. We have solved the complete nonlinear model of [12] by shooting methods and have noticed only a negligible discrepancy between optimal solutions for both models. This justifies our point of view to use the simplified model for the following second order test and sensitivity analysis.

It is also noteworthy to consider the solution to the optimal control problem (1)–(5) where the penalty parameter in the cost functional (5) is set to $c = 0$. All control constraints (4) become active and the optimal control is composed by *bang-bang and singular arcs*. We were not yet able to compute this solution via

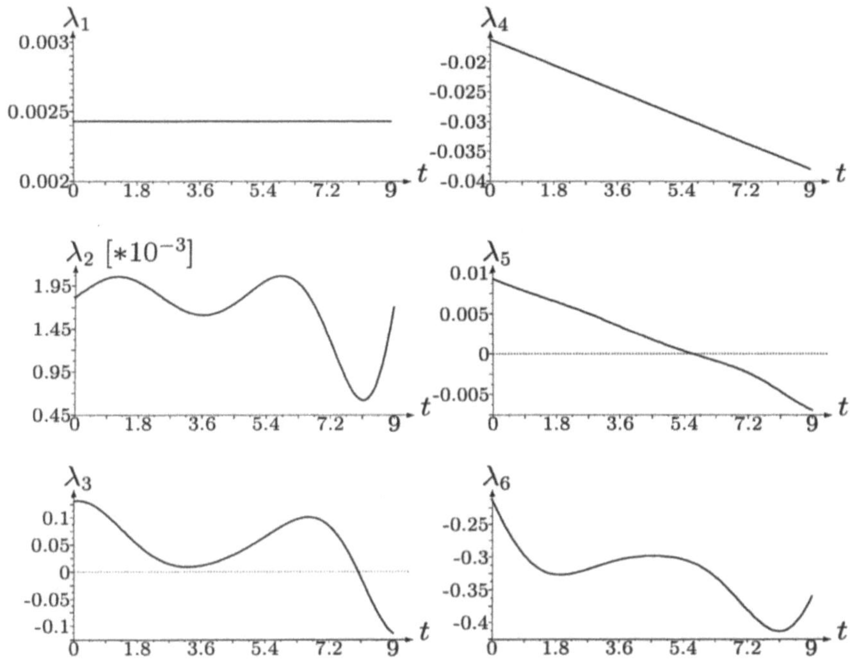

Figure 4. Adjoint variables $\lambda_1(t), \ldots, \lambda_6(t)$

shooting methods. Instead, we have used the nonlinear programming approach and code NUDOCCCS described in [3,4]. The optimal solution shown in Figure 5 is obtained on the basis of $N = 210$ grid points which yields the optimal functional value $F(x_0, u_0) = 5.1504365 \times 10^{-3}$. This solution is substantially different from the one in Figure 3 since, rather inexpectedly, the solution contains *two boundary arcs* with $x_5(t) \equiv -1$. Also, this solution does not agree with the one presented in Teo and Jennings [13] who obtained the functional value $F(x_0, u_0) = 5.361 \times 10^{-3}$. The solution in [13] does not clearly exhibit the bang-bang and singular arcs which is probably due to the fact that the authors used a a rather coarse discretisation.

4 Second Order Sufficient Conditions, Riccati Equations

The last test needed to verify that the solution (12) provides a local optimum is the coercivity condition (AS-5) in [8]. We are able to test this condition in a strong form by showing that the Riccati equation (3.12) in [8], i.e., the equation

$$\begin{aligned}\dot{Q}(t) = \; & -Q(t)f_x[t] - f_x[t]^*Q(t) - H_{xx}[t] \\ & +(H_{xu}[t] + Q(t)f_u[t])(H_{uu}[t])^{-1}(H_{ux}[t] + f_u[t]^*Q(t)), \quad (13)\end{aligned}$$

has a *bounded* solution; cf. also [6]. The jump condition for $Q(t)$ in [8], equation (5.15), does not apply since $t_1 = 0$. Note also that the boundary conditions (3.22)

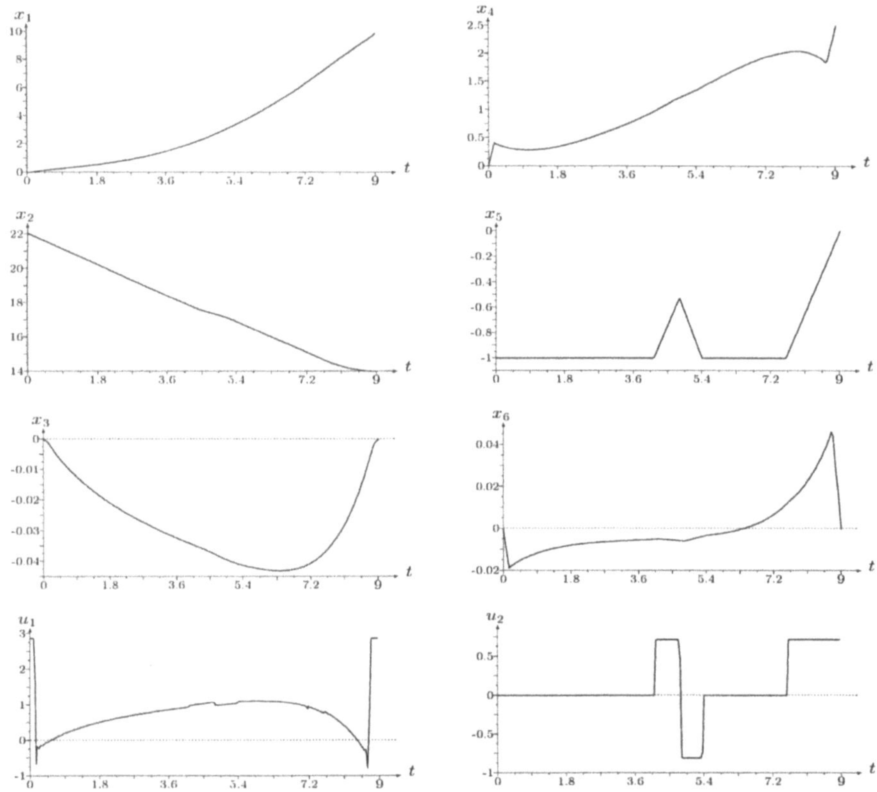

Figure 5. Cost functional (5) with $c = 0$: state variables $x_1(t), \ldots, x_6(t)$ and control variables $u_1(t), u_2(t)$ with bang-bang and sigular arcs

in [8] do not yield any boundary conditions for $Q(0)$ and $Q(t_f = 9)$ since the initial and final states are specified. The crane model has dimension $n = 6$. Hence, for solving the Riccati equation (13) we consider a symmetric (6×6)-matrix $Q(t)$ in the form

$$Q(t) = (q_{ik}(t))_{1 \le i,k \le 6}, \quad q_{ik}(t) = q_{ki}(t).$$

To give the reader an impression of the complexity of the problem, we explicitly provide some of the 21 equations for the elements of the matrix Riccati equation (13). For this purpose it is convenient to introduce the notation

$$c_1 := 17.2656, \quad c_2 := 27.0756, \quad k(x,u) := u_1 + c_2 x_3 + 2x_5 x_6.$$

The first 11 Riccati equations are *homogeneous* in the corresponding variables. Since no boundary conditions are prescribed for $Q(t)$, these equations admit the trivial solutions

$$q_{11} \equiv q_{12} \equiv q_{13} \equiv q_{14} \equiv q_{15} \equiv q_{16} \equiv q_{24} \equiv q_{34} \equiv q_{44} \equiv q_{45} \equiv q_{46} \equiv 0.$$

This allows to simplify the remaining 10 Riccati equations:

$$\dot{q}_{22} = -2k(x,u)\frac{q_{26}}{x_2^2} + 100\left(\left(q_{24} - \frac{q_{26}}{x_2} + \frac{\lambda_6}{x_2^2}\right)^2 + q_{25}^2\right),$$

$$\dot{q}_{23} = \frac{c_2 q_{26}}{x_2} - \frac{k(x,u)q_{36} + c_2\lambda_6}{x_2^2} + 100\left(\left(\frac{q_{26}}{x_2^2} - \frac{\lambda_6}{x_2^3}\right) q_{36} + q_{25}q_{35}\right)$$

$$\dot{q}_{25} = \frac{2x_6 q_{26}}{x_2} - q_{22} - \frac{k(x,u)q_{56} - 2\lambda_6 x_6}{x_2^2} + 100\left(\left(\frac{q_{26}}{x_2^2} - \frac{\lambda_6}{x_2^3}\right) q_{56} + q_{25}q_{55}\right),$$

$$\dot{q}_{26} = \frac{2x_5 q_{26}}{x_2} - q_{23} - \frac{k(x,u)q_{66} + 2\lambda_6 x_5}{x_2^2} + 100\left(\left(\frac{q_{26}}{x_2^2} - \frac{\lambda_6}{x_2^3}\right) q_{66} + q_{25}q_{56}\right),$$

$$\dot{q}_{33} = \frac{2c_2 q_{36}}{x_2} - 1 + \frac{100 q_{36}^2}{x_2^2} + 100 q_{35}^2,$$

$$\dot{q}_{35} = \frac{2x_6 q_{36} + c_2 q_{56}}{x_2} - q_{23} + 100\left(\frac{q_{36}q_{56}}{x_2^2} + q_{35}q_{55}\right),$$

$$\dot{q}_{36} = \frac{2x_5 q_{36} + c_2 q_{66}}{x_2} - q_{33} + 100\left(\frac{q_{36}q_{66}}{x_2^2} + q_{35}q_{56}\right),$$

$$\dot{q}_{55} = \frac{4x_6 q_{56}}{x_2} - 2q_{25} + 100\left(\frac{q_{56}^2}{x_2^2} + q_{55}^2\right),$$

$$\dot{q}_{56} = \frac{2x_5 q_{56} + 2x_6 q_{66}}{x_2} - q_{35} - q_{26} + 100\left(\frac{q_{56}q_{66}}{x_2^2} + q_{55}q_{56}\right),$$

$$\dot{q}_{66} = \frac{4x_5 q_{66}}{x_2} - 2q_{36} - 1 + 100\left(\frac{q_{66}^2}{x_2^2} + q_{56}^2\right).$$

Unfortunately, there does not exist a systematic way to determine appropriate initial values which yield a *bounded solution* of these equations. Our strategy was to set some initial values to zero and to find appropriate initial values for the remaining variables. Finally, a bounded solution of this set of equations is obtained by choosing the initial values

$$q_{33}(0) = -3, \quad q_{36}(0) = 0.3, \quad q_{66}(0) = -2,$$

$$q_{ij}(0) = 0, \quad (i,j) \in \{(2,2),(2,3),(2,5),(2,6),(3,5),(5,5),(5,6)\},$$

which yield the bound $\|Q(t)\|_\infty \leq 4 \;\; \forall\, t \in [0,9]$. Hence we may conclude that the solution characterized by (12) is a local minimum of problem (1)–(5). We point out that the correctness of the Riccati equations can be checked via the routines for symbolic computation in the toolbox MATLAB.

5 Sensitivity Analysis for the Swing Angle and Swing Velocity

The two preceding sections have demonstrated that all assumptions for Theorem 4 in section 5 of [8] are satisfied. Thus we arrive at the following conclusion: if p is any finite-dim. parameter vector which enters the control system (1)–(5) and is sufficiently close to a nominal parameter p_0, then the parametric optimal control problem admits a local optimal solution $x(t,p)$, $u(t,p)$, $\lambda(t,p)$ which is Fréchet-differentiable with respect to both variables (t,p) except at the junction point t_2.

First, we take the initial value $p = x_3(0)$ of the swing angle as a *scalar* perturbation parameter. Clearly, the nominal parameter corresponding to the nominal solution (12) is $p_0 = 0$. The *sensitivity differentials*

$$y(t) := \frac{\partial x}{\partial p}(t;p_0), \quad \gamma(t) := \frac{\partial \lambda}{\partial p}(t;p_0)$$

can be obtained by formally differentiating the BVP (1), (2) and (8)–(11) with respect to $p = x_3(0)$. This leads to the following set of equations that are *linear* in the variables y and γ:

$$\dot{y}_1 = y_4, \quad \dot{y}_2 = y_5, \quad \dot{y}_3 = y_6, \quad \dot{y}_4 = \frac{\partial u_1}{\partial p} + 17.2656\, y_3, \quad \dot{y}_5 = \frac{\partial u_2}{\partial p},$$

$$\begin{aligned} \dot{y}_6 = {} & \frac{y_2}{x_2^2}(u_1 + 27.0756\, x_3 + 2x_5x_6) \\ & - \frac{1}{x_2}\left(\frac{\partial u_1}{\partial p} + 27.0756\, y_3 + 2(y_5x_6 + x_5y_6)\right), \end{aligned}$$

$$\dot{\gamma}_1 = 0, \quad \dot{\gamma}_4 = -\gamma_1,$$

$$\begin{aligned} \dot{\gamma}_2 = {} & \frac{2y_2\lambda_6}{x_2^3}(u_1 + 27.0756\, x_3 + 2x_5x_6) - \frac{\gamma_6}{x_2^2}(u_1 + 27.0756\, x_3 + 2x_5x_6) \\ & - \frac{\lambda_6}{x_2^2}\left(\frac{\partial u_1}{\partial p} + 27.0756\, y_3 + 2(y_5x_6 + x_5y_6)\right), \end{aligned}$$

$$\dot{\gamma}_3 = -y_3 - 17.2656\,\gamma_4 + 27.0756\,\frac{\gamma_6 x_2 - \lambda_6 y_2}{x_2^2},$$

$$\dot{\gamma}_5 = -\gamma_2 + \frac{2}{x_2^2}(x_2(y_6\lambda_6 + x_6\gamma_6) - x_6\lambda_6 y_2),$$

$$\dot{\gamma}_6 = -\gamma_3 - y_6 + \frac{2}{x_2^2}(x_2(y_5\lambda_6 + x_5\gamma_6) - x_5\lambda_6 y_2).$$

In this set of equations, the functions $x_1, \ldots, x_6, \lambda_1, \ldots, \lambda_6$ represent the nominal solution (12). The boundary conditions are

$$y_3(0) = 1, \quad y_i(0) = 0 \ (i = 1,2,4,5,6), \quad y_i(t_f) = 0 \ (i = 1,\ldots,6).$$

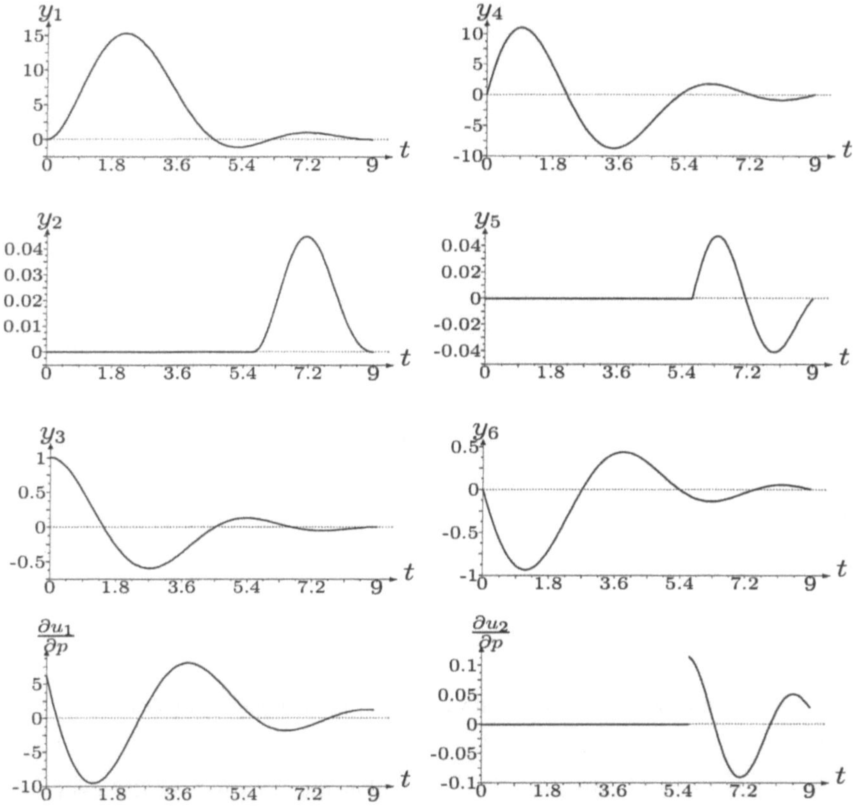

Figure 6. State variables sensitivities $y_1, \ldots, y_6$ and control sensitivities $\frac{\partial u_1}{\partial p}, \frac{\partial u_2}{\partial p}$

The sensitivity differentials $\partial u_i/\partial p$, $i = 1, 2$, are computed by differentiating the relations (9):

$$\frac{\partial u_1}{\partial p} = 100 \left(\frac{\gamma_6}{x_2} - \frac{\lambda_6 y_2}{x_2^2} - \gamma_4 \right) \text{ on } [0, 9], \quad \frac{\partial u_2}{\partial p} = \left\{ \begin{array}{rl} 0, & \text{on } [0, t_2] \\ -100\gamma_5, & \text{on } [t_2, 9] \end{array} \right\}.$$

The multiple shooting code BNDSCO of [9] provides the following solution:

$$\begin{array}{ll}
\gamma_1(0) = -1.755650107 \times 10^{-2}, & \gamma_2(0) = 6.266880594 \times 10^{-3}, \\
\gamma_3(0) = 4.744706998, & \gamma_4(0) = -1.413689234 \times 10^{-1}, \\
\gamma_5(0) = -5.392809798 \times 10^{-3}, & \gamma_6(0) = -1.740429703, \\
\gamma_1(9) = \gamma_1(0), & \gamma_2(9) = -7.343275885 \times 10^{-4}, \\
\gamma_3(9) = -1.725036536 \times 10^{-1}, & \gamma_4(9) = 1.663958622 \times 10^{-2}, \\
\gamma_5(9) = -2.842474846 \times 10^{-4}, & \gamma_6(9) = 4.127712739 \times 10^{-1}.
\end{array}$$

Sensitivity differentials for state and control variables are displayed in Figure 6 and those for the adjoint variables are given in Figure 7.

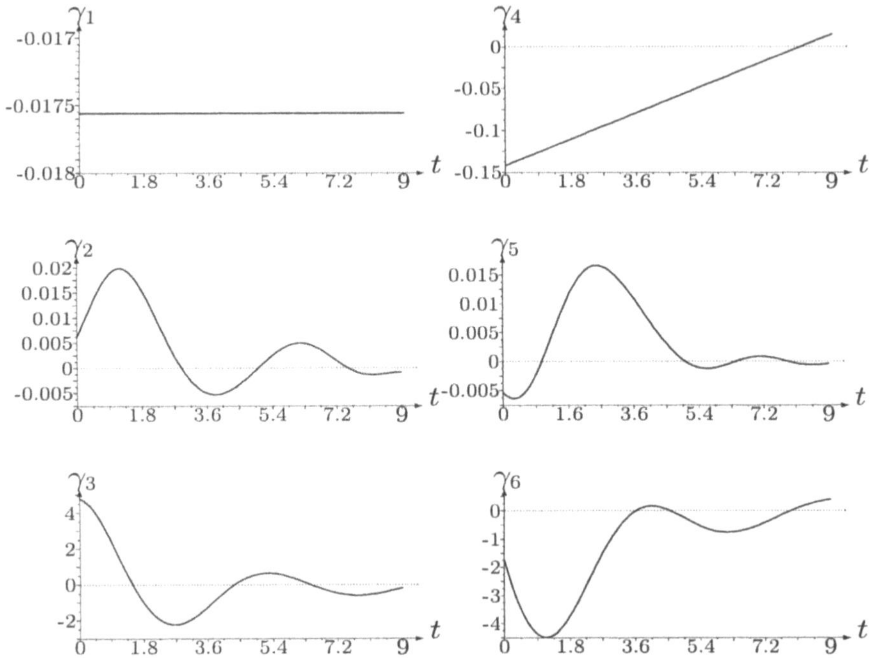

Figure 7. Sensitivity differentials $\gamma_1, \ldots, \gamma_6$ of adjoint variables

To conclude the sensitivity analysis we calculate the sensitivity of the switching point t_2. Due to the junction condition (10) the relation $\lambda_5(t_2(p), p) = 0$ holds for all p near $p_0 = 0$. By differentiation we obtain

$$\dot{\lambda}_5(t_2(p_0), p_0) \frac{dt_2}{dp}(p_0) + \gamma_5(t_2(p_0)) = 0$$

which yields

$$\frac{dt_2}{dp}(p_0) = -\gamma_5(t_2(p_0))/\dot{\lambda}_5(t_2(p_0), p_0) = -0.8340417352. \qquad (14)$$

Another perturbation parameter of practical interest is the initial value $q := x_6(0)$ of the swing velocity for which a sensitivity analysis can be performed in a similar way. We merely give the result for the sensitivity derivative of the switching point $t_2 = t_2(q)$:

$$\frac{dt_2}{dq}(q_0) = -\gamma_5(t_2(q_0))/\dot{\lambda}_5(t_2(q_0), q_0) = -6.655147054. \qquad (15)$$

6 Real-Time Control Approximations

In this section, we consider the two-dim. perturbation $(p, q) = (x_3(0), x_6(0))$ of the initial values for the swing angle and its velocity. The nominal parameter is $(p_0, q_0) = (0,0)$. Having computed the sensitivity differentials $y_p(t) = \frac{\partial x}{\partial p}(t, p_0, q_0)$ and $y_q(t) = \frac{\partial x}{\partial q}(t, p_0, q_0)$ we may proceed to the real-time approximation of perturbed solutions $x(t,p,q)$ by its first-order Taylor expansion

$$x(t,p,q) \approx x_0(t) + y_p(t)p + y_q(t)q. \tag{16}$$

The same procedure applies to control and adjoint variables and to the exit-time t_2. The real-time capacity of this approximation follows from the fact that the right hand side requires only *on-line* multiplications with the particular perturbation (p,q) while all other terms may be computed *off-line*. In order to demonstrate the quality of the approximation provided by the Taylor expansion, we choose the perturbation $(p,q) = (0.05, 0.05)$ whose magnitude reaches the size of the maximum value of the swing angle and velocity; see Figure 3. We find a rather large deviation of the perturbed solution from the unperturbed one,

$$\max_{0 \le t \le 9} \|x(t,p,q) - x_0(t)\|_\infty \le 1.6,$$

whereas the first order Taylor expansion (16 yields the estimate

$$\max_{0 \le t \le 9} \|x(t,p,q) - (x_0(t) + y_p(t)p + y_q(t)q)\|_\infty \le 0.01$$

which confirms the favorable real-time approximation. This is also reflected in the results for the exit-time $t_2 = t_2(p,q)$. The optimal nominal exit-time is

$$t_2(0,0) = 5.672903236$$

whereas the optimal perturbed exit-time is computed as

$$t_2(0.05, 0.05) = 5.29567173.$$

Using the values of sensitivity derivatives (14) and (15) for the junction point, the Taylor expansion yields the approximate value

$$t_2(0,0) + \frac{dt_2}{dp}(0,0)p + \frac{dt_2}{dq}(0,0)q = 5.298443797$$

which represents a good approximation of the exact value in view of the rather large perturbation.

Acknowledgement

We would like to thank an anonymous reviewer for helpful comments.

References

1. A. V. Arutyunov, S. M. Aseev: Investigation of the degeneracy phenomenon of the maximum priciple for optimal control problems with state constraints. SIAM J. Control and Optimization **35** (1997) 930–952

2. D. Augustin, H. Maurer: Second order sufficient conditions and sensitivity analysis for the optimal control of a container crane under state constraints. To appear in Optimization.
3. C. Büskens: Optimierungsmethoden und Sensitivitätsanalyse für optimale Steuerprozesse mit Steuer- und Zustands-Beschränkungen. Dissertation, Institut für Numerische Mathematik, Universität Münster, Germany, 1998.
4. C. Büskens, H. Maurer: Sensitivity analysis and real-time control of parametric optimal control problems using nonlinear programming methods. This volume.
5. R. F. Hartl, S. P. Sethi, and R. G. Vickson: A survey of the maximum principle for optimal control problems with state constraints. SIAM Review **37** (1995) 181–218
6. K. Malanowski: Sufficient optimality conditions for optimal control problems subject to state constraints. SIAM J. on Control and Optimization **35** (1997), 205–227
7. K. Malanowski, H. Maurer: Sensitivity analysis for state constrained optimal control problems. Discrete and Continuous Dynamical Systems **4** (1998) 241–272
8. H. Maurer, D. Augustin: Sensitivity analysis and real-time control of parametric optimal control problems using boundary value methods. This volume.
9. H. J. Oberle, W. Grimm: BNDSCO – A program for the numerical solution of optimal control problems. Institute for Flight Systems Dynamics, DLR, Oberpfaffenhofen, Germany, Internal Report **515–89/22**, 1989
10. R. Pytlak, R. B. Vinter: Feasible direction algorithm for optimal control problems with state and control constraints: Implementation. J. of Optimization Theory and Applications **101** (1999) 623–649
11. F. Rampazzo, R. B. Vinter: Degenerate optimal control problems with state constraints. SIAM J. on Control and Optimization **39** (2000) 989–1007
12. Y. Sakawa, Y. Shindo: Optimal control of container cranes. Automatica **18** (1982), 257–266
13. K. L. Teo, J. L. Jennings: Nonlinear optimal control problems with continuous state inequality constraints. J. of Optimization Theory and Applications **63** (1989), 1–22

Real-Time Control of an Industrial Robot under Control and State Constraints

Christof Büskens[1] and Helmut Maurer[2]

[1] Lehrstuhl für Ingenieurmathematik, Universität Bayreuth, Germany
[2] Institut für numerische Mathematik, Universität Münster, Germany

Abstract The dynamical model in Otter and Türk [10] for the robot Manutec r3 leads to a highly nonlinear optimal control problem with various control and state constraints. The nonlinear programming (NLP) techniques in [1,2,4] are applied to compute the optimal nominal solution for a fixed set of parameters in the system. We consider perturbations in the model which frequently occur in practice: deviations from the precomputed nominal trajectory or perturbations in the mass load. Since the re-optimization of the system for the perturbed set of parameters largely exceeds the running time of the robot, we apply the real-time control techniques developed in [2,4,5]. These methods require the computation of the parameter sensitivity derivatives and implement the first order Taylor expansion of the perturbed optimal solution with respect to the parameters. Real-time computations for the Manutec r3 robot are presented which demonstrate the quality of the real-time approximations.

1 Introduction

Industrial robots play an important role in manufacturing processes like welding or spray painting. International competition as well as increasing quality standards and economic reasons impose high demands on the precision, speed and reliability of industrial robots. Due to nonlinear coupling effects (e.g., coriolis and centrifugal forces) and vibrations in the joints, the control of robots works less accurately at high speed tasks such as glueing or transport jobs.

Relying on the knowledge and the intuition of experienced personnel, on-site teaching of robots is still common practice. This situation has stimulated research in performing a more precise trajectory planning of robots. The two main problem classes (optimal point-to-point trajectories and the so called prescribed-path problem) can be formulated and solved within the framework of optimal control theory. The resulting model constitutes a highly nonlinear optimal control problem subject to various control and state constraints. Up to now, the great complexity of the model has prevented the construction of *feedback* control strategies within a reasonable time frame. As a consequence, the engineer must confine himself to determining *open-loop* control solutions.

During robot motion one may often detect deviations from nominal parameters in the system, e.g., deviations in the load mass or in the coordinates of the trajectory. These perturbations lead to tracking errors and hence require time consuming corrections. Nevertheless, when patched in automated production lines, these robots have to hold the rhythm given by the predecessor and successor. In general, it is by far too time-demanding to compute a highly precise optimal solution for perturbed system parameters. This fact motivates the need for fast and reliable *real-time control approximations* of perturbed optimal solutions.

In [2,4], we have proposed numerical techniques for the real-time optimization of perturbed nonlinear programming (NLP) problems. These methods have been extended in [2,5] to perturbed optimal control problems by using appropriate discretisation techniques. In the present article, these numerical techniques will be applied to the Manutec r3 robot in Otter and Türk [10] which has become a benchmark problem in the optimal control of complex nonlinear systems [2,6,11].

2 THE INDUSTRIAL ROBOT MANUTEC R3

The original model of Otter and Türk [10] for the robot Manutec r3 comprises a robot with 6 links. The first 3 degrees of freedom are responsible for the position of the tool centre, whereas the other 3 degrees of freedom refer to the orientation of the tool itself. The six arms are made of aluminium and are connected by rotational

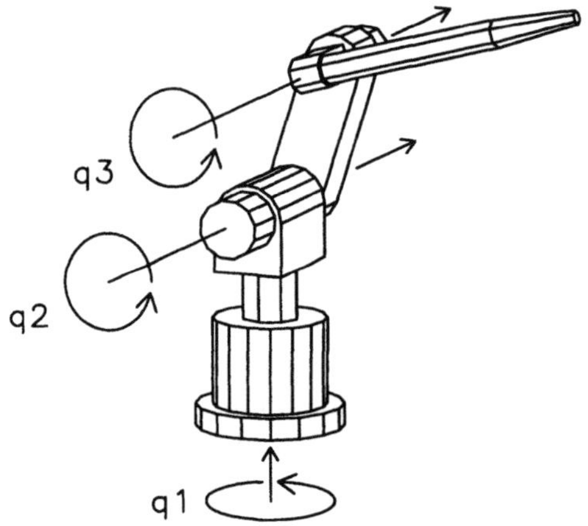

Figure 1. Manutec r3 (3 degrees of freedom)

joints. Each arm is driven by an electronically commuted motor and a gear box which consists of steel gear wheels embedded in the preceding arm. The position and velocity of each motor is measured by an encoder on the motor axis. The angle between two arms can not be measured directly, but it can be calculated from the motor position and the gear ratio of the corresponding gear box. Therefore, the motion of the links is described as a function of the control input signals of the robot drives which are the voltages for controlling the armature current of electronically commuted ac motors. The robot is able to transport loads up to 15 kg. In this article, a reduced model with 3 links is discussed, since the attention is focused more on the motion of the robot than on the orientation of the tool, cf. Figure 1.

Let the vector $q(t) = (q_1(t), q_2(t), q_3(t))^T$ describe the relative angles between the arms and let the vector $u(t) = (u_1(t), u_2(t), u_3(t))^T$ represent the torque controls. For a detailed description of the mechanical foundations of following robot equations, we refer to Vucobratovic [12]. The dynamical model of Otter and Türk [10] is given by the equations

$$M(q(t))\ddot{q}(t) = Du(t) + \chi^d(\dot{q}(t), q(t)) + \chi^g(q(t)), \tag{1}$$

where the control vector $u(t)$ is scaled by the diagonal matrix $D = \mathrm{diag}(-126.0, 252.0, 72.0)$. The terms $\chi^g(q(t))$ in (1) are the moments caused by gravitational forces:

$$\begin{aligned}
\chi_1^g(q) &= 0,\\
\chi_2^g(q) &= b_1 \cos(q_2)\sin(q_3) + b_1 \sin(q_2)\cos(q_3) + b_2 \sin(q_2),\\
\chi_2^g(q) &= b_1 \sin(q_2 + q_3),
\end{aligned} \tag{2}$$

where $g = 9.81$ is the constant of gravity and the constants b_i, $i = 1, 2$, depend on the transport load m,

$$b_1 = (12.1806 + 0.98 \cdot m)g, \qquad b_2 = (41.7325 + 0.50 \cdot m)g. \tag{3}$$

The moments caused by coriolis and centrifugal forces are described by the term $\chi^d(\dot{q}(t), q(t))$ in (1) which has the following structure:

$$\begin{aligned}
\chi_i^d(\dot{q}, q) &= \sum_{j=1}^{3} \left(\sum_{k=1}^{3} \Gamma_{i,j,k}(q)\dot{q}_k \right) \dot{q}_j, \quad i = 1, 2, 3,\\
\Gamma_{i,j,k}(q) &= -\frac{1}{2} \left(\frac{\partial M_{i,j}(q)}{\partial q_k} + \frac{\partial M_{i,k}(q)}{\partial q_j} - \frac{\partial M_{j,k}(q)}{\partial q_i} \right).
\end{aligned} \tag{4}$$

The moments of inertia in (1) and (4) are given by the symmetric and positive definite (3×3) mass-matrix $M(q)$:

$$\begin{aligned}
M_{1,1}(q) &= c_1 \sin^2(q_2 + q_3) + c_2 \sin(q_2 + q_3)\sin(q_2) + c_3 \sin^2(q_2)\\
&\quad + c_4 \cos^2(q_2 + q_3) + c_5 \cos^2(q_2) + c_6,\\
M_{1,2}(q) &= c_7 \cos(q_2 + q_3) + c_8 \cos(q_2),\\
M_{1,3}(q) &= c_7 \cos(q_2 + q_3),\\
M_{2,2}(q) &= c_2 \cos(q_3) + c_9,\\
M_{2,3}(q) &= c_{10} \cos(q_3) + c_{11},\\
M_{3,3}(q) &= c_{12},
\end{aligned} \tag{5}$$

where the constants c_i, $i = 1, \ldots, 12$, are functions of the transport load m which can be recovered from Otter and Türk [10] as

$$\begin{aligned}
c_1 &= 8.0604812 + 0.9604 \cdot m, & c_7 &= -0.0110568,\\
c_2 &= 12.1806000 + 0.9800 \cdot m, & c_8 &= -3.2963900,\\
c_3 &= 20.1794125 + 0.2500 \cdot m, & c_9 &= 85.2298937 + 1.2104 \cdot m,\\
c_4 &= 0.3900000, & c_{10} &= 6.0903000 + 0.4900 \cdot m,\\
c_5 &= 0.6400000, & c_{11} &= 7.9484812 + 0.9604 \cdot m,\\
c_6 &= 17.2112712, & c_{12} &= 12.5504812 + 0.9604 \cdot m.
\end{aligned} \tag{6}$$

Since the mass-matrix $M(q)$ is positive definite, equation (1) is equivalent to

$$\ddot{q} = M(q)^{-1}(Du(t) + \chi^d(\dot{q}, q) + \chi^g(q)). \tag{7}$$

The Fortran subroutine R3M2SI in Otter and Türk [10] offers an efficient method for calculating the right side of (7). The final time t_f is either fixed or free. We consider the following objective functional which is a weighted combination of energy and final time with a weight factor $0 \le \alpha \le 1$:

$$F(u, t_f) = \alpha \int_0^{t_f} \sum_{i=1}^{3} u_i(t)^2 \, dt + (1 - \alpha) t_f. \tag{8}$$

Otter and Türk [10] have introduced the following 18 control and state constraints: *control constraints* defined by the torque voltages,

$$-7.50 \le u_i(t) \le 7.50, \qquad i = 1, 2, 3, \tag{9}$$

state constraints of first order imposed for the angular velocities,

$$\begin{aligned} -3.00 &\le \dot{q}_1(t) \le 3.00, \\ -1.50 &\le \dot{q}_2(t) \le 1.50, \\ -5.20 &\le \dot{q}_3(t) \le 5.20, \end{aligned} \tag{10}$$

and *state constraints of second order* for the angles,

$$\begin{aligned} -2.97 &\le q_1(t) \le 2.97, \\ -2.01 &\le q_2(t) \le 2.01, \\ -2.86 &\le q_3(t) \le 2.86. \end{aligned} \tag{11}$$

The notion of the *order of a state constraint* may be found in Hartl et al. [8] or in [9], section 5. Terminal conditions are given by

$$q(t_f) = (1, -1.95, 1)^T, \qquad \dot{q}(t_f) = (0, 0, 0)^T. \tag{12}$$

In summary, the optimal control problem is the task to determine control functions $u_i : [0, t_f] \to \mathbb{R}$, $i = 1, 2, 3$, that minimize the functional (8) subject to the constraints (1)–(7) and (9)–(12).

In this control problem, we admit perturbations $p = (p_1, p_2)^T \in \mathbb{R}^2$ which either appear in the transport load in the form $m = 15.0 - p_1$ or which represent deviations p_2 in the initial conditions according to

$$q(0) = (0, -1.5 + 0.1 \cdot p_2, 0)^T, \qquad \dot{q}(0) = (0, 0, 0)^T. \tag{13}$$

The factor 0.1 in $q(0)$ is only used for graphical reasons in Figures 5–7. We point out that this special case of perturbations in the initial values also includes the more general case that deviations from the nominal trajectory occur during the motion of the robot. The *nominal* perturbation or unperturbed parameter is $p_0 = ({p_1}_0, {p_2}_0)^T = (0, 0)^T$ to which belongs the nominal optimal control and state function $u_0(t)$ and

$x_0(t)$. As in [5,9] we denote the optimal perturbed solution for any specific perturbation p by $u(t,p)$ and $x(t,p)$. By definition, we have $u(t,p_0) = u_0(t)$ and $x(t,p_0) = x_0(t)$. Our aim is to compute the parameter sensitivity derivatives

$$\frac{\partial u}{\partial p}(t,p_0), \quad \frac{\partial x}{\partial p}(t,p_0), \tag{14}$$

at the nominal parameter p_0. These sensitivity differentials enable us to approximate the perturbed solutions by the following first order Taylor expansion

$$u(t,p) \approx u_0(t) + \frac{\partial u}{\partial p}(t,p_0)(p-p_0), \quad x(t,p) \approx x_0(t) + \frac{\partial x}{\partial p}(t,p_0)(p-p_0). \tag{15}$$

A similar analysis can be performed for the adjoint variables. Let us recall the principle of real-time control developed in [2,5,9]. Since both the nominal solution and the sensitivity differentials can be computed *off-line*, the right-hand sides in (15) represent real-time approximations which can be computed *on-line* in a time which is much smaller than the running time of the robot.

3 Computation of the Nominal Solution

We restrict the discussion to the energy minimal robot with $\alpha = 1$ and a fixed final time $t_f = 0.62$ in the objective (8). Note that the proposed methods work as well for all other values of α. The optimal nominal solution is computed by solving the NLP problem (13) formulated in Büskens, Maurer [5]. We use a linear interpolation of the controls and a 4th order Runge-Kutta approximation for the state variables q and $\dot{q}$. An equidistant grid of $N_t = 101$ discrete time points with step size $h = t_f/(N_t - 1)) = t_f/100$ is used (see equation (7) of [5]) which results in $N_z = 303$ optimization variables, $N_c = 1818$ inequality constraints and $N_e = 6$ equality constraints. Initial estimates for the control functions are chosen as $u^i = 0,\ i = 1,\dots,101$.

All calculations are performed with the subroutine NUDOCCCS developed in Büskens [1,2]. We obtain the optimal value $F(u,t_f) = 38.176716$ for the objective functional which is correct with a relative error of $1.5 \cdot 10^{-7}$. The second order sufficient conditions are satisfied for the nominal problem, since the projected Hessian matrix (16) in [4] is positive definite with smallest eigenvalue $\nu_{min} = 0.309 \cdot h > 0$. Note that this second order test is the discrete analogon for the coercivity condition (AC-5) or (AG-5) in [9]. The strict Legendre-Clebsch condition (AC-4) or (AS-4) in [9] trivially holds due to the choice of $\alpha > 0$ in the objective functional (8). In the time optimal case ($\alpha = 0$) the theoretical treatment of the continuous case is not entirely developed. Nevertheless, the numerical methods work just as well.

Figure 2 shows that the optimal nominal control functions exhibit two boundary arcs for the control u_1 and one boundary arc for the control u_3. We point out that the nominal controls satisfy two assumptions in [9] that are important for sensitivity analysis. Namely, the optimal control functions are *continuous* in $[0,t_f]$ and their derivatives are *discontinuous* at an entry or exit point of the boundary arcs. This

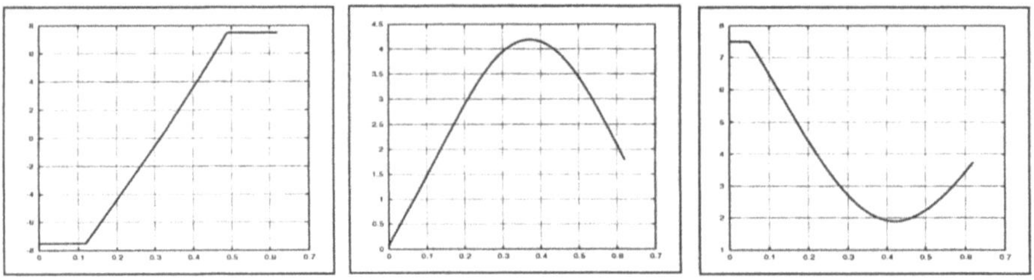

Figure 2. Optimal nominal control variables $u(t, p_0)$

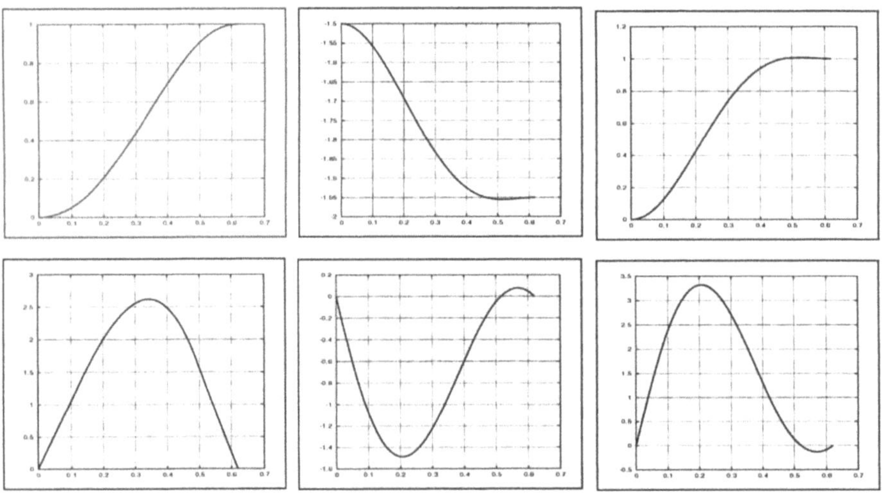

Figure 3. Optimal nominal states variables $q(t, p_0)$, $\dot{q}(t, p_0)$

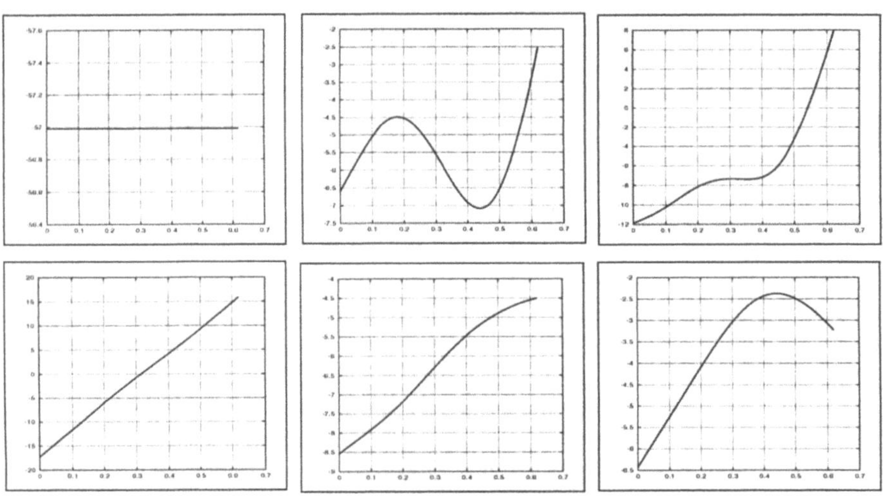

Figure 4. Optimal adjoint variables $\lambda_q(t, p_0)$, $\lambda_{\dot{q}}(t, p_0)$

property is equivalent to the *non-tangential junction condition* (AC-6) in [9]; see also Lemma 2 in [9].

The optimal trajectories are displayed in Figure 3. It is noteworthy that the above state constraints (10) and (11) do *not* become active for the energy minimal solution. Figure 4 depicts the nominal solutions for the six adjoint variables which are calculated from the Lagrange multipliers of the NLP problem; compare the relations (3) and (22) in Büskens, Maurer [5].

4 Computational Sensitivity Analysis and Real-Time Control

For the nominal parameter $p_0 = (0,0)^T$, the sensitivity differentials of the control, state and adjoint variables are obtained from the expressions (24)–(26) in Büskens, Maurer [5]. Figures 5–7 display the respective sensitivity differentials. The thick lines denote the sensitivities with respect to the initial value perturbation p_2 and the thin lines represent the sensitivity differentials with respect to the transport load perturbation p_1. Note that the sensitivities of the controls are zero on the boundary arcs and that the overshooting at each junction point of the control constraints results from the linear interpolation of the control variables.

First and second order sensitivity derivatives of the objective function (8) can be computed from the equations (23), (25) and (26), (27) in [4] which yield

$$\begin{aligned} \frac{dF}{dp}[p_0] &= (-2.17191228, -0.659061661), \\ \frac{d^2F}{dp^2}[p_0] &= \begin{pmatrix} 0.127702046 & 0.118596916 \\ 0.118596916 & 0.079051731 \end{pmatrix}. \end{aligned} \tag{16}$$

Using the relations (27) and (28) in [5] and taking into account the factor 0.1 in (13), we see that the values $\frac{dF}{dp_2}[p_0]$ and $\frac{d^2F}{dp_2^2}[p_0]$ coincide with the values $\lambda_{q_2}(0)$ within 8 digits. Recall from (15) that the sensitivity differentials are needed to evaluate a first order Taylor expansion of the perturbed optimal solution. In order to judge the quality of the real-time approximation (15), we set up the following Table 1 which lists the maximal relative errors in the terminal states for different perturbations p. The notations in Table 1 have the following meaning:

S.Err. denotes the maximal relative error in the terminal states which is obtained by an integration of the perturbed system using the first order real-time control approximation;

W.Err. is the same error resulting from an integration of the perturbed system with the nominal controls.

The computing time for calculating the real-time approximation of the complete controls is about $2 \cdot 10^{-6}$ seconds on a PIII 450MHZ computer. In a practical implementation, the computing time can be reduced drastically by an additional factor of 101 (number of grid points), if the time during the motion of the robot is used for computing the approximation (15). The re-optimization of the perturbed problem by

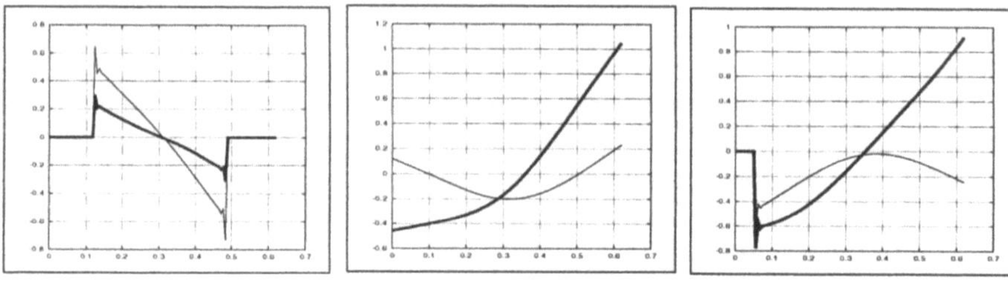

Figure 5. Sensitivity differentials $\frac{du}{dp}(t, p_0)$

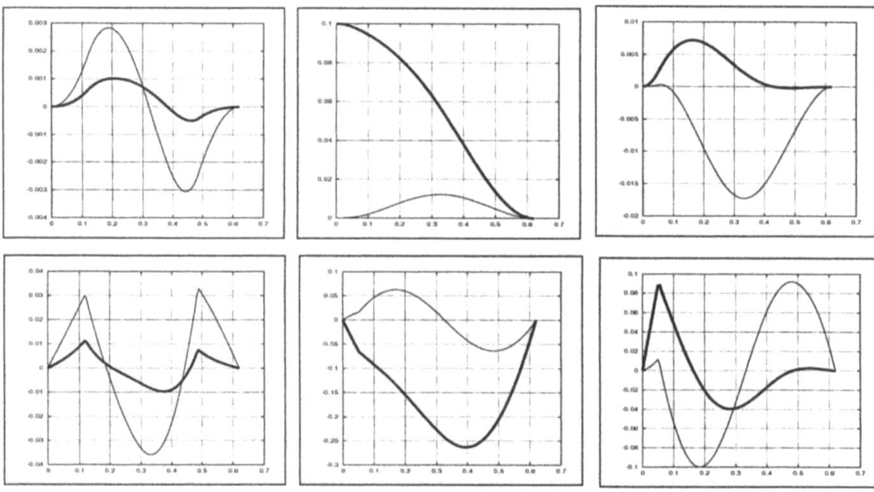

Figure 6. Sensitivity differentials $\frac{dq}{dp}(t, p_0)$ and $\frac{d\dot{q}}{dp}(t, p_0)$

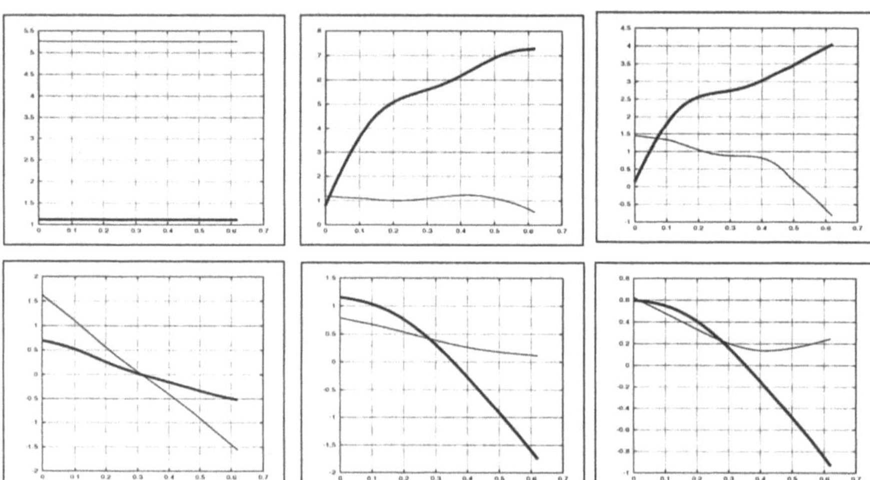

Figure 7. Sensitivity differentials $\frac{d\lambda_q}{dp}(t, p_0)$ and $\frac{d\lambda_q}{dp}(t, p_0)$

	$p_2 = 0.0$		$p_2 = 0.1$		$p_2 = 1.0$	
	S.Err.	W.Err.	S.Err.	W.Err.	S.Err.	W.Err.
$p_1 = 0.00$	$2.8 \cdot 10^{-14}$	$2.8 \cdot 10^{-14}$	$3.8 \cdot 10^{-05}$	$8.4 \cdot 10^{-03}$	$3.7 \cdot 10^{-03}$	$8.7 \cdot 10^{-02}$
$p_1 = 0.01$	$2.9 \cdot 10^{-08}$	$9.6 \cdot 10^{-04}$	$3.9 \cdot 10^{-05}$	$9.3 \cdot 10^{-03}$	$3.7 \cdot 10^{-03}$	$8.8 \cdot 10^{-02}$
$p_1 = 0.05$	$7.3 \cdot 10^{-07}$	$4.8 \cdot 10^{-03}$	$4.4 \cdot 10^{-05}$	$1.3 \cdot 10^{-02}$	$3.8 \cdot 10^{-03}$	$9.2 \cdot 10^{-02}$
$p_1 = 0.10$	$2.9 \cdot 10^{-06}$	$9.7 \cdot 10^{-03}$	$5.1 \cdot 10^{-05}$	$1.8 \cdot 10^{-02}$	$3.8 \cdot 10^{-03}$	$9.7 \cdot 10^{-02}$
$p_1 = 0.50$	$7.3 \cdot 10^{-05}$	$4.9 \cdot 10^{-02}$	$1.8 \cdot 10^{-04}$	$5.7 \cdot 10^{-02}$	$4.3 \cdot 10^{-03}$	$1.4 \cdot 10^{-01}$
$p_1 = 1.00$	$3.0 \cdot 10^{-04}$	$1.0 \cdot 10^{-01}$	$5.1 \cdot 10^{-04}$	$1.1 \cdot 10^{-01}$	$5.1 \cdot 10^{-03}$	$1.9 \cdot 10^{-01}$
$p_1 = 2.00$	$1.2 \cdot 10^{-03}$	$2.1 \cdot 10^{-01}$	$1.6 \cdot 10^{-03}$	$2.1 \cdot 10^{-01}$	$6.7 \cdot 10^{-03}$	$2.9 \cdot 10^{-01}$
$p_1 = 5.00$	$8.0 \cdot 10^{-03}$	$5.8 \cdot 10^{-01}$	$9.2 \cdot 10^{-03}$	$5.9 \cdot 10^{-01}$	$2.1 \cdot 10^{-02}$	$6.8 \cdot 10^{-01}$
$p_1 = 10.0$	$3.6 \cdot 10^{-02}$	$1.6 \cdot 10^{+00}$	$3.9 \cdot 10^{-02}$	$1.6 \cdot 10^{+00}$	$6.7 \cdot 10^{-02}$	$1.8 \cdot 10^{+00}$

Table 1. Maximal relative error in the terminal states

starting from the nominal solution takes about 1–5 seconds depending on the values of the perturbations. This time considerably exceeds the time of motion $t_f = 0.62$ sec. For this specific problem. Hence, the numerical results given in Table 1 clearly indicate that the first order approximations enjoy favourable real-time control properties: the terminal conditions are satisfied with a sufficient high precision and the computing time for the approximations is much smaller than the operation time of the robot. The real-time approximations for $\alpha \in [0, 1)$ yield comparable results, even in the case of time optimal solutions ($\alpha = 0$), with a number of singular subarcs. In this situation the results can be improved taking into account the switching points, cf. Büskens, Pesch and Winderl [7]. A further improvement of the real-time approximations for all $\alpha \in [0, 1]$ with respect to optimality and admissibility (constraints and terminal conditions) can be achieved by applying the mixed open- and closed-loop strategy in Büskens [3] which will not be further discussed here.

References

1. C. Büskens: Direkte Optimierungsmethoden zur numerischen Berechnung optimaler Steuerungen. Diploma thesis, Institut für Numerische Mathematik, Universität Münster, Münster, Germany, (1993)
2. C. Büskens: Optimierungsmethoden und Sensitivitätsanalyse für optimale Steuerprozesse mit Steuer- und Zustands- Beschränkungen. Dissertation, Institut für Numerische Mathematik, Universität Münster, Münster, Germany, (1998)
3. C. Büskens: Real-Time Solutions for Perturbed Optimal Control Problems by a Mixed Open- and Closed-Loop Strategy. This volume.
4. C. Büskens, H. Maurer: Sensitivity Analysis and Real-Time Optimization of Parametric Nonlinear Programming Problems. This volume.
5. C. Büskens, H. Maurer: Sensitivity Analysis and Real-Time Control of Parametric Optimal Control Problems Using Nonlinear Programming Methods. This volume.
6. C. Büskens, H. Maurer: Real-Time Control of Robots with Initial Value Perturbations Via Nonlinear Programming Methods. Optimization **47** (2000) 383–405
7. C. Büskens, H. J. Pesch, S. Winderl: Real-Time Solutions of Perturbed Control Problems with Linear Controls. This volume

8. R. F. Hartl, S. P. Sethi, and R. G. Vickson: A Survey of the Maximum Principle for Optimal Control Problems with State Constraints. SIAM Review **37** (1995) 181–218
9. H. Maurer, D. Augustin: Sensitivity Analysis and Real-Time Control of Parametric Optimal Control Problems Using Boundary Value Methods. This volume.
10. M. Otter, and S. Türk: The DFVLR Models 1 and 2 of the Manutec R3 Robot. DFVLR-Mitteilung 88-13, Institut für Dynamik der Flugsysteme, Oberpfaffenhofen, Germany, (1988)
11. O. von Stryk: Numerische Lösung optimaler Steuerungsprobleme: Diskretisierung, Parameteroptimierung und Berechnung der adjungierten Variablen. Fortschritt-Berichte VDI, Reihe 8, **441**, VDI Verlag, Germany, (1995)
12. M. Vukobratović: Introduction to Robotics. Springer-Verlag, Berlin, (1989)

Real-Time Optimal Control of Shape Memory Alloy Actuators in Smart Structures

Stefan Seelecke[1], Christof Büskens[2], Ingo Müller[3], and Jürgen Sprekels[4]

[1] Department of Mechanical & Aerospace Engineering, North Carolina State University, USA
[2] Lehrstuhl für Ingenieurmathematik, Universität Bayreuth, Germany
[3] Institut für Verfahrenstechnik, Technische Universität Berlin, Germany
[4] Weierstraß-Institut für Angewandte Analysis und Stochastik, Berlin, Germany

Abstract An optimal control method is proposed for shape memory alloy actuators in smart structures. The method is capable of compensating for the hysteretic behavior present in these materials and thus qualifies for high frequency applications where conventional methods fail. Furthermore, a version has been developed, which is fully real-time capable. It is based on a parametric sensitivity analysis of nonlinear programming problems and is demonstrated to work very robustly for a wide range of parameters.

1 Introduction

The advent of multifunctional materials like shape memory alloys (SMAs) has stimulated the design of a completely new generation of structures, capable of automatically adapting to changes in their environment. Examples of these so-called "smart structures" are adaptive airfoils and helicopter blades, adaptive submarine structures or microelectromechanical systems (MEMS), see [15] for an overview of the subject.

In this paper, we will focus on the application of SMA wires as actuators in an adaptive airplane wing. Upon heating, these wires contract and function as "metal muscles" which are able to exert a strong bending moment on the structure. By appropriately controlling the temperature of the wires, say by input of an electric current and convective heat exchange with the surrounding medium, one is able to adapt the shape of the wing to the flow conditions at hand.

Control methods like proportional or integral feedback have long been established and can be considered state of the art at the present time. However, due to the nonlinearity and hysteretic phenomena observed in shape memory alloys, these may by far not be optimal for this type of material. Additionally, the hysteresis induces a time delay in applications with higher frequencies. Usually, this is not a serious problem as shape memory alloys intrinsically prohibit higher frequency applications due to their low cooling speed when exposed to still air at room temperature. However, if one faces an application in an efficient cooling environment, like flight at high altitude or underwater applications, the hysteretic behavior develops into a major obstacle. The same inevitably holds for the important field of MEMS application, where the surface area approaches the order of magnitude of the volume. This allows for actuation frequencies of up to 300 *Hz*, see [9].

A way to overcome the problem is the design of advanced control laws which compensate for the hysteresis. One approach has been given in [8] and [7], another method has been proposed in [18]. For aquatic applications, the removal of heat is so quick that an operational frequency of up to 20 *Hz* can be achieved, see [17], where the authors also demonstrated that a conventional PI control method breaks down already at 4 *Hz*. In order to control the SMA actuator in the frequency range up to 20 *Hz* they introduced an adaptive control method based on Krasnoselskii-Preisach (KP) operators. A strong feature of the method is the ability to perform 7an on-line parameter identification, allowing to adapt to changing hysteresis parameters often observed in cyclic applications.

All of the above methods have in common that they improve standard control techniques by the introduction of model based predictions. However, one aspect that has never been taken into account so far is optimality. It is the scope of the following sections to embed SMA modeling into the framework of optimal control theory, allowing for the inclusion of optimality criteria like speed of adjustment or energy consumption. For the description of the shape memory behavior, we use the Müller-Achenbach model in the improved version by Seelecke [13]. Apart from the good reproduction of experimental results, it is attractive due to its mathematical structure, allowing for a smooth integration into standard optimal control codes. We first discuss the basic procedure for the calculation of an optimal control for SMA actuators in smart structures and then extend the method to real-time capability. A simple smart structure is used for illustration, an elastic beam of which the shape can be adjusted by a SMA wire. Note, however, that the method is by no means restricted to such simple structures or such limited number of actuators, as will be discussed at the end of the section.

2 Basic Equations

A shape memory wire, which is mounted tangentially to the beam axis at a distance h, contracts upon electric heating and exerts a bending moment on the beam. By an appropriate control of the heating current, within certain limits depending on the placement and number of actuators, arbitrary beam shapes can thus be realized, see Figure 1. Other authors have treated similar examples, see [1], [10] and [16].

The mathematical modeling of the beam follows [12] with the usual assumptions of elementary beam theory. As the focus of our investigations is on the implementation of the SMA model into the framework of optimal control theory, we confine attention to small strains and cross sections that remain plane and normal to the beam axis. The beam shape then follows from the well-known Euler-Bernoulli equation, according to which the bending moment is proportional to the beam curvature

$$EIv''(x,t) = h\,P(t)\ [H(x-a) - H(x-b)]\,, \tag{1}$$

with bending rigidity EI and curvature $v''(x,t)$ as second space derivative of the transversal displacement $v(x,t)$. The right hand side (RHS) of (1) is the bending moment. It is given by the lever arm h times the wire force $P(t)$, and $H(x-x')$ is

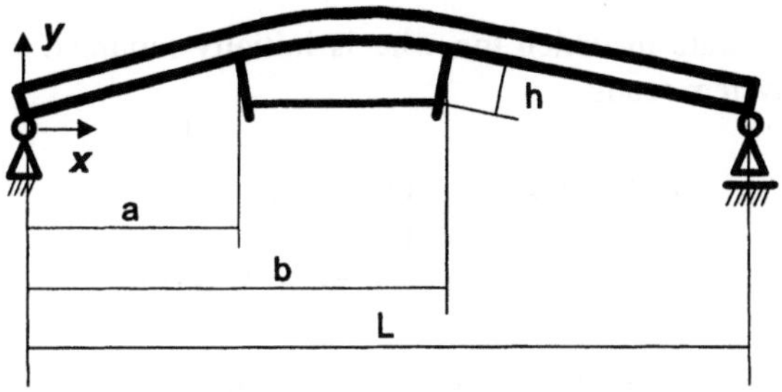

Figure 1. Elastic beam with heated SMA wire actuator

the Heaviside unit step function. This formulation is particularly useful in the case of several actuators, as it avoids the introduction of several domains of integration, but in the following, for simplicity, we will restrict ourselves to just one actuator. The wire force follows from the SMA model equations, which are given below in a concise form,

$$\begin{aligned} \dot{x}_+(t) &= \dot{x}_+(x_\pm(t), T(t), P(t)), \\ \dot{x}_-(t) &= \dot{x}_-(x_\pm(t), T(t), P(t)), \\ \dot{T}(t) &= \dot{T}(x_\pm(t), T(t), P(t), j(t), T_E(t)), \\ P(t) &= P(x_\pm(t), T(t), D(t)), \end{aligned} \tag{2}$$

with $x_\pm(t) := (x_+(t), x_-(t))^T$. The quantities $x_\pm(t)$ are the phase fractions of martensite M_+ or M_-, respectively, which characterize the phase transformation in the crystal lattice structure of a shape memory alloy, see [11] for more details. $T(t)$ is the temperature in the SMA, $P(t)$ and $D(t)$ denote its force and deformation, and $j(t)$ and $T_E(t)$ are the Joule heat of an electric current passing through the SMA and the environmental temperature, respectively. We here encounter the case of a system of degenerate partial differential equations, consisting of one ordinary differential equation in space, three ODEs in time and an algebraic relation (DAE) between the force $P(t)$ and the deformation $D(t)$. The Euler-Bernoulli equation in this case may easily be integrated to yield

$$v(x,t) = \frac{h\,P(t)}{2EI}\left[(x-a)^2 H(x-a) + -(x-b)^2 H(x-b)\right] + C_1 x + C_2. \tag{3}$$

The meaning of the constants a, b, L and h can be seen in Figure 1. With this result another algebraic relation between $P(t)$ and $D(t)$ can be derived, see [14] for details,

$$D_{max} - D(t) = P(t)\frac{h^2 L}{EI}\left[\frac{(a^2-b^2)}{L^2} - 2\frac{a-b}{L}\right], \tag{4}$$

so that both quantities can be completely eliminated from the DAE system (2). In (4), D_{max} denotes the length change in the unloaded state with 100% martensite

M_+, which is the state in which the wire is initially mounted to the beam. The remaining ODE system reads

$$\begin{aligned} \dot{x}_+(t) &= \dot{x}_+(x_\pm(t), T(t)), \\ \dot{x}_-(t) &= \dot{x}_-(x_\pm(t), T(t)), \\ \dot{T}(t) &= \dot{T}(x_\pm(t), T(t), j(t), T_E(t)). \end{aligned} \tag{5}$$

For given initial conditions, this system may be integrated if Joule heating $j(t)$, which appears linearly in (5), and environmental temperature $T_E(t)$ are known as functions of time. Subsequently, $P(t)$ and $D(t)$ can be determined from $(2)_4$ and (4), and the beam shape follows from (3).

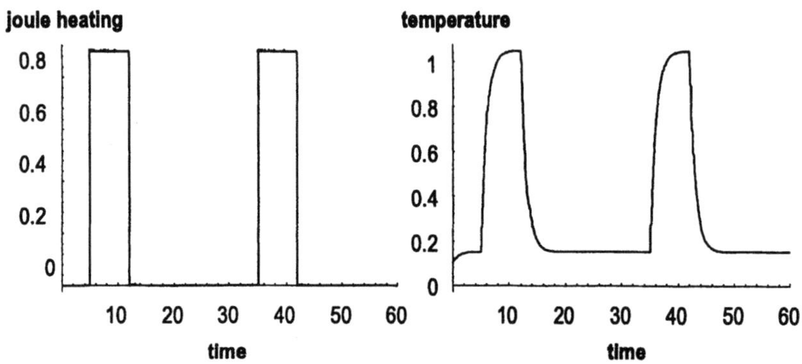

Figure 2. Two heating pulses applied to the wire (left) and the temperature (right)

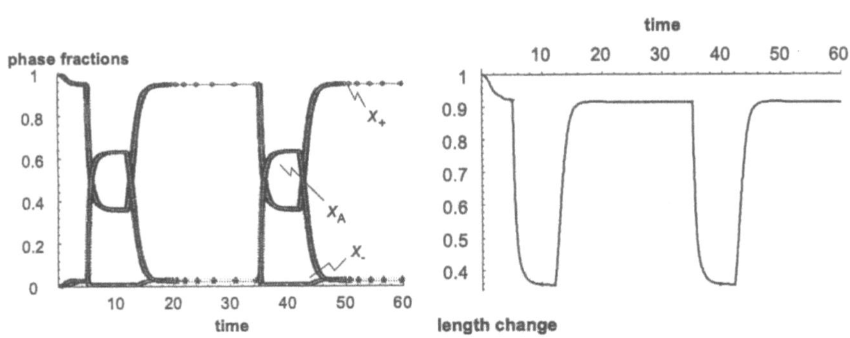

Figure 3. Evolution of the phase fractions during the heating and cooling process (left) and time-dependent wire deformation leading to different beam shapes (right)

Figures 2 and 3 are meant to illustrate the resulting behavior. The left hand side of Figure 2 shows two rectangular heating pulses that are applied to the SMA wire, and the right diagram gives the resulting temperature. The central quantities are the three phase fractions which can be seen on the left hand side of Figure 3. The wire, initially in the pure M_+-phase, transforms into austenite with increasing temperature, and during cooling the reversed process takes place. These phase transitions are the explanation for the length change that can be observed on the right hand side of Figure 3. During the transition to austenite the wire contracts, while during the reformation of martensite the beam acts as a spring and elongates the wire again.

3 Optimal Control Method (A First Model)

The preceding section discussed the so-called forward problem of beam shape adjustment. The heating current was given, and the model calculates the corresponding beam shape. If we are to calculate an optimal control, however, we have to solve the inverse problem.

We look at a certain cost functional, and the control $j(t)$ is determined in such a way as to minimize this functional subject to the ODE system (5) and additional constraints. As in [14], we consider the functional

$$F(x, u, t_f) = \int_0^{t_f} [D(t) - D^*]^2 \, dt. \tag{6}$$

Herein $x(t) := (x_\pm(t), T(t)) \in \mathbb{R}^3$ denotes the state variable and $u(t) := j(t) \in \mathbb{R}$ denotes the control variable. The final time t_f is either fixed or free. Note, that due to the relations in (2) and (4)

$$D(t) = D(x(t), u(t)) = D(x_\pm(t), T(t), j(t)) \tag{7}$$

holds. According to Eq. (3), the shape of the beam from Figure 1 consists of a parabolic part in the region above the SMA wire and two straight parts to the left and right of the wire, respectively. The curvature of the parabola is uniquely determined by the current deformation $D(t)$, so that the prescription of a defined wire deformation D^* determines a desired beam shape. The minimization of (6) thus corresponds to an asymptotic beam shape adaptation, which corresponds approximately to a solution in shortest time.

In a concise form, the optimal control problem is defined by (compare Equation (1) in [6])

$$\begin{array}{lll} \text{Minimize} & F(x, u, t_f) \quad \text{(defined in (6))} & \\ \text{subject to} & \dot{x}(t) = f(x(t), u(t)), & \text{for all } t \in [0, t_f], \\ & \psi(x(0), x(t_f)) = 0, & \\ & C(x(t), u(t)) \le 0, & \text{for all } t \in [0, t_f], \end{array} \tag{8}$$

with $f : \mathbb{R}^3 \times \mathbb{R} \to \mathbb{R}^3$ defined by (5) and $\psi : \mathbb{R}^3 \times \mathbb{R}^3 \to \mathbb{R}^r$, $0 \le r \le 6$ defines initial and terminal conditions. The function $C : \mathbb{R}^3 \times \mathbb{R} \to \mathbb{R}^k$ allows to consider additional path or control constraints. All functions in (1)–(8) are assumed to be sufficiently smooth on appropriate open sets.

The above model has been implemented into the code NUDOCCCS of Büskens [2, 3]. The code is based on a direct method, which transforms the optimal control problem into a nonlinear optimization problem (NLP). Then the resulting NLP problem is solved via a standard sequential quadratic programming (SQP) method. Instead of a continuous solution $u(t)$ the control function is approximated by optimization variables $u^i \approx u(\tau_i)$ at discrete points in time $\tau_i \in [0, t_f]$, $i = 1, \ldots, N_t$, cf. Büskens and Maurer [6] of this issue.

We illustrate the method for the case of an initial beam shape corresponding to a value of nondimensionalized deformation $D(0) = 0.8$. The target value is $D^* = 0.7$, and the control period is chosen to be $t_f = 4$ seconds.

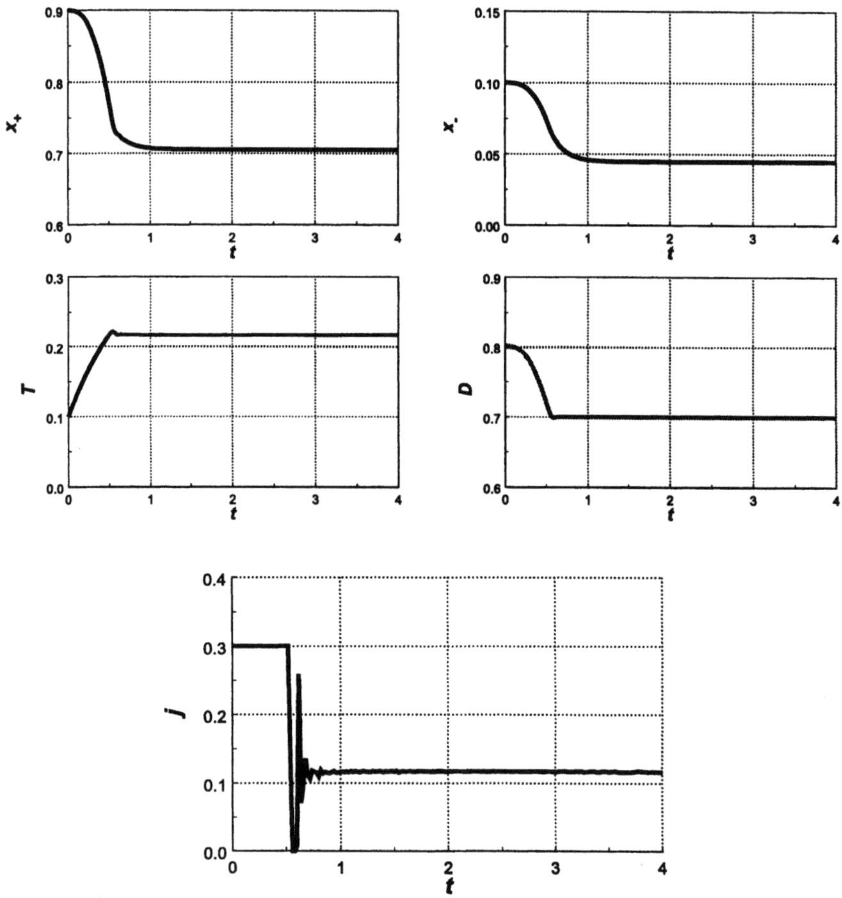

Figure 4. Optimal solution of beam shape adjustment with SMA wires of functional (6)

The upper four diagrams in Figure 4 show the solutions of the three ODEs $x_+(t)$, $x_-(t)$ and $T(t)$ together with the wire deformation $D(t)$ for $N_t = 201$ discrete points in time. The lower diagram shows the corresponding optimal control $j(t)$, which is subject to the constraint $0 \leq j(t) \leq 0.3$. It exhibits a typical bang-bang behavior, before it reaches a stationary value of ~ 0.11. Note, that theoretical investigations of optimal control theory and higher discretizations in terms of the parameter N_t lead us to the speculation, that there exist infinitely many bang-bang arcs, before the optimal control reaches the final singular arc. Since the hysteresis behavior of the SMA is already included in the model and it may thus be efficiently compensated for, the typical overshooting in $D(t)$, which can be observed by, e.g., PI controlers, is avoided.

It is noteworthy, that the computational time for solving the above optimal control problem was about 32.5 seconds on a DEC Alpha 500/333 workstation. Seelecke and Büskens [14] were able to reduce this times to the order of magitude of 1 second. However, this is still not in the range of real-time applications (20 Hz) yet. An algorithm is needed, which provides a considerable time improvement, and it is the scope of the following section to review such a method.

4 Real-Time Optimal Control

The above approach has shown the general suitability of the model (1)–(4) to calculate an optimal control for a SMA actuator in a simple smart structure. However, the computational time required for such a method is prohibitive for an on-line application. In the following, we will use the ideas of [3, 5, 6] in this issue which are perfectly well-suited to overcome this difficulty, see also [11] for an application to SMA actuators.

The two basic requirements of a real-time optimal control application are to provide results in an extremely fast manner and to be capable to cope with unforeseeable changes of the process parameters. A typical example is an adaptive aircraft wing, most simply modeled by the elastic beam treated above. Such a wing, or at least a part of it, is supposed to adapt its shape to changes in the environmental flow conditions measured on-line during the flight. The special real-time feature is that the system measures the data resulting in a new target shape only immediately prior to the control process itself.

A way to an optimal solution accounting for such circumstances is based on the treatment of a parametric optimal control process, with the parameters being represented by measurable disturbances. Of central importance here is the concept of sensitivity analysis, cf., e.g., [5,6] of this issue, determining the sensitivity of the solution with respect to changes in the disturbance parameters.

4.1 *Optimal Control Problems with Parameter Perturbations*

Let us consider the following process: At time $t = 0$ a measurement is performed, from which the target shape results, characterized by D^*. The state of the SMA wire, viz. its phase fractions $x_{\pm}(0)$, its temperature $T(0)$ and, in particular, its deformation $D(0)$ are known. The objective now is to control the beam in such a way as to reach the target shape with a compromise of high adjustment speed and low energy consumption. In contrast to the objective function in (6), this special mathematical formulation requires the minimization of the following weighted convex combination of free final time and a regularized energy consumption.

$$F(x,u,t_f) := \alpha t_f + (1-\alpha)\int_0^{t_f} u^2(t)dt, \qquad \alpha \in [0,1]. \tag{9}$$

Note, that the target shape characterized by D^* now does not appear in the functional itself. However, it enters the problem as a terminal condition

$$D(t_f) = D^*. \tag{10}$$

Hence, the function $\psi(x(0), x(t_f))$ in (8) is defined by

$$\psi(x(0), x(t_f)) := \begin{pmatrix} x(0) - x_0 \\ D(t_f) - D^* \end{pmatrix}. \tag{11}$$

The constant $x_0 \in \mathbb{R}^3$ represents the initial values for $x_{\pm}(0)$ and $T(0)$.

If disturbances can occur during the optimal control problem, problem (8) is given in a more general form by, cf. Eq. (1) in [6] of this issue

$$\begin{array}{lll} \text{Minimize} & F(x,u,t_f,p), & \\ \text{subject to} & \dot{x}(t) = f(x(t),u(t),p), & \text{for all } t \in [0,t_f], \\ & \psi(x(0),x(t_f),p) = 0, & \\ & C(x(t),u(t),p) \le 0, & \text{for all } t \in [0,t_f]. \end{array} \tag{12}$$

Here, $p \in P \subset \mathbb{R}^{N_p}$ denotes the vector of disturbance parameters which accounts for the real-time character of the optimization process. Solutions of (12) for p equal to a reference parameter p_0 ($p = p_0$) will be called *nominal* or *unperturbed* solutions. In the present paper perturbations for the deformation of the wire $D(t_f)$ at the terminal state and perturbations in the environmental temperature $T_E(t)$ will be investigated.

4.2 Real-Time Optimal Control

Let the nominal solution of the discretized optimal control problem (12) satisfy the assumptions of Theorem 3 in [5] of this issue. Then the strategy of real-time optimization can be characterized by the following steps

1. Calculate the nominal solution $u(t, p_0)$ a priori off-line as described in [5].
2. Perform a sensitivity analysis for the nominal solution with respect to the parameter p, cf. [5,6]. This yields sensitivity differentials $\frac{\partial u}{\partial p}(t, p_0)$.

3. Compute a real-time approximation of the perturbed control $u(t, p)$ by means of a truncated Taylor series

$$u(t,p) \approx \tilde{u}(t,p) := u(t,p_0) + \frac{\partial u}{\partial p}(t,p_0)(p - p_0). \tag{13}$$

The real-time approximation in (13) has the advantage that only one matrix multiplication and one vector summation are required. This can be performed in an extremely short time period.

In the following, the real-time optimization of a SMA actuator is investigated under conditions that will typically be encountered in a real-time process. These examples will demonstrate the efficiency and robustness of the proposed method.

Example 1: Unknown Final Value

As a first example, the elementary case of an unknown final shape is considered. The final shape and thus the final length of the SMA wire result from an aerodynamic measurement at the beginning of the control process. Hence the target is modeled by parameter deviations of the form $D(t_f) = D^* + p$ with nominal value $p_0 = 0$. Suppose that the wire has an initial dimensionless deformation $D(0) = 0.97$ and a dimensionless environmental temperature $T_E = 0.1$. The nominal solution $u(t, p_0)$ and its sensitivity differentials $\frac{\partial u}{\partial p}(t, p_0)$ are calculated for a target length $D^* = 0.7$ and weight factor $\alpha = 0.02$.

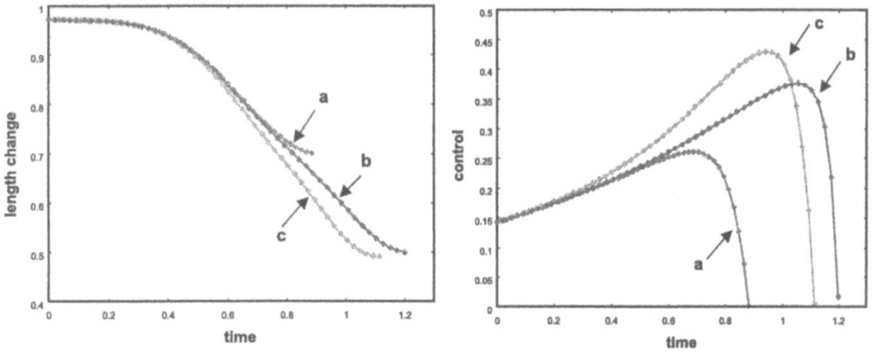

Figure 5. Wire deformation (left) and optimal control (right) for unknown D^*

The resulting nominal deformation $D(t, p_0)$ and the nominal optimal control $j(t, p_0)$ are shown by the curves marked a in Figure 5. Based on this solution and its sensitivity differentials, the real-time approximation (13) is used to calculate the curves marked b for the perturbation $p = -0.2$, which corresponds to a final value $D(t_f) = 0.5$. For comparison, another set of curves, marked c, is shown in Figure 5. These are the "exact" results of the optimal control problem with final value $D(t_f) = 0.5$. Despite the extremely large perturbation of $\sim 30\%$ in the terminal condition for $D(t)$, the error is only about $\sim 2\%$ with respect to the final value.

Note, that in our formulation t_f is assumed to be free and hence it represents an additional optimization variable in the discretized optimal control problem. For this quantity, a sensitivity analysis can be performed as well. Hence a first order Taylor approximation for the terminal time t_f yields

$$t_f(p) \approx \tilde{t}_f(p) := t_f(p_0) + \frac{\partial t_f}{\partial p}(p_0)(p - p_0). \tag{14}$$

Compared to the value of the "exact" c-curve, it is within an error range of only $\sim 4\%$.

The approximation method thus proves to be very robust with respect to disturbances in the target length of the SMA wire. In the following, we proceed to look at other possible disturbances.

Example 2: Disturbance in Environmental Temperature

A second case of practical relevance is the one of possible fluctuations in the environmental dimensionless temperature $T_E(t) = T_E(t, p) := T_E^* + p$. This is not a disturbance of an initial or final condition, rather it is a disturbance acting directly on the level of the model ODEs. The wire is exposed to a lower temperature, and we consider a 50% disturbance $T_E^*(p) = 0.05$, $p = -0.05$ of the nominal value $T_E* = 0.1$, $p_0 = 0$. The (fixed) final value of the length change is $D(t_f) = 0.7$.

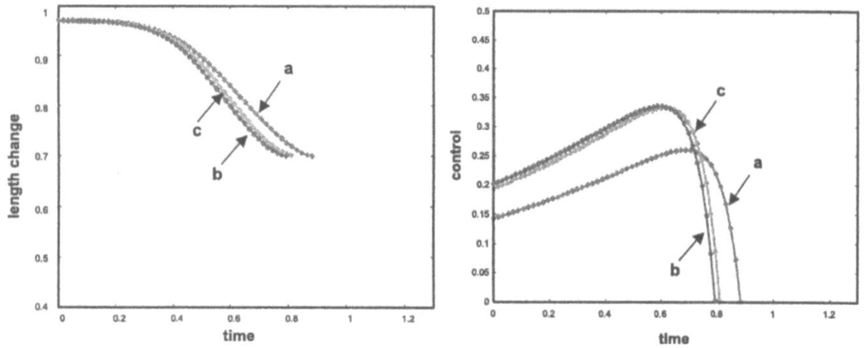

Figure 6. Wire deformation (left) and optimal control (right) in the case of perturbations in the environmental temperature $T_E(t)$

In Figures 6, again, the a-curves show the results of the unperturbed case. This type of perturbation requires a larger amount of heat to be applied, reproduced by the approximation b as well as by the reference solution c. Here, too, the method exhibits only a very small error despite the large disturbance.

Example 3: Coupled Problem

To demonstrate the capabilities of the proposed real-time method, we finally treat the general case of a combination of the two disturbances discussed before. The perturbations correspond to those of the uncoupled problems. The wire deformation and the optimal control in Figure 7 show a very robust behavior for this case as well.

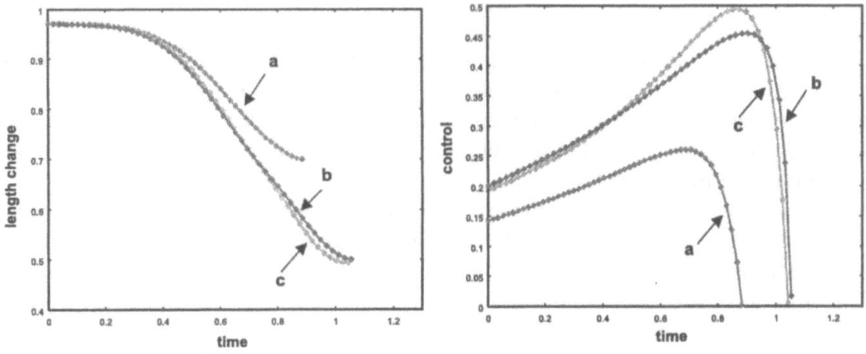

Figure 7. Wire deformation (left) and optimal control (right) for the coupled problem

5 Computing Times and Perspective

The computational time to obtain the "exact" optimal solution by the direct method NUDOCCCS was in the order of magnitude of 1 second for all three cases. The approximate solutions, however, only require time of the order of magnitude of a floating point operation. All computations have been performed on a 533 MHz DEC Alpha workstation with resulting CPU time of ~ 0.5μs. Even an Intel Pentium 200 MHz PC only used about 15μs. Note, that in a practical implementation, the computing times can be reduced drastically by an additional factor of 201 (number of gridpoints), if the time during the control of the SMA actuator is used for computing the needed approximations.

It is clear that computational times in this range make it feasible to treat more realistic problems, e.g., with a considerably larger number of wires and/or more complex structures which have to be modeled by finite elements or comparable methods. The main task is to create data bases with a sufficient number of nominal solutions, which, however, does not have to be very large due to the robustness of the method illustrated above.

Future tasks will cover the incorporation of FE schemes into the real-time optimal control code and an extension of the method to include a feedback of the current state as described in Büskens [4] of this issue. This is necessary in order to compensate for modeling "imperfections" or unforeseen and non-measurable perturbations which would cause a deviation from the actual current state of the system. This feedback can be implemented by predicting the final state as described in [4] and subsequently using this to re-calibrate the initial conditions and the current control. Note, that the mathematical formulation allows for disturbances in the initial conditions as well, see (12), so this feature, too, may be taken care of in real-time.

References

1. L. C. Brinson, M. S. Huang, C. Boller, and W. Brand: Analysis of Controlled Beam Deflections Using SMA Wires, J. Intelligent Mat Syst Struct **8** (1997) 12–25
2. C. Büskens: Direkte Optimierungsmethoden zur numerischen Berechnung optimaler Steuerungen. Diploma thesis, Institut für Numerische Mathematik, Universität Münster, Münster, Germany (1993)
3. C. Büskens: Optimierungsmethoden und Sensitivitätsanalyse für optimale Steuerprozesse mit Steuer- und Zustands-Beschränkungen. Dissertation, Institut für Numerische Mathematik, Universität Münster, Münster, Germany (1998)
4. C. Büskens: Real-Time Solutions for Perturbed Optimal Control Problems by a Mixed Open- and Closed-Loop Strategy. This volume.
5. C. Büskens and H. Maurer: Sensitivity Analysis and Real-Time Optimization of Nonlinear Programming Problems. This volume.
6. C. Büskens and H. Maurer: Sensitivity Analysis and Real-Time Control of Optimal Control Problems Using Nonlinear Programming Methods. This volume.
7. J. M. Cruz-Hernandez and V. Hayward: An Approach to Reduction of Hysteresis in Smart Materials, Proceedings of the 1998 IEEE, International Conference on Robotics & Automation (1998) 1510–1515
8. D. Grant and V. Hayward: Variable Structure Control of Shape Memory Alloy Actuators, IEEE Contr Syst Mag **17** (1997) 80–88
9. K. K. Ho, J. J. Gill, G. P. Carman and P. Jardine: Fabrication and Characterization of Thin Film NiTi for use as a Microbubble for Active Flow Control, M Wuttig, ed, Proc. 6th Ann. Int. Symp. Smart Struct. Mat., 1-5, Newport Beach, CA (1999)
10. D. C. Lagoudas and S. G. Shu: Residual Deformation of Active Structures with SMA Actuators, Int J Mech Sci **41** (1999) 595–619
11. N. Papenfuss, and S. Seelecke: Simulation and Control of SMA Actuators, Proc. 6th Ann. Int. Symp. Smart Struct. Mat., Newport Beach, USA, 1-5 March, 1999, **3667** (1999) 586–595
12. S. Seelecke: Control of Beam Structures by Shape Memory Wires, 2nd Sci. Conf. Smart Mechanical Systems - Adaptronics, Otto-von-Guericke Univ. Magdeburg (1997) 43–52
13. S. Seelecke: Adaptive Structures with SMA Actuators - Modeling and Simulation (in German), Habilitation thesis, TU Berlin, Berlin, Germany (1999)
14. S. Seelecke and C. Büskens: Optimal Control of Beam Structures by Shape Memory Wires, S. Hernandez and C.A. Brebbia, eds, OPTI 97, Computer Aided Optimum Design of Structures, Rome, Italy (1997) 457–466
15. S. Seelecke, and I. Müller: Shape Memory Alloy Actuators in Smart Structures - Modeling and Simulation, Applied Mechanics Review, submitted (2000)
16. S. G. Shu, D. C. Lagoudas, D. Hughes and J. T. Wen: Modeling of Flexible Beam Actuated by Shape Memory Alloy Wires, Smart Mat Struct **6** (1997) 265–277
17. G. V. Webb and D. C. Lagoudas: Control of SMA Actuators Under Dynamic Enviroments, Proc. 6th Ann. Int. Symp. Smart Struct. Mat., Newport Beach, USA, **3667** (1999)
18. M. W. M. van der Wijst, P. J. G. Schreurs, and F. E. Veldpaus: Application of Computed Phase Transformation Power to Control Shape Memory Alloy Actuators, Smart Mat Struct **6** (1997) 190–198

Real-Time Solutions for Perturbed Optimal Control Problems by a Mixed Open- and Closed-Loop Strategy

Christof Büskens

Lehrstuhl für Ingenieurmathematik, Universität Bayreuth, Germany

Abstract Many dynamical processes arising in engineering and natural science can be mathematically modelled by systems of differential equations. Often control functions, by which the process can be influenced, are to be chosen such that a certain objective function is optimized under observation of the differential equations and additional constraints. Today one of the greatest demands on such *optimal control problems* is to solve them in real-time. Open-loop strategies have their advantage with respect to optimality whereas the approximated solutions may not be admissible. Closed-loop strategies guarantee admissibility, but may lead to approximate solutions which are worse in view of optimality. Hence a new mixed strategy is proposed which benefits from the advantages of both methods, without suffering from their disadvantages. Finally an illustrative example from flight mechanics is discussed which shows the efficiency and robustness of the proposed method.

1 Introduction

Modern real-time optimization or control algorithms have to fulfil various requirements. Beyond the properties which can be achieved by usual control algorithms, such as fastness of the algorithms, low memory requirement, and admissibility of the solutions, additional requirements have now to be fulfilled for real-time optimal control methods: (Sub-) Optimality, capability of providing instantaneous information on improved controls, capability of a further improvement of controls if still computing time is left, robustness and stability of the solutions with respect to large and unpredictable perturbations. The real-time algorithm should be able to handle high dimensional problems and should allow a flexible treatment of different types of problems. These requirements can be obtained by methods of high order with *'cheap'* iteration steps, e.g., gradient free algorithms. From the user's viewpoint the resulting software must be simple and the required knowledge of the theoretical background must be low.

The above requirements can be put into a hierarchical order of importance: real-time ability, admissibility and, at last, optimality, which we will call AAO hierarchy. This is why admissibility and feasibility have been more in the focus of interest in the development of real-time control methods in the past, whereas real-time *optimal* control methods are not as well developed. Real-time control methods are usually of closed-loop type, whilst real-time optimal control methods are mostly of open-loop type. In view of the afore mentioned requirements closed-loop methods have their advantages on admissibility and feasibility, open-loop methods on optimality. In particular the advantages of one method are the disadvantages of the other and vice versa. Hence it seems to be obvious to combine the open-loop and the closed-loop ideas.

For this reason a mixed strategy is suggested based on the open-loop approach presented in Büskens and Maurer [3, 4], which leads to an iterative self-correcting and gradient-free closed-loop approach of Newton type.

In Section 2 the general idea of the mixed strategy is presented. Based on the error analysis of the open-loop approach in Büskens and Maurer [3] for perturbed NLP problems, correcting feedback steps in direction of the optimal solution are developed to assure the admissibility of the solution. Section 3 applies these ideas to perturbed optimal control problems. A demanding problem from flight dynamics is discussed in Section 4.

2 A Mixed Method for Perturbed NLP Problems

2.1 *Error Analysis of NLP Constraints*

Consider the perturbed NLP problem

$$\mathbf{NLP}_1(p) \quad \begin{array}{ll} \min\limits_{z} & F(z,p), \\ \text{subject to} & G^1(z,p) = 0, \\ & G^2(z,p) \le 0, \end{array} \tag{1}$$

with sufficiently smooth functions $G = (G_1, \ldots, G_{N_c})^T := (G^1, G^2)^T$, $G_i : \mathbb{R}^{N_z} \times P \to \mathbb{R}$, $i = 1, \ldots, N_c$ and $F : \mathbb{R}^{N_z} \times P \to \mathbb{R}$. Problem (1) is the same problem as (1) in Büskens and Maurer [3] of this issue. It is supposed that the assumptions of Theorem 2 in [3] are fulfilled. Then the real-time approximation of the minimizer $z(p)$ in Eq. (31) in [3] of this issue yields

$$z(p) \approx \tilde{z}^{[1]}(p) := z(p_0) + \frac{dz}{dp}(p_0)\,(p - p_0) \tag{2}$$

with $\frac{dz}{dp}(p_0)$ from Eq. (21) in [3]. Formula (2) represents a linear approximation of $z(p)$ and hence one can expect that, in general,

$$G^a(\tilde{z}^{[1]}(p), p) = \varepsilon_1 \neq 0, \tag{3}$$

with $\varepsilon_1 \in \mathbb{R}^{N_a}$ small. N_a is the number of active components G^a in (1), cf. Eq. (7) of [3]. In general the open-loop expression (2) does not represent an admissible solution of (1), due to the violation ε_1 in Eq. (3). For this reason, the approximation (2) can only be used for perturbed NLP problems, if ε_1 is sufficient small in view of practical requirements. Especially the exponential error growth in initial value problems for ODE systems caused by parameter deviations explains, why the real-time approximation (2) for perturbed optimal control problems cannot be used in general. Hence the following section is concerned with the reduction of those violations in the constraints caused by the real-time approximations (2).

2.2 Real-Time Error Reduction in NLP Constraints

Consider the following modification of (1)

$$\mathbf{NLP}_2(p,q) \quad \begin{array}{ll} \min\limits_{z} & F(z,p), \\ \text{subject to} & G^1(z,p) - q^1 = 0, \\ & G^2(z,p) - q^2 \le 0, \end{array} \tag{4}$$

with perturbation $(p,q)^T \in \mathbb{R}^{N_p+N_c}$, $q = (q^1, q^2)^T$. Moreover let q^a denote the perturbations for the active constraints. Without restriction let $(p_0, q_0) = (p_0, 0) \in \mathbb{R}^{N_p+N_c}$. Hence $\mathbf{NLP}_2(p,0)$ is equivalent to $\mathbf{NLP}_1(p)$.
In case that an actual deviation of the form $(p,0)^T \in \mathbb{R}^{N_p+N_c}$ from the nominal parameter $(p_0,0)^T$ is detected, Eq. (2) provides a very fast *open-loop approximation* for the perturbed solution. It was shown in (3), that this approximation causes an error ε_1 in the active constraints $G^a(p)$. This error is of the form of the new perturbation parameter q in (4). Hence a better approximation in view of optimality and especially admissibility can be found by

$$\begin{aligned} z(p) \approx \tilde{z}^{[2]}(p) &:= z(p_0) + \tfrac{dz}{dp}(p_0)\,(p-p_0) - \tfrac{dz}{dq^a}(0)\ \varepsilon_1 \\ &= \tilde{z}^{[1]}(p) - \tfrac{dz}{dq^a}(0)\ G^a(\tilde{z}^{[1]}(p),p), \end{aligned} \tag{5}$$

with $\tilde{z}^{[1]}(p)$ from (2) and $\frac{dz}{dp}(p_0)$ respectively $\frac{dz}{dq^a}(0)$ from Eq. (21) respectively from the simplified expression (28) in Büskens and Maurer [3] of this issue.

Since the nominal solution $z(p_0)$ as well as the sensitivity differentials $\frac{dz}{dp}(p_0)$ and $\frac{dz}{dq^a}(0)$ can be calculated off-line, Eq. (5) provides also a fast computation of the real-time approximation, since no gradient calculation is needed.
The additional term $\frac{dz}{dq^a}(0)G^a(\tilde{z}^{[1]}(p),p)$ in Eq. (5) can be understood as a correcting *feedback step* for the error caused by Eq. (2). Note that the approximation (5) again causes an error ε_2 in the active constraints $G^a(p)$.

$$G^a(\tilde{z}^{[2]}(p),p) = \varepsilon_2 \neq 0, \tag{6}$$

which is also of the form (4) with the new perturbation parameter q. Hence an additional improvement of (5) is given by

$$z(p) \approx \tilde{z}^{[3]}(p) := \tilde{z}^{[2]}(p) - \frac{dz}{dq^a}(0)\ G^a(\tilde{z}^{[2]}(p),p). \tag{7}$$

Thus we see, that the correction terms in (5) and (7) form an iterative algorithm which can be described as follows:

1. Choose $\varepsilon^\infty \in \mathbb{R}_+$ and initialize $\tilde{z}^{[1]}(p)$ by (2), set $k := 1$.
2. If $\|G^a(\tilde{z}^{[k]}(p),p)\|_2 < \varepsilon^\infty$ then STOP.
3. Calculate
$$\tilde{z}^{[k+1]}(p) := \tilde{z}^{[k]}(p) - \frac{dz}{dq^a}(0)\ G^a(\tilde{z}^{[k]}(p),p), \tag{8}$$
and set $k := k+1$.

4. Goto 2.

Note that the Sensitivity-Theorem 3 in [3] predicts the existence of a neighbourhood where the active constraints remain unchanged. This guarantees the existence of a fixpoint.

2.3 *Higher Order Convergence Rate by the Method of Steffensen*

An analysis of Eq. (8) yields, that this algorithm is basically a simplified Newton method for $G^a(\tilde{z}, p) = 0$ with convergence order one. Anyhow one finds a fast convergence if only small perturbations are given. To improve the rate of convergence one can apply the well known method of Steffensen, cf. Stoer [7], where Formula (8) is replaced by

$$\tilde{z}^{[k+1]}(p) := \tilde{z}^{[k]}(p) - \frac{(\hat{z}^{[1,k]}(p) - \tilde{z}^{[k]}(p))^2}{\hat{z}^{[2,k]}(p) - 2\hat{z}^{[1,k]}(p) + \tilde{z}^{[k]}(p)} \tag{9}$$

with

$$\begin{aligned} \hat{z}^{[1,k]}(p) &:= \tilde{z}^{[k]}(p) - \tfrac{dz}{dq^a}(0)\; G^a(\tilde{z}^{[k]}(p), p), \\ \hat{z}^{[2,k]}(p) &:= \hat{z}^{[1,k]}(p) - \tfrac{dz}{dq^a}(0)\; G^a(\hat{z}^{[1,k]}(p), p). \end{aligned} \tag{10}$$

Note that the method of Steffensen is at least of order 2.
Furthermore both algorithms, i.e., the original algorithm of Section 2.2 and the modification of this section, are also able to handle deviations not taken into account by the parameter sensitivity analysis, for which the sensitivity differentials are not available. The reason is that such perturbations influence the state of the solution and especially violate the constraints indirectly.

3 A Mixed Method for Perturbed Optimal Control Problems

In this section the conceptions developed before are adapted to discretized optimal control problems as described in Büskens and Maurer [4] of this issue. We focus attention on additional artificial perturbations in the constraints. Consider the following modification of optimal control problem (1) in [4], where for the sake of simplicity, only additional perturbations in the terminal constraints are considered

$$\begin{aligned} \text{Minimize} \qquad & F(x,u,p) = g(x(t_f),p) + \int_{t_0}^{t_f} f_0(x(t),u(t),p)\, dt \\ \text{subject to} \qquad & \dot{x}(t) = f(x(t),u(t),p), \\ & x(t_0) = \varphi(p), \\ & \psi(x(t_f),p) - q = 0, \\ & C(x(t),u(t),p) \le 0, \quad t \in [t_0, t_f], \end{aligned} \tag{11}$$

with $q \in \mathbb{R}^r$. By discretisation of (11) we obtain a perturbed NLP problem of form (9) in [4] or, as represented hereafter, of form (13) in the same article:

$$\begin{aligned}
&\text{Minimize} \quad F(z,p) := g(x^{N_t}(z,p),p),\\
&\text{subject to} \quad G(z,p) := \begin{pmatrix} \psi(x^{N_t}(z,p),p) - q \\ C(x^1(z,p),u^1,p) \\ \vdots \\ C(x^{N_t}(z,p),u^{N_t},p) \end{pmatrix} \in \mathbb{R}^{N_c}, \; N_c = kN + r
\end{aligned} \tag{12}$$

For this reason NLP problem (12) is of form (4) and we can apply the algorithm from Section 2.2. Thus, confining algorithm (8) to the artifical perturbation parameter q in (12), we arrive the following algorithm

1. Choose $\varepsilon^\infty \in \mathbb{R}_+$ and initialize $\tilde{z}^{[1]}(p)$ by (2) (respectively by Eqs. (23) and (24) of [4]), set $k := 1$.
2. If $\|\psi(x^{N_t}(\tilde{z}^{[k]}(p),p),p)\|_2 < \varepsilon^\infty$ then STOP.
3. Calculate
$$\tilde{z}^{[k+1]}(p) := \tilde{z}^{[k]}(p) - \frac{dz}{dq^a}(0)\, \psi(x^{N_t}(\tilde{z}^{[k]}(p),p),p), \tag{13}$$
and set $k := k + 1$.
4. Goto 2.

Obviously, this algorithm can also be improved by Steffensen's method.

4 Example: Emergency Landing of a Hypersonic Flight System

A study about the reliability of U.S. launch vehicles since the beginning of human space flight states, that more than $\frac{2}{3}$ of all failures affect the propulsion system. Therefore we investigate the mission abort of a winged two-stage hypersonic flight system due to an ignition failure of the main rocket engine shortly after separation, cf. Mayrhofer and Sachs [5, 6]. The upper stage of the flight system is still able to manoeuvre although the propulsion system is damaged, compare the Emergency Landing Site Landing (ELSL) scenario in Figure 1.

A point mass model consisting of six state functions and two control functions describes the dynamics of the flight system:

$$\begin{aligned}
\dot{v} &= -D(v,h;C_L)\tfrac{1}{m} - g(h)\sin\gamma + \\
&\quad + \omega^2 \cos\Lambda(\sin\gamma\cos\Lambda - \cos\gamma\sin\chi\sin\Lambda)r(h),\\
\dot{\gamma} &= L(v,h;C_L)\tfrac{\cos\mu}{mv} - \left(\tfrac{g(h)}{r(h)} - \tfrac{v}{r(h)}\right)\cos\gamma + \\
&\quad + 2\omega\cos\chi\cos\Lambda + \omega^2\cos\Lambda(\sin\gamma\sin\chi\sin\Lambda + \cos\gamma\cos\Lambda)\tfrac{r(h)}{v},\\
\dot{\chi} &= L(v,h;C_L)\tfrac{\sin\mu}{mv\cos\gamma} - \cos\gamma\cos\chi\tan\Lambda\tfrac{v}{r(h)} + \\
&\quad + 2\omega(\sin\chi\cos\Lambda\tan\gamma - \sin\Lambda) - \omega^2\cos\Lambda\sin\Lambda\cos\chi\tfrac{r(h)}{v\cos\gamma},\\
\dot{h} &= v\sin\gamma,\\
\dot{\Lambda} &= \cos\gamma\sin\chi\tfrac{v}{r(h)},\\
\dot{\Theta} &= \cos\gamma\cos\chi\tfrac{v}{r(h)\cos\Lambda},
\end{aligned} \tag{14}$$

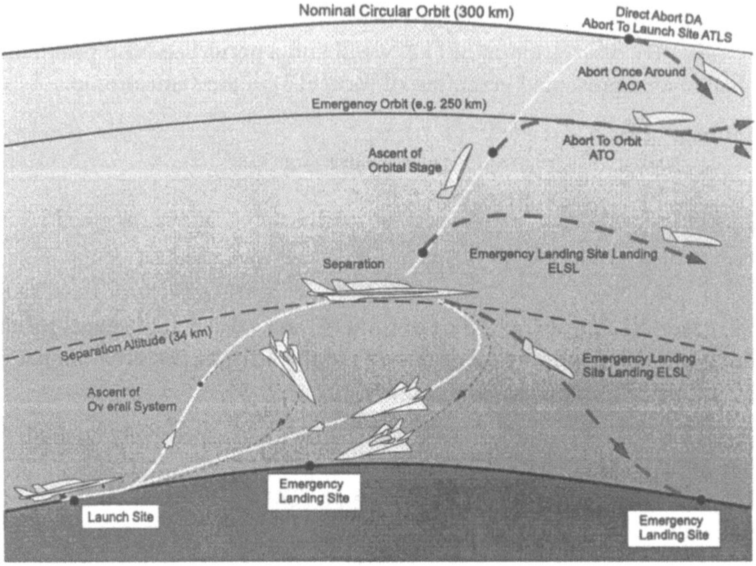

Figure 1. The emergency landing problem (reprint from [5, 6])

with functions

$$
\begin{aligned}
r(h) &= r_0 + h, & g(h) &= g_0\left(\frac{r_0}{r(h)}\right)^2, \\
D(v,h;C_L) &= q(v,h)Fc_D(C_L), & \rho(h) &= \rho_0 e^{-\beta h}, \\
c_D(C_L) &= c_{D_0} + kC_L^2, & q(v,h) &= \tfrac{1}{c}\rho(h)v^2, \\
L(v,h;C_L) &= q(v,h)FC_L
\end{aligned} \tag{15}
$$

and constants

$$
\begin{aligned}
c &= 2.0, & c_{D_0} &= 0.017, & r_0 &= 6.371 \cdot 10^6, \\
F &= 305.0, & g_0 &= 9.80665, & k &= 2.0, \\
\omega &= 7.270 \cdot 10^{-5}, & \beta &= \tfrac{1}{6900.0}.
\end{aligned} \tag{16}
$$

Herein the state variables are defined by the velocity v, the flight path angle γ, the azimuth angle χ, the altitude h, the geographic longitude Λ, the geocentric latitude Θ. The controls are denoted by the lift coefficient C_L and the bank angle μ and are restricted by the box constraints

$$
0 \le C_L \le 1, \qquad 0 \le \mu \le 1. \tag{17}
$$

The mass is assumed to be constant $m = 115000$. Initial values are given by

$$
\begin{pmatrix} v(0) \\ \gamma(0) \\ \chi(0) \\ h(0) \\ \Lambda(0) \\ \Theta(0) \end{pmatrix} = \begin{pmatrix} 2150.5452900 \\ 0.1520181770 \\ 2.2689279889 \\ 33900.000000 \\ 0.8651416299 \\ 0.1980907302 \end{pmatrix}, \tag{18}
$$

which corresponds to a starting point over Bayreuth/Germany (not to be taken seriously). For security reasons an emergency landing trajectory on a rotating earth with maximum range has to be found

$$F(\mu, C_L, t_f) = \left(\frac{\Lambda(t_f) - \Lambda(t_0)}{\Lambda(t_0)}\right)^2 + \left(\frac{\Theta(t_f) - \Theta(t_0)}{\Theta(t_0)}\right)^2 . \tag{19}$$

The admissible set is defined by a final altitude of 500 meters:

$$h(t_f) = 500.0. \tag{20}$$

The perturbations p are deviations in the air density modelled by

$$\rho_0 = 1.249512 \cdot p \tag{21}$$

and additional non-detectable parameters of the system, which is here simulated by deviations in the initial altitude. The terminal time t_f is assumed to be free, thus it has to be considered for the real-time calculations as well.

For $N_t = 101$ equidistant gridpoints the number of optimization variables $z = (\mu^1, C_L^1, \ldots, \mu^{N_t}, C_L^{N_t}, t_f)^T$ is $N_z = 203 = 2 \cdot 101 + 1$ due to two control variables and the free final time. For $p_0 = 1.0$ the nominal objective function is calculated to $F(\mu(p_0), C_L(p_0), t_f(p_0)) = -0.777047$, while the terminal time is found by $t_f(p_0) = 735.242$. Figure 2 shows the nominal control functions on the normalized time interval $[0, 1]$.

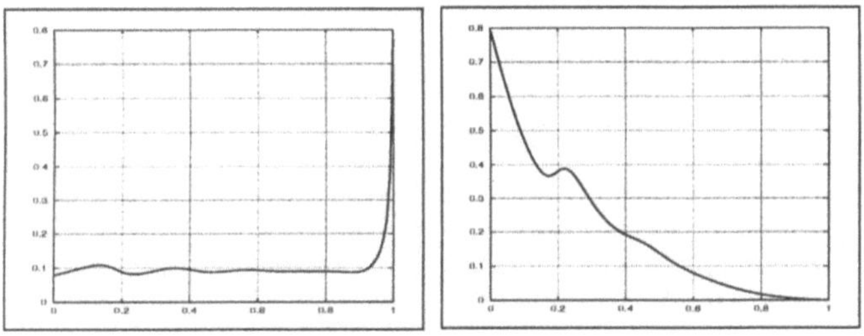

Figure 2. Optimal controls $\mu(t, p_0)$ and $C_L(t, p_0)$

The nominal optimal states are shown in Figure 3.

Figure 4 shows the optimal trajectory as a three dimensional plot and its projection onto ground. The landing site is on the British Island.

Finally, Figure 5 depicts the sensitivity differentials for parameter p and the artifical parameter q for the terminal altitude as described in Eq. (12).

Table 1 shows the first ten iterates for the convergence behaviour of the terminal altitude and the objective function for perturbations between 1% and 100%, calculated

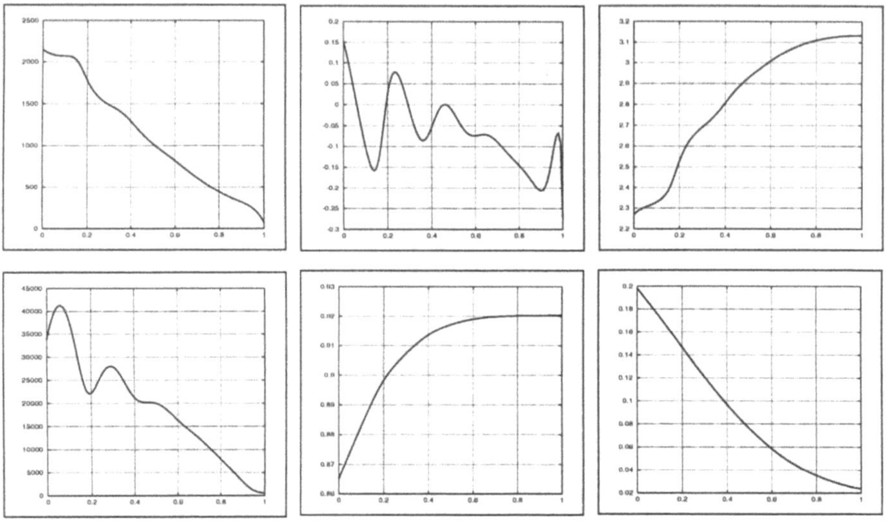

Figure 3. Optimal states $v(t, p_0)$, $\gamma(t, p_0)$, $\chi(t, p_0)$, $h(t, p_0)$, $\Lambda(t, p_0)$ and $\Theta(t, p_0)$

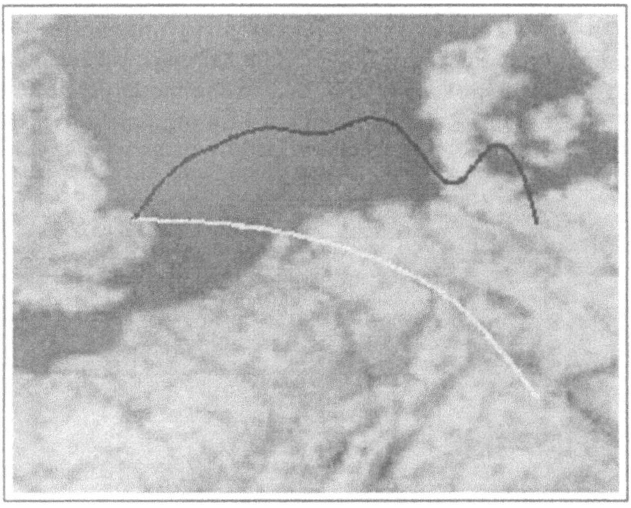

Figure 4. Optimal trajectory

with a fourth-order Runge-Kutta approximation for the state and a linear interpolation of the control variables. **Con.Err.** is the relative error for the constraint (20)

$$\zeta^{[k]}_{err}(p) := \left| \frac{h(\tilde{z}^{[k]}(p); t_f) - 500.0}{500.0} \right|, \tag{22}$$

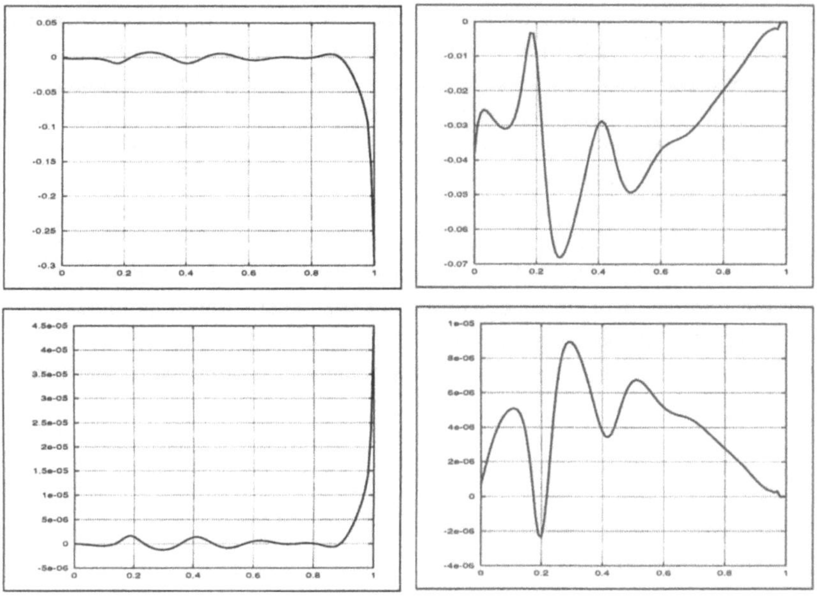

Figure 5. Sensitivity differentials $\frac{\partial \mu}{\partial p}(t)$, $\frac{\partial C_L}{\partial p}(t)$ and $\frac{\partial \mu}{\partial q}(t)$, $\frac{\partial C_L}{\partial q}(t)$

	$p = 1.01$		$p = 1.1$		$p = 2.0$	
Iter.	Con.Err.	Obj.Err.	Con.Err.	Obj.Err.	Con.Err.	Obj.Err.
0	$1.24 \cdot 10^{-01}$	$6.19 \cdot 10^{-04}$	$1.18 \cdot 10^{+00}$	$5.87 \cdot 10^{-03}$	$8.49 \cdot 10^{-00}$	$4.07 \cdot 10^{-02}$
1	$4.97 \cdot 10^{-04}$	$2.48 \cdot 10^{-06}$	$4.62 \cdot 10^{-02}$	$2.27 \cdot 10^{-04}$	$2.80 \cdot 10^{-00}$	$1.06 \cdot 10^{-02}$
2	$1.02 \cdot 10^{-06}$	$4.86 \cdot 10^{-09}$	$9.24 \cdot 10^{-04}$	$4.20 \cdot 10^{-06}$	$3.72 \cdot 10^{-01}$	$7.96 \cdot 10^{-04}$
3	$2.01 \cdot 10^{-09}$	$2.22 \cdot 10^{-10}$	$1.81 \cdot 10^{-05}$	$2.58 \cdot 10^{-07}$	$4.67 \cdot 10^{-02}$	$5.78 \cdot 10^{-04}$
4	$4.28 \cdot 10^{-12}$	$2.32 \cdot 10^{-10}$	$3.56 \cdot 10^{-07}$	$3.45 \cdot 10^{-07}$	$5.80 \cdot 10^{-03}$	$7.52 \cdot 10^{-04}$
5	$1.03 \cdot 10^{-14}$	$2.32 \cdot 10^{-10}$	$7.01 \cdot 10^{-09}$	$3.47 \cdot 10^{-07}$	$7.19 \cdot 10^{-04}$	$7.73 \cdot 10^{-04}$
6	$1.56 \cdot 10^{-14}$	$2.32 \cdot 10^{-10}$	$1.38 \cdot 10^{-10}$	$3.47 \cdot 10^{-07}$	$8.91 \cdot 10^{-05}$	$7.76 \cdot 10^{-04}$
7	$8.64 \cdot 10^{-15}$	$2.32 \cdot 10^{-10}$	$2.72 \cdot 10^{-12}$	$3.47 \cdot 10^{-07}$	$1.11 \cdot 10^{-05}$	$7.76 \cdot 10^{-04}$
8	$2.55 \cdot 10^{-14}$	$2.32 \cdot 10^{-10}$	$8.24 \cdot 10^{-14}$	$3.47 \cdot 10^{-07}$	$1.37 \cdot 10^{-06}$	$7.76 \cdot 10^{-04}$
9	$2.94 \cdot 10^{-14}$	$2.32 \cdot 10^{-10}$	$1.19 \cdot 10^{-14}$	$3.47 \cdot 10^{-07}$	$1.70 \cdot 10^{-07}$	$7.76 \cdot 10^{-04}$
10	$2.75 \cdot 10^{-14}$	$2.32 \cdot 10^{-10}$	$1.44 \cdot 10^{-14}$	$3.47 \cdot 10^{-07}$	$2.11 \cdot 10^{-08}$	$7.76 \cdot 10^{-04}$

Table 1. Convergence behaviour (simplified Newton)

and **Con.Obj.** is the relative error for the objective functional (19)

$$\left| \frac{F(\tilde{z}^{[k]}(p)) - F(z(p))}{F(z(p))} \right| . \tag{23}$$

Herein $z(p)$ denotes the "exact" solution of the perturbed problem as computed by the code NUDOCCCS, cf. Büskens [1,2]. **Iter.** denotes the iteration number in (13). For reasons of comparison the errors for Iter.$= 0$ denote those for the nominal solution $\tilde{z}^{[0]}(p) = z_0$ and for Iter.= 1 those for $\tilde{z}^{[1]}(p)$ which is the *open-loop* initialisation (2). The CPU-time on a Pentium III processor with 450MHZ is about $3.4 \cdot 10^{-7}$ seconds for $\tilde{z}^{[1]}(p)$ and about $1.2 \cdot 10^{-4}$ seconds for each iteration step $\tilde{z}^{[k]}(p)$, $k \geq 2$. Note that the CPU-time can be reduced drastically by a factor which is given by the number of gridpoints (here 101), if the approximate controls are only computed on a small interval necessary to continue the flight. Depending on the magnitude of the perturbation p, the CPU-time for calculating the exact solution $z(p)$ ranges between 6 and 12 seconds. The error in the constraint (representing the admissibility) can be reduced almost to machine precision. For the sake of comparability the relative error in the constraint for the exact solution is given by about $5 \cdot 10^{-7}$. Beyond this, the relative error in the objective is improved, too.

In the left part of Figure 6 the convergence behaviour is depicted for perturbations

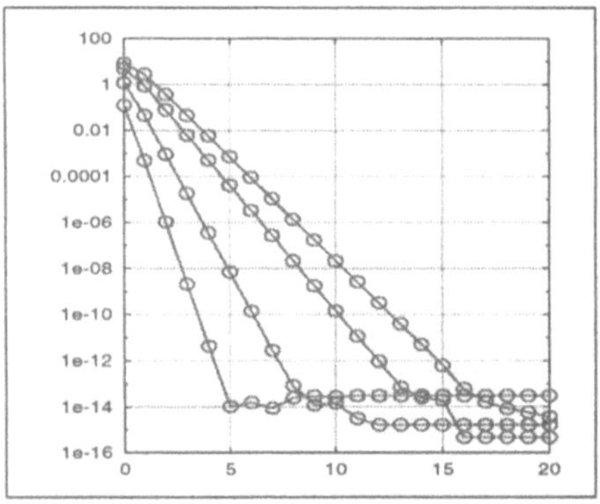

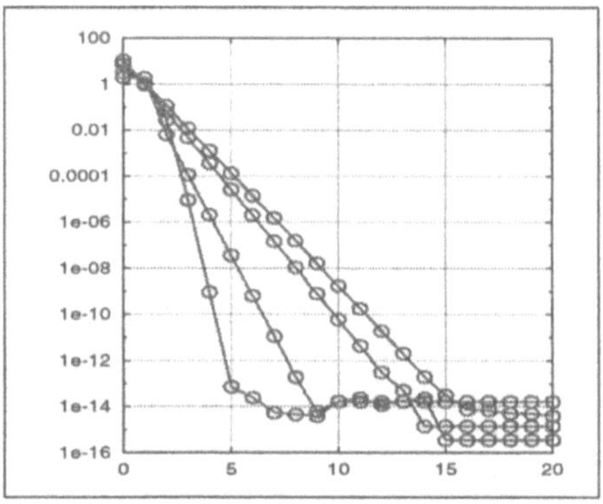

Figure 6. Convergence behaviour of the mixed strategy (simplified Newton)

$p = 1.01$, $p = 1.1$ $p = 1.5$ and $p = 2.0$. The iteration number is on the x–axis while the relative error in the constraint is on the logarithmic y–axis. Similarly, for the same perturbations in the air density the convergence behaviour for an additional deviation of +10000 meters in the initial altitude are shown in the right part of 6. Even though this large perturbation has not be taken into account by the sensitivity analysis the method is able to compensate it due to its influence on the terminal altitude constraint.

Next, we want to determine the *numerical* order α of convergence by estimating α in

$$|\zeta_{err}^{[k+1]}(p)| \approx c(p)|\zeta_{err}^{[k]}(p)|^{\alpha}, \tag{24}$$

for different iteration numbers $k \geq 1$. Since $\frac{\zeta_{err}^{[k+1]}(p)}{\zeta_{err}^{[k]}(p)} = c(p)$ is constant, the order α equals one with a small convergence factor $c(p)$, here $c(1.01) \approx \frac{2}{1000}$, $c(1.1) \approx \frac{2}{100}$, $c(2.0) \approx \frac{1}{10}$.

Table 2 shows the quadratical convergence of the Steffensen method (9) applied to the mixed strategy.

	p = 1.01		p = 1.1		p = 2.0	
Iter.	Con.Err.	Obj.Err.	Con.Err.	Obj.Err.	Con.Err.	Obj.Err.
0	$1.24 \cdot 10^{-01}$	$6.19 \cdot 10^{-04}$	$1.18 \cdot 10^{+00}$	$5.87 \cdot 10^{-03}$	$8.49 \cdot 10^{-00}$	$4.07 \cdot 10^{-02}$
1	$4.97 \cdot 10^{-04}$	$2.48 \cdot 10^{-06}$	$4.62 \cdot 10^{-02}$	$2.27 \cdot 10^{-04}$	$2.80 \cdot 10^{-00}$	$1.06 \cdot 10^{-02}$
2	$3.86 \cdot 10^{-12}$	$5.08 \cdot 10^{-11}$	$3.32 \cdot 10^{-07}$	$3.10 \cdot 10^{-07}$	$3.72 \cdot 10^{-03}$	$7.82 \cdot 10^{-04}$
3	–	–	$4.66 \cdot 10^{-15}$	$3.08 \cdot 10^{-07}$	$8.33 \cdot 10^{-09}$	$7.66 \cdot 10^{-04}$
4	–	–	–	–	$8.64 \cdot 10^{-15}$	$7.66 \cdot 10^{-04}$

Table 2. Convergence behaviour (method of Steffensen)

Again perturbations between 1% and 100% are treated to show the relative errors in the terminal altitude and the objective function. The notations are similar to Table 1. The CPU-time is about $3.4 \cdot 10^{-7}$ seconds for $\tilde{z}^{[1]}(p)$, too. Due to the twofold evaluation of the constraints in (10) the CPU-time for the iteration steps $\tilde{z}^{[k]}(p)$ in (9) rises to $2.4 \cdot 10^{-4}$ seconds for each iteration step $k \geq 2$. Note, that the CPU-times again can be reduced drastically by an additional factor up to 101 (number of gridpoints), if the time during the flight is used for computing the needed approximations. Similar good results, not discussed here, are obtained for perturbations not taken into account for the sensitivity analysis. It could be shown, that the mixed-strategy of section 3 is able to compensate perturbations between −80% and larger than 300%. Although these perturbations are not realistic, the results obtained for the difficult emergency landing manoeuvre shows the enormous potential of the method.

5 Conclusion

This paper presents improved iterative methods for real-time optimization of perturbed optimal control problems. The first essential feature of the methods is to compute an error estimations for the violations of the constraints in the underlying perturbed NLP problem. These error estimations are obtained by analyzing the admissibility of the open-loop approximation of [3, 4] resp. of the different iterates of the methods. The second essential feature is the interpretation of these errors as additional perturbations in the discretized optimal control problem for which a parametric sensitivity analysis, as described in Büskens and Maurer [4], is performed. Herewith, and this is the third essential feature, an iterative feedback strategy of Newton type has been formulated which guarantees the admissibility of the solution, while the optimality is still preserved to first order. The main advantages of the new methods are: the high computational speed, the self-correcting property in view of admissibility, the gradient-free formulation of the methods, the capability for compensating large and unforeseen perturbations and the *any time property*, i.e., the iterates can be continually be improved as long as the real process allows additional computing time. Finally, the efficiency and robustness of the proposed methods have been demonstrated by solving a demanding problem from flight mechanics, the emergency landing of a hypersonic space transportation system.

References

1. C. Büskens: Direkte Optimierungsmethoden zur numerischen Berechnung optimaler Steuerungen. Diploma thesis, Institut für Numerische Mathematik, Universität Münster, Münster, Germany (1993)
2. C. Büskens: Optimierungsmethoden und Sensitivitätsanalyse für optimale Steuerprozesse mit Steuer- und Zustands- Beschränkungen. Dissertation, Institut für Numerische Mathematik, Universität Münster, Münster, Germany (1998)
3. C. Büskens and H. Maurer, Sensitivity Analysis and Real-Time Optimization of Parametric Nonlinear Programming Problems. This volume.
4. C. Büskens and H. Maurer, Sensitivity Analysis and Real-Time Control of Optimal Control Problems Using Nonlinear Programming Methods. This volume.
5. M. Mayrhofer and G. Sachs, Notflugbahnen eines zweistufigen Hyperschall-Flugsystems ausgehend vom Trennmanöver. Seminar des Sonderforschungsbereichs 255: Transatmosphärische Flugsysteme, Technische Universität München, (1996) 109–118
6. M. Mayrhofer and G. Sachs, Abort Trajectory Range Increase by Optimal Fuel Draining, in Optimalsteuerungsprobleme in der Luft- und Raumfahrt, Sonderforschungsbereichs 255: Transatmosphärische Flugsysteme, Technische Universität München (2000) 115–124
7. J. Stoer, Einführung in die Numerische Mathematik I Springer-Verlag, Berlin Heidelberg, (1983) Transatmosphärische Flugsysteme, Technische Universität München (2000) 115–124

Real-Time Optimization of DAE Systems

Christof Büskens[1] and Matthias Gerdts[2]

[1] Lehrstuhl für Ingenieurmathematik, Universität Bayreuth, Germany
[2] Lehrstuhl für Angewandte Mathematik, Universität Bayreuth, Germany

Abstract We consider approaches for real-time optimization of dynamical systems governed by differential-algebraic (DAE) systems. Special attention is turned on the calculation of consistent initial values such that the DAE system is solvable.

The underlying DAE optimal control problem is solved numerically by a direct single shooting method. This method reduces the infinite dimensional optimal control problem by discretization of the control over a suitable chosen grid to a finite dimensional optimization problem.

A real-time optimization method is achieved by performing a parameter sensitivity analysis of the discretized optimal control problem, taking the calculation of consistent initial values into account.

A high dimensional and highly nonlinear example of a car's jink is discussed to demonstrate the capabilities of the proposed methods.

1 Introduction

Subject of our interest is the parameter depending optimal control problem of Mayer type ($\mathbf{OCP}_{\text{DAE}}(p)$):

$$\begin{array}{lrl} \text{Minimize} & J[x,u] = g(x(t_f),t_f,p), & t_0 < t_f \text{ fixed or free,} \\ \text{subject to} & F(x(t),\dot{x}(t),u(t),p) = 0, & \text{for a.e. } t \in [t_0,t_f], \\ & C(x(t),u(t),p) \leq 0, & t \in [t_0,t_f], \\ & \psi(x(t_0),x(t_f),p) = 0, & \end{array} \tag{1}$$

with state variables $x \in \mathbb{R}^n$, control variables $u \in \mathbb{R}^m$, fixed parameters $p \in P \subset \mathbb{R}^{N_p}$, objective $g : \mathbb{R}^n \times \mathbb{R} \times P \to \mathbb{R}$, differential-algebraic (DAE) constraints $F : \mathbb{R}^n \times \mathbb{R}^n \times \mathbb{R}^m \times P \to \mathbb{R}^n$ given in implicit form with *singular* Jacobian $F_{\dot{x}}$, path constraints $C : \mathbb{R}^n \times \mathbb{R}^m \times P \to \mathbb{R}^k$ and boundary conditions $\psi : \mathbb{R}^n \times \mathbb{R}^n \times P \to \mathbb{R}^r$. All functions are assumed to be sufficiently smooth.

DAE optimal control problems pose several problems due to the singular character of DAE systems. One of the main problems is that, in contrast to ordinary differential equation (ODE) systems with regular Jacobian $F_{\dot{x}}$, not any initial value is admissible. Hidden constraints have to be fulfilled, i.e., initial values have to be *consistent* with the DAE system, see Brenan et al. [1] for a more detailed discussion. A broad class of problems is described by *semi-explicit DAE systems* of the form

$$F^d(x^d(t),x^a(t),\dot{x}^d(t),u(t),p) = 0, \tag{2}$$

$$F^a(x^d(t),x^a(t),u(t),p) = 0 \tag{3}$$

with the regularity assumptions

$$\left.\begin{array}{l}\dfrac{\partial F^d}{\partial \dot{x}^d}(x^d(t), x^a(t), \dot{x}^d(t), u(t), p),\\ \dfrac{\partial F^a}{\partial x^a}(x^d(t), x^a(t), u(t), p)\end{array}\right\} \text{ regular, } \forall\, t \in [t_0, t_f]. \tag{4}$$

The implicit DAE system $F := (F^d, F^a)^\top \in \mathbb{R}^{n_d} \times \mathbb{R}^{n_a}$ is separated into a *differential equation* (2) and an *algebraic equation* (3). Similarly, the state x is partitioned a priori in $x := (x^d, x^a)^\top \in \mathbb{R}^{n_d} \times \mathbb{R}^{n_a}$ with *differential variables* x^d and *algebraic variables* x^a.

We call initial values $x^d(t_0)$ and $x^a(t_0)$ *consistent* for (2), (3), if and only if

$$F^a(x^d(t_0), x^a(t_0), u(t_0), p) = 0 \tag{5}$$

holds. For a given initial value $x^d(t_0)$ obviously not all values for $x^a(t_0)$ comply with (5). However, together with the regularity of the Jacobian $\frac{\partial F^a}{\partial x^a}$ in (4), the implicit function theorem provides a locally unique resolving function f^a of (5) w.r.t. $x^a(t_0)$ depending on $x^d(t_0)$, $u(t_0)$ and p:

$$x^a(t_0) = f^a(x^d(t_0), u(t_0), p). \tag{6}$$

Please note, that one differentiation of the algebraic constraint (3) w.r.t. time is needed to receive a differential equation for the algebraic variable x^a. For this reason system (2) and (3) together with (4) is called a semi-explicit DAE system of *index 1*.

As a further consequence of the implicit function theorem, the relationships

$$\begin{aligned}\frac{\partial f^a}{\partial x^d}(x^d, u, p) &= -\left(\frac{\partial F^a}{\partial x^a}(x^d, x^a, u, p)\right)^{-1} \frac{\partial F^a}{\partial x^d}(x^d, x^a, u, p),\\ \frac{\partial f^a}{\partial u}(x^d, u, p) &= -\left(\frac{\partial F^a}{\partial x^a}(x^d, x^a, u, p)\right)^{-1} \frac{\partial F^a}{\partial u}(x^d, x^a, u, p),\\ \frac{\partial f^a}{\partial p}(x^d, u, p) &= -\left(\frac{\partial F^a}{\partial x^a}(x^d, x^a, u, p)\right)^{-1} \frac{\partial F^a}{\partial p}(x^d, x^a, u, p)\end{aligned} \tag{7}$$

hold in t_0. The differentials $\frac{\partial f^a}{\partial x^d}$, $\frac{\partial f^a}{\partial u}$ and $\frac{\partial f^a}{\partial p}$ will be used for gradient calculations and real time computations described in the subsequent sections.

The situation becomes more difficult if we consider semi-explicit DAE systems of type

$$\dot{x}^d(t) = F^d(x^d(t), x^a(t), u(t), p), \tag{8}$$

$$0 = F^a(x^d(t), u(t), p) \tag{9}$$

with

$$\frac{\partial F^a}{\partial x^d}(x^d(t), u(t), p) \cdot \frac{\partial F^d}{\partial x^a}(x^d(t), x^a(t), u(t), p) \quad \text{regular,} \forall\, t \in [t_0, t_f]. \tag{10}$$

Differentiation of (9) w.r.t. time yields the nonlinear algebraic equation

$$0 = \frac{\partial F^a}{\partial x^d}(x^d(t), u(t), p) \cdot F^d(x^d(t), x^a(t), u(t), p) + \frac{\partial F^a}{\partial u}(x^d(t), u(t), p) \cdot \dot{u}(t) \tag{11}$$

which has to be satisfied for all t. In this case initial values $x^d(t_0)$ and $x^a(t_0)$ are called consistent if and only if they satisfy equations (9) and (11). Equation (11) is called *hidden constraint*. With the regularity condition equation (11) can be solved w.r.t. $x^a(t_0)$. Please note, that (5), (9) and (11) depend on u and additionally (11) depends on the derivative of u. Therefore, in contrast to optimal control problems with ODE systems, one can not expect in general, that all state variables are continuous, because the control or its derivative may be discontinuous.

Note, that differentiation of (11) w.r.t. time yields a differential equation for x^a. Therefore DAE systems of type (8) and (9) are called of *index 2*.

It is easy to imagine that the situation gets much more complicated if no special structure of the DAE system is assumed, i.e., the DAE system is given in general implicit form

$$F(x(t), \dot{x}(t), u(t), p) = 0 \tag{12}$$

with singular Jacobian $\frac{\partial F}{\partial \dot{x}}$. For a more detailed characterization of these systems please refer to Campbell and Gear [6], Gear [7] and Leimkuhler et al. [10].

2 Numerical Solution of ($\mathbf{OCP}_{\mathrm{DAE}}(p)$)

In view of the example presented later on and for simplicity reasons, we restrict the following discussion to semi-explicit index-1 systems (2) and (3). Extensions to higher index DAE systems are given in Büskens and Gerdts [5, 8].

The optimal control problem ($\mathbf{OCP}_{\mathrm{DAE}}(p)$) is solved numerically by a direct (single) shooting method, whose basic ideas are to be summarized shortly. Similar to the discretization (10)-(13) in Büskens and Maurer [4] of this issue, the optimal control problem ($\mathbf{OCP}_{\mathrm{DAE}}(p)$) is discretized by introducing a suitable grid

$$t_0 = \tau_1 < \tau_2 < \cdots < \tau_{N_t} = t_f. \tag{13}$$

The control function u is approximated by the continuous and piecewise linear function

$$u_{app}(t) = u^i + \frac{t - \tau_i}{\tau_{i+1} - \tau_i}(u^{i+1} - u^i) \tag{14}$$

for $\tau_i \leq t \leq \tau_{i+1}, i = 1, \ldots, N_t - 1$, or any higher order approximation, where u^i denotes an approximation of $u(\tau_i)$. It is important to notice, that it is recommended to use at least a continuous approximation of the control for this class of DAE systems. If a discontinuous approximation is chosen, the recalculation of consistent values at each grid point becomes necessary, because a continuous state in combination with a discontinuous control approximation will not suffice Equation (5) in general. According to DAE systems of index greater than one, control approximations with even higher smoothness properties have to be chosen.

For a given initial value $x^{d,0} \approx x^d(t_0)$ of the differential part of the state at $t_0 = \tau_1$, the corresponding consistent algebraic part $x^{a,0}$ is calculated by applying Newton's method to (5) evaluated at $(x^{d,0}, x^{a,0}, u_{app}(\tau_1), p)$. Note again, that the nonlinear system is solvable under the assumptions (4).

Let $z = (x^{d,0}, u^1, \dots, u^{N_t})^\top \in \mathbb{R}^{n_d + N_t m}$ denote the vector of variables, then the DAE system (2) and (3) is solved on $[\tau_1, \tau_{N_t}]$ by standard DAE integrators, e.g., DASSL [11]. This yields an approximation

$$x_{app}(t) = x_{app}(t; z, p) \tag{15}$$

of the state $x(t)$ on $[t_0, t_f]$. The notations $u_{app}(t; z)$ and $x_{app}(t; z, p)$ are used to point out the dependency from z and p.

Introduction of u_{app} and x_{app} into (**OCP**$_{\text{DAE}}(p)$) and evaluation of the constraints on the grid results in the finite dimensional nonlinear optimization problem (**NLP**(p)) for z

$$\begin{aligned}
&\text{Minimize } f(z,p) := g(x_{app}(t_f; z, p), t_f, p)\\
&\text{subject to } G(z,p) := \begin{pmatrix} C(x_{app}(\tau_1; z, p), u_{app}(\tau_1; z), p) \\ \vdots \\ C(x_{app}(\tau_{N_t}; z, p), u_{app}(\tau_{N_t}; z), p) \end{pmatrix} \leq 0, \\
&H(z,p) := \psi(x_{app}(\tau_1; z, p), x_{app}(\tau_{N_t}; z, p), p) = 0,
\end{aligned} \tag{16}$$

which can be solved numerically by, e.g., SQP methods. Because SQP methods work iteratively in each iteration consistent initial values have to be computed in the depicted way before the DAE system can be solved. The gradient of the objective and the Jacobian of the constraints in (**NLP**(p)) are computed either by finite differences or by use of the *sensitivity DAE system*

$$\begin{aligned}
\frac{\partial F^d}{\partial x^d} \cdot S^d(t) + \frac{\partial F^d}{\partial x^a} \cdot S^a(t) + \frac{\partial F^d}{\partial \dot{x}_d} \cdot \dot{S}^d(t) + \frac{\partial F^d}{\partial u} \cdot \frac{\partial u_{app}}{\partial z}(t; z) &= 0, \\
\frac{\partial F^a}{\partial x^d} \cdot S^d(t) + \frac{\partial F^a}{\partial x^a} \cdot S^a(t) + \frac{\partial F^a}{\partial u} \cdot \frac{\partial u_{app}}{\partial z}(t; z) &= 0
\end{aligned}$$

corresponding to (2) and (3) with sensitivity matrices $S^d(t) := \partial x^d(t; z, p)/\partial z$, $S^a(t) := \partial x^a(t; z, p)/\partial z$ and consistent initial components

$$\begin{aligned}
S^d(\tau_1) &= \frac{\partial x^{d,0}}{\partial z} = [\, I_{n_d} \mid 0 \mid \cdots \mid 0 \,], \\
S^a(\tau_1) &= \frac{\partial x^{a,0}}{\partial z} = \frac{\partial f^a}{\partial z}(x^{d,0}, u_{app}(\tau_1; z), p)
\end{aligned} \tag{17}$$

computed out of (7) at $t = \tau_1$, compare Gerdts [8] and Hinsberger [9]. The Jacobians of F^d and F^a w.r.t. x^d, x^a and u are evaluated at time t.

3 Real-Time Optimization

Let us assume that solutions of (**OCP**$_{\text{DAE}}(p)$) for different values of p are needed in such short time intervals, that it is not possible to solve (**NLP**(p)) within them.

Under these circumstances, approaches for real-time optimization of $(\mathbf{OCP}_{\mathrm{DAE}}(p))$ become necessary. We follow the ideas of Büskens and Maurer [3,4] of this issue.

Let g, C, ψ, F_d and F_a be twice continuously differentiable w.r.t. its arguments. Furthermore, let $p_0 \in P$ the *nominal parameter* and

$$z_0 := z(p_0) = (x^{d,0}(p_0), u^1(p_0), \dots, u^{N_t}(p_0))^\top \tag{18}$$

a corresponding *nominal solution* of $(\mathbf{NLP}(p_0))$, which satisfies the assumptions of Theorem 3 in Büskens and Maurer [3] in this issue. Then the sensitivity differentials $\frac{dz}{dp}(p_0)$ evaluated at the optimal solution $z(p_0)$ are given by the solution of the linear equation system (21) in the cited article. In particular first order real-time approximations of the differential variables $x^{d,0}(p)$ and the control variables $u^i(p)$, $i = 1, \dots, N_t$, for the perturbed parameter p are given by

$$\begin{aligned} x^{d,0}(p) &\approx x_p^{d,0} := x^{d,0}(p_0) + \frac{dx^{d,0}}{dp}(p_0)(p - p_0), \\ u^i(p) &\approx u_p^i := u^i(p_0) + \frac{du^i}{dp}(p_0)(p - p_0). \end{aligned} \tag{19}$$

Similarly, an approximation of consistent algebraic components is given by

$$x^{a,0}(p) \approx x_p^{a,0} := x^{a,0}(p_0) + \frac{dx^{a,0}}{dp}(p_0)(p - p_0). \tag{20}$$

Since

$$x^{a,0}(p_0) = f_a(x^{d,0}(p_0), u_{app}(\tau_1; z(p_0)), p_0) = f_a(x^{d,0}(p_0), u^1(p_0), p_0) \tag{21}$$

holds in (6), the sensitivity differential $\frac{dx^{a,0}}{dp}(p_0)$ is given by

$$\begin{aligned} \frac{dx^{a,0}}{dp}(p_0) &= \frac{\partial f^a}{\partial x^d}[p_0]\frac{dx^{d,0}}{dp}(p_0) + \frac{\partial f^a}{\partial u}[p_0]\frac{du^1}{dp}(p_0) + \frac{\partial f^a}{\partial p}[p_0] \\ &= -\left(\frac{\partial F^a}{\partial x^a}[p_0]\right)^{-1}\left(\frac{\partial F^a}{\partial x^d}[p_0]\frac{dx^{d,0}}{dp}(p_0) + \frac{\partial F^a}{\partial u}[p_0]\frac{du^1}{dp}(p_0) + \frac{\partial F^a}{\partial p}[p_0]\right) \end{aligned} \tag{22}$$

where in addition the relationships in (7) are incorporated and the notation $[p_0]$ stands for all nominal arguments. The partial derivatives of f^a are evaluated at $(x^{d,0}(p_0), u^1(p_0), p_0)$ and those of F^a at $(x^{d,0}(p_0), x^{a,0}(p_0), u^1(p_0), p_0)$.

Unfortunately we cannot expect that the first order approximation $x_p^{a,0}$ in (20) is consistent in general, i.e.,

$$F_a(x_p^{d,0}, x_p^{a,0}, u_p^1, p) \neq 0, \tag{23}$$

which is an essential requirement for the numerical solution.

To overcome this problematic, a simplified Newton iteration

$$x_{k+1}^{a,0} = x_k^{a,0} - J \cdot F_a(x_p^{d,0}, x_k^{a,0}, u_p^1, p), \qquad (k = 0, 1, 2, \dots), \tag{24}$$

with constant matrix

$$J = \left(\frac{\partial F_a}{\partial x^a}(x^{d,0}(p_0), x^{a,0}(p_0), u^1(p_0), p_0) \right)^{-1}, \tag{25}$$

and good initial starting vector $x_0^{a,0} := x_p^{a,0}$ is used to calculate consistent algebraic components in a fast and accurate manner.

Note, that an exact Newton method might be too time consuming, since expensive gradient calculations are needed. Moreover a further improvement of all real-time approximations in view of optimality and admissibility can be archived by applying the mixed strategy in Büskens [2] of this issue.

4 Example: Virtual Test-Drive of an Automobile

This car model permits the analysis of the planar motion of an automobile with centre of gravity $S = (X(t), Y(t), h) \in \mathbb{R}^3$ w.r.t. a suitably chosen reference system at velocity $v(t) = \|\dot{S}(t)\|$ at constant height h. The car-body is connected to the chassis by spring and damper suspensions such that the investigation of its rolling-, pitching and yawing behaviour with its corresponding angles $\kappa(t)$, $\phi(t)$, $\psi(t)$ and angle velocities $w(t) = (w_x(t), w_y(t), w_z(t))^\top \in \mathbb{R}^3$ becomes possible. The motion of the car-body results in a dynamical vertical tyre load distribution $F_z(t) = (F_{z1}(t), \ldots, F_{z4}(t))^\top \in \mathbb{R}^4$, where $F_{zi}(t)$ denotes the vertical tyre load of tyre i at time t. The longitudinal tyre forces $F_{Rx} = (F_{Rx1}, \ldots, F_{Rx4})^\top \in \mathbb{R}^4$ and the lateral tyre forces $F_{Ry} = (F_{Ry1}, \ldots, F_{Ry4})^\top \in \mathbb{R}^4$ are calculated by the HSRI-tyre model described in Wiegner [14] and Uffelmann [13]. The steering wheel angle δ_R together with the deformation of the tyre and the acting tyre forces cause steering angles $\delta(t) = (\delta_1(t), \ldots, \delta_4(t))^\top \in \mathbb{R}^4$ at the wheels of the vehicle. The angular velocities of the rotating wheels are given by $w_\varphi = (w_{\varphi 1}, \ldots, w_{\varphi 4})^\top \in \mathbb{R}^4$. The drift angle $\beta(t)$ denotes the deviation between the velocity vector of the centre of gravity and the longitudinal axis of the car-body. For illustration purposes some of the mentioned quantities are visualized in Figure 1. The constants J_{xA}, J_{yA}, J_{zA} and J_R denote the moments of inertia of the car body w.r.t. the x-,y- and z-axis and the wheels, respectively. The torques M_{xA}, M_{yA} and M_{zA}, the aerodynamical forces F_{Lx}, F_{Ly} as well as the static and dynamical tyre loads $F_{zst,i}, F_{zdyn,i}$ are functions of the state z. The forces F_i, $i = 1, \ldots, 4$ include friction and tyre forces. The angles $\delta_{S,i}, \delta_K, \delta_{E,i}$ and $\delta_{L,i}$ state the steering behaviour of the car. The constant k_D influences the differential at the front axis. A more detailed description of the model is given in Gerdts [8] and Risse [12]. The resulting equations of motion (26) are given in terms of a highly nonlinear DAE system of index 1 in semi-explicit form. The state of the system is $z = (X, Y, \kappa, \phi, \psi, w, \beta, v, w_\varphi, \delta_R, F_z, \delta)^\top \in \mathbb{R}^{23}$. Herein, the eight algebraic variables are given by the vertical tyre loads F_z and the steering angles δ.

$$
\begin{aligned}
(\dot{X},\dot{Y}) &= v(\cos(\psi+\beta),\sin(\psi+\beta)),\\
(\dot{\kappa},\dot{\phi},\dot{\psi}) &= (w_x,w_y,w_z),\\
J_{xA}\dot{w}_x &= M_{xA}(\kappa,w,\beta,v,\delta,w_\varphi,F_z,\dot{w}_x,\dot{w}_y)+(J_{yA}-J_{zA})w_yw_z,\\
J_{yA}\dot{w}_y &= M_{yA}(\phi,w,\beta,v,\delta,w_\varphi,F_z,\dot{w}_y)+(J_{zA}-J_{xA})w_xw_z,\\
J_{zA}\dot{w}_z &= M_{zA}(w,\beta,v,\delta,w_\varphi,F_z)+(J_{xA}-J_{yA})w_xw_y,\\
\dot{\beta} &= -w_z+\tfrac{1}{mv}\Bigg(F_{Lx}(\beta,v)\sin\beta+F_{Ly}(\beta,v)\cos\beta\\
&\quad+\sum_{i=1}^{4}F_{Rx,i}(w_z,\beta,v,\delta_i,w_{\varphi,i},F_{z,i})\sin(\delta_i-\beta)\\
&\quad+\sum_{i=1}^{4}F_{Ry,i}(w_z,\beta,v,\delta_i,w_{\varphi,i},F_{z,i})\cos(\delta_i-\beta)\Bigg),\\
\dot{v} &= \tfrac{1}{m}\Bigg(F_{Ly}(\beta,v)\sin\beta-F_{Lx}(\beta,v)\cos\beta\\
&\quad+\sum_{i=1}^{4}F_{Rx,i}(w_z,\beta,v,\delta_i,w_{\varphi,i},F_{z,i})\cos(\delta_i-\beta)\\
&\quad-\sum_{i=1}^{4}F_{Ry,i}(w_z,\beta,v,\delta_i,w_{\varphi,i},F_{z,i})\sin(\delta_i-\beta)\Bigg),\\
J_R\left(1+\tfrac{k_D}{2}\right)\dot{w}_{\varphi 1} &= \left(1+\tfrac{k_D}{4}\right)\left(M_1(M)+F_1(w_z,\beta,v,\delta_1,w_{\varphi,1}F_{z,1})r_{st}\right),\\
&\quad-\tfrac{k_D}{4}\left(M_2(M)+F_2(w_z,\beta,v,\delta_2,w_{\varphi,2}F_{z,2})r_{st}\right),\\
J_R\left(1+\tfrac{k_D}{2}\right)\dot{w}_{\varphi 2} &= \left(1+\tfrac{k_D}{4}\right)\left(M_2(M)+F_2(w_z,\beta,v,\delta_2,w_{\varphi,2}F_{z,2})r_{st}\right)\\
&\quad-\tfrac{k_D}{4}\left(M_1(M)+F_1(w_z,\beta,v,\delta_1,w_{\varphi,1}F_{z,1})r_{st}\right),\\
i=3,4:\ J_R\dot{w}_{\varphi,i} &= M_i(M)+F_i(w_z,\beta,v,\delta_i,w_{\varphi,i}F_{z,i})r_{st},\\
\dot{\delta}_R &= w_\delta,\\
i=1,\dots,4:\quad 0 &= F_{z,i}-(F_{zst,i}+F_{zdyn,i}(\kappa,\phi,w,\beta,v,\delta,w_\varphi,F_z,\dot{w}_x,\dot{w}_y)),\\
i=1,2:\quad 0 &= \delta_i-(\delta_{S,i}+\delta_\kappa(\kappa)+\delta_{E,i}(w_z,\beta,v,\delta_i,w_{\varphi,i},F_{z,i})\\
&\quad+\delta_{L,i}(\delta_R,w_z,\beta,v,\delta_1,\delta_2,w_{\varphi,1},w_{\varphi,2},F_{z,1},F_{z,2})),\\
i=3,4:\quad 0 &= \delta_i-(\delta_{S,i}+\delta_\kappa(\kappa)+\delta_{E,i}(w_z,\beta,v,\delta_i,w_{\varphi,i},F_{z,i}))\,.
\end{aligned}
\tag{26}
$$

The actuation system is given by

$$
M_1(M)=M_2(M)=\frac{2i_A}{3}M,\ M_3(M)=M_4(M)=\min\left\{0,\frac{i_A}{3}M\right\} \tag{27}
$$

with parameter i_A. The control M is modelled as a combined speed-up and brake assembly. It brings up a torque, which is distributed only on the front wheels in the case $M \geq 0$ (speed-up) and on all four wheels in the case $M < 0$ (slow down). For the following calculations, M is chosen in such a way that the velocity of the car remains nearly constant during the manoeuvre, e.g., $M = 20$ [N]. The reason for this is that the considered manoeuvre, i.e., the double-lane-change manoeuvre, conventionally is driven at constant speed. Hence, w_δ is the only control variable and

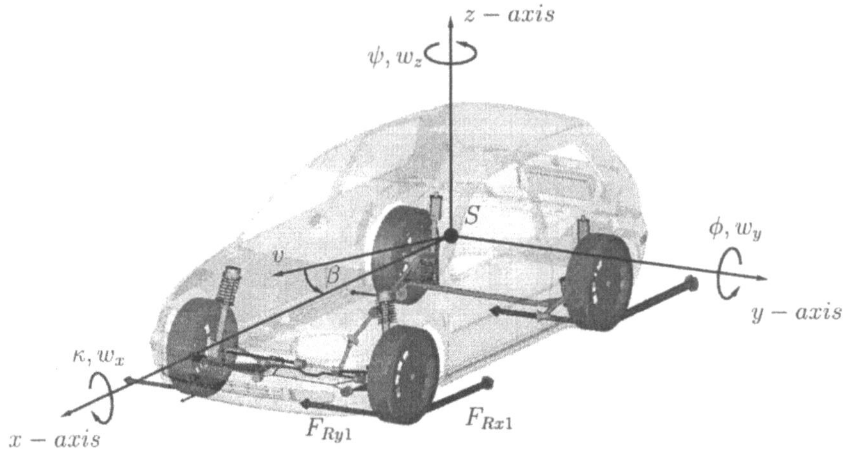

Figure 1. Components of an automobile

prescribes the angular velocity of the steering wheel. Due to the restricted steering capability of the driver w_δ is bounded by $|w_\delta| \leq 5$ [rad/s].

The double-lane-change manoeuvre is a standardized manoeuvre in the automobile industry, compare Zomotor [15]. The driver has to manage an offset of 3.5 [m] and afterwards he has to reach the original track, see Figure 2. Of course, the driver has to avoid a collision with the marking cones.

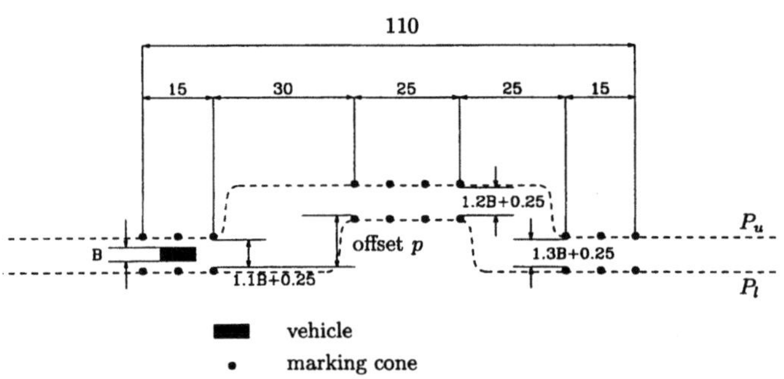

Figure 2. Measurements of the track and boundaries P_l and P_u (dashed)

This task can be reformulated as an optimal control problem, wherein the dynamics of the vehicle are given by (26). The required observation of the marking cones results in two state constraints in (29) given in terms of piecewise defined polynomials $P_l(X, p)$ (lower boundary) and $P_u(X, p)$ (upper boundary) as globally continously differentiable functions of the X-position of the car's centre of gravity, compare Figure 2. Finally restrictions to the initial and final position of the car form boundary conditions. Initial conditions are chosen as follows.

$$\begin{array}{rlrl} X(t_0) = & -30\ [m], & Y(t_0) = & 1\ [m], \\ \kappa(t_0) = & 0\ [rad], & \phi(t_0) = & 0\ [rad], \\ \psi(t_0) = & 0\ [rad], & w_x(t_0) = & 0\ [rad/s], \\ w_y(t_0) = & 0\ [rad/s], & w_z(t_0) = & 0\ [rad/s], \\ \beta(t_0) = & 0\ [rad], & v(t_0) = & 22\ [m/s], \\ \delta_R(t_0) = & 0\ [rad], & w_{\varphi,i}(t_0) = & \frac{v(t_0)}{R_0}\ (i = 1,\dots,4). \end{array} \quad (28)$$

The width of the car is $B = 1.5$ [m]. Hence, the optimal control problem considered is

$$\begin{array}{ll} \text{Minimize} & \int_{t_0}^{t_f} (w_\delta(t))^2 dt, \quad t_0 = 0,\ t_f \text{ free}, \\ \text{subject to} & \text{DAE system (26) with initial conditions (28)}, \\ & \psi(t_f) + \beta(t_f) = 0, \\ & P_l(X(t), p) + \frac{B}{2} \leq Y(t) \leq P_u(X(t), p) - \frac{B}{2}, \\ & |w_\delta| \leq 5. \end{array} \quad (29)$$

The free terminal time t_f is used as an additional optimization variable in (**NLP**(p)). The quantity $\psi(t)+\beta(t)$ is called *track angle*. Thus, the boundary condition $\psi(t_f)+\beta(t_f) = 0$ ensures that the track angle at final time t_f is zero, which means that the car moves parallel to the boundaries of the test-course. Without this condition we might get solutions where the car at final time t_f ends up in a position that does not allow to continue the drive without violation of the boundaries for any $t > t_f$. Please note, that the objective is to minimize the *steering effort* which corresponds to a rather comfortable driver. In Gerdts [8] alternative objective functions are discussed.

The double-lane-change manoeuvre can be understood as a model for a jink caused by a suddenly occurring obstacle on the road. Hence, the offset p to be managed depends on the behaviour of the obstacle. It is desirable that the driver is able to respond to various situations, i.e., various offsets p in Figure 2, in real-time. The optimal control problem depends on the offset p illustrated in Figure 2 and thus can be understood as an parameter perturbed optimal control problem. Figure 3 illustrates the *nominal solution* for $p_0 = 3.5$ [m] and the *real-time approximation* for the perturbed parameter value $p = 3.2$ [m]. Please note that the computational effort to calculate the real-time approximations, $i = 1, \dots, N_t$

$$u^i(p) \approx u^i(p_0) + \frac{du^i}{dp_0}(p_0)(p - p_0), \quad t_f(p) \approx t_f(p_0) + \frac{dt_f}{dp_0}(p_0)(p - p_0), \quad (30)$$

consist only in three basic operations and therefore the computational time can be reduced to a few nanoseconds, if the time during the motion of the car is used to calculate the needed real-time approximations.

The control approximation together with the nominal control is depicted on the left side of Figure 4. The sensitivities $\frac{\partial u}{\partial p}(t, p_0)$ with $\frac{\partial u}{dp}(\tau_i, p_0) \approx \frac{du^i}{dp}(p_0)$, $i =$

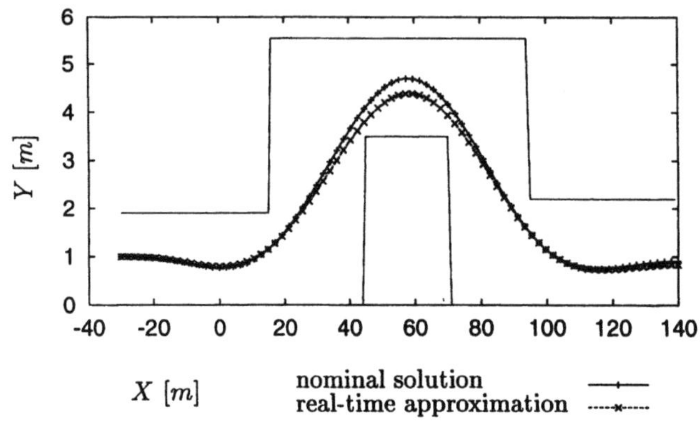

Figure 3. Nominal solution for the double-lane-change manoeuvre with nominal parameter $p_0 = 3.5$ [m] and real-time approximation for perturbation $p = 3.2$ [m]

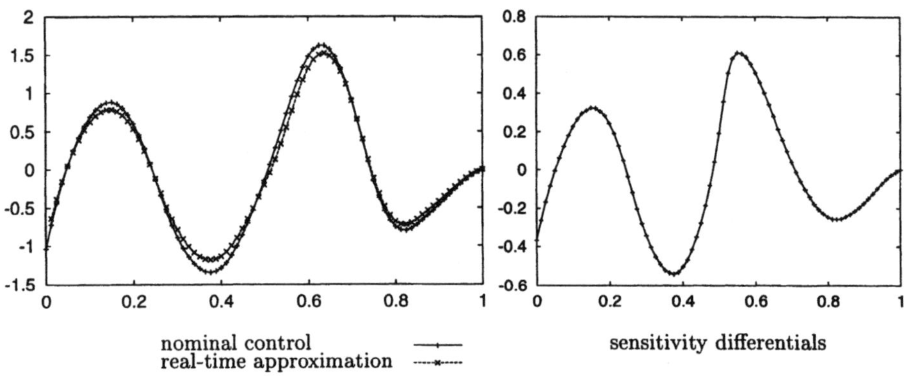

Figure 4. Nominal control and real-time control approximation for parameter $p = 3.2$ [m] (left) and Sensitivities $\frac{\partial u}{\partial p}(t, p_0)$ (right)

$1, \ldots, N_t$ are shown on the right hand side of Figure 4. Finally Figure 5 confirms the excellent conformity between the real-time approximation of the control and the exact optimal control for the perturbed system. The nominal terminal time is calculated to $t_f(p_0) = 7.78349$, while the sensitivity of the terminal time yields $\frac{dt_f}{dp}(p_0) = 0.032124$.

5 Conclusion

A method for real-time optimization of perturbed optimal control problems with differential-algebraic equation systems is investigated. The nominal solution of the optimal control problem is computed numerically by a direct shooting approach. Besides stability, ill-conditioning and stiffness problems within the numerical in-

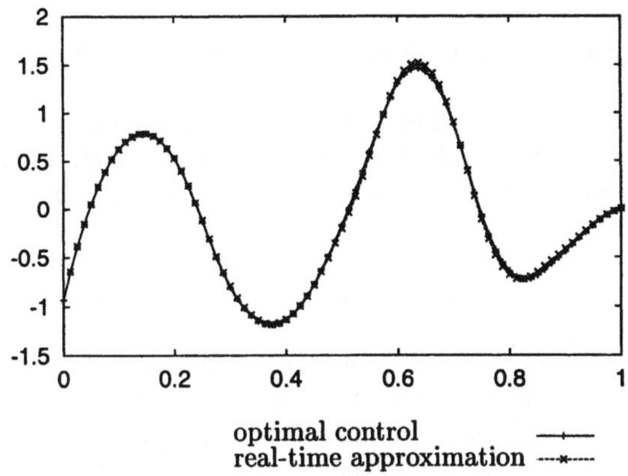

Figure 5. Optimal perturbed control and real-time approximation of the control for parameter $p = 3.2$ [m]

tegration of DAE systems, the calculation of consistent initial values during the iterative solution process is a main problem. For a special class of DAE systems consistent initial values are computed by use of Newton's method.

A sensitivity analysis of the nominal solution allows to state first order real-time approximations for the state and the control. The suggested methods are used to optimize a jink of an automobile in real-time, if an obstacle suddenly occurs on the road. In addition a new method is proposed to compute consistent initial values in real-time.

References

1. K. E. Brenan, S. L. Campbell, L. R. Petzold: Numerical Solution of Initial-Value Problems in Differential-Algebraic Equations. Classics In Applied Mathematics **14**, SIAM, Philadelphia, (1996)
2. C. Büskens, Real-Time Solutions for Perturbed Optimal Control Problems by a Mixed Open- and Closed-Loop Strategy. This volume.
3. C. Büskens and H. Maurer, Sensitivity Analysis and Real-Time Optimization of Parametric Nonlinear Programming Problems. This volume.
4. C. Büskens and H. Maurer, Sensitivity Analysis and Real-Time Control of Optimal Control Problems Using Nonlinear Programming Methods. This volume
5. C. Büskens, M. Gerdts: Numerical Solution of Optimal Control Problems with DAE Systems of Higher Index. Proceedings of the workshop "Optimalsteuerungsprobleme in der Luft- und Raumfahrt" at Greifswald, SFB 255: Transatmosphärische Flugsysteme, München, (2000)
6. S. L. Campbell, C. W. Gear: The Index of General Nonlinear DAEs. Numerische Mathematik **71** (1995)
7. C. W. Gear: Differential-Algebraic Equation Index Transformations. SIAM Journal on Scientific and Statistical Computing **9** (1988) 39–47
8. M. Gerdts: Numerische Methoden optimaler Steuerprozesse mit differential-algebraischen Gleichungssystemen höheren Indexes und ihre Anwendungen in der Kraft-

fahrzeugsimulation und Mechanik. Bayreuther Mathematische Schriften **61**, Bayreuth, (2001).
9. H. Hinsberger: Ein direktes Mehrzielverfahren zur Lösung von Optimalsteuerungsproblemen mit großen, differential-algebraischen Gleichungssystemen und Anwendungen aus der Verfahrenstechnik. Dissertation, Institut für Mathematik, Technische Universität Clausthal, (1997)
10. B. Leimkuhler, L. R. Petzold, C. W. Gear: Approximation Methods for the Consistent Initialization of Differential-Algebraic Equations. SIAM Journal on Numerical Analysis **28**, (1), (1991) 205–226
11. L. R. Petzold: A Description of DASSL: A Differential/Algebraic System Solver. Report Sand **82-8637**, Sandia National Laboratory, Livermore, (1982)
12. H.-J. Risse: Das Fahrverhalten bei normaler Fahrzeugführung. VDI Fortschrittberichte Reihe 12: Verkehrstechnik/Fahrzeugtechnik **160** VDI-Verlag, (1991)
13. F. Uffelmann: Berechnung des Lenk- und Bremsverhaltens von Kraftfahrzeugen auf rutschiger Fahrbahn. Dissertation, Fakultät für Maschinenbau und Elektrotechnik, Technische Universität Braunschweig, (1980)
14. P. Wiegner: Über den Einfluß von Blockierverhinderern auf das Fahrverhalten von Personenkraftwagen bei Panikbremsungen. Dissertation, Fakultät für Maschinenbau und Elektrotechnik, Technische Universität Braunschweig, (1974)
15. A. Zomotor: Fahrwerktechnik: Fahrverhalten. Vogel Buchverlag, Stuttgart, (1991)

Real-Time Solutions of Bang-Bang and Singular Optimal Control Problems

Christof Büskens, Hans Josef Pesch, and Susanne Winderl

Lehrstuhl für Ingenieurmathematik, Universität Bayreuth, Germany

Abstract In many applications of optimal control some or all of the control variables appear linearly in the objective function and the dynamical equations. Therefore, the optimal solutions may exhibit both bang-bang and singular subarcs. Unfortunately, the theory for linear problems of that type is not as well developed as for regular problems, in particular with respect to second order sufficiency conditions. This results in serious problems in developing real-time capable methods to approximate optimal solutions in the presence of data perturbations. In this paper, two discretization methods are presented by which *linear* optimal control problems can be transcribed into *nonlinear* programming problems. Based on a stability and sensitivity analysis of the resulting nonlinear programming problems it is possible to compute sensitivity differentials for the discretized problems, by means of which near-optimal solutions can now be computed in real-time for linear problems, too. The performance of one of these methods is demonstrated for the optimal control of a batch reactor.

1 Introduction

Little is known on second order sufficiency conditions for optimal control problems with control variables entering linearly the objective function and the constraints; see for example Milyutin and Osmolovski [9]. However, all indirect methods for real-time optimal control based on a stability and sensitivity analysis of the optimal control problem, see Maurer, Augustin [8] in this issue and the respective methods cited therein, require that second order sufficient conditions are fulfilled. On the other hand, linear optimal control problems are of utmost importance for practical applications. For example, most of the optimal control problems in robotics are of this type. The thrust in aerospace optimal control problems usually enters the model equations linearly. In chemical process control generally most if not all of the control variables appear linearly.

Moreover, the computation of optimal solutions for linear control problems suffers from the existence of both bang-bang and singular subarcs which are difficultly to detect in general. Therefore, only direct methods are competitive for the solution of complicated and large scale problems. Hereby, the *linear* optimal control problem is transcribed into a *nonlinear* programming problem, whose solution may allow a prediction on the switching structure, i. e. the sequence of bang-bang and singular arcs. Unfortunately, the discretized problem may be of high dimension. On the other hand, the discretized problem is no longer linear so that a sensitivity analysis of the nonlinear programming problem based on second order sufficiency conditions for the discretized problem can now be performed. This circumvents the lack of an appropriate theoretical background for the optimal control problems themselves and leads, in practice, to real-time capable methods for the online computation of optimal solutions also for linear control problems.

In this paper, two new methods of transcribing linear optimal control problems into nonlinear programming problems are presented. First, the optimal switching structure must be guessed from a solution of a direct method such as NUDOCCCS [1,2]. Based on the obtained switching structure, nonlinear programming problems can be formulated where the lengths of the subarcs are the most important discrete optimization variables, cf., e.g., Fraser-Andrews [6,7].

In case of singular subarcs, one has additionally to take into account the discretized control variables on the singular subarcs as optimization variables. Obviously, this transcription method leads to considerably lower dimensional nonlinear programming problems. Moreover, the resulting two direct methods are more robust and allow, as the original method NUDOCCCS, the computation of sensitivity differentials. By means of these sensitivity differentials real-time optimal control of linear control problems can now be practically performed, despite the theoretical lack of the knowledge on sufficiency conditions.

In Section 2 parametric optimal control problems with linear controls are discussed. In Section 3 the two methods for the solution of the transcribed linear optimal control problems are described. The optimal real-time control approach for linear optimal control problems is then presented in Section 4. Section 5 is devoted to a complicated optimal control problem from chemical process control, whose optimal solutions possesses bang-bang and singular subarcs as well a state constraint for an algebraic state variable.

2 Parametric Optimal Control Problems with Linear Controls

We consider parametric control problems with linear controls subject to control and state constraints. Data perturbations are modeled by a parameter $p \in P := \mathbb{R}^{N_p}$. The following parametric control problem of Mayer form will be referred to as problem **OLCP(p)**

$$\begin{aligned}
\text{Minimize} \qquad & F(x,u,p) = g(x(t_f),p) \\
\text{subject to} \qquad & \dot{x}(t) = a(x(t),p) + b(x(t),p)\cdot u(t), \\
& x(t_0) = \varphi(p), \\
& \psi(x(t_f),p) = 0, \\
& c(x(t),p) + d(x(t),p)\cdot u(t) \le 0, \quad t \in [t_0,t_f], \\
& u_{min} \le u(t) \le u_{max}, \quad t \in [t_0,t_f]
\end{aligned} \tag{1}$$

Herein $x(t) \in \mathbb{R}^n$ denotes the state of a system and $u(t) \in \mathbb{R}^m$ the control in a given time interval $[t_0,t_f]$. The final time t_f is either fixed or free. The functions $g : \mathbb{R}^n \times P \to \mathbb{R}$, $a : \mathbb{R}^n \times P \to \mathbb{R}^n$, $b : \mathbb{R}^n \times P \to \mathbb{R}^{n\times m}$, $c : \mathbb{R}^n \times P \to \mathbb{R}^k$, $d : \mathbb{R}^n \times P \to \mathbb{R}^{k\times m}$, $\varphi : P \to \mathbb{R}^n$ and $\psi : \mathbb{R}^n \times P \to \mathbb{R}^r$, $0 \le r \le n$, are assumed to be sufficiently smooth on appropriate open sets. The admissible class of control functions is that of piecewise continuous controls. For reasons of simplicity let the number of control variables be $m = 1$. Note that the formulation of mixed state-control constraints $c(x(t),p)+d(x(t),p)\cdot u(t) \leq 0$ in (1) includes *pure state* constraints $c(x(t),p) \leq 0$, too. *Non autonomous* control problems can be reduced to the *autonomous* problem **OLCP(p)** in a standard way.

Suppose the optimal solution $\bar{u}(t)$ of **OLCP(p)** and hence the switching structure with bang-bang and singular arcs are known. Then there exist sets

$$\begin{aligned} J_{min} &:= \{t \in [t_0,t_f] \mid \bar{u}(t) = u_{min}\}, \\ J_{max} &:= \{t \in [t_0,t_f] \mid \bar{u}(t) = u_{max}\}, \\ J_{sing} &:= \{t \in [t_0,t_f] \mid u_{min} < \bar{u}(t) =: u_{sing}(t) < u_{max}\}, \end{aligned} \tag{2}$$

such that the optimal control $\bar{u}(t)$ is given by

$$\bar{u}(t) = \begin{cases} u_{min}, & t \in J_{min}, \\ u_{max}, & t \in J_{max}, \\ u_{sing}(t), & t \in J_{sing}. \end{cases} \tag{3}$$

Note, that in this paper singular subarcs are understood as arcs, where $\bar{u}(t) \in]u_{min}, u_{max}[$, which includes both the usual singular subarcs from the minimum principle and constrained subarcs. Without restriction let $t_0 = 0$. Let $N_s \in \mathbb{N}$ and let $t_i, i = 1,\dots,N_s$, denote the switching times, at which the optimal control switches from $\bar{u}(t) = u_{min}$ to $\bar{u}(t) = u_{max}$ or vice versa or enters or leaves a singular subarc. Finally let $t_{N_s} := t_f$. Let $N_{sing} \leq N_s$ be the number of singular arcs in the set J_{sing}. Define a function $\iota : \{1,\dots,N_s\} \longrightarrow \{1,\dots,N_{sing}\}$, which maps the switching interval number i of the j'th singular arc to j, $\iota(i) = j$. Then **OLCP(p)** can be replaced by

$$\begin{aligned} &\text{Minimize } F(y,v,p) = g(y_{N_s}(1),p) \\ &\text{subject to } (i = 1,\dots,N_s) \\ &\dot{y}_i(\tau) = \begin{cases} (a(y_i(\tau),p) + b(y_i(\tau),p)\cdot u_{min})(t_i - t_{i-1}), & \text{if } \frac{t_i+t_{i-1}}{2} \in J_{min}, \\ (a(y_i(\tau),p) + b(y_i(\tau),p)\cdot u_{max})(t_i - t_{i-1}), & \text{if } \frac{t_i+t_{i-1}}{2} \in J_{max}, \\ (a(y_i(\tau),p) + b(y_i(\tau),p)\cdot v_{\iota(i)}(\tau))(t_i - t_{i-1}), & \text{if } \frac{t_i+t_{i-1}}{2} \in J_{sing}, \end{cases} \\ &y_1(0) = \varphi(p), \\ &y_{i+1}(0) = y_i(1), \quad i = 1,\dots,N_s - 1, \\ &\psi(y_{N_s}(1),p) = 0, \\ &c(y_i(\tau),p) + d(y_i(\tau),p)\cdot v_{\iota(i)}(\tau) \leq 0, \text{ if } \iota(i) \in \{1,\dots,N_{sing}\}, \end{aligned} \tag{4}$$

with the new normalized time variable τ on the interval $[0,1]$, the new state variable $y(t) := (y_1(t),\dots,y_{N_s}(t))^T \in \mathbb{R}^{nN_s}$ and new control variable $v(t) = (v_1(t),\dots,v_{N_{sing}}(t))^T \in \mathbb{R}^{mN_{sing}}$. The conditions $y_{i+1}(0) = y_i(1)$, $i = 1,\dots,N_s - 1$, in (4) are called *matching conditions*. Note, that the simple box constraints $u_{min} \leq u(t) \leq u_{max}$ in (1) can be neglected in (4), since the switching structure is assumed to be known as well as the sequence of unconstrained and constrained subarcs due to the inequality constraints in (1).

3 Switching Interval Optimization

Suppose that a sufficiently dense grid

$$\tau_i = t_0 + (i-1)h = (i-1)h, \quad i = 1, \ldots, N_t, \quad h := \frac{t_f - t_0}{N_t - 1} \tag{5}$$

with N_t grid points is given. Then a discretized solution $\bar{z} = (\bar{u}_1, \ldots, \bar{u}_{N_t}) \in \mathbb{R}^{mN_t}$ of (1) is obtained by the solution of Eqs. (10)–(13) in Büskens and Maurer [5] of this issue. The values $\bar{u}_i$ denote approximations for the optimal control functions at the discrete times, $\bar{u}(\tau_i) \approx \bar{u}_i$. Hence a discrete approximation of the sets in (2) immediately follows

$$\begin{aligned} \tilde{J}_{min} &:= \{\tau_i \in \{\tau_1, \ldots, \tau_{N_t}\} \mid \bar{u}_i = u_{min}\}, \\ \tilde{J}_{max} &:= \{\tau_i \in \{\tau_1, \ldots, \tau_{N_t}\} \mid \bar{u}_i = u_{max}\}, \\ \tilde{J}_{sing} &:= \{\tau_i \in \{\tau_1, \ldots, \tau_{N_t}\} \mid \tau_i \notin \tilde{J}_{min},\ \tau_i \notin \tilde{J}_{max}\}. \end{aligned} \tag{6}$$

Based on the investigations in Büskens and Maurer [5] of this issue, two methods for discretizing the transformed optimal control problem in (4) are presented in the following. Let $N_t^s > 0$ be a positive integer and for notational simplicity choose equidistant mesh points τ_j^s, $j = 1, \ldots, N_t^s$, for the new time interval $[0, 1]$ with

$$\tau_j^s = (j-1)h^s, \quad j = 1, \ldots, N_t^s, \quad h^s := \frac{1}{N_t^s - 1}. \tag{7}$$

Denote approximations of the values $y(\tau_j^s)$ and $v(\tau_j^s)$ by $y^j = (y_1^j, \ldots, y_{N_s}^j)^T$ and $v^j = (v_1^j, \ldots, v_{N_{sing}}^j)^T$.

3.1 Switching Interval Optimization by Direct Discretization

Define an optimization vector

$$z := (y_1^0, \ldots, y_{N_s}^0, v^1, \ldots, v^{N_t^s}, s_1, \ldots, s_{N_s})^T \in \mathbb{R}^{(n+1)N_s + mN_{sing}N_t^s} \tag{8}$$

where y_i^0 approximates the initial values $y_i^0 \approx y_i(0)$, $i = 1, \ldots, N_s$, for the differential equations and s_i, $i = 1, \ldots, N_s$, approximates the interval lengths $t_i - t_{i-1}$ in (4). Calculating the state variables by a standard integration method for the differential equations in (4) gives the relations

$$\begin{aligned} y_i(\tau) &= y_i(\tau; z, p) = y_i(\tau; y_i^0, v_{\iota(i)}, s_i, p), && \text{if } \iota(i) \in \{1, \ldots, N_{sing}\}, \\ y_i(\tau) &= y_i(\tau; z, p) = y_i(\tau; y_i^0, s_i, p), && \text{else}. \end{aligned} \tag{9}$$

Herewith the optimal control problem (4) can be approximated by the perturbed nonlinear optimization problem

$$
\begin{aligned}
\text{Minimize } & F(z,p) = g(y_{N_s}(1;z,p),p) \\
\text{subject to } & y_1^0 = \varphi(p), \\
& y_{i+1}^0 = y_i(1;z,p), \qquad i = 1,\dots,N_s-1, \\
& \psi(y_{N_s}(1;z,p),p) = 0, \\
& \textstyle\sum_{i=1}^{N_s} s_i = t_f, \\
& c(y_i(\tau_j^s;z,p),p) + d(y_i(\tau_j^s;z,p),p)\cdot \nu_{\iota(i)}^j \le 0, \\
& \text{for } j = 1,\dots,N_t^s \text{ and } i = 1,\dots,N_s : \text{if } \iota(i) \in \{1,\dots,N_{sing}\}.
\end{aligned}
\tag{10}
$$

Herein the constraint $\sum_{i=1}^{N_s} s_i = t_f$ guarantees, that the sum of the lengths s_i of the switching intervals equals the terminal time t_f. Note, that formulations in (9) and (10) include different integration methods for the state variables as well as different approximation method for the control variables.

3.2 Switching Interval Optimization by Sequential Integration

For a further reduction of the number of optimization variables and constraints one can take into account the special structure of the differential equations in (4). Define an optimization vector

$$
z := (\nu^1,\dots,\nu^{N_t^s}, s_1,\dots,s_{N_s-1})^T \in \mathbb{R}^{N_s-1+mN_{sing}N_t^s} \tag{11}
$$

where again s_i approximates the interval lengths $t_i - t_{i-1}$ in (4) for $i = 1,\dots,N_s-1$. The value for s_{N_s} can be calculated from $s_{N_s} = t_f - \sum_{i=1}^{N_s-1} s_i$. Similar to the previous section the state variables can be calculated by standard integration methods, if the matching conditions in (4) are evaluated recursively for $i = 1,\dots,N_s-1$, e.g.,

$$
\begin{aligned}
y_1(\tau) &= y_1(\tau;z,p) &&= y_1(\tau;\varphi(p),\nu_{\iota(1)},s_1,p), \\
y_{i+1}(\tau) &= y_{i+1}(\tau;z,p) &&= y_{i+1}(\tau;y_i(1;z,p),\nu_{\iota(i)},s_i,p).
\end{aligned}
\tag{12}
$$

In this case an approximation of the perturbed control problem (4) is given by the perturbed nonlinear optimization problem

$$
\begin{aligned}
\text{Minimize } & F(z,p) = g(y_{N_s}(1;z,p),p) \\
\text{subject to } & \psi(y_{N_s}(1;z,p),p) = 0, \\
& c(y_i(\tau_j^s;z,p),p) + d(y_i(\tau_j^s;z,p),p)\cdot \nu_{\iota(i)}^j \le 0, \\
& \text{for } j = 1,\dots,N_t^s \text{ and } i = 1,\dots,N_s : \text{if } \iota(i) \in \{1,\dots,N_{sing}\}.
\end{aligned}
\tag{13}
$$

The discretization methods transform the multi-stage process (with continuous matching conditions for the state variables) into a parallel treatment of the stages in (10) or a sequential treatment in (13). A good initial guess for all optimization variables in (8) or (11) can be obtained from the optimal solution of the discretized control problem (1) by the methods described in Büskens and Maurer [5] of this issue.

The decrease of the number of optimization variables when transforming an optimal control problem into a switching interval optimization problem is schematically shown by means of Figure 1.

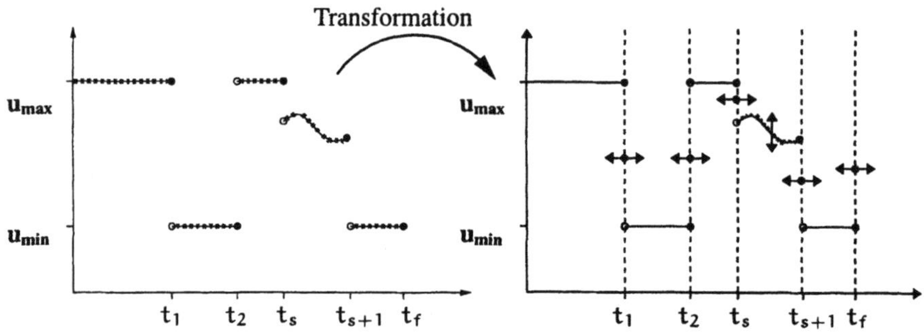

Figure 1. Transformation into a "Switching Interval Optimization Problem" (discrete optimization variables are marked with boxes)

4 Real-Time Optimization of Switching Points and Singular Controls

Formulations (10) and (13) result in a perturbed nonlinear optimization problem

$$\begin{aligned} &\min_z && F(z,p), \\ &\text{subject to } && G_i(z,p) = 0,\ i = 1,\dots,N_e, \\ & && G_i(z,p) \le 0,\ i = N_e+1,\dots,N_c, \end{aligned} \tag{14}$$

with optimization variables z as defined in Eq. (1) of Büskens and Maurer [4] in this issue. Please remember, that the optimal solution $\bar{z}$ includes both the length of the switching intervals and an approximation of the singular controls. Suppose, that for the reference parameter $p = p_0$ an optimal solution

$$\bar{z}(p_0) = (\bar{y}_1^0(p_0),\dots,\bar{y}_{N_s}^0(p_0),\bar{v}^1(p_0),\dots,\bar{v}^{N_t^s}(p_0),\bar{s}_1(p_0),\dots,\bar{s}_{N_s}(p_0))^T$$

of (10) resp.

$$\bar{z}(p_0) = (\bar{v}^1(p_0),\dots,\bar{v}^{N_t^s}(p_0),\bar{s}_1(p_0),\dots,\bar{s}_{N_s-1}(p_0))^T$$

of (13) is given. Under the assumptions of Theorem 3 in Büskens and Maurer [4] we can apply Formulae (21) in the same article and obtain values for the sensitivity differentials of the interval lengths s_i, and the singular controls v^j, for example in case of (13),

$$\begin{aligned}
\frac{dv^j}{dp}(p_0) &= \frac{dz_j}{dp}(p_0) \in \mathbb{R}^{mN_{sing}\times N_p}, \quad j = 1,\dots,N_t^s, \\
\frac{ds_i}{dp}(p_0) &= \frac{dz_k}{dp}(p_0) \in \mathbb{R}^{N_p}, \quad k = mN_{sing}N_t^s + i, \quad i = 1,\dots,N_s - 1,
\end{aligned} \tag{15}$$

In the sequel we restrict the discussion to the switching interval optimization by sequential integration as described in Section 3.2. In case that an actual deviation from p_0 is detected, real-time approximations $\tilde{s}_i(p)$ and $\tilde{v}^j(p)$ of the switching interval length $s_i(p), i = 1,\dots,N_s$, and the singular controls $v^j(p), j = 1,\dots,N_t^s$, can be obtained by its first order Taylor approximation

$$\begin{aligned}
\tilde{s}_i(p) &= \tilde{s}_i(p_0) + \frac{ds_i}{dp}(p_0)(p - p_0), \quad i = 1,\dots,N_s - 1, \\
\tilde{s}_{N_s}(p) &= t_f - \sum_{i=1}^{N_s-1} \tilde{s}_i(p), \\
\tilde{v}^j(p) &= \tilde{v}^j(p_0) + \frac{dv^j}{dp}(p_0)(p - p_0).
\end{aligned} \tag{16}$$

Additionally, real-time approximations $\tilde{t}_i(p)$ of the switching points t_i, $i = 1,\dots,N_s - 1$, are obtained by

$$\begin{aligned}
\tilde{t}_1(p) &= \tilde{s}_1(p), \\
\tilde{t}_j(p) &= \tilde{t}_{j-1}(p) + \tilde{s}_j(p), \quad j = 2,\dots,N_s - 1,
\end{aligned} \tag{17}$$

where $\tilde{s}_j(p)$ are taken from (16). If required, a free final time t_f can be handled as an additional optimization variable in (14). Its sensitivity differential $\frac{dt_f}{dp}(p_0)$ can similarly be calculated. In this case, t_f in (16) is replaced by its first order Taylor approximation $\tilde{t}_f(p) = \tilde{t}_f(p_0) + \frac{dt_f}{dp}(p_0)(p - p_0)$. A further improvement of the real-time approximations in view of optimality and admissibility can be achieved by applying the mixed open- and closed-loop strategy in Büskens [3] in this issue.

5 Example

Now, a model of a batch reactor (cf. [11]) describing a certain chemical reaction

$$A + B \rightarrow C$$

and its side reaction

$$B + C \rightarrow D$$

is considered.

Both reactions are assumed to be strongly exothermic. Thus, direct mixing of the entire necessary amounts of the reactants must be avoided.
Reactant A is charged in the reactor vessel, which is fitted with a cooling jacket to remove the generated heat of the reaction, while reactant B is added. These reactions result in the product C and the undesired byproduct D.

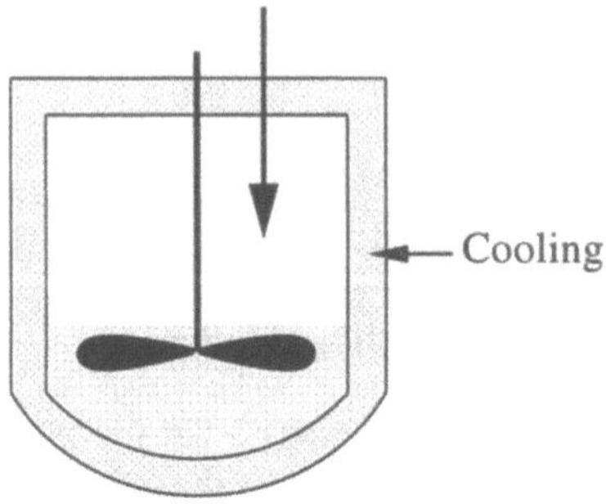

Figure 2: Batch Reactor

Let

$$x = (M_A, M_B, M_C, M_D, H, C_A, C_B, C_C, C_D, T_R, V)^T \in \mathbb{R}^{11} \quad (18)$$

denote the vector of state variables, where $M_i(t)$ [mol] and $C_i(t)$ [mol/m^3] stand for the molar holdups and the molar concentrations of the components, where $i = A, B, C, D$, respectively. $H(t)$ [MJ] denotes the total energy holdup, $T_R(t)$ [K] the reactor temperature and finally $V(t)$ [m^3] the volume of liquid in the system. The two dimensional control vector is given by

$$u = (F_B, Q)^T \in \mathbb{R}^2 \quad (19)$$

where $F_B(t)$ [mol/s] controls the feed rate of component B and $Q(t)$ [kW] controls the cooling load. The objective is to determine a control u that maximizes the molar holdup of the component C. The performance index

$$F(x, u, p) = g(x(t_f), p) = -M_C(t_f) \quad (20)$$

has to be minimized subject to

$$\begin{aligned} \dot{M}_A &= -V \cdot r_1, \\ \dot{M}_B &= F_B - V \cdot (r_1 + r_2), \\ \dot{M}_C &= V \cdot (r_1 - r_2), \\ \dot{M}_D &= V \cdot r_2, \\ \dot{H} &= F_B \cdot h_f - Q - V \cdot (r_1 \cdot \Delta H_1 + r_2 \cdot \Delta H_2), \end{aligned} \quad (21)$$

with the reaction rates r_j, the corresponding Arrhenius rate constants k_j of both reactions ($j = 1, 2$):

$$\begin{aligned} &\mathbf{A + B \to C:} \quad r_1 = k_1 \cdot C_A \cdot C_B \text{ with } k_1 = A_1 \cdot e^{-E_1/T_R}, \\ &\mathbf{C + B \to D:} \quad r_2 = k_2 \cdot C_B \cdot C_C \text{ with } k_2 = A_2 \cdot e^{-E_2/T_R}. \end{aligned} \quad (22)$$

and functions

$$S = \sum_{i=A,B,C,D} M_i \cdot \alpha_i, \qquad W = \sum_{i=A,B,C,D} M_i \cdot \beta_i,$$
$$T_R = \frac{1}{W} \cdot \left(-S + \sqrt{(W \cdot T_{ref} + S)^2 + 2 \cdot W \cdot H} \right), \tag{23}$$
$$C_i = \frac{M_i}{V}, \quad i = A, B, C, D, \qquad V = \sum_{i=A,B,C,D} \frac{M_i}{\rho_i}.$$

In addition a state constraint

$$T_R(t) \leq 520 \tag{24}$$

has to be fulfilled, while the two linear controls are bounded by

$$0 \leq F_B(t) \leq 10 \text{ and } 0 \leq Q(t) \leq 1000. \tag{25}$$

Finally, initial values

$$M_A(0) = 9000, \; M_i(0) = 0, \; i = B, C, D, \; H(0) = 152.50997 \tag{26}$$

and a terminal condition

$$T_R(t_f) = 300, \tag{27}$$

have to be fulfilled. The reaction and component data appearing in (21) and (22) are given in Table 1. Perturbations p in the control problem (18) – (27) are modeled by deviations in the activation energy $E_1 = E_1(p)$ for reaction 1. The nominal value is chosen as $p_0 = 1$.

Abbr.	Reactions		Meaning
	$j = 1$	$j = 2$	
A_j $[\frac{m^3}{mol \cdot s}]$	0.008	0.002	pre-exponential Arrhenius constants
E_j [K]	$3000 \cdot p$	2400	activation energies
ΔH_j $[\frac{kJ}{mol}]$	-100	-75	enthalpies

Abbr.	Components				Meaning
	$i = A$	$i = B$	$i = C$	$i = D$	
ρ_i $[\frac{mol}{m^3}]$	11250	16000	10400	10000	molar density of pure component i
α_i $[\frac{kJ}{mol \cdot K}]$	0.1723	0.2	0.16	0.155	coefficient of the linear (α_i) and quadratic (β_i) term in the pure component specific enthalpy expression
β_i $[\frac{kJ}{mol \cdot K^2}]$	0.000474	0.0005	0.00055	0.000323	

Table 1. Reaction and Component Data

The reference temperature for the enthalpy calculations is given by $T_{ref} = 298$ [K] and the specific molar enthalpy of the reactor feed stream is given by $h_f = 20$

[kJ/mol]. Several calculations show that for increasing t_f the switching structure gets more and more complex. However, the total profit of M_C ($|F(x, u, p)|$) is nearly constant if t_f is greater than a certain value ($t_f \sim 1600$) (cf. Figure 2). This is why we have chosen $t_f = 1600$ for the following considerations.

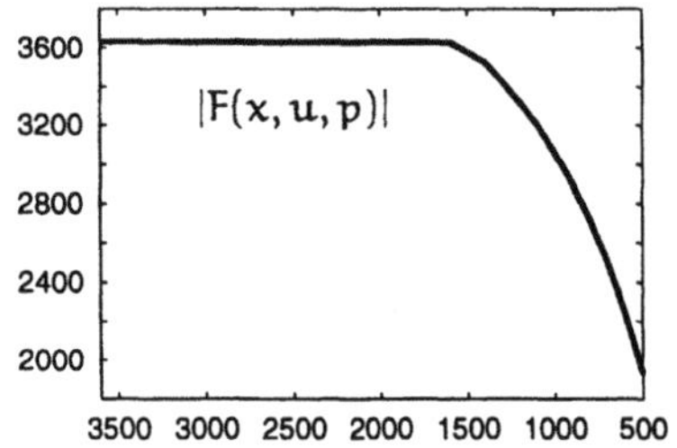

Figure 3. The performance index w.r.t. different t_f

The ODE system has been used for a numerical computation of the optimal solution by the direct method NUDOCCCS [1,2] with $N_t = 256$ discrete points of time. A fourth-order Runge-Kutta-method due to England, cf. [10] and a linear interpolation of the controls are used. The nominal solutions for the optimal controls of the original optimal control problem are depicted in Figure 3. The solution in Figure 3 yields the switching structure given in Table 2.

By applying the transformation into a switching interval optimization problem as described in Section 3.2, a perturbed nonlinear optimization problem of form (13) is obtained with $N_s = 4$, $N_{sing} = 1$. An initial guess of the optimization vector z in (11) can be obtained from the solutions in Figure 3.

Note that the number of grid points N_t^s can be chosen much smaller than the number of grid points N_t for the code NUDOCCCS, because of the considerably lower dimension of the switching interval optimization problem. Since the terminal time t_f is fixed the NLP(p) problem in (13) consists of $3 + N_t^s$ (s_1, s_2, s_3 and the discretized singular control) optimization variables.

The transformed problem is solved with $N_t^s = 64$ discrete points of time and yields the nominal solutions shown in Figure 4. Note, that the optimal controls calculated by the transformed problem (13) coincide with the solutions given in Figure 3. The switching intervals are calculated to $s_1 = 495.65$, $s_2 = 75.09$, $s_3 = 319.86$ which yields $s_4 = 1600 - \sum_{i=1}^{3} s_i \approx 709.4$. Herewith the switching points can be calculated to $t_1 = 495.65$, $t_2 = 570.74$, $t_3 = 890.6$.

Note that the singular subarc (t_2, t_3) splits into two arcs of which the first very short one is not caused by the state constraint, whereas the second one does. Next, a sensitivity analysis, cf. [4,5], is performed to calculated the sensitivity differentials

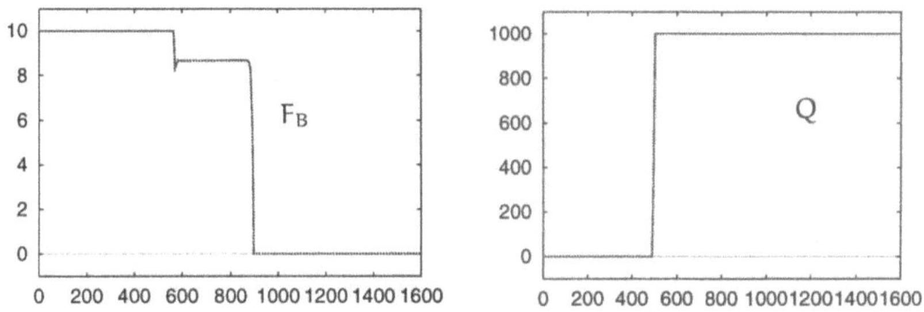

Figure 4. Optimal control variables F_B and Q

	$[0,t_1)$	$[t_1,t_2)$	$[t_2,t_3)$	$[t_3,1600]$
F_B	max	max	singular	min
Q	min	max	max	max

Table 2. Corresponding switching structure, $t_f = 1600$

of the switching intervals $\frac{ds_i}{dp}(p_0)$, $i = 1, \ldots, 3$, in (15):

$$\frac{ds_1}{dp}(p_0) = 0.2132, \quad \frac{ds_2}{dp}(p_0) = -1.4673, \quad \frac{ds_3}{dp}(p_0) = 1.3035. \tag{28}$$

The sensitivity differentials of the controls on the normalized time interval $[0, 1]$ are given in Figure 5. We suppose that the overshooting at the beginning of the interval in the left part of Figure 5 results from additional singular subarc in the optimal solution which is not detectable with $N_t = 256$ by the NUDOCCCS approximation. However note, that the prediction of the additional subarc may be detected by a higher discretization. In the right part of Figure 5 the sensitivity differential of the singular subarc is depicted on the relevant interval $[0.04, 0.85]$ to indicate that the sensitivity differential on the singular subarc is small but by no means equal to zero.

Finally, we apply Eqs. (16) to obtain real-time approximations of the switching intervals and the singular control. Table 3 shows the quality of the real-time approximations for different perturbations $p \in \{1.001, 1.01, 1.03\}$. The values $\zeta^1_{err} := \frac{s_i(p)-\tilde{s}_i(p)}{s_i(p)}$, $i = 1, \ldots, 4$, denote the relative errors between the "exact" solution $s_i(p)$ and the real-time approximation $\tilde{s}_i(p)$ from (16) for the switching intervals, resp. the maximal relative error in the singular control. Here, $s_i(p)$ is obtained by the switching interval optimization method. For reasons of comparability the values $\zeta^2_{err} := \frac{s_i(p)-s_i(p_0)}{s_i(p)}$ denote the relative errors between the "exact" perturbed solution and the nominal solution for the same variables. In addition the relative errors of the switching points t_i, $i = 1, \ldots, 3$, as calculated by Eq. (17) are given in Table 3. The CPU-times for the real-time approximations on a Pentium III 550MHz PC are about $3.4 \cdot 10^{-9}$ seconds for each of the variables in (16). Neglecting data transfer

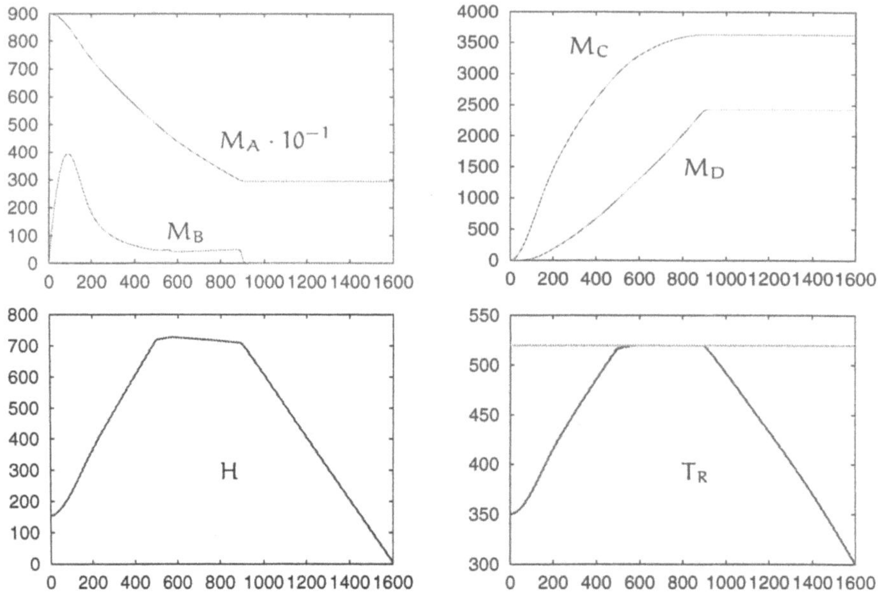

Figure 5. Optimal state variables and the behaviour of the state constraint

the obtained CPU-time for each variable in (16) do also apply for large scale problems. Note that for perturbations $p \gtrapprox 1.03$ the second switching interval vanishes.

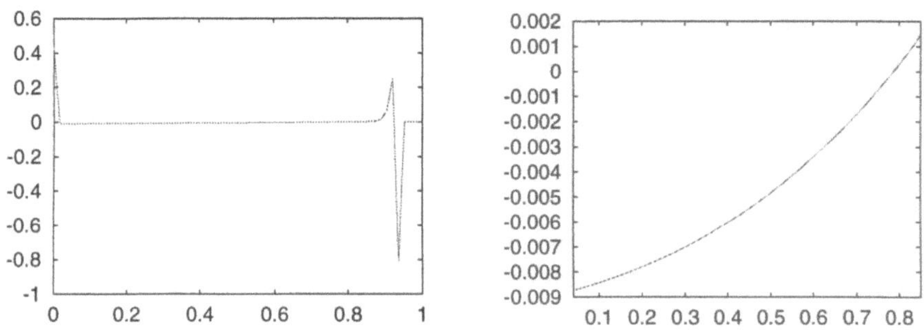

Figure 6. Sensitivity of the control F_B on the singular arc (left) and the relevant part of that (right)

	p=1.001		p=1.01		p=1.02	
	$\lvert\zeta^1_{err}\rvert$	$\lvert\zeta^2_{err}\rvert$	$\lvert\zeta^1_{err}\rvert$	$\lvert\zeta^2_{err}\rvert$	$\lvert\zeta^1_{err}\rvert$	$\lvert\zeta^2_{err}\rvert$
s_1	$9.59 \cdot 10^{-5}$	$7.84 \cdot 10^{-4}$	$8.01 \cdot 10^{-4}$	$7.63 \cdot 10^{-4}$	$1.27 \cdot 10^{-3}$	$1.48 \cdot 10^{-2}$
s_2	$4.88 \cdot 10^{-3}$	$3.73 \cdot 10^{-2}$	$5.99 \cdot 10^{-2}$	$5.42 \cdot 10^{-1}$	$1.90 \cdot 10^{-1}$	$2.18 \cdot 10^{\pm 0}$
s_3	$9.37 \cdot 10^{-4}$	$7.41 \cdot 10^{-3}$	$7.21 \cdot 10^{-3}$	$6.80 \cdot 10^{-2}$	$1.03 \cdot 10^{-2}$	$1.24 \cdot 10^{-1}$
s_4	$5.45 \cdot 10^{-6}$	$1.06 \cdot 10^{-4}$	$5.87 \cdot 10^{-5}$	$1.06 \cdot 10^{-3}$	$1.30 \cdot 10^{-4}$	$2.10 \cdot 10^{-3}$
t_1	$9.59 \cdot 10^{-5}$	$7.84 \cdot 10^{-4}$	$8.01 \cdot 10^{-4}$	$7.63 \cdot 10^{-4}$	$1.27 \cdot 10^{-3}$	$1.48 \cdot 10^{-2}$
t_2	$5.38 \cdot 10^{-4}$	$4.07 \cdot 10^{-3}$	$4.59 \cdot 10^{-3}$	$4.12 \cdot 10^{-2}$	$7.33 \cdot 10^{-3}$	$8.35 \cdot 10^{-2}$
t_3	$4.34 \cdot 10^{-6}$	$8.44 \cdot 10^{-5}$	$4.67 \cdot 10^{-5}$	$8.40 \cdot 10^{-4}$	$1.03 \cdot 10^{-4}$	$1.67 \cdot 10^{-3}$
$v = u_{sing}$	$1.29 \cdot 10^{-6}$	$1.17 \cdot 10^{-5}$	$1.06 \cdot 10^{-5}$	$1.10 \cdot 10^{-4}$	$1.99 \cdot 10^{-5}$	$2.19 \cdot 10^{-4}$

Table 3. Quality of the real-time approximations for different perturbations

6 Conclusion

In this paper, new transcription methods for *linear* optimal control problems are presented which allow the computation of sensitivity differentials despite the missing second order sufficiency conditions for linear optimal control problems. Hence, real-time optimal control becomes, in practice, also possible for the important class of linear control problems. Moreover, the new transcription methods yield also robust solution methods for linear control problems if the correct switching structures are known. In the future, the automatic detection of bang-bang and singular subarcs by use of optimality conditions must be improved. For this purpose, approximations of the adjoint variables are to be taken into account, which can be obtained by NUDOCCCS, too.

References

1. C. Büskens: Direkte Optimierungsmethoden zur numerischen Berechnung optimaler Steuerungen. Diploma thesis, Institut für Numerische Mathematik, Universität Münster, Münster, Germany, (1993)
2. C. Büskens: Optimierungsmethoden und Sensitivitätsanalyse für optimale Steuerprozesse mit Steuer- und Zustands- Beschränkungen. Dissertation, Institut für Numerische Mathematik, Universität Münster, Germany, (1998)
3. C. Büskens: Real-Time Solutions for Perturbed Optimal Control Problems by a Mixed Open- and Closed-Loop Strategy. This volume.
4. C. Büskens and H. Maurer: Sensitivity Analysis and Real-Time Optimization of Nonlinear Programming Problems. This volume
5. C. Büskens and H. Maurer: Sensitivity Analysis and Real-Time Control of Optimal Control Problems Using Nonlinear Programming Methods. This volume.
6. G. Fraser-Andrews: Finding Candidate Singular Optimal Controls: A State of the Art Survey. JOTA **60** (2) (1989) 173 –190
7. G. Fraser-Andrews: Numerical Methods for Singular Optimal Control. JOTA **61** (3) (1989) 377 –401
8. H. Maurer and D. Augustin, Sensitivity Analysis and Real-Time Control of Parametric Optimal Control Problems Using Boundary Value Methods. This volume

9. A. A. Milyutin, N. P. Osmolovskii: Calculus of Variations and Optimal Control. Translations of mathematical Monographs **180**, American Mathematical Society, Providence (1998)
10. H. G. Schwarz: Numerische Mathematik. B.G. Teubner Stuttgart (1988)
11. V. S. Vassiliadis, R. W. H. Sargent, C. C. Pantelides: Solution of a Class of Multistage Dynamic Optimization Problems. 2. Problems with Path Constraints. Ind. Eng. Chem. Res. **33** (9) (1994) 2123–2133

Conflict Avoidance During Landing Approach Using Parallel Feedback Control

Bernd Kugelmann and Wolfgang Weber

Fachbereich Mathematik und Informatik, Ernst Moritz Arndt Universität Greifswald, Germany

Abstract In the last years air traffic density has increased all over the world. This calls for new tools to support the overworked controllers in the air traffic control centers. In this article an approach to compute optimal feedback-controls in realtime will be described. As a prerequisite a previously computed flight path is needed. Because of unforeseeable events like delays or environmental influences the controlled planes will leave this path in almost any case. Therefore it is necessary to update the computed trajectory to adapt it to the present conditions. To get the updates for the controls the known nominal trajectory and the actual state of the controlled object is used. The new idea is to not only adapt the controls, but also the reference path in every correction step. This yields an increase in robustness but on the other hand this raises the computational efforts. To reduce the time needed a parallel algorithm has been developed.

1 Formulation of the Air Traffic Control Problem

Because of the major increase in air traffic, automatic guidance systems to support the air traffic controllers are very interesting research objects. Tools for this kind of tasks would be especially helpful at large airports where a lot of aircraft have to be controlled every day. As an example for the developed algorithm a special flight phase is investigated where the planes must descend from their cruise altitude to a target altitude from where the final landing approach is started. The nominal starting position of each aircraft is defined by the flight plan, the final positions depend on the landing sequence. The criteria used to find this order are not a part of the considered problem. For the algorithm it is assumed that the boundary conditions are known. In the descend phase aircraft coming from different directions have to be merged and necessary passing maneuvers have to be carried out to obtain the desired landing sequence. Also further restrictions have to be met: first the aircraft must stay on fixed air routes given by waypoints. Also the locations where the descend begins respectivly ends are prescribed. As a second point the aircraft must observe a given minimum distance condition. This means that around each aircraft a security envelope exists which no other one is allowed to enter. The term *conflict* is used if this condition is violated. The problem shown in this article is derivated from a problem mentioned in [1].

The following equations describe the dynamical system of the aircraft movement:

$$
\begin{aligned}
\dot{x}_i(t) &= v_i(t)\cos\gamma_i(t)\cos\chi_i(t),\\
\dot{y}_i(t) &= v_i(t)\cos\gamma_i(t)\sin\chi_i(t),\\
\dot{z}_i(t) &= v_i(t)\sin\gamma_i(t),\\
\dot{v}_i(t) &= p_i(t),\\
\dot{\gamma}_i(t) &= q_i(t),\\
\dot{\chi}_i(t) &= r_i(t),
\end{aligned}
\tag{1}
$$

with i as the aircraft-number ($1 \leq i \leq n$; n: total number of aircraft), t as the time and

Name	Type	Description
x_i	state	down range
y_i	state	cross range
z_i	state	altitude
v_i	state	airspeed
γ_i	state	flight path angle
χ_i	state	heading angle
p_i	control	acceleration
q_i	control	angular velocity of γ_i
r_i	control	angular velocity of χ_i.

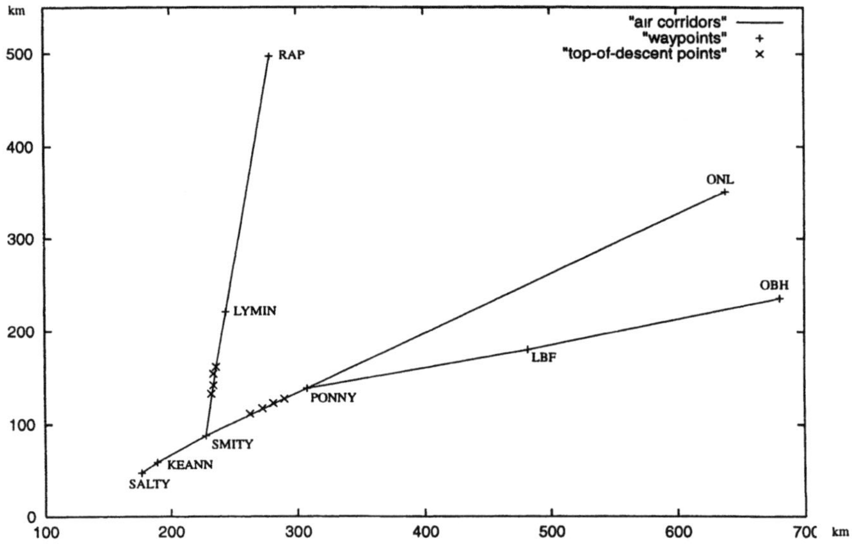

Figure 1. Waypoints and air routes from Denver ARTCC

The previously mentioned air corridors are defined through waypoints. Figure 1 shows the situation for the *Air Route Traffic Control Center* (ARTCC) of Denver. Depicted are the standard waypoints and the top-of-descent points for some cruise flight altitudes (looking from SMITY to PONNY respectively LYMIN the sequence is for the altitudes 8.839 km, 9.449 km, 10.058 km and 10.668 km). The flight direction on these corridors is from the points RAP, ONL and OBH to SALTY. The equations to be met for the point PONNY respectively a top-of-descent point are

$$\begin{aligned} x_i(t_{i,\mathrm{PONNY}}) &= x_{\mathrm{PONNY}} \\ y_i(t_{i,\mathrm{PONNY}}) &= y_{\mathrm{PONNY}} \\ z_i(t_{i,\mathrm{PONNY}}) &= z_{\mathrm{PONNY}} \end{aligned} \qquad (2)$$

$$\begin{aligned} x_i(t_{i,\mathrm{TOD}_j}) &= x_{\mathrm{TOD}_j} \\ y_i(t_{i,\mathrm{TOD}_j}) &= y_{\mathrm{TOD}_j} \\ z_i(t_{i,\mathrm{TOD}_j}) &= z_{\mathrm{TOD}_j} \\ \gamma_i(t_{i,\mathrm{TOD}_j}) &= 0 \end{aligned} \qquad (3)$$

The additional equation for the flight path angle inhibits a change in altitude before reaching the top-of-descent point (the waypoint SALTY is the bottom-of-descent point and satisfies the same conditions (3)).

As mentioned above there is another type of restriction to look at. Aircraft have to stay outside a security envelope of each other. This envelope is chosen as an ellipsoid around each plane. The mathematical representation is an inequality of the following form for every pair of aircraft with numbers i and j, respectively:

$$S_{ij}(t) \geq 0 \qquad (4)$$

with

$$S_{ij}(t) := \left(\frac{x_i(t) - x_j(t)}{d_x}\right)^2 + \left(\frac{y_i(t) - y_j(t)}{d_y}\right)^2 + \left(\frac{z_i(t) - z_j(t)}{d_z}\right)^2 - 1 \qquad (5)$$

According to the ARTCC Denver the values for d_x and d_y are 9.26 km and for d_z it is 0.61 km.

For optimality purposes a minimal-energy approach is used. Therefore the goal is to control the planes with as little acceleration or steering as possible. This leads to a performance index of the form

$$\frac{1}{2}\int_0^{t_f} \sum_{i=1}^{n} \left(a_i p_i(t)^2 + b_i q_i(t)^2 + c_i r_i(t)^2\right) dt \qquad (6)$$

which has to be minimized. a_i, b_i and c_i are weight coefficient and can be chosen for each aircraft individually.

2 Mathematical Basics

In this section the mathematical background for the air traffic control problem from section 1 will be introduced. The notation used here is following the notation used in [2].

The mathematical problem is to find *state functions* $\mathbf{x}(t)$ and *control functions* $\mathbf{u}(t)$ to minimize the functional

$$F(\mathbf{x},\mathbf{u},p) = g(\mathbf{x}(0),\mathbf{x}(T),p) + \int_0^T f_0(\mathbf{x}(t),\mathbf{u}(t),p)\,dt \qquad (7)$$

with respect to the following conditions

$$\begin{aligned} \dot{\mathbf{x}}(t) &= f(\mathbf{x}(t),\mathbf{u}(t),p) && \text{for a.e. } t \in [0,T], && (8)\\ \varphi(\mathbf{x}(0),\mathbf{x}(T),p) &= 0, && && (9)\\ S(\mathbf{x}(t),p) &\le 0 && \text{for all } t \in [0,T], && (10)\\ \mathcal{N}_i(\mathbf{x}(t_i),p) &= 0 && (1 \le i \le j), && (11) \end{aligned}$$

with

$$\mathbf{x} : [0,T] \to \mathbb{R}^n, \quad \mathbf{u} : [0,T] \to \mathbb{R}^m$$

and given functions

$$\begin{gathered} g : \mathbb{R}^n \times \mathbb{R}^n \times P \to \mathbb{R}, \quad f_0 : \mathbb{R}^n \times \mathbb{R}^m \times P \to \mathbb{R}, \\ f : \mathbb{R}^n \times \mathbb{R}^m \times P \to \mathbb{R}^n, \\ \varphi : \mathbb{R}^n \times \mathbb{R}^n \times P \to \mathbb{R}^r, \ 1 \le r \le 2n, \\ \mathcal{N}_i : \mathbb{R}^n \times P \to \mathbb{R}^{k_{\mathcal{N}_i}}, \ S : \mathbb{R}^n \times P \to \mathbb{R}^{k_S} \end{gathered}$$

and the parameter-vector $p \in P \subset \mathbb{R}^q$. The parameter p describes the actual inflight-situation.

While T is fixed, the interior points t_i are subject to the minimization. For simplicity in the following considerations j is set to 1 and (11) changes to

$$\mathcal{N}(\mathbf{x}(t_I),p) = 0. \qquad (11')$$

Another simplification is the assumption that each state constraint S_j has only one junction point $t_{V,j}$.

For p fixed there are several ways to deal with such a problem (see for example ([13])). Here the so called indirect method is considered. Using this method the constraints are coupled directly to the cost functional (see [3]). An important aspect is the *order* l of the state constraints. For the definition of the order of a state constraint and the functions $S^l(\mathbf{x},\mathbf{u})$ see [2]. The augmented Hamiltonian and augmented function $\hat{g}$ are

$$H(\mathbf{x},\mathbf{u},\lambda,\nu^l,p) := f_0(\mathbf{x},\mathbf{u},p) + \lambda^* f(\mathbf{x},\mathbf{u},p) + \nu^{l^*} S^l(\mathbf{x},\mathbf{u},p), \qquad (13)$$

$$\hat{g}(\mathbf{x}(0),\mathbf{x}(t_I),\mathbf{x}(T),p) := g(\mathbf{x}(0),\mathbf{x}(T),p) + \nu_T^*\varphi(\mathbf{x}(0),\mathbf{x}(T),p) + \nu_I^*\mathcal{N}(\mathbf{x}(t_I),p) + \sum_{j=1}^{k_S} \nu_{V,j} S_j(\mathbf{x}(t_{V,j}),p). \tag{14}$$

Optimal control theory (see for example [4]) gives necessary conditions for an optimal solution of this problem. An example is the differential equation for the adjoint variable λ along the optimal solution:

$$\dot{\lambda}(t) = -H_x(\mathbf{x},\mathbf{u},\lambda,\nu^l)^*. \tag{15}$$

These conditions lead to an multipoint boundary value problem for the state variable $\mathbf{x}$ and the adjoint variable λ. The desired control variable $\mathbf{u}$ can be computed from $\mathbf{x}$ and λ. To solve this problem for fixed p in the offline case various methods are available (e.g., [10], [11], [12]).

3 Feedback Control

For realistic problems the computation of an optimal trajectory in advance (for example for $p = 0$) is only one part of the control process. In almost any case the controlled object will leave the precalculated path. So it is absolutly necessary to update the controls to adapt them to the present situation ($p \neq 0$). If a measurement of the state shows a deviation from the nominal trajectory an update has to be executed. The perturbation can be represented by modified initial values in (9):

$$\mathbf{x}(0) = \mathbf{x}_0 + d\mathbf{x}_0. \tag{16}$$

Now the original problem with the new starting values has to be solved online. The base of the approach discussed in this article is the assumption that the solutions of the nominal and the perturbed problem are close together. Therefore the influence of the perturbation parameter p to the solution of the problem is investigated (see [2] for details). The meaning of this parameter in the example problem is the deviation of the actual state of an aircraft to the precomputed nominal state [7]. Then it can be shown that the nominal solution can be embedded into a family of solutions of the perturbed problem [14,15]. An expansion of the equations for the perturbed problem about $p = 0$ and the following neglection of terms of higher order is leading to an approximation for the necessary conditions of the actual problem. Some problems arising in the intervals between the nominal and the actual junction point can be solved by the use of a so called comparison trajectory [5]. As a result we get linear equations for the variations

$$\delta\mathbf{x}(t) := \frac{\partial\mathbf{x}}{\partial p}(t,0)\cdot p, \qquad \delta\lambda(t) := \frac{\partial\lambda}{\partial p}(t,0)\cdot p,$$

$$\delta\mathbf{u}(t) := \frac{\partial\mathbf{u}}{\partial p}(t,0)\cdot p, \qquad \delta\nu^l(t) := \frac{\partial(\nu^l)}{\partial p}(t,0)\cdot p$$

and the differentials

$$dt_I := \frac{\partial t_I}{\partial p}(0) \cdot p, \qquad d\nu_I := \frac{\partial \nu_I}{\partial p}(0) \cdot p,$$
$$dt_{V,j} := \frac{\partial t_{V,j}}{\partial p}(0) \cdot p, \qquad d\nu_{V,j} := \frac{\partial \nu_{V,j}}{\partial p}(0) \cdot p,$$

$$d\nu_T := \frac{\partial \nu_T}{\partial p}(0) \cdot p.$$

Supposing that the perturbated problem has a solution with the same switching structure as the nominal solution the variables defined above are solution of a homogeneous multipoint boundary value problem (BVP). Since the starting conditions and the terminal conditions are uncoupled in the reference problem it is possible to write the BVP as

$$\begin{pmatrix} \delta\dot{\mathbf{x}} \\ \delta\dot{\lambda} \end{pmatrix} = \mathcal{T}(t) \begin{pmatrix} \delta\mathbf{x} \\ \delta\lambda \end{pmatrix}, \tag{17}$$

$$D_0 \begin{pmatrix} \delta\mathbf{x}(0) \\ \delta\lambda(0) \end{pmatrix} = \begin{pmatrix} \delta\mathbf{x}_0 \\ \delta\lambda_0 \end{pmatrix}, \quad D_i \begin{pmatrix} \delta\mathbf{x}(t_i) \\ \delta\lambda(t_i) \end{pmatrix} = 0, \quad D_T \begin{pmatrix} \delta\mathbf{x}(T) \\ \delta\lambda(T) \end{pmatrix} = 0. \tag{18}$$

For simplicity further linear equations for the mentioned differentials were omitted. The switching points t_i in (18) are the collection of the interior and the junction points.

To solve this BVP a multiple shooting approach is used. Therefore the interval $[0, T]$ is divided into subintervals by gridpoints $\tau_j, j = 1, \ldots, m$:

$$0 = \tau_1 < \tau_2 < \cdots < \tau_m = T. \tag{19}$$

With this approach $m - 1$ initial value problems (IVP) of the form

$$\dot{y}(t) = \mathcal{T}(t)\, y(t), \quad y(\tau_j) = s_j \tag{20}$$

have to be solved. The solutions $y(t; s_j)$ of this set of IVPs are a solution of the BVP (17) – (18), if and only if the continuity conditions

$$y(\tau_{j+1}; s_j) = s_{j+1}, \quad j = 1, \ldots, m \tag{21}$$

and the boundary conditions

$$D_0 s_1 = \begin{pmatrix} \delta\mathbf{x}_0 \\ \delta\lambda_0 \end{pmatrix}, \tag{22}$$

$$D_i y(t_i, s_j) = 0, \qquad t_i \in [\tau_j, \tau j + 1), \tag{23}$$

$$D_T s_m = 0 \tag{24}$$

are satisfied. Assuming that there exists only one switching point t_i, this leads to the following linear system of equations:

$$\underbrace{\begin{pmatrix} G_1 & -I & 0 & \cdots & & \cdots & 0 \\ 0 & G_2 & -I & \ddots & & & \vdots \\ \vdots & \ddots & \ddots & \ddots & & & \\ & & & & & \ddots & \vdots \\ \vdots & & & & \ddots & \ddots & 0 \\ 0 & \cdots & & & \ddots & G_{m-1} & -I \\ D_0 & 0 & \cdots & 0 \; D_i\hat{G} \; 0 & \cdots & 0 & D_T \end{pmatrix}}_{M} \underbrace{\begin{pmatrix} s_1 \\ \vdots \\ \vdots \\ s_j \\ \vdots \\ \vdots \\ s_m \end{pmatrix}}_{s} = \underbrace{\begin{pmatrix} 0 \\ \vdots \\ \\ \\ \vdots \\ 0 \\ \delta Z_0 \end{pmatrix}}_{c}, \tag{25}$$

with $\delta Z_0 := (\delta \mathbf{x}_0, \delta\lambda_0, 0, 0)^*$ from equations (22) – (24) and I as the identity. Denoting with $G(t;j)$ the unique fundamental solutions of (17) with $G(\tau_j;j) = I$ the submatrices G_j and $\hat{G}$ are defined by

$$G_j := G(\tau_{j+1};j), \quad \hat{G} := G(t_i;j).$$

A generalization to more than one switching point can be found in [5]. It is obvious that the matrix M depends only on the nominal solution and not on the measured perturbation. Therefore M can be calculated off-line as soon as the nominal solution is known. To get an approximation for the perturbed problem only matrix multiplications have to be done. Because of the linearization used in this approach the correction step must be carried out repeatedly. This procedure is already examined in literature [16–19].

A disadvantage of this method is that each correction step has the same nominal trajectory as the base and the point (time) of the correction step is fixed before the run of the dynamical system. If the deviation of the measured actual state of the observed system is too far from the nominal path, this method cannot be used anymore. The idea to avoid this problem is to use the obtained information from the previous correction step to update the nominal solution. Provided that a correction step has been done, informations about an approximation of the actual solution are available on the gridpoints τ_j. With this a new comparison function $\mathbf{x}^c$, λ^c can be defined piecewisely on the intervals $[\tau_j, \tau_{j+1})$ $(j = 1, \ldots, m-1)$ by using the differential equations (8) and (15) and the initial values

$$\mathbf{x}^c(\tau_j) = \mathbf{x}^n(\tau_j) + \delta\mathbf{x}(\tau_j),$$
$$\lambda^c(\tau_j) = \lambda^n(\tau_j) + \delta\lambda(\tau_j)$$

(the notation $\mathbf{x}^n$ and λ^n is used for the nominal solution). Because of the linearization error the comparison function is discontinuous at the gridpoints:

$$d_{x,j} := \mathbf{x}^c(\tau_j^+) - \mathbf{x}^c(\tau_j^-), \tag{26}$$

$$d_{\lambda,j} := \lambda^c(\tau_j^+) - \lambda^c(\tau_j^-) \tag{27}$$

for $j = 2, \ldots, m$. Since the computed updates are only approximations for the actual solution, it is obvious that the updated trajectory is not an exact solution of the original optimal control problem. But that does not reduce the possibility to use it as a new base for the next correction step. Looking at the linearization used in this method it can be seen that some terms cannot be dropped anymore. As an example the boundary condition (9) is used in the following. Linearisation of this condition and neglecting the terms of higher orders gives

$$\varphi(\mathbf{x}^c(0), \mathbf{x}^c(T)) + \varphi_{\mathbf{x}(0)}[0,T]\,\delta\mathbf{x}(0) + \varphi_{\mathbf{x}(T)}[0,T]\,\delta\mathbf{x}(T) + \varphi_p[0,T]\,p = 0. \tag{28}$$

Now the first term does not disappear like it would in the case of using an exact solution. In equation (28) the partial differentials $\varphi_{\mathbf{x}(0)}$, $\varphi_{\mathbf{x}(T)}$ and φ_p will be evaluated at the approximated nominal solution x^c, λ^c. Doing this for the whole problem leads to a linear multipoint boundary value problem similar to (17), (18) but with different right hand sides. Using the same multiple shooting approach as above the resulting matrix equation is

$$\underbrace{\begin{pmatrix} G_1^c & -I & 0 & \cdots & & \cdots & 0 \\ 0 & G_2^c & -I & \ddots & & & \vdots \\ \vdots & \ddots & \ddots & \ddots & & & \\ & & & & & & \\ & & & & & \ddots & \vdots \\ \vdots & & & & \ddots & \ddots & 0 \\ 0 & \cdots & & & \ddots & G_{m-1}^c & -I \\ D_0 & 0 & \cdots & 0 \; D_i\hat{G} \; 0 \cdots & & 0 & D_T \end{pmatrix}}_{M^c} \underbrace{\begin{pmatrix} s_1^c \\ \vdots \\ \vdots \\ s_j^c \\ \vdots \\ \vdots \\ s_m^c \end{pmatrix}}_{s^c} = \underbrace{\begin{pmatrix} d_2 \\ \vdots \\ \\ \\ \\ \vdots \\ d_m \\ \delta Z_0^c \end{pmatrix}}_{c^c}. \tag{29}$$

The letter "c" used as a high index is a hint that the submatrices depend on the comparison function.

4 Parallel Algorithm

The method described in the previous section can be easily translated into an algorithm:

```
read nominal trajectory
initialize comparison solution as nominal solution
while actual time < T do do
   read perturbation
   compute M^c
   solve linear equation system M^c s^c = c^c
   update actual control
   update comparison solution
end while
```

Because of the dependency of the matrix M^c on the comparison trajectory, which is changed in every correction step, M^c cannot be calculated off-line. This means the computational effort to get solutions for the perturbed problem is rised by a large amount. This is a major drawback for a realtime algorithm. But now the choice for a multiple shooting approach is of advantage because it is known that this method is well suited for parallelisation (see [8] or [9]).

Taking a closer look at the matrix M^c shows a possibility for parallelisation. If the segmentation

$$\begin{array}{ccccccccc}
G_1^c & -I & 0 & \cdots & & & & \cdots & 0 \\
0 & G_2^c & -I & \ddots & & & & & \vdots \\
\vdots & \ddots & \ddots & \ddots & & & & & \\
 & & & & & & & \ddots & \vdots \\
\vdots & & & & & & \ddots & \ddots & 0 \\
0 & \cdots & & & & & \ddots & G_{m-1}^c & -I \\
D_1 & 0 & \cdots & 0 & D_i\hat{G} & 0 & \cdots & 0 & D_T
\end{array}$$

of the matrix M^c is used, the submatrices in each of the L-shaped parts can be calculated independently from the others. Only information from the reference trajectory in the corresponding interval is needed.

Going back to the algorithm it can be seen that it uses a precalculated nominal trajectory as the first comparison solution. In the loop the perturbation is read and as long as the comparison solution is to poor for this perturbation update-vectors will be calculated. In the loop the computation of M^c and the new comparison solution can be done in parallel, while the linear equation system must be solved serially. To minimize the serial part of the algorithm each process can carry out some preparations on its submatrix G_j^c (e.g., a LR-decomposition). Since the computation

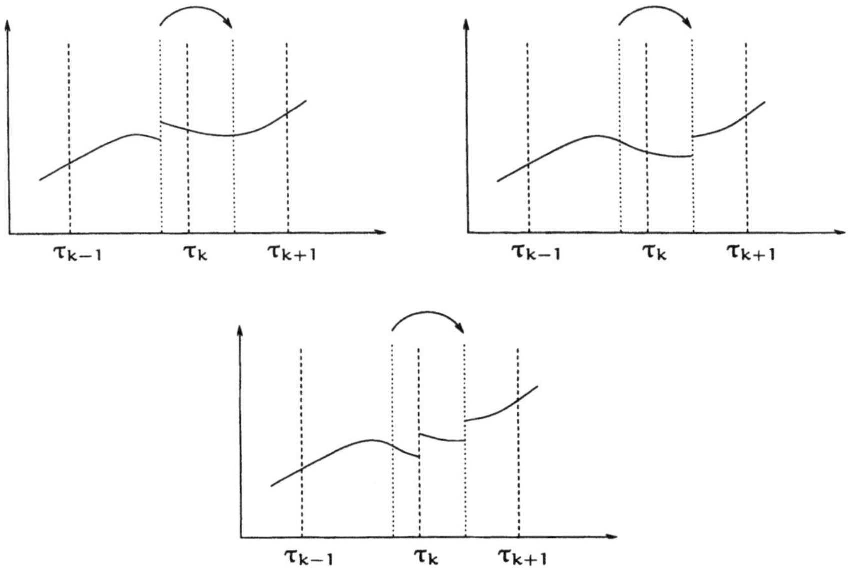

Figure 2. Boundary Problem

of M^c is by far the most time consuming part of the algorithm this serial section is not a big problem (but some additional computational time is lost due to the necessary synchronisation of the parallel processes in this phase).

An additional problem which arises through parallelisation is shown in Figure 2. Suppose that for some value of p there is a switching point in the intervall $[\tau_{k-1}, \tau_k]$ where an adjoint variable jumps (see top left picture). In the flow of computation (changing parameter value p) this point might move into the intervall $[\tau_k, \tau_{k+1}]$. What we want to get is shown in the top right picture. But straight forward computation yields the situation shown in the picture beneath. So the result of such moves are more jumps on the multiple shooting gridpoints, which may lead to more iterations und thus to a longer computation time. The algorithm compensates that by monitoring the motion of the switching points and modifying the starting values if necessary to get the wanted result.

For the reference problem mentioned in the first section, Figure 3 shows some typical flight paths. At the moment up to 10 aircraft are calculated, but considering clearness of the figure only 3 are printed. The steep descent is a consequence of the different scale for the variable z. Figure 4 shows the behaviour of the algorithm. The dotted line is the nominal trajectory. At $t = 300$ [sec] or $t = 650$ [sec] respectively a deviation is measured and the algorithm has to correct the flight path. It can also be seen that the various state variables remain within reasonable bounds so that no additional state constraints are needed. However they could be included if necessary.

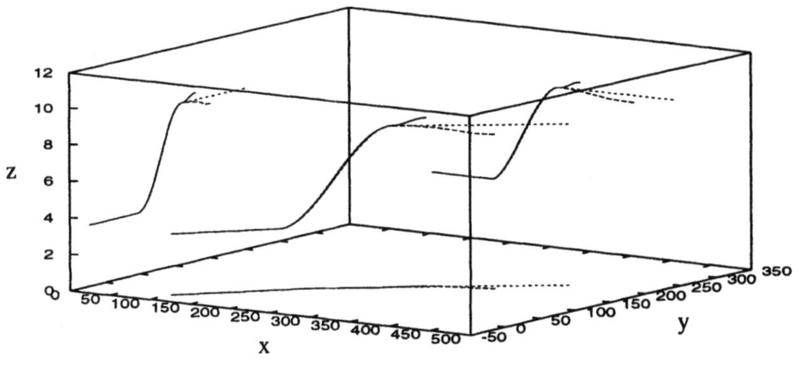

Figure 3. 3D-view of Flight Paths

The parallel feedback algorithm described in this paper has been implemented on a multiple-instruction-multiple-data parallel computer with shared memory. A fast data-exchange between the computational nodes has a favourable effect because of the need to assemble the parallel computed parts of the matrix M^c to solve the linear equation system.

The algorithm is implemented in Fortran90. One of the advantages over the older Fortran77 is the possibility to allocate storage dynamically. So it is not necessary to recompile the program if only the magnitude of the problem changes (only the input-files must be modified).

A graphical user interface is near completion.

References

1. P. K. A. Menon: Control Theoretic Approach to Air Traffic Conflict Resolution. AIAA Report93-3832-CP (1993)
2. H. Maurer, D. Augustin: Sensitivity Analysis and Real-Time Control of Parametric Optimal Control Problems Using Boundary Value Methods. Schwerpunktbuch des DFG Schwerpunktes *Echtzeitoptimierung großer Systeme* (2001)
3. D. H. Jacobson, M. M. Lele, J. L. Speyer: New Necessary Conditions of Optimality for Control Problems with State Variable Inequality Constraints. J. of Math. Anal. and Appl. 35, 255-284 (1971)
4. M. R. Hestenes: Calculus of Variations and Optimal Control Theory. Wiley and Sons, New York (1966)
5. B. Kugelmann: Ein paralleles Rückkopplungsverfahren zur Lösung von Optimalsteuerungsproblemen. Habilitationsschrift, Mathematisches Institut, Technische Universität München (1994)

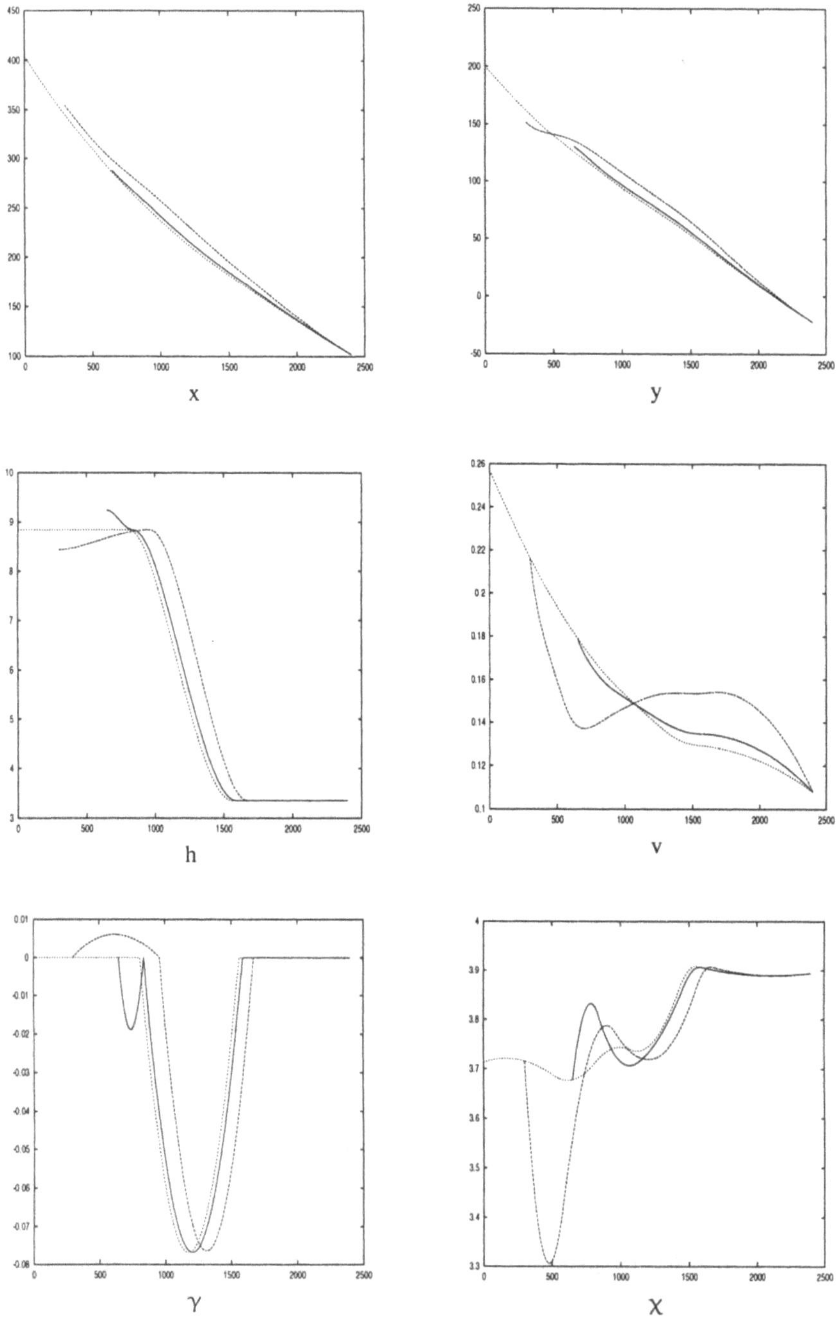

Figure 4. Nominal and Deviated Trajectories of an Aircraft

6. J. Stoer, R. Bulirsch: Numerische Mathematik 2. Springer Lehrbuch, 3. Auflage, Springer Verlag, Berlin, Heidelberg, New York (1990)
7. B. Kugelmann, W. Weber: A parallel Method for Optimal Guidance in Air Traffic. Preprint 99-6, Schwerpunkt *Echtzeitoptimierung grosser Systeme*, DFG (1999)
8. S. J. Wright: Stable Parallel Elimination for Boundary Value ODEs. Preprint MCS-P229-0491 of the Mathematics and Computer Science Division, Argonne National Laboratory (1991)
9. M. Kiehl, R. M. Mehlhorn: Parallel Multiple Shooting for Optimal Control Problems under NX. J. Optimization Methods and Software, 4, 259-271 (1995)
10. R. Bulirsch: Die Mehrzielmethode zur numerischen Lösung von nichtlinearen Randwertproblemen und Aufgaben der optimalen Steuerung. Report der Carl-Cranz Gesellschaft (1971)
11. H. J. Oberle: Numerische Berechnung optimaler Steuerungen von Heizung und Kühlung für ein realistisches Sonnenhausmodell. Habilitationsschrift, Technische Universität München (1982)
12. P. Hiltmann: Numerische Lösung von Mehrpunkt-Randwertproblemen und Aufgaben der optimalen Steuerung mit Steuerfunktionen über endlichdimensionalen Räumen. Dissertation, Institut für Mathematik, Technische Universität München (1989)
13. O. von Stryk: Numerische Lösung optimaler Steuerungsprobleme: Diskretisierung, Parameteroptimierung und Berechnung der adjungierten Variablen. Dissertation, Mathematisches Institut, Technische Universität München (1994). Fortschritt-Berichte VDI, Reihe 8, VDI-Verlag, Düsseldorf, 150 foll.
14. H. Maurer, H. J. Pesch: Solution Differentiability for Parametric Nonlinear Control Problems. Schwerpunktprogramm der Deutschen Forschungsgemeinschaft, Report 316 (1994). SIAM Journal on Control and Optimization 32, Nr. 6 (1994)
15. H. Maurer, H. J. Pesch: Solution Differentiability for Parametric Nonlinear Control Problems with Control-State Constraints. Schwerpunktprogramm der Deutschen Forschungsgemeinschaft, Report 474 (1993). Journal of Optimization Theory & Applications 86, Nr. 2 (1995)
16. H. J. Pesch: Numerische Berechnung optimaler Flugbahnkorrekturen in Echtzeit-Rechnung. Dissertation, Institut für Mathematik der Technischen Universität München (1978)
17. H. J. Pesch: Echtzeitberechnung fastoptimaler Rückkopplungssteuerungen bei Steuerungsproblemen mit Beschränkungen. Habilitationsschrift, Technische Universität München (1986)
18. H. J. Pesch: Real-Time Computation of Feedback Controls for Constrained Optimal Control Problems, Part 1: Neighbouring Extremals. Optimal Control Appl. & Methods 10, 129-145 (1989)
19. H. J. Pesch: Real-Time Computation of Feedback Controls for Constrained Optimal Control Problems, Part 2: A Correction Method Based on Multiple Shooting. Optimal Control Appl. & Methods 10, 147-171 (1989)

II

Optimal Control for Partial Differential Equations

*Optimal Control Problems with a First Order PDE System – Necessary and Sufficient Optimality Conditions

Sabine Pickenhain and Marcus Wagner

Institut für Mathematik, Technische Universität Cottbus, Germany

Abstract This paper gives an overview over the results concerning necessary and sufficient optimality conditions for optimal control problems with multiple integrals and first order partial differential equations. Second order sufficiency conditions are illustrated by the problem of minimal k-energy in an n-dimensional space. It can be shown by the developed theory that a certain cone has strong minimizing properties.

1 Introduction

1.1 Historical Remarks

In the past there where many efforts to extend Pontryagins maximum principle of optimal control theory of one independent variable to the case of multiple integral problems. The investigations where done into two different directions

- The large theory of optimal control problems with distributed parameters, in which one of the variables plays a distinctive leading role.
- The theory of Dieudonné-Rashevsky type problems, where the independent variables have an equal rank.

The second kind of problems are the topic of this paper.

In 1969 Cesari [3] stated a generalized maximum principle, which is a formal extension of Pontryagins maximum principle to Dieudonné-Rashevsky type problems. In 1969 relevant papers followed by Klötzler [12] and Rund [26]. These results use regularity properties of the solution of the corresponding Hamilton-Jacobi-equation, which are not fulfilled in general. In 1993 Klötzer and Pickenhain [15] proved an ϵ-maximum principle with canonical veriables y_ϵ in function spaces. However, the limiting procedure $\epsilon \to 0$ in function spaces is not possible in general. The main subject of the first part of this paper is to formulate assumptions for which a maximum principle can be shown with canonical variables in C^*.

The second part gives a general concept of duality in optimal control and different realizations by Fenchel/Rockafellar, Klötzler and the authors. Second order sufficient optimality conditions are obtained by using duality results. These sufficiency conditions are needed to prove stability and sensitivity for perturbed optimal control problems, see the paper of Maurer/Augustin of this book. On the basis of these sensibility results, perturbed solutions can be calculated in real time.

1.2 Problem Formulation

We consider the following optimal control problem (P) with first-order partial differential equations:

$$J(x,u) = \int_\Omega f_0(t,x(t),u(t))\,dt + \sum_{k=1}^{s} \int_\Omega f_k(t,x(t))\,d\alpha_k(t) \longrightarrow \text{Min!} \quad (1)$$

subject to $x \in W_p^{1,n}(\Omega)$, $u \in L_p^r(\Omega)$ $(p > m)$, satisfying a.e. on $\Omega \subset \mathbb{R}^m$

state equations

$$x_{i;t_j}(t) = g_{ij}(t,u(t)), \qquad i = 1,\dots,n,\ j = 1,\dots,m; \quad (2)$$

control restrictions

$$u(t) \in U \subset \mathbb{R}^r, \qquad U \text{ compact}; \quad (3)$$

boundary conditions

$$x(t) = \varphi(t), \qquad \text{for all } t \in \Gamma \subset \Omega,\ \Gamma \text{ compact},\ \Gamma \neq \emptyset; \quad (4)$$

and *state constraints*

$$x(t) \in \overline{G(t)} \iff h_l(t,x(t)) \leq 0, \quad \text{for all } t \in \Omega,\ l = 1,\dots,w. \quad (5)$$

The formulation of problem (P) includes *state-constrained problems* of *Dieudonné-Rashevsky type* if $f_k = 0$ for all k see [3], [15], as well as *state-constrained deposit problems* for $f_0 = 0$, $f_k(t,x(t)) = x_k(t)$, see [15]. Convexity assumptions are needed for the derivation of existence results as well as necessary and sufficient optimality conditions. If these assumptions are not valid we construct the standard relaxation (or convexification) of (P) by use of Young measures, see [24].

Our basic assumptions for (P) are the following:

(V1) We have $m \geq 2$ and $m < p < \infty$. $\Omega \subset \mathbb{R}^m$ is a compact Lipschitz domain (in strong sense, see [18]). Then functions $x \in W_p^{1,n}(\Omega)$ are continuously representable, and functions $x \in W_\infty^{1,n}(\Omega)$ have Lipschitz representatives on Ω, [1] (p.185, Theorem 5.5.).

(V2) The functions f_0, f_k, g_{ij}, h_l and φ are continuous w. r. to all of their arguments; $f_0(t,\cdot,\cdot)$, $f_k(t,\cdot)$, $g_{ij}(t,\cdot)$ and $h_l(t,\cdot)$ are continuously differentiable w. r. to ξ resp. (ξ,v) for all $t \in \Omega$.

(V3) $\alpha_k \in rca\,(\Omega)$ are signed regular measures on the σ-algebra of the Borel sets on Ω.

(V4) The set of feasible solution (x,u) of (P) is denoted by Z and Z is non empty.

We emphasize *two special types of boundary conditions:*

$$x(t_0) = x_0 \text{ for fixed } t_0 \in \Omega\text{, i. e. } \Gamma = \{t_0\}; \tag{6}$$

$$x(t) = \varphi(t) \text{ for all } t \in \partial\Omega\text{, i. e. } \Gamma = \partial\Omega. \tag{7}$$

1.3 Notations

We abbreviate the m-dimensional Lebesgue measure of A by $|A|$, the closure of A by $\overline{A}$ and the actual zero element by 0. $C^{k,n}(\Omega)$, $L_p^n(\Omega)$ and $W_p^{k,n}(\Omega)$ $(1 \le p \le \infty)$ denote the spaces of n-dimensional vector functions on Ω whose components are k-times continuously differentiable, resp. belong to $L_p(\Omega)$ or to the Sobolev space of $L_p(\Omega)$-functions having weak derivatives up to k^{th} order in $L_p(\Omega)$. The subspace of $C^{k,n}(\Omega)$-functions with compact support is denoted by $\mathring{C}^{k,n}(\Omega)$; instead of $C^{0,1}(\Omega)$ we write shortly $C^0(\Omega)$. For the classical and weak partial derivatives of x_i by t_j we use the same notation: $x_{i;t_j}$. The Banach space of Radon measures (signed regular measures with the total variation $V\mu(\Omega)$ as norm) is denoted by *rca* (Ω). Due to the compactness of Ω, *rca* (Ω) is isomorphical to the dual space $(C^0(\Omega))^*$ [6] (p. 265, Theorem 3) so that each linear, continuous functional on $C^0(\Omega)$ can be represented by an integral w. r. to a Radon measure $\nu \in$ *rca* (Ω). (P) suggests to declare the following sets:

Definition 1.

$$Z_1 = \{z \in L_p^{nm}(\Omega)\,|z_{ij} = g_{ij}(t, u(t)) \text{ a. e. on } \Omega;\ u \in L_p^r(\Omega), u(t) \in U\} \tag{8}$$

$$Z_2 = \{(z, \zeta) \in L_p^{nm}(\Omega) \times W_p^{1,n}(\Omega)|z_{ij} = \zeta_{i;t_j}; \zeta \in W_\infty^{1,n}(\Omega), \zeta|_\Gamma \equiv 0\}. \tag{9}$$

Z_1 is the set of the admissible right-hand sides of (2).

Definition 2. (Baire classification) We call a continuous function ψ defined on the compact set $\Omega \subset \mathbb{R}^m$ *from* 0th *Baire class* and write $\psi \in \mathbf{B}^0(\Omega)$. The limit functions of everywhere pointwise convergent sequences $\{\psi^K\}$, $\psi^K \in \mathbf{B}^0(\Omega)$, form the *first Baire class* $\mathbf{B}^1(\Omega)$; the limit functions of everywhere pointwise convergent sequences $\{\psi^K\}$, $\psi^K \in \mathbf{B}^1(\Omega)$, form the *second Baire class* $\mathbf{B}^2(\Omega)$ and so on.

Obviously, $\mathbf{B}^0(\Omega) \subset \mathbf{B}^1(\Omega) \subset \mathbf{B}^2(\Omega) \subset \ldots$ holds. Note that each finite function contained in any Baire class is measurable [2] (p. 404, Theorem 4); conversely, any measurable, essentially bounded function on Ω agrees a.e. with some function of second Baire class [2] (p. 406, Theorem 5). Each Baire class is closed under (pointwise) addition und multiplication of finite functions [2] (p. 397, Theorems 6 and 7). For more details, [2] (pp. 393 ff).

Defining functionals $H_l: \; C^{0,n}(\Omega) \to \mathbb{R}$ by

$$H_l(x) = \mathrm{Max}_{t \in \Omega}\, h_l(t, x(t)), \;\; 1 \le l \le w,$$

the state constraints (5) can be expressed as follows:

$$h_l(t, x(t)) \le 0 \;\; \forall t \in \Omega \quad \Longleftrightarrow \quad H_l(x) \le 0, \; l = 1, \ldots, w. \tag{10}$$

2 The Maximum Condition as Necessary Optimality Condition

If we prove an appropriate version of Pontryagin's maximum principle for problems (P), considering the inclusion $Z_1 \subseteq L_\infty^{nm}(\Omega)$, it seems to be necessary to take the multipliers corresponding to (2) from the space $\left(L_\infty^{nm}(\Omega)\right)^*$. However, to avoid the difficulties in the practical handling of conditions in presence of $(L_\infty)^*$ multipliers (which can be represented only by finitely additive set functions, see [29] (p. 12 f.), we propose a different way. We do not model \piecewise continuous controls" resp. \piecewise continuous right-hand sides" of (2.2) as (equivalence classes of) L_∞-functions but as functions from *first Baire class* $\mathbf{B}^1(\Omega)$. In [23] (p. 5, Proposition 1.6), it was shown that also \piecewise continuous" functions of several variables belong to this class. Since the class $\mathbf{B}^0(\Omega)$ coincides with the space $C^0(\Omega)$ and, on the other hand, each $L_\infty(\Omega)$-function admits a representation by an element of the class $\mathbf{B}^2(\Omega)$ [2] (p. 406, Theorem 5). This setting generates an intermediate situation which generalizes in a natural way the classical assumptions of the one-dimensional theory. In this frame, we obtain as main result the following theorem.

Theorem 3. *(Maximum principle with multipliers from* $\left(C^{0,nm}(\Omega)\right)^*$ *for solutions of* (P) *which can be strongly varied).*

Let (x^*, u^*) *be a global minimizer of the problem* (P) *under assumptions* (V1) – (V4). *Assume that the weak derivatives* $x^*_{i;t_j}$ *have representatives from first Baire class* $\mathbf{B}^1(\Omega)$. *Further, let us assume:*

(V5) *For each active index* l *(i.e.* $H_l(x^*) = 0$ *) there exists a feasible process* (x^l, μ^l) *with* $H_l'(x^*, x^l - x^*) < 0$ *(i.e. *(x^*, u^*) *can be strongly varied").*

(V6) *There is a continuous function* z^0 *in the interior of* Z_1 *(with respect to the* $L_\infty^{nm}(\Omega)$*-norm topology) which is part of a pair* $(z^0, \zeta^0) \in Z_2$.

(V7) *Each function* $z \in Z_1$ *of first Baire class admits an approximation by an (everywhere) pointwise convergent sequence of continuous functions* $z^K \in Z_1$.

(V8) *The function* $f_0(t, \xi, \cdot)$ *is assumed to be convex for all* $(t, \xi) \in \mathbb{R}^{n+1}$ *and* g_{ij} *is assumed to be linear with respect to the control argument.*

(V9) *The control set* U *is assumed to be convex.*

Then there exist multipliers $\nu \in \left(rca\,(\Omega)\right)^{nm}$ *and* $\sigma_l \in rca\,(\Omega)$, $1 \leq l \leq w$, *satisfying the maximum condition (in integrated form)* $(\mathbf{M})_0$:

$$
\begin{aligned}
&-\int_\Omega f_0(t,x^*(t),u^*(t)) - f_0(t,x^*(t),u(t))]\,dt \,+ \\
&\sum_{i,j}\int_\Omega [g_{ij}(t,u^*(t)) - g_{ij}(t,u(t))]\,d\nu_{ij}(t) \geq 0 \qquad (11)\\
&\forall u \in \mathcal{M}_U := \left\{ u \in L_\infty^{nm}(\Omega),\, u(t) \in U \;\middle|\; g_{ij}(\cdot,u(\cdot)) \in \mathbf{B}^1(\Omega) \right\}
\end{aligned}
$$

as well as the weak canonical equation $(\mathbf{K})_0$:

$$
\begin{aligned}
&\sum_{i,j}\int_\Omega \zeta_{i;t_j}(t)\,d\nu_{ij}(t) \\
&+\int_\Omega \nabla_\xi^T \left[\left[g_{ij}(t,u^*(t)) - g_{ij}(t,u(t))\right]_{\xi=x^*(t)} \zeta(t)\,dt \qquad (12)\\
&+\sum_k \int_\Omega \nabla_\xi^T \left[f_k(t,\xi)\right]_{\xi=x^*(t)} \zeta(t)\,d\alpha_k(t) \\
&-\sum_l \int_\Omega \nabla_\xi^T \left[h_l(t,\xi)\right]_{\xi=x^*(t)} \zeta(t)\,d\sigma_l(t) = 0
\end{aligned}
$$

for all test functions $\zeta \in C^{1,n}(\Omega)$ *with* $\zeta\,|\,\Gamma \equiv 0$. *If the boundary conditions are of type B then* $(\mathbf{K})_0$ *can be restricted on test functions* $\zeta \in \overset{\circ}{C}{}^{\infty,n}(\Omega)$. *Furthermore, the measures* σ_l *satisfy the complementarity condition* $(\mathbf{C})_0$:

$$
\begin{aligned}
&\sigma_l \geq 0,\ \operatorname{supp}\sigma_l \subseteq \{t \in \Omega \mid h_l(t,x^*(t)) = 0\} \\
&\Longrightarrow \int_\Omega h_l(t,x^*(t))\,d\sigma_l(t) = 0,\ 1 \leq l \leq w. \qquad (13)
\end{aligned}
$$

Proof. The proof of a corresponding theorem for relaxed problems was given in [24]. Assumptions (V8) and (V9) garantee that identical methods used in the proof are applicable to the problem (P) of this paper.

Theorem 3 generalizes IOFFE/TICHOMIROV's theorem [11] (p. 207 f., Theorem 1) for one-dimensional problems (in the normal case). Note that the general results of IOFFE/TICHOMIROV [11] (p. 73 f., Theorem 3) and GINSBURG/IOFFE [9] (p. 92 and 96, Theorems 3.3. and 3.6.) concerning \smooth-convex" problems in Banach spaces cannot be applied to (P) since the differential operator in (2) maps onto a subspace of infinite codimension in $L_\infty^{nm}(\Omega)$.

3 DUALITY AND SUFFICIENCY CONDITIONS

3.1 Construction of Dual Problems

In a very general setting, a problem (D) of maximization of an (extended real-valued) functional L over an arbitrary set $S \neq \emptyset$ is said to be a dual problem to

(P), if the *weak duality relation*

$$\sup(\mathrm{D}) \leq \inf(\mathrm{P})$$

is satisfied.

Different notions of duality can be embedded into the following construction scheme:

Step 1 The set of admissible pairs $(x, u) = z \in Z$ is represented by the intersection of two suitable non-empty sets X_0 and X_1.

Step 2 For an (extended real-valued) functional $\breve{}\colon X_0 \times S_0 \to \overline{\mathbb{R}}$ the *equivalence relation*

$$\inf_{z \in Z} J(z) = \inf_{z \in X_0} \sup_{S \in S_0} \breve{}(z, S)$$

holds.

Step 3 Assuming $L_0(S) := \inf_{z \in X_0} \breve{}(z, S)$, each problem (D),

$$\begin{aligned} &\textit{maximize } L(S) \\ &\textit{subject to } S \in S \subseteq S_0 \end{aligned} \tag{D}$$

is a (weak) dual problem to (P) if

$$L(S) \leq L_0(S) \quad \text{forall} \quad S \in S. \tag{14}$$

The proof of the weak duality relation results from the well-known inequality

$$\inf_{z \in X_0} \sup_{S \in S_0} \breve{}(z, S) \geq \sup_{S \in S_0} \inf_{z \in X_0} \breve{}(z, S).$$

3.2 Fenchel-Rockafellar Duality

In the follwing sections we investigate problem (P) with boundary conditions (7) and $f_k = 0$ for $k = 1, \ldots, s$.

In accordance with [7], we transform (P) into a general variational problem (V):

$$\begin{aligned} &\textit{minimize} \int_\Omega l(t, x(t), x_{t_\alpha}(t))\, dt \\ &\textit{subject to } x \in X \end{aligned} \tag{V}$$

where $l\colon \mathbb{R}^n \times \mathbb{R}^n \times \mathbb{R}^{nm} \to \overline{\mathbb{R}}$ is given by

$$l(t, \xi, w) := \begin{cases} \inf\{ r(t, \xi, v) \mid v \in U \textit{ with} \\ w = g(t, v)\} \textit{for } (t, \xi) \in \overline{X} \\ \infty \qquad\qquad\qquad \textit{else} \end{cases}$$

and

$$\bar{X} = \left\{ x \in W_p^{1,n}(\Omega) \mid x(t) \in \overline{G(t)} \textit{ on } \bar{\Omega},\ x(s) = \varphi(s) \textit{ on } \partial\Omega \right\}. \tag{15}$$

Then (P) is called convex if (V) is convex in the sense of [7, p. 113]. In this case both problems are equivalent [21]. The Fenchel-dual problem is obtained by the following settings in the above construction scheme:

1.

$$X_0 = \{ z = (x,u) \in W_p^{1,n}(\Omega) \times L_p^{1,\nu}(\Omega) \mid \; u(t) \in U \textit{ a.e. on } \Omega, \; x(t) \in \overline{G(t)} \textit{ a.e. on } \Omega, \; x(s) = \varphi(s) \textit{ on } \partial\Omega \; \},$$

$$X_1 = \{ z = (x,u) \in W_p^{1,n}(\Omega) \times L_p^{1,\nu}(\Omega) \mid \; x^i_{t_\alpha}(t) = g^i_\alpha(t, x(t), u(t)) \textit{ a.e. on } \Omega, \; (\alpha = 1, \dots, m;\; i = 1, \dots, n) \}$$

2. $S_0 = L_q^{nm)}(\Omega)$ $(p^{-1} + q^{-1} = 1)$, ˇ is the classical Lagrange functional,

$$\check{}(z, y) = J(z) + \sum_{i,\alpha} \langle x^i_{t_\alpha} - g(\cdot, x, u), y^\alpha_i \rangle,$$

where $\langle \cdot, \cdot \rangle$ is the bilinear canonical pairing over $L_p(\Omega) \times L_p^*(\Omega)$, [21].

By use of the Hamiltonian of (P), $\mathcal{H}\colon \mathbb{R}^1 \times \mathbb{R} \times \mathbb{R}^{nm} \to \overline{\mathbb{R}}$,

$$\mathcal{H}(t, \xi, \eta) = \sup\{ H(t, \xi, v, \eta) \mid v \in U \}$$

with

$$H(t, \xi, v, \eta) = -r(t, \xi, v) + \sum_{i,\alpha} \eta^\alpha_i \, g^i_\alpha(t, \xi, v)$$

the dual problem $(D_R)_q$ can be formulated as follows [21]:

$$\textit{maximize} \quad \Big[- \int_\Omega \Big(\sup_{\xi \in G(t)} \big[H(t, \xi, -y(t)) - y_0(t)^T \xi \big] \Big) \, dt - \sup_{\zeta \in X} \Big[\int_\Omega \Big(y_0(t)^T \zeta(t) + \sum_{i,\alpha} y^\alpha_i(t) \, \zeta^i_{t_\alpha}(t) \Big) \, dt \Big] \Big]$$

$$\textit{subject to } (y_0, y) \in L_q^{n(1+m)}(\Omega).$$

3.3 *Duality in the sense of Klötzler*

The duality in the sense of Klötzler is realized by the following settings in the general construction scheme [19]:

1. X_0 and X_1 are chosen as before.
2. $S_0 = W_q^{1,m}(\overline{X})$, and ˇ is an extended Lagrange functional,

$$\check{}(z, S) = J(z) + \sum_{i,\alpha} \langle x^i_{t_\alpha} - g^i_\alpha(\cdot, x, u), S^i_{\xi_\alpha}(\cdot, x) \rangle,$$

where $\langle \cdot, \cdot \rangle$ is again the bilinear canonical pairing over $L_p(\Omega) \times L_p^*(\Omega)$.

By use of Gauss' theorem, the dual problem $(D_K)_q$ reads as follows, [14]

$$\begin{aligned}
&\textit{maximize} \int_{\partial\Omega} S(s,\zeta(s))\, n(s)\, do(s)\\
&\textit{subject to } S \in S_1,\\
&S_1 := \Big\{ S \in S_0 \,\Big|\, \sum_{\alpha=1}^{m} S^{\alpha}_{t_\alpha}(t,\xi) + \mathcal{H}(t,\xi,S_\xi(t,\xi)) \le 0 \textit{ a.e. on } \overline{X} \Big\},
\end{aligned}$$

where $n(\cdot)$ is the exterior unit normal vector to $\partial\Omega$.

In this way we can characterize minimizers of (P) in terms of solutions of the Hamilton-Jacobi inequality or of the Hamilton-Jacobi equation. Since classical solutions of the latter equation may fail to exist, on the one hand technics were developed to construct generalized solutions of this equation (viscosity solutions [16], generalized solutions involving the Clarke generalized gradient [4] or lower Dini derivatives [28]). On the other hand, optimization techniques for parametric problems in finite dimensional spaces are used to minimize the defect in the Hamilton-Jacobi inequality and to get sufficient conditions for (local) optimality [20, 22, 30–32].

3.4 Duality in Measure Spaces

We study problem (P) under class-qualification without state constraints:

Definition 4. The class-qualified problem $(P)_{\mathbf{B}^k}$ is given by problem (P) where each $(x,u) \in Z$ admits one representative $x_{i;t_j} \in \mathbf{B}^k(\Omega)$.

The duality for $(P)_{\mathbf{B}^k}$ in the sense of Pickenhain/Wagner is realized in the case that (P) has a minimizing sequence in $C^{1,n}(\Omega) \times \mathbf{B}^{0,nm}(\Omega)$, see [25]. We use the following setting in the general construction scheme:

1.

$$\begin{aligned}
X_0 = \{(x,u) \in W^{1,n}_{\infty}(\Omega) \times L^r_p(\Omega) \,|\; & x_{i;t_j} \in \mathbf{B}^0(\Omega),\ z_{ij} \in \mathbf{B}^1(\Omega),\\
& z_{ij}(t) = g_{ij}(t,u(t)),\ \forall\,(i,j),\\
& x(t) = \varphi(t) \ \ \forall t \in \Gamma\};\\
X_1 = \{(x,u) \in W^{1,n}_{\infty}(\Omega) \times L^r_p(\Omega) \,|\; & x_{i;t_j}(t) = g_{ij}(t,u(t))\\
& \forall (i,j)\, \forall t \in \Omega\},
\end{aligned}$$

2.

$$S_0 = \big(rca\,(\Omega)\big)^{nm}.$$

$$\Phi(x,u,v) = J(x,u) + \sum_{i,j} \int_{\Omega} \Big[x_{i;t_j}(t) - g_{ij}(t,u(t)) \Big]\, dv_{ij}(t). \qquad (16)$$

Let the problem (P) given and assume that all assumptions of theorem 3 are satisfied. Assume furthermore that

$$\mathrm{int} Z_1 \subset L^{nm}_{\infty}(\Omega) \neq \emptyset.$$

Then it can be shown by a direct estimation that the following problem $(D)_m$ is dual to each for the problems (P) and $(P)_{B^1}$, see [25]

$$\inf_{(x,u)\in X_0}\left[J(x,u)+\sum_{i,j}\int_\Omega\left[x_{i;t_j}(t)-g_{ij}(t,u(t))\right]d\nu_{ij}(t)\right]\to \text{Max}!$$
$$\nu\in Y_0=\left(rca\,(\Omega)\right)^{nm}.$$

3.5 *Sufficient Optimality Conditions*

First and second order sufficient optimality for global minimizers can be derived by means of duality. In the general concept, $(x^*,u^*)\in Z$ is a global minimizer of (P) if

$$J(x^*,u^*)=\inf_{z\in X_0}\ \sup_{S\in S_0}\ \check{}(z,S)=\max_{S\in S_0}\ \inf_{z\in X_0}\ \check{}(z,S)$$

and it exists an $S^*\in S$ with

$$L_0(S^*)=\max_{S\in S}L(S)=\max_{S\in S_0}L_0(S).$$

Following the concept of Klötzler, these equations are satisfied if there is a finite decomposition of Ω into strongly Lipschitz domains $\Omega_1,\dots,\Omega_\nu$ such that $S^*\in C^{1,n}(\overline{X_i})\cap C(\overline{X})$ with

$$\bar{X_i}=\{(t,\xi)\,|\,t\in\bar{\Omega_i},\ \xi\in\overline{G(t)}\}\tag{17}$$

and

a) *the Hamilton-Jacobi inequality*

$$\Lambda(t,\xi):=\sum_\alpha S^{*\alpha}_{t_\alpha}(t,\xi)+\mathcal{H}(t,\xi,S^*_\xi(t,\xi))\le 0\ on\ \overline{X_i}\tag{18}$$

b) *the Hamilton-Jacobi equation*

$$\sum_\alpha S^{*\alpha}_{t_\alpha}(t,x^*(t))+\mathcal{H}(t,x^*(t)),S^*_\xi(t,x^*(t)))=0\ on\ \overline{\Omega}\tag{19}$$

and

c) *the maximum condition*

$$\mathcal{H}(t,x^*(t)),S^*_\xi(t,x^*(t)))=H(t,x^*(t),u^*(t),S^*_\xi(t,x^*(t)))\ a.e.\ on\ \Omega\tag{20}$$

are fulfilled. From conditions a) and b) follows that $x^*(t)$ must be a global maximizer of the parametric optimization problem

$$\begin{aligned}&maximize\ \Lambda(t,\xi)\\&subject\ to\ \xi\in\overline{G(t)}\end{aligned}\tag{(P)$_t$}$$

with parameter $t\in\overline{\Omega_i}$. For this last problem $(P)_t$ first and second order sufficient optimality conditions can be derived with the quadratic setting

$$S^{*\alpha}(t,\xi)=a^\alpha(t)+y^{\alpha^T}(t)\,(\xi-x(t))+\tfrac{1}{2}(\xi-x^*(t))\,Q^\alpha(t)\,(\xi-x^*(t))\tag{21}$$

in the dual problem $(D_K)_\infty$, where

$$y^\alpha \in C^{1,n}(\overline{\Omega_i}) \cap C^{0,n}(\overline{\Omega}) \quad \text{and} \quad Q^\alpha \in C^{1,nn}(\overline{\Omega_i}) \cap C(\overline{\Omega})$$

are symmetric.

The ideas, mentioned above, can be used for identifying *strong* local minimizers of (P) too. In this case $\overline{X}$ is to be replaced by

$$\overline{X}_\varepsilon = \overline{X} \cap \{ (t,\xi) \in \mathbb{R}^{n+1} \mid \| \xi - x(t) \| < \varepsilon,\ t \in \overline{\Omega} \}, \tag{22}$$

see [20,22,30–32]. The second order condition for $(P)_t$ yields a definiteness condition for a Riccati-type expression which generalizes the known theory of conjugated points in the calculus of variations in one independent variable.

We formulate a second order sufficiency result for the problem (P) with the following additional assumptions:

(A1) $\Lambda(t,\cdot)$ is continuously differentiable on $\mathcal{K}_\epsilon(x^*(t))$ for each $t \in \bar{\Omega}_i$ $(i = 1,\dots,\nu)$,

$$\mathcal{K}_\epsilon(x^*(t)) = \{\xi \in \mathbb{R}^n \mid \|\xi - x^*(t)\| \leq \epsilon\} \tag{23}$$

(A2) $\text{grad}_\xi\, \Lambda(t,\cdot)$ is continuous on $\bar{\mathcal{K}}_\epsilon(x^*(t))$.
(A3) $\text{grad}_\xi\, \Lambda(t,\cdot)$ is locally Lipschitz on $\bar{\mathcal{K}}_\epsilon(x^*(t))$ for all $t \in \bar{\Omega}_i$ $(i = 1,\dots,\nu)$.
(A4) The set-valued mapping

$$(t,\xi) \to \partial(\text{grad}_\xi\, \Lambda(t,\xi))$$

is closed and locally bounded on $\bar{\mathcal{K}}_\epsilon$.
(A5) $h_l(t,\cdot)$ are twice continuously differentiable on $\bar{\mathcal{K}}_\epsilon(x^*(t))$, $l = 1,\dots,w$.
(A6) (LICQ) is fulfilled on $\bar{\Omega}$, i.e. the vectors

$$\{\text{grad}_\xi\, h_l(t,x^*(t)) \mid l \in I_0(t)\},$$
$$I_0(t) := \{l \in \{1,\dots,w\} \mid h_l(t,x^*(t)) = 0\},$$

are linearly independent for all $t \in \bar{\Omega}$.

Proposition 5. *Assume that the functions* Λ *in* $(P)_t$, *as well as* $h_1,\dots,h_w$ *in the state constraints satisfy the conditions (A1)–(A6). Moreover, let* $(x^*(t),\lambda(t))$ *be a stationary solution of* $(P)_t$, *i.e., for* $i = 1,\dots,\nu$ *let*

$$\text{grad}_\xi \Big(\Lambda(t,\xi) + \sum_{l=1}^{w} \lambda_l(t) h_l(t,\xi) \Big)_{\xi = x^*(t)} = 0 \tag{26}$$

$$\text{with} \quad \lambda_l(t) \geq 0, \quad \lambda_l(t) h_l(t,x^*(t)) = 0 \quad \text{on} \quad \bar{\Omega}_i. \tag{27}$$

If each matrix of the set

$$\partial^2_{\xi\xi}\Lambda(t,x^*(t)) + \sum_{l \in I^+(t)} \lambda_l(t) \partial^2_{\xi\xi} h_l(t,x^*(t)) \tag{28}$$

is negative definite on

$$W^+(t) := \{h \in \mathbb{R}^l \mid h^T \text{grad}_\xi\, h_l(t,x^*(t)) = 0,\ l \in I^+(t)\} \tag{29}$$

with

$$I^+(t) := \{l \in \{1, \dots, w\} | \lambda_l(t) > 0\}, \tag{30}$$

then $x^*(t)$ *is a strong local maximizer of* $(P)_t$.

Proof. See [20].

Using this result for the parametric problem $(P)_t$, we can formulate now sufficiency conditions for the problem (P).

Theorem 6. *Let* (x^*, u^*) *be feasible for the problem* (P) *and let the functions* y^α $(\alpha = 1, \dots, m)$ *and* λ_l $(j = 1, \dots, w)$*, as well as the matrix functions* Q^α $(\alpha = 1, \dots, m)$ *be chosen in such way that for* Λ *and* h_l, $l = 1, \dots, v$ *assumptions (A1)–(A6) of proposition 4 are satisfied. Moreover if*

$$-y^\alpha_{t_\alpha}(t) = \operatorname{grad}_\xi \left(\mathcal{H}(t, \xi, y(t)) + \sum_{j=1}^{l} \lambda_j(t) h_l(t, \xi) \right)_{\xi = x^*(t)} \tag{31}$$

$$\textit{with} \quad \lambda_l(t) \geq 0, \quad \lambda_l(t) h_l(t, x^*(t)) = 0 \quad \text{on} \quad \bar{\Omega}_i, \tag{32}$$

$$Q^\alpha_{t_\alpha}(t)[x^*_{t_\alpha}(t) - \operatorname{grad}_{\eta_\alpha} \mathcal{H}(t, x^*(t), y(t))] = 0 \quad \textit{on} \quad \bar{\Omega}_i, \tag{33}$$

hold and each matrix $M(t)$,

$$\begin{aligned} M(t) \in\ & \partial^2_{\xi\xi} \mathcal{H}(t, x^*(t), y(t)) + Q^\alpha_{t_\alpha}(t) + \partial^2_{\xi\eta^\alpha} \mathcal{H}(t, x^*(t), y(t)) Q^\alpha(t) \\ & + Q^\alpha(t) \partial^2_{\eta^\alpha \xi} \mathcal{H}(t, x^*(t), p(t)) \\ & + Q^\alpha(t) \partial^2_{\eta^\alpha \eta^\beta} \mathcal{H}(t, x^*(t), p(t)) Q^\beta(t) \\ & + \sum_{l \in I^+(t)} \lambda_l(t) d^2_{\xi\xi} h_l(t, x^*(t)) \end{aligned} \tag{34}$$

is negative definite on $W^+(t)$ *for all* $t \in \bar{\Omega}$*, then* (x^*, u^*) *is a strong local minimizer of the problem* (P).

Proof. See [20].

We apply this second order sufficiency conditions to a well known problem in calculus of variations.

Example (The problem of minimal k–energy) It is well known that the one-dimensional variational problem

$$\mathcal{E}_1(x) = \int_a^b x(t) \sqrt{(1 + |\operatorname{grad}_t x(t)|^2)}\, dt \rightarrow \text{Min!} \tag{35}$$

leads to rotationally symmetric minimal surfaces bounded by two coaxial circles. Fixing the length of the curve this problem can also be formulated as follows: What are the curves, connecting two given points and having the lowest center of gravity?

In this formulation also the n-dimensional analogue of the problem make sense: Which surfaces with fixed boundary conditions minimize the potential energy

$$\mathcal{E}_k(x) = \int_\Omega x^k(t) \sqrt{(1 + |\text{grad}_t x(t)|^2)}\, dt \ ? \tag{36}$$

In the case $k = 0$ this is the problem of minimal surfaces which was widely investigated, e.g., in [10, 17].

We study of the problem for $k \geq 1$ and $n \geq 1$. The essential property of this multidimensional variational problem is the weak ellipticity of the integrand for $k \geq 1$. Because of this weak ellipticity we can not expect that the solutions are analytic and therefore an essential question is in which cases the (non-smooth) solution of the corresponding Euler-equation

$$x^*(t) = \sqrt{\frac{k}{n-1}}[t_1^2 + \cdots + t_n^2]^{1/2} \tag{37}$$

minimizes the functional (36) under fixed boundary conditions

$$x(s) = \sqrt{\frac{k}{n-1}}\rho \quad s \in \partial\Omega, \tag{38}$$

where $\Omega := \{t \in \mathbb{R}^n | t_1^2 + \cdots + t_n^2 \leq \rho^2\}$. Precisely we formulate the following problem (P):

$$\mathcal{E}_k(x) = \int_\Omega x^k(t) \sqrt{(1 + |\text{grad}_t x(t)|^2)}\, dt \ \rightarrow \ \text{Min!} \tag{39}$$

with respect to $x \in W^1_\infty(\Omega)$ *with*

$$x(t) \geq 0 \quad \text{on} \quad \bar{\Omega} \subset \mathbb{R}^n, \tag{40}$$

$$x(s) = \sqrt{\frac{k}{n-1}}\rho \quad s \in \partial\Omega \tag{41}$$

This problem was treaded by Schoen, Simon, and Yau [27] and Dierkes [5] in the class of functions of bounded variations with the following results:

A. For $k + n < 4 + \sqrt{8}$ the extremal (37) has no minimizing properties [27].
B. For $k + n \geq 4 + \sqrt{8}$ the extremal (37) is weak locally minimizing, i.e., the second order variation of the functional (36) is positive, see [5].
C. For $k \geq 2$, $n \geq 3$ and $k + n \geq 7$ or for $k \geq 1$, $n \geq 2$ and $k + n \geq 8$ the extremal (37) is a global minimizer of (P), see [5].

An open question was, if the extremal (42) has strong local minimizing properties for $k + n \geq 4 + \sqrt{8}$. Using Theorem 5 we prove the following propositions:

Proposition 7. *For* $k = 1$, $n \geq 6$ *the solution (35) of the Euler - equation is a strong local minmizer for (P) in* $\bar{\Omega}$.

Proposition 8. *For* $n \geq 2$, $n + k \geq 4 + \sqrt{8}$ *the solution (35) is a strong local minimizer for* (P) *in* $\bar{\Omega} \backslash \mathcal{K}_\delta(0)$ *for each* $\delta > 0$.

The conditions of Theorem 6 for a strong local minimizer for (P) can be verified by a suitable quadratic part Q^{α} in (21):

$$Q^{\alpha}(t) := \gamma(n,k)t_{\alpha}(t_1^2+\cdots+t_n^2)^{(k/2)-1}. \tag{42}$$

4 Conclusion

It is well known from one-dimensional control theory that especially second order sufficiency theorems are needed for the proof of stability results. Up to now for the case of multidimensional control problems of type (P) such results can be shown only under too restrictive assumptions. It should succeed to verify sufficiency conditions which a closer to the necessary conditions and to make to difference between necessary and sufficient optimality conditions smaller.

References

1. H. W. Alt: *Lineare Funktionalanalysis.* Springer, New York – Berlin, 1992
2. C. Carathéodory: *Vorlesungen über reelle Funktionen.* Chelsea, New York, 1968.
3. L. Cesari *Optimization with partial differential equations in Dieudonne-Rashevski form and conjugate problems.* Arch. Rat. Mech. Anal. **33**(1969), pp. 339–357.
4. F. H. Clarke, and R. B. Vinter: *Local optimality conditions and Lipschitz solutions to the Hamilton-Jacobi equation.* SIAM J. Control and Optimization **21** (1983), 856–870.
5. U. Dierkes: *Über singuläre Lösungen gewisser mehrdimensionaler Variationsprobleme.* Habilitationsschrift, Universität des Saarlandes, Saarbrücken, 1989.
6. N. Dunford, J. T. Schwartz: *Linear Operator. Part I: General Theory*, Wiley-Interscience, New York, 1988.
7. I. Ekeland, and R. Temam: *Convex Analysis and Variational Problems.* North Holland Publ. Comp. and New York: Amer. Elsevier Publ. Comp., Inc., 1976.
8. R. V. Gamkrelidze: *Principles of Optimal Control Theory.* Plenum Press, New York, London, 1978.
9. B. Ginsburg, A. D. Ioffe: *The maximum principle in optimal control of systems governed by semilinear equations.* In: B. S. Mordukhovich; H. J. Sussmann (Eds.): Nonsmooth Analysis and Geometric Methods in Deterministic Optimal Control (IMA Volumes in Mathematics and its Applications 78), Springer, New York, Berlin, pp. 81–110, 1996.
10. E. Giusti: *Minimal surfaces and functions of bounded variations.* Birkhäuser, Boston-Basel-Stuttgart. 1984.
11. A. D. Ioffe, V. M. Tichomirov: *Theorie der Extremalaufgaben.* VEB Deutscher Verlag der Wissenschaften, Berlin, 1979.
12. R. Klötzler: *On Pontrjagins maximum principle for multiple integrals.* Beiträge zur Analysis 8 (1996), 67 -75.
13. R. Klötzler: *On a general conception of duality in optimal control.* Proceedings Equadiff 4, Prague 1977. 189–196.
14. R. Klötzler: *Starke Dualität in der Steuerungstheorie.* Math. Nachrichten **95** (1980), 253 – 263.
15. R. Klötzler, and S. Pickenhain: *Pontryagin's Maximum principle for multidimensional control problems.* International Series of Numerical Mathematics **111**, Birkhäuser, Basel (1993), pp. 21–30.

16. P.-L. Lions: *Generalized solutions of Hamilton-Jacobi equations.* Research Notes in Mathematics **69**, Pitman, Boston, 1982.
17. U. Massani and M. Miranda: *Minimal surfaces of codimension one.* North Holland, Amsterdam-New York-Oxford, 1984.
18. C. B. Morrey: *Multiple Integrals in the Calculus of Variations.* (Grundlehren 130), Springer, Berlin, Heidelberg, New York, 1966.
19. S. Pickenhain: *Zum Vergleich verschiedener Dualitätsbegriffe aus der Sicht der Steuerungstheorie.* Z. Anal. Anw. **7**, no. 3 (1988), 277–285.
20. S. Pickenhain, and K. Tammer: *Sufficient conditions for local optimality in multidimensional control problems with state restrictions.* Z. Anal. Anw. **10**, no. 3 (1991), 397 - 405.
21. S. Pickenhain: *Beiträge zur Theorie mehrdimensionaler Steuerungsprobleme.* Habilitationsschrift, Universität Leipzig, 1992.
22. S. Pickenhain: *Sufficiency conditions for weak local minima in multidimensional optimal control problems with mixed control-state restrictions.* Z. Anal. Anw. **11**, no. 4 (1992), 559–568.
23. S. Pickenhain, M. Wagner: *Critical points in relaxed deposit problems.* Proceedings of the conference \Calculus of variations and related topics, Haifa 1998", Pitman Research Notes in Mathematics,1999.
24. S. Pickenhain, M. Wagner: *Pontryagin's principle for state-constrained control problems governed by a first-order PDE system.* JOTA, Vol. 107 (2) (2000), 297–330.
25. S. Pickenhain, M. Wagner: *Minimizing sequences in class-qualified deposit problems.* BTU Cottbus, Preprint-Reihe Mathematik M-06/2001. (submitted)
26. H. Rund: *Pontrjygin functions for multiple integral control problems.* J. Optim. Theory Appl. 18 (1976), 511–520.
27. R. Schoen, L. Simon and S. T. Yau: *Curvature estimates for minimal hypersurfaces,* Acta Math. 134(1975), 276–288.
28. R. B. Vinter, and P. Wolenski: *Hamilton-Jacobi theory for problems with data measurable in time.* SIAM J. Control and Optimization **28** (1990), 1404–1419.
29. M. Wagner: *Pontryagin's maximum principle for Dieudonné-Rashevsky type problems involving Lipschitz functions.* Optimization **44** (1999), 165–184.
30. V. Zeidan: *Sufficient conditions for the generalized problem of Bolza.* Trans. Am. Math. Soc. **275** (1983), 561–586.
31. V. Zeidan: *First- and second-order sufficient conditions for optimal control and the Calculus of Variations.* Applied Mathematics and Optimization **11** (1984), 209–226.
32. V. Zeidan: *Extendend Jacobi sufficiency criterion for optimal control.* SIAM J. Control and Optimization **22** (1984), 294–301.

*Optimal Control Problems for the Nonlinear Heat Equation

Karsten Eppler and Fredi Tröltzsch

[1] Fakultät für Mathematik, Technische Universität Chemnitz, Germany
[2] Fachbereich Mathematik, Technische Universität Berlin, Germany

Abstract Some aspects of numerical analysis are surveyed for the optimal control of the nonlinear heat equation. In the analysis, special emphasis is on second order sufficient optimality conditions. In particular, the case of pointwise state constraints is adressed. Moreover, a numerical technique of instantaneous control type is presented.

1 Control Problem and Optimality Conditions

The optimal control of heating and cooling processes belongs to the core of optimal control theory of parabolic equations. It covers most of the main difficulties of this theory but is not yet overlaid by the technicalities, which are typical for the optimization of other parabolic systems. Therefore, the study of heat control gives also good insight in the methods for the control of other equations such as Burgers equation [14, 23], fuel ignition models [16], Navier-Stokes equations [9, 10, 13], or phase-field models [11, 12].

We report on some applications of control theory to the optimal cooling of steel profiles, which has already been considered in a sequence of papers [20, 24, 30, 31]. Related issues were discussed in [5, 8]. We present the results of our applied research in our second paper in this volume. Here, we give a brief survey on parts of the theory of optimization in semilinear parabolic equations. In real applications to cooling steel, the equation is quasilinear and the results of the semilinear case cannot be applied. However, the study of semilinear problems provides good information on the effects, which should be expected for quasilinear equations as well. To remain simple in the presentation, we begin our short course with the following optimal control problem:

$$\text{(OC)} \quad \min J(y,u) = \beta/2 \int_0^T \int_\Omega (y(x,t) - y_d(x,t))^2 dx dt \qquad (1)$$

$$+ \gamma/2 \int\int_\Omega (y(x,T) - y_\Omega(x))^2 \, dx + \nu/2 \int_0^T \int_\Gamma u(x,t)^2 dS_x dt$$

subject to the heat equation with nonlinear boundary condition

$$\begin{aligned} \frac{\partial y}{\partial t} &= \Delta y && \text{in } Q, \\ y(x,0) &= y_0(x) && \text{in } \Omega, \\ \frac{\partial y}{\partial n} &= b(x,t,y,u) && \text{in } \Sigma, \end{aligned} \tag{2}$$

and subject to the control constraints

$$u_a \le u(x,t) \le u_b \tag{3}$$

to be satisfied a.e. on Σ. Let us consider the state y as the temperature distribution in the bounded domain $\Omega \subset \mathbb{R}^n$ ($n = 1,2,3$ in the applications), while u is the control, acting on $\Sigma = \Gamma \times (0,T)$, where Γ denotes the boundary of Ω and is supposed to be of class $C^{1,s}$.

The control may have various meanings. For instance, it can denote the outer temperature, it may express the intensity of cooling or heating by some surrounding medium, and it might stand for some energy supply. Let us adopt for a while the first view. Then we search an optimal heating strategy $\bar{u} = \bar{u}(x,t)$ such that, starting from the initial temperature y_0, the associated temperature in $Q = (0,T) \times \Omega$ evolves in an optimal way, expressed by the functional J in (1).

Here, $y_d \in L^\infty(Q)$ denotes a desired trajectory of temperature, which has to be followed as closely as possible, and $y_\Omega \in L^\infty(\Omega)$ is a desired final temperature distribution. The constants β, γ are positive weights, and $\nu > 0$ can be interpreted as cost of the control u. Moreover, constant bounds $u_a < u_b$ are given.

We assume $y_0 \in C(\bar{\Omega})$. The function $b = b(x,t,y,u)$ is assumed to be of class C^2 with respect to $(y,u) \in \mathbb{R}^2$ and measurable w.r. to $(x,t) \in Q$ (the other variables fixed, respectively). In general, b and its first and second order derivatives must satisfy certain Lipschitz conditions on bounded sets with respect to (y,u) and the partial derivative of b with respect to y, denoted by b_y, is assumed to be nonpositive. We refer, for instance, to [3], [27]. To shorten the presentation and to have direct access to the literature we assume for simplicity that

$$b_y(x,t,y,u) \le 0 \tag{4}$$

holds a.e. on $Q \times \mathbb{R}^2$, $b(x,t,y,u)$, $b'(x,t,y,u)$, $b''(x,t,y,u)$ are uniformly bounded on $Q \times \mathbb{R}^2$ and uniformly Lipschitz with respect to (y,u) on $Q \times \mathbb{R}^2$. Here, b' and b'' stand for the gradient and the Hessian matrix of the real function b with respect to $(y,u) \in \mathbb{R}^2$. Then the parabolic problem (2) ist well-posed. The assumptions on second order derivatives are not necessary for this. They are needed to establish second order optimality conditions. In the sequel, we fix constants $p > n+1$, $q > n/2 + 1$ and introduce the state space

$$Y = \{y \in W(0,T) \,|\, y_t - \Delta y \in L^q(Q),\ \frac{\partial y}{\partial n} \in L^p(\Sigma),\ y(0) \in C(\bar{\Omega})\}.$$

For the definition of $W(0,T)$ we refer to [25] and the concrete choice in [27]. Y is known to be continuously embedded in $C(\bar{Q})$. Moreover, we define the set of admissible controls $U_{ad} = \{u \in L^\infty(\Sigma) \,|\, u_a \le u(x,t) \le u_b \quad \text{a.e. on } \Sigma.\}$

Theorem 1 ([3,27]). *Let* b *satisfy the assumptions stated above and let a control* $u \in U_{ad}$ *be given. Then the parabolic initial boundary value problem (2) has a unique solution* $y = y(u)$ *in* Y. *There is a positive constant* K *such that* $\|y(u)\|_{C(\bar{Q})} \leq K$ *holds uniformly for all* $u \in U_{ad}$.

The next question concerns the solvability of the optimal control problem, i.e., the existence of a globally optimal control $\bar{u}$ with associated optimal state $\bar{y} = y(\bar{u})$. To give a practicable answer, we need an additional property of b.

Theorem 2. *Suppose that*

$$b(x,t,y,u) = b_1(x,t,y) + b_2(x,t,y)u, \tag{5}$$

i.e., b *is affine-linear with respect to* u. *Then the optimal control problem (OC) admits at least one (globally) optimal control* $\bar{u}$.

The well known proof relies on weak compactness of U_{ad} in $L^p(\Sigma)$, because this permits to select a minimizing subsequence of elements $u_n \in U_{ad}$ such that $u_n \rightharpoonup \bar{u}$ in $L^p(\Sigma)$. By uniform boundedness of $\{y(u_n)\}_{n=1}^{\infty}$, we can select a subsequence of $b_n(x,t) = b(x,t,u_n,y(u_n))$ converging weakly to some function $\bar{b}$ in $L^p(\Sigma)$. Consequently, we have w.l.o.g. $y_n \to \bar{y}$ in $C(\bar{Q})$. The additional assumption (5) is needed to guarantee that $b(\cdot,\cdot,y_n,u_n) \rightharpoonup \bar{b} = b(\cdot,\cdot,\bar{y},\bar{u})$ so that finally $\bar{y} = y(\bar{u})$ holds.

In the numerical analysis, the consideration of global solutions is not the only way to deal with the problem (OC). Iterates, generated by numerical algorithms, will in general converge to *local* solutions only. Hence an alternative way is to consider a triplet $(\bar{y},\bar{u},\bar{p})$ that satisfies the first order necessary conditions and to ensure local optimality by second order sufficient conditions.

Theorem 3 ([3,27]). *Let* $\bar{u}$ *be a locally optimal control of (OC) with associated state* $\bar{y} = y(\bar{u})$. *Then a unique adjoint state* $\bar{p} = \bar{p}(x,t)$ *exists in* $W(0,T)$ *such that the adjoint equation*

$$\begin{aligned} -\frac{\partial p}{\partial t} &= \Delta p + \beta\,(\bar{y} - y_d), \\ p(x,T) &= \gamma\,(\bar{y}(x,T) - y_\Omega(x)), \\ \frac{\partial p}{\partial n} &= b_y(x,t,\bar{y},\bar{u})\,p \end{aligned} \tag{6}$$

is satisfied together with the variational inequality

$$\int_0^T\!\!\int_\Gamma (\nu\bar{u} + b_u(\bar{y},\bar{u})\,\bar{p})(u - \bar{u})\,dS\,dt \geq 0 \quad \forall u \in U_{ad}. \tag{7}$$

This result follows, for instance, from the more general *Pontryagin maximum principle* proved in [27], [3] or directly from [28]. The adjoint state $\bar{p}$ is shown to be in $C(\bar{Q})$. Let us discuss the particular case, where $U_{ad} = L^\infty(\Sigma)$ (unrestricted control) and b satisfies (5). Then (7) implies $\bar{u} = -\nu^{-1} b_2(\cdot, \bar{y})\bar{p}$, and u can be eliminated in (2), (6) to obtain a forward-backward coupled system of two parabolic equations for y and p. This system might be solved, for instance, by the Newton method. It may have multiple solutions.

One of the basic difficulties for the numerical solution is the enormous number of variables the system has after discretization. To give an intuitive estimate for this, assume that $\Omega \subset \mathbb{R}^2$ is the unit square with each edge discretized by 100 node points. Adopt the same simple discretization for the time interval $(0, T)$. Then we have to process $2 \cdot 10^6$ variables. For $\Omega \subset \mathbb{R}^3$ this number increases considerably.

Nevertheless, solving the optimality system (2), (6), (7) for the unconstrained case $U_{ad} = L^\infty(\Sigma)$ is one of the core procedures to solve the constrained case as well.

Formally, Theorem 3 can be derived in the following intuitive way. Define the Lagrange function

$$L = L(y, u, p) = J(y, u) - \int_Q (y_t - \Delta y)p \, dxdt - \int_\Sigma (\frac{\partial y}{\partial n} - b(y, u))p \, dSdt. \quad (8)$$

According to well known Lagrange multiplier rules of mathematical programming in Banach spaces, $(\bar{y}, \bar{u})$ must satisfy, together with $\bar{p}$, the relations

$$L_y(\bar{y}, \bar{u}, \bar{p})y = 0$$

for all $y \in Y$ with $y(0) = 0$ and

$$L_u(\bar{y}, \bar{u}, \bar{p})(u - \bar{u}) \geq 0 \qquad \forall u \in U_{ad}.$$

After some transformations including integration by parts and Greens formulas, these relations imply (6) and (7).

Assume next that $\bar{u} \in U_{ad}$, $\bar{y} = y(\bar{u})$, and $\bar{p}$ satisfy the optimality system (2), (6), (7). What condition can ensure $\bar{u}$ to be optimal, at least locally? To this end, second order sufficient optimality conditions (SSC) can be invoked. We need for their formulation the second order Fréchet-derivative of L w. r. to $(y, u) \in Y \times L^\infty(\Sigma)$,

$$L''(\bar{y}, \bar{u}, \bar{p})[y, u]^2 = \beta \int_Q y^2 dxdt + \gamma \int_\Omega y^2(\cdot, T) \, dx + \int_\Sigma (\nu u^2 + b_{yy}(\bar{y}, \bar{u}) \, \bar{p} \, y^2) \, dSdt.$$

Theorem 4 ([26]). *(SSC) Suppose that* $\bar{u} \in U_{ad}$ *and* $(\bar{y}, \bar{u}, \bar{p})$ *satisfy (2),(6),(7). Assume the existence of* $\delta > 0$ *such that*

$$L''(\bar{y}, \bar{u}, \bar{p})[y, u]^2 \geq \delta \, \|u\|^2_{L^2(\Sigma)}, \quad (9)$$

holds for all $u \in L^\infty(\Sigma)$, $y \in Y$ *satisfying the linearized equation*

$$\begin{aligned} \frac{\partial y}{\partial t} &= \Delta y, \\ y(0,x) &= 0, \\ \frac{\partial y}{\partial n} &= b_y(\bar{y},\bar{u})\,y + b_u(\bar{y},\bar{u})\,u. \end{aligned} \tag{10}$$

Then there exist constants $\varepsilon > 0$, $\sigma > 0$ *such that the quadratic growth condition*

$$J(y,u) \geq J(\bar{y},\bar{u}) + \sigma \|u - \bar{u}\|^2_{L^2(\Sigma)} \tag{11}$$

holds for all $u \in U_{ad}$, $y = y(u)$ *such that* $\|u - \bar{u}\|^2_{L^\infty(\Sigma)} < \varepsilon$. *Hence* $\bar{u}$ *is locally optimal in the norm of* $L^\infty(\Sigma)$.

Remarks: (i) If b admits the form (5), then $L^p(\Sigma)$ can be substituted here for $L^\infty(\Sigma)$. This is an essential advantage, since $\|u - \bar{u}\|_{L^\infty(\Sigma)} < \varepsilon$ requires more or less that jumps of $\bar{u}$, if there are any, must be reproduced by u.

(ii) The second order sufficient conditions can be relaxed by considering active sets, [26]. Then $u = 0$ can be assumed in (10) on so-called strongly active sets.

The theory of (SSC) for problems of the type (OC) is well understood. This refers also to the elliptic case, see [4]. The situation is much more complicated, if state constraints are added. In the case of pointwise state constraints the theory is widely open. For elliptic problems, satisfactory results were obtained in two-dimensional domains Ω, [4], while for parabolic problems the one-dimensional case is considered best, [26].

Let us illustrate by a simple example, where the main difficulty appears. Regard, for instance, (OC) with the additional pointwise state constraint

$$y(x_1,t) - y_2(x_2,t) \leq c \qquad \forall t \in [0,T]. \tag{12}$$

Constraints of this type will occur in our application to cooling steel. They are well formulated, since the choice of Y guarantees $y \in C(\bar{Q})$, hence the functions $y(x_i,t)$ are well defined and continuous on $[0,T]$. In the theory of optimality conditions, the state constraint (12) is considered by another Lagrange multiplier μ, which is a monotone increasing function of bounded variation on [0,T]. We have to introduce the extended Lagrange function

$$\tilde{L}(y,u,p,\mu) = L(y,u,p) + \int_0^T (y(x_1,t) - y(x_2,t))\, d\mu(t).$$

The associated theory of first order *necessary* conditions is well developed, see [3], [27]. The main difficulty in proving *sufficient* conditions is the appearence of measures like $d\mu$ extending the right hand side of the adjoint equation (6). This makes the adjoint $\bar{p}$ state less regular. Therefore, in the general case we do not have the important property $p \in L^\infty(Q)$, which is helpful to estimate $L''(\bar{y},\bar{u},\bar{p},\bar{\mu})[y,u]^2$ with respect to (y,u) in the appropriate norms.

2 Numerical Methods

The numerical solution of optimal control problems for semilinear elliptic and parabolic equations has made considerable progress. Various methods were discussed, and the numerical results provide essential contributions to the fast developing field of large scale optimization. To give the reader an access to further study, we quote [1, 2, 6, 7, 11, 12, 15, 17–19, 21, 22].

Elliptic problems in two-dimensional domains Ω and parabolic problems in domains of dimension one can be solved in short time, since the number of variables after discretization of the problem is still moderate. If the dimension of Ω is larger than one, then the solution of parabolic problems is still time consuming. However, they can be treated succesfully. For the solution of parabolic problems in two-dimensional domains we refer to [6], [11]. One of the favorite techniques is that of (S)equential (Q)uadratic (P)rogramming. Let us briefly describe the main idea for the classical SQP method, which reduces the solution of the nonlinear problem (OC) to a sequence of quadratic optimal control problems with *linear* equation.

Suppose that (y_n, u_n, p_n) has already been determined. Then the next iterate $(y, u) = (y_{n+1}, u_{n+1})$ is found as the solution of the linear-quadratic problem

$$\text{(QP)} \quad \min \quad J'(y,u)[y - y_n, u - u_n] + \frac{1}{2} L''_{(y,u)}(y_n, u_n, p_n)[y - y_n, u - u_n]^2$$

subject to $u \in U_{ad}$ and

$$\begin{aligned} \frac{\partial y}{\partial t} &= \Delta y, \\ y(x,0) &= y_0(x), \\ \frac{\partial y}{\partial n} &= b_y(y_n, u_n)(y - y_n) + b_u(y_n, u_n)(u - u_n) + b(y_n, u_n). \end{aligned}$$

The new Lagrange multiplier p_{n+1} is obtained from (6), where (y_{n+1}, u_{n+1}) is substituted for $(\bar{y}, \bar{u})$. Under natural assumptions, among them second order sufficient conditions are most essential, this method locally converges q-quadratically to $(\bar{y}, \bar{u}, \bar{p})$, if considered in the infinite dimensional setting [6], [7], [29]. For instance, the second order assumptions (2), (6), (7), (9), (10) of Theorem 4 can be used for this purpose. For semilinear elliptic equations, the convergence analysis was presented in [32]. Discretizing the problem, various approximation errors influence the performance of the method. Moreover, modifications of the standard SQP method can be numerically more effective.

Our computational experience shows that the SQP method converges very fast, i.e., only a few steps are needed to gain high precision. However, each single step of the method can be very expensive, in particular for domains of higher dimension. If the parabolic equation is quasilinear rather than semilinear, then the situation is even more complicated.

Therefore, in our problem of cooling steel we did not apply the SQP method. First we applied a method of feasible directions, [20], [24], [30]. Later, a suboptimal technique was implemented – a method of instantaneous control type. These techniques are considerable cheaper than SQP methods and have been sucessfully applied to find suboptimal solutions in the control of fluid flows. We only quote Hinze [13] and refer the reader to the extensive references therein. We also mention [14] for the case of the Burgers equation.

As a preparation of our report on optimal cooling of steel in this volume, here we explain the main idea of our technique for the following simplified control problem with state constraints. Let points $x_i \in \Omega$ be given fixed, $i = 1,2,3$, and assume that $\Gamma = \cup_{i=1}^{P} \bar{\Gamma}_i$, where $\{\Gamma_i\}_{i=1}^{P}$ is a partition of Γ into pairwise disjoint relatively open subsets. Moreover, consider an equidistant partition of $[0,T]$ into subintervals $I_k = (t_{k-1}, t_k]$, $k = 1(1)K$, $0 = t_0 < t_1 < \cdots < t_{K-1} < t_K = T$. Define $\Sigma_{ik} = \Gamma_i \times I_k$. The partition of Γ and $[0,T]$ into subsets should not be viewed as a result of discretization. In our application, it reflects the associated *technical construction*. In cooling steel, Γ_i is the zone influenced by spray nozzle i, and the time interval I_k is associated with passing the cooling segment k. The control function $u = u(x,t)$ is assumed to be constant on Σ_{ik}, i.e., $u(x,t) = u_{ik}$ on Σ_{ik}. Our simplified control problem "steel" is

(OCS) $$\min \; y(x_0, T)$$

subject to

$$\begin{aligned} \frac{\partial y}{\partial t} &= \Delta y && \text{in } Q, \\ y(x,0) &= y_0(x) && \text{in } \Omega, \\ \frac{\partial y}{\partial n} &= u_{ik}\, \alpha(x,y)[y_{fl} - y] && \text{in } \Sigma_{ik}, \end{aligned} \tag{13}$$

$$y(x_1, t_k) - y(x_2, t_k) \le c, \; k = 1(1)K,$$

$$0 \le u_{ik} \le 1, \qquad i = 1(1)P, \; k = 1(1)K.$$

In this setting, $\alpha = \alpha(x,y)$ is the heat exchange coefficient and y_{fl} is the temperature of the cooling fluid. We assume that α is sufficiently smooth with respect to $y \in \mathbb{R}$. The main idea of instantaneous control is as follows: First minimize $y(x_0, t_1)$, i.e., find optimal controls $\bar{u}_{i1}$ on the short time horizon $[t_0, t_1]$. Next insert $y(x, t_1)$ as a new initial condition in (13), to optimize next the process on $[t_1, t_2]$. In this way, we have to solve K single optimal short horizon control problems with P control variables $u_{1k}, \ldots, u_{Pk}$, each. However, the problems are nonlinear, since the boundary condition is nonlinear (notice that $\alpha = \alpha(x,y)$ depends on the state y). Even if the boundary condition would be linear with respect to y, i.e., $\alpha = \alpha(x)$ (or $\alpha = \alpha(x,t)$), the mapping $u \mapsto y$ is still nonlinear (bilinear), because the product $y\,u$ appears in the boundary condition of (13).

We resolve this difficulty by several manipulations. First of all, we introduce the right hand side of the boundary condition in (13) as a new auxiliary control v, i.e., on $[0,T]$ we put

$$v_i(t) := u_i(t)\alpha(x, y(x,t))(y_{fl} - y(x,t)), \tag{14}$$

where $u_i(t)$ denotes the step function being equal to u_{ik} on I_k. From now on, we search controls $v_i(t)$ subject to the linear boundary condition

$$\frac{\partial y}{\partial n} = v_i(t) \quad \text{on } \Gamma_i,$$

$i = 1(1)K$. The $v_i(t)$ are approximated by step functions. To this aim, we consider partitions of $I_k = (t_{k-1}, t_k]$ into L smaller subintervals having equidistant length $\tau = (t_k - t_{k-1})/L$ and define

$$I_{kl} = (t_{k-1} + (l-1)\tau, t_{k-1} + l\tau) = (t_{k-1,l-1}, t_{k-1,l}),\ k = 1(1)K,\ l = 1(1)L.$$

Finally, these are the intervals, where we really apply the idea of instantaneous control. The optimization is started on $I_{11} = (t_0, t_0 + \tau)$ to obtain optimal values $\bar{v}_{i11}$, $i = 1(1)P$. Define $y_{01}(x) = y(x, t_0 + \tau)$ as the new initial temperature for I_{12}. Next the $\bar{v}_{i12}$ are determined, and we put $y_{02}(x) := y(x, t_0 + 2\tau)$. Proceeding in this way, linear short-time optimal control problems (OCS_{kl}) are solved for $k = 1(1)K$, $l = 1(1)L$,

(OCS_{kl}) $$\min\ y(x_0, t_{kl})$$

subject to

$$\begin{aligned} \frac{\partial y}{\partial t} &= \Delta y && \text{in } \Omega \times I_{kl}, \\ y(x, t_{k-1,l-1}) &= y_{k-1,l-1}(x) && \text{in } \Omega, \\ \frac{\partial y}{\partial n} &= v_i && \text{on } \Gamma_i \times I_{kl}, \end{aligned} \tag{15}$$

$$y(x_1, t_{k-1,l}) - y(x_2, t_{k-1,l}) \le c,$$

$$q_{ikl} \le v_i \le 0,\ i = 1(1)P.$$

The optimal controls of (OCS_{kl}) are denoted by $\bar{v}_{ikl}$. Moreover, we put

$$y_{k-1,l}(x) := y(x, t_{k-1,l}), \quad \text{if } l < L,$$

and

$$y_{k,0}(x) := y(x, t_{k-1,L}).$$

It remains to define the values q_{ikl}. We preselect some characteristic points $\hat{x}_i \in \Gamma_i$ (say midpoints of Γ_i in some sense) and regard formula (14) at $y = y_{k-1,l-1}(\hat{x}_i)$ with maximal control value $u = +1$. This should result in the minimum heat flux

$$q_{ikl} := 1 \cdot \alpha(\hat{x}_i, y_{k-1,l-1}(\hat{x}_i))(y_{fl} - y_{k-1,l-1}(\hat{x}_i)). \tag{16}$$

After having exhausted the whole interval $[0, T]$ by the optimization process, we compose the auxiliary controls $\bar{v}_{ikl}$ to suboptimal controls $\bar{u}_{ik}$, $i = 1(1)P$, $k = 1(1)K$, as follows: Motivated by (14), resolving for $u_i(t)$, we define

$$\begin{aligned} u^-_{kli} &= \bar{v}_{kli}/\alpha(y(\hat{x}_i, t_{k-1,l-1}))(y_{fl} - y(\hat{x}_i, t_{k-1,l-1})), \\ u^+_{kli} &= \bar{v}_{kli}/\alpha(y(\hat{x}_i, t_{k-1,l}))(y_{fl} - y(\hat{x}_i, \tilde{t}_{k-1,l})). \end{aligned}$$

Finally, the mean values

$$\bar{u}_{kli} = \frac{u^+_{kli} + u^-_{kli}}{2},$$

are taken to compose

$$\bar{u}_{ki} = \frac{\sum_{l=1}^{L} l\,\bar{u}_{kli}}{\sum_{l=1}^{L} l}, \qquad i = 1(1)P. \tag{17}$$

The principle of superposition can be used to efficiently generate the problems (OCS_{kl}). On I_{kl}, the solution y of (15) is represented in the form

$$y(x,t) = y_I(x,t) + \sum_{i=1}^{P} \nu_{ikl}\, y_i(x,t), \tag{18}$$

where y_I solves the heat equation subject to $\partial y_I/\partial n = 0$ and $y_I(x, t_{k-1,l-1}) = y_{k-1,l-1}(x)$, while the *response functions* y_i solve the heat equation on I_{kl} with homogeneous initial condition and boundary condition $\partial y_i/\partial n = \chi(\Gamma_i)$.

We notice that $y_i(x,t)$ does not depend on k and l, because $y_i(x,t) = z_i(x, t - t_{k-1,l-1})$ holds, where, for $i = 1(1)P$,

$$\begin{aligned} \frac{\partial z_i}{\partial t} &= \Delta z_i && \text{in } \Omega \times (0,\tau), \\ z_i(x,0) &= 0 && \text{on } \Omega, \\ \frac{\partial z_i}{\partial n} &= \chi(\Gamma_i) && \text{on } \Gamma_i \times (0,\tau). \end{aligned} \tag{19}$$

After having solved the P parabolic problems (19) at the beginning of the computations, the functions z_i can be taken to define y_i on all I_{kl}. In this way, (OCS_{kl}) is given by

$$\min \sum_{i=1}^{P} \nu_i\, z_i(x_0, \tau)$$

subject to

$$\sum_{i=1}^{P} \nu_i \left(z_i(x_1,\tau) - z_i(x_2,\tau)\right) \le c + y_I(x_2, t_{k-1,l}) - y_I(x_1, t_{k-1,l}),$$
$$q_{ikl} \le \nu_i \le 0, \; i = 1(1)P.$$

As the $z_i(x_j, t)$, j=1,2,3, have been determined at the beginning, only the values $y_I(x_2, t_{k-1,l})$, $y_I(x_1, t_{k-1,l})$, and q_{ikl} must be updated during the optimization process. This drastically reduces the number of PDE solves.

The application to the concrete example of cooling steel is based on the same type of ideas. However, we need some essential modifications since the heat equation will be nonlinear and the constraints are more complex.

The suboptimal method, despite of all its heuristics, delivered surprisingly precise results, [31].

References

1. N. Arada, J.-P. Raymond, and F. Tröltzsch. On an augmented Lagrangian SQP method for a class of optimal control problems in Banach spaces. *Submitted to Computational Optimization and Applications*, to appear.
2. M. Bergounioux and K. Kunisch. Primal-dual strategy for state-constrained optimal control problems. Submitted.
3. E. Casas. Pontryagin's principle for state-constrained boundary control problems of semilinear parabolic equations. *SIAM J. Control and Optimization*, 35:1297–1327, 1997.
4. E. Casas, F. Tröltzsch, and A. Unger. Second order sufficient optimality conditions for some state-constrained control problems of semilinear elliptic equations. *SIAM J. Control and Optimization*, 38(5):1369–1391, 2000.
5. J. Chen, H. W. Engl, G. Landl, and K. Zeman. Optimal strategies for the cooling of steel strips in hot strip mills. *Inverse Problems in Engineering*, 2:103–118, 1995.
6. H. Goldberg and F. Tröltzsch. On a SQP–multigrid technique for nonlinear parabolic boundary control problems. In W.W. Hager and P.M. Pardalos, editors, *Optimal Control: Theory, Algorithms and Applications*, pages 154–177, Dordrecht, 1997. Kluwer Academic Publishers B.V.
7. H. Goldberg and F. Tröltzsch. On a Lagrange-Newton method for a nonlinear parabolic boundary control problem. *Optimization Methods and Software*, 8:225–247, 1998.
8. W. Grever, A. Binder, H. W. Engl, and K. Mörwald. Optimal cooling strategies in continuous casting of steel with variable casting speed. *Inverse Problems in Engineering*, 2:289–300, 1997.
9. M. Gunzburger and S. Manservisi. The velocity tracking problem for Navier-Stokes flows with bounded distributed controls. *SIAM J. Control and Optimization*, 37:1913–1945, 1999.
10. M. Gunzburger and S. Manservisi. Analysis and approximation of the velocity tracking problem for Navier-Stokes flows with distributed controls. *SIAM J. Numer. Analysis*, To appear.
11. M. Heinkenschloss and E. W. Sachs. Numerical solution of a constrained control problem for a phase field model. In *Control and Estimation of Distributed Parameter Systems*, volume 118 of *Int. Ser. Num. Math.*, pages 171–188. Birkhäuser-Verlag, 1994.
12. M. Heinkenschloss and F. Tröltzsch. Analysis of the Lagrange-SQP-Newton method for the control of a phase field equation. *Control and Cybernetics*, 28(2):178–211, 1999.
13. M. Hinze. *Optimal and instantaneous control of the instationary Navier-Stokes equations*. Habilitation thesis, TU Berlin, 1999.
14. M. Hinze and S. Volkwein. Instantaneous control for the Burgers equation: Convergence analysis and numerical implementation. Tech. Report, Spezialforschungsbereich Optimierung und Kontrolle 170, Inst. f. Mathematik, Karl-Franzens-Universität Graz, 1999.
15. K. Ito and K. Kunisch. Augmented Lagrangian-SQP methods for nonlinear optimal control problems of tracking type. *SIAM J. Control and Optimization*, 34:874–891, 1996.
16. A. Kauffmann. *Optimal control of the solid fuel ignition model*. Thesis, TU Berlin, 1998.
17. C. T. Kelley and E. Sachs. Multilevel algorithms for constrained compact fixed point problems. *SIAM J. Sci. Comput.*, 15:645–667, 1994.
18. C. T. Kelley and E. Sachs. Solution of optimal control problems by a pointwise projected Newton method. *SIAM J. Control and Optimization*, 33:1731–1757, 1995.
19. C. T. Kelley and E. Sachs. A trust region method for parabolic boundary control problems. *SIAM J. Optimization*, 9:1064–1081, 1999.

20. R. Krengel, R. Standke, F. Tröltzsch, and H. Wehage. Mathematisches Modell einer optimal gesteuerten Abkühlung von Profilstählen in Kühlstrecken. Preprint 98–6, TU Chemnitz, Fakultät für Mathematik, 1996.
21. K. Kunisch and E. Sachs. Reduced SQP-methods for parameter identification problems. *SIAM Journal Numerical Analysis*, 29:1793–1820, 1992.
22. K. Kunisch and S. Volkwein. Augmented Lagrangian-SQP techniques and their approximations. *Contemporary Math. 209*, pages 147–159, 1997.
23. K. Kunisch and S. Volkwein. Control of Burgers equation by a reduced order approach using proper orthogonal decomposition. Preprint, Spezialforschungsbereich Optimierung und Kontrolle 138, Inst. f. Mathematik, Karl-Franzens-Universität Graz, 1998.
24. R. Lezius and F. Tröltzsch. Theoretical and numerical aspects of controlled cooling of steel profiles. In H. Neunzert, editor, *Progress in Industrial Mathematics at ECMI 94*, pages 380–388. Wiley-Teubner, 1996.
25. J. L. Lions. *Contrôle optimal de systèmes gouvernès par des équations aux dérivées partielles*. Dunod, Gauthier-Villars, Paris, 1968.
26. J.-P. Raymond and F. Tröltzsch. Second order sufficient optimality conditions for nonlinear parabolic control problems with state constraints. *Discrete and Continuous Dynamical Systems*, 6:431–450, 2000.
27. J.-P. Raymond and H. Zidani. Hamiltonian Pontryagin's principles for control problems governed by semilinear parabolic equations. *Applied Mathematics and Optimization*, 39:143–177, 1999.
28. F. Tröltzsch. *Optimality conditions for parabolic control problems and applications*, volume 62 of *Teubner Texte zur Mathematik*. B.G. Teubner Verlagsgesellschaft, Leipzig, 1984.
29. F. Tröltzsch. An SQP method for the optimal control of a nonlinear heat equation. *Control and Cybernetics*, 23(1/2):267–288, 1994.
30. F. Tröltzsch, R. Lezius, R. Krengel, and H. Wehage. Mathematische Behandlung der optimalen Steuerung von Abkühlungsprozessen bei Profilstählen. In K.H. Hoffmann, W. Jäger, T. Lohmann, and H. Schunck, editors, *Mathematik – Schlüsseltechnologie für die Zukunft, Verbundprojekte zwischen Universität und Industrie*, pages 513–524. Springer-Verlag, 1997.
31. F. Tröltzsch and A. Unger. Fast solution of optimal control problems in the selective cooling of steel. *ZAMM*, 81:447–456, 2001.
32. A. Unger. *Hinreichende Optimalitätsbedingungen 2. Ordnung und Konvergenz des SQP-Verfahrens für semilineare elliptische Randsteuerprobleme*. PhD thesis, Technische Universität Chemnitz, 1997.

Fast Optimization Methods in the Selective Cooling of Steel

Karsten Eppler and Fredi Tröltzsch

[1] Fakultät für Mathematik, Technische Universität Chemnitz, Germany
[2] Fachbereich Mathematik, Technische Universität Berlin, Germany

Abstract We consider the problem of cooling milled steel profiles at a maximum rate subject to given bounds on the difference of temperatures in prescribed points of the steel profile. This leads to a nonlinear parabolic control problem with state constraints in a 2D domain. The controls can admit values from continuous or discrete sets. A method of instantaneous control is applied to establish a fast solution technique. Moreover, continuous and discrete control strategies are compared, and conclusions are given from an applicational point of view.

1 INTRODUCTION

The selective cooling of steel profiles is an important part of the production process in steel mills. Intelligent future strategies aim to combine a reduction of temperature in the rolled profile with an equalization of its interior temperature distribution. An accelerated optimal cooling will reduce the amount of investment in cooling sections. Moreover, it is able to stabilize the interior structure of the steel during phase transitions. Reducing the temperature in the profile as uniformly as possible leads to a higher quality of the steel.

We believe that the intuition of engineers alone is not able to control this process. The mathematical tools of optimal control theory will be helpful to find optimal cooling strategies.

We have reported on this issue in a number of mathematical papers, for instance in Krengel et.al. [7] or Lezius and Tröltzsch [11], where a method of feasible directions was developed to solve the optimal control problem. The numerical tests confirmed the stability and reliability of the method. However, the computing time was high. Therefore, Tröltzsch and Unger [14] dealed with a very fast and precise suboptimal solution method, where, after discretization, the optimization is reduced to a sequence of low-dimensional linear programming problems. A similar problem was discussed by Landl and Engl [9]. In contrast to the setting in [14], where the intensity of the cooling spray nozzles can be chosen *continuously*, in [9] the cooling is controlled by switching on and off the nozzles. This *discrete* strategy seems to be more adequate for the technical process.

In this paper we investigate the application of continuous as well as of discrete control strategies to the model discussed in [14]. It is not realistic to solve the associated large scale mixed integer programming problem up to the optimal solution in the discrete case. This refers also to the continuous problem. Therefore, we decided to extend the suboptimal method of instantaneous control type from the case of continuously controllable nozzles to discrete 0-1-controls. We will show that the extension can be done in a quite simple and straightforward way:

The core of the suboptimal method of [14] consists of small scale linear programming problems, which have been solved by the simplex method. Here, we arrive at small scale linear *integer* programming problems, which are solved by appropriate methods. Combining this main idea with some special techniques to make the method work, we finally achieved computing times of the same order as for the continuous control strategies. This is important to allow interactive work of the engineer and - at least in principle - an online-control of the cooling process.

Obviously, the restriction to discrete strategies shrinks the set of feasible controls. It is quite natural that integer controls are less flexible than continuous ones. Therefore, in our numerical tests, we increased the number of spray nozzles to compensate for this. However, the numerical experience shows that increasing the number of nozzles alone is not the best solution. Using nozzles of smaller size turned out to be more helpful.

The selective cooling of steel profiles is only one of various applications of control theory in metallurgy. Other important issues are the continuous casting of steel, Engl, Langthaler and Mansellio [2], Grever [4], Laitinen and Neittaanmäki [8], Neittaanmäki [12], the firing of kilns, Leibfritz and Sachs [10], or the Laser hardening of steel, Hömberg and Sokołowski [6].

2 The Optimal Control Problem and Iterative Solution

A cooling line consists of a certain number of cooling segments, where water is sprayed on the surface of the hot steel profile. Each cooling segment is followed by a zone of air cooling equalizing the developed temperature differences. The basic scheme is shown in Figure 1.

In the cooling segments, a certain fixed number of spray nozzles is located in groups around the profile. There can be a sequence of groups in each cooling segment. To explain the mathematical model, let us regard one fixed cross section $\Omega \subset \mathbb{R}^2$ of the steel profile. We follow its run through the whole cooling line. This causes an internal time scheme for the reference domain Ω. The cross section Ω enters the first nozzle group of the first cooling segment at time $t_0 = 0$. Now the surface is sprayed on by the p nozzles of the first nozzle group. After leaving this group, Ω reaches the second one at time t_1. (Note that there is a small difference to the notation in [7], therein t_1 denotes the time for passing the first cooling segment.) After r steps, Ω has passed the first cooling segment. Now an area of air cooling follows. At time t_s the next cooling segment is entered. Finally, the cross section reaches the end of the last air cooling area at time $t_K = T$, where the profile has passed M zones of water or air cooling.

To shorten the presentation, we rely on the following simplifications: All cooling segments contain the same number r of nozzle groups with the same number p of nozzles. The time for passing any single nozzle group is equal along the whole cooling line. Moreover, the lengths of all cooling segments and air cooling areas are

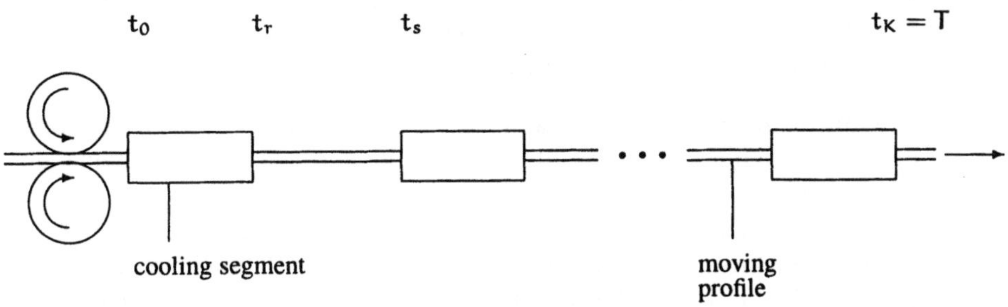

Figure 1. Scheme of a cooling section

assumed to be equal. Therefore, the time to pass an arbitrary segment is constant. These restrictions are not necessary for the computational technique to work. We adopt them only here to simplify the notation. The resulting discretization of the internal time is given by

$$0 = t_0 < t_1 < \cdots < t_r = rt_1 < \cdots < t_s = 2t_r < \cdots < t_K = Mt_r = T. \quad (1)$$

The heat conduction in axial direction is dominated by the heat exchange in Ω. Moreover, the steel profiles are very long, so that we can view them to be endless. This justifies to neglect the heat conduction in axial direction and to regard a 2D heat equation in our reference domain Ω. Related to this and to the real technical situation, we can assume that the intensity of any single spray nozzle is constant i.e., stationary with respect to the (outer) time.

We associate to each nozzle one part of the boundary $\Gamma = \partial\Omega$ standing for its zone of influence. This leads to a partition of Γ into disjoint subdomains Γ_i, $i = 1(1)p$. Denote by u_{ki} the cooling intensity of nozzle i in the group k, $k = 1(1)rM, i = 1(1)p$. Notice that this numbering covers some "phantom" nozzles in the air cooling areas. The numbers u_{ki} will be our *control variables*. In the case of *continuously* controllable nozzles we assume that the constraints $0 \leq u_{ki} \leq 1$ are imposed for all k and i. The value 0 stands for an inactive nozzle, while 1 characterizes a nozzle spraying with maximal intensity. This implies $u_{ki} \in \{0, 1\}$ as the set of admissible controls, if the nozzles can be only switched in (1) or off (0).

Adopting these notations, the mathematical model for the evolution of the temperature admits the following form, which is equivalent to that introduced in [7]: The temperature ϑ in the profile is obtained from the nonlinear heat conduction problem

$$\begin{aligned} c(y)\rho(y)\, y_t &= \operatorname{div}(\lambda(y)\,\operatorname{grad} y) && \text{in } Q, \\ \lambda(y)\,\partial_n y &= \sum_{i,k} u_{ki}\,\chi(\Sigma_{ki})\,\alpha(\cdot, y)(y_{fl} - y) && \text{in } \Sigma, \\ y(x,0) &= y_0(x) && \text{in } \Omega, \end{aligned} \quad (2)$$

where $Q = \Omega \times (0,T)$, $\Sigma = \Gamma \times (0,T)$, $\Sigma_{ki} = \Gamma_i \times (t_{k-1}, t_k)$, and $\chi(\Sigma_{ki})$ is the characteristic function of Σ_{ki}. In this setting, y_t and $\partial_n y$ denote the derivatives $\partial y/\partial t$ and $\partial y/\partial n$ with respect to the time and the outer normal n at Γ, respectively. Moreover, the following quantities are used:

- $y = y(x,t)$ denotes the temperature at $t \in [0,T]$ and $x \in \Omega$. T stands for the fixed terminal time. Ω is a two-dimensional domain, and y_{fl} is the temperature of the cooling fluid.
- $u_{ki} \in \mathbb{R}$ are the control variables mentioned above. Outside the cooling segments the controls u_{ki} are taken zero to model heat isolation in the areas of air cooling. This is expressed by the characteristic function $\chi(\Sigma_{ki})$ in the boundary condition of (2).
- The coefficients c, ρ, and λ are functions of y denoting heat capacity, specific gravity, and heat conductivity, respectively. The function $\alpha = \alpha(x,y)$ models the heat exchange coefficient. To find a good model for α is a nontrivial task. In a simplified setting for cooling of cylindrical rods of steel, this issue was investigated by Zurdel and Brennecke [15]. Moreover, we refer to Rösch, [13].
- Our cooling process starts with the entrance temperature $y_0 = y_0(x)$.

The coefficients c, ρ, λ do not have appropriate properties of smoothness and monotonicity to show the unique solvability of the heat conduction problem. Moreover, the modelling of material changes during the subsequent heating and cooling of the steel is still partially open. The form (2) of the heat equation seems to give only an approximate picture of the temperature changes. Therefore, we do not discuss the question of existence and uniqueness of a solution to (2). Moreover, our computational method will mainly work with linearized versions. For these problems, the existence of a unique solution corresponding to a given vector of controls $u = (u_{ki})$ is clear.

The restrictions on the control variables u_{ki} are alternatively given by

$$0 \le u_{ki} \le u_k \quad \text{or} \quad u_{ki} \in \{0, u_k\}, \tag{3}$$

$k = 1(1)rM$, $i = 1(1)p$, depending on wether we assume a continuous or discrete control strategy. Here, $u_k = 0$ holds for $k = (2j-1)r(1)2jr$ with $j = 1(1)M/2$ (air cooling) and $u_k = 1$ otherwise (cooling segment).

The main aim of the cooling process is to reduce the temperature in the domain. Certainly, this can be expressed in various ways. In our model, the temperature should be minimized in a selection of points $P_n \in \Omega$, $n = 1(1)N$, which characterize the hottest regions. In this way, the objective F is defined by the *linear* functional

$$F(y) = \sum_{n=1}^{N} a_n \, y(P_n, T) \tag{4}$$

with some positive weighting constants a_n.

In the model developed so far, most likely full intensity of all spray nozzles is optimal. However, this strategy is certainly wrong, since very large temperature differences would develop in Ω. This would amount to a low quality of steel and possibly lead to large deformations of the profile. Therefore, we include a finite number of pointwise state constraints in the optimal control problem to bound the temperature differences in Ω. Following [7], these constraints are given by

$$|y(R_\mu, t) - y(Q_\nu, t)| \leq \Theta_{\mu\nu}, \ \mu = 1(1)N_R, \ \nu = 1(1)N_Q. \tag{5}$$

In this setting, R_μ and Q_ν denote points from the closure of Ω. For instance, the minimization points $R_\mu := P_\mu$ can be chosen together with some comparison points Q_ν. The situation of our test example is shown in Figure 2, where the points P_ν and Q_μ are numbered as follows: Q_1 coincides with the origin. Following the boundary of the domain in mathematical positive sense, the next points are $Q_2, \ldots, Q_9$, P_3, P_2, P_1. In this way, Q_9 is located at the top, and P_1 is the lowest among the P_i.

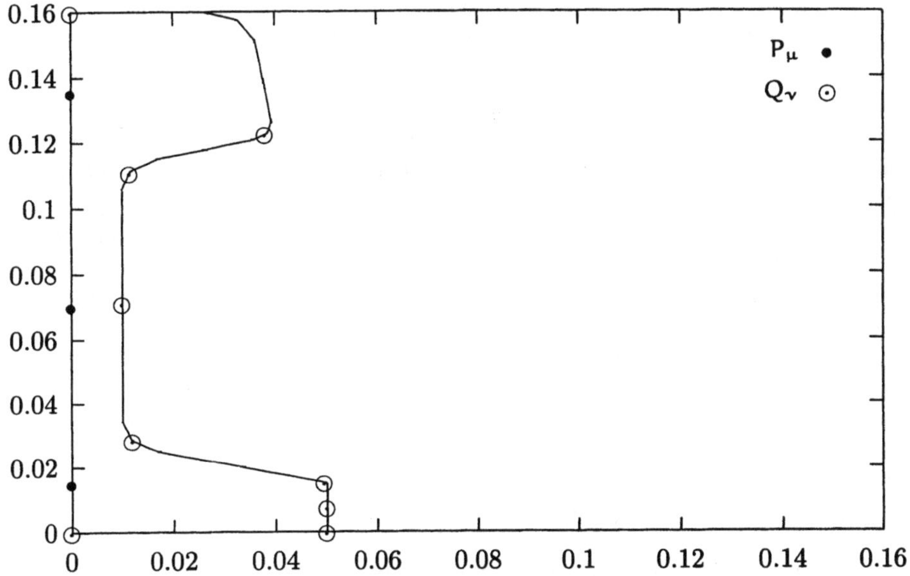

Figure 2. Points of Minimization and of Comparison

Now the definition of the control problem (P) is complete. For the continuous case, it reads as

$$\text{(P)} \qquad \min \ F(y) = \sum_{n=1}^{N} a_n y(P_n, T)$$

subject to the state equation

$$\begin{aligned} c(y)\rho(y)\, y_t &= \operatorname{div}(\lambda(y)\,\operatorname{grad} y) && \text{in } Q,\\ \lambda(y)\,\partial_n y &= \sum_{i,k} u_{ki}\,\chi(\Sigma_{ki})\,\alpha(\cdot,y)(y_{fl}-y) && \text{in } \Sigma,\\ y(x,0) &= y_0(x) && \text{in } \Omega, \end{aligned}$$

and subject to the constraints on control and state

$$\begin{aligned} |y(R_\mu,t)-y(Q_\nu,t)| \le \Theta_{\mu\nu},\quad \mu = 1(1)N_R,\ \nu = 1(1)N_Q,\\ 0 \le u_{ki} \le 1,\qquad i = 1(1)p,\quad k = 1(1)K. \end{aligned}$$

A more detailed motivation can be found, for instance, in [7], [11]. We refer also to these papers for details of the numerical solution of the nonlinear parabolic initial-boundary value problem (2) by a finite-element-multigrid method. Let us briefly recall for convenience the main ideas characterizing the optimization technique of [7], [11].

The optimal control problem is difficult in several respects. The state equation is nonlinear, pointwise constraints on the state are given along with constraints on the controls, and the domain Ω has a curved boundary. Besides the fact that the theory of optimal control problems for nonlinear distributed parameter systems with state-constraints is still far from being complete, the numerical solution is complicated. Readers interested in optimality conditions of first and second order for associated semilinear optimal control problems with state constraints are referred to our first paper in this volume.

Solving the heat equation by a sufficiently precise finite element multigrid method, a huge number of state variables appears in the discretized optimal control problem. However, compared with more academic problems discussed in literature, the technical circumstances of the cooling section show an essential advantage: The number of control variables is very low in comparison with the huge number of state variables. Therefore, we decided to use a direct method, where the controls appear as optimization variables, while the state equation is solved only for a certain number of basis controls. In [7], [11] an iterative method of feasible direction is developed. This algorithm proceeds as follows (below, the *control* u stands for the vector (u_{ki}) of control variables):

1. Choose an admissible starting control vector u^0 and compute the associated state y^0, put $n = 0$. Determine the active state constraints.
2. Linearize the state equation at y^n and u^n, solve it for each standard basis vector of controls. Then the state associated to an arbitrary admissible control can be obtained by superposition, see also the explanations in our first paper of this volume.
3. Express the state in the linearized optimal control problem as a linear image of the standard basis vectors using the results of step 2. Solve the associated linear optimization problem with respect to u by the Simplex method. Only active restrictions are considered in the optimization. The result is a new direction of descent $\tilde{u}$.

4. Put $u^{n+1} = u^n + \gamma(\tilde{u} - u^n)$ and perform a line search with respect to γ while considering all state constraints. Define $n = n + 1$ and go to 2.

This method of feasible direction is of gradient type. Computational tests have shown a quite robust behaviour. We stopped the iteration when the change of the controls was sufficiently small. The convergence rate is quite low. This is the characteristic behaviour of gradient methods. Moreover, the computing time to perform one step of the iteration was very high. Notice that linear partial differential equations are to be solved for each basis vector in step 2. Moreover, we have to solve some nonlinear equations arising from the line search.

Therefore, we propose a suboptimal strategy of instantaneous control type for approximately solving the problem (P). A comparison to the results of the iterative solution method shows a very fast and surprisingly exact behaviour. Furthermore, the method can be easily extended to the case of discrete 0-1 controls.

3 Suboptimal Continuous and Discrete Methods

Let us first explain, how to accelerate the optimization procedure in the case of continuous controls. The main idea is of instantaneous control type and in some sense similar to the method, developed by Choi [1] and Hinze and Kunisch [5].

The first simplification is to *linearize the state equation* during certain intervals of time. Nevertheless, the resulting optimal control problem is still nonlinear. The point is the bilinear coupling of state and control in the boundary condition.

Therefore, we introduce the *heat flux* $v := \lambda\, \partial_n y$ on the boundary as a new *auxiliary control vector*. After having determined the optimal heat flux, we derive an associated original control u by some heuristic formula. Notice that the heat flux has to be nonpositive during a cooling process.

Remark: This approach makes the optimization independent from the working hypothesis on the form of the boundary condition.

Introducing the heat flux as auxiliary control is combined with the idea to *shorten the time horizon* for minimizing the objective functional. This is the core of the instantaneous control technique. In the original formulation of the control problem, we have to achieve the minimal temperature at the final time T. Now we reduce the time horizon to certain small time intervals. The controls associated to the short interval under consideration are chosen to minimize the objective functional at the end of the time interval. In this way, we compute the (sub)optimal solution with respect to a short time horizon regardless of its influence on future times. As a byproduct of linearization, we shall have to solve the state equation only on the associated short time intervals.

Next we shall explain the idea of instantaneous control in more detail. Select an index $k \in \{1, \dots, K\}$ standing for a nozzle group. Suppose that the optimization process has already been performed for the nozzle groups $1, \dots, k-1$, that is up to the time t_{k-1}. Let $y_{k-1} := y(x, t_{k-1})$ denote the temperature distribution obtained at time t_{k-1}. Freeze the coefficients of the heat equation at y_{k-1} on the whole time interval $(t_{k-1}, t_k]$,

$$c = c(x) := c(y_{k-1}(x)), \quad \rho = \rho(x) := \rho(y_{k-1}(x)), \quad \lambda = \lambda(x) := \lambda(y_{k-1}(x)).$$

The associated time interval $[t_{k-1}, t_k]$ is divided in L computational intervals I_{kl} of length $\tau = (t_k - t_{k-1})/L$, $I_{kl} = [t_{k-1} + (l-1)\tau, t_{k-1} + l\tau]$, $l = 1(1)L$. We require constant heat fluxes on I_{kl} and denote them by v_{kli}, $i = 1(1)p$. The situation is shown in Figure 3.

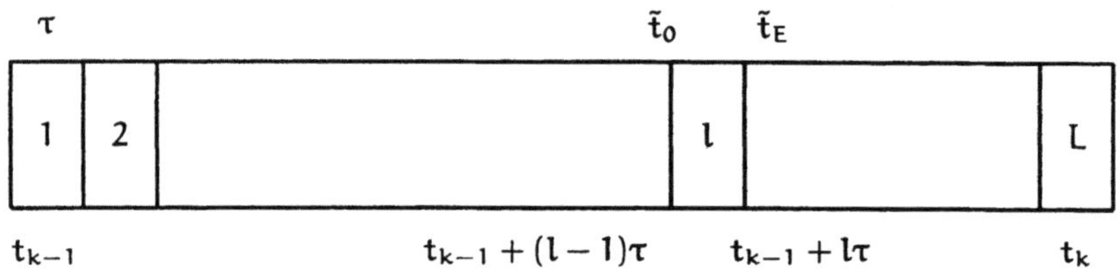

Figure 3. Partition of $[t_{k-1}, t_k]$

Now we solve a *finite sequence of linear optimization problems* (P_{kl}) associated to the small subintervals I_{kl}, $l = 1(1)L$:

Having $k-1$ fixed, regard now the partition of $[t_{k-1}, t_k]$ for $l = 1(1)L$. Assume that the optimization has already delivered the solution up to the subinterval $I_{k(l-1)}$ and regard the next subinterval I_{kl}. Denote by $y^I_{k(l-1)}$ the initial temperature computed at the time $\tilde{t}_0 := t_{k-1} + (l-1)\tau$ (we put $y_{k0} := y_{k-1}$) and solve the following optimal control problem up to the time $\tilde{t}_E := t_{k-1} + l\tau$:

(P_{kl}) *Minimize*

$$F(y(\tilde{t}_E)) = \sum_{n=1}^{N} a_n \, y(P_n, \tilde{t}_E)$$

subject to the state equation

$$\begin{aligned} c(x)\rho(x)\, y_t &= \operatorname{div}(\lambda(x) \operatorname{grad} y) && \text{in } \Omega \\ \lambda(x)\, \partial_n y &= \sum_{i=1}^{p} v_i\, \chi(\Gamma_i) && \text{on } \Gamma \\ y(x, \tilde{t}_0) &= y^I_{k(l-1)}(x) && \text{in } \Omega, \end{aligned} \tag{6}$$

$t \in (\tilde{t}_0, \tilde{t}_E]$, subject to the state constraints

$$|y(R_\mu, \tilde{t}_E) - y(Q_\nu, \tilde{t}_E)| \le \Theta_{\mu\nu}, \tag{7}$$

$\mu = 1(1)N_R$, $\nu = 1(1)N_Q$, and to the restrictions on the control vector $v = (v_i)$

$$q_{kli} \le v_i \le 0.$$

The choice of the bounds q_{kli} will be explained later. To unify the notation, let us consider air cooling areas as cooling segments as well. Here, the restriction $u_{ki} = 0$ should imply $q_{kli} = 0$. We assume this. Then the only admissible control vector $\nu = 0$ is optimal in air-cooling areas. This convention also holds for discrete strategies. We denote the obtained optimal solution by $\bar{\nu_i}$, $i = 1(1)p$, and put $\nu_{kli} := \bar{\nu_i}$, $i = 1(1)p$, to keep the index kl underlying the definition of (P_{kl}).

The solution of the optimal control problems (P_{kl}) is the core of our suboptimal strategy. However, some further ideas are needed to make this strategy work effectively. The following points are still open: In the continuous case, we have to compute the original control vector $u = (u_{ki})$ from the knowledge of the heat fluxes ν_{kli}, $l = 1(1)L$, which served as auxiliary variables. Further, the initial temperatures $y^I_{k(l-1)}(x)$ must be computed in an appropriate way. In particular, we have to control the error caused by the effects of linearization. The bounds q_{kli} must be chosen.

Remark: The state constraints might be required at further instants of time. We check them only at the times $\tilde{t}_E$. Owing to this, small violations of the state constraints may occur inside the cooling areas.

(i) Computation of auxiliary controls u_{kli}*:*

Given the optimal heat fluxes ν_{kli}, we define auxiliary controls u_{kli} as follows: Select some computational points $x_i \in \Gamma_i$. Take the mean value of

$$u^-_{kli} = \nu_{kli}/[\alpha(y(x_i,\tilde{t}_0))(y_{fl} - y(x_i,\tilde{t}_0))]$$

and

$$u^+_{kli} = \nu_{kli}/[\alpha(y(x_i,\tilde{t}_E))(y_{fl} - y(x_i,\tilde{t}_E))],$$

that is

$$u_{kli} = \frac{u^+_{kli} + u^-_{kli}}{2}. \tag{8}$$

(ii) Computation of initial temperatures for $I_{k(l+1)}$*:*

The initial temperature for the next optimization step can be determined on two ways: Solve the heat equation up to time $\tilde{t}_E$ using the linear or nonlinear equation with boundary conditions of third kind inserting the computed controls u_{kli}. We preferred the nonlinear version. After having determined the auxiliary controls u_{kli}, we solve the *nonlinear* heat conduction problem

$$\begin{aligned} c(y)\rho(y)\, y_t &= \operatorname{div}(\lambda(y)\operatorname{grad} y) && \text{in } \Omega \\ \lambda(y)\, \partial_n y &= \sum_{i=1}^{p} u_{kli}\, \chi(\Gamma_i)\alpha(\cdot,y)(y_{fl} - y) && \text{on } \Gamma \\ y(x,\tilde{t}_0) &= y^I_{k(l-1)}(x) && \text{in } \Omega \end{aligned} \tag{9}$$

on $[\tilde{t}_0, \tilde{t}_E]$. Then we put $y^I_{kl}(x) := y(x,\tilde{t}_E)$. In other words, updating of temperatures is performed nonlinearly, while the optimization is done linearly.

(iii) Choice of the bounds q_{kli}*:*

The background to define q_{kli} is the relation

$$\nu_{kli} \approx u_{kli}\, \alpha(x, y(x,t))(y_{fl} - y(x,t)).$$

In view of this, inserting the upper bound 1 for u we define

$$q_{kli} = 1 \cdot \alpha(x_i, y^I_{k(l-1)}(x_i))(y_{fl} - y^I_{k(l-1)}(x_i)) \tag{10}$$

as the lower bound for ν_{kli}.

(iv) Definition of original controls (continuous case):

The optimal control problems (P_{kl}) are solved for $k = 1(1)K$ (outer loop) and $l = 1(1)L$ (inner loop). For each fixed index k, the problems (P_{kl}) deliver the solutions u_{kli}, $l = 1(1)L$, $i = 1(1)p$, on the time intervals I_{kl}. Notice that, according to the given technical construction, only one control vector $u_k = (u_{ki})$ has to be defined on $[t_{k-1}, t_k]$. This is done by the following heuristic formula, which turned out to be very useful:

$$u_{ki} = \frac{\sum_{l=1}^{L} l\, u_{kli}}{\sum_{l=1}^{L} l}. \tag{11}$$

This rule says that a change in the first small intervals of time can be compensated on the last intervals.

Now we explain the modifications of the continuous control strategy to the discrete counterpart. First of all, we have to increase the number of spray nozzles. This is to compensate for the loss of flexibility caused by restricting the controls to $\{0, 1\}$. We assume that each short time interval $[t_{k-1} + (l-1)\tau, t_{k-1} + l\tau]$ corresponds to p spray nozzles located around the profile (= 1 nozzle group). In this way, $L \cdot p$ spray nozzles are associated with the cooling segment passed during $[t_{k-1}, t_k]$, i.e., flexibility w. r. t. the value is substituted by flexibility in time. The corresponding spray intensities are u_{kli}, $l = 1(1)L$, $i = 1(1)p$. The heat fluxes ν_{kli} (auxiliary variables) and the spray intensities u_{kli} are connected by the boundary condition (see (8)). As y varies in time and space, we cannot assume that u_{kli} and ν_{kli} are constant on $\Gamma_i \times (t_{k-1} + (l-1)\tau, t_{k-1} + l\tau)$ at the same time. However, if τ is small we are justified to consider y to be almost constant. This also motivates the choice of the bounds q_{kli} (see (10)) as well as the synthesis rule (8), (11) in the continuous case. Moreover, in the discrete case the choice $u \in \{0, 1\}$ corresponds directly to $\nu \in \{q_{kli}, 0\}$ so that a synthesis rule is not needed.

By (i)–(iv), the whole interval $[t_{k-1}, t_k]$ is processed. Now we proceed with the next interval $[t_k, t_{k+1}]$. In this way, we arrive after finitely many steps at the final time T. Obviously, this procedure requires the numerical solution of many linear and nonlinear partial differential equations. On using the principle of superposition, we are able to considerably reduce the associated numerical effort. These details are explained in the next section.

4 Processing the Subproblems

In each nozzle group, the number of controls is very low in comparison with the number of state variables arising from the finite element discretization. In our test example, we have $p = 9$ control variables per nozzle group (there are 16 nozzles in each nozzle group, see Figure 6, hence, by symmetry, the number of control variables is 9 in each group). In contrast to this, the number of state variables is some thousands. Therefore, in the optimization the state is eliminated by precomputing the response to each standard basis vector for the control, obtained from the linear equation: Regard, for k fixed, the interval $[t_{k-1}, t_k]$. For all $i = 1(1)p$, on $[0, \tau]$ the *response function* $y_{ki} = y_{ki}(x, t)$ is determined by

$$\begin{aligned} c(x)\rho(x)\, y_t &= \text{div}\,(\lambda(x)\,\text{grad}\, y) \\ \lambda(x)\, \partial_n y &= \chi(\Gamma_i) \\ y(x, 0) &= 0. \end{aligned}$$

These p systems have to be solved only once for the whole interval $[t_{k-1}, t_k]$. On the small subintervals $I_{kl} = (t_{k-1} + (l-1)\tau, t_{k-1} + l\tau)$, the temperature y is given by superposition,

$$y(x, t) = y^I(x, t) + \sum_{i=1}^{p} \nu_i\, y_{ki}(x, t - (l-1)\tau).$$

Here, $y^I(x, t)$ is the fixed part, associated to the initial temperature and homogeneous boundary conditions. It is defined by

$$\begin{aligned} c(x)\rho(x)\, y_t &= \text{div}(\lambda(x)\,\text{grad}\, y) \\ \lambda(x)\, \partial_n y &= 0 \\ y(x, 0) &= y^I_{k(l-1)}(x). \end{aligned}$$

The second part represents the contribution associated to the controls ν_i. During the optimization process, only the fixed part has to be updated from one subinterval to the next one. Then the optimization problem on I_{kl} reads for the case of *continuously* controllable nozzles

(P_{kl}) *Minimize*

$$\sum_{n=1}^{N} \sum_{i=1}^{p} c_{in}\, \nu_i$$

subject to

$$\begin{aligned} \sum_{i=1}^{p} \nu_i\, a_{i\mu\nu} &\leq \Theta_{\mu\nu} - b_{\mu\nu} \\ -\sum_{i=1}^{p} \nu_i\, a_{i\mu\nu} &\leq \Theta_{\mu\nu} + b_{\mu\nu} \\ q_i \leq \nu_i &\leq 0, \end{aligned}$$

$i = 1(1)p$, where

$$\begin{aligned} c_{in} &= c_{kin} = y_{ki}(P_n, \tau) \\ a_{i\mu\nu} &= a_{ki\mu\nu} = y_{ki}(R_\mu, \tau) - y_{ki}(Q_\nu, \tau) \\ b_{\mu\nu} &= b_{kl\mu\nu} = y^I(R_\mu, \tilde{t}_E) - y^I(Q_\mu, \tilde{t}_E). \end{aligned}$$

The bounds $q_i = q_{kli}$ are defined according to (10). This linear programming problem is solved by the Simplex method. Its optimal solution $\bar{v} = (\bar{v}_i)$ is denoted by $\bar{v}_{kli}$, $i = 1(1)p$, to preserve the index kl. Notice that the numbers c_{in}, $a_{i\mu\nu}$ have to be computed only once on $[t_{k-1}, t_k]$, while the $b_{\mu\nu}$ and q_i depend on l, hence they must be updated on all subintervals.

For discrete strategies, we arrive at linear *integer* programming problems of the same structure like above, which have to be solved by appropriate methods.

$$\min \quad \sum_{n=1}^{N} \sum_{i=1}^{p} c_{in} \, v_i,$$

$$\text{subject to}$$

$$\sum_{i=1}^{p} v_i \, a_{i\mu\nu} \le \Theta_{\mu\nu} - b_{\mu\nu}$$

$$-\sum_{i=1}^{p} v_i \, a_{i\mu\nu} \le \Theta_{\mu\nu} + b_{\mu\nu}$$

$$v_i \in \{q_i, 0\}, \; i = 1, \dots, p.$$

Strictly speaking, this is not a binary problem, since $q_m \neq 1$ in general. If particular discrete optimization methods are based on a binary structure, the problem must be transformed appropriately. Since q_m is updated after each time step τ, this means changing the matrix of constraints and the coefficients of the objective in each step. If the discrete method needs a special preprocessing of these data, this has to be repeated for each subproblem.

During the computations we observed effects of ill-posedness for small values of τ close to the time step for solving the PDEs. For instance, even in the continuous case, we observed that some controls were switching from 0.4 to 1.0 and reverse by changing the discretization of time. To overcome this problem, instead of using the original linear objective functional given above, we minimized in all cases the *linearly regularized objective*

$$\text{Min} \quad \sum_{n=1}^{N} \sum_{i=1}^{p} c_{in} v_i + \varepsilon \sum_{i=1}^{p} v_i$$

subject to the constraints given above. This trick stabilized the computed optimal controls.

5 Numerical Results

5.1 The Test Example

One of our standard test examples is the cooling of rail profiles. Following [7, 11], we consider the domain shown in Figure 5(a) with a moderate discretization. The concrete formulas for the coefficients c, ρ, λ, α are adopted from these papers. All other data were transmitted by the Mannesmann-Demag-Sack GmbH.

We restrict ourselves to the situation of [7]. That is, we consider 3 minimization points P_n on the axis of symmetry and take them as comparison points too, that is $R_\mu := P_\mu$. 9 points of comparison are chosen on the boundary. Their location is shown in Figure 2. The temperature at these points is compared with the temperature at the minimization points according to the table below.

Point	compared with
P_1	Q_1, Q_2, Q_3, Q_4
P_2	$Q_4, Q_5, Q_6,$
P_3	Q_6, Q_7, Q_8, Q_9

Table 1. Comparison points

In the test example, we regard a cooling line composed of one cooling segment followed by one air cooling area both with length equivalent to 15 seconds. Hence our cross section Ω passes the whole plant in 30 seconds. The cooling segment contains two blocks. In the continuous case, each block is identified with one nozzle group, whereas in the discrete case we have essentially more nozzle groups and each block corresponds to the time interval for freezing the coefficients c, ρ and λ. Each nozzle group consists of 16 spray nozzles, hence by symmetry we have 9 control variables, see Figure 6. Following the notation of Section 2 we have $m = 2$, $r = 2$, $p = 9$. The geometry is shown in Figure 4. According to the general setting, for t_1 we get the value 7.5 seconds.

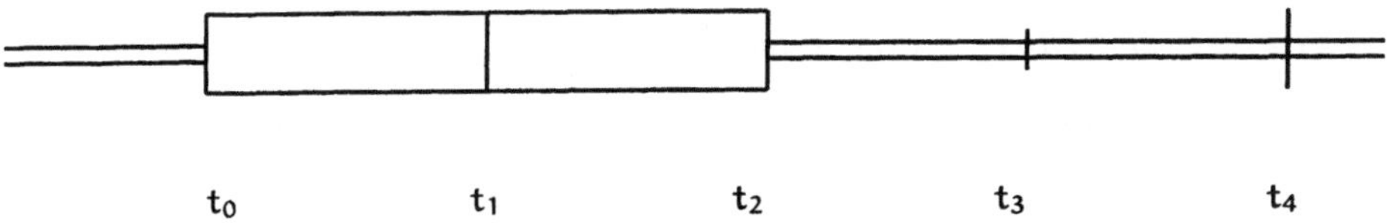

Figure 4. Test geometry

The partition of the boundary Γ into parts Γ_i is roughly indicated in Figure 5(b). For the exact geometry of the rail profile we refer to Figure 2.

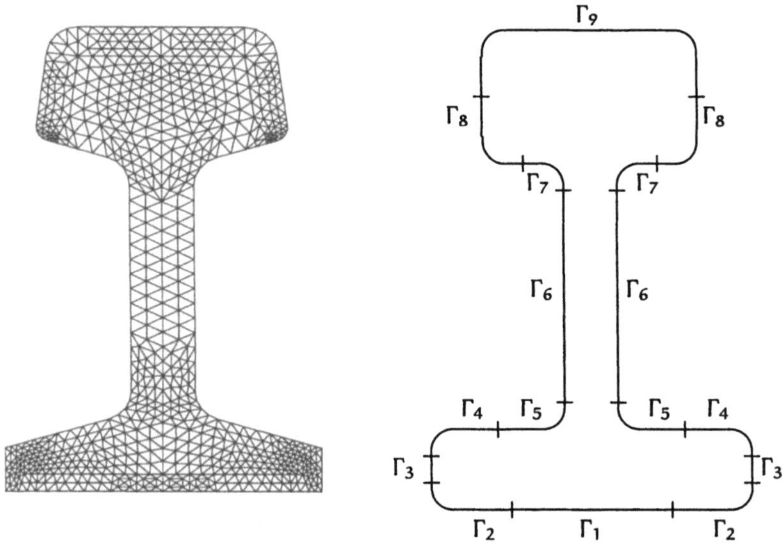

(a) The rail profile (b) Partition of Γ

Figure 5.

The parabolic state equation was solved by a Crank-Nicholson scheme in time and a 3-step FEM multigrid algorithm for the elliptic subproblems. The initial value is chosen as in [7] assuming constant temperatures in 3 areas.

For presenting test results, we proceed as follows: In a first part we compare the suboptimal strategy with the method of feasible directions in the case of continuous controls. In a second part we compare between continuous and discrete cooling strategies, in particular we discuss the lack of efficiency using discrete 0-1 nozzles. Moreover, we present results for improving efficiency of discrete strategies by applying nozzles of *lower* size, i.e. of *lower* maximal cooling intensity.

5.2 Feasible Directions versus Suboptimal Strategy

According to subsection 5.1, we consider the following test problem **(E)**:

$$\min \quad F(y) = \sum_{n=1}^{3} a_n \, y(P_n, T) \tag{12}$$

subject to

$$\begin{aligned} c(y)\rho(y)\, y_t &= \operatorname{div}(\lambda(y) \operatorname{grad} y) \\ \lambda(y)\, \partial_n y &= \textstyle\sum_{i,k} u_{ki}\, \chi(\cdot, \Gamma_i)\, \alpha(y)(y_{fl} - y) \\ y(x,0) &= y_0(x), \end{aligned} \tag{13}$$

to the control constraints

$$0 \leq u_{ki} \leq 1, \tag{14}$$

$k = 1, 2, i = 1(1)9$, (cooling segment), $u_{ki} = 0$, $k = 3, 4, i = 1(1)9$ (air cooling area), and subject to the state constraints

$$|y(P_\mu, t) - y(Q_\nu, t)| \leq \Theta_{\mu\nu}, \ \mu = 1(1)3, \ \nu = 1(1)9, \tag{15}$$

where $T = 30$ sec, and we have $18 = 2 \cdot 9$ controls acting for 7.5 seconds on different time intervals and on different boundary parts. We have chosen the values $\Theta_{\mu\nu} = |P_\mu - Q_\nu| \cdot 8000$ K/m, if the point P_μ is compared with the point Q_ν according to Table 1, and $\Theta_{\mu\nu} = \infty$ otherwise. In the computations we omit the constraints, where $\Theta_{\mu\nu} = \infty$. Furthermore, we take the weights $a_1 = a_3 = 3$ and $a_2 = 1$.

In the test runs of this subsection we worked with a time step of 0.75 seconds to solve the PDE. Therefore, we splitted each interval $[t_{k-1}, t_k]$ into 10 parts having just this length $\tau = 0.75$ sec. Obviously, this is the smallest length we can use for computational intervals in our case. In this way, we got the discretization of time $0 = t_0 < t_0 + \tau < \cdots < t_0 + 10\tau = t_1 < \cdots < t_1 + 10\tau = t_2 < \cdots < T$, where $T = 30$ sec and $\tau = 0.75$ sec.

The fast suboptimal strategy determines the solution in a very short time. Moreover, it is a direct method. In particular, no admissible initial control u_0 is needed. We list the computational results for the values $L = 1, \ 5, \ 10$ in Table 2. The CPU time was about 2 minutes on a workstation HP Apollo 9000. Table 2 contains the computed controls and the corresponding values of the cost functional together with the temperature in the minimization points. Our suboptimal method was applied for different numbers of computational intervals. Our values show the surprising effect that more (but smaller) computational intervals increase the precision of our method while decreasing the computational time. The reason for the gain of speed is that computing the response functions is cheaper on shorter intervals.

The results are compared with those obtained by the method of feasible directions in [7]. To that aim this slow iterative method was started at controls computed by our suboptimal method for the largest number of computational intervals. Nevertheless, to get the marginal improved "optimal" values in the last column of the table, the iterative method required 177 iteration steps. Hence we needed 2.5 days to get this slightly better result. *One* iteration by the method of [7] needs a between five and ten times longer computational time than our whole method. Altogether, accuracy and running time of the fast method of instantaneous control are very convincing. However, there appear small problems with violating the state constraints.

Therefore, after the first iterations the method of [7] still delivered a solution with considerably larger value than that of our fast approximate solution. The level of violation is low (0.3 K at most). In our opinion, this is sufficiently small to accept the computed control. Nevertheless, one should carefully observe this problem in more complicated situations. For more details we refer to [14].

τ_c	1	5	10	Method of [7]
u_{11}	0.351402	0.354894	0.357386	0.361015
u_{12}	1.000000	0.798217	1.000000	1.000000
u_{13}	0.118034	0.588312	0.462750	0.510719
u_{14}	1.000000	1.000000	1.000000	1.000000
u_{15}	1.000000	1.000000	1.000000	1.000000
u_{16}	0.336405	0.368480	0.376729	0.383220
u_{17}	0.562985	0.606946	0.612365	0.644712
u_{18}	0.572915	0.661840	0.686257	0.720357
u_{19}	0.434374	0.458182	0.454216	0.488881
u_{21}	0.398711	0.407742	0.413433	0.422056
u_{22}	1.000000	0.896221	0.927565	0.935165
u_{23}	0.114328	0.079897	0.036162	0.077026
u_{24}	1.000000	0.957728	0.977744	0.982834
u_{25}	1.000000	1.000000	1.000000	1.000000
u_{26}	0.378044	0.387829	0.386588	0.384954
u_{27}	0.581319	0.569530	0.565634	0.578580
u_{28}	0.516497	0.456389	0.449471	0.469555
u_{29}	0.437347	0.429484	0.428212	0.436120
$y(T,P_1)$	781.909	781.855	781.055	780.630
$y(T,P_2)$	755.870	752.048	751.399	750.944
$y(T,P_3)$	854.393	853.167	853.106	851.579
$F(y)$	5664.776	5657.115	5653.881	5647.572
CPU	208 sec	106 sec	100 sec	1200 sec / It.

Table 2. Performance of the suboptimal strategy

5.3 Continuous versus Discrete Suboptimal Strategies

In the continuous case, $[t_{k-1}, t_k]$ contains *one* nozzle group, where the nozzles can admit all intensities in $[0, 1]$. The interval $[t_{k-1}, t_k]$ was splitted into L subintervals of length τ, which served as *auxiliary* subintervals. Now we compare the results with the following discrete situation: To *each* subinterval we associate one nozzle group, i.e. we have $L = 10$ times more nozzles in $[t_{k-1}, t_k]$. In other words, 18 *continuously* controllable nozzles are replaced by 180 nozzles. However, these nozzles can only admit the intensities 0 (off) and 1 (on), and they influence the profile for a shorter time, i.e., only for $\tau = (t_k - t_{k-1})/10$.

The integer programming subproblems were solved by complete enumeration. The size of the subproblems is so small that this method was faster than standard branch and bound algorithms. We obtained computational times close to the ones for the continuous case. This shows that the solution of the integer problem needs approximately the same time as the simplex method, essentially faster than the time needed by the method of feasible directions. Optimal values and computing times are compared in Table 3. We observed that data generation, in particular computation of response functions, is stronger sensitive with respect to the length of time step than the solution of the state equation. We had to find a reasonable compromise

between accuracy and computing time. In Table 3, different choices of time steps are compared for solving the instantaneous control problem:
In all cases I-III (dis=discrete, con=continuous), the optimization subproblems are solved on time horizons of length τ. However, we do not update all coefficients of the problem after each time step τ. The coefficients c_{in} of the objective and the matrix $A = (a_{i\mu\nu})$ of the constraints are updated only after larger intervals of time (in our test example after 10τ), while the right hand sides $\Theta_{\mu\nu} \pm b_{\mu\nu}$ of the constraints are updated after each time step τ. In case I, τ is used as the step length to solve the parabolic equation for updating the state and the $b_{\mu\nu}$ as well as for the generation of the optimization data c_{in}, $a_{i\mu\nu}$ via response functions. In case II we used a smaller time step $\tau_{II} = \tau/5$ for all computations. Case III proceeds as case I with respect to the update of the state, but the coefficients of the matrix and the objective are computed with higher precision by the time step $\tau/5$.

	I dis	II dis	III dis	I con	II con	III con	Method of [7]
$F(y)$	5841.1	5848.8	5853.2	5641.8	5643.3	5653.7	5647.6
CPU	122. sec	431 sec	177 sec	124 sec	433 sec	180 sec	59 h (177 It.)

Table 3. Optimal values and computing times

Table 4 contains the associated optimal controls. In the continuous cases I-III con, the values u_{ki}, $k = 1, 2$, $i = 1(1)9$, express the intensity of nozzle i in group k. In the discrete cases, the reader would expect $K \cdot p \cdot L = 2 \cdot 9 \cdot 10$ values $u_{kli} \in \{0, 1\}$. To avoid the associated large table and to make the results comparable with the continuous case, the columns I dis - III dis contain the mean values, defined by formula (11),

$$u_{ki} = \frac{\sum_{l=1}^{10} l u_{kli}}{\sum_{l=1}^{10} l}, \quad i = 1(1)9.$$

It is quite natural that integer controls are not so flexible as the continuous ones. Even by using ten times more nozzles as in the continuous case, there is an essential lack of efficiency causing a temperature difference of almost 30 K in every "hot spot" P_n selected for the objective. Due to the state constraints, some of the nozzles (u_{16}, u_{26}) must kept switched off during the whole process, which is the main reason for the gap. Nevertheless, using the same number of nozzles of *lower* size can improve the efficiency of discrete cooling. A careful comparison of the optimal controls in Table 4 shows the main reason for the lack of efficiency in the discrete case: In particular, nozzle 6 is never active in all variants, while in the continuous case a moderate and almost *uniform* cooling takes place. This can be observed from the solution of the subproblems, because it holds $0.365 \leq u_{1l6} \leq 0.375$, and $0.38 \leq u_{2l6} \leq 0.39$ for all $l = 1(1)10$. Obviously, the constraints for nozzle 6

r_c	I dis	II dis	III dis	I con	II con	III con
u_{11}	0.14545	0.18182	0.18182	0.35739	0.35719	0.34315
u_{12}	0.74545	0.78182	0.78182	1.00000	1.00000	1.00000
u_{13}	0.45454	0.32727	0.32727	0.46275	0.41935	0.39777
u_{14}	1.00000	1.00000	1.00000	1.00000	1.00000	1.00000
u_{15}	1.00000	1.00000	1.00000	1.00000	1.00000	1.00000
u_{16}	0.00000	0.00000	0.00000	0.37673	0.36948	0.35291
u_{17}	0.47273	0.34545	0.40000	0.61237	0.60966	0.58313
u_{18}	0.61818	0.43636	0.43636	0.68626	0.67156	0.66524
u_{19}	0.27273	0.20000	0.20000	0.45422	0.45254	0.43632
u_{21}	0.18182	0.23636	0.25455	0.41343	0.41091	0.39926
u_{22}	0.98182	0.80000	0.80000	0.92756	0.91320	0.91700
u_{23}	0.01818	0.01818	0.01818	0.03616	0.03209	0.02941
u_{24}	1.00000	0.81818	0.81812	0.97774	0.96000	0.96622
u_{25}	1.00000	1.00000	1.00000	1.00000	1.00000	1.00000
u_{26}	0.00000	0.00000	0.00000	0.38659	0.38420	0.36980
u_{27}	0.54545	0.45455	0.54545	0.56569	0.56630	0.55394
u_{28}	0.38182	0.30909	0.41818	0.44947	0.44810	0.44394
u_{29}	0.29091	0.27273	0.27273	0.42821	0.42644	0.42035

Table 4. Optimal controls (all variants)

are too strong for cooling with intensity 1, even if the time of cooling is very short. Consequently, no cooling is the only admissible control. This seems to be a typical difficulty for discrete strategies. However, the comparison with the continuous problem shows the possibility of an almost uniform cooling, if the maximal intensity of the nozzle is reduced to $i_6 = 0.35$. In Table 5 we present two different cases (data generation and update by refined stepsize $\tau/3$): The first column contains the results, where the maximal cooling intensity of nozzle 6 was reduced to 0.35, while the others still had the size 1.0 as before. The values in the second column correspond to the following maximal cooling intensities $i_1, \dots, i_9$ for the nozzle 1(1)9: $i_1 = 0.5,\ i_2 = 0.8,\ i_3 = 0.4,\ i_4 = 1,\ i_5 = 1,\ i_6 = 0.35,\ i_7 = i_8 = i_9 = 0.5$. This version is called the refined strategy. Notice that these intensities are fixed in advance by our experience from the continuous case. The last two columns are added for a comparison with the standard choice of maximal intensity 1.0 for all nozzles.

The refined strategy essentially improves the standard discrete method, although the efficiency of the continuous strategy cannot be reached completely. Moreover, we observed the following: Nozzles with reduced intensities prevent the controls from chattering - the number of switches between consecutive nozzles is reduced. Some of the nozzles were active all the time. For some more details, including resulting different temperature distributions of the profile, we refer to [3].

	$i_6 = 0.35$	refined	con	dis
u_{11}	0.18182	0.30175	0.35733	0.18182
u_{12}	0.78182	0.80000	1.00000	0.78182
u_{13}	0.32727	0.40000	0.42245	0.32727
u_{14}	1.00000	1.00000	1.00000	1.00000
u_{15}	1.00000	1.00000	1.00000	1.00000
u_{16}	0.35000	0.35000	0.36983	0.00000
u_{17}	0.40000	0.50000	0.60945	0.34544
u_{18}	0.43636	0.50000	0.67217	0.43636
u_{19}	0.18182	0.43778	0.45205	0.18182
u_{21}	0.21818	0.36526	0.41118	0.21818
u_{22}	0.80000	0.80000	0.91482	0.80000
u_{23}	0.01818	0.10964	0.03246	0.01818
u_{24}	1.00000	1.00000	0.96173	0.81818
u_{25}	1.00000	1.00000	1.00000	1.00000
u_{26}	0.35000	0.35000	0.38444	0.00000
u_{27}	0.36364	0.50000	0.56625	0.45454
u_{28}	0.49091	0.50000	0.44818	0.49091
u_{29}	0.27273	0.35566	0.42653	0.27273
$F(y)$	5746.41	5689.95	5642.87	5847.74

Table 5. Optimal controls for the refined strategy

6 Conclusions

The instantaneous control technique is successful for continuous and discrete control strategies. In our example, it is able to deal with discrete cooling strategies in almost the same time as for the continuous method. The application of fixed maximum intensity 1.0 turned out to be insufficient: Even essentially more nozzles cannot deliver the same final temperature as the continous strategy. Using nozzles of lower size can overcome this problem. The solution of the continuous problem is helpful to design the size of nozzles.

References

1. H. Choi. Suboptimal control of turbulent flow using control theory. In *Proceed. Sympos. Math. Modelling of Turbulent Flows*, Tokio, 1995.
2. H. W. Engl, T. Langthaler, and P. Mansellio. An inverse problem for a nonlinear heat equation connected with continuos casting of steel. In K.-H. Hoffmann and W. Krabs, editors, *Optimal control of partial differential equations*, pages 67–90, Basel, 1987. Birkhäuser.
3. K. Eppler and F. Tröltzsch. Discrete and continuous optimal control strategies in the selective cooling of steel. Preprint 01–3, DFG-Forschungsschwerpunkt "Real-time optimization of large systems", 2001.
4. W. Grever. A nonlinear parabolic initial boundary value problem modelling the continuous casting of steel. *ZAMM*, 78:109–119, 1998.

5. M. Hinze and K. Kunisch. On suboptimal strategies for the Navier-Stokes equations with continuos casting of steel. In *Contrôle et Équations aux Derivées Partielles*, pages 181–198. ESAIM: proceedings, vol. 4, 1998.
6. D. Hömberg and J. Sokołowski. Optimal control of laser hardening. Preprint 315, Weierstraß Institute of Applied Analysis and Stochastics Berlin, 1997.
7. R. Krengel, R. Standke, F. Tröltzsch, and H. Wehage. Mathematisches Modell einer optimal gesteuerten Abkühlung von Profilstählen in Kühlstrecken. Preprint 98–6, TU Chemnitz, Fakultät für Mathematik, 1996.
8. E. Laitinen and P. Neittaanmäki. On numerical solution of the continuous casting process. *J. of Engineering Mathematics*, 22:335–354, 1988.
9. G. Landl and H. W. Engl. Optimal strategies for the cooling of steel strips in hot strip mills. *Inverse Problems in Engineering*, 2:102–118, 1995.
10. F. Leibfritz and E. W. Sachs. Numerical solution of parabolic state constraint control problems using SQP- and interior-point-methods. In W.W. Hager, editor, *Large scale optimization*, pages 245–258, Dordrecht, 1994. Kluwer Academic Publishers B.V.
11. R. Lezius and F. Tröltzsch. Theoretical and numerical aspects of controlled cooling of steel profiles. In H. Neunzert, editor, *Progress in Industrial Mathematics at ECMI 94*, pages 380–388. Wiley-Teubner, 1996.
12. P. Neitaanmäki. On the control of the secondary cooling in the continuos casting of steel. In K.-H. Hoffmann and W. Krabs, editors, *Optimal control of partial differential equations*, pages 161–178, Basel, 1987. Birkhäuser.
13. A. Rösch. Identification of nonlinear heat transfer laws by optimal control. *Num. Funct. Analysis and Optimization*, 15(3&4):417–434, 1994.
14. F. Tröltzsch and A. Unger. Fast solution of optimal control problems in the selective cooling of steel. *ZAMM*, 81:447–456, 2001.
15. K. Zurdel and N. Brennecke. *Untersuchungen zum Wärmeübergang bei der Wasserkühlung von Feinstahl und Walzdraht.* PhD thesis, Technische Hochschule Magdeburg, 1974.

Real-Time Optimization and Stabilization of Distributed Parameter Systems with Piezoelectric Elements.

Karl-Heinz Hoffmann and Nikolai D. Botkin

Stiftung caesar, Center of Advanced European Studies and Research, Bonn, Germany

Abstract The investigation of this paper is related to the design of active real-time controls which provide the desired performance of flexible constructions subjected to varying disturbances. Physical controllers are piezoelectric elements that transform applied voltages into mechanical forces and moments. Control voltages are computed on the base of signals measured on piezoelectric sensors. The underlying structure is a thin plate or shell. The piezoelectric elements are either surface mounted or embedded within the structure. Conventional averaging procedures are used to eliminate the thickness in mathematical models. A homogenization procedure is used to reduce structures with large number of piezoelectric elements to the case of continuously distributed input or output signals.

1 Conception of Active Control

The idea of active real-time control is intensively developed during several last years. The acute interest to this method of elimination of parasitic disturbances is caused by the rapid development of the material science, electronics, and computer technology. This approach can be characterized as an intelligent one by contrast with routine approaches based on the reenforcement of the construction, which makes it heavier and more expensive.

Consider a precise structure subjected to varying acoustic or thermal conditions. Even though carefully designed, it will disturbed as a result of unpredictable thermal gradients or acoustic pressures. One way to prevent this is to build the structure from massive components and to provide very good isolation from external influences. An alternative way is to use a set of actuators and sensors connected by a feedback loop. In this case, we exploit the main virtue of the feedback which is to attenuate the effect of disturbances within the bandwidth of the control system. Active structures may be cheaper or lighter than passive structures with comparable performances, or they can sufficiently improve the performance. Here, we want to refer to the following classical example (see [1]). The telescope at ESO in La Silla, Chili, uses adaptive optics for the compensation of atmospheric turbulences. The primary mirror of the telescope is connected at the back to a set of hundred actuators. The control system uses an image analyzer to evaluate the distributed amplitude of the perturbation. The correction is computed to minimize the effect of the perturbation and is applied to the actuators. The computation of the correcting forces is based on the influence matrix describing the relation between the actuator forces and the wave front changes. This matrix is determined experimentally from the analysis of images. Note that such a relatively simple relation between the controls and objectives is not always possible. Very often, this relation is described by a very complicated mathematical model based on nonlinear partial differential equations. This approach is carefully developed in the monograph [2]. The schema of the active control design is shown on Figure 1. We study the real-time control design with the use of piezoelectric elements as actuators and sensors. Piezoelectric materials exhibit significant deformations in response to an applied electric field as well as produce polarization in response to mechanical strains.

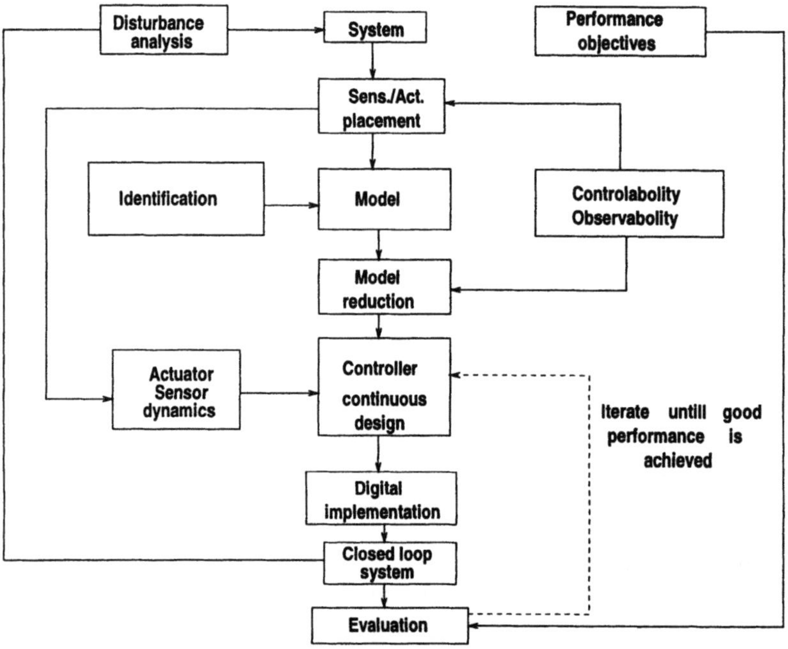

Figure 1. Steps of the control design

The field-strain relations are nearly linear for small electric fields and deformations, which is an advantage when employing piezoelectric elements in control systems. The linear direct and converse constitutive relationships for piezoelectric materials (see, e.g., [3] and compare with [4] concerning more general models) are given by

$$\sigma_{ij} = C_{ijkl}\varepsilon_{kl} - e_{kij}E_k,$$
$$D_i = \epsilon_{ij}E_j + e_{ikl}d_{kl}. \tag{1}$$

Here, D and E denote the electric displacement and field, respectively, while ϵ, e, and C denote the material dielectric tensor, the stress piezoelectric tensor and the elastic stiffness tensor, respectively. As usually, the repeated subscripts indicate summation; Latin subscripts run from 1 to 3 whereas Greek subscripts run from 1 to 2. To cover all possible cases of piezoelectric materials, assume that all coefficients may be nonzero and different. This is really the case for the triclinic crystal systems (see [3]).

Note that relations (1) are nonlinear with respect to the displacements $u_i = y_i(x_1, x_2, x_3, t) - x_i$, $i = 1, 2, 3$, because the strain tensor contains quadratic terms, namely

$$\varepsilon_{ij} = \frac{1}{2}\left(\frac{\partial u_i}{\partial x_j} + \frac{\partial u_j}{\partial x_i} + \frac{\partial u_k}{\partial x_i} \cdot \frac{\partial u_k}{\partial x_j}\right), \quad i, j = 1, 2, 3. \tag{2}$$

Therefore, the energy density of a piezoelectric medium,

$$\chi = \frac{1}{2}(\sigma_{ij}\varepsilon_{ij} - E_i D_i), \tag{3}$$

is not quadratic and may be not convex, which may violate the uniqueness of solutions.

The underlying structure is a plate or shell in our case. The piezoelectric elements are either surface mounted or embedded within the structure. Therefore, we have to describe the coupling between two media with different elastic, electric and piezoelectric properties. Then, we have to apply an averaging to exclude the thickness and obtain a thin composite structure. Moreover, the interaction between the piezoelectric elements and the substrate is time-dependent with various transient effects; hence dynamic models are necessary.

We derive our models using the following variational principle: the work of inertia forces plus the variation of the free energy must be equal to the work of external forces. The free energy is the sum of the free energies of the piezoelectric elements and the substrate. One advantage of this approach is that the right interface conditions are being obtained automatically.

The design of a control law must be done in accordance with that the actuators and sensors yield unbounded input and output operators in the mathematical problem formulation. An extension of H^∞ theory (see [5]) to this case is developed in [6]. A way to avoid unbounded input and output operators is related to the homogenization of model equations. If the number of actuators and sensors is large enough, we can assume that it goes to infinity, whereas the size of each element tends to zero. Some special procedure based on two-scale convergence is applied to the model equations in oder to obtain limiting equations. It is remarkable that the limiting equations are much better than the original ones: they do not contain unbounded operators and discontinuous coefficients because the interface is smeared when applying the homogenization. Numerical simulations prove that the control law designed using the limiting equations provides some good performance of the original controlled system; the number of actuators does not have to be to large: 30-40 elements are sufficient to provide some good consistence.

Summarizing, we can outline the frameworks of our study as follows:

- Derivation of mathematical models of thin plates and shells with actuators and sensors in various configurations.
- Mathematical investigation of the model equations concerning their solubility and correctness including nonlinear cases.
- Estimation of unknown parameters.
- Homogenization.
- Feedback control design using H^∞ theory or the theory of differential games.

2 Mathematical Models of Thin Plates and Shells with Surface Mounted Piezoelectric Elements

2.1 *Variational Principle*

We use a variational principle (see [7]) for the derivation of dynamic equations:

$$\text{The work of the inertia forces} + \delta F = 0, \tag{4}$$

where δF is the variation of the total free energy. The total free energy F can be computed trough the densities of the free energy for the substrate, piezoelectric material and surrounding substance (say air). Thus,

$$F = \int_{V_S} \mathcal{F}_S + \int_{V_P} \mathcal{F}_P + \int_{V_A} \mathcal{F}_A. \tag{5}$$

Here, $\mathcal{F}_P$, $\mathcal{F}_S$ and $\mathcal{F}_A$ are the densities of the free energy; V_P, V_S and V_A are the corresponding volumes. Note that each of volumes V_P, V_S and V_A can be divided into several parts to simplify the computation of the free energy or to account the discontinuity of material parameters that appear due to a multi piece structure of the system.

2.2 *Elimination of the Thickness*

To obtain two-dimensional structures, we have to average along the thickness. This can be archived using the Kirchhoff-Love-Koiter hypothesis that allow us to express all components of the strain tensor through the displacement of points lying on some reference surface (say middle surface). An expansion of the electric field with respect to the deviation from the reference surface is also necessary.

Let Q be a point of the plate, M orthogonal projection of Q onto the middle surface, u_1, u_2 longitudinal displacements of M and w the transversal displacement of M. According to the Kirchhoff-Love-Koiter hypothesis (see, e.g., [8]), the components of the strain tensor at the point Q are given by

$$\begin{aligned} &\varepsilon_{\alpha\beta} = 1/2(u_{\alpha x_\beta} + u_{\beta x_\alpha} + w_{x_\alpha} w_{x_\beta}) - x_3 \cdot w_{x_\alpha x_\beta}, \\ &\varepsilon_{3\alpha} = \varepsilon_{\alpha 3} = 0, \quad \varepsilon_{33} \quad \text{is found from} \quad \sigma_{33} = 0. \end{aligned} \tag{6}$$

The assumption $\varepsilon_{3\alpha} = \varepsilon_{\alpha 3} = 0$, $\alpha = 1, 2$, means the absence of transverse shear strains, which implies the conservation of the normal. The component ε_{33} is found from the hypothesis of plain stresses. Therefore, ε_{33} is material and field dependent. The first approximation of the electric field within a thin piezoelectric element whose longitudinal faces are covered by a metal is given by

$$E_1 = E_2 = 0, \quad E_3 = V/h, \tag{7}$$

where V is a voltage applied between the longitudinal faces, and h is the thickness of the piezoelectric element. Now we are in position to compute the total free energy and exclude the thickness using (3), (5), (6) and (7). The application of (4) yields dynamic equations. Note that the Kirchhoff-Love-Koiter hypotheses can be replaced by more complex assumptions permitting transverse shear strains. This demands two additional variables describing the change of the normal. Some more accurate approximation of the electric field can be archived through the hypothesis of linear dependence of E_i, $i = 1, 2, 3$, on x_3.

2.3 A Fully Coupled Nonlinear Model

We begin with a very general fully coupled model of a plate with surface mounted piezoelectric elements (see [9] and [10]). The inverse piezoelectric effect (i.e., the excitation of the electric field due to strains), electric properties of the surrounding substance (say air), and the quadratic terms in the strain tensor are taken into account.

Consider a plate with symmetrically mounted piezoelectric elements.

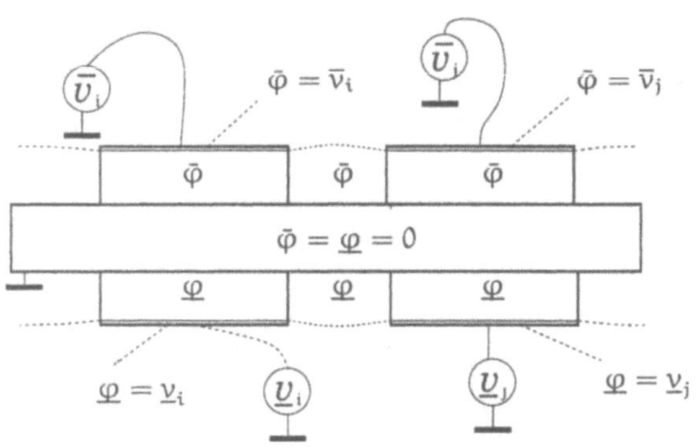

Figure 2. Fragment of a plate with many piezoelectric elements; $\bar{\varphi}$ and $\underline{\varphi}$ are potential functions, $\bar{v}_i$ and $\underline{v}_i$ are applied voltages

Let Ω be the region of the plate; Ω_{p_i} the projection of the i-th piezoelectric pair onto the middle plain of the plate; $\Omega_p = \bigcup_i \Omega_{p_i}$; $\Omega_s = \Omega \setminus \Omega_p$; u_1 and u_2 longitudinal displacements of points laying on the middle plain; w the transversal deflection; $\bar{v}_i$ and $\underline{v}_i$ voltages applied to the i-th piezoelectric pair; d the thickness of the substrate; h the thickness of the piezoelectric patches; $\bar{E}_i = \partial\bar{\varphi}/\partial x_i$ and $\underline{E}_i =$

$\partial\underline{\varphi}/\partial x_i$ the electric fields, where $\bar{\varphi}$ and $\underline{\varphi}$ are the potential functions. Assuming that $\bar{E}_i$ and $\underline{E}_i$ are linear in x_3 and taking into account conventional boundary conditions for $\bar{\varphi}$ and $\underline{\varphi}$, we can compute quadratic polynomials $\bar{f}_1(x_3)$, $\bar{f}_2(x_3)$, $\underline{f}_1(x_3)$ and $\underline{f}_2(x_3)$ such that the potentials within the i-th piezoelectric pair are given by

$$\bar{\varphi}(x_1, x_2, x_3) = \bar{f}_1(x_3)\bar{v}_i + \bar{f}_2(x_3)\bar{\phi}(x_1, x_2),$$

$$\underline{\varphi}(x_1, x_2, x_3) = \underline{f}_1(x_3)\underline{v}_i + \underline{f}_2(x_3)\underline{\phi}(x_1, x_2),$$

where $\bar{\phi}(x_1, x_2)$ and $\underline{\phi}(x_1, x_2)$ are unknown distributions of the electric field above and below the plate along the $\{x_1, x_2\}$-plane.

Using relations (1), (2), (3) and Kirchhoff-Love-Koiter hypothesis (6) yields the energy density within the piezoelectric patches and the substrate:

$$\bar{\chi}^p = \frac{1}{2}c^p_{\alpha\beta\gamma\mu}s_{\alpha\beta}s_{\gamma\mu} + \frac{1}{2}x_3^2 c^p_{\alpha\beta\gamma\mu}w_{x_\alpha x_\beta}w_{x_\gamma x_\mu} - e_{i\alpha\beta}s_{\alpha\beta}\bar{E}_i$$
$$-\frac{1}{2}\epsilon_{ij}\bar{E}_i\bar{E}_j + x_3 c^p_{\alpha\beta\gamma\mu}s_{\alpha\beta}w_{x_\gamma x_\mu} - x_3 e_{i\alpha\beta}w_{x_\alpha x_\beta}\bar{E}_i;$$

$$\underline{\chi}^p = \frac{1}{2}c^p_{\alpha\beta\gamma\mu}s_{\alpha\beta}s_{\gamma\mu} + \frac{1}{2}x_3^2 c^p_{\alpha\beta\gamma\mu}w_{x_\alpha x_\beta}w_{x_\gamma x_\mu} - e_{i\alpha\beta}s_{\alpha\beta}\underline{E}_i$$
$$-\frac{1}{2}\epsilon_{ij}\underline{E}_i\,\underline{E}_j + x_3 c^p_{\alpha\beta\gamma\mu}s_{\alpha\beta}w_{x_\gamma x_\mu} - x_3 e_{i\alpha\beta}w_{x_\alpha x_\beta}\underline{E}_i;$$

$$\chi^s = \frac{1}{2}c^s_{\alpha\beta\gamma\mu}s_{\alpha\beta}s_{\gamma\mu} + \frac{1}{2}x_3^2 c^s_{\alpha\beta\gamma\mu}w_{\alpha x_\beta}w_{x_\gamma x_\mu} + x_3 c^s_{\alpha\beta\gamma\mu}s_{\alpha\beta}w_{x_\gamma x_\mu};$$

$$F = \iint_{\Omega_p}\int_{d/2}^{d/2+h}\bar{\chi}^p\,dx_3 + \iint_{\Omega_p}\int_{-d/2-h}^{-d/2}\underline{\chi}^p\,dx_3 + \iint_{\Omega}\int_{-d/2}^{d/2}\chi^s\,dx_3$$
$$+\iint_{\Omega_s}\int_{d/2}^{d/2+h}\epsilon_0\bar{E}_i^2\,dx_3 + \iint_{\Omega_s}\int_{-d/2-h}^{-d/2}\epsilon_0\,\underline{E}_i^2\,dx_3.$$

Here, ϵ_0 is the material dielectric constant of the air, while

$$c^s_{\alpha\beta\gamma\mu} = C'_{\alpha\beta\gamma\mu} - \frac{C'_{33\alpha\beta}C'_{33\gamma\mu}}{C'_{3333}}$$

with C' being the elastic stiffness tensor of the substrate. Moreover,

$$c^p_{\alpha\beta\gamma\mu} := C_{\alpha\beta\gamma\mu} - \frac{C_{33\alpha\beta}C_{33\gamma\mu}}{C_{3333}}, \quad e_{i\alpha\beta} := e_{i\alpha\beta} - \frac{C_{33\alpha\beta}e_{i33}}{C_{3333}},$$
$$\epsilon_{ij} := \epsilon_{ij} + \frac{e_{i33}e_{j33}}{C_{3333}},$$

where C, e and ϵ are the elastic stiffness tensor, piezoelectric stress tensor and material dielectric tensor involved in (1). Note that the old notations for e and ϵ are kept. The in-plain strain tensor is given by

$$s_{\alpha\beta} = 1/2(u_{\alpha x_\beta} + u_{\beta x_\alpha} + w_{x_\alpha} \cdot w_{x_\beta}). \tag{8}$$

The application of the variational principle [7] leads to the equations

$$\rho w_{tt} + \frac{\partial}{\partial x_\beta}(\sigma_{\alpha\beta} w_{x_\alpha}) + \Delta(\gamma \Delta w) - \theta \frac{\partial}{\partial x_\beta}(e_{\gamma\alpha\beta} w_{x_\alpha}(\bar{\phi}_{x_\gamma} + \underline{\phi}_{x_\gamma}))+$$
$$\theta \frac{\partial^2}{\partial x_\alpha \partial x_\beta}(e_{3\alpha\beta}(\bar{\phi} + \underline{\phi})) - r \frac{\partial^2}{\partial x_\alpha \partial x_\beta}(e_{\gamma\alpha\beta}(\bar{\phi}_{x_\gamma} - \underline{\phi}_{x_\gamma}))-$$
$$q \frac{\partial}{\partial x_\beta}(e_{3\alpha\beta} w_{x_\alpha}(\bar{v} - \underline{v})) - \ell \frac{\partial^2}{\partial x_\alpha \partial x_\beta}(e_{3\alpha\beta}(\bar{v} + \underline{v})) = 0;$$

$$\rho u_{\alpha tt} - \frac{\partial}{\partial x_\beta}\sigma_{\alpha\beta} + \theta \frac{\partial}{\partial x_\beta}(e_{\gamma\alpha\beta}(\bar{\phi}_{x_\gamma} + \underline{\phi}_{x_\gamma}))+$$
$$q \frac{\partial}{\partial x_\beta} e_{3\alpha\beta}(\bar{v} - \underline{v}) = 0, \qquad \alpha = 1, 2;$$
$$a \frac{\partial}{\partial x_\alpha}(\epsilon_{\alpha\beta} \bar{\phi}_{x_\beta}) + b\epsilon_{33}\bar{\phi} + \theta \frac{\partial}{\partial x_\gamma}(e_{\gamma\alpha\beta} s_{\alpha\beta})+$$
$$r \frac{\partial}{\partial x_\gamma} e_{\gamma\alpha\beta} w_{x_\alpha x_\beta} - \theta \frac{\partial}{\partial x_\alpha} e_{3\alpha\beta} w_{x_\beta} - g\epsilon_{33}\bar{v} = 0;$$
$$a \frac{\partial}{\partial x_\alpha}(\epsilon_{\alpha\beta} \underline{\phi}_{x_\beta}) + b\epsilon_{33}\underline{\phi} + \theta \frac{\partial}{\partial x_\gamma}(e_{\gamma\alpha\beta} s_{\alpha\beta})-$$
$$r \frac{\partial}{\partial x_\gamma} e_{\gamma\alpha\beta} w_{x_\alpha x_\beta} - \theta \frac{\partial}{\partial x_\alpha} e_{3\alpha\beta} w_{x_\beta} - g\epsilon_{33}\underline{v} = 0;$$

where

$$\sigma_{\alpha\beta} = \frac{1}{2}\ell_{\alpha\beta\mu\nu}(u_{\mu x_\nu} + u_{\mu x_\nu} + w_{x_\mu} w_{x_\nu}) \tag{9}$$

is the in-plain stress tensor; $\rho, \gamma, e_{i\alpha\beta}, \epsilon_{\alpha\beta}$ and $\ell_{\alpha\beta\mu\nu}$ are piecewise constant discontinuous functions with jumps on the boundary of Ω_p.

The constants $a, b, \theta, r, \ell, q, g$ are computed from the problem's data. Moreover,

$$\underline{v} = \sum \underline{v}_i(t) I_{\Omega_{p_i}}(x_1, x_2), \qquad \bar{v} = \sum \bar{v}_i(t) I_{\Omega_{p_i}}(x_1, x_2),$$

where $I_{\Omega_{p_i}}$ is the indicator function of Ω_{p_i}. Introduce the notation:

$$\vec{u} = (u_1, u_2, w), \quad \vec{\phi} = (\bar{\phi}, \underline{\phi}), \quad \vec{v} = (\bar{v}, \underline{v}),$$

$$H = \left(L_2(\Omega)\right)^3, \quad V = \left(H_0^1(\Omega)\right)^2 \times H_0^2(\Omega), \quad \Phi = \left(H^1(\Omega)\right)^2.$$

Theorem 1 (Solvability: see [9, 10]).
If $\vec{u}(0) \in V$, $\vec{u}_t(0) \in H$ *and* $\vec{v} \in H^1(0, T; R^2)$, *then the system has a solution such that:*

$$\vec{u} \in L_\infty(0, T; V), \quad \vec{u}_t \in L_\infty(0, T; H), \quad \vec{\phi} \in L_\infty(0, T; \Phi). \quad \square$$

The uniqueness can also be proved, if some additional smoothness of solutions is assumed. Such an assumption is not realistic since the smoothness can not be improved because of discontinuity of the coefficients.

2.4 Simplified Models

Electrically Decoupled Nonlinear Models

Nonlinear models without the inverse piezoelectric effect involve coupled equations for the transversal and longitudinal displacements, whereas the equations for the electric field vanish because the electric field is defined immediately through the applied voltages (see (7)). It was assumed that piezoelectric and substrate materials are nearly isotropic, and voltages on the lower piezoelectric patches are equal to zero. These assumptions lead to the equations:

$$\begin{aligned} \rho w_{tt} - \Delta\Big(\gamma \Delta w\Big) - \tfrac{\partial}{\partial x_\alpha}\Big(\sigma_{\alpha\beta} w_{x_\beta}\Big) &= \\ K\Delta\Big(\nu_i(t) I_{\Omega_{p_i}}\Big) + G\tfrac{\partial}{\partial x_\alpha}\Big(w_{x_\alpha}\nu_i(t) I_{\Omega_{p_i}}\Big), \qquad (10) \\ \rho u_{\alpha tt} - \tfrac{\partial}{\partial x_\beta}\sigma_{\alpha\beta} = G\tfrac{\partial}{\partial x_\alpha}\Big(\nu_i(t) I_{\Omega_{p_i}}\Big), \quad \alpha = 1, 2. \end{aligned}$$

Here Δ is the Laplace operator; K and G are constants; ρ and γ are piecewise constant discontinuous functions with jumps on the boundary of Ω_p; $\nu_i(t)$ is a voltage applied to the i-th piezoelectric element.

The solvability follows from the previous theorem (see also [11]) but the uniqueness is not proved because of the discontinuous coefficients that violate the regularity of solutions.

Linear Models

The simplest mathematical model is related to the case, where the nonlinear terms in (6) and the inverse piezoelectric effect, the second equation of (1), are omitted. The equation for w is independent from u_1, u_2 and given by

$$\rho(x_1, x_2) w_{tt} + \Delta\Big(\gamma(x_1, x_2)\Delta w\Big) = K\Delta\Big(\nu_i(t) \cdot I_{\Omega_{p_i}}(x_1, x_2)\Big). \qquad (11)$$

Note that discontinuous coefficients and unbounded input operators are inherent even in this simple model. Taking into account the inverse piezoelectric effect yields electrically coupled linear models that involve additional equations for u_1, u_2 and the electric field (see, e.g., [13]). The linear case is good from the mathematical point of view because the corresponding equations are uniquely solvable and solutions are differentiable with respect to model parameters (see [12] and [13]).

2.5 Shells with Piezoelectric Actuators

The theory of usual thin shells is summarized very good in [8]. Consider a shell with two piezoelectric patches mounted symmetrically with respect to the middle surface. Let $(\xi^1, \xi^2) \in \Omega$ be curvilinear coordinates of the middle surface of the shell; Ω_p the region of the piezoelectric pair; $\Omega_s = \Omega \setminus \Omega_p$ the region of the uncovered substrate; ξ^3 the transversal coordinate along the normal; r the radius-vector of shell points; a the metric tensor of the middle surface; b the curvature tensor of the middle surface; g the global metric tensor related to the shell coordinate system; u_1, u_2 and u_3 displacements of points of the middle surface; d the thickness of the substrate; h the thickness of the piezoelectric patches; $v_1(t)$ and $v_2(t)$ voltages applied to the piezoelectric patches. Let the vertical bar before a subscript denotes the corresponding covariant derivative. The strain tensor of the middle surface is given by the formula

$$\gamma_{\alpha\beta}(\vec{u}) = 1/2(u_{\alpha|\beta} + u_{\beta|\alpha}) - b_{\alpha\beta}u_3.$$

In the case of the Koiter model, where the conservation of the normal is assumed, the curvature change is given by

$$\wp_{\alpha\beta}(\vec{u}) = -(u_{3|\alpha\beta} - b^{\lambda}_{\alpha} b_{\lambda\beta} u_3 + b^{\lambda}_{\alpha|\beta} u_{\lambda} + b^{\lambda}_{\alpha} u_{\lambda|\beta} + b^{\lambda}_{\beta} u_{\lambda|\alpha}).$$

The strain tensor is of the form

$$\begin{aligned}
\varepsilon_{\alpha\beta}(\vec{U}) &= \gamma_{\alpha\beta}(\vec{u}) + \xi^3 \wp_{\alpha\beta}(\vec{u}) \\
\varepsilon_{3\alpha}(\vec{U}) &= 0 \\
\varepsilon_{33}(\vec{U}) &= -\frac{\nu}{1-\nu} a^{\alpha\beta} \varepsilon_{\alpha\beta}(\vec{U}).
\end{aligned}$$

The variation of the energy is given by the formula

$$a^K[\vec{u}, \vec{\varphi}; v_1, v_2] =$$

$$\int_{\Omega} d\sqrt{a} E^{\alpha\beta\lambda\mu} \left[\gamma_{\alpha\beta}(\vec{u}) \gamma_{\lambda\mu}(\vec{\varphi}) + \frac{d^2}{12} \wp_{\alpha\beta}(\vec{u}) \wp_{\lambda\mu}(\vec{\varphi})\right] d\xi^1 d\xi^2 +$$

$$\int_{\Omega_p} \sqrt{a} \left\{ \frac{v_1(t)}{h} \int_{d/2}^{d/2+h} e^{3kl} \varepsilon_{kl}(\vec{\varphi}) d\xi^3 + \frac{v_2(t)}{h} \int_{-d/2-h}^{-d/2} e^{3kl} \varepsilon_{kl}(\vec{\varphi}) d\xi^3 \right\} d\xi^1 d\xi^2. \tag{12}$$

The bilinear form a^K is defined on the space

$$\vec{V}^K(\Omega) = \left\{\vec{u} : u_1, u_2 \in H^1(\Omega);\ u_3 \in H^2(\Omega);\ \vec{u}|_{\Gamma_0} = \partial u_3/\partial n|_{\Gamma_0} = 0\right\}.$$

The coefficients $E^{\alpha\beta\lambda\mu}$ are discontinuous functions of ξ^1 and ξ^2. In the case of shells with actuators made of piezoelectric ceramics belonging to the hexagonal class (6m m), the coefficients are of the form

$$E^{\alpha\beta\lambda\mu} = \begin{cases} E_s^{\alpha\beta\lambda\mu}(\xi^1, \xi^2), & (\xi^1, \xi^2) \in \Omega_s; \\ E_p^{\alpha\beta\lambda\mu}(\xi^1, \xi^2), & (\xi^1, \xi^2) \in \Omega_p; \end{cases} \qquad \nu = \begin{cases} \nu_s, (\xi^1, \xi^2) \in \Omega_s; \\ \hat{\nu}_p, (\xi^1, \xi^2) \in \Omega_p; \end{cases}$$

$$E_s^{\alpha\beta\lambda\mu} = \frac{E_s}{2(1+\nu_s)} \left(a^{\alpha\lambda} a^{\beta\mu} + a^{\alpha\mu} a^{\beta\lambda} + \frac{2\nu_s}{1-2\nu_s} a^{\alpha\beta} a^{\lambda\mu} \right),$$

$$E_p^{\alpha\beta\lambda\mu} = \frac{\hat{E}_p}{2(1+\hat{\nu}_p)}\left(a^{\alpha\lambda}a^{\beta\mu} + a^{\alpha\mu}a^{\beta\lambda} + \frac{2\hat{\nu}_p}{1-2\hat{\nu}_s}a^{\alpha\beta}a^{\lambda\mu}\right),$$

$$\hat{E}_p = x + xy/(x+y), \quad \hat{\nu}_p = y/(x+y),$$

$$x = \frac{d}{(2h+d)}\frac{E_s}{1+\nu_s} + \frac{2h}{2h+d}\frac{E_p}{1+\nu_p},$$

$$y = \frac{d}{2h+d}\frac{\nu_s E_s}{1+\nu_s} + \frac{2h}{2h+d}\frac{\nu_p E_p}{1+\nu_p},$$

$$e^{fhs}(\xi^1,\xi^2,\xi^3) = g^{fi}g^{hj}g^{sk}e^{00}_{mnl}\frac{\partial r^m}{\partial \xi^i}\frac{\partial r^n}{\partial \xi^j}\frac{\partial r^l}{\partial \xi^k}.$$

The subscripts s and p point out to substrate and piezoelectric materials, respectively. The symbols ν_s, ν_p, E_s, E_p and e^{00}_{mnl} denote the Poisson ratios, Young moduli and piezoelectric stress tensor in the Euclidean coordinate system, while $\hat{\nu}_p$ and $\hat{E}_p$ denote the effective Poisson ratio and Young module of the substrate-piezoelectric structure. It is convenient to introduce the following seven parameters

$$p_1 = \frac{E_s}{1+\nu_s}, \quad p_2 = \frac{E_s\nu_s}{(1+\nu_s)(1-2\nu_s)},$$

$$p_3 = \frac{\hat{E}_p}{1+\hat{\nu}_p}, \quad p_4 = \frac{\hat{E}_p\hat{\nu}_p}{(1+\hat{\nu}_p)(1-2\hat{\nu}_p)},$$

$$p_5 = e^{00}_{311}, \quad p_6 = e^{00}_{333}, \quad p_7 = e^{00}_{113}$$

that define all of the coefficients. Besides, the form a^K is linear in $\vec{p} = (p_1, \dots, p_7)$.

According to the variational principle, the dynamic equations are given by

$$\int_\Omega \sqrt{a}\rho\vec{u}_{tt}\vec{\varphi}\,d\xi^1 d\xi^2 + a^K[\vec{u},\vec{\varphi};\nu_1,\nu_2] = 0, \quad \vec{\varphi} \in \vec{V}^K(\Omega). \tag{13}$$

The initial and boundary conditions look like that

$$\vec{u}|_{t=0} = \vec{u}^0, \quad \vec{u}_t|_{t=0} = \vec{u}^{0\prime}, \quad \vec{u}|_{\Gamma_0} = 0, \quad \partial u_3/\partial n|_{\Gamma_0} = 0.$$

The following theorem states the existence and the uniqueness of solutions for models of the Koiter's type. The proof follows from the properties of the elliptic part, the first integral, of (12) stated in [8]. The input operator, the second integral, is similar to that in [11].

Theorem 2. *If* $\vec{u}^0 \in \vec{V}^K(\Omega)$, $\vec{u}^{0\prime} \in (L_2(\Omega))^3$, $\nu_1(\cdot), \nu_2(\cdot) \in H^1(0,T)$, *then the system (13) is uniquely solvable. The solution* $\vec{u}$ *possesses the regularity:*

$$\vec{u} \in C([0,T];\vec{V}^K(\Omega)), \quad \vec{u}_t \in C([0,T];(L_2(\Omega))^3). \quad \square$$

3 Identification of Parameters for Shells

We propose two procedures appropriate for the treatment of composite structures. The first one is related to the minimization of a residual of dynamic equations (see [14]). The second procedure is the conventional least squares method supported by the "strong" differentiability of solutions with respect to parameters (see [12]).

3.1 Minimization of Residual

A solution $\vec{u}$ is measured at few points that lie sufficiently dense in a subset Ω_m of Ω so that they form a triangulation of Ω_m. Then, an approximation $\vec{u}^\varepsilon$ of the solution and its derivatives is computed from the measured data using Finite Elements. Here ε denotes the error due to the approximation and measurement. The residual on the region Ω_m is a functional that is equal to zero on the exact solution and positive for other functions. We use the following form of the residuum:

$$\ell(\vec{u}^\varepsilon, \vec{\psi}, \vec{p}) = \int\limits_{\Omega_m} \sqrt{a}\rho \vec{u}^\varepsilon_{tt} \vec{\psi} d\xi^1 d\xi^2 + a^K_m[\vec{u}^\varepsilon, \vec{\psi}; \vec{p}], \quad \vec{\psi} \in \vec{V}^K(\Omega),$$

where a^K_m is defined in the same way that a^K but with the integration over Ω_m instead of Ω. The dependence of a^K_m on $\vec{p}$ is indicated, whereas the dependence on ν_1 and ν_2 is omitted. Let Π be a set that a priory bounds the exact parameter vector $\vec{p}$. An approximation $\vec{p}^{\,\varepsilon}$ is being found as the minimizer of the functional ℓ, that is

$$\vec{p}^{\,\varepsilon} = \arg\min_{\vec{p} \in \Pi} \int\limits_0^T \ell(\vec{u}^\varepsilon, \vec{\psi}^{t,\vec{p},\varepsilon}, \vec{p}) dt, \tag{14}$$

where the auxiliary function $\vec{\psi}^{t,\vec{p},\varepsilon}$ satisfies the equation

$$(\vec{\psi}^{t,\vec{p},\varepsilon}, \omega)_{\vec{V}^K_\varepsilon} = \ell(\vec{u}^\varepsilon, \vec{\omega}, \vec{p}), \quad \forall \vec{\omega} \in \vec{V}^K_\varepsilon, \quad t \in [0,T].$$

Here, $\vec{V}^K_\varepsilon$ is a finite element approximation of the space $\vec{V}^K(\Omega)$. Because of the linearity of both $\ell(\vec{u}^\varepsilon, \vec{\psi}, \vec{p})$ and $\vec{\psi}^{t,\vec{p},\varepsilon}$ with respect to $\vec{p}$, the integral in (14) is a quadratic form, i.e.

$$\int\limits_0^T \ell(\vec{u}^\varepsilon, \vec{\psi}^{t,\vec{p},\varepsilon}, \vec{p}) dt = \vec{p}\,' Q_\varepsilon \vec{p} - 2 q'_\varepsilon \vec{p} + d_\varepsilon =: G^\varepsilon_{\Omega_m}(\vec{p}). \tag{15}$$

The matrix Q_ε has the following structure:

$$Q_\varepsilon = \int\limits_0^T R'_\varepsilon(t) D_\varepsilon^{-1} R_\varepsilon(t) dt, \tag{16}$$

where R_ε is some matrix and D_ε^{-1} is a positive definite matrix.

Proposition 3 (see [14]). *If solutions of (13) possess the following additional regularity:* $\int_0^T \|\vec{u}_{tt}\|^2_{L_2(\Omega_m)} \leq \infty$, *then the approximation* $\vec{u}^\varepsilon$ *can be chosen such that* $\int_0^T \left(\|\vec{u}^\varepsilon_{tt} - \vec{u}_{tt}\|^2_{L_2(\Omega_m)} + \|\vec{u}^\varepsilon - \vec{u}\|^2_{\vec{V}^K(\Omega_m)}\right) dt \to 0$ *as* $\varepsilon \to 0$.

Proposition 4. (see [14]) *Let* Π^0 *be the set of the parameters that are compatible with the exact solution. There is a function* $Z(\varepsilon)$ *such that* $Z(\varepsilon) \to 0$ *as* $\varepsilon \to 0$, *and*

$$\sup_{\vec{p} \in \Pi^0} \left|G^\varepsilon_{\Omega_m}(\vec{p}) - G^\varepsilon_{\Omega_m}(\vec{p}^{\,\varepsilon})\right| \leq Z(\varepsilon). \quad \square$$

We assume in addition that Ω_m depends on ε and may shrink with ε. Suppose that the function $G^\varepsilon_{\Omega_m}(\vec{p})$ defined by (15) has the following properties:

G1. There is a unique minimizing element $\vec{p}^{\,\varepsilon} \in \Pi$ of the function $G^\varepsilon_{\Omega_m}(\cdot)$.
G2. There is $\nu > 0$ and a positive function $X(\Omega_m, \varepsilon)$ such that, for any $\vec{p} \in \Pi$,

$$G^\varepsilon_{\Omega_m}(\vec{p}) - G^\varepsilon_{\Omega_m}(\vec{p}^{\,\varepsilon}) \geq X(\Omega_m, \varepsilon)\|\vec{p} - \vec{p}^{\,\varepsilon}\|^\nu.$$

Theorem 5 (see [14]). *If* $Z(\varepsilon)/X(\Omega_m, \varepsilon) \to 0$ *as* $\varepsilon \to 0$, *then the set of all parameters compatible with the exact solution consists of a unique element* $\vec{p}^{\,0}$, *and* $\vec{p}^{\,\varepsilon} \to \vec{p}^{\,0}$ *as* $\varepsilon \to 0$.

Remark 6. Properties G1 and G2 express the positive definiteness of the matrix Q_ε (this matrix depends on the domain Ω_m). As a rule, the function X decreases with Ω_m and is almost insensitive w.r.t. ε. Therefore, if Ω_m decreases not too quick with ε, the conditions of the Theorem are expected to be satisfied. Note that the matrix under the integral in (16) is positive semi-definite for each t. The resulting matrix Q_ε has a good chance to be positive definite and well conditioned, if the zero spaces of $R'_\varepsilon(t)D_\varepsilon^{-1}R_\varepsilon(t)$ vary with t. This corresponds to the accumulation of the information during the observation.

3.2 Last Squares Method

The formal differentiation of (13) w.r.t. the parameters yields the so-called variational equations

$$\int_\Omega \sqrt{a}\rho\vec{u}_{j\,tt}\vec{\varphi} + a^K[\vec{u}_j, \vec{\varphi}; \vec{p}] + a^K[\vec{u}, \vec{\varphi}; \vec{1}_j] = 0. \tag{17}$$

Here, the following notation is used:

$$\vec{u}_j = \partial\vec{u}/\partial p_j, \quad \vec{1}_j = (0, \ldots, \overset{j}{1}, \ldots, 0)\,.$$

Theorem 7 (see [12]). *Let* $\vec{u}^0 = 0$, $\vec{u}^{0'} = 0$, $v_\alpha(\cdot) \in H^3(0,T)$, $v_\alpha(0) = 0$, $v'_\alpha(0) = 0$, $\alpha = 1, 2$. *Then the solution* $\vec{u}$ *of (13) is continuously differentiable w.r.t.* $\vec{p}$. *The partial derivatives can be computed from the variational equations (17).* $\square$

The parameters fitting can be done using a gradient descent method, where the gradient is computed using (17). Moreover, higher derivatives exist and can be computed using a sequence of variational equations similar to (17). This makes possible the application of SQP methods.

3.3 *Simulation*

To illustrate theoretical results, consider a cylindric shell with a piezoelectric pair as in Figure 3.

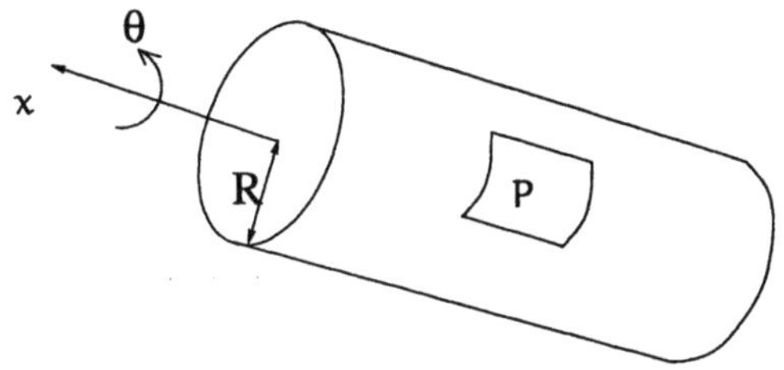

Figure 3. Cylindric shell; $s = x/R$ and θ is the polar angle

The coordinates of the middle surface are defined through the variable $s = x/R$ and the polar angle θ.

The equations read:

$$\rho u_{1tt} + (a u_{1s} + b u_{2\theta} + b u_3)_s + (c u_{2s} + c u_{1\theta})_\theta = m(t) K (I_{\Omega_p})_\theta,$$

$$\rho u_{2tt} + (b u_{1s} + a u_{2\theta} + a u_3)_\theta + (c u_{2s} + c u_{1\theta})_s = m(t) K (I_{\Omega_p})_\theta + n(t) G (I_{\Omega_p})_\theta,$$

$$\rho u_{3tt} + b u_{1s} + a u_{2\theta} + a u_3 + \Delta(k \Delta u_3) = n(t) G \Delta I_{\Omega_p} - n(t) G - m(t) K.$$

Here, the notation is used:

$$m(t) = \nu_1(t) + \nu_2(t), \quad n(t) = \nu_1(t) - \nu_2(t).$$

The coefficients ρ, a, b, c, k are of the form:

$$\rho(s,\theta) = \rho_p I_{\Omega_p} + (1 - I_{\Omega_p})\rho_s, \quad a(s,\theta) = a_p I_{\Omega_p} + (1 - I_{\Omega_p}) a_s,$$

$$b(s,\theta) = b_p I_{\Omega_p} + (1 - I_{\Omega_p}) b_s, \quad c(s,\theta) = c_p I_{\Omega_p} + (1 - I_{\Omega_p}) c_s,$$

$$k(s,\theta) = k_p I_{\Omega_p} + (1 - I_{\Omega_p}) k_s,$$

Assume that ρ_s and ρ_p are known parameters because they can be easily measured. Thus, the parameters $a_p, b_p, c_p, k_p, a_s, b_s, c_s, k_s, K, G$ are to estimate.

Results

Here, the process of estimation based on the differentiability of solution w.r.t. parameters is shown. The transversal deflection is measured at 20 spatial points for 100 times with the error equal to 5%. The least squares problem is solved using a gradient descent method. The gradient is computed from the corresponding variational equations. In Figure 4, the horizontal axes represent the number of steps of the gradient descent method.

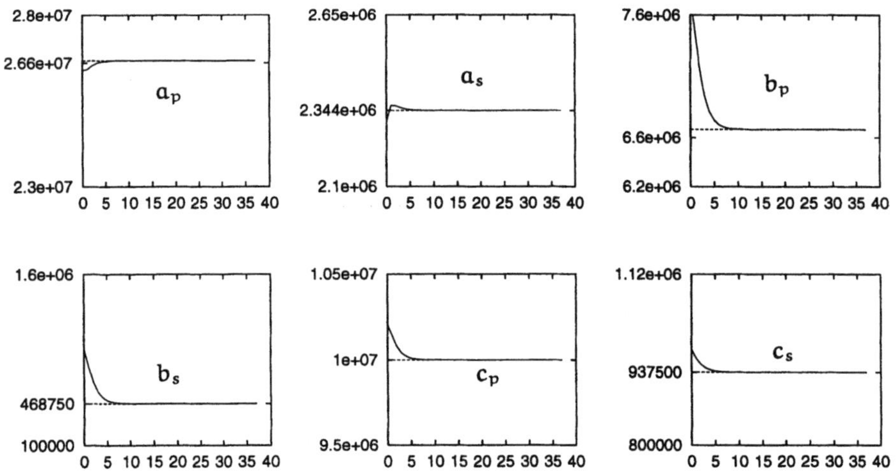

Figure 4. Estimations of several parameters. Error is equal to 5%

4 HOMOGENIZATION

Consider a von Kármán plate with surface mounted numerous piezoelectric actuators that form a self similar periodic structure (see Figure 5).

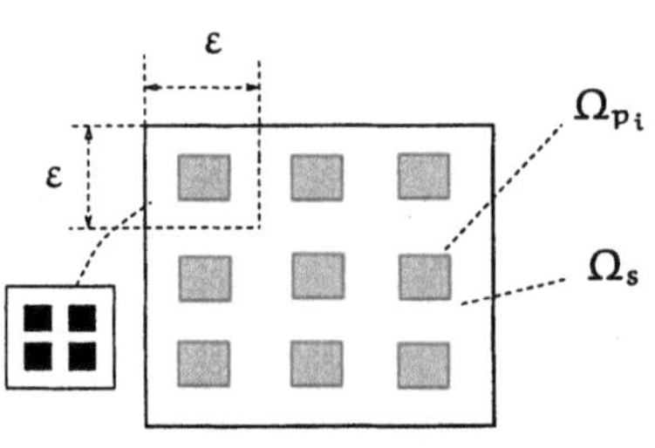

Figure 5. Self similar structure. Refinement of each cell if $\varepsilon := \varepsilon/2$

The inverse piezoelectric effect and the interaction with the surrounding substance are omitted but the geometrical nonlinearities are accounted. The dynamic equations can be written (compare with 10) as follows.

PROBLEM P_ε.

$$\rho\left(\tfrac{x}{\varepsilon}\right) w^\varepsilon_{tt} - \operatorname{div}\left(\mu\left(\tfrac{x}{\varepsilon}\right)\nabla w^\varepsilon_{tt}\right) + \Delta\left(\gamma\left(\tfrac{x}{\varepsilon}\right)\Delta w^\varepsilon\right) - \tfrac{\partial}{\partial x_\alpha}\left(\sigma^\varepsilon_{\alpha\beta} w^\varepsilon_{x_\beta}\right) =$$

$$K\Delta\left(V(t,x)I\left(\tfrac{x}{\varepsilon}\right)\right) + G\tfrac{\partial}{\partial x_\alpha}\left(w^\varepsilon_{x_\alpha} V(t,x) I\left(\tfrac{x}{\varepsilon}\right)\right).$$

$$\rho\left(\tfrac{x}{\varepsilon}\right) u^\varepsilon_{\alpha tt} - \tfrac{\partial}{\partial x_\beta}\sigma^\varepsilon_{\alpha\beta} = G\tfrac{\partial}{\partial x_\alpha}\left(V(t,x)I\left(\tfrac{x}{\varepsilon}\right)\right), \quad \alpha = 1,2.$$

Here, $\sigma^\varepsilon_{\alpha\beta} = \ell_{\alpha\beta\lambda\eta}\left(\tfrac{x}{\varepsilon}\right)\left(u^\varepsilon_{\lambda x_\eta} + u^\varepsilon_{\eta x_\lambda} + w^\varepsilon_{x_\lambda} w^\varepsilon_{x_\eta}\right)$ is the in-plane stress tensor, $\ell_{\alpha\beta\lambda\eta}$, ρ, μ, γ, I are 1×1-periodic piecewise-constant discontinuous functions, $V(t,x)$ is a distributed applied voltage. The inertia of mechanical moments is taken into account through the term $\operatorname{div}\left(\mu\left(\tfrac{x}{\varepsilon}\right)\nabla w^\varepsilon_{tt}\right)$. The boundary and initial conditions are imposed,

$$\begin{aligned} &w|_{\partial\Omega} = 0, \quad \left.\frac{\partial w}{\partial \vec{n}}\right|_{\partial\Omega} = 0, \quad w|_{t=0} = w_0, \quad w_t|_{t=0} = w'_0, \\ &u_\alpha|_{\partial\Omega} = 0, \quad u_\alpha|_{t=0} = u_{\alpha 0}, \quad u_{\alpha t}|_{t=0} = u'_{\alpha 0}. \end{aligned} \tag{18}$$

A homogenization procedure based on two-scale convergence (see [15] and [16]) yields the following limiting equations with constant coefficients.

PROBLEM P.

$$\begin{aligned} \hat{\rho} w_{tt} - \hat{\mu}\Delta w_{tt} + \hat{\gamma}\Delta^2 w - \tfrac{\partial}{\partial x_\alpha}\left(\hat{\sigma}_{\alpha\beta} w_{x_\beta}\right) \\ = K\hat{I}\Delta V(t,x) + G\hat{J}_{\alpha\beta}\tfrac{\partial}{\partial x_\alpha}\left(w_{x_\beta} V(t,x)\right), \end{aligned}$$

$$\hat{\rho} u_{\alpha tt} - \tfrac{\partial}{\partial x_\beta}\hat{\sigma}_{\alpha\beta} = G\hat{J}_{\alpha\beta}\tfrac{\partial}{\partial x_\beta} V(t,x), \quad \alpha = 1,2.$$

The stress tensor contains only constant coefficients:

$$\hat{\sigma}_{\alpha\beta} = \hat{\ell}_{\alpha\beta\lambda\eta}\left(u_{\lambda x_\eta} + u_{\eta x_\lambda} + w_{x_\lambda} w_{x_\eta}\right).$$

Boundary and initial values are the same, see (18). The coefficients are given by

$$\hat{\rho} = \langle\rho(y)\rangle, \quad \hat{\mu} = \langle\mu(y)\rangle, \quad \hat{\gamma} = \langle 1/\gamma(y)\rangle^{-1}, \quad \hat{I} = -\langle 1/\gamma(y)\rangle^{-1}\langle I(y)/\gamma(y)\rangle,$$

$$\hat{\ell}_{\alpha\beta\,\lambda\eta} = \left\langle \ell_{\alpha\beta\,\tau\theta}(y)\left(\delta_{\tau\lambda}\delta_{\theta\eta} + \frac{\partial N_{\tau\lambda\eta}(y)}{\partial y_\theta} + \frac{\partial N_{\theta\lambda\eta}(y)}{\partial y_\tau}\right)\right\rangle,$$

$$\hat{J}_{\alpha\beta} = \left\langle \ell_{\alpha\beta\,\tau\theta}(y)\left(\frac{\partial M_\tau(y)}{\partial y_\theta} + \frac{\partial M_\theta(y)}{\partial y_\tau}\right) + \delta_{\alpha\beta} I(y)\right\rangle.$$

The auxiliary functions $N_{1\lambda\eta}(y)$, $N_{2\lambda\eta}(y)$, $(\lambda,\eta) = (1,1),(2,2),(1,2),(2,1)$, $M_1(y)$, and $M_2(y)$ are defined by elliptic systems which are to be solved on the unit square $[0,1] \times [0,1]$.

$$\begin{cases} \frac{\partial}{\partial y_\beta}\left[\ell_{\mu\nu1\beta}(y)\left(\delta_{\mu m}\delta_{\nu n} + \frac{\partial N_{\mu\lambda\eta}}{\partial y_\nu} + \frac{\partial N_{\nu\lambda\eta}}{\partial y_\mu}\right)\right] = 0, \\ \frac{\partial}{\partial y_\beta}\left[\ell_{\mu\nu2\beta}(y)\left(\delta_{\mu m}\delta_{\nu n} + \frac{\partial N_{\mu\lambda\eta}}{\partial y_\nu} + \frac{\partial N_{\nu\lambda\eta}}{\partial y_\mu}\right)\right] = 0. \end{cases}$$

$$\begin{cases} \frac{\partial}{\partial y_\beta}\left[\ell_{\mu\nu1\beta}(y)\left(\frac{\partial M_\mu}{\partial y_\nu} + \frac{\partial M_\nu}{\partial y_\mu}\right) + \delta_{1\beta}I(y)\right] = 0, \\ \frac{\partial}{\partial y_\beta}\left[\ell_{\mu\nu2\beta}(y)\left(\frac{\partial M_\mu}{\partial y_\nu} + \frac{\partial M_\nu}{\partial y_\mu}\right) + \delta_{2\beta}I(y)\right] = 0. \end{cases}$$

The boundary conditions must be $[0,1]\times[0,1]$-periodic, which yields the solvability and the uniqueness up to a constant.

The following theorem (see [17]) states the relation between problems $\mathbf{P}_\varepsilon$ and $\mathbf{P}$.

Theorem 8. *If* $w_0 = 0$, $w_0' = 0$, $u_{\alpha 0} = 0$, $u_{\alpha 0}' = 0$, $V \in H^2(0,T;L_2(\Omega))$, $V(0,\cdot) = 0$, *then problem* $\mathbf{P}$ *has a unique strong solution* $(w,\ u_\alpha)$ *that possesses the following regularity:*

$$w \in C([0,T];H_0^2(\Omega)) \cap C^1([0,T];H_0^1(\Omega)) \cap \\ W_\infty^2(0,T;H_0^1(\Omega)) \cap L_\infty(0,T;H^3(\Omega)).$$

$$u_\alpha \in C([0,T];H_0^1(\Omega)) \cap C^1([0,T];L_2(\Omega)) \cap \\ W_\infty^2(0,T;L_2(\Omega)) \cap L_\infty(0,T;H^2(\Omega)).$$

The set G_ε *of all limits of Galerkin approximations of* $\mathbf{P}_\varepsilon$ *shrinks to* $(w,\ u_\alpha)$ *as* $\varepsilon \to 0$, *that is all sequences* $(w^\varepsilon, u_\alpha^\varepsilon) \in G_\varepsilon$ *converge to* (w, u_α) *in*

$$C^1([0,T];H_0^2(\Omega)) \times C^1([0,T];H_0^1(\Omega))$$

as $\varepsilon \to 0$. □

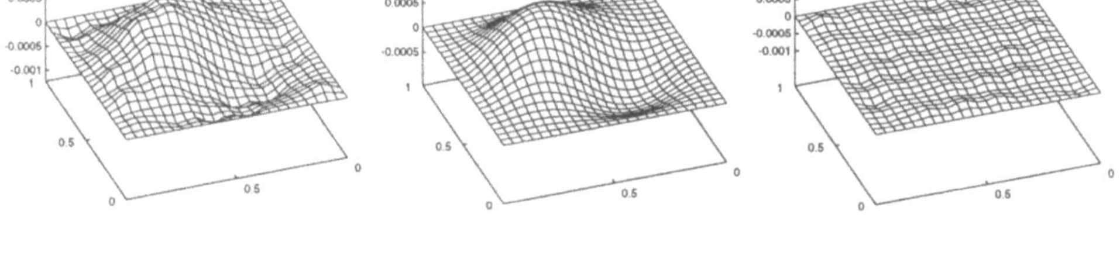

Figure 6. Snapshot of solutions to $\mathbf{P}_\varepsilon$ and $\mathbf{P}$ and their difference. The number of uniformly distributed piezoelectric actuators is equal to 36

In conclusion of this subsection, we note that the general fully coupled system can also be homogenized. Nevertheless, the result obtained in [10] is weaker then the claim of the previous theorem: the relation between the original and limiting equations is very weak so that the L_2-convergence holds for u_α^ε, and the H^1-convergence holds for w^ε. Another feature consist in arising new nonlinear terms in the limiting equations. This leads to the idea to control the creations of such terms to improve the structure of the limiting equations with respect to their controllability and observability.

5 Using Homogenization for Control Design

For brevity, consider the control design for equation (11). Assume that the number of piezoelectric actuators is controlled by the refinement parameter ε (see Figure 6). Then, the equation reads

$$\rho w_{tt}^\varepsilon + \Delta\Big(\gamma \Delta w^\varepsilon\Big) = K\Delta\Big(\sum_{i=1}^{n_p(\varepsilon)} v_i(t) \cdot I_{\Omega_{p_i}}\Big). \tag{19}$$

We begin with a distributed voltage $V \in H^1(0,T;L_2(\Omega))$ that defines voltages applied to the piezoelectric actuators as follows:

$$v_i(t) = \frac{1}{\text{meas}(\Omega_{p_i})} \int_{\Omega_{p_i}} V(t,x)dx, \quad i = 1,\dots,n_p(\varepsilon). \tag{20}$$

The voltages $v_i(t), i = 1,\dots,n_p(\varepsilon)$, define the function

$$V^\varepsilon(t,x) = \begin{cases} V(t,x), & x \in \Omega_s, \\ \text{meas}(\Omega_{p_i})^{-1} \int_{\Omega_{p_i}} V(t,x)dx, & x \in \Omega_{p_i}. \end{cases}$$

V^ε is constant on each piezoelectric actuator and approximates the distributed control V. Now, we can rewrite equation (19) as follows

$$r\left(\tfrac{x}{\varepsilon}\right) w_{tt}^\varepsilon + \Delta(g\left(\tfrac{x}{\varepsilon}\right) \Delta w^\varepsilon) = K\Delta\big(V^\varepsilon I\left(\tfrac{x}{\varepsilon}\right)\big), \tag{21}$$

where $r(y)$, $g(y)$, and $I(y)$ are appropriate 1×1 periodic functions. The homogenization procedure yields a limiting equation of the form

$$\hat{r} w_{tt} + \hat{g}\Delta^2 w = K\hat{I}\Delta V. \tag{22}$$

One can prove that $w^\varepsilon \to w$, in $C([0,T];\mathcal{E}^{0,\nu}(\bar{\Omega}))$, $\nu < 1$, where $\mathcal{E}^{0,\nu}(\bar{\Omega}) \subset C(\bar{\Omega})$ is the Hölder space. Using the notation, $v(t,x) = K\hat{I}\Delta V(t,x)$, we rewrite (22) as follows

$$\hat{r} w_{tt} + \hat{g}\Delta^2 w = v(t,x). \tag{23}$$

Note that (23) does not contain unbounded operators on the right-hand-side. Let $\nu(t,x)$ be the optimal control for (23), then the optimal control for (22) is the solution of the Laplace equation

$$\Delta V = (K\hat{I})^{-1}\nu(t,x), \quad V|_{\partial\Omega} = 0. \tag{24}$$

Theorem 9. *Let $\nu(t,x)$ be the optimal control in (23), and $\nu_i(t)$, $i = 1,\dots,m_p$, are defined through (24) and (20). Then*

$$\lim_{\varepsilon\to 0} \|w^\varepsilon - w\|_{C([0,T];\mathcal{E}^{0,\nu}(\bar{\Omega}))} = 0,$$

where w^ε and w are solutions of (19) and (23), respectively. □

The proof follows from the fact that $V^\varepsilon \rightharpoonup V$ in $H(0,T;L_2(\Omega))$ and from the results of [17] about the convergence rate of w^ε to a limiting solution.

6 Stabilization of Shells with Piezoelectric Actuators

A method for stabilizations of shells is described in [18]. An auxiliary open-loop control problem is stated and the dependence of the adjoint variable on the initial state vector of the shell equations is computed.

For a shell with two symmetrically mounted piezoelectric patches, the variation (12) of the total energy is of the form

$$a^K[\vec{u},\vec{\varphi};\nu_1(t),\nu_2(t)] = a_0^K[\vec{u},\vec{\varphi}] + \vec{\nu}(t)\cdot\vec{L}[\vec{\varphi}],$$

with $\vec{\nu} = (\nu_1,\nu_2)$ and $\vec{L} = (L_1, L_2)$, where L_1 and L_2 are clearly to obtain from (12). This representation holds in the case of several pairs of actuators, if the appropriate dimension of the vectors $\vec{\nu}$ and $\vec{L}$ is meant. The auxiliary functional is defined as follows:

$$J_T = \frac{\ell}{2}\int_\Omega |\vec{u}(T)|^2 d\xi_1 d\xi_2 + \frac{b}{2}\int_\Omega |\vec{u}_t(T)|^2 d\xi_1 d\xi_2 + \frac{c}{2}\int_0^T |\vec{\nu}(t)|^2 dt.$$

The coupled system of the shell and adjoint equations reads:

$$\int_\Omega \sqrt{a}\rho\vec{u}_{tt}\vec{\varphi}\,d\xi^1 d\xi^2 + a_0^K[\vec{u},\vec{\varphi}] = -c^{-1}\vec{L}^K[\vec{\pi}]\cdot\vec{L}^K[\vec{\varphi}], \quad \forall\vec{\varphi}\in\vec{V}^K(\Omega),$$

$$\int_\Omega \sqrt{a}\rho\vec{\pi}_{tt}\vec{\psi}\,d\xi^1 d\xi^2 + a_0^K[\vec{\pi},\vec{\psi}] = 0, \quad \forall\vec{\psi}\in\vec{V}^K(\Omega),$$

$$\vec{u}(0) = \vec{u}^0, \quad \vec{u}_t(0) = \vec{u}^{0\prime},$$

$$\vec{\pi}(T) = b\,\vec{u}_t(T), \quad \vec{\pi}_t(T) = -\ell\,\vec{u}(T).$$

The optimal control $\vec{v}_T^0$ of the auxiliary problem is defined through the adjoint variable as follows

$$\vec{v}_T^0(t;\vec{u}^0,\vec{u}^{0\prime}) = c^{-1}\,\vec{L}^K[\vec{\pi}].$$

An approximation of the optimal stabilizing control is given by the formula

$$\vec{v}_{stb}(\vec{u}^0,\vec{u}^{0\prime}) = \vec{v}_T^0(0;\vec{u}^0,\vec{u}^{0\prime}). \tag{25}$$

Proof (Sketch of the proof). The following representation holds

$$\vec{v}_T(t;\vec{u}^0,\vec{u}^{0\prime}) = c^{-1}\mathcal{B}\mathcal{P}(t,T)\begin{pmatrix}\vec{u}(t;\vec{u}^0,\vec{u}^{0\prime})\\ \vec{u}_t(t;\vec{u}^0,\vec{u}^{0\prime}\end{pmatrix},$$

where the input operator $\mathcal{B}$ is defined by the relation: $\mathcal{B}\vec{\pi} = \vec{L}^K[\vec{\pi}]$ for all $\vec{\pi} \in \vec{V}^K(\Omega)$, while $\mathcal{P}(t,T)$ is the solution of the corresponding operator Ricatti equation (see, e.g., [6]). According to (25), we obtain

$$\vec{v}_{stb}(\vec{u}^0,\vec{u}^{0\prime}) = c^{-1}\mathcal{B}\mathcal{P}(0,T)\begin{pmatrix}\vec{u}^0\\ \vec{u}^{0\prime}\end{pmatrix}.$$

It is known that the optimal stabilizing control is defined by

$$\vec{v}_{opt}(\vec{u}^0,\vec{u}^{0\prime}) = c^{-1}\mathcal{B}\mathcal{P}\begin{pmatrix}\vec{u}^0\\ \vec{u}^{0\prime}\end{pmatrix},$$

where $\mathcal{P}$ is a solution of the corresponding stationary Ricatti equation. Moreover, $\mathcal{P}(0,T) \to \mathcal{P}$ as $T \to \infty$. Thus, $\vec{v}_{stb}(\vec{u}^0,\vec{u}^{0\prime})$ approximates $\vec{v}_{opt}(\vec{u}^0,\vec{u}^{0\prime})$ and, therefore, stabilizes the system, whenever T is sufficiently large.

7 Active Sound Reduction

The control design that we want to discuss in this section differs from that discussed in Section 6 because the control actions must be computed in real-time from some incomplete information about the state of the object. This information is the history of the signal measured on sensors up to the current time. This section is closely related to the investigation [2] and [19].

7.1 Physical Model

The physical model is a container with a sound source within them. The problem consists in the reduction of the sound that is radiated by the structure. The control circuit includes piezoelectric actuators and sensors delivering information about the current state (see Figure 7).

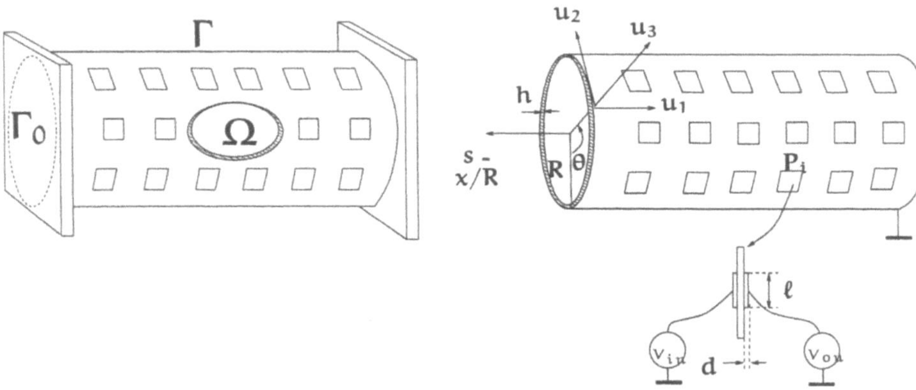

Figure 7. Physical Model

7.2 *Coupled Acoustic Structure*

This system describes the interface between acoustic waves and a cylindric shell

$$\phi_{tt} = c^2 \Delta \phi \quad \text{on} \quad \Omega,$$

$$\nabla \phi \vec{n}|_{\Gamma_0} = 0, \quad \phi_r|_\Gamma = u_{3t}.$$

$$\rho u_{1tt} + (a u_{1s} + b u_{2\theta} + b u_3)_s + (c u_{2s} + c u_{1\theta})_\theta = K v_i(t)(I_{p_i})_s,$$

$$\rho u_{2tt} + (b u_{1s} + a u_{2\theta} + a u_3)_\theta + (c u_{2s} + c u_{1\theta})_s = L v_i(t)(I_{p_i})_\theta, \quad (\theta, s) \in \Gamma,$$

$$\rho u_{3tt} + b u_{1s} + a u_{2\theta} + a u_3 + \Delta k \Delta u_3 = G v_i(t) \Delta I_{p_i} - \hat{\rho} \phi_t(t, R, \theta, s).$$

Boundary conditions read: $u_1 = u_2 = u_3 = u_{3s} = 0$ on Γ_0; u_1, u_2, u_3 are periodic in θ. Here, ϕ is the acoustic potential; $u_i, i = 1, 2, 3$, are the tangential and transversal displacements; θ is the polar angle; $s = x/R$ is the specific length. The coefficients a, b, c, k are of the form: $a(s, \theta) = a_{piez} I_p + a_{subs}(1 - I_p)$, where I_p is the indicator function of the region related to the piezoelectric actuators.

7.3 *Reduced Basis Method and Proper Orthogonal Decomposition*

First, the coupled acoustic system is approximated very carefully to obtain a sufficient number of "snapshots" of the solution or its time derivatives at various times:

$$y_j(\theta, s) = \begin{pmatrix} u_1(t_j, \theta, s) \\ u_2(t_j, \theta, s) \\ u_3(t_j, \theta, s) \end{pmatrix} \text{ or } y_j(\theta, s) = \begin{pmatrix} u_{1t}(t_j, \theta, s) \\ u_{2t}(t_j, \theta, s) \\ u_{3t}(t_j, \theta, s) \end{pmatrix}, \quad j = 1, \ldots, N_s$$

where $u_k(t,\theta,s), k = 1,2,3$, is a solution of the coupled elastic-acoustic system. Then, the covariation matrix,

$$C_{kl} = \frac{1}{N_s}\langle y_k, y_l\rangle,$$

is computed along with the eigenvalues and eigenvectors,

$$\lambda_1 \geq \lambda_2 \geq \ldots \lambda_{N_s} \geq 0; \quad \alpha^1, \alpha^2, \ldots, \alpha^{N_s}; \quad \alpha^k \cdot \alpha^l = \begin{cases} 0, & k \neq l \\ 1/(\lambda_k N_s), & k = l. \end{cases}$$

New basis functions Φ_j, $j = 1, \ldots, N_r$, are chosen in such a way:

$$\Phi_j = \sum_{i=1}^{N_s} \alpha_i^j y_i, \quad j = 1, \ldots, N_r << N_s, \quad (N_s \approx 100, \quad N_r \approx 3 \div 6).$$

Thus, the reduced basis yields a low dimensional Galerkin approximation,

$$\dot{x}(t) = A\,x(t) + B\,V_{inp}(t) + g(t), \quad V_{out}(t) = C\,x(t), \quad x \in R^{2N_r}, \tag{26}$$

of the coupled acoustic system. Here, N_r is the dimension of the reduced basis; V_{inp} and V_{out} are the vectors of applied and measured voltages on the piezoelectric actuators and sensors, respectively; $g(t)$ is an unpredictable disturbance due to the acoustic excitation.

It is known from H^∞ theory that the feedback control is given by the relation

$$V_{inp}(t) = -Kx_c(t),$$

where $x_c(t)$ is the solution of the so called compensator equation,

$$\dot{x}_c(t) = A_c x_c(t) - FV_{out}(t),$$

$$A_c = A - FC - BK + \gamma^{-2}\hat{Q}\Pi,$$

$$K = R^{-1}B'\Pi, \quad F = \left[I - \gamma^{-2}\hat{\Pi}\Pi\right]^{-1}\hat{\Pi}C'\hat{R}^{-1}.$$

Here, the matrices Π and $\hat{\Pi}$ satisfy two Riccati equations,

$$\Pi A + A'\Pi - \Pi\left[BR^{-1}B' - \gamma^{-2}\hat{Q}\right]\Pi + Q = 0,$$

$$\hat{\Pi}A' + A\hat{\Pi} - \hat{\Pi}\left[B\hat{R}^{-1}B' - \gamma^{-2}Q\right]\hat{\Pi} + \hat{Q} = 0,$$

related to (26) and to a quadratic functional of the form

$$J = \int_0^\infty \{\langle Qx, x\rangle + \langle RV_{inp}, V_{inp}\rangle - \gamma^2|g(t)|^2\}\,dt. \tag{27}$$

Thus, the control circuit is implemented as follows:

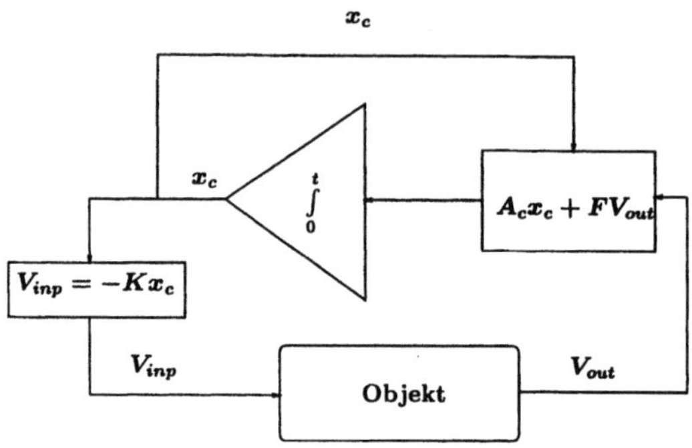

7.4 Results of the Simulation

The dotted curve denotes a free multi-frequency oscillation. The solid curve represents damped oscillations. The duration is equal to 0.001 s. The number of actuators and sensors is equal to 36

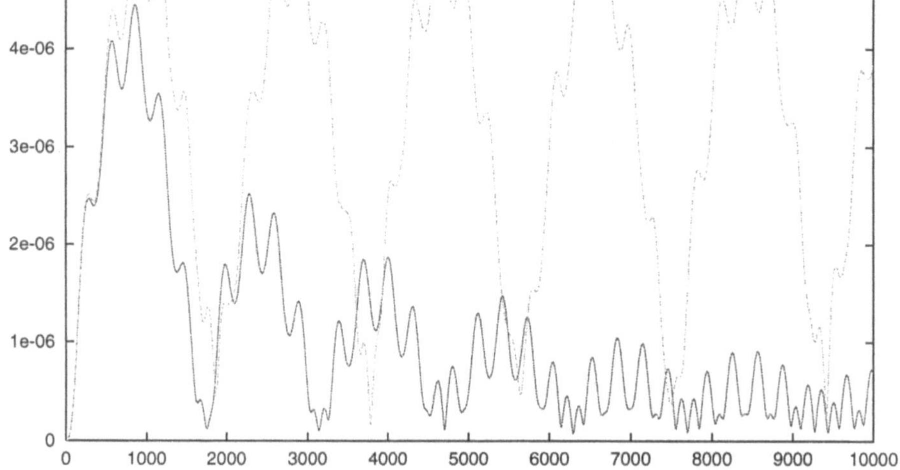

7.5 Another Auxiliary Functional

The utilization of other auxiliary functionals for the feed-back control design in (26) is possible and, maybe, preferable. For example, the functional

$$J = \int_0^\infty \left\{ \sqrt{\langle Qx, x\rangle} + \langle R\, V_{\mathrm{inp}}, V_{\mathrm{inp}}\rangle - \gamma^2 |g(t)|^2 \right\} dt \qquad (28)$$

provides a slower grows with respect to the state vector than (27). This guarantees the existence of the value function of the control problem independently from the disturbance g in (26) even though the observation contains arbitrary large disturbances. The computation of the value function is based on some stochastic procedures (see [20] and [21]). Nevertheless the final formulae are pure deterministic an can be implemented numerically. Roughly speaking, the optimal control at the state $x(t)$ and subject to the measured signal $\{V_{\mathrm{out}}(\tau), 0 \le \tau \le t\}$ is computed through the solution of the problem

$$h^0 \to \min_{|h|=1} \{\langle \mathcal{D}h, h\rangle - \langle x(t), h\rangle\},$$

where $\mathcal{D}$ is a positive semidefinite matrix that is easily computed using the problem data and the signal $\{V_{\mathrm{out}}(\tau), 0 \le \tau \le t\}$.

REFERENCES

1. A. Preumont, *Vibration Control of Active Structures.* Solid Mechanics and its Applications. **50**, Kluver Academic Publisher (Dordrecht-Boston-London) 1997.
2. H. T. Banks, R. C. Smith and Y. Wang, *Smart material structures: modeling, estimation and control.* Wiley, Chichester 1996.
3. J. Zelenka, *Piezoelectric Resonators and their Applications.* Studies in Electrical and Electronic Engineering. **24**, Elsevier 1986.
4. G. A. Maugin, *Continuum mechanics of electromagnetic solids.* North-Holland series in Applied Mathematics and mechanics. **33**, North-Holland 1987.
5. T. Basar and P. Bernhard, H^∞*-Optimal Control and Related Minimax Design Problems.* Birkhäuser, Boston 1987.
6. I. Lasiecka and R. Trigiani, *Control theory for partial equations: continuous and approximation theories. II. Abstract hyperbolic-like systems over a finite time horizon.* Encyclopedia of mathematics and its applications **75**, Edited by G.-C.Rota, Cambridge University Press 2000.
7. L. D. Landau and E. M. Lifschitz, *Elastizitätstheorie.* Akademie-Verlag (Berlin) 1975.
8. M. Bernadou, *Finite Element Methods for Thin Shell Problems.* Wiley (Chichester) 1996.
9. K.-H. Hoffmann and N. D. Botkin, *A Fully coupled model of a nonlinear thin plate excited by piezoelectric actuators.* Smart materials. Proceedings of the first caesarium, Bonn, Nov. 17–19, 1999. Edited by K.-H. Hoffmann, Springer 2001, available on the preprint server[1].
10. K.-H. Hoffmann and N. D. Botkin, *Homogenization of a fully coupled model of a nonlinear thin plate excited by piezoelectric actuators.* International conference: Optimal

[1] http://www.zib.de/dfg-echtzeit/Publikationen/index.html

Control of Complex Structures, Mathematical Research Center Oberwolfach, June, 4–10, 2000 (to appear in proceedings, available on the preprint server[2].
11. K.-H. Hoffmann and N. D. Botkin, *Oscillations of nonlinear thin plates excited by piezoelectric patches,* ZAMM: Z. angew. Math. Mech., **78** (1998), 495–503.
12. N. D. Botkin, *Estimation of parameters of a linear thin plate.* Analysis **17** (1997) 367-378.
13. N. D. Botkin, W. G. Litvinov, Y. I. Rubezhansky, *Models and problems for piezoelectric elastic bodies and thin shells.* DFG-Project "Echtzeit-Optimierung großer Systeme", Preprint 99-5. Preprintserver[3]
14. N. D. Botkin, *Estimation of parameters of thin plates excited by piezoelectric patches.* Preprint of the Technical University of Munich. M9704, 1997.
15. G. Allaire, *Homogenization and two-scale Convergence,* SIAM J. Math. Anal., **23 (6)** (1992), 1482-1518.
16. H. Haller, *Verbundwerkstoffe mit Formgedächtnislegierung – Mikromechanische Modellierung und Homogenisierung,* Dissertation, TU-München 1997.
17. K.-H. Hoffmann and N. D. Botkin, *Homogenization of von Kármán plates excited by piezoelectric patches,* ZAMM: Z. angew. Math. Mech., **133** (1999), 191–200.
18. K.-H. Hoffmann and N. D. Botkin, *Adaptive Materialien in der Echtzeitoptimierung.* DFG-Projekt "Echtzeitoptimierung großer Systeme", Arbeitsbericht über den ersten Förderungszeitraum von 1995-1997. München 1997 (available on the preprint server[4]).
19. H. T. Banks, R. C. H. Rosario and R. C. Smith, *Reduced order model feedback control design: computational studies for thin cylindric shells.* Report of the Center for Research in Scientific Computation. North Carolina State University. CRSC-TR98-24, June, 1998.
20. N. N. Krasovskii, *Control for dynamic system (in Russian).* Nauka (Moscow), 1985.
21. N. N. Krasovskii, *Control problems with incomplete information (in Russian).* Academy of Sciences of USSR, Ural Scientific Center, Institute of Mathematics and Mechanics (Sverdlovsk) 1984.

[2] `http://www.zib.de/dfg-echtzeit/Publikationen/index.html`

[3] `http://www.zib.de/dfg-echtzeit/Publikationen/index.html`

[4] `http://www.zib.de/dfg-echtzeit/Publikationen/index.html as Project report. I.`

Instantaneous Control of Vibrating String Networks

Ralf Hundhammer and Günter Leugering

Fachbereich Mathematik, Technische Universität Darmstadt, Darmstadt

Abstract We introduce an algorithm for instantaneous control of vibrating string networks, where, after semidiscretization of the time variable, the resulting elliptic optimization systems are solved for each single time step. The elliptic network problems themselves are solved by domain decomposition. An interpretation of the algorithm as a local optimality system is given, as well as an interpretation as a local discrete feedback law. Numerical results are presented for simulation and control of several string networks.

1 Introduction

We consider dynamic network structures in the spirit of [10] and look for algorithms to steer these networks from some initial state to the zero state. For this purpose we will briefly discuss our model problem and introduce some notation.

A network structure is described either by a planar or a threedimensional graph $G = (V, E)$, where V denotes the set of nodes, E the set of edges and $n_v := \#V$, $n_e := \#E$. In the case of a planar graph we have to distinguish the cases in which the evolution is onedimensional "out of plane" or twodimensional "in plane". Therefore we have to deal with a function $y(x,t)$, which is either scalar, two– or threedimensional. In the latter two cases local coordinate systems $(e_{i_j})_{i=1:n_e}^{j=1:d}$, $d \in \{2,3\}$ are needed. For instance, y on edge i has in the 2d-case the form

$$y_i(x,t) = u_i(x,t)e_i + w_i(x,t)e_i^{\perp},$$

where e_i denotes the local unit vector in direction of the edge. So u_i describes the longitudinal part of the displacement, while w_i gives the lateral one.

Multiple nodes in the graph are denoted by $v_J \in V_M \subset V$, while simple nodes are distinguished by their boundary conditions, which can be clamped (Dirichlet) $v_J \in V_D \subset V$, free (Neumann) $v_J \in V_N \subset V$ or controlled $v_J \in V_C \subset V$, where J is a node's index with $1 \leq J \leq n_v$. We have $V = V_D \cup V_N \cup V_C \cup V_M$. Let $\mathcal{E}_J$ be the index set of all edges adjacent at node v_J and $d(v_J) = d_J$ the number of these adjacent edges. Then for the inner nodes in the graph V_M the equation $V_M = \{v \in V \mid d(v) > 1\}$ holds.

Let furthermore K_i denote the local stiffness matrix of the system, l_i the length of the edge i and ϵ_{iJ} the orientation of the outer normal vector on the boundary (the nodes of the edge), i.e. $\epsilon_{iJ} := -1$ if the edge starts at node v_J and $\epsilon_{iJ} := 1$ if it ends there in the local coordinate system. We define $\epsilon_{iJ} := 0$ if edge i is not adjacent at node v_J.

With this notations, our model problem takes the following form:

$$\left.\begin{array}{lll}
\ddot{y}_i = K_i y_i'', & (0,l_i)\times(0,T) =: Q_i,\ i=1:n_e & (1.1)\\
y_d(v_D) = 0, & v_D \in V_D,\ d\in\mathcal{E}_D,\ t\in(0,T) & (1.2)\\
y_i(v_J) = y_j(v_J), & \forall\, i,j\in\mathcal{E}_J,\ v_J\in V_M,\ t\in(0,T) & (1.3)\\
\sum_{i\in\mathcal{E}_J} \epsilon_{iJ} K_i y_i'(v_J) = 0 & v_J\in V_M\cup V_N,\ t\in(0,T) & (1.4)\\
\epsilon_{iJ} K_i y_i'(v_C) = u_C & v_C\in V_C,\ i\in\mathcal{E}_C,\ t\in(0,T) & (1.5)\\
y_i(x,0) = \tilde{y}_i^0 & x\in(0,l_i),\ i=1:n_e & (1.6)\\
\dot{y}_i(x,0) = \tilde{y}_i^1 & x\in(0,l_i),\ i=1:n_e & (1.7)
\end{array}\right\} \quad (1)$$

where dots and primes refer to time and spatial derivatives, respectively.

The first equation describes the evolution of the system, the third and forth one the continuity and balance of forces conditions at inner nodes of the network, the remaining equations constitute the initial and boundary conditions. This system has been shown to be wellposed in the space setting

$$\dot{y}, \tilde{y}^1 \in H := \prod_{i=1}^{n_e} L^2(0,l_i)^d;$$

$$y, \tilde{y}^0 \in V := \left\{ y \in \prod_{i=1}^{n_e} H^1(0,l_i)^d \middle| y \text{ satisfies } (1.2), (1.3) \right\}$$

see [9].

2 Instantaneous Control

2.1 *Suboptimal Control*

The task of finding optimal- or LQR-controls for system (1) is hard from the numerical point of view. One has to deal with a global adjoint system running backwards in time, strongly coupled with the primal equation, which leads to many unknowns, at least if the network geometry becomes complicated and/or the discretization scheme becomes very fine.

In online applications one has to compute appropriate real time controls in the following sense: once a vibration (caused by some external force) is quantified at a certain measuring point of the network, one has to compute the control just in the time, the wave needs to propagate to the closest adjacent actuator in the network. Only a real time capable algorithm can ensure the system to be able to apply a reasonable control, when the wave reaches at the actuators. Consequently, the computational time needed, is given through the wave propagation velocity in the network's components.

We are interested in calculating stabilizing control laws immediately at each single time step, taking into account the geometry of the network (unlike absorbing control laws do, for instance). The basic idea of instantaneous control was, to our knowledge, first mentioned in [3] and successfully applied in [7] for Navier-Stokes equations. The principle is the following. In the first step the time variable is discretized by an appropriate discretization scheme. Afterwards the resulting elliptic problems have to be solved at each single time step by some algorithm. The calculated controls are then used to steer the system from a certain time step to the next.

Note, that after semidiscretization in time we still have to deal with network problems, but on the elliptic level. This will be done by domain decomposition.

We have successfully applied instantaneous control to the wave equation (also in the 2d case), see [2] for implementation details and numerical results for strings and membranes. In the meanwhile, we also combined domain decomposition and instantaneous control for some twodimensional examples.

2.2 Semidiscretization of the System

We consider the network problem (1) on a certain edge. Sometimes we will drop the index i, when this causes no confusion.

We have chosen Newmark's scheme for time-discretization, see [12] for details. In this method two parameters β and γ are introduced, which can be used to make the scheme either explicit, implicit or mixed. The usual choice of these parameters is $\beta = \frac{1}{4}$ and $\gamma = \frac{1}{2}$, in which case the method can be shown to conserve the discrete energy, see [12]. Therefore we shall not have any algorithmical damping effects, as they arise, e.g., in the implicit Euler scheme, which can be applied to the network equations rewritten as a first order system.

Let y^n denote the approximation of the solution of system (1) at time step n, say $t = n \cdot \delta t$, where δt denotes the step size of the time discretization. Then Newmark's scheme has the form

$$y_i^{n+1} = y_i^n + \delta t \cdot z_i^n + \delta t^2\big(\beta K_i(y_i^{n+1})'' + (\frac{1}{2} - \beta)K_i(y_i^n)''\big) \tag{1}$$

$$z_i^{n+1} = z_i^n + \delta t\big(\gamma K_i(y_i^{n+1})'' + (1-\gamma)K_i(y_i^n)''\big). \tag{2}$$

In addition at each time step, the following boundary and nodal conditions apply:

$$\begin{aligned}
y_i^{n+1}(v_D) &= 0 && \text{for } v_D \in V_D \\
\epsilon_{iC}K_i(y_i^{n+1})'(v_N) &= 0 && \text{for } v_N \in V_N \\
\epsilon_{iC}K_i(y_i^{n+1})'(v_C) &= u^{n+1} && \text{for } v_C \in V_C \\
y_i(v_J) &= y_j(v_J) && \forall\, i,j \in \mathcal{E}_J,\ v_J \in V_M, \\
\sum_{i\in\mathcal{E}_J} \epsilon_{iJ}K_i y_i'(v_J) &= 0 && v_J \in V_M \cup V_N.
\end{aligned} \tag{3}$$

Here, the artificial variable z^n gives an approximation of the time derivative of y^n.

From equations (1) and (2) we can now derive the desired elliptic systems.

$$\begin{aligned} y_i^{n+1} - \delta t^2 \beta K_i (y_i^{n+1})'' &= y_i^n + \delta t \cdot z_i^n + \delta t^2 (\frac{1}{2} - \beta) K_i (y_i^n)'' =: f_i^n \\ z_i^{n+1} - \delta t \gamma K_i (y_i^{n+1})'' &= z_i^n + \delta t (1-\gamma) K_i (y_i^n)'' =: g_i^n \end{aligned} \tag{4}$$

in addition to the boundary conditions (3). Note that the second equation in (4) is just an update for z, so that we have to deal with only one elliptic equation on the network in each time step.

In order to derive an optimality system for the elliptic problem, we have to define a cost functional. For the sake of simplicity let J be defined as

$$J(u^{n+1}) := \frac{1}{2} \sum_{v_C \in V_C} \sum_{i \in \mathcal{E}_C} (u_i^{n+1})^2 + \frac{1}{2} \cdot \sum_{i=1}^{n_e} \int_0^{l_i} k (y_i^{n+1})^2 + \rho_0 (z_i^{n+1})^2 dx. \tag{5}$$

In some sense, at each time step, we are looking for a control u, which brings the system to the zero state as far as possible at reasonable control costs. We will focus on the choice of the regularization parameters k and ρ_0 later.

Now, equations (1) and (4) allow to express z^{n+1} in terms of y^{n+1}, f^n and g^n only. Note that there are no derivatives included anymore, which is of some importance from the numerical point of view:

$$z^{n+1} = \frac{\gamma}{\delta t \beta} (y^{n+1} - f^n) + g^n \tag{6}$$

Analogous computations for f^n and g^n result in

$$\begin{aligned} f^n &= \delta t z^n + ((\frac{1}{2} - \beta)\beta^{-1} + 1) y^n - (\frac{1}{2} - \beta)\beta^{-1} f^{n-1} \\ g^n &= z^n + \delta t^{-1} (1-\gamma) \beta^{-1} (y^n - f^{n-1}). \end{aligned} \tag{7}$$

From (6) we get

$$\begin{aligned} J(u^{n+1}) = &\frac{1}{2} \sum_{v_C \in V_C} \sum_{i \in \mathcal{E}_C} (u_i^{n+1})^2 + \frac{k}{2} \cdot \sum_{i=1}^{n_e} \int_0^{l_i} (y_i^{n+1})^2 dx \\ &+ \frac{\rho}{2} \cdot \sum_{i=1}^{n_e} \int_0^{l_i} (y_i^{n+1} \underbrace{- f_i^n + \frac{\delta t \beta}{\gamma} g_i^n}_{=: -\hat{z}_i^n})^2 dx, \end{aligned}$$

where

$$\rho := \rho_0 \cdot \frac{\gamma}{\delta t \beta}.$$

Finally the cost functional takes the following standard form

$$\begin{aligned} J(u^{n+1}) = &\frac{1}{2} \sum_{v_C \in V_C} \sum_{i \in \mathcal{E}_C} (u_i^{n+1})^2 + \frac{k}{2} \cdot \sum_{i=1}^{n_e} \int_0^{l_i} (y_i^{n+1})^2 dx + \\ &\frac{\rho}{2} \cdot \sum_{i=1}^{n_e} \int_0^{l_i} (y_i^{n+1} - \hat{z}_i^n)^2 dx. \end{aligned} \tag{8}$$

We are now ready to derive the elliptic optimality system. Therefore we need to calculate the variational derivative of J. Let $\phi = (\phi_i)_{i=1:n_e}$ solve the elliptic equations with homogeneous right hand sides and $\epsilon_{iC}K_i\phi_i'(\nu_c) = \nu_i \quad \forall \nu_C \in V_C$. Since the problem reduces to the solution of an elliptic system in each single time step, we drop the time iteration index for convenience. Then the variational derivative of $J(u^{n+1})$ in direction ν is

$$J'(u)(\nu) = \sum_{\nu_C \in V_C} \sum_{i \in \mathcal{E}_C} (u_i)(\nu_i) + k \sum_{i=1}^{n_e} \int_0^{l_i} y_i \phi_i \, dx + \rho \sum_{i=1}^{n_e} \int_0^{l_i} (y_i - \hat{z}_i^1) \phi_i \, dx. \tag{9}$$

The adjoint equation is given by

$$p_i = \delta t^2 \beta K_i p_i'' + k y_i + \rho(y_i - \hat{z}_i), \tag{10}$$

where p_i fulfils the same boundary conditions as the primal system (1), with exception of controlled nodes, where $\epsilon_{iC}K_i p_i'(\nu_C) = 0 \quad \forall \nu_C \in V_C$ is enforced. Then

$$\begin{aligned}
0 &= \sum_{i=1}^{n_e} \int_0^{l_i} (\phi_i - \delta t^2 \beta K_i \phi_i'') p_i \, dx \\
&= \sum_{i=1}^{n_e} \int_0^{l_i} \phi_i p_i \, dx - \sum_{i=1}^{n_e} \int_0^{l_i} \delta t^2 \beta K_i \phi_i'' p_i \, dx \\
&= \sum_{i=1}^{n_e} \int_0^{l_i} \phi_i p_i \, dx - \sum_{i=1}^{n_e} \int_0^{l_i} \delta t^2 \beta K_i \phi_i p_i'' \, dx - \sum_{i=1}^{n_e} \delta t^2 \beta K_i \phi_i' p_i \big|_0^{l_i} \\
&\quad + \sum_{i=1}^{n_e} \delta t^2 \beta K_i \phi_i p_i' \big|_0^{l_i} \\
&\underset{1}{=} \sum_{i=1}^{n_e} \int_0^{l_i} (\phi_i p_i - \delta t^2 \beta K_i p_i'') \, dx - \sum_{i=1}^{n_e} \delta t^2 \beta K_i \phi_i' p_i \big|_0^{l_i} \\
&\underset{2}{=} \sum_{i=1}^{n_e} \int_0^{l_i} (\phi_i p_i - \delta t^2 \beta K_i p_i'') \, dx - \sum_{\nu_C \in V_C} \sum_{i \in \mathcal{E}_C} \delta t^2 \beta \epsilon_{iC} K \phi_i'(\nu_C) p_i(\nu_C) \\
&= \sum_{i=1}^{n_e} \int_0^{l_i} \phi_i (k y_i + \rho((y_i - \hat{z}_i)) \, dx - \sum_{\nu_C \in V_C} \sum_{i \in \mathcal{E}_C} \delta t^2 \beta g_i p_i(\nu_C).
\end{aligned}$$

Together with (9) we finally derive

$$\begin{aligned}
J'(u)(\nu) &= \sum_{\nu_C \in V_C} \sum_{i \in \mathcal{E}_C} (u_i)(\nu_i) + \sum_{\nu_C \in V_C} \sum_{i \in \mathcal{E}_C} \delta t^2 \beta \nu_i p_i(\nu_C) \overset{!}{=} 0 \\
&\Longrightarrow \quad u_i = -\delta t^2 \beta p_i(\nu_C) \quad \forall \nu_C \in V_C,
\end{aligned}$$

[1] $\phi_i(\nu_D) = 0 \; \forall \nu_D \in V_D$, $p_i'(\nu) = 0 \; \forall \nu \in V_N \cup V_C$, $\sum_{\nu_J \in V_M} \sum_{i \in \epsilon_J} \delta t^2 \beta K p_i'(\nu_J) = 0 \; \forall \nu_J \in V_M$

[2] $\delta t^2 \beta \epsilon_{iJ} K_i \phi_i' p_i(\nu) = 0 \; \forall \nu \in V_D \cup V_N \cup V_M$

which leads to the following corollary.

Corollary 1 (Optimality system). *The optimality system with respect to the cost functional* (8) *at the discrete time level* $n \cdot \delta t$ *has the form*

$$
\begin{aligned}
& y_i^n = \delta t^2 \beta K_i (y_i^n)'' + f_i^{n-1} && p_i^n = \delta t^2 \beta K_i (p_i^n)'' + k y_i^n + \rho (y_i^n - \hat{z}_i^{n-1}) \\
& y_d^n(\nu_D) = 0 && p_d^n(\nu_D) = 0 \\
& y_i^n(\nu_J) = y_j^n(\nu_J) && p_i^n(\nu_J) = p_j^n(\nu_J) \\
& \sum_{i \in \mathcal{E}_J} \epsilon_{iJ} K_i (y_i^n)' = 0 && \sum_{i \in \mathcal{E}_J} \epsilon_{iJ} K_i (p_i^n)' = 0 \\
& \epsilon_{iJ} K_i (y_i^n)'(\nu_C) = -\delta t^2 \beta p_i(\nu_C) && \epsilon_{iJ} K_i (p_i^n)'(\nu_C) = 0.
\end{aligned}
\tag{11}
$$

We emphasize, that this optimality system is globally defined on the whole network and has to be solved at each single time step. In the numerical examples this was done by a domain decomposition method (ddm) and finite differences. We will focus on the details of the ddm in the next section. Finally, we can formulate the instantaneous control algorithm.

Algorithm 2 (Instantaneous control of the string network).

$y_i^0 := \bar{y}_i^0$
$z_i^0 := \bar{z}_i^0$
$f_i^0 := y_i^0 + \delta t z_i^0 + \delta t^2 (\frac{1}{2} - \beta) K_i (y_i^0)''$
$g_i^0 := z_i^0 + \delta t (1 - \gamma) K_i (y_i^0)''$
solve the global optimality system (11) for $n = 1$
compute $z_i^1 = \frac{\gamma}{\delta t \beta}(y_i^1 - f_i^0) + g_i^0$
for $n = 1 : N_t - 1$ **do**
  $f_i^n = \delta t z_i^n + ((\frac{1}{2} - \beta)\beta^{-1} + 1) y_i^n - (\frac{1}{2} - \beta)\beta^{-1} f_i^{n-1}$
  $g_i^n = z_i^n + \delta t^{-1}(1 - \gamma)\beta^{-1}(y_i^n - f_i^{n-1})$
  solve the global optimality system (11) for time step $n + 1$
  compute $z_i^{n+1} = \frac{\gamma}{\delta t \beta}(y_i^{n+1} - f_i^n) + g_i^n$
end for

In the next section we will focus on the problem, how to solve the global optimality system (11) numerically. Before we do this, we will give an interpretation of the instantaneous control algorithm as a discrete closed loop system.

2.3 *Interpretation as Discrete Feedback System*

We recall Newmark's scheme on a certain edge

$$
\begin{aligned}
y^{n+1} - \delta t^2 \beta (K y^{n+1})'' &= y^n + \delta t \cdot z^n + \delta t^2 (\frac{1}{2} - \beta)(K y^n)'' \\
z^{n+1} - \delta t \gamma (K y^{n+1})'' &= z^n + \delta t (1 - \gamma)(K y^n)''
\end{aligned}
\tag{12}
$$

with the boundary- and initial conditions (3) and the discrete unknowns y^{n+1} and z^{n+1}. Let the matrices A and B be defined through the spatial discretization of an edge, where A is the coefficient matrix of a finite differences/elements scheme and B acts basically as a projection of the control onto the boundaries. Then

$$\begin{aligned} y^{n+1} + \delta t^2 \beta A y^{n+1} &= \delta t^2 \beta B u^{n+1} + y^n + \delta t z^n - \delta t^2 (\frac{1}{2} - \beta) A y^n \\ &\quad + \delta t^2 (\frac{1}{2} - \beta) B u^n \\ z^{n+1} + \delta t \gamma A y^{n+1} &= \delta t \gamma B u^{n+1} + z^n - \delta t (1-\gamma) A y^n \\ &\quad + \delta t (1-\gamma) B u^n . \end{aligned}$$

Furthermore define $x := \binom{y}{z}$, which is an approximation of the continuous system's phase space variable $\binom{y}{y_t}$. Then

$$\underbrace{\begin{pmatrix} I + \delta t^2 \beta A & 0 \\ \delta t \gamma A & I \end{pmatrix}}_{=:\mathcal{A}_1} \begin{pmatrix} y^{n+1} \\ z^{n+1} \end{pmatrix} = \underbrace{\begin{pmatrix} I - \delta t^2 (\frac{1}{2} - \beta) A & \delta t I \\ -\delta t (1-\gamma) A & I \end{pmatrix}}_{=:\mathcal{A}_0} \begin{pmatrix} y^n \\ z^n \end{pmatrix} + \underbrace{\begin{pmatrix} \delta t^2 \beta B \\ \delta t \gamma B \end{pmatrix}}_{=:\mathcal{B}_1} u^{n+1} + \underbrace{\begin{pmatrix} \delta t^2 (\frac{1}{2} - \beta) B \\ \delta t (1-\gamma) B \end{pmatrix}}_{=:\mathcal{B}_0} u^n ,$$

i.e.

$$\mathcal{A}_1 x^{n+1} = \mathcal{A}_0 x^n + \mathcal{B}_1 u^{n+1} + \mathcal{B}_0 u^n . \tag{13}$$

From the adjoint system, we derive the control law.

$$\begin{aligned} (I + \delta t^2 \beta A) p^{n+1} &= k y^{n+1} + \rho_0 z^{n+1} \\ u^{n+1} &= -\delta t^2 \beta (I + \delta t^2 \beta A)^{-1} (k y^{n+1} + \rho_0 z^{n+1}), \end{aligned}$$

and therefore

$$\begin{pmatrix} \delta t^2 \beta B \\ \delta t \gamma B \end{pmatrix} u^{n+1} = \underbrace{- \delta t^3 \beta B (I + \delta t^2 \beta A)^{-1} \begin{pmatrix} \delta t \beta k I & \delta t \beta \rho_0 I \\ \gamma k I & \gamma \rho_0 I \end{pmatrix}}_{=:\mathcal{C}_1} \begin{pmatrix} y^{n+1} \\ z^{n+1} \end{pmatrix}$$

$$\begin{pmatrix} \delta t^2 (\frac{1}{2} - \beta) B \\ \delta t (1-\gamma) B \end{pmatrix} u^n = \underbrace{-\delta t^3 \beta B (I + \delta t^2 \beta A)^{-1} \begin{pmatrix} \delta t (\frac{1}{2} - \beta) k I & \delta t (\frac{1}{2} - \beta) \rho_0 I \\ (1-\gamma) k I & (1-\gamma) \rho_0 I \end{pmatrix}}_{=:\mathcal{C}_0} \begin{pmatrix} y^n \\ z^n \end{pmatrix} ,$$

i.e.

$$\begin{aligned} \mathcal{B}_1 u^{n+1} &= -\mathcal{C}_1 x^{n+1} \\ \mathcal{B}_0 u^n &= \mathcal{C}_0 x^n , \end{aligned}$$

and finally

$$x^{n+1} = (\mathcal{A}_1 + \mathcal{C}_1)^{-1}(\mathcal{A}_0 + \mathcal{C}_0)x^n. \tag{14}$$

Equation (14) is a discrete closed loop system. A spectral analysis of the feedback operator above is part of our current investigations.

We also made similar calculations, where we used the implicit Euler scheme (applied to the network equations rewritten as a first order system) instead of Newmark's scheme. In this case we can show that the discrete iteration matrix has spectral radius smaller than one, such that the system is guaranteed to be damped to the zero state. Unfortunately, due to algorithmical damping effects, the numerical results do not properly represent our philosophy.
Since Newmark's scheme preserves the discrete energy, we do not have these algorithmical damping effects. Therefore we clearly see, that the energy is steered out of the system only at the boundaries, where the controls apply. Evidently, this does not hold true for the implicit Euler scenario.

3 A Domain Decomposition Method for the Elliptic Optimality System

Obviously, the edges of the network are coupled through the conditions of continuity and balance of forces. To make computations as simple (and fast) as possible, it seems to be reasonable to split the network into its physical components, the edges. Domain decomposition is applied as an iterative method, which makes it possible to handle specific domains (here: edges) independently from the others at each iteration level. Taking into account the possibilities of parallel computing, we have a tool to decrease the necessary computing time. For further reading we refer to [9], [10].

3.1 The Local Optimality System

In the sequel we specify the domain decomposition method for the elliptic network equations. We also give an interpretation of the resulting local equations as a local optimality system of a control problem with artificial controls at the inner nodes of the network.

Proposition 3 (Local optimality systems).
We solve the global optimality system at a certain time step in parallel on each single edge by the following iterative domain decomposition method ($m \triangleq$ iteration index, time step index dropped):

$$\begin{aligned}
y_i^{m+1} &= \delta t^2 \beta K_i (y_i^{m+1})'' + f_i^m & p_i^{m+1} &= \delta t^2 \beta K_i (p_i^{m+1})'' + k y_i^{m+1} \\
& & &\quad + \rho(y_i^{m+1} - \hat{z}_i^1) \\
y_d^{m+1}(\nu_D) &= 0 & p_d^{m+1}(\nu_D) &= 0 \\
\epsilon_{ij} K_i (y_i^{m+1})'(\nu_C) &= -\delta t^2 \beta p_i^{m+1}(\nu_C) & \epsilon_{ij} K_i (p_i^{m+1})'(\nu_C) &= 0
\end{aligned} \tag{15}$$

To enforce the continuity- and balance of forces conditions at the inner nodes, we use the following transmission conditions on these boundaries:

$$\begin{aligned}\epsilon_{iJ}\delta t^2\beta K_i(y_i^{m+1})'(\nu_J) + \sigma_J p_i^{m+1}(\nu_J) &= \frac{2}{d_J}\sum_{j\in\mathcal{E}_J} \sigma_J p_j^m(\nu_J) - \sigma_J p_i^m(\nu_J) \\ &-\Big(\frac{2}{d_J}\sum_{j\in\mathcal{E}_J} \epsilon_{jJ}\delta t^2\beta K_j(y_j^m)'(\nu_J) - \epsilon_{iJ}\delta t^2\beta K_i(y_i^m)'(\nu_J)\Big),\end{aligned} \tag{16}$$

$$\begin{aligned}\epsilon_{iJ}\delta t^2\beta K_i(p_i^{m+1})'(\nu_J) - \sigma_J y_i^{m+1}(\nu_J) &= -\Big(\frac{2}{d_J}\sum_{j\in\mathcal{E}_J} \sigma_J y_j^m(\nu_J) - \sigma_J y_i^m(\nu_J)\Big) \\ &-\Big(\frac{2}{d_J}\sum_{j\in\mathcal{E}_J} \epsilon_{jJ}\delta t^2\beta K_j(p_j^m)'(\nu_J) - \epsilon_{iJ}\delta t^2\beta K_i(p_i^m)'(\nu_J)\Big),\end{aligned} \tag{17}$$

where σ_J is an arbitrary positive parameter.

The ddm-iteration above follows the ideas of Benamou [1] and was extended by Leugering [10] to fit to context of networks. Note, that the right hand sides of the transmission conditions depend on the iteration history only. On the iteration level $m+1$ of the ddm, first the right hand sides of (16) and (17) are evaluated and then used as boundary conditions for the local optimality system (15). We emphasize, that the equations for each single edge are independent from the other edges, such that the network is decomposed into its physical components.

We will now give an interpretation of system (15), (16), (17) as an optimality system of a local control problem on a single edge.
To this end we consider the most complicated case of an edge with an inner node at $\nu_J = 0$ and a controlled node at $\nu_C = l$ and the following optimal control problem on edge i with the control u_C and the artificial control u_J:

$$\begin{aligned}J_{loc}(u) := \frac{1}{2}(u_C)^2 + \frac{1}{(2\sigma_J)}(u_J)^2 + \frac{k}{2}\int_0^l (y_i^{m+1})^2 dx \\ + \frac{\rho}{2}\int_0^l (y_i^{m+1} - \hat{z}_i^1)^2 dx + \frac{1}{2}(\sigma_J y_i^{m+1}(\nu_J) + \mu_{iJ}^m)^2.\end{aligned} \tag{18}$$

s.t.

$$\begin{aligned} y_i^{m+1} &= \delta t^2\beta K_i(y_i J^{m+1})'' + f_i^m \\ \epsilon_{iC}K_i(y_i^{m+1})'(\nu_C) &= u_C \\ \epsilon_{iJ}\delta t^2\beta K_i(y_i^{m+1})'(\nu_J) &= u_J + \lambda_{iJ}^m \end{aligned} \tag{P_{loc}}$$

where λ_{iJ} and μ_{iJ} are defined as

$$\mu_{iJ} = -\Big(\frac{2}{d_J}\sum_{j\in\mathcal{E}_J} \sigma_J y_j^m(\nu_J) - \sigma_J y_i^m(\nu_J)\Big)$$
$$- \Big(\frac{2}{d_J}\sum_{j\in\mathcal{E}_J} \epsilon_{jJ}\delta t^2\beta K_j (p_j^m)'(\nu_J) - \epsilon_{iJ}\delta t^2\beta K_i (p_i^m)'(\nu_J)\Big)$$
$$\lambda_{iJ} = \frac{2}{d_J}\sum_{j\in\mathcal{E}_J} \sigma_J p_j^m(\nu_J) - \sigma_J p_i^m(\nu_J)$$
$$- \Big(\frac{2}{d_J}\sum_{j\in\mathcal{E}_J} \epsilon_{jJ}\delta t^2\beta K_j (y_j^m)'(\nu_J) - \epsilon_{iJ}\delta t^2\beta K_i (y_i^m)'(\nu_J)\Big).$$

We compute again the variational derivative of J with respect to $u = (u_C, u_J)$ in direction $v = (v_C, v_J)$.

$$J'_{loc}(u)(v) = u_C v_C + \sigma_J^{-1} u_J v_J + k\cdot\int_0^l y_i^{m+1}\phi dx$$
$$+ \rho\cdot\int_0^l (y_i^{m+1} - \hat{z}_i^1)\phi dx + (\sigma_J y_i(0) + \mu_{iJ}^m)\phi(0) \stackrel{!}{=} 0, \tag{19}$$

where ϕ solves the differential equation above with homogeneous right hand side and fulfils $\epsilon_{iC} K_i \phi_i'(l) = v_C$ as well as $\epsilon_{iJ}\delta t^2 \beta K_i \phi_i'(0) = v_J$. Thus

$$0 = \int_0^l (\phi_i - \delta t^2\beta K_i \phi_i'')p_i dx$$
$$= \int_0^l \phi_i p_i + \int_0^l \delta t^2 \beta K_i \phi_i' p_i' dx - \delta t^2\beta K_i \phi_i' p_i|_0^l$$
$$= \int_0^l (p_i - \delta t^2\beta K_i p_i'')\phi_i dx - \delta t^2\beta K_i\phi_i' p_i|_0^l + \delta t^2\beta K_i \phi_i p_i'|_0^l$$
$$\underset{(P_{loc})}{=} \int_0^l (p_i - \delta t^2\beta K_i p_i'')\phi_i + \epsilon_{iJ} v_J p_i(0) - \delta t^2 \beta\epsilon_{iC} v_C p_i(l) + \delta t^2\beta K_i\phi_i p_i'|_0^l.$$

We recall the definition of the adjoint equation (10) and obtain

$$\int_0^l \big(k(y_i^{m+1} - \tilde{z}_i^0) + \rho(y_i^{m+1} - \hat{z}_i^1)\big)\phi_i dx =$$
$$v_J p_i(0) + \delta t^2\beta v_C p_i(l) - \delta t^2\beta K_i \phi_i p_i'|_0^l.$$

Together with (19) we conclude

$$-u_C v_C - \sigma_J^{-1} u_J v_J - (\sigma_J y_i(0) + \mu_{ij})\phi_i(0)$$
$$= p_i(0) v_J + \delta t^2 \beta p_i(l) v_C - \delta t^2 \beta K_i \epsilon_{iJ} p_i'(0)\phi_i(0) - \delta t^2\beta K_i \epsilon_{iC} p_i'(l)\phi_i(l).$$

Therefore the boundary conditions are

$$\begin{aligned}
u_C &= -\delta t^2 \beta p_i^{m+1}(l) \\
u_J &= -\sigma p_i^{m+1}(0) \\
\epsilon_{iJ}\delta t^2 \beta K_i (p_i^{m+1})'(0) - \sigma_J y_i^{m+1}(0) &= \mu_{iJ}^m \\
\epsilon_{iC} K_i (p_i^{m+1})'(l) &= 0,
\end{aligned}$$

which are exactly the transmission conditions (16) and (17) in Proposition 3.

3.2 An Alternative Iteration Scheme

A disadvantage of the ddm in the above form is the necessity of calculating the derivatives at the inner nodes in order to compute the right hand sides of the transmission conditions. In [4] Q. Deng proposed a simple scheme for the evaluation of the right hand sides of the equations without any explicit calculation of derivatives. We adapted this scheme to the network and control context in the manner described by the following proposition.

Proposition 4.
The transmission conditions in Proposition 3 are replaced by

$$\epsilon_{iJ}\delta t^2 \beta K_i (y_i^{m+1})'(v_J) + \sigma_J p_i^{m+1}(v_J) = g_{iJ}^{m+1}$$

and

$$\epsilon_{iJ}\delta t^2 \beta K_i (p_i^{m+1})'(v_J) - \sigma_J y_i^{m+1}(v_J) = h_{iJ}^{m+1},$$

where g_{iJ} *and* h_{iJ} *are defined as*

$$\begin{aligned}
g_{iJ}^{m+1} &:= 2\sigma_J \Big(\frac{2}{d_J}\sum_{j\in\mathcal{E}_J} p_j^m(v_J) - p_i^m(v_J)\Big) - \Big(\frac{2}{d_J}\sum_{j\in\mathcal{E}_J} g_{jJ}^m - g_{iJ}^m\Big), \\
h_{iJ}^{m+1} &:= -2\sigma_J \Big(\frac{2}{d_J}\sum_{j\in\mathcal{E}_J} y_j^m(v_J) - y_i^m(v_J)\Big) - \Big(\frac{2}{d_J}\sum_{j\in\mathcal{E}_J} h_{jJ}^m - h_{iJ}^m\Big).
\end{aligned}$$

To initialize the iteration, we chose g_{iJ}^0 *and* h_{iJ}^0 *arbitrary (e.g.* $g_{iJ}^0 = h_{iJ}^0 = 0$*).*

At any time level $m > 1$, there holds

$$
\begin{aligned}
&\epsilon_{iJ}\delta t^2\beta K_i(y_i^{m+1})'(v_J) + \sigma_J p_i^{m+1}(v_J) = g_{iJ}^{m+1} \\
&= 2\sigma_J\Big(\frac{2}{d_J}\sum_{j\in\mathcal{E}_J} p_j^m(v_J) - p_i^m(v_J)\Big) - \Big(\frac{2}{d_J}\sum_{j\in\mathcal{E}_J} g_{jJ}^m - g_{iJ}^m\Big) \\
&= 2\sigma_J\Big(\frac{2}{d_J}\sum_{j\in\mathcal{E}_J} p_j^m(v_J) - p_i^m(v_J)\Big) - \Bigg(\frac{2}{d_J}\sum_{j\in\mathcal{E}_J} \big(\epsilon_{jJ}\delta t^2\beta K_j(y_j^m)'(v_J) + \\
&\qquad \sigma_J p_j^m(v_J)\big) - \big(\epsilon_{jJ}\delta t^2\beta K_i(y_i^m)'(v_J) + \sigma_J p_i^m(v_J)\big)\Bigg) \\
&= \sigma_J\Big(\frac{2}{d_J}\sum_{j\in\mathcal{E}_J} p_j^m(v_J) - p_i^m(v_J)\Big) - \Big(\frac{2}{d_J}\sum_{j\in\mathcal{E}_J} \epsilon_{jJ}\delta t^2\beta K_j(y_j^m)'(v_J) \\
&\qquad - \epsilon_{jJ}\delta t^2\beta K_i(y_i^m)'(v_J)\Big),
\end{aligned}
$$

i.e. in the limit the same transmission conditions are fulfilled as in the classical scheme. The computation for h_{iJ}^m are analogous.

4 Numerical Examples

4.1 *Choice of Penalty Parameters*

It is not always easy to find good parameter sets for the instantaneous control algorithms above. We are looking for appropriate choices of the three parameters k, ρ_0 and σ_J, which can be interpreted as weights in a Tykchonov regularization formulation. Obviously, the penalization terms have to guarantee a balance between a regularization effect and accuracy. One can find several methods for adjusting optimal regularization parameters in the literature, see [5] and [6] for details. Another problem is the interrelation of the penalty parameters k, ρ_0 and σ_J with the discretization schemes in time and space, say the step sizes δt and δx. From this point of view it seems reasonable to use a global optimization routine for finding good parameter sets, which allows to optimize both the penalty- and discretization parameters in between given ranges. We used a MATLAB implementation of a genetic algorithm to compute appropriate parameter sets for our problem. The basic properties of genetic algorithms as well as a description of the optimization routine can be found in [8]. Also see [2] for numerical results of instantaneous control applied to membranes.

We emphasize, that the parameter problem has to be solved off-line only once by any global optimization algorithm or by one of the methods in [5] and [6]. All future applications of the instantaneous control will work with this special parameter set on-line.

4.2 *Instantaneous Control of a String*

We used a finite difference scheme for the spatial discretization of the string of length $l = 1.0$ with 35 grid points with controls on both boundaries. The genetic algorithm toolbox gave us the following parameter set: $\delta t = 0.018$, $k = 3.04$, $\rho_0 = 1.7$. The control was computed up to the final time $T = 1.0$. At $t = 0$ the initial condition gives a discrete maximum norm of $\|y^0\| = 0.1$. After application of the instantaneous control algorithm we found $\|y^{N_T}\| = 0.0004$ in the same norm at $T = 1.0$. Figure 1 shows the evolution of the string in the z-axis plotted versus time and space. The MATLAB code took 0.072 seconds on a Pentium II / 166 MHz, running Linux.

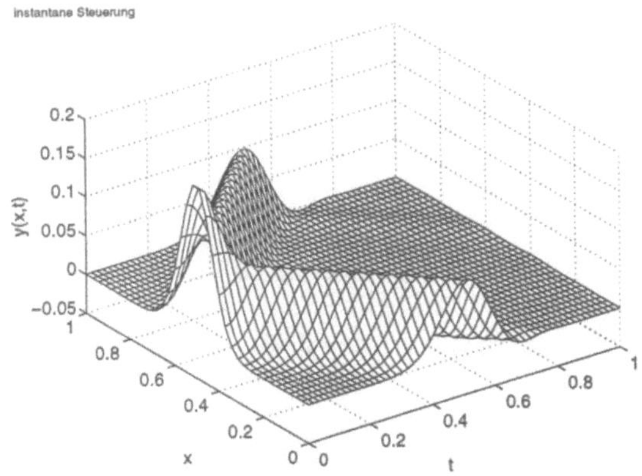

Figure 1. Instantaneous control of a string

In order to check the quality of the instantaneous control, we compared the results to the approximation of the optimal control. For this, we used the same setting, as in the example above but with the string clamped at $x = 0$. The approximation of the optimal control was computed by optimizing the continuous cost functional

$$J_{cont}(u) := \frac{1}{2}\int_0^T \|u(t)\|^2 dt + \frac{\sigma}{2}\int_0^T \|y(T)\|_{L^2(0,1)} dt + \frac{\sigma}{2}\int_0^T \|y_t(T)\|_{L^2(0,1)} dt,$$

such that 1 is fulfilled, with a high penalty parameter σ. For this computation we have chosen $T_{opt} = 2.0$, which is the smallest possible time interval, in which controllability of the wave equation holds in this setting (for $t > 2.0$ just the zero function was plotted). The instantaneous control was computed up to $T = 4.0$. Figure 2 shows the evolution of the optimally controlled system, plotted against the evolution of the instantaneously controlled one (solid line) at several time steps.

4.3 Network Simulation

We consider uncontrolled systems at first, i.e. $k = \rho = 0$, with homogeneous Dirichlet boundary conditions at single nodes, in order to check the domain decomposition

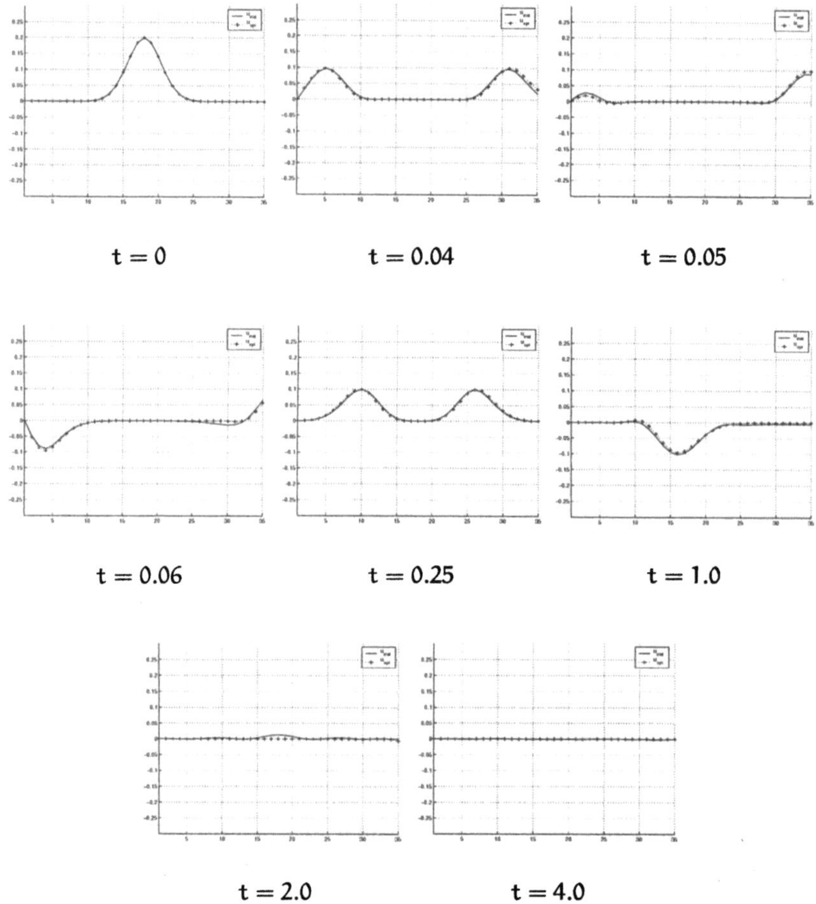

Figure 2. Instantaneous versus optimal control

algorithm. In these examples, we have chosen $\delta x = 1/19$ and $\delta t = \delta x/2$. The iteration history plots contain in the top window the number of necessary ddm-iteration steps, to get the differences of the evolution on the adjacent edges of a node, as well as the deviations in the transmission conditions $||g_{iJ}^{n+1} - g_{iJ}^{n}||$, below a certain threshold, here $\varepsilon = 10^{-4}$. In the middle and bottom windows, we see the evolution- and transmission condition error, plotted against the time steps, respectively. For the simulations, we have chosen $\sigma_J = 2 \cdot \frac{\delta t^2 \beta}{dx}$ for the ddm-parameter.

The first example shows the simulation of a serial string network. Figure 3 shows its iteration history, Figure 4 the evolution of the system.

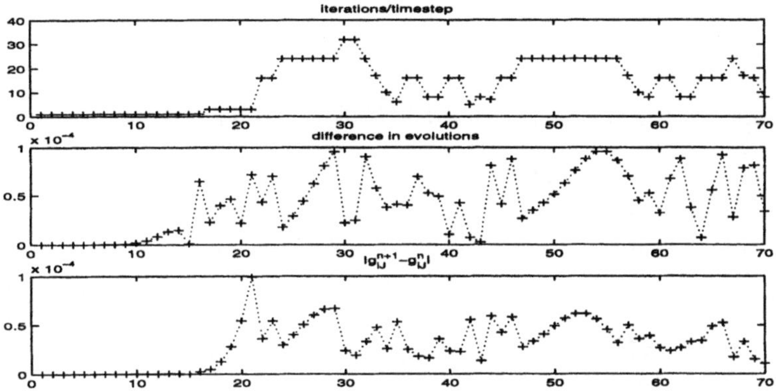

Figure 3. Iteration history of a serial string network

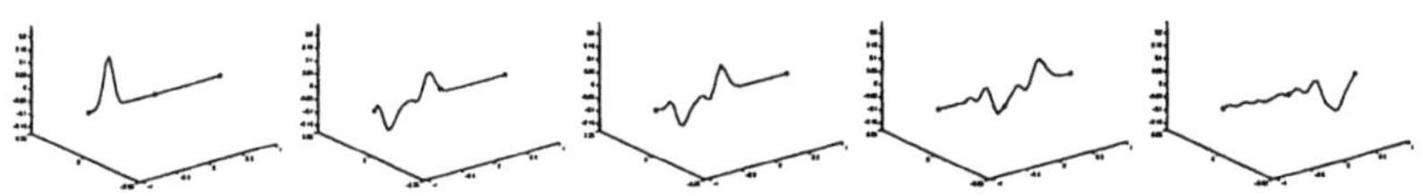

0.0 0.58 0.76 1.29 2.0

Figure 4. Simulation of a serial string network

As an example for a network, which contains a node with edge degree $d_J > 1$, we present the simulation of a string trihedron. Figures 5 and 6 show again the iteration history and the evolution of the system.

Finally, Figure 7 shows the simulation of a network with 40 edges and 32 nodes. We applied free (Neumann) boundary conditions at the single nodes.

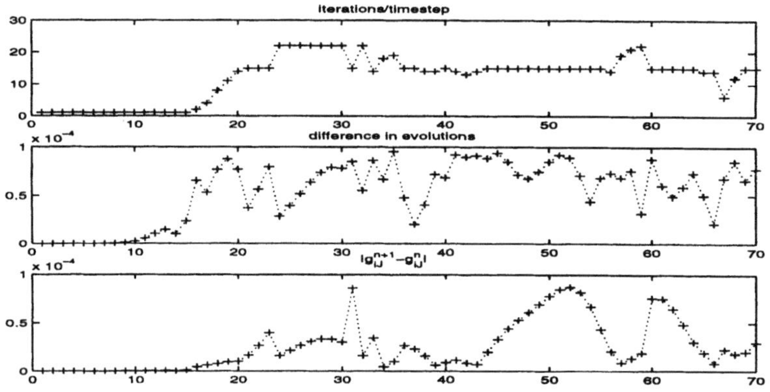

Figure 5. Iteration history of a trihedron string network

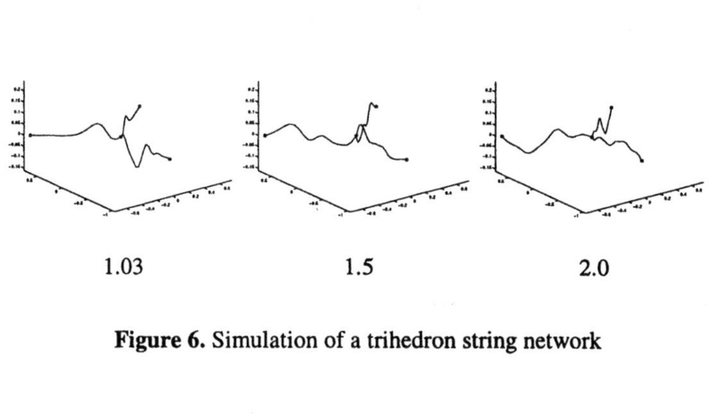

Figure 6. Simulation of a trihedron string network

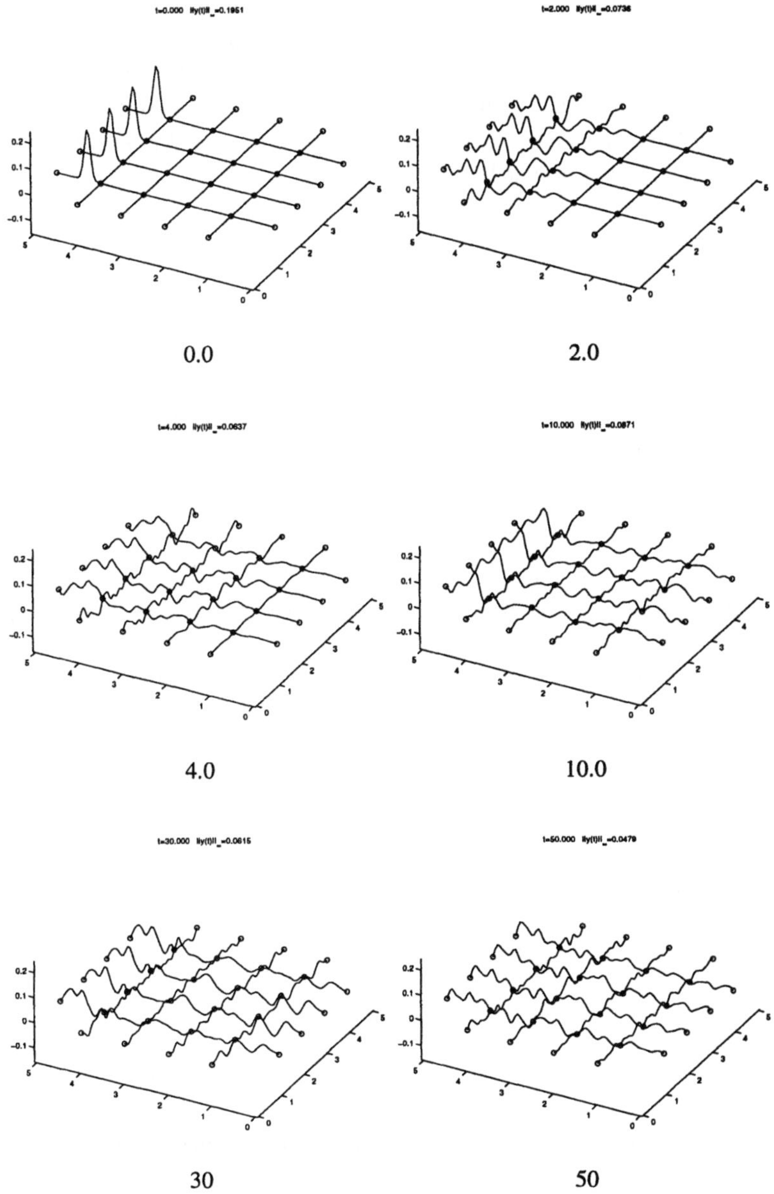

Figure 7. Simulation of a network

4.4 Instantaneous Control of Networks

We present numerical results for the last two examples, with instantaneous controls applied at single nodes. For the trihedron example, we have chosen $k = 10000$ and $\rho_0 = 7500$ for the control penalty parameters and $\sigma_J = k \cdot \frac{\delta t^2 \beta}{dx}$ for the ddm-parameter, i.e. the difference of the displacements on the different edges is penalized by $\frac{\sigma_J}{\delta t^2 \beta} = \frac{k}{dx} = 1.9 \cdot 10^5$. Figure 8 shows the evolution history, 9 the evolution of the trihedron. We achieved the zero state with a precision smaller than $0.5 \cdot 10^{-3}$ in the discrete maximum norm at $T = 4.0$.

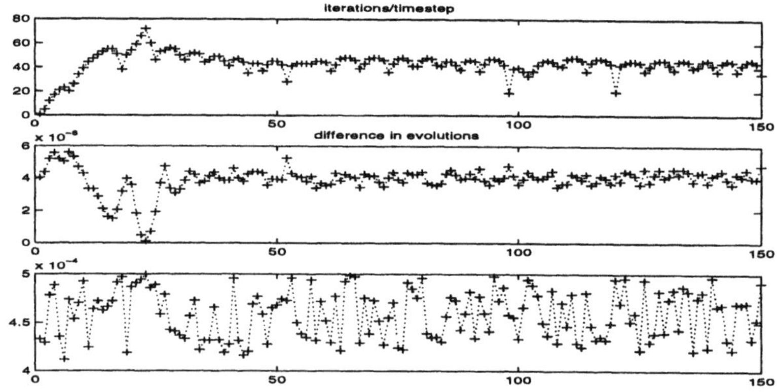

Figure 8. Iteration history of a trihedron string network

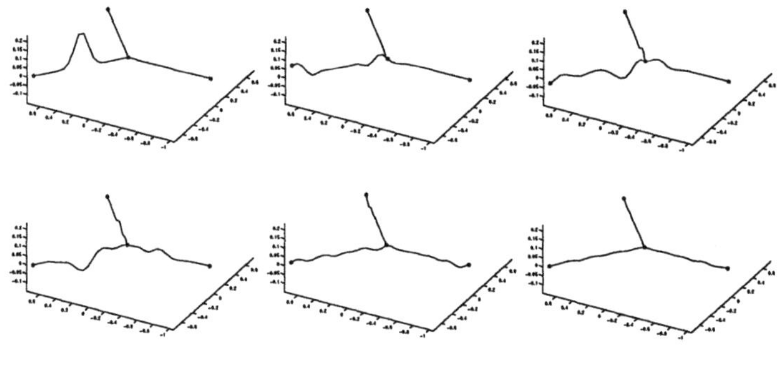

Figure 9. Instantaneous control of a trihedron string network

Finally, Figure 10 shows again the evolution of the 40-edges network from the precedent example, but with instantaneous control applied on the single nodes. Note, that we have an uncontrollable situation, due to the fact, that the network contains circles. For $T = 50.0$ we found for the discrete maximum norm of the evolution over the whole network $||y^{Nt}|| = 0.0055$.

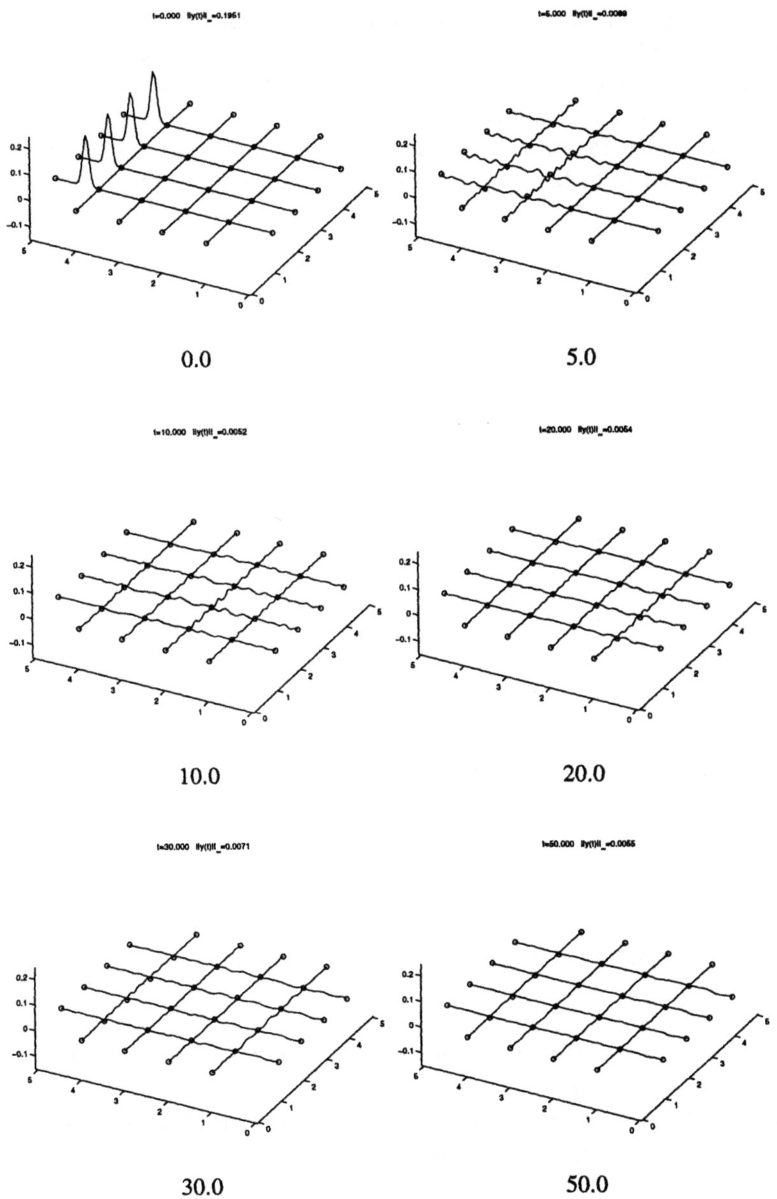

Figure 10. Instantaneous control of a network

5 Remarks

Evidently, the instantaneous control framework can be extended to the two dimensional case. Instantaneous control was successfully applied to vibrating membranes [2]. We also combined a different instantaneous control strategies ("one gradient step stabilisation", see [7] for details) and domain decomposition to the so-called "L-shape problem" and obtained encouraging numerical results, see Figure 11. In this case, the L-shaped domain was subdivided into three quadratic sections. Controls were applied at the long edges of the "L" while the membrane was clamped at the short ones.

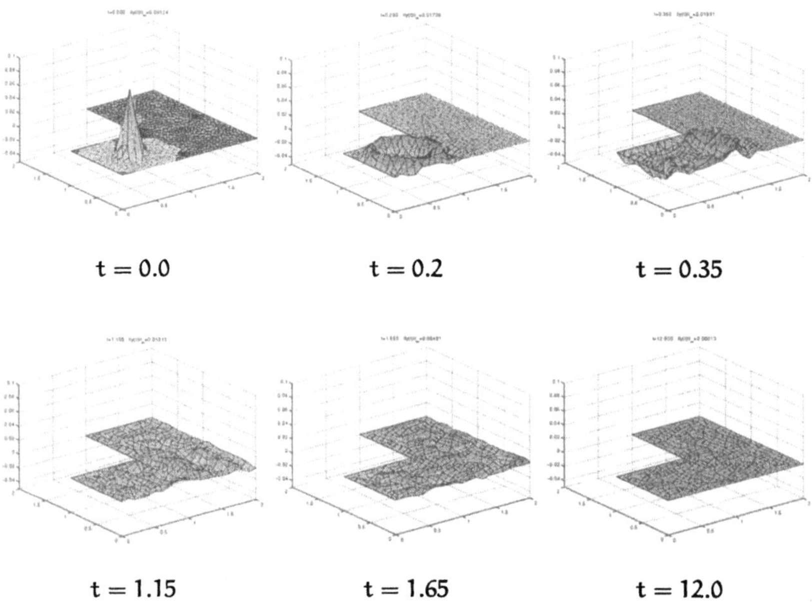

Figure 11. Instantaneous control of the "L-shape problem"

We note in closing that we have not yet considered the implementation of other domain decomposition techniques.

Notably, the methods by J. L. Lions and O. Pironneau [11] appear to be promising with respect to instantaneous controls.

References

1. J.-D. Benamou, (1995), "A domain decomposition method for the optimal control of systems governed by the Helmholtz equation", *Mathematical and numerical aspects of wave propagation*, (Cohen, G. Ed.), SIAM, pp. 653-662

2. F. Bourquin, R. Hundhammer, G. Leugering (2000), "Instantaneous Control of Vibrating Strings and Membranes", *Proceedings of the Second European Conference on Structural Control*, to appear.
3. H. Choi, R. Temam, P. Moin, J. Kim (1993), "Feedback control for unsteady flow and its applications to the stochastic Burgers equation", *J. Fluid Mech.*, pp. 253–509.
4. Q. Deng, "An Analysis for a Nonoverlapping Domain Decomposition Iterative Procedure", *SIAM J. SCI. COMPUT.*, Vol. 18, No 5, pp. 1517-1525
5. M. Hanke, P. C. Hansen (1993), "Regularization methods for large-scale problems", *Surv. Math. Ind.*, pp. 253-315
6. P. C. Hansen (1990), "Truncated singular value decomposition solutions to discrete ill-posed problems with ill-determined numerical rank", *SIAM J. Sci. Comp.*, Vol. 11, No. 3, pp. 503-518
7. M. Hinze, K. Kunisch (1997), "On Suboptimal Control Strategies for the Navier-Stokes Equations", *TU Berlin*, Preprint No. 568/1997.
8. C. R. Houck, J. A. Joines, M. G. Kay, "A Genetic Algorithm for Function Optimization: A Matlab Implementation", *North Carolina State Univ.*, Preprint.
9. J. E. Lagnese and G. Leugering and Schmidt E.J.P.G. (1994), *Modeling, Analysis and Control of Dynamic Elastic Multi-Link Structures*, Birkhäuser, Boston.
10. G. Leugering (2000), "Domain decomposition of optimal control problems for dynamic networks of elastic strings.", *Comp. Opt. and Appl. 16/1* pp. 5-28.
11. J. L. Lions, O. Pironneau (1998), "Algorithmes parallèles pour la solution de problèmes aux limites", *C. R. Acad. Sci. Paris*, t. 327, Série I, p. 947-952
12. P. A. Raviart, J. M. Thomas (1992), "Introduction à l'analyse numérique des équations aux dérivées partielles", *Masson.*

Modelling, Stabilization, and Control of Flow in Networks of Open Channels

Martin Gugat[1], Günter Leugering[1], Klaus Schittkowski[2], and E. J. P. Georg Schmidt[3]

[1] Fachbereich Mathematik, Technische Universität Darmstadt, Germany
[2] Institut für Mathematik, Universität Bayreuth, Germany
[3] Department of Mathematics and Statistics, McGill University, Montreal, Canada

Abstract In this paper we present a model for the controlled flow of a fluid through a network of channels using a coupled system of St Venant equations. We generalize in a variety of ways recent results of Coron, d'Ándrea-Novel and Bastin concerning the stabilizability around equilibrium of the flow through a single channel to serially connected channels and finally to networks of channels. The work presented here is entirely based on the theory of quasilinear hyperbolic systems. We also consider open-loop optimal control problems and provide numerical schemes both for the simulation and the control of such systems.

1 Introduction

The problem which we are going to consider in this paper is related to the real-time optimal control of sewer- or irrigation systems. Such systems are described by planar graphs representative of the channel system along with one-dimensional flow equations, namely the 1-d shallow water- or the 'de St.Venant' equations. These equations constitute a non linear hyperbolic system first introduced in [14]. They have, in particular, become a standard tool for hydraulic engineers used in the modelling of the dynamics of channels and rivers. The books [5] and [6] provide useful engineering references to this topic.

In a typical real-time application, of course, one has to consider various model reductions in order to comply with the requirements of the computing time available. As a matter of fact, in the application to a particular sewer-system which we have in mind, the real-time requirements strongly depend on the velocity of the flow in the network. We consider a nominal flow in that network which is essentially at an equilibrium level. Now, once heavy rain falls cause a significant discharge at the boundary of our network of channels, we can estimate the velocity of flow within the reaches (links, channels) and, therefore, we can estimate the arrival time of the 'flood' at a given node in that network. Now, as the network stretches from a mountainous region into a flat region, those channels which are close to the boundary will be much steeper than those far from the boundary. The water in the steep channels is transported essentialy without 'backwater effects' and can be modelled as transport links without resorting to partial differential equations. Those channels which are less steep but still have a significant bed slope can be approximated by some nonlinear diffusion advection equations or stiff ordinary differential equations. We will report on these approximations and their numerical treatment in a different note.

Finally, the channels which are located in the flat region, i.e., those with moderate or small bed slope, will be modelled by the St. Venant equations.

Assume now that the flow of water through such a network of channels is controlled at certain points (junctions, serial joints etc.) by the action of sluice gates, weirs and pumps into reservoirs. The control actions are related to measurements that are taken at certain locations both at the boundary and within the network. The final goal of this project then is to compute optimal controls that keep the water-level at given 'down-stream' locations in some given bounds. In the system under consideration there is purification plant at the root of the (tree-like) network where constraints on the height of the water are to be respected. While the modelling and control of such systems has been considered in the framework of ordinary differential equations in the engineering literature, the treatment of partial differential equations in this context is still in its infancy. However, backwater effects and other effects that occur when controlling the flow through the openings of, say, sluice gates necessitate the consideration of continuous models. For that reason, in this paper, we proceed to take into account such effects by considering the St. Venant model. We emphazise that this report is preliminary in the sense that we do not consider real-time implementations of numerical algorithms. Rather, at this point we take into account software which we think will turn out to be real-time capable. The final implementation will be reported on later. As has become clear from this description of the problem, we will be faced with a hierarchy of models and controls according to the real-time requirements. As the continuum models appear in the region with small bed slope and wide channels with simple geometry, we will concentrate on exemplaric situations as serially linked channels and multiply linked channels. Only the simplest situations have been discussed in the mathematical literature so far, namely a single channel bounded by two drowned sluice gates, or by a series of such channels. In particular, in a recent paper Coron et.al. [3] have considered the flow along a channel between two large bodies of water controlled by sluice gates at the ends of the channel. The settings of these gates determine the fluid velocity at the two ends and the main result shows that suitable feedback boundary conditions can be used to exponentially stabilize a given operational state of the channel. This result depends on a subtle theorem of Greenberg and Ta Tsien [7] which guarantees the exponential decay of solutions to certain hyperbolic systems in two variables subject to boundary conditions which impose damping. These results have been extended by Leugering and Schmidt [11] to more general boundary conditions and to controlled star-like networks. Also the geometry of the reaches can be taken more general. We will restate the results below. As was the case for the previously cited papers the methods are entirely non linear but do remain within the realm of classical, shock free, solutions.

2 The model

We consider first a single channel parametrized lengthwise by $x \in [0, L]$. Let $Y_b(x)$ denote the altitude above sea level of the bed of the channel at x. The variable

$y = y(x) \in [0, d(x)]$ denotes the elevation above the channel bed where $d(x)$ denotes the depth of the channel. Let $\sigma(x, y)$ denote the width of the channel cross section at x and elevation y. Let $A(x, t)$ denote the area of the crosssection at x occupied by water at time t. We assume that the water level is constant across the channel at height $h(x, t) \in [0, d(x)]$. Clearly $A = A(x, h)$ and $h = h(x, A)$. In particular, leaving aside the t- dependence for the moment,

$$A(x, h) = \int_0^h \sigma(x, y)\, dy \quad \text{and} \quad A = \int_0^{h(x,A)} \sigma(x, y)\, dy. \tag{1}$$

It follows that

$$\partial_h A(x, h) = b(x, h) \quad \text{and} \quad \partial_A h(x, A) = \frac{1}{b(x, A)}, \tag{2}$$

where $b(x, h) \triangleq \sigma(x, h)$ (or $b(x, A) \triangleq b(x, h(x, A))$) is the width of the water surface at x. See Figure 2.

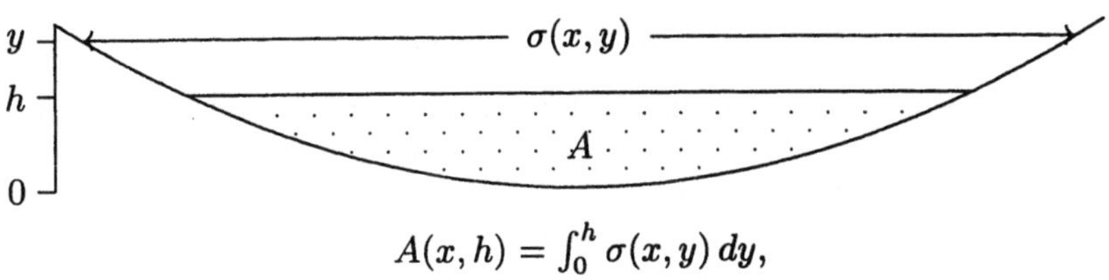

Figure 1. Channel cross section at x

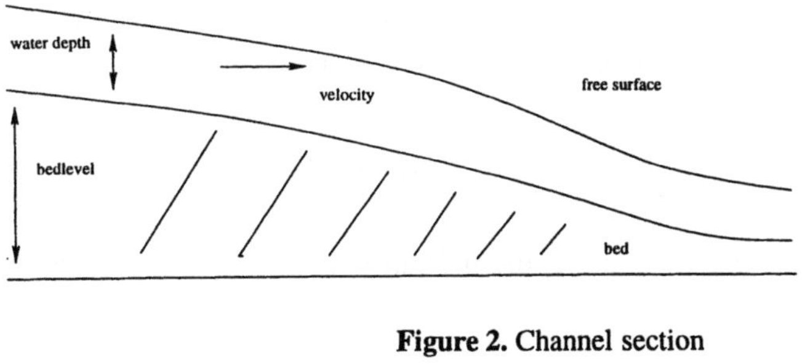

Figure 2. Channel section

It turns out to be convenient to choose $A(x,t)$, rather than $h(x,t)$ as our geometric state variable describing the distribution of water along the channel at a given time since it conveniently leads to a system of conservation laws. The derivation of the St Venant equations depends on the assumption that the flow of water along the channel can be represented by a scalar velocity function $V(x,t)$ in the direction of the channel from 0 to L. This can be thought of either as a constant, or as an average, velocity over the crosssection of the channel. For a more detailed physical discussion of the underlying assumptions of the St Venant model see [5] page 8.

Assuming a constant density ($\rho \equiv 1$, say) the mass flow rate of the liquid along the channel is given by $Q(x,t) \triangleq A(x,t)V(x,t)$. Conservation of mass is then expressed by the conservation equation

$$\partial_t A + \partial_x [AV] = 0, \tag{3}$$

The St. Venant system is completed by the balance of moments

$$\partial_t V + \partial_x \left[\tfrac{1}{2}V^2 + gh(x,A) + gY_b(x)\right] = 0. \tag{4}$$

Remark 1. One can add an empirically motivated resistance, or friction, term to the left hand side of the last equation. Various alternatives occur in the engineering literature (see, for example, [5] pages 19-22). Generically these are of the form $F(x,A,V)$ satisfying

$$F(x,A,0) = 0 \quad \text{and} \quad VF(x,A,V) \geq 0. \tag{5}$$

Now we consider networks of channels. We use notation similar to that introduced in [10] for networks of strings and beams. We index the channels, and the quantities associated with the channels, by $i \in \mathcal{I} = \{1,\dots,n\}$. We label the locations of the end points of the channels, which we shall refer to as nodes, by $j \in \{1,\dots,m\}$. We distinguish between multiple nodes, indexed by $j \in \mathcal{J}_M$, at which various channels come together and the simple nodes, indexed by $j \in \mathcal{J}_S$ which are endpoints of a single channel. For $j \in \mathcal{J}$ we introduce

$$\mathcal{I}_j = \{i \in \mathcal{I} : \text{ the } i\text{-th channel meets the } j\text{-th node}\}.$$

For $i \in \mathcal{I}_j$ we set $x_{ij} = 0$ or L_i corresponding to the end which meets the other channels at the j-th node. We the also set $\epsilon_{ij} = 1$, if $x_{ij} = L_i$ or $\epsilon_{ij} = -1$, if $x_{ij} = 0$.

At simple nodes we shall later impose boundary conditions through which controls can be imposed on the network. At multiple nodes we shall impose first the following condition expressing conservation of fluid in the flow through the node indexed by $j \in \mathcal{J}_M$:

$$\sum_{i\in\mathcal{I}_j} \epsilon_{ij} Q_i(x_{ij},t) = \sum_{i\in\mathcal{I}_j} \epsilon_{ij} A_i(x_{ij},t)V_i(x_{ij},t) = 0, \quad \text{for } j \in \mathcal{J}_M. \tag{6}$$

We can derive a second dynamic node condition from Hamilton's principle. To this end we introduce

$$S_i \triangleq \tfrac{1}{2}V_i^2 + gh_i(x,A_i) + gY_{bi}(x)$$

(a quantity which, divided by g, is called the *specific energy*). Then applying Hamilton's principle we obtain

$$S_k(x_{kj}, t) \equiv S_l(x_{lj}, t),$$

for all incident edges. The analysis can be modified to deal with the various cases in which x_{kj} and x_{lj} are not both zero. So we end up with the following dynamic node condition for each $j \in \mathcal{J}_M$

$$\left[\tfrac{1}{2}V_i^2 + gh_i(\cdot, A_i) + gY_{bi}\right](x_{ij}, t) \text{ coincide for } i \in \mathcal{I}_j \tag{7}$$

In addition we have to consider boundary conditions to be imposed at the simple nodes and initial conditions prescribing $A(x,0)$ and $V(x,0)$. The discussion of the boundary conditions will be given later taking into account issues concerning hyperbolic systems. It should be noted that the continuity of the water level at auch a junction is also a reasonable condition, if one is willing to sacrifice conservation of energy for an entropy inequality. This simpler condition, which is considered by some engineers, will also be used in some of the numerical simulations. To be more precise, let

$$\begin{aligned} E(A,V) &\triangleq \tfrac{1}{2}AV^2 + g\int_0^{h(x,A)} [Y_b(x) + y]\sigma(x,y)\,dy \\ &= \tfrac{1}{2}AV^2 + gAY_b(x) + \int_0^{h(x,A)} y\sigma(x,y)\,dy, \end{aligned} \tag{8}$$

where A and V satisfy (3) and (4). Then one easily verifies

$$\partial_t E + \partial_x[QS] = 0. \tag{9}$$

This of course also holds in indexed form for each channel and if one introduces the total energy

$$\mathcal{E}(t) = \mathcal{E}(\mathbf{A}(\cdot,t), \mathbf{V}(\cdot,t)) \triangleq \sum_{i\in\mathcal{I}} \int_0^{L_i} E_i(A_i, V_i)(x,t)\,dx$$

one obtains, using the multiple node conditions,

$$D_t\mathcal{E}(t) = \sum_{j\in\mathcal{J}_S} Q(x_{ij}, t)S(x_{ij}, t),$$

so that energy is conserved if, for example, there is no flow through the simple nodes of the channel network.

Remark 2. Friction can be introduced with a friction term $F_i(x, A_i, V_i)$ in the left hand side of the local St. Venant equations.

$$\partial_t E + \partial_x[QS] + F(x,A,V) = 0.$$

as well as

$$D_t\mathcal{E}(t) = \sum_{j\in\mathcal{J}_S} Q(x_{ij},t)S(x_{ij},t) - \sum_{i\in\mathcal{I}} \int_0^{L_i} A_iV_iF_i(x,A_i,V_i)\,dx.$$

The friction terms lead to energy decay because of the assumption (5).

3 Equilibrium Flows and their Perturbations

We seek solutions $\overline{\mathbf{A}}$ and $\overline{\mathbf{V}}$ of the network system described in the previous section which depend only on x and not on t. Explicitly they are required to satisfy

$$\begin{cases} \partial_x[\overline{A}_i\overline{V}_i] = 0 \ \text{ on } [0,L_i] & \text{for } i \in \mathcal{I}, \\ \partial_x[\frac{1}{2}\overline{V}_i^2 + gh_i(x,\overline{A}_i) + gY_{bi}(x)] = 0 \ \text{ on } [0,L_i] & \text{for } i \in \mathcal{I}, \\ \sum_{i\in\mathcal{I}_j} \epsilon_{ij}\overline{A}_i(x_{ij})\overline{V}_i(x_{ij}) = 0 & \text{for } j \in \mathcal{J}_M, \\ \frac{1}{2}\overline{V}_i(x_{ij})^2 + gh_i(x_{ij},\overline{A}_i(x_{ij})) + gY_{bi}(x) \text{ coincide} & \text{for } j \in \mathcal{J}_M, i \in \mathcal{I}_j. \end{cases} \tag{10}$$

It is easy to deduce that one must have

$$\overline{A}_i(x)\overline{V}_i(x) \equiv \overline{Q}_i \ \text{ with } \sum_{i\in\mathcal{I}_j} \epsilon_{ij}\overline{Q}_{ij} = 0, \tag{11}$$

$$\overline{S}_i(x) = \frac{1}{2}\overline{V}_i(x)^2 + gh_i(x,\overline{A}_i(x)) + gY_{bi}(x) \equiv \overline{S}, \tag{12}$$

where $\overline{Q}_i$ and $\overline{S}_i$ are constants.

We say that the fluid in the channels is *still* if $V = 0$. In that case the components of $A(x)$ are determined from

$$h_i(x,\overline{A}_i(x)) = \frac{1}{g}[\overline{S} - gY_{bi}(x)],$$

which will have a solution respecting the depth restriction on each channel if and only if

$$0 \leq \frac{1}{g}[\overline{S} - gY_{bi}(x)] \leq d_i(x) \ \text{ for all } i \in \mathcal{I}.$$

The equilibria which are not still are more difficult to determine. For the purposes of this paper we now make two restrictive assumptions, namely

- the channels are *prismatic* which means that the crossections of the channels do not depend on x (so d_i, h_i, σ_i and b_i do not depend directly on x);
- the system of channels is *level*, the beds of the channels all lying at the same constant elevation Y_b.

In this case equilibria A, V will not depend on x. One can think of channels designed with certain operating conditions in mind. For example one can specify a standard water height of $\overline{h}$ for the whole channel with $\overline{h} < d_i$ for each $i \in \mathcal{I}$. For each channel one can fix the flow direction, encoding this by $\epsilon_i = 1$ or -1 depending on whether $\overline{V}_i$ is to be positive or negative. This is to be done in such a way that at each multiple nodes flows occur both into and out of the node. We then try to design the crosssections of each channel in such a way that $\overline{A}_i = A_i(\overline{h})$ satisfy

$$\sum_{i \in \mathcal{I}_j} \epsilon_{ij} \epsilon_i \overline{A}_i = 0. \tag{13}$$

If this is possible we can set $\overline{V}_i = \epsilon_i \overline{V}$ where $\overline{V}$ is any given positive velocity and easily check that the conditions for an equilibrium are satisfied. To clarify the meaning of the above condition we note that when $\epsilon_{ij}\epsilon_i = 1$ the flow in the i-th channel is into the j-th node while when $\epsilon_{ij}\epsilon_i = -1$ that flow is away from the node. For channels of rectangular crosssection with b_i the breadth of the i-th channel one can require

$$\sum_{i \in \mathcal{I}_j} \epsilon_{ij} \epsilon_i b_i = 0, \tag{14}$$

independently of the value $\overline{h}$. It is almost obvious that the above process always works in the following particular network configurations:

- star configurations in which n channels all meet at one multiple node;
- tree configurations in which the direction of flow is always towards the trunk, or always away from the trunk.

Remark 3. One can easily experiment with a variety of particular networks and direction assignments to find many other configurations in which the above process works. These may include closed paths.

A general goal now is to stabilize the flow around such an equilibrium flow by means of suitable feedback boundary conditions at the simple nodes, which lie at the extremities of the channel network i.e., at the points where water flows into or out of the system of channels. At present we can do this only for star configurations. In fact we first consider a single channel for which our results are also in large part new.

4 Stabilization and Null Controllability for a Single Channel

We consider one level, prismatic channel and study perturbations of constant equilibrium conditions.

First we make use of the standard method of Riemann invariants for hyperbolic systems to be found, for example, in Taylor [13], Chapter 16. We begin with a single channel. Evaluating the x derivatives in (3) and(4) we can rewrite these equations as a system

$$\partial_t \begin{pmatrix} A \\ V \end{pmatrix} + \begin{pmatrix} V & A \\ g/b(A) & V \end{pmatrix} \partial_x \begin{pmatrix} A \\ V \end{pmatrix} = \begin{pmatrix} 0 \\ 0 \end{pmatrix}. \tag{15}$$

The eigenvalues of the matrix are

$$\lambda_\pm(A,V) = V \pm \gamma(A) \quad \text{where } \gamma(A) = \sqrt{gA/b(A)} \tag{16}$$

with $\lambda_+ > 0$ and $\lambda_- < 0$ in the *subcritical* case that

$$V^2 < gA/b(A). \tag{17}$$

The corresponding left eigenvectors are

$$\mathbf{l}_\pm(A) = \frac{1}{2}(\pm\sqrt{g/[Ab(A)]}, 1).$$

Riemann invariants are then obtained by solving

$$\nabla\xi_\pm(A,V) = (\partial_A, \partial_V)\xi_\pm(A,V) = \mathbf{l}_\pm(A).$$

We can take

$$\xi_\pm(A,V) = \frac{1}{2}\left(V \pm \int_0^A q(\alpha)\,d\alpha\right) \quad \text{with } q(\alpha) = \sqrt{\frac{g}{\alpha b(\alpha)}}.$$

The system (15) is now equivalent to

$$\partial_t\xi_\pm(A,V) + \lambda_\pm(A,V)\partial_x\xi_\pm(A,V) = 0.$$

For solutions A and V the *Riemann invariants* $\xi_\pm$ are constant along characteristic curves $(x_\pm(t), t)$ with

$$D_t x_\pm(t) = \lambda_\pm(A(x_\pm(t),t), V(x_\pm(t),t)).$$

Now we consider perturbations $A = \overline{A} + a$ and $V = \overline{V} + v$ of an equilibrium state. We assume that the equilibrium flow is subcritical, so that this will continue to be the case for small perturbations. The system becomes

$$\partial_t \begin{pmatrix} a \\ v \end{pmatrix} + \begin{pmatrix} \overline{V}+v & \overline{A}+a \\ g/b(\overline{A}+a) & \overline{V}+v \end{pmatrix} \partial_x \begin{pmatrix} a \\ v \end{pmatrix} = \begin{pmatrix} 0 \\ 0 \end{pmatrix}. \tag{18}$$

In terms of the perturbation variables a and v the eigenvalues of the matrix are

$$\lambda_\pm(a,v) = \overline{V} + v \pm \beta(a) \quad \text{where } \beta(a) = \sqrt{g\frac{\overline{A}+a}{b(\overline{A}+a)}}, \tag{19}$$

corresponding to Riemann invariants

$$\xi_\pm(a,v) = \frac{1}{2}\left(v \pm \int_0^a p(\alpha)\,d\alpha\right) \quad \text{with } p(\alpha) = \sqrt{\frac{g}{[\overline{A}+\alpha]b(\overline{A}+\alpha)}}. \tag{20}$$

The characteristic curves $(x_\pm(t), t)$ are now determined by solving

$$D_t x_\pm(t) = \lambda_\pm(a(x_\pm(t),t), v(x_\pm(t),t)). \tag{21}$$

We next turn to the feedback stabilization of a single channel, which should drive perturbations a and v to zero exponentially in time. The system has to be complemented by initial conditions

$$a(x,0) = a^0(x), \quad v(x,0) = v^0(x). \tag{22}$$

In terms of the Riemann invariants it is well known that one can impose boundary conditions of the form

$$\xi_+(0,t) = g^0(\xi_-(0,t)), \quad \text{and } \xi_-(L,t) = g^L(\xi_+(L,t)).$$

In particular one could impose the absorbing boundary conditions

$$\xi_+(0,t) = 0, \quad \text{and } \xi_-(L,t) = 0, \tag{23}$$

equivalent to the following feedback boundary conditions in terms of a and v,

$$v(0,t) = -P(a(0,t)), \; v(L,t) = P(a(L,t)) \;\text{ with } P(a) = \int_0^a p(\alpha)\, d\alpha. \tag{24}$$

One could also replace $P(a)$ by Taylor approximations $P_1(a) = p(0)a$ or $P_2(a) = p(0)a + p'(0)a^2/2$, say, or alternatively by other functions of a.

Theorem 4. *Consider the perturbed St Venant system with boundary conditions*

$$v(0,t) = f^0(a(0,t)), \quad v(L,t) = f^L(a(L,t)),$$

where f^0, f^L are continuously differentiable in a neighbourhod of the origin and satisfy

$$f^0 = 0, \quad Df^0(0) \neq p(0), \quad f^L = 0, \quad Df^L(0) \neq -p(0)$$

We assume that initial conditions are given satisfying the compatibility conditions

$$\begin{aligned}
&v^0(0) = f^0(a^0(0)), \; v^0(L) = f^L(a^0(L)), \\
&\frac{g}{b(\overline{A}+a^0(0))} Da^0(0) + (\overline{V}+v^0(0))Dv^0(0) \\
&\qquad = Df^0(a^0(0)) \left[(\overline{V}+v^0(0))Da^0(0) + (\overline{A}+a^0(0))Dv^0(0)\right], \\
&\frac{g}{b(\overline{A}+a^0(L))} Da^0(L) + (\overline{V}+v^0(L))Dv^0(L) \\
&\qquad = Df^L(a^0(L)) \left[(\overline{V}+v^0(L))Da^0(L) + (\overline{A}+a^0(L))Dv^0(L)\right].
\end{aligned} \tag{25}$$

Suppose that

$$|Dg^0(0)Dg^L(0)| = \left|\frac{Df^0(0)+p(0)}{Df^0(0)-p(0)}\,\frac{Df^L(0)-p(0)}{Df^L(0)+p(0)}\right| < 1. \tag{26}$$

Then, if $\| (a^0, v^0) \|_1$ *is sufficiently small, there exists a unique continuously differentiable solution* $(a(x,t), v(x,t))$ *to the problem which is defined for all positive* t *and satisfies an estimate*

$$\| (a(\cdot,t), v(\cdot,t)) \|_1 < Ce^{-\alpha t} \| (a^0, v^0) \|,$$

where C *and* α *are suitable positive constants.*

For a proof see [11].

Remark 5. In the notation of the paper [3] the boundary conditions (16) of that paper can be written

$$u_0 = -\left(\frac{\overline{V}}{2} + \lambda_0\right) \frac{y_0 - \overline{y}}{\overline{y} + (y_0 - \overline{y})} \triangleq f^0(y_0 - \overline{y}),$$
$$u_L = -\left(\frac{\overline{V}}{2} - \lambda_L\right) \frac{y_L - \overline{y}}{\overline{y} + (y_L - \overline{y})} \triangleq f^L(y_L - \overline{y}).$$

The variables y_0, y_L and $\overline{y}$ correspond to $A(0,t)$, $A(L,t)$ and $\overline{A}$, with $y_0 - \overline{y}$ corresponding to $a(0,t)$, etc. In our notation $u_0 = u(0,t)$ and $u_L = u(L,t)$. The quantities calculated in the proof of their Theorem 1 correspond directly to our calculation of $Dg^0(0)$ and $Dg^L(0)$. Our result is more general in that it allows for a broad class of boundary conditions and does not require rectangular crosssections.

It is also possible to treat feedback-boundary controllability problems using Riemann invariants. See [11] and [8].

5 Stabilization and Null Controllability for a Star Configuration of Channels

Now we consider a star configuration of channels and perturbations $A = \overline{A} + a$ and $V = \overline{V} + v$ of equilibrium states $\overline{A}, \overline{V}$ constructed at the end of the previous section. We assume that the equilibrium flow is subcritical on each channel, so that this will continue to be the case for small perturbations. The flow in each channel is governed by the St.Venant equations with all the quantities indexed by i and $x \in [0, L_i]$. It is convenient to suppose that for each i the parameter value $x = 0$ corresponds to the end of the channel at the single multiple node. It is then useful to parametrize all the channels with a parameter x over a common interval $[0, L]$, where L could, for example, be the average channel length. For the i-th channel this would entail a parameter change $x \mapsto L_i/L\, x$ and the system corresponding to that channel becomes

$$\partial_t \begin{pmatrix} a_i \\ v_i \end{pmatrix} + \frac{L_i}{L} \begin{pmatrix} \overline{V}_i + v_i & \overline{A}_i + a_i \\ g/b_i(\overline{A}_i + a_i) & \overline{V}_i + v_i \end{pmatrix} \partial_x \begin{pmatrix} a_i \\ v_i \end{pmatrix} = \begin{pmatrix} 0 \\ 0 \end{pmatrix}. \tag{27}$$

The corresponding eigenvalues of the matrix are now

$$\lambda^i_{\pm}(a_i, v_i) = \frac{L_i}{L}(\overline{V}_i + v_i \pm \beta_i(a_i)).$$

The Riemann invariants $\xi_{\pm}^{i}$ are unaffected by the parameter change.

This gives us a hyperbolic system of 2n equations on $[0, L] \times [0, T]$. Such systems are discussed in the appendix, which uses a different indexing of the Riemann invariants, setting

$$\xi_i = \begin{cases} \xi_{+}^{i}, & \text{for } i = 1, \dots, n, \\ \xi_{-}^{i-n}, & \text{for } i = n+1, \dots, 2n. \end{cases}$$

In our case the equations are pairwise decoupled and it is convenient to stay with the indexing $\xi_{\pm}^{i}$. Initial conditions are given by

$$a(x,0) = a^0(x), \;\; v(x,0) = v^0(x). \tag{28}$$

At $x = L$ we can introduce decoupled boundary conditions acting independently on each channel:

$$v_i(L,t) = f_i(a_i(L,t)), \quad \text{or} \quad \xi_{-}^{i}(L,t) = g_i^L(\xi_{+}^{i}(L,t)). \tag{29}$$

The coupling between the variables occurs through the multiple node conditions which translate into a boundary condition at $x = 0$. Let

$$\begin{aligned} S_i(a_i, v_i) &\triangleq \frac{1}{2}(\overline{V}_i + v_i)^2 + g h_i(\overline{A}_i + a_i), \\ Q_i(a_i, v_i) &\triangleq (\overline{A}_i + a_i)(\overline{V}_i + v_i). \end{aligned} \tag{30}$$

Then we have the following set of n multiple node conditions in $2n$ variables holding at $(0,t)$

$$\begin{cases} S_i(a_i, v_i) - S_n(a_n, v_n) = 0, & \text{for } i = 1, \dots, n-1, \\ \sum_{i=1}^{n} Q_i(a_i, v_i) = 0. \end{cases} \tag{31}$$

We have the following

Theorem 6. *Consider the systems* (27) *with boundary conditions* (30) *and* (29) *holding at* $x = 0$ *and* $x = L$ *respectively and initial conditions* (28) *with data satisfying the appropriate compatibility conditions as in* (25). *Suppose that*

$$\begin{aligned} &\| \nabla_{x i_-} g^0(0) \|_\infty \max_{1 \le i \le n} |D g_i^L(0)| \\ &= \| \overline{\Lambda}_+^{-1} [G_+^{-1} G_-] \overline{\Lambda}_- \|_\infty \max_{1 \le i \le n} \left| \frac{D f_i^L(0) + p_i(0)}{D f_i^L(0) - p_i(0)} \right| < 1. \end{aligned} \tag{32}$$

Then, for $\| (a^0, v^0) \|_1$ *sufficiently small, there exists a unique continuously differentiable solution to the problem which is defined for all positive* t *and satisfies*

$$\| (a(\cdot,t), v(\cdot,t)) \|_1 < C e^{-\alpha t} \| (a^0, v^0) \|,$$

where C *and* α *are suitable positive constants.*

Proof. (See [11])

It is also possible to adapt the proof of this Theorem to prove a result on null controllability for the star system.

6 Numerical Simulations

We now consider a flow process in open rectangular channels, to simplify the analysis. Hence, $A(x,t) = b \cdot h(x,t)$, where again b is the constant width and $h(x,t)$ the height or water level, respectively. Let C denote the Manning number and R the fraction of cross section versus wetted boundary, i.e.,

$$R(x,t) = \frac{A(x,t)}{b + 2h(x,t)},$$

We apply the so-called Manning formula for computing the friction slope

$$S_{fric}(x,t) = \frac{C^2 |V(x,t)| V(x,t)}{R(x,t)^{4/3}}. \tag{33}$$

Initial values describe the discharge and water level distribution at $t = 0$, $V(x,0) = V_0(x)$ and $h(x,0) = h_0(x)$. Boundary values are chosen to model a specific situation, where a time dependent inflow at one side is given, say $x = 0$, with an input function $s(t)$, and an outflow controlled by an underflow gate opening subject to a control function $u(t)$, leading to

$$\begin{aligned} Q(0,t) &= V(0,t)bh(0,t) = s(t), \\ Q(L,t) &= V(L,t)bh(L,t) = \alpha u(t)\sqrt{g(h(L,t) - H_f)}, \end{aligned} \tag{34}$$

see Graf [6], where H_L is the right water level outside the reach and $\alpha > 0$ a constant, see Figure 3. These conditions are considered in [3] and fit into the framework of Riemann invariants outlined above. In many practical situations, channels are connected and form networks with different topologies. To give two simple examples, consider three channels connected at one end, with one inflow and two outflows, see Figure 4(a), or the two serial channels of Figure 4(b).

In the second case, two serial channels are given with the same width for simplicity, and are connected at $x = L_1$. Water height can be controlled by two underflow gates, one between the two channels, one at the right boundary $x = L_2$. Both

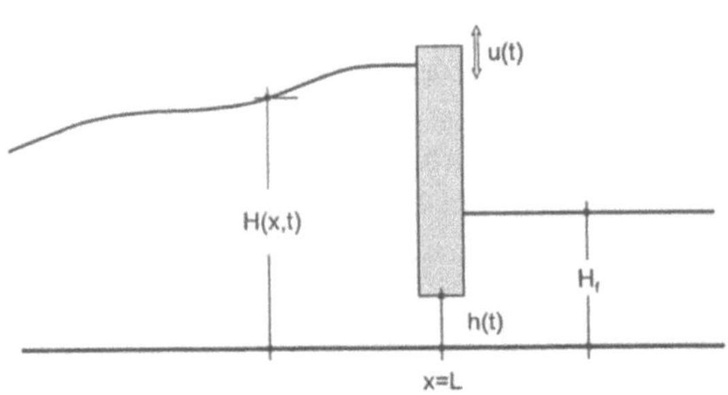

Figure 3. Underflow Gate

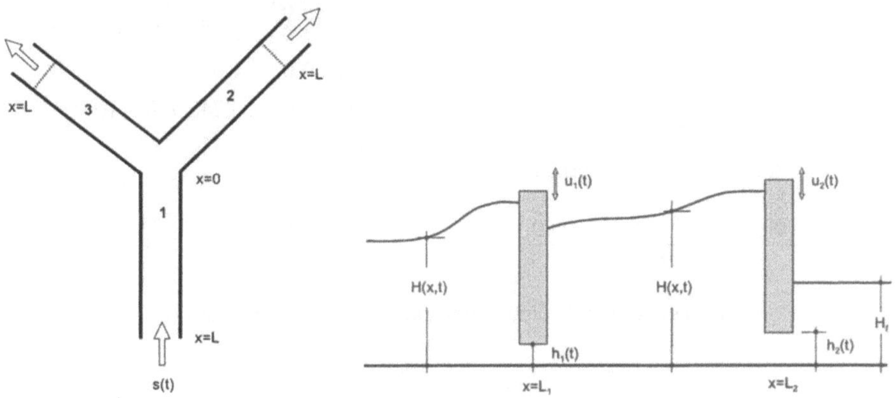

Figure 4. Three-star channel node (a) and two serial channels (b)

are controllable by time-dependent, smooth functions $u_1(t)$ and $u_2(t)$. Now the dynamic system is defined in two different integration areas from $x = 0$ to $x = L_1$, then again from $x = L_1$ to $x = L_2$, leading boundary and transition conditions of the form

$$\begin{aligned} Q(0,t) &= s(t), \\ Q^-(L_1,t) &= Q^+(L_1,t), \\ Q^+(L_1,t) &= \alpha u_1(t)\sqrt{g(h^-(L_1,t) - h^+(L_1,t))}, \\ Q(L_2,t) &= \alpha u_2(t)\sqrt{g(h(L_2,t) - H_f)}, \end{aligned} \tag{35}$$

Here the minus and plus signs denote the corresponding limits from the left and right side at transition point $x = L_1$. The flow is assumed to go from upstream to downstream.

In order to present a numerical example, we consider two serial channels as outlined above, with constant widths $b = 5$, gravitational constant $g = 9.81$, Boussinesq coefficient $\beta = 1$, Manning number $C = 0.1$, constant bed slope $S_{floor} = 0.001$, initial discharge $Q(x,0) = 1.65$, initial water level $h(x,0) = 1.2$, outer water level $H_f = 0.5$, and boundary coefficient $\alpha = 2$. The length of both channels is 500, i.e., $L_1 = 500$ and $L_2 = 1000$. The inflow $s(t)$ is given by linear interpolation of some data, see Table 1(a).

We are interested in the question, whether the flow in the channel can be controlled at the two underflow gates, so that the water levels are $H_{goal}(x) = 1$ for $x \in [0, L_1]$ and $H_{goal}(x) = 0.9$ for $x \in (L_1, L_2]$. Controlled are the openings of the two underflow gates by simple polynomial control functions of the type

t	s(t)
0.0	1.65
20.0	50.00
80.0	1.65
500.0	0.00

(a)

	initial	final
a_{11}	1.0	2.5550385
a_{12}	0.0	$-4.0682183 \cdot 10^{-3}$
a_{13}	0.0	$6.2459060 \cdot 10^{-6}$
a_{21}	1.0	1.1101733
a_{22}	0.0	$4.587581 \cdot 10^{-3}$
a_{23}	0.0	$-6.7988864 \cdot 10^{-6}$

(b)

Table 1. Inflow (a) and initial and final parameter values (b)

$u_1(t) = a_{11} + a_{12}t + a_{13}t^2$ and $u_2(t) = a_{21} + a_{22}t + a_{23}t^2$. The final time horizon is $t = 500$.

The system of hyperbolic equations is discretized by 19 lines in the first and 15 lines in the second integration area. The ENO method is applied, where the fixed stepsize of the Runge-Kutta method is 1. A least squares problem is formulated to reach the given goal values at $t = 500$, measured at spatial values $x = 50$, $x = 100$, $x = 150, \ldots, x = 1000$. The code DFNLP [12] computes a solution within 25 steps with termination accuracy $1.0 \cdot 10^{-6}$, see Table 1(b).

Corresponding discharge and water level surface plots are shown in Figures 5 to 8 for the initial, then the controlled situation.

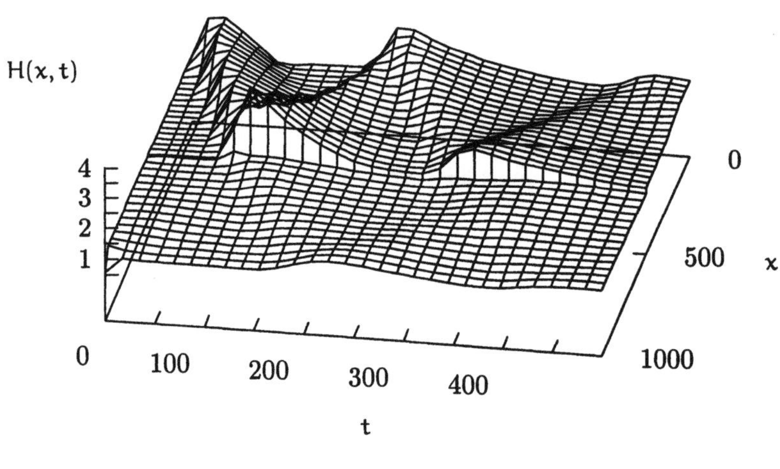

Figure 5. Initial Water Level Distribution

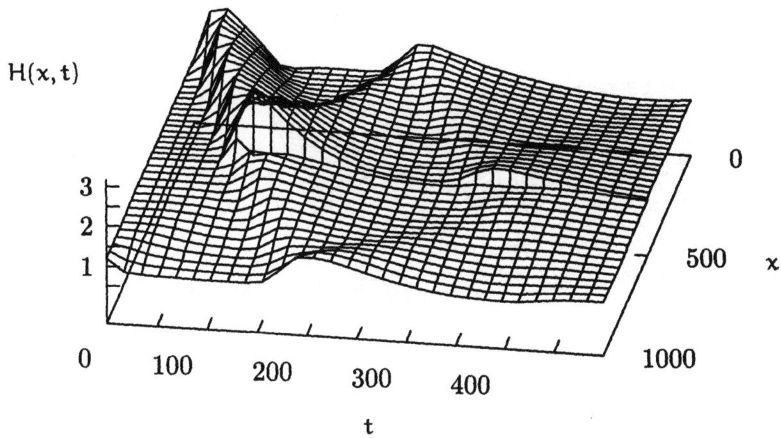

Figure 6. Final Water Level Distribution

6.1 Methods of Characteristics

The discussions in Section 4 and 5 are based upon the Riemann invariants and characteristic curves. In particular, the absorbing boundary conditions are given in terms of the Riemann invariants. In fact, in terms of the characteristic variables the absorbing boundary conditions are linear, whereas with other sets of variables they are nonlinear. Therefore, it is natural to make also numerical experiments with a method of characteristics. In the engineering literature, such methods are well represented, see, e.g., [1]. They are also discussed in [4]. An advantage of these methods is that they accurately model the domains of determinacy and the ranges of influence which are decisive for control problems. We have performed simulations with such a method that will be reported on in the paper [9].

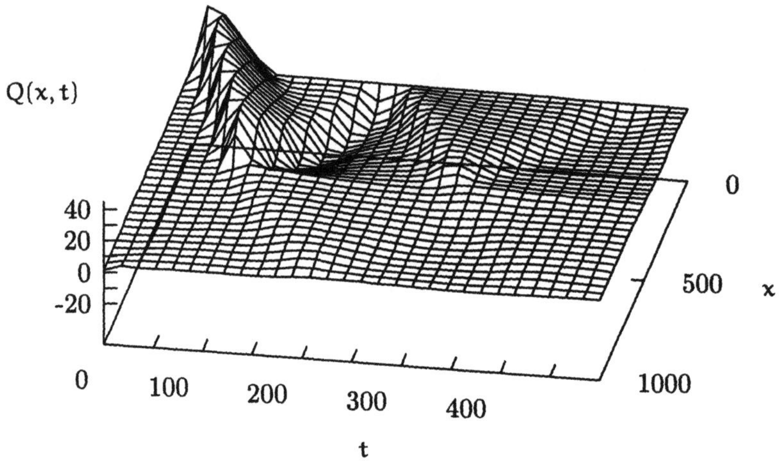

Figure 7. Initial Discharge Distribution

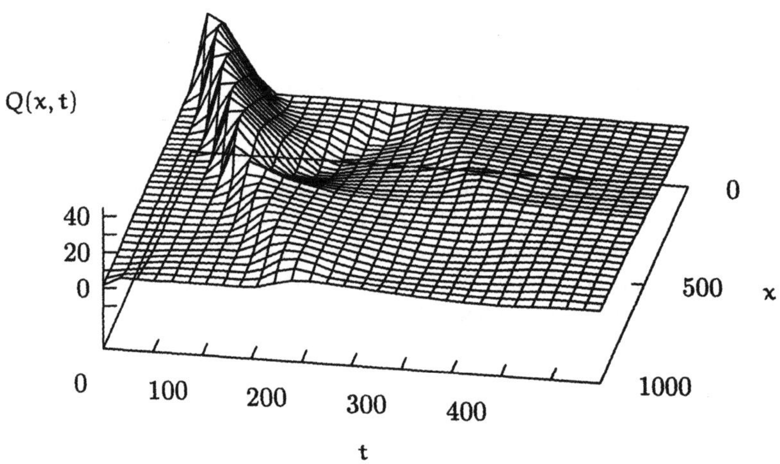

Figure 8. Final Discharge Distribution

Example 7. We consider a junction point where three channels of length $L = 10m$ meet. The channels are rectangular hence for the wave celerity c we have $c^2 = gh$. The junction point is the end L of channel 1 and channel 2 which go from zero to L and the end zero of channel 3 that goes form zero to L.

The channels have constant slope −0.0125. For the friction slope we use the Chezy formula $S_{fric} = V|V|/(C^2h)$ with $C^2 = 5g$.

The width of channel 1 and channel 2 is $2m$ and the width of channel 3 equals $4m$.

At $t = 0$ we have the constant initial values $V = 0.5m/s$ and $c = 2m/s$.

At the end zero of channel 1 we have the boundary condition

$$\xi_+(t) = V(t) + 2c(t) = \begin{cases} (t-1)^2(t-20)^2 * 10^{-5} + 4.5m/s & \text{if } t \in (1,2) \\ 4.5m/s & \text{else.} \end{cases}$$

At the end zero of channel 2 we have the constant boundary condition $\xi_+(t) = V(t) + 2c(t) = 4.5m/s$.

In channel 3 at the end L, we have the constant boundary condition $\xi_-(t) = V(t) - 2c(t) = -3.5m/s$.

In the junction point we use as junction conditions the continuity equation

$$h_1 b_1 V_1 + h_2 b_2 V_2 = h_3 b_3 V_3,$$

where V_i is the velocity in channel i, h_i is the corresponding water height and b_i is the width of channel i. So we have $b_1 + b_2 = b_3$ and $h_i b_i V_i$ is the discharge corresponding to channel i in the junction point. Moreover, we require that the water surface is continuous in the junction point:

$$h_1 = h_2 = h_3.$$

At $t = 0$ we started with 64 equidistant points for each channel.

The following pictures show the computed Froude numbers V/c (corresponding to Mach's number in gas dynamics) for each of the three channels.

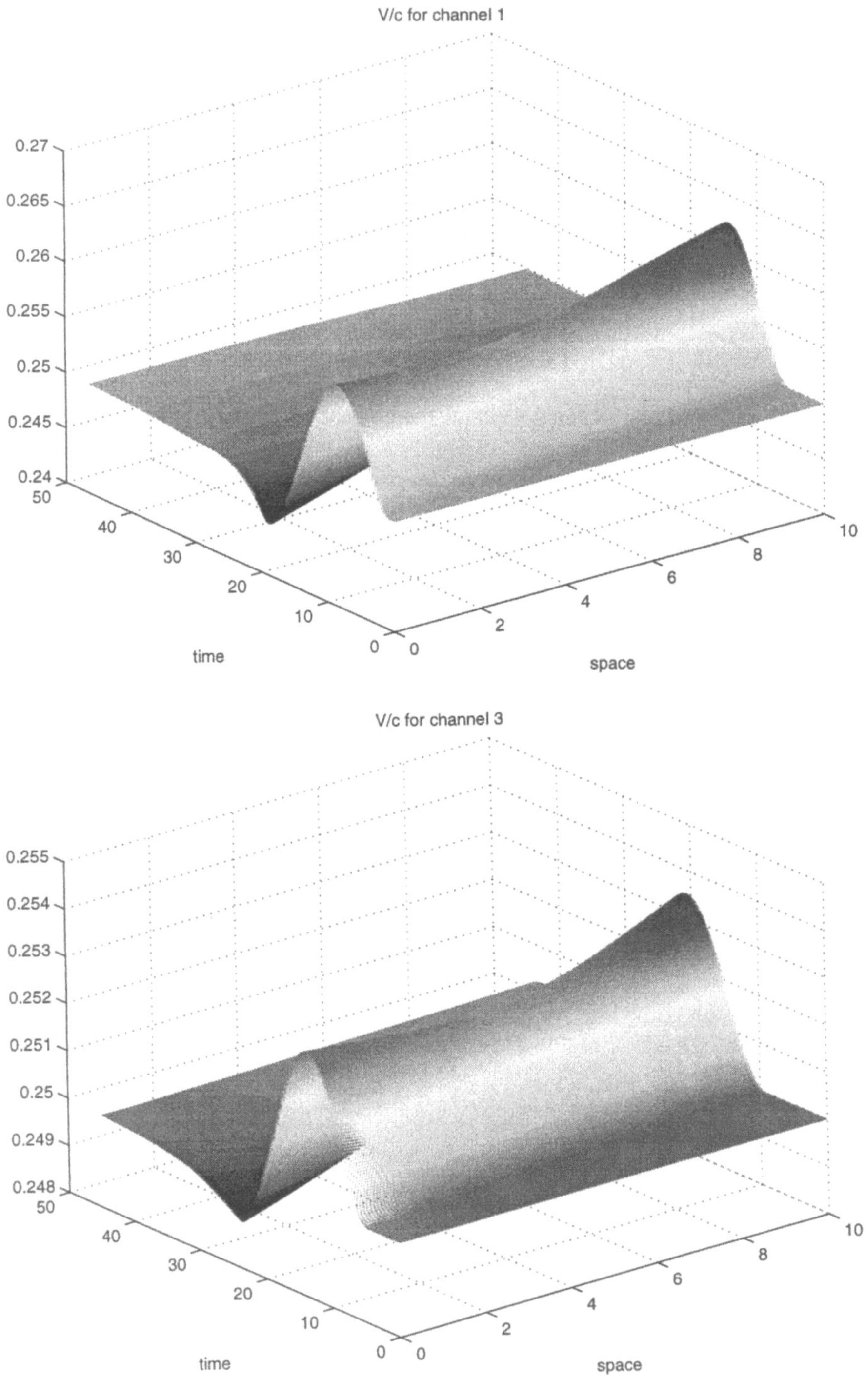

Figure 9. V/c for channel 1 and channel 3

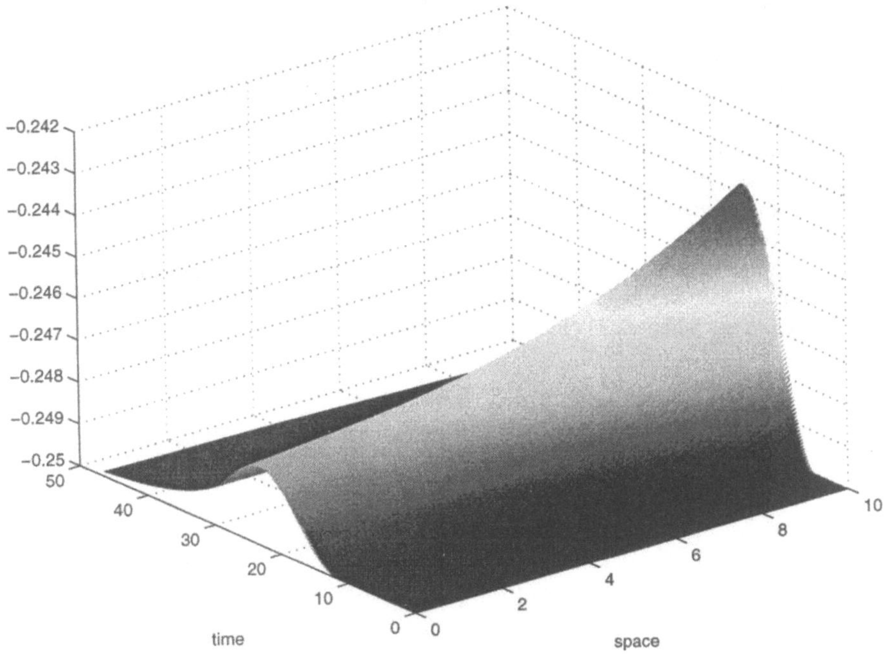

Figure 10. $-V/c$ for channel 2

REFERENCES

1. M. B. Abbott, *Computational Hydraulics*, Pitman, London, 1979.
2. M. Cirinà, *Nonlinear Hyperbolic problems with solutions on preassigned sets*, Mich. Math J., 17 (1970), pp.193-209.
3. J. M. Coron, B. d'Ándréa-Novel and G. Bastin, *A Lyapunov approach to control irrigation channels modeled by Saint-Venant equations*, to appear in Proceedings of ECC Karlsruhe, 1999.
4. R. Courant and D. Hilbert *Methods of Mathematical Physics*, Volume II, Interscience Publishers, 1962.
5. J. A. Cunge, F. M. Holly, A. Verwey, *Practical aspects of computational river hydraulics*, Pitman, Boston 1980.
6. W. H. Graf *Fluvial Hydraulics*, J. Wiley and Sons, Chichester, 1998.
7. J. M. Greenberg and Li Ta Tsien, *The effect of boundary damping for the quasilinear wave equation*, J.D.E., 52 (1984), pp.66–75.
8. M. Gugat and G. Leugering, *Global boundary controllability of the de St. Venant equations*, submitted 2000.
9. M. Gugat *A method of characteristics for the control of networks governed by the de St. Venant equations*, preprint, 2001.

10. J. E. Lagnese, G. Leugering and E. J. P. G. Schmidt, *Modeling, Analysis and Control of Dynamic Elastic Multi-Link Structures*, Birkhauser, Boston/Basel/Berlin, 1994.
11. G. Leugering and E. J. P. G. Schmidt, *On the modelling and stabilization of flows in networks of open channels*, submitted to SIAM J Control and Optimzation, July 2000.
12. K. Schittkowski, *Solving nonlinear least squares problems by a general purpose SQP-method*, in: Trends in Mathematical Optimization, K.-H. Hoffmann, J.-B. Hiriart-Urruty, C. Lemarechal, J. Zowe eds., International Series of Numerical Mathematics, Vol. 84, Birkhaeuser, 1988.
13. M. E. Taylor, *Partial Differential Equations Vol III*, Applied Math.Series **117**, Springer, New York, 1996.
14. B. de Saint-Venant, *Théorie du mouvement non-permanent des eaux avec application aux crues des rivières et à l'íntroduction des marées dans leur lit*, Comptes Rendus Academie des Sciences 73 (1871), pp.148-154, 237-240.

Optimal Control of Distributed Systems with Break Points

Michael Liepelt and Klaus Schittkowski

Fachbereich Mathematik, Universität Bayreuth, Germany

Abstract We consider optimal control of distributed parameter systems, that are frequently used in chemical engineering to model for example tubular reactors. Break points are introduced to take also cleaning operations into account, where a reactor is shut down for a while and then restarted again. The goal is to minimize a cost function over a fixed time horizon, where the number of cleaning operations, the length of reactor operation between successive cleanings, and the reactor feed rates for each time interval are to be computed. We assume that product prices and consumer demands are time-dependent. It must be guaranteed that the decrease of the free cross-sectional area of the tube caused by coke deposition never exceeds a certain limit. Moreover, there are time and position dependent constraints for the state and control variables such as a maximum bound for the temperature. The general mathematical model and the applied discretization schemes are outlined in detail. We present two different approaches, one based on the method of lines, the other on full discretization where discretized state variables are treated as additional optimization variables. Numerical results are presented for a case study, where optimal input feeds and maintenance times of an acetylene reactor are computed.

1 Introduction

The computation of optimal feed controls for chemical reactors, especially for tubular reactors, is a well-known technique, see Edgar and Himmelblau [11], Nishida et al. [17], and Buzzi-Ferraris et al. [5, 6]. Our intention is to extend the underlying mathematical model structure with the aim to determine also the number of reactor cleanings and the operation length between successive maintenance times under the assumption that time-dependent cost and consumer demand functions are known over the whole time horizon.

The mathematical model is given as a distributed parameter system with break points, that is in form of a set of first order partial differential equations in one space dimension. At each break point, initial values with respect to the time variable are reset. The right-hand side of the distributed system depends in addition on a discrete, i.e., time and space independent optimization parameter. The system is controlled at the left boundary of the spatial interval. There are bounds for control and state variables, dynamic constraints for a state variable at the right boundary of the spatial interval, and additional constraints for the parameter vector. The cost functional to be minimized is given in form of an integral over the time variable.

In case of a tubular reactor the chemical reactions and the temperature depend dynamically on the space variable, whereas the dynamic decrease of the cross-sectional area caused by coke deposition is time-dependent. In both cases we know initial values either in the form of time-dependent feed control functions or a constant tube diameter. The break points are needed to schedule the reactor maintenance, i.e., we want to determine a subdivision into operation intervals, where the reactor is cleaned at the end of each interval. Then we want to compute dynamic feed rates for each interval such that the overall profit of the reactor is maximized. It must be guaranteed that the free cross-sectional area never falls below a given lower bound, and that the temperature never exceeds a maximum value. As soon as the cross-sectional area reaches this minimum level, the reaction is stopped, the reactor has to be cleaned, and the process is restarted. Moreover, there are bounds for the input controls, mass flows, and operation lengths. The cleaning times represent additional optimization variables, since the price and demand functions may change over time. Time-dependent alterations of these functions are the only reasons for differences between the operation intervals, since we always start with the same initial distribution of the cross-sectional area after reactor cleanings.

Two possibilities are outlined to discretize the dynamical system. In the first case the system equations are discretized w.r.t. the time variable only, and the resulting system of ordinary differential equations in the space variable is solved using standard algorithms, see for example of Hairer et al. [13] for non-stiff and of Hairer and Wanner [14] for stiff equations. Stiffness can be introduced by a mixture of fast and slow chemical reactions. This approach is called the method of lines, see Schiesser [19]. The dynamic constraints are discretized using uniform grids over the space and time interval. Since the input control functions can be described by a finite number of variables, we obtain a finite dimensional nonlinear programming problem. The resulting optimization problem is solved by the sequential quadratic programming (SQP) code NLPQL of Schittkowski [20, 21]. A special advantage of SQP methods is that they can efficiently handle a relatively large number of inequality restrictions. Gradients are approximated by finite differences, where the special sparsity structure of the Jacobian of the problem is exploited.

The second approach consists of a full discretization of the state variables with respect to both space and time. Partial derivatives are substituted by suitable finite difference approximations. The discretized state variables are treated as optimization parameters, where the dynamic equations are transformed into a large and sparse system of nonlinear equations. The resulting large scale nonlinear optimization problem is efficiently solved by a partially reduced SQP algorithm of Dennis et al. [8]. State constraints are added to the objective function using a modified barrier functions (MBF) of Ben-Tal and Zibulevsky [1]. Particular advantages of this approach are the exact evaluation of gradients obtained by automatic differentiation, see Dobmann et al. [9], and the possibility to overcome instabilities in case of non-feasible programs.

However the assumption that time-dependent price and consumer demand functions are a priori known is somewhat unrealistic. In a real-life environment these functions will permanently change, and even the technical conditions or the chemical process can vary. Thus, one has to guarantee that maintenance times and feed rates can be adapted under real-time conditions, i.e., they must be re-evaluated as fast as possible whenever model data are altered. Hence the implemented numerical algorithms allow restarts, where the known solution and additional iteration data of the numerical algorithm are exploited to get a new solution within very few additional iterations.

In Section 2 we give an outline of the considered mathematical model in form of a general distributed system, and the optimal control problem is formulated. The discretization of the distributed parameter system by the numerical method of lines is described in Section 3, whereas the full discretization approach and the application of the RSQP algorithm are found in Section 4. In Section 5 we introduce the dynamic equations of an acetylene reactor that serves as a case study. Some numerical results are presented in Section 6.

2 The Optimal Control Problem

We consider distributed systems of the form

$$\begin{aligned} u_x &= f(u,v), \\ v_t &= g(u,v) \end{aligned} \tag{1}$$

with initial values $u(t,0) = u_0(t)$, $v(0,x) = v_0(x)$. The state variables u and v are vector-valued functions, $u \in \mathbb{R}^{n_u}$, $v \in \mathbb{R}^{n_v}$, and the equations are to be satisfied for all $t \in [0, t_f]$ and $x \in [0, L]$ with given time horizon t_f and spatial length L. Obviously the system is completely symmetric, that is we obtain exactly the same structure when interchanging time variable t and spatial variable x.

Equations (1) are quite general and can be applied to a large number of real applications. In our case we are mainly interested in deriving a general scheme that can be used then to model for example tubular chemical reactors. Having this goal in mind, we obtain an optimal control problem by the following steps:

a) First we add an additional parameter vector $p \in \mathbb{R}^{n_p}$ to the right-hand side of (1) that does not depend on time or spatial variable,

$$\begin{aligned} u_x &= f(p,u,v), \\ v_t &= g(p,u,v). \end{aligned} \tag{2}$$

Since the solution then depends also on p, we use the notation $u(p,t,x)$ and $v(p,t,x)$, respectively, and write the initial conditions in the form

$$\begin{aligned} u(p,t,0) &= u_0(t), \\ v(p,0,x) &= v_0(x). \end{aligned} \tag{3}$$

b) Break points $0 =: \tau_0 < \tau_1 < \cdots < \tau_{n_c} < t_f =: \tau_{n_c+1}$ are introduced where the integration is restarted always at the same initial values, i.e., the dynamical system (2) must be satisfied for all $t \in [\tau_i, \tau_{i+1})$ and $x \in [0, L]$ with

$$\begin{aligned} u(p,t,0) &= u_0(t), \\ v(p,\tau_i,x) &= v_0(x) \end{aligned} \tag{4}$$

for $i = 0, \ldots, n_c$. We will see later that these break points are considered also as optimization parameters. However, when applying a linear scaling of the time variable, we obtain the same formulation over constant time intervals, where the break points become now part of the parameter vector p. Thus, we assume without loss of generality that they are constant values.

c) Since we want to control the system at the left boundary of the spatial interval, i.e., at $x = 0$, we consider $u_0(t)$ as control variable. If only a few of the available control variables are needed in a practical situation, we suppose that the remaining ones are simply known functions of t. It is required that all control functions are bounded by constants

$$\underline{u}_0 \leq u_0(t) \leq \overline{u}_0 \tag{5}$$

for all $t \in [0, t_f]$.

d) State variables are bounded by

$$\begin{aligned} \underline{u} \leq u(p,t,x) \leq \overline{u}, \\ \underline{v} \leq v(p,t,x) \leq \overline{v} \end{aligned} \tag{6}$$

for all $t \in [0, t_f]$ and $x \in [0, L]$.

e) There are additional restrictions for the state variables u at the right boundary of the spatial area

$$\underline{h}(t) \leq h(u(p,t,L)) \leq \overline{h}(t) \tag{7}$$

for all $t \in [0, t_f]$, where the lower und upper bounds are given time-dependent functions and where $h(u)$ is a nonlinear, smooth function of u.

f) Parameters indicated by the vector p are also restricted in the form

$$r(p) \geq 0 \tag{8}$$

with a given smooth function $r(p)$.

g) The objective function to be minimized is given in form of an integral over time depending on the parameters p, the time variable t, the control function $u_0(t)$, and the output function, i.e., the state variable u evaluated at the right boundary

$$\int_0^{t_f} c(p,t,u_0(t),u(p,t,L))dt \tag{9}$$

with a given cost function $c(p,t,u_0,u)$.

The dynamical system and the optimal control problem can be formulated in more general terms. But for the underlying application in mind, the above optimal control problem is sufficiently general. It should be noted that there do not exist any exactly formulated and theoretically verified optimality conditions, at least to the knowledge of the authors. A similar problem based on a one-dimensional partial differential equation, is considered in Pickenhain and Wagner [18]. However a distributed control is assumed in this case.

The role of the additional parameter vector p is important. Since the dynamical system is always restarted without exploiting any data from a previous integration interval, the problem would be completely separable in the other case.

In case of a tubular reactor model the state variable u consists for example of molar concentrations of some chemical components and the temperature, where the control functions are input feeds. A time-dependent equation could describe the dynamic decrease of the cross-sectional area of the reactor. Break points are cleaning times where the reaction is stopped and define also the operation intervals. Objective function could consist of the total profit of the reactor over the whole operation interval $[0, t_f]$ to be maximized. In a natural way there are bounds for the state variables, in particular for temperature and cross sectional area. From the reactor yields time-dependent constraints for the customer demands can be derived. Minimum runtime of the reactor and monotonicity requirements for the cleaning times lead to the additional parameter constraints.

Thus, the model is quite general and reflects the underlying real-life application. As soon as the cross-sectional area reaches this minimum level, the reaction is stopped, the reactor has to be cleaned, and the process is restarted.

3 Semi-Discretization by the Numerical Method of Lines

3.1 *Discretization of State Equations*

Our first attempt to solve the optimal control problem for the distributed system uses the numerical method of lines, see, e.g., Schiesser [19]. For simplicity, let us consider only the basic equations (1),

$$\begin{aligned} u_x &= f(u, v), \\ v_t &= g(u, v). \end{aligned}$$

The idea is to discretize the system with respect to one variable, x or t, to replace the corresponding derivatives by a difference formula and to formulate a system of ordinary differential equations subject to the other variable.

The application problems we have in mind, possess different dynamical behaviour. Along the x-variable chemical reactions are described. Since slow and fast reactions can appear and must be taken into account simultaneously, the resulting differential equation can be stiff. On the other hand the dynamical behaviour of v is expected to be quite harmless. We know that the cross-sectional area is a monotonously decreasing function with respect to the time variable for a fixed spatial position in the reactor. In other words, the dynamical structure of the system in the time variable t is completely different from the behaviour in the space variable x.

Thus, we perform an equidistant discretization subject to the time variable with grid points $t_k = k\Delta t$, $\Delta t = \frac{t_f}{n_t}$, $k = 0, \ldots, n_t$, and replace the time derivative of v by a simple difference formula, e.g., the explicit Euler formula. When using the notation $u_k(x) = u(t_k, x)$ and $v_k(x) = v(t_k, x)$ for each grid point t_k, we are able to compute $v_k(x)$ successively from available data by

$$v_k(x) = v_{k-1}(x) + \Delta t \; g(u_{k-1}(x), v_{k-1}(x)) \tag{10}$$

for each $x \in [0, L]$. The process is started at $k = 1$ with known initial value $v_0(x)$. Since start-up conditions are not taken into account, we set $u_0(x) = 0$. We obtain then a system of $n_t n_u$ ordinary differential equations

$$\frac{d}{dx} u_k(x) = f(u_k(x), v_k(x)), \tag{11}$$

with initial conditions $u_k(0) = u_0(x)$, $k = 1, \ldots, n_t$. Note again that each $u_k(x)$ is a vector-valued variable of dimension n_u.

The proposed procedure does not depend on the particular choice of the difference formula for $v_t(x, t_k)$. Any other explicit method is applicable as well. Moreover, we may insert also an implicit formula, for example the implicit Euler formula, leading to a set of additional algebraic equations

$$v_k(x) - v_{k-1}(x) - \Delta t \; g(u_k(x), v_k(x)) = 0. \tag{12}$$

In this case we get a system of $n_t(n_u + n_v)$ ordinary differential algebraic equations that must be solved by an implicit integration scheme. If the same semi-discretization of the distributed system is applied to all $n_c + 1$ integration intervals $0 < \tau_1 < \cdots < \tau_{n_c} < t_f$, and if we apply the same number of grid points in each interval, we obtain either $(n_c + 1)n_t n_u$ or $(n_c + 1)n_t(n_u + n_v)$ equations, respectively.

3.2 The Discretized Optimization Problem

Control functions $u_0(t)$ are approximated by a finite number of parameters, for example in form of a piecewise linear function. In order to simplify the notation and to avoid additional interpolation, we suppose that the grid points that are used for these approximations, are identical to those that are used for the discretization of $v(x, t)$. Together with the additional parameter vector $p \in \mathbb{R}^{n_p}$, the discretized problem possesses $n_p + (n_c + 1)n_t$ optimization variables.

In order to discretize the constraints, we also need a discretization of the space interval, for example in equidistant form $x_j = j\Delta x$, $\Delta x = \frac{L}{n_x}$, $j = 0, \ldots, n_x$ Together with the time discretization which must be applied for each integration interval separately, we get the following finite dimensional approximations for the dynamic control and state constraints:

$$\begin{aligned} \underline{u}_0 \leq & \quad u_0(t_k) && \leq \overline{u}_0, \\ \underline{u} \leq & \quad u_k(p, x_j) && \leq \overline{u}, \\ \underline{v} \leq & \quad v_k(p, x_j) && \leq \overline{v}, \\ \underline{h}(t_k) \leq & \; h(u_k(p, x_{n_x})) && \leq \overline{h}(t_k), \\ & \quad r(p) && \geq 0, \end{aligned} \tag{13}$$

where $k = 1, \ldots, n_t, j = 1, \ldots, n_x$, and where the additional parameter vector p is introduced again. To summarize, we get a set of $2(n_c + 1)n_t(n_u + n_x(n_u + n_v) + n_h) + n_r$ bounds and nonlinear inequality constraints, where n_h is the dimension of $h(u)$ and n_r the dimension of $r(p)$.

Finally the integral of the objective function (9) is approximated by numerical integration subject to grid points t_k over all $n_c + 1$ runtime intervals.

3.3 Numerical Implementation

Our goal is to achieve a highly modular and flexible numerical implementation. Control functions $u_0(t)$ are approximated by piecewise constant functions, piecewise linear functions, or piecewise cubic splines. After a suitable discretization as outlined above, we have to solve large and eventually stiff systems of ordinary differential equations by some of the routines published in Hairer et al. [13] (DOPRI5, DOP853, ODEX) and in Hairer and Wanner [14] (RADAU5, SEULEX, SDIRK4). The integral of the objective function is evaluated by Simpson's rule.

As a result the optimal control problem is transformed into a finite dimensional parameter optimization problem, that is solved by the sequential quadratic programming (SQP) algorithm NLPQL of Schittkowski [21]. A particular advantage of these methods is that they can handle a large number of inequality constraints without severe increase of the number of iterations or computing time.

Gradients of objective function and constraints are evaluated by the forward scaled difference formula. However the special structure of the Jacobian of the constraints is exploited based on the observation that the state variables in different time intervals are linked only by the parameter p. Moreover the time and space discretization procedures lead to lower triangular sub-matrices for the sensitivities of the control parameters.

There is no necessity to exploit this special structure within the SQP algorithm NLPQL. The sparsity pattern belongs to inequality constraints, that are passed to an algorithm for solving certain quadratic programming problems obtained by a quadratic approximation of the Lagrangian function and a linearization of the constraints. However the full Jacobian of all constraints is evaluated only at the starting point. Afterwards only gradients belonging to a few active constraints are re-evaluated. Moreover the quadratic programming problem is solved by the dual method of Goldfarb and Idnani [12], where only active inequality constraints are added to a so-called working set, i.e., very few in our case.

4 Full Discretization

4.1 Discretization of State Equations

Now we perform a full discretization of the distributed system (4). Again we consider first the most simple case

$$\begin{aligned} u_x &= f(u,v), \\ v_t &= g(u,v). \end{aligned}$$

Proceeding now from a time grid $t_k = k\Delta t$, $\Delta t = \frac{t_f}{n_t}$, $k = 0, \ldots, n_t$, and a spatial grid $x_j = j\Delta x$, $\Delta x = \frac{L}{n_x}$, $j = 0, \ldots, n_x$, we denote by $u_{kj} = u(t_k, x_j)$ and $v_{kj} = v(t_k, x_j)$ certain approximations of the state variables for each pair of grid points t_k and x_j. The partial derivatives can be replaced by any reasonably accurate difference formula, see for example Thomas [22]. Since the dynamical behaviour of the system is supposed to be different in both directions, we apply the simple implicit Euler formula in t direction, and get

$$v_{kj} - v_{k-1,j} - \Delta t\, g(u_{kj}, v_{kj}) = 0. \tag{14}$$

see also (12).

On the other hand, we can, e.g., use a symmetric fourth order approximation for the spatial derivatives, i.e.

$$2v_{k,j-2} - 16v_{k,j-1} + 16v_{k,j+1} - 2v_{k,j+2} - 4!\,\Delta x\, f(u_{kj}, v_{kj}) = 0. \tag{15}$$

The formula must be adapted at the boundaries, see, e.g., Schiesser [19] for details. The advantage of this formulation is that we can ensure a sufficient accuracy of the approximation by a rather coarse discretization, which prevents the number of state and control variables from becoming prohibitively large.

4.2 The Discretized Optimization Problem

By the full discretization procedure as outlined above, we obtain a system of $(n_c + 1)n_t n_x(n_u + n_v)$ nonlinear equations in $(n_c + 1)n_t n_x(n_u + n_v)$ unknowns u_{kj} and v_{kj}, respectively, i.e., the fully discretized state variables. Now we take the individual integration intervals as defined by break points, into account. The remaining control variables, objective function and constraints are discretized in the same manner shown in Section 3. Thus, we finally get a large, sparse finite dimensional optimization problem in $(n_c + 1)n_t n_x(n_u + n_v) + (n_c + 1)n_t + n_p$ variables and $2(n_c + 1)n_t(n_u + n_x(n_u + n_v) + n_h) + n_r$ bounds and nonlinear inequality constraints, cf. (13).

The discretized state variables are now considered as additional optimization variables, and the discretized state equations (14) and (15) as additional nonlinear equality constraints. Thus, we get a very large optimization problem with a specific sparsity pattern in the Jacobians, and also a very large number of nonlinear inequality constraints. However the inequality constraints are not sharp, at least for the applications we have in mind. It can be assumed that minor violation of a state restriction will not lead to dramatic negative effects on the overall process. This holds especially since some of the parameters of the model can only be estimated and the whole model is flawed with certain inaccuracies.

Beyond that, an unfortunate choice of the model parameters, for example a too low number of break points, may result in an empty feasible set of the optimization problem. In this case, the algorithm should not stop with an error message, but calculate an optimal control that minimizes the violation of these inequality constraints in addition.

Therefore we incorporate inequality constraints in a modified barrier function (MBF), which leads to a sequence of optimization problems with discretized state equations as equality constraints and simple bounds for the discretized control variables and the additional parameters. In order to motivate the MBF method, we consider the discretized optimal control problem in form of the nonlinear program

$$\begin{array}{ll} \min & F(y,z) \\ & H_j(y,z) = 0, \quad j = 1,\dots,n_z, \\ y \in \mathbb{R}^{n_y}, z \in \mathbb{R}^{n_z}: & G_i(y,z) \geq 0, \quad i = 1,\dots,n_G, \\ & y_l \leq y \leq y_u, \end{array} \tag{16}$$

where F, H_j, and G_i are real valued differentiable functions defined for $y \in \mathbb{R}^{n_y}$ and $z \in \mathbb{R}^{n_z}$. y contains the discretized control variables and the parameter vector p, and z the discretized state variables. H_j describes now the discretized state equations, and G_i the bounds and nonlinear inequality constraints for state and control variables. To simplify the subsequent notation, bounds for state variables are not handled separately.

The proposed MBF method transforms the constrained minimization problem (16) into a sequence of equality constrained minimization problems by defining the auxiliary objective function

$$F_{MBF}(y,z,\mu,\lambda,\beta,s) = F(y,z) - \mu \sum_{i=1}^{n_G} \lambda_i \phi(G_i(y,z),\mu,\beta). \tag{17}$$

The function ϕ is a variation of the mixed quadratic-logarithmic penalty function of Ben-Tal and Zibulevsky [1],

$$\phi(G_i(y,z),\mu,\beta) = \begin{cases} \log(s_i + G_i(y,z)/\mu), & \text{if } G_i(y,z) \geq -\beta\mu s_i, \\ \frac{1}{2}a_i G_i^2(y,z) + b_i G_i(y,z) + c_i, & \text{if } G_i(y,z) < -\beta\mu s_i, \end{cases} \tag{18}$$

where the coefficients a_i, b_i, and c_i are chosen so that the function is twice continuously differentiable, and λ_i are estimates of the Lagrange multipliers.

The penalty parameter μ is monotonously decreasing, the parameter β determines how close to the singularity of the logarithm the barrier function is extrapolated, and s_i are suitable scaling factors. For a detailed description of these parameters and a more rigorous treatment of the MBF algorithm, we refer to Breitfeld and Shanno [3], or Liepelt [15].

The resulting subproblems we have to solve, are of the form

$$\begin{array}{ll} \min & F_{MBF}(y,z) \\ y \in \mathbb{R}^{n_y}, z \in \mathbb{R}^{n_z} : & H(y,z) = 0, \\ & y_l \le y \le y_u, \end{array} \tag{19}$$

where z are the discretized state variables, y are the discretized control variables, and $H(y,z)$ are the discretized state equations. Note that the number of state equations is the same as the number of state variables, i.e., the Jacobian matrix $H_z(y,z)$ is quadratic.

The solution of the nonlinear optimization problem is computed by solving a sequence of reduced quadratic programming problems, where each problem is reduced to the tangent space of the state equations. Trust-regions are used to guarantee that convergence is attained from any starting point and to regularize the subproblems arising in the SQP framework. The bound constraints are handled by an affine scaling interior-point strategy with accurate dual approximation. For more details on this algorithm, we refer to Dennis et al. [8].

4.3 Numerical Implementation

The fundamental advantage of a method based on a complete discretization of the distributed system is that the derivatives of the objective function and constraints with respect to all variables can be computed without using approximations by divided differences. This holds especially in our case where the underlying equations are highly nonlinear. All problem functions are modeled using the automatic differentiation program PCOMP, see Dobmann et al. [9] or Liepelt and Schittkowski [16] for a reference, from which a Fortran code for function and gradient evaluations is generated.

For the successive solution of the nonlinear subproblems (19) of (16), we use the code TRICE that was implemented by Heinkenschloss, see Dennis et al. [8]. The user has to provide subroutines for solving the linearized state equations

$$H_z(y,z)s_z + H_y(y,z)s_y + H(y,z) = 0.$$

Furthermore we need to program matrix-vector products of the form

$$z = H_z(y,z)^{-1}H_y(y,z)v, \quad z = H_y(y,z)^T(H_z(y,z)^T)^{-1}v$$

with suitable vectors v and z. In both cases we have to solve a linear system of equations with Jacobian $H_z(y,z)$.

Due to the finite difference schemes that are used for the approximation of the partial derivatives of the distributed system, the matrix $H_z(y,z)$ is very sparse. Since this matrix is not symmetric, we cannot apply the usual sparse Cholesky factorization and use an algorithm instead that has been proposed by Davis and Duff [7].

5 Case Study: Acetylene Reactor

We consider an existing reactor producing acetylene (C_2H_2), reacting the methane (CH_4) in natural gas with oxygen. This reaction requires less oxygen compared to complete combustion. The products are quickly quenched to keep the acetylene from being converted entirely to coke, see Wansbrough [23].

During the reaction process, a small part of the carbon is deposited in the reactor as coke. The quantity and its distribution in the reactor depend on the reaction equations. Since it is impossible to measure the cross-sectional area directly, we need a mathematical model that describes the functional dependence of the cross-sectional area on other system parameters. If the deposition of coke reaches a certain limit, the reactor must be stopped and the tube is cleaned.

There are six reactions to be taken into account. Reactions 1 through 5 are the main ones that produce acetylene, but also undesirable byproducts such as coke. Reaction 6 is included only to balance the hydrogen stoichiometry, see Birk et al. [2]. The chemical reactions can be described by a system of 8 ordinary differential equations, where C_i denotes the molar concentration of the i-th component. Together with initial values

$$\begin{aligned} C_1(0,t) &= C_1^0(t), \\ C_2(0,t) &= C_2^0(t), \\ C_i(0,t) &= 0, \quad i = 3,\ldots,8, \end{aligned} \tag{20}$$

and a reaction parameter ε, we have

$$\begin{aligned} \tfrac{\partial}{\partial x} C_1(x,t) &= (-r_1(x,t) - r_2(x,t))/v(x,t), \\ \tfrac{\partial}{\partial x} C_2(x,t) &= (-r_2(x,t) - \tfrac{1}{2} r_3(x,t) - \tfrac{1}{2} r_5(x,t))/v(x,t), \\ \tfrac{\partial}{\partial x} C_3(x,t) &= (\tfrac{1}{2} r_1(x,t) - r_4(x,t))/v(x,t), \\ \tfrac{\partial}{\partial x} C_4(x,t) &= r_3(x,t)/v(x,t), \\ \tfrac{\partial}{\partial x} C_5(x,t) &= (\tfrac{3}{2} r_1(x,t) + r_2(x,t) + r_4(x,t) \\ &\qquad - r_5(x,t) - n(1-\varepsilon) r_4(x,t))/v(x,t), \\ \tfrac{\partial}{\partial x} C_6(x,t) &= (r_2(x,t) - r_3(x,t))/v(x,t), \\ \tfrac{\partial}{\partial x} C_7(x,t) &= (r_2(x,t) + r_5(x,t))/v(x,t), \\ \tfrac{\partial}{\partial x} C_8(x,t) &= 2(1-\varepsilon) r_4(x,t)/v(x,t). \end{aligned} \tag{21}$$

These eight material balance equations depend on the rates of the various reactions and on the velocity of the mixture in the reactor, because this speed determines the time that the components spent in the reactor. The reaction rates are expressed by

$$\begin{aligned}
r_1(x,t) &= k_1 \exp\left(-\tfrac{E_1}{R}\left(\tfrac{1}{T(x,t)} - \tfrac{1}{T_r}\right)\right) C_1^{a_1}(x,t),\\
r_2(x,t) &= k_2 \exp\left(-\tfrac{E_2}{R}\left(\tfrac{1}{T(x,t)} - \tfrac{1}{T_r}\right)\right) C_1(x,t) C_2^{a_2}(x,t),\\
r_3(x,t) &= k_3 \exp\left(-\tfrac{E_3}{R}\left(\tfrac{1}{T(x,t)} - \tfrac{1}{T_r}\right)\right) C_6(x,t) C_2^{0.5}(x,t),\\
r_4(x,t) &= k_4 \exp\left(-\tfrac{E_4}{R}\left(\tfrac{1}{T(x,t)} - \tfrac{1}{T_r}\right)\right) C_3^{a_4}(x,t),\\
r_5(x,t) &= k_5 \exp\left(-\tfrac{E_5}{R}\left(\tfrac{1}{T(x,t)} - \tfrac{1}{T_r}\right)\right) C_5(x,t) C_2^{0.5}(x,t),
\end{aligned} \tag{22}$$

with five reaction constants $k_1, \ldots, k_5$, five activation energies $E_1, \ldots. E_5$, and three reaction orders a_1, a_2, and a_4. n denotes the average number of H atoms in CH_n. For the smaller and less important reactions, the stoichiometric order can be used as an estimate for the reaction order. For the other reactions these parameters have to be derived from the real reactor that is going to be examined. The average temperature T_r is used to scale the exponential functions and R denotes the gas constant. Normalized reaction values used for the numerical tests, are given in the Appendix.

The velocity of the mixture in the reactor depends on the cross-sectional area $A(x,t)$, the total mass flow $\dot{m}(t)$ in the reactor, and the density $\rho(x,t)$, and is given by

$$v(x,t) = \frac{\dot{m}(t)}{\rho(x,t)A(x,t)}. \tag{23}$$

Because the amount of carbon that is deposited in the reactor as coke, is very small, we may assume that the total mass flow

$$\dot{m}(t) = \dot{m}_1(0,t) + \dot{m}_2(0,t) \tag{24}$$

is constant during the reaction. The density of the mixture is given by

$$\rho(x,t) = \sum_{j=1}^{8} C_j(x,t) M_j, \tag{25}$$

where M_j denotes the molar weight of the j-th component.

Since the acetylene reactor in controlled by the feeds of natural gas and oxygen, these are the only components with non-vanishing initial values. The initial molar concentrations are given by

$$\begin{aligned}
C_1^0(t) &= \frac{\dot{m}_1(0,t)\rho(0,t)}{\dot{m}(t)M_1},\\
C_2^0(t) &= \frac{\dot{m}_2(0,t)\rho(0,t)}{\dot{m}(t)M_2}.
\end{aligned} \tag{26}$$

Just like the molar concentrations of the individual components, also the temperature in the reactor can be described by an ordinary differential equation,

$$\frac{\partial}{\partial x} T(x,t) = \frac{1}{\rho(x,t)v(x,t)c_p(x,t)} \sum_{i=1}^{5} r_i(x,t)\Delta H_i \tag{27}$$

with the initial condition $T(0,t) = T_0$. The incremental change of the temperature is determined by the rate of heat release for all reactions, which depends on the total heat capacity

$$c_p(x,t) = \frac{\sum_{j=1}^{8} c_{pj} M_j C_j(x,t)}{\sum_{j=1}^{8} M_j C_j(x,t)}. \tag{28}$$

The individual heat capacities c_{pj} are considered to be constant, and the parameters ΔH_i denote the known heats of reaction.

If we neglect the deposition of coke, the underlying partial differential equation is stationary, i.e., does not depend on the time. But a decrease of the cross-sectional area $A(x,t)$ increases the velocity $v(x,t)$ of the mixture in the reactor, which influences the incremental change of the concentrations $C(x,t)$ and the temperature $T(x,t)$, see (21), (23), and (27). The coke deposition is modeled by the time-dependent differential equation

$$\frac{\partial}{\partial t} A(x,t) = -\beta r_4(x,t) \tag{29}$$

with the initial condition $A(x,0) = A_0$ and a reaction parameter $\beta \in \mathbb{R}$.

Now we want to maximize the profit of the reactor over an operation interval $[0, t_f]$, i.e., we are looking for optimal input mass feeds

$$u_j(t) := \dot{m}_j(0,t), \quad j = 1,2, \tag{30}$$

and additional break points τ_i for $i = 1, \ldots, n_c$, at which the reaction is stopped and the reactor is cleaned,

$$t_0 < \tau_1 < \cdots < \tau_{n_c} < t_f. \tag{31}$$

Note that the number of cleaning times corresponds to the number of break points n_c discussed before. In order to avoid a nonlinear mixed integer optimization problem, we assume that the number n_c of reactor cleanings is given, but their positions within the whole operation interval can vary. The total profit of the acetylene reactor over the whole time horizon is given by the integral over given time-dependent prices multiplied by corresponding mass flows over all runtime intervals, where cleaning and control costs are subtracted, see again Birk et al. [2] for details.

We do not assume that the reactor has been cleaned at the initial time t_0 of the process, i.e., we allow for an arbitrary initial distribution $A^*(x)$ of the cross-sectional area,

$$\begin{aligned} A(x,t_0) &= A^*(x), \quad x \in [0,L], \\ A(x,\tau_i) &= A_0, \quad x \in [0,L], \quad i = 1,\ldots,n_c. \end{aligned} \tag{32}$$

This is especially important in real-time applications, when the model or some data change and we have to compute a new optimal control, using the known solution as an initial value.

Besides of the state equations that are implied by the distributed system described above, the considered optimal control problem has also a series of additional inequality constraints. Some of them are process related, such as a maximum temperature, which may never be exceeded,

$$T(x,t) \leq T^{max}(t), \quad x \in [0,L], \quad t \in [t_0,t_f] \tag{33}$$

and a lower bound for the free cross-sectional area,

$$A(x,t) \geq A^{min}(t), \quad x \in [0,L], \quad t \in [t_0,t_f]. \tag{34}$$

Note that $A(x,t)$ is monotonously decreasing with t for every fixed x and that the reaction is stopped and the reactor is cleaned, whenever this limit has been reached.

Although the primary goal is to maximize the total profit of the reactor over the entire operation interval, it is also necessary to define lower and upper bounds for the output of the individual component mass flows for technical reasons. The reactor is controlled by the initial feeds of methane and oxygen, for which we also have additional lower and upper bounds. There is a minimum runtime Δt^{min} of the reactor leading to the additional restrictions

$$\tau_i - \tau_{i-1} \geq \Delta t^{min}, \quad i = 1, \ldots, n_c, \tag{35}$$

and the cleaning times must remain in the interior of the operation interval. Variables τ_i correspond to the additional parameter vector p introduced in Section 2.

6 Numerical Results

In this section we present some numerical results that are obtained for the acetylene reactor described in the previous section. Computational experiments are performed on a PC equipped with an Intel Pentium III processor running with 750 MHz under Microsoft Windows 2000. The discretized system of ordinary differential equations was integrated using the explicit Runge-Kutta method of Dormand and Prince [10], i.e., the code DOPRI5 of Hairer et al. [13]. It turns out that a quite crude discretization yields acceptable results and that the ordinary differential equations are not stiff in the present case study.

To give a first impression of the the reactions in this model, we start with a simulation using constant input feeds of oxygen and methane, i.e.,

$$\dot{m}_1(0,t) = \dot{m}_2(0,t) = 500, \quad t \in [0,t_f], \tag{36}$$

where we assume that the following process data are given:

name	*value*	*description*
A_0	0.1	cross-sectional area of the tube
A^{min}	0.08	minimum cross-sectional area
L	1	length of the reactor
T_0	873.15	initial reaction temperature
T^{max}	1,300	maximum permissible temperature
t_f	200	duration of the process
Δt^{min}	60	minimum runtime of the reactor
P_c	5,000	costs of a reactor cleaning
l	20	number of space discretizations

The corresponding surface plot of the cross-sectional area is shown in Figure 1. We observe a drastic violation of the minimum admissible cross sectional area of the tube A^{min}.

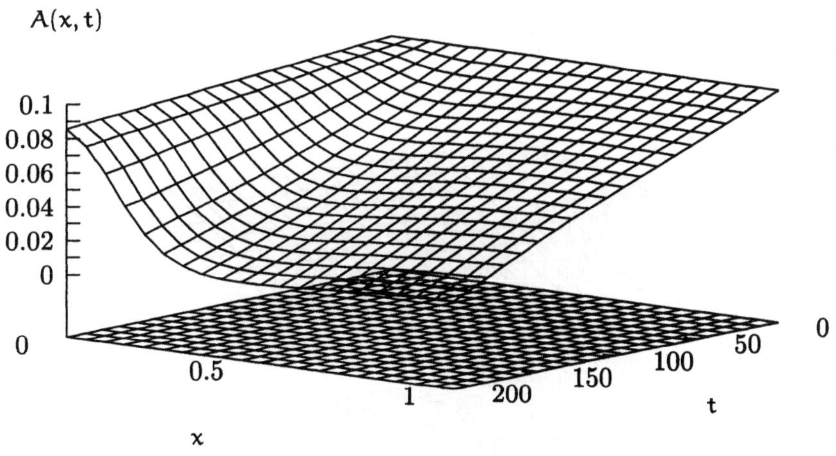

Figure 1. Cross-sectional area for constant feed

Now we want to compute optimal control functions of the acetylene reactor, i.e., input feeds that yield feasible values of the state and control variables and maximize the overall profit of the process. Throughout this section we use piecewise linear approximations for the feeds of O_2 and CH_4. For starting the optimization algorithm we choose constant values of the control functions, $\dot{m}_1(0,t) = 700$ and $\dot{m}_2(0,t) = 400$. The expected time-dependent price and demand data are given in the form of piecewise linear approximations, see Birk et al. [2] for details. Moreover we assume that the following bounds for the feed controls are given:

$$200 \leq \dot{m}_1(0,t) \leq 800, \quad t \in [0,t_f],$$
$$200 \leq \dot{m}_2(0,t) \leq 800, \quad t \in [0,t_f].$$

Our goal is to show the influence of a different number of operation periods. Thus, we compute the optimal control for zero to five maintenance breaks, i.e., for $n_c = 0$ to $n_c = 5$. To be able to compare some performance data, we vary the number of time grid points to get discretized optimization problems of approximately the same size, i.e., we require that $(n_c + 1)n_t \approx 36$.

First we consider the semi-discretization approach based on the numerical method of lines. Depending on the number of break points, the resulting discretized nonlinear programming problem has between 72 and 77 variables and between 2,034 and 2,093 nonlinear inequality constraints. The size of the system of ordinary differential equations to be solved for each function call, is 324. A typical surface plot of the cross-sectional area for two operation intervals satisfying all constraints, is shown in Figure 2. The optimal solution for the CH_4 feed attains its upper bound in all test runs, i.e., $\dot{m}_1(0,t) = 800$ for $t \in [0,t_f]$.

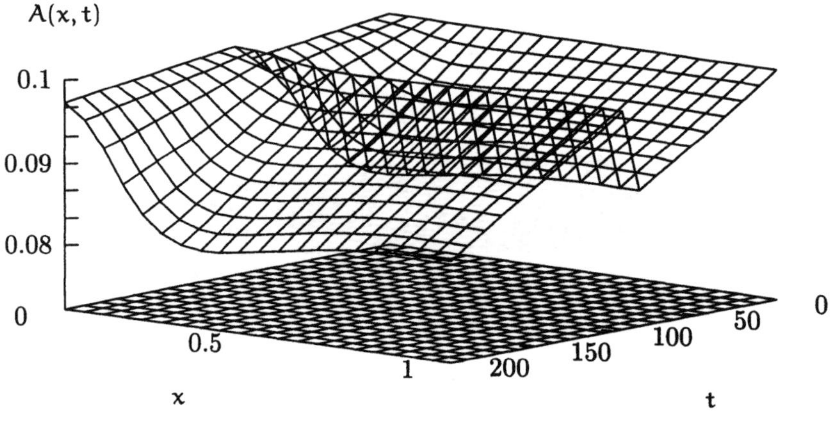

Figure 2. Cross-sectional area for optimal feed

The termination accuracy for the nonlinear programming code NLPQL is set to 10^{-5}, and the relative error of the ODE solver DOPRI5 is set to 10^{-6}. Since the starting trajectories are poor and the objective function is very flat in a neighborhood of an optimal solution, NLPQL requires a large number of iterations to reach a solution. The number of iterations, the obtained objective function values, and some other convergence results are listed in Table 1. Using some terms defined in Schittkowski [20, 21], we get the following abbreviations:

n_c – number of cleaning times
f – objective function value
it – number of iterations
cv – constraint violation
ac – number of active constraints
oc – optimality criterion
time – calculation time in minutes

n_c	f	*it*	*cv*	*ac*	*oc*	*time*
0	44,419	106	$0.40 \cdot 10^{-10}$	3	$0.44 \cdot 10^{-5}$	13.4
1	47,004	110	$0.19 \cdot 10^{-14}$	4	$0.12 \cdot 10^{-5}$	15.2
2	47,958	104	$0.41 \cdot 10^{-12}$	6	$0.61 \cdot 10^{-5}$	10.5
3	45,383	140	$0.37 \cdot 10^{-9}$	8	$0.42 \cdot 10^{-5}$	11.9
4	40,747	316	$0.12 \cdot 10^{-9}$	12	$0.43 \cdot 10^{-5}$	31.7
5	35,941	286	$0.74 \cdot 10^{-9}$	14	$0.83 \cdot 10^{-5}$	21.1

Table 1. Performance results for semi-discretization approach

The overall profit increases with the number of reactor cleanings if the corresponding costs are neglected. For maintenance costs of 5,000 units, we maximize the profit for two breaks. The corresponding control function for the O_2 feed of two maintenance intervals is shown in Figure 3(a). It is interesting to note that the SQP algorithm NLPQL reaches a quite accurate solution despite of the numerical errors in the gradient approximation. Since the assumed bounds for consumer demands and the prices do not vary too much over the time horizon, the cleaning times are not altered drastically. The input data are defined only at 5 grid points, where linear interpolation is used to access intermediate values.

The second approach based on a full discretization is executed for the same discretization accuracy and the same starting values for the control variables and cleaning times. The number of optimization variables of the discretized nonlinear program ranges from 7,272 to 7,277, and the number of nonlinear inequality constraints is of the same size as before, i.e., about 2,000. Note that the equality constrained subproblem to be solved by TRICE, has 7,200 nonlinear equality constraints. TRICE is executed with error tolerance 10^{-3} and at most 20 iterations. We prefer an approximate solution of the subproblem to get a faster update of penalty and multiplier values. The termination tolerance of the outer MBF method is set to 10^{-4}. Corresponding control function for two maintenance breaks leading to the best objective function value, is plotted in Figure 3(b). The corresponding performance results are shown in Table 2, where we use the following abbreviations:

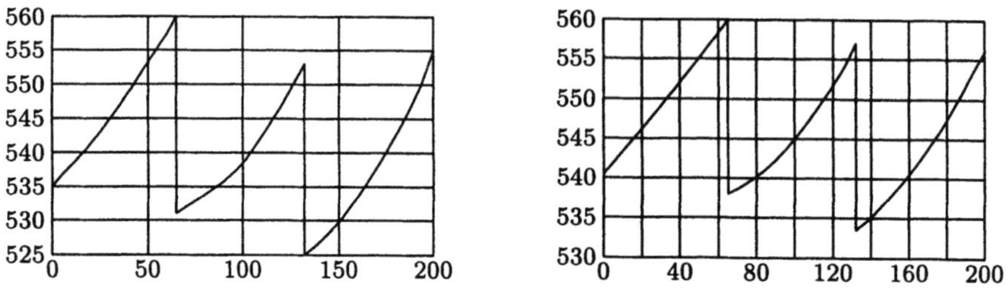

(a) O_2 feed (method of lines)

(b) O_2 feed (full discretization)

Figure 3. O_2 feed

n_c – number of cleaning times
f – objective function value
it_out – number of outer iterations
it – total number of TRICE iterations
cv – norm of constraint violation
ac – number of active constraints
cs – complementary slackness
time – calculation time in minutes

n_c	f	*it_out*	*it*	*cv*	*ac*	*cs*	*time*
0	45,951	44	575	$0.90 \cdot 10^{-6}$	1	$0.89 \cdot 10^{-4}$	49.3
1	47,389	20	416	$0.66 \cdot 10^{-7}$	2	$0.21 \cdot 10^{-4}$	33.5
2	47,488	20	401	$0.45 \cdot 10^{-6}$	3	$0.22 \cdot 10^{-4}$	29.4
3	44,466	21	317.	$0.43 \cdot 10^{-4}$	4	$0.55 \cdot 10^{-4}$	23.2
4	39,864	10	210	0.0	0	$0.88 \cdot 10^{-4}$	16.4
5	35,112	16	210	0.0	0	$0.13 \cdot 10^{-5}$	20.3

Table 2. Performance results for full discretization approach

Taking into account the different discretization methods and numerical realization, the results are comparable to the results obtained for the first approach, see Table 1. In both cases the calculation times are by far too big to apply the proposed strategies on-line. Obviously the initial guesses for the optimization variables are too far away from the optimal solution to get a better performance. In a real-life environment, however, we may assume that there exists an optimal solution of a running process, and that we need to recompute another one after some minor changes of the model data, e.g., of prices or consumer demands.

For simplicity, we consider only the semi-discretization approach. Starting from a given optimal control of the tubular reactor with three intervals, we perform some small changes of data to demonstrate the possibility of computing a new optimal feed control under real-time conditions. For this purpose we use the solution of the original problem, the approximations of the Lagrange multipliers and the BFGS matrix for a restart. The results of these tests are listed in Table 3, where we use the same abbreviations as before.

It turns out that alterations of active or nearly active bounds for cross sectional area and temperature lead to some significant alterations in the cost function. In particular the increase of the admissible tube diameter by 10 % leads to a profit loss of about 20 %. An increase of the CH_4 prices leads also to a decrease of the objective function, in contrast to an increase of the C_2H_2 prices. Nevertheless the much lower number of iterations and calculation times indicate that the implemented algorithm is capable to run under real-time conditions.

changes	*it*	f	*time*
set $T^{max} = 1287$ (−1 %)	14	48,089	1.41
set $A^{min} = 0.0808$ (+1 %)	18	47,671	1.82
set $A^{min} = 0.088$ (+10 %)	50	41,274	5,05
increase price of CH_4 by 0.02	27	46,369	2,73
increase price of C_2H_2 by 0.02	17	56,993	1.72

Table 3. Real-time simulations

7 Conclusions

We extended an optimal control model for tubular reactors with the aim to take into account also the number and position of cleaning times. Two different approaches are presented to discretize the distributed system and to formulate the optimal control problem. In the first case the partial differential equations are discretized using the method of lines, where the resulting system of ordinary differential equations is integrated by a standard ODE solver. Since control functions are represented by a finite number of variables and since dynamic constraints are discretized at given time and space grid points, we get a finite dimensional nonlinear programming problem, for which a standard SQP method is applied. In the second case we apply finite difference formulae to get a full discretization of the state variables and equations, where the optimization problem is extended by additional variables and equality constraints. The large subproblem is solved by a reduced SQP method, where additional state constraints are taken into account by a modified barrier function.

Proceeding from a case study in form of an acetylene reactor, is is shown that operating intervals and optimal feed controls can be simultaneously computed. Both approaches outlined above, have their advantages and disadvantages. The first semi-discretization method is more robust and more efficient in terms of calculation time, whereas the full discretization procedure is quite sensible w.r.t. choice of input parameters and yields less accurate solutions. The main advantage of the second approach is that inconsistent constraints can be detected.

The underlying model is quite complex. The implementation and numerical solution of both approaches is so different that we cannot claim which one is preferable in general. The obvious advantage of full discretization, to get exact derivatives within machine precision, is somehow cancelled by the much larger and *harder* nonlinear programming subproblems to be solved in each step. Moreover the solution methods for the discretized optimization problems differ drastically and prevent a direct numerical comparison.

The first approach we presented, applies the method of lines to get a system of ordinary differential equations. For solving this semi-discretized problem under real-time conditions, alternative approaches are available, see for example Büskens and Maurer [4]. It should be possible to exploit also the post-optimal sensitivity analysis as proposed there, by successive linearization at perturbed solutions.

References

1. A. Ben-Tal, M. Zibulevsky: Penalty/barrier multiplier methods for convex programming problems. SIAM J. Optim., **7** (1997) 347–366
2. J. Birk, M. Liepelt, K. Schittkowski, F. Vogel: Computation of optimal feed rates and operation intervals for tubular reactors. J. Process Control **9** (1999) 325–336
3. M. G. Breitfeld, D. F. Shanno: Computational experience with modified log-barrier methods for nonlinear programming. Technical Report RRR 17-93, RUTCOR, Rutgers University, New Brunswick, New Jersey (1993)
4. C. Büskens, H. Maurer: Sensitivity analysis and real-time control of parametric optimal control problems using nonlinear programming methods. This volume (2001)
5. G. Buzzi-Ferraris, G. Facchi, P. Forzetti, E. Troncani: Control optimization of tubular catalytic decay. Industrial Engineering in Chemistry **23** (1984) 126–131
6. Buzzi-G. Ferraris, M. Morbidelli, P. Forzetti, S. Carra: Deactivation of catalyst – mathematical models for the control and optimization of reactors. International Chemical Engineering **24** (1984) 441–451
7. T. A. Davis, I. S. Duff: An unsymmetric-pattern multifrontal method for sparse LU factorization. SIAM J. Matrix Analysis and Applications **18** (1997) 140–158
8. J. E. Dennis, M. Heinkenschloss, L. M. Vicente: Trust-region interior-point algorithms for a class of nonlinear programming problems. SIAM J. Control and Optimization **36** (1998) 1750–1794
9. M. Dobmann, M. Liepelt, K. Schittkowski: Algorithm 746: PCOMP: A Fortran code for automatic differentiation. ACM Trans. Math. Software **21** (1995) 233–266
10. R. J. Dormand, P. J. Prince: High order embedded Runge-Kutta formulae. J. Comp. Appl. Maths. **7** (1981) 67–75
11. T. F. Edgar, D. M. Himmelblau: Optimization of Chemical Processes. McGraw-Hill, New York (1988)
12. D. Goldfarb, A. Idnani: A numerically stable method for solving strictly convex quadratic programs. Mathematical Programming **27** (1983) 1–33

13. E. Hairer, S. P. Nørsett, G. Wanner: Solving Ordinary Differential Equations I: Non-stiff Problems. Springer Series in Computational Mathematics **8**, Springer-Verlag, Berlin, Heidelberg, New York (1983)
14. E. Hairer, G. Wanner: Solving Ordinary Differential Equations II: Stiff Problems. Springer Series in Computational Mathematics **14**, Springer-Verlag, Berlin, Heidelberg, New York (1991)
15. M. Liepelt: Optimierung großer Systeme mit Penalty-Barrier-Verfahren. Dissertation, Department of Mathematics, University of Bayreuth, D - 95440 Bayreuth (1997)
16. M. Liepelt, K. Schittkowski: Algorithm 746: New features of PCOMP: A Fortran code for automatic differentiation. ACM Trans. Math. Software **26** (2000) 335–362
17. N. Nishida, A. Ichikawa, E. Tazaki: Optimal design and control in a class of distributed parameter systems under uncertainty. AIChE J. **18** (1972) 561–568
18. S. Pickenhain, M. Wagner: Pontryagin principle for state-constrained control problems governed by a first-order PDE. J. Optimization Theory and Applications **107** (2000) 297–330
19. W. E. Schiesser: The Numerical Method of Lines. Academic Press, San Dieago (1991)
20. K. Schittkowski: On the convergence of a sequential quadratic programming method with an augmented Lagrangian line search function. Optimization **14** (1983) 1–20
21. K. Schittkowski: NLPQL: A Fortran subroutine solving constrained nonlinear programming problems. Annals of Operations Research **5** (1985) 485–500
22. J. W. Thomas: Numerical Partial Differential Equations. Texts in Applied Mathematics **22**, Springer, New York, Berlin, Heidelberg (1995)
23. R. W. Wansbrough: Modeling chemical reactors. Chemical Engineering **5** (1985) 95–102

APPENDIX: REACTION DATA

$a_1 = 0.72$	$E_3 = 3.4 \cdot 10^4$	$k_5 = 3.6 \cdot 10^4$
$a_2 = 0.56$	$E_4 = 3.2 \cdot 10^4$	$M_1 = 16.0$
$a_4 = 0.56$	$E_5 = 3.0 \cdot 10^4$	$M_2 = 32.0$
$\beta = 1.4 \cdot 10^{-5}$	$\varepsilon = 1.0 \cdot 10^{-3}$	$M_3 = 26.0$
$c_{p1} = 2.219$	$\Delta H_1 = -263.0 \cdot 10^3$	$M_4 = 44.0$
$c_{p2} = 0.917$	$\Delta H_2 = 277.5 \cdot 10^3$	$M_5 = 2.0$
$c_{p3} = 1.683$	$\Delta H_3 = 283.0 \cdot 10^3$	$M_6 = 28.0$
$c_{p4} = 0.837$	$\Delta H_4 = 226.7 \cdot 10^3$	$M_7 = 18.0$
$c_{p5} = 14.32$	$\Delta H_5 = 241.8 \cdot 10^3$	$M_8 = 14.0$
$c_{p6} = 1.042$	$k_1 = 1.8 \cdot 10^4$	$n = 2.0$
$c_{p7} = 2.500$	$k_2 = 9.0 \cdot 10^6$	$R = 8.314$
$c_{p8} = 1.595$	$k_3 = 2.6 \cdot 10^5$	$\rho_0 = 0.047$
$E_1 = 2.7 \cdot 10^4$	$k_4 = 8.9 \cdot 10^2$	$T_r = 873.15$
$E_2 = 3.8 \cdot 10^4$		

III

Moving Horizon Methods in Chemical Engineering

*Introduction to Model Based Optimization of Chemical Processes on Moving Horizons

Thomas Binder[1], Luise Blank[2], H. Georg Bock[3], Roland Bulirsch[4], Wolfgang Dahmen[2], Moritz Diehl[3],
Thomas Kronseder[4], Wolfgang Marquardt[1], Johannes P. Schlöder[3], and Oskar von Stryk[5]

[1] Lehrstuhl für Prozesstechnik, Rheinisch-Westfälische Technische Hochschule Aachen,
[2] Institut für Geometrie und praktische Mathematik, Rheinisch-Westfälische Technische Hochschule Aachen, Germany
[3] Interdisziplinäres Zentrum für wissenschaftliches Rechnen, Universität Heidelberg, Germany
[4] Lehrstuhl für Höhere Mathematik und Numerische Mathematik, Technische Universität München, Germany
[5] Fachgebiet Simulation und Systemoptimierung, Technische Universität Darmstadt, Germany

Abstract Dynamic optimization problems are typically quite challenging for large-scale applications. Even more challenging are on-line applications with demanding real-time constraints. This contribution provides a concise introduction into problem formulation and standard numerical techniques commonly found in the context of moving horizon optimization using nonlinear differential algebraic process models.

1 Introduction

Safe and economical process operation is of crucial importance for the success of chemical companies. Model based optimization is a promising technique to increase the operational profit in process operation. Moving horizon optimization includes model predictive control (MPC) and receding horizon estimations (RHE) and requires on-line dynamic optimization (see, e.g., Helbig et al. [61]). MPC regulates processes whereas RHE is used to estimate unaccessible process states and parameters (see Allgöwer et al. [3] for an excellent survey on both problems). In these applications a multi-variable optimization problem restricted to a large scale mathematical process model has to be solved on-line. The large scale nature as well as the real-time requirement of the problem is a clear challenge where the cutting edge of currently commercially available technology needs to be pushed further forward. It is the intention of this article to provide a concise introduction into the exciting field of dynamic optimization applied on moving horizons and to summarize the available numerical techniques. However, due to the limitations in space we only focus on techniques which from our point of view are considered as standard technologies currently applied. Recent results developed within the Schwerpunktprogramm are covered elsewhere in this book by additional contributions of each research group (Binder et al., [21]; Diehl et al., [47]; Kronseder et al., [85]).

This introductory article is organized as follows. In Section 2 we introduce the generic problem formulation of control and estimation problems which are closely related. However, we will also illuminate the differences between both. Special attention is given to the mathematical process model in Section 2.1 whereas in Sections 2.2 and 2.3 the optimization problems on fixed and moving horizons are defined respectively. Furthermore, we emphasize the particular real-time character intrinsic to moving horizon optimization.

Closed loop stability of the applied algorithm is of crucial importance. A short introduction into the terminology and basic concepts is given in Section 3. Here, we motivate how stability problems arise when a finite dimensional horizon instead of an infinite dimensional one is chosen.

Section 4 summarizes the basic techniques to solve the dynamic optimization problem on a fixed horizon. We start with a short introduction into optimal feedback controls which are described by the Hamilton-Jacobi-Carathéodory-Bellman partial differential equation, and briefly introduce techniques based on the so called Maximum Principle in Section 4.2.

In Section 5 we focus on the so called direct methods for the numerical solution of optimal control problems. Here, the infinite dynamic optimization problem is transformed into a nonlinear program (NLP) by parameterizing the controls. Special room is given to direct single shooting (Section 5.1), direct multiple shooting (Section 5.2) and direct collocation (Section 5.3), techniques which are commonly used so solve large scale problems. The numerical solution of the NLP by sequential quadratic programming (SQP) is outlined in Section 5.4, and the three presented direct techniques are compared in Section 5.5.

Extensions of fixed horizon optimization to moving horizon optimization are discussed in Section 6. We start with the well-known recursive solution approaches for regulation (Section 6.1) and estimation (Section 6.2) available for unconstrained linear quadratic optimization problems. The simplicity and power of these recursive techniques motivate extensions to nonlinear models. Therefore Section 6.3 discusses optimal feedback control obtained by linearization along a specific reference solution. In moving horizon optimization the numerical cost can be lowered substantially using appropriate initialization techniques. The various options of commonly applied approaches are outlined Section 6.4.

The article is concluded in Section 7 by a short summary.

2 Problem Formulation

2.1 Model

Mathematical process models are an abstraction of real process systems and aim to capture the essential features of concern. In general, the process models are either based on fundamental principles or empirical observations or in the hybrid case on a mixture of both. The basis for virtually all fundamental process models are the general conservation principles of mass, momentum and energy. As long as the underlying assumptions remain valid, fundamental models can be expected to extrapolate to new operating regions where no data sets are available. However, it is a rather difficult and time consuming task to construct and validate good fundamental process models (see, e.g., Aris, [4]; Bauer et al., [13]; Marquardt [95]). An empirical model built from available process data might be more convenient in some instances since a detailed process understanding is not required for the model development, although a suitable model structure has to be selected as well. Artificial neural networks are the most popular framework for empirical model development (Su and McAvoy, [130]), but other techniques based on Hammerstein and Wiener models (Norquay et al., [107]; Pearson and Pottmann, [111]; Wellers and Rake, [144]), Volterra models (Maner et al., [94]), and polynomial ARMAX models (Sriniwas and Arkun, [127]) might be considered alternatively. In this contribution a detailed discussion of the particular advantages and disadvantages of fundamental or empirical modeling are off focus since from an optimization point of view we only need a sufficiently good process model. The underlying principles of the building process are of minor importance, although they very well might affect the applicability of the model and the particular choice of the numerical solution method. We assume that a fundamental process model is available, but we keep in mind that other model types might be used as well. There is a wide variety of phenomena in chemical process systems such that we have various types of process models which vary over a large range starting from simple algebraic equation systems, to ordinary (ODE) or differential-algebraic (DAE) equations systems, and to more complicated (partial-) integro-differential equations. Despite this richness our discussion is limited to mathematical process models which can be represented as DAE systems given by

$$\mathbf{0} = \mathbf{f}(\dot{\mathbf{x}}(t), \mathbf{x}(t), \mathbf{z}(t), \mathbf{u}(t), \mathbf{w}(t), \mathbf{p}, t), \quad \forall t \in I, \tag{1}$$

$$\mathbf{0} = \mathbf{g}(\mathbf{x}(t), \mathbf{z}(t), \mathbf{u}(t), \mathbf{w}(t), \mathbf{p}, t), \quad \forall t \in I, \tag{2}$$

Here, $\mathbf{x}(t) \in \mathbb{R}^{n_x}$ and $\mathbf{z}(t) \in \mathbb{R}^{n_z}$ denote the differential and the algebraic system state vectors, respectively. $\mathbf{u}(t) \in \mathbb{R}^{n_u}$ are operational variables which can be directly manipulated by process operators. Modeling uncertainties and disturbances are concatenated without further specification into a vector function $\mathbf{w}(t) \in \mathbb{R}^{n_w}$. $\mathbf{p} \in \mathbb{R}^{n_p}$ denotes a vector of time-invariant system parameters. The function $\mathbf{f}$ (with $\frac{\partial \mathbf{f}}{\partial \dot{\mathbf{x}}}$ invertible) describes the differential portion while the function $\mathbf{g}$ represents the algebraic portion of the process model. In general the Jacobian $\frac{\partial \mathbf{g}}{\partial \mathbf{z}}$ might be singular such that the DAE could be of a higher index, where roughly spoken, the index denotes the minimum number the system has to be differentiated with respect to time to be able to transform the DAE system into an ODE system. Details on the theory of DAE's can be found for example in Brenan et al. [32] or Unger et al. [133]. Because of the richness of phenomena occuring in higher index problems we limit ourselves to problems of index one. The time interval of interest is denoted in the sequel by $I := [t_0, t_f]$ where t_0, t_f are starting and final times respectively. The process model (1) and (2) might be used for simulation. Given particular values of $\mathbf{u}^\star(t)$[1], $\mathbf{w}^\star(t)$, $t \in I$, $\mathbf{p}^\star$ and appropriate initial conditions the process model

[1] The superscript $\star$ denotes specific but arbitrary values.

(1) and (2) can be solved using a suitable integration routine. For notational convenience we assume in the remainder that initial conditions are provided for the differential states, i.e., $\mathbf{x}(t_0) = \mathbf{x}_0$. A more general discussion on the specification of initial conditions can be found in e.g., Kröner et al. [83], Brenan et al. [32] or Unger et al. [133] for index one and higher index problems.

Some functions of the system states are measurable, therefore we augment the model (1) and (2) by the sensor model

$$\mathbf{y}(t) = \mathbf{h}(\mathbf{x}(t), \mathbf{z}(t), \mathbf{u}(t), \mathbf{p}, t) \in \mathbb{R}^{n_y}, \quad \forall t \in I, \tag{3}$$

that determines the output variables $\mathbf{y}$ as a function of the other system variables.

2.2 *Off-line Optimization on a Fixed Horizon*

Before we start to outline moving horizon dynamic optimization we consider first an off-line problem on a fixed horizon such as the optimization of batch processes. These problems require the minimization of an objective function by adjusting the free operational variables $\mathbf{u}$, also referred to as controls, in an appropriate manner within the finite interval $I^r = [t_0^r, t_f^r]$ which denotes an operational phase of the process, such as the time required for a grade change of a continuous process or the reaction phase in a batch process. The final time may be fixed or subject to optimization.

The controls $\mathbf{u}$ cannot be adjusted arbitrarily since they might be restricted by constraints which are typically associated with physical limits such as,, e.g., restrictions on valve position or rate of change.

Further (mixed) constraints on controls *and* states comprise, e.g., limits on capacity of production units and quality specifications on the product, as well as safety constraints. For notational simplicity, both types of restrictions are concatenated in a general constraint vector function $\mathbf{c}(\mathbf{x}, \mathbf{z}, \mathbf{u}, \mathbf{p}, t)$. The constraints $\mathbf{c}$ have to be enforced during process operation at any time $t \in I$.

Optimal operation of the process with respect to the specified cost functional could be achieved if a perfect process model (1), (2) and (3) of the process would be available and if the initial state $\mathbf{x}_0$ at t_0, the parameters $\mathbf{p}$, and the disturbances $\mathbf{w}$ were known exactly. Then the controls and therefore the operational trajectory could be determined entirely off-line through the solution of the following dynamic optimization problem (provided that it is solvable)

$$\min_{\substack{\mathbf{x}^r(\cdot), \mathbf{z}^r(\cdot), \\ \mathbf{u}^r(\cdot), t_f^r}} \quad E^r(\mathbf{x}^r(t_f^r), \mathbf{z}^r(t_f^r), \mathbf{p}^r) + \int_{t_0^r}^{t_f^r} L^r(\mathbf{x}^r, \mathbf{z}^r, \mathbf{u}^r, \mathbf{p}^r, \tau)\, d\tau \tag{4}$$

$$\begin{aligned}
\text{s.t.} \qquad 0 &= \mathbf{f}(\dot{\mathbf{x}}^{\tau}(t), \mathbf{x}^{\tau}(t), \mathbf{z}^{\tau}(t), \mathbf{u}^{\tau}(t), \mathbf{w}^{\tau}(t), \mathbf{p}^{\tau}, t), \quad t \in I^{\tau}, \\
\mathbf{x}^{\tau}(t_0^{\tau}) &= \mathbf{x}_0^{\tau}, \\
0 &= \mathbf{g}(\mathbf{x}^{\tau}(t), \mathbf{z}^{\tau}(t), \mathbf{u}^{\tau}(t), \mathbf{w}^{\tau}(t), \mathbf{p}^{\tau}, t), \quad t \in I^{\tau}, \\
0 &\le \mathbf{c}^{\tau}(\mathbf{x}^{\tau}(t), \mathbf{z}^{\tau}(t), \mathbf{u}^{\tau}(t), \mathbf{p}^{\tau}, t), \quad t \in I^{\tau}, \\
0 &= \mathbf{r}^{\tau}(\mathbf{x}^{\tau}(t_f^{\tau}), \mathbf{z}^{\tau}(t_f^{\tau}), \mathbf{p}^{\tau}).
\end{aligned}$$

In the case of tracking problems the Lagrange term L^{τ} may be given by an appropriate norm of the difference between the output trajectory $\mathbf{y}$ and a given reference trajectory $\boldsymbol{\eta}^{\tau}(t)$, such as a weighted Euclidean norm with the particular weighting $\mathbf{S}$:

$$L^{\tau}(\mathbf{x}^{\tau}, \mathbf{z}^{\tau}, \mathbf{u}^{\tau}, \mathbf{p}^{\tau}, t) := \|\mathbf{h}(\mathbf{x}^{\tau}(t), \mathbf{z}^{\tau}(t), \mathbf{u}^{\tau}(t), \mathbf{p}^{\tau}, t) - \boldsymbol{\eta}^{\tau}(t)\|_{\mathbf{S}}^2.$$

E^{τ} is then the penalty for the final states. In a more general case L^{τ} and E^{τ} may denote an economical cost function. The vector function $\mathbf{r}^{\tau}$ is used to account for endpoint constraints. The superscript τ in all quantities indicates that (4) is typically an optimal control problem which aims to determine the regulating optimal trajectory for the control $\mathbf{u}$. Furthermore, problem (4) is commonly referred to as an open loop optimal control problem, since no feedback from the process enters the problem formulation.

A similar dynamic optimization problem can be formulated if one aims to determine unknown or hardly accessible process quantities such as initial conditions, process parameters, process disturbances, or model uncertainty from process measurements on $I^e = [t_0^e, t_f^e]$. We can formulate the following dynamic optimization problem for the off-line estimation of the unknown quantities using data that have been collected during an operational phase I^e.

$$\min_{\substack{\mathbf{x}_0^e, \mathbf{x}^e(\cdot), \mathbf{z}^e(\cdot), \\ \mathbf{w}^e(\cdot), \mathbf{p}^e}} \quad E^e(\mathbf{x}^e(t_0^e), \mathbf{z}^e(t_0^e), \mathbf{p}^e) + \int_{t_0^e}^{t_f^e} L^e(\mathbf{x}^e, \mathbf{z}^e, \mathbf{w}^e, \mathbf{p}^e, \tau)\, d\tau \tag{5}$$

$$\begin{aligned}
\text{s.t.} \qquad 0 &= \mathbf{f}(\dot{\mathbf{x}}^{e}(t), \mathbf{x}^{e}(t), \mathbf{z}^{e}(t), \mathbf{u}^{e}(t), \mathbf{w}^{e}(t), \mathbf{p}^{e}, t), \quad t \in I^{e}, \\
\mathbf{x}^{e}(t_0^{e}) &= \mathbf{x}_0^{e}, \\
0 &= \mathbf{g}(\mathbf{x}^{e}(t), \mathbf{z}^{e}(t), \mathbf{u}^{e}(t), \mathbf{w}^{e}(t), \mathbf{p}^{e}, t), \quad t \in I^{e}, \\
0 &\le \mathbf{c}^{e}(\mathbf{x}(t), \mathbf{z}^{e}(t), \mathbf{u}^{e}(t), \mathbf{p}^{e}, t), \quad t \in I^{e}.
\end{aligned}$$

Now, the superscript e is used to indicate the estimation. Here, the Lagrange term is typically given as a weighted Euclidean norm of the difference between the measurements $\boldsymbol{\eta}^e(t)$ and the model response $\mathbf{y}$

$$L^{e}(\mathbf{x}^{e}, \mathbf{z}^{e}, \mathbf{w}^{e}, \mathbf{p}^{e}, t) := \|\mathbf{h}(\mathbf{x}^{e}(t), \mathbf{z}^{e}(t), \mathbf{u}^{e}(t), \mathbf{p}^{e}, t) - \boldsymbol{\eta}^{e}(t)\|_{\mathbf{S}}^2.$$

A typical weighting matrix $\mathbf{S}$ is the inverse of the covariance matrix of the measurement error. Nevertheless, more general weights like, e.g., time dependent operators are possible, too (Binder et al. [20]). The measurement function $\boldsymbol{\eta}^e(t)$ has to be

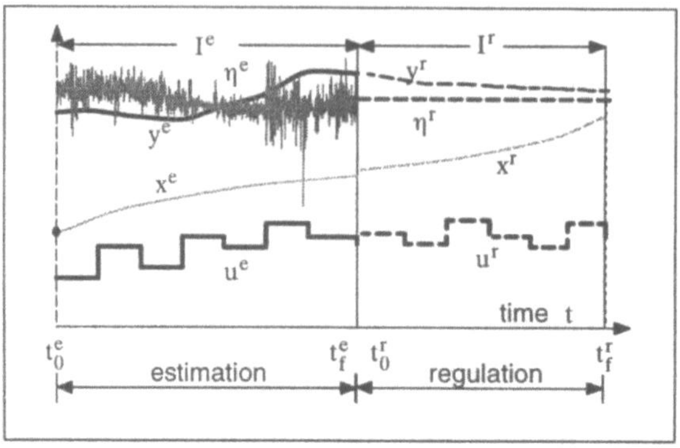

Figure 1. Moving horizon approach

generated appropriately from the measurements taken from the process at discrete sampling times. The measurements might be as well included pointwise by substituting the integral by a finite sum. The initial conditions $\mathbf{x}_0^e$, the parameters $\mathbf{p}^0$, and the disturbances $\mathbf{w}^e$ are free variables to be determined by the optimizer.

A reference value $\bar{\mathbf{x}}_0$ which could be close to the true initial conditions can be incorporated into the initial penalty E^e, e.g., by

$$E^e(\mathbf{x}^e(t_0^e), \mathbf{z}^e(t_0^e), \mathbf{p}^e) := \|\mathbf{x}^e(t_0^e) - \bar{\mathbf{x}}_0^e\|_S^2,$$

where the particular weight **S** reflects the confidence in such a reference value. The controls $\mathbf{u}^e$ are typically accessible and therefore assumed to be known.

So far we considered the regulating and estimation problems independently of each other, each formulated on a fixed horizon. It is obvious that both problems can be also coupled. For example, consider an operational phase of a production process where first process data is collected in some interval to estimate unknown quantities. Then, based on these estimates an optimal operational trajectory is determined on the remaining time interval. Let's assume the time interval $[t_0, t_f]$ is split in two parts, i.e., $I^e = [t_0^e, t_f^e]$, $I^r = [t_0^r, t_f^r]$ where $t_0^e = t_0$, $t_f^e = t_0^r$, $t_0^r = t_f$; see Figure 1. The optimal solution of (5) provides estimates on I^e of parameters **p**, disturbances $\mathbf{w}(t)$, and measured or unmeasured states $\mathbf{x}(t), \mathbf{z}(t)$ which are consistent with the process model. The estimates can then be used in the regulator problem (4) on I^r. In (4) the uncertainty and disturbance vector $\mathbf{w}^r$ is assumed to be fixed and known. Typically its values are suitable predictions based upon the estimates $\mathbf{w}^e$ computed on I^e, e.g., $\mathbf{w}^r = \mathcal{P}_w(\mathbf{w}^e)$ where $\mathcal{P}_w$ denotes a prediction operator. The predictions are computed by extrapolation or by use of simple disturbance models as discussed, e.g., in Ricker [121]. Furthermore, the problems (4) and (5) are coupled by the initial condition $\mathbf{x}^r(\mathbf{t}_0^r) = \mathbf{x}^e(\mathbf{t}_f^e)$ and the parameters $\mathbf{p}^r = \mathcal{P}_p(\mathbf{p}^e)$ which have been determined from (5) for further use in (4) using the prediction operator $\mathcal{P}_p$. Note that in this example the solution to (4) and (5) cannot be computed off-line anymore.

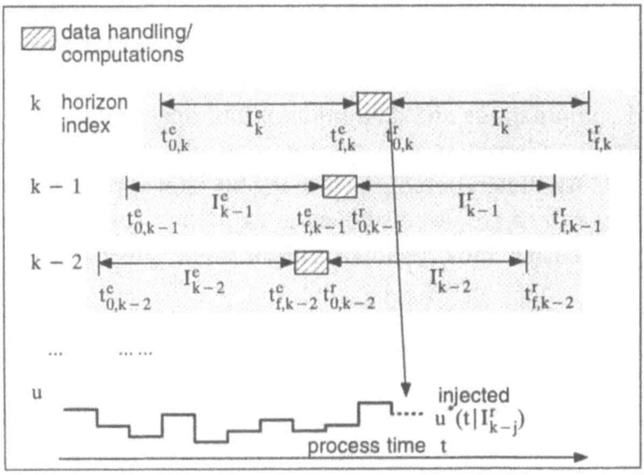

Figure 2. On-line optimization

The concept of estimation and regulation has been outlined for a simple setting. It is clear that the methodology also applies in more general situations, to better deal with the uncertainty in the model and the disturbances acting on the real process. Next, the horizons are repetitively shifted with time by a sampling time interval ΔT such that a moving horizon optimization problem is obtained.

2.3 *On-line Optimization on Moving Horizons*

In moving horizon optimization problems (4) and (5) are solved repeatedly. Unknown process quantities are estimated from the collected process measurements using (5). Based on this estimates an optimal trajectory $\mathbf{u}^{r*}(t)\ t \in I^r$ is determined by solving (4)[2], but $\mathbf{u}^{r*}(t)\ t \in I^r$, is applied to the process only during an interval ΔT, i.e., $\mathbf{u}^{r*}(t),\ t_0^r \leq t < t_0^r + \Delta T$. Then, new measurement information is collected, the estimation and regulation horizons I^e, I^r are shifted by ΔT, and (4), (5) are resolved. We now have several horizons which typically overlap such that we introduce a horizon index k which is also used as subscript in notation, i.e., the horizons are denoted by $I_k^e := [t_{0,k}^e, t_{f,k}^e]$ and $I_k^r := [t_{0,k}^r, t_{f,k}^r]$. Furthermore, we include I_k^e, I_k^r as a second argument in all quantities appearing in (4) and (5) to distinguish the solutions computed on different horizons. Therefore $\mathbf{u}^{r*}(t|I_k^r), t \in I_k^r$, denotes the optimal solution $\mathbf{u}^{r*}$ in problem (4) obtained on horizon I_k^r. Similar notation applies for all other quantities in problems (4) and (5). Note, that in this problem setting we estimate all quantities simultaneously. In practice the unknowns might live on different time-scales such that there is no need to estimate slowly varying quantities in each time step (Helbig et al., [61]).

[2] The superscript $*$ denotes the optimal values.

So far, the framework introduced is quite idealistic since it assumes that problems (4) and (5) can be solved instantly. For practical applications the estimation and prediction horizons have to be separated by some time to perform the necessary numerical computations and data input/output operations. Therefore, the estimation and prediction horizons are separated typically by one sampling time, i.e., $t^r_{0,k} = t^e_{f,k} + \Delta T$. An illustration of the moving horizon approach is given in Figure 2, where however $\mathbf{w}, \mathbf{z}, \mathbf{p}$ are not displayed to avoid an overload of the graph. With these definitions the basic moving horizon algorithm is given by:

1. While $t^e_{f,k} \leq t < t^r_{0,k}$:
   - Apply $\mathbf{u}^r(t|I^r_{k-1})$ to the process.
   - Access the measurement values $\boldsymbol{\eta}^e(t)$, $t \in I^e_k$.
   - Solve the estimation problem (5), use the so far injected controls as $\mathbf{u}^e(t)$ for $t \in I^e_k$, i.e., compute $\mathbf{x}^e(t|I^e_k)$, $\mathbf{z}^e(t|I^e_k)$, $\mathbf{w}^e(t|I^e_k)$, $\mathbf{p}^e_{I^e_k}$.
   - Compute $\mathbf{x}^r(t^r_{0,k})$ using the model $\mathbf{f}, \mathbf{g}$ in (5) as prediction model with initial condition $\mathbf{x}^e(t^e_{f,k})$, control $\mathbf{u}^r(t|I^r_{k-1})$ and suitable prediction models $\mathcal{P}_w, \mathcal{P}_p$ to extrapolate $\mathbf{w}(t^e|I^e_k)$ and $\mathbf{p}^e_{I_k}$ on $[t^e_{f,k}, t^r_{0,k}]$. Extend the extrapolation to I^r_k such that $\mathbf{w}^r$, $\mathbf{p}^r_{I_k}$ is obtained.
   - Solve the control problem (4) with the extrapolated quantities and determine $\mathbf{u}^r(t|I^r_k)$ to be injected into the process in the upcoming step $t^e_{f,k+1} < t < t^r_{0,k+1}$.
2. $k := k+1$.
3. Goto 1.

The underlying assumption of the algorithm is that the prediction models are of sufficient quality such that the initial guess $\mathbf{x}^r(t^r_{0,k})$ and the extrapolations $\mathbf{w}^e(t)$, $t^e_{f,k} \leq t \leq t^r_{0,k}$, $\mathbf{w}^r(t)$, $t \in I^r_k$, are close to the true values.

Obviously, the time ΔT should be as small as possible where at least ΔT has to be sufficiently smaller than the dominating process time constants. These time constants depend on a number of factors such as for example on the particular chemical species involved or on the particular unit operations used. While distillation column time constants with regard to product concentration are in the range of hours the product concentration of chemical reactors can change in seconds. The computational complexity of an algorithm to solve (4) and (5) depends in addition on a number of other factors such as the used model (type, structure, dimension), numerical solution approaches (optimization method, discretization), choice of cost functional and the horizon length. While the time constant is given by the process, the computational complexity is affected by engineer and mathematician through modeling and algorithmic design decisions. The designed algorithm has to prompt in any event the optimal values (or at least suitable approximations) of (4) and (5) within the available time span ΔT since otherwise proper function of the on-line optimization scheme cannot be guaranteed. This is an important real-time requirement which should be addressed by the design of the algorithms.

The functionality of the process is further affected by the closed loop stability properties (e.g., Bitmead, Gevers and Wertz, [22]) of the moving horizon approach which will be addressed in the next section.

3 REMARKS ON CLOSED LOOP STABILITY

First, we discuss closed loop stability for the regulator problem (4) assuming fully accessible differential states, i.e., $\mathbf{y}^r = \mathbf{x}^r$, and given $\mathbf{p}^r$, $\mathbf{w}^r(t)$. Let's assume that we want to find an optimal control $\mathbf{u}^r(t), t_0^r \leq t \leq \infty$ which moves the process state $\mathbf{x}^r$ from some given initial conditions $\mathbf{x}^r(t_{0,k}^r)$ to a target state which for simplicity is chosen to be the origin, i.e., $\boldsymbol{\eta}^r(t) = 0, \forall t$. Suppose, that the considered system is controllable (see, e.g, Ogunnaike and Ray, [109], for an introduction to the concept of controllability) and that no unknown disturbances, unknown parameters and model uncertainties are present. Furthermore, we assume to know the true initial conditions $\mathbf{x}(t_{0,k}^r)$.

It follows from Bellman's Principle of Optimality (e.g., Anderson and Moore, [2]), that in each horizon k the predicted state and control trajectories $\mathbf{x}^r(t, I_k^{r,\infty})$, $\mathbf{z}^r(t, I_k^{r,\infty})$, $\mathbf{u}^r(t, I_k^{r,\infty})$ of problem (4), where $I^{r,\infty} := [t_{0,k}^r, \infty]$, are equal to the optimal process trajectories $\mathbf{x}^r(t, I_0^{r,\infty}), \mathbf{z}^r(t, I_0^{r,\infty}), \mathbf{u}^r(t, I_0^{r,\infty})$ of the process system determined on $[t_0^r, \infty]$. This holds only if the first problem $k = 0$ is feasible and if the initial conditions $\mathbf{x}^r(t_{0,k}^r)$ are known for all k (Keerthi and Gilbert, [73]). Therefore, for infinite horizons there is no difference between the subsequent control sequences determined at certain time steps and the control trajectory obtained by solving a single problem. This implies closed loop stability, as any feasible optimized trajectory goes to the origin (Keerthi and Gilbert, [73]).

When instead a (small) finite horizon I_k^r, $t_{f,k} < \infty$, is chosen the actual closed loop control and state trajectories will differ in general from the predicted open loop trajectories even if no model uncertainty and unknown disturbances are present which is nicely illustrated by Bitmead, Gevers, and Wertz [22]. The solutions computed on $I_k^{r,\infty}$ and I_k^r may differ significantly the shorter I_k^r is chosen. Since from a theoretical perspective the minimum requirement of a model based controller is that it yields a stable closed-loop system if a perfect model of the plant is available and if the state is completly accessible by measurements (Henson, [64]), intense research has been undertaken in the last decade to develop schemes with guaranteed nominal stability properties. The major developments are summarized in excellent surveys given by Mayne [96], Morari and Lee [103], Allgöwer et al. [3], and Mayne et al. [97]. However, the developed approaches are yet computationally expensive, difficult to design and therefore limited to processes with low state dimensions. So far, moving horizon schemes with guaranteed stability have been only applied in academia. Besides the inherent drawbacks of the approaches with guaranteed stability this might as well be due to the fact that it is typically not difficult for practical problems to find long enough horizons by trial and error such that closed loop stability is obtained. However, it should be admitted that it is difficult to come up with a generally applicable horizon design procedure, which, given a specific problem, determines stabilizing prediction and control horizons based on the process model and the cost functional chosen (Allgöwer et. al, [3]).

Similar stability considerations apply to the estimation problem. Here, stability of the estimator is defined as the convergence of the estimated states to the true states for $t \rightarrow \infty$ for arbitrarily specified initial conditions, if the measurements contain no errors, the model is correct and the disturbances and parameters are known. Stability is trivially obtained by dropping the Mayer term in the objective ($E^e = 0$), because the minimization of the cost immediately moves the initial state to the correct value. However, such a strategy would lead to poor estimation quality if measurement noise and model uncertainty would be present as in any real situation. In these cases one should include an appropriate guess of the initial condition, say $\bar{\mathbf{x}}^e_{0,k}$, to improve estimation quality. This can be accomplished by introducing the Mayer term $E^e(\mathbf{x}^e(t^e_{0,k}) - \bar{\mathbf{x}}^e_{0,k})$ where E^e typically reflects some kind of least-squares formulation of the error. Alternatively, one could account for all available measurement information for the current estimation by keeping $t^e_{0,k} = t_{0,0}$, $\forall k$. The resulting problems are not computationally tractable since the problem dimension grows as the estimation horizon grows. Instead in the k-th horizon past data in $[t^e_{0,0}, t^e_{f,k}]$ are indirectly accounted for by $\bar{\mathbf{x}}^e_{0,k}$ which is used to reflect the past estimate and thus indirectly the information content of the past measurement data. Thus the weights in E^e reflect the confidence in the past estimates. E^e has to be chosen rather carefully to ensure proper weighting of the old data. Estimator divergence may result if the initial penalty E^e biases the old data by too strongly weighting the past estimates, while performance may suffer if the initial penalty neglects the old data by not sufficiently weighting them. Stability and performance implications for several choices of E^e for a number of problems are discussed in a rigorous manner by a number of contributions, e.g., see Michalska and Mayne [101], Muske and Rawlings [104], Robertson et al. [122], and Rao and Rawlings [120].

So far stability has been illuminated separately for the regulator and estimation problem. For linear time invariant models with quadratic cost functionals and no inequality restrictions present the separation principle holds and closed loop stability of the combined problem follows if the estimation and regulation problem are stable independently. Furthermore, it can be proven for general systems (Meadows and Rawlings, [98]) that if an exponentially converging estimator is combined with a stable control algorithm where all states are measurable, then this observer-controller system is stable. This holds even for nonlinear regulators where the separation principle obviously does not hold.

4 Overview of Solution Methods for Optimal Control Problems on Fixed Horizon

Next, we discuss the numerical techniques which are commonly applied to solve dynamic optimization problems. First we review available methods to solve optimal control problems on a fixed horizon before we examine particular extensions towards a moving horizon.

Many methods for the on-line solution of optimal control problems on moving horizons are based on algorithms designed for the off-line computation of solutions to optimal control problems on a fixed interval $I := [t_0, t_f]$ in time (including problems where t_f is as well a degree of freedom in the optimization problem).

Therefore, in this section we give an overview of the most common off-line optimal control methods which will form the core of any receding horizon strategy.

In the previous sections the necessity to distinguish the estimation and regulator problems required an extended notation which we drop here for convenience, since both types of optimal control problems can be solved using similar numerical techniques. Additionally, we restrict our attention to ODE models as the DAE case introduced above poses additional theoretical and practical difficulties which are beyond the scope of this general discussion. However, for further information on dynamic optimization with DAE systems we refer to Pytlak [119] as well as to the articles by Büskens et al. [38], Diehl et al. [47], and Kröner et al. [84] within this book as a starting point.

We consider a deterministic optimal control problem in Bolza form on a fixed horizon $I := [t_0, t_f]$ with

$$\min_{\mathbf{u}(\cdot),\mathbf{x}(\cdot)} J[\mathbf{u}(\cdot),\mathbf{x}(\cdot)] := E\left(\mathbf{x}(t_f)\right) + \int_{t_0}^{t_f} L\left(\mathbf{x}(t),\mathbf{u}(t),t\right) dt \tag{6}$$

subject to

$$\dot{\mathbf{x}}(t) = \mathbf{f}\left(\mathbf{x}(t),\mathbf{u}(t),t\right), \quad t \in I \tag{7}$$

$$\mathbf{x}(t_0) = \mathbf{x}_0, \tag{8}$$

$$\mathbf{0} \leq \mathbf{c}\left(\mathbf{x}(t),\mathbf{u}(t),t\right), \quad t \in I \tag{9}$$

$$\mathbf{0} = \mathbf{r}\left(\mathbf{x}(t_f)\right), \tag{10}$$

where $\mathbf{x} : I \to \mathbb{R}^{n_x}$, $n_x \geq 1$, and $\mathbf{u} : I \to \mathbb{R}^{n_u}$, $n_u \geq 1$, denote the state and control variables. The model ODE is denoted by $\mathbf{f} : \mathbb{R}^{n_x} \times \mathbb{R}^{n_u} \times I \to \mathbb{R}^{n_x}$, $\mathbf{c} : \mathbb{R}^{n_x} \times \mathbb{R}^{n_u} \times I \to \mathbb{R}^{n_c}$, $n_c \geq 1$, is a general nonlinear inequality constraint function, and $\mathbf{r} : \mathbb{R}^{n_x} \to \mathbb{R}^{n_r}$, $n_r \geq 0$ describes the end point constraints. The objective incorporates a Mayer term $E : \mathbb{R}^{n_x} \to \mathbb{R}$ and a Lagrange term with $L : \mathbb{R}^{n_x} \times \mathbb{R}^{n_u} \times I \to \mathbb{R}$.

For simplicity, the final time $t_f > t_0$ as well as the initial conditions $\mathbf{x}_0$ and the model parameters are assumed to be known and fixed, but an extension of the solution methods presented towards a free end time and unknown initial initial conditions and model parameters can be obtained straightforwardly. For convenience, the model parameters have been suppressed in Eqns. (6)-(10).

The functions E, L, $\mathbf{f}$, $\mathbf{c}$, and $\mathbf{r}$ are assumed to be twice continuously differentiable with respect to their arguments.

There are three basic approaches to solving optimal control problems of the form (6)-(10):

(I) Hamilton-Jacobi-Carathéodory-Bellman (HJCB) partial differential equations (PDEs) and Dynamic Programming,
(II) Calculus of Variations, Euler-Lagrange differential equations (EL-DEQ), and the Maximum Principle (indirect methods), and
(III) direct methods based on a finite dimensional parameterization of the controls.

We will briefly comment on the first two approaches in Subsections 4.1 and 4.2. The direct methods will be presented in detail in Section 5, as they have proven to be most successful for the treatment of real life large scale optimal control problems.

4.1 Hamilton-Jacobi-Carathéodory-Bellman Partial Differential Equation, Dynamic Programming

In HJCB the optimal *feedback* control $\mathbf{u}^*(\mathbf{x}, t)$ is obtained by solving a PDE for a so-called value function (e.g., Pesch and Bulirsch, [113]). In practice, however, the PDE can be solved numerically for very small state dimensions only. A further severe drawback is that inequality constraints on the state variables as well as dynamical systems with switching points usually lead to discontinuous partial derivatives and cannot easily be included. Discretization methods to compute numerical approximations of the value function by solving a first order PDE with dynamic programming are described by Bardi and Dolcetta [10], Falcone and Ferretti [49], and Lions [92] (*viscosity solutions* of the HJCB equation). It is worth mentioning here that for the subclass of *linear-quadratic regulator* problems, the HJCB-PDE can be solved analytically or numerically by solving either an algebraic or dynamic matrix Riccati equation. This approach is described in more detail in Section 6.2.

A similar solution methodology is obtained by *dynamic programming* (Bellman [14]), which provides the *global* optimal control. Unfortunately, its application is severely restricted in the case of continuous states systems – at most three state dimensions seem feasible so far because of the *curse of dimensionality*. Recently, the application of neural network approximations has been investigated to handle the curse of dimensionality and the *curse of modeling* if dynamic programming is applied to higher dimensional, nonlinear and also stochastic problems (*neuro-dynamic programming*, Bertsekas and Tsitsiklis [15]). Another new development is the adaptive critic method which relies on neural network approximations, reinforcement learning strategies and dynamic programming (e.g., Naumer [105]; Werbos [143]). However, these approaches are still restricted to problems with small state dimension.

4.2 Calculus of Variations, Euler-Lagrange Differential Equations, Maximum Principle (Indirect Methods)

A common approach to compute the optimal control is based on the Maximum Principle, that we will sketch for the case of optimal control problems with the control constrained to the (nonempty) set $\mathcal{U}(t) := \{\mathbf{u} \in \mathbb{R}^{n_u} | \mathbf{0} \le \mathbf{c}(\mathbf{u}, t)\}$

First, a Hamiltonian is defined as

$$\mathcal{H}(\mathbf{x}, \mathbf{u}, \boldsymbol{\lambda}, t) := -L(\mathbf{x}, \mathbf{u}, t) + \boldsymbol{\lambda}(t)^T \cdot \mathbf{f}(\mathbf{x}, \mathbf{u}, t), \tag{11}$$

where the vector $\boldsymbol{\lambda}(t) : I \to \mathbb{R}^{n_x}$ denotes the so-called adjoint variables. Necessary conditions for optimality of solution trajectories $\mathbf{x}^*(t)$ and $\mathbf{u}^*(t)$, $t \in I$, can then be given by the following boundary value problem in the states $\mathbf{x}^*(t)$ and in the

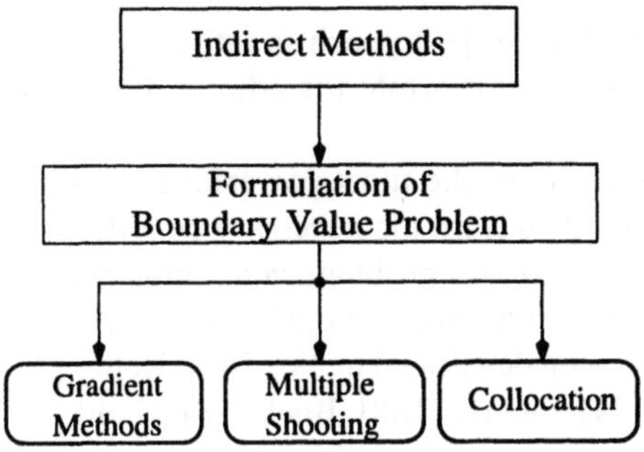

Figure 3. Overview of numerical methods based on the indirect approach

adjoints $\boldsymbol{\lambda}^*(t)$ [3], which form the EL-DEQ for the situation considered

$$\begin{aligned} \mathbf{x}^*(t_0) &= \mathbf{x}_0 \\ \mathbf{0} &= \mathbf{r}(\mathbf{x}^*(t_f)) \\ \boldsymbol{\lambda}^*(t_f) &= \nabla_{\mathbf{x}} E(\mathbf{x}^*(t_f)) - \nabla_{\mathbf{x}} \mathbf{r}(\mathbf{x}^*(t_f))\boldsymbol{\alpha} \\ &\text{and for almost all } t \in [t_0, t_f] \\ \dot{\mathbf{x}}^*(t) &= \nabla_{\boldsymbol{\lambda}} \mathcal{H}(\mathbf{x}^*(t), \mathbf{u}^*(t), \boldsymbol{\lambda}^*(t), t) \\ \dot{\boldsymbol{\lambda}}^*(t) &= -\nabla_{\mathbf{x}} \mathcal{H}(\mathbf{x}^*(t), \mathbf{u}^*(t), \boldsymbol{\lambda}^*(t), t). \end{aligned} \tag{12}$$

The vector $\boldsymbol{\alpha} \in \mathbb{R}^{n_r}$ denotes Lagrange multipliers for the end point constraints. The optimal controls are obtained by a *pointwise* maximization of the Hamiltonian, which may lead to discontinuities:

$$\mathbf{u}^*(t) = \arg \max_{\mathbf{u} \in \mathcal{U}(t)} \mathcal{H}(\mathbf{x}^*(t), \mathbf{u}, \boldsymbol{\lambda}^*(t), t) \tag{13}$$

Early developments of the Maximum Principle have been carried out by Pontryagin et al. [116], Isaacs [68], and Hestenes [65]. The approach has been extended to handle general constraints (9) on the control and state variables (for an overview see, e.g., Hartl, Sethi, and Vickson [60]). Then the EL-DEQ form an intricate multipoint boundary value problem (MPBVP) with a priori unknown interior switching points denoting the times when one of the constraints becomes active or inactive. Activation or deactivation of a state constraint generally leads to jumps in the adjoint variables.

Several families of numerical methods are based on the EL-DEQ and the Maximum Principle, some of which are listed in Figure 3.

Gradient methods are intended to iteratively improve an approximation of the optimal control by minimizing the Hamiltonian subject to a boundary value problem (Cauchy [41]; Kelley [74]; Tolle [132]; Bryson and Ho [33]; Miele [102];

[3] $\nabla_{\boldsymbol{\lambda}} \mathcal{H}(\mathbf{x}, \mathbf{u}, \boldsymbol{\lambda}, t) := \left[\frac{\partial \mathcal{H}(\cdot)}{\partial \lambda_1}, \ldots, \frac{\partial \mathcal{H}(\cdot)}{\partial \lambda_{n_x}}\right]^T$ for $\boldsymbol{\lambda} = [\boldsymbol{\lambda}_1, \ldots, \boldsymbol{\lambda}_{n_x}]^T$

Chernousko and Luybushin [43]). In each iteration step, the model (7) is numerically integrated forward in time while the adjoint differential equations are integrated backwards in time.

Multiple shooting is one of the most powerful numerical methods for solving the resulting MPBVP derived from the necessary conditions of optimality of a constrained nonlinear optimal control problem, generating highly accurate and verified (with respect to necessary conditions of optimality) solutions. Numerical multiple shooting methods have been developed by Fox [51], Keller [75], Bulirsch [35], Deuflhard [44], Bock [23, 24], Oberle [108], Bock [28], Kiehl [76], Hiltmann [66], and Callies [39]. For an introduction into multiple shooting we refer to Ascher et al. [7] or Stoer and Bulirsch [129].

Collocation methods have also been investigated to solve the boundary value problem of the EL-DEQ (e.g., Dickmanns and Well, [45]; Bär, [9]; Ascher et al. [6]) but they have been applied more successfully in the context of direct methods (Section 5).

The practical drawbacks of indirect methods are:

- Proper formulations of the necessary conditions (EL-DEQ etc.) in a numerically suitable way must be derived. The application of automatic differentiation (e.g., Griewank, [58]) may help to partly reduce the efforts to formulate the MBPVP (e.g., Mehlhorn and Sachs, [99]). In spite of this, significant knowledge and experience in optimal control is still required by the user of an indirect method.
- In order to handle active constraints properly, their switching structure must be guessed.
- Suitable initial guesses of the state and adjoint trajectories must be provided to start the iterative methods.
- Changes in the problem formulation (e.g., by a modification of the model equations), or low differentiability properties of the model functions (e.g., by low order interpolation of tabular data), are difficult to include in the solution procedure.

5 Introduction into Direct Solution Algorithms

The basic idea of direct methods for the solution of optimal control problems introduced above is to transcribe the original infinite dimensional problem (6)-(9) into a finite dimensional Nonlinear Programming problem (NLP) (Kraft, [79]; Bock and Plitt, [27]; Biegler [19]; Betts [16]; von Stryk and Bulirsch [140]), which has been pushed by the progress in nonlinear optimization (Han, [59]; Powell, [117]; Barclay, Gill, and Rosen [11]; Betts [16]). Two basically different solution strategies for the reformulated problem exist (see Pytlak, [119], for a survey):

(i) Sequential simulation and optimization:
In every iteration step of the optimization method, the model equations (7) are solved "exactly" by a numerical integration method for the current guess of control parameters. This method is also referred to as control vector parameterization.

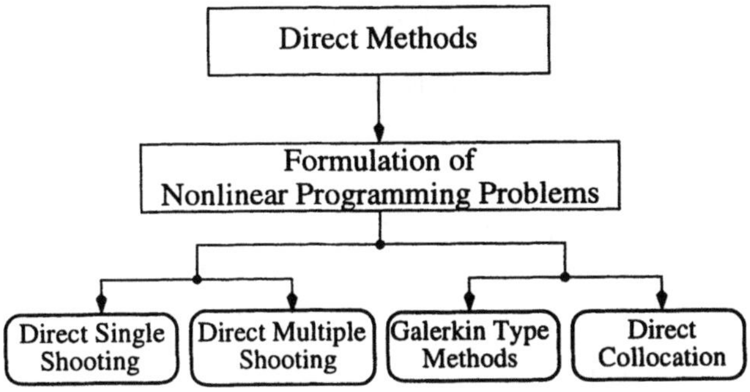

Figure 4. Overview of numerical methods based on the direct approach

(ii) Simultaneous simulation and optimization:
The discretized differential equations (7) enter the transcribed optimization problem as nonlinear constraints that can be violated during the optimization procedure. At the solution, however, they have to be satisfied.

Figure 4 outlines four particular methods which differ in the way the transcription is achieved. Collocation methods arise from general Galerkin type approaches by an appropriate choice of the approximation spaces and quadrature rules (see, e.g., Fletcher, [50]). Therefore, we will not comment any further on Galerkin type methods since the statements made for direct collocation apply as well for the more general Galerkin type methods. In this section we will only elaborate on direct single shooting, direct multiple shooting, and direct collocation.

Direct single shooting represents a pure sequential approach, whereas collocation is a pure simultaneous approach; direct multiple shooting may be considered a hybrid method, as the model equations are solved "exactly" only on intervals during the solution iterations.

5.1 Direct Single Shooting

In the direct single shooting method (e.g., Kraft [79], [80]), the infinitely many degrees of freedom $\mathbf{u}(t)$ for $t \in I$ are reduced by a control parameterization $\tilde{\mathbf{u}}(t, \mathbf{q})$ that depends on a finite dimensional vector $\mathbf{q} \in \mathbb{R}^{n_q}$. The parameterization of the control can be based on general functions with local or global support or a mixture of both. An example based on a parameterization using functions with global support, e.g., a polynomial with N coefficients $\mathbf{q}_0, \ldots, \mathbf{q}_{N-1}$, is given by

$$\tilde{\mathbf{u}}(t, \mathbf{q}_0, \ldots, \mathbf{q}_{N-1}) := \sum_{i=0}^{N-1} \mathbf{q}_i t^i, \quad t \in I.$$

A second example (see Figure 5) employing a localized parameterization is obtained using a piecewise constant control representation on a partition of the interval I into

N subintervals I_i, $i = 0, 1, \ldots, N-1$, such that

$$\tilde{\mathbf{u}}(t, \mathbf{q}_0, \ldots, \mathbf{q}_{N-1}) := \mathbf{q}_i, \quad t \in I_i.$$

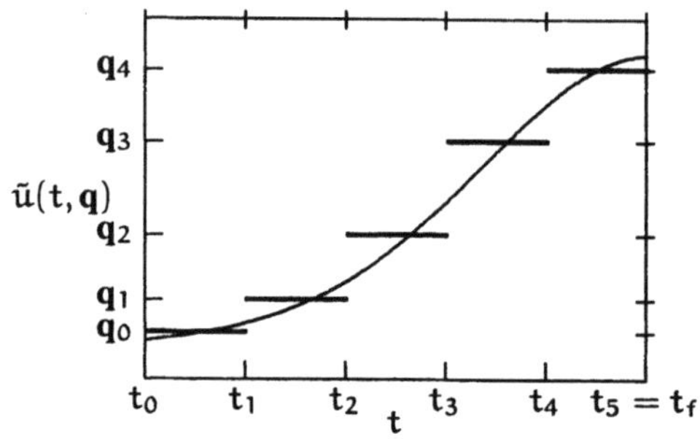

Figure 5. Piecewise constant representation of a control ($N = 5$). The control intervals are given as $I_i = [t_i, t_{i+1}]$ for $i = 0, \ldots, 4$ with intermediate time points $t_1, \ldots, t_4$

Besides these two explicit parameterizations of the controls one can also define controls implicitly via additional parameterized ODEs (or – if DAE models are admissible – by additional algebraic equations containing the so called shape parameters), e.g.,

$$\begin{aligned} \dot{\tilde{\mathbf{u}}}(t; \mathbf{q}) &= \tilde{\mathbf{f}}\left(\mathbf{x}(t), \tilde{\mathbf{u}}(t; \mathbf{q}), t, \mathbf{q}\right), \quad t \in I \\ \tilde{\mathbf{u}}(t_0; \mathbf{q}) &= \tilde{\mathbf{u}}_0(\mathbf{q}). \end{aligned}$$

The additional equations can be added to the model equations Eq. (7). In this case, the parameterized controls $\tilde{\mathbf{u}}$ are reinterpreted as (parameter dependent) states.

Given an initial value $\mathbf{x}_0$ and a parameter vector $\mathbf{q}$, the following Initial Value Problem (IVP) can be solved:

$$\begin{aligned} \dot{\mathbf{x}}(t) &= \mathbf{f}\left(\mathbf{x}(t), \tilde{\mathbf{u}}(t, \mathbf{q}), t\right), \quad t \in I, \\ \mathbf{x}(t_0) &= \mathbf{x}_0. \end{aligned}$$

The solution of this problem is a trajectory $\mathbf{x}(t)$ which is a function of $\mathbf{q}$ only. To keep this dependency in mind we will denote this solution by $\tilde{\mathbf{x}}(t; \mathbf{q})$ in the following. By substituting this trajectory into the objective functional defined in (6) we can define the cost function $\tilde{J} : \mathbb{R}^{n_q} \to \mathbb{R}$ as

$$\tilde{J}(\mathbf{q}) := E\left(\tilde{\mathbf{x}}(t_f; \mathbf{q})\right) + \int_{t_0}^{t_f} L\left(\tilde{\mathbf{x}}(t; \mathbf{q}), \tilde{\mathbf{u}}(t, \mathbf{q}), t\right) dt$$

In order to incorporate the path inequality constraints $\mathbf{c}$ into the NLP, different methods have been developed (cf. Vassiliadis, Sargent, and Pantelides [135]). Two popular methods are

1. Introduction of a penalty term in the objective function:

$$\hat{J}[\mathbf{u}(\cdot),\mathbf{x}(\cdot)] := J[\mathbf{u}(\cdot),\mathbf{x}(\cdot)] + \sum_{j=1}^{n_h} \kappa_j \cdot \int_{t_0}^{t_f} \left(\max(0, -c_j(\cdot))\right)^2 dt$$

   where $\kappa_j \in \mathbb{R}^+$, $j = 1, \dots, n_h$ are large positive constants. A difficulty with the max operator is that it hides all information about a constraint as long as it is inactive, and that its smoothness is limited.
2. Using a time grid $t_0 < t_1 < \cdots < t_N = t_f$ the infinite dimensional path inequality constraints (9) are reformulated into $N+1$ vector inequality constraints

$$\mathbf{0} \leq \tilde{\mathbf{c}}_i(\mathbf{q}) := \mathbf{c}\left(\tilde{\mathbf{x}}(t_i;\mathbf{q}), \tilde{\mathbf{u}}(t_i,\mathbf{q}), t_i\right), \quad i = 0, \dots, N.$$

   By construction, this method enforces the path inequality constraints at the points on the time grid only. A sufficiently good approximation of the original constraint can be obtained by a sufficiently fine grid. Also a combination with the first method is possible.

In the sequel we adopt the second approach.

The endpoint constraint is similarly reformulated as

$$\mathbf{0} = \tilde{\mathbf{r}}(\mathbf{q}) := \mathbf{r}\left(\tilde{\mathbf{x}}(t_N;\mathbf{q})\right).$$

In summary, the finite dimensional NLP in the direct single shooting parameterization is given as

$$\min_{q \in \mathbb{R}^{n_q}} \tilde{J}(\mathbf{q})$$

subject to

$$\begin{aligned} \mathbf{0} &\leq \tilde{\mathbf{c}}_i(\mathbf{q}), \quad i = 0, \dots, N, \\ \mathbf{0} &= \tilde{\mathbf{r}}(\mathbf{q}). \end{aligned} \tag{15}$$

The numerical effort to solve the NLP (15) is determined to a large extent by the complexity of the parameterization of the control vector. Clearly, a piecewise constant parameterization with a uniform mesh length might not be the best for general problems such that adaptive parameterization schemes should be employed to resolve the trajectory at the right place. However, it is by no means trivial to generate such problem adapted meshes a-priori, i.e., before the actual optimal solution is known (see Waldraff et al. [142], Betts and Huffmann [18], Binder et al. [20]).

The solution of the NLP (15) requires sensitivity information of the states with respect to the control parameters $\mathbf{q}$. The computation of these sensitivities should be done according to the principle of *Internal Numerical Differentiation (IND)* (Bock, [25]) and *not* by trying to generate derivates by finite differences of independently computed approximations of the solution of disturbed initial value problems [4]. Many

[4] This is also valid in the context of multiple shooting which we will introduce in the next section.

ODE and DAE solvers exist that can efficiently compute sensitivities according to the principle of IND; see, e.g., Caracotsios and Stewart [40], Leis and Kramer [91], Heim [62], Buchauer, Hiltmann, and Kiehl [34], Bock, Schlöder, and Schulz [30], Maly and Petzold [93], Kiehl [77], Engl et al. [48], Bauer [12].

In many practical applications the problem functions have only low, local differentiability properties, i.e., discontinuities in the first or second order derivatives occur.

In these cases obtaining a useful gradient approximation is rather involved, since a numerical sensitivity analysis for initial value problems with switching points must be carried out, e.g., Rozenvasser [123], Bock [28], von Schwerin, Winckler, and Schulz [136], Galán, Feehery, and Barton [52].

5.2 Direct Multiple Shooting

In the direct multiple shooting method (Plitt, [115]; Bock and Plitt, [27]), the transcription of the optimal control problem (6)-(9) into an NLP starts similar to the direct (single) shooting method with a local control representation. First, the time horizon $I = [t_0, t_f]$ is divided into N subintervals $I_i := [t_i, t_{i+1}]$, $i = 0, 1, \ldots, N-1$, with $t_0 < t_1 < \cdots < t_N = t_f$. Then, the control trajectory is parameterized by a piecewise representation

$$\tilde{\mathbf{u}}_i(t, \mathbf{q}_i) \text{ for } t \in [t_i, t_{i+1}]$$

with N *local* control parameter vectors $\mathbf{q}_0, \mathbf{q}_1, \ldots \mathbf{q}_{N-1}$, $\mathbf{q}_i \in \mathbb{R}^{n_q}$. The trivial example for such a parameterization is again the piecewise constant representation shown in Figure 5.

In a crucial second step, $N+1$ additional vectors $\mathbf{s}_0, \mathbf{s}_1, \ldots, \mathbf{s}_N$ of the same dimension n_x as the system state are introduced, to which we will refer to as the multiple shooting *node values*. All but the last serve as initial values for N independent IVPs on the intervals I_i:

$$\begin{aligned} \dot{\mathbf{x}}_i(t) &= \mathbf{f}(\mathbf{x}_i(t), \tilde{\mathbf{u}}_i(t, \mathbf{q}_i), t), \quad t \in [t_i, t_{i+1}] \\ \mathbf{x}_i(t_i) &= \mathbf{s}_i. \end{aligned}$$

The solutions of these problems are N independent trajectories $\mathbf{x}_i(t)$ on $[t_i, t_{i+1}]$, which are a function of $\mathbf{s}_i$ and $\mathbf{q}_i$ only. To keep this dependency in mind, we will denote these solutions by $\tilde{\mathbf{x}}_i(t; \mathbf{s}_i, \mathbf{q}_i)$ in the following. For an illustration, see Figure 6.

By substituting the independent trajectories $\tilde{\mathbf{x}}_i(t; \mathbf{s}_i, \mathbf{q}_i)$ into the Lagrange term L in Eq. (6) we can calculate the objective contributions $\tilde{J}_i : \mathbb{R}^{n_x} \times \mathbb{R}^{n_q} \rightarrow \mathbb{R}$ for $i = 0, \ldots, N-1$ as

$$\tilde{J}_i(\mathbf{s}_i, \mathbf{q}_i) := \int_{t_i}^{t_{i+1}} L(\tilde{\mathbf{x}}_i(t; \mathbf{s}_i, \mathbf{q}_i), \tilde{\mathbf{u}}_i(t, \mathbf{q}_i), t)\, dt.$$

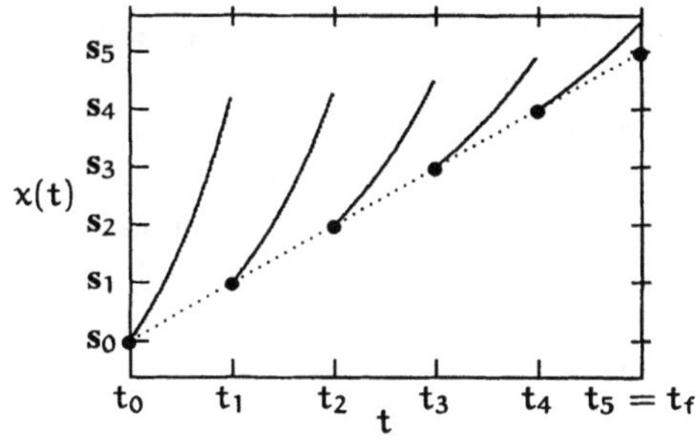

Figure 6. Five trajectories in the multiple shooting parameterization ($N = 5$)

The decoupled IVPs are connected by matching conditions which require that each node value should equal the final value of the preceding trajectory:

$$\mathbf{s}_{i+1} = \tilde{\mathbf{x}}_i(t_{i+1}; \mathbf{s}_i, \mathbf{q}_i), \quad i = 0, \ldots, N-1. \tag{16}$$

The first multiple shooting node variable $\mathbf{s}_0$ is required to be equal to the initial value $\mathbf{x}_0$ of the optimization problem:

$$\mathbf{s}_0 = \mathbf{x}_0. \tag{17}$$

Together, the constraints (16) and (17) remove the additional degrees of freedom which were introduced with the parameters $\mathbf{s}_i$, $i = 0, \ldots, N$. It is by no means necessary that the constraints (16) and (17) are satisfied *during* the optimization iterations – on the contrary, it is a crucial feature of the direct multiple shooting method that it can deal with *infeasible* initial guesses of the variables $\mathbf{s}_i$ and $\mathbf{q}_i$.

Using for notational convenience the same time grid as for the multiple shooting parameterization (finer or coarser grids are equally possible), the infinite dimensional path inequality constraints (9) are transcribed into $N + 1$ vector inequality constraints

$$0 \le \tilde{\mathbf{c}}_i(\mathbf{s}_i, \mathbf{q}_i) := \mathbf{c}\left(\mathbf{s}_i, \tilde{\mathbf{u}}_i(t_i, \mathbf{q}_i), t_i\right), \quad i = 0, \ldots, N.$$

Summarizing, the finite dimensional NLP in the direct multiple shooting parameterization is given as

$$\min_{\mathbf{s}_0, \ldots, \mathbf{s}_N, \mathbf{q}_0, \ldots, \mathbf{q}_{N-1}} E(\mathbf{s}_N) + \sum_{i=0}^{N-1} \tilde{J}_i(\mathbf{s}_i, \mathbf{q}_i)$$

subject to

$$\begin{aligned}
\mathbf{s}_{i+1} &= \tilde{\mathbf{x}}_i(t_{i+1}; \mathbf{s}_i, \mathbf{q}_i), \quad i = 0, \ldots, N-1, \\
\mathbf{s}_0 &= \mathbf{x}_0, \\
0 &\le \tilde{\mathbf{c}}_i(\mathbf{s}_i, \mathbf{q}_i), \\
0 &= \mathbf{r}(\mathbf{s}_N).
\end{aligned}$$

An important feature of the direct multiple shooting method is the sparse structure of this large scale NLP. Its Lagrangian function $\mathcal{L}$ is *partially separable*, i.e., its Hessian matrix $\nabla^2_{\mathbf{s},\mathbf{q}}\mathcal{L}$ is block diagonal with non-zero blocks $\nabla^2_{\mathbf{s}_i,\mathbf{q}_i}\mathcal{L}$ that correspond to *local* variables $\mathbf{s}_i, \mathbf{q}_i$ only (Bock and Plitt, [27]).

Extensions of the Direct Multiple Shooting Method to treat DAE systems efficiently are described by Bock, Eich, Schlöder [29], Schulz, Bock, Steinbach [126], and Heim and von Stryk [63] for the case of parameter estimation and for very general multistage optimal control problems by Leineweber [89].

5.3 Direct Collocation

We now consider a general direct collocation discretization of the optimal control problem Eqs. (6)-(10). For ease of notation, we assume that the functional Eq. (6) is in Mayer form $J[\mathbf{u},\mathbf{x}] = E(\mathbf{x}(t_f))$. This is no restriction of generality, as the transformation of the Bolza functional (6) to Mayer form is easily done: As a first step, an additional state $\mathbf{x}_{n_x+1}$ and an additional differential equation

$$\dot{\mathbf{x}}_{n_x+1}(t) = L(\mathbf{x}(t),\mathbf{u}(t),t), \quad \mathbf{x}_{n_x+1}(t_0) := 0$$

are introduced. In the second step, the objective $E(\mathbf{x}(t_f))$ is redefined as $E(\mathbf{x}(t_f)) + \mathbf{x}_{n_x+1}(t_f)$ (in order to keep notation at a minimum, no new symbol for the *redefined* objective is introduced).

Both state and control variables are approximated by piecewise defined functions $\tilde{\mathbf{x}}(t;\cdot)$ and $\tilde{\mathbf{u}}(t;\cdot)$ on the time grid

$$t_0 < t_1 < \cdots < t_{N+1} = t_f.$$

Within each collocation interval $[t_i, t_{i+1}[$, $0 \le i \le N$, these functions are chosen as parameter dependent polynomials of order $k, l \in \mathbb{N}$ respectively:

$$\tilde{\mathbf{x}}(t;\mathbf{s})\big|_{[t_i,t_{i+1}[} := \tilde{\mathbf{x}}_i(t;\mathbf{s}_i) := \pi_i^x(t;\mathbf{s}_i) \in \Pi_k^{n_x},$$
$$\tilde{\mathbf{u}}(t;\mathbf{q})\big|_{[t_i,t_{i+1}[} := \tilde{\mathbf{u}}_i(t;\mathbf{q}_i) := \pi_i^u(t;\mathbf{q}_i) \in \Pi_l^{n_c}.$$

Here, Π_μ^ν denotes the space of ν-dimensional vectors of polynomials up to degree μ. The coefficients of the polynomials (*shape parameters*) are collected in the vectors

$$\mathbf{s} := (\mathbf{s}_0^T,\ldots,\mathbf{s}_N^T)^T \in \mathbb{R}^{N\cdot(k+1)\cdot n_x}, \quad \mathbf{s}_i \in \mathbb{R}^{(k+1)\cdot n_x}, \ i = 0,\ldots,N,$$
$$\mathbf{q} := (\mathbf{q}_0^T,\ldots,\mathbf{q}_N^T)^T \in \mathbb{R}^{N\cdot(l+1)\cdot n_c}, \quad \mathbf{q}_i \in \mathbb{R}^{(l+1)\cdot n_c}, \ i = 0,\ldots,N.$$

Matching conditions of the form

$$\pi_i(t_{i+1}^-,\cdot) = \pi_{i+1}(t_{i+1}^+,\cdot), \quad i = 0,\ldots,N-1$$

have to be imposed at the boundaries of the subintervals to enforce continuity of the approximating functions in $[t_0, t_f]$. Additionally, higher order differentiability may be imposed by

$$\frac{d^\kappa}{dt^\kappa}\pi_i(t_{i+1}^-,\cdot) = \frac{d^\kappa}{dt^\kappa}\pi_{i+1}(t_{i+1}^+,\cdot), \quad \begin{cases} \kappa = 1,\ldots,J \\ i = 0,\ldots,N-1 \end{cases}$$

where J denotes the desired order of differentiability.

In order to formulate a nonlinear optimization problem, the model equations and the continuous constraints are explicitly discretized:

1. The model equations (7) are only to be satisfied at the *collocation points* $t_{i\,\mu}$, $\mu = 1, \ldots, M$, within each subinterval $[t_i, t_{i+1}[$, $i = 0, \ldots, N-1$, and within $[t_N, t_{N+1}]$:

$$t_i \le t_{i0} < \cdots < t_{iM} < t_{i+1}, \qquad i = 0, \ldots, N-1$$
$$t_N \le t_{N0} < \cdots < t_{NM} \le t_{N+1},$$

2. The inequality constraints $\mathbf{c}(\cdot)$ are sampled on a second grid within $[t_0, t_f]$:

$$t_0 \le t_1^c < \cdots < t_L^c \le t_f$$

Altogether, this leads to the formulation of the discretized optimal control problem derived from (6)-(10) (in Mayer form) by collocation:

$$\min_{\mathbf{s},\mathbf{q}} \tilde{E}(\mathbf{s}) = E\left(\tilde{\mathbf{x}}(t_f; \mathbf{s})\right) \tag{19}$$

subject to the nonlinear (point) constraints

$$\mathbf{f}\left(\tilde{\mathbf{x}}(t_{il}; \mathbf{s}), \tilde{\mathbf{u}}(t_{il}; \mathbf{q}), t\right) - \dot{\tilde{\mathbf{x}}}(t_{il}; \mathbf{s}) = \mathbf{0}, \quad \begin{cases} i = 0, \ldots, N \\ l = 0, \ldots, M \end{cases} \tag{20}$$

$$\mathbf{c}\left(\tilde{\mathbf{x}}(t_\gamma^c; \mathbf{s}), \tilde{\mathbf{u}}(t_\gamma^c; \mathbf{q}), t_\gamma^c\right) \ge \mathbf{0}, \quad \gamma = 1, \ldots, L \tag{21}$$

$$\tilde{\mathbf{x}}(t_0; \mathbf{s}) - \mathbf{x}_0 = \mathbf{0}, \tag{22}$$

$$\mathbf{r}\left(\tilde{\mathbf{x}}(t_f; \mathbf{s})\right) = \mathbf{0}. \tag{23}$$

If the solution is restricted to (higher order) continuously differentiable state and control variables, the matching conditions have to be fulfilled additionally:

$$\frac{d^\kappa}{dt^\kappa}\pi_i^x(t_{i+1}^-; \mathbf{s}_i) - \frac{d^\kappa}{dt^\kappa}\pi_{i+1}^x(t_{i+1}^+; \mathbf{s}_{i+1}) = 0, \quad \begin{cases} \kappa = 0, \ldots, J_s \\ i = 0, \ldots, N-1 \end{cases} \tag{24a}$$

$$\frac{d^\kappa}{dt^\kappa}\pi_i^u(t_{i+1}^-; \mathbf{q}_i) - \frac{d^\kappa}{dt^\kappa}\pi_{i+1}^u(t_{i+1}^+; \mathbf{q}_{i+1}) = 0, \quad \begin{cases} \kappa = 0, \ldots, J_c \\ i = 0, \ldots, N-1 \end{cases} \tag{24b}$$

where J_s is the order of differentiability in the state variables and J_c is the order of differentiability in the control variables.

The constrained nonlinear optimization problem Eqs. (19), (20)-(22), (24a)-(24b) can be efficiently solved using SQP algorithms that will be discussed in Section 5.4. SQP methods are based on the availability of gradient information. This gradient information can be obtained very easily, e.g.,

$$\frac{d}{d\mathbf{s}_i}(\mathbf{f} - \dot{\tilde{\mathbf{x}}})\bigg|_{t = t_{il}; \mathbf{s}, \mathbf{q}} = \frac{\partial \mathbf{f}}{\partial \mathbf{x}} \cdot \frac{\partial \tilde{\mathbf{x}}_i}{\partial \mathbf{s}_i}(t_{il}; \mathbf{s}_i) - \frac{\partial \dot{\tilde{\mathbf{x}}}_i}{\partial \mathbf{s}_i}(t_{il}; \mathbf{s}_i).$$

Due to the full discretization of both control and state space, the NLPs generated by direct collocation tend to become very large for practically interesting problems. Thus, special care has to be taken in the implementation of a collocation algorithm to account for the special structure and the high sparsity of the Jacobian of the constraints Eqs. (20)-(22), (24a)-(24b); see, e.g., Betts [16], von Stryk [139].

5.4 *Numerical Solution of the NLP by Sequential Quadratic Programming*

Sequential quadratic programming (SQP)(Han, [59]; Powell, [117]) is a very efficient iterative method for the solution of NLP arising from the discretization of optimal control problems by direct transcription methods as described above. Now, let ξ be the set of *parameters* introduced by the discretization of an infinite dimensional optimal control problem. In each SQP iteration a current guess of the optimal set of *optimization variables* ξ^* is improved by the solution of a quadratic subproblem derived from a quadratic approximation of the Lagrangian of the NLP subject to the linearized constraints (for a description see, e.g., Barclay, Gill, and Rosen [11]; Gill, Murray, and Saunders [54]).

In the sequel, we consider an NLP of the form

$$\min_{\xi} \varphi(\xi) \tag{25}$$
$$\text{subject to} \quad \mathbf{a}(\xi) = 0, \quad \mathbf{b}(\xi) \geq 0,$$

and their solution by SQP methods equipped with a relaxation strategy based on line search.

For the class of SQP methods considered, the vector of optimization variables $\xi_k \in \mathbb{R}^{n_\nu}$ itself and the vector of multipliers $\mathbf{v}_k := (\boldsymbol{\mu}, \boldsymbol{\sigma})_k \in \mathbb{R}^{n_a+n_b}$ are changed from (the major SQP) iteration number k to iteration number k + 1 by

$$\begin{pmatrix} \xi_{k+1} \\ \mathbf{v}_{k+1} \end{pmatrix} = \begin{pmatrix} \xi_k \\ \mathbf{v}_k \end{pmatrix} + \alpha_k \begin{pmatrix} \mathbf{d}_k \\ \mathbf{u}_k - \mathbf{v}_k \end{pmatrix}, \quad k = 0, 1, 2, \ldots$$

where the *search direction* $(\mathbf{d}_k, \mathbf{u}_k)$ is obtained as the solution of a linearly constrained quadratic problem (QP) resulting from a quadratic approximation of the Lagrangian $\mathcal{L}$

$$\mathcal{L}(\xi, \boldsymbol{\mu}, \boldsymbol{\sigma}) := \varphi(\xi) - \sum_{i=1}^{n_a} \mu_i a_i(\xi) - \sum_{j=1}^{n_b} \sigma_j b_j(\xi), \quad \boldsymbol{\mu} \in \mathbb{R}^{n_a}, \quad \boldsymbol{\sigma} \in \mathbb{R}^{n_b} :$$

$$\min_{\mathbf{d} \in \mathbb{R}^{n_\nu}} \frac{1}{2} \mathbf{d}^T \mathbf{C}_k \mathbf{d} + \nabla \varphi(\xi_k)^T \mathbf{d} \tag{26}$$
$$\text{subject to} \quad \nabla a_i(\xi_k)^T \mathbf{d} + a_i(\xi_k) = 0, \; i = 1, \ldots, n_a,$$
$$\nabla b_j(\xi_k)^T \mathbf{d} + b_j(\xi_k) \geq 0, \; j = 1, \ldots, n_b.$$

Usually, $\mathbf{C}_k$ is a positive definite approximation of the Hessian $\mathbf{H}_k$ of the Lagrangian $\mathcal{L}(\xi_k, \mu_k, \sigma_k)$. The search direction $\mathbf{d}_k$ is the solution of the QP (26) and $\mathbf{u}_k$ is the corresponding multiplier. The quadratic (sub-)problem Eqs. (26) itself is solved by an iterative method (usually, an active set strategy or an interior point method is employed).

The *step size* $\alpha_k \in \mathbb{R}$ is obtained by a (approximate) one-dimensional minimization of a merit function (*line search*)

$$\psi_r\left(\begin{pmatrix}\boldsymbol{\xi}\\ \mathbf{v}\end{pmatrix} + \alpha\begin{pmatrix}\mathbf{d}\\ \mathbf{u}-\mathbf{v}\end{pmatrix}\right)$$

with respect to α. A suitable merit function is, e.g., the Lagrangian augmented by penalty terms (*augmented Lagrangian*, e.g., Gill, Murray, Saunders, and Wright [55])

$$\begin{aligned}\psi_r(\boldsymbol{\xi}, \boldsymbol{\mu}, \boldsymbol{\sigma}) = \varphi(\boldsymbol{\xi}) &- \sum_{i=1}^{n_a}\left(\mu_i a_i(\boldsymbol{\xi}) - \frac{1}{2} r_i a_i^2(\boldsymbol{\xi})\right)\\ &- \sum_{j\in J}\left(\sigma_j b_j(\boldsymbol{\xi}) - \frac{1}{2} r_{n_a+j} b_j^2(\boldsymbol{\xi})\right) - \frac{1}{2}\sum_{j\in K}\frac{\sigma_j^2}{r_{n_a+j}}.\end{aligned}$$

The index sets J and K are chosen according to $J = \{j | 1 \leq j \leq n_b, b_j(y) \geq \sigma_j / r_{n_a+j}\}$, $K = \{1, \ldots, n_b\} \backslash J$, where $r_i > 0$, $i = 1, \ldots, n_a + n_b$.

A widely used and robust general-purpose line search based SQP method is NPSOL (Gill, Murray, Saunders, and Wright [56]), which is suitable for small to medium sized problems. The new, sparse SQP method SNOPT is a successor of NPSOL and one of the most advanced, efficient and robust, general-purpose SQP methods currently available for large-scale problems (Gill, Murray, and Saunders [54]; Gould and Toint [57]).

A discussion of other SQP methods, e.g., of the trust-region method, can be found in Gould and Toint [57] or in Nocedal and Wright [106].

5.5 Comparison of Direct Methods

We will try to develop the advantages and disadvantages of the previously described three methods – direct single shooting, direct multiple shooting, and direct collocation – when the resulting NLPs are solved by appropriately designed SQP methods. A brief summary of this discussion is given in Table 1. Additional background information and a broad list of references regarding direct methods can be found, e.g., in von Stryk [139].

Direct (Single) Shooting

- In each major SQP iteration an initial value problem is numerically solved with high solution accuracy (even though the controls may be far from from their optimal solution values).
- Possible use of existing dynamic simulation facilities (Engl, Kröner, Kronseder, and von Stryk [48]) can increase the confidence of users not deeply familiar with optimization techniques.

- Use of efficient state-of-the-art ODE and DAE solvers allows to profit from recent developments in the field.
- Small size of NLP facilitates the use of off-the-shelf NLP/QP solvers.
- Only initial guesses for the control parameters (and if free, for the initial values) are needed.
- For highly unstable systems (i.e., initial value problems with a strong dependence on the initial values) the optimization algorithm inherits the ill-conditioning of the initial value problem, even if the optimization problem itself is well-conditioned (this well-conditioning may, e.g., be due to end point constraints or an objective function penalizing trajectory deviations as, e.g., in tracking/estimation problems).
- The dynamic model is fulfilled during all SQP iterations (up to integrator accuracy), so that in time critical cases a premature stop with a physical system trajectory is possible. However, state and end point constraints (9), (10) may still be violated – roughly spoken, they have only second priority in the single shooting formulation.
- If the initial value is fixed (as in the optimal control problem (4), but not in the estimation problem (5)), the number of derivatives corresponds to the number of control parameters – this may limit the numerical effort very efficiently for large scale systems with few control parameters.
- The single shooting algorithm can, for instance, be found in the software packages gOPT (Process Systems Enterprise, [118]), DYNOPT (Abel et al., [1]), OPTISIM® (Engl et al., [48], Kröner et al., [84]). These packages have been successfully applied to solve large scale industrial problems.

Direct Multiple Shooting

- Similar to single shooting, the underlying initial value problems are numerically solved with prespecified accuracy in each SQP iteration.
- Use of existing dynamic simulation facilities, and of efficient state-of-the-art DAE solvers is possible, as for single shooting.
- The relatively large number of variables requires specially tailored NLP/QP algorithms. On the other hand, the structure can be exploited to yield even faster convergence than for direct single shooting ("high rank updates", Bock and Plitt, [27]), which is especially useful in the case of long horizons with many control parameters. For the QP solution, recursive schemes allow to reduce the linear algebra effort to essentially the same as for single shooting ("condensing", Bock and Plitt, [27]). Alternatively, an efficient QP solution based on dynamic programming (Steinbach, [128]) is possible which is *linear* in the number of control intervals.
- Initial guesses for the whole state trajectory are needed. This is an advantage, if a-priori knowledge about the state trajectory is available, as, e.g., in tracking problems, where it can damp the influence of poor initial guesses for the controls (which are usually much less known).
- The optimization of highly unstable or even chaotic systems can be possible (cf. Baake et al. [8]; Kallrath et al. [70]). A detailed numerical stability analysis for the case of parameter estimation is given by Bock [28].

- The method is well suited for parallel computation, since the IVP solutions and derivative computations are decoupled on different multiple shooting intervals (Gallitzendörfer and Bock, [53]).
- The continuity (Eq. (16)) of the system trajectory is only fulfilled after successful termination of the SQP solution procedure (up to the solution tolerance). At premature stops, both, continuity conditions (16) and state and end point constraints (9), (10) may be equally violated.
- An implementation of the multiple shooting method is found, e.g, within the highly advanced optimal control package MUSCOD-II (Leineweber, [89]), or in Petzold et al. [114].

Direct Collocation

- The ODE simulation (7) and the control optimization problems (6)-(10) are solved simultaneously, which leads to potentially faster computations compared to shooting techniques.
- Existing dynamic simulation facilities and DAE solvers cannot be reused directly.
- The very large number of variables requires tailored NLP/QP algorithms. On the other hand, similar as for the direct multiple shooting method, a careful exploitation of the structure can lead to excellent convergence behaviour and very efficient QP solutions. Furthermore, sparsity can be exploited at all levels.
- As for multiple shooting, initial guesses for the whole state trajectory are needed, which may be an advantage, if a-priori knowledge about the state trajectory is available.
- The optimization of highly unstable systems is also possible.
- The discretized DAE model equations (7) are only fulfilled after succesful termination of the SQP solution procedure (up to the solution tolerance). At premature stops, all constraints (20)-(24b) are equally violated.
- A reliable estimation of the adjoint variables is available on the entire state variable discretisation grid. Moreover, the estimates are also valid along arcs with active state constraints. The estimation of the adjoint variables from the Lagrange multipliers at the solution of the NLP corresponding to the infinite dimensional optimal control problem has been described, e.g., in von Stryk [137] for the case without state constraints and in von Stryk [138] for problems including state constraints.

  In this way, collocation can be used within a *hybrid* approach (von Stryk and Bulirsch, [140]) to provide information required for a highly accurate indirect multiple shooting method (see Section 4.2), i.e., good start estimates for all optimal trajectories *including* the adjoint states (e.g., Bulirsch et al., [31]), as well as for the switching structure (e.g., von Stryk and Schlemmer [141]).
- Highly advanced collocation algorithms have been implemented by Betts and Huffmann [18] (SOCS), by Cervantes and Biegler [42], by Schulz [124, 125] (OCPRSQP), and by von Stryk [139] (DIRCOL).

	Direct Single Shooting	Direct Multiple Shooting	Direct Collocation
general solution approach	sequential	hybrid	simultaneous
use of (state of the art) DAE solvers	yes	yes	no
number of variables / size of NLP	small	intermediate	large
initial guess for system states	initial state	all node values	all node values
applicable to highly unstable sytems	no	yes	yes
DAE model fullfilled in each iteration step	yes	partially	no

Table 1. Comparison of direct methods

The development of these pieces of software has been facilitated by the advent of new optimization methods which allow the solution of very large scale NLP. [5]

6 Optimization Techniques on Moving Horizons

When a sequence of moving horizon optimization problems is solved on-line, several questions regarding the employed numerical algorithm arise:

- Can the solution of each optimization problem be computed in a time ΔT that is known a-priori?
- If not so, what are suitable approximations of the feedback control that can be used instead?
- What can in advance be computed off-line, what has necessarily to be done on-line?
- How can the similarity of subsequent optimization problems be exploited to reduce computation times?

As the approaches to address these questions vary broadly and are not easily classified, we will here only mention some classical approaches which we consider a

[5] OCPRSQP uses a partially reduced SQP method.
DIRCOL employs SNOPT (Gill et al., [54]). SNOPT approximates the Hessian of the NLP Lagrangian by limited-memory quasi-Newton updates and uses a reduced Hessian algorithm for solving the QP subproblems. The null-space matrix of the working set in each iteration is obtained from a sparse LU factorization.
In the code SOCS of Betts and Frank [17] a Schur-complement QP method is implemented instead of a reduced-Hessian QP method.

useful basis for understanding current developments. We explicitly encourage the reader to consult the research articles of Binder et al., Kronseder et al. and Diehl et al. in this book for some recent approaches.

Before briefly introducing some classical approaches, let us first go a step backwards and formulate what is the aim of numerical moving horizon optimization algorithms. Further, we will distinguish between those problem specific data that are known a priori, and those that are only available on-line. For simplicity, we will here only treat the optimal control problem (4). In Subsection 6.1, however, we will briefly address the estimation problem (5) for linear systems and introduce the Kalman filter algorithm.

Optimal Moving Horizon Feedback Control

Let us recall that the task of on-line optimization on moving horizons is to compute an open-loop control $\mathbf{u}(t|I_k^r)$ for all $t \in I_k^r$. Only the first part on the time interval $t \in [t_{0,k}^r, t_{0,k}^r + \Delta T]$ is applied to the process. In the limit of negligible computation times the sampling time ΔT could be set to zero, so that the only essentially needed output of the algorithm is the first value of the open-loop control, i.e., the vector $\mathbf{u}^r(t_{0,k}|I_k^r) \in \mathbb{R}^{n_u}$.

On the other hand, what data are necessary to specify the k-th optimal control problem (4)? First, a DAE model, constraint functions and an objective functional have to be given a priori – however, some of the model parameters $\mathbf{p}^r$ and similarly the disturbance prediction $\mathbf{w}^r(t), t \in I_k^r$ may not be known before the process runs. In practice, we have to provide in advance a disturbance model that provides explicitly the predicted disturbance trajectory $\mathbf{w}^r(t)$, depending on some additional parameters. We will assume that this parameterized disturbance model is contained in the model equations, and that the vector $\mathbf{p}^r$ of a priori unknown parameters is suitably enlarged. Secondly, the objective function, or more precisely, the reference trajectory $\eta^r(t), t \in I_k^r$, may be changed during process operation – e.g., due to a change in the desired operating point. Again, we have to assume that a parameterization of all possible reference trajectories $\eta^r(t)$ exists, and that the additional parameters are again added to the general parameter vector $\mathbf{p}^r$.

Thus, the only quantities that are on-line inputs to our optimization algorithm are

- the parameter vector $\mathbf{p}_k^r$,
- the initial value $\mathbf{x}_{0,k}^r$, and
- the starting time $t_{0,k}^r$.

In summary, the purpose of idealized on-line optimization on moving horizons is to compute the *optimal moving horizon feedback control* function, that we define as follows:

$$\begin{aligned} \mathbf{u} : D \subset \mathbb{R}^{n_p} \times \mathbb{R}^{n_x} \times \mathbb{R} &\to \mathbb{R}^{n_u} \\ (\mathbf{p}_k^r, \mathbf{x}_{0,k}^r, t_{0,k}^r) &\mapsto \mathbf{u}(\mathbf{p}_k^r, \mathbf{x}_{0,k}^r, t_{0,k}^r) := \mathbf{u}^r(t_{0,k}^r|I_k^r), \end{aligned} \tag{27}$$

where we have introduced the bounded domain $D \subset \mathbb{R}^{n_p} \times \mathbb{R}^{n_x} \times \mathbb{R}$ to account for the fact that all inputs are expected to vary in a finite range only. Note that the control vector $\mathbf{u}(\mathbf{p}_k^r, \mathbf{x}_{0,k}^r, t_{0,k}^r)$ is computed as the first value of an open-loop optimal control, but that the idea of optimal moving horizon feedback control is to apply exactly this value to the real system. If $\mathbf{p}_k^r, \mathbf{x}_{0,k}^r$ were directly accessible (and not the result of on-line estimation), the optimal moving horizon feedback control function alone would define the closed-loop system behaviour.

In principle, this function could be precalculated off-line on a sufficiently fine grid on its domain D, thus eliminating the need for any on-line calculations. In practice, even for moderate state and parameter dimensions n_p and n_x, the necessary off-line calculation time and the storage requirements would be excessive, thus creating the need for on-line optimization.

For notational convenience, we go back to the problem formulation (6)-(10) introduced at the beginning of Section 4, and therefore omit the parameters $\mathbf{p}_k^r$ in the rest of this section. In the presented framework they can be treated in the same way as the initial values $\mathbf{x}_{0,k}^r$.

Time Dependence of Moving Horizon Problems

We can divide the possible moving horizon problem formulations into three major classes:

Finite Moving Horizon Problems. In this class of problem, the initial and the final time of the horizon move simultaneously, i.e., the horizon length $T = t_{f,k}^r - t_{0,k}^r$ is constant for all k. If the model equations and objective function are time independent, the output of the optimization algorithm looses its direct dependence on $t_{0,k}^r$. This can be exploited in the numerical solution of subsequent problems.

Shrinking Horizon Problems. This class comprises problems with a finite horizon length $t_{f,k}^r - t_{0,k}^r$ which is typically decreasing with growing k. Two cases are distinguished:

a) Fixed end time problems, where $t_{f,k}^r = t_{f,k-1}^r = t_f^r$. This may, e.g., occur in batch processes with a prespecified delivery time. Even when the system model and objective are time independent, the optimal control problems differ in the horizon length $t_{f,k} - t_{0,0}$, so that the resulting feedback control $\mathbf{u}(t_{0,k}^r | I_k^r)$ usually has a time dependence.
b) Open end time problems, which leave the final time $t_{f,k}^r$ as a degree of freedom of the optimization (or restrict them by a state dependent constraint). This may occur, e.g., in batch processes that should stop when the product or conversion specifications are attained. This formulation leads again to a *time independent* feedback control, if the system model and objective is time invariant.

Infinite Horizon Problems. Though so far not numerically tractable for general systems, it is worth mentioning here that they again lead to time independent control laws if the problem formulation is time invariant. This can so far only be exploited in the linear quadratic regulator problem investigated in Subsection 6.2.

Unconstrained optimal control problems for linear systems with quadratic cost can be solved very elegantly by dynamic programming techniques that will be reviewed in the first two subsections for both, the estimation and the regulator problem. It will be seen that subsequent problems can essentially be solved with negligible computational cost. Many textbooks consider this topic in far greater detail (see, e.g., Anderson and Moore, [2]). Let us first treat the estimation problem.

6.1 Linear Quadratic Estimation Problem

The typical problem formulation of such systems is given by

$$\min_{\mathbf{x}^e(\cdot),\mathbf{y}^e(\cdot),\mathbf{w}^e(\cdot)} \ \tfrac{1}{2}\,(\mathbf{x}^e(t^e_{0,k}) - \bar{\mathbf{x}}^e_{0,k})^T \mathbf{E}^e (\mathbf{x}^e(t^e_{0,k}) - \bar{\mathbf{x}}^e_{0,k}) \tag{28}$$

$$+ \tfrac{1}{2}\int_{t^e_{0,k}}^{t^e_{f,k}} (\boldsymbol{\eta}^e(\tau) - \mathbf{y}^e(\tau))^T \mathbf{Q}^e (\boldsymbol{\eta}^e(\tau) - \mathbf{y}^e(\tau)) + \mathbf{w}^e(\tau)^T \mathbf{R}^e \mathbf{w}^e(\tau)\, d\tau$$

$$\text{s.t.} \quad \dot{\mathbf{x}}^e(t) = \mathbf{A}\mathbf{x}^e(t) + \mathbf{B}\mathbf{u}^e(t) + \mathbf{w}^e(t), \quad \forall\, t \in I^e_k,$$
$$\mathbf{y}^e(t) = \mathbf{C}\mathbf{x}^e(t), \quad \forall\, t \in I^e_k.$$

The matrices $\mathbf{A}, \mathbf{B}, \mathbf{C}$ reflect the time-invariant model matrices and $\mathbf{E}^e, \mathbf{Q}^e, \mathbf{R}^e$ are time-invariant positive semi-definite weighting matrices. $\bar{\mathbf{x}}^e_{0,k}$ refers to a reference value of the initial state. The control $\mathbf{u}^e(t)$ is assumed to be known. Since we have a linear quadratic problem the optimal solution $\mathbf{x}^e(t|I^e_k)$ of (28) can also be written as (Kailath, [72])

$$\mathbf{x}^e(t) = \boldsymbol{\xi}^e(t) + \mathbf{P}^e(t)\boldsymbol{\lambda}^e(t), \ t \in I^e_k, \tag{29}$$

where $\mathbf{P}^e(t)$ and $\boldsymbol{\lambda}^e(t), \boldsymbol{\xi}^e(t)$ denote differentiable time dependent matrix and vector functions, respectively.

Explicit equations for $\mathbf{P}^e(t), \boldsymbol{\lambda}^e(t)$, and $\boldsymbol{\xi}^e(t)$ can be derived exploring the necessary optimality conditions commonly referred to as Euler-Lagrange Equations, which have been discussed in Section 4.2 (cf. Eqs. (12) and (13), and more specifically Kailath, [72]):

$$\dot{\mathbf{P}}^e(t) = \mathbf{A}\mathbf{P}^e(t) + \mathbf{P}^e(t)\mathbf{A}^T + \mathbf{R}^{e-1} - \mathbf{P}^e(t)\mathbf{C}^T\mathbf{Q}^e\mathbf{C}\mathbf{P}^e(t), \quad t \in I^e_k, \tag{30}$$

$$\dot{\boldsymbol{\lambda}}^e(t) = (\mathbf{C}^T\mathbf{Q}^e\mathbf{C}\mathbf{P}^e(t) - \mathbf{A}^T)\boldsymbol{\lambda}^e(t) - \mathbf{C}^T\mathbf{Q}^e(\boldsymbol{\eta}^e(t) - \mathbf{C}\boldsymbol{\xi}^e(t)), \ t \in I^e_k, \tag{31}$$

$$\dot{\boldsymbol{\xi}}^e(t) = \mathbf{A}\boldsymbol{\xi}^e(t) + \mathbf{B}\mathbf{u}^e(t) + \mathbf{P}^e(t)\mathbf{C}^T\mathbf{Q}(\boldsymbol{\eta}^e(t) - \mathbf{C}\boldsymbol{\xi}^e(t)), \quad t \in I^e_k. \tag{32}$$

The initial conditions arise from transversality conditions and are given by

$$\mathbf{P}^e(t_{0,k}) = \mathbf{E}^{e-1}, \tag{33}$$

$$\boldsymbol{\lambda}^e(t_{0,k}) = \mathbf{0}, \tag{34}$$

$$\boldsymbol{\xi}^e(t_{0,k}) = \bar{\mathbf{x}}^e_{0,k}. \tag{35}$$

Equation (30) is commonly referred to as the matrix Riccati equation, (31) is the governing equation for the dual variable $\boldsymbol{\lambda}^e(t)$ and (32) denotes a filter equation which will be further discussed at the end of this section. Note that initial conditions (33) and (35) are specified at $t^e_{0,k}$ while (34) is a final condition at $t^e_{f,k}$. $\boldsymbol{\xi}^e(t)$ and $\mathbf{P}^e(t)$ can be solved by forward integration using the derived initial conditions. If the trajectories for $\mathbf{P}^e, \boldsymbol{\xi}^e$ are available, $\boldsymbol{\lambda}^e(t)$ can be computed by an integration backwards in time starting at $t^e_{f,k}$. Thus forward and backward integration are necessary to determine $\mathbf{x}^e(t|I_k)$, $t \in I^e_k$. However, the integration of (31) becomes unnecessary if only the end value $\mathbf{x}(t_{f,k}|I^e_k)$ is of interest. It refers to a filtered state estimate using all preceeding data collected in the interval I^e_k. On the other hand, the estimates $\mathbf{x}^e(t|I^e_k)$, $t^e_{0,k} \leq t < t^e_{f,k}$ in the interior of I^e_k, which require a backward integration of (31), are referred to as smoothed states.

The Kalman Filter

The solution of (28) has been outlined using a deterministic problem formulation. The equations can also be derived using a stochastic approach. Then, $\mathbf{w}^e(t)$ and $\mathbf{v}^e(t) := \boldsymbol{\eta}^e(t) - \mathbf{y}^e(t)$ are assumed to follow an uncorrelated zero-mean Gaussian statistic with covariances $E\{\mathbf{w}^e(t)\mathbf{w}^{e^T}(\tau)\} = \mathbf{R}^{e^{-1}}\delta(t-\tau)$ and $E\{\mathbf{v}^e(t)\mathbf{v}^{e^T}(\tau)\} = \mathbf{Q}^{e^{-1}}\delta(t-\tau)$ where δ denotes the Dirac distribution and E is the expected value. Furthermore, let $E\{(\mathbf{x}^e(t^e_{0,k})\} = \bar{\mathbf{x}}^e_{0,k}$ and $E\{(\mathbf{x}^e(t^e_{0,k}) - \bar{\mathbf{x}}^e_{0,k})(\mathbf{x}^e(t^e_{0,k}) - \bar{\mathbf{x}}_{0,k})^T\} = \mathbf{E}^{e^{-1}}$. Then, (28) defines a maximum likelihood problem. Bias free estimates of minimal variance are obtained and $\mathbf{P}^e$ can be interpreted to be the covariance matrix of the state estimation error. However, one should be aware that the statistical assumptions might not be justified in practical applications. The problem (28) based on a statistical formulation was originally formulated and solved by Kalman (1960).

If subsequent estimation problems differ only by an increasing end time, i.e., if $t^e_{f,k} > t^e_{f,k-1}$, but $t^e_{0,k} = t^e_{0,k-1} (= t_{0,0})$, and if only the filtered state estimates $\mathbf{x}^e(t_{f,k}|I^e_k)$ are of interest, a solution can be obtained efficiently as follows: Starting with the end values $\mathbf{P}^e(t^e_{f,k-1})$ and $\boldsymbol{\xi}^e(t^e_{f,k-1})$ of the previous problem, Eqs. (30) and (32) have to be integrated on the appended part $[t^e_{f,k-1}, t^e_{f,k}]$ of the interval only. The end value $\boldsymbol{\xi}^e(t^e_{f,k})$ provides already the new filtered state estimate because of Eq. (34) and (29) evaluated at $t^e_{f,k}$. In fact, the integration of (30) and (32) can be performed simultaneously with the data acquisition, providing a continuous stream of filtered state estimates. Equation (32) is commonly referred to as the Kalman filter equation for continuous problems where $\mathbf{K}^e(t) := \mathbf{P}^e(t)\mathbf{C}^{e^T}\mathbf{Q}^e$ denotes the filter gain.

For $t \to \infty$, the matrix $\mathbf{P}^e(t)$ approaches a constant steady state $\bar{\mathbf{P}}^e$ that can be calculated a priori as the solution of the algebraic Riccati equation that is obtained by setting $\dot{\mathbf{P}}^e(t) = \mathbf{0}$ in Eq. (30). In this case, the (relatively expensive) integration of the matrix Riccati equation (30) can be omitted, and only a constant gain matrix $\bar{\mathbf{K}}^e = \bar{\mathbf{P}}^e\mathbf{C}^{e^T}\mathbf{Q}^e$ has to be kept for use in the Kalman filter equation (32).

The Extended Kalman Filter (EKF)

The Kalman filter algorithm for linear systems can be extended to non-linear systems to obtain a heuristic algorithm that is known as the *Extended Kalman Filter*. Though successful in practical applications, this algorithm does neither provide a solution to a general non-linear optimization problem of a similar form as (28), nor does it have a statistical interpretation. However, for a discrete-time system and a horizon length of one time step, the extended Kalman-filter can be related to moving horizon estimation (Robertson et al., 1996).

The extension of the Kalman filter equations is as follows: Assuming a nonlinear ODE system with outputs

$$\begin{aligned}\dot{\mathbf{x}}(t) &= \mathbf{f}(\mathbf{x}(t), \mathbf{u}(t)),\\ \mathbf{y}(t) &= \mathbf{h}(\mathbf{x}(t), \mathbf{u}(t)),\end{aligned}$$

the matrix Riccati and Kalman filter equations (30) and (32) can be generalized to obtain nonlinear analogues: Eq. (30) can directly be used with the substitutions

$$\begin{aligned}\mathbf{A}(t) &:= \frac{\partial \mathbf{f}}{\partial \mathbf{x}}(\boldsymbol{\xi}(t), \mathbf{u}(t)),\\ \mathbf{C}(t) &:= \frac{\partial \mathbf{h}}{\partial \mathbf{x}}(\boldsymbol{\xi}(t), \mathbf{u}(t)),\end{aligned}$$

and Eq. (32) is modified to

$$\dot{\boldsymbol{\xi}}^e(t) = \mathbf{f}(\boldsymbol{\xi}^e(t), \mathbf{u}(t)) + \mathbf{P}^e(t)\mathbf{C}(t)^T\mathbf{Q}\left(\boldsymbol{\eta}^e(t) - \mathbf{h}(\boldsymbol{\xi}^e(t), \mathbf{u}(t))\right).$$

The initial conditions (33) and (35) are the same.

Note that the EKF, when applied to linear systems, coincides with the Kalman filter.

6.2 *Linear Quadratic Regulator Problem*

Similar analysis as for the linear quadratic estimation problem holds for the Linear Quadratic Regulator (LQR) problem which is given by

$$\begin{aligned}\min_{\mathbf{x}^r(\cdot),\mathbf{u}^r(\cdot)} \quad & \frac{1}{2}\mathbf{x}^r(t^r_{f,k})^T\mathbf{E}^r\mathbf{x}^r(t^r_{f,k}) \qquad (36)\\ & + \frac{1}{2}\int_{t^r_{0,k}}^{t^r_{f,k}} \left(\mathbf{x}^r(\tau)^T\mathbf{Q}^r\mathbf{x}^r(\tau) + 2\mathbf{u}^r(\tau)^T\mathbf{S}^r\mathbf{x}^r(\tau) + \mathbf{u}^r(\tau)^T\mathbf{R}^r\mathbf{u}^r(\tau)\right) d\tau\\ \text{s.t.} \quad & \dot{\mathbf{x}}^r(t) = \mathbf{A}\mathbf{x}^r(t) + \mathbf{B}\mathbf{u}^r(t), \quad \forall\, t \in I^r_k \quad ,\\ & \mathbf{x}^r(t_{0,k}) = \mathbf{x}^r_{0,k}.\end{aligned}$$

The optimal control $\mathbf{u}^r(t|I_k^r)$, $t \in I_k^r$, has to be determined by the optimizer. For the sake of simplicity, $\mathbf{w}$ has been left out in (36) but extensions to include *known* forcing functions are only a matter of notation. The optimal solution to (36) for any initial state is a linear function of the state $\mathbf{u}(t) = -\mathbf{K}^r(t)\mathbf{x}(t)$ where the time-variant gain $\mathbf{K}^r(t)$ is given by

$$\mathbf{K}^r(t) = \mathbf{R}^{r^{-1}} \left(\mathbf{B}^T\mathbf{P}^r(t) + \mathbf{S}\right). \tag{37}$$

Similar to the solution of the estimation problem, $\mathbf{P}^r(t)$ denotes a differentiable time dependent matrix which has to satisfy a matrix Riccati equation given by

$$\begin{aligned} \dot{\mathbf{P}}^r(t) &= -\mathbf{P}^r(t)\mathbf{A} - \mathbf{A}^T\mathbf{P}^r(t) - \mathbf{Q}^r \\ &\quad + \left(\mathbf{P}^r(t)\mathbf{B} + \mathbf{S}^T\right)\mathbf{R}^{r^{-1}}\left(\mathbf{B}^T\mathbf{P}^r(t) + \mathbf{S}\right), \quad t \in I_k^r, \end{aligned} \tag{38}$$

$$\mathbf{P}^r(t_{f,k}^r) = \mathbf{E}^r. \tag{39}$$

Equation (38) can be solved by integration backwards in time starting at $t_{f,k}^r$.

Three interesting cases of moving horizons allow very efficient on-line schemes to calculate the optimal moving horizon feedback control $\mathbf{u}(\mathbf{x}_{0,k}^r, t_{0,k})$ for problem k. All of them make use of the fact that the solution of the matrix Riccati equation (38) is independent of the initial value $\mathbf{x}_{0,k}^r$ and can thus be solved before $\mathbf{x}_{0,k}^r$ is specified.

Shrinking Horizon

For a sequence of problems with fixed end time $t_{f,k}^r = t_{f,k-1}^r$, but $t_{0,k}^r > t_{0,k-1}^r$, e.g., for batch problems with a fixed end time, we can use the fact that the backwards integration of the matrix Riccati equation (39) starts at the identical "initial" condition (39) and thus gives identical trajectories $\mathbf{P}^r(t)$, but on shrinking time intervals. It is possible to perform the computation of $\mathbf{P}^r(t)$ on the interval $[t_{0,0}^r, t_{f,0}^r]$ off-line, and to store just the gain matrix trajectory $\mathbf{K}^r(t)$, $t \in [t_{0,0}^r, t_{f,0}^r]$. This allows to obtain the optimal feedback control law $\mathbf{u}(\mathbf{x}_{0,k}^r, t_{0,k}^r) := -\mathbf{K}^r(t_{0,k}^r)\mathbf{x}_{0,k}^r$, that can be evaluated in negligible time.

Note that this method is equally applicable to linear time-variant systems. It also provides the basis for the linearized neighboring feedback control method for non-linear systems presented in Subsection 6.3.

Moving Horizon

A second interesting simplification arises in the case that the initial and the final time of the horizon move simultaneously, i.e., that $T = t_{f,k}^r - t_{0,k}^r$ is constant for all k. An inspection of Eqs. (38) and (39) shows that the solution $\mathbf{P}^r(t)$ of problem k does *not* depend on the index k. In particular, $\mathbf{P}^r(t_{0,k}^r)$ is identical for different problems k, and therefore also the gain matrix $\mathbf{K}^r(t_{0,k}^r) = \bar{\mathbf{K}}^r$. The optimal moving horizon feedback control is therefore simply given by a matrix multiplication $\mathbf{u}(\mathbf{x}_{0,k}^r) = -\bar{\mathbf{K}}^r\mathbf{x}_{0,k}^r$. In contrast to the shrinking horizon case, this feedback law is time independent and requires storage of one matrix $\bar{\mathbf{K}}^r$ only. Unfortunately, this method cannot be generalized to time variant linear systems, because Eq. (38) would loose its time invariance.

Infinite Horizon

A third and very prominent case arises when $t^r_{f,k} = \infty$ for all k. Here, the matrix $\mathbf{P}(t)$ is simply constant for all times in all problems; it is the solution of an algebraic Riccati equation that can be obtained by requiring $\dot{\mathbf{P}}^r(t) = 0$ in Eq. (38). As in the moving horizon case, the gain matrix is constant, $\mathbf{K}^r(t_{0,k}) = \bar{\mathbf{K}}^r$ for all k, and can be computed off-line. The resulting linear controller is commonly referred to as the *Linear Quadratic Regulator (LQR).*

For both problems, the linear state estimation (28) and the linear quadratic regulator (36), efficient and robust numerical techniques have been developed which are also applicable to large scale processes (see, e.g. Mehrmann [100]; Jacobson et al. [69]). However, the problem formulations are restricted to linear process models and general inequality restrictions cannot be considered.

6.3 Linearized Neighboring Feedback Control along Reference Solutions

The simplicity and the power of the recursive techniques that are applicable to linear systems with quadratic cost motivates the question how they can help to provide an approximation to the optimal moving horizon control for non-linear systems. One such technique will be briefly described in this subsection. The method is applicable to a much wider class of problems than considered. Numerical techniques to solve them have been developed, e.g., by Pesch [112], Krämer-Eis et al., [81, 82], and Kugelmann and Pesch [86, 87]. Linearized neighboring techniques have also been used in similar approaches, e.g., by Terwiesch and Agarwal [131] and de Oliveira and Biegler [110].

Let us assume that we have found an optimal solution to the problem (6)-(10) for some $\mathbf{x}_0$ and t_0 by the indirect approach. The result are trajectories $\mathbf{x}^*(t)$, $\mathbf{u}^*(t)$ and $\boldsymbol{\lambda}^*(t)$, which have to satisfy the necessary conditions for optimality stated in Eqs. (12) and (13). We rephrase these equations here for a slightly simplified problem:

$$\begin{aligned}
&\mathbf{0} = \mathbf{x}^*(t_0) - \mathbf{x}_0\\
&\mathbf{0} = \boldsymbol{\lambda}^*(t_f) - \nabla_{\mathbf{x}} E(\mathbf{x}^*(t_f))\\
&\text{and for almost all } t \in [t_0, t_f]\\
&\mathbf{0} = \mathbf{f}(\mathbf{x}^*(t), \mathbf{u}^*(t), t) - \dot{\mathbf{x}}^*(t)\\
&\mathbf{0} = \nabla_{\mathbf{x}} \mathcal{H}(\mathbf{x}^*(t), \mathbf{u}^*(t), \boldsymbol{\lambda}^*(t), t) + \dot{\boldsymbol{\lambda}}^*(t)\\
&\mathbf{0} = \nabla_{\mathbf{u}} \mathcal{H}(\mathbf{x}^*(t), \mathbf{u}^*(t), \boldsymbol{\lambda}^*(t), t).
\end{aligned} \tag{40}$$

The final state constraint (10) and all path constraints (9) are omitted and we assume that the Hamiltonian $\mathcal{H}$ from Eq. (11) depends twice continuously differentiable on $\mathbf{x}$ and $\mathbf{u}$ and is concave in $\mathbf{u}$, so that the last equation is equivalent to the maximization of $\mathcal{H}(\mathbf{x}^*(t), \mathbf{u}, \boldsymbol{\lambda}^*(t), t)$ with respect to $\mathbf{u}$.

Let us now investigate how the solution trajectories change if the initial value changes to a slightly disturbed value $\mathbf{x}_0' = \mathbf{x}_0 + \epsilon$. Under mild regularity assumptions, the solution trajectories depend continuously differentiable on $\mathbf{x}_0$; let us introduce the shorthands

$$\begin{aligned}
\mathbf{x}^\epsilon(t) &= \tfrac{d\mathbf{x}^*(t;\mathbf{x}_0)}{d\mathbf{x}_0}(\mathbf{x}_0' - \mathbf{x}_0),\\
\mathbf{u}^\epsilon(t) &= \tfrac{d\mathbf{u}^*(t;\mathbf{x}_0)}{d\mathbf{x}_0}(\mathbf{x}_0' - \mathbf{x}_0),\\
\boldsymbol{\lambda}^\epsilon(t) &= \tfrac{d\boldsymbol{\lambda}^*(t;\mathbf{x}_0)}{d\mathbf{x}_0}(\mathbf{x}_0' - \mathbf{x}_0).
\end{aligned}$$

We can apply the implicit function theorem to compute these derivatives. A linearization of system (40) along the reference trajectories $\mathbf{x}^*(t)$, $\mathbf{u}^*(t)$ and $\boldsymbol{\lambda}^*(t)$ yields[6]:

$$\begin{aligned}
\mathbf{0} &= \mathbf{x}^\epsilon(t_0) - (\mathbf{x}_0' - \mathbf{x}_0),\\
\mathbf{0} &= \boldsymbol{\lambda}^\epsilon(t_f) - \frac{\partial^2 E}{\partial \mathbf{x}^2}(t_f)\mathbf{x}^\epsilon(t_f),
\end{aligned}$$

and for all $t \in [t_0, t_f]$,

$$\begin{aligned}
\mathbf{0} &= \frac{\partial \mathbf{f}}{\partial \mathbf{x}}(t)\mathbf{x}^\epsilon(t) + \frac{\partial \mathbf{f}}{\partial \mathbf{u}}(t)\mathbf{u}^\epsilon(t) - \dot{\mathbf{x}}^\epsilon(t),\\
\mathbf{0} &= \frac{\partial^2 \mathcal{H}}{\partial \mathbf{x}^2}(t)\mathbf{x}^\epsilon(t) + \frac{\partial^2 \mathcal{H}}{\partial \mathbf{x}\partial \mathbf{u}}(t)\mathbf{u}^\epsilon(t) + \frac{\partial \mathbf{f}}{\partial \mathbf{x}}(t)^T\boldsymbol{\lambda}^\epsilon(t) + \dot{\boldsymbol{\lambda}}^\epsilon(t),\\
\mathbf{0} &= \frac{\partial^2 \mathcal{H}}{\partial \mathbf{u}\partial \mathbf{x}}(t)\mathbf{x}^\epsilon(t) + \frac{\partial^2 \mathcal{H}}{\partial \mathbf{u}^2}(t)\mathbf{u}^\epsilon(t) + \frac{\partial \mathbf{f}}{\partial \mathbf{u}}(t)^T\boldsymbol{\lambda}^\epsilon(t).
\end{aligned}$$

It turns out that this system of linear equations is nothing else than the indirect approach applied to a time variant linear quadratic regulator problem of the same form as (36). This problem can be formulated as follows:

$$\begin{aligned}
\min_{\mathbf{x}^\epsilon(\cdot),\mathbf{u}^\epsilon(\cdot)} \quad & \frac{1}{2}\mathbf{x}^\epsilon(t_f)^T\frac{\partial^2 E}{\partial \mathbf{x}^2}\mathbf{x}^\epsilon(t_f)\\
&+ \frac{1}{2}\int_{t_0}^{t_f}\left(\mathbf{x}^{\epsilon T}\frac{\partial^2 \mathcal{H}}{\partial \mathbf{x}^2}\mathbf{x}^\epsilon + 2\mathbf{u}^{\epsilon T}\frac{\partial^2 \mathcal{H}}{\partial \mathbf{u}\partial \mathbf{x}}\mathbf{x}^\epsilon + \mathbf{u}^{\epsilon T}\frac{\partial^2 \mathcal{H}}{\partial \mathbf{u}^2}\mathbf{u}^\epsilon\right) dt \qquad (42)
\end{aligned}$$

[6] For the Jacobian of a vector valued function $\mathbf{f}(\mathbf{x})$ we write $\frac{\partial \mathbf{f}}{\partial \mathbf{x}}$ which denotes the matrix with entries $\left(\frac{\partial \mathbf{f}}{\partial \mathbf{x}}\right)_{ij} := \frac{\partial f_i}{\partial x_j}$. The second derivative matrix of a scalar function $\mathcal{H}(\mathbf{x}, \mathbf{u})$ is denoted, e.g., by $\frac{\partial^2 \mathcal{H}}{\partial \mathbf{x}\partial \mathbf{u}}$ with $\left(\frac{\partial^2 \mathcal{H}}{\partial \mathbf{x}\partial \mathbf{u}}\right)_{ij} := \frac{\partial^2 \mathcal{H}}{\partial x_i \partial u_j}$. For brevity, we do not repeat all function arguments, but only the time t, and implicitly assume that the derivatives are evaluated at the corresponding point of the trajectories $\mathbf{x}^*(t)$, $\mathbf{u}^*(t)$ and $\boldsymbol{\lambda}^*(t)$.

$$\text{s.t.} \qquad \dot{\mathbf{x}}^\epsilon(t) = \frac{\partial \mathbf{f}}{\partial \mathbf{x}}\mathbf{x}^\epsilon(t) + \frac{\partial \mathbf{f}}{\partial \mathbf{u}}\mathbf{u}^\epsilon(t), \quad \forall\, t \in [t_0, t_f],$$
$$\mathbf{x}^\epsilon(t_0) = (\mathbf{x}_0' - \mathbf{x}_0).$$

The matrix Riccati equation (38) can be solved on the horizon $[t_0, t_f]$ for the linearized problem (42) along the reference trajectory with initial value $\mathbf{x}_0$ (cf. Eq. (40)). Then the feedback matrix $\mathbf{K}(t_0)$ can be precalculated to provide a first order approximation $\tilde{\mathbf{u}}$ to the optimal feedback for a system state $\mathbf{x}_0'$ at time t_0:

$$\tilde{\mathbf{u}}(\mathbf{x}_0', t_0) := \mathbf{u}^*(t_0) - \mathbf{K}(t_0)(\mathbf{x}_0' - \mathbf{x}_0). \tag{43}$$

Shrinking Horizon

For shrinking horizon problems, the matrix function $\mathbf{K}(t)$ can be precomputed along the reference solution for $t \in [t_0, t_f]$ and can serve to provide an immediate feedback analogous to the shrinking horizon method described in Section 6.2. For given $\mathbf{x}_{0,k}$ and $t_{0,k} \in [t_0, t_f]$ we compute a first order approximation $\tilde{\mathbf{u}}$ of the optimal moving horizon feedback control that is given by

$$\tilde{\mathbf{u}}(\mathbf{x}_{0,k}, t_{0,k}) := \mathbf{u}^*(t_{0,k}) - \mathbf{K}(t_{0,k})(\mathbf{x}_{0,k} - \mathbf{x}^*(t_{0,k})).$$

The motivation for this approximation is the expectation that the real system trajectory stays sufficiently close to the reference trajectory. In particular, we assumed that no model uncertainties and disturbances have been present. However, if severe model uncertainty and disturbances are present the approach will encounter difficulties.

6.4 *Initialization Techniques for Direct Methods*

A straightforward approach to moving horizon optimization is to apply one of the direct methods described in Section 5 to solve the moving horizon optimization problems. Though they are originally designed for off-line use, their on-line application can lead to good results, depending on the real-time requirements of the problem, as the advantages of direct methods (flexibility, robustness, handling of constraints) can be fully exploited (see, e.g., Leineweber, [90]). It should be kept in mind, however, that no general run-time guarantees can be given for these methods as the number of SQP iterations is not limited (an interesting approach that requires only one iteration per sampling time can be found in the research article by Diehl et al. [47] in this book).

The computing times for the subsequent NLP solutions depend considerably on the initial guess ξ_0 for the optimization variables and the initial setup of the SQP algorithm (in particular the Hessian). We will present some apparent approaches to find a good initial guess ξ_0^k for the optimization variables in the NLP (25) that arises after the discretization of the k-th optimal control problem (6)-(10). We will briefly discuss them for moving and shrinking horizon problems.

Moving Horizon Problems

For time independent moving horizon problems, three possibilities seem to suggest themselves for the initialization ξ_0^k of the NLP (25):

- *Set-point initialization:* If the optimization problem is formulated with the objective to steer the system into a desired steady state (the setpoint state $\mathbf{x}_{ss}$ and controls $\mathbf{u}_{ss}$), the (constant) setpoint trajectory is the solution of an optimization problem (6)-(10) with initial value $\mathbf{x}_0 = \mathbf{x}_{ss}$. The NLP solution ξ_*^{ss} of this optimization problem in the chosen transcription may be used to serve as an initial guess for the NLP solution iterations (5.4): $\xi_0^k := \xi_*^{ss}$. As long as the real system state $\mathbf{x}_{0,k}$ stays close to $\mathbf{x}_{ss}$ this may be a good initial guess. The setpoint initialization provides every optimization problem with the same initial guess.
- *Simple warm start:* This strategy is based on the conjecture that the solution ξ_*^{k-1} of the previous optimization problem $k-1$ would provide a good initial guess for the current problem k: $\xi_0^k := \xi_*^{k-1}$. This may be justified if the new initial state $\mathbf{x}_{0,k}$ has not changed much compared to $\mathbf{x}_{0,k-1}$, as can be expected if the sampling time ΔT is short relative to the time constant of the system.
- *Shift strategy:* The third strategy is motivated by the following observation: for a fictitious *undisturbed* system controlled by a moving horizon algorithm with *infinite* horizon, the (open-loop) solution of the first optimization problem on $[t_{0,0}, \infty]$ would already provide the whole *closed-loop* control trajectory – thanks to the dynamic programming property, the part of the precalculated control strategy that remains at problem k on the horizon $[t_{0,k}, \infty]$ is still optimal (this is similar for shrinking horizon problems). In the finite moving horizon framework the dynamic programming property does no longer hold strictly, but the idea to shift the problem in time may still be advantageous if the horizon is chosen to be sufficiently long. We will illustrate this strategy in the context of the direct single shooting method described in Section 5.1; we choose a piecewise constant control representation with N intervals I_i each of length ΔT. Using the $(k-1)$st solution $\xi_*^{k-1} = (\mathbf{q}_{*,0}^{k-1}, \ldots, \mathbf{q}_{*,N-1}^{k-1})$, the initial guess ξ_0^k of the kth problem would be determined by a "shift" in the controls

$$\mathbf{q}_{0,i}^k := \mathbf{q}_{*,i+1}^{k-1} \quad \text{for} \quad i = 0, 1, \ldots, N-2.$$

  The new initial value for the last control variable cannot be obtained by the shift and must be extrapolated; a convenient initialization is, e.g.: $\mathbf{q}_{0,N-1}^k := \mathbf{q}_{*,N-1}^{k-1}$. This method is applicable to general time-variant nonlinear systems.

The setpoint initialization provides every optimization problem with the same initialization and thus leads to optimization outcomes that are independent of the optimization history. In practice, however, both the warm start and shift strategy perform clearly faster (cf. Diehl et al., [46], for a test in the context of the direct multiple shooting method). From the programmer's point of view, the warm start technique can often easier be incorporated into existing off-line optimization software and may therefore be preferable.

Shrinking Horizon Problems

An initialization method very similar to the shift strategy can be applied for shrinking horizon problems with fixed end time $t_{f,k} = t_{f,0}$. Here, only the part of the old solution ξ_*^{k-1} that corresponds to the new horizon $[t_{0,k}, t_{f,0}] \subset [t_{0,k-1}, t_{f,0}]$ is used to initialize the (reduced) optimization variable vector ξ_0^k of the new problem. For the direct single shooting method this would, e.g., mean that the reduced new piecewise control vector $\xi_0^k = \xi_0^k = (\mathbf{q}_{0,0}^k, \ldots, \mathbf{q}_{0,N^k}^k)$, where $N^k = N^{k-1} - 1$, is initialized by

$$\mathbf{q}_{0,i}^k := \mathbf{q}_{*,i+1}^{k-1} \quad \text{for} \quad i = 0, 1, \ldots N^k - 1.$$

As the shift strategy this method is applicable to general time-variant nonlinear systems.

7 Summary

An introduction has been given to dynamic optimization on moving horizons. We first focused on the generic problem formulation for both, control and estimation problems where we illuminated the special on-line character of the problem. Secondly, we reviewed standard numerical techniques to solve the problem on a fixed horizon. Special emphasis has been given to direct optimization methods which are typically used in practise. Furthermore we discussed basic extensions of the fixed horizon approaches to the moving horizon case. An extended discussion of more advanced concepts to solve these demanding dynamic optimization problems, proposed by the authoring research groups, can be found in this book elsewhere.

In particular, a methodology using a multiscale approach is suggested by Binder et al. [21] where a hierarchy of successively refined finite dimensional problems are constructed and solved as long as time permits. Therefore an approximate solution is provided at any time where the approximation quality of the solution scales with the used computation time.

Diehl et al. (2001) develop a real-time iteration scheme for the direct multiple shooting method that is aimed for large-scale real-time optimization problems arising in nonlinear MPC. They perform closed-loop experiments with a high purity distillation column that is described by a DAE model involving 164 state equations; sampling times of a few seconds are feasible with this approach.

In Kronseder et al. [85] a concept for model predictive control of very large-scale dynamical systems that arise in the control of air separation plants and consist in thousands of DAEs is developed. The concept considers the different time scales prescribed by the nature of the process. Emphasis is put on mid term and short term computations, which are here represented by online computation of parameterized optimal set point trajectories on a moving horizon and by update of set point trajectories via linearization of neighboring parameterized extremals respectively. Additionally, fundamental issues of the notion of real-time optimality are discussed.

REFERENCES

1. O. Abel, A. Helbig, W. Marquardt: DYNOPT User Manual, Release 2.4, Lehrstuhl für Prozesstechnik, RWTH Aachen (1999)
2. B. D. O. Anderson, J. B. Moore: Optimal Control: Linear Quadratic Methods. Prentice-Hall (1989)
3. F. Allgöwer, T. A. Badgwell, J. S. Qin, J. B. Rawlings, S. J. Wright: Nonlinear Predictive Control and Moving Horizon Estimation - An introductory overview. In: P.M. Frank (ed.): Advances in Control. Highlights of ECC'99, Springer (1999) 391-449
4. R. Aris: Mathematical Modeling Techniques. Dover Publications (1994)
5. J. A. Arwell, B. B. King: Proper orthogonal decomposition for reduced basis feedback controllers for parabolic equations. Report 99-01-01, Interdisciplinary Center for Applied Mathematics, Virginia Tech (1999)
6. U. Ascher, J. Christiansen, R. D. Russell: A collocation solver for mixed order systems of boundary value problems. Math. Comp. **33**, (1979) 659–679
7. U. Ascher, R. Mattheij, R. D. Russell: Numerical Solution of Boundary Value Problems for Differential Equations. SIAM (1988)
8. E. Baake, M. Baake, H. G. Bock, K. Briggs: Fitting Ordinary Differential Equations to Chaotic Data. Phys. Rev. A, **45** (1992)
9. V. Bär: Ein Kollokationsverfahren zur numerischen Lösung allgemeiner Mehrpunktrandwertaufgaben mit Schalt- und Sprungbedingungen mit Anwendungen in der Optimalen Steuerung und der Parameteridentifizierung. Diplomarbeit, Bonn (1983)
10. M. Bardi, I. C. Dolcetta: Optimal Control and Viscosity Solutions of Hamilton-Jacobi-Bellman Equations. Birkhäuser (1996)
11. A. Barclay, P. E. Gill, J. B. Rosen: SQP methods and their application to numerical optimal control. In: W. H. Schmidt, K. Heier, L. Bittner, R. Bulirsch (eds.): Variational Calculus. Optimal Control and Application (1998)
12. I. Bauer: Numerische Verfahren zur Lösung von Anfangswertaufgaben und zur Generierung von ersten und zweiten Ableitungen mit Anwendungen bei Optimierungsaufgaben in Chemie und Verfahrenstechnik. PhD thesis, University of Heidelberg (2000).
13. I. Bauer, H. G. Bock, S. Körkel, J. P. and Schlöder: Numerical Methods for Optimum Experimental Design in DAE Systems. J. Comp. Appl. Math. Vol. **120** (2000) 1-25
14. R. E. Bellman: Dynamic Programming. Princeton University Press (1957)
15. D. P. Bertsekas, J. N. Tsitsiklis: Neuro-Dynamic Programming. Athena Scientific (1996)
16. J. T. Betts: Survey of numerical methods for trajectory optimization. AIAA J. Guidance, Control, and Dynamics **21**, 2 (1998) 193-207.
17. J. T. Betts, P. D. Frank: A sparse nonlinear optimization algorithm. J. Optimization Theory and Applications **82**, 3 (1994) 519-541.
18. J. T. Betts, W. P. Huffmann: Mesh refinement in direct transcription methods for optimal control. Optim. Control Appl. Meth. **19**, (1998) 1-21
19. L. T. Biegler: Solution of dynamic optimization problems by successive quadratic programming and orthogonal collocation. Comput. chem. Engng. **8**, 3/4 (1984) 243-248.
20. T. Binder, L. Blank, W. Dahmen, W. Marquardt: Towards Multiscale Dynamic and Data Reconciliation. In: R. Berber, C. Kravaris (eds.): Nonlinear Model Based Process Control Kluwer Academic Publishers, (1998), 623-665
21. T. Binder, L. Blank, Dahmen W., Marquardt W.: Multiscale Concepts for Moving Horizon Optimization. In: M. Groetschel, S.O. Krumke, J. Rambau (eds.): Online Optimization of Large Scale Systems: State of the Art, Springer (2001)
22. R. R. Bitmead, M. Gevers, V. Wertz: Adaptive optimal Control - The Thinking Man's GPC. Prentice Hall (1990)

23. H. G. Bock: Numerische Berechnung zustandsbeschränkter optimaler Steuerungen. Carl-Cranz-Gesellschaft, Tech. Rep 1.06/78 Heidelberg (1978)
24. H. G. Bock: Numerical Solution of Nonlinear Multipoint Boundary Value Problems with Applications to Optimal Control. ZAMM **58**, T407 (1978)
25. H. G. Bock: Numerical treatment of inverse problems in chemical reaction kinetics. In: K. H. Ebert, P. Deuflhard, and W. Jäger, (eds.): Modelling of Chemical Reaction Systems. volume 18 of Springer Series in Chemical Physics, Springer (1981)
26. H. G. Bock: Recent advances in parameter identification techniques for O.D.E. In: P. Deuflhard und E. Hairer (eds.): Numerical Treatment of Inverse Problems in Differential and Integral Equations. Birkhäuser (1983)
27. H. G. Bock, K. J. Plitt: A multiple shooting algorithm for direct solution of optimal control problems. In Proc. 9th IFAC World Congress Budapest, July 2-6, Pergamon Press (1984), 242-247
28. H. G. Bock: Randwertproblemmethoden zur Parameteridentifizierung in Systemen nichtlinearer Differentialgleichungen. Bonner Mathematische Schriften 183, University of Bonn (1987)
29. H. G. Bock, E. Eich, J. P. Schlöder: Numerical solution of constrained least squares boundary value problems in differential-algebraic equations. In: K. Strehmel (ed.): Numerical Treatment of Differential Equations. Teubner (1988)
30. H. G. Bock, J. P. Schlöder, V. Schulz: Numerik großer Differentiell-Algebraischer Gleichungen — Simulation und Optimierung. In: H. Schuler (ed.): Prozeßsimulation. VCH Verlagsgesellschaft mbH (1995) 35-80
31. R. Bulirsch, E. Nerz, H. J. Pesch, O. von Stryk: Combining direct and indirect methods in optimal control: range maximization of a hang glider. In: R. Bulirsch, A. Miele, J. Stoer, K.-H. Well (eds.): Optimal Control - Calculus of Variations, Optimal Control Theory and Numerical Methods, International Series of Numerical Mathematics **111**, Birkhäuser (1993) 273-288.
32. K. E. Brenan, S. L. Campbell, L. R. Petzold: Numerical Solution of Initial-Value Problems in Differential-Algebraic Equations. SIAM (1996)
33. A. E. Bryson, Y. C. Ho: Applied Optimal Control. Ginn and Company, (1969), Rev. printing, Hemisphere (1975)
34. O. Buchauer, P. Hiltmann, M. Kiehl: Sensitivity analysis of initial-value problems with application to shooting techniques. Numerische Mathematik **67** (1994) 151-159
35. R. Bulirsch: Die Mehrzielmethode zur numerischen Lösung von nichtlinearen Randwertproblemen und Aufgaben der optimalen Steuerung. Report of the Carl-Cranz-Gesellschaft e.V., DLR, Oberpfaffenhofen (1971)
36. R. Bulirsch, D. Kraft (eds.): Computational Optimal Control. International Series in Numerical Mathematics **115** Birkhäuser-Verlag (1994)
37. J. A. Burns, B. B. King: A reduced basis approach to the design of low order feedback controllers for nonlinear continuous systems. To appear in: Journal of Vibration and Control
38. C. Büskens, M. Gerdts: Real-time optimization of DAE-systems. In: M. Groetschel, S.O. Krumke, J. Rambau (eds.): Online Optimization of Large Scale Systems: State of the Art. Springer (2001)
39. R. Callies: Habilitationsschrift. Technische Universität München, submitted
40. M. Caracotsios, W. E. Stewart: Sensitivity analysis of initial value problems with mixed ODEs and algebraic equations. Computers & Chemical Engineering **9**, 4 (1985) 359-365
41. A. L. Cauchy: Méthode générale pour la résolution systémes d'équations simultanées. Compt. rend. acad. sci. **25** (1847) 536-538

42. A. Cervantes, L. T. Biegler: Large-scale DAE optimization using a simultaneous NLP formulation. AIChE Journal **44**, 5 (1998) 1038-1050.
43. F. L. Chernousko, A. A. Luybushin: Method of successive approximations for optimal control problems (survey paper). Opt. Contr. Appl. and Meth. **3**, (1982) 101-114
44. P. Deuflhard: A modified Newton method for the solution of ill-conditioned systems of nonlinear equations with application to multiple shooting. Numer. Math. **22** (1974) 289-315
45. E. D. Dickmanns, K. H. Well: Approximate solution of optimal control problems using third order Hermite polynomial functions. Lec. Notes in Comp. Sc., Vol. 27, Springer (1975) 158-166
46. M. Diehl, H. G. Bock, D. B. Leineweber, J. P. Schlöder: Efficient direct multiple shooting in nonlinear model predictive control. In: F. Keil, W. Mackens, H. Voß, and J. Werther (eds.): Scientific Computing in Chemical Engineering II, volume 2, Springer (1999), 218-227
47. M. Diehl, I. Disli-Uslu, S. Schwarzkopf, F. Allgöwer, H. G. Bock, T. Bürner, R. Findeisen, E. D. Gilles, A. Kienle, J. P. Schlöder, E. Stein: Real-Time Optimization of Large Scale Process Models: Nonlinear Model Predictive Control of a High Purity Distillation Column. In: M. Groetschel, S.O. Krumke, J. Rambau (eds.): Online Optimization of Large Scale Systems: State of the Art. Springer (2001)
48. G. Engl, A. Kröner, T. Kronseder, O. von Stryk: Numerical simulation and optimal control of air separation plants. In: H.-J. Bungartz, F. Durst, Chr. Zenger (eds.): High Performance Scientific and Engineering Computing. Lecture Notes in Computational Science and Engineering **8**, Springer (1999) 221-231
49. M. Falcone, R. Ferretti: Discrete time high-order schemes for viscosity solutions of Hamilton-Jacobi-Bellman equations. Numerische-Mathematik **67**, 3 (1994) 315-344
50. C. A. Fletcher: Computational Galerkin Methods. Springer (1984)
51. L. Fox: Some numerical experiments with eigenvalue problems in ordinary differential equations. In: Langer R.E. (ed): Boundary Value Problems in Differential Equations. (1960)
52. S. Galán, W. F. Feehery, P. I. Barton: Parametric sensitivity functions for hybrid discrete/continuous systems. Applied Numerical Mathematics **31**, 1 (1999) 17-47
53. J. V. Gallitzendörfer, H. G. Bock: Parallel algorithms for optimization boundary value problems in DAE. In: H. Langendörfer (ed.): Praxisorientierte Parallelverarbeitung. Hanser (1994)
54. P. E. Gill, W. Murray, M. A. Saunders: SNOPT: An SQP algorithm for large-scale constrained optimization. Report NA 97-2, Department of Mathematics, University of California, San Diego (1997)
55. P. E. Gill, W. Murray, M. A. Saunders, M. H. Wright: Some theoretical properties of an augmented Lagrangian merit function. In: P. M. Pardalos (ed.): Advances in Optimization and Parallel Computing. Elsevier Science Publishers (1992) 101-128
56. P. E. Gill, W. Murray, M. A. Saunders, M. H. Wright: User's Guide for NPSOL (Version 5.0): a Fortran package for nonlinear programming. Numerical Analysis Report 98-2, Department of Mathematics, University of California, San Diego (1998)
57. N. I. M. Gould, P. L. Toint: SQP Methods for Large-Scale Nonlinear Programming. Proceedings of the 19th IFIP TC7 Conference on System Modelling and Optimization, Cambidge, England, July 12th to 16th 1999, also available as: Rutherford Appleton Laboratory Technical Report RAL-TR-1999-055
58. A. Griewank: Evaluating Derivatives: Principles and Techniques of Algorithmic Differentiation. Frontiers in Applied Mathematics **19** SIAM (2000)

59. S. P. Han: Superlinearly convergent variable-metric algorithms for general nonlinear programming problems. Math. Progr. **11** (1976) 263-282
60. R. F. Hartl, S. P. Sethi, R. G. Vickson: A survey of the Maximum Principles for optimal control problems with state constraints. SIAM Review **37**, 2 (1995) 181-218
61. A. Helbig, O. Abel, W. Marquardt: Structural concepts for optimization based control of transient processes, In: F. Allgöwer, A. Zheng (eds.): Nonlinear Predictive Control. Birkhäuser (2000) 295-312
62. A. Heim: Parameteridentifizierung in differential-algebraischen Gleichungssystemen. Diploma thesis, Department of Mathematics, Technische Universität München (1992)
63. A. Heim, O. von Stryk: Documentation of PAREST — A multiple shooting code for optimization problems in differential-algebraic equations. Report TUM-M9616, Mathematisches Institut, Technische Universität München (1996)
64. M. A. Henson: Nonlinear model predictive control: Current status and future directions. Comp. Chem. Eng. **23** (1998) 187-202
65. M. R. Hestenes: Calculus of Variations and Optimal Control Theory. Wiley (1966)
66. P. Hiltmann: Numerische Lösung von Mehrpunkt-Randwertproblemen und Aufgaben der optimalen Steuerung mit Steuerfunktionen über endlichdimensionalen Räumen. Dissertation, Fakultät für Mathematik und Informatik, Technische Universität München (1990)
67. H. Hinsberger: Ein direktes Mehrzielverfahren zur Lösung von Optimalsteuerungsproblemen mit großen, differential-algebraischen Gleichungssystemen und Anwendungen aus der Verfahrenstechnik. Dissertation, Mathematisch-Naturwissenschaftliche Fakultät, Technische Universität Clausthal (1997)
68. R. Isaacs: Differential Games: A Mathematical Theory with Applications to Warfare and Pursuit, Control and Optimization. J. Wiley & Sons (1965)
69. D. H. Jacobson, D. H. Martin, M. Pachter, T. Geveci: Extensions of Linear-Quadratic Control Theory. Lecture Notes in Control and Inform. Sciences **27**, Springer (1980)
70. J. Kallrath, H. G. Bock, J. P. Schlöder: Least Squares Parameter Estimation in Chaotic Differential Equations. Celestial Mechanics and Dynamical Astronomy, **56** (1993)
71. R. E. Kalman: A new approach to linear filtering and prediction problems. Trans. ASME, J. Basic Engineering (1960) 35-45
72. T. Kailath: Linear systems. Prentice Hall (1980)
73. S. S. Keerthi, E. G. Gilbert: Optimal infinite-horizon feedback laws for a general class of constraint discrete-time systems: Stability and moving horizon approximations. J. Opt. Theory and Appl. **57**(2) (1988) 265-293
74. H. J. Kelley: Gradient theory of optimal flight paths. Journal of the American Rocket Society **30** (1960) 947-953
75. H. J. B. Keller: Numerical Methods for Two-Point Boundary Value problems. Waltham: Blaisdell (1968)
76. M. Kiehl: Vektorisierung der Mehrzielmethode zur Lösung von Mehrpunkt-Randwertproblemen und Aufgaben der optimalen Steuerung. PhD Thesis, Mathematisches Institut, Technische Universität München (1989)
77. M. Kiehl: Sensitivity analysis of ODEs and DAEs — theory and implementation guide. Optimization Methods and Software **10**, 6 (1999) 803-821
78. B. B. King, E. W. Sachs: Semidefinite programming techniques for reduced order systems with guaranteed stability margins. Submitted to: Computational Optimization and Applications
79. D. Kraft: On converting optimal control problems into nonlinear programming problems. In: K. Schittkowski (ed.): Computational Mathematical Programming. NATO ASI Series, Vol. **F15**, Springer (1985) 261-280.

80. D. Kraft: Algorithm 733: TOMP – Fortran modules for optimal control calculations. ACM Transactions on Mathematical Software **20**, 3 (1994) 262-281.
81. P. Krämer-Eis: Ein Mehrzielverfahren zur numerischen Berechnung optimaler Feedback-Steuerungen bei beschränkten nichtlinearen Steuerungsproblemen. Bonner Mathematische Schriften 166, University of Bonn (1985)
82. P. Krämer-Eis, H. G. Bock: Numerical Treatment of State and Control Constraints in the Computation of Feedback Laws for Nonlinear Control Problems. In: P. Deuflhard et al. (eds.): Large Scale Scientific Computing. Birkhäuser (1987), 287-306
83. A. Kröner, W. Marquardt, E. D. Gilles: Computing consistent initial conditions for differential algebraic process models. Comp. Chem. Eng., **16** (1992) 131-138
84. A. Kröner, T. Kronseder, G. Engl, O. von Stryk: Dynamic optimization for air separation plants. Proceesdings of the European Symposium on Computer Aided Process Engineering (ESCAPE-11), Kolding, Denmark, May 27-30, 2001 (2001)
85. T. Kronseder, O. von Stryk, R. Bulirsch: Towards Nonlinear Model Based Predictive Optimal Control of Large-Scale Process Models with Application to Air Separation Plants. In: M. Groetschel, S.O. Krumke, J. Rambau, (eds.): Online Optimization of Large Scale Systems: State of the Art. Springer (2001)
86. B. Kugelmann, H. J. Pesch: New general guidance method in constrained optimal control, Part 1: Numerical method. J. Optimization Theory and Applications **67**, 3 (1990) 421-435.
87. B. Kugelmann, H. J. Pesch: New general guidance method in constrained optimal control, Part 2: Application to space shuttle guidance. J. Optimization Theory and Applications **67**, 3 (1990) 437-446.
88. K. Kunisch, S. Volkwein: Control of Burger's equation by a reduced order approach using proper orthogonal decomposition. Karl-Franzens-Universität Graz, Spezialforschungsbereich F003, Bericht Nr. 138 (Sep. 1998)
89. D. B. Leineweber: Efficient reduced SQP methods for the optimization of chemical processes described by large sparse DAE models. volume 613 of Fortschr.-Ber. VDI Reihe 3, Verfahrenstechnik. VDI Verlag (1999)
90. D. B. Leineweber, H. G. Bock, J. P. Schlöder: Fast direct methods for real-time optimization of chemical processes. Proc. 15th IMACS World Congress on Scientific Computation, Modelling and Applied Mathematics Berlin, Wissenschaft- und Technik-Verlag (1997)
91. J. R. Leis, M. A. Kramer: Sensitivity analysis of systems of differential and algebraic equations. Comput. Chem. Eng. **9** (1985) 93-96
92. P. L. Lions: Generalized Solutions of Hamilton-Jacobi Equations. Pittman (1982)
93. T. Maly, L. R. Petzold: Numerical methods and software for sensitivity analysis of differential-algebraic equations. Applied Numerical Mathematics **20** (1996) 57-79
94. B. R. Maner, F. J. Doyle, B. A. Ogunnaike, R. K. Pearson: Nonlinear model predictive control of a simulated multivariable polymerization reactor using second-order Volterra models. Automatica **32**, (1996) 1285-1301
95. W. Marquardt: Nonlinear model reduction for optimization based control of transient chemical processes. Proceedings of Chemical Process Control-6, Tuscon USA (2001)
96. D. Q. Mayne: Optimization in model based control. In: J. B. Rawlings (ed.): The 4-th IFAC symposium on dynamics and control of chemical reactors, distillation columns, and batch processes. DYCORD 95 (1995) 229-242
97. D. Q. Mayne, J. B. Rawlings, C. V. Rao, P. O. M. Scokaert: Constrained model predictive control: Stability and optimality. Automatica **36** (2000) 789-814
98. E. S. Meadows, J. B. Rawlings: Topics in model predictive control. In: R. Berber (ed.): Methods of Model-Based Control. NATO-ASI Series, Kluwer Press (1995) 331-347

99. R. Mehlhorn, G. Sachs: A new tool for efficient optimization by automatic differentiation and program transparency. Optimization Methods and Software **4** (1994) 225-242
100. V. L. Mehrmann: The Autonomous Linear Quadratic Control Problem. Lecture Notes in Control and Information Sciences **163** Springer (1991)
101. H. Michalska, D. Mayne: Moving horizon observers-based control. IEEE Transactions on Automatic Control **40**(6) (1995) 995-1006
102. A. Miele: Gradient algorithms for the optimization of dynamic systems. In: C.T. Leondes (ed.): Control and Dynamic Systems **16** (1980) 1-52
103. M. Morari, J. H. Lee: Model predictive control: past, present and future. Computers and Chemical Engineering **23** (1999) 667-682
104. K. R. Muske, J. B. Rawlings: Nonlinear receding horizon state estimation. In: R. Berber (ed.): Methods of Model-Based Control. NATO-ASI Series. Kluwer Press (1995) 489-504
105. B. Naumer: Über approximative Methoden der Dynamischen Programmierung in der Optimalen Steuerung. Utz (1999)
106. J. Nocedal, S. J. Wright: Numerical optimization. Springer (1999)
107. S. J. Norquay, A. Palazoglu, J. A. Romagnoli: Application of Wiener model predictive control (WMPC) to an industrial C2-splitter. J. Process Control **9** (1999) 461-473
108. H. J. Oberle: Numerische Berechnung optimaler Steuerungen von Heizung und Kühlung für ein realistisches Sonnenhausmodell. Habilitationsschrift, Technische Universität München, Germany (1982)
109. B. A. Ogunnaike, W. H. Ray: Process Dynamics, Modelling and Control. Oxford University Press (1994)
110. N. M. C. de Oliveira, L. T. Biegler An extension of Newton-type algorithms for nonlinear process control. Automatica **31** (2) (1995) 281-286
111. R. K. Pearson, M. Pottmann: Gray-box identification of block-oriented nonlinear models. J. Process Control **10** (2000) 301-314
112. H. J. Pesch: Numerische Berechnung optimaler Flugbahnkorrekturen in Echtzeitrechnung. Ph.D. thesis, TU München (1978)
113. H. J. Pesch, R. Bulirsch: The Maximum Principle, Bellman's equation, and Caratheodory's work. J. Optimization Theory and Applications **80**, 2 (1994) 199-225
114. L. Petzold, J. B. Rosen, P. E. Gill, L. O. Jay, K. Park: Numerical optimal control of parabolic PDEs using DASOPT. In: Biegler, Coleman, Conn, Santosa (eds.): Large Scale Optimization with Applications, Part II. Springer (1997), 271-299
115. K. J. Plitt: Ein superlinear konvergentes Mehrzielverfahren zur direkten Berechnung beschränkter optimaler Steuerungen, Diplomarbeit, Bonn (1981)
116. L. S. Pontryagin, V. G. Boltyanski, R. V. Gamkrelidze, E. F. Miscenko: The Mathematical Theory of Optimal Processes. Wiley (1962)
117. M. J. D. Powell: A fast algorithm for nonlinearly constrained optimization calculations. in G.A. Watson (Hrsg.), Numerical Analysis, Dundee 1977, Lecture Notes in Mathematics **630**, Springer (1978).
118. Process Systems Enterprise Ltd: gPROMS Advanced User Guide (Release 1.8), London (2000)
119. R. Pytlak: Numerical Methods for Optimal Control Problems with State Constraints. Springer (1999)
120. C. V. Rao, J. B. Rawlings: Nonlinear Horizon State Estimation, In: F. Allgöwer, A. Zheng (eds.): Nonlinear Predictive Control. Birkhäuser (2000) 45-70
121. N. L. Ricker: Model predictive control: State of the art. In: Y. Arkun and W. H. Ray (eds.): Chemical Process Control - CPC IV, CACHE. Elsevier (1991) 271-296

122. D. Robertson, K. H. Lee, J. B. Rawlings: A moving horizon-based approach for least-squares estimation. AIChE Journal **42**(8) (1996) 2209-2223
123. E. N. Rozenvasser: General sensitivity equations of discontinuous systems. Automat. Remote Control (1967) 400-404
124. V. Schulz: Reduced SQP methods for large-scale optimal control problems in DAE with application to path planning problems for satellite mounted robots. Dissertation, Naturwiss.-Math. Gesamtfakultät, Universität Heidelberg (1996).
125. V. H. Schulz: Solving discretized optimization problems by partially reduced SQP methods. Comput. Visual. Sci. **1** (1998) 83-96
126. V. H. Schulz, H. G. Bock, and M. C. Steinbach: Exploiting invariants in the numerical solution of multipoint boundary value problems for DAEs. SIAM J. Sci. Comput. **19** (1998) 440-467
127. G. R. Sriniwas, Y. Arkun: A global solution to the nonlinear model predictive control algorithms using polynomial ARX models. Comp. Chem. Engng **21**, (1997) 431-439
128. M. C. Steinbach: Fast recursive SQP methods for large-scale optimal control problems. PhD thesis. University of Heidelberg (1995)
129. J. Stoer, R. Bulirsch: Introduction to Numerical Analysis. 2nd ed., Springer (1993)
130. H. T. Su, T. J. McAvoy: Artificial neural networks for nonlinear process identification and control. In: M. A. Henson, D. E. Seborg (eds.): Nonlinear Process Control. Prentice-Hall (1997) 371-428
131. P. Terwiesch, M. Agarwal: On-line corrections of pre-optimized input profiles for batch reactors. Comp. Chem. Eng. **18** (1994) S433-S437
132. H. Tolle: Optimization Methods. Springer (1975)
133. J. Unger, A. Kröner, W. Marquardt: Structural analysis of DAE-systems - Theory and applications. Comput. Chem. Eng. **19**, 8 (1995) 867-882
134. V. S. Vassiliadis, R. W. H. Sargent, C. C. Pantelides: Solution of a class of multistage dynamic optimization problems. 1. problems without path constraints. Ind. Eng. Chem. Res. **33** (1994) 2111-2122
135. V. S. Vassiliadis, R. W. H. Sargent, C. C. Pantelides: Solution of a class of multistage dynamic optimization problems. 2. problems with path constraints. Ind. Eng. Chem. Res. **33** (1994) 2123-2133
136. R. von Schwerin, M. J. Winckler, V. H. Schulz: Parameter estimation in discontinuous descriptor models. In: Bestle and Schiehlen (eds.): IUTAM Symposium on Optimization of Mechanical Systems. Kluwer Academic Publishers (1996) 269-276
137. O. von Stryk: Numerical solution of optimal control problems by direct collocation. In: R. Bulirsch, A. Miele, J. Stoer, K. H. Well (eds.): Optimal Control — Calculus of Variations, Optimal Control Theory and Numerical Methods. International Series of Numerical Mathematics **111** Birkhäuser (1993) 129-143
138. O. von Stryk: Numerische Lösung optimaler Steuerungsprobleme: Diskretisierung, Parameteroptimierung und Berechnung der adjungierten Variablen. Fortschritt-Berichte VDI, Reihe 8, Nr. 441, VDI-Verlag (1995).
139. O. von Stryk: Numerical Hybrid Optimal Control and Related Topics. Habilitation, Department of Mathematics, Technische Universität München (2000), submitted
140. O. von Stryk, R. Bulirsch: Direct and indirect methods for trajectory optimization. Annals of Operations Research **37** (1992) 357-373
141. O. von Stryk, M. Schlemmer: Optimal control of the industrial robot Manutec r3. In: R. Bulirsch, D. Kraft (eds.): Computational Optimal Control. International Series of Numerical Mathematics **115**, Basel (1994) 367-382
142. W. Waldraff, R. King, E. D. Gilles: Optimal feeding strategies by adaptive mesh selection for fed-batch bioprocesses. Bioprocess Engineering **17** (1997) 221-227

143. P. J. Werbos: Approximate dynamic programming for real-time control and neural modeling. In: D. A. White and D. A. Sofge (eds.): Handbook of Intelligent Control. Van Nostrand Reinhold (1992) 493-559
144. M. Wellers, H. Rake: Nonlinear model predictive control based on stable Wiener and Hammerstein models. In: F. Allgöwer, A. Zheng (eds.): Nonlinear Predictive Control. Birkhäuser (2000) 357-368

Multiscale Concepts for Moving Horizon Optimization

Thomas Binder[1], Luise Blank[2], Wolfgang Dahmen[2], and Wolfgang Marquardt[1]

[1] Lehrstuhl für Prozesstechnik, Rheinisch-Westfälische Technische Hochschule Aachen, Germany
[2] Institut für Geometrie und Praktische Mathematik, Rheinisch-Westfälische Technische Hochschule Aachen, Germany

Abstract In chemical engineering complex dynamic optimization problems formulated on moving horizons have to be solved on-line. In this work, we present a multiscale approach based on wavelets where a hierarchy of successively, adaptively refined problems are constructed. They are solved in the framework of nested iteration as long as the real-time restrictions are fulfilled. To avoid repeated calculations previously gained information is extensively exploited on all levels of the solver when progressing to the next finer discretization and/or to the moved horizon. Moreover, each discrete problem has to be solved only with an accuracy comparable to the current approximation error. Hence, we suggest the use of an iterative solver also for the arising systems of linear equations. To facilitate fast data transfer the necessary signal processing of measurements and setpoint trajectories is organized in the same framework as the treatment of the optimization problems. Moreover, since the original estimation problem is potentially ill-posed we apply the multiscale approach to determine a suitable regularization without a priori knowledge of the noise level.

1 Introduction

The numerical solution of dynamic optimization problems is quite challenging for large-scale problems. The challenge becomes even more severe when real-time applications formulated on moving horizons such as model predictive control (MPC) or receding horizon estimation (RHE) are envisaged, since the response time where a solution has to be prompted is fixed. For a more detailed introduction into moving horizon optimization we refer to [2] in this book and references therein. Both, the regulator problem and the estimation problem have to be solved repetitively within a fixed time span ΔT which is dictated by the dynamics of the process. An acceptable algorithm has to prompt the optimal values within ΔT since proper process operation cannot be guaranteed otherwise.

The majority of known implementations of dynamic optimization algorithms for large-scale systems are based on the direct approach [2] since it does not require any analytic expression for the necessary optimality conditions which can be quite cumbersome to determine especially in the presence of inequality constraints. For this class of methods the dynamic optimization problem is transformed into a finite-dimensional nonlinear programming problem (NLP) by either discretizing the states and control profiles (simultaneous approach) [18] or by parameterizing the control variables only (sequential approach) [38]. It is common practice to approximate the time-variant quantities on a discretization mesh of fixed resolution.

This applies to sequential as well as to simultaneous approaches, although in the sequential approach a stepsize adaption is used in the integrator to control the error in the state equations. The particular discretization mesh is typically chosen based on a rather conservative estimate of the computing time needed to solve the optimization problem. In recent years sophisticated algorithms have been developed which are capable to solve dynamic optimization problems reliably and fast, see, e.g., [12,14,38], but they employ a fixed mesh and are therefore of given complexity. However, due to the nonlinear nature of the problem it is impossible to estimate exactly the necessary computation time. Hence, either a certain fraction of the available time span remains unused or the algorithm may even fail to prompt the solution in due time. Furthermore, an accurate estimate of the available computation time is hindered by competing software processes which run simultaneously on multitasking process control systems. As a consequence the available computation time is a priorly unknown and ΔT only gives an upper limit. This suggests a new view on *real time requirements* in on-line computations: *Provide an approximate solution at any time in* ΔT *with increasing approximation quality.*

The conceptual framework we want to present now hinges on this real time requirement. Instead of keeping the degrees of freedom fixed we propose the solution of a suitable *hierarchy of optimization problems* of increasing resolution using the simultaneous approach. We start with a very coarse approximation of states and control profiles in the optimization problem. Thus already after a hopefully very short period of time at least the coarsest approximate solution can serve as a minimal response. During the remaining time this initial solution is then *successively refined* by an *adaptive strategy* so that the full available time span ΔT is exploited in an optimal way. Moreover, for any discretization level the corresponding discrete problem has to be solved only with an accuracy that is comparable to the corresponding discretization error. Such a concept can only offer significant advantages over simply using a hierarchy of conventional discretizations if the work needed to compute an approximate solution on a coarser level need not be repeated when progressing to the next finer approximation. Hence, in the very spirit of classical *nested iteration* [13], we exploit the current approximation as initial guess and reduce the current error only by a fixed factor when progressing to the next discretization level. An extension of the refinement concept to the sequential solution approach is given in [11].

Figure 1 sketches our approach tailored to the receding horizon estimation problem. The discrete measurements are processed by denoising, compression and data fitting schemes to produce functions. This step provides also an initial mesh, respectively a first set of basis functions Λ^0, for discretizing the optimization problem. A NLP solver determines a solution of the optimization problem on the coarse initial mesh. Then, if time permits, this solution is updated in an adaptive way. The coarse mesh solution of the optimization problem as well as the potentially denoised inputs function as indicators for the new, refined mesh, respectively for the new basis index set $\Lambda^{\ell+1}$ where ℓ denotes the refinement counter. The optimization problem is resolved on this refined mesh with additional input information and by exploiting any available information from previous stages. As long as time permits this

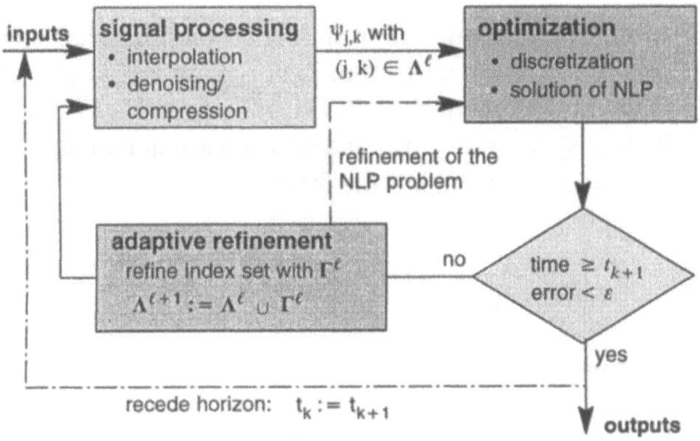

Figure 1. Receding Horizon Estimation

refinement procedure is repeated, until the time horizon is moved eventually after an elapsed time of ΔT. The subsequent calculations exploit again the information of the previous time horizon.

A *unified framework* for signal (input) processing and optimization is to facilitate a fast data exchange between these two conceptual blocks. We propose to realize the above concept with the aid of *wavelet methods* for the following reasons. Firstly, wavelets constitute a very efficient and established tool for denoising and compression in signal processing [15, 16, 24]. Secondly, a multiscale representation of the problem variables in the numerical solution avoids a change of bases in the updating procedure. Moreover, wavelet properties can be exploited for example for preconditioning and for adaptive refinement of the solution. For the reader who is unfamiliar with wavelets we include in Section 3 a short introduction to some of their basic properties and indicate their significance in the present context. For a more detailed treatment of wavelets we refer to [16, 17, 19, 21].

Before we proceed with the details of the refinement concept we introduce briefly a general problem formulation of receding horizon optimization problems. More details with an emphasis on state estimation can be found in [7, 8]. The dynamic behavior of the plant is often modeled by a system of differential-algebraic equations which, together with bounds on selected variables, form the constraints of the optimization. However, this work is only based on models where the state equations are described by ordinary differential equations. For a first step towards an extension to differential-algebraic equations we refer to [29]. The control functions are denoted by $\mathbf{u}$ and the parameters by $\mathbf{p}$. For *state estimation* both, $\mathbf{u}$ and $\mathbf{p}$ are given. The goal is to estimate the output functions $\mathbf{y}$, which correspond to the signals, and the state functions $\mathbf{x}$ on the receding fixed time interval $[t_{k-m}, t_k]$. Here t_k denotes the current time. The underlying measurements are typically discrete and noisy. So they have to be transformed into denoised functions $\mathbf{z}$. Additive model correction terms are introduced into the model equations as functions $\mathbf{v}$ and $\mathbf{w}$ which have to be estimated as well. The introduction of model correction terms actually may enhance ill-posedness as discussed in Section 2. For given $z \in (L_2)^{n_y}$, $p \in \mathbb{R}^{n_p}$ and $u \in (L_2)^{n_u}$ and unknown functions $x \in (H^1)^{n_x}$, $y \in (L_2)^{n_y}$, $v \in (L_2)^{n_v}$ and $w \in (L_2)^{n_w}$ with $n_w \leq n_x, n_v \leq n_y$, and time-invariant indicator matrices W, V the resulting optimization problem has the form

$$\min_{x,y,v,w} \int_{t_{k-m}}^{t_k} \{(y-z)^T Q(y-z)\} \, d\tau \tag{1}$$

subject to the constraints

$$\dot{x} - f(x, u, p) - Ww = 0, \tag{2}$$

$$y - g(x, u, p) - Vv = 0, \tag{3}$$

$$l_x(t) \leq x(t) \leq s_x(t), \tag{4}$$

$$l_y(t) \leq y(t) \leq s_y(t), \tag{5}$$

$$t \in [t_{k-m}, t_k].$$

The size of the problem is dominated by the number of state functions n_x.

In the sequel, we focus on the estimation problem only, but want to emphasize the similarity to general dynamic optimization problems such as the *model predictive control* problem. There, typically the parameters p and the initial conditions $x_0 = x(t_k)$ are known and the controls u are to be determined, while model correction terms are not present. The signals z represent in this case the given reference trajectories to be tracked by control. A very similar optimization problem to (1)–(5) is to be solved to determine the controls u.

2 Problem Regularization

The goal of the problem formulation (1)–(5) is to provide good estimates for x, y, v, w from noisy measurements. If no v and no inequalities are present a minimal prerequisite for a unique solution is a sufficient number n_y of measured model outputs. Moreover, observability of the model has to be guaranteed, i.e., given y, v, w, u, p there has to exist a unique solution x of the model equations (2)–(3). For a general discussion of necessary and sufficient conditions concerning the invertibility of the process model we refer the reader to [28, 36]. Even if a unique optimum exists, the inverse problem (1)–(5) to determine from given measurements z the quantities x, y, v, w might not be well-posed in the sense of Hadamard [25,30] with respect to continuity in L_2. In fact, a small perturbation of z with respect to the L_2-norm with high frequency oscillations would force y to oscillate as well. In view of (2), (3) this may cause x to have large derivatives which gives rise to arbitrarily large variation of w in L_2 [7]. Consequently, the solution operator $T : (L_2)^{n_y} \longrightarrow (H^1)^{n_x} \times (L_2)^{n_y+n_v+n_w}$ induced by (1)–(5) may not be continuous.

Regularization can be used to guarantee continuity as well as uniqueness. For example, regularization of Tikhonov type includes v and w as quadratic terms in the cost functional [37]. Then, the extended problem formulation

$$\min_{x,y,v,w} \int_{t_{k-m}}^{t_k} \left\{ (y-z)^T Q (y-z) + v^T R_v v + w^T R_w w \right\} \, d\tau \tag{6}$$

with constraints (2)–(5) is well-posed. Here, the regularization parameters R_v, R_w are time-invariant penalty matrices. Nevertheless, more general weights like, e.g., time dependent operators are possible, too. Here, the weights are interpreted in a deterministic sense and do not depend on statistical assumptions. Obviously, an increase of the weights improves stability of the estimates but causes also a growing approximation error in case of non-vanishing v, w due to bias. Regularization can also be achieved by projection of (1)–(5) into finite dimensional subspaces, e.g., using a Petrov-Galerkin scheme [34]. No regularization parameter need to be present explicitly. However, there is hidden regularization introduced by the associated discretization. The data error increases when refining the discretization (weak regularization) while a too coarse problem discretization (strong regularization) leads to large regularization errors [25, 30]. Of course, regularization by projection might be combined with direct regularization methods based on penalty terms such as in (6).

Immediately the question arises how to choose the regularization parameters in order to achieve a good compromise between data and regularization error. Optimal parameter selection strategies are available if the noise level of the measurements and the smoothness of the exact solution is known [32]. However, since this information is not available for most practical problems we examined in [7] a strategy, namely the L-curve criterion [27], that does not require this type of knowledge. The approximately best compromise is determined by relating the residual norms under a systematic variation of parameters to a (semi)-norm of the approximation itself. The procedure seems to be well suited for a low number of corrections w. However, the problem of quantitatively assessing the best compromise becomes increasingly difficult for a growing number of unknown functions w.

In [7] we explore the performance of regularization by projection of (1)–(5) for a simple linear model. The presented refinement concept developed in the context of real-time estimation is here successfully applied to automatically construct sequences of discretization meshes based on upgrades of the approximation spaces. Using the L-curve criterion the refinement process is stopped when a good compromise between data and regularization errors in the estimates is accomplished. Uniform as well as non-uniform problem adapted upgrades of the approximation spaces are employed using techniques outlined in more detail in Section 4. In that particular example it turned out, that non-uniform approximations lead to estimates whose quality cannot be achieved by employing only a uniform mesh. The analysis of ill-posedness presented in [7] can be also extended to problems with nonlinear models.

3 Wavelets

Apart from boundary adaptation wavelet functions $\psi_{j,k}$ are obtained by dilation and translation of a suitable *mother wavelet* ψ, i.e., $\psi_{j,k} = 2^{j/2}\psi(2^j \cdot -k)$, where the normalizing factor $2^{j/2}$ keeps the L_2-norm of the $\psi_{j,k}$ independent of j, k. Thus when ψ is chosen to have bounded support, as it will always be the case in our applications, one has $\mathrm{diam}\,\mathrm{supp}\,\psi_{j,k} \sim 2^{-j}$, $j \to \infty$. The key feature is that for suitable ψ the collection $\Psi = \{\psi_{j,k}\}$ forms a *Riesz basis* of L_2, i.e., every $f \in L_2$ has a unique expansion

$$f = \Psi^T \mathbf{d}_f = \sum_{j,k} d_{j,k}\psi_{j,k}$$

and the norm equivalence

$$c_1 \parallel \mathbf{d}_f \parallel_{\ell_2} \leq \parallel f \parallel_{L_2} \leq c_2 \parallel \mathbf{d}_f \parallel_{\ell_2} \tag{7}$$

holds for some positive constants c_1, c_2 independent of f. In the special case when the $\psi_{j,k}$ form an orthonormal basis, equality holds between coefficient and function norm. In general, (7) implies the existence of a *dual basis* $\tilde{\Psi}$ in L_2 which is *biorthogonal* to Ψ, i.e.

$$\langle\langle \Psi, \tilde{\Psi} \rangle\rangle = \mathbf{I} \tag{8}$$

where we use the shorthand notation $\langle\langle \Psi, \tilde{\Psi} \rangle\rangle = (\langle \psi_{j,k}, \psi_{i,l} \rangle)_{(j,k),(i,l)}$ and $\langle \cdot, \cdot \rangle$ is the standard L_2 inner product. Thus the wavelet coefficients $d_{j,k}$ are given by $\langle f, \tilde{\psi}_{j,k} \rangle$.

Clearly, (7) means that small perturbations of the expansion coefficients $\mathbf{d}_f$ cause only small changes of f in the L_2-norm and vice versa. In particular, retaining only the first N largest coefficients in modulus provides – up to a uniform constant – the best approximation to f that can be composed from any selection of N basis functions. This is the basis for nonlinear compression techniques (*best* N-*term approximation*). Accordingly the objective of *adaptive* approximation is to successively track the most significant coefficients of the function to be determined.

Furthermore, depending on the smoothness of the employed wavelets also norm equivalences with respect to other function spaces such as Sobolev and Besov spaces hold with appropriate scale dependent weights on the wavelet coefficients. They are essential for signal analysis (see Section 5) and for the preconditioning of discretized differential and integral operators (see Section 6.4).

The starting point for the construction of wavelet bases is usually a so-called *multiresolution sequence* of nested spaces $S_{j_0} \subset \cdots \subset S_j \subset S_{j+1} \subset \ldots$ whose union is dense in L_2. Here j_0 stands for some coarsest discretization level. The spaces S_j are spanned by scaling functions, $S_j = \mathrm{span}\{\varphi_{j,k} : k \in \mathcal{I}_j\}$, which are obtained by a suitable, compactly supported function φ. Due to the compact support of φ we can view 2^{-j} as the uniform mesh size of S_j. Given a suitable dual sequence $\tilde{S}_j$, a successive decomposition of S_j leads to the multiscale splitting $S_J = S_{j_0} \bigoplus_{j=j_0}^{J-1} W_j$ where $W_j = \mathrm{span}\{\psi_{j,k} : k \in \mathcal{I}'_j\} \perp \tilde{S}_j$. Analogous splittings for the spaces $\tilde{S}_j$ yield complement spaces $\tilde{W}_j \perp S_j$. The bases Ψ and $\tilde{\Psi}$ for the spaces W_j respectively $\tilde{W}_j$ satisfy (8).

All properties above can be realized when choosing φ as the cardinal B-spline of order m. Moreover, one can construct for any $\tilde{m} \in \mathbb{N}$, $\tilde{m} \geq m$, $m+\tilde{m}$ even, a compactly supported dual scaling function $\tilde{\varphi}$ such that the dual multiresolution spaces $\tilde{S}_j$ contain all polynomials up to order $\tilde{m}$. As a consequence of biorthogonality, one has the *moment conditions*

$$\langle (\cdot)^r, \psi \rangle = 0, \qquad r = 0, \dots, \tilde{m} - 1. \tag{9}$$

Likewise one has $\langle (\cdot)^r, \tilde{\psi} \rangle = 0$ for $r = 0, \dots, m-1$. Therefore, smooth functions have wavelet expansions with rapidly decaying wavelet coefficients so that levelwise truncation of wavelet expansions provide good approximations. However, as soon as sharp transitions or even singularities occur different nonuniform selections of basis functions are expected to provide more economical approximations. In fact, combining the above statements with the locality of wavelets, large wavelet coefficients reflect a large local change of the function.

4 Discretization: Wavelets as Trial and Test Functions

For the discretization of the problem on a finite time horizon $[t_{k-m}, t_k]$ we scale the horizon to $[0, 1]$ and formulate the equality constraints (2), (3) in a weak sense. We obtain the equality constraints:

$$\langle \dot{x} - f(x, u, p) - Ww, v_1 \rangle = 0 \qquad \text{for all } v_1 \in L_2^{n_x} \tag{10}$$

$$\langle y - g(x, u, p) - Vv, v_2 \rangle = 0 \qquad \text{for all } v_2 \in L_2^{n_y}. \tag{11}$$

Discretization is then given by the representation of each function with respect to an appropriately chosen wavelet basis Ψ, i.e., we represent x as $x = \Psi_{\Lambda_x}^T d_x$, y as $y = \Psi_{\Lambda_y}^T d_y$ etc.. We obtain an equivalent, *infinite dimensional* but *discretized* optimization problem for the wavelet coefficients $\zeta_\Lambda = (d_x^T, d_y^T, d_v^T, d_w^T)^T$:

$$\min_{\zeta_\Lambda} \; \zeta_\Lambda^T H_\Lambda \zeta_\Lambda + b_\Lambda^T \zeta_\Lambda \tag{12}$$

$$\text{s.t.} \qquad F_\Lambda(\zeta_\Lambda) = 0 \tag{13}$$

$$n_\Lambda(\zeta_\Lambda) \leq 0. \tag{14}$$

In our particular case we use piecewise linear, continuous wavelets, (i.e the scaling function is a B-spline of order $m = 2$ fulfilling the moment conditions with $\tilde{m} = 2$), as trial functions for x, y and v and for the test functions v_2, while we employ piecewise constant wavelets ($m = \tilde{m} = 1$) for w and the test functions v_1. This corresponds to the minimal regularity for a conforming discretization.

While the above formulation yields still the exact solution, for the numerical treatment we have to choose finite dimensional approximations. Hence, for each function involved we have to choose a finite set of basis functions, described by their index set Λ_x etc.. As mentioned before, we will start with some initial possibly small index sets Λ_x^0 etc. and *adaptively* enlarge these index sets to $\Lambda_x^{\ell+1} = \Lambda_x^{\ell} \cup \Gamma_x$ in each refinement step. Due to the piecewise linear basis functions the inequality constraints are guaranteed to be satisfied exactly, by enforcing their validity at the corresponding mesh points.

5 SIGNAL PROCESSING

The tasks to be performed in the signal processing part are transformation of the discrete measurement samples into a continuous representation, possibly continuation of missing signal parts, denoising and outlier removal, as well as data compression. Recent developments show that wavelets are very well suited for dealing with these tasks [15, 22–24]. Hence, we restrict ourselves to mentioning only the basic properties of wavelets employed for signal analysis, namely, the norm equivalences with respect to Besov spaces and the vanishing polynomial moments mentioned in Section 3. They give rise to nonlinear approximation techniques based on efficient threshold and/or shrinkage algorithms. However, the choice of the involved parameters depends typically on statistical model assumptions, such as white Gaussian noise with a given noise level. To our knowledge, theoretical results under less stringent assumptions do not exist. The current state of our algorithm is based on data fitting on dyadic meshes $\{2^{-J}k\}$ and a shrinkage algorithm based on the results of [15]. The data are considered as scaling function coefficients of a correspondingly high level of resolution. To this data format one can then apply the wavelet transform. In case of nonuniform sampling rates a more sophisticated fitting procedure is needed. Considering a single function z_i and omitting the index i, we obtain the wavelet representation $z^\delta = \Psi^T \mathbf{d}_z^\delta$. Corresponding to the trial functions for the outputs y, we choose piecewise linear, continuous wavelets for this purpose. Then, to obtain the denoised signal functions $z = \Psi^T \mathbf{d}_z$ we apply the shrinkage algorithm (omitting the index z) $d_{j,k} := d_{j,k}^\delta - \mathrm{sign}(d_{j,k}^\delta)\,\varepsilon$ for $|d_{j,k}^\delta| \geq \varepsilon$ and $d_{j,k} := 0$ otherwise, where ε depends only on the noise level. If little or nothing is known about the noise, ε has to be tuned in a rather heuristic manner, usually depending on the magnitude of the wavelet coefficient, the location k and the scale j. For a more detailed description we refer the reader to [10].

The representation of the measurements in terms of wavelets is fully compatible with our multiscale discretization of the constrained optimization problem. In particular, since the best N-term wavelet approximation is essentially determined by the first N largest (in modulus) wavelet coefficients, the initial index set Λ_0 for discretizing the optimization problem is easily identified. Only these coefficients will be used in the first step. In further refinement steps of the optimization problem additional wavelet coefficients of z will be processed. For the refinement of the employed basis index set, which is based on the measurements as well as on the current approximation of the estimates, we refer to Section 6.1.

6 The Optimization Problem on a Fixed Horizon

To realize the overall concept of multiscale moving horizon optimization under the particular real time requirement, we encounter many new problems on each level of the solution process even for an optimization problem on a fixed horizon. The sketch in Figure 2 indicates the main conceptual blocks of our approach on a fixed horizon, respectively its current state of development.

Dynamic optimization problem
adaptive refinement employing
- sensitivity of the cost functional with respect to $\boldsymbol{w}$ and $\boldsymbol{v}$
- local error estimator for the differential equations

> **NLP**
> *treatment of the nonlinearities*
> - general purpose algorithm based on a SQP-method
> - efficient evaluation of $\langle f(\Psi_\Lambda^T \mathbf{d_x}), \psi_{j,k}\rangle$ for given $\mathbf{d_x}$
>
> > **QP**
> > *treatment of the inequality constraints*
> > - interior-point methods for the first approximations
> > - active-set methods with initialization of the active sets for the refinement steps
> >
> > > **Linear system of equations**
> > > *iterative solver (simplified Uzawa)*
> > > - employing previous solution as initial value
> > > - using adapted Jacobi preconditioner motivated by wavelet theory

Figure 2. Outline of the conceptual blocks on a fixed horizon

In the subsequent subsections we shall briefly describe the main ingredients listed in Figure 2. The outermost level corresponds to the discretization of the whole optimization problem and the adaptive refinement. The goal is to spend minimal computational effort for realizing a fixed decay rate of the current error in each refinement step. The next subsection will briefly present our algorithm and some numerical results. Then, the treatment of nonlinearities will be presented. The optimization problem arising on each level of the resolution hierarchy is currently solved by a general purpose algorithm based on a SQP-method. Some algorithmic details are given in 6.2. Ultimately, the general purpose SQP-solver is to be replaced by a fully problem adapted scheme. A core ingredient is the treatment of the corresponding linearized problems. Here, the outer loop is concerned with the treatment of the inequality constraints. We address the question whether interior-point methods or active-set methods can exploit the information of the previous refinement step in a more efficient way. Results are given in Section 6.3. Finally, the solution process for the arising linear systems is outlined in 6.4. The computational cost for a fixed error reduction should remain proportional to the current size of the problem. This immediately suggests the idea to use an iterative solver rather than a direct solver. The current state of development is briefly described. Also preconditioning and stopping criteria are discussed in the context of nested iteration.

6.1 Adaptive Refinement

An adaptive discretization strategy is essential for an efficient treatment of the optimal control problem. In our refinement approach we start with a coarse approximation with relatively few trial functions collected in Λ^0. Then, based on the information of the previously computed approximation $\mathbf{d}_\Lambda^{*\ell}$ at refinement step ℓ an improved index set $\Lambda^{\ell+1}$ is generated and (12)–(14) is resolved. Algorithmic details on the adaptive refinement strategy are given in [5, 9]. Therefore, we present here only the underlying ideas.

The *refinement of the known quantities* $\mathbf{z}$ *and* $\mathbf{u}$ is only discussed for a single function z_i, omitting the index i for convenience, since everything applies to any component in the same way. We are interested in compressed approximations of the measurements z_Λ^ℓ satisfying

$$\| z - z_\Lambda^\ell \|_{L_2} \leq \varepsilon_\ell' \| z \|_{L_2} \tag{15}$$

with given tolerance ε_ℓ'. As mentioned in Section 3 the N largest wavelet coefficients of z give up to a fixed constant factor the best N-term approximation. Hence, given the wavelet coefficients $\mathbf{d}_z$ of the full but finite expansion $z = \mathbf{\Psi}_\Lambda^T \mathbf{d}_z$, we simply have to neglect the elements of $\mathbf{d}_z$ whose moduli are below a certain threshold ε which uniquely depends on ε_ℓ'. The remaining entries are the *significant* coefficients and their indices form the index set Λ_z^ℓ. Typically the number of significant wavelet coefficients is by far smaller than the number of discrete measurement samples. During the refinement sequence the approximation quality is increased according to $\varepsilon_{\ell+1}' < \varepsilon_\ell'$.

The *refinement of states* $\mathbf{x}$ *and outputs* $\mathbf{y}$ is based on an error analysis of $\mathbf{x}_\Lambda^\ell$ only, since errors in $\mathbf{y}_\Lambda$ are directly linked to errors in $\mathbf{x}_\Lambda$. Residual based error analysis is computationally inexpensive and gives usually a qualitatively good grasp on the error behavior. However, sharp error estimates which are needed to efficiently treat index problems arising from active state inequality constraints are difficult to obtain [29]. Therefore, the refinement is based on local error estimation where $\mathbf{x}_\Lambda^{*\ell}$ is compared to *locally* refined solutions keeping $\mathbf{w}_\Lambda^{*\ell}$ at their current optimal values. In particular, we use a local trapezoidal rule and evaluate the error at the midpoints of the current mesh determined by Λ_x^ℓ. Where the error bounds are violated, we increase the set of wavelet basis functions by adding to the current set those basis functions with neighboring indices in the time frequency plane.

In contrast to the local error estimator used for the states $\mathbf{x}$ and output functions $\mathbf{y}$, the problem formulation allows us to employ the sensitivity of the Lagrange functional for the *refinement of unknown inputs* $\mathbf{v}$ *and* $\mathbf{w}$. Hence, for an improved approximation of w_i we evaluate the gradients of the Lagrange functional with respect to *potentially new* trial functions for w_i determined by the index set $\Pi^\ell_{w_i}$. Local neighbors of the indices in $\Lambda^\ell_{w_i}$ are excellent *candidates* for $\Pi^\ell_{w_i}$ where we require $\Pi^\ell_{w_i} \cap \Lambda^\ell_{w_i} = \emptyset$. Once the gradients are computed, the trial functions associated with the larger absolute gradients are interpreted as those basis functions which have large impact on the solution and therefore should be used for refinement. The smaller ones are neglected. The same approach is applied to $\mathbf{v}$.

The approximations for $\mathbf{x}, \mathbf{y}, \mathbf{v}, \mathbf{w}$ might change during the cycles of refinement such that previously needed trial functions may become obsolete. The *elimination of unnecessary trial functions* is based on a compression technique as outlined before. Basis functions corresponding to significant wavelet coefficients are kept while the small ones are discarded.

Finally the different index sets for the various functions have to be combined to take their close interactions into account.

In summary, the overall *adaptive refinement algorithm* proceeds as follows denoting by Γ^ℓ the set of newly added indices:

Algorithm 1 (Adaptive Refinement Algorithm).

1: Solve problem (12)–(14) with the index set Λ^ℓ.
2: Refine $\mathbf{z}, \mathbf{u}$ by *thresholding* $\rightarrow \Gamma^\ell_z, \Gamma^\ell_u$.
3: Apply a *local error* estimator for $\mathbf{x}, \mathbf{y} \longrightarrow \Gamma^\ell_{x_i}, \Gamma^\ell_{y_i}$.
4: Use the *sensitivity* of Lagrange functional for $\mathbf{v}, \mathbf{w} \longrightarrow \Gamma^\ell_{v_i}, \Gamma^\ell_{w_i}$.
5: *Compress* current approximation by thresholding $\longrightarrow \tilde{\Lambda}^\ell_{x_i}$ etc.
6: Take *interactions* into account $\longrightarrow \Gamma^\ell$
7: Form $\Lambda^{\ell+1} = \tilde{\Lambda}^\ell \cup \Gamma^\ell$.
8: Set $\ell := \ell + 1$, go to 1.

Figure 3 shows for a typical example the comparison between different choices of the index sets Λ^ℓ, i.e., a uniform refinement which adds all basis functions of the next scale and an adaptive refinement which is based on the considerations above. The L_2 error is plotted versus the number of trial functions. We see, for example, that for an error 0.105 we need roughly 550 basis functions with an adaptive approach while approximately 1000 degrees of freedom arise for a uniform mesh (mesh size 2^{-10}). Obviously adaptivity outperforms uniform discretization with regard to the cost for computing an approximation for a given target accuracy.

6.2 Solving the NLP

In principle, problem (12)–(14) can be solved with any available NLP method like sequential quadratic programming (SQP) or generalized Gauss-Newton methods. The latter ones are particularly attractive when the cost functional is close to zero [35]. However, to achieve highest possible efficiency the structure of the problem formulation, i.e., the sparsity pattern of the model equations system, as well as the

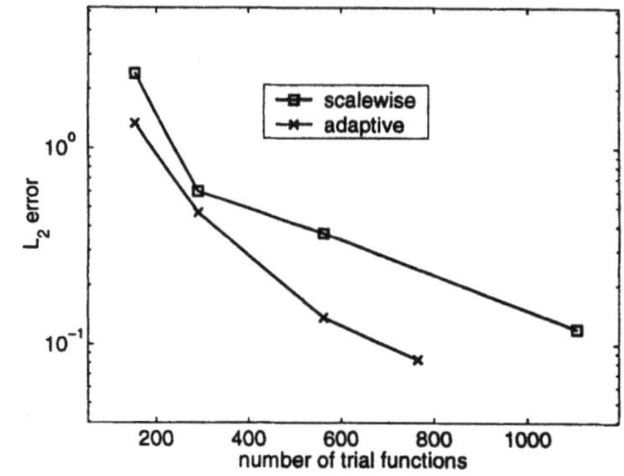

Figure 3. Scalewise versus adaptive refinement

structure of the discretization should be exploited to their full extent. Nevertheless, for nonlinear problems we currently employ only a general purpose SQP-method designed for solving large sparse NLP problems based on an active set strategy. For the specific case of linear dynamic models a tailored strategy exploiting the discretization structure is presented in more detail in Subsection 6.4.

A general purpose SQP-method such as SNOPT [26] typically requires routines to evaluate the cost functional in (12) and the residuals of the restrictions (13), (14) at a current iterate $\zeta^{\ell,i}$ as well as their first order derivatives with respect to $\zeta^{\ell,i}$. In order to provide these data, inner products of wavelet functions have to be computed. Inner products of linear terms in the equations (13), (14) do not depend on current iterates and therefore can be computed prior to the optimization run with known numerical techniques for a sufficiently large index set Λ. Then, in each refinement step ℓ those rows and columns of the precomputed and stored inner product matrix whose indices are contained in Λ^ℓ have to be retrieved only from these files.

The computation of nonlinear terms in the optimization problem is more difficult and costly since we have to evaluate inner products of the form $\langle f(x_\Lambda^{\ell,i}), \psi_{j,k}\rangle$ where $x_\Lambda^{\ell,i}$ is given in a wavelet expansion and f is a nonlinear function. Straightforward quadrature would spoil the complexity gain by adaptive refinements such that more sophisticated efficient evaluation schemes are needed. Recently, Dahmen *et al.* [20] developed a methodology to efficiently evaluate such inner products which avoids the complexity of the finest uniform mesh and requires only a computational effort proportional to the dimension of the adapted wavelet basis.

A simple but efficient warmstart of the SQP-method is obtained by reusing the solution in refinement step $\ell - 1$ as initial guess for the solver in step ℓ. Due to the hierarchical structure of the wavelet basis a change of basis is not necessary. The initial values for the unknowns $\zeta^{\ell,0}$ as well as for the Lagrange multipliers are efficiently provided by $d_{j,k}^{\ell,0} = d_{j,k}^{*\ell-1}$ for $(j,k) \in \Lambda^{\ell-1} \cap \Lambda^\ell$ and $d_{j,k}^{\ell,0} = 0$ otherwise. Additional warm start strategies like updating the Hessian matrices or their approximations may also reduce the numerical cost but have not been investigated yet for the nonlinear problem.

6.3 Solving the QP

For the treatment of the quadratic programming problems, which arise either as subproblems of the nonlinear programming problem or directly for dynamic optimization problems with linear process model and quadratic cost functional, several decisions concerning the solution strategy have to be made. Typically, a large number of algebraic inequalities is present due to the discretization of the inequality restrictions (14). These inequality restrictions have to be handled efficiently and a method is required which takes full advantage of the refinement concept.

Two common approaches based on active sets (AS) and on an interior point (IP) method have been compared. On one hand, IP methods are often applied to systems with a very large number of inequalities. They usually need a fixed number of iterations independently of the dimension of the system [40]. Hence they suggest themselves in the present context. On the other hand, AS methods need to solve smaller linear systems in each iteration step. Nevertheless, the number of iterations typically increases with the number of inequality constraints. This typically happens if one solves the optimization problem in a single run. In our refinement algorithm, though, this will be the case only in the first loop. For the next refinement we can employ the information from the previous step. In case of the IP method we can use the solution as initial guess for the refined optimization problem. However, the central idea of the IP is to drive all quantities simultaneously to the bounds at approximately the same rate [40]. Hence, even if a good initial guess is taken, some of the variables might be too close to the boundary and a centering step becomes necessary. The next iterate presumably moves away from the optimum. Therefore, the warmstart potential of IP methods seems to be limited in our context. In contrast, in case of AS methods we can initialize the active set on refinement level $\ell+1$ with the set for which the current solution ζ^ℓ is active on the refined mesh. Since the current solution should be a good approximation of the refined solution, we do not expect significant changes of the active set and, therefore, we do not expect a large increase of iteration numbers.

These considerations are confirmed by numerical experiments reported in [6, 39]. Therefore, our strategy is to use an *interior-point* method for the *first* approximate solution of the optimization problem and for all *following* refinements we use the *active-set* method *with initialization* based on the previous solution.

6.4 Solving the Linear System

Since even for nonlinear problems the same issues would arise after linearization we confine the discussion here for the sake of clarity to a linear model problem. Moreover, the application of the active-set method leads to saddle point problems of the form:

$$\mathcal{L}_\Lambda \begin{pmatrix} \Upsilon_\Lambda \\ \lambda_\Lambda \end{pmatrix} := \begin{pmatrix} \mathcal{A}_\Lambda & \mathcal{B}_\Lambda{}^T \\ \mathcal{B}_\Lambda & \mathbf{0} \end{pmatrix} \begin{pmatrix} \Upsilon_\Lambda \\ \lambda_\Lambda \end{pmatrix} = \begin{pmatrix} \mathbf{h}_\Lambda \\ \mathbf{g}_\Lambda \end{pmatrix}, \tag{16}$$

which are structurally the same as if only equality constraints are present in the optimization problem. Hence, in order to focus on the treatment of such linear systems we consider in a first step the following optimization problem

$$\min_{\mathbf{x},\mathbf{y},\mathbf{v},\mathbf{w}} \int_0^1 \{(\mathbf{y}-\mathbf{z})^T \mathbf{Q} (\mathbf{y}-\mathbf{z}) + \mathbf{v}^T \mathbf{R}_v \mathbf{v} + \mathbf{w}^T \mathbf{R}_w \mathbf{w}\}\, d\tau \tag{17}$$

$$\text{subject to} \quad \dot{\mathbf{x}} - \mathbf{A}\mathbf{x} - \mathbf{W}\mathbf{w} = \mathbf{B}\mathbf{u} \tag{18}$$

$$\mathbf{y} - \mathbf{C}\mathbf{x} - \mathbf{V}\mathbf{v} = \mathbf{0}. \tag{19}$$

However, the ideas can be extended to the nonlinear case with inequality constraints. Figure 4(a) shows the typical structure of the Karush-Kuhn-Tucker matrix $\mathcal{L}_\Lambda$ for an example of fourth order. It exhibits the block structure, where each block corresponds to entries of the regularization and model matrices. The blocks are either diagonal or have finger structure arising from wavelet scalar products.

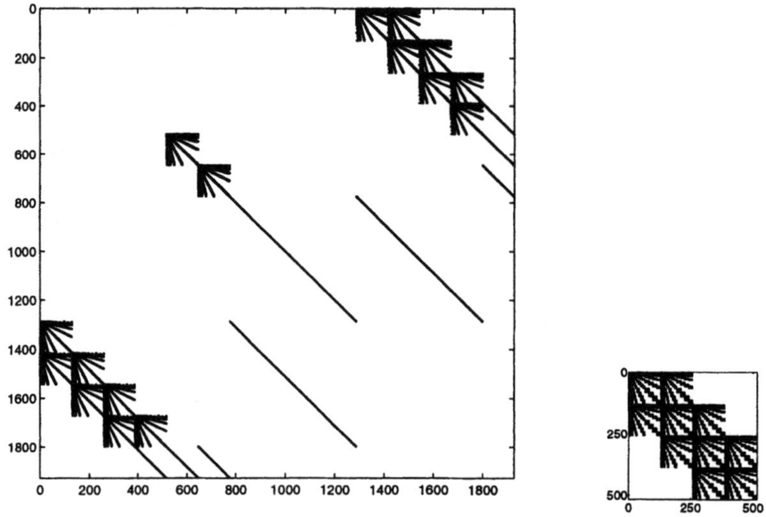

(a) KKT-matrix $\mathcal{L}_\Lambda$

(b) Schur complement $\mathcal{S}_\Lambda$

Figure 4. KKT-matrix and Schur complement

The discretization error is bounded from below by the approximation error of the functions in the adaptively chosen wavelet space. Hence, it is sufficient to solve the above system of equations only up to an error of this order. The goal is now to use an efficient solver up to an *accuracy* which *corresponds* to the *discretization error on each refinement level.* For efficiency reasons, it is important to exploit the previous approximation. Furthermore, successive error reduction is in the spirit of the overall concept. Hence, while most optimization solvers use direct methods in this context we focus on *iterative solvers.* Of course, then the discretization matrices do not have to be assembled but only their application to a vector has to be guaranteed. Due to the indefiniteness of the system one could think of using *gmres*, or, exploiting the symmetry, of using *symmlq* or *minres* [1,31]. Nevertheless, to exploit the problem structure as much as possible, we *reduce the system* and obtain a matrix $\tilde{\mathcal{L}}_\Lambda$, where we can apply the Uzawa algorithm (see [13]). This algorithm is particularly designed for saddle point problems with positive definite $\mathcal{A}_\Lambda$. Conceptionally it applies the *pcg*-method to the Schur complement and, to avoid inversion of $\mathcal{A}_\Lambda$, also to $\mathcal{A}_\Lambda$. In our case the inner application of the *pcg*-method is not necessary, since $\mathcal{A}_\Lambda^{-1}$ can actually be set up very efficiently. This means, that we apply essentially the *pcg*-method to the Schur complement $\mathcal{S}_\Lambda$ of the reduced system. For a detailed discussion of this *adapted Uzawa algorithm* we refer to [8]. In Figure 4(b) we can see the structure of the linear system after a reduction of $\mathcal{L}_\Lambda$ to the Schur complement $\mathcal{S}_\Lambda$. The problem size is reduced from a size corresponding to all functions x, y, v, w and the Lagrange parameter functions λ_1, λ_2 to the size of the discretization of x. Furthermore, $\mathcal{S}_\Lambda$ is still very sparse and has finger structure. Nevertheless, it will never be set up explicitly to avoid matrix multiplications.

As for the application of iterative methods there arise, of course, two essential questions: preconditioning and stopping criteria. These two questions will be discussed next in the context of nested iteration, which is applied to the sequence of equations

$$\tilde{\mathcal{L}}_{\Lambda^\ell} \begin{pmatrix} \tilde{\lambda}_{\Lambda^\ell} \\ \tilde{\Upsilon}_{\Lambda^\ell} \end{pmatrix} = \begin{pmatrix} \tilde{h}_{\Lambda^\ell} \\ \tilde{g}_{\Lambda^\ell} \end{pmatrix}, \quad \ell = \ell_0, \ell_0 + 1, \dots, \tag{20}$$

with increasing dimension corresponding to the refined optimization problems.

The corresponding operator on the infinite dimensional function spaces

$$\tilde{\mathcal{L}} : (L_2)^{n_y+n_w} \times (H^1)^{n_x} \longrightarrow (L_2)^{n_y+n_w} \times ((H^1)^{n_x})' \tag{21}$$

is bounded and has a bounded inverse with respect to these appropriate norm. In order to obtain well conditioned discretizations in the Euclidean metric l_2 one exploits the fact that suitably weighted l_2-norms of the wavelet coefficients are equivalent to Sobolev norms, i.e., $\{2^{-sj}\psi_{j,k}\}$ form a Riesz basis of H^s, here for $s \in \{0,1\}$. This corresponds to a symmetric diagonal scaling of the wavelet representation $\tilde{\mathcal{L}}_\Lambda$ of $\tilde{\mathcal{L}}$ with infinite index set Λ. This scaling can also be viewed as preconditioning of $\tilde{\mathcal{L}}_\Lambda$. For our particular case, this leads to a symmetric preconditioning of the Schur complement $\mathcal{S}_\Lambda$ with a diagonal matrix $M_{\Lambda^\ell}^{1/2}$ with scale dependent entries 2^{-j}. The boundedness of the preconditioned $\tilde{\mathcal{L}}_\Lambda$ and its inverse remain valid for finite index sets Λ^ℓ independently of ℓ provided that the Galerkin scheme associated with these trial spaces is stable, i.e., the trial spaces satisfy the LBB-condition [13]. Consequently, we can deduce that the condition numbers fulfill

$$\kappa_\ell := \mathrm{cond}_2(\mathbf{M}_{\Lambda^\ell}^{1/2} \boldsymbol{S}_{\Lambda^\ell} \mathbf{M}_{\Lambda^\ell}^{1/2}) \leq c \tag{22}$$

independently of ℓ. The convergence rate of the adapted Uzawa algorithm can be estimated in terms of κ_ℓ and, therefore, is also bounded independently of the refinement step.

Instead of the diagonal entries 2^{-2j} in $\mathbf{M}$ we use an *approximate Jacobi preconditioner*, i.e., an approximation of the diagonal of the Schur complement $\boldsymbol{S}_{\Lambda^\ell}$ which can be set up very efficiently. The entries reflect the necessary scale dependent scaling. In addition this preconditioner corresponds to normalizing the wavelet basis in the energy norm and usually gives better results. Numerical experiments confirm the superiority to only scaling by powers of two [8].

The *stopping criteria* for the iterative solver hinges on three properties: (i) the maximal approximation order of the discretization spaces in the energy norm giving a reduction rate α, (ii) the norm equivalences relating the energy norm to the wavelet coefficients, and (iii) the convergence order of the underlying *pcg*-method, which can be estimated in terms of the condition numbers κ_ℓ. These properties can be used to determine an upper bound for the number of iterations

$$\max_{\mathrm{it}} \approx \log(2\bar{c}/\alpha) / \left(-\log\left((\sqrt{\kappa}-1)/(\sqrt{\kappa}+1)\right)\right) \tag{23}$$

and a bound for the maximal relative residual error

$$\alpha_{\mathrm{rel}} \approx \alpha/(\bar{c}\sqrt{\kappa}) \tag{24}$$

needed to produce on each level approximate solutions with discretization error accuracy (see [8]). The constant $\bar{c}$ is given by the norm equivalences. For our particular discretization we expect at best a first order convergence in the energy norm. Hence, in terms of the number of degrees of freedom we take $\alpha = \#\Lambda^{\ell-1}/\#\Lambda^\ell$, which is $1/2$ for uniform mesh refinement. Moreover, to determine α_{rel} and $\max_{\mathrm{it}}$ an upper bound for κ has to be estimated. As a pragmatic choice we used $\alpha_{\mathrm{rel}} = 10^{-4}$ and $\max_{\mathrm{it}} = 20$ in our experiments, which was sufficient in most cases.

In spite of the accomplished asymptotic boundedness, large condition numbers may still arise due to system inherent features as for example, a poor observability measure, stiff differential equations or inadequate regularization. The effect of the choice of the regularization parameter is studied in [8] by means of numerical experiments. A rather weak regularization of the model error functions $\boldsymbol{w}$ motivated by stabilizing the original ill-posed problem of state estimation turns out to be favorable for the condition numbers of the systems considered here. Nevertheless, conclusive statements about the quantitative effect of regularization and of system inherent features on the iterative schemes can not be made, yet.

While the condition numbers provide an upper bound for the iteration numbers, the actual eigenvalue distributions give rise to considerably smaller iteration numbers. Superconvergence, known for the cg-method, can be observed for the adapted

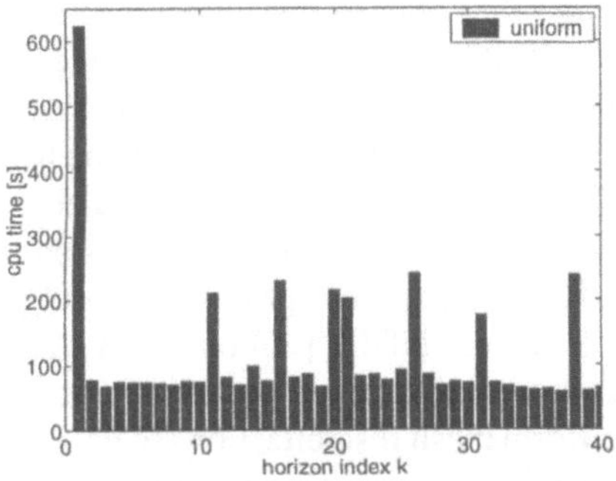

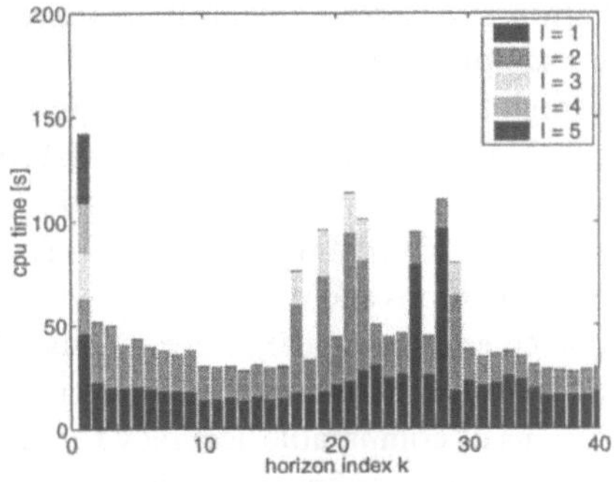

Figure 5. CPU times in each horizon. Uniform and adaptive discretization

Uzawa algorithm, too. Nevertheless, more sophisticated methods have to be developed to affect the smallest eigenvalues in a more drastic way. Currently, a Schur complement method based on the adaptive refinement algorithm is under investigation.

First numerical results for the algorithm above can be found in [8]. It turns out that depending on the system inherent features, the nested iteration process is more efficient than solving the fully discretized system directly. Perhaps more important than the expected higher efficiency of nested iteration is the fact it provides successively improved approximations at a much earlier stage. We can see also that it is not necessary to determine on each scale the exact solution. The discretization error on each refinement level is nearly reached, although the residuals are still quite large, in full agreement with the size of the condition numbers.

7 Moving Horizon Optimization

So far we have outlined the refinement approach for the optimization on a fixed horizon. In a moving horizon framework quantitative information about the solution, its structure and the corresponding mesh is available from the last horizon $[t_{k-m-1}, t_{k-1}]$. This information is to be exploited on the next horizon. The approximate solution on the $k-1$-th horizon is extrapolated in step k to the non-overlapping window interval $[t_{k-1}, t_k]$ and employed as initial guess ($\ell = 0$) for the optimization problem. For simplicity we employ linear extrapolation. Then, the approximation on the k-th horizon is expressed in its wavelet expansion. In order to ensure that the refinement procedure has time to prompt an approximate solution, signal based compression (see Section 5) of all quantities with a suitable compression rate provides an initial index set $\Lambda^{0,k}$ for the optimization problem on the k-th horizon.

We applied the refinement approach presented to a well known literature example (see, e.g., [33]). The process consists of an ideal continuous stirred tank reactor where a reversible exothermic reaction $A \rightleftharpoons R$ is carried out. The objective in the example is to move the system from an equilibrium point of low conversion and high temperature to a target equilibrium point of low temperature and significantly higher conversion. This optimal control problem is approached using an MPC framework and employing a quadratic cost functional. The horizon length is chosen to be $T = 400\ min$ and the horizon is shifted every $\Delta T = 10\ min$. Details on the model equations and parameters used can be found in [6]. The optimization problem is solved employing the active set based NLP solver SNOPT [26] on a Sun Ultra 2 (167 MHz) workstation. The numerical effort on each horizon needed to compute solutions of comparable accuracy for one uniform mesh discretization (scale 7) and for the adaptive refinement approach is shown in Figure 5. For the latter we used a mesh relaxation factor of $\eta = 0.7$. Note, that both figures are scaled differently. The numerical work with and without warm start favor the adaptive refinement approach. The gain is particularly large in the very first horizon where no solution is available for initialization. Moreover, the intermediate solutions given by the refinement approach provide backup estimates for the real time restrictions. They might also be directly applied to the real process at an earlier time.

8 Conclusions

We have introduced a multiscale concept for moving horizon optimization based on wavelets in order to meet real-time requirements. Of course, the development and validation of the presented concept is a rather complex task. The information obtained on the current level of discretization has to be exploited on the next hierarchy level. It is used to identify a suitable refinement, to possibly adjust the regularization, and to speed up the solution of the refined discretization of the estimation or control problem. So far we investigated, although on different levels of depth, the following components of the problem: regularization, input processing, optimization on a fixed horizon as well as moving the horizon. The relevant ingredients are now essentially available and the results indicate the potential of this approach.

However, we had to face in the development of these components a diversity of problems, some of which do only occur due to this particular real-time requirement and the multiscale ansatz whereas others are problem inherent to process monitoring. Not all of these obstructions could be overcome in a satisfactory way yet. Further investigations should include for example the combinations of direct and indirect regularization techniques presented, as well as level dependent Tikhonov-type regularization and additional regularization of the states. Also the influence of the length of the horizon on the estimation quality for given noisy data has to be investigated more thoroughly. The signal processing part has to be extended to non uniform sampling and colored noise. For the treatment of the optimization on a fixed horizon, for example, the used general purpose SQP-method should be replaced by a fully problem adapted method. A necessary improvement of the preconditioner for the arising linear systems is currently under investigations. It exploits the multiscale setting and is based on a Schur complement technique.

The developed ingredients still have to be combined to one single software implementation. Moreover, while the numerical examples have been mostly of model character, the optimization software with all its necessary extensions should be finally examined for more realistic industrial problems of large scale under real-time conditions. Of course, the listed future directions are not exhaustive.

REFERENCES

1. O. Axelsson: *Iterative Solution Methods.* Cambridge University Press, New York, 1994.
2. T. Binder, L. Blank, H. G. Bock, R. Bulirsch, W. Dahmen, M. Diehl, T. Kronseder, W. Marquardt, J. P. Schlöder, O. von Stryk: *Introduction to Model Based Optimization of Chemical Processes on Moving Horizons.* In M. Grötschel, S. O. Krumke, J. Rambau (editors): Online Optimization of Large Scale Systems: State of the Art, Springer-Verlag, Berlin, 2001.
3. T. Binder, L. Blank, W. Dahmen, W. Marquardt: *On the Regularization of Dynamic Data Reconciliation Problems.* RWTH Aachen, LPT-Report, LPT-2000-29, (submitted to J. of Proc. Contr.), 2000.
4. T. Binder, L. Blank, W. Dahmen, W. Marquardt: *Iterative Multiscale Methods for Process Monitoring.* RWTH Aachen, IGPM-Report 196, (to be published in: Proc. to "Fast Solution of Discretized Optimization Problems", WIAS, Berlin, Germany, 8.–12.5.2000, ISNM, Birkäuser Verlag), 2000.
5. T. Binder, L. Blank, W. Dahmen, W. Marquardt: *Grid Refinement in Multiscale Dynamic Optimization.* In S. Pierucci (editor): Proc. European Symposium on Computer Aided Process Engineering, ESCAPE-10, Florence, Italy, 7.–10.5.2000, 31–36, 2000.
6. T. Binder, L. Blank, W. Dahmen, W. Marquardt: *An Adaptive Multiscale Method for Real-Time Moving Horizon Optimization.* Proc. American Control Conference 2000, Chicago, USA, 28.–30.6.2000, Omnipress, 4234–4238, 2000.
7. T. Binder, L. Blank, W. Dahmen, W. Marquardt: *Regularization of Dynamic Data Reconciliation Problems by Projection.* In L. T. Biegler, A. Brambilla, C. Scali (editors): Proc. IFAC International Symposium on Advanced Control of Chemical Processes, ADCHEM 2000, Pisa, Italy, 14.–16.6.2000, Vol.2, 689–694, 2000.
8. T. Binder, L. Blank, W. Dahmen, W. Marquardt: *Iterative Algorithms for Multiscale Dynamic Data Reconciliation.* RWTH Aachen, IGPM-Report 186, (submitted to J. Opt. Th. Appl.), 2000.
9. T. Binder, L. Blank, W. Dahmen, W. Marquardt: *Adaptive Multiscale Approach to Real-Time Dynamic Optimization.* RWTH Aachen, LPT-Report, LPT–1999–14, 1999.
10. T. Binder, L. Blank, W. Dahmen, W. Marquardt: *Towards Multiscale Dynamic and Data Reconciliation.* In R. Berber, C. Kravaris (editors): Nonlinear Model Based Process Control, NATO ASI, Kluwer Academic Publishers, 623–665, 1998.
11. T. Binder, A. Cruse, C. Villar, W. Marquardt: *Optimization using a Wavelet based Adaptive Control Vector Parametrization Strategy.* Comput. Chem. Eng. 24, 1201–1207, 2000.
12. H. G. Bock, M. M. Diehl, D. B. Leineweber, J. P. Schlöder: *A Direct Multiple Shooting Method for Real-Time Optimization of Nonlinear DAE Processes.* In F. Allgöwer, A. Chen (editors): Nonlinear Model Predictive Control, Basel, Birkhäuser Verlag, 245–268, 2000.
13. D. Braess: *Finite Elemente.* Springer-Verlag, Berlin, 1992.
14. A. Cervantes, L. T. Biegler: *Large-scale DAE Optimization using a Simultaneous NLP Formulation.* AIChE J. 44(5), 1038–1050, 1998.

15. A. Chambolle, R. DeVore, N. Y. Lee, J. Lucier: *Nonlinear Wavelet Image Processing: Variational Problems, Compression, and Noise Removal through Wavelet Shrinkage.* IEEE Image Processing 7, 319–335, 1998.
16. C. K. Chui: *An Introduction to Wavelets.* Academic Press, Boston, 1992.
17. A. Cohen, I. Daubechies, J. C. Feauveau: *Biorthogonal Bases of Compactly Supported Wavelets.* Comm. Pure and Appl. Math. 45, 485–560, 1992.
18. J. E. Cuthrell, L. T. Biegler: *On the Optimization of Differential-Algebraic Process Systems.* AIChE J. 33(8), 1257–1270, 1987.
19. W. Dahmen: *Wavelet and Multiscale Methods for Operator Equations.* Acta Numerica, 55–228, 1997.
20. W. Dahmen, R. Schneider, Y. Xu: *Nonlinear Functionals of Wavelet Expansions – Adaptive Reconstruction and Fast Evaluation.* Numer. Math. 86(1), 49–101, 2000.
21. I. Daubechies: *Ten Lectures on Wavelets.* Society for Industrial and Applied Math., Philadelphia, 1992.
22. R. A. DeVore, B. Jawerth, V. A. Popov: *Compression of Wavelet Decompositions.* Amer. J. Math. 114, 719–746, 1992.
23. D. Donoho, I. Johnstone: *Ideal Spatial Adaption by Wavelet Shrinkage*, Biometrika 81, 425–455, 1994.
24. D. Donoho, I. Johnstone, G. Kerkyacharian, D. Picard: *Wavelet Shrinkage: Asymptopia?* J. Roy. Statist. Assoc. 90, 301–369, 1995.
25. H. W. Engl, M. Hanke, A. Neubauer: *Regularization of Inverse Problems* Kluwer, Dordrecht, The Netherlands, 1996.
26. P. E. Gill, M. Murray, M. A. Saunders: *User's Guide for SNOPT 5.3: a Fortran Package for Large-Scale Nonlinear Programming.* Internal report, Department of Operations Research, Stanford University, 1997.
27. C. Hansen: *Rank-Deficient and Discrete Ill-posed Problems: Numerical Aspects of Linear Inversion.* SIAM, Philadelphia, 1998.
28. R. M. Hirschhorn: *Invertability of Multivariable Nonlinear Control Systems.* IEEE Trans. Auto. Cont., AC-24(6), 855–865, 1979.
29. J. V. Kadam: *Index Analysis and Adaptive Refinement in Multiscale Dynamic Optimization.* Master Thesis, Lehrstuhl für Prozesstechnik, RWTH Aachen, 2000.
30. A. Kirsch: *An Introduction to the Mathematical Theory of Inverse Problems.* Springer-Verlag, 1996.
31. G. Meurant: *Computer Solution of Large Linear Systems.* Elsevier Science B.V., Amsterdam, The Netherlands, 1999.
32. A. Neumaier: *Solving Ill-Conditioned and Singular Linear Systems: A Tutorial on Regularization.* SIAM Rev. 40, 636–666, 1998.
33. S. de Oliveira, M. Morari: *Contractive Model Predictive Control with Local Linearization for Nonlinear Systems.* In R. Berber, Kravaris C. (editors): Nonlinear Model Based Process Control, NATO ASI, Kluwer Academic Publishers, 403–432, 1998.
34. R. Plato, G. Vainikko: *On the Regularization of Projection Methods for Solving Ill-posed Problems.* Numer. Math. 57, 63–79, 1990.
35. K. Schittkowski: *Solving Constraint Nonlinear Least Squares Problems by a General Purpose SQP-method.* International Series of Numerical Mathematics 84, 295–309, 1988.
36. L. M. Silverman: *Inversion of Multivariable Linear Systems.* Automatica, AC–14(3), 270–276, 1969.
37. A. N. Tikhonov, A. V. Goncharsky, V. V. Stepanov, A. G. Yagola: *Numerical Methods for the Solution of Ill-posed Problems.* Kluwer Academic Publishers, New York, 1996.

38. V. S. Vassiliadis, R.W.H. Sargent, C. C. Pantelides: *Solution of a Class of Multistage Dynamic Optimization Problems. 1. Problems without Path Constraints.* Ind. Eng. Chem. Res. 33(9), 2111–2122, 1994.
39. C. Walter: *Implementation and Evaluation of Interior-Point-Methods for the Solution of Wavelet Discretized Quadratical Optimal Control Problems.* Senior project, Lehrstuhl für Prozesstechnik, RWTH Aachen, 1999.
40. S. Wright: *Primal-Dual Interior-Point Methods.* SIAM, 1997.

Real-Time Optimization for Large Scale Processes: Nonlinear Model Predictive Control of a High Purity Distillation Column

Moritz Diehl[1], Ilknur Uslu[2], Rolf Findeisen[3], Stefan Schwarzkopf[2], Frank Allgöwer[3], H. Georg Bock[1], Tobias Bürner[1], Ernst Dieter Gilles[2,4], Achim Kienle[4], Johannes P. Schlöder[1], and Erik Stein[4]

[1] Interdisziplinäres Zentrum für wissenschaftliches Rechnen, Universität Heidelberg, Germany
[2] Institut für Systemdynamik und Regelungstechnik (ISR), Universität Stuttgart, Germany
[3] Institut für Systemtheorie technischer Prozesse (IST), Universität Stuttgart, Germany
[4] Max-Planck-Institut für Dynamik komplexer technischer Systeme, Magdeburg, Germany

Abstract The purpose of this paper is an experimental proof-of-concept of the application of NMPC for large scale systems using specialized dynamic optimization strategies. For this aim we investigate the application of modern, computationally efficient NMPC schemes and real-time optimization techniques to a nontrivial process control example, namely the control of a high purity binary distillation column. All necessary steps are discussed, from formulation of a DAE model with 164 states up to the final application to the experimental apparatus. Especially an efficient real-time optimization scheme based on the direct multiple shooting method is introduced. It is characterized by an *initial value embedding* strategy, that allows to immediately respond to disturbances, and *real-time iterations*, that dovetail the optimization iterations with the real process development. Using this scheme, sampling times of 10 seconds are feasible on a standard PC. This shows that an efficient NMPC scheme based on large scale DAE models is feasible for the real-time control of a pilot scale distillation column.

Introduction

Over the last two decades linear model predictive control has emerged as a powerful and widely used control technique, especially in the process industry. Recently there is growing interest in model predictive control for *nonlinear* systems in academia and in the industrial process control community, and the properties of a variety of NMPC schemes have been investigated theoretically (see e.g., [1, 16] for a review). In addition, there has been significant progress in the area of dynamic process optimization that made on-line optimization for NMPC feasible [5, 8, 22] (compare also to the overview article on optimization on moving horizon in this book [4]), and simulation studies have shown the real-time feasibility even for large scale process models [6, 13], as considered in this paper.

The main purpose of the paper is an experimental proof-of-concept of the application of NMPC for large scale systems using specialized dynamic optimization strategies. In particular, we consider the experimental application of NMPC to a high purity binary distillation column. We want to show that NMPC can be applied to large scale chemical processes, if well suited optimization strategies are used, and that it leads to a reasonable control performance without much tuning. This is experimentally validated, after addressing the challenges of the practical realization as parameter estimation, on-line state estimation and reliable data transfer with an existing process control system. Considering the optimization based control of distillation columns, we also refer to the research article [18] which deals with the problem of probabilistic constraints.

The paper is structured as follows: In Section 1 we describe the considered distillation column as well as the used DAE model. Section 2 contains the utilized formulation of the NMPC optimization problem, and in Section 3 we present the employed real-time optimization algorithm. In Section 4 we describe the experimental setup for the closed-loop experiments, and comment on state estimation. Section 5 contains experimental closed-loop results and the observed computation times, and gives a short comparison with a conventional PI controller.

1 Distillation Column and Equations

The experimental implementation of NMPC was carried out on a pilot plant distillation column for the separation of a binary mixture of Methanol and n-Propanol. The desired product compositions are minimum 0.99 mol/mol (low boiling component) for the top product and maximum 0.01 mol/mol for the bottom product.

The column has a diameter of 0.10 m and a height of 7 m and consists of $N = 40$ bubble cap trays. The overhead vapour is totally condensed in a water cooled condenser which is open to atmosphere. The reboiler is heated electrically. Several variables are measured and monitored on-line during each experiment, such as temperatures of feed and reflux streams, at the reboiler and condenser and on each tray of the column, volumetric flow rates of feed, reflux, distillate and bottom product streams and the column pressure. Fluid dynamic stable operation of the column is checked by the pressure drop along the column for all operating conditions presented in this study. The nominal operating conditions of the plant are listed in Table 1.

The flowsheet of the distillation system is shown in Figure 1. The preheated feed stream F_{vol} enters the column at the feed tray as saturated liquid. It can be switched automatically between two feed tanks in order to introduce well defined disturbances in feed concentrations.

Process inputs available for control purposes are the heat input to the reboiler, Q, and the reflux flow rate L_{vol}. Although the main control purpose is to maintain the product purity specifications for a distillation column, product composition measurements are often expensive, unreliable and with delays. Therefore in this study temperatures T_{14} and T_{28} on trays 14 and 28 are selected as two controlled variables (cf. Sec. 2).

A distributed control system (DCS) is used for data acquisition and the basic control loops of the flow rates, the heat input, the liquid levels in the reboiler and the condenser. To implement more advanced control schemes the DCS is connected to a PC from and to which direct access from UNIX workstations is possible. The

NMPC scheme and the state estimator are implemented on UNIX workstations, i.e., the DCS is only used for data acquisition and the basic control loops.

Feed rate, F_{vol} [l/h]	14.0
Feed composition, x_F	0.32
Feed temperature, T_F [°C]	70.0
Top composition, x_{41}	0.99
Bottom composition, x_0	0.0006
Temperature tray 28 T_{28} [°C]	70.0
Temperature tray 14 T_{14} [°C]	88.0
Reflux flow, L_{vol} [l/h]	4.3
Heat input, Q [kW]	2.5
Top pressure [bar]	0.97
Reboiler holdup [l]	8.5
Condenser holdup [l]	0.17

Table 1. Nominal operating conditions

1.1 Differential Algebraic Model

Depending on the model simplifications different kinds of models can be obtained for the dynamics of the distillation column. For the predictions in the NMPC controller we use a (simple) equilibrium stage model, which is considered to capture the main features of the column dynamics. The presented nonlinear DAE model is based on the following assumptions:

- total condenser
- negligible vapor holdup
- constant molar liquid holdup
- perfect mixing
- the mixture is at equilibrium temperature
- Murphree efficiency is applied for each tray

The model consists principally of overall and component material balances and energy balance for each tray ℓ where $\ell = 1, 2, \ldots, N$ and N is the total number of trays (compare also Figure 2). For notational convenience the index $\ell = 0$ is used for the reboiler and $\ell = N + 1$ for the condenser. The following balance equations are each written for the trays, the reboiler and the condenser, in this order. Since the molar liquid holdup, n_ℓ, is constant, overall material balances become:

$$0 = L_{\ell+1} - L_\ell + V_{\ell-1} - V_\ell + F_\ell \quad (1)$$

$$0 = L_1 - L_0 - V_0 \quad (2)$$

$$0 = V_N - L_{N+1} - D \quad (3)$$

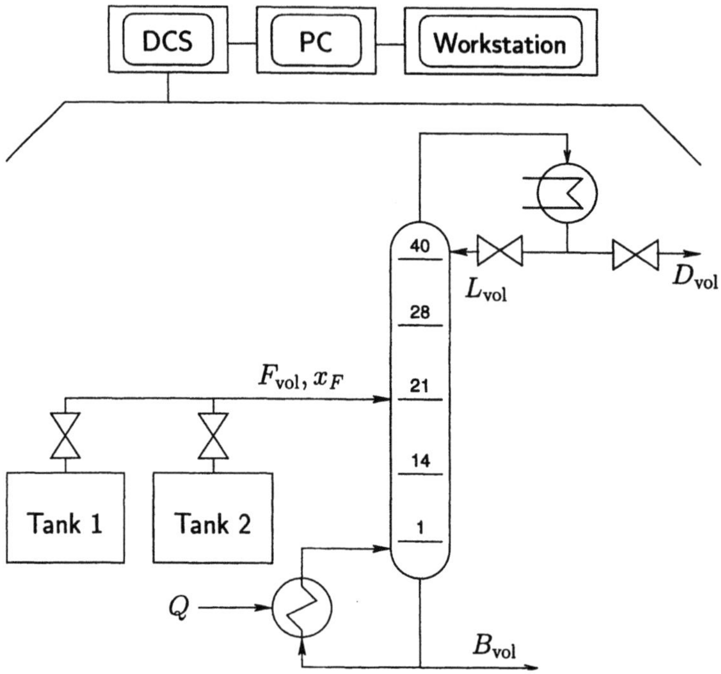

Figure 1. Flowsheet of the distillation column

Component material balance for components $k = 1, N_c - 1$:

$$n_\ell \dot{x}_{\ell,k} = L_{\ell+1} x_{\ell+1,k} - L_\ell x_{\ell,k} + V_{\ell-1} y_{\ell-1,k} - V_\ell y_{\ell,k} + F_\ell x_{F,\ell,k} \quad (4)$$

$$n_0 \dot{x}_{0,k} = L_1 x_{1,k} - L_0 x_{0,k} - V_0 y_{0,k} \quad (5)$$

$$n_{N+1} \dot{x}_{N+1,k} = V_N y_{N,k} - (L_{N+1} + D) x_{N+1,k} \quad (6)$$

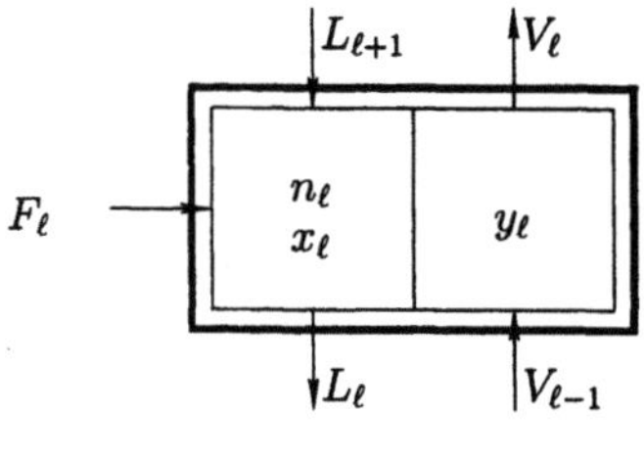

Figure 2. Distillation tray, ℓ

Energy balances:

$$
\begin{aligned}
n_\ell \dot{h}_\ell^L &= n_\ell \sum_{i=1}^{N_c-1} (h_{\ell,k}^L - h_{\ell,N_c}^L)\dot{x}_{\ell,k} + n_\ell c_{p,\ell}^L \dot{T}_\ell \\
&= L_{\ell+1} h_{\ell+1}^L - L_\ell h_\ell^L + V_{\ell-1} h_{\ell-1}^V - V_\ell h_\ell^V + F_\ell h_{F,\ell}^L
\end{aligned} \tag{7}
$$

$$
\begin{aligned}
n_0 \dot{h}_0^L &= n_0 \sum_{i=1}^{N_c-1} (h_{0,k}^L - h_{0,N_c}^L)\dot{x}_{0,k} + n_0 c_{p,0}^L \dot{T}_0 \\
&= L_1 h_1^L - L_0 h_0^L - V_0 h_0^V + Q - Q_{loss}
\end{aligned} \tag{8}
$$

In the equations above, the molar concentrations of the liquid and vapour phases are represented by $x_{\ell,k}$ and $y_{\ell,k}$; k is the index for components and N_c is the total number of components. For the process of interest in this study $N = 40$ and $N_c = 2$. The molar vapour and liquid fluxes are denoted by V_ℓ and L_ℓ, the molar feed and distillate flows by F_ℓ and D. F_ℓ becomes 0 if the tray ℓ has no feed stream. A single feed stream is introduced to the column on tray 21. The energy balance for the reboiler includes terms for the heat input, Q, and the possible heat loss, Q_{loss}. There are only $N_c - 1$ independent component balances since by definition

$$\Sigma_k x_{\ell,k} = 1 \tag{9}$$

$$\Sigma_k y_{\ell,k} = 1 \tag{10}$$

Assuming an ideal mixture, the vapour phase composition in equilibrium with the liquid phase, y_ℓ^*, is described by Rault's law:

$$y_{\ell,k}^* = \frac{P_k^s(T_\ell)x_{\ell,k}}{P_\ell} \tag{11}$$

Here $P_\ell^s(T_\ell)$ is the equilibrium vapour pressure of the pure components and determined by the Antoine equation in terms of the temperature, T_ℓ, and constant parameters A, B, and C:

$$P_k^s(T_\ell) = \exp\left(A_k - \frac{B_k}{T_\ell + C_k}\right) \tag{12}$$

It should be noted that $\dot{T}$ in the energy balances (7) and (8) is obtained by implicit differentiation of the combination of (10) - (12).

To account for the deviation from thermodynamic equilibrium due to finite mass transfer resistance the definition of Murphree efficiency, α_ℓ, is applied for each tray:

$$\alpha_\ell = \frac{y_{\ell,k} - y_{\ell-1,k}}{y_{\ell,k}^* - y_{\ell-1,k}} \quad \ell = 1, \ldots, N \tag{13}$$

To determine the (constant) pressures we assume that the condenser pressure is fixed to the outside pressure, i.e., $P_{N+1} = P_{top}$, whereas the pressures P_ℓ on the trays and the reboiler are calculated under the assumption of a constant pressure drop, ΔP_ℓ, from tray to tray, i.e.,

$$P_\ell = P_{\ell+1} + \Delta P_\ell \quad \ell = N, N-1, \ldots, 2, 1, 0. \tag{14}$$

Additionally to the molar flow rates, which cannot be measured, also the *volumetric* flow rates of the feed and reflux streams, F_{vol} and L_{vol}, are required. They can be measured and controlled by the DCS. For example L_{vol} is determined from $L_{vol} = L_{N+1}v^m$, where v^m is the molar volume of the reflux stream.

The enthalpies of the liquid and vapour phases, h_ℓ^L and h_ℓ^V, partial molar volumes, v^m, and heat capacities, $c_{p,\ell}^L$, are defined as functions of temperature and composition; the effect of pressure on h_ℓ and v^m are neglected in the model. For details of the correlations the reader is refered to [14].

Summarizing the DAE

We can subsume all system states in two vectors **x** and **z** which denote the differential and the algebraic state vectors, respectively.

The (molar) Methanol concentrations in reboiler, on the 40 trays, and in the condenser x_ℓ for $\ell = 0, 1, \ldots, N+1$ are the 42 components of the differential state vector **x**. The liquid and vapor (molar) fluxes L_ℓ and V_ℓ ($\ell = 1, 2, \ldots, N$) out of the 40 trays as well as the 42 temperatures T_ℓ ($\ell = 0, 1, 2, \ldots, N+1$) of reboiler, trays and condenser form the 122 components of the algebraic state vector $\mathbf{z} = (L_1, \ldots, L_N, V_1, \ldots, V_N, T_0, \ldots, T_{N+1})^T$ [1]. Note that those algebraic variables that can easily be eliminated (as, e.g., h_ℓ^L, y_ℓ, $P_k^s(T_\ell)$, etc.) do not count as algebraic variables in this formulation.

The two components of the control vector $\mathbf{u} = (L_{vol}, Q)^T$ are the volumetric reflux flow, L_{vol}, out of the condenser, and the heat input, Q, determining the molar vapour flux out of the reboiler. All system parameters can be subsumed in a vector **p**. The resulting model has 42 differential equations **f**, and 122 algebraic equations **g**.

We can write the DAE system, which has index one, in the following summarized form:

$$\dot{\mathbf{x}}(t) = \mathbf{f}(\mathbf{x}(t), \mathbf{z}(t), \mathbf{u}(t), \mathbf{p}) \tag{15}$$

$$\mathbf{0} = \mathbf{g}(\mathbf{x}(t), \mathbf{z}(t), \mathbf{u}(t), \mathbf{p}). \tag{16}$$

1.2 Estimation of the Model Parameters

In the actual application, the performance of NMPC depends on the quality of the model. Considering this fact, steady state and and open-loop dynamic experiments have been performed. To obtain measurements of the dynamic behaviour of the column step changes in the feed rate F_{vol} and composition x_F, the reflux rate L_{vol}, and heat input Q were performed.

Measurements of *all* temperatures $T_0, \ldots, T_{N+1}$ were taken for least squares fitting of the simulated to the observed behaviour. The assumptions for this fit are that the tray efficiencies are constant on each of the two column sections, i.e. $\alpha_1 =$

[1] The equilibrium temperature of the condenser mixture helps to define the temperature of the reflux, when not specified. Otherwise, this last algebraic variable could be eliminated without changing the dynamics.

$\cdots = \alpha_{N_F}$ and $\alpha_{N_F+1} = \cdots = \alpha_N$, that the pressure losses are equal on all trays: $\Delta P_0 = \cdots = \Delta P_N$, and that the molar tray and condenser holdups coincide: $n_1 = \cdots = n_N$. Therefore, the parameters to be adjusted to the dynamic experimental data were: Q_{loss}, α_1, α_{N_F+1}, ΔP_0, and n_1. The parameter estimation was performed with the off-line version of the multiple shooting method that is described below in the context of real-time optimization. For details we refer to [14].

2 Nonlinear Model Predictive Controller Setup

As usual in distillation control, the product purities x_B and x_D at reboiler and condenser are not controlled directly – instead, an inferential control scheme which controls the deviation of the temperatures on tray 14 and 28 from a given setpoint is used. Earlier investigations have shown that the temperatures (respectively concentrations) on these trays are much more sensitive to changes in the inputs of the system than the product concentrations [2]. It can be expected, that if these concentrations are kept constant, the product purities are safely maintained for a large range of process conditions. Since the tray temperatures correspond directly to the concentrations via the Antoine equation and the temperatures on tray 14 and 28 are measured directly, we refine the controll objective to keep these temperatures as close to their setpoint values as possible. In the following we will use $\tilde{\mathbf{T}}(\mathbf{z}) := (T_{14}, T_{28})^T$ for the controlled temperatures and $\tilde{\mathbf{T}}_{ref} := \left(T_{14}^{ref}, T_{28}^{ref}\right)^T$ for the desired setpoints.

A desired steady state $\mathbf{x}_s$, $\mathbf{z}_s$, and the corresponding control $\mathbf{u}_s$ as well as the steady state temperatures can be determined as the solution of the steady state equation

$$\begin{aligned} \mathbf{f}(\mathbf{x}_s, \mathbf{z}_s, \mathbf{u}_s, \mathbf{p}) &= \mathbf{0}, \\ \mathbf{g}(\mathbf{x}_s, \mathbf{z}_s, \mathbf{u}_s, \mathbf{p}) &= \mathbf{0}, \\ \tilde{\mathbf{T}}(\mathbf{z}_s) - \tilde{\mathbf{T}}_{ref} &= \mathbf{0},. \end{aligned} \tag{17}$$

In the following we will refer to this set of equations as $\mathbf{r}(\mathbf{x}_s, \mathbf{z}_s, \mathbf{u}_s, \mathbf{p}) = \mathbf{0}$. Notice that last equation restricts the steady state to satisfy the inferential control aim of keeping the temperatures at the fixed reference values. The necessary degrees of freedom are the two components of the steady state controls $\mathbf{u}_s$.

The open-loop objective is formulated as the integral of a least squares term $||\mathbf{l}(\mathbf{x}, \mathbf{z}, \mathbf{u}, \mathbf{u}_s, \mathbf{p})||_2^2$ with

$$\mathbf{l}(\mathbf{x}, \mathbf{z}, \mathbf{u}, \mathbf{u}_s, \mathbf{p}) := \begin{pmatrix} \tilde{\mathbf{T}}(\mathbf{z}) - \tilde{\mathbf{T}}_{ref} \\ \mathbf{R}(\mathbf{u} - \mathbf{u}_s) \end{pmatrix}. \tag{18}$$

The second component is introduced for regularization, with a small diagonal weighting matrix $\mathbf{R} = \mathrm{diag}(0.05\,°\mathrm{C\,h\,l^{-1}}, 0.05\,°\mathrm{C\,kW^{-1}})$.

To ensure nominal stability of the closed loop-system, we follow here a somewhat practical approach based on results given in [11, 15]. We append an additional prediction interval $[t_0 + T_c, t_0 + T_p]$ to the control horizon $[t_0, t_0 + T_c]$, with the controls fixed to the setpoint values $\mathbf{u}_s$ determined by (17). If the control horizon is sufficiently large, the closed-loop system will be stable. A horizon length of $T_p - T_c = 3600$ seconds proved to be sufficient in all performed experiments.

The Optimal Control Problem

The resulting optimal control problem is formulated as follows:

$$\min_{\mathbf{u}(\cdot),\mathbf{x}(\cdot),\mathbf{p}} \int_{t_0}^{t_0+T_p} \|\mathbf{l}(\mathbf{x}(t),\mathbf{z}(t),\mathbf{u}(t),\mathbf{u}_s,\mathbf{p})\|_2^2 \, dt \tag{19}$$

subject to the model DAE

$$\begin{aligned} \dot{\mathbf{x}}(t) &= \mathbf{f}(\mathbf{x}(t),\mathbf{z}(t),\mathbf{u}(t),\mathbf{p}) \\ \mathbf{0} &= \mathbf{g}(\mathbf{x}(t),\mathbf{z}(t),\mathbf{u}(t),\mathbf{p}) \end{aligned} \quad \text{for } t \in [t_0, t_0 + T_p]$$

with

$$\mathbf{u}(t) = \mathbf{u}_s \quad \text{for } t \in [t_0 + T_c, t_0 + T_p].$$

Furthermore the initial values for the differential states and values for the system parameters are given by:

$$\begin{aligned} \mathbf{x}(t_0) &= \mathbf{x}_0, \\ \mathbf{p} &= \mathbf{p}_0 = \text{constant}. \end{aligned}$$

All state and control inequality constraints are combined to:

$$\tilde{\mathbf{c}}(\mathbf{x}(t),\mathbf{z}(t),\mathbf{u}(t),\mathbf{p}) \geq \mathbf{0} \quad t \in [t_0, t_0 + T_p].$$

In particular, we require that the bottoms product and distillate streams L_0 and V_{N+1} cannot become negative, which implicitly leads to "natural" upper limits on the inputs Q and L_{vol}. Additionally, explicit lower and upper bounds for Q and L_{vol} are given.

The steady state control $\mathbf{u}_s$ is determined by the steady state equation

$$\mathbf{0} = \mathbf{r}(\mathbf{x}_s, \mathbf{z}_s, \mathbf{u}_s, \mathbf{p}).$$

3 Real-Time Solution of the NMPC Optimization Problems

In this section we will describe the newly developed *real-time iteration* scheme that we employed for the on-line computations in this study. For a more detailed description of the algorithm we refer to [14]; cf. also [6, 13, 17]

The scheme is based on the direct multiple shooting method [10, 20], which is introduced in the overview article [4] in this book (Section 5, "Introduction into Direct Solution Algorithms"), to which we explictly refer here. We stay close to the notation used in this article, with one important difference: in contrast to direct multiple shooting for ODEs, we need to account for the algebraic states in the DAE model equations.

Direct Multiple Shooting for DAE. In addition to the *differential* node values s_i^x (which are denoted s_i in [4]), we also introduce *algebraic* node values s_i^z. For simplicity we use a piecewise constant control representation with control parameters $q_0, \ldots, q_{N-1}$ on the N multiple shooting intervals. The prediction horizon $[t_0 + t_c, t_0 + T_p]$ is chosen to be the last interval, so that $t_{N-1} = t_0 + T_c$ and $t_N = t_0 + T_p$, with constant steady state control $q_{N-1} := u_s$.

On each subinterval $[t_i, t_{i+1}]$ we compute the independent trajectories $x_i(t)$ and $z_i(t)$ as the solution of a *relaxed* initial value problem:

$$\dot{x}_i(t) = f(x_i(t), z_i(t), q_i, p) \tag{24}$$

$$0 = g(x_i(t), z_i(t), q_i, p) - e^{-\beta \frac{t - t_i}{t_{i+1} - t_i}} g(s_i^x, s_i^z, q_i, p) \tag{25}$$

$$x_i(t_i) = s_i^x, \quad z_i(t_i) = s_i^z \tag{26}$$

The decaying subtrahend in (25) with $\beta > 0$ is deliberately introduced to allow an efficient DAE solution for initial values s_i^z that may violate temporarily the consistency conditions (16) (cf. Bock et al. [7], Schulz et al. [21]). Note that the i-th multiple shooting trajectories $x_i(t), z_i(t)$ on $[t_i, t_{i+1}]$ are functions of s_i^x, s_i^z, q_i, and p, so that we will write: $x_i(t; s_i^x, s_i^z, q_i, p)$, and $z_i(t; s_i^x, s_i^z, q_i, p)$.

To remove the freedom introduced by the DAE relaxation, additional equalities $g(s_i^x, s_i^z, q_i, p) = 0$ have to be added to the NLP formulation given in [4], and taking account of the additional variables x_s, z_s, u_s, p as well of the steady state constraint (17), we formulate the following structured **Nonlinear Program (NLP)** in the unknowns $\xi := (s_0^x, \ldots, s_N^x, s_0^z, \ldots, s_{N-1}^z, q_0, \ldots, q_{N-2}, x_s, z_s, u_s, p)$:

$$\min_{\xi} \sum_{i=0}^{N-1} \int_{t_i}^{t_{i+1}} \|l(x_i(t; s_i^x, s_i^z, q_i, p), z_i(t; s_i^x, s_i^z, q_i, p), q_i, u_s, p)\|_2^2 \, dt \tag{27}$$

subject to

$$s_0^x = x_0, \quad p = p_0, \tag{28}$$

$$s_{i+1}^x = x_i(t_{i+1}; s_i^x, s_i^z, q_i, p), \quad i = 0, \ldots, N-1, \tag{29}$$

$$0 = g(s_i^x, s_i^z, q_i, p), \quad i = 0, \ldots, N-1, \tag{30}$$

$$0 \leq \tilde{c}(s_i^x, s_i^z, q_i, p), \quad i = 0, \ldots, N-1, \tag{31}$$

$$0 = r(x_s, z_s, u_s, p). \tag{32}$$

3.1 Real-Time Iterations and Initial Value Embedding

In the real-time context, the above NLP (27)-(32) has to be solved several times; a crucial observation is that the optimization problems differ *only in the values* x_0 *and* p_0, which enter the problem through the *linear* constraints (28). Instead of considering a sequence of unrelated optimization problems $P(x_0, p_0)$, each of which is solved independently by an iterative SQP type method, we shift the focus towards the solution iterations themselves: the real-time iteration scheme can be considered as a sequence of Newton type iterates towards the solution of the above problem, with the particularity that the values for $\mathbf{x_0}$ and $\mathbf{p_0}$ are changed *during* the iterations.

The variant of the algorithm that we used in this study is based on the constrained Gauss-Newton method. For the current iterate ξ_k and the current values $(\mathbf{x}_0)_{k+1}$ and $(\mathbf{p}_0)_{k+1}$, all problem functions are linearized to yield the following, specially structured **Quadratic Program (QP)** in the variables $\Delta\xi = (\Delta s_0^x, \ldots, \Delta s_N^x, \Delta s_0^z, \ldots, \Delta s_{N-1}^z, \Delta \mathbf{q}_0, \ldots, \Delta \mathbf{q}_{N-2}, \Delta \mathbf{x}_s, \Delta \mathbf{z}_s, \Delta \mathbf{u}_s, \Delta \mathbf{p})$:

$$\min_{\Delta\xi} \sum_{i=0}^{N-1} \int_{t_i}^{t_{i+1}} \|\mathbf{l}_i(t) + \mathbf{L}_i(t)\, ({\Delta s_i^x}^T, {\Delta s_i^z}^T, {\Delta \mathbf{q}_i}^T, {\Delta \mathbf{u}_s}^T, \Delta \mathbf{p}^T)^T\|_2^2 \, dt \tag{33}$$

subject to

$$\Delta s_0^x = (\mathbf{x}_0)_{k+1} - s_0^x, \quad \Delta \mathbf{p} = (\mathbf{p}_0)_{k+1} - \mathbf{p}, \tag{34}$$

$$\Delta s_{i+1}^x = \mathbf{x}_i + \mathbf{X}_i\, ({\Delta s_i^x}^T, {\Delta s_i^z}^T, {\Delta \mathbf{q}_i}^T, \Delta \mathbf{p}^T)^T, \qquad i = 0, \ldots, N-1, \tag{35}$$

$$\mathbf{0} = \mathbf{g}_i + \mathbf{G}_i^z \Delta s_i^z + \mathbf{G}_i\, ({\Delta s_i^x}^T, {\Delta \mathbf{q}_i}^T, \Delta \mathbf{p}^T)^T, \quad i = 0, \ldots, N-1, \tag{36}$$

$$\mathbf{0} \le \tilde{\mathbf{c}}_i + \tilde{\mathbf{C}}_i\, ({\Delta s_i^x}^T, {\Delta s_i^z}^T, {\Delta \mathbf{q}_i}^T, \Delta \mathbf{p}^T)^T, \qquad i = 0, \ldots, N-1, \tag{37}$$

$$\mathbf{0} = \mathbf{r} + \mathbf{R}\, ({\Delta \mathbf{x}_s}^T, {\Delta \mathbf{z}_s}^T, {\Delta \mathbf{u}_s}^T)^T + \mathbf{R}^p \Delta \mathbf{p}. \tag{38}$$

The solution $\Delta\xi_k$ of this quadratic program is first used to determine the control $(\mathbf{q}_0)_{k+1} := (\mathbf{q}_0)_k + (\Delta \mathbf{q}_0)_k$ which is immediately given to the plant, and secondly to compute the next real-time iterate:

$$\xi_{k+1} := \xi_k + \Delta\xi_k. \tag{39}$$

The iterations *never terminate*. Instead, the values $(\mathbf{x}_0)_{k+1}$ and $(\mathbf{p}_0)_{k+1}$ are changed from one iteration to the next, according to the current state and parameter estimates.

Remark 1. Note that the objective function in (33) can, neglecting a constant, equivalently be written as

$$\sum_{i=0}^{N-1}\{({\Delta s_i^x}^T, {\Delta s_i^z}^T, {\Delta \mathbf{q}_i}^T, {\Delta \mathbf{u}_s}^T, \Delta \mathbf{p}^T)\, \mathbf{h}_i$$
$$+\tfrac{1}{2}({\Delta s_i^x}^T, {\Delta s_i^z}^T, {\Delta \mathbf{q}_i}^T, {\Delta \mathbf{u}_s}^T, \Delta \mathbf{p}^T)\, \mathbf{H}_i\, ({\Delta s_i^x}^T, {\Delta s_i^z}^T, {\Delta \mathbf{q}_i}^T, {\Delta \mathbf{u}_s}^T, \Delta \mathbf{p}^T)^T\}.$$

with Hessian blocks and gradient vectors

$$\mathbf{H}_i := 2\int_{t_i}^{t_{i+1}} \mathbf{L}_i(t)^T \mathbf{L}_i(t)\, dt \quad \text{and} \quad \mathbf{h}_i := 2\int_{t_i}^{t_{i+1}} \mathbf{L}_i(t)^T \mathbf{l}_i(t)\, dt.$$

This formulation explicitly shows that the linear least squares problem (33)-(38) is nothing else than a finite dimensional quadratic programming problem. In fact, the Gauss-Newton matrices $\mathbf{H}_i$ can be regarded a cheap approximation of the *exact Hessian* blocks $\mathbf{H}_i^{\text{exact}}$, as they arise in the exact Hessian SQP method.

Remark 2. Instead of using $\xi_k + \Delta\xi_k$ directly as the new iterate (we call this the *warm start* strategy) it is alternatively possible to account for the movement of the horizon in time, if the multiple shooting intervals are equally spaced with interval lengths that correspond to a sampling time. To achieve this, a *shift* in the problem variables is performed before the new iterate is defined, i.e., Eq. (39) is replaced by $\xi_{k+1} := S(\xi_k + \Delta\xi_k)$, where S is a shift operator which removes the variables of the first interval, shifts all variables by one interval, and appends new guesses on the last interval. For details, see [14].

Remark 3. For the related class of *shrinking horizon* problems (as defined in [4]), we can prove contractivity of the real-time algorithm under mild conditions, if plant and model coincide after an initial disturbance [14]. The contractivity result can conceptually be generalized to the shift strategy on long horizons, but in practice warm start and shift strategy show very similar performance [12].

Remark 4. The formulation of the constraints (28) in the NLP (27)-(32) can be considered an *initial value embedding* of each optimization problem into the manifold of perturbed problems. It allows a very efficient transition from one optimization problem to the next: let us for a moment assume that ξ_k is equal to the *exact* solution ξ_k^* of the optimization problem $P((\mathbf{x}_0)_k, (\mathbf{p}_0)_k)$, and that the above QP (33)-(38) is formulated with the *exact* Hessian blocks $\mathbf{H}_i^{\text{exact}}$. Then it can be shown under mild conditions that the next real-time iterate $\xi_{k+1} = \xi_k^* + \Delta\xi_k$ is a first order prediction for the solution ξ_{k+1}^* of the optimization problem $P((\mathbf{x}_0)_{k+1}, (\mathbf{p}_0)_{k+1})$, i.e.

$$\|\xi_{k+1} - \xi_{k+1}^*\| = O\left(\left\| \begin{matrix} (\mathbf{x}_0)_{k+1} - (\mathbf{x}_0)_k \\ (\mathbf{p}_0)_{k+1} - (\mathbf{p}_0)_k \end{matrix} \right\|^2\right),$$

even at points where the active set changes [14]. In practice, the initial value embedding strategy ensures that the real-time iterates ξ_k stay close to the exact solutions ξ_k^*, even if a Gauss-Newton Hessian is employed instead of the exact one.

Real-Time QP Generation and Solution

The generation and solution of the structured quadratic program (33)-(38) are dovetailed in the real-time iteration scheme. The initial value embedding turns out to be crucial for the real-time performance, as it allows to prepare large parts of the QP solution without knowledge of $(\mathbf{x}_0)_{k+1}, (\mathbf{p}_0)_{k+1}$. The following steps are performed during each real-time iteration:

1. Reduction: Generate the equality constraints (36) by linearizing the consistency conditions (30), and resolve (36) to eliminate $\Delta\mathbf{s}_i^z$ from the problem ($\mathbf{G}_i^z$ is invertible due to the index one assumption). Similarly, generate (38) and resolve it to eliminate $\Delta\mathbf{x}_s, \Delta\mathbf{z}_s, \Delta\mathbf{u}_s$ from the problem (assuming that the square matrix $\mathbf{R}$ is invertible).
2. DAE solution and derivative generation: Solve the relaxed initial value problems (24)-(26) and compute simultaneously the directional derivatives of the trajectories $\mathbf{x}_i(t)$ $\mathbf{z}_i(t)$ in the reduced directions, using the principle of *internal numerical differentiation* (IND) as described by Bock [9]. This yields the

reduced version of (35).[2] Linearize also the constraints (31) and generate a reduced version of (37).

3. Gradient and Hessian generation: Compute a reduced version of the gradient integrals $\mathbf{h}_i$ and of the Gauss-Newton Hessian blocks $\mathbf{H}_i$. The sparse DAE solver DAESOL [3] has been adapted to compute numerical approximations of reduced versions of $\mathbf{h}_i$ and $\mathbf{H}_i$ *simultaneously* with the sensitivity calculations from step 2, with negligible additional costs [14].
4. Condensing: Using the (reduced) linearized continuity conditions (35), eliminate the variables $\Delta\mathbf{s}_1^x, \ldots, \Delta\mathbf{s}_N^x$. Project the objective gradient and Hessian, as well as the linearized path constraints (37) onto the space of the remaining variables $\Delta\mathbf{s}_0^x, \Delta\mathbf{q}_0, \ldots, \Delta\mathbf{q}_{N-2}, \Delta\mathbf{p}$.
5. Step generation: take the current values $(\mathbf{x}_0)_{k+1}$, $(\mathbf{p}_0)_{k+1}$ from the state estimator, and eliminate $\Delta\mathbf{s}_0^x := (\mathbf{x}_0)_k - \mathbf{s}_0^x$ and $\Delta\mathbf{p} := (\mathbf{p}_0)_k - \mathbf{p}$, and generate a fully condensed QP in the variables $(\Delta\mathbf{q}_0, \ldots, \Delta\mathbf{q}_{N-2})$. Solve this QP with an efficient dense QP solver using an active set strategy. Give the value $\mathbf{q}_0 + \Delta\mathbf{q}_0$ immediately as a control to the plant.
6. Expansion: Expand the solution to yield values for all variables $\Delta\xi_k$ and perform the iteration $\xi_{k+1} = \xi_k + \Delta\xi_k$. Go to 1.

Note that the computations of step 5 are in typical applications orders of magnitude shorter than the overall computations of one cycle. This means that the response delay of one sampling time, that is present in all previous NMPC optimization schemes, is practically avoided. It is interesting to observe that step 5 corresponds to the solution of a QP in *linear* MPC, which is based on a system linearization along the currently best predicted trajectory. Note, however, that this trajectory is updated after each iteration, and that the real-time iteration scheme maintains all advantages of a fully nonlinear treatment of the optimization problems.

The major computational costs for each real-time iteration, those of step 2, scale roughly linear with the number N of predicted control intervals.

A comparison of the closed-loop behaviour of the real-time iteration strategy with a full iteration scheme, where each optimization problem is iterated to convergence, can be seen in Figure 7. In the shown example scenario even the full iteration scheme (that was started using the initial value embedding) was nearly always able to meet the limit of 10 seconds sampling time; the comparison shows no significant differences in the closed loop behaviour. Note, however, that bounds on the number of iterations for the full iteration scheme are difficult to establish.

4 Experimental Setup

In Section 5 we will demonstrate the real-time feasibility of the presented NMPC scheme. Furthermore, we give a short comparison of the achieved results with a conventional PI-controller. The purpose of the performance comparison is to show

[2] Note that the full matrix $\mathbf{X}_i$ is *never* computed in this *partial reduction approach* that was developed by Leineweber [19].

that NMPC does lead to reasonable performance without much tuning required. In the following, we shortly outline the two used controller setups.

4.1 NMPC Controller Setup

For the real-time application we implemented the presented NMPC scheme using the real-time optimization strategy given. For the NMPC setup we assume that the column is given in LV configuration, i.e., we use L_{vol} and the heat input Q into the boiler (which corresponds to the vapor flow out of the boiler) as manipulated variable. State estimates are obtained using a variant of an extended Kalman filter. Figure 3 shows the overall controller/plant/estimator setup. As described in Sec-

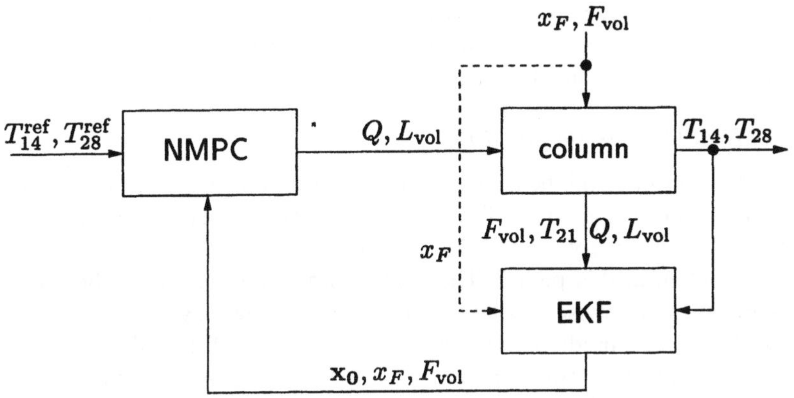

Figure 3. Closed loop NMPC setup

tion 2, we use a inferential control scheme, i.e., the product concentrations are not directly controlled, but instead the temperatures T_{14} and T_{28} which are directly measured at the column. The deviation of these temperatures from the desired reference temperatures T_{14}^{ref} and T_{28}^{ref} are weighted in the stage cost function, cf. Eq. (18).

State Estimation

To obtain an estimate of the 42 differential system states and of the model parameter x_F we have implemented a variant of an Extended Kalman Filter (EKF). To improve the performance of the estimator, the temperature T_{21} is fed into the state estimator additionally to the controlled temperatures T_{14} and T_{28}. Together with the implemented controls (Q and L_{vol}), the measured feed flow rate F_{vol} is also given to the estimator. The usage of the *measured* heat input and reflux flow (in contrast to the optimization output) is necessary to overcome input disturbances, since both values can only be controlled indirectly giving the setpoints to the low-level control loops in the DCS.

The EKF is based on subsequent linearizations of the system model at each current estimate; each measurement is compared with the prediction of the nonlinear model, and the estimated state is corrected according to the deviation. The weight of past measurement information is kept in a weighting matrix, which is updated according to the current system linearization.

In contrast to an ordinary EKF the implemented estimator can incorporate additional knowledge about states and parameters in form of bounds. This is especially useful as the tray concentrations have to be in the interval [0, 1] to make a reasonable DAE solution possible. For details, we refer to [14].

In Section 5 we will consider two scenarios for the "estimates" of the feed concentration disturbance. In the first scenario it is assumed that the time and value of the disturbance is known exactly and directly fed into the EKF. In the second scenario we consider the disturbance in the feed concentration x_F as unknown, i.e., the value of x_F is also estimated by the EKF.

Tuning of NMPC Controller and EKF

The NMPC controller and the EKF were tuned independently based on simulations and measured data, respectively. The EKF receives new measurements and provides new state estimates every 10 seconds, and the NMPC optimizer solves one optimization problem in this period. The control inputs on the control horizon were parameterized as piecewise constant, with 10 control intervals each of 120 seconds length, followed by a prediction interval of $T_p - T_c = 3600$ seconds with the inputs fixed to the steady state values $\mathbf{u}_s$ (cf. Sec. 3). Note that it does not cause any difficulty that the interval length of 120 seconds in the control horizon is *not* equal to the sampling time of 10 seconds.

Implementation of the NMPC Setup

The NMPC controller was implemented on a Unix workstation running under Linux. The open-loop optimization problem was solved on-line using our specially tailored version of MUSCOD2 as outlined in Section 3. The EKF was implemented using MATLAB and the integration routine DAESOL [3] to obtain the necessary derivatives. The file transfer between the DCS and the workstation was done using a PC connected to the DCS from which the measured and manipulated variables could be read and written to via ftp, which was done every 10 seconds.

4.2 Configuration of the PI Control Loops

The conventional control scheme used consists of two decoupled single-input/single-output PI loops which where already implemented on the column. In contrast to the NMPC setup a L/D,V configuration is used. The controlled variables are, as in the NMPC case the temperatures on trays 14 and 28. The manipulated variables are the heat input Q to the boiler (corresponding to the liquid flow V out

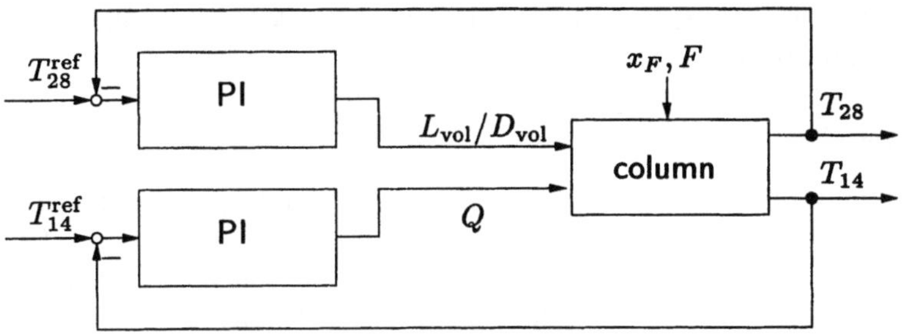

Figure 4. Closed loop PI setup

of the boiler) and the reflux ratio L_{vol}/D_{vol}. However, for comparisons we plot in section 5 the reflux flow rate for the decoupled PI as well as the NMPC controller.

PI Controller Tuning

To achieve good control performance, the PI controllers are tuned as follows: In a first step, a Ziegeler-Nichols tuning based on the process reaction curve method is performed for each loop. This method was chosen since it is easy to apply and does only require single step tests. In a second step, the resulting PI controllers were detuned to compensate for the interactions between the control loops.

Implementation of the PI Setup

The PI controllers were already implemented using the basic control function in the used DCS. The data collection was done using a PC connected to the DCS and the Unix workstations, compare Figure 1.

5 Experimental Results

In this section we present results on the computational demand and performance of the PI control setup and the NMPC controller using the dynamic optimization strategy outlined in Section 3. The main result here is an experimental proof-of-concept that NMPC can be used in real-time even for large scale models. The performance results given are only supposed to show that NMPC, without much further tuning, does lead to satisfying control performance.

5.1 Considered Disturbance Scenarios

The NMPC scheme and the decoupled PI controllers were tested on various scenarios. As scenarios we used a series of step changes in the feed flow rate (F_{vol}); a step change in the feed composition (x_F); and a short reflux breakdown ($L_{vol} = 0.5$). In the following figures the reflux flow rate, L_{vol}, is plotted for both NMPC and PI controllers although the corresponding manipulated variable is reflux ratio, (L_{vol}/D_{vol}), for the PI controller setup.

Feed Flow Change

Figure 5 shows the controlled outputs (T_{28} and T_{14}) and input responses (L_{vol} and Q) when the feed flow rate, F_{vol}, is changed stepwise first -10 % at t= 0.57 h, then +20 % at t=1.16 h, finally -10 % at t=1.52 h. The plots on the left hand side show the results of NMPC and those on the right hand side belong to the PI controller. As can be seen, while there is no obvious difference between the two schemes in the first phase, the performance of NMPC is better than that of PI in the second and third phases. In the final phase, NMPC can handle with the corresponding load change in about 20 min. whereas the PI controller requires 40 min. In the case of PI controllers,

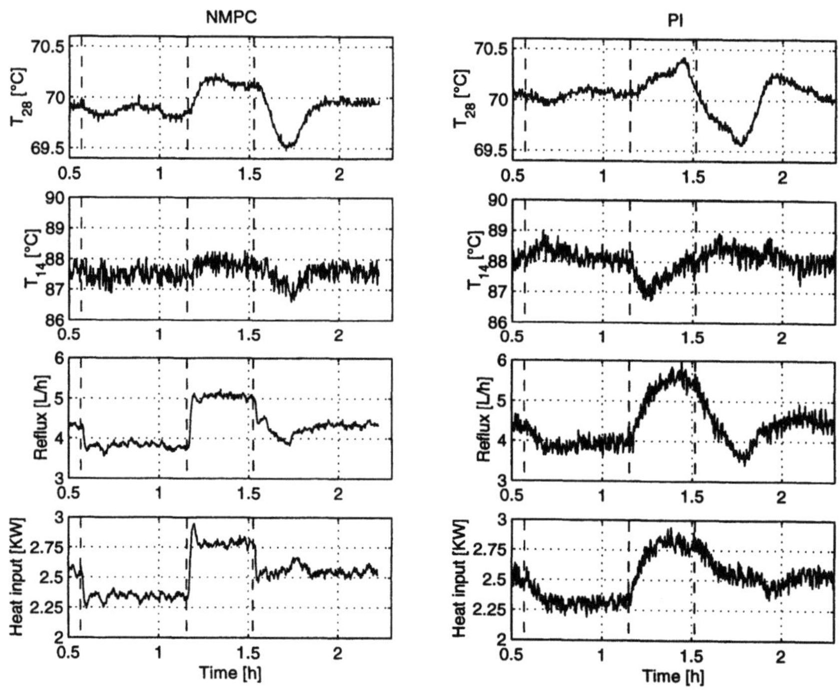

Figure 5. Comparison of closed loop NMPC and conventional PI controller performances for step disturbances in F_{vol}

the slightly higher deviation of the controlled variables from their set points can be explained by the fact that no feedforward disturbance information is used, whereas NMPC uses the disturbance information. This advantage of NMPC results in finding the optimum values for L_{vol} and Q at once while the manipulated variables move very slowly for PI controllers following the change in feedback measurements.

Feed Concentration Change

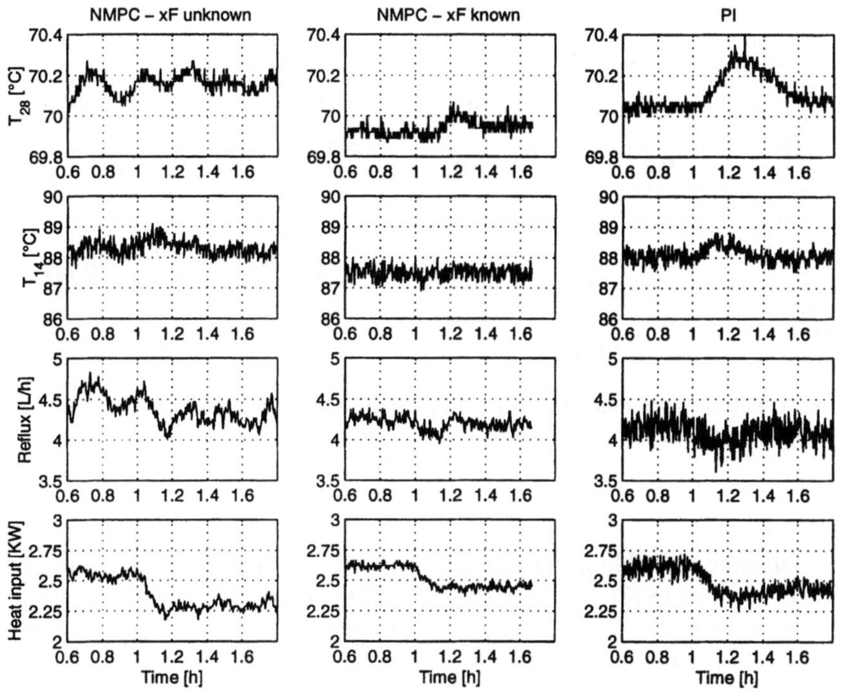

Figure 6. Step change in x_F: Comparison of closed loop NMPC – for unknown and known x_F – with the conventional PI controller performances

For the next test, a step change in the feed composition is considered ; x_F is decreased from 0.320 to 0.272 at t=1.0 h. In Figure 6, the first column on the left hand side illustrates the results of NMPC which incorporates EKF estimating x_F. As can be seen, the controlled outputs (T_{14} and T_{28}) oscillate with small deviations from the set points but NMPC performance is still acceptable.

From the last two columns of Figure 6 it is evident that the performance of NMPC with known disturbance in x_F outperforms the PI controller in maintaining the controlled variable T_{28}.

Comparison of Real-Time Iterations with a Full-Iteration Scheme. Figure 7 compares the computational and control performance of the real-time iteration scheme, that we used in all presented experiments, with a full iteration scheme, for the same scenario as in Figure 6 (x_F known). The full iteration scheme was initialized by

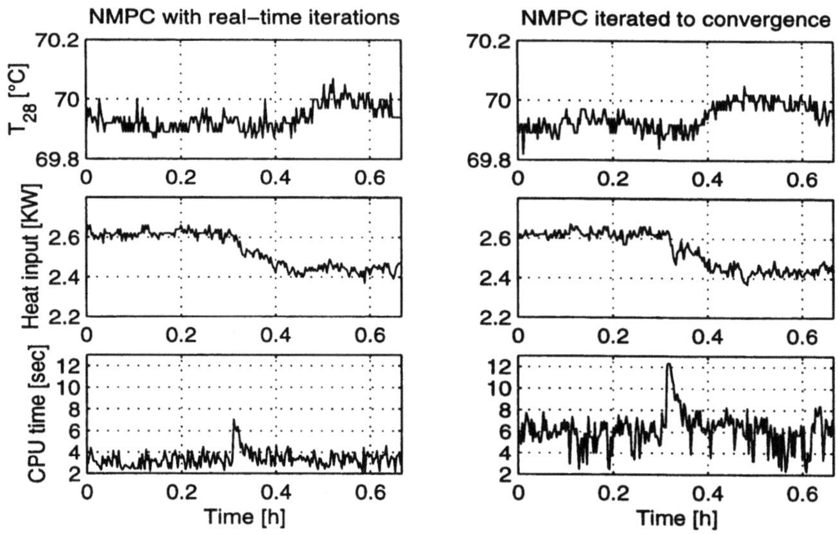

Figure 7. Performance and CPU time comparison of real-time iterations (cf.Figure 6, x_F known) with full iterations to convergence, on a standard PC

an initial value embedding, but then iterated until a prespecified convergence criterion was satisfied (KKT tolerance of 10^{-3}). If the optimization time exceeded the sampling time of 10 seconds, the old control was used for another sampling period.

It can be seen that the control performance is nearly identical in our example, but that the CPU times differ significantly. Note that the real-time iteration scheme had unused CPU capacity as the sampling time was fixed to 10 seconds in both schemes.

The CPU time variations in the real-time iterations are due to integrator adaptivity. The largest computation times occur for both schemes at the moment when the step change in x_F happens, which makes large changes in the predicted trajectory necessary.

Due to the unused capacity of the real-time iteration scheme, more detailed process models are computationally feasible for the same sampling time. First experimental tests with a larger and stiffer DAE model (82 differential and 122 algebraic states) have shown that the corresponding NMPC controller is still real-time implementable [14].

Short Reflux Breakdown

In the previous two cases the disturbing effects of load changes (in the feed flow and composition) on controlled variables are reasonably small. In order to have large disturbance effects on T_{14} and T_{28} we applied a reflux flow breakdown on the system starting at 0.22 h. for a short period of time (approximately 4 minutes). After the reflux flow is switched on again, NMPC and PI controller are able to bring the

T_{14} and T_{28} back to their set points within 1 hour. However, both controllers lead to slight oscillations in the closed loop.

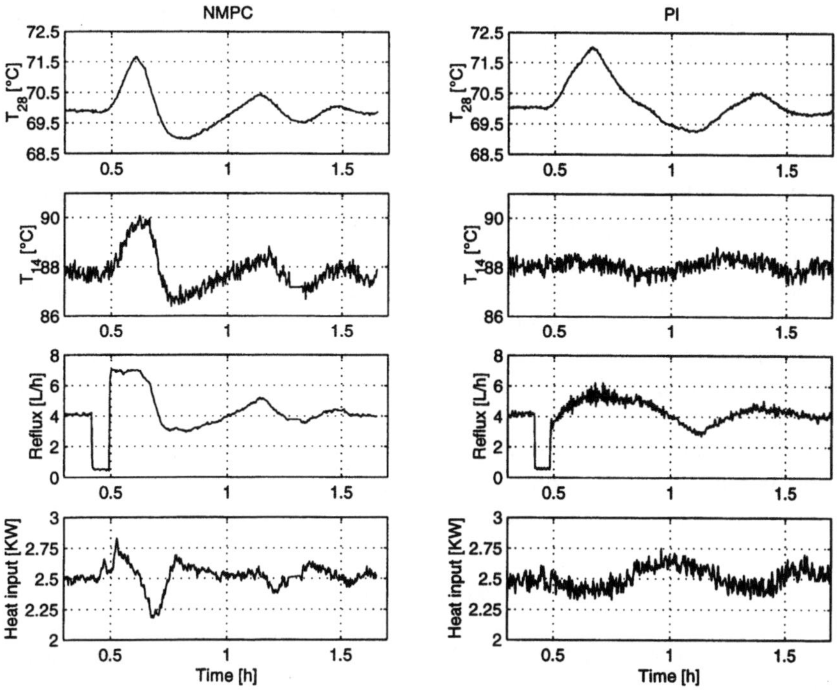

Figure 8. Comparison of closed loop NMPC and conventional PI controller performances for a short time reflux (L_{vol}) breakdown

It is interesting to note that the PI controller of T_{28} shows an aggressive response during the initial transition period. We have experienced that if the reflux breakdown is applied for a longer period of time, the PI controller of T_{28} becomes completely unstable.

5.2 Discussion of Computational Demand and Closed-Loop Performance

The presented experiments show that NMPC can handle the control problem satisfactorily while being real-time implementable. The results are comparable to the performance of (existing) PI controllers for a moderate range of disturbances.

Due to the efficiency of our real-time optimization scheme a more complex model would still be feasible for on-line implementation.

One of the challenges to be solved is to improve the state estimation and disturbance detection, as load changes which are unlikely to be measured in an industrial application, like x_F, need to be estimated for a realistic NMPC scheme. As can be seen in Figure 6, the closed loop performance when x_F is estimated by the EKF could still be improved.

It is the aim of future work to show the benefits of real-time NMPC applications for more complex processes or for a wide range of operating conditions like start-up periods of chemical processes.

6 Conclusions

We have presented an experimental proof-of-concept of the application of NMPC to the control of a high purity binary distillation column.

An efficient real-time optimization scheme based on the direct multiple shooting method is described. Among its features are an initial value embedding strategy, that allows to immediately respond to disturbances, and real-time iterations, that dovetail the optimization iterations with the real process development. This approach makes sampling times of 10 seconds for a system model of 164th order possible on a standard PC.

Our study shows that real-time implementation of NMPC using large scale DAE models is feasible for the control of a pilot scale distillation column, if efficient numerical optimization techniques are used.

References

1. F. Allgöwer, T. A. Badgwell, J. S. Qin, J. B. Rawlings, and S. J. Wright. Nonlinear predictive control and moving horizon estimation – An introductory overview. In P. M. Frank, editor, *Advances in Control, Highlights of ECC'99*, pages 391–449. Springer, 1999.
2. F. Allgöwer and J. Raisch. Multivariable controller design for an industrial distillation column. In N.K Nichols and D.H. Owens, editors, *The Mathematics of Control Theory*. Clarendon Press, Oxford, 1992.
3. I. Bauer. *Numerische Verfahren zur Lösung von Anfangswertaufgaben und zur Generierung von ersten und zweiten Ableitungen mit Anwendungen bei Optimierungsaufgaben in Chemie und Verfahrenstechnik.* PhD thesis, University of Heidelberg, 2000.
4. T. Binder, L. Blank, H. G. Bock, R. Bulirsch, W. Dahmen, M. Diehl, T. Kronseder, W.Marquardt, J. P. Schlöder, and O. von Stryk. Introduction to model based optimization of chemical processes on moving horizons. In *to be announced.* Springer, 2001.
5. H. G. Bock, M. Diehl, D. Leineweber, and J. P. Schlöder. A direct multiple shooting method for real-time optimization of nonlinear DAE processes. In F. Allgöwer and A. Zheng, editors, *Nonlinear Predictive Control*, pages 245–268. Birkhäuser, 2000.
6. H. G. Bock, M. Diehl, J. P. Schlöder, F. Allgöwer, R. Findeisen, and Z. Nagy. Real-time optimization and nonlinear model predictive control of processes governed by differential-algebraic equations. In *ADCHEM2000 - International Symposium on Advanced Control of Chemical Processes*, volume 2, pages 695–703, Pisa, 2000.
7. H. G. Bock, E. Eich, and J. P. Schlöder. Numerical solution of constrained least squares boundary value problems in differential-algebraic equations. In K. Strehmel, editor, *Numerical Treatment of Differential Equations*. Teubner, Leipzig, 1988.

8. L. Biegler. Efficient solution of dynamic optimization and NMPC problems. In F. Allgöwer and A. Zheng, editors, *Nonlinear Predictive Control*. Birkhäuser, 2000.
9. H. G. Bock. Numerical treatment of inverse problems in chemical reaction kinetics. In K. H. Ebert, P. Deuflhard, and W. Jäger, editors, *Modelling of Chemical Reaction Systems*, volume 18 of *Springer Series in Chemical Physics*. Springer, Heidelberg, 1981.
10. H. G. Bock and K. J. Plitt. A multiple shooting algorithm for direct solution of optimal control problems. In *Proc. 9th IFAC World Congress*, Budapest, 1984.
11. H. Chen and F. Allgöwer. A quasi-infinite horizon nonlinear model predictive control scheme with guaranteed stability. *Automatica*, 34(10):1205–1218, 1998.
12. M. Diehl, H. G. Bock, D. B. Leineweber, and J. P. Schlöder. Efficient direct multiple shooting in nonlinear model predictive control. In F. Keil, W. Mackens, H. Voß, and J. Werther, editors, *Scientific Computing in Chemical Engineering II*, volume 2, pages 218–227, Berlin, 1999. Springer.
13. M. Diehl, H. G. Bock, J. P. Schloeder, R. Findeisen, Z. Nagy, and F. Allgoewer. Real-time optimization and nonlinear model predictive control of processes governed by large-scale DAE. *J. Proc. Contr.*, 2001.
14. M. Diehl. *Real-Time Optimization for Large Scale Nonlinear Processes*. PhD thesis, University of Heidelberg, 2001.
15. G. De Nicolao, L. Magni, and R. Scattolini. Stabilizing nonlinear receding horizon control via a nonquadratic terminal state penalty. In *Symposium on Control, Optimization and Supervision, CESA'96 IMACS Multiconference*, pages 185–187, Lille, 1996.
16. G. de Nicolao, L. Magni, and R. Scattolini. Stability and robustness of nonlinear receding horizon control. In F. Allgöwer and A. Zheng, editors, *Nonlinear Predictive Control*, pages 3–23. Birkhäuser, 2000.
17. R. Findeisen, F. Allgöwer, M. Diehl, H. G. Bock, J. P. Schlöder, and Z. Nagy. Efficient nonlinear model predictive control. Submitted to 6th International Conference on Chemical Process Control – CPC VI, 2000.
18. R. Henrion, P. Li, A. Moeller, M. Wendt, and G. Wozny. Optimal control of a continuous distillation process under probabilistic constraints. In M. Groetschel, S. O. Krumke, and J. Rambau, editors, *Online Optimization of Large Scale Systems: State of the Art*. Springer, 2001.
19. D. B. Leineweber. *Efficient reduced SQP methods for the optimization of chemical processes described by large sparse DAE models*, volume 613 of *Fortschr.-Ber. VDI Reihe 3, Verfahrenstechnik*. VDI Verlag, Düsseldorf, 1999.
20. K. J. Plitt. Ein superlinear konvergentes Mehrzielverfahren zur direkten Berechnung beschränkter optimaler Steuerungen. Master's thesis, University of Bonn, 1981.
21. V. H. Schulz, H. G. Bock, and M. C. Steinbach. Exploiting invariants in the numerical solution of multipoint boundary value problems for daes. *SIAM J. Sci. Comp.*, 19:440–467, 1998.
22. S. J. Wright. Applying new optimization algorithms to model predictive control. In J. C. Kantor, C. E. Garcia, and B. Carnahan, editors, *Fifth International Conference on Chemical Process Control – CPC V*, pages 147–155. American Institute of Chemical Engineers, 1996.

Towards Nonlinear Model-Based Predictive Optimal Control of Large-Scale Process Models with Application to Air Separation Plants

Thomas Kronseder[1], Oskar von Stryk[2], Roland Bulirsch[1], and Andreas Kröner[3]

[1] Lehrstuhl für Höhere Mathematik und Numerische Mathematik, Technische Universität München, Germany
[2] Fachgebiet Simulation und Systemoptimierung, Technische Universität Darmstadt, Germany
[3] Linde AG, Process Engineering and Contracting Division, Höllriegelskreuth, Germany

Abstract We propose a concept for model predictive control of large-scale dynamical systems. This concept has been designed for the optimal control of chemical engineering processes, in particular for cryogenic air separation plants which are modelled by large systems of coupled differential and algebraic equations of (differential) index two with state dependent discontinuities. Our concept considers different time scales for various tasks which are prescribed by the real-time nature of the process of interest. In this paper (items refer to Figure 4) the components (a)–(d) and (f) of the general concept are considered in detail. Until now there has been a lack of a clear concept for real-time optimality. Therefore, we conclude by discussing some fundamental issues of the notion of real-time optimality.

1 Introduction

We start with a description of the investigated application, namely the control of the load change of air separation plants. Modelling approaches to the dynamic process models and their resulting properties are discussed next.

1.1 An Industrial Optimal Control Challenge

Gases like nitrogen (N_2), oxygen (O_2), and argon (Ar) are used in industry as raw materials or as auxiliary substances. E.g., nitrogen is the basis of fertilisers, oxygen is required for the refinement of steel, and argon is needed for welding. At the same time, these three gases are the main components of air (see Table 1). Thus it appears natural to obtain such industrial gases by the separation of *air* which is available at any place on earth in sufficient amounts.

Nitrogen	N_2	78.084 [vol%]
Oxygen	O_2	20.946 [vol%]
Argon	Ar	0.932 [vol%]
Carbon dioxide	CO_2	≈ 335 [vppm]
Neon	Ne	18.18 [vppm]
Helium	He	5.239 [vppm]
Krypton	Kr	1.14 [vppm]
Xenon	Xe	0.086 [vppm]

Table 1: Average composition of air (w/o variable constituents; Rohde [50])

In 1902 Dr. Carl von Linde, the founder of the Linde AG, built the world's

name	min [mol/mol]	max [mol/mol]	description
O_2 LOX	0.997	1.0	O_2 fraction in liquid O_2 product
O_2 GOX	0.997	1.0	O_2 fraction in gaseous O_2 product
O_2 DLIN	0.0	$5.0 \cdot 10^{-6}$	O_2 fraction in liquid N_2 product
O_2 GAN	0.0	$5.0 \cdot 10^{-6}$	O_2 fraction in gaseous N_2 product
Ar Prod	0.965	1.0	Ar fraction in Ar product
O_2 feed ArC	0.90	1.0	O_2 fraction in feed of argon column

Table 2. Example data for constraints to be satisfied during load changes of an existing air separation plant of the type shown in Figure 2 (Engl et al. [19])

first commercially viable cryogenic air separation plant. The idea was to liquefy ambient air and use rectification (which can be regarded as a highly efficient form of distillation) in order to obtain the various pure fractions contained in the feed. In industrial applications the energy consumption required for cooling to the prescribed deep temperatures governs the operational costs of such plants. Therefore, the different processing parts in such air separation plants are strongly interconnected in order to keep as much of the energy within the process as possible.

Figure 1: An air separation plant (picture by Linde AG)

Nowadays, air separation plants are often integral parts of other industrial production facilities. Their typical capacities range from 25 [t/d] (tons per day) to 15000 [t/d] of processed air. E.g., the plant depicted in Figure 1 has a capacity of 2 × 2150 [t/d] gaseous O_2 in 99.5 [%] purity and 2 × 600 [t/d] gaseous N_2 with a maximum impurity of 10 [ppm] O_2.

A critical phase in the control of these plants occurs when a load change, i.e., a transition from one operational point to another one, must be performed. It is of utmost importance that various constraints on gas concentrations at certain points in the plant are not violated in order to guarantee safe operation of the plant and purity of the products also during the load change (the data for an existing air separation plant are given in Table 2). Other objectives such as maximisation of product gain or minimisation of energy consumption are of minor importance. During a load change

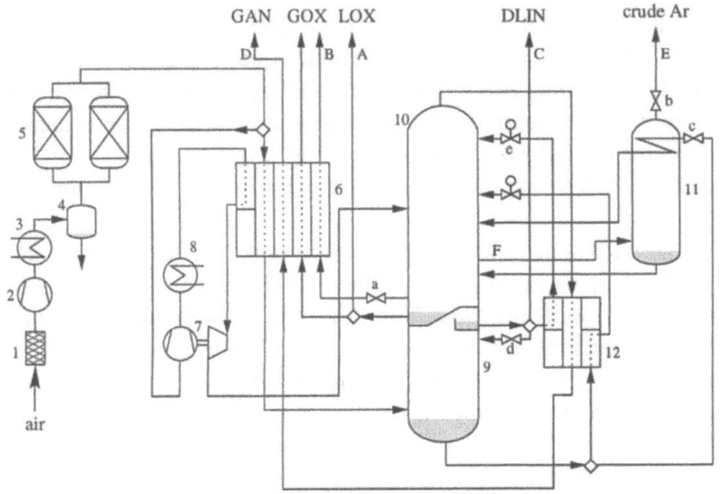

Figure 2. A simplified flowsheet of a low-pressure cryogenic air separation plant according to Baldus et al. [1], Eich-Söllner et al. [17], Zapp [57]
main units: 1: filter, 2: compressor, 3: cooling, 4: flash, 5: molecular sieves, 6: main heat exchanger, 7: compressor/expander, 8: cooling, 9: high pressure column, 10: low pressure column, 11: argon column, 12: heat exchanger; **products**: GAN: gaseous N_2, DLIN: high pressure liquid N_2, GOX: gaseous O_2, LOX: liquid O_2, crude Ar; **controls**: a: GOX drain, b: crude Ar drain, c: Ar condenser turnover, d: reflux HPC, e: reflux LPC; **constraints**: (an example set of constraints is specified in Table 2) A: O_2 LOX, B: O_2 GOX, C: O_2 DLIN, E: Ar Prod, F: O_2 feed ArC

the controls of the process essentially consist of continuously working valves (see Figure 2).

The numerical treatment of this difficult problem using the powerful tools of numerical optimal control (for an overview see, e.g., Binder et al. [3]) provides the basic motivation for the work presented in this paper; though, the techniques presented here also apply to other optimal control problems.

1.2 Modelling of Chemical Engineering Processes

Dynamical process models of chemical engineering plants are generally described by systems of coupled differential and algebraic equations (DAEs) in *linear implicit* form

$$\begin{bmatrix} \mathbf{A}_1(t,\mathbf{x},\mathbf{y},\mathbf{u},\operatorname{sign}\mathbf{q}) & \mathbf{A}_2(t,\mathbf{x},\mathbf{y},\mathbf{u},\operatorname{sign}\mathbf{q}) \\ 0 & 0 \end{bmatrix} \begin{bmatrix} \dot{\mathbf{x}} \\ \dot{\mathbf{y}} \end{bmatrix} = \begin{bmatrix} \mathbf{f}(t,\mathbf{x},\mathbf{y},\mathbf{u},\operatorname{sign}\mathbf{q}) \\ \mathbf{g}(t,\mathbf{x},\mathbf{y},\mathbf{u},\operatorname{sign}\mathbf{q}) \end{bmatrix} \qquad (1)$$

where the *differential variables* are denoted by $\mathbf{x} : \mathbb{R} \longrightarrow \mathbb{R}^{n_x}$, the *algebraic variables* are denoted by $\mathbf{y} : \mathbb{R} \longrightarrow \mathbb{R}^{n_y}$, and the *control variables* are denoted as $\mathbf{u} : \mathbb{R} \longrightarrow \mathbb{R}^{n_u}$. Furthermore, we have a regular matrix $\mathbf{A}_1 : \mathbb{R}^{1+n_x+n_y+n_u+n_q} \longrightarrow \mathbb{R}^{n_x \times n_x}$, a general matrix $\mathbf{A}_2 : \mathbb{R}^{1+n_x+n_y+n_u+n_q} \rightarrow \mathbb{R}^{n_x \times n_y}$, the differential equations $\mathbf{f} : \mathbb{R}^{1+n_x+n_y+n_u+n_q} \rightarrow \mathbb{R}^{n_x}$ and the algebraic equations $\mathbf{g} : \mathbb{R}^{1+n_x+n_y+n_u+n_q} \rightarrow \mathbb{R}^{n_y}$. Typically, the model exhibits discontinuities. Advanced techniques for the efficient and reliable treatment of discontinuities are based on the availability of an additional vector of *switching functions* $\mathbf{q} = \mathbf{q}(t, \mathbf{x}, \mathbf{y}, \mathbf{u}) : \mathbb{R}^{1+n_x+n_y+n_u} \rightarrow \mathbb{R}^{n_q}$ which describe each discontinuity as a zero pass in one of its components (e. g., Eich [16]).

Due to the enormous system dimensions found in industrial applications (see, e.g., the example given in Section 5.1), the models have to be generated by means of computer-aided modelling tools. This large size is caused by both the complexity of the plants and the high accuracy requirements (Table 2).

A mathematical model of a chemical engineering plant is based on the *flowsheet* of the process (as an example see Figure 2). A flowsheet is a graph-oriented description which consists in the parts of a plant (the *units*) in its nodes while the edges between the nodes represent physical flows or flows of information (the *streams*). This description allows automated compilation of the entire mathematical model of the plant using standard models of the single units. In turn, the standard models are provided in unit libraries.

1.3 Special Properties of the Models Considered

OPTISIM® – an in-house developed modelling, simulation, and optimisation tool of the Linde AG (Burr [8]) – uses models of the form

$$\dot{\mathbf{x}} = \mathbf{f}(t, \mathbf{x}, \mathbf{y}, \mathbf{u}, \operatorname{sign} \mathbf{q}) \tag{2a}$$

$$0 = \mathbf{g}(t, \mathbf{x}, \mathbf{y}, \mathbf{u}, \operatorname{sign} \mathbf{q}) \tag{2b}$$

which are a special case of Eq. (1). In the applications of interest, the models show several properties that become important factors in the design and practical applicability of off-line and on-line optimal control algorithms.

Size of the Models Considered

As already noted, dynamic models of the processes of interest are of very large scale ($n_x + n_y \approx \mathcal{O}(1000) \ldots \mathcal{O}(10000)$).

Currently, the numerical simulation of such models is still only possible by exploiting the sparsity of the Jacobian matrices (about 0.01 to 0.1 percent nonzero entries) (e. g., Eich et al. [17]). Although the Jacobians neither possess any special structure nor in general exhibit definiteness, their sparsity allows the application of special solvers designed for very large and sparse linear systems of equations (Duff [15] discusses various methods).

Differential Index 2

The *differential index* of Eqs. (2a)–(2b) is even more important from both a theoretical as well as a numerical point of view. Simply speaking, the *differential index* of a DAE is the number of differentiations required in order to transform the DAE into an ordinary differential equation (ODE) (Brenan et al. [6]). In practice, the index of the DAE can be regarded as a measure for the numerical (and theoretical) difficulties connected with the treatment of the plain DAE.

The large-scale models of real-world air separation plants (an example is discussed in Section 5.1) treated so far exhibit an index of two. This index has been verified in a structural sense (cf. Unger et al. [54]) by an analysis facility which was recently added to the plant simulator OPTISIM® (Kronseder [32]). Based on a closer inspection of the DAEs, the *higher index* can be attributed in part to necessary simplifications in the model equations. From a more general point of view the *connection* of several index-1 unit models can produce an index of the overall system exceeding one (Lefkopoulos and Stadtherr [36], [37]). Therefore, higher index problems are an active field of research, e.g., Gritsis et al. [46], Feehery [22], Engl et al. [19].

General index reduction to index-1 in order to circumvent the problems connected with the higher index is outweighed by its high numerical costs when applied to the dynamic simulation of the large DAE models considered.

Discontinuities

Finally, the DAE models Eqs. (2a)–(2b) of chemical engineering processes exhibit discontinuous changes in the model equations. The discontinuities are commonly classified into purely *time dependent* discontinuities (where the time of the discontinuity is fixed, e.g., in ramp forcing functions) and *state dependent* discontinuities which in general occur at a-priori unknown times. State dependent discontinuities arise, e.g., due to the physical nature of the process (such as a batch process), by the necessity for different models in different domains in state space (e.g., the model of an open valve differs from that of a closed valve), or by necessary model simplifications (say, continuously interpolated physical property measurements).

In order to provide an adequate treatment for these discontinuities, switching functions $\mathbf{q}(t, \mathbf{x}, \mathbf{y}, \mathbf{u})$ are employed. These functions indicate a discontinuity in the model by a zero in at least one of $\mathbf{q}_i(t, \mathbf{x}, \mathbf{y}, \mathbf{u})$, $i = 1, \ldots, n_\mathbf{q}$. This additional modelling effort allows for an efficient and exact location of discontinuities by root finding algorithms (e.g., Eich [16]).

After the detection of a discontinuity the model equations are switched and the numerical integration procedure is restarted at the switching point. In contrast to the ODE case, in the DAE case the restart is far more complicated. Especially in the context of higher index DAEs the determination of *consistent initial values* for a theoretically satisfying and numerically flawless (re-)start of an integration is a difficult problem and still subject to current research (e.g., Kröner et al. [29], Kronseder [32]).

Yet until recently, in OPTISIM® the effort required for consistent initialisation could be circumvented utilizing a special property of the BDF method (see Section 4.1 and Sincovec et al. [52], Brenan et al. [6]). In most cases encountered it allows to continue integration without any analysis of the DAE Eqs. (2a)–(2b). However, the situation changes as the integration method must be extended to parametric sensitivity calculations in the case of *state dependent* discontinuities (see Section 4).

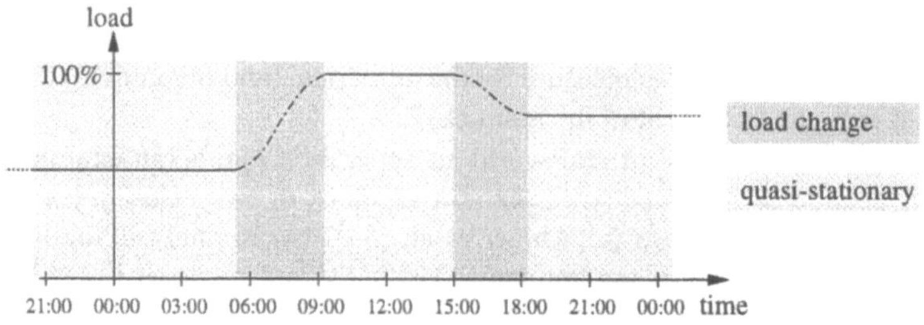

Figure 3. load change, and quasi-stationary operation at constant load (air input)

Consistent initialisation is *required* in order to perform the correct transfer of the sensitivities across discontinuities (e.g., Galán, Feehery, and Barton [23]).

Due to its importance and level of difficulty we will treat the problem of sensitivity transfer at discontinuities and the consistent initialisation problem in detail within Section 4.

2 A Concept for Model Predictive Optimal Control

2.1 *Problem Setting*

In common, cryogenic air separation plants are operated quasi-stationarily at a constant load over many periods of time, i.e., the amount of processed air is kept constant. From time to time, a transition of the process to a different load has to be performed (see Figure 3); such a transition is termed a *load change*. For the problems treated here, a load change requires about 1–3 hours. In our framework (see Figure 4) the operator has to specify the type of the load change, the optimality criterion to be minimised as well as the constraints to be satisfied during the load change. As discussed in Section 1.1 the strict fulfilment of the constraints is of highest priority during the load change.

We now consider the optimal control of an air separation plant in permanent operation, e.g., during a year ($[t_0, t_f] = [0, 365]$ days). The load history (see Figure 3) admits application of the Bernoulli/Bellman principle of optimality so that we may split the long term optimal control problem into optimal control problems on smaller horizons without loss of optimality. Each of these horizons covers *quasi-stationary* operation at load A — *instationary* load change — and *quasi-stationary* operation at load B.

The quasi-stationary points of operation can be computed efficiently, given the specifications of the customer. Therefore we assume that the values of the state variables $\mathbf{x}^*_{\mathrm{loadA/B}}$, and $\mathbf{y}^*_{\mathrm{loadA/B}}$ and of the control variables $\mathbf{u}^*_{\mathrm{loadA/B}}$ at both beginning and end of the load change are known.

In contrast to the quasi-stationary case, the determination of load change strategies was until recently almost entirely based on human expertise and experimental data. The data needed to be obtained from expensive tests from the ready-to-use air

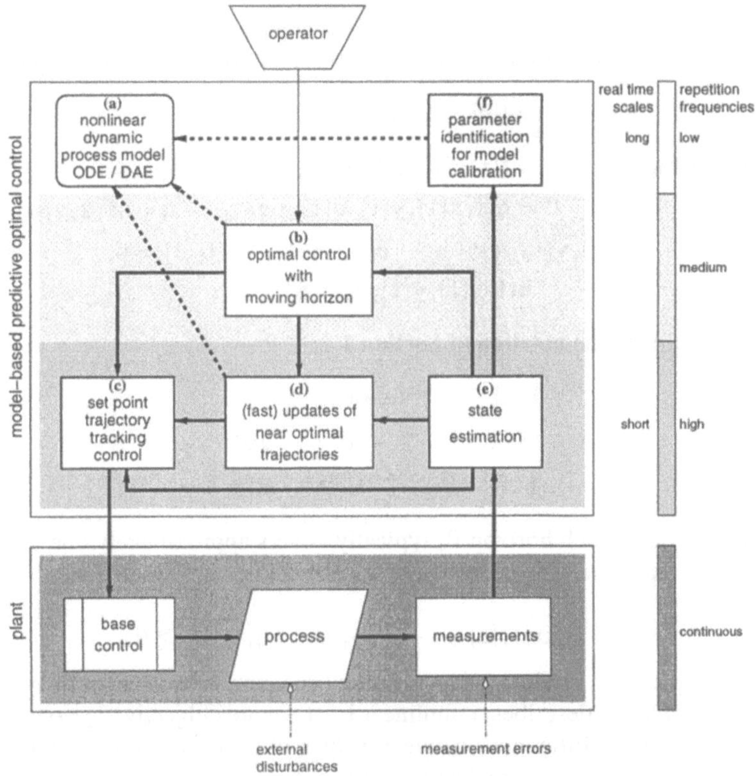

Figure 4. Overall concept of a model predictive control algorithm for real-time optimal control of large-scale dynamical systems

separation plant. Nijsing [40] proposed a direct transcription method suitable for index-1 DAE models in the context of air separation plants. A problem, however, was the approximation of the gradients required for the sequential quadratic programming (SQP) solver using finite differences of perturbed trajectories via external numerical differentiation which is numerically costly, of limited accuracy, and to be handled with great care (Kiehl [27]). In Engl et al. [19] a method for the numerical off-line determination of load change strategies for air separation plants modelled by large scale index-2 DAEs has been developed. This method is based on the direct transcription of the optimal control problem into a direct single shooting approach where reliable gradient information is obtained via the computation of the sensitivities from the efficient solution of the sensitivity DAE corresponding to the model DAE (see Section 4 and Binder et al. [3] for details). However, only time-dependent discontinuities in the models have been considered there.

Applying the Bernoulli/Bellman principle of optimality we now split the long term optimal control problem into a sequence of load change problems, each of which can be solved separately. If we consider the jth load change on a prediction horizon $P_j=[t_{0,j}, t_{f,j}]$ covering the entire load change, the corresponding optimal control problem incorporating the DAE model Eqs. (2a)–(2b) can be formulated as

$$\begin{aligned} &\min_{\mathbf{u}} J[\mathbf{u}] = E(t_{f,j}, \mathbf{x}(t_{f,j}), \mathbf{y}(t_{f,j})) \\ \text{subject to} \quad &\dot{\mathbf{x}}(t) = \mathbf{f}(t, \mathbf{x}(t), \mathbf{y}(t), \mathbf{u}(t), \operatorname{sign} \mathbf{q}), \quad t_{0,j} \le t \le t_{f,j}, \\ &0 = \mathbf{g}(t, \mathbf{x}(t), \mathbf{y}(t), \mathbf{u}(t), \operatorname{sign} \mathbf{q}), \quad \mathbf{q} = (t, \mathbf{x}, \mathbf{y}, \mathbf{u}), \\ &\mathbf{x}(t_{0,j}), \mathbf{y}(t_{0,j}) \text{ given}, \quad \mathbf{r}(t_{f,j}, \mathbf{x}(t_{f,j}), \mathbf{y}(t_{f,j})) = 0, \\ &\mathbf{h}(t, \mathbf{x}(t), \mathbf{y}(t), \mathbf{u}(t), \mathbf{q}) \ge 0. \end{aligned} \tag{3}$$

where in addition to the notation of Section 1.2,

$$\begin{aligned} E &: \mathbb{R}^{1+n_x+n_y} \to \mathbb{R}, \\ \mathbf{r} &: \mathbb{R}^{1+n_x+n_y} \to \mathbb{R}^{n_r}, \quad \text{and} \\ \mathbf{h} &: \mathbb{R}^{1+n_x+n_y+n_u+n_q} \to \mathbb{R}^{n_h}. \end{aligned}$$

In our applications, each horizon P_j typically covers approximately one up to several hours of real-time. The objective is constructed as a weighted sum of partial objectives

$$J[\mathbf{u}] = \omega_1 J_1[\mathbf{u}] + \cdots + \omega_{n_J} J_{n_J}[\mathbf{u}], \quad \omega_i \ge 0,$$

corresponding to, e.g., product gain and energy consumption. In case of a parameter estimation problem J describes a nonlinear least squares objective l_2 consisting of a sum of squares of differences between values obtained from measurement and simulation as, e.g., (f) in Figure 4 and Section 5.2.

2.2 *Offline-Optimisation*

The deterministic optimal control problem (3) is solved by control parameterisation (direct single shooting) and nonlinear optimisation (SQP). Currently, we investigate two types of parameterisation for the control variables:

a) full (open loop) parameterisation of the controls by piecewise polynomial shape functions $\widehat{\mathbf{u}}(t, \mathbf{p}) = \sum_{i=1}^{n_l} \mathbf{p}_i \mathbf{u}_i(t)$, $\mathbf{p} = (\mathbf{p}_1^T, \ldots, \mathbf{p}_{n_l}^T)^T \in \mathbb{R}^{n_u \cdot n_l}$, $\mathbf{p}_i \in \mathbb{R}^{n_u}$, $i = 1, \ldots, n_l$, and
b) a parameterisation according to control functions $\widehat{\mathbf{u}}(\mathbf{x}, \mathbf{p})$, $\mathbf{p} \in \mathbb{R}^{n_p}$ that are used in conventional (closed loop) automated control.

By substitution of the controls $\widehat{\mathbf{u}}$ with their parameterised counterparts and enforcement of the path inequality constraints $\mathbf{h}$ on a mesh $\tau_j^{0,\mathbf{h}} < \cdots < \tau_j^{M,\mathbf{h}} \in [t_{0,j}; t_{f,j}]$ (Vassiliadis et al. [55,56]) the infinite dimensional optimal control problem (3) is converted into the finite dimensional optimisation problem

$$\begin{aligned} &\min_{\mathbf{p}} \widehat{J}[\mathbf{p}] = \widehat{E}(t_{f,j}, \widehat{\mathbf{x}}(t_{f,j}), \widehat{\mathbf{y}}(t_{f,j})) \\ \text{subject to} \quad &\dot{\widehat{\mathbf{x}}}(t; \mathbf{p}) = \widehat{\mathbf{f}}(t, \widehat{\mathbf{x}}(t; \mathbf{p}), \widehat{\mathbf{y}}(t; \mathbf{p}), \mathbf{p}, \operatorname{sign} \widehat{\mathbf{q}}), \quad t_{0,j} \le t \le t_{f,j}, \\ &0 = \widehat{\mathbf{g}}(t, \widehat{\mathbf{x}}(t; \mathbf{p}), \widehat{\mathbf{y}}(t; \mathbf{p}), \mathbf{p}, \operatorname{sign} \widehat{\mathbf{q}}), \quad \widehat{\mathbf{q}} = (t, \widehat{\mathbf{x}}, \widehat{\mathbf{y}}, \mathbf{p}), \end{aligned} \tag{4}$$

$$\widehat{\mathbf{x}}(t_{0,j};\mathbf{p}),\widehat{\mathbf{y}}(t_{0,j};\mathbf{p})\text{ given},\quad \widehat{\mathbf{r}}(t_{f,j},\widehat{\mathbf{x}}(t_{f,j};\mathbf{p}),\widehat{\mathbf{y}}(t_{f,j});\mathbf{p})=0,$$
$$\widehat{\mathbf{h}}(\tau_j^{\nu,h},\widehat{\mathbf{x}}(\tau_j^{\nu,h};\mathbf{p}),\widehat{\mathbf{y}}(\tau_j^{\nu,h};\mathbf{p}),\mathbf{p},\widehat{\mathbf{q}})\geq 0,\quad \nu=0,\ldots,M.$$

Here, we use $\widehat{\cdot}$ as a qualifier in order to denote the functions and variables which are derived by the parameterisation of the respective terms in (3).

Most other numerical optimal control methods (see, e.g., Binder et al. [3]) are (currently) not suitable for the special class of problems treated in this paper. Obstacles are given by, e.g., the inavailability of data required, or the size and complexity of the processes to be considered.

2.3 A Real-Time Capable Method for Optimisation and Disturbance Rejection

In Section 2.2, the numerical *off-line* solution of a deterministic optimisation problem gives approximations for parameterised optimal controls $\widehat{\mathbf{u}}^*$ as well as for corresponding optimal state variable trajectories $\mathbf{x}^*$ and $\mathbf{y}^*$.

Real-time control of the load change problem requires the compensation of disturbances. Due to inevitable disturbances a load change implementing off-line (open loop) computed controls without compensation by state feedback is likely to fail as the constraints may be violated or the value of the objective may become inacceptable. Disturbances arise, e.g., from

- the deviation between the real-life process and its mathematical/numerical model (limited precision of modelling),
- uncertainties in the model, or
- errors in the measurement (or estimation) of actual state variables, or
- time lags between measurement, numerical computation, and the implementation of a control decision[1].

Small disturbances can be compensated by (conventional) setpoint trajectory tracking control, i.e., by enforcing the open loop computed parameterised optimal trajectory. Using this technique, the larger the disturbances and thus the deviations from the (parameterised) optimal reference trajectory are, the more the trajectory obtained in real-time will differ from the optimal trajectory. Apart from the loss in optimality the constraints will be increasingly violated as the disturbances are growing. In the context of the application envisaged this is a serious drawback of the method.

On the other hand, a perpetual complete recomputation of the optimal reference trajectory according to Section 2.2 is prohibitive due to the required computation times. As long as the disturbances are not "too large", a possible remedy is given by the method of *linearisation of neighbouring parameterised extremals*, applied here to the parameterised optimal control problem introduced in Section 2.2. The idea is to modify *on-line* the optimal trajectory which is used as a reference trajectory

[1] Bellman [2] has stated the *Principle of Macroscopic Uncertainty* according to which perfect control of a large system is not possible due to the time required for the essential steps of control.

according to the deviations in order to approximate an optimal feedback control. The parameterised extremals are then updated iteratively in "short" time steps.

As we will see in the following Section 3, this technique requires the *computation of sensitivity functions*. Until now, fully satisfactory methods for the computation of the sensitivity functions have only been developed for ODE and index-1 DAE models exhibiting state dependent discontinuities (Galán, Feehery, and Barton [23]). Currently, we extend our algorithm in order to compute the sensitivities of large-scale index-2 DAEs Eqs. (2a)–(2b) with state dependent discontinuities (see Section 4).

3 Neighbouring Extremals

The method of linearisation of neighbouring parameterised extremals has been developed in the 70s and 80s originally for indirect methods and for the coupled boundary value system of equations of motion and adjoint differential equations arising from the optimal control of ODE systems (e.g., Pesch [48,49], Kugelmann, Pesch [33], Krämer-Eis [28]). The linearisation of neighbouring extremals for parameterised optimal control problems derived from large-scale discontinuous index-2 DAE process models Eqs. (2a)–(2b) in Section 1.3 has not previously been treated.

3.1 *Solution Differentiability*

In Maurer and Pesch [43,44] solution differentiability of (infinite dimensional) optimal control problems

$$
\begin{aligned}
\min_{\tilde{\mathbf{u}}, t_f} \tilde{J}[\tilde{\mathbf{u}}, t_f] &= \tilde{E}(\tilde{\mathbf{x}}(t_f;\boldsymbol{\pi}),\boldsymbol{\pi}) + \int_{t_0}^{t_f} \tilde{\Lambda}(\tilde{\mathbf{x}}(t;\boldsymbol{\pi}),\tilde{\mathbf{u}}(t;\boldsymbol{\pi}),\boldsymbol{\pi}) \\
\dot{\tilde{\mathbf{x}}}(t;\boldsymbol{\pi}) &= \tilde{\mathbf{f}}(\tilde{\mathbf{x}}(t;\boldsymbol{\pi}),\tilde{\mathbf{u}}(t;\boldsymbol{\pi}),\boldsymbol{\pi}); \quad t \in [t_0, t_f] \\
0 &\geq \tilde{\mathbf{h}}(\tilde{\mathbf{x}}(t;\boldsymbol{\pi}),\tilde{\mathbf{u}}(t;\boldsymbol{\pi}),\boldsymbol{\pi}) \\
0 = \tilde{\mathbf{x}}(0;\boldsymbol{\pi}) &- \tilde{\mathbf{r}}^{\text{ini}}(\boldsymbol{\pi}) = 0, \quad 0 = \tilde{\mathbf{r}}^{\text{end}}(\tilde{\mathbf{x}}(t_f;\boldsymbol{\pi}),\boldsymbol{\pi})
\end{aligned} \tag{5}
$$

with respect to the *disturbance parameters* $\boldsymbol{\pi} \in \mathbb{R}^{n_\pi}$ is examined.

The basic task is to find the control $\tilde{\mathbf{u}}(t;\boldsymbol{\pi}) \in \mathbb{R}^{n_{\tilde{u}}}$ minimising the objective functional $\tilde{J}$ subject to the path inequality constraints $\tilde{\mathbf{h}} : \mathbb{R}^{n_{\tilde{x}}+n_{\tilde{u}}+n_\pi} \to \mathbb{R}^{m_{\tilde{h}}}$ and the final point constraints $\tilde{\mathbf{r}}^{\text{end}} : \mathbb{R}^{n_{\tilde{x}}+n_\pi} \to \mathbb{R}^{m_{\tilde{r}\text{end}}}$ given a nominal disturbance vector $\boldsymbol{\pi} := \boldsymbol{\pi}_0$. The system dynamics $\tilde{\mathbf{x}}(t;\boldsymbol{\pi}) \in \mathbb{R}^{n_{\tilde{x}}}$ is determined by the ODE initial value problem $\tilde{\mathbf{f}} : \mathbb{R}^{n_{\tilde{x}}+n_{\tilde{u}}+n_\pi} \to \mathbb{R}^{n_{\tilde{x}}}$, $\tilde{\mathbf{r}}^{\text{ini}} : \mathbb{R}^{n_\pi} \to \mathbb{R}^{n_{\tilde{x}}}$.

The nominal solution $(\tilde{\mathbf{x}}_0(t), \tilde{\mathbf{u}}_0(t)) := (\tilde{\mathbf{x}}(t;\boldsymbol{\pi}_0), \tilde{\mathbf{u}}(t;\boldsymbol{\pi}_0))$ is defined as the solution of the optimal control problem (5) for the reference parameter $\boldsymbol{\pi} := \boldsymbol{\pi}_0$. Associated with the nominal solution are the adjoint variables $\boldsymbol{\lambda}_0(t) := \boldsymbol{\lambda}(t;\boldsymbol{\pi}_0) \in \mathbb{R}^{n_{\tilde{x}}}$ and the Lagrange multipliers $\boldsymbol{\mu}_0(t) := \boldsymbol{\mu}(t;\boldsymbol{\pi}_0) \in \mathbb{R}^{m_{\tilde{h}}}$.

Based on the BVP derived from the description of the optimal solution of problem (5) by variational calculus, second order sufficient conditions (SSC), and a Riccati ODE related to the SSC solution differentiability with respect to the disturbance parameters around the nominal solution is shown. Büskens and Maurer [12] summarise the results in a theorem on solution differentiability. This theorem provides the theoretical justification for the application of neighbouring extremal approaches in order to calculate near optimal approximations to disturbed ($\boldsymbol{\pi} \neq \boldsymbol{\pi}_0$) optimal control problems Eq. (5).

The idea of neighbouring extremals is that – given differentiability of the solution with respect to the disturbance parameters – corrections to the nominal solution can be found according to truncated Taylor's series expansions. E.g., for a first-order correction of the state variable trajectory one has

$$\tilde{\mathbf{x}}(t;\boldsymbol{\pi}) \doteq \tilde{\mathbf{x}}(t;\boldsymbol{\pi}_0) + \frac{\partial \tilde{\mathbf{x}}(t;\boldsymbol{\pi}_0)}{\partial \boldsymbol{\pi}}(\boldsymbol{\pi} - \boldsymbol{\pi}_0).$$

The quantities

$$\frac{\partial \tilde{\mathbf{x}}(t;\boldsymbol{\pi}_0)}{\partial \boldsymbol{\pi}}, \frac{\partial \tilde{\mathbf{u}}(t;\boldsymbol{\pi}_0)}{\partial \boldsymbol{\pi}}, \frac{\partial \boldsymbol{\lambda}(t;\boldsymbol{\pi}_0)}{\partial \boldsymbol{\pi}}, \frac{\partial \boldsymbol{\mu}(t;\boldsymbol{\pi}_0)}{\partial \boldsymbol{\pi}}, \text{ and } \frac{\partial t_f(\boldsymbol{\pi}_0)}{\partial \boldsymbol{\pi}}$$

are the *sensitivity differentials* of the various unknowns with respect to the *disturbance parameter* $\boldsymbol{\pi}$ at the nominal solution. The question is on how to obtain the sensitivity differentials.

3.2 Sensitivity of Parameterised ODE Optimal Control Problems

In Büskens and Maurer [11, 12] perturbed optimal control problems of the type Eq. (5) with an eventually free final time $t_f \in \mathbb{R}_+$ are considered.

Starting point is the NLP obtained by the parameterisation of the controls in the infinite dimensional optimal control problem Eq. (5)

$$\begin{aligned} \min_{\mathbf{p}} \bar{J}(\mathbf{p},\boldsymbol{\pi}) := \bar{E}(\bar{\mathbf{x}}_N(\mathbf{p},\boldsymbol{\pi}), t_N, \boldsymbol{\pi}) + \sum_{\nu=0}^{N-1} (t_{\nu+1} - t_\nu)\bar{\Lambda}(\bar{\mathbf{x}}_\nu(\mathbf{p},\boldsymbol{\pi}), t_\nu, \boldsymbol{\pi}), \\ \text{subject to} \quad 0 = \bar{G}_\nu(\mathbf{p},\boldsymbol{\pi}); \quad 1 \le \nu \le r, \\ 0 \le \bar{G}_\nu(\mathbf{p},\boldsymbol{\pi}); \quad r+1 \le \nu \le m_{\bar{G}}. \end{aligned} \tag{6}$$

$\mathbf{p} \in \mathbb{R}^{n_p}$ contains the optimisation parameters, i.e., the shape parameters for the parameterised control functions $\tilde{\mathbf{u}}(t;\boldsymbol{\pi}) \curvearrowright \bar{\mathbf{u}}(t,\mathbf{p};\boldsymbol{\pi})$ (see Section 2.2), and $\bar{\mathbf{x}}_\nu(\mathbf{p},\boldsymbol{\pi})$ are values of the state variables on the mesh $t_0 \le t_1 \le \cdots \le t_N = t_f$. $\bar{G}_\mu(\mathbf{p},\boldsymbol{\pi})$ are a collection of both point equality and inequality constraints; the explicit dependency from state and control variables is dropped as both are identified by their respective parameterised and discretised counterparts.

Of primary interest are the parametric sensitivities $\partial \mathbf{p}(\boldsymbol{\pi})/\partial \boldsymbol{\pi}$. They are obtained following the sensitivity analysis for NLPs given in Fiacco [21]:

Associated with the NLP (6) is a Lagrangian function

$$\bar{L}(\mathbf{p},\bar{\boldsymbol{\mu}},\boldsymbol{\pi}) := \bar{J}(\mathbf{p},\boldsymbol{\pi}) + \bar{\boldsymbol{\mu}}^T \cdot \bar{G}(\mathbf{p},\boldsymbol{\pi}),$$

where $\bar{\boldsymbol{\mu}} \in \mathbb{R}^{m_{\bar{G}}}$ are the Lagrangian multipliers. Now let $(\mathbf{p}_0, \bar{\boldsymbol{\mu}}_0)$ be the solution of problem (6) at the nominal value $\boldsymbol{\pi} = \boldsymbol{\pi}_0$. Further let $\bar{G}^a$ be the vector of all active constraints $\bar{G}_\nu(\mathbf{p},\boldsymbol{\pi}) = 0$, $\nu = 1,\ldots,m_{\bar{G}}$, and let $\bar{\boldsymbol{\mu}}^a$ be the vector of associated multipliers. Then, under the strict complementary condition (the multipliers for active inequality constraints are positive), the nominal solution can be embedded into a C^1 family of solutions $(\mathbf{p}(\boldsymbol{\pi}),\bar{\boldsymbol{\mu}}(\boldsymbol{\pi}))$, with the 1st order sensitivities given by

$$\begin{bmatrix} \frac{\partial \mathbf{p}}{\partial \boldsymbol{\pi}} \\ \frac{\partial \bar{\boldsymbol{\mu}}^a}{\partial \boldsymbol{\pi}} \end{bmatrix}_{\boldsymbol{\pi}_0} = - \begin{bmatrix} \frac{\partial^2}{\partial \mathbf{p}^2}\bar{L} & \left[\frac{\partial}{\partial \mathbf{p}}\bar{G}^a\right]^* \\ \frac{\partial}{\partial \mathbf{p}}\bar{G}^a & 0 \end{bmatrix}^{-1}_{\substack{\mathbf{p}_0,\bar{\boldsymbol{\mu}}_0,\\ \boldsymbol{\pi}_0}} \begin{bmatrix} \frac{\partial^2}{\partial \mathbf{p}\partial \boldsymbol{\pi}}\bar{L} \\ \frac{\partial}{\partial \boldsymbol{\pi}}\bar{G}^a \end{bmatrix}_{\substack{\mathbf{p}_0,\bar{\boldsymbol{\mu}}_0,\\ \boldsymbol{\pi}_0}}.$$

If Eq. (6) is solved with SQP methods, we can assume that the constraint Jacobian $\partial\bar{G}(\mathbf{p},\boldsymbol{\pi})/\partial\mathbf{p}$ and the gradient of the objective function $\partial\bar{J}(\mathbf{p},\boldsymbol{\pi})/\partial\mathbf{p}$ is provided by the user, as well as $\partial\bar{G}(\mathbf{p}_0,\boldsymbol{\pi}_0)/\partial\boldsymbol{\pi}$. Still, we require the Hessians of the Lagrangian $\partial^2\bar{L}(\mathbf{p}_0,\bar{\boldsymbol{\mu}}_0,\boldsymbol{\pi}_0)/\partial\mathbf{p}^2$ and $\partial^2\bar{L}(\mathbf{p}_0,\bar{\boldsymbol{\mu}}_0,\boldsymbol{\pi}_0)/\partial\mathbf{p}\partial\boldsymbol{\pi}$. Unfortunately, the approximate Hessian $\partial^2\bar{L}/\partial\mathbf{p}^2$ generated by the SQP method cannot be used due to its lack of accuracy. Therefore, Büskens and Maurer [11] propose to compute the Hessian explicitly after the NLP has been solved either by approximation via finite differences (Büskens [10]) or by solution of an ODE (Büskens [9]).

First order updates to the nominal solution are then given by, e.g.,

$$\bar{\mathbf{u}}(t,\mathbf{p};\boldsymbol{\pi}) \doteq \bar{\mathbf{u}}(t,\mathbf{p}_0;\boldsymbol{\pi}_0) + \left[\frac{\partial\bar{\mathbf{u}}}{\partial\mathbf{p}}\cdot\frac{\partial\mathbf{p}}{\partial\boldsymbol{\pi}} + \frac{\partial\bar{\mathbf{u}}}{\partial\boldsymbol{\pi}}\right]_{\substack{\mathbf{p}_0,\\ \boldsymbol{\pi}_0}} (\boldsymbol{\pi}-\boldsymbol{\pi}_0).$$

4 Sensitivities of Very Large Scale Index-2 DAEs with Discontinuities

As we have seen in Section 2.2 and Section 3.2, parametric sensitivity functions are a fundamental ingredient of both the off-line and the on-line optimal control methods subject to our interest:

- In the direct optimisation method they are used in order to provide gradient information for the underlying SQP solver.
- In the algorithm for updating optimal parameterised trajectories according to the theory of neighbouring extremals they contain derivative information required in order to set up a truncated Taylor's series expansion of the trajectories.

Currently, there are three basic methods for the numerical computation of parametric sensitivity functions which are applied in the context of DAE models: Finite difference approximations (e.g., Buchauer et al. [7], Schwerin et al. [53]), backward integration of the adjoint equations (e.g., Gritsis et al. [46], Morrison and Sargent [45]), and integration of the sensitivity equations (e.g., Heim and von Stryk [26], Maly and Petzold [41]). Each of these methods has its special advantages and disadvantages. Unless dictated by the availability of software, in non-standard applications the choice of the method to be used is strongly problem dependent. Due to reliability and efficiency considerations, an implementation should use an IND (internal numerical differentiation) approach instead of an END (external numerical differentiation) scheme whenever possible (Bock et al. [4], Kiehl [27]).

4.1 Integration of the Sensitivity System by a Staggered Direct Method

In the context of the simulation and optimisation package OPTISIM®, the (parameterised) optimal control problem (4), and the chemical engineering application (cf. Section 1.1: load change of a cryogenic air separation plant), we are faced with the following setting:

a) The model is a large scale index-2 DAE with discontinuities.
b) The inequality constraints are enforced on a grid.
c) The dimension of the space of optimisation parameters is typically small.
d) The source code of the (BDF-)integrator can be accessed (Burr [8]).

Based upon this data, our method of choice (Engl et al. [19], Kronseder [31]) is the integration of the sensitivity DAE of the model by *differentiation of the integrator* in a *staggered direct method* implementation (Caracotsios and Stewart [13], Leis and Kramer [35], Li and Petzold [38]).

Consider the parameterised optimal control problem (4) and focus on the model DAE initial value problem

$$\begin{aligned} \dot{\hat{\mathbf{x}}}(t;\mathbf{p}) &= \hat{\mathbf{f}}(t,\hat{\mathbf{x}}(t;\mathbf{p}),\hat{\mathbf{y}}(t;\mathbf{p}),\mathbf{p},\operatorname{sign}\hat{\mathbf{q}}), \quad t_{0,j} \le t \le t_{f,j}, \\ 0 &= \hat{\mathbf{g}}(t,\hat{\mathbf{x}}(t;\mathbf{p}),\hat{\mathbf{y}}(t;\mathbf{p}),\mathbf{p},\operatorname{sign}\hat{\mathbf{q}}), \quad \hat{\mathbf{q}} = (t,\hat{\mathbf{x}},\hat{\mathbf{y}},\mathbf{p}), \\ &\hat{\mathbf{x}}(t_{0,j};\mathbf{p}),\hat{\mathbf{y}}(t_{0,j};\mathbf{p}) \text{ given.} \end{aligned} \tag{7}$$

Given a fixed set of parameters $\mathbf{p}$, the model equations in OPTISIM® are integrated using a method based on backward difference formulas (BDF) (Burr [8]). BDF methods are implicit linear multi-step methods with variable order and variable step-size suited for solving index-1 DAEs numerically. With a modified error control criterion they can also be used for the direct integration of semi-explicit index-2 DAEs (Brenan et al. [6]).

In the nth integration step of a BDF method the interpolating polynomial of the $k-1$ previously computed points $(\hat{\mathbf{x}}_{n-k+1},\hat{\mathbf{y}}_{n-k+1}), \ldots, (\hat{\mathbf{x}}_n,\hat{\mathbf{y}}_n)$ and of the next point $(\hat{\mathbf{x}}_{n+1},\hat{\mathbf{y}}_{n+1})$ is formally constructed. The new point $(\hat{\mathbf{x}}_{n+1},\hat{\mathbf{y}}_{n+1})$ is determined by the condition that the interpolating polynomial has to fulfil the DAE at t_{n+1}. By extrapolation of the interpolating polynomial of $(\hat{\mathbf{x}}_{n-k},\hat{\mathbf{y}}_{n-k}), \ldots, (\hat{\mathbf{x}}_n,\hat{\mathbf{y}}_n)$ for t_{n+1}, estimates (*predictors*) $\hat{\mathbf{x}}_{n+1}^{\text{pred}}$, $\hat{\mathbf{y}}_{n+1}^{\text{pred}}$, and $\dot{\hat{\mathbf{x}}}_{n+1}^{\text{pred}}$ are obtained. The BDF method with fixed leading coefficient results in the system of nonlinear equations

$$\begin{aligned} 0 &= \hat{\mathbf{f}}(t_{n+1},\hat{\mathbf{x}}_{n+1},\hat{\mathbf{y}}_{n+1},\mathbf{p},\operatorname{sign}\hat{\mathbf{q}}) - \dot{\hat{\mathbf{x}}}_{n+1}^{\text{pred}} + \frac{\alpha_k}{\eta_{n+1}}(\hat{\mathbf{x}}_{n+1} - \hat{\mathbf{x}}_{n+1}^{\text{pred}}) \\ 0 &= \hat{\mathbf{g}}(t_{n+1},\hat{\mathbf{x}}_{n+1},\hat{\mathbf{y}}_{n+1},\mathbf{p},\operatorname{sign}\hat{\mathbf{q}}), \quad \hat{\mathbf{q}} = (t_{n+1},\hat{\mathbf{x}}_{n+1},\hat{\mathbf{y}}_{n+1},\mathbf{p}), \end{aligned} \tag{8}$$

where the *leading coefficient* $\alpha_k \in \mathbb{R}$ depends on the order k of the method and $\eta_{n+1} = t_{n+1} - t_n$ is the step-size. Eq. (8) is solved by a modified Newton algorithm

$$\hat{\mathbf{x}}_{n+1}^{[0]} := \hat{\mathbf{x}}_{n+1}^{\text{pred}}, \quad \hat{\mathbf{y}}_{n+1}^{[0]} := \hat{\mathbf{y}}_{n+1}^{\text{pred}},$$

$$\begin{bmatrix} \frac{\alpha_k}{\eta_{n+1}} \mathbf{Id} + \widehat{\mathbf{f}}_{\widehat{\mathbf{x}}} & \widehat{\mathbf{f}}_{\widehat{\mathbf{y}}} \\ \widehat{\mathbf{g}}_{\widehat{\mathbf{x}}} & \widehat{\mathbf{g}}_{\widehat{\mathbf{y}}} \end{bmatrix} \begin{bmatrix} \Delta\widehat{\mathbf{x}}^{[m]} \\ \Delta\widehat{\mathbf{y}}^{[m]} \end{bmatrix} = - \begin{bmatrix} \widehat{\mathbf{f}} - \dot{\widehat{\mathbf{x}}}^{\text{pred}}_{n+1} + \frac{\alpha_k}{\eta_{n+1}} (\widehat{\mathbf{x}}^{[m]}_{n+1} - \widehat{\mathbf{x}}^{\text{pred}}_{n+1}) \\ \widehat{\mathbf{g}} \end{bmatrix} \tag{9}$$

$$\widehat{\mathbf{x}}^{[m+1]}_{n+1} := \widehat{\mathbf{x}}^{[m]}_{n+1} + c\Delta\widehat{\mathbf{x}}^{[m]}, \quad \widehat{\mathbf{y}}^{[m+1]}_{n+1} := \widehat{\mathbf{y}}^{[m]}_{n+1} + c\Delta\widehat{\mathbf{y}}^{[m]}, \quad 0 < c \leq 1$$

($\widehat{\mathbf{f}}_{\widehat{\mathbf{x}}} := \partial\widehat{\mathbf{f}}/\partial\widehat{\mathbf{x}}, \widehat{\mathbf{f}}_{\widehat{\mathbf{y}}} := \partial\widehat{\mathbf{f}}/\partial\widehat{\mathbf{y}}, \ldots$). In each step of the Newton iteration, the linear system Eq. (9) is solved by a hierarchical algorithm efficiently applying modern direct sparse matrix techniques to the large, sparse and unstructured system matrix (Eich-Söllner et al. [17])

$$\mathsf{D} = \begin{bmatrix} \frac{\alpha_k}{\eta_{n+1}} \mathbf{Id} + \widehat{\mathbf{f}}_{\widehat{\mathbf{x}}} & \widehat{\mathbf{f}}_{\widehat{\mathbf{y}}} \\ \widehat{\mathbf{g}}_{\widehat{\mathbf{x}}} & \widehat{\mathbf{g}}_{\widehat{\mathbf{y}}} \end{bmatrix}.$$

Now consider the total differentiation of the corrector system Eq. (8) with respect to the parameters $\mathbf{p}$. This yields the *linear* system of equations

$$0 = \mathsf{D} \begin{bmatrix} \rho_{n+1} \\ \sigma_{n+1} \end{bmatrix} + \begin{bmatrix} \widehat{\mathbf{f}}_{\mathbf{p}} - (\dot{\rho}^{\text{pred}} + \frac{\alpha_k}{\eta_{n+1}} \rho^{\text{pred}}_{n+1}) \\ \widehat{\mathbf{g}}_{\mathbf{p}} \end{bmatrix} \tag{10}$$

where $\rho = \rho(t;\mathbf{p}) := \left[\frac{\partial\widehat{\mathbf{x}}(t;\mathbf{p})}{\partial\mathbf{p}}\right] \in \mathbb{R}^{n_{\widehat{x}} \times n_p}$ and $\sigma = \sigma(t;\mathbf{p}) := \left[\frac{\partial\widehat{\mathbf{y}}(t;\mathbf{p})}{\partial\mathbf{p}}\right] \in \mathbb{R}^{n_{\widehat{y}} \times n_p}$ denote the *sensitivity matrices*. On the other hand, application of the BDF scheme to the sensitivity equations related to the original DAE Eq. (7) results in the *same* linear system Eq. (10).

Thus, the sensitivity matrices can be computed after each integration step for the state variable trajectories by (direct) solution of Eq. (10) (*staggered direct method*). This step is computationally cheap, since (an approximation of) the matrix D is already available in decomposed form in the modified Newton iteration Eq. (9). Moreover, by the equivalence noted above this method corresponds to the (separate) integration of the sensitivity equations with the same step-sizes and order sequence as used for the original DAE which is important in the context of IND.

4.2 Sensitivity Transfer at Discontinuities

The method for the computation of sensitivity functions for very large DAE models of index 2 described in Section 4.1 has been successfully applied to the solution of real-life optimal control problems from chemical engineering with purely time dependent discontinuities (see Section 5.1). In our special problem setting, the computation of sensitivity functions in the general case of time- *and* state-dependent discontinuities shows to be very demanding.

In the context of ODEs, the formulae for the sensitivity transfer at discontinuities have been clearly elaborated by Rozenvasser [51]. Feehery [22] and Gálan et al. [23] report the corresponding results for index ≤ 1 DAEs. Adapted and simplified, the model considered consists in a sequence of DAEs

$$\widehat{\mathbf{F}}_n(t, \widehat{\mathbf{x}}(t;\mathbf{p}), \widehat{\mathbf{y}}(t;\mathbf{p}), \dot{\widehat{\mathbf{x}}}(t;\mathbf{p}), \mathbf{p}) = 0; \quad t_{n-1} < t < t_n, \ n = 1, 2, \ldots \tag{11}$$

$$\text{with} \qquad \text{rank}\left(\left[\frac{\partial \widehat{\mathbf{F}}_n}{\partial \widehat{\mathbf{y}}}\ \frac{\partial \widehat{\mathbf{F}}_n}{\partial \dot{\widehat{\mathbf{x}}}}\right]\right) = n_{\widehat{\mathbf{x}}} + n_{\widehat{\mathbf{y}}} \quad \text{(along the solution)} \tag{12}$$

where $\widehat{\mathbf{x}} \in \mathbb{R}^{n_{\widehat{x}}}$ is the vector of differential variables, $\widehat{\mathbf{y}} \in \mathbb{R}^{n_{\widehat{y}}}$ is the vector of algebraic variables, $\mathbf{p} \in n_{\mathbf{p}}$ is the vector of parameters, $t \in \mathbb{R}$ is the independent variable and $\widehat{\mathbf{F}}_n \in \mathcal{C}^1\left(\mathbb{R}^{1+n_{\widehat{x}}+n_{\widehat{y}}+n_{\widehat{x}}+n_p}, \mathbb{R}^{n_{\widehat{x}}+n_{\widehat{y}}}\right)$, $n = 1, 2, \dots$, are the different models. The rank criterion Eq. (12) restricts the range of problems covered to ODEs and most index-1 DAEs. In order to tackle with higher index problems Feehery [22] applies index reduction according to the method of dummy derivatives introduced by Mattsson and Söderlind [42]. The necessary differentiations are carried out symbolically by automatic differentiation which is a feature of the underlying simulation package ABACUSS that does not apply to OPTISIM®.

Furthermore we have the (a-priori adequately defined) sets of initial conditions and jump conditions

$$\widehat{k}_0(t_0, \widehat{\mathbf{x}}_0, \widehat{\mathbf{y}}_0, \dot{\widehat{\mathbf{x}}}_0, \mathbf{p}) = 0$$

$$\widehat{k}_n(t_n, \widehat{\mathbf{x}}_n^+, \widehat{\mathbf{y}}_n^+, \dot{\widehat{\mathbf{x}}}_n^+, \widehat{\mathbf{x}}_n^-, \widehat{\mathbf{y}}_n^-, \dot{\widehat{\mathbf{x}}}_n^-, \mathbf{p}) = 0; \quad n = 1, 2, \dots$$

where $\widehat{k}_0 \in \mathcal{C}^1(\mathbb{R}^{1+n_{\widehat{x}}+n_{\widehat{y}}+n_{\widehat{x}}+n_p}, \mathbb{R}^{n_{\widehat{x}}})$, $\widehat{k}_n \in \mathcal{C}^1(\mathbb{R}^{1+2(n_{\widehat{x}}+n_{\widehat{y}}+n_{\widehat{x}})+n_p}, \mathbb{R}^{n_{\widehat{x}}})$, $n = 1, 2, \dots$, and with $\mathbf{z} \in \{\widehat{\mathbf{x}}, \widehat{\mathbf{y}}, \dot{\widehat{\mathbf{x}}}\}$, $\mathbf{z}_0 = \mathbf{z}_0(t_0; \mathbf{p}) := \mathbf{z}(t; \mathbf{p})|_{t=t_0}$, $\mathbf{z}_n^- = \mathbf{z}_n^-(t_n; \mathbf{p}) := \lim_{t \nearrow t_n} \mathbf{z}(t; \mathbf{p})$, $\mathbf{z}_n^+ = \mathbf{z}_n^+(t_n; \mathbf{p}) := \lim_{t \searrow t_n} \mathbf{z}(t; \mathbf{p})$, $n = 1, 2, \dots$. The times of the discontinuity events are given as the zeros of the scalar valued switching functions $\widehat{\mathbf{q}}_n \in \mathcal{C}^1\left(\mathbb{R}^{1+n_{\widehat{x}}+n_{\widehat{y}}+n_{\widehat{x}}+n_p}, \mathbb{R}\right)$:

$$\widehat{\mathbf{q}}_n(t_n, \widehat{\mathbf{x}}(t_n; \mathbf{p}), \widehat{\mathbf{y}}(t_n; \mathbf{p}), \dot{\widehat{\mathbf{x}}}(t_n; \mathbf{p}), \mathbf{p}) = 0.$$

The roots of the switching functions must not coincide. Later on the higher order time derivatives $\ddot{\widehat{\mathbf{x}}}$ and $\dot{\widehat{\mathbf{y}}}$ will be required. Given Eq. (12) these derivatives can be obtained at a *consistent* point $[t, \widehat{\mathbf{x}}, \widehat{\mathbf{y}}, \dot{\widehat{\mathbf{x}}}, \mathbf{p}]$ from the definition of index-1 DAEs according to

$$\frac{d}{dt}\widehat{\mathbf{F}}_n(t, \widehat{\mathbf{x}}, \widehat{\mathbf{y}}, \dot{\widehat{\mathbf{x}}}, \mathbf{p}) = 0 \overset{\text{Eq. (12)}}{\Leftrightarrow} \left[\frac{\partial \widehat{\mathbf{F}}_n}{\partial \widehat{\mathbf{y}}}\ \frac{\partial \widehat{\mathbf{F}}_n}{\partial \dot{\widehat{\mathbf{x}}}}\right] \begin{bmatrix} \dot{\widehat{\mathbf{y}}} \\ \ddot{\widehat{\mathbf{x}}} \end{bmatrix} = -\left[\frac{\partial \widehat{\mathbf{F}}_n}{\partial t} + \frac{\partial \widehat{\mathbf{F}}_n}{\partial \widehat{\mathbf{x}}}\dot{\widehat{\mathbf{x}}}\right]. \tag{13}$$

In addition to the notation introduced up to now we use for $\mathbf{z} \in \{\boldsymbol{\rho}, \boldsymbol{\sigma}, \dot{\boldsymbol{\rho}}\}$ $\mathbf{z}_n^- = \mathbf{z}_n^-(t_n; \mathbf{p}) := \lim_{t \nearrow t_n} \mathbf{z}(t; \mathbf{p})$, $\mathbf{z}_n^+ = \mathbf{z}_n^+(t_n; \mathbf{p}) := \lim_{t \searrow t_n} \mathbf{z}(t; \mathbf{p})$, as well as $\widehat{\mathbf{F}}_n^+ := \lim_{t \searrow t_n} \widehat{\mathbf{F}}_{n+1}$ and $\widehat{\mathbf{q}}_n^- := \lim_{t \nearrow t_n} \widehat{\mathbf{q}}_n$. By definition, at each discontinuity t_n, $n = 1, 2, \dots$, the system of equations

$$\widehat{\mathbf{F}}_{n+1}(t_n, \widehat{\mathbf{x}}_n^+, \widehat{\mathbf{y}}_n^+, \dot{\widehat{\mathbf{x}}}_n^+, \mathbf{p}) = 0 \tag{14a}$$

$$\widehat{k}_n(t_n, \widehat{\mathbf{x}}_n^+, \widehat{\mathbf{y}}_n^+, \dot{\widehat{\mathbf{x}}}_n^+, \widehat{\mathbf{x}}_n^-, \widehat{\mathbf{y}}_n^-, \dot{\widehat{\mathbf{x}}}_n^-, \mathbf{p}) = 0 \tag{14b}$$

$$\widehat{\mathbf{q}}_n(t_n, \widehat{\mathbf{x}}_n^-, \widehat{\mathbf{y}}_n^-, \dot{\widehat{\mathbf{x}}}_n^-, \mathbf{p}) = 0 \tag{14c}$$

holds. Therefore, sensitivity of the switching time $t_n = t_n(\mathbf{p})$ with respect to the parameters $\mathbf{p}$ can be obtained from total differentiation of the switching condition Eq. (14c)

$$\left[\frac{\partial\widehat{q}_n^-}{\partial t} + \frac{\partial\widehat{q}_n^-}{\partial\widehat{x}}\dot{\widehat{x}}^- + \frac{\partial\widehat{q}_n^-}{\partial\widehat{y}}\dot{\widehat{y}}^- + \frac{\partial\widehat{q}_n^-}{\partial\widehat{x}}\ddot{\widehat{x}}^-\right]\left[\frac{\partial t_n}{\partial p}\right] = \\ -\left[\frac{\partial\widehat{q}_n^-}{\partial\widehat{x}}\rho^- + \frac{\partial\widehat{q}_n^-}{\partial\widehat{y}}\sigma^- + \frac{\partial\widehat{q}_n^-}{\partial\widehat{x}}\dot{\rho}^- + \frac{\partial\widehat{q}_n^-}{\partial p}\right]$$

given $\left[\frac{\partial\widehat{q}_n^-}{\partial t} + \frac{\partial\widehat{q}_n^-}{\partial\widehat{x}}\dot{\widehat{x}}^- + \frac{\partial\widehat{q}_n^-}{\partial\widehat{y}}\dot{\widehat{y}}^- + \frac{\partial\widehat{q}_n^-}{\partial\widehat{x}}\ddot{\widehat{x}}^-\right] \neq 0$ (transversality condition). Accordingly, total differentiation of Eqs. (14a)–(14b) with respect to the parameters under consideration of the parametric dependency of the switching time gives the sensitivities of the *dynamic consistent initialisation problem*

$$\begin{bmatrix} \frac{\partial\widehat{F}_n^+}{\partial\widehat{x}} & \frac{\partial\widehat{F}_n^+}{\partial\widehat{y}} & \frac{\partial\widehat{F}_n^+}{\partial\widehat{x}} \\ \frac{\partial\widehat{k}_n}{\partial\widehat{x}^+} & \frac{\partial\widehat{k}_n}{\partial\widehat{y}^+} & \frac{\partial\widehat{k}_n}{\partial\widehat{x}^+} \end{bmatrix} \begin{bmatrix} \rho^+ \\ \sigma^+ \\ \dot{\rho}^+ \end{bmatrix} = - \begin{bmatrix} \frac{\partial\widehat{F}_n^+}{\partial\widehat{x}} & \frac{\partial\widehat{F}_n^+}{\partial\widehat{y}} & \frac{\partial\widehat{F}_n^+}{\partial\widehat{x}} \\ \frac{\partial\widehat{k}_n}{\partial\widehat{x}^+} & \frac{\partial\widehat{k}_n}{\partial\widehat{y}^+} & \frac{\partial\widehat{k}_n}{\partial\widehat{x}^+} \end{bmatrix} \begin{bmatrix} \dot{\widehat{x}}^+ \\ \dot{\widehat{y}}^+ \\ \ddot{\widehat{x}}^+ \end{bmatrix} \cdot \left[\frac{\partial t_n}{\partial p}\right]$$
$$- \begin{bmatrix} \frac{\partial\widehat{F}_n^+}{\partial t} & 0 & 0 & 0 & \frac{\partial\widehat{F}_n^+}{\partial p} \\ \frac{\partial\widehat{k}_n}{\partial t} & \frac{\partial\widehat{k}_n}{\partial\widehat{x}^-} & \frac{\partial\widehat{k}_n}{\partial\widehat{y}^-} & \frac{\partial\widehat{k}_n}{\partial\widehat{x}^-} & \frac{\partial\widehat{k}_n}{\partial p} \end{bmatrix} \begin{bmatrix} \frac{\partial t_n}{\partial p} \\ \rho^- + \dot{\widehat{x}}^- \cdot \frac{\partial t_n}{\partial p} \\ \sigma^- + \dot{\widehat{y}}^- \cdot \frac{\partial t_n}{\partial p} \\ \dot{\rho}^- + \ddot{\widehat{x}}^- \cdot \frac{\partial t_n}{\partial p} \\ \mathbf{Id} \end{bmatrix}. \tag{15}$$

In the OPTISIM® environment symbolic or automatic differentiation cannot be applied in order to derive an index-1 system from the index-2 DAE given. This, in connection with the lack of a-priori given adequate initial conditions, has initiated the development of our algorithms for consistent initialisation and sensitivity transfer at discontinuities for index-2 DAEs.

In Kronseder [32] attempts of various authors to tackle with the problem of sensitivity transfer in discontinuous DAEs are discussed. One of these methods is to build finite differences of the jump function, i.e, of the jump conditions Eq. (14b) in explicit form (Ertel and Arnold [20]), which has been implemented and successfully applied to the solution of a parameter identification problem from chemical engineering in Kröner et al. [30]; see also in Section 5.2 below. A central point in our implementation is the evaluation of the unknown jump function by *back-tracing* according to Sincovec et al. [52]. In our case, back-tracing utilises a special property of BDF integrators: starting from "not too inconsistent" initial values they can arrive at a consistent solution manifold after a number of integration steps. The idea is to reverse the direction of integration as soon as the (consistent) solution manifold is reached; arrived back at the start of the integration the integrator should then give the set numerically consistent initial values corresponding to the current solution manifold.

By numerical experiments we have found that numerical differentiation of the jump function in connection with back-tracing can give satisfactory results suitable for the solution of "small scale" optimal control problems (cf. Section 5.2). But when applied to problems from chemical engineering with higher levels of difficulty this approach has shown a lack of reliability; especially we encountered severe problems during (disturbed) back-tracing, frequently resulting in aborts during the backward integration phase.

Therefore, we have examined a second approach starting from the results of Feehery [22] discussed above in closer detail. Our approach uses structural analysis of the DAE model (*Algorithm of Pantelides*, Pantelides [47], Unger et al. [54]) and numerical approximation of the total time derivatives required for local index reduction (Leimkuhler et al. [34]). A suitable set of dynamic degrees of freedom is determined automatically by application of a structural algorithm, detail knowledge from chemical engineering, and a heuristic method. For the solution of the nonlinear system of reduced consistency equations several different algorithms can be applied in combination (SQP (Gill et al. [24]), Newton, Levenberg-Marquardt and Dog-Leg (Chen and Stadtherr [14]). In order to address the problem of obtaining suitable initial estimates (which may be a serious problem with chemical engineering models) back-tracing (Sincovec et al. [52]) may be employed again. Further numerical results for this second approach to sensitivity transfer across discontinuities are given in Kronseder [32].

5 NUMERICAL RESULTS

Numerical results are presented that demonstrate the efficiency of the solution procedures developed for our general concept for model-based predictive optimal control (Figure 4). The interplay between the components within the general framework and the balancing of the different time scales as well as a comparison of numerical results provided by the method of neighbouring parameterised extremals with the ones obtained using a direct update of optimal controls whenever the state vector is disturbed is a subject of ongoing investigations.

5.1 *Off-Line Solution of an Industrial Optimal Control Problem*

We consider a load change process of an existing air separation plant (Kröner et al. [30]) of the type introduced in Section 1.1. The actual task is to decrease the load of the plant from 100 % air input to 60 %. In real life, the load change itself (i.e., the reduction of the feed air) takes about one hour. The time horizon considered in the mathematical optimal control problem (3) is from $t_0 = 0$ [s] to $t_f = 6000$ [s] in order to be able to take into account the long range dynamics of the process especially in the treatment of constraints.

The air separation plant is modelled by an index-2 DAE system of the form Eqs. (2a)–(2b) consisting of about $n_x = 900$ differential and $n_y = 2600$ algebraic equations. The plant model is the basis of one of the largest, nonlinear DAE optimal control problems that has ever been solved numerically. However, the approach described in this paper aims at even larger plant models.

The purity restrictions result in lower and upper path constraints for six of the state variables, i.e., $n_h = 12$; the constraints and their bounds are specified in Table 2. The $n_u = 5$ control variables describe the valve settings (items (a) - (e) in Figure 2). The control variables are parameterised implicitly based on functions used

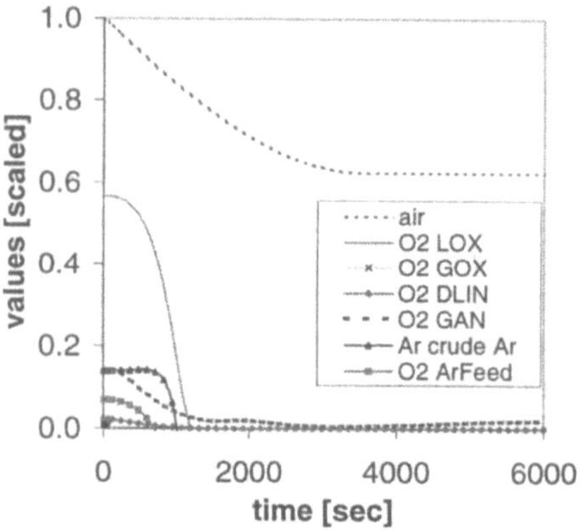

Figure 5. Purities and air flow for original parameter setting (values are scaled to lower and upper bounds from Table 2)

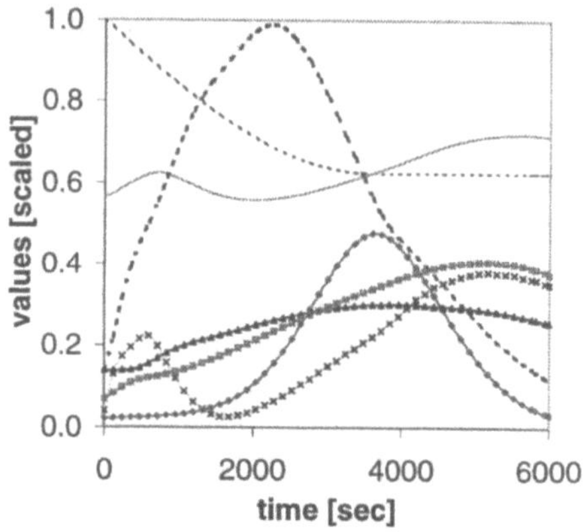

Figure 6. Purities and air flow for optimised parameter setting (for legend see Figure 5)

in conventional load control; altogether, $n_{\mathbf{p}} = 9$ parameters are used. The path inequality constraints are discretised in time on an equidistant mesh with 10 nodes yielding $10 \cdot n_{\hat{\mathbf{h}}} = 120$ nonlinear point inequality constraints for the NLP. The objective is to maximise an integral term describing product gain. Such an objective of Lagrange form is easily brought to Mayer form (3) by adding an additional differential equation. However, the primary interest is in finding a feasible control leading to the satisfaction of all constraints for such a highly complex plant.

Figure 5 shows the trajectories of the relevant purities of the process, simulated with the starting values for the optimisation parameters **p**. This set leads to a breakdown of the air separation process as several variables severely violate their bounds. After successful solution of the parameterised optimal control problem (4) with the state-of-the art SQP algorithm SNOPT (Gill et al. [24]) all purities are feasible within their lower and upper bounds during the entire time horizon, as displayed in Figure 6.

We have already solved the same optimal control problem using the NAG Fortran Library implementation (NAG [39]) of NPSOL, the predecessor of SNOPT. In the direct comparison of the two methods, SNOPT could improve the final objective by approximately 16%. Additionally, the user interaction necessary in our NPSOL-based implementation in order to obtain a set of optimisation parameters giving a first *feasible* state trajectory for this difficult problem (Kronseder [31], Engl et al. [19]) was almost eliminated due to the sophisticated methods used by SNOPT for generating a feasible solution starting from an infeasible point.

5.2 *Parameter Identification and Sensitivity Functions*

As a numerical illustration for the correct transfer of sensitivities across state-dependent discontinuities using our first algorithm based on back-tracing and finite

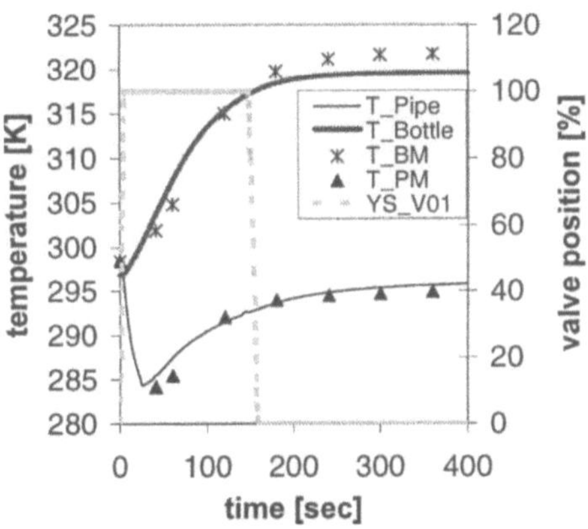

Figure 7. Non-insulated bottling process, fitted and measured data for bottle (T_Bottle, T_BM) and piping (T_Pipe, T_PM) temperature

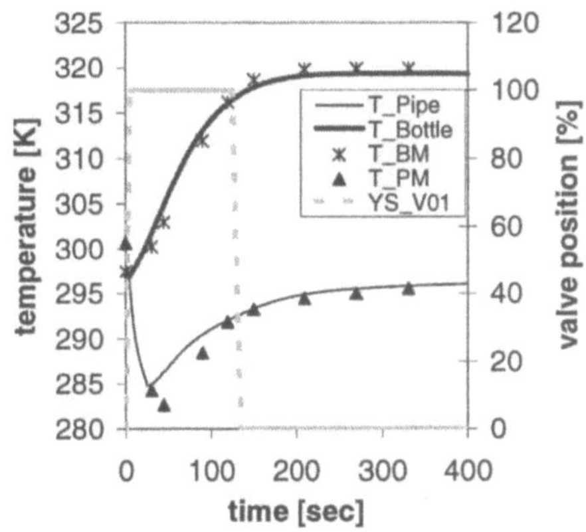

Figure 8. Insulated bottling process, simulated and measured data for bottle (T_Bottle, T_BM) and piping (T_Pipe, T_PM) temperature

differencing, we now consider a parameter identification problem solved by the direct shooting method described in Section 2.2 (Kröner et al. [30]).

The underlying problem arising from industrial application is the filling of industrial gases (such as N_2, O_2, or Ar) into high pressure gas bottles for retail. Initially, the bottles are depressurised to vacuum in order to avoid contamination of the pure product with residue in the bottle. The process in view is the subsequent filling of the bottles with the gas from a high pressure gas tank up to a final pressure of 200 [bar].

An important point in the investigation of potentials for the acceleration of the filling process is the heat balance of the bottle system, which requires knowledge on the coefficients for the gas-bottle heat transfer in order to determine the heat flows in the system. As no model for prediction of heat transfer coefficients in a fast pressurised gas volume is available from the literature, heat transfer coefficients must be determined from measurements. The newly implemented algorithms are applied to a model tuning problem, where in a first approximation constant heat transfer coefficients are identified.

In the experiment, a single depressurised gas bottle is filled from a bottle battery. The valve between bottle battery and bottle to be filled is opened at time zero and closed after 120 seconds when pressure equilibrium has been achieved. Measured observables are the temperature at the filling pipe and the temperature at the outside of the single gas bottle. Based on measurements from an experiment with a non-insulated bottle, $n_p = 3$ heat transfer coefficients within the bottle are to be determined by our dynamic parameter identification algorithm. The optimised results (solid lines) and the 8 sampling points ($*$, ▲) entering the optimisation are shown in Figure 7. YS_V01 denotes the valve position. The simulation model for the entire process, i.e., bottle battery, the single gas bottle, piping, and valves, has

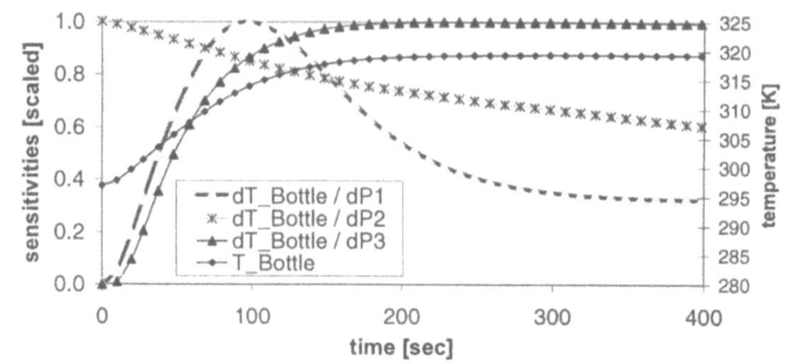

Figure 9. Sensitivity functions of the state variables depicted in Figure 7

$n_{\hat{x}} + n_{\hat{y}} = 144$ equations and a differential index of 1. Optimisation parameters are three heat transfer coefficients. The parameter identification algorithm finally reduces the value of the weighted l_2 objective function (weighted sum of squares of deviations between simulation and measurement) by 17%.

A comparison of measurements at 9 points in time originating from an experiment with an insulated bottle with simulation results using the heat transfer coefficients determined above show the applicability of the fitted parameters even to a not too closely related case (Figure 8).

6 Discussion of Real-Time Optimality

To our knowledge, until now there has been a lack of a clear concept of real-time optimality. Some reasoning has been done in the working paper of Engell et al. [18]. Based on this paper, we propose the following terms concerning *real-time optimisation* of large scale dynamical systems as a first step towards a more rigorous definition of various aspects of real-time optimality:

Let there be a *real-causal process*[2] P with time-dependent observable variables **x** and **y**, manipulable inputs **u** (continuous control functions, parameters, or discrete decision variables) and an objective function depending on the state variable and on the control variable trajectories within a finite or infinite interval in time. The objective may include requirements depending on time. Causality means that inputs can influence present and future values of the observable variables only, i.e., the same inputs up to a point in time t and the same initial conditions generate the same output trajectories up to t.

[2] i.e., a process with a time dependent evolution determined by physical laws

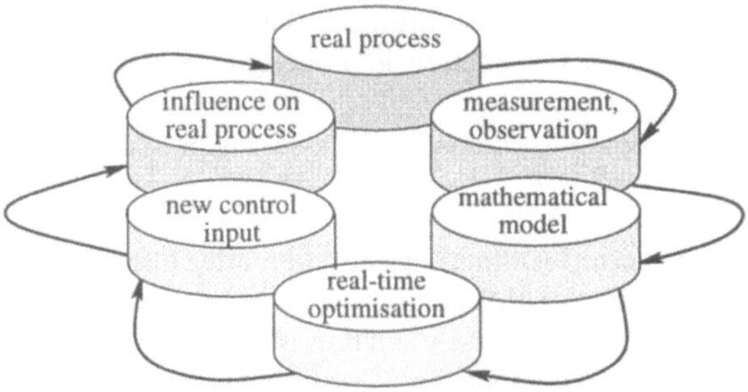

Figure 10. control loop structure of real-time optimisation algorithms

The structure of information is such that the inputs at a certain time can only be determined on the basis of the observable variables and of the requirements until that time. These observable variables and requirements may only be available with a delay due to measurement, transmission, or computation times. Furthermore, they may be biased due to errors in measurement and transmission. The value of the objective function as a function of the inputs may possess uncertainty, e.g., due to unknown or not exactly predictable disturbances on the system of interest, imprecise knowledge of the system, or since requirements for the process performance may change unpredictably with time. A *characteristic property* of real-time optimisation problems is the presence of incomplete information and uncertainty (e.g., current air density or wind force at the re-entry of a spacecraft (Breitner [5])).

The task of real-time optimisation is to determine the control variables aiming to optimise the objective variables while observing constraints if they are present. The determination of the controls depends on the data, observable variables, and requirements up to the respective point in time. In general, a *model* of the process and of the unknown disturbances and requirements is used. The notion of "real-time optimality" implies

1. a real process,
2. a facility in order to (one time, several times, repeatedly) interact (online) with the process during its evolution in order to modify or control it.
3. a facility for the determination of the interactions with the process, i.e., for the determination of the control variables, and
4. a time limit for the determination of the interactions, which is due to the intention of the obtained controls to influence the real process evolving in time (if possible so as to obtain an optimum performance of the process for the overall process time).

In a more general sense one could think of a control loop (Figure 10).

A critical point in the discussion of real-time optimal control is the *assessment* of the performance of a process evolution. The "classical" notion of optimality means an optimum in the objective function during the duration of the process (in the off-line sense) provided that complete information is available. But, as motivated above, incomplete information is a characteristic property of real-time optimisation problems.

There are several possible criteria for the assessment of a real-time optimisation algorithm:

1. If possible, a posteriori (offline) optimisation after termination of the process, i.e., under the assumption that all information is known finally. (However this is not the case with many processes, and thus often neither practicable nor reasonable.) The a-posteriori computed off-line optimal solution gives a bound on the best possible performance of the real-time optimal control algorithm, which is equivalent to the solution of the optimal control problem with complete model and arbitrary amount of computation time.
   We call this notion of real-time optimality "*a-posteriori optimality*".
2. Consider a deterministic mathematical model incorporating all (or at least the most important) unknowns, and use this *meta-* or *super-model* for the analysis of robustness-, stability-, or optimality properties.
   Due to the possible applications of this notion of optimality, we call it *robust optimality*.
3. It appears to be reasonable to demand from a real-time optimal control algorithm that it achieves only the best possible performance that can be expected in view of the "circumstances" present. "Circumstances" or criteria are, e.g.,
   - the available information,
   - the (real-)time interval available for the generation of a reaction,
   - and the size of the system (i. e., the process or its model, respectively).

   While the first criterion is of more fundamental nature, the second and the third criterion are strongly related to the hard- and software available. However, these criteria are difficult to state and define in a general sense. We call this notion of optimality "*real-time optimality*".
   This notion of optimality appears to suit best with Bellman's idea of "on-line [optimal] control", cf. Bellman [2, Section 14.5]: "Closely associated ... is that [problem] of 'on-line' control. Here the constraint is a novel one mathematically – one not previously encountered in scientific research. We are required to render a decision, perhaps supply a numerical answer, within a specified period of time. It is no longer a question of devising a computationally feasible algorithm; instead we must obtain the best approximation within a specified time."

If we are interested in a *physical* process or plant, the availability of an exact a-posteriori optimal solution depends on the model of the process. Mathematical statements can only be made on mathematical models. In common, an assessment of the optimality of the real process cannot be made exactly. At the best one may judge its performance based on empirical knowledge, e.g., that a method gave an improvement of at least X % on a set of examples in relation to another known method (Engell et al. [18]).

7 Summary

A framework for model-based on-line optimal control of air separation plants described by very large scale DAE models of index 2 incorporating state dependent switchings has been suggested. The method consists of various tasks to be performed on different time scales. Among them are the mid-term computation of parameterised optimal controls, the short-term updates of optimal open loop reference trajectories through linearisation, and the long-term re-calibration of dynamic model parameters. As the key problem towards a numerical solution of these tasks the efficient computation of sensitivity functions for large scale DAE models of index 2 in the presence of state dependent switchings has been identified and discussed in detail. After the presentation of several numerical results, the paper is concluded with some general thoughts about various notions of real-time optimality.

References

1. H. Baldus, K. Baumgärtner, H. Knapp, M. Streich: Verflüssigung und Trennung von Gasen. In: K. Winnacker, Küchler, L.: Chemische Technologie Band 3: Anorganische Technologie II. Carl Hanser Verlag (1983)
2. R. E. Bellman: Introduction to the Mathematical Theory of Control Processes: Volume II: Nonlinear Processes. Academic Press (1971)
3. T. Binder, L. Blank, H. G. Bock, R. Bulirsch, W. Dahmen, M. Diehl, T. Kronseder, W. Marquardt, J. P. Schlöder, O. von Stryk: Introduction to Model Based Optimization of Chemical Processes on Moving Horizons. This volume.
4. H. G. Bock, J. P. Schlöder, V. Schulz: Numerik großer Differentiell-Algebraischer Gleichungen — Simulation und Optimierung. In: H. Schuler (ed.), Prozeßsimulation VCH Verlagsgesellschaft mbH (1995) 35-80
5. M. H. Breitner: Robust optimal on-board reentry guidance of an European space shuttle: dynamic game approach and guidance synthesis with neural networks. Schwerpunktprogramm der Deutschen Forschungsgemeinschaft Echtzeitoptimierung großer Systeme, Preprint 99-4 (1999)
6. K. E. Brenan, S. L. Campbell, L. R. Petzold: Numerical Solution of Initial-Value Problems in Differential-Algebraic Equations. Classics in Applied Mathematics **14**, SIAM (1996)
7. O. Buchauer, P. Hiltmann, M. Kiehl: Sensitivity Analysis of Initial-Value Problems with Application to Shooting Techniques. Numer. Math. **67** (1994) 151-159
8. P. S. Burr: The Design of Optimal Air Separation and Liquefaction Processes with the OPTISIM equation-oriented Simulator and its Application to on-line and off-line plant Optimization. AIChE Symposium Series **89**, 294 (1993) 1-7
9. C. Büskens: Optimierungsmethoden und Sensitivitätsanalyse für optimale Steuerprozesse mit Steuer- und Zustandsbeschränkungen. Ph.D. thesis, Westfälische Wilhelms-Universität Münster (1998)
10. C. Büskens.: private communications (1999)
11. C. Büskens, H. Maurer: Real-Time Control of an Industrial Robot Using Nonlinear Programming Methods. Schwerpunktprogramm der Deutschen Forschungsgemeinschaft Echtzeitoptimierung großer Systeme, Preprint 98-12 (1998)
12. C. Büskens, H. Maurer: SQP-Methods for Solving Optimal Control Problems with Control and State Constraints: Adjoint Variables, Sensitivity Analysis and Real-Time Con-

trol. Schwerpunktprogramm der Deutschen Forschungsgemeinschaft Echtzeitoptimierung großer Systeme, Preprint 00-03 (2000)
13. M. Caracotsios, W. E. Stewart: Sensitivity Analysis of Initial Value Problems with Mixed ODEs and Algebraic Equations. Computers Chem. Engng. **9**, 4 (1985) 359-365
14. H.-S. Chen, M. A. Stadtherr: A Modification of Powell's Dogleg Method for Solving Systems of Nonlinear Equations. Computers Chem. Engng. **5**, 3 (1981) 143-150
15. I. S. Duff: Matrix Methods. Rutherford Appleton Laboratory Technical Report RAL-TR-1998-076 (1998)
16. E. Eich: Projizierende Mehrschrittverfahren zur numersichen Lösung von Bewegungsgleichungen technischer Mehrkörpersysteme mit Zwangsbedingungen und Unstetigkeiten. Fortschritt-Berichte VDI, Reihe 18, Nr. 109 (1992)
17. E. Eich-Söllner, P. Lory, P. Burr, A. Kröner: Stationary and dynamic flowsheeting in the chemical engineering industry. Surv. Math. Ind. **7** (1997) 1-28
18. S. Engell, O. von Stryk, M. H. Breitner, U. Zimmermann: Definitionsversuch Echtzeitoptimierung. Internal working paper of the DFG Schwerpunktprogramm "Echtzeitoptimierung großer Systeme" (1997)
19. G. Engl, A. Kröner, T. Kronseder, O. von Stryk: Numerical simulation and optimal control of air separation plants. In: H.-J. Bungartz et al. (eds.): High Performance Scientific and Engineering Computing. Lecture Notes in Computational Science and Engineering **8**, Springer (1999) 221-231
20. S. Ertel, M. Arnold: Die Berechnung von Sensitivitätsmatrizen unter Berücksichtigung unstetiger Zustandsänderungen. Manuscript, DLR, Oberpfaffenhofen (March 1998)
21. A. V. Fiacco: Introduction to Sensitivity and Stability Analysis in Nonlinear Programming. Mathematics in Science and Engineering, Vol. 165, Academic Press, New York (1983)
22. W. F. Feehery: Dynamic Optimization with Path Constraints. Ph.D. thesis, Massachusetts Institute of Technology March 5 (1998)
23. S. Galán, W. F. Feehery, P. I. Barton: Parametric Sensitivity Functions for Hybrid Discrete/Continuous Systems. Appl. Numer. Math. **31**, 1 (1999) 17-47
24. P. E. Gill, W. Murray, M. A. Saunders: SNOPT: An SQP Algorithm for Large-Scale Constrained Optimization. University of California, Department of Mathematics, Numerical Analysis Report 97-2, San Diego (USA) (1997)
25. P. E. Gill, W. Murray, M. A. Saunders, M. H. Wright: User's Guide for NPSOL (Version 5.0): A FORTRAN Package for Nonlinear Programming. University of California, Department of Mathematics, Numerical Analysis Report 98-2, San Diego (USA), (1998)
26. A. Heim, O. von Stryk: Documentation of PAREST — A Multiple Shooting Code for Optimization Problems in Differential-Algebraic Equations. Report TUM-M9616, Mathematisches Institut, Technische Universität München (1996)
27. M. Kiehl: Sensitivity Analysis of ODEs and DAEs — Theory and Implementation Guide. Optimization Methods and Software **10**, 6 (1999) 803-821
28. P. Krämer-Eis: Ein Mehrzielverfahren zur numerischen Berechnung optimaler Feedback-Steuerungen bei beschränkten nichtlinearen Steuerungsproblemen. Brieskorn, E. et al. (eds.), Bonner Mathematische Schriften **166** (1985)
29. A. Kröner, W. Marquardt, E. D. Gilles: Getting Around Consistent Initialization of DAE Systems? Computers Chem. Engng. **21**, 2 (1996) 145-158
30. A. Kröner, T. Kronseder, G. Engl, O. von Stryk: Dynamic optimization for air separation plants. Paper accepted for ESCAPE-11, European Symposium on Computer Aided Process Engineering, May 27-30, 2001, Kolding, Denmark
31. T. Kronseder: Optimal Control of Chemical Engineering Processes. Diploma thesis, Mathematisches Institut der Technischen Universität München (1998)

32. T. Kronseder: Towards model-based online optimal control of chemical engineering plants: Parameterised controls and sensitivity functions for very large-scale index-2 DAE systems with state dependent discontinuities. Manuscript, to be sumitted as Ph.D. dissertation (2001).
33. B. Kugelmann, H. J. Pesch: New general guidance method in constrained optimal control. Part 1: Numerical method. J. Opt. Th. Appl. **67**, 3 (1990) 421-435
34. B. Leimkuhler, L. R. Petzold, C. W. Gear: Approximation Methods for the Consistent Initialization of Differential-Algebraic Equations. SIAM Journal on Numerical Analysis **28**, 1 (1991) 205-226
35. J. R. Leis, M. A. Kramer: Sensitivity Analysis of Systems of Differential and Algebraic Equations. Computers Chem. Engng. **9**, 1 (1985) 93-96
36. A. Lefkopoulos, M. A. Stadtherr: Index Analysis of Unsteady-State Chemical Process Systems I: An Algorithm for Problem Formulation. Computers Chem. Engng. **17**, 4 (1993) 399-413
37. A. Lefkopoulos, M. A. Stadtherr: Index Analysis of Unsteady-State Chemical Process Systems II: Strategies for Determining the Overall Flowsheet Index. Computers Chem. Engng. **17**, 4 (1993) 415-430
38. S. Li, L. R. Petzold: Design of New DASPK for Sensitivity Analysis. Computer Science Technical Report **TRCS99-28**, Department of Computer Science, University of California, Santa Barbara (1999)
39. NAG Fortran Library Mark 18. The Numerical Algorithms Group Ltd., Oxford (1997).
40. J. A. R. Nijsing: Increasing the ASU Controllability by the use of Dynamic Modeling & Simulation. Proc. Munich Meeting of Air Separation Technology MUST'96, Linde AG, Höllriegelskreuth (1996) 93-111
41. T. Maly, L. R. Petzold: Numerical Methods and Software for Sensitivity Analysis of Differential-Algebraic Systems. Appl. Numer. Math. **20** (1996) 57-79
42. S. E. Mattsson, G. Söderlind: Index Reduction in Differential-Algebraic Equations Using Dummy Dervatives. SIAM J. Sci. Comput. **14**, 3 (1993) 677-692
43. H. Maurer, H. J. Pesch: Solution Differentiability for Nonlinear Parametric Control Problems. SIAM Journal on Control and Optimization **32**, 6 (1994) 1542-1554
44. H. Maurer, H. J. Pesch: Solution Differentiability for Parametric Nonlinear Control Problems with Control-State Constraints. Journal of Optimization Theory and Applications **86**, 3 (1995) 649-667
45. K. R. Morison, R. W. H. Sargent: Optimization of Multistage Processes Described by Differential-Algebraic Equations. In: Lecture Notes in Mathematics 1230, A. Dold, Eckmann, B. (eds.), Springer (1986) 86-102
46. D. M. Gritsis, C. C. Pantelides, R. W. H. Sargent: Optimal Control of Systems Described by Index Two Differential-Algebraic Equations. SIAM J. Sci. Comput. **16** (1995) 1349-1366.
47. C. C. Pantelides: The Consistent Initialization of Differential-Algebraic Systems. SIAM Journal on Scientific and Statistical Computing **9**, 2 (1988) 213-231
48. H. J. Pesch: Real-Time Computation of Feedback Controls for Constrained Optimal Control Problems, Part 1: Neighbouring Extremals. Optimal Control, Applications & Methods **10**, 2 (1989) 129-145
49. H. J. Pesch: Real-Time Computation of Feedback Controls for Constrained Optimal Control Problems, Part 2: A Correction Method Based on Multiple Shooting. Optimal Control, Applications & Methods **10**, 2 (1989) 147-171
50. W. Rohde: Production of pure argon from air. Linde Reports on Science and Technology **54** (1994) 3-7

51. E. Rozenvasser: General Sensitivity Equations of Discontinuous Systems. Automation and Remote Control **28**, 3 (1967) 400-404
52. R. F. Sincovec, A. M. Erisman, E. L. Yip, M. A. Epton: Analysis of Descriptor Systems Using Numerical Algorithms. IEEE Transactions on Automatic Control **AC-26**, 1 (1981) 139-147
53. R. von Schwerin, M. J. Winckler, V. H. Schulz: Parameter estimation in discontinuous descriptor models. In: Bestle and Schiehlen (eds.): IUTAM Symposium on Optimization of Mechanical Systems Kluwer Acad. Publ. (1996) 269-276
54. J. Unger, A. Kröner, W. Marquardt: Structural Analysis of Differential-Algebraic Equations — Theory and Applications. Computers Chem. Engng. **19**, 8 (1995) 867-882
55. V. S. Vassiliadis, R. W. H. Sargent, C. C. Pantelides: Solution of a class of multistage dynamic optimization problems. 1. Problems without path constraints. Industrial & Engineering Chemistry Research **33** (1994) 2111-2122
56. V. S. Vassiliadis, R. W. H. Sargent, C. C. Pantelides: Solution of a class of multistage dynamic optimization problems. 2. Problems with path constraints. Industrial & Engineering Chemistry Research **33** (1994) 2123-2133
57. G. Zapp: Dynamic simulation of air separation plants and the use of "Relative Gain Analysis" for design of control systems. Linde Reports on Science and Technology **54** (1994) 13-18

IV

Delay Differential Equations in Medical Decision Support Systems

*Differential Equations with State-Dependent Delays

Eberhard P. Hofer[1], Bernd Tibken[2], and Frank Lehn[1]

[1] Abteilung Meß-, Regel- und Mikrotechnik, Universität Ulm, Germany
[2] Fachgruppe für Automatisierungstechnik und Technische Kybernetik (atk) Bergische Universität—Gesamthochschule Wuppertal, Germany

Abstract In this article an introduction to the wide field of retarded or delay differential equations with state-dependent delays is given. Hereby, the most important features of this type of differential equations which are profoundly different from ordinary differential equations are discussed. The presence of discontinuities in higher derivatives of the solution of delay differential equations require suitable integration methods. Thus, an efficient numerical code for the integration of delay differential equations is presented. This code is based on a fourth order RK one step algorithm. Furthermore, the presence of discontinuities may lead to severe problems when parameters are to be estimated using higher order optimization techniques. These problems are pointed out and a systematic way for the analysis of the smoothness of a least squares objective function is presented.

1 Introduction

The main goal in this section is to give the reader an overview and some insight into the field of retarded or delay differential equations (DDE). The main differences between DDEs and ordinary differential equations (ODE) and the resulting problems concerning numerical simulation and parameter estimation will be pointed out. Therefore, the remainder of this article is splitted into three parts: The first, dealing with an introduction to DDEs and discussing special properties of DDEs, the second, presenting requirements for the numerical simulation of DDEs, and the third, where difficulties within parameter estimation for DDEs are discussed. In the Sections 2 and 4 the most important parts of [5,8], [12,13] and [1,2] are cited.

2 Delay Differential Equations

Delay differential equations are equations of the form

$$\dot{x}(t) = f(t, x(t), x(\alpha_1(t, x(t))), \dots, x(\alpha_r(t, x(t)))), \quad t \in [t_0, t_e], \tag{1}$$

where $\dot{x}(t)$ denotes the time derivative of x. The functions $\alpha_i(t, x(t)) \leq t$, $i = 1, \dots, r$ are called the *retarding arguments, retarding functions* or *lag functions* [5]. In the literature the *delayed terms* $x(\alpha_i(t, x(t)))$, $i = 1, \dots, r$, in (1) are often given in the form $\alpha_i(t, x(t)) = t - \sigma_i(t, x(t))$, $i = 1, \dots, r$, where the functions $\sigma_i(\cdot)$, $i = 1, \dots, r$, are called the *delays*. The simpliest case for a delay is a constant delay, when $\sigma_i = T_i$ holds. In this case the delay is called *constant*. In the case, that the delay depends on the time t it is called *time dependent* and if it further depends on the state $x(t)$ it is called *state dependent* [5]. With these definitions (1) is called a delay differential equation with multiple state dependent delays. For the rest of this section the case $r = 1$ is considered for simplicity, thus, equations of the form

$$\dot{x}(t) = f(t, x(t), x(\alpha(t, x(t)))), \quad t \in [t_0, t_e] \tag{2}$$

are investigated. The results presented in the following can easily be extended to multiple delays and the restriction to $r = 1$ is no loss of generality.

Delay differential equations frequently show properties which cannot be observed in ordinary differential equations. For illustration consider the equation

$$\dot{x}(t) = f(t, x(t), x(t-1)), \quad t \in [t_0, t_e]. \tag{3}$$

Due to the term $x(t-1)$ it is obvious that (3) is a delay differential equation. Furthermore, it is evident that an initial function over $[t_0 - 1, t_0]$ has to be given in order to define the solution to (3) uniquely.

Thus, in order to define an initial value problem, a function

$$x(t) \equiv \Phi(t), \quad t \in [\hat{t}, t_0] \tag{4}$$

where $\hat{t} = \min\{\alpha(t, x(t))\}$, $t \in [t_0, t_e]$, has to be given. Hereby, the set $[\hat{t}, t_0]$ is called the *initial set*, $\Phi(t)$ is called the *initial function* and $\Phi(t_0)$ is called the *initial value* [5, 14]. The term *solution* is defined in the following.

Definition 1 (Feldstein and Neves [5]). A function $x(t)$ is said to be a *solution* of (2) and (4) if the following conditions.

(i) $x(t)$ is a continuous extension of Φ on $[t_0, t_e]$,
(ii) $\alpha(t, x(t)) \leq t$ for $t \in [t_0, t_e]$, and
(iii) $x(t)$ satisfies (2) on $[t_0, t_e]$, where the right hand derivative is used at $t = t_0$

are fulfilled.

The existence and uniqueness of solutions have been studied by Driver [3, 4]. In these publications the author gives sufficient conditions for the existence and uniqueness of a solution which are resumed in the following theorem.

Theorem 2 (Driver [3]). *Let the following conditions hold.*

(i) f, α *and* Φ *are continuous with respect to their respective arguments and* f *is bounded.*
(ii) f *satisfies a Lipschitz condition with respect to the last two arguments,* α *satisfies a Lipschitz condition with respect to the last argument and* Φ *satisfies a Lipschitz condition with respect to* t.

Then, there exists an unique solution for the problem (2) and (4).

The function f is said to satisfy a Lipschitz condition in a region D, if the inequality

$$\|f(t, x_2, y_2) - f(t, x_1, y_1)\| \leq L_1\|x_2 - x_1\| + L_2\|y_2 - y_1\|$$

with Lipschitz constants L_1 and L_2 for all (t, x_1, y_1), (t, x_2, y_2) in D holds [3].

For delay differential equations with only constant delay $\dot{x}(t) = f(t, x(t), x(t-\tau))$ it is easier to answer the question of existence of a solution. Given an initial function $\Phi(t)$ on the initial set $[t_0 - \tau, t_0]$ it is clear that the delayed term $x(t-\tau)$ is a known function of t for $[t_0, t_0 + \tau]$. Thus, the delay differential equation becomes an ordinary differential equation. Consequently, the question of existence on this interval can be treated using well known existence theories for ordinary differential equations. By calculating the solution on $[t_0, t_0+\tau]$ the delayed term is known for $[t_0 + \tau, t_0 + 2\tau]$ and the solution can be computed on this time interval. By repetition of this procedure the solution may be calculated up to any arbitrary time. This method is known in literature as the *method of steps* [6].

It is well known in the literature that the extension of initial conditions for a single point (which has to be given for ODEs) to an initial set (which has to be given for DDEs) is very significant for the continuity of the solution. In order to analyze the solution's continuity at the point $t = t_0$ the left hand derivatives $\Phi^{(r)}(t_0^-)$, defined by the initial function, and the right hand derivatives $x^{(r)}(t_0^+)$, defined by the differential equation, have to be considered for $r = 0, 1, 2, \ldots$. Regarding these derivatives it is evident that a jump discontinuity may emerge at the *initial point* t_0 in some derivative of the solution if $\Phi^{(r)}(t_0^-) \neq x^{(r)}(t_0^+)$ for some r. Due to delayed terms in the right-hand side of a DDE, this jump discontinuity may propagate to subsequent times within the solution. This effect consequently leads to piecewise continuous differentiable solutions. To illustrate the above effect the following example is considered.

Example 3 (Willé and Baker [12]).

$$\dot{x}(t) = x(t-1), \quad t \geq 0 \tag{5}$$

$$x(t) = 1, \quad t \in [-1, 0). \tag{6}$$

By differentiating (6) it is clear that $\dot{x}(t) = 0$ holds on $[-1, 0)$ but using (5) leads to $\dot{x}(t) = 1$ on $[0, 1)$. Thus, $\dot{x}(t)$ has a jump discontinuity at $t_0 = 0$. The effect of this discontinuity propagates to the points $t = 1, 2, 3, \ldots$ due to the delayed term $x(t-1)$ in (5). Differentiating (5) k times leads to $x^{(k+1)}(t) = x^{(k)}(t-1)$, which can be written as $x^{(k+1)}(t) = \dot{x}(t-k)$. Thus, $x^{(k+1)}$ has a jump discontinuity at $t = k$. In this case x is said to have a $(k+1)$th-*order discontinuity* at $t = k$.

Derivative discontinuities of this type are a common feature in delay differential problems. These discontinuities may be of importance in numerical algorithms. If points which have insufficient continuity are disregarded in a numerical algorithm, this may lead to the invalidity of the integration schemes used within the method. Thus, in a numerical code it is necessary to know the positions and the orders of the derivative discontinuities. It can be shown [4, 14], that all derivative discontinuities originate from the initial set or point. Thus, it is sufficient to understand how discontinuities propagate through the solution. In the following, the main results from [5, 8] concerning the propagation of jump discontinuities within the solution of a delay differential equation are presented.

In order to analyze the mentioned propagation of derivative discontinuities the following definitions are given and will be used repeatedly. The following definition introduces the terms *compatibility* and *incompatibility* for the initial function $\Phi(t)$. Throughout the following text the standard notation $C^p[I]$ is used to denote the set of all functions whose first p derivatives exist and are continuous on the closed set I.

Definition 4 (Feldstein and Neves [5]). The initial function $\Phi(t)$ of (4) is called J-*incompatible*, provided $J \geq 1$ is the least integer such that $\Phi^{(J)}(t_0^-) \neq x^{(J)}(t_0^+)$. If no such finite J exists, then Φ is called *compatible*.

In the following a definition of a multiple zero is given, which can be regarded as a *left-handed multiplicity*.

Definition 5 (Feldstein and Neves [5]). Let Y be a real number in the range of $\alpha(t, x(t))$ for $t \in [t_0, t_e]$. Denote $g(t) \equiv \alpha(t, x(t)) - Y$. Z will be called a *(left-handed) zero of multiplicity* m of $g(t)$ provided:

(i) $g(Z) = 0$,
(ii) $g^{(m)}(Z^-)$ exists from the left,
(iii) $m \geq 1$ is the least integer such that $g^{(m)}(Z^-) \neq 0$.

Definition 6 establishes the term of a C^p *extension* of a function.

Definition 6 (Feldstein and Neves [5]). Let $t_0 \leq t_k \leq t_e$ and $u(t) \in C^p[t_0, t_k]$. The function $u^*(t)$ defined on $[t_0, t_e]$ will be said to be a C^p *extension* of $u(t)$ from $[t_0, t_k]$ to $[t_0, t_e]$ provided

(i) $u^*(t) = u(t)$ for all $t \in [t_0, t_k]$,
(ii) $u^*(t) \in C^p[t_0, t_e]$.

The following Definition 7 gives a precise definition of the term *changing sign.*

Definition 7 (Feldstein and Neves [5]). A function $u(t)$ is said to *change sign* at a point $t = Z$ provided there exists an open interval U containing Z on which for all $t \in U$ either

$$u(t) \begin{cases} < 0 \text{ for } t < Z, \\ = 0 \text{ for } t = Z, \\ > 0 \text{ for } t > Z \end{cases} \quad \text{or} \quad u(t) \begin{cases} > 0 \text{ for } t < Z, \\ = 0 \text{ for } t = Z, \\ < 0 \text{ for } t > Z \end{cases}$$

It should be mentioned that Definition 7 excludes functions like $\sin\left(\frac{1}{t}\right)$ at $t = 0$ [5]. In the following definition 8 smoothness properties of the problem (2) and (4) are specified.

Definition 8 (Feldstein and Neves [5]). Problem (2) and (4) has *continuity class* $p \geq 1$, if and only if the following hold over the appropriate domains:

(i) All of the mixed partial derivatives $f_{i,j,k}$ are continuous for all $i + j + k \leq p$,

(ii) All of the mixed partial derivatives $\alpha_{i,j}$ are continuous for all $i + j \le p$,
(iii) $\Phi \in C^p[\hat{t}, t_0]$

In Definition 8 the quantities $f_{i,j,k}$ and $\alpha_{i,j}$ are abbreviations for the partial derivatives

$$f_{i,j,k} := \frac{\partial^{i+j+k} f(x,y,z)}{\partial^i x \partial^j y \partial^k z}, \tag{7}$$

$$\alpha_{i,j} := \frac{\partial^{i+j} \alpha(x,y)}{\partial^i x \partial^j y} \tag{8}$$

where $i, j, k = 0, 1, 2, \ldots$. In the following definition some notation conventions on the location and the order of discontinuities and their relation are given.

Definition 9 (Feldstein and Neves [5]). Derivatives are denoted by lower case letters. Y and Z denote points of discontinuities in derivatives of x, while y and z are the lowest derivatives of x that jump at these points, respectively. Specifically, let $Z > t_0$ denote a point of a jump discontinuity of some derivative of x. Then

1. Z and z are related as follows: z is the positive integer such that
$$x \in C^{z-1}(Z) \quad \text{and} \quad x \notin C^z(Z).$$
2. Y and Z are related as follows: $Z > Y$ and
$$Y = \alpha(Z, x(Z)). \tag{9}$$
3. Y and y are related as follows: y is the positive integer such that
$$x \in C^{y-1}(Y) \quad \text{and} \quad x \notin C^y(Y).$$
4. Let y and p be integers such that $0 \le y \le p$. For $Y \in [t_0, t_e]$ and $\zeta > 0$, denote
$$C^p_y[Y-\zeta, Y+\zeta] = C^p[Y-\zeta, Y] \cap C^p[Y, Y+\zeta] \cap C^y[Y-\zeta, Y+\zeta]. \tag{10}$$
For simplicity, when Y and ζ are understood, denote this by C^p_y.
5. Since $C^p_y \subset C^p_{y-1}$, we consider the set differences
$$D^p_y \equiv C^p_{y-1} - C^p_y. \tag{11}$$
If $u(t) \in D^p_y[Y-\zeta, Y+\zeta]$, then both of the limits $u^{(y)}(Y-)$ and $u^{(y)}(Y+)$ exist and are finite but
$$u^{(y)}(Y^-) \neq u^{(y)}(Y^+). \tag{12}$$
6. For $u \in C(I) \equiv C^0[I]$ denote $\|u\|^I_\infty = \max |u(t)|$ for $t \in I$.

Due to the incompatibility of the initial function Φ jumps in $x^{(q)}$ for $q \le p+1$ can be propagated. If $x^{(y)}$ jumps at Y, one would expect that $x^{(y+1)}$ jumps at $Z > Y$, but not $x^{(y)}$, where Z satisfies (9). In other words, one would expect that $z = y + 1$. However, the connection between y and z is a function of m, the multiplicity of the solution to (9) which was proved in [8]. The following theorem establishes this connection precisely.

Theorem 10 (Feldstein and Neves [5]). *Let the problem given by (2) and (4) have continuity class* $p \geq 1$. *For* $Y \in [t_0, t_e]$, *let the integer* $y \in [1, p]$ *be such that* $x \in C^p_{y-1}[Y - \zeta, Y + \zeta]$ *for some* $\zeta > 0$. *Assume that there exists a least number* $Z \in (Y, t_e)$ *such that* Z *is a zero of integer multiplicity* $m \geq 1$ *of* $\alpha(t, x(t)) - Y$. *Then* $x \in C^p_q[Z - \xi, Z + \xi]$ *for some* $\xi > 0$ *(thus* $z \geq q + 1$*) where*

(i) $q = p$ *if* m *is even,*
(ii) $q = \min(p, my)$ *if* m *is odd.*

The solution of (2) and (4) has $p + 1$ continuous derivatives except at the various derivative jump points if the problem has continuity class p. Furthermore, if, for example, α is piecewise monotone these jump points are isolated. However, it may appear that the jump points have a finite cluster point. This appears, if the following equation holds true for a clustering sequence of propagated jump points $\{Y_i\}$.

$$Y_i = \alpha(Y_{i+1}, x(Y_{i+1})) \quad \text{and} \quad \lim_{i \to \infty} Y_i = W. \tag{13}$$

Using the continuity of x and α this results in

$$W = \alpha(W, x(W))$$

if the limits are taken in (13). This relationship leads to the following proposition.

Proposition 11 (Feldstein and Neves [5]). *Let the problem given by (2) and (4) have continuity class* $p \geq 1$. *Then either the jump points are isolated or they cluster at a fixed point of* α.

If it is assumed that W is a fixed point of α then either W is the limit of an infinite sequence of jump points of $x^{(q)}$ for various q, or it is not. If W is not such a limit, then the right-hand side of (2) belongs to C^p which is implied by Definition 8. Furthermore, this means that $x \in C^{p+1}$ in a neighborhood of W which follows by differential equations theory. If W is the limit of such a sequence of jump points, then $x \in C^{p+1}$ in a neighborhood of W is true again because $x \in C^{p+1}$ between jump points and the fact F4 (generalized smoothing) given below implies that the smoothness of $x(t)$ increases by at least one at each successive jump point. Thus, the following proposition is formulated.

Proposition 12 (Feldstein and Neves [5]). *Let the problem (2) and (4) have continuity class* $p \geq 1$. *If* W *is a fixed point of* α, *then* $x \in C^{p+1}$ *in a neighborhood of* W.

In order to develop a pth order numerical method, it is of extreme importance to locate points where $x^{(q)}$, for $q \leq p$, actually jumps. By the Propositions 11 and 12 such jump points are isolated, and neighborhoods of fixed points of α may be ignored without loss of generality. This setting is stated in the following hypothesis 13 which expresses that one isolated jump point is propagated from another.

Hypothesis 13 (Feldstein and Neves [5]). *Let the problem given by (2) and (4) have continuity class* $p \geq 1$. *Let* z *be an integer satisfying* $z \in [1, p]$. *Let*

$$x(t) \in D_z^{p+1}[Z - \xi, Z + \xi] \cap D_y^{p+1}[Y - \zeta, Y + \zeta], \tag{14}$$

for some integer $y \in [1, p]$, *where* $Y = \alpha(Z, x(Z))$ *and* $t_0 \leq Y - \zeta < Y + \zeta < Z - \xi < Z + \xi \leq t_e$.

Additionally, the following theorems hold true.

Theorem 14 (Feldstein and Neves [5]). *Let* m *be the multiplicity of the zero* Z *of* $\alpha(t, x(t)) - Y$. *Then Hypothesis 13 implies that* m *is odd and* $my \leq z + 1$.

Theorem 15 (Feldstein and Neves [5]). *Let* x^* *and* x^{**} *be any two* C^{p+1} *extensions of* x, *the first from* $[Z - \xi, Z]$ *forward to* $[Z - \xi, Z + \xi]$, *and the second from* $[Z, Z + \xi]$ *back to* $[Z - \xi, Z + \xi]$. *Similarly, let* x^o *and* x^{oo} *be any two* C^{p+1} *extensions of* x, *the first from* $[Y - \zeta, Y]$ *to* $[Y - \zeta, Y + \zeta]$, *and the second from* $[Y, Y + \zeta]$ *back to* $[Y - \zeta, Y + \zeta]$. *Let Hypothesis 13 hold. Then, for all* $\rho \leq \min(\zeta, \xi)$,

$$\|x^{**} - x\|_\infty^{[Z-\rho, Z]} = O(\rho^z), \quad \|x^* - x\|_\infty^{[Z, Z+\rho]} = O(\rho^z),$$
$$\|x^{oo} - x\|_\infty^{[Y-\rho, Y]} = O(\rho^y), \quad \|x^o - x\|_\infty^{[Y, Y+\rho]} = O(\rho^y).$$

Subsequently, these and other results from [5, 8] are summarized by the use of a directed graph given in Figure 1, which shows a typical jump discontinuity situation. Each node, denoted by a dot in the graph represents a zero Z of odd multiplicity of $\alpha(t, x(t)) - Y$ for some appropriate Y. By Theorem 14 zeros of even multiplicity of $\alpha(t, x(t)) - Y$ do not lead to jump points and for this reason they are excluded from the graph. The propagation of a jump discontinuity from Y to Z is denoted by an arrow where Z is the tip of the arrow, and $Y = \alpha(Z, x(Z))$ is at the tail of the arrow. It is clear that each Z has only one immediate ancestor Y, but however, each Y could be the immediate ancestor to many different nodes Z where the number of such nodes may be infinite. A linearly ordered, connected collection of nodes and arrows beginning at level 0 is called a chain of jump points. Theorem 10 implies that each node is a possible jump point for some derivative of x. For the scenario given in Figure 1, level 1 consits of four nodes which represent all zeros of odd multiplicity of $\alpha(t, x(t)) - t_0$ for $t \in [t_0, t_e]$. In terms of the graph the facts F1 and F2 given below represent the Theorems 10 and 14, respectively. The remaining facts belong to theorems which have been proved in [8]. For the following nine facts (F1) - (F9) cited from [5] it is assumed that Hypothesis 13 holds and that Φ is J-incompatible with $J \geq 1$.

F1. Let Z be a node at level k and Y be its unique ancestor in level $k-1$. Then $x^{(q)}$ is continuous at $t = Z$ for $q \leq \min(p, my)$, where the odd integer m is the multiplicity of the zero Z of $\alpha(t, x(t)) - Y$. Each chain terminates when $q = p$.

F2. If Z is a jump point for the zth derivative of x, then Z is a root of $\alpha(t, x(t)) - Y = 0$, and Z has odd multiplicity $m \leq (z-1)/y$. Equivalently, if Y is the immediate ancestor of Z and if m is the multiplicity of the zero Z, then $x \in D_q^{p+1}[Y - \zeta, Y + \zeta]$ for some $q \leq (z-1)/m$.

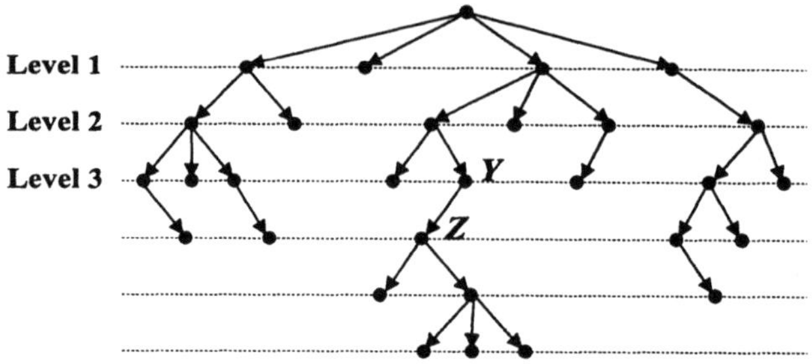

Figure 1. Z is a multiple root of $\alpha(t, x(t)) - Y = 0$ (with odd multiplicity), from [5]

F3. The graph contains all possible jump points of the first p derivatives of the solution x. Each node represents the location of a simple jump in some derivative of x, and $\alpha(t, x(t))$ changes sign at each node of the graph.

F4. Let N_k be the highest derivative of x that is continuous at all nodes in level k. Then $N_{k+1} \geq N_k + 1$. (This means that descending a chain increases the degree of smoothness of x by at least 1. This is called *generalized smoothing*.)

F5. If $\alpha(t, x(t))$ is a strictly increasing function, the graph degenerates to a linearly ordered chain.

F6. If p is finite, then so is the chain and it has no more than p levels. If p is infinite and α is strictly increasing, the linearly ordered chain is infinite, and the nodes along any chain can cluster at a fixed point at a fixed point of $\alpha(t, x(t))$.

F7. If for all $t \in [t_0, t_e]$ either $\alpha(t, x(t)) \leq t_0$ or $\alpha(t, x(t)) \geq t_0$, then the graph degenerates to the singleton node t_0.

F8. If α is piecewise monotone, then the graph cannot be infinitely broad without being infinitely long, and if, in addition, p is finite (i.e., the graph is not infinitely long), then the graph has a finite number of nodes.

F9. For every node on the graph, there exists a unique finite chain of ancestors connecting it to t_0. Further, the graph has no cycles (closed chains); that is, the graph is a tree.

2.1 *Propagation of Discontinuities in Systems of DDEs*

The aim in this subsection is to give some insight into the propagation of discontinuities within systems of delay differential equations, denoted by

$$\dot{x}(t) = f(t, x(t), x(\alpha_1(t, x(t))), \ldots, x(\alpha_k(t, x(t)))), \quad t \in [t_0, t_e] \tag{15}$$

where f is a known function and

$$x(t) = \Phi(t) \tag{16}$$

for $\hat{t} \leq t \leq t_0$ is given. Here $x(t) \in \mathbb{R}^n$ and Φ and f are known functions with appropriate dimension. The delays $\{\alpha_i\}$ are again required to satisfy

$$\hat{t} \leq \alpha_i(t, x(t)) \leq t \quad \forall i, \ t \in [t_0, t_e].$$

Subsequently, results of [12, 13] are presented concerning the propagation and the tracking of discontinuities in systems of delay differential equation of the form (15) and (16). The latter citations should be referred for further information. Apart from scalar delay differential equations ($n = 1$), where discontinuities may propagate between various time instances, for systems of delay differential equations discontinuities additionally may propagate between various components of the respective state vector $\mathbf{x}$. Although the vector representation (15) is a very compact description of the system, it neglects large parts of the system's relevant structure. This structure, however, is of high importance for the analysis of the propagation and the tracking of discontinuities within the solution to the system. In analogy to systems of ordinary differential equations, any vector representation of delay differential equations can be expressed as a system of coupled scalar equations.

If, for instance, the ith component of f depends on k_i components of the current solution $\mathbf{x}(t)$ and $\hat{k}_i$ components at retarded times $\{\alpha_{\hat{m}_{ij}}\}$ then the ith component can be written as

$$\frac{dx_i}{dt} = f_i\big(t, x_{n_{i1}}, \dots, x_{n_{ik_i}}, x_{m_{i1}}(\alpha_{\hat{m}_{i1}}(t,\mathbf{x})), \dots, x_{m_{i\hat{k}_i}}(\alpha_{\hat{m}_{i\hat{k}_i}}(t,\mathbf{x}))\big).$$

In the above equation the sequences n_{ij} and m_{ij} index the required solution components. It should be mentioned that for each i the number of arguments $1 + k_i + \hat{k}_i$ in general is not the same. Thus, the whole system (15) can be written as

$$\frac{dx_1}{dt} = f_1\big(t, x_{n_{11}}, \dots, x_{n_{1k_1}}, x_{m_{11}}(\alpha_{\hat{m}_{11}}(t,\mathbf{x})), \dots, x_{m_{1\hat{k}_1}}(\alpha_{\hat{m}_{1\hat{k}_1}}(t,\mathbf{x}))\big),$$
$$\frac{dx_2}{dt} = f_2\big(t, x_{n_{21}}, \dots, x_{n_{2k_2}}, x_{m_{21}}(\alpha_{\hat{m}_{21}}(t,\mathbf{x})), \dots, x_{m_{2\hat{k}_2}}(\alpha_{\hat{m}_{2\hat{k}_2}}(t,\mathbf{x}))\big),$$
$$\vdots$$
$$\frac{dx_n}{dt} = f_n\big(t, x_{n_{n1}}, \dots, x_{n_{nk_n}}, x_{m_{n1}}(\alpha_{\hat{m}_{n1}}(t,\mathbf{x})), \dots, x_{m_{n\hat{k}_n}}(\alpha_{\hat{m}_{n\hat{k}_n}}(t,\mathbf{x}))\big)$$

subject to appropriate initial conditions. In order to analyze the way in which discontinuities are propagated through systems of delay differential equations, two new terms have to be introduced. These terms are those of *strong* and *weak coupling* whose definitions [12] are given below. In [12] it is shown, that the ith component of the solution x_i can only have a discontinuous derivative if

(i) it inherits a discontinuity from some other component of $\mathbf{x}(t)$ or
(ii) some of its delays $\alpha_{\hat{m}_{ij}}(t, \mathbf{x}(t))$ lie on a past discontinuity

provided that f_i and $\{\alpha_i\}$ are sufficiently smooth.

These two modes of propagation (i) and (ii) given above correspond to the definitions of strong and weak coupling which have already been mentioned. The following two definitions are precise statements for these terms.

Definition 16 (Willé and Baker [12]). A component $x_i(t)$ is *strongly coupled* to a component $x_j(t)$ if $x_j(t)$ appears and is referenced in either the argument list of f_i

or in a lag function α referenced by f_i. This means that a discontinuity in $x_j(t)$ is directly inherited to $x_i(t)$. Then one says that i is strongly coupled to j, which is denoted by $j\,\mathsf{S}\,i$.

Definition 17 (Willé and Baker [12]). A component i is *weakly coupled* to j if f_i depends on the component x_j at some previous time. This is denoted by $j\,\mathcal{W}\,i$.

Graphical representations of the strong ($j\,\mathsf{S}\,i$) and weak ($j\,\mathcal{W}\,i$) coupling relation are given in Figure 2.

Figure 2. Graphical representation of strong and weak coupling.

The following example clearly illustrates the coupling properties.

Example 18 (Willé and Baker [12]). Consider the system

$$\begin{aligned} \dot{x}_1(t) &= f_1(t, x_3(t), x_1(t-1)), \\ \dot{x}_2(t) &= f_2(x_1(t)), \\ \dot{x}_3(t) &= f_3(t, x_4(\alpha(t, x_2(t)))), \\ \dot{x}_4(t) &= f_4(x_3(t)). \end{aligned} \tag{17}$$

given with appropriate initial conditions. Then, $3\mathsf{S}1$, $1\mathsf{S}2$, $2\mathsf{S}3$, $3\mathsf{S}4$, $1\mathcal{W}1$, and $4\mathcal{W}3$ holds.

A graphical representation of these coupling dependencies can be given in a so called *dependency network*, where the strong coupling is indicated by solid and the weak coupling by dashed lines. The collection of all dashed lines herein is then called *weak dependency network* and the collection of all solid lines analogously *strong dependency network*. In this way the dependency network consists of two parts, the weak and the strong dependency network. Thus, the dependency network for the example above is given in Figure 3.

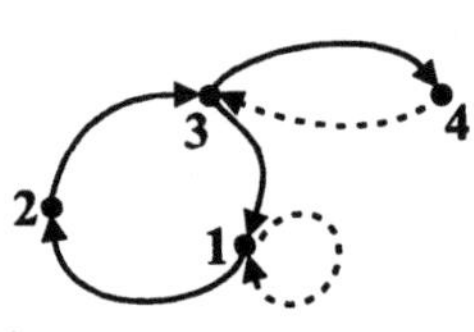

Figure 3. Dependency Network for Example 18, from [12].

The main result from [12] is given in the following theorem.

Theorem 19 (Willé and Baker [12]). *In terms of the dependency network, the order of a propagated discontinuity increases by at least one for each node of the network it crosses. Extending this principal across many nodes the minimum degree of smoothing between any two points is equal to the minimum distance between them taken along the directed graph.*

In order to illustrate this statement assume, for example, that a discontinuity of order k occurs in x_1 of (17). In terms of the dependency network given in Figure 3, this discontinuity may propagate to a discontinuity of order at least $k+1$ in x_2 and to a discontinuity of order at least $k+2$ in x_3 and so on. Theorem 10 clarifies that there is no way, of course, of getting information from the dependency network wether these minimum bounds are in fact achieved. This will in general depend on the precise nature of $\{f_i\}$ and $\{\alpha_i\}$ (also see fact F4, generalized smoothing, in the preceeding section for the case of scalar equations).

The main difference between the coupling structure in systems of ordinary differential equations and systems of delay differential equations consists in the effect of weak coupling. While strong coupling is common to systems of ordinary and delay differential systems the weak coupling sets delay equations apart from ordinary equations since it cannot appear in ordinary systems. In order to gain some insight to the structure of delay differential systems it has to be annotated that the strong coupling governs the propagation of discontinuities within a given time instant, whereas weak coupling is responsible for the propagation directly from one time instant to another. For illustration consider the following example.

Example 20 (Willé and Baker [12]).

$$\begin{aligned} \dot{x}_1 &= f_1(x_2, x_3(t-1)),\\ \dot{x}_2 &= f_2(x_3),\\ \dot{x}_3 &= f_3(x_1) \end{aligned} \tag{18}$$

where $\{f_i\}$ are analytic, with its associated dependency network shown in Figure 4.

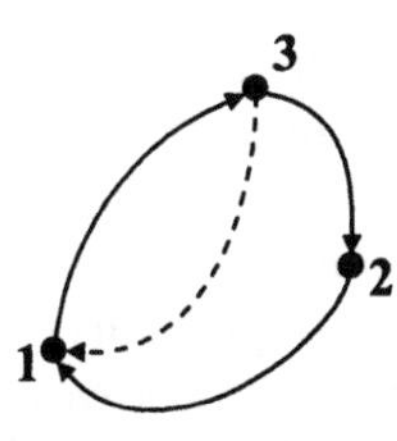

Figure 4. Dependency Network for Example 18, from [12]

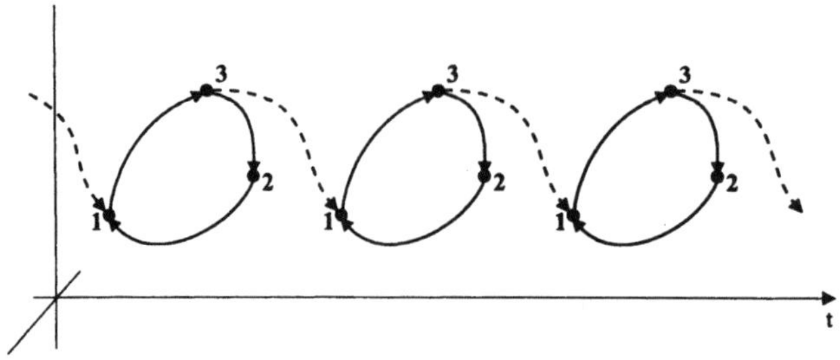

Figure 5. The role of strong and weak coupling, from [12]

Since the weak coupling propagates discontinuities between various time instants the dependency network shown in Fig. 4 can be expanded along the time axis. The corresponding figure is given in Fig. 5. It shows the strong dependence network lying in a plane perpendicular to the t-axis which represents that the strong coupling is responsible for the propagation of discontinuities within a specific time instance. Furthermore, one observes the weak coupling network expanded along the time axis representing the propagation of discontinuities between different times. From this figure it is also clear that the order of an inherited discontinuity does not only depend on the overall order of the solution at the previous time, but on the order of a specific component. From Figure 5, for example, it is clear that the minimum order of smoothing between successive discontinuities is of order two and not of order one which would have been expected.

3 Numerical Simulation

The main goal in this section is to point out the basic requirements in order to develop a code for the numerical integration of delay differential equations. For simplicity of exposition scalar delay differential equations with respective initial functions in the form given by (2) and (4) are considered. The delay differential equation may be solved by combinig an interpolation method for evaluating delayed solution values with an integration method for ordinary differential equations. However, due to the presence of derivative discontinuities within the solution of a delay differential equation this approach may not be adequate since the presence of discontinuities may invalidate the integration and interpolation schemes. Thus, in a numerical code it is necessary to know the positions of derivative discontinuities [5, 11, 14]. In [5] the authors show that the insertion of positions of discontinuities into the time grid for the numerical integration ensures the order of the applied integration method. In Figure 6, taken from [5], a flow diagram for the numerical integration of delay differential equations is printed. In this figure one observes an index j, counting the current position on the time grid $[t_0, t_1, \ldots, t_N]$ where $t_N = t_e$ holds and an index i, counting the positions of already detected discontinuities $Z_{h,i}$ where $Z_{h,1} = t_0$ holds. Within the time grid a fixed step size h such that $h := t_{j+1} - t_j$ holds

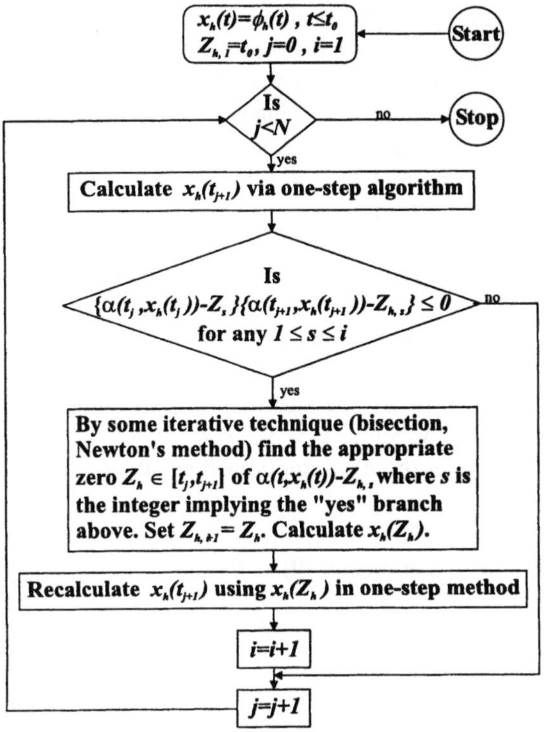

Figure 6. Flow diagram for the numerical integration of DDEs, from [5].

is assumed. Subsequently, the steps within the flow diagram are explained, where quantities indexed with h, e.g., $x_h(t)$ or $Z_{h,1}$, denote approximations to the exact quantities which are not indexed. The main loop stops if $j \geq N$ holds, that is, the integration has finished over the integration interval $[t_0, t_e]$ since $t_j = t_N = t_e$ holds. For $j < N$, that is, the integration has not finished yet, we further consider an integration step from t_j to t_{j+1} where t_j and t_{j+1} are given and $x_h(t_j)$ has already been calculated in the previous integration step. The first action then is to calculate $x_h(t_{j+1})$ via an one step method, which is used for the integration, e.g., a Runge-Kutta method. Thereafter, it is verified if a discontinuity occured within the considered time step by checking if

$$(\alpha(t_j, x_h(t_j)) - Z_{h,s}) \cdot (\alpha(t_{j+1}, x_h(t_{j+1})) - Z_{h,s}) \leq 0 \tag{19}$$

holds for any index $s \in [1, i]$. In the literature the above equation is called *tracking equation* [5, 14] since it checks if $g_h(t) := \alpha(t, x_h(t)) - Z_{h,s}$ changes sign on $t \in [t_j, t_{j+1}]$ which characterizes the occurence of a discontinuity in this interval and therefore allows to track their propagation. If this condition is not satisfied the already calculated solution $x_h(t_{j+1})$ is valid and the integration proceeds. If not, that is a discontinuity is detected within $[t_j, t_{j+1}]$ for some position of a past discontinuity $Z_{h,s}$ with $s \in [1, i]$, then the position of the propagated discontinuity $Z_{h,i+1}$ in $[t_j, t_{j+1}]$ is calculated using some iterative method by solving $g_h(Z_{i+1}) = 0$ and accordingly the index i is incremented by one. Once calculated the position $Z_{h,i+1}$ it is included into the time grid, that is $t_j \leq Z_{h,i+1} \leq t_{j+1}$ and accordingly $N = N + 1$, and the solution at $Z_{h,i+1}$ is calculated in a step from t_j to $Z_{h,i+1}$. Finally, the required solution at t_{j+1} is calculated in a step from $Z_{h,i+1}$ to t_{j+1} and the main loop is proceeded. In this way one is enabled to integrate any delay differential equation with state dependent delay. The extension to multiple delays is obvious by evaluating the tracking equation for all possible delay functions $\{\alpha(\cdot)\}$. Subsequently, it is assumed that $Z_{h,i+1}$ is a root of $\alpha(t, x_h(t)) - Z_{h,s} = 0$ where $x_h(t)$ was a pth-order approximation to $x(t)$ and where $Z_{h,s}$ is an approximation to the position of the discontinuity Z_s. The position Z_s itself is the ancestor of Z_{i+1}. Because Z_s and Z_{i+1} are the locations of discontinuities it is assumed that they are zeros of odd multiplicities y and z, respectively, of equations of the form $\alpha(t, x(t)) - Y = 0$. Under these assumptions the following theorem proved in [9] holds.

Theorem 21 (Neves and Thompson [9]). *If*

$$|Z_s - Z_{h,s}| = O(h^n), \quad |Z_{i+1} - Z_{h,i+1}| = O(h^r), \tag{20}$$

then

$$|x(t) - x_h(t)| = O(h^Q), \quad \textit{where } Q = \min(p, rz, ny)$$

and $x_h(t)$ *is the* p*th-order numerical solution to the problem (2) and (4).*

The above theorem states that if the products, rz and ny, are equal to p or exceeding p, the pth-order convergence will be preserved in crossing approximate locations of discontinuities when using a suitably modified pth-order integration method. Thus, the theorem sets a standard for "how accurate" the approximate locations of a discontinuity must be calculated in order to maintain the pth-order convergence of the underlying integration method.

Furthermore, if a numerical interpolation method for the calculation of past solution values $x_h(\alpha(t, x_h(t)))$ is considered the overall order $\hat{p}$ of an integration method implemented according to Figure 6 is given by

$$\hat{p} = \min\{p, q\} \tag{21}$$

whereby p is the order of the applied one step integration method and q is the order of the applied interpolation method [10].

A modified 4th-order Runge-Kutta method ($p = 4$) for the use with delay differential equations has been presented in [7] where the solution at t_{j+1} is calculated by

$$x_h(t_{j+1}) = x_h(t_j) + \frac{h}{6}(k_1 + 2k_2 + 2k_3 + k_4) \tag{22}$$

where $h = t_{j+1} - t_j$ holds and the quantities k_1, k_2, k_3, and k_4 are given by

$$k_1 = f\left(t_j, x_h(t_j), x_h(\alpha(t_j, x_h(t_j)))\right), \tag{23}$$

$$k_2 = f\left(t_j + \frac{h}{2}, x_h(t_j) + \frac{h}{2}k_1, x_h(\alpha(t_j + \frac{h}{2}, x_h(t_j) + \frac{h}{2}k_1))\right), \tag{24}$$

$$k_3 = f\left(t_j + \frac{h}{2}, x_h(t_j)) + \frac{h}{2}k_2, x_h(\alpha(t_j + \frac{h}{2}, x_h(t_j)) + \frac{h}{2}k_2))\right), \tag{25}$$

$$k_4 = f\left(t_j + h, x_h(t_j) + hk_3, x_h(\alpha(t_j + h, x_h(t_j) + hk_3))\right). \tag{26}$$

A possible 4th-order interpolation method ($q = 4$) for the calculation of past solution values $x_h(\tilde{t})$ with $\tilde{t} \in [t_j, t_{j+1}]$ is the two point hermite interpolation method where

$$x_h(\tilde{t}) = x_{h,j} + d_1(\dot{x}_{h,j} + q_1(s_1 - q_2(s_1 + s_2))) \tag{27}$$

using the notations $x_{h,j} = x_h(t_j)$, $x_{h,j+1} = x_h(t_{j+1})$, $\dot{x}_{h,j} = \frac{d}{dt}x_h(t_j)$, $\dot{x}_{h,j+1} = \frac{d}{dt}x_h(t_{j+1})$ and

$$\Delta t = t_{j+1} - t_j, \quad d_1 = \tilde{t} - t_j, \quad d_2 = \tilde{t} - t_{j+1},$$
$$q_1 = \frac{d_1}{\Delta t}, \quad q_2 = \frac{d_2}{\Delta t},$$
$$s_1 = \frac{x_{h,j+1} - x_{h,j}}{\Delta t} - \dot{x}_{h,j}, \quad s_2 = \frac{x_{h,j+1} - x_{h,j}}{\Delta t} - \dot{x}_{h,j+1}$$

holds. Combining these two methods to an integration procedure according to Figure 6 consequently leads to an integration method of order $\hat{p} = 4$ according to (21) which can easily be extended to systems of delay differential equations. An implementation of this method is described in [7]. For detailed information about the development of alternative integration methods for delay differential equations refer to [5,9–11].

4 Parameter Estimation

In this section an introduction to problems that may arise when estimating parameters in delay differential equations is given. The results presented here are based on [1,2]. The typical parameter estimation problem may be summarized as, "*Given values of a solution, what was the problem?*" [2]. A parameter estimation problem normally is solved by minimizing a suitable (least-squares) objective function. The problem that may arise when estimating parameters in delay differential equations is, that the objective function is not sufficiently smooth. This observation then has practical consequences in the optimization process, since it may restrict the usability of higher order optimization techniques.

Subsequently, a scalar delay differential equation depending on the parameter vector $\mathbf{p} = (p_1, \ldots, p_q)^T$ given by

$$\begin{aligned} \dot{x}(t,\mathbf{p}) &= f(t, x(t,\mathbf{p}), x(\alpha(t, x(t,\mathbf{p}), \mathbf{p}), \mathbf{p}), \mathbf{p}) \quad && t \geq t_0, \\ x(t,\mathbf{p}) &= \Phi(t,\mathbf{p}) && t \leq t_0 \end{aligned} \tag{28}$$

and a parameter estimation problem

$$J(\mathbf{p}^*) = \min_{\mathbf{p}}\{J(\mathbf{p})\} \tag{29}$$

with

$$J(\mathbf{p}) := \sum_{i=1}^{k} (X(\tau_i) - x(\tau_i, \mathbf{p}))^2 \tag{30}$$

is considered. In (30), $X(\tau_i)$ are given measurements taken at times τ_i and $x(\tau_i, \mathbf{p})$ is the solution of (28) evaluated at the measurement times τ_i.

The following definition gives a notation of left- and right hand derivatives of a function which will be used later.

Definition 22 (Baker and Paul [1]). The notation $\dot{\varphi}(t)$ usually denotes the derivative $\frac{d\varphi(t)}{dt}$, and the implication is that the left- and right-hand derivatives

$$\left(\frac{d\varphi(t)}{dt}\right)_- \equiv \dot{\varphi}_-(t) = \lim_{\delta\to 0}\left[\frac{\varphi(t) - \varphi(t-\delta)}{\delta}\right]$$
$$\left(\frac{d\varphi(t)}{dt}\right)_+ \equiv \dot{\varphi}_+(t) = \lim_{\delta\to 0}\left[\frac{\varphi(t+\delta) - \varphi(t)}{\delta}\right]$$

both exist and are equal. In addition,

$$\varphi_\pm^{(s)}(t) := \left(\frac{d\varphi_\pm^{(s-1)}(t)}{dt}\right)_\pm$$

for $s = 1, 2, \ldots$ and $\varphi_\pm^{(0)} = \varphi(t)$ is written.

According to the above definition, the derivative $\varphi^{(r)}(t)$ has a jump at $t = s$ if the left- and right hand derivatives both exist at $t = s$ and $0 < |\varphi_+^{(r)}(s) - \varphi_-^{(r)}(s)| < \infty$. In the preceeding section it was indicated how discontinuities in the solution of a delay differential equation may arise and propagate through the solution. Such discontinuities, when arising from the initial point t_0 may propagate into $J(\mathbf{p})$ via the solution values $\{x(\tau_i, \mathbf{p})\}$ in (30). The main aspect of the remainder of the section is to give the reader some insight how this can occur.

In the following it is supposed that the solution of (28) has discontinuities with respect to t occuring at the points

$$\Sigma(\mathbf{p}) := \{Z_1(\mathbf{p}), Z_2(\mathbf{p}), \ldots\} \tag{31}$$

Furthermore, the following theorem from [1] holds true.

Theorem 23 (Baker and Paul [1]). *By elementary differentiation of (30) we have the results*

$$\left(\frac{\partial J(\mathbf{p})}{\partial p_l}\right)_{\pm} = -2\sum_{i=1}^{k} (X(\tau_i) - x(\tau_i, \mathbf{p}))\left(\frac{\partial x(\tau_i, \mathbf{p})}{\partial p_l}\right)_{\pm},$$

$$\left(\frac{\partial^2 J(\mathbf{p})}{\partial p_l \partial p_m}\right)_{\pm\pm} = 2\sum_{i=1}^{k}\left(\left(\frac{\partial x(\tau_i, \mathbf{p})}{\partial p_l}\right)_{\pm}\left(\frac{\partial (\tau_i, \mathbf{p})}{\partial p_m}\right)_{\pm} - (X(\tau_i) - x(\tau_i, \mathbf{p}))\left(\frac{\partial^2 x(\tau_i, \mathbf{p})}{\partial p_l \partial p_m}\right)_{\pm\pm}\right).$$

Regarding the equations in the above theorem it is clear that a jump in a first partial derivative of $J(\mathbf{p})$ can occur if

$$x_{p_l}(t, \mathbf{p}) := \frac{\partial x(t, \mathbf{p})}{\partial p_l}$$

has a jump at $t = \tau_i$ for at least one i. However, the jump does not propagate into $\frac{\partial J(\mathbf{p})}{\partial p_l}$ if $X(\tau_i) = x(\tau_i, \mathbf{p})$ [1]. Sufficient insight on jumps in the second partial derivatives of $J(\mathbf{p})$ is obtained for the case $m = l$. A jump in

$$x_{p_l p_l}(t, \mathbf{p}) := \frac{\partial^2 x(t, \mathbf{p})}{\partial p_l^2}$$

at $t = \tau_i$ can propagate in $\frac{\partial^2 J(\mathbf{p})}{\partial p_l^2}$. In analogy to the first partial derivative the jump will not propagate if $X(\tau_i) = x(\tau_i, \mathbf{p})$. By Theorem 23, jumps in the partial derivatives of $J(\mathbf{p})$ can arise from corresponding jumps in the partial derivatives of $x(t, \mathbf{p})$. In order to analyze the relationship between jumps in the derivatives of $x(t, \mathbf{p})$ with respect to t and those in $J(\mathbf{p})$ with respect to some p_l the behaviour of $x_l(t, \mathbf{p})$ and its derivatives with respect to the variables p_l is investigated. In the following it is supposed that all derivatives are all right-hand derivatives and that

$$\lim_{t \to Z_r(\mathbf{p})} x_{p_l}(t, \mathbf{p}) \neq x_{p_l}(Z_r(\mathbf{p}), \mathbf{p})$$

for some r, then $x_{p_l}(t, \mathbf{p})$ has a jump at the point $Z_r(\mathbf{p}) \in \Sigma(\mathbf{p})$. However, if $x(Z_r(\mathbf{p}), \mathbf{p})$ varies smoothly with respect to p_l when the other components of $\mathbf{p}$ are fixed, then $x_{p_l}(Z_r(\mathbf{p}), \mathbf{p})$ does not have jumps. Differentiation of $x(Z_r(\mathbf{p}), \mathbf{p})$ with respect to some parameter p_l leads to

$$\begin{aligned}\frac{\partial x(Z_r(\mathbf{p}), \mathbf{p})}{\partial p_l} &= \left.\frac{dx(t, \mathbf{p})}{dt}\right|_{t=Z_r(\mathbf{p})} \cdot \frac{\partial Z_r(\mathbf{p})}{\partial p_i} + \left.\frac{\partial x(t, \mathbf{p})}{\partial p_l}\right|_{t=Z_r(\mathbf{p})} \\ &= \dot{x}(Z_r(\mathbf{p}), \mathbf{p})\frac{\partial Z_r(\mathbf{p})}{\partial p_l} + x_{p_l}(Z_r(\mathbf{p}), \mathbf{p}) \qquad (32)\end{aligned}$$

where $\dot{x}(t, \mathbf{p}) := \frac{dx(t,\mathbf{p})}{dt}$ and $x_{p_l}(t, \mathbf{p}) := \frac{\partial x(t,\mathbf{p})}{\partial p_l}$ hold. Analogously, the second partial derivatives read as

$$\begin{aligned}\frac{\partial^2 x(Z_r(\mathbf{p}), \mathbf{p})}{\partial p_l \partial p_m} &= (\ddot{x}(Z_r(\mathbf{p}), \mathbf{p}) + x_{p_m}(Z_r(\mathbf{p}), \mathbf{p}))\,\frac{\partial Z_r(\mathbf{p})}{\partial p_l} \\ &\quad + \dot{x}(Z_r(\mathbf{p}), \mathbf{p})\frac{\partial^2 Z_r(\mathbf{p})}{\partial p_l \partial p_m} \\ &\quad + \dot{x}_{p_l}(Z_r(\mathbf{p}), \mathbf{p})\frac{\partial Z_r(\mathbf{p})}{\partial p_m} + x_{p_l p_m}(Z_r(\mathbf{p}), \mathbf{p}) \qquad (33)\end{aligned}$$

where $\ddot{x}(t, \mathbf{p}) := \frac{d^2 x(t,\mathbf{p})}{dt^2}$ and $x_{p_l p_m}(t, \mathbf{p}) := \frac{\partial^2 x(t,\mathbf{p})}{\partial p_l \partial p_m}$ hold.

From (32) it is seen that a jump must occur in $x_{p_l}(t, \mathbf{p})$ at $t = Z_r(\mathbf{p})$ if $\dot{x}(t, \mathbf{p})$ has a jump at $t = Z_r(\mathbf{p})$ and $\frac{\partial Z_r(\mathbf{p})}{\partial p_l}$ does not vanish. Equivalently, from (33) it follows that, if $\ddot{x}(t, \mathbf{p})$ has a jump at $t = Z_r(\mathbf{p})$ and $\frac{\partial Z_r(\mathbf{p})}{\partial p_l}$ does not vanish, then $x_{p_l p_m}(t, \mathbf{p})$ must also have a jump at $t = Z_r(\mathbf{p})$. Thus, $\frac{\partial x(\tau_i,\mathbf{p})}{\partial p_l}$ has a jump provided that $\tau_i = Z_r(\mathbf{p})$ for some r and $Z_r(\mathbf{p})$ varies as p_l varies.

The above results are illustrated by the following example taken from [1].

Example 24 (Baker and Paul [1]). Consider the parameter estimation problem

$$\dot{x}(t, \mathbf{p}) = \lambda x(\nu t, \mathbf{p}), \quad t \geq 1, \qquad (34)$$

$$x(t, \mathbf{p}) = 2, \quad t < 1, \quad x(1, \mathbf{p}) = 0 \qquad (35)$$

where $\mathbf{p} = (\lambda, \nu)^T$. This is a delay differential equation only if $\nu \leq 1$ since $\alpha(\cdot) = \nu t$ must remain smaller or equal t for all $t \geq 1$.

The data $\{X(\frac{3}{2}), X(\frac{4}{2}), X(\frac{5}{2}), X(\frac{6}{2}), X(\frac{7}{2}), X(\frac{8}{2})\}$ are obtained from (34) using $\mathbf{p}^* = (0.5, 0.5)^T$. Simple analysis shows that jumps occur in $\dot{x}(t, \mathbf{p})$ at $t = \frac{1}{\nu} =: Z_1(\mathbf{p})$ and in $\ddot{x}(t, \mathbf{p})$ at $t = \frac{1}{\nu^2} =: Z_2(\mathbf{p})$. In the above example the locations of discontinuities are depemdent on the parameter ν but not with λ. It follows, that a jump may occur in $\frac{\partial J(\mathbf{p})}{\partial \nu}$ when $\nu = \frac{1}{\tau_i}$ for any i, that is, if $\nu \in \{\frac{2}{3}, \frac{2}{4}, \frac{2}{5}, \frac{2}{6}, \frac{2}{7}, \frac{2}{8}\}$. Setting $\lambda = \frac{3}{4}$ ensures that there is no smoothing of the discontinuity in $\frac{\partial x(t,\mathbf{p})}{\partial \nu}$ as it propagates to $\frac{\partial J(\mathbf{p})}{\partial \nu}$ at $\nu = \frac{1}{2}$, and such jumps are clearly visible in Figure 7 (a). Jumps in $\frac{\partial^2 J(\mathbf{p})}{\partial \nu^2}$ that occur at $\nu = \sqrt{\frac{1}{\tau_i}}$ can also been deduced from sharp bends in Figure 7 (b).

5 Conclusion

In this report it is shown that delay differential equations significantly differ from ordinary differential equations. The most substantial difference consists in the occurence of derivative discontinuities within the solution of a delay differential equation. In general, these discontinuities originate from the non-smooth transition at the initial point t_0 of the integration interval. The presence of these discontinuities

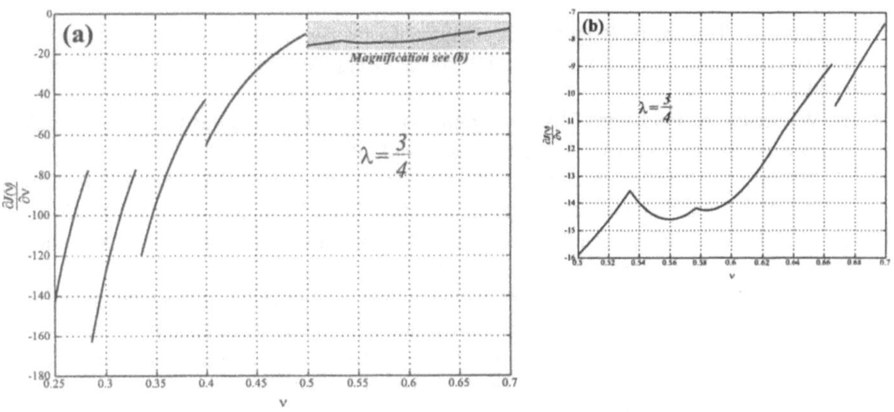

Figure 7. Gradient of the objective function, from [1]

does not only require adequate integration schemes for the numerical integration of delay differential equations but may also lead to problems when parameters in a delay differential equation are subject to be estimated. Therefore, a fundamental theory which delivers deep insight in the occurence and the propagation of discontinuities within the solution of a delay differential equation was given. Additionally, an analysis of the propagation of such discontinuities within systems of delay differential equations has been presented. The basic requirements for the development a pth order solver for the numerical integration of delay differential equations have been pointed out. Finally, the difficulties which may occur when parameters within delay differential equation or its respective initial function are subject to be estimated have been discussed. These difficulties may severly restrict the applicability of higher order optimization techniques since the considered objective function may not be sufficiently smooth.

References

1. C. T. H. Baker, C. A. H. Paul: Pitfalls in parameter estimation for delay differential equations. NAREP 267, Manchester Centre for Computational Mathematics (1995)
2. C. T. H. Baker, C. A. H. Paul: Pitfalls in parameter estimation for delay differential equations. SIAM J. Sci. Comp. **18,1** (1997) 305–314
3. R. D. Driver: Existence Theory for a Delay Differential System. Contrib. Diff. Eq. **1,3** (1963) 317–336
4. R. D. Driver: Ordinary and Delay Differential Equations. Applied Mathematical Sciences **20**, Springer Verlag, New York (1977)
5. A. Feldstein, K. W. Neves: High order methods for state dependent delay differential equations with nonsmooth solutions. SIAM J. Numer. Anal. **21,5** (1984) 844–863
6. E. Hairer, S. P. Nørsett, G. Wanner: Solving Ordinary Differential Equations. Springer, Berlin (1987)

7. F. Lehn, B. Tibken, E. P. Hofer: Development of a Simulation Environment for Time Delay Differential Equations., Preprint 98-29, 'Echtzeit-Optimierung großer Systeme' (1998)
8. K. W. Neves, A. Feldstein: Characterization of jump discontinuities for state dependent delay differential equations. J. Math. Anal. Appl. **56** (1976) 689–707
9. K. W. Neves, S. Thompson: Software for the numerical solution of systems of functional-differential equations with state-dependent delay. Appl. Num. Math. **9** (1992) 385–401
10. C. A. H. Paul: Developing a delay differential equation solver. NAREP 204, Manchester Centre for Computational Mathematics (1991)
11. C. A. H. Paul: Designing efficient software for solving delay differential equations. NAREP 368, Manchester Centre for Computational Mathematics (2000)
12. D. R. Willé, C. T. H. Baker: The tracking of derivative discontinuities in systems of delay differential equations. Appl. Num. Math. **9** (1992) 209–222
13. D. R. Willé, C. T. H. Baker: The propagation of derivative discontinuities in systems of delay-differential equations. NAREP 160, Manchester Centre for Computational Mathematics (1988)
14. D. R. Willé, C. T. H. Baker: Delsol – a numerical code for the solution of systems of delay differential equations. Appl. Num. Math. **9** (1992) 223–224

Biomathematical Models with State-Dependent Delays for Granulocytopoiesis

Eberhard P. Hofer[1], Bernd Tibken[2], and Frank Lehn[1]

[1] Abteilung Meß-, Regel- und Mikrotechnik, Universität Ulm, Germany
[2] Fachgruppe für Automatisierungstechnik und Technische Kybernetik (atk) Bergische Universität—Gesamthochschule Wuppertal, Germany

Abstract The development of a biomathematical model of the human granulopoiesis builds the core of this article. Hereby, the term granulopoiesis specifies the dynamical process of the generation of granulocytes, a subclass of white blood cells. The modeling of this process is based on delay differential equations with state-dependent delays in order to describe non-constant cell maturation times. This model is used to estimate the severeness of radiation damage and, furthermore, to predict the recovery of an irradiated person satisfying real-time requirements. This task is solved by the estimation of the initial conditions of the time delay model since they directly represent the degree of damage to the granulopoietic system harmed by acute radiation exposure. Since it is known that the integration and the parameter estimation for delay differential equations may pose severe difficulties, these problems are analyzed in detail for the model of granulopoiesis and suitable integration and optimization techniques are deduced. Several prediction results for real patient data are presented and show the power of our system in order to estimate an irradiation damage and to predict the recovery of granulopoiesis.

1 Introduction

At the University of Ulm engineers in cooperation with medical doctors have modeled cell renewal systems especially the human granulopoiesis [8], whereby the term granulopoiesis describes the cell renewal system of the granulocytes, a subclass of white blood cells. The goal of this modeling was the development of a decision support system which assists the medical doctor in the treatment of the acute radiation syndrome. The cell renewal system of the granulocytes hereby represents an excellent indicator, to be used in such a system, for the severeness of a damage, since the life time of granulocytes in the blood is very short and therefore reactions on a radiation exposure occur in this cell system. Various nuclear accidents, e.g., in Chernobyl/Russia (1986), show the demand for such a system. In this project we followed up a novel kind of modeling the human granulopoiesis, whereby a biomathematical model of the human granulopoiesis, based on time-delay differential equations with state-dependent delays has been developed. The foundation of this new model is an existing model of Fliedner and Steinbach [7]. The model is used to perform an estimation of its initial conditions based on the minimization of a quadratic distance measure. The distance measure hereby consists of the difference between real patient granulocyte measures out of the blood and the equivalent model output, parameterized by the choice of its initial conditions. Hereby, the initial cell contents are subject to be estimated, which corresponds to the degree of damage caused by radiation exposure for an individual patient. The estimation of the degree of damage then aids the medicating doctor to find appropriate treatment of the patient. Subsequently, we present the medical background by an introduction to the human granulopoiesis followed by the derivation of a new time delay model of human granulopoiesis. Furthermore, we discuss the estimation of the initial conditions for the derived model and show typical estimation results.

2 Medical Background

In this section we give an overview of the generation of granulocytes and its dynamics. A typical sample of human blood consists of approximately 50% blood plasma and 50% blood cells. One can separate three main groups of blood cells: The red blood cells (erythrocytes) which represent the largest group with approximately 4.5 million cells per cubic millimeter, the platelets (thrombocytes) with approximately 250.000 cells per cubic millimeter and the white blood cells (leucocytes) with approximately 7.000 cells per cubic millimeter. The granulocytes are a subclass of white blood cells. All of these cell types have different functions within the human body. The erythrocytes are mainly responsible for the transport of oxygen and carbon dioxide through the body while the main function of the platelets is to initiate the blood-clotting mechanism which prevents blood loss. The white blood cells themselves consist of three main subgroups: the granulocytes, the monocytes, and the lymphocytes. These white blood cells are responsible for the protection of the body against disease and infection. All of the mentioned blood cells have their specific life span and therefore have to be reproduced. This reproduction is subject to a specific dynamical behaviour in each case. To understand the cell renewal system of the granulocytes, the granulopoiesis, it is helpful to study Figure 1 which is taken from [4]. It shows the human haemopoiesis, the cell renewal system of all the blood cells. The granulopoiesis for neutrophil granulocytes is underlayed gray. At the lower end of the branches one observes the fully matured blood cells, namely platelets, erythrocytes, granulocytes, and lymphocytes. The granulopoiesis starts with the pluripotent stem cell. Until a fully matured granulocyte emerges several different intermediate stages have to be considered: myeloic stem cell, CFU-GEMM, CFU-GM, CFU-G, myeloblast, neutrophil myelocyte, and neutrophil granulocyte. During these stages the maturation and the proliferation occur. For further information about the human granulopoiesis we refer to [4]. The human body is able to satisfy the demand for granulocytes and is also able to react on perturbations of this cell renewal system. In our case we are interested in the regeneration of the granulopoiesis after perturbations caused by radiation.

3 Modeling

In this section we present the derivation of a novel time delay model of human granulopoiesis. Since this new model is based on an existing model of Fliedner and Steinbach, we present this model as introduction to the modeling of cell systems.

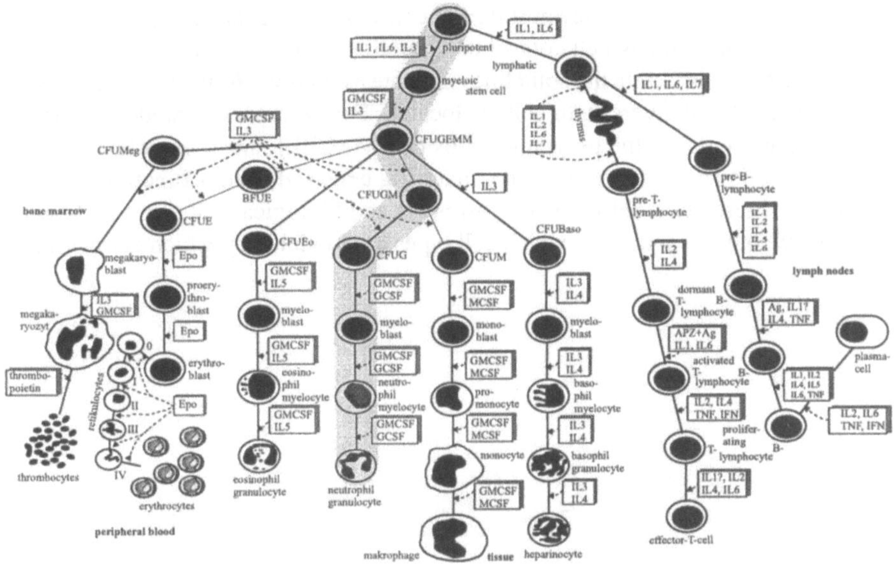

Figure 1. Human haemopoiesis; translated from [4].

3.1 Model of Fliedner and Steinbach

In this section we will present the model of Fliedner and Steinbach [7] which allows us to simulate the human granulopoiesis. The biomathematical model of the human granulopoiesis is given by the scheme which is shown in Figure 2.

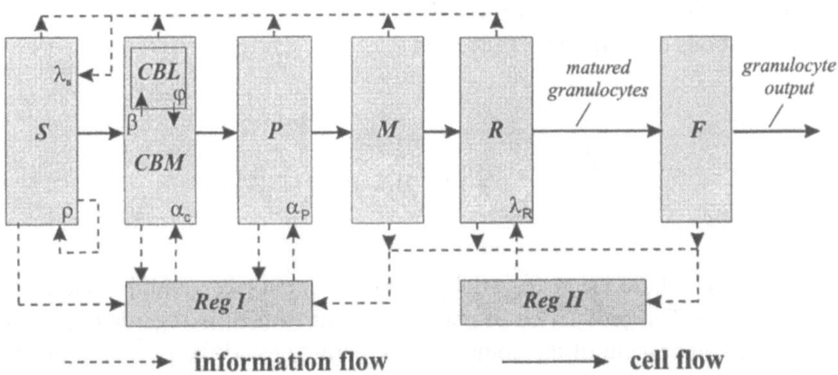

Figure 2. Fliedner-Steinbach model of granulopoiesis.

This model consists of seven different cell compartments and additionally two hormon compartments which control the granulopoiesis. The dashed lines indicate an information flow while the solid lines indicate a cell flow. At the beginning a stem cell pool (S) that delivers stem cells is located. Furthermore the model consists of two compartments for the progenitor cells (CBM/CBL), one compartment for the precursor cells (P), reserve cells (R), maturing cells (M) and a function compartment (F). The relation between compartments and biological cell types is shown in Figure 3 where it is indicated which cell types are modeled by which compartment.

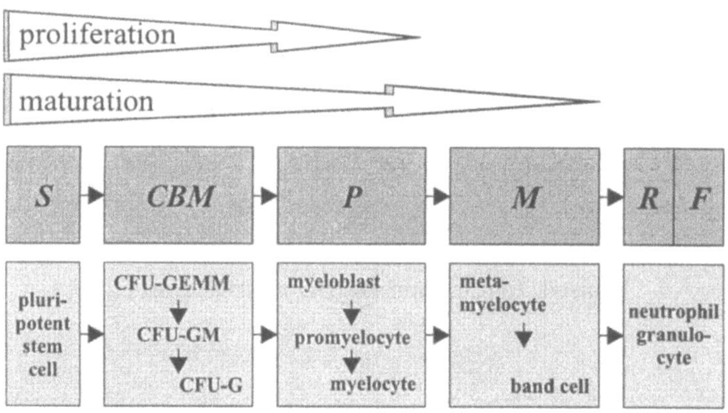

Figure 3. Relation between compartments and cell types.

The basic idea for the derivation of a dynamical biomathematical model from this compartment structure is to introduce state variables for each compartment. As state variables the number of cells of each compartment are chosen. Using cell balance equations the differential equation

$$\frac{d}{dt}x(t) = u(t) - y(t) + \alpha(t)x(t) \tag{1}$$

is computed [1] where the state variable $x(t)$, the input $u(t)$, and the output $y(t)$ have been used. The production of cells in a compartment is regarded as proportional to the cell content of the compartment, namely as $\alpha(t)x(t)$. Using this type of modeling, Fliedner and Steinbach arrive at a biomathematical model of the human granulopoiesis consisting of 37 ordinary differential equations (ODE).

The model equations are given by

$$\frac{dS}{dt} = \lambda_S(2\rho - 1)S, \tag{2}$$

$$\frac{dS_{inj}}{dt} = \lambda_{inj}(2\rho_{inj} - 1)S_{inj}, \tag{3}$$

$$\begin{aligned}\frac{dCBM_1}{dt} &= 2\lambda_S(1-\rho)S + 2\lambda_{inj}(1-\rho_{inj})S_{inj} - (\lambda_C - \alpha_C)CBM_1 \\ &\quad -\beta CBM_1 + \Phi CBL_1,\end{aligned} \tag{4}$$

$$\begin{aligned}\frac{dCBM_i}{dt} &= \lambda_C CBM_{i-1} - (\lambda_C - \alpha_C)CBM_i \\ &\quad -\beta CBM_i + \Phi CBL_i,\ i = 2,\ldots,10,\end{aligned} \tag{5}$$

$$\frac{dCBL_i}{dt} = \beta CBM_i - \Phi CBL_i,\ i = 1,\ldots,10, \tag{6}$$

$$\frac{dP_1}{dt} = \lambda_C CBM_{10} - (\lambda_P - \alpha_P)P_1, \tag{7}$$

$$\frac{dP_i}{dt} = \lambda_P P_{i-1} - (\lambda_P - \alpha_P)P_i,\ i = 2,\ldots,10, \tag{8}$$

$$\frac{dM}{dt} = \lambda_P P_{10} - \lambda_M M, \tag{9}$$

$$\frac{dR}{dt} = \lambda_M M - \lambda_R R, \tag{10}$$

$$\frac{dF}{dt} = \lambda_R R - \lambda_F F, \tag{11}$$

$$\frac{dRegI}{dt} = \omega_1 e^{(-\Theta_1 Z)} - \lambda_{RegI} RegI, \tag{12}$$

$$\frac{dRegII}{dt} = \omega_2 e^{(-\Theta_2 F)} - \lambda_{RegII} RegII, \tag{13}$$

where the regulation functions

$$\lambda_S = \gamma_1 e^{-\nu_1 S} + \gamma_2 e^{-\nu_2 ZK} + \gamma_3, \tag{14}$$

$$\alpha_C = \gamma_4 - \gamma_5 e^{-\nu_3 RegI}, \tag{15}$$

$$\alpha_P = \gamma_6 - \gamma_7 e^{-\nu_4 RegI}, \tag{16}$$

$$\lambda_M = \gamma_8 - \gamma_9 e^{-\nu_5 RegII}, \tag{17}$$

$$\lambda_R = \gamma_{10} - \gamma_{11} e^{-\nu_6 RegII} \tag{18}$$

and the abbreviations

$$CBM = \sum_{i=1}^{10} CBM_i, \quad P = \sum_{i=1}^{10} P_i, \quad ZK = CBM + P + M + R + F,$$
$$Z = g_1 S + g_2 CBM + g_3(P + M + R + F)$$

have been used. All other parameters occuring in these equations are constant and are derived from biological experiments except for β and ρ which are piecewise linear functions of CBM and S, respectively. For the model parameters refer to [7, 17]. The compartments CBM, CBL, and P are each modeled using ten subcompartments in order to reproduce the maturation times occuring in these compartments. The transfer times and the cell proliferation rates are given in Tab. 1, which is taken from [17].

compartment	transfer time	cellular fission	static cell content
S (pluripotential stem cells)	generation time 20h	10% active in fission process	$1,25 \cdot 10^9$ cells
CBM (progenitor cells in the bone marrow)	$125h \pm 30\%$	5 (4,5–15)	$8,27 \cdot 10^9$ cells
CBL (progenitor cells in the blood)	$1h \pm 30\%$ (then returning to the bone marrow)	–	$7,52 \cdot 10^6$ cells
P (precursorcells)	$100h \pm 30\%$	4	$1,24 \cdot 10^{11}$ cells
M (maturation)	70h	–	$2,24 \cdot 10^{11}$ cells
R (reserve)	70h	–	$2,24 \cdot 10^{11}$ cells
F (function)	12,5h	–	$4 \cdot 10^{10}$ cells
Reg I (regulator I)	half-time 10h	–	10
Reg II (regulator II)	half-time 10h	–	10

Table 1. Compartments used in the Fliedner-Steinbach model

3.2 *Partial Differential Equation Approach*

Regarding the Fliedner-Steinbach model one observes that subcompartment structures within the CBM, CBL and P compartments have been used in order to model cell maturation times, which led to the introduction of 30 dynamic equations. Consequently, the dimension of the whole model is extremely high because the integration of this system leads to realtime problems when the estimation of initial values has to be carried out on a small laptop computer in case of an accident.

A different way of modeling maturation times is to introduce time delays in the differential equations which both have the advantage of reducing the number of model equations and also to include knowledge about cell maturation times in a very direct way without detour via subcompartments. Since cell proliferation by cell division occurs sychronously together with the cell maturation a suitable approach to describe this is the utilization of a partial differential transport equation including a source term.

The main idea to derive such an equation is to consider a mechanical conveyor belt where the mass located on the belt is able to proliferate. Regarding the mass on the belt as the number of cells in a compartment, the mass proliferation as a cell proliferation by mitoses, and the conveyor belt as the compartment itself the analogy between the mechanical belt and the biological cell compartment is obvious.

In the following section the problem of modeling such a conveyor belt and solving the resulting partial differential equation is treated [12, 14]. Later the resulting equations are used to derive a new time delay description of the above mentioned compartments which finally results in a novel time delay model of granulopoiesis conserving the basic structure of the Fliedner-Steinbach model.

3.3 Model introduction

Within this section we regard a conveyor belt, on which at the beginning we have an input and at the end an output of mass. During the transport over the belt a mass amplification is allowed. If we introduce a finite velocity of the belt the transport will take a finite time which represents the transport time.

The connection between these terms and the biological background consists in regarding the belt as a compartment, the mass as cells and the transport time as maturation time. Since the cell maturation times of the human granulopoiesis are

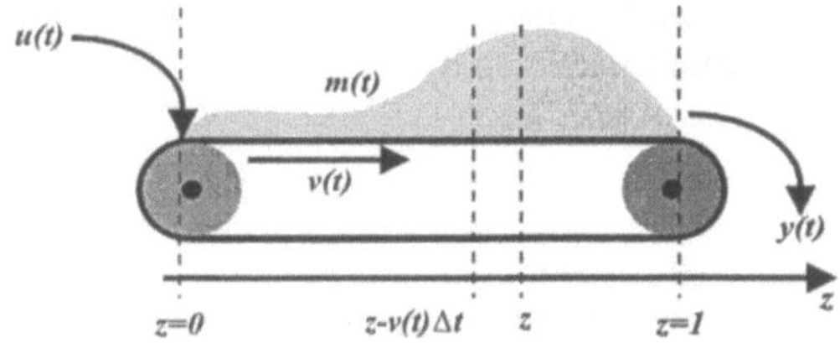

Figure 4. Conveyor belt.

subject to the regulation of the human body one has to consider a non-constant velocity $v(t)$ of the mechanical conveyor belt. It will be no restriction to set the length of the belt to the normalized value 1 because the generation of any arbitrary time delay is possible by an appropriate choice of the velocity $v(t)$. Analogously we introduce a non-constant amplification function $a(t)$ in order to be compatible with the ability of the human body to regulate cell division in our model approach. In the next step we will describe the dynamics of the conveyor belt shown in Figure 4.

We consider a mass $m(t)$ on this belt, a mass input $u(t)$, and a mass output $y(t)$. For the dynamical description of this belt we regard a non-constant mass density function $c(z,t)$ on the belt where the independent variable z measures the position along the belt. For the mass density function $c(z,t)$ the following partial differential equation

$$\frac{\partial}{\partial t}c(z,t) + \nu(t)\frac{\partial}{\partial z}c(z,t) = a(t)c(z,t) \tag{19}$$

is valid [12]. The input $u(t)$ and the cell output $y(t)$ are connected with the mass density function $c(z,t)$ by

$$u(t) = c(0,t)\nu(t) \tag{20}$$

and

$$y(t) = c(1,t)\nu(t). \tag{21}$$

Equation (19) together with (20) defines a boundary value problem which will be solved in the next section.

3.4 *Solution of the Boundary Value Problem*

The boundary value problem given by (19) and (20) will be solved using the method of characteristics [18, 19]. We parameterize the characteristics of (19) by

$$t = t(s), \quad z = z(s), \quad \text{and} \quad c(z(s),t(s)) = \phi(s) \tag{22}$$

with the real parameter s. Using this parameterization we have to solve the system of ordinary differential equations given by

$$\frac{dt(s)}{ds} = 1, \tag{23}$$

$$\frac{dz(s)}{ds} = \nu(t(s)), \tag{24}$$

$$\frac{d\phi(s)}{ds} = a(t(s))\phi(s) \tag{25}$$

in order to compute the characteristics of (19). The integration of (23) from s_0 to s leads to

$$t(s) = s - s_0 + t(s_0). \tag{26}$$

Using (26) we integrate (24) and (25) which results in

$$z(s) = z(s_0) + \int_{s_0}^{s} \nu(\tau - s_0 + t(s_0))d\tau \tag{27}$$

and

$$\phi(s) = \phi(s_0) \cdot \exp\left\{\int_{s_0}^{s} a(\tau - s_0 + t(s_0))d\tau\right\}. \tag{28}$$

Subsequently, we calculate the characteristics that satisfy the boundary condition (20), namely,

$$c(0,t) = \frac{u(t)}{\nu(t)}, \quad \nu(t) > 0 \; \forall t.$$

This means that for any given parameter value s^* with $z(s^*) = 0$ and $t(s^*) =: t^*$ the boundary condition

$$\phi(s^*) = c(0,t^*) = \frac{u(t^*)}{\nu(t^*)} \tag{29}$$

must hold. The characteristics that satisfy the boundary value problem therefore are given by

$$t(s) = s - s^* + t^*, \tag{30}$$

$$z(s) = \int_{s^*}^{s} \nu(\tau - s^* + t^*)d\tau, \tag{31}$$

$$\phi(s) = \frac{u(t^*)}{\nu(t^*)} \cdot \exp\left\{\int_{s^*}^{s} a(\tau - s^* + t^*)d\tau\right\}. \tag{32}$$

In the next step we calculate the general solution $c(z,t)$ of the boundary value problem. A parameter value $\hat{s}$ is chosen, such that a characteristic meets an arbitrarily given point (z,t). Thus, we require

$$t(\hat{s}) = \hat{s} - s^* + t^* \stackrel{!}{=} t, \tag{33}$$

$$z(\hat{s}) = \int_{s^*}^{\hat{s}} \nu(\tau - s^* + t^*)d\tau \stackrel{!}{=} z. \tag{34}$$

From (33) follows $t^* = s^* - \hat{s} + t$ and (34) reads as

$$z = \int_{s^*}^{\hat{s}} \nu(\tau - \hat{s} + t)d\tau.$$

Now, we substitute $\sigma = \tau - \hat{s} + t$ which leads to

$$z = \int_{\lambda(z,t)}^{t} \nu(\sigma)d\sigma \tag{35}$$

where the abbreviation $\lambda(z,t) = s^* - \hat{s} + t$ has been used. Then, the general solution $c(z,t)$ of the boundary value problem is given by

$$c(z,t) = \phi(\hat{s}) = \frac{u(\lambda(z,t))}{v(\lambda(z,t))} \cdot \exp\left\{ \int_{\lambda(z,t)}^{t} a(\sigma) d\sigma \right\}, \tag{36}$$

where $\lambda(z,t)$ has to be calculated from (35) for any given value of z and t.

Another function of interest is the output $y(t)$ given by (21). In order to calculate the output function we need $c(1,t)$. Therefore, we consider $z = 1$ in (35) and rename $\lambda(1,t)$ by $\gamma(t)$. Thus,

$$1 = \int_{\gamma(t)}^{t} v(\sigma) d\sigma \tag{37}$$

holds. Differentiating both sides of (37) with respect to time t results in the delay differential equation

$$\frac{d\gamma(t)}{dt} = \frac{v(t)}{v(\gamma(t))} \tag{38}$$

for the delay function $\gamma(t)$. For convenience we define

$$\alpha(t) := \int_{\gamma(t)}^{t} a(\sigma) d\sigma \tag{39}$$

which is the exponent of (36) for $z = 1$ and not equal to $\alpha(t)$ used in (1). By differentiating both sides of (39) the delay differential equation

$$\frac{d\alpha(t)}{dt} = a(t) - \frac{v(t)}{v(\gamma(t))} a(\gamma(t)) \tag{40}$$

follows which describes the mass growth. Then, together with (38) and (40), the output of the conveyor belt reads as

$$y(t) = c(1,t)v(t) = \frac{v(t)u(\gamma(t))}{v(\gamma(t))} \exp\{\alpha(t)\}. \tag{41}$$

3.5 Derivation of a Time-Delay Model

In this section we will use the results of the preceeding section in order to derive a new time delay model of human granulopoiesis where we will describe the CBM and P compartments by one time delay equation, respectively. In our new model we neglect the CBL compartment introduced within the Fliedner-Steinbach model since simulations show only small impact of this compartment to the overall dynamical behavior of granulopoiesis. The same holds true for the regulation compartments RegI and RegII which are neglected while the regulation in our new model is implemented by direct feedback of cell numbers.

In the following we derive time delay equations for the mentioned CBM and P compartments and, by conserving the basic structure of the Fliedner-Steinbach model, we will arrive at a new model of human granulopoiesis.

To derive dynamical equations for the compartments CBM and P an expression for the dynamical behavior of the number of cells in these compartments is needed. The mass density of the cells on the belt is given by (36). Thus, the mass on the belt is given by

$$m(t) = \int_{z=0}^{z=1} c(z,t)dz = \int_{0}^{1} \frac{u(\lambda(z,t))}{v(\lambda(z,t))} \cdot \exp\left\{\int_{\lambda(z,t)}^{t} a(\sigma)d\sigma\right\} dz \tag{42}$$

where $\lambda(z,t)$ has to be calculated from (35). Substituting $x = \lambda(z,t)$ we get $\frac{dz}{dx} = -v(x)$ from (35) and the expression for the mass $m(t)$ reads as

$$m(t) = \int_{x=\gamma(t)}^{x=t} u(x) \cdot \exp\left\{\int_{x}^{t} a(\sigma)d\sigma\right\} dx \tag{43}$$

where $\gamma(t)$ satisfies (37). Differentiating equation (43) with respect to time t leads to

$$\frac{dm(t)}{dt} = u(t) + a(t) \cdot m(t) - \frac{v(t)}{v(\gamma(t))} \exp\{\alpha(t)\} u(\gamma(t)) \tag{44}$$

where $\gamma(t)$ and $\alpha(t)$ are subject to (38) and (40), respectively.

In (44) we recognize the input function $u(t)$, an amplification term $a(t) \cdot m(t)$, and the output function $y(t)$ given by (41). The required dynamic equations which describe the CBM compartment therefore are given by

$$\frac{dCBM(t)}{dt} = u(t) + a_{CBM}(t)CBM(t) - \frac{v_{CBM}(t)u(\gamma_{CBM}(t))}{v_{CBM}(\gamma_{CBM}(t))} \exp\{\alpha_{CBM}(t)\}, \tag{45}$$

$$\frac{d\alpha_{CBM}(t)}{dt} = a_{CBM}(t) - \frac{v_{CBM}(t)}{v_{CBM}(\gamma_{CBM}(t))} a_{CBM}(\gamma_{CBM}(t)), \tag{46}$$

$$\frac{d\gamma_{CBM}(t)}{dt} = \frac{v_{CBM}(t)}{v_{CBM}(\gamma_{CBM}(t))}. \tag{47}$$

Analogously, a set of delay differential equations to describe the dynamics of the P compartment is computed. Since in the P compartment a constant cell proliferation occurs there is no need to introduce a dynamical equation for $\alpha_P(t)$. Thus, $\exp\{\alpha_P(t)\}$ is chosen as constant $2^4 = 16$ which corresponds to four cell divisions and is achieved by setting $a_P(t) = 4\ln(2) \cdot v_P(t)$. Then, the complete set of model equations is given by

$$\frac{dS}{dt} = \lambda_S\,(2\rho - 1)\,S, \quad y_S = 2\lambda_S\,(1 - \rho)\,S, \tag{48}$$

$$\frac{dS_{inj}}{dt} = \lambda_{inj}\,(2\rho_{inj} - 1)\,S_{inj}, \quad y_{inj} = 2\lambda_{inj}\,(1 - \rho_{inj})\,S_{inj}, \tag{49}$$

$$\frac{dCBM_1}{dt} = y_S + y_{inj} + a_{CBM}CBM_1 - \left(y_S^{(C)} + y_{inj}^{(C)}\right)\frac{\nu_{CBM}}{\nu_{CBM}^{(C)}}\exp\{\alpha_{CBM}\}, \tag{50}$$

$$\frac{dCBM_2}{dt} = \left(y_S^{(C)} + y_{inj}^{(C)}\right)\frac{\nu_{CBM}}{\nu_{CBM}^{(C)}}\exp\{\alpha_{CBM}\} - \lambda_{CBM}CBM_2, \tag{51}$$

$$\frac{dP_1}{dt} = \lambda_{CBM}CBM_2 + a_P P_1 - \lambda_{CBM}CBM_2^{(P)}\frac{\nu_P}{\nu_P^{(P)}}\cdot 2^4, \tag{52}$$

$$\frac{dP_2}{dt} = \lambda_{CBM}CBM_2^{(P)}\frac{\nu_P}{\nu_P^{(P)}}\cdot 2^4 - \lambda_P P_2, \tag{53}$$

$$\frac{dM}{dt} = \lambda_P P_2 - \lambda_M M, \quad \frac{dR}{dt} = \lambda_M M - \lambda_R R, \quad \frac{dF}{dt} = \lambda_R R - \lambda_F F, \tag{54}$$

$$\frac{d\gamma_{CBM}}{dt} = \frac{\nu_{CBM}}{\nu_{CBM}^{(C)}}, \quad \frac{d\gamma_P}{dt} = \frac{\nu_P(t)}{\nu_P^{(P)}}, \quad \frac{d\alpha_{CBM}}{dt} = a_{CBM} - \frac{\nu_{CBM}}{\nu_{CBM}^{(C)}}a_{CBM}^{(C)}. \tag{55}$$

The division of each of the compartments CBM and P into two subcompartments CBM_1, CBM_2, and P_1, P_2, respectively, is necessary in order to model the biologically observed dynamics of these compartments by a certain smoothing of the input-output behavior. Further, it should be mentioned that the superscripts $(\cdot)^{(C)}$ and $(\cdot)^{(P)}$ indicate that the corresponding functions have to be evaluated at the retarded arguments $\gamma_{CBM}(t)$ and $\gamma_P(t)$, respectively, while all other time-dependent functions are evaluated at the current time instant t. The regulation functions λ_S, ρ, λ_M, λ_R, ν_{CBM}, ν_P, a_{CBM} and a_P are computed in analogy to the Fliedner-Steinbach model and are listed in [12].

3.6 Analysis of the Time Delay Equations

As already mentioned the novelty of our time delay model in comparison to the Fliedner-Steinbach model is the description of the compartments CBM and P using time delay differential equations instead of a set of ordinary equations as introduced by Fliedner and Steinbach. The new model is based on the solution of the general partial differential transport equation, which is used to derive a set of time delay differential equations for the compartments CBM and P. The new model has been implemented and simulated but showed numerical instability, which caused us to investigate the new model equations with regard to their stability.

Subsequently, we regard time delay differential equations of type (45) - (47) for $\nu(t) = \nu_0 = \text{constant} > 0$ and $a(t) = a_0 = \text{constant} > 0$. Under these assumptions equations (47) and (46) do have a closed form solution which can be calculated using their integral representations

$$1 = \int_{\gamma(t)}^{t} \nu(\sigma)d\sigma \stackrel{!}{=} \int_{\gamma(t)}^{t} \nu_0 d\sigma \quad \text{and} \quad \alpha(t) = \int_{\gamma(t)}^{t} a(\sigma)d\sigma \stackrel{!}{=} \int_{\gamma(t)}^{t} a_0 d\sigma, \tag{56}$$

respectively. It follows directly that $\gamma(t) = t - 1/\nu_0 = t - T$, where $T := 1/\nu_0$, must hold. With this result $\alpha(t) = a_0 T$ follows and we arrive at a relatively simple differential equation given by

$$\frac{dm(t)}{dt} = u(t) + a_0 m(t) - \exp\{a_0 T\} u(t-T), \tag{57}$$

where $u(t)$ is an inflow into the compartment and $m(t)$ is the state which is the content of the compartment. After Laplace-Transformation (57) results in

$$G(s) := \frac{M(s)}{U(s)} = \frac{1 - \exp\{-(s - a_0)T\}}{s - a_0}, \tag{58}$$

where $M(s)$ and $U(s)$ are the Laplace transforms functions of $m(t)$ and $u(t)$, respectively, using the Laplace variable s. From (58) it is obvious that there exists a pole-zero cancellation in the right half of the complex plane at $s = a_0$ which explains the observed numerical instabilities during simulation [14].

3.7 Model Reduction

Despite the problems described in the preceeding section, it is possible to perform an approximation avoiding these problems for the modeling. This approximation consists in splitting the compartment under consideration into two new compartments connected sequentially, namely, the first which models a pure maturation using a time delay differential equation which can be obtained by setting $a(t) \equiv 0$ in (19), and the second which models the amplification using an ordinary differential equation. Each of the respective compartments then can be described using equations of the form

$$\frac{d}{dt}x_1(t) = u(t) - \frac{\nu(t)}{\nu(\gamma(t))}u(\gamma(t)), \quad \frac{d}{dt}x_2(t) = \frac{\nu(t)}{\nu(\gamma(t))}u(\gamma(t)) \cdot \alpha - \lambda x_2(t)$$

together with a state dependent amplifying regulation function α and

$$\frac{d}{dt}\gamma(t) = \frac{\nu(t)}{\nu(\gamma(t))}$$

describing the maturation time of the respective compartment. Further details about this approach can be obtained from [13]. Since the time delay effect occurs in both of the above equations through the term $\frac{\nu(t)}{\nu(\gamma(t))}u(\gamma(t))$ it is feasible to neglect the first, pure transport equation. This leads to a compartment description for the CBM and P compartment using a set of equations of the form

$$\frac{d}{dt}x(t) = \frac{\nu(t)}{\nu(\gamma(t))}u(\gamma(t)) \cdot \alpha - \lambda x(t), \quad \frac{d}{dt}\gamma(t) = \frac{\nu(t)}{\nu(\gamma(t))}$$

for each compartment [14]. Hereby, we introduce a new amplification function α and a new parameter λ which are used to represent the dynamics of the respective compartment in analogy to the Fliedner-Steinbach model.

Following this approach we derive a set of time delay differential equations describing the human granulopoiesis given by

$$\frac{dS}{dt} = \lambda_S(2\rho - 1)S, \quad y_S := 2\lambda_S(1-\rho)S, \tag{59}$$

$$\frac{dS_{inj}}{dt} = \lambda_{inj}(2\rho_{inj} - 1)S_{inj}, \quad y_{inj} := 2\lambda_{inj}(1-\rho_{inj})S_{inj}, \tag{60}$$

$$\frac{d\,CBM}{dt} = \frac{\nu_{CBM}}{\nu_{CBM}^{(C)}}\left(y_S^{(C)} + y_{inj}^{(C)}\right)\alpha_{CBM} - \lambda_{CBM}CBM, \tag{61}$$

$$\frac{dP}{dt} = \lambda_{CBM}\frac{\nu_P}{\nu_P^{(P)}}CBM^{(P)}\alpha_P - \lambda_P P, \tag{62}$$

$$\frac{dM}{dt} = \lambda_P P - \lambda_M M, \quad \frac{dR}{dt} = \lambda_M M - \lambda_R R, \tag{63}$$

$$\frac{dF}{dt} = \lambda_R R - \lambda_F F, \tag{64}$$

$$\frac{d\gamma_{CBM}}{dt} = \frac{\nu_{CBM}}{\nu_{CBM}^{(C)}}, \quad \frac{d\gamma_P}{dt} = \frac{\nu_P}{\nu_P^{(P)}} \tag{65}$$

where we use the regulation functions

$$\lambda_S = \gamma_1 e^{-\nu_1 S} + \gamma_2 e^{-\nu_2 ZK} + \gamma_3, \tag{66}$$

$$\nu_{CBM} = \gamma_4 - \gamma_5 e^{-\nu_3 Z}, \tag{67}$$

$$\nu_P = \gamma_6 - \gamma_7 e^{-\nu_4 Z}, \tag{68}$$

$$\lambda_M = \gamma_8 + \gamma_9 e^{-\nu_5 F}, \tag{69}$$

$$\lambda_R = \gamma_{10} + \gamma_{11} e^{-\nu_6 F}, \tag{70}$$

$$\alpha_{CBM} = 2^{\Omega_{CBM}}, \quad \Omega_{CBM} = \gamma_{12} + \gamma_{13} e^{-\nu_7 Z}, \tag{71}$$

$$\rho = \begin{cases} 0.63 & \text{for } S < 5 \cdot 10^5 \\ 0.50 & \text{for } S > 1.25 \cdot 10^9 \\ \gamma_{14} + \gamma_{15} e^{-\nu_8 (S - 5 \cdot 10^5)} & \\ +\gamma_{16} e^{-\nu_9 (S - 5 \cdot 10^5)} & \text{else.} \end{cases} \tag{72}$$

in accordance with the Fliedner-Steinbach model. The parameters in the above equations are computed by conserving the steady-state values and the respective ranges of the regulation functions of the Fliedner-Steinbach model and are given in [15]. Regarding the structure of the model equations, we can fomally write the system as

$$\frac{d}{dt}x(t) = f(x(t), x(\alpha_1(x(t))), x(\alpha_2(x(t)))) \tag{73}$$

using the state vector

$$x(t) = (S(t), S_{inj}(t), CBM(t), P(t), M(t), R(t), F(t), \gamma_{CBM}(t), \gamma_P(t))^T.$$

For the numerical integration of systems of time delay differential equations we refer to [11].

4 Estimation of Initial Conditions

As mentioned in the introduction the initial conditions of the new time delay model are subject to be estimated by minimization of a quadratic distance measure. The distance measure depends on the real blood granulocyte cell counts of an irradiated person and the content of the compartment F of the time delay model, since it represents the amount of granulocytes in the blood. Therefore, we define the optimization problem $J(p^*) = \min_p\{J(p)\}$ with

$$J(p) := \sum_{i=1}^{k} \left(\hat{F}(\tau_i) - F(\tau_i, p)\right)^2 \tag{74}$$

where $\hat{F}(\tau_i)$ are real blood granulocyte measurements from an irradiated person at times τ_i and $F(\tau_i, p)$ represents the models blood granulocyte compartment (64) evaluated at the time τ_i. The parameters enter the model via the initial conditions $\Phi(t, p)$ for $t \leq 0$. In our case, we assume constant initial functions for the cell compartments such that

$$\Phi(t, p) = (p_1, \ldots, p_6, \beta, t - T_{CBM}, t - T_P)^T \quad t \leq 0 \tag{75}$$

is chosen. Within (75) β represents the constant initial function for the F compartment which is not estimated but either chosen as the steady state value of the F compartment if no measurement is available at $t = 0$ or as the value of the measurement if available at $t = 0$. The initial functions for γ_{CBM} and γ_P are chosen as functions of time t since they represent maturating delay times.

Using this initial function for the integration of the model equations makes the whole solution a function not only of time, but also on the parameters p_i, $i = 1, \ldots, 6$. Since it is known [2,3], that the estimation of parameters in time delay differential equations may pose difficulties concerning the smoothness of the objective function this problem has to be investigated. Because the time delays in the model equations are a function of the state, which itself is a function of the parameters, the position of derivative discontinuities will be a function of the parameters. To get information about the smoothness of the objective function, the smoothness of the state $F(t,p)$ has to be considered because only this state enters the objective function. If the state $F(t,p)$ is sufficiently smooth with respect to the time, then according to [2, 3] also the objective function will be accordingly smooth. In order to analyze the smoothness of the state F we use the so called network dependency graph theory presented in [20–22]. Subsequently, we will derive the network dependency graph for our model of granulopoiesis, which then should allow insight in the propagation of discontinuities in this system. We investigate the model equations and remind, that the regulation functions within this model are functions of the states. For better understanding we rename the states as $x_1(t) = S(t)$, $x_2(t) = S_{inj}(t)$, $x_3(t) = CBM(t)$, $x_4(t) = P(t)$, $x_5(t) = M(t)$, $x_6(t) = R(t)$, $x_7(t) = F(t)$, $x_8(t) = \gamma_{CBM}(t)$, and $x_9(t) = \gamma_P(t)$. The regulation function λ_S, for instance, is a function of the state S and of $ZK = CBM + P + M + R + F$. Using the above nomenclature λ_S is a function $\lambda_S = \lambda_S(x_1, x_2, x_4, x_5, x_6, x_7)$. For the remaining regulation functions one obtains equivalently

$$\begin{aligned} &\rho = \rho(x_1), \quad \nu_{CBM} = \nu_{CBM}(x_1, x_3, x_4, x_5, x_6, x_7), \\ &\alpha_{CBM} = \nu_{CBM}(x_1, x_3, x_4, x_5, x_6, x_7), \quad \nu_P = \nu_P(x_1, x_3, x_4, x_5, x_6, x_7), \\ &\lambda_M = \lambda_M(x_7), \quad \lambda_R = \lambda_R(x_7). \end{aligned}$$

The differential equations for $S = x_1$ and $S_{inj} = x_2$ read as

$$\frac{d}{dt}x_1 = f_1(x_1, x_3, x_4, x_5, x_6, x_7), \quad \frac{d}{dt}x_2 = f_1(x_2)$$

Functions evaluated at retarded times $\gamma_{CBM} = x_8$, for instance the function $\nu_{CBM}(\gamma_{CBM})$, are occuring in (61). With the above nomenclature this dependence represents a dependence on $x_1(x_8)$, $x_3(x_8)$, $x_4(x_8)$, $x_5(x_8)$, $x_6(x_8)$, and $x_7(x_8)$. The formal representation for (61) is given by

$$\frac{dx_3}{dt} = f_3(x_1, x_3, \ldots, x_7, x_1(x_8), x_2(x_8), x_3(x_8), \ldots, x_7(x_8))$$

For the remaining states one calculates

$$\begin{aligned} &\frac{d}{dt}x_4 = f_4(x_1, x_3, \ldots, x_7, x_1(x_9), x_3(x_9), \ldots, x_7(x_9)), \\ &\frac{d}{dt}x_5 = f_5(x_4, x_5, x_7), \quad \frac{d}{dt}x_6 = f_6(x_5, x_6, x_7), \quad \frac{d}{dt}x_7 = f_7(x_6, x_7), \\ &\frac{d}{dt}x_8 = f_8(x_1, x_3, \ldots, x_7, x_1(x_8), x_3(x_8), \ldots, x_7(x_8)), \\ &\frac{d}{dt}x_9 = f_9(x_1, x_3, \ldots, x_7, x_1(x_9), x_3(x_9), \ldots, x_7(x_9)). \end{aligned}$$

Thus, in terms of strong and weak coupling, the dependencies for the model of granulopoiesis are 1S1, 3S1, 4S1, 5S1, 6S1, 7S1, 2S2, 1S3, 3S3, 4S3, 5S3, 6S3,

7S3, 8S3, 1W3, 2W3, 3W3, 4W3, 5W3, 6W3, 7W3, 1S4, 3S4, 4S4, 5S4, 6S4, 7S4, 9S4, 1W4, 3W4, 4W4, 5W4, 6W4, 7W4, 4S5, 5S5, 6S5, 5S6, 6S6, 7S6, 6S7, 7S7, 1S8, 3S8, 4S8, 5S8, 6S8, 7S8, 1W8, 3W8, 4W8, 5W8, 6W8, 7W8, 1S9, 3S9, 4S9, 5S9, 6S9, 7S9, 1W9, 3W9, 4W9, 5W9, 6W9, and 7W9. This coupling is depicted in Figure 5 where only the most relevant joints with regard to our discussion are shown.

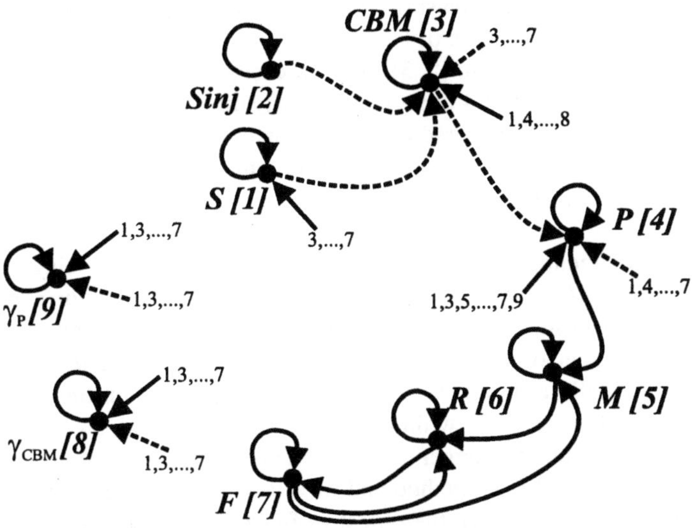

Figure 5. Dependency network for the model of granulopoiesis.

Due to the regulation functions there exist more joints of strong (solid arrow) and weak (dashed arrow) coupling between the different states, which are denoted by solid and dashed arrows followed by the respective state numbers. The state names $S, \ldots, \gamma_P$ are printed together with their assigned component number in square brackets.

By the choice of the initial function (75) a discontinuity of order one occurs at $t = 0$ in all components of the solution of the model equations since in general $\dot{\Phi}(0, p) \neq \dot{x}(0, p)$. Thus, all components themselves are continuous but their first derivatives have jumps at $t = 0$. But this is not a severe feature with regard to the estimation problem since the locations of these discontinuities do not vary with p but are fixed. Therefore, they do not lead to discontinuities within the objective function (74). The first and lowest order discontinuities, whose locations depend on the parameter vector p, do occur in CBM and in P because only in these compartments delayed terms enter directly. The locations of these discontinuities are fixed by $\gamma_{CBM}(t_{1,CBM}) = 0$ and $\gamma_P(t_{1,P}) = 0$ where $t_{1,CBM}$ and $t_{1,P}$ are the locations of the discontinuities in the CBM and P compartment, respectively. The order of these discontinuities is at least two because they appear due to a retarded argument passing over the location of the first order discontinuity at $t = 0$, which has the order one. All subsequent discontinuities have higher orders. Consequently, the minimum order discontinuities depending on p are occuring in CBM and P and have an order of two. Now using the dependency network given in Figure 5, it can be seen that a discontinuity occuring in F has at least order 5 (as the discontinuity in P of order 2 propagates to F). This consequently represents the minimum order of discontinuities which may appear in F. Summarizing this paragraph we showed that discontinuities in F are at least of order 5, which means that jumps may occur at least in the 5th time derivative of F.

Using the approaches from [2, 3] one can easily reproduce that the objective function (74) must be at least 4 times continuously differentiable with respect to p. Thus, it is possible to calculate gradients and hessians of the objective function without encountering difficulties concerning the smoothness of the objective function. This means, that standard optimization methods can be used for the minimization of (74).

We have implemented the model and an appropriate solver for delay differential equations based on [6] under C++. In order to solve the optimization problem, we use the Nelder-Mead method to calculate a rough estimate $p^{(0)}$ and subsequently we switch to a higher order method, where $p^{(0)}$ is used as the initial parameter vector. We aditionally estimate T_{CBM} and T_P from (75) within the Nelder-Mead method [16] which afterwards, within the higher order method, are assumed to be known and therefore are not longer estimated. This means, we have $m = 8$ parameters within the Nelder-Mead method and $m = 6$ within the higher order method. This strategy is necessary since the objective function is very insensitive with respect to T_{CBM} and T_P after a few optimization steps. It should be mentioned, that the parameter space is restricted to positive values of the parameters since negative values would correspond to negative cell numbers which are not feasible in the biological sense. Thus we have to deal with optimization techniques for box constrained optimization problems. Well known classical methods are the constrained Newton method, the Levitin-Poljak method or the Wilson method [9], which have also been implemented. The results in the following section have been computed using the Wilson method. The numerical calculation of the gradient and the hessian of the objective function themselves are carried out using the automatic differentiation method [10] in forward mode. The QP subproblems are solved using a standard method [5].

5 Results

To demonstrate the excellent qualification of our decision support system in view of the prediction of the human granulopoiesis, we give some typical prediction results for real granulocyte datasets of Chernobyl victims.

It has to be mentioned that only the patient data denoted by linked circles have been used to perform the prediction. The remaining data denoted by linked triangles have not been used for the prediction but are printed in order to compare the prediction

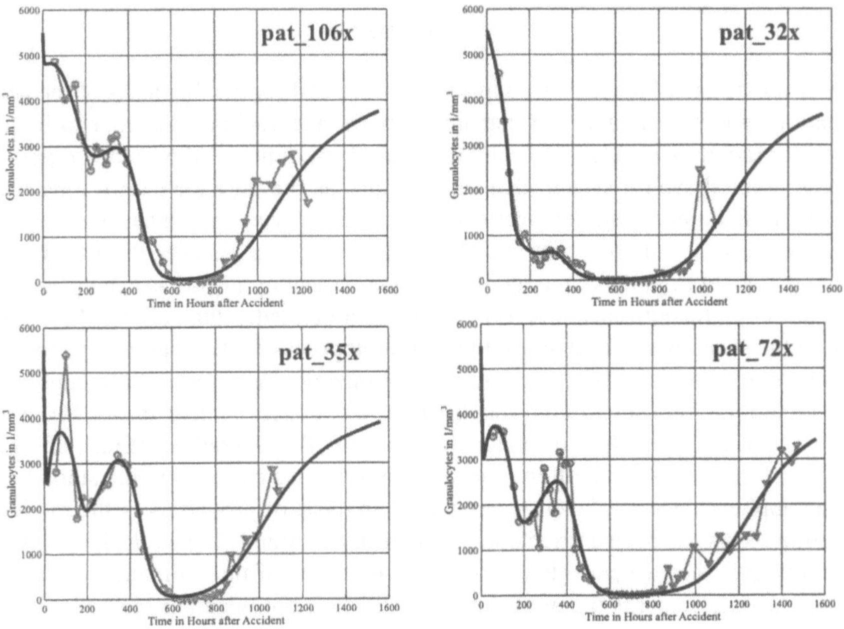

Figure 6. Prediction results for real patient data.

result with the real patient data. In Tab. 2, the number of steps for the Wilson method, the norms of the gradients of the optimal solution, and the computation time are presented for the respective data sets.

Data Set	Nelder-Mead steps	Wilson steps	$\|\nabla J(p^*)\|$	Time
pat106x	89	6	6.29597e-009	151.147 sec
pat32x	194	4	1.72508e-009	97.2 sec
pat35x	241	4	3.05051e-005	106.783 sec
pat72x	76	3	4.87781e-005	110.919 sec

Table 2. Optimization Results

The computations have been carried out on a standard PC with Pentium-II 400 MHz CPU and 128 MB RAM. Regarding the computation times it gets clear that our system is able to satisfy real-time requirements since it is able to compute predictions within a time period of a few minutes. This complies with the demands from the medical part.

6 Conclusion

The development of a decision support system (DSS) for the diagnosis of the acute radiation syndrome is a strong demand from the medical part in order to be prepared for radiation accidents like the Chernobyl accident in 1986. In this report we have presented such a system that incorporates as the core element a time-delay model of the human granulopoiesis since the granulocytes are the best indicator for the severeness of radiation injuries. The model derivation has been discussed in detail and a novel time-delay model of the human granulopoiesis has been presented. Furthermore, the optimization problem has been defined, which allows the estimation of the severeness of a damage to the granulopoietic system caused by radiation exposure of a human being. Hereby, the optimization task specifically consists in an estimation of constant initial functions for the models compartments. Since it is known, that the estimation of parameters in delay differential equations, of which our novel model consists, often poses severe problems concerning the smoothness of the applied objective function, a detailed analysis of properties of the model of granulopoiesis has been carried out. As the result it turned out that the least squares objective function we considered for the formulation of the estimation problem is sufficiently smooth for the application of standard optimization methods. Within the implementation of our DSS, called MODRAT, we have used the Wilson SQP method to calculate the optimal solutions coupled with Automatic Differentiation techniques in order to calculate the gradients and hessians of the objective function needed. Hereby, MODRAT is a C++ implementation with graphical user interface of the mentioned techniques designed for 32 bit Microsoft Windows operating systems. Typical prediction results show the excellent qualification of our system in order to predict the recovery of radiation exposed persons, whereby the computation times lie in the range of a few minutes. Regarding these times our system fulfilles the realtime requirements as they are formulated by the medical part.

References

1. A. Asachenkov, G. Marchuk, R. Mohler, S. Zuev: Disease Dynamics. Birkhäuser, Boston, Basel, Berlin (1994)
2. C. T. H. Baker, C. A. H. Paul: Pitfalls in parameter estimation for delay differential equations. NAREP 267, Manchester Centre for Computational Mathematics (1995)
3. C. T. H. Baker, C. A. H. Paul: Pitfalls in parameter estimation for delay differential equations. SIAM J. Sci. Comp. **18,1** (1997) 305–314
4. H. Begemann, J. Rastetter: Klinische Hämatologie. Georg Thieme Verlag (1993)
5. T. Coleman, M. A. Branch, A. Grace: Optimization Toolbox User's Guide. The MathWorks, Inc., Natick (1999)
6. A. Feldstein, K. W. Neves: High order methods for state-dependent delay differential equations with nonsmooth solutions. SIAM J. Numer. Anal. **21,5** (1984) 844–863
7. T. M. Fliedner, K.-H. Steinbach: Simulationsmodelle von Perturbationen des granulozytären Zellerneuerungssystems. in W. Doerr: 'Modelle der pathologischen Physiologie' (1987) 89–105

8. T. M. Fliedner, B. Tibken, E. P. Hofer, W. Paul: Stem cell responses after radiation exposure: A key to the evaluation and prediction of its effects. Health Physics **70,6** (1996) 787–797
9. Ch. Großmann, J. Terno: Numerik der Optimierung. Teubner Studienbücher. B. G. Teubner, Stuttgart (1993)
10. J. Herzberger: Wissenschaftliches Rechnen. Akademie Verlag, Berlin (1995)
11. E. P. Hofer, B. Tibken, F. Lehn: Differential Equations with State-Dependent Delays. in Grötschel, M., S. O. Krumke, J. Rambau: 'Online Optimization of Large Scale Systems', Springer Verlag, Berlin, Heidelberg, New York (2001)
12. F. Lehn, B. Tibken, E. P. Hofer: Development of a Decision Support System for the Treatment of Irradiated Persons - Modeling the Human Granulopoiesis. Preprint 98-30, 'Echtzeit-Optimierung großer Systeme' (1998)
13. F. Lehn, B. Tibken, E. P. Hofer: Human Granulopoiesis – A New Time Delay Model. Proc. of the European Control Conference ECC 1999, Karlsruhe, Germany (1999) 1–5, CP–15
14. F. Lehn, E. P. Hofer: A comparison of different approaches in modelling human granulopoiesis., in E. P. Hofer, O. Sawodny: 'Synergies in Engineering Research' (2000) 237–250
15. F. Lehn, B. Tibken, E. P. Hofer: Estimation of Initial Conditions for a Time Delay Model of Human Granulopoiesis. Preprint 00-24, 'Echtzeit-Optimierung großer Systeme' (2000)
16. J. A. Nelder, R. Mead: A simplex method for function minimization. Computer Journal, **7** (1965) 308–313
17. W. Paul: Die Analyse der Regeneration der Granulopoese nach Stammzelltransplantationen mit Hilfe einer regelungstechnischen Implementation eines biomathematischen Modells. PhD thesis, Universität Ulm (1997)
18. R. Sauer, I. Szabó: Mathematische Hilfsmittel des Ingenieurs, Vol. II. Springer-Verlag, Berlin, Heidelberg, New York (1969)
19. F. Treves: Basic Linear Partial Differential Equations. Academic Press, New York, San Francisco, London (1975)
20. D. R. Willé, C. T. H. Baker: The tracking of derivative discontinuities in systems of delay differential equations. Appl. Num. Math. **9** (1992) 209–222
21. D. R. Willé, C. T. H. Baker: The propagation of derivative discontinuities in systems of delay-differential equations. NAREP 160, Manchester Centre for Computational Mathematics (1988)
22. D. R. Willé, C. T. H. Baker: The tracking of derivative discontinuities in systems of delay-differential equations. NAREP 185, Manchester Centre for Computational Mathematics (1990)

V

Stochastic Optimization in Chemical Engineering

*Stochastic Optimization for Operating Chemical Processes under Uncertainty

René Henrion[1], Pu Li[2], Andris Möller[1], Marc C. Steinbach[3], Moritz Wendt[2], and Günter Wozny[2]

[1] Weierstraß-Institut für Angewandte Analysis und Stochastik (WIAS), Berlin, Germany
[2] Institut für Prozess- und Anlagentechnik, Technische Universität Berlin, Germany
[3] Konrad-Zuse-Zentrum für Informationstechnik Berlin (ZIB), Germany

Abstract Mathematical optimization techniques are on their way to becoming a standard tool in chemical process engineering. While such approaches are usually based on deterministic models, uncertainties such as external disturbances play a significant role in many real-life applications. The present article gives an introduction to practical issues of process operation and to basic mathematical concepts required for the explicit treatment of uncertainties by stochastic optimization.

1 Operating Chemical Processes

Chemical industry plays an essential role in the daily life of our society. The purpose of a chemical process is to transfer some (cheap) materials into other (desired) materials. Those materials include any sorts of solids, liquids and gas and can be single components or multicomponent mixtures. Common examples of chemical processes are reaction, separation and crystallization processes usually composed of operation units like reactors, distillation columns, heat exchangers and so on. Based on market demands, those processes are designed, set up and put into operation. From the design, the process is expected to be run at a predefined operating point, i.e., with a certain flow rate, temperature, pressure and composition [22].

Distillation is one of the most common separation processes which consumes the largest part of energy in chemical industry. Figure 1 shows an industrial distillation process to separate a mixture of methanol and water to high purity products (methanol composition in the distillate and the bottom should be $x_D \geq 99.5\,\text{mol\%}$ and $x_B \leq 0.5\,\text{mol\%}$, respectively). The feed flow F to the column is from outflows of different upstream plants. These streams are first accumulated in a tank (a middle buffer) and then fed to the column. The column is operated at atmospheric pressure. From the design, the diameter of the column, the number of trays, the reboiler duty Q and the reflux flow L will be defined for the given product specifications.

For an existing chemical process, it is important to develop flexible operating policies to improve its profitability and reducing its effect of pollution. The ever-changing market conditions demand a high flexibility for chemical processes under different product specifications and different feedstocks. On the other hand, the increasingly stringent limitations to process emissions (e.g., $x_B \leq 0.5\,\text{mol\%}$ in the above example) require suitable new operating conditions satisfying these constraints. Moreover, the properties of processes themselves change during process

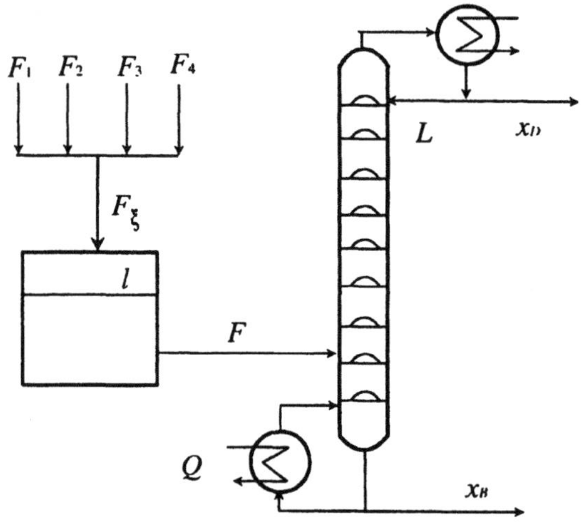

Figure 1. An industrial distillation column with a feed tank

operation, e.g., tray efficiencies and fouling of the equipment, which leads to reduction of product quality if the operating point remains unchanged. Therefore, keeping a constant operating point given by the process design is nowadays an out-dated concept. That is to say, optimal and robust operating policies should be searched for and implemented online, corresponding to the real-time process situations.

In the past, heuristic rules were used for improving process operation in chemical industry. However, since most chemical processes behave nonlinear, time-dependent and possess a large number of variables, it was impossible to find the optimal solutions or even feasible solutions by heuristic rules. Therefore, systematic methods including modeling, simulation and optimization have been developed in the last two decades for process operation. These methods are model-based deterministic approaches and have been more and more used in chemical industry [10].

1.1 Process Modeling

Conservation laws are used for modeling chemical processes. A balance space is first chosen, for which model equations will be established by balancing mass, momentum and energy input into and output from the space [3]. Thus variables of a space can be classified into independent and dependent variables. Independent variables are input variables including manipulated variables and disturbance variables. For instance, the reflux flow and the reboiler duty are usually manipulated variables for a distillation column, while the feed flow and composition are disturbance variables. Dependent variables are output variables (usually called state variables) which depend on the input variables. The compositions and temperatures on the trays inside the column are dependent variables. Besides conservation laws, correlation equations based on physical and chemical principles are used to describe relations between state variables. These principles include vapor-liquid equilibrium if two phases exist in the space, reaction kinetics if a reaction takes place and fluid dynamics for describing the hydraulics influenced by the structure of the equipment.

Let us consider modeling a general tray of a distillation column, as shown in Figure 2, where i and j are the indexes of components ($i = 1, NK$) and trays (from the condenser to the reboiler), respectively. The dependent variables on each tray are the vapor and liquid compositions $y_{i,j}, x_{i,j}$, vapor and liquid flow rate V_j, L_j, liquid molar holdup M_j, temperature T_j and pressure P_j. The independent variables are the feed flow rate and composition $F_j, z_{Fi,j}$, heat flow Q_j and the flows and compositions from the upper as well as lower tray. The model equations include component and energy balances, vapor-liquid equilibrium equations, a liquid holdup equation as well as a pressure drop equation (hydraulics) for each tray of the column:

- Component balance:

$$\frac{d(M_j x_{i,j})}{dt} = L_{j-1}x_{i,j-1} + V_{j+1}y_{i,j+1} - L_j x_{i,j} - V_j y_{i,j} + F_j z_{Fi,j} \quad (1)$$

- Phase equilibrium:

$$y_{i,j} = \eta_j K_{i,j}(x_{i,j}, T_j, P_j)x_{i,j} + (1 - \eta_j)y_{i,j+1} \quad (2)$$

- Summation equation:

$$\sum_{i=1}^{NK} x_{i,j} = 1, \qquad \sum_{i=1}^{NK} y_{i,j} = 1 \quad (3)$$

- Energy balance:

$$\frac{d(M_j H_j^L)}{dt} = L_{j-1}H_{j-1}^L + V_{j+1}H_{j+1}^V - L_j H_j^L - V_j H_j^V + F_j H_{F,j}^L + Q_j \quad (4)$$

- Holdup correlation:

$$M_j = \varphi_j(x_{i,j}, T_j, L_j) \quad (5)$$

- Pressure drop equation:

$$P_j = P_{j-1} + \psi(x_{i,j-1}, y_{i,j}, L_{j-1}, V_j, T_j) \quad (6)$$

In addition to the equations (1)–(6), there are auxiliary relations to describe the vapor and liquid enthalpy H_j^V, H_j^L, phase equilibrium constant $K_{i,j}$, holdup correlation φ_j and pressure drop correlation ψ_j which are functions of the dependent variables. Parameters in these correlations can be found in chemical engineering handbooks like [9, 19]. Murphree tray efficiency η_j is introduced to describe the nonequilibrium behavior. This is a parameter that can be verified by comparing the simulation results with the operating data.

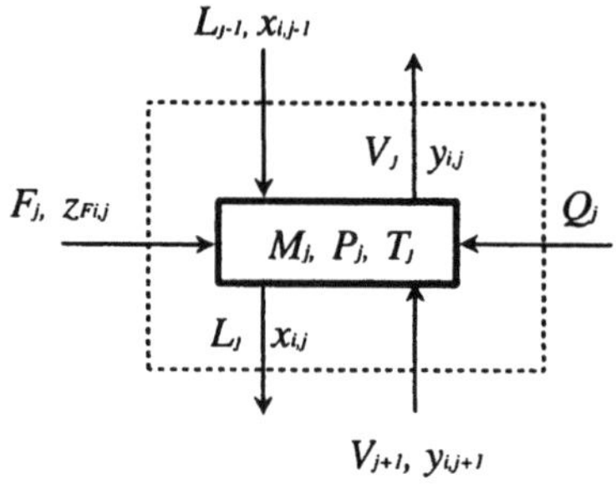

Figure 2. A general tray of the distillation column

Equations of all trays in the column lead to a complicated nonlinear DAE system. Moreover, some dependent variables are required to be kept at a predefined value (e.g., the bottom liquid level of the column). This will be realized by feedback control loops usually with PID (proportional-integral-derivative) controllers. Thus controller equations have to be added to the model equation system, if closed loop behaviors will be studied. Process simulation means, with given independent variables, to solve the DAE so as to gain the profiles of the dependent variables. In the framework of optimization, an objective function will be defined (e.g., minimizing the energy consumption during the operation). The above DAE system will be the equality constraints. The inequality constraints consist of the distillate and bottom product specifications as well as the physical limitations of vapor and liquid flow rates. Thus a dynamic nonlinear optimization problem is formulated. Approaches to solve dynamic optimization problems use a discretization method (either multiple-shooting or orthogonal collocation) to transform the dynamic system to a NLP problem. They can be classified into simultaneous approaches, where all discretized variables are included in a huge NLP problem, and sequential approaches, where a simulation step is used to compute the dependent variables and thus only the independents will be solved by NLP. Solution approaches to such problems can be found in [15, 23]. As a result, optimal operating policies for the manipulated variables can be achieved. It should be noted that some processes may have zero degree of freedom. In the above example, when the product specifications become equalities, it implies that the independent variables at the steady state must be fixed for fulfilling these specifications.

1.2 Uncertainties in Process Operation

Although deterministic approaches have been successfully applied to many complex chemical processes, their results are only applicable if the real operating conditions are included in the problem formulation. To deal with the unknown operating reality a priori, optimization under uncertainty has to be considered [13]. From the viewpoint of process operation there are two general types of uncertainties.

Internal Uncertainties

These uncertainties represent the unavailability of the knowledge of a process. The process model is only an approximation and thus can not describe the real behavior of the process exactly. Internal uncertainties include the model structure and the parameter uncertainty. For the description of a chemical or a thermodynamic phenomenon several representations always exist. The selection of a representation for the model leads to a structure uncertainty. Model parameters (such as parameters of reaction kinetics and vapor-liquid equilibrium) are usually estimated from a limited number of experimental data and hence the model may not be able to predict the actual process [28].

External Uncertainties

These uncertainties, mainly affected by market conditions, are from outside but have impacts on the process. These can be the flow rate and composition of the feedstock, product specifications as well as the supply of utilities. The outlet stream from an upstream unit and the recycle stream from a downstream unit are usually uncertain streams of the considered operating unit. For some processes which are sensitive to the surrounding conditions, the atmospheric temperature and pressure will be considered as external uncertain variables.

While some uncertain variables are treated as constants during the process operation, there are some time-dependent uncertain variables which are dependent on the process operating conditions. For instance, the tray efficiency of a distillation column often changes with its vapor and liquid load. Another example is the uncertainty of the feed streams coming from the upstream plants. In these cases a dynamic stochastic optimization problem will be formulated. For such problems, rather than individual stochastic parameters, continuous stochastic processes should be considered. Approximately, most of them can be considered as normal distributed stochastic processes. There may exist correlation between these variables. Operation data from historic records can be used to estimate these stochastic properties.

In deterministic optimization approaches the expected values of uncertain variables are usually employed. In the reality the uncertain variables will deviate from their expected values. Based on the realized uncertain variables a reoptimization can be carried out to correct the results from the last iteration. For dynamic optimization, a moving horizon with N time intervals will be introduced. Figure 3 shows the implementation of the three consecutive paces of the moving horizon. At the current horizon k only the values of the available policies for the first time interval u_1 which were developed in the past horizon $k-1$, will be realized to the process. During this time interval a reoptimization is carried out to develop the operating policies for the future horizon $k+1$. The method in which the expected values of the uncertainties are used in the problem formulation is the so-called wait-and-see strategy. The shortcoming of this strategy is that it can not guarantee holding inequality constraints.

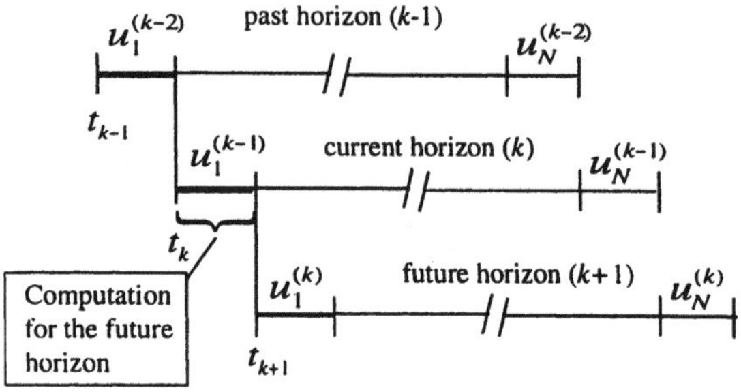

Figure 3. Reoptimization over a moving horizon

1.3 Distillation Column Operation under Uncertain Feed Streams

Now we consider again the industrial distillation process. The flows from the upstream plants often change considerably due to the varying upstream operation. We may have high flow rates of the feed during the main working hours and decreased flow rates during the night hours or at the weekend. Figure 4 shows the measured profiles of the total feed flow, composition and temperature for 24 hours. Here we only focus on the impact of the variation of the flow rate. One consequence resulted from the fluctuating feed streams is that the tank level $\mathfrak{l}$ may exceed the upper bound $\mathfrak{l}^{max}$ (then a part of the liquid must be pumped out to an extra tank) or fall below the lower bound $\mathfrak{l}_{min}$ (then a redundant feed stream must be added to the feed flow). Since the appearance of these cases will lead to considerable extra costs, a careful planning for the operation should be made to prevent these situations.

Another consequence of a large feed change is that it causes significant variations of the operating point of the distillation column. To guarantee the product quality ($\mathtt{x}_D, \mathtt{x}_B$), a conservative operating point is usually used for a higher purity than the required specification. This leads, however, to more energy consumption than required. The growth of energy requirement for a column operation is very sensitive to the product purity, especially for a high purity distillation.

Conventionally a feedback control loop is used in process industry to keep the level of the feed tank, using the outflow as the manipulated variable. The drawback of this control loop is that it can not guarantee the output constraints and it will propagate the inflow disturbance to the downstream distillation column.

To describe the continuous uncertain inflow this stochastic process will be discretized as multiple stochastic variables in fixed time intervals. We assume they have a multivariate normal distribution with an available expected profile and a covariance matrix in the considered time horizon. The reason for this assumption is that the total feed of the tank is the sum of several independent streams from the upstream plants. According to the central limit theorem [16], if a random variable is generated as the sum of effects of many independent random parameters, the distribution of the variable approaches a normal distribution, regardless of the distribution

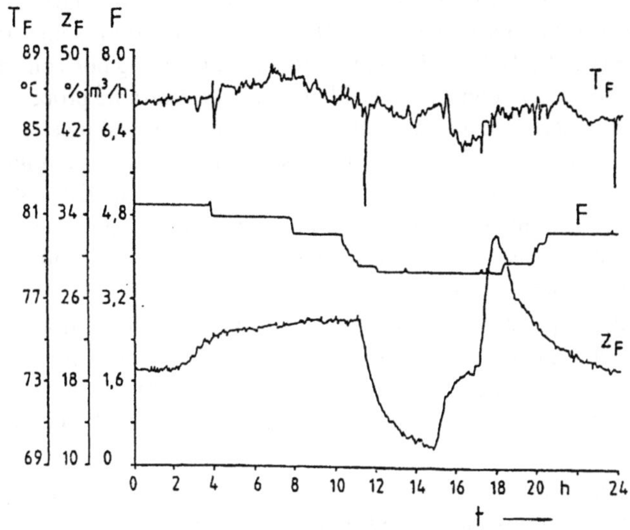

Figure 4. Measured feed profiles of an industrial methanol-water distillation process

of each individual parameter. These parameters can be readily obtained by analyzing daily measured operating data. It is obvious that a wait-and-see strategy is not appropriate to be used in this process. Setting the feed flow with its expected profile in a deterministic optimization can not guarantee holding the tank level in the desired region. The product specification will also be easily violated by the drastically changing real feed flow. Therefore, a here-and-now strategy, which includes the uncertainties in the optimization problem, should be used. This will be discussed in the next sections.

2 Modeling Uncertainty

As discussed in the previous section, a common technique of correcting random disturbances in chemical processes is *moving horizon* control (or *model predictive control*): states are measured (or estimated) in relatively small intervals, and optimal open-loop strategies are computed over a given planning horizon—"optimal" under the simplifying assumption that no further disturbances occur. In effect, the frequent repetition of this process implicitly generates a (possibly nonlinear) feedback controller that reacts to the measured disturbances.

The stochastic approaches described here are naturally applicable within such a moving horizon framework but differ in a fundamental aspect: rather than just reacting they *look ahead* by taking stochastic information on future events *explicitly* into account. This is possible if it is known which random events may occur and how likely they are. In other words, a stochastic model of the disturbances is required, taking the form of a random process $\xi = (\xi_t)_{t\in[0,T]}$ defined on some underlying probability space $(\Omega, \mathcal{F}, P)$. Here T is the length of the planning horizon and 0 is the current time. In the present context, only $\mathbb{R}^k$-valued *discrete-time* processes for $t = 0, 1, \ldots, T$ are considered, and it is assumed that ξ_t is observed just before time t so that ξ_0 is known at $t = 0$. Thus, the processes can be seen as random variables $\xi = (\xi_1, \ldots, \xi_T)$ in $\mathbb{R}^{kT}$. Moreover, we consider either *discrete* distributions P_ξ or distributions with a continuous *density function* on $\mathbb{R}^{kT}$. (More details will be given below.) For a comprehensive treatment of the measure-theoretic and probability-theoretic foundations see, e.g., Bauer [1,2].

Apparently the explicit modeling of uncertainty adds information to the optimization model and allows for more robust process control. The price one has to pay is the necessity of solving a *stochastic* optimization problem whose complexity may exceed the complexity of the underlying *deterministic* problem by orders of magnitude.

The precise nature of uncertainties (such as the time dependence and the significance in objective and constraints) leads to different classes of stochastic optimization models; we will describe two of them. The first approach yields a multistage recourse strategy consisting of optimal reactions to every observable sequence of random events. It minimizes expected costs while satisfying all constraints. This is appropriate if feasible solutions exist for every possible disturbance, or if costs for the violation of soft constraints can be quantified (as penalty terms). The second approach yields a single control strategy that does not react to random events but is guaranteed to satisfy the constraints with a prescribed probability. This is appropriate if constraint violations are unavoidable in certain extreme cases, or if they cause significant costs that cannot be modeled exactly. For detailed discussions of stochastic modeling aspects and problem classes we refer to the textbooks [5] and [14].

3 Scenario-Based Stochastic Optimization

In scenario-based optimization, uncertainty is modeled as a finite set of possible realizations of the future with associated positive probabilities. Each realization is called a *scenario* and represents a certain event or, in our case, history of events. In precise probabilistic terms this corresponds to a discrete distribution given by a finite probability space $(\Omega, \mathcal{F}, P)$, $|\Omega| = N$. One may simply think of ω as the "number" of a scenario, which is often emphasized by using index notation. Thus, each elementary event $\omega \in \Omega$ labels a possible realization $\xi_\omega = (\xi_{\omega 1}, \ldots, \xi_{\omega T})$, and the distribution is given by N probabilities $(p_\omega)_{\omega\in\Omega}$, that is, $P_\xi(\xi_\omega) = P(\omega) = p_\omega$. (The σ-field is then simply the power set of the sample space, $\mathcal{F} = 2^\Omega$.)

3.1 Scenario Trees

As indicated, we have to deal with event histories rather than single events. This means that there is a finite number of realizations of ξ_1, each of which may lead to

$L_T \equiv L_2 = \{3,4,5\}$ $\quad$ $\mathcal{F}_2 \equiv \mathcal{F}_T = \sigma(\{3\},\{4\},\{5\})$

$L_1 = \{1,2\}$ $\quad$ $\mathcal{F}_1 = \sigma(\{3,4\},\{5\})$

$L_0 = \{0\}$ $\quad$ $\mathcal{F}_0 = \sigma(\{3,4,5\})$

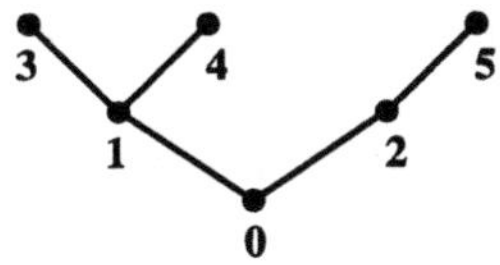

Figure 5. A small scenario tree with level sets and corresponding σ-fields

a different group of realizations of ξ_2, and so on. The repeated branching of partial event histories $\xi^t := (\xi_0, \ldots, \xi_t)$, called *stage* t *scenarios*, defines a *scenario tree* (or *event tree*) whose root represents ξ_0, the known observation at $t = 0$, and whose leaves represent the complete set of scenarios. Thus any node represents a group of scenarios that share a partial history ξ^t. We denote by V the set of nodes (or *vertices*) of the tree, by $L_t \subseteq V$ the level set of nodes at time t, and by $L \equiv L_T$ the set of leaves; further by $0 \in L_0$ the root, by $j \in L_t$ the "current" node, by $i \equiv \pi(j) \in L_{t-1}$ its unique predecessor (if $t > 0$), and by $S(j) \subseteq L_{t+1}$ its set of successors. The scenario probabilities are $p_j > 0$, $j \in L$. All other nodes also have a probability p_j satisfying $p_j = \sum_{k \in S(j)} p_k$. Hence, $\sum_{j \in L_t} p_j = 1$ holds for all t, and $p_0 = 1$.

Seen as a partitioning of the scenarios into groups, each level set L_t consists of *atoms* generating a sub-σ-field $\mathcal{F}_t = \sigma(L_t) \subseteq \mathcal{F}$ (where $\mathcal{F}_0 = \{\emptyset, \Omega\}$ and $\mathcal{F}_T = \mathcal{F}$), and ξ_t is measurable with respect to $\mathcal{F}_t$. The tree structure is thus reflected by the fact that these σ-fields form a *filtration* $\mathcal{F}_0 \subseteq \cdots \subseteq \mathcal{F}_T$ to which the process $(\xi_t)_{t=0}^T$ is *adapted*. For instance, in Figure 5 the nodes represent scenario sets as follows: $0 \leftrightarrow \{3,4,5\}$, $1 \leftrightarrow \{3,4\}$, $2 \leftrightarrow \{5\}$, $3 \leftrightarrow \{3\}$, $4 \leftrightarrow \{4\}$, and $5 \leftrightarrow \{5\}$. Since these abstract probability-theoretic notions are unnecessarily general for our purposes, we will use the more natural concept of scenario trees in the following. The notation $\xi_t = (\xi_j)_{j \in L_t}$ or $\xi = (\xi_j)_{j \in V}$ refers to the distinct realizations of ξ on level t or on the entire tree, respectively. (Here we include the deterministic initial event ξ_0 in ξ.)

3.2 *Multistage Stochastic Programs*

The main topic of this section are *multistage decision processes*, that is, sequences of alternating decisions and observations over the given planning horizon. The initial decision must be made without knowledge of the actual realizations of future events; hence it is based solely on ξ_0 and the probability distribution of ξ. As the future unfolds, the decision maker observes realizations ξ_j of random events ξ_t, thus collecting additional information which he or she takes into account from then on. The resulting sequence of decisions is therefore called *nonanticipative*. For instance, in controlling the feed tank of the distillation column in Section 1, we have to decide in each time step how much liquid to extract during the next period based on observations of the inflow during all previous periods and taking into account the probability distribution of future inflows. (For the initial decision, past observations do not appear explicitly in the problem but are implicitly modeled in the distribution.)

The specific class of problems considered here are (convex quadratic) *multistage recourse problems* on scenario trees, with decision vectors $y_j \in \mathbb{R}^n$, $j \in V$. Given are convex quadratic objective functions

$$\phi_j(y_j) := \frac{1}{2} y_j^* H_j y_j + f_j^* y_j,$$

and polyhedral feasible sets depending on the previous decision $y_i \equiv y_{\pi(j)}$,

$$Y_0 := \{ y_0 \geq 0 : W_0 y_0 = h_0 \}, \tag{7}$$

$$Y_j(y_i) := \{ y_j \geq 0 : W_j y_j = h_j - T_j y_i \}, \quad j \in V^* := V \setminus \{0\}. \tag{8}$$

These are the realizations of random costs $\phi_t(y_t)$ and random sets $Y_t(y_{t-1})$, that is, we take as random events the problem matrices and vectors

$$\xi_j = (H_j, f_j, W_j, h_j, T_j)$$

or, more generally, the functions and sets $\xi_j = (\phi_j, Y_j)$. (Conceptually we are thus allowing entirely random problem data. In practice, however, only a subset of matrix and vector elements will usually depend on an even smaller number of random influences.) Decisions y_j are to be made so as to minimize the immediate costs plus the expected costs of (optimal) future decisions; this is expressed in the general multistage recourse problem

$$\min_{y_0 \in Y_0} \phi_0(y_0) + \mathbb{E}_{L_0} \Big[\min_{y_1 \in Y_1(y_0)} \phi_1(y_1) + \cdots + \mathbb{E}_{L_{T-1}} \Big[\min_{y_T \in Y_T(y_{T-1})} \phi_T(y_T) \Big] \cdots \Big]. \tag{9}$$

Here $\mathbb{E}_{L_t}$ denotes the conditional expectation with respect to L_t,

$$\mathbb{E}_{L_t}(X_{t+1}) = \Big\{ \sum_{k \in S(j)} \frac{p_k}{p_j} X_k \Big\}_{j \in L_t}.$$

The recourse structure of this class of stochastic programs is induced by the stage-coupling equations in (8); it is best seen in the *deterministic equivalent* form. Defining $\mathcal{Q}_{T+1}(y_T) := 0$ and then recursively

$$\mathcal{Q}_t(y_{t-1}) := \mathbb{E}_{L_{t-1}} \Big[\min_{y_t \in Y_t(y_{t-1})} \phi_t(y_t) + \mathcal{Q}_{t+1}(y_t) \Big], \quad t = T(-1)1,$$

or, in terms of the realizations,

$$\mathcal{Q}_t(y_i) := \sum_{j \in S(i)} \frac{p_j}{p_i} \Big[\min_{y_j \in Y_j(y_i)} \phi_j(y_j) + \mathcal{Q}_{t+1}(y_j) \Big], \quad t = T(-1)1, \quad i \in L_{t-1},$$

the deterministic equivalent problem reads

$$\min_{y_0 \in Y_0} \phi_0(y_0) + \mathcal{Q}_1(y_0).$$

This has the form of a deterministic optimization problem (hence the name), but $\mathcal{Q}_1$ is nonlinear and in general non-smooth, so it is not necessarily an appropriate formulation for numerical computations. In the case of interest one can actually unwrap the nesting of minimizations to obtain a single objective; the deterministic equivalent then takes the form of a large but structured convex quadratic program in the decision variables $y = (y_j)_{j \in V}$,

$$\min_y \sum_{j \in V} p_j \left[\frac{1}{2} y_j^* H_j y_j + f_j^* y_j\right], \tag{10}$$

$$\text{s.t. } W_0 y_0 = h_0, \tag{11}$$

$$W_j y_j = h_j - T_j y_i \quad \forall j \in V^*, \tag{12}$$

$$y_j \geq 0 \quad \forall j \in V. \tag{13}$$

This is called the *extensive form*. In stochastic notation with $y = (y_0, \ldots, y_T)$ the same problem reads

$$\min_y \sum_{t=0}^{T} \mathbb{E}\left[\frac{1}{2} y_t^* H_t y_t + f_t^* y_t\right] \tag{14}$$

$$\text{s.t. } W_0 y_0 = h_0, \tag{15}$$

$$W_t y_t = h_t - T_t y_{t-1} \quad \forall t \in \{1, \ldots, T\}, \tag{16}$$

$$y_t \geq 0 \quad \forall t \in \{0, \ldots, T\}. \tag{17}$$

Problem (14–17) and its deterministic equivalent (10–13) represent a standard problem class in stochastic programming. Especially the *linear* case with *fixed recourse* (i.e., objective $\sum_t \mathbb{E}[f_t^* y_t]$ and *deterministic* W_t) is very well-understood and widely used in practice. An important property of the deterministic equivalent is that, except for the recourse *sub*-structure, it has the form of a standard mathematical program (LP, QP, CP, or NLP). Thus, even though the scenario tree may cause exponential growth of the problem size, standard solution approaches are applicable when combined with suitable techniques that exploit the sparsity induced by the stochastic nature. The most prominent such techniques are *decomposition* approaches which split the large stochastic program into smaller problems associated with clusters of nodes (or scenarios). For a discussion of these techniques we refer the reader to the excellent survey articles [4, 21]; our own approach combines interior point methods with specially developed sparse-matrix techniques.

3.3 Dynamic Structure

The stage-coupling equations (16) or (12) define (implicitly) an underlying dynamic process, usually combined with further equality constraints. More precisely, the rows of conditions (16) can be categorized into *dynamic equations* and certain types of constraints which possess natural interpretations and satisfy associated regularity conditions. In [25, 26] we have developed complete such categorizations for several formulations of stochastic programs, accompanied by solution algorithms that employ natural pivot orders resulting from the refined sparse structure.

In processes governed by differential (or difference) equations there is typically also a natural partitioning of the decision variables $y = (x, u)$ into (independent) *control* variables u and (dependent) *state* variables x, the former representing the actual degrees of freedom available to the decision maker. The dynamic equations are then often given in explicit form,

$$x_j = G_j x_i + E_j u_i + h_j, \tag{18}$$

which is equivalent to (12) if we define $W_j := (I \;\; 0)$ and $T_j := -(G_j \;\; E_j)$. In this notation (and with the convention $x_{\pi(0)}, u_{\pi(0)} \in \mathbb{R}^0$), the multistage stochastic program of interest takes the form

$$\min_{(x,u)} \sum_{j \in V} p_j \left[\frac{1}{2} \begin{pmatrix} x_j \\ u_j \end{pmatrix}^* \begin{pmatrix} H_j & J_j^* \\ J_j & K_j \end{pmatrix} \begin{pmatrix} x_j \\ u_j \end{pmatrix} + \begin{pmatrix} f_j \\ d_j \end{pmatrix}^* \begin{pmatrix} x_j \\ u_j \end{pmatrix} \right], \tag{19}$$

$$\text{s.t.} \;\; x_j = G_j x_i + E_j u_i + h_j \quad \forall j \in V, \tag{20}$$

$$x \in [x^{\min}, x^{\max}], \tag{21}$$

$$u \in [u^{\min}, u^{\max}], \tag{22}$$

$$\sum_{j \in V} p_j (F_j x_j + D_j u_j + e_j) = 0. \tag{23}$$

Apart from the form of dynamic equations, the major difference to the standard formulation consists in the additional equality constraint (23). This condition represents a sum of expectations; we call it a *global* constraint since it may couple all nodes of the tree. In the standard formulation (10–13), such a condition cannot be modeled directly; it would require surplus variables and additional constraints.

The natural interpretation of the dynamics (18) is that the decision u_i at time $t-1$ controls all subsequent states $x_j, j \in S(i)$, at time t. This is the typical situation in discretized continuous-time processes: actually u_i determines a control action for the entire interval $(t-1, t)$ which becomes effective in x_j one period later. In other application contexts (particularly in the financial area), decisions become effective immediately, leading to dynamic equations

$$x_j = G_j x_i + E_j u_j + h_j. \tag{24}$$

Here each state x_j has "its own" control u_j rather than sharing u_i with the siblings $S(i)$.

The problem classes and solution algorithms associated with the possible formulations of dynamics are closely related; we refer to them collectively as *tree-sparse*. (For details see [24–26].) Applications are not only in discrete-time deterministic and stochastic optimal control but also in other dynamic optimization problems with an underlying tree topology; extensions to network topologies with "few" cycles are straightforward. A very general related problem class is investigated in [20] using a similar formulation of dynamics but σ-fields and probability spaces rather than scenario trees.

3.4 Convex Programs

Since we are concerned with convex quadratic stochastic programs, we recall here some basic definitions and facts of convex optimization. A convex optimization problem has the general form

$$\min_{y\in Y} f(y) \tag{25}$$

where Y is a convex set and $f\colon Y \to \mathbb{R}$ is a convex function, that is,

$$(1-t)y_0 + ty_1 \in Y \quad \text{and} \quad f((1-t)y_0 + ty_1) \leq (1-t)f(y_0) + tf(y_1)$$

for all $y_0, y_1 \in Y$ and $t \in [0,1]$. A *convex program* (CP) is the special case

$$\min_{y} f(y) \quad \text{s.t.} \quad g(y) = 0, \quad h(y) \geq 0, \tag{26}$$

where $g\colon \mathbb{R}^n \to \mathbb{R}^m$ is an affine mapping, $g(y) \equiv Ay + a$, and $h\colon \mathbb{R}^n \to \mathbb{R}^k$ is a (component-wise) concave mapping. If f, g, h are twice continuously differentiable, this means that the Hessians of f and $-h_i$ are positive semidefinite, $D^2 f(y) \geq 0$ and $D^2 h_i(y) \leq 0$. The convex quadratic case (with $H \geq 0$) reads

$$\min_{y} \frac{1}{2} y^* H y + f^* y \quad \text{s.t.} \quad Ay + a = 0, \quad By + b \geq 0. \tag{27}$$

The feasible set Y is a *polyhedron* if and only if it is given by finitely many linear equalities and inequalities, as in (27). It is easily seen that all level sets $N_c := \{y \in Y : f(y) \leq c\}$ of (25) are convex. Moreover, every local solution is automatically a global solution, and the set $\mathcal{S}$ of all solutions is convex. In the general case $\mathcal{S}$ may be empty even if feasible solutions exist. This happens either if f is unbounded below, $\inf_{y\in Y} = -\infty$, or if f is bounded below but the level sets N_c are unbounded for $c \downarrow \inf_{y\in Y} > -\infty$. Both situations are impossible in the convex quadratic case (27): existence of a solution $\hat{y} \in Y$ is then always guaranteed (unless the problem is infeasible). Uniqueness of a solution $\hat{y}$ holds under standard conditions. For the convex QP (27), a sufficient condition is positive definiteness of H on the null space $N(A)$ or, more generally, on its intersection with the null spaces of the rows of B associated with *strictly active* inequalities at $\hat{y}$. All this applies in particular to the stochastic problems (10–13) and (19–23). For an exhaustive treatment of the theory and numerical aspects of convex and (nonconvex) nonlinear programming we refer the reader to standard textbooks, such as [7, 8, 17].

4 Stochastic Optimization under Probabilistic Constraints

An important instance of optimization problems with uncertain data occurs if the constraints depend on a stochastic parameter, such as the inequality system

$$h(x, \xi) \geq 0, \tag{28}$$

where $h: \mathbb{R}^n \times \mathbb{R}^s \to \mathbb{R}^m$, ξ is an s-dimensional random variable defined on some probability space $(\Omega, \mathcal{A}, P)$ and the inequality sign has to be understood componentwise. Written as such, the constraint set is not a well-defined part of an optimization problem since, usually, the decision on the variables x has to be taken before ξ can be observed. It is clear that, in order to arrive at an implementable form of the constraints, one has to remove in an appropriate way the dependence of h on specific outcomes of ξ. The most prominent approaches to do so are

- the expected value approach
- the compensation approach
- the worst case approach
- the approach by probabilistic constraints

Using expected values, the system (28) is replaced by $\mathbb{E}\, h(x, \xi) \geq 0$, which now can be understood as an inequality system depending on x only, as the expectation operator acts as an integrator over ξ. An even simpler form is obtained when the random variable itself is replaced by its expectation: $h(x, \mathbb{E}\xi) \geq 0$ (both forms coincide in case that h depends linearly on ξ). The last form corresponds to the naive idea of substituting random parameters by average values. It seems obvious (and will be demonstrated later) that such reduction to first moment information ignores substantial information about ξ. Indeed, the expectation approach guarantees the inequality system to be satisfied on the average only, but a decision x leading to a failure of a system for about half of the realizations of ξ is usually considered as unacceptable. On the other extreme, the worst case approach enforces a decision to be feasible under all possible outcomes of ξ: $h(x, \xi) \geq 0\ \forall \xi$. This puts emphasis on absolute safety which is frequently either not realizable in the strict sense or is bought by extreme increase of costs. Although diametrically opposed in their modeling effects, both the expected value and worst case approach share some ignorance of the stochastic nature of ξ.

The basic idea of compensation relies on the possibility to adjust constraint violations in the system (28) after observation of ξ by later compensating actions. Accordingly, the set of variables splits into first stage decisions x (to be fixed before realization of ξ) and second stage decisions y (to be fixed after realization of ξ). As an example, one may think of power scheduling where an optimal load pattern of power generating units has to be designed prior to observing the unknown demand, and, where possible later gaps between supply and demand can be corrected by additional resources (e.g., hydro-thermal units, contracts etc.). The adjustment of constraint violation is modeled by an inequality system $H(x, \xi, y) \geq 0$, connecting all three types of variables and it causes additional costs $g(y, \xi)$ for the second stage decisions. Of course, given x and ξ, y should be chosen as to minimize second stage costs among all feasible decisions. Summarizing, compensation models replace the original problem

$$\min\{ f(x) \mid h(x,\xi) \geq 0\}$$

of minimizing the costs of first stage decision under stochastic inequalities by a problem where the sum of first stage costs and expected optimal second stage costs is minimized:

$$\min\{f(x) + Q(x)\}, \; Q(x) = \mathbb{E}q(x,\xi), \; q(x,\xi) = \min\{ g(y,\xi) \mid H(x,\xi,y) \geq 0\}.$$

The compensation approach, however, requires that compensating actions exist at all and can be reasonably modeled. In many situations this is not the case. For instance, operating an abundance of inflow in a continuous distillation process may cause adjusting actions which are inconvenient to carry out or the costs of which are hard to specify. In such circumstances, emphasis is shifted towards the reliability of a system by requiring a decision to be feasible at high probability. More precisely, (28) is replaced by the probabilistic constraint

$$P(h(x,\xi) \geq 0) \geq p.$$

Here, P is the probability measure of the given probability space and $p \in (0,1]$ is some probability level. Of course, the higher p the more reliable is the modeled system. On the other hand, the set of feasible x is more and more shrunk with $p \uparrow 1$ which makes increase the optimal value of the objective function at the same time. The extreme case $p = 1$ is similar to the worst case approach mentioned before. Fortunately, in a typical application, considerable increase in reliability can be obtained—for instance when contrasted to the expected value approach—at a small expense of the objective function and it is only for requirements close to certainty that the optimal value of the objective function worsens critically. This makes the use of probabilistic constraints a good compromise between the afore-mentioned methods. For a detailed introduction into various models of stochastic optimization the reader is referred to the monographs [5], [14] and [18].

4.1 Types of Probabilistic Constraints

Both for theoretical and practical reasons it is a good idea to identify different types of probabilistic constraints. First let us recall, that (28) is a system of inequalities given in components by $h_1(x,\xi) \geq 0, \ldots, h_m(x,\xi) \geq 0$. Now, when passing to probabilistic constraints as described before, one has the choice of integrating or separating these components with respect to the probability measure P:

$$P[h_1(x,\xi) \geq 0, \ldots, h_m(x,\xi) \geq 0] \geq p \qquad \text{or}$$
$$P[h_1(x,\xi) \geq 0] \geq p, \quad \ldots, \quad P[h_m(x,\xi) \geq 0] \geq p.$$

These alternatives are referred to as joint and individual probabilistic constraints, respectively. It is easily seen that feasibility in the first case entails feasibility in the second case while the reverse statement is false. In the context of control problems, the components of ξ may relate to a discretization of the time interval. Then, joint probabilistic constraints express the condition that at minimum probability p certain trajectories satisfy the given constraints over the whole interval whereas individual ones mean that this statement holds true for each fixed time of the discretized interval. From the formal point of view, passing from joint to individual constraints may appear as a complication as a single inequality (with respect to the decision variables x) is turned into a system of m inequalities. However, introducing one-dimensional random variables (depending on x) $\eta_i(x) := h_i(x, \xi)$, it can be seen that the joint constraints involve all components η_i simultaneously, whereas in each of the individual constraints just one specific component η_i figures as a scalar random variable. Taking into account that the numerical treatment of probability functions involving high-dimensional random vectors is much more delicate than in dimension one, where typically a reduction to quantiles of one-dimensional distributions can be carried out, the increase in the number of inequalities is more than compensated by a much simpler implementation. Of course, the choice between both formulations is basically governed by the modeling point of view.

Another important structure of probabilistic constraints occurs if in the constraint function h decision and random variables are separated in the sense that $h(x, \xi) = \tilde{h}(x) - \hat{h}(\xi)$. Using the distribution function $F_\eta(z) := P(\eta \leq z)$ for the transformed random variable $\eta = \hat{h}(\xi)$, the resulting (joint) probabilistic constraint may be equivalently written as

$$P(h(x, \xi) \geq 0) \geq p \iff P(\tilde{h}(x) \geq \hat{h}(\xi)) \geq p \iff F_\eta(\tilde{h}(x)) \geq p.$$

In this way, the originally implicit constraint function on x has been transformed into a composed function $F_\eta \circ \tilde{h}$. Taking into account that $\tilde{h}$ is analytically given from the very beginning and that there exist satisfactory approaches of evaluating distribution functions (in particular multivariate normal distribution), one has arrived at an explicit, implementable constraint. Thus it makes sense to speak of explicit probabilistic constraints here.

In the general implicit case, the evaluation of probabilities $P(h(x, \xi) \geq 0)$ as well as of their gradients with respect to x may become very difficult and efficient only in lower dimension. Nevertheless, there is some good chance for special cases like $h_i(\cdot, \xi)$ concave. Another option for solution is passing from joint to individual constraints.

4.2 *Storage Level Constraints*

An important instance of probabilistic constraints arises with the control of stochastic storage levels. Here it is assumed that some reservoir storing water or energy or anything similar is subject to lower and upper capacity levels $l^{\min}$ and $l^{\max}$. The reservoir is continuously fed and emptied. The feed ξ is assumed to be stochastic whereas extraction x is carried out in a controlled way. We consider this process over a fixed time horizon $[t_a, t_b]$ and discretize ξ and x according to subintervals of time as $(\xi_1, \ldots, \xi_s)$ and $(x_1, \ldots, x_s)$, where ξ_i and x_i denote the amount of substance directed to or extracted from the reservoir, respectively, during the i-th time subinterval. Accordingly, the current capacity level after the i-th interval amounts to $l^0 + \xi_1 + \cdots + \xi_i - x_1 - \cdots - x_i$, where l^0 refers to the initial capacity at t_a. Thus, the stochastic storage level constraints may be written as

$$l^{\min} - l^0 \le \xi_1 + \cdots + \xi_i - x_1 - \cdots - x_i \le l^{\max} - l^0 \quad (i = 1, \ldots, s)$$

or more compactly as the system $l^1 \le L\xi - Lx \le l^2$, where L is a lower left triangular matrix filled with '1'. Obviously, decision and random variables are separated here and, according to the preceding section, the resulting probabilistic constraints become explicit and can basically be reduced to level sets of s-dimensional distribution functions in case of joint constraints. The problem becomes particularly simple if the constraints are considered individually both with respect to the upper and lower level and to time index i. For instance, the i-th upper level constraint writes as

$$\begin{aligned} P(\eta_i \le l_i^2 + x_1 + \cdots + x_i) \ge p &\iff F_{\eta_i}(l_i^2 + x_1 + \cdots + x_i) \ge p \\ &\iff x_1 + \cdots + x_i \ge (F_{\eta_i})^{-1}(p) - l_i^2, \end{aligned}$$

where $\eta_i = \xi_1 + \cdots + \xi_i$, F_{η_i} refers to the 1-dimensional distribution function of η_i and $(F_{\eta_i})^{-1}(p)$ denotes the (usually tabulated) p-quantile of this distribution. Consequently, the probabilistic constraints can be transformed to a system of simple linear inequalities in the decision variable x then. Storage level constraints will be considered later in the context of controlling a continuous distillation process where the role of the reservoir is played by the so-called feed tank which acts as a buffer between stochastic inflows and the operating distillation unit.

4.3 Numerical Treatment

The solution of optimization problems involving probabilistic constraints requires at least the ability of evaluating the function $\varphi(x) = P(h(x, \xi) \ge 0)$. Thinking of discretized control problems which are typically large dimensional, efficient methods like SQP have to be employed. Then, of course, the gradient of φ has to be provided as well if not even second partial derivatives.

Assuming ξ to have a density f_ξ, the function φ is formally defined as the parameter-dependent multivariate integral

$$\varphi(x) = \int\limits_{h(x,z) \ge 0} f_\xi(z)dz, \tag{29}$$

where integration takes place over an s-dimensional domain. Thinking of discretized control problems again, the dimension s of the random variable may correspond to the discretization of a time interval, hence values of $s = 20$ are more than moderate. In such dimension, however, an 'exact' evaluation of the above integral by numerical integration is far from realistic. Rather, two principal 'inexact' strategies have proven powerful in the past, namely bounding and simulation. Some rough ideas can be illustrated for the example of distribution functions, i.e., the special case where the domain of integration becomes a cell $h(x) + \mathbb{R}^s_-$. As mentioned in the previous sections, the evaluation of distribution functions is crucial for the important special case of explicit probabilistic constraints.

The generic representatives of the bounding and simulation procedures are the Bonferroni bounds and the crude Monte-Carlo estimator. The Bonferroni bounds refer to the determination of the probability $P(\bigcup_{k=1}^s A_k)$ of the union of s abstract probability events A_k, and they are based on the inequalities

$$\sum_{k=1}^{2m}(-1)^{k-1}S_k \leq P(\bigcup_{k=1}^{s} A_k) \leq \sum_{k=1}^{2m+1}(-1)^{k-1}S_k,$$

where $m = 1, \ldots, \lfloor s/2 \rfloor$ on the left hand side and $m = 0, \ldots, \lfloor (s-1)/2 \rfloor$ on the right hand side, and

$$S_k = \sum_{1 \leq i_1 < \cdots < i_k \leq s} P(A_{i_1} \cap \cdots \cap A_{i_k})$$

denotes the summarized probability of all possible intersections of order k. In case of $s = 2$, for instance, the very properties of a measure yield

$$P(A_1 \cup A_2) = P(A_1) + P(A_2) - P(A_1 \cap A_2) \leq P(A_1) + P(A_2) = S_1,$$

so we have recovered the first Bonferroni upper bound in a trivial case. For the evaluation of a distribution function one has

$$F(z) = P(\xi_1 \leq z_1, \ldots, \xi_s \leq z_s) = P(A_1 \cap \cdots \cap A_s) = 1 - P(\bigcup_{k=1}^{s} A_k),$$

hence the Bonferroni bounds can be applied to the last expression. Specifying these bounds for m up to 2, one gets

$$1 - S_1 \leq F(z) \leq 1 - S_1 + S_2.$$

Increasing m, these bounds become sharper and sharper until the maximum possible value of m exactly realizes the desired probability. On the other hand, the determination of S_k becomes increasingly complex. For instance, in the context of F being a multivariate normal distribution, the determination of probabilities $P(A_{i_1} \cap \cdots \cap A_{i_k})$ leads to k-dimensional integration of that distribution. This can be efficiently done for $k = 1, 2$ but gets quickly harder with higher dimension. At the same time, the number of such probability terms to be summed up in the determination of S_k equals $\binom{s}{k}$ and thus makes the numerical effort soon explode. That is why in the determination of distribution functions, one has to be content with the very few first terms S_k. Often, the gap between the resulting Bonferroni bounds is too large for practical purposes then.

Fortunately, sharper bounds can be derived on the basis of appropriate linear programs (see [18]). For $k \leq 4$ there even exist explicit expressions for these improved bounds, for instance $1 - S_1 + \frac{2}{s}S_2 \leq F(z)$ provides a much better lower bound on the basis of S_k with $k \leq 2$ than the Bonferroni counterpart $1 - S_1 \leq F(z)$ (where S_2 does not figure at all in the first lower bound). Still the gap may remain unsatisfactory. Another strategy of deriving bounds relies on graph-theoretical arguments. The prominent Hunter bound (see [12]), for instance, is based on finding a maximum weight spanning tree in a graph the vertices of which are represented by the single events A_k and the edges of which correspond to pairwise intersections of events $A_k \cap A_l$. The weight of an edge is given by the probability $P(A_k \cap A_l)$ which is easily calculated for all edges. The Hunter bound can be shown to be at least as good as, but frequently much better than the (improved) lower Bonferroni bound $1 - S_1 + \frac{2}{s}S_2$ mentioned above, although calculated with basically the same effort. The idea behind the Hunter bound has been continuously generalized towards more complex graph structures (hypertrees defined by hyperedges) in the last few years resulting in amazingly efficient lower and upper bounds. Excellent results for the multivariate normal distribution are reported in [6] with dimension up to $s = 40$.

The simplest scheme of Monte-Carlo simulation for evaluating (29) consists in generating a sample of N realizations $z_1, \ldots, z_N$ of ξ and to take then the ratio k/N as an estimate for the desired probability, where $k = \#\{i \mid h(x, z_i) \geq 0\}$. For larger dimension s, the variance of this estimate becomes quite large which makes it unsatisfactory soon. Similar to the starting point of Bonferroni bounds, more efficient simulation schemes have been developed as well. At this point, we may refer to Szántai's simulation scheme (see [27], related approaches are described in [18]) which is based on the knowledge of the first two terms S_1, S_2 of probabilities of single events and pairwise intersections. Using the same sample as already generated for the crude Monte-Carlo estimator, these terms allow immediately to calculate two additional Monte-Carlo estimators, the reason behind being a simple cancellation rule of binomial expressions. Now, the main idea is to convexly combine these three Monte-Carlo estimators (including the crude one) and to exploit correlations between them in order to minimize the variance of the combined estimator. In this way, simulation results become considerably more precise. Finally, an extension to incorporating Hunter's and the other mentioned graph-theoretical bounds into this scheme has been successfully carried out.

The procedures described so far are related to the evaluation of functional values of φ in (29) with special emphasis on distribution functions. As for gradients or higher order derivatives, these can be reduced analytically to the determination of functional values again at least in case of a multivariate normal distribution (for details see [18]). Hence, the same basic strategies apply although with repeated effort now (n components for the gradient and $n(n+1)/2$ components for the Hessian if wanted).

4.4 Probability Maximization

As already mentioned above, increasing the probability level p in a probabilistic constraint shrinks the feasible set. Typically, the feasible set becomes empty starting from a critical value $\bar{p}$ which may be less than 1. In particular, a user of some implemented solution method dealing with probabilistic constraints might unintentionally have chosen a value of p above that critical value. Then, for instance, SQP codes working with infeasible iterates and enforcing feasibility in convergence only, will consume a lot of computing time in vain due to operating on an empty constraint set. This effect is particularly undesirable in an environment of on-line optimization. Therefore, one has good reason prior to the optimization problem itself to determine $\bar{p}$ by probability maximization over the constraints:

$$\max\{p \mid P(h(x,\xi) \geq 0) \geq p\}.$$

As long as probabilistic constraints are considered alone in this auxiliary problem, it can be solved rather quickly as compared to the original optimization problem. However, one has to take into account that the obtained maximum value of p is just an upper bound for $\bar{p}$ since the other constraints of the optimization problem (usually related to the dynamics of the underlying control problem) are not involved here. At least, this bound gives an indication for a probability level which cannot be exceeded at all. In order to calculate the exact bound, one would have to include all constraints which, of course, is almost as time consuming as the original problem.

4.5 Structural Properties

For an efficient treatment of probabilistic constraints, it is crucial to have some insight into their analytical, geometrical and topological structure. While corresponding statements are well-known and immediate for usual (analytical) constraints of the type $g(x) \leq 0$ (e.g., when g is linear, convex or differentiable), there are no obvious relations between the quality of data and the structure of probabilistic constraints. Most results in this direction are concerned with convexity issues which have direct consequences for numerics and theoretical analysis. A corresponding important statement in simplified form is the following one (cf. [18]):

Theorem 1. *In (28), let the components* h_i *of* h *be convex and assume that* ξ *has a density the logarithm of which is concave. Then, the function* $\varphi(x) = P(h(x,\xi) \geq 0)$ *is concave and, hence, the corresponding probabilistic constraint may be convexly described, i.e.,* $P(h(x,\xi) \geq 0) \geq p \iff -\varphi(x) \leq p$.

Many but not all of the prominent multivariate distribution share the property of having a log-concave density as required in the last theorem (e.g., multivariate normal distribution or uniform distribution on bounded convex sets, cf. [18]).

An alternative structural characterization relates to the weaker property of connectedness (cf. [11]):

Theorem 2. *The constraint set* $\{x \mid P(\tilde{h}(x) \geq \hat{h}(\xi)) \geq p\}$ *of an explicit probabilistic constraint is connected whenever the components* $\tilde{h}_i$ *are concave and the constraint qualification*

$$\mathrm{Im}(\tilde{h}) \cap (t \cdot (1, \ldots, 1) + \mathbb{R}^m_+) \neq \emptyset \quad \forall t \in \mathbb{R}.$$

In the affine linear case $\tilde{h}(x) = Ax + b$, *this constraint qualification reduces to the positive linear independence of the rows of* A.

Note that this last result does not require any assumptions on the distribution of the random variable. Applying the previous theorems to the specific situation of joint storage level constraints to be considered later on in the context of a distillation process, one may infer that the feasible set is convex for many and connected for all distributions of the random variable ξ.

REFERENCES

1. H. Bauer. *Maß- und Integrationstheorie. (Measure and integration theory).* de Gruyter Lehrbuch. De Gruyter, Berlin, 2nd, revised edition, 1992. (In German.).
2. H. Bauer. *Probability Theory.* de Gruyter Studies in Mathematics 23. de Gruyter, Berlin, 1996. (Translated from the German by Robert B. Burckel.).
3. L. T. Biegler, I. E. Grossmann, and A. W. Westerberg. *Systematic Methods of Chemical Process Design.* Prentice-Hall, Englewood Cliffs, NJ, 1997.
4. J. R. Birge. Stochastic programming computations and applications. *INFORMS J. Comput.*, 9(2):111–133, 1997.
5. J. R. Birge and F. Louveaux. *Introduction to Stochastic Programming.* Springer-Verlag, New York, 1997.
6. J. Bukszár and A. Prékopa. Probability bounds with cherry trees. *Math. Oper. Res.*
7. A. V. Fiacco and G. P. McCormick. *Nonlinear Programming: Sequential Unconstrained Minimization Techniques.* Wiley, New York, 1968. Reprinted by SIAM Publications, 1990.
8. P. E. Gill, W. Murray, and M. H. Wright. *Practical Optimization.* Addison Wesley, 1981.
9. J. Gmehling, U. Onken, and W. Arlt. *VLE Data Collection.* DECHEMA, 1997.
10. I. E. Grossmann and A. W. Westerberg. Research challenges in process systems engineering. *AIChE J.*, 46:1700–1703, 2000.
11. R. Henrion. A note on the connectedness of chance constraints. Preprint 21, Stochastic Programming Eprint Series (SPEPS), 2000. Submitted to J. Optim. Theory Appl.
12. D. Hunter. Bounds for the probability of a union. *J. Appl. Probab.*, 13:597–603, 1976.
13. M. G. Ierapetritou, J. Acevedo, and E. N. Pistikopoulos. An optimization approach for process engineering problems under uncertainty. *Comput. Chem. Eng.*, 20:703–709, 1996.
14. P. Kall and S. W. Wallace. *Stochastic Programming.* Wiley, New York, 1994.
15. P. Li, H. A. Garcia, G. Wozny, and E. Reuter. Optimization of a semibatch distillation process with model validation on the industrial site. *Ind. Eng. Chem. Res.*, 37:1341–1350, 1998.
16. M. Loeve. *Probability Theory.* Van Nostrand-Reinhold, Princeton, NJ, 1963.
17. J. Nocedal and S. J. Wright. *Numerical Optimization.* Springer-Verlag, New York, 1999.
18. A. Prékopa. *Stochastic Programming.* Kluwer Academic Publishers, Dordrecht, The Netherlands, 1995.

19. R. C. Reid, J. M. Prausnitz, and B. E. Poling. *The Properties of Gases and Liquids*. McGraw-Hill, New York, 1987.
20. R. T. Rockafellar and R. J.-B. Wets. Generalized linear-quadratic problems of deterministic and stochastic optimal control in discrete time. *SIAM J. Control Optim.*, 28(4):810–822, 1990.
21. A. Ruszczyński. Decomposition methods in stochastic programming. *Math. Programming*, 79(1-3):333–353, 1997.
22. W. D. Seider, J. D. Seader, and D. R. Lewin. *Process Design Principles: Synthesis, Analysis and Evaluation*. Wiley, New York, 1998.
23. M. C. Steinbach. *Fast Recursive SQP Methods for Large-Scale Optimal Control Problems*. Ph. D. dissertation, University of Heidelberg, Germany, 1995.
24. M. C. Steinbach. Recursive direct algorithms for multistage stochastic programs in financial engineering. In P. Kall and H.-J. Lüthi, editors, *Operations Research Proceedings 1998*, 241–250, New York, 1999. Springer-Verlag.
25. M. C. Steinbach. Hierarchical sparsity in multistage convex stochastic programs. In S. P. Uryasev and P. M. Pardalos, editors, *Stochastic Optimization: Algorithms and Applications*, 363–388, Kluwer Academic Publishers, 2001. Dordrecht, The Netherlands.
26. M. C. Steinbach. Tree-sparse convex programs. Technical Report ZR-01-08, ZIB, 2001. Submitted for publication.
27. T. Szántai and A. Habib. On the k-out-of-r-from-n : F system with unequal element probabilities. In F. Gianessi et al., editor, *New Trends in Mathematical Programming*, 289–303. Kluwer Academic Publishers, Dordrecht, The Netherlands, 1998.
28. W. B. Whiting. Effects of uncertainties in thermodynamic data and models on process calculations. *J. Chem. Eng. Data*, 41:935–941, 1996.

A Multistage Stochastic Programming Approach in Real-Time Process Control

Izaskun Garrido and Marc C. Steinbach

Konrad-Zuse-Zentrum für Informationstechnik Berlin (ZIB), Germany

Abstract Standard model predictive control for real-time operation of industrial production processes may be inefficient in the presence of substantial uncertainties. To avoid overly conservative disturbance corrections while ensuring safe operation, random influences should be taken into account explicitly. We propose a multistage stochastic programming approach within the model predictive control framework and apply it to a distillation process with a feed tank buffering external sources. A preliminary comparison to a probabilistic constraints approach is given and first computational results for the distillation process are presented.

1 Introduction

The work reported here is part of a joint research effort aiming at real-time control of chemical processes under uncertainty. Two different stochastic optimization approaches are studied, with the intention to explore and compare their respective general properties and their usefulness in certain practical situations. A specific distillation process serves as a prototypical application example which is investigated under various aspects.

Distillation processes are used to separate liquid or vapor mixtures of several substances into products with different compositions of a desired purity, by the application and removal of heat. Distillation is the most widely used separation process in chemical industry; it consumes large amounts of energy.

The specific process under investigation is the separation of a binary mixture of methanol and water in a continuously running system of two energetically coupled distillation columns. The process is fed from a buffer tank that collects several external sources. In practice, uncertainty occurs when the inflow into the tank may vary at random due to disturbances in the upstream processes. A robust extraction strategy is then required to prevent the tank from running dry or spilling over while keeping the process in favorable operating conditions. Specifically, we assume that the total energy consumption is to be minimized over a given planning horizon; cf. [11, §1].

A pilot system of the process just described is installed at the Institute of Process Dynamics of the Technical University of Berlin. Optimization results have been obtained in the partner project for a simplified one-stage column model called a *flash unit* [13]. In that work, the composition of the inflow mixture and its temperature are assumed to be deterministic whereas the inflow *rate* may vary at random. The rate is modeled as an autocorrelated Gaussian process representing the superposition of many independent inflows, as indicated in Figure 1. A rectangular inflow profile modeling a single event with known rate and duration but random starting time is

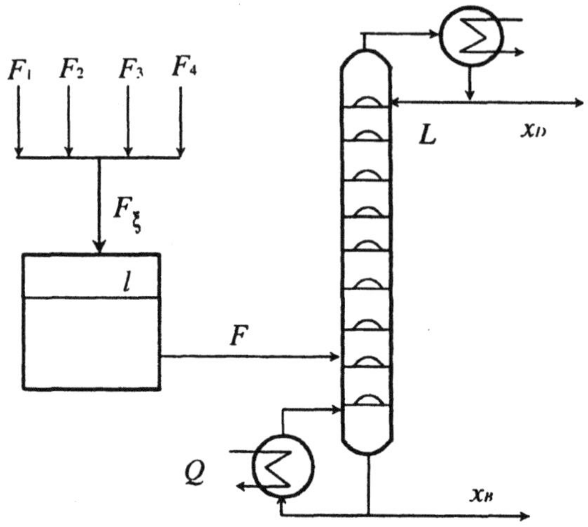

Figure 1. Distillation column with feed tank. F_ξ: random inflow rate; F: feed rate; L: reflux flow rate; Q: reboiler duty; x_D: distillate; x_B: bottom

also considered in [13]. Each of these stochastic models represents disturbances that occur in usual process operation, as opposed to exceptional events like failures.

The stochastic optimization approach presented here is based on a scenario tree model; cf. [11, §§2, 3]. Possible deviations from the expected inflow profile over the entire planning horizon are thus represented as a discrete-time stochastic process with finitely many realizations. An optimal solution in this framework minimizes the mathematical expectation of the cost (energy consumption) over all scenarios. The optimal control strategy is itself a *nonanticipative* stochastic process: it determines a different extraction profile for each scenario, thus specifiying *a priori* how to *react* to future measurements of actual inflows. This approach requires that (within a linearized model) the constraints can be satisfied for any possible sequence of random disturbances (the tank filling level can be kept feasible for any sequence of inflows). More generally, the approach is applicable if *hard* constraints can always be satisfied and costs for potential violations of *soft* constraints can be quantified. In the latter case, soft constraints will be satisfied only if this is possible and economic.

The probabilistic constraints approach pursued in the partner project also models the inflow history as a discrete-time process but allows continuous probability distributions given by a density function; cf. [11, §§2, 4]. The optimal control strategy in this case is *deterministic*: it does not react to actual disturbances. Instead, it minimizes the cost under the restriction that constraints will be satisfied with high probability (for instance, in at least 95% of all cases). This approach is appropriate if it is not possible to satisfy all constraints with certainty (in certain extreme cases one cannot prevent the tank from running dry or spilling over), or if such events cause substantial costs for which no precise model is available.

Both stochastic optimization approaches are naturally applicable within a moving horizon framework. In the present context, *real-time* process control means response times in the order of 10 to 15 minutes. This is appropriate for the distillation process with a planning horizon of about a day and reoptimizations every 2 or 3 hours.

As indicated in [11, §1], uncertainty may influence the planning and operation of chemical production processes in various ways. For instance, [22, 23] study the *design* of chemical plants under uncertainty, with the aim of guaranteeing the existence of feasible control strategies after observing the random event(s). Process *operation* under uncertainty is investigated, e.g., in [15, 24] and, for the case of random feed streams, in [7] and in the partner project [12, 13]. A stochastic integer programming approach to online scheduling of batch processes is given in [8]. The rigorous treatment of exceptional events by scenario-based DAE models is described in [1]. A coarse classification of relevant types of uncertainty and a general discussion of the topic can be found in [16]. The specific area of distillation processes is particularly well-studied under various aspects. See, e.g., [19] for a recent general survey, [5] for optimal control in the presence of random feed, or [2] for a large-scale industrial application. For the background in stochastic optimization required in this paper we refer to [11, §§2, 3] and standard textbooks [3, 14].

The current investigation treats the same general situation as [13], using a tracking approach presented in §2. The discretization of the Gaussian inflow process in a scenario tree framework is described in §3, and a new, straightforward technique of evaluating integrals of the multivariate normal density is proposed in §4. Finally we present first computational results in §5 and give some conclusions in §6.

2 OPTIMIZATION MODEL

A schematic diagram of a distillation column with a buffer tank is shown in Figure 1. Uncertainty occurs in the tank inflow F_ξ; control variables are the reboiler duty Q, the feed extraction F directed from the tank to the column, and the reflux flow L. For more detailed descriptions ot the system we refer to [11, §1] and [9, 12].

Our first investigations, as reported here, are aimed at answering the following question: given a desired extraction profile $\hat{F}$, under which inflow conditions is it possible to satisfy the level constraints in the tank, and how difficult is it? Difficulty is measured as the expectation of the accumulated quadratic deviation between actual extraction and target profile. That is, we solve on-line a stochastic *tracking problem* where the target profile typically results from an *off-line* process optimization. Uncertainty is thus effectively decoupled from the process dynamics. Of course, such a simplified approach will be practically useful only in situations where the distillation column is always capable of processing the extracted amount of liquid without violating the purity constraints, and the total energy consumption $\int Q\,dt$ is not too sensitive to the deviations.

2.1 Continuous Time

Since only the basic model structure is of interest here, we formulate a deterministic tracking problem for simplicity (without uncertainty in the inflow rate F_ξ). Given a target feed rate $\hat{F}$ and denoting by v the liquid volume in the tank, the model reads

$$\min_{F} \int_0^T \frac{1}{2}[F(t) - \hat{F}(t)]^2\, dt \tag{1}$$

$$\text{s.t.}\quad \dot{v}(t) = F_\xi(t) - F(t), \tag{2}$$

$$v(0) = \hat{v}_0, \tag{3}$$

$$v(T) = \hat{v}_T, \tag{4}$$

$$v(t) \in [v^{\min}, v^{\max}], \tag{5}$$

$$F(t) \in [F^{\min}, F^{\max}]. \tag{6}$$

Here $\hat{v}_0$ is the known initial volume, and $\hat{v}_T$ is a prespecified final level.

Note that some terminal condition on the liquid volume is always required in the original problem since minimizing the total energy consumption of the reboiler would otherwise result in processing as little liquid as possible, and hence yield a full tank at the end of the planning horizon. Here we simply fix $v(T)$; the nature of the condition (called a *cycling constraint*) will be discussed in more detail in the following section.

2.2 Discrete Time

Considering T periods (not necessarily of equal physical length) in discrete time $t = 0, 1, \ldots, T$, we denote by v_t the liquid volume in the tank at time t, by f_t the feed volume extracted during $(t, t+1)$, by $\hat{f}_t$ the associated target extraction volume, and by ξ_t the random inflow volume during $(t-1, t)$. In our approach, the latter is assumed to vary only within some known finite interval $[\xi_t^{\min}, \xi_t^{\max}]$. We also assume that the volume f_t is extracted at a constant rate, which is consistent with standard practice in process operation. The tank filling volume now evolves according to

$$v_t = v_{t-1} - f_{t-1} + \xi_t, \qquad t = 1, \ldots, T.$$

Note that the time index on f_t and $\hat{f}_t$ refers to the following period whereas on ξ_t it refers to the previous period. This reflects the fact that decisions on extraction volumes must be made at the beginning of each period whereas inflow volumes are only measured at the end. An obvious consequence is that the control f_t will always lag behind one period in compensating for undesired inflows. It can never steer the state v_t to a precise value but only into some range determined by the inflow variation: $v_t \in (v_{t-1} - f_{t-1}) + [\xi_t^{\min}, \xi_t^{\max}]$.

In particular, the cycling constraint $v_T = \hat{v}_T$ cannot be satisfied with certainty but only in an average sense. It is thus replaced by

$$\mathbb{E}(v_T) = \hat{v}_T \tag{7}$$

in [13]. This is the best one can do in a probabilistic constraints framework, but the condition is actually quite weak in our approach: final values of ν_T may vary over the entire feasible range rather than being clustered around $\hat{\nu}_T$ (as intended). In fact we can do better: it is possible to satisfy a similar condition *independently* for every realization of the preceding state ν_{T-1}, which amounts to prescribing the *conditional expectation*

$$\mathbb{E}(\nu_T | \nu_{T-1}) = \hat{\nu}_T. \tag{8}$$

(This condition can also be interpreted as a limiting case of (7) when the feasible range $[\nu_T^{min}, \nu_T^{max}]$ is continuously reduced in an appropiate manner.) We will present optimization results for both alternatives.

The discrete-time stochastic tracking problem minimizes the *expected* tracking error. In variables $\nu = (\nu_0, \ldots, \nu_T)$ and $f = (f_0, \ldots, f_{T-1})$ it reads

$$\min_{(\nu,f)} \sum_{t=0}^{T-1} \frac{1}{2}\mathbb{E}[(f_t - \hat{f}_t)^2] \tag{9}$$

$$\text{s.t.}\quad \nu_t = \nu_{t-1} - f_{t-1} + \xi_t, \quad t = 0, \ldots, T, \tag{10}$$

$$\mathbb{E}(\nu_T) = \hat{\nu}_T \quad \text{or} \quad \mathbb{E}(\nu_T | \nu_{T-1}) = \hat{\nu}_T, \tag{11}$$

$$\nu_t \in [\nu^{min}, \nu^{max}], \quad t = 1, \ldots, T, \tag{12}$$

$$f_t \in [f^{min}, f^{max}], \quad t = 0, \ldots, T-1. \tag{13}$$

Uncertainty occurs only in the right-hand side of the dynamic equations (10). The special case $t = 0$ is formally handled by setting $\nu_{-1} := f_{-1} := 0$ and $\xi_0 := \hat{\nu}_0$, which is equivalent to using the physical quantities ν_{-1}, f_{-1}, and ξ_0 from the actual, continously running process.

Given a scenario tree with vertex set V, we denote by $L_t \subseteq V$ the level set of nodes at time t and by $L \equiv L_T$ the set of leaves; further by $0 \in L_0$ the root, by $j \in L_t$ the "current" node, by $i \equiv \pi(j) \in L_{t-1}$ its unique predecessor (if $t > 0$), and by $S(j) \subseteq L_{t+1}$ its set of successors. The node probabilities are $p_j > 0$, $j \in V$. For further details see [11, §2], or [17, §2.4] where alternative scenario models (with explicit nonanticipativity constraints) are discussed.

In the numerical formulation, the state variable is $x_t := \nu_t$ and the control is defined as the tracking error, $u_t := f_t - \hat{f}_t$ (with limits $u_t^{min} := f^{min} - \hat{f}_t$ and $u_t^{max} := f^{max} - \hat{f}_t$). Correspondingly, we define $h_t := \xi_t - \hat{u}_{t-1}$. Objective terms then simplify to $\mathbb{E}(u_t^2)$, and the remaining equations and constraints retain their original form with proper variable replacements—except that the control bounds are now time-dependent. On the scenario tree, stochastic quantities x_t, u_t, h_t are represented by their realizations x_j, u_j, h_j, $j \in L_t$. Letting $V^* := V \setminus \{0\}$ and recalling $i \equiv \pi(j)$, the first optimization problem (with cycling constraint $\mathbb{E}(\nu_T) = \hat{\nu}_T$)

then reads

$$\min_{(x,u)} \sum_{j\in V\setminus L} \frac{1}{2} p_j u_j^2 \tag{14}$$

$$\text{s.t. } x_j = x_i - u_i + h_j \quad \forall j \in V, \tag{15}$$

$$x_j \in [x^{\min}, x^{\max}] \quad \forall j \in V^*, \tag{16}$$

$$u_j \in [u_t^{\min}, u_t^{\max}] \quad \forall j \in V \setminus L, \tag{17}$$

$$\sum_{j\in L} p_j x_j = \hat{x}_T. \tag{18}$$

The second problem can be written in the same form where the cycling constraint $\mathbb{E}(\nu_T|\nu_{T-1}) = \hat{\nu}_T$ replacing (18) translates to

$$\sum_{k\in S(j)} \frac{p_k}{p_j} \nu_k = \hat{\nu}_T \quad \forall j \in L_{T-1}.$$

However, this (set of) condition(s) is not explicitly specified in the problem formulation. Instead, we use it to pre-eliminate the final period entirely as follows. Substituting the dynamic equation (10) into (8) yields

$$\hat{\nu}_T = \mathbb{E}(\nu_{T-1} - f_{T-1} + \xi_T|\nu_{T-1}) = \nu_{T-1} - f_{T-1} + \mathbb{E}(\xi_T|\nu_{T-1}).$$

Hence, the final-period feed extraction is uniquely determined as

$$f_{T-1} = \nu_{T-1} - \hat{\nu}_T + \hat{\xi}_{T-1} \tag{19}$$

where $\hat{\xi}_{T-1} := \mathbb{E}(\xi_T|\nu_{T-1})$ is the conditional expectation of the final-period inflow. The uniqueness of f_{T-1} shows that (8) is actually the strongest possible cycling constraint in our framework.

It remains to clarify the roles of conditions (12) at $t = T$ and (13) at $t = T - 1$. The bounds on ν_T translate to a restriction of the problem *data*,

$$\xi_T - \hat{\xi}_{T-1} \in [\nu^{\min} - \hat{\nu}_T, \nu^{\max} - \hat{\nu}_T].$$

This yields an *a priori* feasibility check: no feasible solution can exist if the final inflow ξ_T (conditioned on ν_{T-1}) varies too much. On the other hand, since

$$\xi_T - \hat{\xi}_{T-1} \in [\xi_T^{\min} - \xi_T^{\max}, \xi_T^{\max} - \xi_T^{\min}],$$

the restriction is always satisfied if the latter range is contained in the former.

The bounds on f_{T-1} simply imply a further restriction of ν_{T-1},

$$\nu_{T-1} \in (\hat{\nu}_T - \hat{\xi}_{T-1}) + [f^{\min}, f^{\max}].$$

In the numerical formulation, the final state variable is now defined as the final-period tracking error, $x_{T-1} := (\nu_{T-1} - \hat{\nu}_T + \hat{\xi}_{T-1}) - \hat{f}_{T-1}$ by (19). Accordingly, we have $h_{t-1} := \xi_{t-1} - \hat{\nu}_T + \hat{\xi}_{T-1} - \hat{f}_{T-1}$ and limits

$$x_{T-1}^{\min} := \max(f^{\min}, \nu^{\min} - \hat{\nu}_T + \hat{\xi}_{T-1}) - \hat{f}_{T-1},$$
$$x_{T-1}^{\max} := \min(f^{\max}, \nu^{\max} - \hat{\nu}_T + \hat{\xi}_{T-1}) - \hat{f}_{T-1},$$

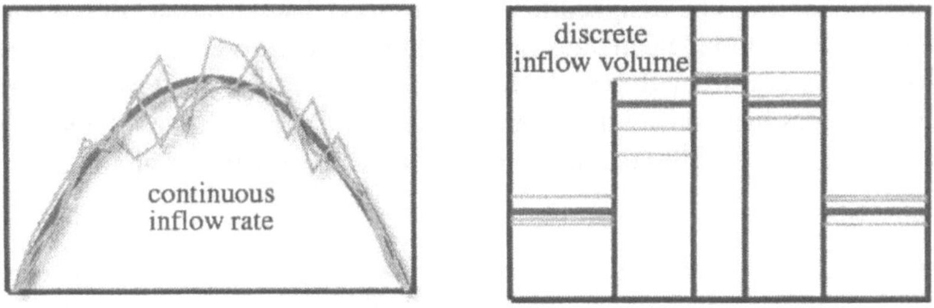

Figure 2. Expected inflow profiles (bold) and realizations. Left: continuous time. Right: discrete time

yielding a possibly empty interval (which is also checked a priori). At times $t = 0, \dots, T-2$ we use the same variables as before. That is, $x_t := v_t$, $u_t := f_t - \hat{f}_t$, (with limits $u_t^{\min}, u_t^{\max}$), and $h_t := \xi_t - \hat{u}_{t-1}$. The state bounds for $0 < t < T-1$ are $x_t^{\min} := v^{\min}$ and $x_t^{\max} := v^{\max}$.

Due to the eliminations, the leaves of the original scenario tree are now obsolete (typically a drastical reduction in size!), and the optimization problem is defined on the subtree with vertex set $V_c := V \setminus L$ and leaf set $L_c := L_{T-1}$. It reads

$$\min_{(x,u)} \sum_{j \in V_c \setminus L_c} \frac{1}{2} p_j u_j^2 + \sum_{j \in L_c} \frac{1}{2} p_j x_j^2 \tag{20}$$

$$\text{s.t. } x_j = x_i - u_i + h_j \quad \forall j \in V_c, \tag{21}$$

$$x_j \in [x_t^{\min}, x_t^{\max}] \quad \forall j \in V_c^*, \tag{22}$$

$$u_j \in [u_t^{\min}, u_t^{\max}] \quad \forall j \in V_c \setminus L_c. \tag{23}$$

This is still very similar to problem (14–18), but instead of a terminal condition we now have objective terms in the final period, and the state bounds in $T-1$ are now defined by the (possibly empty) intersection of two intervals.

3 Discretizing the Gaussian Process

The autocorrelated Gaussian process model for the inflow rate F_ξ leads immediately to an autocorrelated discrete-time Gaussian process of inflow volumes ξ_t, see Figure 2. The latter is given by a general multivariate normal distribution $\mathcal{N}(\bar{\xi}, \Sigma)$ whose dimension is the number of time periods, T. The density function φ of the normal distribution is positive on the entire space, that is, its support is $\mathbb{R}^T$. Thus, although with small probability, it allows arbitrarily large inflows and even negative ones (which are physically impossible unless the tank leaks). On the other hand, a scenario tree representation necessarily corresponds to a *discrete* probability distribution, with *compact* support. Since we intend to compare the two stochastic optimization approaches, this raises several nontrivial questions: How should the support be chosen, how should the scenario tree be constructed, and how should the node probabilities be assigned to obtain a "good discretization" of the continuous normal distribution? Such questions would be less relevant in practice: scenario trees would be constructed directly from measurements, and a Gaussian process would be seen as just one possible approximation of the real data, with the property of being particularly well tractable in the probabilistic constraints approach.

Compact Support

Due to the absence of real data we make the following assumptions that seem reasonable for a basic investigation: the actual inflow rate may only vary within a (possibly variable) symmetric bandwith around the expected rate, and "negative inflows" are impossible,

$$F_\xi(t) \in [F_\xi^{\min}(t), F_\xi^{\max}(t)] \equiv [\bar{F}_\xi(t) - \Delta F_\xi(t), \bar{F}_\xi(t) + \Delta F_\xi(t)] \subset \mathbb{R}_+. \tag{24}$$

Integration over the subintervals then yields a similar relation for the inflow volumes,

$$\xi_t \in [\xi_t^{\min}, \xi_t^{\max}] \equiv [\bar{\xi}_t - \Delta\xi_t, \bar{\xi}_t + \Delta\xi_t] \subset \mathbb{R}_+. \tag{25}$$

This means that the discrete distribution will be supported by a T-dimensional compact box centered at the mean and lying entirely in the positive orthant,

$$\Xi := [\bar{\xi} - \Delta\xi, \bar{\xi} + \Delta\xi] \equiv \prod_{t=1}^{T} [\bar{\xi}_t - \Delta\xi_t, \bar{\xi}_t + \Delta\xi_t] \subset \mathbb{R}_+^T. \tag{26}$$

We now define a new density function (whose support is precisely this box) by restricting the given normal density to Ξ and renormalizing the weight,

$$\varphi_\Xi(\xi) := \frac{1}{P(\Xi)} \chi_\Xi(\xi)\varphi(\xi), \qquad P(\Xi) = \int_\Xi \varphi(\xi)\, d\xi. \tag{27}$$

Probabilities are thus replaced by *conditional* probabilities with respect to Ξ. Obviously the construction leaves expected inflows invariant by symmetry. The correlations of inflows, however, will deviate from their original values, giving increasingly inaccurate approximations with decreasing weight $P(\Xi)$. Moreover, assumption (24) on the inflow rate guarantees that all constraints are satisfied in continuous time if this is true in the discrete-time model.

Scenario Tree

Currently we construct the scenario tree from a uniform recursive partitioning of Ξ. Each scenario corresponds to an elementary box of full dimension. The stage-t scenarios correspond to unions of elementary boxes having the same geometry in the first t dimensions (i.e., identical projections into the associated subspace). This means that the same nodes are traversed up to level t in the scenario tree or, equivalently, that inflow volumes are identical during the first t periods, see Figure 3. The number k_t of partitions in dimension t is the number $|S(j)|$ of successors of each node j on level $t-1$. The resulting total number of boxes (scenarios) is $N = \prod_{t=1}^{T} k_t$.

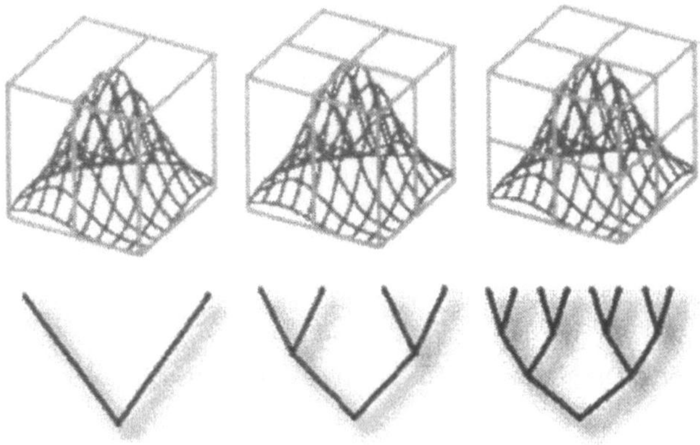

Figure 3. Construction of a scenario tree by recursive partitioning of the compact support Ξ. Node probabilities correspond to the weights of (unions of) sub-boxes

Scenario Probabilites

As scenario probability we define the weight of the associated elementary box (with respect to the renormalized density φ_Ξ). This weight is assumed to be concentrated in the geometric center of the box, whose coordinates represent the sequence of associated inflow volumes. Although the center of gravity would yield a more accurate approximation, we prefer the geometric center since this allows to choose an exact range of inflow variations *a priori*: realizations of the discrete distribution will be evenly spaced at a distance of $2\Delta\xi_t/k_t$ between limits

$$\xi^{\min} + \frac{\Delta\xi_t}{k_t}, \quad \xi^{\max} - \frac{\Delta\xi_t}{k_t}. \tag{28}$$

(With the center of gravity, minimal and maximal realizations would depend on the density.)

For a standard univariate normal distribution, Figure 4 shows the probability density function, the renormalized density for the interval $\Xi = [-2, 2]$ (having weight 0.9545), an approximation by a piecewise constant density for $k_1 = 5$ scenarios, and the weights of the five subintervals concentrated in their respective midpoints.

4 Calculating Scenario Probabilities

Consider without loss of generality a centralized normal distribution $\mathcal{N}(0, \Sigma)$ (with mean $\bar{\xi} = 0$) in $\mathbb{R}^T$. The density function reads

$$\varphi(\xi) = \frac{1}{\sqrt{(2\pi)^T \det(\Sigma)}} \exp\left(-\frac{1}{2}\xi^* \Sigma^{-1} \xi\right). \tag{29}$$

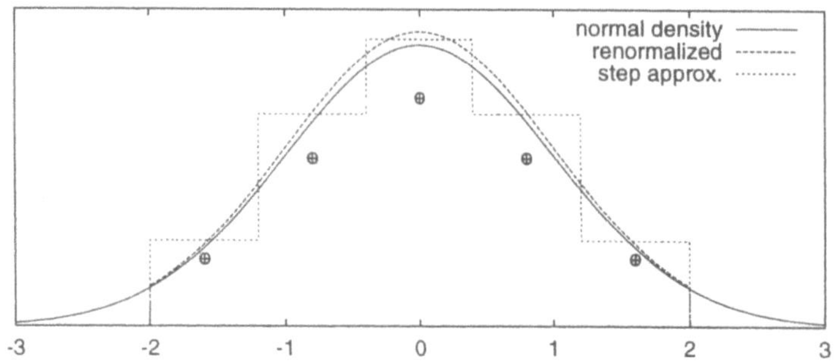

Figure 4. Discretization of a univariate normal distribution on $[-2, 2]$

In our optimization model with equidistant time discretization we make the same assumptions as our colleagues [12, 13]: random inflow volumes ξ_t have the same variance σ^2 in all periods, and their correlations c_{st} decrease linearly with the distance $|s-t|$ such that the elements of the covariance matrix are

$$\Sigma_{st} = \sigma^2 c_{st}, \quad c_{st} = 1 - \frac{1}{T}|s-t|, \qquad s, t = 1, \dots, T. \tag{30}$$

We have to calculate the scenario probabilities which are defined as multivariate integrals of the density function over rectangular domains. Multi-dimensional numerical integration is generally hard since the required effort in direct generalizations of univariate integration methods grows exponentially with the dimension. Thus, Monte Carlo techniques are often applied.

In the special case of normal distributions, some alternative approaches are reported in the literature. Schervish [18] employs an adaptive quadrature routine using an error estimate based on the Newton-Cotes approximation with non-local modifications. Deák [4] combines a transformation to spherical coordinates with Monte-Carlo techniques. The method of Genz [10] transforms the integration domain to the unit cube and applies either Monte-Carlo, adaptive subregions, or lattice rules to the transformed integral.

Here we propose a straightforward, easily implementable approach based on direct integration of the second order Taylor series expansion in combination with adaptive refinement. This appears to be a new method; it seems appropriate in our situation since we have to evaluate integrals over comparatively small domains: the elementary boxes representing scenarios.

Taylor Approximation

In scalar product notation $\langle \xi, \eta \rangle := \xi^* \Sigma^{-1} \eta$, the first three derivatives of the density function (29) are

$$\begin{aligned} D\varphi(\xi)[h] &= -\varphi(\xi)\langle \xi, h \rangle, \\ D^2\varphi(\xi)[h,h] &= +\varphi(\xi)\left[\langle \xi, h \rangle^2 - \langle h, h \rangle\right], \\ D^3\varphi(\xi)[h,h,h] &= -\varphi(\xi)\left[\langle \xi, h \rangle^3 - 3\langle \xi, h \rangle \langle h, h \rangle\right]. \end{aligned}$$

Hence, with some $\theta \in [0, 1]$, the third-order Taylor series expansion reads

$$\varphi(\xi+h) = \varphi(\xi) - \varphi(\xi)\langle\xi, h\rangle + \frac{1}{2}\varphi(\xi)\left[\langle\xi, h\rangle^2 - \langle h, h\rangle\right] \\ - \frac{1}{6}\varphi(\xi)\left[\langle\xi, h\rangle^3 - 3\langle\xi, h\rangle\langle h, h\rangle\right] + \frac{1}{24}D^4\varphi(\xi+\theta h)[h, h, h, h].$$

We need to integrate φ over rectangular boxes $\xi + A = \{\xi + h\colon h \in A\}$ where

$$A = [-a, a] := \prod_{t=1}^{T}[-a_t, a_t], \qquad \mathrm{Vol}(A) = \prod_{t=1}^{T} 2a_t. \tag{31}$$

For the expansion above, this integration is easily evaluated in closed form. In terms of the Hessian

$$H(\xi) := \varphi(\xi)(\Sigma^{-1}\xi\xi^*\Sigma^{-1} - \Sigma^{-1})$$

one obtains

$$\int_{\xi+A}\varphi(h)\,dh = \varphi(\xi)\,\mathrm{Vol}(A)\left[1 + \frac{1}{24}\sum_{t=1}^{T} H_{tt}(\xi)a_t^2 + O(\|a\|^4)\right]. \tag{32}$$

Observe that, due to symmetry, all odd-order terms vanish in the integration. Thus we get an asymptotic error of order four by adding just one correction (of second order) to the trivial approximation $P(\xi + A) \approx \varphi(\xi)\,\mathrm{Vol}(A)$.

Asymptotic Error Control

To ensure sufficient accuracy, a simple adaptive strategy is employed. After evaluating (32), A is partitioned into a "left" half A_l and a "right" half A_r, and the same weight approximation is applied to A_l and A_r, yielding values w, w_l, w_r. The bisection procedure is recursively repeated with each box until the relative difference falls below a given tolerance,

$$\frac{|w_l + w_r - w|}{w_l + w_r} < \epsilon.$$

Two heuristic strategies have been tested for determining the side s in which to cut the box:

- the largest side, $a_s = \max_t a_t$;
- the side giving the largest second order term, $H_{ss}(\xi)a_s^2 = \max_t H_{tt}(\xi)a_t^2$.

The second strategy was found to perform better and was used in numerical calculations, with the tolerance set to $\epsilon = 10^{-3}$.

Matrix Determinant and Inverse

It turns out that both the determinant and the inverse of the specific correlation matrix (30) have closed-form representations,

$$\det(\Sigma) = \frac{T+1}{4}\left(\frac{2\sigma^2}{T}\right)^T$$

and (for $T \geq 3$)

$$\Sigma^{-1} = \frac{T}{2\sigma^2} \begin{bmatrix} 1 & -1 & & & \\ -1 & 2 & -1 & & \\ & \ddots & \ddots & \ddots & \\ & & -1 & 2 & -1 \\ & & & -1 & 1 \end{bmatrix} + \frac{T}{2\sigma^2(T+1)} \begin{bmatrix} 1 & 0 & \dots & 0 & 1 \\ 0 & 0 & & & 0 \\ \vdots & & \ddots & & \vdots \\ 0 & & & 0 & 0 \\ 1 & 0 & \dots & 0 & 1 \end{bmatrix}.$$

Using these formulae in (32) yields significant savings in the numerical computation, especially for higher dimensions. The computational effort is still immense (several hours for a complete discretization), but it should be kept in mind that the weight calculation is an *off-line* task which is required just once for a given inflow distribution.

5 Computational Results

In this section we report on computational experiments with the stochastic tracking problems stated in §2.2, where either the expectation or the conditional expectation of the final liquid volume are prescribed, $\mathbb{E}(\nu_T) = \hat{\nu}_T$ or $\mathbb{E}(\nu_T | \nu_{T-1}) = \hat{\nu}_T$. All problems are solved by a primal-dual interior point method combined with a tree-sparse KKT solver [20, 21].

5.1 Prescribed Expectation of ν_T

Problem Data

To allow a comparison of optimization approaches, our computations are based on the following problem data for which optimization runs have also been performed in the partner project. They are slight modifications of the computations described in [13].

The planning horizon has a length of 16 hours and is equidistantly partitioned into eight discretization intervals of two hours each, $\Delta t = 2\,\mathrm{h}$. Only the flow rate into the buffer tank is assumed to vary at random, whereas its temperature and the respective concentrations of methanol and water remain constant. A parabolic profile of the expected inflow rate is assumed, starting and ending with 11.0 ml/h and reaching a maximum of 127.6 ml/h after eight hours, at $t = 4$. The associated variance of the inflow volume in each two-hour period is $\sigma^2 = 20.0\,\mathrm{ml}^2$, yielding by (30) the covariance matrix

$$\Sigma_{st} = 20 c_{st}, \quad c_{st} = 1 - \frac{1}{8}|s - t|, \qquad s, t = 1, \dots, 8.$$

The liquid volume in the tank is to be kept between 440 ml and 1320 ml, with an initial filling level of 1210 ml. This value is also specified as the final level so that the distillation process can be repeated periodically if there are no disturbances. In the presence of disturbances, preventing violations of the upper limit $\nu^{\max}$ is a major concern since the initial level and desired final level are quite close to that limit.

Table 1. Target extraction volumes for each two-hour period

Target type	1	2	3	4	5	6	7	8
Expected inflow	75.44	162.9	221.2	250.3	250.3	221.2	162.9	75.44
Deterministic	139.8	186.3	186.3	186.3	186.3	186.3	186.3	162.0
Probabilistic	126.2	193.0	196.9	198.0	198.0	198.0	198.0	111.7

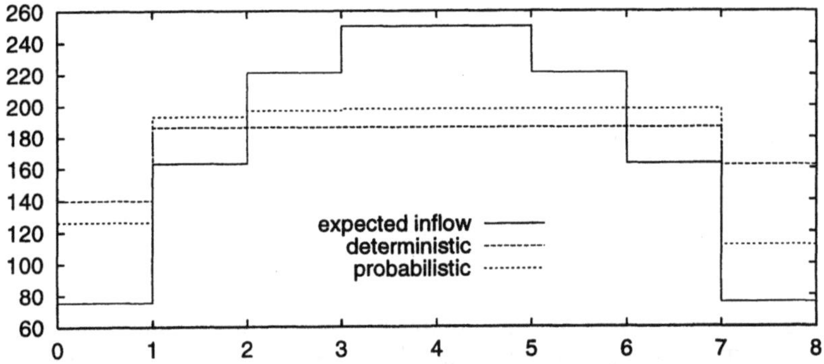

Figure 5. Target extraction profiles over eight two-hour periods

Problem Types

We consider three problem types corresponding to the following target extraction profiles:

1. the expected inflow;
2. the optimal extraction strategy of a deterministic optimization based on the expected inflow ("deterministic solution");
3. the optimal extraction strategy under probabilistic constraints as obtained in the partner project ("probabilistic solution").

Tracking the expected inflow is a very simplistic approach; we include this case only for comparison purposes. The deterministic and probabilistic cases are discussed in [13], where both are solved for a DAE model of a flash unit, using a finer discretization of 32 time periods. The extraction limit is $f^{\max} = 186.34$ (average inflow plus 5%) in the deterministic case, and $f^{\max} = 198$ in the probabilistic case. For the problem data given above (with 8 periods), the expected inflow and optimal profiles are displayed in Table 1 and Figure 5. Here the probabilistic solution satisfies the lower (less critical) level constraint with certainty, and the upper level constraint with a probability of 0.95.

Inflow and Extraction Bounds

For a given inflow limit $\xi^{\max}$, we choose as support Ξ (on which the normal distribution is discretized) the largest cube centered at the mean $\bar{\xi}$ and lying entirely in $[0, \xi^{\max}]^{\mathsf{T}}$,

$$\Xi := \bar{\xi} + [-\Delta, \Delta]^{\mathsf{T}}, \qquad \Delta := \min_{t} \min(\bar{\xi}_t, \xi^{\max} - \bar{\xi}_t). \tag{33}$$

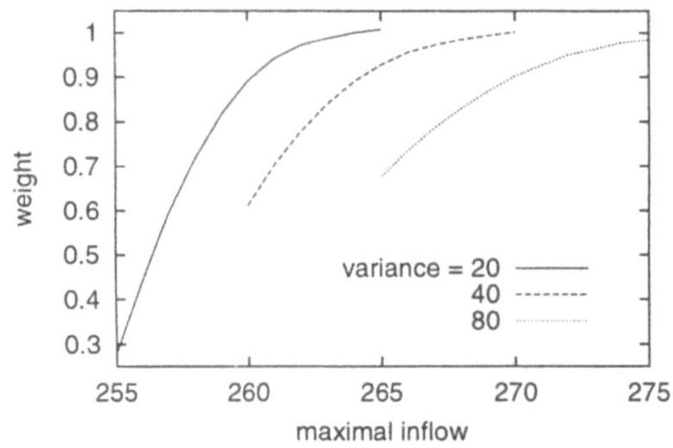

Figure 6. Weight of support versus inflow limit

Recalling that $\xi^{\max} = F_\xi^{\max}\Delta t$, this obviously models a fixed bandwidth of the inflow rate F_ξ. To study the feasibility question stated in §2, we vary the inflow limit $\xi^{\max}$ in a suitable range of values slightly larger than the largest expected inflow $\max_t \bar{\xi}_t = 250.3$ ml, namely $\xi^{\max} \in \{255, 256, \ldots, 265\}$. The discrete bandwidth is then determined by the largest expected inflow as $\Delta = \xi^{\max} - \max_t \bar{\xi}_t$, covering the range $[4.7, 14.7]$ and yielding cubes Ξ of different sizes with weights roughly between 0.3 and 1; see the solid line in Figure 6. (For our standard deviation $\sigma \approx 4.5$, this allows to compare "good" and "poor" discretizations of the normal distribution.)

Obviously, feasibility is harder to achieve when $\xi^{\max}$ is increased (giving larger inflow variations) or when $f^{\max}$ is decreased (giving a smaller control range). For each $\xi^{\max}$ value we therefore set $f^{\max} = \xi^{\max}$ first and then decrease $f^{\max}$ until the problem becomes infeasible. For all combinations of $\xi^{\max}$ and $f^{\max}$ we solve problem (14–18). Data for variances $\sigma^2 = 40$ and $\sigma^2 = 80$ appearing in the plots are associated with problems in § 5.2 and will be discussed below.

Expected Inflow

First we consider optimal solutions for the expected inflow as target profile. Figure 7 plots the optimal objective value, i.e., the (expected) "tracking error", versus extraction limit $f^{\max}$ for selected inflow limits $\xi^{\max}$ covering the entire range $[255, 265]$. One observes that the tracking error increases with decreasing extraction limit; closer inspection of the data reveals that its value is actually zero in all problems with $f^{\max} \geq 251$. The first observation confirms precisely the expected behavior: since the target profile has a peak inflow of 250.3 in the middle, large extractions are required in earlier and later periods when $f^{\max}$ is reduced below that peak value. The second observation says that (accumulated) inflow deviations can never violate a limit if precisely the expected inflow is extracted. This is easily verified; it indicates that inflow variations are moderate even in the case $\xi^{\max} = 265$ which covers about 99.9% of the distribution.

It is also observed that the tracking error is almost independent of the inflow limit $\xi^{\max}$ for large values of $f^{\max}$ whereas there are significant differences for small

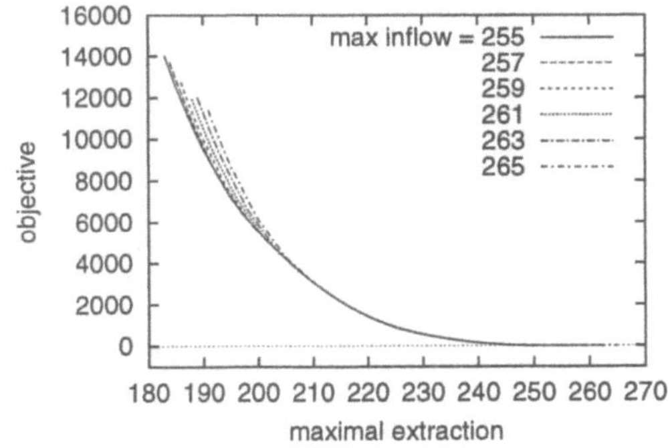

Figure 7. Tracking error versus extraction limit for expected inflow

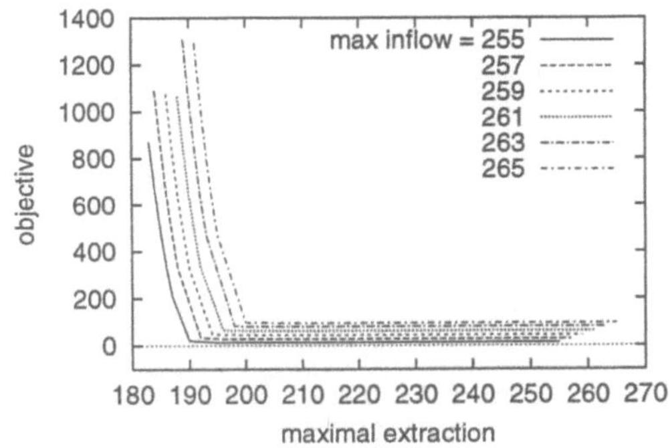

Figure 8. Tracking error versus extraction limit for deterministic solution

values. This is also easily explained: A large extraction limit is not a severe restriction in any case, whereas a small limit will be binding in most periods until the problem becomes eventually infeasible.

Deterministic Solution

Next we consider the deterministic solution as target profile. Figure 8 plots the tracking error versus extraction limit f^{max} for the same ξ^{max} values as before. Here it is observed that in all cases the tracking error remains constant (on a low level) over a wide range of f^{max} values but increases rapidly when infeasibility is approached. Smaller inflow limits yield smaller tracking errors for all extraction limits. This might come as a surprise but can be explained when the target profile is inspected: the deterministically optimal extraction is constantly at its upper limit 186.34, except for the first and last interval. Again, only small f^{max} values are a severe restriction, and a violation of the upper level constraint by accumulated inflow deviations can be avoided with comparatively small corrections. This is consistent with the results in [13]: although many trajectories violate the upper level constraint for the given profile, the limit is only slightly exceeded.

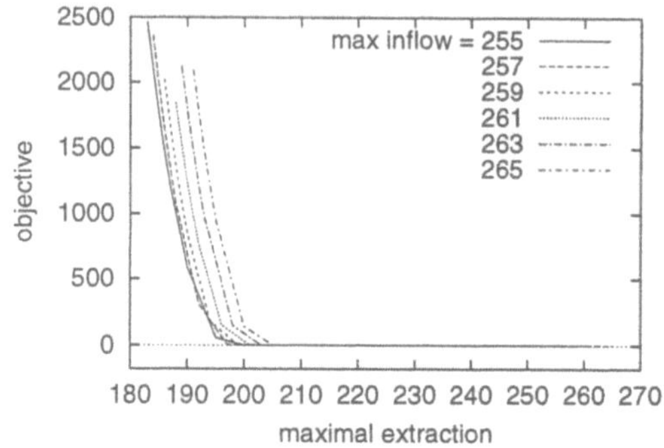

Figure 9. Tracking error versus extraction limit for probabilistic solution

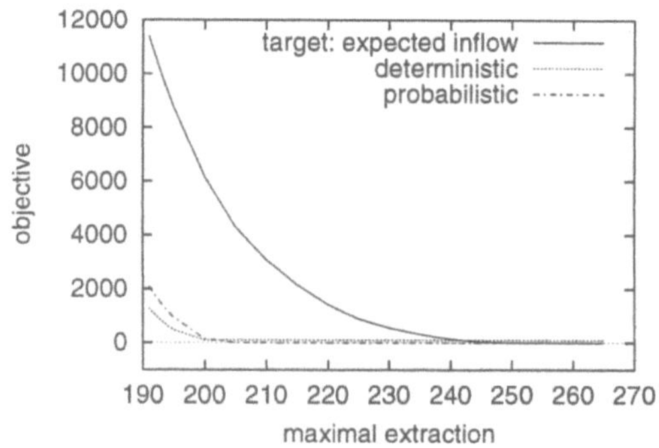

Figure 10. Tracking error versus extraction limit for all target profiles, with inflow limit $\xi^{\max} = 265$

Probabilistic Solution

The probabilistic solution as target profile yields almost identical results, as shown in Figure 9. However, the constant level for large $f^{\max}$ values is consistently smaller than in the deterministic case: at most 1.86, but often exactly zero. This confirms the robustness of the probabilistic solution (certain feasibility is achieved with negligible extra effort) and is again a consequence of the precise shape of the target profile. On the other hand, for small $f^{\max}$ values the tracking error is *larger* than in the deterministic case, which can be seen in the direct comparison of all three target profiles displayed in Figure 10 (where $\xi^{\max} = 265$). The latter fact indicates that reducing the *probability* of constraint violations may go along with an increase of their *size*.

Feasibility

Figure 11 shows how the smallest feasible extraction limit $f^{\max}$ depends on the inflow limit $\xi^{\max}$. (There is obviously no dependence on the target profile.) At $\xi^{\max} = 265$, a feasible solution can still be obtained with $f^{\max} = 191$, and at

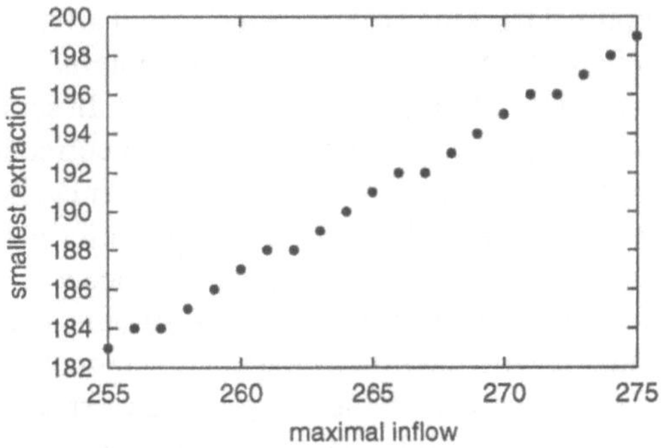

Figure 11. Smallest feasible extraction limit versus inflow limit

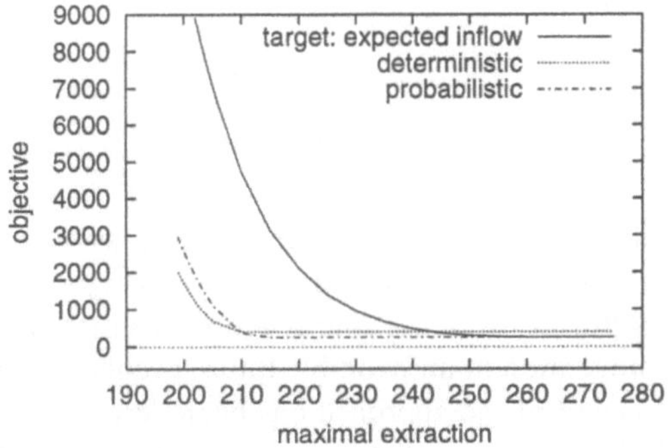

Figure 12. Tracking error versus extraction limit for all target profiles, with inflow limit $\xi^{\max} = 275$ and conditional expectation in cycling constraint

$\xi^{\max} = 274$ with $f^{\max} = 198$. This demonstrates the flexibility of the stochastic programming approach: with reasonable extraction limits, constraint violations can be avoided with certainty by suitable predetermined reactions to inflow measurements.

5.2 *Prescribed Conditional Expectation of* ν_T

Feasibility was comparatively easy to achieve in the previous problem due to the moderate inflow variance and weak cycling constraint. Therefore we also investigate problem (20–23), using the variance values $\sigma^2 \in \{20, 40, 80\}$ and respective $\xi^{\max}$ ranges 255–265, 260–270, and 265–275. Associated weights of the support Ξ and smallest feasible extraction limits are included in Figures 6 and 11. A comparison of all three target profiles for $\sigma^2 = 80$ and $\xi^{\max} = 275$ is given in Figure 10. As expected, increasing the variance (and accordingly the inflow limit, to cover sufficient weight of the distribution) requires larger extraction limits to achieve feasibility. Moreover, the optimal tracking error is never zero in all these cases. This shows that, even if no level constraints are violated by the nominal extraction, corrective action is required to meet the stronger cycling constraint. In some of the higher variance problems, part of the tracking error is also due to corrections preventing violation of the upper level constraint. However, the stronger cycling constraint has no significant impact on feasibility. This is because the prescribed final level is close to the critical upper level constraint: if no violations occur as long as inflows are large, the control will be able to extract a sufficient amount of liquid during the last period so that the final level is met on average. (If accumulated inflow deviations are negative, the control just extracts less liquid than expected.) In contrast, test runs show that certain feasibility *is* harder to achieve if the final level is close to the middle of the feasible range. To sum up, results for the stronger cycling constraint are similar to the previous case, and they scale in some sense for the larger variance values. Differences for the three target profiles are slightly accentuated; cf. Figures 10 and 12.

6 Conclusions

Multistage stochastic programming has been proposed as a new approach in real-time control of chemical processes. This can be seen as a generalization of standard model predictive control in the sense that reactions to measured disturbances are combined with the prevention of unfavorable future events by means of a stochastic model. The basic concept of the approach has been demonstrated for the problem of controlling the buffer tank of a distillation column with random inflows, and a preliminary comparison with a probabilistic constraints approach has been given. Future research should extend this work in different directions. For instance, a more realistic, integrated treatment of stochasticity and process dynamics in the application example is intended. Further, scenario reduction techniques as in [6] appear promising in obtaining better discrete distributions and at the same time allowing finer time discretizations. Finally, warm start techniques and other algorithmic improvements may increase the efficiency of the approach so that faster processes can be controlled in real time.

Acknowledgments

We would like to express our gratitude to G. Wozny, P. Li, and M. Wendt (Technische Universität Berlin) and to R. Henrion and A. Möller (Weierstraß-Institut Berlin) for numerous vital discussions and for providing the process model and optimization data used in the comparison. We also wish to thank L. T. Biegler and G. Rodriguez (Carnegie Mellon University) who suggested this research and introduced the second author to basic robust control issues during two enjoyable research visits. Finally, we gratefully acknowledge financial support by the Deutsche Forschungsgemeinschaft.

REFERENCES

1. O. Abel and W. Marquardt. Scenario-integrated optimization of dynamic systems. *AIChE J.*, 46(4):803–823, 2000.
2. O. E. Agamennoni, J. L. Figueroa, G. W. Barton, and J. A. Romagnoli. Advanced controller design for a distillation column. *Int. J. Control*, 59:817–839, 1994.
3. J. R. Birge and F. Louveaux. *Introduction to Stochastic Programming*. Springer-Verlag, New York, 1997.
4. I. Deák. Three digit accurate multiple normal probabilities. *Numer. Math.*, 35:369–380, 1980.
5. U. M. Diwekar and J. R. Kalagnanam. Efficient sampling technique for optimization under uncertainty. *AIChE J.*, 43:440–447, 1997.
6. J. Dupačová, N. Gröwe-Kuska, and W. Römisch. Scenario reduction in stochastic programming: An approach using probability metrics. Preprint 00-09, Institut für Mathematik, Humboldt-Universität Berlin, 2000.
7. A. M. Eliceche, M. Sanchez, and L. Fernandez. Feasible operating region of natural gas plants under feed perturbations. *Comput. Chem. Eng.*, 22:S879–S882, 1998.
8. S. Engell, A. Märkert, G. Sand, R. Schultz, and C. Schulz. Online scheduling of multiproduct batch plants under uncertainty. Preprint SM-DU-494, Universität Duisburg, 2001. Submitted to this volume.
9. H. A. Garcia, R. Henrion, P. Li, A. Möller, W. Römisch, M. Wendt, and G. Wozny. A model for the online optimization of integrated distillation columns under stochastic constraints. Preprint 98-32, DFG Research Center "Echtzeit-Optimierung großer Systeme", Nov. 1998.
10. A. Genz. Numerical computation of the multivariate normal probabilities. *J. Comput. Graph. Statist.*, 1:141–150, 1992.
11. R. Henrion, P. Li, A. Möller, M. C. Steinbach, M. Wendt, and G. Wozny. Stochastic optimization for operating chemical processes under uncertainty. Technical Report ZR-01-04, ZIB, 2001. Submitted to this volume.
12. R. Henrion, P. Li, A. Möller, M. Wendt, and G. Wozny. Optimization of a continuous distillation process under probabilistic constraints. Submitted to this volume.
13. R. Henrion and A. Möller. Optimization of a continuous distillation process under random inflow rate. Preprint 00-4, DFG Research Center "Echtzeit-Optimierung großer Systeme", Mar. 2000. Submitted to Comput. Math. Appl.
14. P. Kall and S. W. Wallace. *Stochastic Programming*. Wiley, New York, 1994.
15. S. Orçun, I. K. Altinel, and Ö. Hortaçsu. Scheduling of batch processes with operational uncertainty. *Comput. Chem. Eng.*, 20:S1191–S1196, 1996.
16. E. N. Pistikopoulos. Uncertainty in process design and operation. *Comput. Chem. Eng.*, 19:S553–S563, 1995.
17. W. Römisch and R. Schultz. Multistage stochastic integer programs: An introduction. Preprint SM-DU-496, Universität Duisburg, 2001. Submitted to this volume.
18. M. Schervish. Multivariate normal probabilities with error bound. *J. Appl. Statist.*, 33:81–87, 1984.
19. S. Skogestad. Dynamics and control of distillation columns—a critical survey. *Model. Identif. Control*, 18:177–217, 1997.
20. M. C. Steinbach. Recursive direct algorithms for multistage stochastic programs in financial engineering. In P. Kall and H.-J. Lüthi, editors, *Operations Research Proceedings 1998*, pages 241–250, New York, 1999. Springer-Verlag.

21. M. C. Steinbach. Hierarchical sparsity in multistage convex stochastic programs. In S. Uryasev and P. M. Pardalos, editors, *Stochastic Optimization: Algorithms and Applications*, Kluwer Academic Publishers, 2001. Dordrecht, The Netherlands.
22. D. A. Straub and I. E. Grossmann. Design optimization of stochastic flexibility. *Comput. Chem. Eng.*, 17:S339–S354, 1993.
23. R. Swaney and I. E. Grossmann. An index for operational flexibility in chemical process design. *AIChE J.*, 31:621–630, 1985.
24. P. Terwiesch, D. Ravemark, B. Schenker, and D. W. T. Rippin. Semi-batch process optimization under uncertainty: Theory and experiments. *Comput. Chem. Eng.*, 22:201–213, 1998.

Optimal Control of a Continuous Distillation Process under Probabilistic Constraints

René Henrion[1], Pu Li[2], Andris Möller[1], Moritz Wendt[2], and Günter Wozny[2]

[1] Weierstraß-Institut für Angewandte Analysis und Stochastik (WIAS), Berlin, Germany
[2] Institut für Prozess- und Anlagentechnik, Technische Universität Berlin, Germany

Abstract A continuous distillation process with random inflow rate is considered. The aim is to find a control (feed rate, heat supply, reflux rate) which is optimal with respect to energy consumption and which is robust at the same time with respect to the stochastic level constraints in the feed tank. The solution approach is based on the formulation of probabilistic constraints. An overall model including the dynamics of the distillation process and probabilistic constraints under different assumptions on the randomness of inflow is developed and numerical results are presented.

1 Introduction

As noted in [9], Section 1.2, continuous distillation processes are frequently characterized by uncertainties of their inflow. These may relate to the flow rate, to the composition of the mixture to be separated or to its temperature. Typically, the uncertainties are not completely irregular but follow a certain pattern caused by the operation of upstream units. Then it makes sense to model uncertainty as a stochastic parameter, the distribution of which can be estimated from history but the realization of which in the coming period of optimization is unknown. In the following, we are going to consider the rate of inflow as the only random parameter. As a consequence of possible unpredictable peaks, the inflow cannot be processed immediately but has to be stored in a feed tank before being directed at a controlled rate to the distillation unit (see [9], Figure 1). For technological reasons, one has to impose upper and lower level constraints for the feed tank preventing it from running full or empty. Both cases would require unpleasant compensating actions which are desirable to avoid (see [9], Section 1.3). Therefore, a problem will be formulated which reflects the objective to find a feed control being robust with respect to level constraints yet optimal in the sense of minimum energy consumption subject to product specifications.

From the stochastic nature of the inflow rate it is clear that the level constraints are stochastic too, and there is a choice to apply any of the methods briefly sketched in [9]. As costs of compensating actions for possible level violations are difficult to model on the one hand and a worst case approach is much too conservative or even impossible on the other hand, it is proposed to rely on probabilistic constraints. The gain over simply using typical profiles (or expected values) for the inflow rate will be illustrated later on. The aim of the subsequent analysis is to present a model of optimal control for continuous distillation with probabilistic feed tank constraints and to present numerical results for different assumptions on the randomness of inflow rate. As an example serves the separation of a methanol/water mixture.

2 THE DISTILLATION PROCESS

2.1 *Process description*

The process considered relates to a distillation column, as shown in Figure 1, for separating a binary mixture of water and methanol. The column has a diameter of 100 mm and 20 bubble-cap trays with a central downcomer. Isolation coat is mounted to reduce the heat loss from the column wall. The boilup is provided by an electrical thermal device, while the condensation is carried out by a total condenser with cooling water. The plant is equipped with temperature, pressure, level and flow rate measurements and electrical valves for the flow control. All input/output signals are treated by a process control system. Several control loops have been configured and implemented on the plant. The control system is connected to the local area network to manage experimental data. The composition of the feed stream, distillate and bottom product will be measured off-line by gas chromatography.

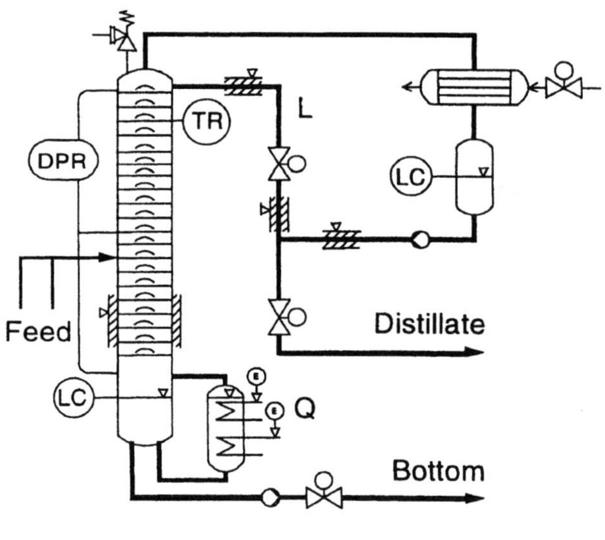

Figure 1. A pilot plant for distillation

From a feed tank, the feed stream with a given composition is fed to the column. The column is operated with atmospheric pressure. Operation of the plant means to run the column for keeping the desired distillate and bottom product purity under an uncertain feed flow profile. This leads to a control problem. Conventionally, to carry out this task two control loops have been used to manipulate the reflux rate and the reboiler duty. However, due to the changing feed stream, the operating point will also be changed frequently. Since distillation is a nonlinear dynamic process, the conventional control loops can not follow the desired changing operating point. Thus a conservative setpoint value, which is higher than the purity specifications, has to be used. But a conservative operation leads to more energy consumption than necessary, since the energy requirement for a column operation increases sensitively to the product purity, especially for a high purity distillation. Therefore, we propose to use a stochastic optimization approach to solve the column operation problem. This has the advantage over conventional feedback control that it employs the non-linear rigorous model and includes uncertainties.

2.2 *The Model of Dynamics*

The dynamics of the distillation process are summarized in Table 1. Here we use a rigorous model details of which can be found in [1, 11, 15]. In the table $i = 1, \ldots, I$ and $j = 1, \ldots, J$ are the indices of components and trays (counting from the condenser over the internal trays to the reboiler), respectively. The control variables are the feed flow rate F, the reboiler duty Q and the reflux rate from the condenser L^1. The dependent variables on each tray are the vapor and liquid mole fractions Y_i^j, X_i^j, vapor and liquid flow rate V^j, L^j, molar liquid holdup M^j, temperature T^j and pressure P^j. The Murphree tray efficiency η^j is included in order to describe the nonequilibrium behavior. The inclusion of the tray hydraulics is necessary to arrive at an index-one DAE system [3]. Additionally, it reflects the reality much better than other commonly used models, where molar or volumetric tray holdups and pressure drops are fixed. Really, the expressions containing F are not included in the differential equations for all internal trays but only for the feed tray. The wet pressure drop ϕ_w is not applied to the tray 2 at the column head.

The DAE system also includes column specific parameters, applicable only to the individual construction and design of the pilot plant. The values of those parameters are listed in Table 2.

2.3 *Model Validation*

Before carrying out the optimization a model validation has to be made to ensure the accuracy of the computational results. In the model, the parameters of the vapour liquid equilibrium are taken from [6] while the parameters of the pure components are taken from [14]. The parameters for the tray holdup and the pressure drop calculated by the gas and liquid fluid dynamics are based on the tray geometric sizes. Several parameters (coefficients) in the model are strongly dependent on the column structure and the operating point and thus should be matched to the experimental results. We consider the tray efficiency in the rectifying section η_R and stripping section η_S of the column which are the most important parameters for the model validation. The weir constant C_w, a parameter in the Francis weir equation needed

Reboiler
total mass balance:
$\frac{d}{dt}(M^J) = L^{J-1} - V^J - B$
component balance:
$\frac{d}{dt}(M^J X_1^J) = L^{J-1} X_1^{J-1} - V^J Y_1^J - B X_1^J$
energy balance:
$\frac{d}{dt}(M^J H^{L,J}) = L^{J-1} H^{L,J-1} - V^J H^{V,J} - B H^{L,J} + Q$
enthalpy calculation:
$H^{L,J} = \sum_{i=1}^{2} X_i^J h_i^L(T^J) \qquad H^{V,J} = \sum_{i=1}^{2} Y_i^J h_i^V(T^J)$
sum balances:
$\sum_{i=1}^{2} X_i^J = 1 \qquad \sum_{i=1}^{2} Y_i^J = 1$
vapor liquid equilibrium:
$Y_i^J = K_i(T^J, P^J, X_1^J, X_2^J) X_i^J \quad i = 1,2$
pressure drop equation:
$P^{J-1} - P^J = \varphi_d(V^J, T^J, Y_1^J, Y_2^J) + \varphi_w(M^{J-1}, T^{J-1}, X_1^{J-1}, X_2^{J-1})$
Internal Trays
total mass balance:
$\frac{d}{dt}(M^j) = F + L^{j-1} + V^{j+1} - L^j - V^j$
component balance:
$\frac{d}{dt}(M^j X_1^j) = F Z_1 + L^{j-1} X_1^{j-1} + V^{j+1} Y_1^{j+1} - L^j X_1^j - V^j Y_1^j$
energy balance:
$\frac{d}{dt}(M^j H^{L,j}) = F H^F + L^{j-1} H^{L,j-1} + V^{j+1} H^{V,j+1} - L^j H^{L,j} - V^j H^{V,j}$
enthalpy calculation:
$H^{L,j} = \sum_{i=1}^{2} X_i^j h_i^L(T^j) \qquad H^{V,j} = \sum_{i=1}^{2} Y_i^j h_i^V(T^j)$
sum balances:
$\sum_{i=1}^{2} X_i^j = 1 \qquad \sum_{i=1}^{2} Y_i^j = 1$
vapor liquid equilibrium:
$Y_i^j = \eta^j K_i(T^j, P^j, X_1^j, X_2^j) X_i^j + (1 - \eta^j) Y_i^{j+1} \quad i = 1,2$
Francis weir equation:
$L^j = \psi(M^j, T^j, X_1^j, X_2^j)$
pressure drop equation:
$P^{j-1} - P^j = \varphi_d(V^j, T^j, Y_1^j, Y_2^j) + \varphi_w(M^{j-1}, T^{j-1}, X_1^{j-1}, X_2^{j-1})$
Total Condenser
total mass balance:
$\frac{d}{dt}(M^1) = V^2 - L^1 - D$
component balance:
$\frac{d}{dt}(M^1 X_1^{L,1}) = V^2 X_1^{V,2} - L^1 X_1^{L,1} - D X_1^{L,1}$
enthalpy calculation:
$H^{L,1} = \sum_{i=1}^{2} X_i^1 h_i^L(T^1) \qquad (T^1 = \text{const.})$

Table 1. Dynamics of the distillation process

	Construction Parameters		
N	Number of trays	20	-
A	Tray area	0.008992	m^2
A_F	Free area	0.00045	m^2
l_W	Weir length	0.015	m
h_W	Weir height	0.03	m
	Operation Parameters		
ε	Vol. fraction of liquid in the bubble area	0.5	-
T_R	Reflux Temp.	60	°C
T_F	Feed Temp.	60	°C
V_D	Vol. of Dist. Vessel	1.5	m^3
V_B	Vol. of bottom Vessel	6.0	m^3
N_F	Feed tray	14	-

Table 2. Column specific data

for describing the dynamic behavior of the column is also estimated from experimental data. The friction factor ζ_w, a parameter in the pressure drop equation used for calculating the dry pressure loss, has to be verified as well. Moreover, the total heat loss Q_V from the column, reboiler and the pipelines has been found at a level of about 10% of the reboiler duty. This result is important for the implementation of the computed reboiler duty values Q_{eff} on the real plant.

To accomplish the above verifications, 10 steady-state operating points have been gained through experiment. The parameters are adjusted by comparing the measured data (temperature, pressure as well as pressure drop, flow rates, compositions and reboiler duty) with the simulation results for these operating points. The comparison was evaluated by the least square function related to the temperature profiles. Moreover,the measured data of the dry pressure drop of the rectifying section and the total pressure drop of the stripping section during the startup phase were also used for estimating the parameters ζ_w and C_w. finally, the adjusted values for those parameters are $\eta_R = 0.85$, $\eta_S = 0.80$, $C_w = 7.614$, $\zeta_w = 2.3046$, $Q_V = 0.6KW$.

To verify these values two tests were done by step changes of the feed rate, reflux rate and reboiler duty from a steady-state operating point of the real plant. Figure 2 (left) shows the temperature responses along the column in the first run, where the black lines with noises are the measured curves and the grey lines are the simulated results. It can be seen that a very good agreement has been received for the top and bottom temperature. The temperature on tray 17 and 18, which are the two most sensitive trays in the column, has a 2-3 C difference between the measured and computed results. The reason for this is the existence of slight disturbances from the atmospheric pressure and temperature during the experiment which led to the oscillation of the column pressure and heat loss from the column. The fact that both the measured and simulated curves have the same tendency demonstrates that the inclusion of the hydraulic computation leads to a successful description of the dynamic behavior for the column operation. The second test was made for a step change in the opposite direction. The temperature profiles are shown in Figure 2 (right). It can be noted that the temperature on the trays in the rectifying section is not sensitive to the step changes, especially for the top of the column. But the temperature in the stripping section has a significant sensitivity. This phenomenon can be identified from the form of the x-y-diagram of the methanol-water mixture.

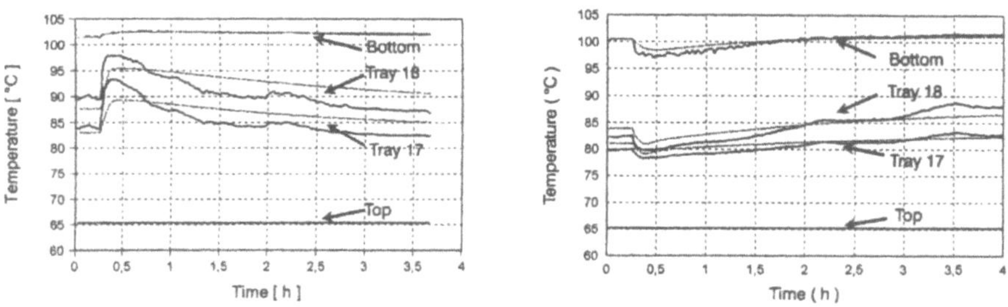

Figure 2. Simulated and measured temperature profiles: experiment 1 (left) and 2 (right)

2.4 *Relation between Feed Rate and Reboiler Duty*

Often one observes a tight relation between the feed supply and the required energy (reboiler duty). In general (especially for dynamic processes), however, this relation is not strong enough to substitute one of these control variables by the others which would lead to simpler mathematical models. In particular, the consideration of probabilistic constraints to be described below could be separated then from the dynamics of the distillation process. The following computational results demonstrate that the total energy consumption (our goal function) is not only a function of the total amount of feed but also of the shape of the feed rate profile.

Figure 3 shows two profiles for the feed flow rate having extremely different shapes. If the required purity constraints are exactly satisfied (here: 0.995 for the distillate and bottom products) during the time horizon, the theoretical energy consumption in the two cases should have almost the same value. It turns out, however, that the total energy consumption for the smooth profile is less than for the irregular profile at a factor of around 0.97. The reason can be figured out from the respective purity profiles in Figure 4: The purity constraints can be almost exactly satisfied for the smooth feed trajectory, whereas in the second case a higher purity than required has to be realized to ensure the constraints to be held under the drastically changing

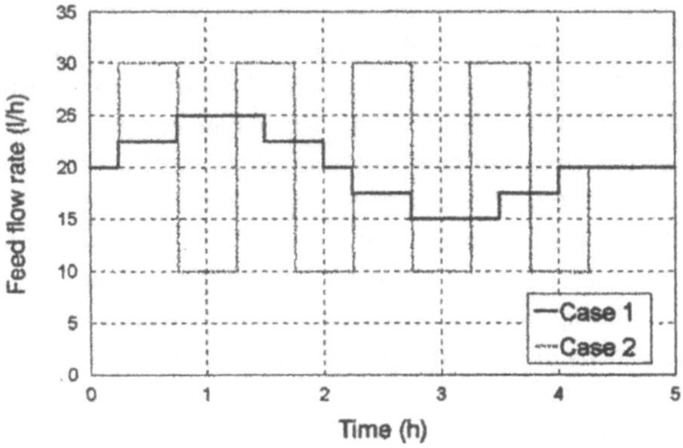

Figure 3. Two different profiles for the feed flow rate

feed stream. This behavior of the purity profiles results from the piecewise constant profiles of the control functions on the discretization scheme.

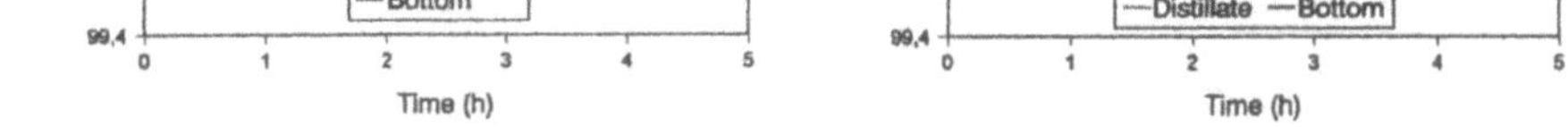

Figure 4. Optimal purity profiles for the feed trajectory from Figure 3: smooth case (left) and irregular case (right)

3 Probabilistic Feed Tank Constraints

We assume that our optimization horizon is given by the interval $[t_0, t_1]$. As mentioned in the introduction, the filling level $l(t)$ of the feed tank has to satisfy the simple constraints

$$l^{\min} \leq l(t) \leq l^{\max} \quad \text{for all } t \in [t_0, t_1]. \tag{1}$$

In view of the contradiction between the finite optimization horizon and the infinite nature of our continuous distillation process, one has to impose an additional so-called cycling constraint on the filling level. Otherwise, an optimal strategy with respect to energy consumption would certainly consist in running the unit on the lowest possible feed extraction level. This would result in a high filling level $l(t_1)$ at the end of the optimization horizon. If, for instance, $l(t_1) = l^{max}$, then it may well happen that there does not exist any feasible control of feed extraction for the next horizon $[t_1, t_2]$ due to the risk of violating the upper level constraint (when the inflow runs faster than the maximum possible rate of feed extraction). In other words, one has to make sure that optimization during any time period is not done at the expense of coming periods. This can be realized by the requirement, that all inflow has to be processed over the interval, or in equivalent terms, the final and initial filling levels coincide: $l(t_1) = l(t_0)$. For a higher operational flexibility it is reasonable to formulate the following slight generalization of the last constraint:

$$l(t_1) = l' \tag{2}$$

Here, $l' \in [l^{min}, l^{max}]$ is any pre-defined end-level which may be in the middle of l^{min} and l^{max} in the regular case, but which could also be appropriate to choose closer to l^{max}, for instance, in front of a week-end where no or few amounts of inflows only can be expected.

Next, we are going to take into account the random character of the filling level caused by the stochastic nature of inflows. To this aim, denote by ξ the inflow rate which we assume to be a one-dimensional stochastic process. Now, the filling level as a function of inflow rate ξ, of feed extraction rate F and of time t writes as

$$l(\xi, F, t) = l_0 + \int_{t_0}^{t} (\xi(\tau) - F(\tau))\, d\tau.$$

Here, l_0 denotes the initial level at t_0. Accordingly, (1) and (2) turn into *uncertain* constraints of the type

$$l^{min} \leq l(\xi, F, t) \leq l^{max} \quad \text{for all } t \in [t_0, t_1] \quad \text{and } l(\xi, F, t_1) = l'.$$

According to [9], uncertain constraints can be turned into implementable constraints by expected value substitution or by probabilistic constraints. Taking expected values seems to be justified for the cycling constraint, since violations of (3) by single inflow realizations do not cause serious harms and we may content ourselves with the fact that over many repeated optimization periods there will be no systematic gain on average by violating (3) in a specific period at the expense of later periods. Accordingly, (2) becomes

$$l(\mathbb{E}\xi, F, t_1) = l', \tag{3}$$

where '$\mathbb{E}$' denotes expectation. On the other hand, frequent violations of the level constraints (1), as they have to be expected when feasibility is reached only for an average profile of inflow rate, do cause troubles for the compensation of which is difficult to measure in terms of costs. Consequently, one may be interested in controls F being feasible with high probability according to the distribution of ξ:

$$P(l^{min} \leq l(\xi, F, t) \leq l^{max} \quad \text{for all } t \in [t_0, t_1]) \geq p. \tag{4}$$

Here, P refers to the probability measure associated with ξ and $p \in (0,1)$ is some probability level at which we require the inequalities inside parentheses to be fulfilled. According to [9], Section 4.1 there is much degree of modelling freedom concerning a joint or individual treatment of probabilistic constraints. We shall focus on a formulation in which time dependence appears as joint constraints whereas lower and upper levels are considered individually. More precisely, the following two probabilistic constraints shall be imposed with respect to filling levels of the feed tank:

$$\begin{aligned} P(l^{\min} &\le l(\xi, F, t) \quad \text{for all } t \in [t_0, t_1]) \ge p \\ P(l^{\max} &\ge l(\xi, F, t) \quad \text{for all } t \in [t_0, t_1]) \ge p. \end{aligned} \tag{5}$$

4 The Overall Problem and its Numerical Solution

Combining the model of dynamics described in section 2.2 and the stochastic feed tank constraints, we are now in a position to set up our control problem in the following compact way:

$$\min J(u) \qquad (P)$$

$$\begin{aligned} &\text{subject to} \\ \dot{x} &= f(x, y, u) \\ g(x, y) &= 0 \\ x(t_0) &= x_0 \\ l(\mathbb{E}\xi, u, t_1) &= l' \\ P(l^{\min} \le l(\xi, u, t) \quad \forall t \in [t_0, t_1]) &\ge p \\ P(l^{\max} \ge l(\xi, u, t) \quad \forall t \in [t_0, t_1]) &\ge p \\ u \in U, x \in X, y &\in Y \end{aligned}$$

Here, u, x, y, ξ are functions defined on our optimization interval $[t_0, t_1]$ and representing control, differential state, algebraic state and random variables, respectively. For instance, $u = (F, Q, L^1)$ comprises the extraction rates for feed, heat supply and reflux rate. The state variables consist of all the remaining physical quantities in the dynamics of the distillation system. Finally, ξ refers to the stochastic inflow rate. The goal function to be minimized is the total heat consumption

$$J(u) = \int_{t_0}^{t_1} Q(t)dt.$$

We assume that the condenser cooling duty is negligible since it will not influence our solution approach and, furthermore, it depends on the specific circumstances whether or not additional costs arise by cooling.

The first three equalities in the constraints above contain the differential and algebraic equations of the dynamics as well as initial conditions. The following three inequalities relate to the cycling constraint (3) and to the probabilistic feed level constraints (5). The abstract constraints at the end represent simple bounds on the control and state variables; in particular, $u \in U$ incorporates limitations on the feed rate, heat supply and flow rates for bottom and distillate products, $x \in X$ includes purity conditions on the bottom and distillate products as well as constraints for the holdup in the reboiler and condenser while $y \in Y$ defines bounds on physical quantities like temperature, pressure etc.

A common approach for solving control problems like (P) is the so-called direct method [2, 13, 16–18], where the differential algebraic equations are discretized and the corresponding functions are finitely parametrized in order to yield a (possibly large scale) nonlinear optimization problem in finite dimensions. In our case, it may be written as

$$\min \varphi(u) \qquad (\mathrm{NLP}) \tag{6}$$

$$\begin{aligned} &\text{subject to} \\ &G(x, y, u) = 0 \\ &h_1(\mathbb{E}\xi, u) = l' \\ &P(h_2(\xi, u) \leq 0) \geq p \\ &P(h_3(\xi, u) \leq 0) \geq p \\ &u \in U, x \in X, y \in Y, \end{aligned}$$

where now u, x, y, ξ are finite-dimensional vectors.

A direct simultaneous approach based on collocation will be used here to discretize the differential algebraic equation (DAE). The obtained optimization problem is then solved by the SQP method *SNOPT* [4], [5]. From technological requirements, the controls have to be piecewise constant on a given grid here. For instance, the operator of the system may be able to only tune constant values once an hour. The grid for the DAE is chosen equal to or as a refinement of the grid for the controls to prevent that a jump of the controls is within an integration interval of the DAE. On each subinterval of the refined grid the DAE is treated by a 3 stage Radau IIa (collocation) scheme [7]. The resulting collocation conditions are formulated in terms of the algebraic variables and in terms of the derivatives of the differential state variables at the collocation points [16, 17]. Additional conditions are needed to ensure continuity of the state variables between the end and the beginning of two succeeding collocation intervals. For consistency purposes these conditions can not be imposed on all variables but only for a part of the variables depending on the index. Since the differential index of our DAE is 1, it is sufficient to restrict the continuity condition to the differential state variables [12]. That means that for the whole problem as well as for the integration intervals initial values are given only for the differential state variables. In convergence, consistency is then automatically fullfiled since the Radau IIa scheme includes the endpoint but excludes the initial point of each collocation interval and for given differential state variables the algebraic variables are uniquely determined by the algebraic constraints of the Index-1-DAE.

In order to solve the overall problem, one has to incorporate numerical techniques for dealing with the probabilistic constraints into the framework described before. The concrete realization depends on the structure of probabilistic constraints as it results from different modelling assumptions. For instance, assuming ξ originally to represent a Gaussian process, the treatment of constraints (5) leads to the evaluation of multivariate normal distribution functions after passing to a time discretization. For implementation, we linked Szantai's simulation scheme described in [9], Section 4.3 and realized in his code *BERNOR* to the SQP code *SNOPT*. As a preliminary step of numerical solution, probability maximization was carried out (see [9], Section 4.4).

5 Stochastic Models for the Inflow Rates

We shall consider two basically different models for the stochastic inflow rate: a model describing some elementary single process of inflow generation and a model reflecting the superposition of numerous such elementary processes. Both situations may be relevant in practice depending on the nature of production processes prior to distillation. We start with the model relating to a lot of independent elementary processes of equal structure. According to the law of large numbers, it is reasonable to assume that the rate ξ_t of overall inflows is a Gaussian process, which means that each finite selection $(\xi_{\tau_1}, \ldots, \xi_{\tau_n})$ of random variables with $\tau_1, \ldots, \tau_n \in [t_0, t_1]$ has a multivariate normal distribution. The second model considers a fixed inflow function the realization of which takes place at a random starting time. The difference between both models is illustrated in Figure 5.

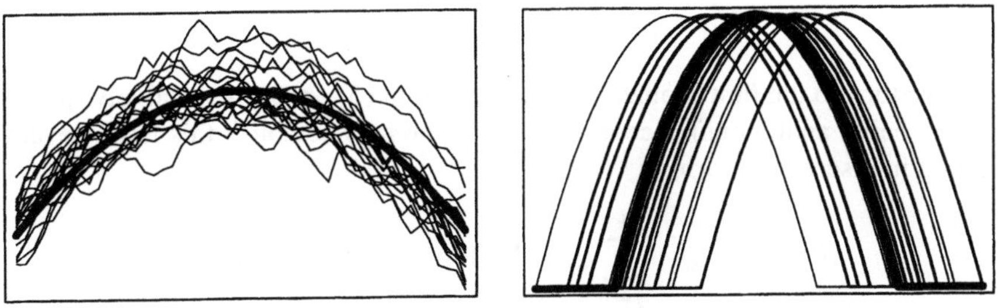

Figure 5. Comparison of two different stochastic models for the inflow rate. Left: Gaussian process; right: fixed profile with random initial time. Thick curves represent expected values of the corresponding processes and thin curves illustrate possible samples (observations) thereof

5.1 *Inflow Rate with Multivariate Normal Distribution*

According to the discretized optimization problem (6) let $\tau_0 < \cdots < \tau_N$ be a subdivision of the time interval $[t_0, t_1]$ with $\tau_0 = t_0$ and $\tau_N = t_1$. For such fixed subdivision, we may identify the average inflow rate ξ_i during the interval $[\tau_{i-1}, \tau_i]$ with the total amount of inflow over this period $(i = 1, \dots, N)$. According to the statement above, it is assumed that the N- dimensional random vector $\xi = (\xi_1, \dots, \xi_N)$ has a multivariate normal distribution, i.e., $\xi \sim \mathcal{N}(\mu, \Sigma)$, where μ and Σ are the expectation and covariance matrix, respectively, of ξ. In the numerical experiments described below, we have supposed a covariance structure with decreasing positive values at increasing distance in time. In practice, μ and Σ have to be estimated from observed data of the inflow process. In the first diagram of Figure 7 a set of sample paths of an exemplary inflow process is illustrated. Here, loosely speaking, the typical profile corresponds to μ, whereas Σ accounts for the scattering around μ and the smoothness of the sample paths.

In order to specify the functions h_1, h_2, h_3 in (NLP), we recall that our feed tank constraints are special cases of the storage level constraints introduced in [9], Section 4.2. Accordingly, the filling level in the feed tank at time τ_i calculates as

$$l_0 + \sum_{j=1}^{i} \xi_j - \sum_{j=1}^{i} F_j, \tag{7}$$

where l_0 is the filling level at $\tau_0 = t_0$ and ξ_j, F_j refer to the amounts of inflow and feed extraction, respectively during $[\tau_{j-1}, \tau_j]$. Now, the cycling constraint (3) requires that the expectation of the final filling level (at $\tau_N = t_1$) equals l'. Taking into account that $\mathbb{E}\xi_j = \mu_j$, the cycling constraint writes as the following simple linear restriction in the control variable F:

$$\sum_{j=1}^{N} F_j = l_0 - l' - \sum_{j=1}^{N} \mu_j.$$

Next, we turn to the first probabilistic constraint in (6) or (5), respectively (the second one being analogous). According to (7), one may establish this lower level restriction in discretized form as

$$P(l^{\min} \leq l_0 + \sum_{j=1}^{i} \xi_j - \sum_{j=1}^{i} F_j; \quad i = 1, \dots, N) \geq p.$$

After some transformations detailed in [10], this last relation can be equivalently written in the explicit form $\Phi(\alpha(F)) \geq p$, where $\alpha(F)$ is a simple affine linear mapping and Φ refers to a N-dimensional standard normal distribution with suitable correlation matrix. Consequently, the whole issue of coping with probabilistic feed tank constraints hinges upon the ability of calculating distribution functions of multivariate normally distributed random vectors. Some possible approaches to do so have been presented in [9], Section 4.3.

5.2 *Fixed Inflow Profiles with Stochastic Initial Time*

Next we turn to an inflow process with a deterministic profile which is realized at a random starting time. In a non-discretized setting, the profile is given by some nonnegative function $\nu \in L_+^\infty([0,d])$ (determined, e.g., by a fixed operating pattern of some machine), where d denotes the duration of the inflow. In the simplest case, the profile is constant over $[0, d]$. Defining ξ' as the random initial time, the inflow process becomes

$$\xi = \begin{cases} \nu(t-\xi') & \text{if } t \in [\xi', \xi'+d] \\ 0 & \text{else} \end{cases}.$$

In [8] it is shown that the probabilistic constraints (5) can be transformed into the explicit linear (functional) constraints

$$\varphi^1(t) \le \int_{t_0}^{t} F(\tau)d\tau \le \varphi^2(t) \quad \forall t \in [t_0, t_1]$$

for the feed extraction rate F. Here,

$$\begin{aligned}
\varphi^{1(2)}(t) &= l_0 - l^{\max}(l^{\min}) + \min\{\max\{0, V(t-q_1(q_2))\}, V(d)\}; \\
V(t) &= \int_0^t \nu(\tau)d\tau; \\
q_1 &= \sup\{t \mid P(\xi' \le t) \le 1-p\} \\
q_2 &= \inf\{t \mid P(\xi' \le t) \ge p\}
\end{aligned}$$

This means that the constraining functions φ^1, φ^2 are easily obtained from the data of the problem. In particular, q_1 and q_2 are related with appropriate quantiles of the distribution of the random initial time ξ'. In discretized form, one arrives at the linear restrictions

$$\varphi^1(\tau_i) \le \sum_{j=1}^{i} F_j \le \varphi^2(\tau_i) \quad i = 1, \dots, N.$$

It is worth mentioning (cf. [8, Thm. 5]) that exactly the same constraints result if, in contrast to (5), the probabilistic constraints are considered individually with respect to time, i.e.,

$$P(l^{\min} \le l(\xi, F, t)) \ge p \quad \text{and} \quad P(l^{\max} \ge l(\xi, F, t)) \ge p \quad \text{for all } t \in [t_0, t_1].$$

This means that the distinction between individual and joint constraints is no longer relevant in this specific model (see discussion in [9], Section 4.1).

6 Numerical Results

In the following, we describe numerical results obtained for the model developed above. A time horizon of 16 hours is considered. In order to illustrate the gain of using stochastic information, the results of models with probabilistic constraints shall be opposed to those obtained when simply assuming the expected inflow profile. We start with the model of a (discretized) Gaussian process for the inflow rate. The first diagram of Figure 7 shows 100 possible inflow realizations for an example. As a distillation unit, a column consisting of a reboiler, a total condenser and three internal trays, is considered. The required purities are defined by methanol concentrations of $\geq 80\%$ and $\leq 10\%$ in the condenser and reboiler, respectively.

If the calculations are based on the expected inflow profile then the solid curves in the first three diagrams of Figure 6 result as optimal - in the sense of minimum heat consumption - profiles for feed extraction rate, reflux rate and heat supply. Not surprisingly, heat consumption and feed extraction run almost in parallel. The last three diagrams illustrate the profiles of a selection of associated state variables, namely purity, pressure and liquid flow rate. The purity diagram confirms in particular the satisfaction of required purities in reboiler and condenser, respectively. These solutions are now contrasted with the solutions based on probabilistic feed tank constraints when imposing a probability level of $p = 0.9$. The optimal profiles for control variables are represented by dotted curves in the first three diagrams of Figure 6. Although the differences appear to be negligible, they have a significant impact on the robustness of the process as shall be seen next. Corresponding plots of state variable profiles are similar to the previous ones and omitted here for the sake of brevity.

Assuming for a moment that the inflow profile realizes indeed its expected values, the filling levels plotted in the second diagram of Figure 7 result from applying the respective feed extraction controls of the first diagram of Figure 6. Again, the solid curve represents the solution based on the expected inflow profile whereas the dotted curve relates to probabilistic constraints. Both trajectories are feasible with respect to satisfying a filling level between the upper value of 1500 and the lower value of 500. Note that the lower level is attained by the solid curve after around 4 hours whereas the dotted curve remains slightly above. The starting level was supposed to be at 1000 and a cycling constraint was set up in order to realize the same level at the end of the time horizon which means that eventually all incoming substance was separated by the distillation column.

In practice, however, it is very unlikely that the expected value is realized. Rather, one is likely to observe one out of the 100 sample profiles plotted in the first diagram of Figure 7 which scatter more or less around the expected profile. Therefore, the filling level obtained from the optimal feed extraction profiles has to be verified with regard to these samples instead. The associated 100 profiles of filling levels are given in the third (application of the feed extraction control based on assuming the expected inflow profile) and fourth (application of the feed extraction control based on probabilistic constraints) diagrams of Figure 7. In both cases it becomes obvious how uncertainty evolves over time with ever increasing variances of filling levels. Furthermore, in both cases the average filling level at the end coincides with the initial value which illustrates satisfaction of the (stochastic) cycling constraint imposed in (3).

In contrast, and more important, a frequent violation of the lower level constraint can be observed in the first case. As this is hardly visible from the total diagram, the interesting section is zoomed in the diagrams below. Indeed, 51 out of the 100 samplesare found to violate the lower level at some time between three or five hours after initial time. This means that undesirable compensating actions have to be carried out with probability of around 50% upon choosing a feed extraction control based on assuming the expected inflow profile and thus ignoring information on the distribution of the random process. This result of 50% violations is not surprising if one recalls that in the expected case the lower level is exactly attained (solid curve in the second diagram of Figure 7).

In contrast, application of the feed extraction control based on probabilistic constraints yields significantly less violations of the lower level as can be easily verified from the corresponding zoomed plot again. This time, just 12 out of the 100 samples fall below l^{min} which is in good accordance with the chosen probability level of $p = 0.9$. The upper level constraint (not zoomed) is violated for less than 10 samples in both cases. Summarizing, there is a considerable gain in robustness when passing from expected value solutions to probabilistic constraints solutions. The value of stochastic information is reflected by the probability of violation dropping from 50% to 10%. In general, one would expect to obtain such gain in robustness only at the expense of worsened values of the objective function. However, there is no measurable difference in heat consumption for the two discussed controls here (compare area below curves of the respective diagram in Figure 6).

Now we turn to corresponding results when assuming the alternative stochastic model of a fixed inflow profile with random initial time. As a fixed profile, we consider an inflow process with constant rate and with a duration of 9 hours. The initial time is assumed to be uniformly distributed in the interval $[1, 7]$. As mentioned above, any other bounded fixed profile and any other random distribution could be chosen equally well as long as the required quantiles are available.

Figure 8 shows the obtained results relating to feed tank constraints and omitting other plots related to the dynamics of the distillation process. In the first diagram, the optimal feed extraction rates are plotted with the meanings of the two curves being analogous to the discussion before. The two bottom diagrams represent the realized filling levels when applying the two different controls to 100 samples of the inflow process. Thick curves refer to the expected inflow. Again, the lower level is violated by almost one half of the samples for the expected value solution whereas just a few violations occur for the probabilistic constraints solution. In the right top diagram the violation probabilities are plotted as functions of time.

It can be seen that the expected value solution reaches a peak of around 45% in the period between 4 and 10 hours. This is in contrast to the probabilistic constraints solution with a peak of $5 - 10\%$ (in good accordance with the level $p = 0.9$) in the period between 6 and 8 hours.

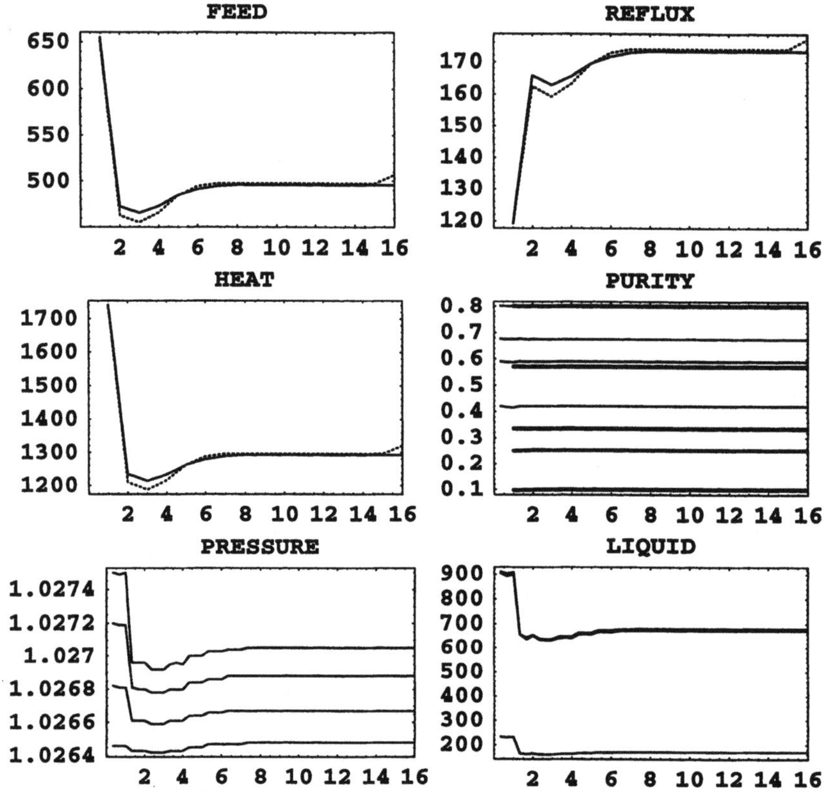

Figure 6. Results for the optimal control of continuous distillation with the inflow rate being a Gaussian process. The first three diagrams plot the optimal profiles for the three control variables (feed extraction rate, reflux and heat supply). The solid curves relate to the assumption of the expected inflow profile whereas the dotted curve relates to the model with probabilistic feed tank constraints. The last three diagrams illustrate the profiles of some selected state variables (purity, pressure, liquid flow rate). Different curves correspond to subsequent trays (possibly including reboiler and condenser) of the distillation column. The purity diagram distinguishes methanol concentrations in the liquid phase (thick lines) from those in the vapour phase (thin lines)

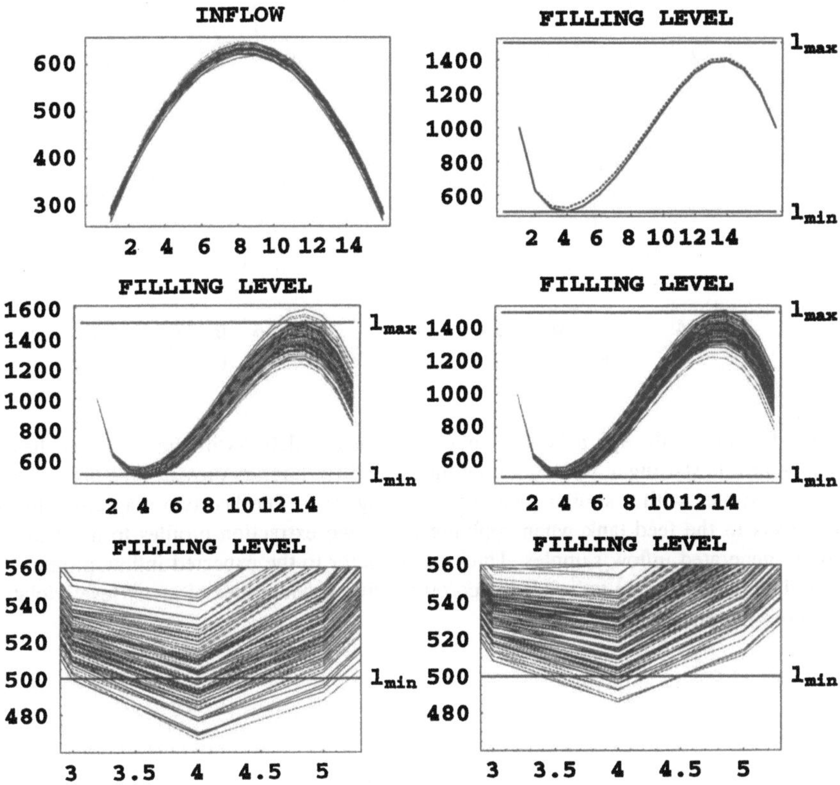

Figure 7. Filling levels in feed tank upon applying the optimal The first diagram illustrates 100 samples of a possible inflow process. In the second diagram, the filling levels in the feed tank are plotted under the assumption that the expected inflow profile is observed and the feed extraction profiles from Figure 6 are applied. The next two diagrams show the resulting filling levels for the 100 inflow samples of the first diagram. The left diagram relates to the expected value solution whereas the right one relates to the probabilistic constraints solution. The critical section with respect to lower level violation is zoomed in the respective diagrams at the bottom

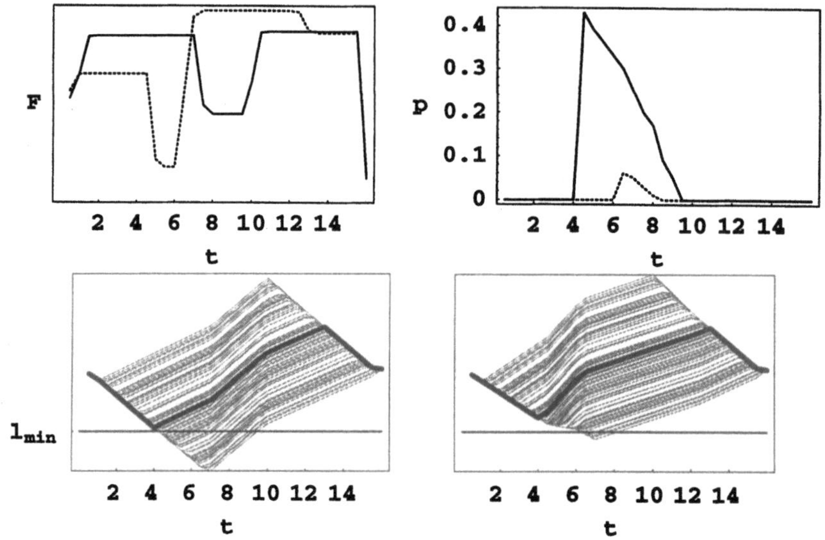

Figure 8. Results for the optimal control of continuous distillation with the inflow rate having a constant profile starting at random initial time. The first diagram provides the optimal feed extraction profiles with a similar meaning as in Figure 6. The last two diagrams plot the filling levels in the feed tank when applying these two extraction profiles to a set of 100 randomly generated inflow samples. Thick curves refer to the expected inflow profile. The second diagram (top right) indicates the corresponding probabilities of lower level violation as functions of time

References

1. H. Arellano-Garcia, R. Henrion, P. Li, A. Möller, W. Römisch, M. Wendt, G. Wozny: A model for the online optimization of integrated distillation columns under stochastic constraints. DFG-Schwerpunktprogramm 'Echtzeit-Optimierung grosser Systeme', Preprint 98-32, 1998.
2. A. Barclay, P. E. Gill, J. B. Rosen: SQP Methods and their Application to Numerical Optimal Controll Report NA 97-3, Department of Mathematics, University of California, San Diego 1997.
3. E. Eich-Soellner, P. Lory, P. Burr, A. Kröner: Stationary and Dynamic Flowsheeting in the Chemical Engineering Industry. Surv. Math. Ind. **7** (1997), 1-28.
4. P. E. Gill, W. Murray, M. A. Saunders: SNOPT: An SQP Algorithm for Large-Scale Constrained Optimization. Report NA 97-2, Department of Mathematics, University of California, San Diego 1997.
5. P. E. Gill, W. Murray, M. A. Saunders: User's guide for SNOPT 5.3: A FORTRAN package for large-scale nonlinear programming. Report NA 97-2, Department of Mathematics, University of California, San Diego, 1997.

6. J. Gmehling, U. Onken, W. Arlt: Vapour Liquid Equilibrium Data Collection, DECHEMA, Frankfurt, 1997.
7. E. Hairer, G. Wanner: Solving Ordinary Differential Equations II: Stiff and Differential-Algebraic Problems. Springer, Berlin, 1996.
8. R. Henrion, Structure and Stability of Probabilistic Storage Level Constraints. Preprint No. 23 (2000) of Stochastic Programming Eprint Series (SPEPS), submitted to: Proc. 4th GAMM/IFIP-Workshop on "STOCHASTIC OPTIMIZATION: Numerical Methods and Technical Applications", held at UniBw Munich, June 27–29, 2000.
9. R. Henrion, P. Li, A. Möller, M. C. Steinbach, M. Wendt, G. Wozny: Stochastic Optimization for Operating Chemnical Processes under Uncertainty. This volume.
10. R. Henrion, A. Möller: Optimization of a continuous distillation process under random inflow rate. Preprint No. 00-4 des DFG-Schwerpunktprogramms 'Echtzeit-Optimierung großer Systeme', Berlin, 2000, submitted to: Computers & Mathematics with Applications.
11. H. Z. Kister: Distillation Design. McGraw-Hill, New York, 1992.
12. R. Lamour: A Well-Posed Shooting Method for Transferable DAE's. Numer. Math. **59** (1991), 815-829.
13. D. B. Leineweber, H. G. Bock and J. P. Schlöder: Fast Direct Methods for Real-Time Optimization of Chemical Processes. in: (A. Sydow ed.) *Proceedings of the 15th IMACS World Congress on Scientific Computation, Modelling and Applied Mathematics* **6**, 1997, pp. 451-456.
14. R. C. Reid, J. M. Prausnitz, B. E. Poling: The Properties of Gases and Liquids. McGraw-Hill, New York, 1987.
15. A. Rix: Modellierung und Prozeßführung wärmeintegrierter Destillationskolonnen. PhD Thesis, TU Berlin, 1997.
16. V. Schulz: Reduced SQP Methods for Large-Scale Optimal Control Problems in DAE with Application to Path Planning Problems for Satellite Mounted Robots, Dissertation, Universität Heidelberg, 1995.
17. M. C. Steinbach: Fast Recursive SQP Methods for Large-Scale Optimal Control Problems, Dissertation, Universität Heidelberg, 1995.
18. O. von Stryk: Numerische Lösung optimaler Steuerungsprobleme: Diskretisierung, Parameteroptimierung und Berechnung der adjungierten Variablen, Dissertation, *VDI Fortschritt-Berichte* Reihe 8, Nr. 441, VDI-Verlag, Düsseldorf, 1995.

VI

Stochastic Trajectory Planning in Robot Control

*Adaptive Optimal Stochastic Trajectory Planning

Numerics and Real-time Application

Andreas Aurnhammer and Kurt Marti

Institut für Mathematik und Informatik, Universität der Bundeswehr München, Germany

Abstract In Optimal Stochastic Trajectory Planning of industrial or service robots the problem can be modelled by a variational problem under stochastic disturbances that compared to ordinary deterministic engineering techniques also accounts for stochastic model parameters. Using stochastic optimisation theory, this variational problem is transformed into a nonlinear mathematical program, that can be solved by means of standard optimisation routines like SQP. However, these methods are not applicable in the on-line control process of robots, since they are not capable of solving mathematical programs in real-time. Hence, Neural Networks are trained based on solutions obtained from a standard optimisation algorithm.

1 Adaptive Optimal Stochastic Trajectory Planning

1.1 Prescribed-Path and Point-to-Point Problems

In optimal trajectory planning of robots usually two main problem classes are discussed. The first, called prescribed-path problem (PP), is the task of following a desired path in work or configuration space as exactly as possible, where additionally a performance index like minimum-time, minimum energy consumption or a combination of both has to be optimised. These sort of tasks appear, e.g. in laser cutting or glueing. The second, called point-to-point problem (PTP), is having only two given points in work or configuration space that have to be connected by an optimal trajectory subject to the performance index and taking into account several state and control constraints. Typical applications here-fore are, e.g., spot welding in the automotive industry.

Mathematically the trajectory planning problem for a prescribed path in work or configuration space can be described, after a nondecreasing time-path transformation $s : [t_0, t_f] \mapsto [s_0, s_f]$ onto a given fixed s-parameter domain, see, e.g., [9, 23], by a variational problem. This is necessary to incorporate a free final time t_f like in minimum-time problems. Then, the Optimal Trajectory Planning Problem (PP) for prescribed-path problems reads:

$$\min_{\beta(.)} \int_{s_0}^{s_f} f_0(s, q_e(s), q_e'(s), q_e''(s), \beta(s), \beta'(s), p_J)\, ds \tag{1a}$$

subject to

$$\beta(s_0) = 0, \ \beta(s_f) = 0, \tag{1b}$$

$$u_i^{\min} \le a_i \beta'(s) + b_i \beta(s) + c_i \le u_i^{\max}, \ s_0 \le s \le s_f, \ i = 1, 2, \ldots, n \tag{1c}$$

$$\dot{q}^{\min} \le q_e'(s)\sqrt{\beta(s)} \le \dot{q}^{\max}, \ s_0 \le s \le s_f, \tag{1d}$$

$$\beta(s) \ge 0, \ s_0 \le s \le s_f, \tag{1e}$$

$$T(q_e(s), p_K) = x_e(s), \ s_0 \le s \le s_f. \tag{1f}$$

Remark 1. If the path is already given in configuration space, equation (1f) can be replaced by

$$q = q_e(s), \ s_0 \le s \le s_f. \tag{1f'}$$

Here, s, $s_0 \le s \le s_f$, denotes the path parameter, $q = q_e(s)$ is the vector of configuration or robot coordinates, $\beta = \beta(s) := \dot{s}^2(t)$ denotes the velocity profile, and $p := (p_D, p_K, p_J)^T$ is the vector of unknown dynamic, kinematic and objective model parameters. $T = T(q, p_K)$, the kinematic operator, describes the relation between configuration coordinates q and the coordinates x in work space. The coefficients $a_i = a_i(q_e, q_e', p_D)$, $b_i = b_i(q_e, q_e', q_e'', p_D)$ and $c_i = c_i(q_e, p_D)$, $i = 1, \ldots, n$, are obtained by putting $q(t) := q_e(s(t))$ into the dynamic equation of the robot [21], where n is the number of degrees of freedom. Moreover, $x_e(.)$ and $q_e(.)$ denote the prescribed path in work or configuration space, (1b) are initial and terminal conditions for the velocity profile $\beta = \beta(s)$, (1c) represent control constraints, which restrict the forces and moments in the robot joints, and (1d) are restrictions for the joint velocities. Uncertainties in the selection of appropriate constraints are represented by a further parameter vector p_C. Finally, the objective function (1a) can describe different optimisation criteria:

- minimum time

$$f_0(s, q_e, q_e', q_e'', \beta, \beta', p_J) = \frac{1}{\sqrt{\beta(s)}} \tag{2}$$

- minimum joint forces and torques

$$f_0(s, q_e, q_e', q_e'', \beta, \beta', p_J) = \frac{1}{\sqrt{\beta(s)}} \sum_{i=1}^{n} (u_i(s))^2, \tag{3}$$

  where $u_i(s) = a_i \beta'(s) + b_i \beta(s) + c_i$
- minimum energy consumption

$$f_0(s, q_e, q_e', q_e'', \beta, \beta', p_J) = \frac{1}{\sqrt{\beta(s)}} \sum_{i=1}^{n} (q_{ei}'(s)\sqrt{\beta(s)} u_i(s))^2. \tag{4}$$

Having the optimal geometric path $q_e(s)$ in configuration space and the optimal velocity profile $\beta(s)$ from (1a-f) and then also the time parameter transformation $s = s(t)$, the trajectory $q(t)$ in time domain is defined by

$$q(t) := q_e(s(t)),\ t \geq t_0. \tag{5}$$

A more detailed description can be found in [3, 9, 13, 17–21, 23–26].

Using the same techniques, point-to-point problems can also be represented by variational problems. In this case the condition (1f) or (1f') of (PP) is replaced by initial and terminal conditions only for the position in work or configuration space. Moreover, position constraints for the configuration coordinates $q = q_e(s)$ have to be added, to prevent the now free path between initial and terminal position from violating the physical boundaries of the robot joints. This leads to a variational problem of the following form (PTP):

$$\min_{\beta(.),\ q_e(.)} \int_{s_0}^{s_f} f_0(s, q_e(s), q_e'(s), q_e''(s), \beta(s), \beta'(s), p_J)\, ds \tag{6a}$$

subject to

$$\beta(s_0) = 0,\ \beta(s_f) = 0, \tag{6b}$$

$$u_i^{\min} \leq a_i \beta'(s) + b_i \beta(s) + c_i \leq u_i^{\max},\ s_0 \leq s \leq s_f,\ i = 1, 2, \ldots, n, \tag{6c}$$

$$q^{\min} \leq q_e(s) \leq q^{\max},\ s_0 \leq s \leq s_f, \tag{6d}$$

$$\dot{q}^{\min} \leq q_e'(s)\sqrt{\beta(s)} \leq \dot{q}^{\max},\ s_0 \leq s \leq s_f, \tag{6e}$$

$$\beta(s) \geq 0,\ s_0 \leq s \leq s_f, \tag{6f}$$

$$T(q_e(s_0), p_K) = x_0,\ T(q_e(s_f), p_K) = x_f \tag{6g}$$

Remark 2. Here, equation (6g) can be replaced by

$$q_e(s_0) = q_0,\ q_e(s_f) = q_f, \tag{6g'}$$

if the initial and terminal position are given in configuration space.

1.2 Stochastic Parameters

A basic drawback is that the parameter vector $p = (p_D, p_K, p_J, p_C)^T$ is not a given fixed quantity, but has to be considered as a random vector with a certain probability distribution due to

- stochastic variations of the material,
- manufacturing errors,
- modelling errors,
- stochastic variations of the work space environment (e.g. stochastic payload).

Hence, to reduce the on-line measurement and correction expenses [16], available a priori and statistical information about the stochastic variations of $p = p(\omega)$ should already be considered in the modelling phase as well as in certain future time-points whenever new information about $p = p(\omega)$ is available. Assume in the following, that the new information at times $t_0, t_1, \ldots$ can be represented by information σ-algebras $\mathcal{A}_{t_0}, \mathcal{A}_{t_1}, \ldots$, which can be determined by on-line estimation methods like, e.g., recursive least square algorithms [8,22]. This leads to an adaptive stochastic trajectory planning problem:

initial time-point, main correction time-points	t_0	t_1	$\ldots$	t_j	$\ldots$
information vector	$\begin{pmatrix} s_0 \\ (q_0, \dot{q}_0) \text{ or} \\ (x_0, \dot{x}_0) \\ \mathcal{A}_{t_0} \end{pmatrix}$	$\begin{pmatrix} s_1 \\ (q_1, \dot{q}_1) \text{ or} \\ (x_1, \dot{x}_1) \\ \mathcal{A}_{t_1} \end{pmatrix}$	$\ldots$	$\begin{pmatrix} s_j \\ (q_j, \dot{q}_j) \text{ or} \\ (x_j, \dot{x}_j) \\ \mathcal{A}_{t_j} \end{pmatrix}$	$\ldots$
$\Downarrow$					
opt. velocity profile, opt. geom. path in configuration space	$\begin{pmatrix} \beta^{(0)}(s) \\ q_e^{(0)}(s) \end{pmatrix}$	$\begin{pmatrix} \beta^{(1)}(s) \\ q_e^{(1)}(s) \end{pmatrix}$	$\ldots$	$\begin{pmatrix} \beta^{(j)}(s) \\ q_e^{(j)}(s) \end{pmatrix}$	$\ldots$

AOSTP (**A**daptive **O**ptimal **S**tochastic **T**rajectory **P**lanning)

1.3 Substitute Problems with Probability Constraints

Since the parameters p in (1a-f) or (6a-g) are random, (PP) and (PTP) can not be solved directly. Instead we use deterministic substitute problems, that are available from stochastic optimisation: In order to get robust optimal trajectories, random parameter variations are incorporated [12–15] into the optimisation process [12–15]

- by taking expectations in (weighted) objective functions,
- by evaluating violations of constraints by the resulting expected penalty costs, as, e.g., the separated or joint probability(ies) of violation of the constraints,

and

- by imposing upper bounds for the expected penalty costs or including them into the objective function (sum of the expected process and penalty costs).

If we consider the conditional expectation of the objective function (1a) with respect to the information available up to time t_j and demand that the control constraints (1c) are fulfilled separately at least with given fixed probabilities α_{u_i}, $i = 1, \ldots, n$, and the joint velocity restriction (1d) with a reliability of $\alpha_{\dot{q}}$, then a basic substitute problem for (1a-f) at stage j can be written as (SPP):

$$\min_{\beta(.)} \int_{s_j}^{s_f} \mathcal{E}(f_0(s, q_e(s), q_e'(s), q_e''(s), \beta(s), \beta'(s), p_J) \mid \mathcal{A}_{t_j})\, ds \tag{7a}$$

subject to

$$q_e'(s_j)\sqrt{\beta(s_j)} = \dot{q}_j, \beta(s_f) = 0, \tag{7b}$$

$$P(u_i^{min} \leq a_i\beta'(s) + b_i\beta(s) + c_i \leq u_i^{max} \mid \mathcal{A}_{t_j}) \geq \alpha_{u_i},$$
$$s_j \leq s \leq s_f,\ i = 1, 2, \ldots, n, \tag{7c}$$

$$P(\dot{q}^{min} \leq q_e'(s)\sqrt{\beta(s)} \leq \dot{q}^{max} \mid \mathcal{A}_{t_j}) \geq \alpha_{\dot{q}},\ s_j \leq s \leq s_f, \tag{7d}$$

$$\beta(s) \geq 0,\ s_j \leq s \leq s_f, \tag{7e}$$

$$q = q_e(s),\ s_j \leq s \leq s_f, \tag{7f}$$

where equation (7f) is replaced by

$$\mathcal{E}(T(q_e(s), p_K) \mid \mathcal{A}_{t_j}) = x_e(s),\ s_j \leq s \leq s_f, \tag{7f'}$$

if the prescribed path is given in work space.

Remark 3. Since $\dot{q}_0 = q_e'(s_0)\sqrt{\beta(s_0)} = 0$ the simpler condition $\beta(s_0) = 0$ is used in (7b) at the initial stage.

The corresponding substitute problem for (PTP) is given by (SPTP):

$$\min_{\beta(.), q_e(.)} \int_{s_j}^{s_f} \mathcal{E}(f_0(s, q_e(s), q_e'(s), q_e''(s), \beta(s), \beta'(s), p_J) \mid \mathcal{A}_{t_j})\, ds \tag{8a}$$

subject to

$$q_e'(s_j)\sqrt{\beta(s_j)} = \dot{q}_j, \beta(s_f) = 0, \tag{8b}$$

$$P(u_i^{min} \leq a_i\beta'(s) + b_i\beta(s) + c_i \leq u_i^{max} \mid \mathcal{A}_{t_j}) \geq \alpha_{u_i},$$
$$s_j \leq s \leq s_f,\ i = 1, 2, \ldots, n, \tag{8c}$$

$$P(q^{min} \leq q_e(s) \leq q^{max} \mid \mathcal{A}_{t_j}) \geq \alpha_q,\ s_j \leq s \leq s_f, \tag{8d}$$

$$P(\dot{q}^{min} \leq q_e'(s)\sqrt{\beta(s)} \leq \dot{q}^{max} \mid \mathcal{A}_{t_j}) \geq \alpha_{\dot{q}},\ s_j \leq s \leq s_f, \tag{8e}$$

$$\beta(s) \geq 0,\ s_j \leq s \leq s_f, \tag{8f}$$

$$q_e(s_j) = q_j,\ q_e(s_f) = q_f, \tag{8g}$$

where equation (8g) is replaced by

$$\mathcal{E}(T(q_e(s_j), p_K) \mid \mathcal{A}_{t_j}) = x_e(s_j),\ \mathcal{E}(T(q_e(s_f), p_K) \mid \mathcal{A}_{t_j}) = x_e(s_f), \tag{8g'}$$

if the initial and terminal position are given in work space. Furthermore, $\alpha_q, \alpha_{\dot{q}}$ are prescribed minimum probabilities for the separated state chance constraints.

Let $\beta^{(j)} = \beta^{(j)}(s)$ in case of (SPP) and $(\beta^{(j)}, q_e^{(j)}) = (\beta^{(j)}(s), q_e^{(j)}(s))$ in case of (SPTP), denote the optimal solution at stage j. Finally, the initial conditions in (7a-f) and (8a-g) can be defined recursively as follows:

a) position q

$$q = q_e^{(j-1)}(s_j) \text{ or } q = \hat{q}(t_j),\ j = 1, 2, \dots, \tag{9}$$

b) joint velocity $\dot{q}_j$

$$\dot{q} = q_e^{(j-1)\prime}(s_j)\sqrt{\beta^{(j-1)}(s_j)} \text{ or } \dot{q} = \hat{\dot{q}}(t_j),\ j = 1, 2, \dots, \tag{10}$$

c) path parameter s_j: See Section 1.5.

In (9),(10) $\hat{q}(t_j), \hat{\dot{q}}(t_j)$ denotes an observation of $q(t_j), \dot{q}(t_j)$, respectively.

1.4 A Path Parameter Transformation

In order to work on a fixed parameter interval $[\tilde{s}_0, \tilde{s}_f]$, $\tilde{s}_0 < \tilde{s}_f$, let

$$s^{[j]} = \frac{s_f - s_j}{\tilde{s}_f - \tilde{s}_0}, \tag{11}$$

and consider the second parameter transformation [14, 15],

$$s = s(\tilde{s}) = s_j + s^{[j]} \cdot (\tilde{s} - \tilde{s}_0),\ \tilde{s}_0 \le \tilde{s} \le \tilde{s}_f \tag{12}$$

having the inverse transformation

$$\tilde{s} = \tilde{s}(s) = \tilde{s}_0 + \frac{1}{s^{[j]}} \cdot (s - s_j),\ s_j \le s \le s_f. \tag{13}$$

At every stage $j = 0, 1, 2, \ldots$, problem (7a-f) can then be replaced by:

$$\min_{\tilde{\beta}(.)} \int_{\tilde{s}_0}^{\tilde{s}_f} \mathcal{E}\Big(f_0\Big(\tilde{s}, \tilde{q}_e(\tilde{s}), \tilde{q}_e'(\tilde{s})\frac{1}{s^{[j]}}, \tilde{q}_e''(\tilde{s})\left(\frac{1}{s^{[j]}}\right)^2,$$

$$\tilde{\beta}(\tilde{s}), \tilde{\beta}'(\tilde{s})\frac{1}{s^{[j]}}, p_J\Big) \mid \mathcal{A}_{t_j}\Big) \cdot s^{[j]}\, d\tilde{s} \tag{14a}$$

subject to

$$\tilde{q}_e'(\tilde{s}_0)\frac{1}{s^{[j]}}\sqrt{\tilde{\beta}(\tilde{s}_0)} = \dot{q}_j, \tilde{\beta}(\tilde{s}_f) = 0, \tag{14b}$$

$$P(u_i^{\min} \le \tilde{a}_i\tilde{\beta}'(\tilde{s})\frac{1}{s^{[j]}} + \tilde{b}_i\tilde{\beta}(\tilde{s}) + \tilde{c}_i \le u_i^{\max}|\mathcal{A}_{t_j}) \ge \alpha_{u_i},$$

$$\tilde{s}_0 \le \tilde{s} \le \tilde{s}_f,\ i = 1, 2, \ldots, n, \tag{14c}$$

$$P(\dot{q}^{\min} \le \tilde{q}_e'(\tilde{s})\frac{1}{s^{[j]}}\sqrt{\tilde{\beta}(\tilde{s})} \le \dot{q}^{\max} \mid \mathcal{A}_{t_j}) \ge \alpha_{\dot{q}},\ \tilde{s}_0 \le \tilde{s} \le \tilde{s}_f, \tag{14d}$$

$$\tilde{\beta}(\tilde{s}) \ge 0,\ \tilde{s}_0 \le \tilde{s} \le \tilde{s}_f \tag{14e}$$

$$q = \tilde{q}_e(\tilde{s}),\ \tilde{s}_0 \le \tilde{s} \le \tilde{s}_f, \tag{14f}$$

where $\tilde{q}_e(\tilde{s}) = q_e(s(\tilde{s}))$, $\tilde{\beta}(\tilde{s}) = \beta(s(\tilde{s}))$ and $\tilde{a}_i = a_i(\tilde{q}_e, \tilde{q}_e', s^{[j]}, p_D)$, $\tilde{b}_i = b_i(\tilde{q}_e, \tilde{q}_e', \tilde{q}_e'', s^{[j]}, p_D)$, $\tilde{c}_i = c_i(\tilde{q}_e, s^{[j]}, p_D)$.

Similarly, the second path parameter transformation can be applied to (8a–g), yielding also a fixed parameter interval $[\tilde{s}_0, \tilde{s}_f]$.

1.5 Time-Path Parameter Transformation

According to Section 1.1, the velocity profile in Optimal Trajectory Planning is defined [9, 23] as

$$\beta(s) = \dot{s}^2(t(s)). \tag{15}$$

Hence, for calculating the time-path parameter transformation at stage j, the following initial value problem has to be solved:

$$\dot{s}(t) = \sqrt{\beta^{(j)}(s(t))},\ t \ge t_j, \tag{16a}$$

$$s(t_j) = s_j, \tag{16b}$$

where $s = s(t)$ is a strictly increasing function.

Let

$$s^{(j)} = s^{(j)}(t),\ t_j \le t \le t_f^{(j)}, \tag{17}$$

denote the unique solution of (16a,b) on the remaining time interval $[t_j, t_f] = [t_j, t_f^{(j)}]$. The terminal time-point $t_f = t_f^{(j)}$, valid for the j-th stage of the process, is the unique solution of the equation

$$s^{(j)}(t_f^{(j)}) = s_f. \tag{18}$$

Finally, the initial path parameters s_j, $j = 1, 2, \ldots$, can be given recursively as follows:

$$s_j := s^{(j-1)}(t_j),\ j = 1, 2, \ldots. \tag{19}$$

2 Numerical Solution of the Substitute Problems

2.1 Numerical Calculation of the Expected Objective Function

Let

$$\tilde{f}_0(s, p_J) := f_0(s, q_e(s), q_e'(s), q_e''(s), \beta(s), \beta'(s), p_J) \tag{20}$$

denote the objective function in the Optimal Stochastic Trajectory Planning problem.

Hence, using a Taylor series expansion of $\tilde{f}_0(s, p_J)$ with respect to the stochastic parameter p_J at $\bar{p}_J^{(j)} = \mathcal{E}(p_J(\omega)|\mathcal{A}_{t_j})$, we are able to approximate the expected objective function by

$$\mathcal{E}(\tilde{f}_0(s, p_J)|\mathcal{A}_{t_j}) \approx \tilde{f}_0(s, \bar{p}_J^{(j)}) + \sum_{k=1}^{K} \frac{1}{k!} \frac{\partial^k \tilde{f}_0}{\partial p_J^k}(s, \bar{p}_J^{(j)}) \cdot \mu_k^{(j)}, \tag{21}$$

where $\mu_k^{(j)}$ is the system of the k-th conditional central moments.

Remark 4. Any other expectations which are involved in the substitute problem (SPP) or (SPTP) can be approximated using the same method.

If $\tilde{f}_0(s, p_J)$ is at least (K+1)-times continuous differentiable subject to p_J, the error in the Taylor expansion of $\tilde{f}_0(s, p_J)$ with respect to p_J at $\bar{p}_J^{(j)}$ can be evaluated as:

$$\frac{1}{(K+1)!} \cdot \frac{\partial^{K+1}}{\partial p_J^{K+1}} \tilde{f}_0(s, \hat{p}_J)(p_J - \bar{p}_J^{(j)})^{K+1}, \tag{22}$$

where $\hat{p}_J$ denotes a value between p_J and $\bar{p}_J^{(j)}$. Now, if the system of the $(K+1)$ derivatives is uniformly bounded on the support of $p_J(\omega)$, that is

$$\|\frac{\partial^{K+1}}{\partial p_J^{K+1}} \tilde{f}_0(s, p_J)\| \leq M_0 \text{ a.s.(almost sure)},\ s_0 \leq s \leq s_f, \tag{23}$$

with a fixed bound M_0, the expected approximation error is less than

$$\frac{M_0}{(K+1)!} \cdot \mathcal{E}(\|p_J - \bar{p}_J^{(j)}\|^{K+1}|\mathcal{A}_{t_j}). \tag{24}$$

Under the standard assumption in robotics, that the parameter p_J is nested in a bounded domain, we finally get as upper bound for the approximation error:

$$\frac{M_0 \rho^{K+1}}{(K+1)!}, \tag{25}$$

where we used $\|p_J - \bar{p}_J^{(j)}\| \leq \rho$ with probability 1. Hence, for $\rho < 1$, small values of K already provide good approximations.

2.2 Numerical Evaluation of the Probability Constraints

Assume first that only a single dynamic parameter $\tilde{p}_D$ of p_D is modelled stochastic. Then, evaluating, e.g., the probabilistic control constraint in (7a-f) and (8a-g), we observe that due to the linear parameterisation property of robots [21,28], the coefficients $a_i(q_e, q_e', p_D)$, $b_i(q_e, q_e', q_e'', p_D)$ and $c_i(q_e, p_D)$, $i = 1, \dots, n$, resulting from the t-s-transformation, can be represented by

$$a_i(q_e, q_e', p_D) = a_{i0}(q_e, q_e', \bar{p}_D) + \tilde{p}_D(\omega) a_{i1}(q_e, q_e'), \tag{26a}$$
$$b_i(q_e, q_e', q_e'', p_D) = b_{i0}(q_e, q_e', q_e'', \bar{p}_D) + \tilde{p}_D(\omega) b_{i1}(q_e, q_e', q_e''), \tag{26b}$$
$$c_i(q_e, p_D) = c_{i0}(q_e, \bar{p}_D) + \tilde{p}_D(\omega) c_{i1}(q_e), \tag{26c}$$

where $\tilde{p}_D = \tilde{p}_D(\omega)$ is the stochastic dynamic parameter and $\bar{p}_D$ contains the remaining deterministic dynamic parameters. Hence, introducing deterministic functions

$$u_{i0} := a_{i0}(q_e, q_e', \bar{p}_D)\beta' + b_{i0}(q_e, q_e', q_e'', \bar{p}_D)\beta + c_{i0}(q_e, \bar{p}_D), \tag{27a}$$
$$u_{i1} := a_{i1}(q_e, q_e')\beta' + b_{i1}(q_e, q_e', q_e'')\beta + c_{i1}(q_e), \tag{27b}$$

we get

$$u_i = u_{i0} + \tilde{p}_D(\omega) u_{i1}, \quad i = 1, 2, \dots, n, \tag{28}$$

and the separated chance constraints (7c) and (8c) may be represented by

$$P(u_i^{\min} \le u_{i0} + \tilde{p}_D(\omega) u_{i1} \le u_i^{\max} \mid \mathcal{A}_{t_j}) \ge \alpha_{u_i}, \quad i = 1, 2, \dots, n. \tag{29}$$

Splitting up (29) for further simplification into the one-sided inequalities

$$P(u_{i0} + \tilde{p}_D(\omega) u_{i1} \le u_i^{\max} \mid \mathcal{A}_{t_j}) \ge \alpha_{u_i}, \quad i = 1, 2, \dots, n, \tag{30a}$$
$$P(u_i^{\min} \le u_{i0} + \tilde{p}_D(\omega) u_{i1} \mid \mathcal{A}_{t_j}) \ge \alpha_{u_i}, \quad i = 1, 2, \dots, n, \tag{30b}$$

we find the following conditions, where

$$F_{\tilde{p}_D}^{(t_j)}(z) := P(\tilde{p}_D(\omega) \le z \mid \mathcal{A}_{t_j}), \quad z \in \mathcal{R}, \tag{31}$$

denotes the conditional distribution function of the random parameter $\tilde{p}_D$ given $\mathcal{A}_{t_j}$, and we suppose that the bounds $u_i^{\min}$ and $u_i^{\max}$, $i = 1, \dots, n$, are fixed:

$$u_i^{\min} \le u_{i0} + p_D^{(t_j,1)} u_{i1}, \quad i = 1, \dots, n, \tag{32a}$$
$$u_{i0} + p_D^{(t_j,2)} u_{i1} \le u_i^{\max}, \quad i = 1, \dots, n. \tag{32b}$$

Here, $p_D^{(t_j,1)}$ and $p_D^{(t_j,2)}$ are defined as:

$$p_D^{(t_j,1)} = p_D^{(t_j,1)}(u_{i1}) = \begin{cases} F_{\tilde{p}_D}^{(t_j)^{-1}}(1 - \alpha_{u_i}), & u_{i1} > 0, \\ 0, & u_{i1} = 0, \\ F_{\tilde{p}_D}^{(t_j)^{-1}}(\alpha_{u_i}), & u_{i1} < 0, \end{cases} \tag{33a}$$

$$p_D^{(t_j,2)} = p_D^{(t_j,2)}(u_{i1}) = \begin{cases} F_{\tilde{p}_D}^{(t_j)^{-1}}(\alpha_{u_i}), & u_{i1} > 0, \\ 0, & u_{i1} = 0, \\ F_{\tilde{p}_D}^{(t_j)^{-1}}(1 - \alpha_{u_i}), & u_{i1} < 0. \end{cases} \tag{33b}$$

For a more general case, represent u_i, cf. (26a-c) by

$$u_i = u_{io} + (U\tilde{p}_D(\omega))_i,\ i = 1,\dots,n, \tag{34}$$

where $\tilde{p}_D$ is the vector of random dynamic parameters and $U = U(q_e, q_e', q_e'')$ is a deterministic matrix (linear parametrisation property of robots); moreover, suppose in the following that $u_i^{\min} = -u_i^{\max}$. Using a well known Chebycheff type probability inequality, see [11], the conditional probability in (7c) or (8c) can be approximated from below as follows:

$$\begin{aligned} & P(u_i^{\min} \le u_{io} + (U\tilde{p}_D(\omega))_i \le u_i^{\max} \mid \mathcal{A}_{t_j}) \\ &= P(|u_{io} + (U\tilde{p}_D(\omega))_i| \le u_i^{\max} \mid \mathcal{A}_{t_j}) \\ &= 1 - P(|u_{io} + (U\tilde{p}_D(\omega))_i| > u_i^{\max} \mid \mathcal{A}_{t_j}) \\ &\ge 1 - \frac{\mathcal{E}(g(u_{io} + (U\tilde{p}_D(\omega))_i) \mid \mathcal{A}_{t_j})}{g(u_i^{\max})},\ i = 1,\dots,n, \end{aligned} \tag{35}$$

where $g = g(t)$ is an arbitrary positive function such that $g(-t) = g(t)$, and $g(t)$ is nondecreasing on the interval $[0, +\infty)$. Appropriate examples are $g(t) = |t|^p$ for $p \ge 1$. The chance constraint (29) may be guaranteed then by the condition

$$\mathcal{E}(g(u_{io} + (U\tilde{p}_D(\omega))_i) \mid \mathcal{A}_{t_j}) \le (1 - \alpha_{u_i}) \cdot g(u_i^{\max}), \tag{36}$$

where the expectation can be calculated, in the general case, by the methods described in Section 2.1.

2.3 *Approximation of the Solution Curves by Spline-Functions*

Approximating the solution $\beta^{(j)} = \beta^{(j)}(s)$ or $(\beta^{(j)}, q_{ei}^{(j)}) = (\beta^{(j)}(s), q_{ei}^{(j)}(s))$, $s_j \le s \le s_f$, $i = 1, 2, \dots, n$, of (SPP) or (SPTP) by certain linear combinations of known basis functions B_k^β, $k = 1, \dots, K_\beta$, and B_k^q, $k = 1, \dots, K_q$, they can be represented by

$$\beta^{(j)}(s) = \sum_{k=1}^{K_\beta} \gamma_k^\beta B_k^\beta(s),\ s_j \le s \le s_f, \tag{37a}$$

$$q_{ei}^{(j)}(s) = \sum_{k=1}^{K_q} \gamma_k^{q_i} B_k^q(s),\ s_j \le s \le s_f, i = 1, 2, \dots, n, \tag{37b}$$

with unknown coefficients γ_k^β, $1 \le k \le K_\beta$, and $\gamma_k^{q_i}$, $1 \le k \le K_q$, $i = 1, 2, \dots, n$, where the representation (37a,b) may depend on the stage j.

Hence, we can reduce the Stochastic Trajectory Planning Problem to a parameter optimisation problem of following type:

$$\min_{\gamma} F(\gamma) \tag{38a}$$

subject to

$$h(\gamma) = 0, \tag{38b}$$

$$g(s,\gamma) \leq 0, \ s_j \leq s \leq s_f, \tag{38c}$$

where γ is the vector of all unknown coefficients in linear combinations (37a,b) and $F(\gamma)$, $h(\gamma)$ and $g(s,\gamma)$ are obtained by substituting (37a,b) into (SPP) or (SPTP). Since $g = g(s,\gamma)$ depends on the path parameter s, (38a-c) defines a semi-infinite optimisation problem [7].

If we choose a finite number of path parameters $s_1, s_2, \ldots, s_N$ in $[s_j, s_f]$ and demand that the equation $g(s,\gamma) \leq 0$ is fulfilled in all s_r, $r = 1, 2, \ldots, N$, problem (38a-c) can be approximated by an ordinary finite parameter optimisation problem under equality and inequality constraints.

The solution of the problem therefore depends on the selection of the basis functions B_k^{β}, $k = 1, \ldots, K_{\beta}$, and B_k^q, $k = 1, \ldots, K_q$. Due to their well known good properties, we use spline basis functions [4,27].

3 Neural Network Approximation

In Sections 1 and 2 we have seen, how Adaptive Optimal Stochastic Trajectory Planning problems can be modelled and how they could be solved using nonlinear programming techniques. Since industrial robot applications demand, due to the short cycle times, real-time processing ability, the methods described in the previous section can not be used directly as they are computationally too expensive. However, since the substitute problem at every stage j only depends on

i) s_j, the initial path parameter,
ii) $\mathcal{A}_{t_j}$, the available information about the parameter distribution, represented, e.g., by certain conditional moments of the random model parameters arising in the computation of conditional expectations by Taylor expansions, see (21) and (35),(36),
iii) q_j, the initial position in configuration space,
iv) $\dot{q}_j$, the initial joint velocities,

we can solve (7a-f) or (8a-g) for a large amount of inputs $\zeta := (s_j, \mathcal{A}_{t_j}, q_j, \dot{q}_j)^T$. By collecting these solutions we get large data sets $(\zeta, \gamma^*)_\pi$, $\pi = 1, \ldots, \Pi$, where $\gamma^* = (\gamma_k^{\beta^*}, \gamma_l^{q_1{}^*}, \ldots, \gamma_l^{q_n{}^*})^T$, $k = 1, \ldots, K_\beta, l = 1, \ldots, K_q$, denotes a set of optimal spline coefficients for the input ζ calculated by means of nonlinear programming:

$$(\zeta, \gamma^*)_\pi = \begin{pmatrix} (s_j, \mathcal{A}_{t_j}, q_j, \dot{q}_j)^T \\ (\gamma_1^{\beta^*}, \ldots, \gamma_{K_\beta}^{\beta^*}, \gamma_1^{q_1{}^*}, \ldots, \gamma_1^{q_n{}^*}, \ldots, \gamma_{K_q}^{q_1{}^*}, \ldots, \gamma_{K_q}^{q_n{}^*})^T \end{pmatrix}_\pi, \tag{39}$$

$\pi = 1, \ldots, \Pi$. Splitting up this data set into $K_\beta + n \cdot K_q$ sets

$$(\zeta, \gamma_1^{\beta^*})_\pi, \ldots, (\zeta, \gamma_{K_\beta}^{\beta^*})_\pi, \pi = 1, \ldots, \Pi, \tag{40a}$$

$$(\zeta, \gamma_1^{q_i^*})_\pi, \ldots, (\zeta, \gamma_{K_q}^{q_i^*})_\pi, \pi = 1, \ldots, \Pi, i = 1, \ldots, n, \tag{40b}$$

we get a pattern set for each unknown spline coefficient in linear combination (37a,b). By means of a Neural Network Simulator for the training process, we can finally interpolate the solution coefficients $\gamma^* = \gamma^*(\zeta)$, for inputs ζ not included in the pattern sets using its approximation $\hat{\gamma} = (\hat{\gamma}_k^\beta, \hat{\gamma}_l^{q_1}, \ldots, \hat{\gamma}_l^{q_n})^T$, $k = 1, \ldots, K_\beta, l = 1, \ldots, K_q$. In detail we get:

$$\gamma_k^{\beta^*} \approx \hat{\gamma}_k^\beta = \hat{\gamma}_k^\beta(\zeta, w_1, \ldots, w_r), k = 1, \ldots, K_\beta, \tag{41a}$$

$$\gamma_k^{q_i^*} \approx \hat{\gamma}_k^{q_i} = \hat{\gamma}_k^{q_i}(\zeta, w_1, \ldots, w_r), k = 1, \ldots, K_q, i = 1, \ldots, n, \tag{41b}$$

where the w_d, $d = 1, \ldots, r$, denote the weights of the trained neural networks. In general optimal weights can be obtained by minimising the summed squared error $E(w_1, \ldots, w_r)$ over all training patterns for the neural net subject to its weights $w_1, \ldots, w_r$.

Hence, if γ_π^* denotes the optimal coefficients related to the input ζ_π, $\pi = 1, \ldots, \Pi$, obtained by optimal stochastic trajectory planning and

$$\hat{\gamma} = \hat{\gamma}(\zeta_\pi, w_1, \ldots, w_r)$$

represents the coefficients given by the neural network, we minimise

$$E(w_1, \ldots, w_r) = \sum_{\pi=1}^{\Pi} \|\gamma_\pi^* - \hat{\gamma}(\zeta_\pi, w_1, \ldots, w_r)\|^2. \tag{42}$$

For a more detailed description of Neural Nets and their training we refer to [1,2,6, 29].

Remark 5. By means of B-spline basis functions, the necessary amount of neural nets can be reduced to $(K_\beta - 2) + n \cdot (K_q - 2)$, since in this case at every stage j we put $B_1^\beta(s_j) = B_1^q(s_j) = 1$ and $B_{K_\beta}^\beta(s_f) = B_{K_q}^q(s_f) = 1$, cf. [4]. Hence, the initial and terminal conditions for the geometric path and velocity profile can be guaranteed by choosing $\gamma_1^\beta = \beta^{(j-1)}(s_j), \gamma_{K_\beta}^\beta = 0$ and $\gamma_1^{q_i} = (q_j)_i, \gamma_{K_q}^{q_i} = (q_f)_i$, $i = 1, \ldots, n$, in case of a continuous stage transition. In case of a transition based on observations in (9, 10) the same changes can be made accordingly.

4 Numerical Examples

4.1 *Industrial Robot Manutec r3*

To test the previous theoretical models we consider a 6 joint Manutec r3 industrial robot. Since the first 3 joints are mainly responsible for placing the end-effector in workspace and the last 3 joints for orientating the end-effector we used a simplified 3 d.o.f. model. In this model the last joints remain in a fixed locked position. A detailed mathematical model for the Manutec r3 can be obtained from [28]. Furthermore, we suppose that the payload mass m_l is the only uncertain parameter. Therefore, we get at stage j for (SPP) the following restrictions:

$$P(-7.5 \le u_i(s) \le 7.5 \mid \mathcal{A}_{t_j}) \ge \alpha_{u_i},\ i=1,2,3,\ s_j \le s \le s_f, \tag{43}$$

$$-3.0\ \tfrac{\text{rad}}{\text{sec}} \le q_1{}'(s)\sqrt{\beta(s)} \le 3.0\ \tfrac{\text{rad}}{\text{sec}},\ s_j \le s \le s_f, \tag{44a}$$

$$-1.5\ \tfrac{\text{rad}}{\text{sec}} \le q_2{}'(s)\sqrt{\beta(s)} \le 1.5\ \tfrac{\text{rad}}{\text{sec}},\ s_j \le s \le s_f, \tag{44b}$$

$$-5.2\ \tfrac{\text{rad}}{\text{sec}} \le q_3{}'(s)\sqrt{\beta(s)} \le 5.2\ \tfrac{\text{rad}}{\text{sec}},\ s_j \le s \le s_f, \tag{44c}$$

where $\alpha_{u_i} = 0.99$, $i = 1,2,3$. For (SPTP) additionally the following position constraints have to be fulfilled:

$$-2.97\ \text{rad} \le q_1(s) \le 2.97\ \text{rad},\ s_j \le s \le s_f, \tag{45a}$$

$$-2.01\ \text{rad} \le q_2(s) \le 2.01\ \text{rad},\ s_j \le s \le s_f, \tag{45b}$$

$$-2.86\ \text{rad} \le q_3(s) \le 2.86\ \text{rad},\ s_j \le s \le s_f. \tag{45c}$$

4.2 *Probability Distributions*

To study the influence of the probability distribution on (SPP) or (SPTP), consider without loss of generality only the stage $j = 0$ and suppose that the stochastic payload mass can be modelled by an uniform, an exponential or a Gaussian distribution. In order to compare the different distributions, we assume that the payload mass has an expectation of $\mathcal{E}(m_l) = 5$ kg and a variance of $V(m_l) = 25$. Additionally, the case of a deterministic payload mass of 5 kg was considered. Hence for a reliability $\alpha_{u_i} = 0.99$, $i = 1,2,3$, we get according to (32a,b) the following deterministic substitutes for control constraint (7c) or (8c):

$$u_i^{\min} \le u_{i0} + p_D^{(t_j,1)} u_{i1},\ i = 1,\dots,n, \tag{46a}$$

$$u_{i0} + p_D^{(t_j,2)} u_{i1} \le u_i^{\max},\ i = 1,\dots,n, \tag{46b}$$

where for the different distributions $p_D^{(t_j,1)}$ and $p_D^{(t_j,2)}$ read as follows, cf. (33a,b):

a) uniform distribution

$$p_D^{(t_j,1)} = \begin{cases} 2(1-\alpha_{u_i})\sqrt{3V(m_l)} + \mathcal{E}(m_l) - \sqrt{3V(m_l)}\,, & u_{i1} > 0, \\ = -3.4870 & \\ 0 & u_{i1} = 0, \\ 2\alpha_{u_i}\sqrt{3V(m_l)} + \mathcal{E}(m_l) - \sqrt{3V(m_l)} & u_{i1} < 0 \\ = 13.4870, & \end{cases} \tag{47a}$$

$$p_D^{(t_j,2)} = \begin{cases} 13.4870 & , u_{i1} > 0, \\ 0 & , u_{i1} = 0, \\ -3.4870 & , u_{i1} < 0. \end{cases} \tag{47b}$$

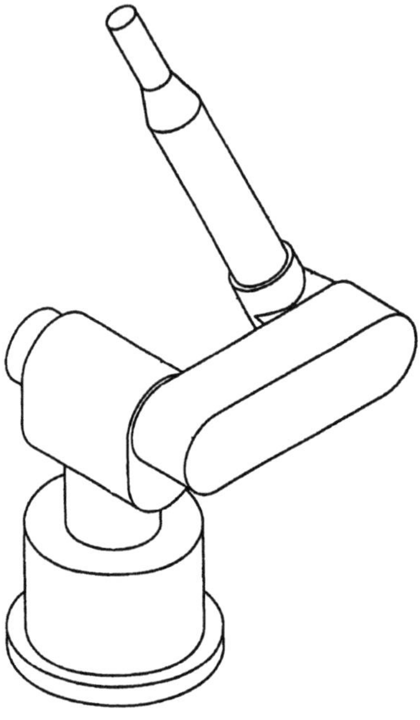

Figure 1. Manutec r3

b) exponential distribution

$$p_D^{(t_j,1)} = \begin{cases} -\ln(\alpha_{u_i}) \cdot \mathcal{E}(m_l) = 0.0502 & , u_{i1} > 0, \\ 0 & , u_{i1} = 0, \\ -\ln(1-\alpha_{u_i}) \cdot \mathcal{E}(m_l) = 23.0258 & , u_{i1} < 0, \end{cases} \tag{48a}$$

$$p_D^{(t_j,2)} = \begin{cases} 23.0258 & , u_{i1} > 0, \\ 0 & , u_{i1} = 0, \\ 0.0502 & , u_{i1} < 0. \end{cases} \tag{48b}$$

c) Gaussian distribution

$$p_D^{(t_j,1)} = \begin{cases} F_{p_D}^{(t_j)^{-1}}(1-\alpha_{u_i}) = -6.6317 & , u_{i1} > 0, \\ 0 & , u_{i1} = 0, \\ F_{p_D}^{(t_j)^{-1}}(\alpha_{u_i}) = 16.6317 & , u_{i1} < 0, \end{cases} \tag{49a}$$

$$p_D^{(t_j,2)} = \begin{cases} 16.6317 & , u_{i1} > 0, \\ 0 & , u_{i1} = 0, \\ -6.6317 & , u_{i1} < 0. \end{cases} \tag{49b}$$

Here, in case of the uniform and the exponential distribution, the inverse conditional distribution function can easily be computed analytically. For the Gaussian distribution it was calculated numerically.

Remark 6. Though in reality not possible, in case of the uniform and Gaussian distribution negative payload masses are included here in the modelling of the probabilistic constraints. However, the main goal in this example was to compare the different distributions. Hence, we assumed the same expectation and variance for all three cases, which led to this results. For applications its is of course more favourable to use a cut off uniform or Gaussian distribution.

Now, in case of (SPTP), the robot should perform a time-optimal movement between initial position

$$q_0 = (-2.4, 1.2, 0.6)^T$$

and terminal position

$$q_f = (-1.3, 0.2, -1.0)^T,$$

where the SQP solver SNOPT [5] was used to solve (SPTP) on a SUN Ultra-Sparc II with a 200 MHz processor.

Distribution	t_f [sec]	**CPU-time** [sec]
deterministic	0.7857	5.25
uniform	0.9113	4.74
exponential	1.2049	9.97
Gaussian	1.0026	4.32

Table 1. Run-time and CPU-time

Table 1 clearly indicates, that the solution of (8a-g) computed by mathematical programming can not be used directly in the on-line control process of the robot, since the time to solve (SPTP) is much longer than the final time t_f of a optimal control. Thus, no adaptions with new information about the payload mass uncertainty are possible without much faster (approximate) solution procedures. A way out is presented in Section 4.3.

Now, using inverse dynamics, that is, substituting a calculated optimal velocity profile and geometric path in configuration-space into the dynamic equation of the robot, also allows the computation of the related optimal feed-forward controls. Finally, applying the inverse time-path parameter transformation, obtained by any ODE-solver (see Section 1.5), the optimal solution, e.g., the third joint, can be plotted in time-domain. Additionally, by means of the robots kinematic equation the optimal trajectories in workspace are given.

In case of an exponential distribution the most cautious controls are obtained, see Table 1. Comparing, e.g., the support of the uniform and exponential distribution, we have that:

$$\begin{aligned} \mathrm{supp}_{\mathrm{uni}}(m_l) &= [\mathcal{E}(m_l) - \Delta m_l, \mathcal{E}(m_l) + \Delta m_l] \\ &= [5 \cdot (1 - \sqrt{3}), 5 \cdot (1 + \sqrt{3})], \end{aligned} \tag{50}$$

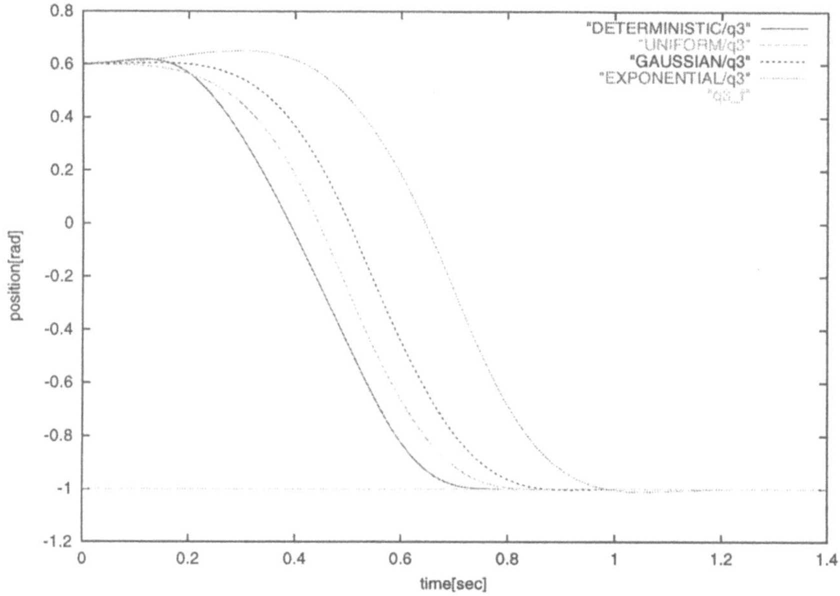

Figure 2. Position $q_3(t)$ of the third joint

where $\Delta m_l = \sqrt{3 \cdot V(m_l)}$, and

$$\text{supp}_{\text{exp}}(m_l) = \mathcal{R}^+. \tag{51}$$

Hence, in case of the exponential distribution a payload mass greater than $5\cdot(1+\sqrt{3})$ contributes still with positive probability to the calculation of the optimal control. Additionally, examining the distribution function

$$F_{m_l}^{\text{exp}}(z) = P(m_l \leq z) = 1 - e^{-\lambda z}, \text{ for all } z \in \text{supp}_{\text{exp}}(m_l), \tag{52}$$

where here $\lambda = 0.2$, we find that the payload can exceed the bound with probability

$$\begin{aligned} P(m_l \geq 5 \cdot (1+\sqrt{3})) &= 1 - P(m_l \leq 5 \cdot (1+\sqrt{3})) \\ &= 1 - F_{m_l}^{\text{exp}}(5 \cdot (1+\sqrt{3})) \approx 0.065. \end{aligned} \tag{53}$$

Thus, the exponential distribution incorporates a much wider range of possible payload masses (given that expectation and variance are equal for both distributions) and therefore we get a greater final time and more cautious controls (see Figure 4). This enables us to preserve the robustness against the bigger payload uncertainty for the exponential distribution. Similar relations between the other distributions can be obtained.

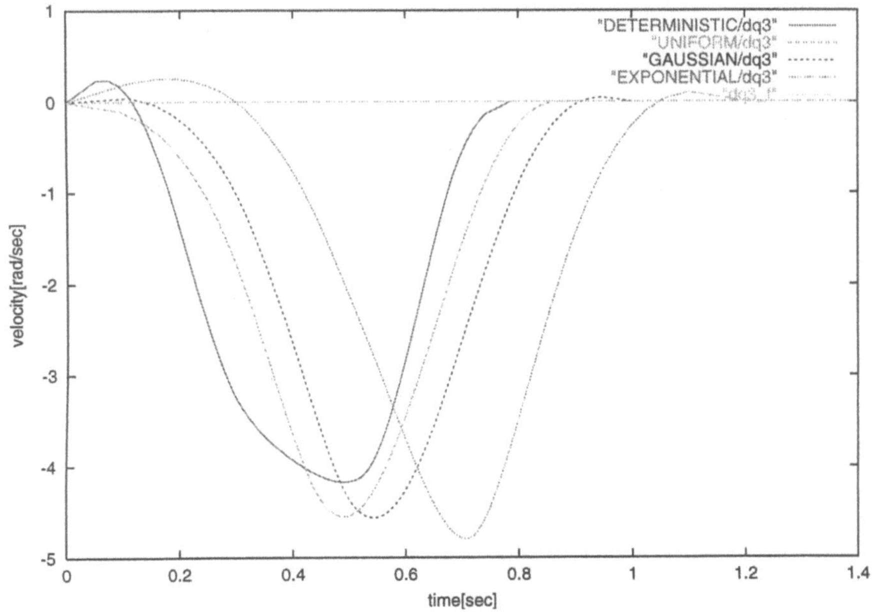

Figure 3. Velocity $\dot{q}_3(t)$ of the third joint

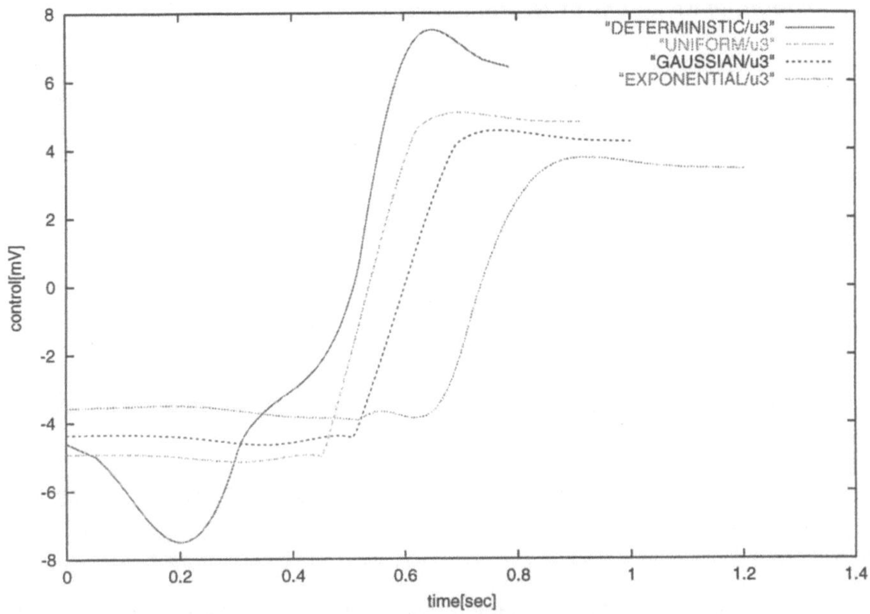

Figure 4. Control $u_3(t)$ of the third joint

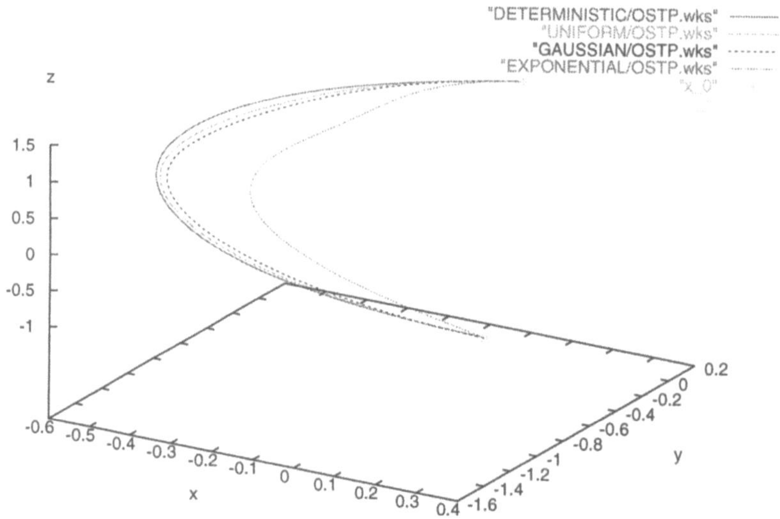

Figure 5. Optimal trajectories in workspace

4.3 *Solutions in Real-Time by Neural Networks*

As already discussed in Section 3, optimal solutions of the substitute problems in real-time may be obtained by means of Neural Networks Approximation. Some first results are presented in the following.

Prescribed-Path Problem

Consider the prescribed path in configuration space

$$q_e(s) = \begin{pmatrix} s \\ s^2+1 \\ s^3+2 \end{pmatrix}, \quad 0 \leq s \leq 1, \tag{54}$$

in (7a-f) and assume at stage 0 we have the following initial information for a time-optimal solution:

- uniform distribution on $[m_l - \Delta m_l, m_l + \Delta m_l]$, where $m_l = 7.5$ kg and $\Delta m_l = 4.0$,
- $\alpha_u = 0.99$,

Furthermore, suppose the path is adapted twice at $t_1 = 0.1484$ sec with new information $m_l = 6.0$ kg and $\Delta m_l = 2.0$ and at $t_2 = 0.3615$ with exact information $m_l = 5.0$ kg. Hence, at stage 2 there remains no uncertainty and we arrive at a

STAGE	METHOD	t_j	s_j	t_f [sec]	CPU-time [sec]
0	SNOPT	0.0	0.0	0.9453	0.58
1	SNOPT	0.1484	0.2273	0.9264	0.43
1	Neural Networks	0.1484	0.2273	0.9278	0.02
2	SNOPT	0.3615	0.5949	0.9120	0.20
2	Neural Networks	0.3615	0.5949	0.9182	0.01

Table 2. Run-time and CPU-time for prescribed path problems

deterministic problem. Under this assumption we get the numerical results on a 200 MHz SUN Ultra-Sparc II shown in Figure 2.

Table 2 shows that, due to the reduced complexity of a prescribed path problem compared to a point-to-point problem, the CPU-time for SNOPT is much smaller here (see Table 1), but still not small enough to be applied for on-line control purposes. However, according to Table 2, Neural Network approximation can be calculated about 20 times faster than a solution using the SQP-solver SNOPT. Moreover, to demonstrate the accuracy of the Neural Network approximation, we plot the velocity profile in time domain, where the curves have the following meanings: The solid line in Figure 6 denotes the result for stage 0, obtained off-line by means of SNOPT. Then, for stage 1 and as well for stage 2, the dashed and the dotted lines which coincide almost exactly represent the results obtained from exact computation using SNOPT and by using Neural-Network-Approximation, respectively.

Point-to-Point Problem

For the point-to-point problem consider the initial and terminal positions $q_0 = (0.0, -1.5, 0.0)^T$ and $q_f = (1.0, -1.95, 1.0)^T$. Moreover, a uniform distribution is supposed, where again $m_l = 7.5$, $\Delta m_l = 4.0$ and $\alpha_u = 0.99$. The path is adapted once at $t_1 = 0.2220$ with new information $m_l = 5.0$ and $\Delta m_l = 2.66$. Then the results shown in Table 3 and Figures 7-10 are obtained.

STAGE	METHOD	t_j	s_j	t_f [sec]	CPU-time [sec]
0	SNOPT	0.0	0.0	0.5817	2.340
1	SNOPT	0.2220	0.3339	0.5697	0.440
1	Neural Networks	0.2220	0.3339	0.5712	0.013

Table 3. Run-time and CPU-time for point-to-point problems

Figures 7-10 show again that the Neural Network solution can approximate the optimal path $q(t)$ and the optimal velocity profile $\beta(t)$ with high accuracy. Additionally, according to Table 3, the Neural Network Approximation can be computed much faster than a solution using the SQP-Solver SNOPT. Hence, this approach is useful in the on-line control process of the robot.

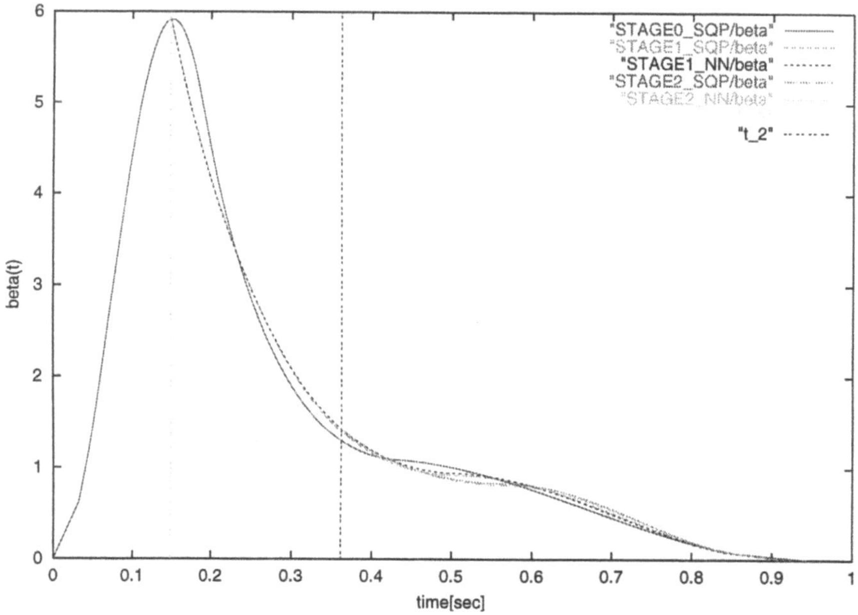

Figure 6. Velocity profile $\beta(t)$

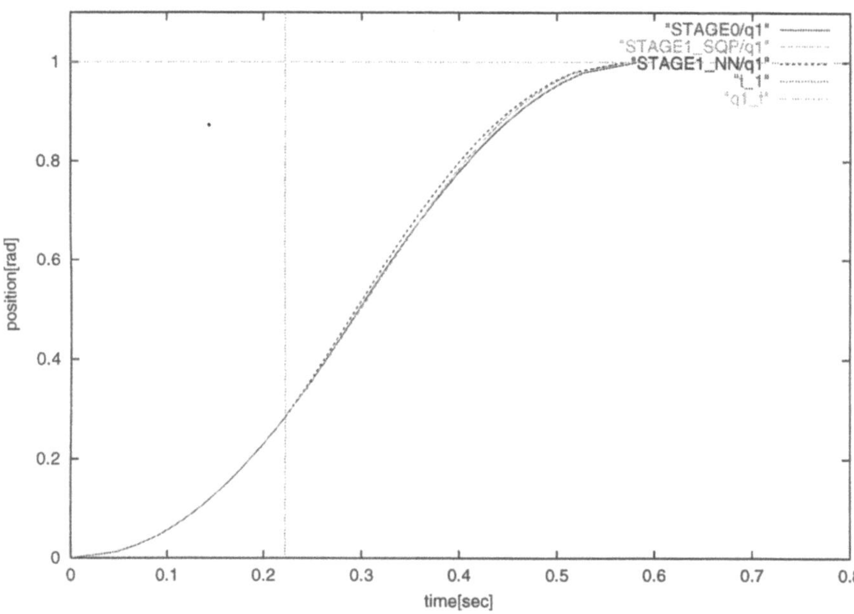

Figure 7. Position $q_1(t)$ of the first joint

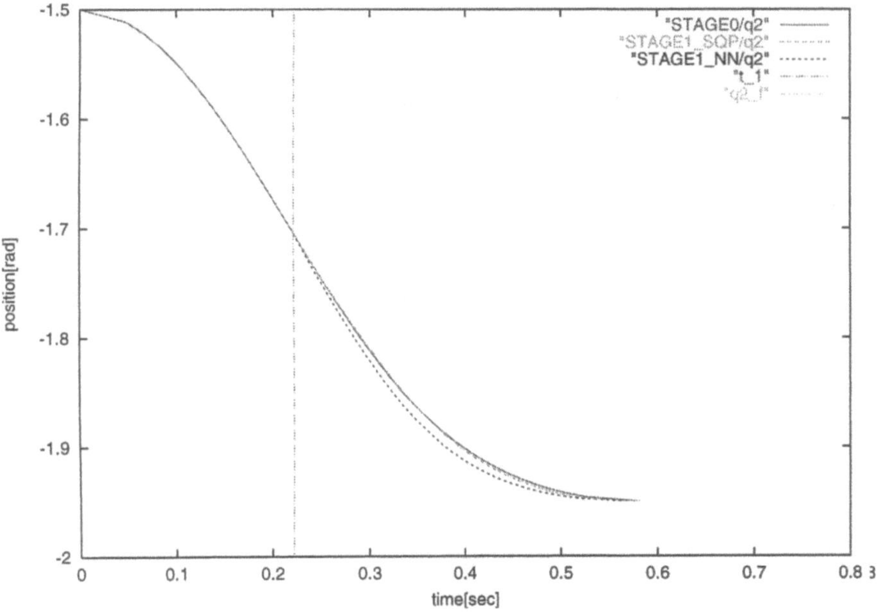

Figure 8. Position $q_2(t)$ of the second joint

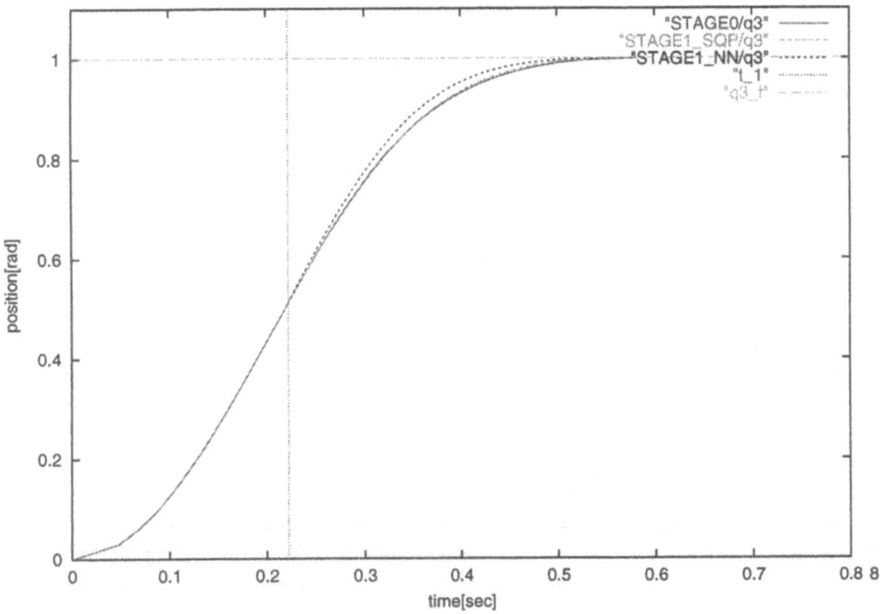

Figure 9. Position $q_3(t)$ of the third joint

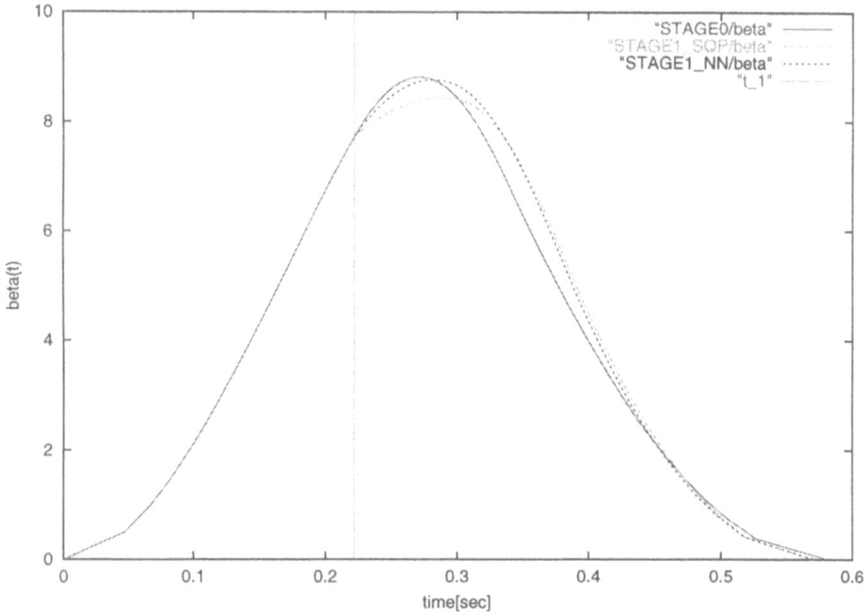

Figure 10. Velocity profile $\beta(t)$

REFERENCES

1. J. A. Anderson, E. Rosenfeld: *Neurocomputing: Foundations of Research*, MIT Press, 1988.
2. G. A. Bekey, K. Y. Goldberg: *Neural Networks in Robotics*, Kluwer Academic Puplishers, 1993.
3. J. E. Bobrow. S. Dubowsky, J. S. Gibson: *Time-optimal Control of Robotic Manipulators*, Int. J. Robot. Res. 4(3), 1985, pp. 3-17.
4. C. de Boor: *A Practical Guide to Splines*, Springer-Verlag, New York, 1978.
5. P. E. Gill, W. Murray, M. A. Saunders: *User's Guide for SNOPT 5.3: A FORTRAN Package for Large-Scale Nonlinear Programming*, University California/Stanford University, March, 1998.
6. R. Hecht-Nielsen: *Neurocomputing*, Addison-Wesley, Reading, MA, 1990.
7. R. Hettich, P. Zencke: *Numerische Methoden der Approximation und semi-infiniten Optimierung*, B.G.Teubner, Stuttgart, 1982.
8. R. Isermann: *Identifikation dynamischer Systeme*, Springer Verlag, Berlin, 1992.
9. R. Johanni: *Optimale Bahnplanung bei Industrierobotern*, VDI Fortschrittberichte, Reihe 18, Nr. 51, Düsseldorf, 1988.
10. P. Kall, St. Wallace: *Stochastic Programming*, Wiley, Chichester, 1995.
11. M. Loeve: *Probability Theory I*, Springer-Verlag, New York-Heidelberg-Berlin 1977.
12. K. Marti: *Approximation stochastischer Optimierungsprobleme* Hain, Königstein/Ts., 1979.
13. K. Marti: *Path Planning for Robots under Stochastic Uncertainty*, Optimization, 45, 1999, pp. 163-195.

14. K. Marti: *Robust Adaptive Control of Robots by Stochastic Optimization Methods*, Proceedings 2nd Int. Conf. on Engineering Computational Technology, 6-8 September 2000, Leuwen, Belgium, Civil-Comp Press 2000.
15. K. Marti: *Adaptive Optimal Stochastic Trajectory Planning and Control (AOSTPC) for Robots*, Preprint 00-21, DFG-priority program 'Real-time Optimization of Large Scale Systems', available for download[1], October 2000.
16. K. Marti, A. Aurnhammer: *Adaptive Stochastic Path Planning and Control (ASPPC)*. In: G. I. Schuëller, P. Kafka (eds.), Safety and Reliability, A. A. Balkeema, Rotterdam 1999, Volume 2, pp. 1609-1614.
17. K. Marti, S. Qu: *Optimal Trajectory Planning for Robot Considering Stochastic Parameters and Disturbances*. In: K. Marti, P. Kall (eds.), Stochastic Programming: Numerical Techniques and Engineering Application, LNEMS, Vol. 423, Springer-Verlag, New York-Berlin-Heidelberg 1995, pp.268-288.
18. K. Marti, S. Qu: *Optimal Trajectory Planning for Robot Considering Stochastic Parameters and Disturbances - Computation of an Efficient Open-Loop Strategy*, J. Intelligent and Robotic Systems 15, 1996, pp. 19-23.
19. K. Marti, S. Qu: *Path Planning for Robots by Stochastic Optimization Methods*, J. Intelligent and Robotic Systems 22, 1998, pp. 117-127.
20. F. Pfeiffer, R. Johanni: *A Concept for Manipulator Trajectory Planning*, IEEE J.Robot. Automat. RA-3(3), 1987, pp. 115-123.
21. F. Pfeiffer, E. Reithmeier: *Roboterdynamik*, B.G.Teubner, Stuttgart, 1987.
22. M. Prüfer: *Reibungsanalyse und Identifikation von Dynamikparametern bei direktangetriebenen und getriebebehafteten Robotern*, Fortschritte in der Robotik 1, Shaker, Aachen, 1996.
23. S. Qu: *Optimale Bahnplanung für Roboter unter Berücksichtigung stochastischer Parameterschwankungen*, VDI Verlag, Düsseldorf, 1995.
24. S. Qu: *Stochastic Trajectory planning and its application for Robot Manutec r3*, In: K. Marti, P. Kall (eds.), Stochastic Programming Methods and Technical Applications, LNEMS, Vol. 458, Springer-Verlag, New York-Heidelberg-Berlin 1998, pp. 382-393.
25. Z. Schiller, S. Dubowsky: *Robot Path Planning with Obstacles, Actuator, Gripper and Payload Constraints*, Int. J. Robot. Res. 8(6),1989, pp. 3-18.
26. K. G. Shin, N. D. McKay: *Minimum-time Control of Robotic Manipulator with Geometric Path Constraints*, IEEE Trans. Automat. Control AC-30(6), 1985, pp. 370-375.
27. L. L. Schumaker: *Spline Functions: Basic Theory*, John Wiley & Sons, New York, 1981.
28. M. Türk: *Zur Modellierung der Dynamik von Robotern mit rotatorischen Gelenken*, VDI-Verlag, Düsseldorf, 1990.
29. A. Zell: *Simulation neuronaler Netze*, Addison-Wesley, Bonn, 1996.

[1] `http://www.zib.de/dfg-echtzeit`

Stochastic Optimization Methods in Robust Adaptive Control of Robots

Kurt Marti

Institut für Mathematik und Informatik, Universität der Bundeswehr München, Germany

Abstract In the optimal control of industrial or service robots, the standard procedure is to determine first off-line a feedforward control and a reference trajectory based on some nominal values of the model parameters, and to correct then the resulting inevitable deviation of the trajectory or performance of the system from the prescribed values by on-line (local) measurement and control actions. Due to stochastic parameter variations, increasing correction actions are then needed during the process.

By adaptive optimal stochastic trajectory planning and control (AOSPTC), the a priori and sample information available about the robot is incorporated into the control process by using stochastic optimization techniques. Moreover, the feedforward control and the reference trajectory are updated somewhat later in order to maintain a high quality of the reference functions.

As a consequence, the deviation between the actual and prescribed trajectory or performance of the robot is reduced. Hence, the on-line correction expenses can be reduced, and more reliable, robust controls are obtained. Analytical estimates for the reduction of the on-line correction expenses are given.

1 Introduction

An industrial or service robot is modelled mathematically by its *dynamic equation*, being a system of second order differential equations for the robot or configuration coordinates $q = (q_1, \ldots, q_n)'$ (rotation angles in case of revolute links, length of translations in case of prismatic links), and the *kinematic equation*, relating the space $\{q\}$ of robot coordinates to the work space $\{x\}$ of the robot. Thereby one meets [5,9,10,14,40,44,48] several model parameters, such as length of links, $l_i(m)$, location of center of gravity of links, $l_{ci}(m)$, mass of links, $m_i(kg)$, payload (N), moments of inertia about centroid, $I_i(kgm^2)$, (Coulomb-)friction coefficients, $R_{ij0}(N)$, etc.. Let p_D, p_K denote the vector of model parameters contained in the dynamic, kinematic equation, respectively. A further vector p_C of model parameters occurs in the formulation of several constraints, especially initial and terminal conditions, control and state constraints of the robot, as, e.g., maximum, minimum torques or forces in the links, bounds for the position, maximum joint, path velocities. Moreover, certain parameters p_J, e.g., cost factors, may occur also in the objective (performance, goal) functional J.

Due to stochastic variations of the material, manufacturing errors, measurement (identification) errors, stochastic uncertainty of the payload, errors in the selection of appropriate bounds arising in state and control constraints, errors in the selection of cost factors, more general modelling errors, etc., the total vector

$$p = \begin{pmatrix} p_D \\ p_K \\ p_C \\ p_J \end{pmatrix} \tag{1a}$$

of model parameters of the robot and its working environment is not a given, fixed quantity. The vector p but must be represented therefore by a random vector

$$p = p(\omega), \quad \omega \in (\Omega, \mathcal{A}, P) \tag{1b}$$

on a certain probability space $(\Omega, \mathcal{A}, P)$, see [5, 16, 33, 43, 44, 50].

Having to control a robotic or a more general dynamical system, the control input $u = u(t)$, is represented [10, 14, 19, 37, 46, 47, 51] usually by the sum

$$u(t) := u^{(0)}(t) + \Delta u(t), t_0 \leq t \leq t_f, \tag{2}$$

of a feedforward control (open-loop control) $u_0(t), t_0 \leq t \leq t_f$, and an on-line (local) control correction (feedback control) $\Delta u(t)$.

Replacing the unknown parameter vector p by a certain vector $p^{(0)}$ of nominal parameter values, as, e.g., the expectation $p^{(0)} := \overline{p} = Ep(\omega)$, in actual engineering practice [19, 46, 47, 50] the feedforward control $u_0(t)$ is determined off-line based on a certain reference trajectory $q^{(0)}(t), t_0 \leq t \leq t_f$, in confi-guration space. The resulting, and mostly growing deviation of the actual state of the robot from the prescribed state, caused by the deviation of the actual parameter values $p(\omega)$ from the chosen nominal values $p^{(0)}$, must be compensated then by on-line control corrections $\Delta u(t)$ at each time point $t > t_0$. This usually requires extensive, growing on-line state observations (measurements) and feedback control actions.

In order to determine a reference trajectory $q^{(0)} = q^{(0)}(t), t_0 \leq t \leq t_f$, in configuration space, being robust with respect to stochastic parameter variations, the a priori information (given by certain moments or parameters of the probability distribution of $p(\cdot)$) about the random variations of the vector $p(\omega)$ is taken into account already at the planning phase. Thus, instead of solving a deterministic trajectory planning problem with a fixed nominal parameter vector $p^{(0)}$, here, an optimal velocity profile $\beta^{(0)}(s), s_0 \leq s \leq s_f$, and - in case of point-to-point control problems - also an optimal geometric path $q_e^{(0)}(s), s_0 \leq s \leq s_f$, in configuration space is determined by using a stochastic optimization approach, see [26–29, 36]. By means of $\beta^{(0)}(s)$ and $q_e^{(0)}(s), s_0 \leq s \leq s_f$, we then find a more reliable, robust reference trajecetory $q^{(0)}(t), t_0 \leq t \leq t_f^{(0)}$, in configuration space. Applying now the so-called "inverse dynamics approach" [10, 14, 31, 37, 40], also a more reliable, robust feedforward control $u^{(0)}(t), t_0 \leq t \leq t_f^{(0)}$, is obtained. Finally, by Taylor expansion of the dynamic equation of the robot in a neighbourhood of $\left(u^{(0)}(t), q^{(0)}(t), E\left(p_D(\omega)|\mathcal{A}_{t_0}\right)\right), t \geq t_0$, where $\mathcal{A}_{t_0}$ denotes the σ-algebra of information up to the initial time point t_0, a control correction $\Delta u^{(0)}(t), t \geq t_0$, is obtained which is related to the so-called feedback linearization of a system [37, 46].

At later moments (main correction time points) t_j,

$$t_0 < t_1 < t_2 < \cdots < t_{j-1} < t_j < \ldots, \tag{3}$$

further information on the parameters of the control system and its environment is available, e.g., by process observation, identification, calibration procedures etc.. Improvements $q^{(j)}(t), u^{(j)}(t), \Delta u^{(j)}(t), t \geq t_j, j = 1,2,\ldots,$ of the preceding reference trajectory $q^{(j-1)}(t)$, feedforward control $u^{(j-1)}(t)$, and control correction (feedback control) $\Delta u^{(j-1)}(t)$ can be determined then by *replanning*, i.e., by optimal stochastic trajectory planning (OSTP) for the remaining time interval $t \geq t_j, j = 1,2,\ldots,$ and by using the information $\mathcal{A}_{t_j}$ on the robot and its working environment available up to the moment $t_j > t_0, j = 1,2,\ldots,$ see [15,41,42].

2 Optimal Trajectory Planning for Robots

According to [6,7,9,10,14,19,34,40], the dynamic equation for a robot is given by the following system of second order differential equations

$$M\Big(p_D, q(t)\Big)\ddot{q}(t) + h\Big(p_D, q(t), \dot{q}(t)\Big) = u(t), t \geq t_0, \tag{4a}$$

for the n-vector $q = q(t)$ of the robot or configuration coordinates $q_1, q_2, \ldots, q_n$. Here, $M = M(p_D, q)$ denotes the $n \times n$ inertia (or mass) matrix, and the vector function $h = h(p_D, q, \dot{q})$ is given by

$$h(p_D, q, \dot{q}) := C(p_D, q, \dot{q})\dot{q} + F_R(p_D, q, \dot{q}) + G(p_D, q), \tag{4b}$$

where $C(p_D, q, \dot{q}) = C(p_D, q)\dot{q}$, and $C(p_D, q) = \Big(C_{ijk}(p_D, q)\Big)_{1 \leq i,j,k \leq n}$ is the tensor of Coriolis and centrifugal terms, $F_R = F_R(p_D, q, \dot{q})$ denotes the vector of frictional forces and $G = G(p_D, q)$ is the vector of gravitational forces. Moreover, $u = u(t)$ is the vector of controls, i.e., the vector of torques/forces in the joints of the robot. Standard representations of the friction term F_R are given [19,45] by

$$F_R(p_D, q, \dot{q} := R_\nu(p_D, q)\dot{q}, \tag{4c}$$

$$F_R(p_D, q, \dot{q}) := R(p_D, q)\,\mathrm{sgn}(\dot{q}), \tag{4d}$$

where $\mathrm{sgn}(\dot{q}) := \Big(\mathrm{sgn}(\dot{q}_1), \ldots, \mathrm{sgn}(\dot{q}_n)\Big)'$. In the first case (4c), $R_\nu = R_\nu(p_D, q)$ is the viscous friction matrix, and in the Coulomb approach (4d), $R = R(p_D, q) = \Big(R_i(p, q)\delta_{ij}\Big)$ is a diagonal matrix.

Remark 1. **Inverse dynamics**. Reading the dynamic equation (4a) from the left to the right hand side, hence, by inverse dynamics [10,14,31,38,40], the control function $u = u(t)$ may be described in terms of the trajectory $q = q(t)$ in configuration space.

The relationship between the so-called configuration space $\{q\}$ of robot coordinates $q = (q_1, \ldots, q_n)'$ and the work space $\{x\}$ of world coordinates (position and orientation of the end-effector) $x = (x_1, \ldots, x_n)'$ is represented by the kinematic equation

$$x = T(p_K, q). \tag{5}$$

As mentioned already in the introduction, p_D, p_K denote the vectors of dynamic, kinematic parameters arising in the dynamic and kinematic equation (4a–d), (5).

Remark 2. **Linear parametrization of robots.** Note that the parametrization of a robot can be chosen, cf. [3, 14], so that the dynamic and kinematic equation depend linearly on the parameter vectors p_D, p_K.

The objective of optimal trajectory planning is to determine [6, 7, 9, 19, 49] a control function $u = u(t), t \geq t_0$, so that the cost functional

$$J\Big(u(\cdot)\Big) := \int_{t_0}^{t_f} L\Big(p_J, q(t), \dot{q}(t), u(t)\Big)\, dt + \phi\Big(p_J, q(t_f), \dot{q}(t_f)\Big) \tag{6}$$

is minimized, where the terminal time t_f may be given explicitly or implicitly, as, e.g., in minimum-time problems [6, 7, 19]. Besides the minimization of (6), an optimal control function $u^* = u^*(t)$ and the related optimal trajectory $q^* = q^*(t), t \geq t_0$, in configuration space must satisfy, of course, the dynamic equation (4a–d) and the following constraints [6, 7, 9, 19, 34, 40, 45, 49]:

i) The initial conditions

$$q(t_0) = \mathbf{q_0}(\omega), \dot{q}(t_0) = \dot{\mathbf{q}}_\mathbf{0}(\omega) \tag{7a}$$

ii) The terminal conditions

$$\psi\Big(t_f, p, q(t_f), \dot{q}(t_f)\Big) = 0, \tag{7b}$$

e.g.

$$q(t_f) = q_f(\omega), \dot{q}(t_f) = \dot{q}_f(\omega). \tag{7c}$$

iii) Control constraints

$$u^{\min}(p) \leq u(t) \leq u^{\max}(p), t_0 \leq t \leq t_f \tag{8a}$$

$$g_I\Big(p, q(t), \dot{q}(t), u(t)\Big) \leq 0, t_0 \leq t \leq t_f \tag{8b}$$

$$g_{II}\Big(p, q(t), \dot{q}(t), u(t)\Big) = 0, t_0 \leq t \leq t_f. \tag{8c}$$

iv) State constraints

$$S_I\Big(p, q(t), \dot{q}(t)\Big) \leq 0, \quad t_0 \leq t \leq t_f \tag{9a}$$

$$S_{II}\Big(p, q(t), \dot{q}(t)\Big) = 0, \quad t_0 \leq t \leq t_f. \tag{9b}$$

In robotics [9, 19] often the following state constraints are used:

$$q^{min}(p_C) \le q(t) \le q^{max}(p_C), \quad t_0 \le t \le t_f \tag{9c}$$

$$\dot{q}^{min}(p_C) \le \dot{q}(t) \le \dot{q}^{max}(p_C), \quad t_0 \le t \le t_f, \tag{9d}$$

with certain vectors $q^{min}, q^{max}, \dot{q}^{min}, \dot{q}^{max}$ of (random) bounds.

A special constraint of the type (9b) occurs if the trajectory in work space

$$x(t) := T\Big(p_K, q(t)\Big) \tag{10}$$

should follow as precise as possible a geometric path in work space

$$x_e = x_e(p_x, s), s_0 \le s \le s_f \tag{11}$$

which is known up to a certain random parameter vector $p_x = p_x(\omega)$, which then is added to the total vector p of model parameters, cf. (1a,b).

Remark 3. In the following we suppose that the functions M, h, L, ϕ and T arising in (4a–d), (5), (6) as well as the functions $\psi, g_I, g_{II}, S_I, S_{II}$ arising in the constraints (7b-9b) are sufficiently smooth.

3 Problem Transformation

Since the terminal time t_f may be given explicitly or implicitly, the trajectory $q(\cdot)$ in configuration space may have a varying domain $[t_0, t_f]$. Hence, in order to work with a given fixed domain of the unknown functions, the reference trajectory $q = q(t), t \ge t_0$, in configuration space is represented, cf. [19], by

$$q(t) := q_e\Big(s(t)\Big), t \ge t_0. \tag{12a}$$

Here

$$s = s(t), t_0 \le t \le t_f, \tag{12b}$$

is a strictly monotoneous increasing transformation from the possibly varying time domain $[t_0, t_f]$ into a given fixed parameter interval $[s_0, s_f]$. E.g., $s \in [s_0, s_f]$ may be the path parameter of a given path in work space, cf. (11). Moreover,

$$q_e = q_e(s), \quad s_0 \le s \le s_f, \tag{12c}$$

denotes the so-called geometric path in configuration space.

Assuming that the transformation $s = s(t)$ is differentiable on $[t_0, t_f]$ with the exception of at most a finite number of points, we introduce now the so-called velocity profile $\beta = \beta(s), s_0 \le s \le s_f$, along the geometric path $q_e(\cdot)$ in configuration space by

$$\beta(s) := \dot{s}^2\Big(t(s)\Big) = \left(\frac{ds}{dt}\Big(t(s)\Big)\right)^2, \tag{13}$$

where $t = t(s), s_0 \leq s \leq s_f$, is the inverse of $s = s(t), t_0 \leq t \leq t_f$. Thus, we have that

$$dt = \frac{1}{\sqrt{\beta(s)}} ds, \tag{14a}$$

and the time $t \geq t_0$ can be represented by the integral

$$t = t(s) := t_0 + \int_{s_0}^{s} \frac{d\sigma}{\sqrt{\beta(\sigma)}}. \tag{14b}$$

Obviously, the terminal time t_f is given, cf. (14b), by

$$t_f = t(s_f) = t_0 + \int_{s_0}^{s_f} \frac{d\sigma}{\sqrt{\beta(\sigma)}}. \tag{15}$$

3.1 *Transformation of the Dynamic Equation*

Because of (12a,b), we find that

$$\dot{q}(t) = q_e'(s)\dot{s} \quad \left(\dot{s} := \frac{ds}{dt}, q_e'(s) := \frac{dq_e}{ds}\right) \tag{16a}$$

$$\ddot{q}(t) = q_e'(s)\ddot{s} + q_e''(s)\dot{s}^2. \tag{16b}$$

Moreover, according to (13) we have that

$$\dot{s}^2 = \beta(s), \dot{s} = \sqrt{\beta(s)}, \tag{16c}$$

and the differentiation of (16c) with respect to time t yields

$$\ddot{s} = \frac{1}{2}\beta'(s). \tag{16d}$$

Hence, (16a–d) yields the following representation

$$\dot{q}(t) = q_e'(s)\sqrt{\beta(s)} \tag{17a}$$

$$\ddot{q}(t) = q_e'(s)\frac{1}{2}\beta'(s) + q_e''(s)\beta(s) \tag{17b}$$

of $\dot{q}(t), \ddot{q}(t)$ in terms of the new unknown functions $q_e(\cdot), \beta(\cdot)$.

Inserting now (17a,b) into the dynamic equation (4a), we find the equi-valent relation

$$u_e\,(p_D, s; q_e(\cdot), \beta(\cdot)) = u(t) \text{ with } s = s(t), t = t(s), \tag{18a}$$

where the function u_e is defined by

$$u_e\Big(p_D, s; q_e(\cdot), \beta(\cdot)\Big) := M\Big(p_D, q_e(s)\Big)\left(\frac{1}{2}q_e'(s)\beta'(s) + q_e''(s)\beta(s)\right) \tag{18b}$$

$$+ h\left(p_D, q_e(s), q_e'(s)\sqrt{\beta(s)}\right). \tag{18c}$$

The initial and terminal conditions (7a–c) are transformed, see (12a,b) and (17a), as follows

$$q_e(s_0) = \mathbf{q}_0(\omega),\ q_e'(s_0)\sqrt{\beta(s_o)} = \dot{\mathbf{q}}_0(\omega) \tag{19a}$$

$$\psi\left(p, q_e(s_f), q_e'(s_f)\sqrt{\beta(s_f)}\right) = 0 \tag{19b}$$

or

$$q_e(s_f) = \mathbf{q}_f(\omega),\ q_e'(s_f)\sqrt{\beta(s_f)} = \dot{\mathbf{q}}_f(\omega). \tag{19c}$$

Remark 4. In most cases we have that the robot is at rest at time $t = t_0$ and $t = t_f$, i.e., $\dot{q}(t_0) = \dot{q}(t_f) = 0$, hence,

$$\beta(s_0) = \beta(s_f) = 0. \tag{19d}$$

3.2 *Transformation of the Control Constraints*

Using (12a,b), the control constraints (8a–c) read in s-form as follows:

$$u^{\min}(p_C) \leq u_e\left(p_D, s; q_e(\cdot), \beta(\cdot)\right) \leq u^{\max}(p_C), \quad s_0 \leq s \leq s_f \tag{20a}$$

$$g_I\left(p_C, q_e(s), q_e'(s)\sqrt{\beta(s)}, u_e\left(p_D, s; q_e(\cdot), \beta(\cdot)\right)\right) \leq 0, \quad s_0 \leq s \leq s_f \tag{20b}$$

$$g_{II}\left(p_C, q_e(s), q_e'(s)\sqrt{\beta(s)}, u_e\left(p_D, s; q_e(\cdot), \beta(\cdot)\right)\right) = 0, \quad s_0 \leq s \leq s_f. \tag{20c}$$

3.3 *Transformation of the State Constraints*

Applying the transformations (12a,b), (16a) and (14b) to the state constraints (9a,b), we find the following s-form of the state constraints:

$$S_I\left(p_C, q_e(s), q_e'(s)\sqrt{\beta(s)}\right) \leq 0, \quad s_0 \leq s \leq s_f \tag{21a}$$

$$S_{II}\left(p_C, q_e(s), q_e'(s)\sqrt{\beta(s)}\right) = 0, \quad s_0 \leq s \leq s_f. \tag{21b}$$

Obviously, the s-form of the special state constraints (9c,d) read

$$q^{\min}(p_C) \leq q_e(s) \leq q^{\max}(p_C), \quad s_0 \leq s \leq s_f, \tag{21c}$$

$$\dot{q}^{\min}(p_C) \leq q_e'(s)\sqrt{\beta(s)} \leq \dot{q}^{\max}(p_C), \quad s_0 \leq s \leq s_f. \tag{21d}$$

In the case that the end-effector of the robot has to follow a given path (11) in work space, the equation (21b) reads

$$T\left(p_K, q_e(s)\right) - x_e(p_x, s) = 0, s_0 \leq s \leq s_f. \tag{21e}$$

In (21e) the parameter vector p_x describes possible uncertainties in the selection of the path to be followed by the roboter in work space.

3.4 Transformation of the Objective Function

Applying the integral transformation $t = t(s), dt = \frac{ds}{\sqrt{\beta(s)}}$ to the integral in the representation (6) of the objective function $J = J\Big(u(\cdot)\Big)$, and transforming also the terminal costs, we find the following s-form of the objective function:

$$J\Big(u(\cdot)\Big) = \int_{s_0}^{s_f} L\left(p_J, q_e(s), q_e'(s)\sqrt{\beta(s)}, u_e\Big(p_D, s; q_e(\cdot), \beta(\cdot)\Big)\right) \frac{ds}{\sqrt{\beta(s)}}$$

$$+ \phi\left(p_J, q_e(s_f), q_e'(s_f)\sqrt{\beta(s_f)}\right). \tag{22a}$$

For the class of time-minimum problems we have that

$$J\Big(u(\cdot)\Big) := t_f - t_0 = \int_{t_0}^{t_f} dt = \int_{s_0}^{s_f} \frac{ds}{\sqrt{\beta(s)}}. \tag{22b}$$

Optimal deterministic trajectory planning (ODTP). By means of the $t-s$-transformation onto the fixed s-parameter domain $[s_0, s_f]$, the optimal control problem (4a–d),(6)-(11) is transformed into a variational problem for finding, see (12a–c) and (13), an optimal velocity profile $\beta(s)$ and an optimal geometric path $q_e(s)$, $s_0 \leq s \leq s_f$. In the deterministic case, i.e. if the parameter vector p is assumed to be known, for the numerical solution of the resulting *optimal deterministic trajectory planning* problem several solution techniques are available, cf. [6,7,9,19,34,35,49].

4 OSTP – Optimal Stochastic Trajectory Planning

In the following we suppose that the initial and terminal conditions (19d) hold, i.e., $\beta_0 = \beta(s_0) = \beta_f = \beta(s_f) = 0$ or $\dot{q}(t_0) = \dot{q}(t_f) = 0$.

Based on the $(t-s)$-transformation described in Section 3, and relying on the inverse dynamics approach, the *robot control problem* (4a–d),(6)-(11) can be represented now by a *variational problem* for $\Big(q_e(\cdot), \beta(\cdot)\Big), \beta(\cdot)$, resp., given in the following. Having $\Big(q_e(\cdot), \beta(\cdot)\Big), \beta(\cdot)$, resp., a reference trajectory and a feedforward control can then be constructed.

After the $(t-s)$-transformation described before, the optimal control problem takes the following equivalent s-form:

$$\min \int_{s_0}^{s_f} L^J\Big(p_J, q_e(s), q_e'(s), q_e''(s), \beta(s), \beta'(s)\Big)\, ds + \phi^J\Big(p_J, q_e(s_f)\Big) \tag{23a}$$

s.t.

$$f^u\Big(p, q_e(s), q_e'(s), q_e''(s), \beta(s), \beta'(s)\Big) \le 0, s_0 \le s \le s_f \tag{23b}$$

$$g^u\Big(p, q_e(s), q_e'(s), q_e''(s), \beta(s), \beta'(s)\Big) = 0, s_0 \le s \le s_f \tag{23c}$$

$$f^S\Big(p, q_e(s), q_e'(s), \beta(s)\Big) \le 0, s_0 \le s \le s_f \tag{23d}$$

$$g^S\Big(p, q_e(s), q_e'(s), \beta(s)\Big) = 0, s_0 \le s \le s_f \tag{23e}$$

$$\beta(s) \ge 0, s_0 \le s \le s_f \tag{23f}$$

$$q_e(s_0) = \mathbf{q}_0(\omega), q_e'(s_0)\sqrt{\beta(s_0)} = \dot{\mathbf{q}}_0(\omega) \tag{23g}$$

$$q_e(s_f) = \mathbf{q}_f(\omega), \beta(s_f) = \beta_f. \tag{23h}$$

Under condition (19d), a more general version of the terminal condition (23h) reads, cf. (19b),

$$\psi\Big(p, q_e(s_f)\Big) = 0, \ \beta(s_f) = \beta_f := 0. \tag{23h'}$$

Here,

$$L^J = L^J\Big(p_J, q_e, q_e', q_e'', \beta, \beta'\Big), \phi^J = \phi^J(p_J, q_e) \tag{24a}$$

$$f^u = f^u(p, q_e, q_e', q_e'', \beta, \beta'), g^u = g^u(p, q_e, q_e', q_e'', \beta, \beta') \tag{24b}$$

$$f^S = f^S(p, q_e, q_e', \beta), g^S = g^S(p, q_e, q_e', \beta) \tag{24c}$$

are the functions representing the s-form of the objective function (22a), the constraint functions in the control constraints (20a–c), and in the state constraints (21a–d), respectively.

In order to get a reliable optimal geometric path $q_e^* = q_e^*(s)$ in configuration open and a reliable optimal velocity profile $\beta^* = \beta^*(s), s_0 \le s \le s_f$, being robust with respect to random parameter variations of $p = p(\omega)$, the variational problem (23a–h,h') under stochastic uncertainty must be replaced by an appropriate *deterministic substitute variational problem.* Depending on the decision theoretical point of view, different approaches are possible.

Assume first that the a priori information about the robot and its environment up to time t_0 is described by means of a σ-algebra $\mathcal{A}_{t_0}$, and let then

$$P^{(0)}_{p(\cdot)} = P_{p(\cdot)}\Big(\cdot|\mathcal{A}_{t_0}\Big) \tag{25}$$

denote the a priori distribution of the random vector $p = p(\omega)$ given $\mathcal{A}_{t_0}$.

Using stochastic optimization methods [20, 21, 23–25, 30], the following two basic classes of reliability-based deterministic substitute problems are considered, see [13, 26–29, 36]:

I) Risk-based minimum expected cost problems
II) Expected total cost-minimum problems.

The first class I of deterministic substitute is related to the chance- or expected cost-constrained approach [20, 21, 30], while the second class II is related to the so-called 2-stage or recourse approach [20, 21, 30] of stochastic optimization.

Substitute problems are constructed now by selecting certain scalar or vectorial loss or cost functions

$$\gamma_f^u, \gamma_g^u, \gamma_f^S, \gamma_g^S, \gamma^\psi, \ldots \tag{26a}$$

for the cost evaluation of the random constraints (23b,c), (23d,e), (23h'), respectively.

In the following all expectations are conditional expectations with respect to the a priori distribution $P_{p(\cdot)}^{(0)}$ of the random parameter vector $p(\omega)$; all expectations are assumed to exist. Moreover, the following compositions are introduced:

$$f_\gamma^u := \gamma_f^u \circ f^u, g_\gamma^u := \gamma_g^u \circ g^u \tag{26b}$$

$$f_\gamma^S := \gamma_f^S \circ f^S, g_\gamma^S := \gamma_g^S \circ g^S \tag{26c}$$

$$\psi_\gamma := \gamma^\psi \circ \psi. \tag{26d}$$

Now the two basic types of substitute problems are described.

I) *Risk-based minimum expected cost problems*
Minimizing the expected (primal) costs $E\left(J(u(\cdot)|\mathcal{A}_{t_0}\right)$, and demanding that the risk, i.e., the expected costs arising from the violation of the constraints of the variations problem (23a–h,h') do not exceed given upper bounds, we find the following substitute problem:

$$\min \int_{s_0}^{s_f} E\left(L^J\left(p_J, q_e(s), q_e'(s), q_e''(s), \beta(s), \beta'(s)\right)|\mathcal{A}_{t_0}\right) ds + E\left(\phi^J\left(p_J, q_e(s_f)\right)|\mathcal{A}_{t_0}\right) \tag{27a}$$

s.t.

$$E\left(f_\gamma^u\left(p, q_e(s), q_e'(s), q_e''(s), \beta(s), \beta'(s)\right)|\mathcal{A}_{t_0}\right) \le \Gamma_f^u, s_0 \le s \le s_f \tag{27b}$$

$$E\left(g_\gamma^u\left(p, q_e(s), q_e'(s), q_e''(s), \beta(s), \beta'(s)\right)|\mathcal{A}_{t_0}\right) \le \Gamma_g^u, s_0 \le s \le s_f \tag{27c}$$

$$E\left(f_\gamma^S\left(p, q_e(s), q_e'(s), \beta(s)\right)|\mathcal{A}_{t_0}\right) \le \Gamma_f^S, s_0 \le s \le s_f \tag{27d}$$

$$E\left(g_\gamma^S\left(p, q_e(s), q_e'(s), \beta(s)\right)|\mathcal{A}_{t_0}\right) \le \Gamma_g^S, s_0 \le s \le s_f \tag{27e}$$

$$\beta(s) \ge 0, s_0 \le s \le s_f \tag{27f}$$

$$q_e(s_0) = q_0,\ q_e'(s_0)\sqrt{\beta(s_0)} = \dot{q}_0 \tag{27g}$$

$$q_e(s_f) = q_f \text{ (if } \phi^J = 0),\ \beta(s_f) = \beta_f \tag{27h}$$

and the more general terminal condition (23h') is replaced by

$$\beta(s_f) = \beta_f := 0, E\left(\psi_\gamma\left(p, q_e(s_f)\right)|\mathcal{A}_{t_0}\right) \le \Gamma_\psi. \tag{27h'}$$

Here,

$$\Gamma_f^u = \Gamma_f^u(s), \Gamma_g^u = \Gamma_g^u(s), \Gamma_f^S = \Gamma_f^S(s), \Gamma_g^S = \Gamma_g^S(s), \Gamma_\psi = \Gamma_\psi(s) \tag{27i}$$

denote scalar or vectorial upper risk bounds which may depend on the path parameter $s \in [s_0, s_f]$. Furthermore, the initial, terminal values $q_0, \dot{q}_0, q_f$ in (27g,h) are determined according to one of the following relations: a)

$$q_0 := \hat{q}(t_0), \dot{q}_0 := \hat{\dot{q}}(t_0), q_f := \hat{q}(t_f), \tag{27j}$$

where $\left(\hat{q}(t), \hat{\dot{q}}(t)\right)$ denotes an estimate, observation, etc., of the state in configuration space at time t; b)

$$\begin{aligned} q_0 &= \overline{q}_0 := E\left(\mathbf{q}_0(\omega)|\mathcal{A}_{t_0}\right), \dot{q}_0 = \overline{\dot{q}}_0 := E\left(\dot{\mathbf{q}}_0(\omega)|\mathcal{A}_{t_0}\right), \\ q_f &= \overline{q_f}^{(0)} := E\left(\mathbf{q}_f(\omega)|\mathcal{A}_{t_0}\right), \end{aligned} \tag{27k}$$

where $\mathbf{q}_0(\omega), \dot{\mathbf{q}}_0(\omega)$ is a random initial position, and $\mathbf{q}_f(\omega)$ is a random terminal position.
Having corresponding information about initial, terminal values $x_0, \dot{x}_0, x_f$ in work space, related equations for $q_0, \dot{q}_0, q_f$ may be obtained by means of the kinematic equation (5).

Remark 5. Problems with Chance Constraints
Substitute problems having chance constraints are obtained if the loss functions γ_f^u, γ_f^S for evaluating the violation of the inequality constraints in (23a–h) are 0–1–functions, cf. [21,22,30].

To give a characteristic example, we demand that the control, state constraints (20a), (21c), (21d), resp., have to be fulfilled at least with probability α_u, α_q, $\alpha_{\dot{q}}$, hence,

$$P\left(u^{\min}(p_C) \le u_e\left(p_D, s; q_e(\cdot), \beta(\cdot)\right) \le u^{\max}(p_C)|\mathcal{A}_{t_0}\right) \ge \alpha_u,$$
$$s_0 \le s \le s_f, \tag{28a}$$
$$P\left(q^{\min}(p_C) \le q_e(s) \le q^{\max}(p_C)|\mathcal{A}_{t_0}\right) \ge \alpha_q, s_0 \le s \le s_f, \tag{28b}$$
$$P\left(\dot{q}^{\min}(p_C) \le q_e'(s)\sqrt{\beta(s)} \le \dot{q}^{\max}(p_C)|\mathcal{A}_{t_0}\right) \ge \alpha_{\dot{q}}, s_0 \le s \le s_f. \tag{28c}$$

Sufficient conditions for the complicated chance constraints (28a–c) can be obtained by applying certain probability inequalities, see [28]. Defining

$$u^c(p_C) := \frac{u^{\max}(p_C) + u^{\min}(p_C)}{2}, \rho_u(p_C) := \frac{u^{\max}(p_C) - u^{\min}(p_C)}{2}, \tag{28d}$$

then a sufficient conditions for (28a) reads, cf. [28],

$$E\left(\mathrm{tr}B\rho_u(p_C)_d^{-1}\left(u_e-u^c(p_C)\right)\left(u_e-u^c(p_C)\right)'\rho_u(p_C)_d^{-1}|\mathcal{A}_{t_0}\right)$$
$$\leq 1-\alpha_u, s_0\leq s\leq s_f, \quad (28e)$$

where $u_e = u_e\left(p_D, s; q_e(\cdot), \beta(\cdot)\right)$, and $\rho_u(p_C)_d$ denotes the diagonal matrix containing the elements of $\rho_u(p_C)$ on its diagonal. Moreover, B denotes a positive definite matrix such that $z'Bz \geq 1$ for all vectors z such that $\|z\|_\infty \geq 1$. Taking, e.g., $B = I$, (28e) reads

$$E\left(\|\rho_u(p_C)_d^{-1}\left(u_e\left(p_D, s; q_e(\cdot), \beta(\cdot)\right)-u^c(p_C)\right)\|^2|\mathcal{A}_{t_0}\right)$$
$$\leq 1-\alpha_u, s_0\leq s\leq s_f. \quad (28f)$$

Obviously, similar sufficient conditions may be derived for (28b,c).

We observe that the above class of risk-based minimum expected cost problems for the computation of $\left(q_e(\cdot), \beta(\cdot)\right), \beta(\cdot)$, resp., is represented completely by the following set of

$$\textit{initial parameters } \zeta_0 : t_0, s_0, q_0, \dot{q}_0, P^{(0)}_{p(\cdot)} \text{ or } \nu_0 \quad (29a)$$

and

$$\textit{terminal parameters } \zeta_f : t_f, s_f, \beta_f, q_f. \quad (29b)$$

In case of problems with a given geometric path $q_e = q_e(s)$ in configuration space, the values q_0, q_f may be deleted. Moreover, approximating the expectations in (27a–h,h') by means of Taylor expansions with respect to the parameter vector p at the conditional mean

$$\overline{p}^{(0)} := E\left(p(\omega)|\mathcal{A}_{t_0}\right), \quad (29c)$$

the a priori destribution $P^{(0)}_{p(\cdot)}$ may be replaced by a certain vector

$$\nu_0 := \left(E\left(\prod_{k=1}^{r} p_l(\omega)|\mathcal{A}_{t_0}\right)_{(l_1,\ldots,l_r)\in\Lambda}\right) \quad (29d)$$

of a priori moments of $p(\omega)$ with respect to $\mathcal{A}_{t_0}$. Here, Λ denotes a certain finite set of multiple indices $(l_1, \ldots, l_\tau), r \leq 1$.

II) *Expected total cost-minimum problem*
Here, the costs arising from violations of the constraints in the variational problem (23a–h,h') are added to the (primary) costs arising along the trajectory, to

the terminal costs, respectively. Of course, corresponding weight factors may be included in the cost functions (26a). Taking expectations with respect to $\mathcal{A}_{t_0}$, the following substitute problem ist obtained:

$$\min \int_{s_0}^{s_f} E\left(L_\gamma^J\left(p, q_e(s), q_e'(s), q_e''(s), \beta(s), \beta'(s)\right)|\mathcal{A}_{t_0}\right) ds$$
$$+E\left(\phi_\gamma^J\left(p, q_e(s_f)\right)|\mathcal{A}_{t_0}\right) \quad (30a)$$

s.t.

$$\beta(s) \geq 0, s_0 \leq s \leq s_f \quad (30b)$$
$$q_e(s_0) = q_0, q_e'(s_0)\sqrt{\beta(s_0)} = \dot{q}_0 \quad (30c)$$
$$q_e(s_f) = q_f(\text{ if } \phi_\gamma^J = 0), \beta(s_f) = \beta_f, \quad (30d)$$

where $L_\gamma^J, \phi_\gamma^J$ are defined by

$$L_\gamma^J := L^J + \lambda_f^u f_\gamma^u + \lambda_g^u g_\gamma^u + \lambda_f^S f_\gamma^S + \lambda_g^S g_\gamma^S \quad (30e)$$
$$\phi_\gamma^J := \phi^J \text{ or } \phi_\gamma^J := \phi^J + \lambda_\psi \psi_\gamma, \quad (30f)$$

and $\lambda_f^u, \lambda_g^u, \lambda_f^S, \lambda_g^S, \lambda_\psi > 0$ are certain positive scale factors.

We observe that the initial/terminal parameters characterizing the above class of substitute problems (30a–f) are given again by (29a,b).

4.1 Computational Aspects

Two main techniques are available for solving substitute problems of type (I), (II):

a) *Reduction to a finite dimensional parameter optimization problem*
Here, the unknown functions $\left(q_e(\cdot), \beta(\cdot)\right)$ or $\beta(\cdot)$ are approximated by a linear combination

$$q_e(s) := \sum_{l=1}^{l_q} \hat{q}_l B_l(s), s_0 \leq s \leq s_f \quad (31a)$$
$$\beta(s) := \sum_{l=1}^{l_\beta} \hat{\beta}_l B_l^\beta(s), s_0 \leq s \leq s_f, \quad (31b)$$

where $B_l = B_l(s), B_l^\beta = B_l^\beta(s), s_0 \leq s \leq s_f, l = 1, \ldots, l_q(l_\beta)$, are given basis functions, e.g., B-splines, and $\hat{q}_l, \hat{\beta}_l, l = 1, \ldots, l_q(l_\beta)$, are vectorial, scalar coefficients. Putting (31a,b) into (27a–h,h'), (30a–f), resp., a semiinfinite optimization problem is obtained. If the inequalities involving explicitly the path parameter $s, s_0 \leq s \leq s_f$, are required for a finite number N of parameter values $s_1, s_2, \ldots, s_N$ only, then this problem is reduced finally to a finite dimensional parameter optimization problem which can be solved now numerically by standard mathematical programming routines or search techniques. Of course, a major problem is the approximative computation of the conditional expectations which is done essentially by means of Taylor expansion with respect to the parameter vector p at $\overline{p}^{(0)}$. Consequently, several conditional moments have to be determined (on-line, for stage $j \geq 1$). For details, see [29].

b) *Variational techniques*
Using methods from calculus of variations, necessary and – in some cases – also sufficient conditions in terms of certain differential equations may be derived for the optimal solutions $\left(q_e^{(0)}, \beta^{(0)}\right), \beta^{(0)}$, resp., of the variational problems (27a–h,h'), (30a–f). For more details, see [36].

An interesting variant of the above mentioned techniques (a,b) is obtained [19] by applying the following bilevel optimization routine:

i) For a given geometric path $\tilde{q}_e$ contained in a certain finite or infinite collection $\{\tilde{q}_{e,\lambda} : \lambda \in \Lambda\}$ of admissible geometric path in configuration space, an optimal velocity profile $\tilde{\beta}^* = \tilde{\beta}^*_\lambda$ is determined. Then
ii) in a second or outer optimization loop the parameter $\lambda \in \Lambda$ is optimized.

4.2 Optimal Reference Trajectory, Optimal Feedforward Control

Assume now that, at least approximatively, the optimal geometric path $q_e^{(0)} = q_e^{(0)}(s)$ and the optimal velocity profile $\beta^{(0)} = \beta^{(0)}(s), s_0 \leq s \leq s_f$, i.e., the optimal solution $\left(q_e^{(0)}, \beta^{(0)}\right) = \left(q_e^{(0)}(s), \beta^{(0)}(s)\right), s_0 \leq s \leq s_f$, of one of the stochastic path planning problems (27a–h, h'), (30a–f), resp., is given. Then, according to (12a–c), (13), the optimal reference trajectory in configuration space $q^{(0)} = q^{(0)}(t), t \geq t_0$, is defined by

$$q^{(0)}(t) := q_e^{(0)}\left(s^{(0)}(t)\right), t \geq t_0, \tag{32a}$$

where the optimal $(t \leftrightarrow s)$-transformation $s^{(0)} = s^{(0)}(t), t \geq t_0$, is given by the initial value problem

$$\dot{s}(t) = \sqrt{\beta^{(0)}(s)},\ t \geq t_0, s(t_0) := s_0. \tag{32b}$$

By means of the kinematic equation (5), the corresponding reference trajectory $x^{(0)} = x^{(0)}(t), t \geq t_0$, in workspace may be defined by

$$x^{(0)}(t) := E\left(T\left(p_K(\omega), q^{(0)}(t)\right)|\mathcal{A}_{t_0}\right) = T\left(\overline{p}_K^{(0)}, q^{(0)}(t)\right), t \geq t_0, \tag{32c}$$

where

$$\overline{p}_K^{(0)} := E\left(p_K(\omega)|\mathcal{A}_{t_0}\right). \tag{32d}$$

Based on the *inverse dynamics approach*, see Remark 1, the optimal reference trajectory $q^{(0)} = q^{(0)}(t), t \geq t_0$, is inserted now into the left hand side of the dynamic equation (4a). This yields next to the random optimal control function

$$\nu^{(0)}\left(t, p_D(\omega)\right) := M\left(p_D(\omega), q^{(0)}(t)\right) \ddot{q}^{(0)}(t) + h\left(p_D(\omega), q^{(0)}(t), \dot{q}^{(0)}(t)\right), t \geq t_0. \tag{33}$$

Starting at the initial state $(q_0, \dot{q}_0) := (q^{(0)}(t_0), \dot{q}^{(0)}(t_0))$, this control obviously keeps the robot exactly on the optimal trajectory $q^{(0)}(t), t \geq t_0$, provided that $p_D(\omega)$ is the true vector of dynamic parameters.

An optimal feedforward control law $u^{(0)} = u^{(0)}(t), t \geq t_0$, related to the optimal reference trajectory $q^{(0)} = q^{(0)}(t), t \geq t_0$, can be obtained therefore by applying a certain averaging or estimating operator $\Psi = \Psi\left(\cdot|\mathcal{A}_{t_0}\right)$ to (33), hence,

$$u^{(0)}(t) := \Psi\left(\nu^{(0)}\left(t, p_D(\cdot)\right)|\mathcal{A}_{t_0}\right), t \geq t_0. \tag{34}$$

If $\Psi(\cdot|\mathcal{A}_{t_0})$ is the conditional expectation, then we find the optimal feedforward control law

$$\begin{aligned} u^{(0)} &:= E\left(M\left(p_D(\omega), q^{(0)}(t)\right)\ddot{q}^{(0)}(t) + h\left(p_D(\omega), q^{(0)}(t), \dot{q}^{(0)}(t)\right)|\mathcal{A}_{t_0}\right), \\ &= M\left(\overline{p}_D^{(0)}, q^{(0)}(t)\right)\ddot{q}^{(0)}(t) + h\left(\overline{p}_D^{(0)}, q^{(0)}(t), \dot{q}^{(0)}(t)\right), t \geq t_0. \end{aligned} \tag{35a}$$

In (35a) $\overline{p}_D^{(0)}$ denotes the conditional mean of $p_D(\omega)$ defined by (29c), and the second equation in formula (35a) holds since the dynamic equation of a robot depends linearly on the parameter vector p_D, see Remark 2.

Inserting, instead of the conditional mean $\overline{p}_D^{(0)}$ of $p_D(\omega)$ given $\mathcal{A}_{t_0}$, another estimator $p_D^{(0)}$ of the true parameter vector p_D or a certain realization $p_D^{(0)}$ of $p_D(\omega)$ at the moment t_0, we obtain the optimal feedforward control law

$$u^{(0)}(t) := M\left(p_D^{(0)}, q^{(0)}(t)\right)\ddot{q}^{(0)}(t) + h\left(p_D^{(0)}, q^{(0)}(t), \dot{q}^{(0)}(t)\right), t \geq t_0. \tag{35b}$$

5 AOSTP – Adaptive Optimal Stochastic Trajectory Planning

As already mentioned in the introduction, by means of direct or indirect measurements, observations of the robot and its environment, as, e.g., by observations of the state $(x, \dot{x}), (q, \dot{q})$, resp., of the robot in work or configuration space, further information about the unknown parameter vector $p = p(\omega)$ is available at each moment $t > t_0$. Let denote, cf. Sections 1, 4,

$$\mathcal{A}_t (\subset \mathcal{A}), t \geq t_0, \tag{36a}$$

the σ-algebra of all information about the random parameter vector $p = p(\omega)$ up to time t. Hence, $(\mathcal{A}_t)$ is an increasing family of σ-algebras. Note that the flow of information in this control process can be described also by means of the stochastic process

$$p_t(\omega) := E\left(p(\omega)|\mathcal{A}_t\right),\ t \geq t_0, \tag{36b}$$

see [4]. By parameter identification [16,48] or robot calibration techniques [43,44] we may then determine the conditional distribution

$$P_{p(\cdot)}^{(t)} = P_{p(\cdot)|\mathcal{A}_t} \tag{36c}$$

of $p(\omega)$ given $\mathcal{A}_t$. Alternatively, we may determine the vector of conditional moments

$$\nu^{(t)} := \left(E\left(\prod_{k=1}^{r} p_{l_k}(\omega)|\mathcal{A}_t\right)\right)_{(l_1,\dots,l_r)\in\Lambda} \tag{36d}$$

arising in the approximate computation of conditional expectations in (OSTP) with respect to $\mathcal{A}_t$, cf. (29c,d).

The increase of information about the unknown parameter vector $p(\omega)$ from one moment t to the next $t+dt$ may be rather low, and the determination of $P_{p(\cdot)}^{(t)}$ or $\nu^{(t)}$ at each time point t may be very expensive, though identification methods in real-time exist [43,44]. Hence, as already mentioned briefly in Section 1, the conditional distribution $P_{p(\cdot)}^{(t)}$ or the vector of conditional moments $\nu^{(t)}$ is determined/updated at discrete moments (t_j):

$$t_0 < t_1 < t_2 < \dots < t_j < t_{j+1} < \dots. \tag{37}$$

The optimal functions $q_e^{(0)}(s), \beta^{(0)}(s), s_0 \leq s \leq s_f$, based on the a priori information $\mathcal{A}_{t_0}$, loose in course of time more or less their qualification to provide a satisfactory pair of guiding functions $\left(q^{(0)}(t), u^{(0)}(t)\right), t \geq t_0$.

This functions can be renewed by using the updated information σ-algebras $\mathcal{A}_{t_j}$ and then the updated a posteriori probability distributions $P_{p(\cdot)}^{(t_j)}$ or conditional moments $\nu^{(t_j)}$ of $p(\omega)$ available at the main correction moments $t_j, j \geq 1$. Adopting an adaptive stochastic control procedure, see [2, 3, 14], the pair of guiding functions $\left(q^{(0)}(t), u^{(0)}(t)\right), t \geq t_0$, is replaced by a sequence of renewed pairs $\left(q^{(j)}(t), u^{(j)}(t)\right), t \geq t_j, j = 1, 2, \dots$, of guiding functions determined by replanning, i.e., by repeated (OSTP) for the remaining time intervals $[t_j, t_f^{(j)}]$ and by using the new information given by $\mathcal{A}_{t_j}$.

The resulting substitute problem at a stage $j \geq 1$ follows from the corresponding substitute problem for the previous stage $j-1$ just by updating $\zeta_{j-1} \to \zeta_j, \zeta_f^{(j-1)} \to \zeta_f^{(j)}$, the initial and terminal parameters, see (29a,b). The renewed

$$\textit{initial parameters}\ \zeta_j : t_j, s_j, q_j, \dot{q}_j, P_{p(\cdot)}^{(j)} \text{ or } \nu_j \tag{38a}$$

for the j-th stage, $j \geq 1$, are determined recursively as follows:

$$s_j := s^{(j-1)}(t_j) \qquad (1-1-\text{transformation } s = s(t)) \tag{38b}$$

$$q_j := \hat{q}(t_j) \qquad (\text{estimate or observation of } q(t_j)) \tag{38c}$$

$$\dot{q}_j := \hat{\dot{q}}(t_j) \qquad (\text{estimate or observation of } \dot{q}(t_j)) \tag{38d}$$

$$P^{(j)}_{p(\cdot)} := P^{(t_j)}_{p(\cdot)} = P_{p(\cdot)|\mathcal{A}_{t_j}} \tag{38e}$$

$$\nu_j := \nu^{(t_j)}. \tag{38f}$$

The renewed

$$\textit{terminal parameters } \zeta_f^{(j)} : t_f^{(j)}, s_f, q_f^{(j)}, \beta_f \tag{39a}$$

for the j-th stage, $j \geq 1$, are defined by

$$s_f \text{ given} \tag{39b}$$

$$q_f^{(j)} := \hat{q}(t_f), q_f^{(j)} = \overline{q}_f^{(j)} \qquad (\text{estimate of } q(t_f)), \text{ cf. (27j,k)} \tag{39c}$$

$$\beta_f = 0 \tag{39d}$$

$$s^{(j)}\left(t_f^{(j)}\right) = s_f. \tag{39e}$$

As already mentioned above, the (OSTP) for the j-th stage, $j \geq 1$, is obtained from the substitute problems (27a–h,h'), (30a–f), resp., formulated for the 0-th stage, $j = 0$, just by substituting

$$\zeta_0 \to \zeta_j \text{ and } \zeta_f \to \zeta_f^{(j)}. \tag{40}$$

Let then denote

$$\left(q_e^{(j)}, \beta^{(j)}\right) = \left(q_e^{(j)}(s), \beta^{(j)}(s)\right), s_j \leq s \leq s_f, \tag{41}$$

the corresponding pair of optimal solutions of the resulting substitute problem for the j-th stage, $j \geq 1$.

The pair of guiding functions $\left(q^{(j)}(t), u^{(j)}(t)\right), t \geq t_j$, for the j-th stage, $j \geq 1$, is then defined as described in Section 4.2 for the 0-th stage. Hence, for the j-th stage, the reference trajectory in configuration space $q^{(j)}(t), t \geq t_j$, reads, cf. (32a),

$$q^{(j)}(t) := q_e^{(j)}\left(s^{(j)}(t)\right), t \geq t_j, \tag{42a}$$

where the transformation $s^{(j)} : \left[t_j, t_f^{(j)}\right] \to [s_j, s_f]$ is defined by the initial value problem

$$\dot{s}(t) = \sqrt{\beta^{(j)}(s)}, \ t \geq t_j, s(t_j) = s_j. \tag{42b}$$

The terminal time $t_f^{(j)}$ for the j-th stage is defined by the equation

$$s^{(j)}\left(t_f^{(j)}\right) = s_f. \tag{42c}$$

Moreover, again by the inverse dynamics approach, the feedforward control $u^{(j)} = u^{(j)}(t), t \geq t_j$, for the j-th stage is given, (33), (34), (35a,b), by

$$u^{(j)}(t) := \Psi\left(v^{(j)}\left(t, p_D(\omega)\right)|\mathcal{A}_{t_j}\right), \tag{43a}$$

where

$$v^{(j)}(t, p_D) := M\left(p_D, q^{(j)}(t)\right)\ddot{q}^{(j)}(t) + h\left(p_D, q^{(j)}(t), \dot{q}^{(j)}(t)\right), t \geq t_j. \tag{43b}$$

Using the conditional expectation $\Psi(\cdot|\mathcal{A}_{t_j}) := E(\cdot|\mathcal{A}_{t_j})$, we find the feedforward control

$$u^{(j)}(t) := M\left(\overline{p}_D^{(j)}, q^{(j)}(t)\right)\ddot{q}^{(j)}(t) + h\left(\overline{p}_D^{(j)}, q^{(j)}, \dot{q}^{(j)}(t)\right), t \geq t_j, \tag{43c}$$

where, cf. (29c),

$$\overline{p}_D^{(j)} := E\left(p_D(\omega)|\mathcal{A}_{t_j}\right). \tag{43d}$$

Corresponding to (32c,d), the reference trajectory in work space $x^{(j)} = x^{(j)}(t)$, $t \geq t_j$, for the remaining time interval $t_j \leq t \leq t_f^{(j)}$ reads

$$x^{(j)}(t) := E\left(T\left(p_K(\omega), q^{(j)}(t)\right)|\mathcal{A}_{t_j}\right) = T\left(\overline{p}_K^{(j)}, q^{(j)}(t)\right), t_j \leq t \leq t_f^{(j)}, \tag{44a}$$

where

$$\overline{p}_K^{(j)} := E\left(p_K(\omega)|\mathcal{A}_{t_j}\right). \tag{44b}$$

5.1 (OSTP)-Transformation

The variational problems (OSTP) at the different stages $j = 0, 1, 2 \ldots$ are determined uniquely by the set of initial and terminal parameters $(\zeta_j, \zeta_f^{(j)})$, cf. (38a–f), (39a–e). Thus, these problems can be transformed to a reference problem depending on $\left(\zeta_j, \zeta_f^{(j)}\right)$ and having a certain fixed reference s-interval.

Theorem 6. *Let* $[\tilde{s}_0, \tilde{s}_f], \tilde{s}_0 < \tilde{s}_f := s_f$, *be a given, fixed reference* s*-interval, and consider for a certain stage* $j, j = 0, 1, \ldots$, *the transformation*

$$\tilde{s} = \tilde{s}(s) := \tilde{s}_0 + \frac{\tilde{s}_f - \tilde{s}_0}{s_f - s_j}(s - s_j), s_j \leq s \leq s_f, \tag{45a}$$

from $[s_j, s_f]$ *onto* $[\tilde{s}_0, s_f]$ *having the inverse*

$$s = s(\tilde{s}) = s_j + \frac{s_f - s_j}{s_f - \tilde{s}_0}(\tilde{s} - \tilde{s}_0), \tilde{s}_0 \leq \tilde{s} \leq \tilde{s}_f. \tag{45b}$$

Represent then the geometric path in work space $q_e = q_e(s)$ *and the velocity profile* $\beta = \beta(s), s_j \leq s \leq s_f$, *for the* j*-th stage by*

$$q_e(s) := \tilde{q}_e\left(\tilde{s}(s)\right), s_j \leq s \leq s_f \tag{46a}$$

$$\beta(s) := \tilde{\beta}\left(\tilde{s}(s)\right), s_j \leq s \leq s_f, \tag{46b}$$

where $\tilde{q}_e = \tilde{q}_e(\tilde{s}), \tilde{\beta} = \tilde{\beta}(\tilde{s}), \tilde{s}_0 \leq \tilde{s} \leq \tilde{s}_f$, *denote the corresponding functions on* $[\tilde{s}_0, \tilde{s}_f]$. *Then the (OSTP) for the* j*-th stage is transformed into a reference variational problem (stated in the following) for* $(\tilde{q}_e, \tilde{\beta})$ *depending on the parameters*

$$(\zeta, \zeta_f) = (\zeta_j, \zeta_f^{(j)}) \tag{47}$$

and having the fixed reference s*-interval* $[\tilde{s}_0, \tilde{s}_f]$. *Moreover, the optimal solution* $\left(q_e^{(j)}, \beta^{(j)}\right) = \left(q_e^{(j)}(s), \beta^{(j)}(s)\right), s_j \leq s \leq s_f$, *may be represented by the optimal (control) law*

$$q_e^{(j)}(s) = \tilde{q}_e^*\left(\tilde{s}(s); \zeta_j, \zeta_f^{(j)}\right), s_j \leq s \leq s_f, \tag{48a}$$

$$\beta^{(j)}(s) = \tilde{\beta}^*\left(\tilde{s}(s); \zeta_j, \zeta_f^{(j)}\right), s_j \leq s \leq s_f, \tag{48b}$$

where

$$\tilde{q}_e^* = \tilde{q}_e^*(\tilde{s}; \zeta, \zeta_f), \tilde{s}_0 \leq \tilde{s} \leq \tilde{s}_f, \tag{48c}$$

$$\tilde{\beta}^* = \tilde{\beta}^*(\tilde{s}; \zeta, \zeta_f), \tilde{s}_0 \leq \tilde{s} \leq \tilde{s}_f, \tag{48d}$$

denotes the optimal solution of the above mentioned reference variational problem.

Proof. According to (46a,b) and (45a,b), the derivatives of the functions $q_e(s)$, $\beta(s)$, $s_j \leq s \leq s_f$, are given by

$$q_e'(s) = \tilde{q}_e'\left(\tilde{s}(s)\right)\frac{\tilde{s}_f - \tilde{s}_0}{s_f - s_j}, \quad s_j \leq s \leq s_f, \tag{49a}$$

$$q_e''(s) = \tilde{q}_e''\left(\tilde{s}(s)\right)\left(\frac{\tilde{s}_f - \tilde{s}_0}{s_f - s_j}\right)^2, s_j \leq s \leq s_f, \tag{49b}$$

$$\beta'(s) = \tilde{\beta}'\left(\tilde{s}(s)\right)\frac{\tilde{s}_f - \tilde{s}_0}{s_f - s_j}, \quad s_j \leq s \leq s_f. \tag{49c}$$

Now putting the transformation (45a,b) and the representation (46a,b), (49a–c) of $q_e(x), \beta(s), s_j \leq s \leq s_f$, and their derivatives into one of the substitute problems (27a–h,h'), (30a–f), the chosen substitute problem is transformed into a corresponding reference variational problem (stated in the following Section 5.2) having the fixed reference interval $[\tilde{s}_0, \tilde{s}_f]$ and depending on the parameter vectors $\zeta_j, \zeta_f^{(j)}$. Moreover, according to (46a,b), the optimal solution $\left(q_e^{(j)}, \beta^{(j)}\right)$ of the substitute problem for the j-th stage may be represented then by (48a–d). □

Remark 7. Based on the above theorem, the stage-independent functions $\tilde{q}_e^*, \tilde{\beta}^*$ can now be determined off-line by using an appropriate numerical procedure.

5.2 *The Reference Variational Problem*

After the (OSTP)-transformation described in Section 5.1 for the problems of type (27a–h,h'), we find

$$\min \int_{\tilde{s}_0}^{\tilde{s}_f} E\left(L^J\left(p_J, \tilde{q}_e(\tilde{s},), \tilde{q}_e'(\tilde{s})\frac{\tilde{s}_f - \tilde{s}_0}{s_f - s_j}, \tilde{q}_e''(\tilde{s})\left(\frac{\tilde{s}_f - \tilde{s}_0}{s_f - s_j}\right)^2,\right.\right.$$
$$\left.\left.\tilde{\beta}(\tilde{s}), \tilde{\beta}'(\tilde{s})\frac{\tilde{s}_f - \tilde{s}_0}{s_f - s_j}\right)|\mathcal{A}_{t_j}\right)\frac{s_f - s_j}{\tilde{s}_f - \tilde{s}_0}\,d\tilde{s} + E\left(\phi^J\left(p_J, \tilde{q}_e(\tilde{s}_f)\right)|\mathcal{A}_{t_j}\right) \quad (50a)$$

s.t.

$$E\left(f_\gamma\left(p, \tilde{q}_e(\tilde{s}), \tilde{q}_e'(\tilde{s})\frac{\tilde{s}_f - \tilde{s}_0}{s_f - s_j}, \tilde{q}_e''(\tilde{s})\left(\frac{\tilde{s}_f - \tilde{s}_0}{s_f - s_j}\right)^2,\right.\right.$$
$$\left.\left.\tilde{\beta}(\tilde{s}), \tilde{\beta}'(\tilde{s})\frac{\tilde{s}_f - \tilde{s}_0}{s_f - s_j}\right)|\mathcal{A}_{t_j}\right) \le \Gamma_f, \tilde{s}_0 \le \tilde{s} \le \tilde{s}_f \quad (50b)$$

$$\tilde{\beta}(\tilde{s}) \ge 0, \tilde{s}_0 \le \tilde{s} \le \tilde{s}_f \quad (50c)$$

$$\tilde{q}_e(\tilde{s}_0) = q_j,\ \tilde{q}_e'(\tilde{s}_0)\frac{\tilde{s}_f - \tilde{s}_0}{s_f - s_j}\sqrt{\tilde{\beta}(\tilde{s}_0)} = \dot{q}_j \quad (50d)$$

$$\tilde{q}_e(\tilde{s}_f) = q_f^{(j)}\ (\text{if } \phi^J = 0),\ \tilde{\beta}(\tilde{s}_f) = 0 \quad (50e)$$

$$\tilde{\beta}(\tilde{s}_f) = 0, E\left(\psi\left(p, \tilde{q}_e(\tilde{s}_f)\right)|\mathcal{A}_{t_j}\right) \le \Gamma_\psi, \quad (50e')$$

where f_γ, Γ_f are defined by

$$f_\gamma := (f_\gamma^u, g_\gamma^u, f_\gamma^s, g_\gamma^s)',\ \Gamma_f := (\Gamma_f^u, \Gamma_g^u, \Gamma_f^S, \Gamma_g^S)'. \quad (50f)$$

Moreover, for the problem type (30a–f) we get

$$\min \int_{\tilde{s}_0}^{\tilde{s}_f} E\left(L_\gamma^J\left(p, \tilde{q}_e(\tilde{s}), \tilde{q}_e'(\tilde{s})\frac{\tilde{s}_f - \tilde{s}_0}{s_f - s_j}, \tilde{q}_e''(\tilde{s})\left(\frac{\tilde{s}_f - \tilde{s}_0}{s_f - s_j}\right)^2,\right.\right.$$
$$\left.\left.\tilde{\beta}(\tilde{s}), \tilde{\beta}'(\tilde{s})\frac{\tilde{s}_f - \tilde{s}_0}{s_f - s_j}\right)|\mathcal{A}_{t_j}\right)\frac{s_f - s_j}{\tilde{s}_f - \tilde{s}_0}\,d\tilde{s} + E\left(\phi_\gamma^J\left(p, \tilde{q}_e(\tilde{s}_f)\right)|\mathcal{A}_{t_j}\right) \quad (51a)$$

s.t.

$$\tilde{\beta}(\tilde{s}) \ge 0, \tilde{s}_0 \le \tilde{s} \le \tilde{s}_f \quad (51b)$$

$$\tilde{q}_e(\tilde{s}_0) = q_j, \tilde{q}_e'(\tilde{s}_0)\frac{\tilde{s}_f - \tilde{s}_0}{s_f - s_j}\sqrt{\tilde{\beta}(\tilde{s}_0)} = \dot{q}_j \quad (51c)$$

$$\tilde{q}_e(\tilde{s}_f) = q_f^{(j)}\ (\text{if } \phi_\gamma^J = 0),\ \tilde{\beta}(\tilde{s}_f) = 0. \quad (51d)$$

6 Online Control Corrections

We now consider the control of the robot at the j-th stage, i.e., for time $t \geq t_j$, see [1–3, 7, 12, 14, 15]. In practice we have random variations of the vector p of the model parameters of the robot and its environment. Moreover, there are possible deviations of the true initial state $(q_j, \dot{q}_j) := \left(q(t_j), \dot{q}(t_j)\right)$ in configuration space from the corresponding initial values $(\overline{q}_j, \dot{\overline{q}}_j) = \left(\overline{q}_j, q_e^{(j)\prime}(s_j)\sqrt{\beta_j}\right)$ of the (OSTP) at stage j. Thus, the actual trajectory

$$q(t) = q\left(t, p_D, q_j, \dot{q}_j, u(\cdot)\right), t \geq t_j \tag{52a}$$

in configuration space of the robot will deviate more or less from the optimal reference trajectory

$$q^{(j)}(t) = q_e^{(j)}\left(s^{(j)}(t)\right) = q\left(t, \overline{p}_D^{(j)}, \overline{q}_j, \dot{\overline{q}}_j, u^{(j)}(\cdot)\right), \tag{52b}$$

see (38a–f), (42a,b) and (43c).

In the following we assume that the state $\left(q(t), \dot{q}(t)\right)$ in configuration space may be observed for $t > t_j$. Now, the control correction (feedback control law) is represented, see [10, 12, 14, 17, 18, 47, 51], by

$$\Delta u^{(j)}(t) = u(t) - u^{(j)}(t) := \varphi^{(j)}\left(t, \Delta z^{(j)}(t)\right), \quad t \geq t_j, \tag{53a}$$

where

$$\Delta z^{(j)}(t) := z(t) - z^{(j)}(t), z(t) := \begin{pmatrix} q(t) \\ \dot{q}(t) \end{pmatrix}, \quad z^{(j)}(t) := \begin{pmatrix} q^{(j)}(t) \\ \dot{q}^{(j)}(t) \end{pmatrix} \tag{53b}$$

and $\varphi^{(j)} = \varphi^{(j)}(t, \Delta q, \Delta\dot{q})$ is such a function that

$$\varphi^{(j)}(t, 0, 0) = 0 \quad \text{for all } t \geq t_j. \tag{53c}$$

For the construction of an appropriate function $\varphi^{(j)}$, the trajectories $q(t)$ and $q^{(j)}(t)$, $t \geq t_j$, are embedded into a one-parameter family of trajectories $q = q(t, \epsilon), t \geq t_j, 0 \leq \epsilon \leq 1$, in configuration space which are defined as follows:

Consider first the following initial data for stage j:

$$q_j(\epsilon) := \overline{q}_j + \epsilon\Delta q_j, \Delta q_j := q_j - \overline{q}_j \tag{54a}$$

$$\dot{q}_j(\epsilon) := \dot{\overline{q}}_j + \epsilon\Delta\dot{q}_j, \Delta\dot{q}_j := \dot{q}_j - \dot{\overline{q}}_j \tag{54b}$$

$$p_D(\epsilon) := \overline{p}_D^{(j)} + \epsilon\Delta p_D, \Delta p_D := p_D - \overline{p}_D^{(j)}, \ 0 \leq \epsilon \leq 1. \tag{54c}$$

Moreover, define the control input $u(t), t \geq t_j$, by (53a), hence,

$$u(t) = u^{(j)}(t) + \Delta u^{(j)}(t) \tag{54d}$$

$$= u^{(j)}(t) + \varphi^{(j)}\left(t, q(t) - q^{(j)}(t), \dot{q}(t) - \dot{q}^{(j)}(t)\right), \quad t \geq t_j. \tag{54e}$$

Let then denote

$$q(t,\epsilon) = q\left(t, p_D(\epsilon), q_j(\epsilon), \dot{q}_j(\epsilon), u(\cdot)\right),\ 0 \le \epsilon \le 1,\ t \ge t_j, \tag{55}$$

the solution of the following initial value problem consisting of the dynamic equation (4a) with the initial values, the vector of dynamic parameters and the total control input $u(t)$ given by (54a–d):

$$F\left(p_D(\epsilon), q(t,\epsilon), \dot{q}(t,\epsilon), \ddot{q}(t,\epsilon)\right) = u(t,\epsilon), 0 \le \epsilon \le 1, t \ge t_j, \tag{56a}$$

where

$$q(t_j,\epsilon) = q_j(\epsilon), \dot{q}(t_j,\epsilon) = \dot{q}_j(\epsilon), \tag{56b}$$

$$u(t,\epsilon) := u^{(j)}(t) + \varphi^{(j)}\Big(t, q(t,\epsilon) - q^{(j)}(t), \dot{q}(t,\epsilon) - \dot{q}^{(j)}(t)\Big), \tag{56c}$$

and $F = F(p_D, q, \dot{q}, \ddot{q})$ is defined, cf. (4a), by

$$F(p_D, q, \dot{q}, \ddot{q}) := M(p_D, q)\ddot{q} + h(p_D, q, \dot{q}). \tag{56d}$$

In the following we suppose that the initial value problem (56a–d) has a unique solution $q = q(t,\epsilon),\ t \ge t_j$, for each parameter value $\epsilon, 0 \le \epsilon \le 1$, see [11,22].

Note that $(t,\epsilon) \to q(t,\epsilon), t \ge t_j, 0 \le \epsilon \le 1$, can be interpreted as a *homotopy* from the reference trajectory $q^{(j)}(t)$ to the actual trajectory $q(t), t \ge t_j$, cf. [39].

6.1 *Basic Properties of the Embedding* $q(t,\epsilon)$

$\epsilon = \epsilon_0 := 0$

Because of condition (53c) of the feedback control law $\varphi^{(j)}$ to be determined, and due to the unique solvability assumption of the initial value problem (56a–d) at the j-th stage, for $\epsilon = 0$ we have that

$$q(t,0) = q^{(j)}(t), t \ge t_j. \tag{57a}$$

$\epsilon = \epsilon_1 := 1$

According to (52a), (53a–c) and (54a–d),

$$q(t,1) = q(t) = q\Big(t, p_D, q_j, \dot{q}_j, u(\cdot)\Big), t \ge t_j, \tag{57b}$$

is the actual trajectory in configuration space under the total control input $u(t) = u^{(j)}(t) + \Delta u^{(j)}(t), t \ge t_j$, given by (54d).

Taylor-Expansion with Respect to ϵ

Let $\Delta\epsilon = \epsilon_1 - \epsilon_0 = 1$, and suppose that the following property known from parameter-dependent differential equations, cf. [11,22], holds:

Assumption 7. The solution $q = q(t,\epsilon), t \geq t_j, 0 \leq \epsilon \leq 1$, of the initial value problem (56a–d) has continuous derivatives with respect to ϵ up to order $\nu > 1$ for all $t_j \leq t \leq t_j + \Delta t_j, 0 \leq \epsilon \leq 1$, with a certain $\Delta t_j > 0$.

Based on the above assumption and (57a,b), by Taylor expansion with respect to ϵ at $\epsilon = \epsilon_0 = 0$, the actual trajectory of the robot can be represented by

$$\begin{aligned} q(t) &= q\Big(t, p_D, q_j, \dot{q}_j, u(\cdot)\Big) = q(t,1) = q(t, \epsilon_0 + \Delta\epsilon) \\ &= q(t,\epsilon_0) + \sum_{l=1}^{\nu-1} \frac{1}{l!} d^l q(t) (\Delta\epsilon)^l + \frac{1}{\nu!}\frac{\partial^\nu q}{\partial \epsilon^\nu}(t,\vartheta)(\Delta\epsilon)^\nu \\ &= q^{(j)}(t) + \sum_{l=1}^{\nu-1} \frac{1}{l!} d^l q(t) + \frac{1}{\nu!}\frac{\partial^\nu q}{\partial \epsilon^\nu}(t,\vartheta), t_j \leq t \leq t_j + \Delta t_j, \end{aligned} \tag{58a}$$

where $\vartheta = \vartheta(t,\nu), 0 < \vartheta < 1$, and

$$d^l q(t) := \frac{\partial^l q}{\partial \epsilon^l}(t,0), t_j \leq t \leq t_j + \Delta t_j, l = 1,2,\ldots,\nu-1, \tag{58b}$$

denote the l-th order differentials of $q = q(t,\epsilon)$ with respect to ϵ at $\epsilon = \epsilon_0 = 0$. Obviously, differential equations for the differentials $d^l q(t), l = 1,2,\ldots$, may be obtained, cf. [22], by successive differentiation of the initial value problem (56a–d) with respect to ϵ at $\epsilon_0 = 0$.

7.1 *The 1st Order Differential* dq

Next to we have to introduce some definitions. Corresponding to (53b) and (55) we put

$$z(t,\epsilon) := \begin{pmatrix} q(t,\epsilon) \\ \dot{q}(t,\epsilon) \end{pmatrix}, t \geq t_j, 0 \leq \epsilon \leq 1; \tag{59a}$$

then, we define the following Jacobians of the function F given by (56d):

$$K(p_D, q, \dot{q}, \ddot{q}) := \frac{\partial F}{\partial q}(p_D, q, \dot{q}, \ddot{q}) \tag{59b}$$

$$D(p_D, q, \dot{q}) := \frac{\partial F}{\partial \dot{q}}(p_D, q, \dot{q}, \ddot{q}) = \frac{\partial h}{\partial \dot{q}}(p_D, q, \dot{q}). \tag{59c}$$

Moreover, it is

$$M(p_D, q) = \frac{\partial F}{\partial \ddot{q}}(p_D, q, \dot{q}, \ddot{q}), \tag{59d}$$

and due to the linear parametrization property of robots, see Remark 2, F may be represented by

$$F(p_D, q, \dot{q}, \ddot{q}) = C(q, \dot{q}, \ddot{q}) p_D \tag{59e}$$

with a certain matrix function $C = C(q, \dot{q}, \ddot{q})$.

By differentiation of (56a–d) with respect to ϵ, for the partial derivative $\frac{\partial q}{\partial \epsilon}(t,\epsilon)$ of $q = q(t,\epsilon)$ with respect to ϵ we find, cf. (53b), the following linear initial value problem [11,22]

$$\begin{aligned} & C\Big(q(t,\epsilon),\dot{q}(t,\epsilon),\ddot{q}(t,\epsilon)\Big)\Delta p_D + K\Big(p_D(\epsilon),q(t,\epsilon),\dot{q}(t,\epsilon),\ddot{q}(t,\epsilon)\Big)\frac{\partial q}{\partial \epsilon}(t,\epsilon) \\ & + D\Big(p_D(\epsilon),q(t,\epsilon),\dot{q}(t,\epsilon)\Big)\frac{d}{dt}\frac{\partial q}{\partial \epsilon}(t,\epsilon) + M\Big(p_D(\epsilon),q(t,\epsilon)\Big)\frac{d^2}{dt^2}\frac{\partial q}{\partial \epsilon}(t,\epsilon) \\ & \qquad = \frac{\partial u}{\partial \epsilon}(t,\epsilon) = \frac{\partial \varphi^{(j)}}{\partial z}\Big(t,\Delta z^{(j)}(t)\Big)\frac{\partial z}{\partial \epsilon}(t,\epsilon) \end{aligned} \tag{60a}$$

with the initial values, see (54a,b),

$$\frac{\partial q}{\partial \epsilon}(t_j,\epsilon) = \Delta q_j, \frac{d}{dt}\frac{\partial q}{\partial \epsilon}(t_j,\epsilon) = \Delta\dot{q}_j. \tag{60b}$$

Putting now $\epsilon = \epsilon_0 = 0$, because of (53a,b) and (57a), system (60a,b) yields then this system of 2nd order differential equations for the 1st order differential $dq(t) = \frac{\partial q}{\partial \epsilon}(t,0)$:

$$\begin{aligned} & C^{(j)}(t)\Delta p_D + K^{(j)}(t)dq(t) + D^{(j)}(t)d\dot{q}(t) + M^{(j)}(t)d\ddot{q}(t) \\ & \quad = du(t) = \frac{\partial \varphi^{(j)}}{\partial z}(t,0)dz(t) \\ & \quad = \frac{\partial \varphi^{(j)}}{\partial q}(t,0)dq(t) + \frac{\partial \varphi^{(j)}}{\partial \dot{q}}(t,0)d\dot{q}(t), t \geq t_j, \end{aligned} \tag{61a}$$

with the initial values

$$dq(t_j) = \Delta q_j, d\dot{q}(t_j) = \Delta\dot{q}_j. \tag{61b}$$

Here,

$$du(t) := \frac{\partial u}{\partial \epsilon}(t,0), \tag{61c}$$

$$dz(t) := \begin{pmatrix} dq(t) \\ d\dot{q}(t), \end{pmatrix}, d\dot{q} := \frac{d}{dt}dq, d\ddot{q} := \frac{d^2}{dt^2}dq, \tag{61d}$$

and the matrices $C^{(j)}(t), K^{(j)}(t), D^{(j)}(t)$ and $M^{(j)}(t)$ are defined, cf. (59b–e), by

$$C^{(j)}(t) := C\Big(q^{(j)}(t),\dot{q}^{(j)}(t),\ddot{q}^{(j)}(t)\Big) \tag{61e}$$

$$K^{(j)}(t) := K\Big(\overline{p}_D^{(j)},q^{(j)}(t),\dot{q}^{(j)}(t),\ddot{q}^{(j)}(t)\Big) \tag{61f}$$

$$D^{(j)} := D\Big(\overline{p}_D^{(j)},q^{(j)}(t),\dot{q}^{(j)}(t)\Big), M^{(j)}(t) := M\Big(\overline{p}_D^{(j)},q^{(j)}(t)\Big). \tag{61g}$$

Local (PD-)control corrections $du = du(t)$ stabilizing system (61a,b) can now be obtained by the following definition of the Jacobian of $\varphi^{(j)}(t,z)$ with respect to z at $z = 0$:

$$\begin{aligned}\frac{\partial\varphi^{(j)}}{\partial z}(t,0) &:= \frac{\partial F}{\partial z}\left(\overline{p}_D^{(j)}, q^{(j)}(t), \dot{q}^{(j)}(t), \ddot{q}^{(j)}(t)\right) - M^{(j)}(t)(K_p, K_d)\\ &= \left(K^{(j)}(t) - M^{(j)}(t)K_p, D^{(j)}(t) - M^{(j)}(t)K_d\right), \end{aligned} \tag{62}$$

where $K_p = (\gamma_{pk}\delta_{kv}), K_d = (\gamma_{dk}\delta_{kv})$ are positive definite diagonal matrices with positive diagonal elements $\gamma_{pk}, \gamma_{dk} > 0, k = 1, \dots, n$.

Inserting (62) into (61a), due to the regularity [3] of $M^{(j)} = M^{(j)}(t)$, for $t \geq t_j$ we find the following linear system of 2nd order differential equations for $dq = dq(t)$:

$$\ddot{d}q(t) + K_d\dot{d}q(t) + K_p dq(t) = -M^{(j)}(t)^{-1}\, C^{(j)}(t)\Delta p_D, t \geq t_j, \tag{63a}$$

$$dq(t_j) = \Delta q_j, \dot{d}q(t_j) = \dot{\Delta}q_j. \tag{63b}$$

Considering the right hand side of (63a), according to (61e), (59e) and (56d) we have that

$$\begin{aligned}C^{(j)}(t)\Delta p_D &= C\left(q^{(j)}(t), \dot{q}^{(j)}(t), \ddot{q}^{(j)}(t)\right)\Delta p_D\\ &= F\left(\Delta p_D, q^{(j)}(t), \dot{q}^{(j)}(t), \ddot{q}^{(j)}(t)\right)\\ &= M\left(\Delta p_D, q^{(j)}(t)\right)\ddot{q}^{(j)}(t) + h\left(\Delta p_D, q^{(j)}(t), \dot{q}^{(j)}(t)\right). \end{aligned} \tag{64a}$$

Using then definition (42a,b) of $q^{(j)}(t)$ and the representation (17a,b) of $\dot{q}^{(j)}(t)$, $\ddot{q}^{(j)}(t)$, we get

$$\begin{aligned}C^{(j)}(t)\Delta p_D &= M\left(\Delta p_D, q_e^{(j)}\left(s^{(j)}(t)\right)\right)\\ &\times\left(q_e^{(j)\prime}\left(s^{(j)}(t)\right)\frac{1}{2}\beta^{(j)\prime}\left(s^{(j)}(t)\right) + q_e^{(j)\prime\prime}\left(s^{(j)}(t)\right)\beta^{(j)}\left(s^{(j)}(t)\right)\right)\\ &+ h\left(\Delta p_D, q_e^{(j)}\left(s^{(j)}(t)\right), q_e^{(j)\prime}\left(s^{(j)}(t)\right)\sqrt{\beta^{(j)}\left(s^{(j)}(t)\right)}\right). \end{aligned} \tag{64b}$$

From (18b) now we obtain the following important representations, where we suppose that the feedforward control $u^{(j)}(t)$, $t \geq t_j$, is given by (43c,d).

Lemma 8. *The following representations hold:*

$$C^{(j)}(t)\Delta p_D = u_e\left(\Delta p_D, s^{(j)}(t); q_e^{(j)}(\cdot), \beta^{(j)}(\cdot)\right), t \geq t_j \tag{65a}$$

$$u^{(j)}(t) = u_e\left(\overline{p}_D^{(j)}, s^{(j)}(t); q_e^{(j)}(\cdot), \beta^{(j)}(\cdot)\right), t \geq t_j \tag{65b}$$

$$u^{(j)}(t) + C^{(j)}(t)\Delta p_D = u_e\left(p_D, s^{(j)}(t); q_e^{(j)}(\cdot), \beta^{(j)}(\cdot)\right), t \geq t_j. \tag{65c}$$

Proof. The first equation follows from (64b) and (18b). Equations (43c), (17a,b) and (18b) yield (65b). Finally, (65c) follows from (65a,b) and the linear parametrization of robots, cf. Remark 2.

Remark 9. Note that according to the transformation (18a) of the dynamic equation onto the s-domain, for the control input $u(t)$ we have the representation

$$\begin{aligned} u(t) &= u_e\left(p_D, s; q_e(\cdot), \beta(\cdot)\right) \\ &= u_e\left(\overline{p}_D^{(j)}, s; q_e(\cdot), \beta(\cdot)\right) + u_e\left(\Delta p_D, s; q_e(\cdot), \beta(\cdot)\right) \end{aligned} \tag{65d}$$

with $s = s(t)$.

Using (61d), it is easy to see that (63a,b) can be described also by the 1st order initial value problem

$$\dot{dz}(t) = A dz(t) + \begin{pmatrix} 0 \\ \psi^{(j)}(t) \end{pmatrix}, t \geq t_j \tag{66a}$$

$$dz(t_j) = \Delta z_j = \begin{pmatrix} q_j - \overline{q}_j \\ \dot{q}_j - \dot{\overline{q}}_j \end{pmatrix}, \tag{66b}$$

where A is the stability or Hurwitz matrix

$$A := \begin{pmatrix} 0 & I \\ -K_p & -K_d \end{pmatrix}, \tag{66c}$$

and $\psi^{(j)}(t)$ is defined, cf. (65a), by

$$\begin{aligned} \psi^{(j)}(t) &:= -M^{(j)}(t)^{-1} C^{(j)}(t) \Delta p_D \\ &= -M^{(j)}(t)^{-1} u_e\left(\Delta p_D, s^{(j)}(t); q_e^{(j)}(\cdot), \beta^{(j)}(\cdot)\right). \end{aligned} \tag{66d}$$

Remark 10. Proceeding this way, also the higher order differentials $d^l q(t), l = 2, 3, \ldots$, may be obtained. This also enables to determine the higher order (tensorial) coefficients of the Taylor expansion of the feedback control law $\varphi^{(j)} = \varphi^{(j)}(t, \Delta z)$. Here we mention that related constructions of feedback controllers by using series expansion techniques are studied in [32].

For the first order expansion term $dz(t)$ of the deviation $\Delta z^{(j)}(t)$ between the actual state $z(t) = \begin{pmatrix} q(t) \\ \dot{q}(t) \end{pmatrix}$ and the prescribed state $z^{(j)}(t) = \begin{pmatrix} q^{(j)}(t) \\ \dot{q}^{(j)}(t) \end{pmatrix}, t \geq t_j$, we have now the representation [11, 22]

$$dz(t) = dz^{(j)}(t) = e^{A(t-t_j)} \Delta z_j + \int_{t_j}^{t} e^{A(t-\tau)} \begin{pmatrix} 0 \\ \psi^{(j)}(\tau) \end{pmatrix} d\tau. \tag{67a}$$

Because of $\mathrm{E}\left(\Delta p_D(\omega)|\mathcal{A}_{t_j}\right) = 0$, see (54c) and (43d), we have that

$$\mathrm{E}\left(\psi^{(j)}(t)|\mathcal{A}_{t_j}\right) = 0, \tag{67b}$$

$$\mathrm{E}\left(dz(t)|\mathcal{A}_{t_j}\right) = e^{A(t-t_j)}\Delta z_j, t \geq t_j. \tag{67c}$$

It is easy to see that the diagonal elements $\gamma_{dk}, \gamma_{pk} > 0, k = 1, \ldots, n$, of the positive definite diagonal matrices K_d, K_p, resp., can be chosen in such a way that the fundamental matrix $\Phi(t,\tau) = e^{A(t-\tau)}, t \geq \tau$, is exponentially stable, i.e.

$$||\Phi(t,\tau)|| \leq a_0 e^{-\lambda_0(t-\tau)}, t \geq \tau, \tag{68a}$$

with positive constants a_0, λ_0. A sufficient condition for (68a) reads

$$\gamma_{dk}, \gamma_{pk} > 0, k = 1, \ldots, n, \text{ and } \gamma_{dk} > 2 \quad \text{in case of double eigenvalues of } A. \tag{68b}$$

For the behaviour of $dz(t), t \geq t_j$, we may formulate now the following result:

Theorem 11. *Suppose that the diagonal matrices* K_d, K_p *are selected such that (68a) holds. Moreover, apply the local (i.e., first order) control correction (PD-controller)*

$$du(t) := \frac{\partial \varphi^{(j)}}{\partial z}(t, 0)\, dz(t), \tag{69a}$$

where $\frac{\partial \varphi^{(j)}}{\partial z}(t, 0)$ *is defined by (62). Then, the following relations hold:*

a) Asymptotic local stability in the mean:

$$\mathrm{E}\left(dz(t)|\mathcal{A}_{t_j}\right) \to 0, t \to \infty; \tag{69b}$$

b) Mean absolute deviation $\delta\left(dz(t)|\mathcal{A}_{t_j}\right) := \mathrm{E}\left(||dz(t) - \mathrm{E}\left(dz(t)|\mathcal{A}_{t_j}\right)||\ |\mathcal{A}_{t_j}\right)$ *of* $dz(t)$*:*

$$\delta\left(dz(t)|\mathcal{A}_{t_j}\right) \leq a_0 \int_{t_j}^{t} \frac{\sqrt{\mathrm{E}\left(||\psi^{(j)}(\tau)||^2|\mathcal{A}_{t_j}\right)}}{e^{\lambda_0(t-\tau)}}\, d\tau, \tag{69c}$$

c) Covariance of $dz(t)$*:*

$$||\operatorname{cov}\left(dz(t)|\mathcal{A}_{t_j}\right)|| \leq a_0^2 \left(\int_{t_j}^{t} e^{-\lambda_0(t-\tau)} \sqrt{\mathrm{E}\left(||\psi^{(j)}(\tau)||^2|\mathcal{A}_{t_j}\right)}\, d\tau\right)^2, \tag{69d}$$

where

$$E\left(\|\psi^{(j)}(t)\|^2|\mathcal{A}_{t_j}\right) \leq \|M^{(j)}(t)^{-1}\|^2\sigma_{u_e}^{(j)2}\left(s^{(j)}(t)\right)$$
$$\leq \|M^{(j)}(t)^{-1}\|^2\|C^{(j)}(t)\|^2 \operatorname{var}\left(p_D(\cdot)|\mathcal{A}_{t_j}\right). \quad (69e)$$

Here, $\sigma_Z^{(j)2} = \operatorname{var}(Z|\mathcal{A}_{t_j}) := E\left(\|Z - E(Z|\mathcal{A}_{t_j})\|^2|\mathcal{A}_{t_j}\right)$ *for any random vector* Z*, and* $\sigma_{u_e}^{(j)}(s)$ *is defined by*

$$\sigma_{u_e}^{(j)^2}(s) := \operatorname{var}\left(u_e\left(p_D(\cdot), s; q_e^{(j)}, \beta^{(j)}(s\cdot)\right)|\mathcal{A}_{t_j}\right), s_j \leq s \leq s_f. \quad (69f)$$

Proof. Using the PD-controller $du(t), t \geq t_j$, defined by (69a) with the gain matrix (62), according to (61a–g) we find that the first order differential $dq(t)$ has to fulfill the system of second order linear differential equations (63a,b) which can be represented then by the system of first order linear differential equation (66a,b), cf. (61d). Its solution $dz = dz(t)$ can be represented in the integral form (67a). Taking then conditional expectations, because of (66d) and $E\left(\Delta p_D|\mathcal{A}_{t_j}\right) = 0$, we get relation (67c) which yields now the first assertion (69b) since K_d, K_p are chosen such that (68a) holds. From (67a,c) and (68a) we obtain

$$\|dz(t) - E\left(dz(t)|\mathcal{A}_{t_j}\right)\| \leq \int_{t_j}^{t} a_i e^{-\lambda_0(t-\tau)}\|\psi^{(j)}(\tau)\|d\tau. \quad (70a)$$

Taking expectations in (70a), by means of Hölder's inequality we get (69c). Using again (67a) and (67c), we have that

$$\operatorname{cov}\left(dz(t)|\mathcal{A}_{t_j}\right) = E\left(\left(dz(t) - E\left(dz(t)|\mathcal{A}_{t_j}\right)\right)(dz(t)\right.$$
$$\left.-E\left(dz(t)|\mathcal{A}_{t_j}\right)\right)'\Big|\mathcal{A}_{t_j}\Big)$$
$$= \int_{t_j}^{t}\int_{t_j}^{t} e^{A(t-\tau)}E\left(\left(\begin{matrix}0\\ \psi^{(j)}(\tau)\end{matrix}\right)\left(\begin{matrix}0\\ \psi^{(j)}(\theta)\end{matrix}\right)'\Bigg|\mathcal{A}_{t_j}\right)e^{A(t-\theta)'}d\tau d\theta. \quad (70b)$$

Using again (68a), the norm of the covariance matrix can then be estimated from above by

$$\left\|\operatorname{cov}\left(dz(t)|\mathcal{A}_{t_j}\right)\right\| \leq \int_{t_j}^{t}\int_{t_j}^{t} a_0^2 e^{-\lambda_0(t-\tau)}e^{-\lambda_0(t-\theta)}$$
$$\times E\left(\|\psi^{(j)}(\tau)\|\cdot\|\psi^{(j)}(\theta)\|\Big|\mathcal{A}_{t_j}\right)d\tau d\theta. \quad (70c)$$

Applying now Hölder's inequality to the conditional expectation in (70c), we get (69d). Finally, inequalities (69e) are a consequence of equations (65a), (66d), the linearity of u_e with respect to p_D, cf. Remark 2, and the definition of the (generalized) variance $\sigma_Z^{(j)2} = \mathrm{var}(Z|\mathcal{A}_{t_j})$ of a random vector Z. □

Remark 12. $\sigma_{u_e}^{(j)2}\left(s^{(j)}(t)\right)$ can be interpreted as the risk of the feedforward control $u^{(j)}(t),\ t \geq t_j$.

Using (52b), (61g), (69e) and then changing variables $\tau \longrightarrow s$ in the integral in (69c,d), we obtain the following result:

Theorem 13. *Let denote* $t^{(j)} = t^{(j)}(s), s \geq s_j$, *the inverse of the parameter transformation* $s^{(j)} = s^{(j)}(t), t \geq t_j$. *Under the assumptions of Theorem 11, the following inequality holds for* $t_j \leq t \leq t_f^{(j)}$*:*

$$\delta\left(dz(t)|\mathcal{A}_{t_j}\right) \leq \int\limits_{s_j}^{s^{(j)}(t)} \frac{a_0 e^{-\lambda_0\left(t-t^{(j)}(s)\right)} \left\|M\left(\overline{p}_D^{(j)}, q_e^{(j)}(s)\right)^{-1}\right\|}{\sqrt{\beta^{(j)}(s)}} \sigma_{u_e}^{(j)}(s)\, ds, \tag{71}$$

and $\|\mathrm{cov}\left(dz(t)|\mathcal{A}_{t_j}\right)\|$ *can be estimated from above by the square of the right hand side of (71).*

Proof. Having the basic inequality (69c) of Theorem 11, by means of the first inequality in (69e) and defintion (69f) of the generalized variance of u_e we get

$$\delta\left(dz(t)|\mathcal{A}_{t_j}\right) \leq \int\limits_{t_j}^{t} a_0 e^{-\lambda_0(t-\tau)} \|M^{(j)}(\tau)^{-1}\| \sigma_{u_e}^{(j)}\left(s^{(j)}(\tau)\right) d\tau. \tag{72}$$

Changing now variables $\tau \longrightarrow s := s^{(j)}(\tau)$ in the above integral, because of definition (61g) of $M^{(j)}(t)$ and the transformation (14a) of the corresponding differentials $d\tau$, ds, inequality (72) yields the assertion (71). □

The meaning of the above results is indicated by the following important *minimality/bounding properties* depending on the chosen substitute problem for the trajectory planning problem under stochastic uncertainty:

i) the factor λ_0 can be decreased by an appropriate selection of the damping matrix K_d;

ii)
$$\underline{c}_M \leq \|M\left(p_D^{(j)}, q_e^{(j)}(s)^{-1}\right)\| \leq \overline{c}_M,\ s_j \leq s \leq s_f, \tag{73a}$$

with positive constants $\underline{c}_M, \overline{c}_M > 0$. This follows from the fact that the mass matrix is always positive definite, cf. [3].

iii)

$$\int_{s_j}^{s^{(j)}(t)} \frac{ds}{\sqrt{\beta^{(j)}(s)}} \leq \int_{s_j}^{s_f} \frac{ds}{\sqrt{\beta^{(j)}(s)}} = t_f^{(j)} - t_f, \tag{73b}$$

where according to (OSTP), for minimum-time and related substitute problems, the right hand side takes a minimum.

iv) Depending on the chosen substitute problem, the generalized variance $\sigma_{u_e}^{(j)}(s)$, $s_j \leq s \leq s_f$, is bounded point by point by an appropriate upper risk level, or $\sigma_{u_e}^{(j)}(\cdot)$ is minimized in a certain weighted mean sense. Two important cases are treated in the following.

iv.1) Working with the chance constraints discussed in Remark 5, we find that the probabilistic constraints (28f) can be guaranteed by

$$\sigma_{u_{:}e}^2(s) + E\left(\|u^e\left(\overline{p}_D^{(j)}, s; q_e^{(j)}(\cdot), \beta^{(j)}(\cdot)\right) - u^c(p_C)\|^2 \Big| \mathcal{A}_{t_j}\right) \leq (1-\alpha_u)\rho^{\min 2}, s_j \leq s \leq s_f, \tag{73c}$$

where $\rho^{\min} > 0$ is a fixed positive lower bound of the components of the vector $\rho_u(p_C)$ defined in (28d). However, (73c) implies the variance constraints

$$\sigma_{u_e}^2(s) \leq (1-\alpha_u)\rho^{\min 2}, s_j \leq s \leq s_f. \tag{73d}$$

iv.2) Define the cost function $L = L(p_J, q, \dot{q}, u)$ along the trajectory, cf. (6) by

$$L(p_J, q, \dot{q}, u) := \|u - u^c\|^2, \tag{73e}$$

where $u^c = u^c(p_C)$ is the center point of the control domain (8a) defined in (28d). According to (22a) we find that the expected total costs along the trajectory are given by

$$E\left(\int_{t_j}^{t_f} L\left(p_J(\omega), q(t), \dot{q}(t), u(t)\right) dt \,\Big|\mathcal{A}_{t_j}\right) = \int_{s_j}^{s_f} \sigma_{u_e}^2(s) \frac{ds}{\sqrt{\beta(s)}}$$
$$+ \int_{s_j}^{s_f} E\left(\left\|u_e\left(\overline{p}_D^{(j)}, s; q_e(\cdot), \beta(\cdot)\right) - u^c\left(p_C(\omega)\right)\right\|^2 \Big|\mathcal{A}_{t_j}\right) \frac{ds}{\sqrt{\beta(s)}}. \tag{73f}$$

Hence, a weighted mean of the generalized variance $\sigma_{u_e}^2(s), s_j \leq s \leq s_f$, is minimized.

8 Conclusions

Due to several stochastic disturbances and random variations, most of the model parameters of an industrial or service robot and its working environment must be modelled in practice by random variables. Instead of working with some fixed nominal parameter values and facing then fast increasing tracking errors, using a stochastic optimization approach, more robust guiding functions, i.e., reference trajectories and feedforward controls, are obtained. Moreover, having then much smaller tracking errors, the online correction expenses are reduced. Obtaining further information about the control process at later time instants, by a replanning procedure the guiding functions can be updated. Since the optimal stochastic trajectory planning problems arising at later stages have the same basic form, they can be transformed to a unique reference variational problem depending on the varying initial/terminal values only.

ACKNOWLEDGEMENTS

The author would like to thank two anonymous referees for their helpful comments and suggestions for improvement. Moreover, the author thanks Ms. Elisabeth Lößl for providing the LaTeX-file of the manuscript.

REFERENCES

1. S. Arimoto. Control theory of non-linear mechanical systems. Clarendon Press, Oxford (etc.), 1996.
2. K. J. Astroem, B. Wittenmark. Adaptive Control. Addison-Wesley, Reading, Mass. (etc.), 1995.
3. G. Bastian, et al. (eds.). Theory of Robot Control. Springer-Verlag, Berlin (etc.), 1996.
4. H. Bauer. Wahrscheinlichkeitstheorie und Grundz"uge der Masstheorie. Walter de Gruyter & Co., Berlin, 1968.
5. R. Bernhardt, S. L. Albright. Robot Calibration. Chapman and Hall, London (etc.), 1993.
6. J.E. Bobrow, S. Dubowsky, J. S. Gibson. Time-Optimal Control of Robotic Manipulators Along Specified Paths. *The Int. J. of Robotics Research* 4, No. 3, pages 3-17, 1985.
7. J. E. Bobrow. Optimal Robot Path Planning Using the Minimum-Time Criterion. *IEEE J. of Robotics and Automation 4*, No. 4, pages 443-450, 1988.
8. E.F. Camacho, C. Bordons. Model predictive control. Springer-Verlag, London, 1999.
9. Y.-Ch. Cheng. Solving Robot Trajectory Planning Problems with Uniform Cubic B-Splines. *Optimal Control Applications and Methods* 12: 247-262, 1991.
10. J. J. Craig. Adaptive Control of Mechanical Manipulators. Reading, Mass. (etc.). Addison-Wesley Publ. Company, 1988.
11. J. Dieudonné. Foundations of modern analysis. Academic Press, New York-London, 1960.
12. J. C. Doyle, B. A. Francis, A. R. Tannenbaum. Feedback Control Theory. Macimillan, New York (etc.), 1992.
13. C. Haubach-Lippmann. Stochastische Strukturoptimierung flexibler Roboter. Fortschritberichte VDI, Reihe 8, Nr. 706. VDI Verlag, Düsseldorf, 1998.
14. D. Holtgrewe. Adaptive Regelung flexibler Roboter. Igel Verlag, Paderborn, 1996.
15. R. Hoppen. Autonome mobile Roboter. B.I. Wissenschaftsverlag, Mannheim (etc.), 1992.
16. R. Isermann. Identifikation dynamischer Systeme. Springer-Verlag, Berlin, 1988.
17. A. Isidori. Nonlinear control systems. Springer-Verlag, Berlin (etc.), 1995.
18. A. Isidori. Nonlinear control systems II. Springer-Verlag, London, 1999.

19. R. Johanni. Optimale Bahnplanung bei Industrierobotern. Fortschrittsberichte VDI, Reihe 18, Nr. 51. VDI-Verlag, Düsseldorf, 1988.
20. P. Kall. Stochastic Linear Programming. Springer-Verlag, Berlin/Heidel-berg/New York, 1976.
21. P. Kall, S. W. Wallace. Stochastic Programming. J. Wiley, Chichester, etc., 1994.
22. H. K. Khalil. Nonlinear Systems. Macmillan Publ. Comp., New York (etc.), 1992.
23. K. Marti. Aproximation stochastischer Optimierungsprobleme. Hain, Königstein/Ts., 1979.
24. K. Marti, P. Kall (eds.). Stochastic Programming: Numerical Methods and Engineering Applications. LNEMS, Vol. 423, Springer-Verlag, Berlin (etc.), 1995.
25. K. Marti, P. Kall (eds.). Stochastic Programming Methods and Technical Applications. LNEMS, Vol. 458, Springer-Verlag, Berlin-Heidelberg, 1998.
26. K. Marti, S. Qu. Optimal Trajectory Planning for Robots under the Consideration of Stochastic Parameters and Disturbances. *J. of Intelligent and Robotic Systems 15*, pages 19-23, 1996.
27. K. Marti, S. Qu. Path Planning for Robots by Stochastic Optimization Methods. *J. of Intelligent and Robotic Systems 22*, pages 117-127, 1998.
28. K. Marti. Path Planning for Robots under Stochastic Uncertainty. Optimization 45, pages 163-195, 1999.
29. K. Marti, S. Qu. Adaptive Stochastic path planning for robots - Real-time optimization by means of neural networks. In M.P. Pohis *et al*, editors, *Systems Modelling and Optimization, Proc. 18th TC7 Conference,* pages 486-494, Chapmann and Hall/CRC, Research Notes in Mathematics, Boca Raton, 1999.
30. J. Mayer: Stochastic Linear Programming Algorithms. Gordon and Breach, Amsterdam, 1998.
31. S. Miesbach. Bahnführung von Robotern mit Neuronalen Netzen. Dissertation, TU München, Fakultät für Mathematik, 1995.
32. J. A. O'Sullivan, M. K. Sain. Optimal Control by Polynomial Approximation: The Discrete Time Case. In A. Isidori, editor, *Nonlinear Control Systems Design*, IFAC Symposia Series, No. 2, pages 303-308, Pergomon Press, Oxford (etc.), 1990.
33. D. K. Pai, M. C. Leu. Uncertainty and Compliance of Robot Manipulators with Application to Task Feasibility. *The Int. J. of Robotic Research 10*, No. 3, pages 200-212, 1991.
34. F. Pfeiffer, R. Johanni. A Concept for Manipulator Trajectory Planning *IEEE J. of Robotics and Automation RA-3*, pages 115-123, 1987.
35. F. Pfeiffer, K. Richter. Optimal Path Planning Including Forces at the Gripper. *J. of Intelligent and Robotic Systems 3*, pages 251-258, 1990.
36. S. Qu. Optimale Bahnplanung unter Berücksichtigung stochastischer Parameterschwankungen. Fortschrittberichte VDI, Reihe 8, Nr. 472. VDI Verlag, Düsseldorf, 1995.
37. Z. Qu, D. M. Dawson. Robust Tracking Control of Robot Manipulators. IEEE Press, New York, 1996.
38. D. A. Redfern, C. J. Goh. Feedback control of state constrained optimal control problems. In J. Dolezal, J. Fidler, editors, *System Modelling and Optimization*, pages 442-449, Chapman and Hall, London (etc.), 1996.
39. K. Reif. Steuerung von nichtlinearen Systemen mit Homotopieverfahren. Fortschrittberichte VDI, Reihe 8, Nr. 631. VDI Verlag, Düsseldorf, 1997.
40. R. J. Schilling. Fundamentals of Robotics; Analysis and Control. Prentice Hall, London (etc.), 1990.
41. K. Schilling, W. Flury. Autonomy and On-Board Mission Management Aspects for the Cassini Titan Probe. Acta Astronautica 21, No. 1, pages 55-68, 1990.

42. K. Schilling et al.. The European Development of Small Planetary Mobile Vehicle. Space Technology 17, No. 3/4, pages 151-162, 1997.
43. J. P. Schlöder. Numerische Methoden zur Behandlung hochdimension- aler Aufgaben der Parameteridentifizierung. Dissertation, Universität Bonn, Math.-Naturwissenschaftliche Fakultät, 1988.
44. K. Schröer. Identifikation von Kalibrierungsparametern kinematischer Ketten. Carl Hanser, München-Wien, 1993.
45. K. G. Shin, N. D. McKay. Minimum-Time Control of Robotic Manipulators with Geometric Path Constraints. *IEEE Trans. on Automatic Control*, AC-30, pages 531-541, 1985.
46. J.-J. Slotine, W. Li. Applied Nonlinear Control. Prentice-Hall Int., Inc., Englewood Cliffs, N.J., 1991.
47. R. F. Stengel. Stochastic Optimal Control - Theory and Application. J. Wiley, New York (etc.), 1986.
48. H. W. Stone. Kinematic Modeling, Identification and Control of Robotic Manipulators. Kluwer Acad. Publ., Boston (etc.), 1987.
49. I. Troch. Time-Optimal Path Generation for Continuous and Quasi-Continuous Path Control of Industrial Robots. *J. Intelligent and Robotic Systems 2*, pages 1-28, 1989.
50. A. Weinmann. Uncertain Models and Robust Control. Springer-Verlag, Wien - New York, 1991.
51. K. Weinzierl. Konzepte zur Steuerung und Regelung nichtlinearer Systeme auf der Basis der Jacobi-Linearisierung. Verlag Shaker, Aachen, 1995.

VII

Integer Stochastic Programming

*Multistage Stochastic Integer Programs: An Introduction

Werner Römisch[1] and Rüdiger Schultz[2]

[1] Institut für Mathematik, Humboldt-Universität zu Berlin, Germany
[2] Fachbereich Mathematik, Gerhard-Mercator-Universität Duisburg, Germany

Abstract We consider linear multistage stochastic integer programs and study their functional and dynamic programming formulations as well as conditions for optimality and stability of solutions. Furthermore, we study the application of the Rockafellar-Wets dualization approach as well as the structure and algorithmic potential of corresponding dual problems. For discrete underlying probability distributions we discuss possible large scale mixed-integer linear programming formulations and three dual decomposition approaches, namely, *scenario*, *component* and *nodal decomposition*.

1 Introduction

Stochastic programming deals with the optimization of decision making under uncertainty over time. Typical objects of study are random optimization problems where outcomes of random data are unveiled over time, and the decisions to be optimized must not anticipate future outcomes (non-anticipativity). The latter provides a tight link to real-time optimization seen as the need for optimal "here-and-now" decision in an incomplete (or uncertain) data environment. Provided that probabilistic information on the uncertain data is available, operational models suitable for real-time optimization often may be formulated as multi-stage stochastic programs. Basic references for theory, algorithmics, and application of stochastic programming are the textbooks [7, 24, 34]. The edited volume [50] provides insight into recent research in the field.

Indispensability of integer requirements is a basic modeling experience in practical optimization. Like in other branches of mathematical optimization this has considerable consequences on structural properties and algorithm design in stochastic programming, too. The models best understood so far are (purely) linear stochastic programs. This is mainly due to the fact that the optimal value of a linear minimization problem is a convex function of the right-hand side and a concave function of the objective function vector. This enables application of the machinery of convex analysis in various contexts, such as duality, stability, and subgradient minimization. For an impression on these developments we refer to [12, 14, 36, 49], with accent on theory, and to [5, 42], with accent on computation.

With integer requirements, the above convexity/concavity observation is not valid anymore, and the mentioned functions become discontinuous. Thus, comparatively little is known on theory and algorithms for mixed-integer linear stochastic programs. A recent survey is provided in [26]. Impressions on developments in theory can be obtained from [2,43,46] and on algorithm design from [9,11,19,27,30, 33,47], see also the Ph.D. thesises [8,32,48].

The present paper aims at a short introduction into some essential theoretical and algorithmic issues in multi-stage stochastic integer programming. Accent is placed on introducing approaches. Proofs are omitted, with references to the original sources instead. The main topics will be modeling, approximation, and algorithmics.

2 Multistage Stochastic Integer Programs

2.1 Modeling

We consider a finite horizon sequential decision process under uncertainty, in which a decision made at stage t is based only on information available at t ($1 \leq t \leq T$). We assume that the information is given by a discrete time stochastic process $\{\xi_t\}_{t=1}^T$ defined on some probability space $(\Omega, \mathcal{F}, P)$ and with ξ_t taking values in $\mathbb{R}^{s_t}$. The information available at stage t consists of the random vector $\xi^t := (\xi_1, \ldots, \xi_t)$, and the stochastic decision x_t at stage t varying in $\mathbb{R}^{m_t}$ is assumed to depend only on ξ^t. The latter property is called *nonanticipativity* and is equivalent to the measurability of x_t with respect to the σ-algebra $\mathcal{F}_t \subseteq \mathcal{F}$ which is generated by ξ^t. Clearly, we have $\mathcal{F}_t \subseteq \mathcal{F}_{t+1}$ for $t = 1, \ldots, T-1$ and, with no loss of generality, we may assume that $\mathcal{F}_1 = \{\emptyset, \Omega\}$, i.e., ξ_1 and x_1 are deterministic, and that $\mathcal{F}_T = \mathcal{F}$.

More precisely, we consider a decision model where the objective is given by expected linear costs and the constraints consist of three groups: the measurability constraints on x_t, a linear constraint describing the relation between decisions at different stages, and constraints characterizing feasibility of the t-th stage decision x_t. The latter constraints consist of a linear inequality constraint and of the general constraint $x_t \in X_t$ where the (fixed) set X_t has the property that its convex hull $\text{conv}(X_t)$ is polyhedral, allowing for mixed-integer decisions in all stages. Furthermore, the data ξ_t at stage t may enter all corresponding cost coefficients, matrices and right-hand sides. This leads to the following stochastic decision model:

$$\min\{E[\sum_{t=1}^{T} c_t\ (\xi_t)x_t] \ :\ x_t \text{ is measurable with respect to } \mathcal{F}_t, \tag{1}$$

$$x_t \in X_t,\ B_t(\xi_t)x_t \geq d_t(\xi_t),\ P-\text{a.s.},\ t = 1, \ldots, T, \tag{2}$$

$$\sum_{\tau=1}^{t} A_{t\tau}(\xi_t)x_\tau \geq g_t(\xi_t),\ P-\text{a.s.},\ t = 2, \ldots, T\} \tag{3}$$

Throughout, the following is imposed: The sets X_t are nonempty and closed. The matrices $A_{t\tau}(\cdot)$, $B_t(\cdot)$ as well as the coefficients $c_t(\cdot)$ and the right-hand sides $d_t(\cdot)$, $g_t(\cdot)$ are affine linear functions of the corresponding component of ξ, for each $\tau = 1, \ldots, t$, $t = 1, \ldots, T$. In order to have the model (1)–(3) well defined, we need that the scalar products $c_t(\xi_t)x_t$ are integrable. The latter property is implied by the integrability of $||\xi_t||||x_t||$ and by the conditions $\xi_t \in L_{q_t}(\Omega, \mathcal{F}_t, P; \mathbb{R}^{s_t})$ and

$x_t \in L_{r_t}(\Omega, \mathcal{F}_t, P; \mathbb{R}^{m_t})$ where $q_t, r_t \in [1, \infty]$ with $\frac{1}{q_t} + \frac{1}{r_t} = 1$. Since it is desirable to impose only weak conditions on the data process ξ and since we assume later on that the set X_t is bounded, we may restrict our attention to decisions $x_t \in L_\infty(\Omega, \mathcal{F}_t, P; \mathbb{R}^{m_t})$ and to the first order moment condition $\xi_t \in L_1(\Omega, \mathcal{F}_t, P; \mathbb{R}^{s_t})$ on the data at stage t for each $t = 1, \ldots, T$. Then the nonanticipativity constraint (1) may be expressed equivalently as

$$x_t \in L_\infty(\Omega, \mathcal{F}, P; \mathbb{R}^{m_t}) \quad \text{and} \quad x_t = E[x_t|\mathcal{F}_t],\ t = 1, \ldots, T, \tag{4}$$

by using the conditional expectation $E[\cdot|\mathcal{F}_t]$ with respect to the σ-algebra $\mathcal{F}_t$. Condition (4) describes a linear subspace $\mathcal{N}_{na}$ of the space $\times_{t=1}^T L_\infty(\Omega, \mathcal{F}, P; \mathbb{R}^{m_t})$. This combination of functional and (P-a.s.) pointwise constraints in our model, i.e., the functional condition $x \in \mathcal{N}_{na}$ and the P-a.s. constraints (2) and (3), forms the theoretical and algorithmic challenge of multistage stochastic programs. A special role is played by the two-stage case (i.e., T=2) where $\mathcal{N}_{na}$ takes the specific form $\mathcal{N}_{na} = \mathbb{R}^{m_1} \times L_\infty(\Omega, \mathcal{F}, P; \mathbb{R}^{m_2})$. An additional complication of the model (1)–(3) is caused by the mixed-integer constraints hidden in the condition $x_t \in X_t$, $t = 1, \ldots, T$.

2.2 *Multistage Models, Dynamic Programming and Optimality*

We adopt the setting of the previous section and assume that X_t is compact and $\xi_t \in L_1(\Omega, \mathcal{F}, P; \mathbb{R}^{s_t})$ for $t = 1, \ldots, T$. For each $\omega \in \Omega$ we define the subset $\mathcal{Y}(\omega)$ of $\mathcal{X} := \times_{t=1}^T \mathbb{R}^{m_t}$ by

$$\begin{aligned}\mathcal{Y}(\omega) := \{y \in \mathcal{X} : y_t \in X_t,\ B_t(\xi_t(\omega))y_t \geq d_t(\xi_t(\omega)),\ t = 1, \ldots, T, \\ \sum_{\tau=1}^t A_{t\tau}(\xi_t(\omega))y_\tau \geq g_t(\xi_t(\omega)),\ t = 2, \ldots, T\}\end{aligned} \tag{5}$$

and the extended real-valued function φ

$$\varphi(y_1, \ldots, y_T, \omega) := \begin{cases} \sum_{t=1}^T c_t(\xi_t(\omega))y_t\ , & (y_1, \ldots, y_T) \in \mathcal{Y}(\omega), \\ +\infty & , \text{otherwise} \end{cases} \tag{6}$$

from $\mathcal{X} \times \Omega$ to $(-\infty, +\infty]$. With these notations, the model (1)–(3) is equivalent to the optimization problem

$$\min\{E[\varphi(x_1, \ldots, x_T, \omega)] : x_t \text{ is measurable w.r.t. } \mathcal{F}_t,\ t = 1, \ldots, T\}. \tag{7}$$

The real-valued function $(y, \omega) \mapsto \sum_{t=1}^T c_t(\xi_t(\omega))y_t$ is continuous in y for each $\omega \in \Omega$ and measurable in ω for each $y \in \mathcal{X}$, and the set-valued mapping $\mathcal{Y}$ from Ω to $\mathcal{X}$ is closed-valued and measurable (cf. Theorem 14.36 in [40]). Hence, the function φ is $\mathcal{B}(\mathcal{X}) \bigotimes \mathcal{F}$-measurable (cf. Example 14.32 in [40]). Furthermore, the following estimate is valid for each $y \in \times_{t=1}^T X_t$ and $\omega \in \Omega$:

$$|\varphi(y_1, \ldots, y_T, \omega)| \leq \sum_{t=1}^T \|c_t(\xi_t(\omega))\| \sup_{y_t \in X_t} \|y_t\| \tag{8}$$

Hence, $\mathbb{E}[\varphi(x_1,\dots,x_T,\omega)]$ is finite for each decision $x = (x_1,\dots,x_T)$ such that $x(\omega) \in \mathcal{Y}(\omega)$ for P-almost all $\omega \in \Omega$. As in [16], we construct recursively two sequences of functions by putting $\psi_{T+1} := \varphi$ and

$$\varphi_t(y_1,\dots,y_t,\omega) := \mathbb{E}^r[\psi_{t+1}(y_1,\dots,y_t,\cdot)|\mathcal{F}_t](\omega), \qquad (9)$$

$$\psi_t(y_1,\dots,y_{t-1},\omega) := \inf_y \varphi_t(y_1,\dots,y_{t-1},y,\omega), \qquad (10)$$

for $t = T,\dots,1$, and for each $\omega \in \Omega$ and $y_\tau \in X_\tau$, $\tau = 1,\dots,T$. Here, $\mathbb{E}^r[\cdot|\mathcal{F}_t]$ denotes the regular conditional expectation with respect to $\mathcal{F}_t$.

We recall that the *regular conditional expectation* is a version of the conditional expectation (i.e., $\mathbb{E}^r[\cdot|\mathcal{F}_t] = \mathbb{E}[\cdot|\mathcal{F}_t]$, P-a.s.) having the property that the mapping $(z,\omega) \mapsto \Phi(z,\omega) := \mathbb{E}^r[\Psi(z,\cdot)|\mathcal{F}_t](\omega)$ from $Z_t \times \Omega$ to $(\infty,+\infty]$ is $\mathcal{B}(\mathcal{Z}_t) \bigotimes \mathcal{F}_t$-measurable if Ψ is $\mathcal{B}(\mathcal{Z}_t) \bigotimes \mathcal{F}$-measurable. Here, Z_t denotes a closed subset of a Euclidean space. The regular conditional expectation exists if Ψ is $\mathcal{B}(\mathcal{Z}) \bigotimes \mathcal{F}$-measurable and uniformly integrable, i.e., there exists a (real) random variable ζ with finite first moment such that $|\Psi(z,\omega)| \le \zeta(\omega)$ for $z \in Z_t$ and $\omega \in \Omega$ (see [15]). Due to condition (8), relation (9) is well defined for $t = T$ and leads to a $\mathcal{B}(\mathcal{Z}) \bigotimes \mathcal{F}_T$-measurable function ϕ_T, where $Z := \times_{t=1}^T X_t$. It is shown in [16] that the relations (9) and (10) are well defined for all $t = T,\dots,1$. Furthermore, the following optimality criterion and existence result for (7) or, equivalently, for (1)–(3) are valid.

Theorem 1. *Let the general assumptions be satisfied and assume that there exists a feasible solution of (1)–(3). Then $\{\bar{x}_t\}_{t=1}^T$ is a solution of (1)–(3) iff*

$$\varphi_t(\bar{x}^t(\omega),\omega) = \psi_t(\bar{x}^{t-1}(\omega),\omega), \; P-a.s., \; t = 1,\dots,T. \qquad (11)$$

Moreover, there exists a solution $\bar{x}_1$ of the first-stage optimization problem

$$\min\{\varphi_1(x_1) = \mathbb{E}[\psi_2(x_1,\omega)] \; : \; x_1 \in X_1, \; B_1(\xi_1)x_1 \ge d_1(\xi_1)\}, \qquad (12)$$

and, given $\mathcal{F}_\tau$-measurable functions $\bar{x}_\tau$ for $\tau = 1,\dots,t-1$, there exists an $\mathcal{F}_t$-measurable function $\bar{x}_t$ such that $\varphi_t(\bar{x}^t(\omega),\omega) = \psi_t(\bar{x}^{t-1}(\omega),\omega)$, $P-a.s.$

The theorem is a special case of the more general results (Theorems 1 and 2) in [16]. Theorem 1 implies the existence of a solution to (1)–(3) and justifies the solution approach (11) which is usually called *dynamic programming approach*. Due to measurable selection arguments (cf. Chapter 14 in [40]), a feasible solution of (1)–(3) exists if the model (1)–(3) has *relatively complete recourse*, i.e., if $\mathcal{Y}(\omega) \neq \emptyset$ P-a.s.

2.3 Structure and Stability

We adopt the setting of the previous sections, denote by $\mathcal{P}(\Xi)$ the set of all Borel probability measures on some closed subset Ξ of $\mathbb{R}^s$ with $s = \sum_{t=1}^T s_t$, which is chosen such that it contains the support of ξ. By $\mu \in \mathcal{P}(\Xi)$ we denote the probability distribution of ξ. We consider the probability space $(\Xi, \mathcal{B}(\Xi), \mu)$ as the underlying probability space $(\Omega, \mathcal{F}, P)$ in Section 1, and define a function f from $X_1 \times \Xi$ to the extended real numbers $\overline{\mathbb{R}}$ by $f(x_1, \xi) := \psi_2(x_1, \xi)$, where ψ_2 is defined by (9) and (10). Then the first-stage optimization problem (12) can be rewritten in the following form:

$$\min\{\int_{\Xi} f(x_1, \xi)\mu(d\xi) \, : \, x_1 \in X_1, \, B_1(\xi_1)x_1 \geq d_1(\xi_1)\} \tag{13}$$

The techniques exploited in [16] and used in the previous section imply that the integrand f is $\mathcal{B}(X_1) \bigotimes \mathcal{B}(\Xi)$-measurable. The recursions (9) and (10) together with the Fatou Lemma for (conditional) expectations as well as lower semicontinuity properties of infima in parametric minimization (e.g., Theorem 1.17 of [40]) imply lower semicontinuity of f with respect to x_1 and of the objective function $x_1 \mapsto \int_{\Xi} f(x_1, \xi)\mu(d\xi)$. If the multistage model (1)–(3) has relatively complete recourse, it holds that $|f(x_1, \xi)| \leq K(1 + \max_{t=1,\ldots,T} \|\xi_t\|)$ for each feasible x_1, each $\xi \in \Xi$ and some constant $K > 0$. Hence, the integrand has a uniform and integrable upper bound, and the objective function is finite at all feasible $\bar{x}_1$. By Lebesgue's theorem, the objective function is continuous at some feasible $\bar{x}_1$ if $\mu(\{\xi \in \Xi : f(\cdot, \xi)$ is not continuous at $\bar{x}_1\}) = 0$. Such discontinuity sets of the integrand f have been studied in [43] for the two-stage situation with fixed recourse matrix A_{22} and recourse costs c_2.

When developing approximation schemes and algorithmic approaches for solving the model (1)–(3), the behaviour of its optimal value val(μ) and of the set Sol(μ) of first-stage solutions to (1)–(3) is important when perturbing or approximating the underlying distribution μ. We say that the model (1)–(3) is *stable* if val($\cdot$) and Sol($\cdot$) satisfy certain continuity properties at μ with respect to some suitable convergence of probability measures. Here, we follow the presentation in [35] and consider the distance

$$d_f(\mu, \nu) := \sup\{|\int_{\Xi} f(x_1, \xi)(\mu - \nu)(d\xi)| \, : \, x_1 \in X_1\} \tag{14}$$

of probability measures μ and ν belonging to the set $\mathcal{P}_1(\Xi) := \{\nu \in \mathcal{P}(\Xi) : \int_{\Xi} \|\xi\| \nu(d\xi) < \infty\}$. Then it holds for any perturbation ν of the original underlying probability distribution μ that

$$|\text{val}(\mu) - \text{val}(\nu)| \leq d_f(\mu, \nu) \tag{15}$$

$$\emptyset \neq \text{Sol}(\nu) \subseteq \text{Sol}(\mu) + \Psi(d_f(\mu, \nu))B_{m_1} \tag{16}$$

where B_{m_1} denotes the closed unit ball in $\mathbb{R}^{m_1}$ and Ψ is some monotonically increasing function on $\mathbb{R}_+$ with $\Psi(0) = 0$, which is related to the growth behaviour of the objective function near the set Sol(μ) (see [35]). While (15) represents a Lipschitz type estimate for the optimal value at μ, the relation (16) says that the sets of first-stage solutions behave upper semicontinuously at μ with respect to d_f. In general, the distance d_f is rather involved and difficult to handle. Hence, it is of considerable interest to derive estimates of d_f in terms of simpler probability metrics and to expose relations to the classical concept of weak convergence of probability measures. For two-stage models with fixed recourse matrix and costs, such results are obtained in [44] and [35]. We also refer to relevant stability studies in [2,18,49]. Altogether, such stability results justify the approximation of the underlying distribution μ by simpler measures and provide techniques for designing approximation schemes. Next, we show that approximations by discrete measures having finitely many atoms or *scenarios* play a prominent role since they lead to specially structured large-scale mixed-integer linear programs.

2.4 Scenario Based Models

We assume throughout this section that Ω is finite, i.e., $\Omega = \{\omega_s\}_{s=1}^S$, $\mathcal{F}$ is the power set of Ω and $P(\{\omega_s\}) = p_s$, $s = 1, \dots, S$. We denote by $\xi_t^s := \xi_t(\omega_s)$ the value of the data scenario s at stage t and by x_t^s the value of the decision scenario s at t for $s = 1, \dots, S$, $t = 1, \dots, T$. Since Ω is finite, there exists a finite subset $\mathcal{E}_t$ of the σ-algebra $\mathcal{F}_t$, for each $t = 1, \dots, T$, such that $\mathcal{E}_t$ is a partition of Ω and that the smallest σ-algebra containing $\mathcal{E}_t$ is just $\mathcal{F}_t$. Then the conditional expectation w.r.t. $\mathcal{F}_t$ in the nonanticipativity condition (4) takes the form

$$\begin{aligned} E[x_t|\mathcal{F}_t] &= \sum_{C \in \mathcal{E}_t} \frac{1}{P(C)} \int_C x_t(\omega) P(d\omega) \chi_C \\ &= \sum_{C \in \mathcal{E}_t} \Big(\sum_{\substack{s=1 \\ \omega_s \in C}}^S p_s \Big)^{-1} \Big(\sum_{\substack{s=1 \\ \omega_s \in C}}^S p_s x_t^s \Big) \chi_C \end{aligned} \tag{17}$$

where χ_C denotes the characteristic function of the set $C \in \mathcal{E}_t$. Hence, the nonanticipativity condition (4) is equivalent to the following equality constraints

$$x_t^\sigma = \sum_{\substack{C \in \mathcal{E}_t \\ \omega_\sigma \in C}} \Big(\sum_{\substack{s=1 \\ \omega_s \in C}}^S p_s \Big)^{-1} \sum_{\substack{s=1 \\ \omega_s \in C}}^S p_s x_t^s, \ \sigma = 1, \dots, S, \ t = 1, \dots, T. \tag{18}$$

Clearly, for $t = 1$ we have $\mathcal{E}_1 = \{\Omega\}$ and, hence, condition (18) is equivalent to the equations $x_1^\sigma = \sum_{s=1}^S p_s x_1^s$, $\sigma = 1, \dots, S$, i.e, to $x_1^1 = \dots = x_1^S$.

Hence, the multistage stochastic program (1)–(3) takes the following form which will be called its scenario formulation:

$$\begin{aligned} \min \Big\{ & \sum_{s=1}^S \sum_{t=1}^T p_s c_t(\xi_t^s) x_t^s \ : \ x \text{ satisfies the constraints (18)}, \\ & x_t^s \in X_t, \ B_t(\xi_t^s) x_t^s \geq d_t(\xi_t^s), \ s = 1, \dots, S, \ t = 1, \dots, T, \\ & \sum_{\tau=1}^t A_{t\tau}(\xi_t^s) x_\tau^s \geq g_t(\xi_t^s), \ s = 1, \dots, S, \ t = 2, \dots, T \Big\} \end{aligned} \tag{19}$$

Since $\mathcal{F}_t \subseteq \mathcal{F}_{t+1}$, every element of $\mathcal{E}_{t+1}$ can be represented as the union of certain elements of $\mathcal{E}_t$. Furthermore, formula (17) shows that the number of elements in

$\mathcal{E}_t$ coincides with the number of realizations of ξ and x at period t, respectively. Hence, representing the relations between the elements of $\mathcal{E}_t$ and $\mathcal{E}_{t+1}$ for $t = 1, \ldots, T-1$, leads to a tree having the same structure as the sets of scenarios of ξ and x, respectively. Therefore, such a tree is called *scenario tree*. It is based on a finite set $\mathcal{N} \subseteq \mathbb{N}$ of nodes. Figure 1 shows an example of a scenario tree where the t_k denote the branching points of the tree.

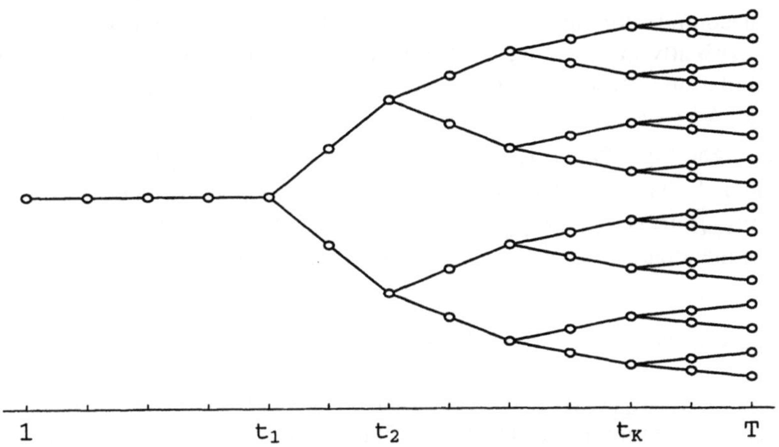

Figure 1. Example of a (binary) scenario tree

The root node $n = 1$ stands for period $t = 1$. Every other node n has a unique predecessor node n_- and a transition probability $\pi_{n/n_-} > 0$, which is the probability of n being the successor of n_-. The probability π_n of each node n is given recursively by $\pi_1 = 1$, $\pi_n = \pi_{n/n_-}\pi_{n_-}$, $n > 1$. We denote by $\mathcal{N}_+(n)$ the set of successors to node n, by $\mathrm{path}(n)$ the path from the root to node n and by $t(n)$ its length, i.e., $t(n) := \mathrm{card}(\mathrm{path}(n))$. $\mathcal{N}_t$ denotes the set $\{n \in \mathcal{N} : t(n) = t\}$, and it holds $\sum_{n \in \mathcal{N}_t} \pi_n = 1$ for each period t. Nodes n with $\mathcal{N}_+(n) = \emptyset$ are called leaves; they constitute the terminal set $\mathcal{N}_T$. A scenario corresponds to a path from the root node to a leaf. Clearly, it holds that $\mathrm{card}(\mathcal{N}_T) = S$ and $\{\pi_n\}_{n \in \mathcal{N}_T} = \{p_s\}_{s=1}^S$. Conversely, given these scenario probabilities, the remaining node and transition probabilities are generated recursively by $\pi_n := \sum_{n_+ \in \mathcal{N}_+(n)} \pi_{n_+}$, $\pi_{n_+/n} := \pi_{n_+}/\pi_n$ for $n_+ \in \mathcal{N}_+(n)$. We use the following notation for the sequence of predecessors of any node $n \in \mathcal{N}$: $n_0 := n$, $n_{-1} := n_-$ if $n > 1$, $n_{-(\kappa+1)} := (n_{-\kappa})_-$ if $t(\kappa) > 1$. Note that $t(n_{-\kappa}) = t(n) - \kappa$ for $\kappa = 1, \ldots, t(n) - 1$. Furthermore, we denote by $\{\xi^n\}_{n \in \mathcal{N}_t}$ the realizations of ξ_t and by $\{x^n\}_{n \in \mathcal{N}_t}$ the realizations of x_t. After these preparations the scenario tree formulation of the multistage stochastic program reads:

$$\min\{\sum_{n\in\mathcal{N}} \pi_n c_{t(n)}(\xi^n)x^n : x^n \in X_{t(n)},\ B_{t(n)}(\xi^n)x^n \geq d_{t(n)}(\xi^n), \quad (20)$$
$$\sum_{\kappa=0}^{t(n)-1} A_{t(n),t(n)-\kappa}(\xi^n)x^{n-\kappa} \geq g_{t(n)}(\xi^n),\ n\in\mathcal{N}\}$$

Both formulations of the multistage stochastic program will be used for the description of decomposition approaches. We recall that the nonanticipativity condition appears explicitly in the scenario formulation (19), but disappears in the scenario tree formulation (20) because it is incorporated into the tree construction. Since it holds that $|\mathcal{N}| := \text{card}(\mathcal{N}) << TS$, the dimensions of both model formulations are quite different. More precisely, the model (19) contains $(\sum_{t=1}^{T} m_t)S$ decision variables and $\sum_{t=1}^{T}(m_t+k_t+r_t)S$ linear constraints, whereas the model (20) contains $\sum_{n\in\mathcal{N}} m_{t(n)}$ decisions and $\sum_{n\in\mathcal{N}}(k_{t(n)}+r_{t(n)})$ linear constraints. Here, k_t and r_t denote the dimensions of $d_t(\cdot)$ and $g_t(\cdot)$, respectively, for $t=1,\ldots,T$.

2.5 *Dualization and the Convex Case*

We assume $\xi \in \times_{t=1}^{T} L_\infty(\Omega,\mathcal{F},P;\mathbb{R}^{s_t})$ and consider the multistage stochastic integer program of Section 1 as an abstract (infinite) optimization problem in the Banach space $\times_{t=1}^{T} L_\infty(\Omega,\mathcal{F},P;\mathbb{R}^{m_t})$, i.e., in the form

$$\min\{E\,[\sum_{t=1}^{T} c_t(\xi_t)x_t] : x \in \times_{t=1}^{T} L_\infty(\Omega,\mathcal{F},P;\mathbb{R}^{m_t}),\ x\in\mathcal{N}_{na}, \quad (21)$$
$$x_t \in X_t,\ B_t(\xi_t)x_t \geq d_t(\xi_t),\ P-\text{a.s.},\ t=1,\ldots,T, \quad (22)$$
$$\sum_{\tau=1}^{t} A_{t\tau}(\xi_t)x_\tau \geq g_t(\xi_t),\ P-\text{a.s.},\ t=2,\ldots,T\}. \quad (23)$$

Let $F(\cdot)$ denote the objective function, i.e., $F(x) := E[\sum_{t=1}^{T} c_t(\xi_t)x_t]$.

Our aim is to introduce a Lagrangian associated with the essential groups of constraints of problem (21)–(23), namely, the (functional) nonanticipativity constraint $x \in \mathcal{N}_{na}$, the k_t coupling constraints $B_t(\xi_t)x_t \geq d_t(\xi_t)$ and r_t dynamic constraints (23). We make use of the concepts and results of [38] and introduce the following sets $\Lambda_1 := \{\lambda_1 \in \times_{t=1}^{T} L_1(\Omega,\mathcal{F},P;\mathbb{R}^{m_t}) : E[\lambda_{1t}|\mathcal{F}_t] = 0,\ P-\text{a.s.},\ t=1,\ldots,T\}$, $\Lambda_2 := \{\lambda_2 \in \times_{t=1}^{T} L_1(\Omega,\mathcal{F},P;\mathbb{R}^{k_t}) : \lambda_2 \geq 0,\ P-\text{a.s.}\}$ and $\Lambda_3 := \{\lambda_3 \in \times_{t=1}^{T} L_1(\Omega,\mathcal{F},P;\mathbb{R}^{r_t}) : \lambda_3 \geq 0,\ P-\text{a.s.}\}$ of Lagrange multipliers. The sets Λ_2 and Λ_3 are convex cones and Λ_1 is a linear space which is complementary to the nonanticipativity subspace $\mathcal{N}_{na}$ with respect to the dual pairing $\langle\cdot,\cdot\rangle$ of L_1 and L_∞, i.e., it holds $\langle\lambda_1, x\rangle := E[\sum_{t=1}^{T}\lambda_{1t}x_t] = 0$ for all $\lambda_1 \in \Lambda_1$ and $x \in \mathcal{N}_{na}$.

The Lagrangian is defined to be the function

$$L(x,\lambda) := E[\sum_{t=1}^{T} \{ c_t(\xi_t)x_t - \lambda_{1t}x_t + \lambda_{2t}(d_t(\xi_t) - B_t(\xi_t)x_t)\} \tag{24}$$
$$+ \sum_{t=2}^{T} \lambda_{3t}(g_t(\xi_t) - \sum_{\tau=1}^{t} A_{t\tau}(\xi_t)x_\tau)]$$

from $\times_{t=1}^{T} L_\infty(\Omega, \mathcal{F}, P; \mathbb{R}^{m_t}) \times \Lambda$ to $\mathbb{R}$, where $\Lambda := \times_{i=1}^{3} \Lambda_i$. The dual function D from Λ to $\mathbb{R}$ is defined by

$$D(\lambda) := \inf\{L(x,\lambda) : x \in \times_{t=1}^{T} L_\infty(\Omega, \mathcal{F}, P; R^{m_t}), \tag{25}$$
$$x_t \in X_t,\ P - \text{a.s.},\ t = 1, \ldots, T\},$$

and the *dual problem* associated with (21)–(23) is

$$\max\{D(\lambda) : \lambda \in \Lambda\}. \tag{26}$$

We assume again that the sets X_t, $t = 1, \ldots, T$, are compact. Then the Lagrangian L and the dual function D are well-defined, D is concave and the *weak duality* estimate

$$D(\lambda) \leq F(x) \quad \text{for all } \lambda \in \Lambda \text{ and all } x \text{ satisfying (21)–(23).} \tag{27}$$

is valid. In the following, we say that the model (21)–(23) is *strictly feasible* if there exist $\tilde{x} \in \mathcal{N}_{na}$ and $\varepsilon > 0$ such that

$$\tilde{x}_t + \varepsilon B_{m_t} \subseteq \text{conv}(X_t),\ B_t(\xi_t)\tilde{x}_t \geq d_t(\xi_t) + \varepsilon,\ P - \text{a.s.},\ t = 1, \ldots, T,$$
$$\sum_{\tau=1}^{t} A_{t\tau}(\xi_t)\tilde{x}_\tau \geq g_t(\xi_t) + \varepsilon,\ P - \text{a.s.},\ t = 2, \ldots, T,$$

where B_m denotes the closed unit ball in $\mathbb{R}^m$. Then we conclude from Theorem 1 and from Theorem 3 of [39] that the following holds.

Theorem 2. *Assume that the general assumptions are satisfied, that the sets* X_t, $t = 1, \ldots, T$, *are convex compact and that the model (21)–(23) has relatively complete recourse and is strictly feasible. Then there exist optimal solutions* $\bar{\lambda}$ *to (26) and* $\bar{x}$ *to (21)–(23), and it holds* $D(\bar{\lambda}) = F(\bar{x})$.

Since the sets X_t, $t = 1, \ldots, T$, fail to be convex, such a duality result is not available in our setting and, due to (27), we are faced with a *duality gap*

$$DG := F(\bar{x}) - \sup_{\lambda \in \Lambda} D(\lambda) \geq 0. \tag{28}$$

This inequality is strict, in general. On the other hand, in case of a discrete underlying probability distribution, the theory of Lagrangian relaxation in mixed-integer linear programming (cf., e.g., Chapter II.3.6 of [31]) implies that the optimal value of (26) is greater than or equal to the optimal value of the linear programming relaxation to (19) or (20). In other words, the lower bound obtained by dualizing constraints is never worse the bound obtained by relaxing the integer requirements.

So far we have associated Lagrange multipliers with nonanticipativity, coupling as well as dynamic constraints. Of course, it is also of interest to consider restricted Lagrangians and restricted duals by associating multipliers with one or with two of these three groups of constraints, only. For such restricted dualization schemes, duality results for the convex case that are similar to Theorem 2 may be derived as well (see [37] for dualizing the nonanticipativity constraints and [38] for other inequality constraints). It is worth recalling that the duality gap increases when dualizing additional constraints (see Section 3.1 in [29]). Since small duality gaps are of algorithmic interest, we take a closer look at dualization schemes where either nonanticipativity or coupling or dynamic constraints are associated with Lagrange multipliers. We denote the corresponding dual functions from Λ_i to $\mathbb{R}$ by D_i for $i = 1, 2, 3$ and start with dualizing nonanticipativity constraints, i.e.,

$$
\begin{aligned}
D_1(\lambda_1) := \inf\{&E[\sum_{t=1}^{T}(c_t(\xi_t)x_t - \lambda_{1t}x_t)] \;:\; x \in \times_{t=1}^{T} L_\infty(\Omega, \mathcal{F}, P; R^{m_t}),\\
&x_t \in X_t,\; B_t(\xi_t)x_t \geq d_t(\xi_t),\; P-\text{a.s.},\; t = 1, \ldots, T,\\
&\sum_{\tau=1}^{t} A_{t\tau}(\xi_t)x_\tau \geq g_t(\xi_t),\; P-\text{a.s.},\; t = 2, \ldots, T\}\\
= E[\inf\{&\sum_{t=1}^{T}(c_t(\xi_t)x_t - \lambda_{1t}x_t) \;:\; x_t \in X_t,\; B_t(\xi_t)x_t \geq d_t(\xi_t),\\
&t = 1, \ldots, T,\; \sum_{\tau=1}^{t} A_{t\tau}(\xi_t)x_\tau \geq g_t(\xi_t),\; t = 2, \ldots, T\}],
\end{aligned}
$$

where the infimum and expectation may be interchanged since the minimization problem only contains P-a.s. pointwise constraints (see e.g., Theorem 14.60 of [40]). Hence, the multistage stochastic program defining D_1 decomposes into pathwise minimization problems. This effect becomes more transparent if the underlying probability distribution of ξ is discrete, i.e., if $\Omega = \{\omega_1, \ldots, \omega_S\}$. Adopting the notation of Section 2.4, the dual function takes the form

$$
\begin{aligned}
D_1(\lambda_1) := \sum_{s=1}^{S} p_s \inf\{&\sum_{t=1}^{T}[c_t(\xi_t^s)x_t^s - \lambda_{1t}^s x_t^s] \;:\; x_t^s \in X_t,\\
&B_t(\xi_t^s)x_t^s \geq d_t(\xi_t^s),\; t = 1, \ldots, T,\\
&\sum_{\tau=1}^{t} A_{t\tau}(\xi_t^s)x_\tau^s \geq g_t(\xi_t^s),\; t = 2, \ldots, T\},
\end{aligned}
\tag{29}
$$

where $\lambda_1 \in \Lambda_1$ has the scenarios $\{\lambda_{1t}^s\}_{t=1}^T$ with probabilities p_s for $s = 1, \dots, S$ and Λ_1 is given by the linear subspace

$$\Lambda_1 = \{\lambda_1 : E[\lambda_{1t}|\mathcal{F}_t] = \sum_{C \in \mathcal{E}_t} (\sum_{\substack{s=1 \\ \omega_s \in C}}^{S} p_s)^{-1} \sum_{\substack{s=1 \\ \omega_s \in C}}^{S} p_s \lambda_{1t}^s \chi_C = 0, t = 1, \dots, T\}$$

$$= \{\lambda_1 : \sum_{\substack{s=1 \\ \omega_s \in C}}^{S} p_s \lambda_{1t}^s = 0, C \in \mathcal{E}_t, t = 1, \dots, T\} \tag{30}$$

of the Euclidean space of dimension $(\sum_{t=1}^T k_t + \sum_{t=2}^T r_t)S$. Since $\mathcal{E}_1 = \{\Omega\}$ and $\mathcal{E}_T = \{\{\omega_1\}, \dots, \{\omega_S\}\}$, the conditions for $t = 1$ and $t = T$ in (30) are equivalent to $\sum_{s=1}^S p_s \lambda_{11}^s = 0$ and $\lambda_{1T}^s = 0$, $s = 1, \dots, S$, respectively. We note that the constraint $\lambda_1 \in \Lambda_1$ means that each $\lambda_1 \neq 0$ is *anticipative*, i.e., λ_{1t} is not $\mathcal{F}_t$-measurable for some t (see also the example in [22]). Sometimes, one might find it more convenient that the dual function is defined and maximized on the whole space, i.e, without regard to the subspace constraint $\lambda_1 \in \Lambda_1$. This can be done be replacing λ_{1t} in the right-hand side of (29) by $\lambda_{1t} - E[\lambda_{1t}|\mathcal{F}_t]$ for $t = 1, \dots, T$. Then the subspace constraint for the multiplier is automatically satisfied and the dual maximization problem is unconstrained.

Next we consider dualizations of certain inequality constraints by some multiplier, but leave the nonanticipativity constraint untouched. In contrast to the anticipativity of multipliers in the previous case, the multipliers may now be chosen nonanticipative, i.e., as elements of $\times_{t=1}^T L_1(\Omega, \mathcal{F}_t, P)$. This is due to the linear separability properties of (21)–(23) (Theorem 7 of [38]). In particular, when dualizing the coupling constraints, the restricted dual function

$$D_2(\lambda_2) := \inf\{E[\sum_{t=1}^{T} (c_t(\xi_t)x_t + \lambda_{2t}(d_t(\xi_t) - B_t(\xi_t)x_t))] : x \in \mathcal{N}_{na}, \tag{31}$$
$$x_t \in X_t, \sum_{\tau=1}^{t} A_{t\tau}(\xi_t)x_\tau \geq g_t(\xi_t), P - a.s., t = 2, \dots, T\},$$

has to be maximized on the convex cone $\Lambda_2 := \{\lambda_2 \in \times_{t=1}^T L_1(\Omega, \mathcal{F}_t, P; \mathbb{R}^{k_t}) : \lambda_{2t} \geq 0, P - a.s., t = 1, \dots, T\}$. Dualizing the dynamic constraints leads to maximizing the restricted dual

$$D_3(\lambda_3) := \inf\{E[\sum_{t=1}^{T} c_t(\xi_t)x_t + \sum_{t=2}^{T} \lambda_{3t}(g_t(\xi_t) - \sum_{\tau=1}^{t} A_{t\tau}(\xi_t)x_\tau)] : \tag{32}$$
$$x \in \mathcal{N}_{na}, x_t \in X_t, B_t(\xi_t)x_t \geq d_t(\xi_t), P - a.s., t = 1, \dots, T\}.$$

subject to the convex cone $\Lambda_3 := \{\lambda_3 \in \times_{t=1}^T L_1(\Omega, \mathcal{F}_t, P; \mathbb{R}^{r_t}) : \lambda_{3t} \geq 0, P - a.s., t = 1, \dots, T\}$. Clearly, both optimization problems on the right-hand sides of (31) and (32), respectively, are stochastic integer programs. While the program in

(31) exhibits the typical multistage structure, the specific feature of the program in (32) is the lack of a dynamic constraint. In Sections 3.2 and 3.3 we gain further information on these programs in case of a discrete underlying probability distribution, i.e., when the data, decisions and multipliers form scenario trees.

3 Decomposition Methods

Due to the enormous size of scenario based models in multi-stage stochastic programming, decomposition is the method of choice when it comes to numerical solution. This is further enhanced by special structures met, both in the scenario formulation (19) and in the scenario tree formulation (20) of multi-stage stochastic programs. If integer requirements are missing in (21) - (23), powerful convexity and duality results (cf. Theorem 2) are the basis of efficient decomposition methods. These methods can be subdivided into primal and dual ones.

Primal decomposition methods employ the scenario tree formulation (20). Starting from the root node, primal proposals are passed down the tree where they are used to compute so called feasibility and optimality cuts that are passed upward to be included into convex optimization problems whose solutions lead to updated primal proposals that are again passed down the tree, and so on. This procedure (nested decomposition) is enhanced by regularization and cut deletion. Its mathematical backbone is convexity, in particular ideas from the area of bundle-trust and proximal point methods.

Dual decomposition circles around duality results such as Theorem 2. The approaches discussed in Section 2.5 then all benefit from a zero duality gap. Particular attention has been paid to dualizing nonanticipativity in the framework of augmented Lagrangians and related proximal point algorithms (progressive hedging, cf. [39]). The survey papers [5, 42] provide further insights into both primal and dual decomposition of multi-stage stochastic linear programs.

With integer requirements in (21)–(23) the mentioned powerful convexity and duality results are lost. Approaches to decomposition, that have proven efficient for purely linear models, have to be rethought from their very beginnings.

The impact of integrality on *primal decomposition* is twofold: Feasibility and optimality cuts can no longer be obtained as linear functionals but as merely subadditive functionals instead. Primal proposals can no longer be obtained via convex programs but via merely lower semicontinuous (discontinuous) nonconvex programs instead. For algorithmic realization this leads to obstacles impossible to overcome with existing methods, [8, 11]. Two-stage models have been tackled with limited success by solving the mentioned lower semicontinuous programs via enumeration [45] or branch-and-bound [1] and exploiting problem similarities in the second stage.

The impact of integrality on *dual decomposition* has already been mentioned in Section 2.5: Theorem 2 is no longer valid, and we face a non-zero duality gap (28). Although progressive hedging then is no longer formally justified, quite satisfactory results have been observed empirically for specific applications, [30, 47].

In what follows, we will return to the dualization schemes introduced in Section 2.5 in case that the underlying probability distribution is discrete. We will discuss the solution of the corresponding dual maximization problems

$$\max\{D_i(\lambda_i) : \lambda_i \in \Lambda_i\} \quad (i = 1, 2, 3)$$

by subgradient type methods and examine the decoupling potential of the different dualizations. Under the conditions imposed in Section 2.5 the dual functions D_i are finite, concave and polyhedral. They have the form

$$D_i(\lambda_i) = \inf_x\{F(x) + \langle \lambda_i, G_i(x) \rangle\}, \tag{33}$$

where F is the objective function, G_i is some affine linear function from L_∞ to L_∞, and $\langle \cdot, \cdot \rangle$ denotes the dual pairing of L_1 and L_∞. Hence, $G_i(x_i(\lambda_i))$ is a subgradient of D_i if $x_i(\lambda_i)$ is a solution to the minimization problem (33) defining D_i. Furthermore, the solution sets of the dual problems are nonempty since their objectives are polyhedral and their suprema finite. Therefore, *subgradient bundle methods* may be used for solving the duals, [23,25,28]. Let us consider the *proximal bundle method* [17,23,25] in some more detail. Starting from an arbitrary point $\lambda_i^1 = \bar{\lambda}_i^1 \in \Lambda_i$, this method generates a sequence $\{\lambda_i^k\}_{k \in \mathbb{N}}$ in Λ_i converging to some solution of the dual problem, and trial points $\bar{\lambda}_i^k$ for evaluating the solutions $x_i^k = x_i(\bar{\lambda}_i^k)$ of (33), the subgradients $G_i(x_i^k)$ of D_i and its linearizations

$$D_i^k(\cdot) := D_i(\lambda_i^k) + \langle \cdot - \bar{\lambda}_i^k, G_i(x_i^k) \rangle \geq D_i(\cdot).$$

Iteration k uses the polyhedral model $D_{ik}(\cdot) := \min_{l \in N^k} D_i^l(\cdot)$ with $k \in N^k \subset \{1, \ldots, k\}$ for finding the next trial point $\bar{\lambda}_i^{k+1}$ as a solution of the quadratic subproblem

$$\max\{D_{ik}(\lambda) - \frac{1}{2}u_k|\lambda - \lambda_i^k|^2 : \lambda \in \Lambda_i\}, \tag{34}$$

where the proximity weight $u_k > 0$ and the penalty term $|\cdot|^2 := \langle \cdot, \cdot \rangle$ should keep $\bar{\lambda}_i^{k+1}$ close to the prox-center λ_i^k. An ascent step to $\lambda_i^{k+1} = \bar{\lambda}_i^{k+1}$ occurs if $D_i(\bar{\lambda}_i^{k+1}) \geq D_i(\lambda_i^k) + \kappa\delta_k$, where $\kappa \in (0, 1)$ is a fixed Armijo-like parameter and $\delta_k := D_{ik}(\bar{\lambda}_i^{k+1}) - D_i(\lambda_i^k) \geq 0$ is the predicted ascent (if $\delta_k = 0$ then λ_i^k is a solution and the method may stop). Otherwise, a null step $\lambda_i^{k+1} = \lambda_i^k$ improves the next model $D_{i,k+1}$ with the new linearization D_i^{k+1}. The choices of the weights u_k and of the index set N^{k+1} are dicussed in [17,25] (see also Section 3.4 of [19]). The quadratic subproblem (34) is essentially influenced by the dual pairing $\langle \cdot, \cdot \rangle$. The latter reads $\langle \lambda_i, y \rangle = \sum_{s=1}^S p_s \sum_{t=1}^T \lambda_{it}^s y_t^s$ and $\langle \lambda_i, y \rangle = \sum_{n \in \mathcal{N}} \pi_n \lambda_i^n y^n$ for the scenario and the node formulations, respectively.

3.1 Scenario Decomposition

Scenario decomposition rests on the dualization of nonanticipativity constraints if the probability distribution of ξ is discrete. This leads to the dual maximization problem

$$\max\{D_1(\lambda_1) : \lambda_1 \in \Lambda_1\} \tag{35}$$

where D_1 and Λ_1 are defined as in (29), (30) of Section 2.5. Since the computation of D_1 decomposes into solving pathwise minimization problems, function values and subgradients of D_1 are obtained by solving the single-scenario problems

$$\begin{aligned}\min\{\sum_{t=1}^{T}\{c_t(\xi_t^s)x_t^s - \lambda_{1t}^s x_t^s\} \; : \; & x_t^s \in X_t, \\ & B_t(\xi_t^s)x_t^s \geq d_t(\xi_t^s),\ t = 1,\ldots,T, \\ & \sum_{\tau=1}^{t} A_{t\tau}(\xi_t^s)x_\tau^s \geq g_t(\xi_t^s),\ t = 2,\ldots,T\}\end{aligned}$$

for all $s = 1,\ldots,S$.
Indeed, if $\bar{x}^s$, $s = 1,\ldots,S$, denote optimal solutions to these problems, then

$$D_1(\lambda_1) = \sum_{s=1}^{S} p_s(\sum_{t=1}^{T}\{c_t(\xi_t^s)\bar{x}_t^s - \lambda_{1t}^s \bar{x}_t^s\}),$$

and $G_1(\bar{x}) = \bar{x}$ is a subgradient of D_1 at λ_1, where $\bar{x}$ has the scenarios $\bar{x}^s$, $s = 1,\ldots,S$. Compared with the scenario formulation (19) of the multi-stage stochastic program (1)–(3), which is a mixed-integer linear program in dimension $S \cdot \sum_{t=1}^{T} m_t$, the above single-scenario problems are S mixed-integer linear programs each of dimension $\sum_{t=1}^{T} m_t$, only. In view of (28), solving (35) provides a lower bound to the optimal value of the multi-stage stochastic integer program (19).

If the single-scenario solutions $\bar{x}_1^s,\ldots,\bar{x}_T^s$ for the optimal λ_1 in (35) fulfilled the nonanticipativity constraints then $\bar{x}$ would be optimal to (19). In general, however, one faces a non-zero duality gap (28). Therefore the lower bounding has to be accompanied by upper bounding procedures resting on the generation of "promising" feasible solutions. This can be accomplished by primal heuristics starting from the results of the dual optimization, i.e., from single-scenario solutions $\bar{x}_1^s,\ldots,\bar{x}_T^s$ corresponding to optimal or nearly optimal λ_1.

An algorithmic realization of scenario decomposition for the case $T = 2$, i.e., for two-stage stochastic integer programs, has been proposed in [8–10]. The nonanticipativity constraints then read $x_1^\sigma = \sum_{s=1}^{S} p_s x_1^s$, $\sigma = 1,\ldots,S$. In [8–10], the equivalent representation $x_1^1 = \cdots = x_1^S$ is employed, and the scenario formulation (19) is set up with (18) replaced by $x_1^1 = \cdots = x_1^S$. Then, the usual Lagrangian relaxation of mixed-integer linear programming is performed with respect to the constraints $x_1^1 = \cdots = x_1^S$. In particular, this leads to a non-probabilistic Lagrangian, in contrast to the probabilistic Lagrangian (24) introduced in Section 2.5. As a consequence, the Lagrangian dual of [8–10] is unconstrained and lives in dimension $(S-1)\cdot m_1$. In the setting of Section 2.5, cf. (30), we obtain a dual in dimension $S \cdot m_1$ constrained by $\sum_{s=1}^{S} p_s \lambda_{11}^s = 0$, i.e., essentially an unconstrained program in dimension $(S-1)\cdot m_1$ as well.

In [8–10], the scheme of lower and upper bounding outlined above is further enhanced by embedding into a branch-and-bound algorithm in the spirit of global optimization. As stated in (13), the stochastic program can be rewritten as a nonconvex global optimization problem. In the branching part of the algorithm, the feasible region of (13) is subdivided. On each member of the subdivision, the bounding part employs dualization of nonanticipativity for the lower and a primal heuristic for the upper bounds. For further details on scenario decomposition for two-stage stochastic integer programs we refer to [21].

Only little is known about algorithmic realizations of scenario decomposition for multi-stage stochastic integer programs with $T > 2$. First experiences on extending the approach of [8–10] will be reported in [4].

3.2 *Component Decomposition*

Dualization of component coupling constraints results in the dual maximization problem

$$\max\{D_2(\lambda_2) : \lambda_2 \in \Lambda_2\},$$

where D_2 and Λ_2 are defined in Section 2.5. We assume that the underlying probability distribution of the data process ξ is discrete and, hence, given in form of a scenario tree $\{\xi^n\}_{n\in\mathcal{N}}$, where $\mathcal{N}$ denotes the finite set of nodes. The notation of Section 2.4 is used, and we denote by $x = \{x^n\}_{n\in\mathcal{N}}$ the decision scenario tree and by $\lambda_2 = \{\lambda_2^n\}_{n\in\mathcal{N}}$ the multiplier scenario tree. Then the dual function (31) may be rewritten in the following form (see also (20)):

$$D_2(\lambda_2) := \inf\{\sum_{n\in\mathcal{N}} \pi_n\{c_{t(n)}(\xi^n)x^n + \lambda_2^n(d_{t(n)}(\xi^n) - B_{t(n)}(\xi^n)x^n)\} : \quad (36)$$
$$x^n \in X_{t(n)}, \sum_{\kappa=0}^{t(n)-1} A_{t(n),t(n)-\kappa}(\xi^n)x^{n-\kappa} \geq g_{t(n)}(\xi^n), n \in \mathcal{N}\}$$

where $\lambda_2 \in \Lambda_2 = \{\{\lambda_2^n\}_{n\in\mathcal{N}} : \lambda_2^n \geq 0, n \in \mathcal{N}\}$. In order to demonstrate the component decoupling potential hidden in D_2, we assume that X_t has the specific structure $X_t = \times_{i=1}^{m_t} X_{ti}$, where the X_{ti} are closed subsets of $\mathbb{R}$, that $m_t = m$, $k_t = k$ and $r_t = mr$ for $t = 1, \ldots, T$ and some $r \in \mathbb{N}$, and that the matrices $A_{t\tau}(\cdot)$ are block-diagonal with m blocks $a_{t\tau}^i(\cdot) \in \mathbb{R}^r$ for $i = 1, \ldots, m$. In particular, this condition means that the constraints in (36) are expressible as componentwise constraints. We denote by $c_t^i(\cdot)$ the i-th component of $c_t(\cdot)$, by $g_t^i(\cdot) \in \mathbb{R}^r$ the i-th component vector of $g_t(\cdot)$, and by $b_t^i(\cdot)$ the i-th column of the matrix $B_t(\cdot)$. With x_i^n denoting the i-th component of x^n, we obtain by exchanging summation w.r.t.

n and i

$$
\begin{aligned}
D_2(\lambda_2) = \inf\{ \sum_{n\in\mathcal{N}} \pi_n\{\sum_{i=1}^{m} [c^i_{t(n)}(\xi^n) - \lambda_2^n b^i_{t(n)}(\xi^n)]x_i^n + \lambda_2^n d_{t(n)}(\xi^n)\} : \\
x_i^n \in X^i_{t(n)}, \sum_{\kappa=0}^{t(n)-1} a^i_{t(n),t(n)-\kappa}(\xi^n) x_i^{n-\kappa} \geq g^i_{t(n)}(\xi^n), \\
i = 1, \ldots, m,\ n \in \mathcal{N}\} \\
= \sum_{i=1}^{m} D_{2i}(\lambda_2) + \sum_{n\in\mathcal{N}} \pi_n \lambda_2^n d_{t(n)}(\xi^n)
\end{aligned}
$$

where the functions D_{2i}, $i = 1, \ldots, m$, from Λ_2 to $\mathbb{R}$ are defined by

$$
\begin{aligned}
D_{2i}(\lambda_2) = \inf\{ \sum_{n\in\mathcal{N}} \pi_n [c^i_{t(n)}(\xi^n) - \lambda_2^n b^i_{t(n)}(\xi^n)]x_i^n : x_i^n \in X^i_{t(n)}, \\
\sum_{\kappa=0}^{t(n)-1} a^i_{t(n),t(n)-\kappa}(\xi^n) x_i^{n-\kappa} \geq g^i_{t(n)}(\xi^n),\ n \in \mathcal{N}\}.
\end{aligned} \tag{37}
$$

By specifying (33) we obtain that $G_2(\bar{x}) = \{d_{t(n)}(\xi^n) - \sum_{i=1}^{m} b^i_{t(n)}(\xi^n)\bar{x}_i^n\}_{n\in\mathcal{N}}$ is a subgradient of D_2 at λ_2, where $\bar{x}_i = \{\bar{x}_i^n\}_{n\in\mathcal{N}}$ is a solution of (37). The dual function (36), which is defined by a multistage stochastic integer program of dimension $m|\mathcal{N}|$, decomposes into m functions each given by a multistage stochastic integer program of dimension $|\mathcal{N}|$. Since the dimension of the dual problem is $k|\mathcal{N}|$, the computational potential of this dualization approach takes effect in situations, where the number k of coupling constraints to be dualized is much smaller than the decision dimension m (i.e., $k << m$) and where the m subproblems (37) of dimension $|\mathcal{N}|$ can be solved much faster than the original multistage model of dimension $m|\mathcal{N}|$. The latter could appear, for example, if complex mixed-integer models decompose into pure integer and pure linear programs.

Component decomposition has been applied successfully under the label *Lagrangian relaxation* of coupling constraints to solving hydro-thermal power management models under data uncertainty. Lagrangian relaxation has a long tradition for solving (deterministic) unit commitment problems of power systems operation. Recently, this technique has been extended to stochastic power management models, where the stochasticity enters the model, for example, via the electric load, streamflows to hydro units, and electricity prices. When letting the production decisions of individual power units play the role of components, the above dualization scheme leads to a decomposition into single (thermal or hydro) power unit models. Such approaches for determining lower bounds have been proposed and implemented in [3, 13, 19, 33, 41]. In [19, 20, 32] encouraging numerical results and computing times have been reported for both solving the dual and determining a nearly optimal primal solution by a Lagrangian based heuristic.

3.3 Nodal Decomposition

Finally, we return to the dualization of the dynamic constraints of (21) - (23) in case of a discrete underlying probability distribution and show that the dual function exhibits a nodewise decoupling structure. We let D_3 and Λ_3 be defined as in Section 2.5 and consider the corresponding dual problem

$$\max\{D_3(\lambda_3) \,:\, \lambda_3 \in \Lambda_3\}.$$

Let $\{\xi^n\}_{n\in\mathcal{N}}$ be the scenario tree representing the data process ξ, $\mathcal{N}$ the finite set of nodes, $\{\pi_n\}_{n\in\mathcal{N}}$ the node probabilities, and $\{x^n\}_{n\in\mathcal{N}}$ and $\{\lambda_3^n\}_{n\in\mathcal{N}}$ the corresponding scenario trees of the decision and of the multiplier process, respectively. Using the notation of Section 2.4, the dual function D_3 takes the following scenario tree representation

$$\begin{aligned} D_3(\lambda_3) = \inf\{c_1(\xi^1)x^1 + &\sum_{n\in\mathcal{N}\setminus\{1\}} \pi_n[c_{t(n)}(\xi^n)x^n + \lambda_3^n(g_{t(n)}(\xi^n) \\ &- \sum_{\kappa=0}^{t(n)-1} A_{t(n),t(n)-\kappa}(\xi^n)x^{n-\kappa})] \,: \\ &x^n \in X_{t(n)},\, B_{t(n)}(\xi^n)x^n \geq d_{t(n)}(\xi^n),\, n \in \mathcal{N}\}, \end{aligned} \tag{38}$$

where $\lambda_3 \in \Lambda_3 = \{\{\lambda_3^n\}_{n\in\mathcal{N}} \,:\, \lambda_3^n \geq 0,\, n \in \mathcal{N}\}$. Since the minimization problem in (38) contains only node constraints for the decision tree, we rearrange its objective function with respect to the decision nodes and obtain

$$\begin{aligned} D_3(\lambda_3) = \inf\{&\sum_{n\in\mathcal{N}} \pi_n(c_{t(n)}(\xi^n) - \sum_{\ell\in\mathrm{Tr}(n)} \pi_\ell\lambda_3^\ell A_{t(\ell),t(n)}(\xi^\ell))x^n \\ &+ \sum_{n\in\mathcal{N}\setminus\{1\}} \pi_n\lambda_3^n g_{t(n)}(\xi^n) \,: \\ &x^n \in X_{t(n)},\, B_{t(n)}(\xi^n)x^n \geq d_{t(n)}(\xi^n),\, n \in \mathcal{N}\}, \end{aligned}$$

where $\mathrm{Tr}(1) := \mathcal{N}\setminus\{1\}$, and $\mathrm{Tr}(n)$ for $n > 1$ denotes the set of all nodes belonging to the subtree with root node n, i.e., $\mathrm{Tr}(n) := \cup_{n_T\in\mathcal{N}_T}\{\mathrm{path}(n_T) \,:\, n \in \mathrm{path}(n_T)\}\setminus \mathrm{path}(n_-)$. Now, we may interchange summation and minimization and arrive at the node decomposed formulation

$$D_3(\lambda_3) = \sum_{n\in\mathcal{N}} D_{3n}(\lambda_3) + \sum_{n\in\mathcal{N}\setminus\{1\}} \pi_n\lambda_3^n g_{t(n)}(\xi^n) \tag{39}$$

of D_3, where the functions D_{3n}, $n \in \mathcal{N}$, are defined on Λ_3 and given by

$$\begin{aligned} D_{3n}(\lambda_3) := \inf\{(\pi_n c_{t(n)}(\xi^n) - &\sum_{\ell\in\mathrm{Tr}(n)} \pi_\ell\lambda_3^\ell A_{t(\ell),t(n)}(\xi^\ell))x^n \,: \\ &x^n \in X_{t(n)},\, B_{t(n)}(\xi^n)x^n \geq d_{t(n)}(\xi^n)\}. \end{aligned} \tag{40}$$

Hence, the representation (39) of D_3 provides a decomposition of the original mixed-integer program of dimension $\sum_{n \in \mathcal{N}} m_{t(n)}$ into $|\mathcal{N}|$ subproblems (40) of dimension $m_{t(n)}$ for $n \in \mathcal{N}$. Formulas for computing subgradients of D_3 may be derived similarly to the previous section. Computational experience of such nodal decomposition schemes for determining lower bounds of multistage stochastic integer programs is not available yet.

Acknowledgement

We wish to thank Krzysztof C. Kiwiel (Systems Research Institute, Polish Academy of Sciences, Warsaw) for helpful discussions on the subject of this paper.

References

1. S. Ahmed, M. Tawarmalani, N. V. Sahinides: A finite branch and bound algorithm for two-stage stochastic integer programs, preprint, University of Illinois, 2000
2. Z. Artstein, R. J.-B. Wets: Stability results for stochastic programs and sensors, allowing for discontinuous objective functions. SIAM J. Optim. **4** (1994) 537–550
3. L. Bacaud, C. Lemaréchal, A. Renaud, C. Sagastizábal: Bundle methods in stochastic optimal power management: A disaggregated approach using preconditioners. Comput. Optim. Appl. (to appear)
4. H. Bachmann, R. Schultz: Scenario grouping for multi-stage stochastic programs with integer requirements (in preparation)
5. J. R. Birge: Stochastic programming computation and applications. INFORMS J. Comput. **9** (1997) 111–133
6. J. R. Birge, M. A. H. Dempster: Stochastic programming approaches to stochastic scheduling, J. Glob. Optim. **9** (1996) 417–451
7. J. R. Birge, F. Louveaux: Introduction to Stochastic Programming. Springer, New York, 1997
8. C. C. Carøe: Decomposition in stochastic integer programming. Ph.D. thesis, Institute of Mathematical Sciences, University of Copenhagen, 1998
9. C. C. Carøe, R. Schultz: Dual decomposition in stochastic integer programming. Oper. Res. Lett. **24** (1999) 37–45
10. C. C. Carøe, R. Schultz: A two-stage stochastic program for unit commitment under uncertainty in a hydro-thermal system, Schwerpunktprogramm "Echtzeit-Optimierung großer Systeme" der Deutschen Forschungsgemeinschaft, Preprint 98-13, 1998, revised 1999
11. C. C. Carøe, J. Tind: L-shaped decomposition of two-stage stochastic programs with integer recourse, Math. Progr. **83** (1998) 451-464
12. M. A. H. Dempster: On stochastic programming II: Dynamic problems under risk. Stochastics **25** (1988) 15–42
13. D. Dentcheva, W. Römisch: Optimal power generation under uncertainty via stochastic programming. In: Stochastic Programming Methods and Technical Applications (K. Marti, Kall, P. eds.), Lecture Notes in Economics and Mathematical Systems Vol. 458, Springer-Verlag, Berlin 1998, 22–56
14. J. Dupačová: Multistage stochastic programs: The state-of-the-art and selected bibliography. Kybernetika **31** (1995) 151–174

15. E. B. Dynkin, I. V. Evstigneev: Regular conditional expectation of correspondences. Theory Probab. Appl. **21** (1976) 325–338
16. I. Evstigneev: Measurable selection and dynamic programming. Math. Oper. Res. **1** (1976) 267–272
17. S. Feltenmark, K. C. Kiwiel: Dual applications of proximal bundle methods, including Lagrangian relaxation of nonconvex problems. SIAM J. Optim. **10** (2000) 697–721
18. O. Fiedler, W. Römisch: Stability in multistage stochastic programming. Ann. Oper. Res. **56** (1995) 79–93
19. N. Gröwe-Kuska, K. C. Kiwiel, M. P. Nowak, W. Römisch, I. Wegner: Power management in a hydro-thermal system under uncertainty by Lagrangian relaxation. Preprint 99-19, Institut für Mathematik, Humboldt-Universität Berlin, 1999 and in IMA Volume "Decision Making under Uncertainty: Energy and Environmental Models", Springer-Verlag (to appear)
20. N. Gröwe-Kuska, M. P. Nowak, I. Wegner: Modeling of uncertainty for the real-time management of power systems. This volume.
21. R. Hemmecke, R. Schultz: Decomposition methods for two-stage stochastic integer programs. This volume.
22. J. L. Higle, S. Sen: Duality in multistage stochastic programs. In: Prague Stochastics '98 (M. Hušková, P. Lachout, J.Á. Višek, eds.), JČMF, Prague 1998, 233-236
23. J.-B. Hiriart-Urruty, C. Lemaréchal: Convex Analysis and Minimization Algorithms II, Springer, Berlin, 1993
24. P. Kall, S. W. Wallace: Stochastic Programming. Wiley, Chichester, 1994
25. K. C. Kiwiel: Proximity control in bundle methods for convex nondifferentiable minimization. Math. Progr. **46** (1990) 105–122
26. W. K. Klein Haneveld, M. H. van der Vlerk: Stochastic integer programming: General models and algorithms. Ann. Oper. Res. **85** (1999) 39–57
27. G. Laporte, F. V. Louveaux: The integer L-shaped method for stochastic integer programs with complete recourse, Oper. Res. Lett. **13** (1993) 133-142
28. C. Lemaréchal: Lagrangian decomposition and nonsmooth optimization: Bundle algorithm, prox iteration, augmented Lagrangian. In: Nonsmooth Optimization, Methods and Applications (Giannessi, F. ed.), Gordon & Breach, Philadelphia, 1992, 201–216
29. C. Lemaréchal, A. Renaud: A geometric study of duality gaps, with applications. Math. Progr. (to appear)
30. A. Løkketangen, D. L. Woodruff: Progressive hedging and tabu search applied to mixed integer (0,1) multi-stage stochastic programming, J. of Heuristics **2** (1996) 111-128
31. G. L. Nemhauser, L. A. Wolsey: Integer and Combinatorial Optimization, Wiley, New York, 1988
32. M. P. Nowak: Stochastic Lagrangian relaxation in power scheduling of a hydro-thermal system under uncertainty. Ph.D. thesis, Institute of Mathematics, Humboldt-University Berlin, 2000
33. M. P. Nowak, W. Römisch: Stochastic Lagrangian relaxation applied to power scheduling in a hydro-thermal system under uncertainty. Ann. Oper. Res. **100** (2001) (to appear)
34. A. Prékopa: Stochastic Programming. Kluwer, Dordrecht, 1995
35. S. T. Rachev, W. Römisch: Quantitative stability in stochastic programming: The method of probability metrics. Preprint 00-22, Institut für Mathematik, Humboldt-Universität Berlin, 2000
36. R. T. Rockafellar: Duality and optimality in multistage stochastic programming. Ann. Oper. Res. **85** (1999) 1–19
37. R. T. Rockafellar, R. J.-B. Wets: Nonanticipativity and L^1-martingales in stochastic optimization problems. Math. Progr. Study **6** (1976) 170–187

38. R. T. Rockafellar, R. J.-B. Wets: The optimal recourse problem in discrete time: L^1-multipliers for inequality constraints. SIAM J. Contr. Optim. **16** (1978) 16–36
39. R. T. Rockafellar, R. J.-B. Wets: Scenarios and policy aggregation in optimization under uncertainty. Math. Oper. Res. **16** (1991) 119–147
40. R. T. Rockafellar, R. J.-B. Wets: Variational Analysis. Springer-Verlag, Berlin, 1997
41. W. Römisch, R. Schultz: Decomposition of a multi-stage stochastic program for power dispatch, Zeitschr. Angew. Math. Mech. 76 (1996) Suppl. 3, 29–32
42. A. Ruszczyński: Decomposition methods in stochastic programming. Math. Progr. **79** (1997) 333–353
43. R. Schultz: On structure and stability in stochastic programs with random technology matrix and complete integer recourse. Math. Progr. **70** (1995), 73–89
44. R. Schultz: Rates of convergence in stochastic programs with complete integer recourse. SIAM J. Optim. **6** (1996), 1138–1152
45. R. Schultz, L. Stougie, M. H. van der Vlerk: Solving stochastic programs with integer recourse by enumeration: A framework using Gröbner basis reductions, Math. Progr. **83** (1998) 229-252
46. L. Stougie: Design and Analysis of Algorithms for Stochastic Integer Programming, CWI Tract 37, Centrum voor Wiskunde en Informatica, Amsterdam, 1987
47. S. Takriti, J. R. Birge, E. Long: A stochastic model for the unit commitment problem, IEEE Trans. Power Syst. **11** (1996), 1497–1508
48. M. H. van der Vlerk: Stochastic programming with integer recourse, Ph.D. thesis, University of Groningen, 1995
49. R. J.-B. Wets: Stochastic programming. In: Handbooks in Operations Research and Management Science, Vol. 1, Optimization (G. L. Nemhauser, A. H. G. Rinnooy Kan, M. J. Todd eds.), North-Holland, Amsterdam 1989, 573–629
50. R. J.-B. Wets, W. T. Ziemba: Stochastic programming. State of the Art, 1998, Ann. Oper. Res. **85** (1999)

Decomposition Methods for Two-Stage Stochastic Integer Programs

Raymond Hemmecke and Rüdiger Schultz

Fachbereich Mathematik, Gerhard-Mercator-Universität Duisburg, Germany

Abstract Stochastic programs are proper tools for real-time optimization if real-time features arise due to lack of data information at the moment of decision. The paper's focus is at two-stage linear stochastic programs involving integer requirements. After a discussion of basic structural properties, two decomposition approaches are developed. While the first approach is directed to decomposition of the stochastic program itself, the second deals with decomposition of the related Graver test set.

1 INTRODUCTION

The two-stage stochastic program is the optimization problem

$$\min_{x}\{c^{\mathsf{T}}x + \mathbb{E}_{\omega}(\min_{y}\{q(\omega)^{\mathsf{T}}y : W(\omega)y = h(\omega) - T(\omega)x, y \in Y\} : x \in X\}. \quad (1)$$

Here, $x \in \mathbb{R}^n, y \in \mathbb{R}^m, \xi(\omega) := (q(\omega), W(\omega), h(\omega), T(\omega))$ denotes a random vector on some probability space $(\Omega, \mathcal{A}, \mathbb{P})$, and the symbol $\mathbb{E}_{\omega}$ is used for expectation. The sets $X \subseteq \mathbb{R}^n, Y \subseteq \mathbb{R}^m$ are polyhedra, possibly involving integer requirements to components of x and y. Accordingly, two-stage linear and two-stage linear mixed-integer stochastic programs are distinguished. Model (1) is a two-stage version of the more general multi-stage stochastic program discussed in [23].

For real-time optimization, model (1) offers the following features. Consider the random optimization problem

$$\min_{x,y}\{c^{\mathsf{T}}x + q(\omega)^{\mathsf{T}}y : T(\omega)x + W(\omega)y = h(\omega), x \in X, y \in Y\} \quad (2)$$

where decisions on x have to be taken under incomplete information on $\xi(\omega) = (q(\omega), W(\omega), h(\omega), T(\omega))$, and where recourse actions y are permitted after decision on x and observation of $\xi(\omega)$. Model (1) then optimizes the here-and-now decision x such that x is feasible ($x \in X$) and the sum of the direct costs $c^{\mathsf{T}}x$ and the expected future costs becomes minimal. The future costs depend on both x and $\xi(\omega)$, and the recourse action y is selected best possible.

As a two-stage *linear* stochastic program, the model (1) is well understood, both structurally and algorithmically. This is different when including integer requirements to components of x and (mainly) y. In the present paper we start out from the roots of these difficulties, present some structural results, and place the main accent on two recently developed decomposition approaches to solving (1).

2 From Linear to Mixed-Integer Linear Stochastic Programs

For ease of exposition we assume that q, W, T in (1) are deterministic such that $\xi(\omega) = h(\omega)$. So far, most of the subsequent results on stochastic integer programs were obtained for that special case. Where appropriate, we will point to existing results under more general randomness in the vector (q, W, h, T). It is convenient to reformulate (1) as follows

$$\min\{c^T x + Q(x) \ : \ x \in X\} \tag{3}$$

where

$$Q(x) := \int_{\mathbb{R}^s} \Phi(h - Tx)\, \mu(dh) \tag{4}$$

and

$$\Phi(t) := \min\{q^T y \ : \ Wy = t, \ y \in Y\}.$$

Here we assume that $h(\omega) \in \mathbb{R}^s$ and that μ denotes the image measure $\mu := \mathbb{P} \circ h^{-1}$ on $\mathbb{R}^s$. The reason for (3) being so well understood in the purely linear situation, i.e., where $Y := \mathbb{R}^m_+$, is the convexity of Q in that case. Indeed, by linear programming duality, we have $\Phi(t) = \max_u \{t^T \ : \ W^T u \leq q\}$. Assuming that $\{u \ : \ W^T u \leq q\}$ has vertices $d_1, \ldots, d_J$, this yields $\Phi(t) = \max_{j=1,\ldots,J} d_j^T t$, for all $t \in \mathbb{R}^s$ such that $\Phi(t)$ is finite. Therefore, $\Phi(h - Tx)$ is convex in x, and, if the integral in (4) is finite, the function Q is convex, too. This can be summarized into the following standard result of stochastic linear programming.

Proposition 1. *Assume that $W(\mathbb{R}^m_+) = \mathbb{R}^s, \{u \in \mathbb{R}^s \ : \ W^T u \leq q\} \neq \emptyset$, and $\int_{\mathbb{R}^s} \|h\|\, \mu(dh) < \infty$. Then $Q : \mathbb{R}^n \longrightarrow \mathbb{R}$ is a real-valued convex function.*

For further reading on stochastic linear programming we refer to the textbooks [2, 15, 21].

The situation changes if components of y are restricted to the integers. Consider the following examples:

$$\Phi_1(t) = \min\{y^+ + y^- \ : \ y^+ - y^- = t, \ y^+ \in \mathbb{R}_+, \ y^- \in \mathbb{R}_+\},$$
$$\Phi_2(t) = \min\{\tfrac{1}{2}v + y^+ + y^- \ : \ y^+ - y^- = t, \ v \in \mathbb{Z}_+, \ y^+ \in \mathbb{R}_+, \ y^- \in \mathbb{R}_+\}$$
$$= \min\{\tfrac{1}{2}v + |t - v| \ : \ v \in \mathbb{Z}_+\},$$
$$\Phi_3(t) = \min\{v^+ + v^- \ : \ y + v^+ - v^- = t, \ y \in \mathbb{R}_+, \ v^+ \in \mathbb{Z}_+, \ v^- \in \mathbb{Z}_+\}$$
$$= \begin{cases} 0 & : \ t \geq 0 \\ \lceil -t \rceil & : \ t < 0. \end{cases}$$

Whereas $\Phi_1(t) = |t|$ is convex, Φ_2 is neither convex nor concave but still (Lipschitz) continuous, and Φ_3 is discontinuous but still lower semicontinuous. If $\Phi(t)$ in (3) is given by

$$\Phi(t) := \min\{q^T y + q'^T y' \ : \ Wy + W'y' = t, \ y' \in \mathbb{R}^{m'}_+, \ y \in \mathbb{Z}^{\bar{m}}_+\}, \tag{5}$$

with rational matrices W, W' of conformal dimensions, then the following is known from parametric integer optimization [1,3].

Proposition 2. *Assume that* $W(\mathbb{Z}_+^{\bar{m}}) + W'(\mathbb{R}_+^{m'}) = \mathbb{R}^s$ *and* $\{u \in \mathbb{R}^s : W^T u \leq q, W'^T u \leq q'\} \neq \emptyset$. *Then it holds*

(i) Φ *is real-valued and lower semicontinuous on* $\mathbb{R}^s$,
(ii) there exists a countable partition $\mathbb{R}^s = \cup_{i=1}^{\infty} \mathcal{T}_i$ *such that the restrictions of* Φ *to* $\mathcal{T}_i$ *are piecewise linear and Lipschitz continuous with a uniform constant* $L > 0$ *not depending on* i,
(iii) each of the sets $\mathcal{T}_i$ *has a representation* $\mathcal{T}_i = \{t_i + \mathcal{K}\} \setminus \cup_{j=1}^{N} \{t_{ij} + \mathcal{K}\}$ *where* $\mathcal{K}$ *denotes the polyhedral cone* $W'(\mathbb{R}_+^{m'})$ *and* t_i, t_{ij} *are suitable points from* $\mathbb{R}^s$, *moreover,* N *does not depend on* i,
(iv) there exist positive constants β, γ *such that* $|\Phi(t_1) - \Phi(t_2)| \leq \beta \|t_1 - t_2\| + \gamma$ *whenever* $t_1, t_2 \in \mathbb{R}^s$.

Returning to (4), we observe that the integrand is measurable by the lower semicontinuity of Φ. Moreover, the growth property in (iv) and the assumption that $\int_{\mathbb{R}^s} \|h\| \, \mu(dh) < \infty$ imply that the integral in (4) is finite. In addition, (iv) then provides integrable lower and upper bounds for Φ, enabling the application of Fatou's Lemma and Lebesgue's Dominated Convergence Theorem. This leads to the following (see [26] for details).

Proposition 3. *Assume that* $W(\mathbb{Z}_+^{\bar{m}}) + W'(\mathbb{R}_+^{m'}) = \mathbb{R}^s$, $\{u \in \mathbb{R}^s : W^T u \leq q, W'^T u \leq q'\} \neq \emptyset$, *and* $\int_{\mathbb{R}^s} \|h\| \, \mu(dh) < \infty$. *Then it holds*

(i) $Q : \mathbb{R}^n \longrightarrow \mathbb{R}$ *is a real-valued lower semicontinuous function,*
(ii) if μ *has a density, then* Q *is continuous on* $\mathbb{R}^n$.

It is still possible to find verifiable and indispensable, but rather technical conditions ensuring Lipschitz continuity of Q on bounded subsets of $\mathbb{R}^n$, [25]. From an algorithmic viewpoint, however, it has to be stated:

- Although a well-defined nonlinear program, (3) lacks essential smoothness and convexity properties for employing the algorithmic machinery of nonlinear programming.
- Computing exact function values of Q requires multidimensional integration of implicitly given integrands. With continuous measure μ, this is beyond existing numerical capabilities and hence motivates discrete approximation of μ.
- For discrete μ, the function Q becomes computable, but is discontinuous, on the other hand.

In conclusion, approximation of the underlying probability measure is a key issue of numerical methods in stochastic programming. Its justification leads to analyzing the stability of (3) under perturbations of μ. Without going into details, we mention that optimal value and optimal solutions to (3), viewed as mappings acting on some topological space of probability measures, fulfil natural continuity requirements. For details see the survey [27] and the references therein. Therefore, "small" perturbations of μ imply only "small" perturbations of the optimal value and the set of optimal solutions to (3).

3 Scenario Decomposition

From now on we assume that the probability measure μ in (3) is discrete, with realizations h_j and probabilities $\pi_j, j = 1, \ldots, N$. To circumvent the discontinuity of Q we reformulate (3) as a mixed-integer linear program:

$$\min_{x,y_j}\{c^\top x + \sum_{j=1}^{N} \pi_j q^\top y_j \; : \; Tx + Wy_j = h_j, \; y_j \in Y, \; x \in X\}. \tag{6}$$

Here, the variables $y_j, j = 1, \ldots, N$, are copies of the vector y according to the number N of realizations (or scenarios) of μ. N being big in general, (6) quickly becomes large-scale and intractable by general purpose algorithms and software in mixed-integer linear programming.

On the other hand, the constraints interlinking the variables x and y_j, $j = 1, \ldots, N$, have a block-angular structure giving rise to decomposition. Fixing x results in decoupling of these constraints into N independent blocks. Since the objective is separable, problem (6) then decomposes into N independent problems. The latter forms the basis of an algorithmic approach referred to as L-shaped or primal decomposition in the literature [7, 22, 32]. The idea is to iterate x in an outer master problem, to avoid solving the full problem by resorting to the mentioned subproblems, and to stop the process when some optimality condition is fulfilled sufficiently accurately. This works nicely provided there are no integer requirements in (6). Then the master problem is a tractable non-smooth convex program. With integer requirements among the y−variables, however, the master problem turns into a discontinuous minimization problem with lower semicontinuous objective. This problem is far less tractable than the mentioned convex counterpart [7, 28].

As an alternative, we propose the following. Introduce copies $x_j, j = 1, \ldots, N$, according to the number of scenarios, and add the constraints $x_1 = \cdots = x_N$ (or an equivalent system), for which we use the notation $\sum_{j=1}^{N} H_j x_j = 0$ with proper (l, n)−matrices $H_j, j = 1, \ldots, N$. Problem (6) then becomes

$$\min\{\sum_{j=1}^{N} \pi_j (c^\top x_j + q^\top y_j) \; : \; Tx_j + Wy_j = h_j, \; x_j \in X, \; y_j \in Y, \; \sum_{j=1}^{N} H_j x_j = 0\}. \tag{7}$$

This model is amenable to Lagrangian relaxation of the constraints $\sum_{j=1}^{N} H_j x_j = 0$. For $\lambda \in \mathbb{R}^l$ we consider the functions

$$L_j(x_j, y_j, \lambda) := \pi_j (c^\top x_j + q^\top y_j) + \lambda^\top H_j x_j, \quad j = 1, \ldots, N,$$

and form the Lagrangian

$$L(x, y, \lambda) := \sum_{j=1}^{N} L_j(x_j, y_j, \lambda).$$

The Lagrangian dual of (7) then is the optimization problem

$$\max\{D(\lambda) \; : \; \lambda \in \mathbb{R}^l\} \tag{8}$$

where

$$D(\lambda) = \min\{\sum_{j=1}^{N} L_j(x_j, y_j, \lambda) : Tx_j + Wy_j = h_j,\ x_j \in X,\ y_j \in Y\}. \quad (9)$$

For separability reasons we have

$$D(\lambda) = \sum_{j=1}^{N} D_j(\lambda) \quad (10)$$

where

$$D_j(\lambda) = \min\{L_j(x_j, y_j, \lambda) : Tx_j + Wy_j = h_j,\ x_j \in X,\ y_j \in Y\}. \quad (11)$$

$D_j(\lambda)$ being the pointwise minimum of affine functions in λ, it is piecewise affine and concave. Hence, (8) is a non-smooth concave maximization problem, or equivalently, a non-smooth convex minimization problem. Non-differentiable optimization [14,16] offers advanced bundle methods for tackling (8). At each iteration, these methods require the objective value and one subgradient of D. Here, the separability in (10) is most beneficial. Altogether, the optimal value φ_{LD} of (8) provides a lower bound to the optimal value φ of problem (6). This is made precise by the following well-known result [20].

Proposition 4. *It holds* $\varphi \geq \varphi_{LD}$. *If for some multiplier* $\lambda \in \mathbb{R}^l$ *the optimal solutions* $(x_j, y_j), j = 1, \ldots, N$, *to the optimization problem in (9) fulfil* $\sum_{j=1}^{N} H_j x_j = 0$, *then* $\varphi = \varphi_{LD}$ *and* $(x_j, y_j), j = 1, \ldots, N$, *are optimal for (7).*

The lower bound φ_{LD} is obtained by decomposing the stochastic program (6) into scenario-specific subproblems that are very close to the non-stochastic version of (6), where $N = 1$, see (9), (11). Therefore the name scenario decomposition. It is well-known that $\varphi_{LD} \geq \varphi_{LP}$ where φ_{LP} denotes the optimal value to the LP relaxation of (6), [20].

The total computational effort is distributed. The single-scenario problems in (11), which differ in their objectives and right-hand sides but have common constraint matrix, have to be solved repeatedly to provide the input for the concave maximization in (8). Powerful software is available for these tasks. In our computational experiments we resorted to CPLEX [9] for solving the single-scenario problems, and to the bundle method NOA 3.0 developed and implemented by K.C. Kiwiel [16,17]. With more complex stochasticity (cf. (1)), the above bounding scheme is still working. The only difference is that single-scenario problems may also differ in their constraint matrices.

In Lagrangian relaxation, feasible points for the original problem are often obtained by suitable heuristics starting from the results of the dual optimization. Our relaxed constraints being very simple ($x_1 = \cdots = x_N$), ideas for such heuristics come up straightforwardly. For example, examine the x_j–components, $j = 1, \ldots, N$, of solutions to (11) for optimal or nearly optimal λ, and decide for the most frequent value arising or average and round if necessary.

If the heuristic results in a feasible solution to (7), then the objective value of the latter provides an upper bound $\bar{\varphi}$ for φ. Together with the lower bound φ_{LD} this yields a quality certificate (gap) $\bar{\varphi} - \varphi_{LD}$. If necessary, this certificate can be improved by embedding the procedure described so far into the following branch-and-bound scheme. Let $\mathcal{P}$ denote the list of current problems and $\varphi_{LD} = \varphi_{LD}(P)$ the Lagrangian lower bound for $P \in \mathcal{P}$.

Step 1 *Initialization:* Set $\bar{\varphi} = +\infty$ and let $\mathcal{P}$ consist of problem (7).

Step 2 *Termination:* If $\mathcal{P} = \emptyset$ then the solution $\hat{x}$ that yielded $\bar{\varphi} = c^T\hat{x} + Q(\hat{x})$, cf. (3), is optimal.

Step 3 *Node selection:* Select and delete a problem P from $\mathcal{P}$ and solve its Lagrangian dual. If the optimal value $\varphi_{LD}(P)$ hereof equals $+\infty$ (infeasibility of a subproblem) then go to Step 2.

Step 4 *Bounding:* If $\varphi_{LD}(P) \geq \bar{\varphi}$ go to Step 2 (this step can be carried out as soon as the value of the Lagrangian dual rises above $\bar{\varphi}$).

(i) The scenario solutions $x_j, j = 1, \ldots, N$, are identical: If $c^T x_j + Q(x_j) < \bar{\varphi}$ then let $\bar{\varphi} = c^T x_j + Q(x_j)$ and delete from $\mathcal{P}$ all problems P' with $\varphi_{LD}(P') \geq \bar{\varphi}$. Go to Step 2.

(ii) The scenario solutions $x_j, j = 1, \ldots, N$ differ: Compute the average $\bar{x} = \sum_{j=1}^{N} \pi_j x_j$ and round it by some heuristic to obtain $\bar{x}^R$. If $c^T\bar{x}^R + Q(\bar{x}^R) < \bar{\varphi}$ then let $\bar{\varphi} = c^T\bar{x}^R + Q(\bar{x}^R)$ and delete from $\mathcal{P}$ all problems P' with $\varphi_{LD}(P') \geq \bar{\varphi}$. Go to Step 5.

Step 5 *Branching:* Select a component $x_{(k)}$ of x and add two new problems to $\mathcal{P}$ obtained from P by adding the constraints $x_{(k)} \leq \lfloor \bar{x}_{(k)} \rfloor$ and $x_{(k)} \geq \lfloor \bar{x}_{(k)} \rfloor + 1$, respectively (if $x_{(k)}$ is an integer component), or $x_{(k)} \leq \bar{x}_{(k)} - \varepsilon$ and $x_{(k)} \geq \bar{x}_{(k)} + \varepsilon$, respectively, where $\varepsilon > 0$ is a tolerance parameter to have disjoint subdomains.

In the subsequent section we will report about some practical experience with the above method. For the moment we state that the method is obviously finite in case X is bounded and all x–components have to be integers. If x is mixed-integer some stopping criterion to avoid endless branching on the continuous components has to be employed. For further details see [5, 6].

4 Unit Commitment with Incomplete Information

Unit commitment aims at finding a fuel cost optimal scheduling of start-up/shut-down decisions and operation levels for power generation units over some time horizon. This is a central task in reliable and efficient operation of power systems. Here we consider a hydro-thermal system as met with the German power company VEAG Vereinigte Energiewerke AG Berlin. This system comprises conventional coal and gas fired thermal units as well as pumped-storage plants. At the beginning of the optimization horizon, information on essential unit commitment data such as electrical load or availability of generating units is incomplete. Furthermore, starting up a coal fired block involves some time delay before the block becomes available for electricity generation. Therefore, switching decisions for these units have to be taken well in advance and cannot be employed as short-term corrective actions. Gas turbines, however, have a sufficiently small delay in this respect such that switching is feasible for short-term corrections. For further technological details of the power system we refer to [6].

As an example on how to employ the modeling of Section 1 and the algorithmics of Section 3 we deal with the problem of finding a cost optimal weekly hedging policy for start-up decisions of coal fired units under incomplete information on the electrical load to be covered. Grouping of decisions follows the operational characteristics above. Start-up decisions for coal fired units will become first-stage, whereas start-up decisions for gas turbines and output levels for all units will be the second-stage decisions. The optimization aims at minimizing the sum of direct costs for first-stage decisions and the expected value of the costs induced by the first- together with the second-stage decisions.

Let $t = 1, \dots, T$ denote the subintervals (e.g., hours) of the optimization horizon. Suppose that there are $i = 1, \dots, I$ coal fired thermal units, $k = 1, \dots, K$ gas turbines, and $j = 1, \dots, J$ pumped storage plants. The stochastic behavior of electrical load is represented by a random variable d on some probability space $(\Omega, \mathcal{A}, \mathbb{P})$ with values in $\mathbb{R}^T$. We assume that d follows a discrete distribution with finite support and use the symbol d^ω, $\omega \in \Omega$, to denote its realizations. (For clarity of exposition we modify the notation from Section 3 by indexing the scenarios by ω and placing the index in superscript throughout.)

By $u_{it} \in \{0,1\}$, $i = 1, \dots, I$ and $t = 1, \dots, T$ we denote the first-stage variables indicating whether the coal fired unit i is on or off at time t. Correspondingly, $u_{kt}^\omega \in \{0,1\}$, $k = 1, \dots, K$, $t = 1, \dots, T$, $\omega \in \Omega$, denotes the on/off decision for gas turbine k in time interval t under scenario d^ω. Along with the on/off decisions we have output levels $p_{it}^\omega, p_{kt}^\omega$ of the mentioned units. The output limitations of thermal units then read as follows

$$p_i^{\min} u_{it} \leq p_{it}^\omega \leq p_i^{\max} u_{it}, \quad i = 1, \dots, I, \; t = 1, \dots, T, \; \omega \in \Omega \tag{12}$$

$$p_k^{\min} u_{kt}^\omega \leq p_{kt}^\omega \leq p_k^{\max} u_{kt}^\omega, \quad k = 1, \dots, K, \; t = 1, \dots, T, \; \omega \in \Omega. \tag{13}$$

Here, $p_i^{\min}$, $p_k^{\min}$ and $p_i^{\max}$, $p_k^{\max}$ are the minimal and maximal outputs of the respective units. By s_{jt}^ω and w_{jt}^ω, $j = 1, \dots, J$, $t = 1, \dots, T$, $\omega \in \Omega$, we denote the levels of generation and pumping in pumped storage plant j at time t under scenario d^ω. With upper bounds $s_j^{\max}, w_j^{\max}$ we have the following box constraints on these variables

$$0 \leq s_{jt}^\omega \leq s_j^{\max}, \quad j = 1, \dots, J, \; t = 1, \dots, T, \; \omega \in \Omega \tag{14}$$

$$0 \leq w_{jt}^\omega \leq w_j^{\max}, \quad j = 1, \dots, J, \; t = 1, \dots, T, \; \omega \in \Omega \tag{15}$$

The variables l_{jt}^ω, $j = 1, \dots, J$, $t = 1, \dots, T$, $\omega \in \Omega$, are introduced for the fills (in energy and at the end of the time interval) of the upper dams. Water balances in the pumped storage plants can then be expressed as follows

$$0 \le l_{jt}^{\omega} \le l_j^{\max}, \qquad t = 1,\ldots,T, \tag{16}$$

$$l_{jt}^{\omega} = l_{jt-1}^{\omega} - s_{jt}^{\omega} + \eta_j w_{jt}^{\omega}, \quad t = 2,\ldots,T, \tag{17}$$

$$l_{j1}^{\omega} = l_j^{\text{ini}} - s_{j1}^{\omega} + \eta_j w_{j1}^{\omega}, \tag{18}$$

$$l_{jT}^{\omega} = l_j^{\text{end}} \tag{19}$$

for $j = 1,\ldots,J$, $\omega \in \Omega$. The constants η_j, $0 < \eta_j < 1$, are the pumping efficiencies, and $l_j^{\max}, l_j^{\text{ini}}, l_j^{\text{end}}$ denote the maximal, initial and final fills, respectively. Inequalities (16) state that the fills must not exceed certain bounds and equations (17) - (19) display the fill dynamics with (19) avoiding empty dams at the end of the time horizon by prescribing proper end conditions. The equilibrium between total generation and electrical load reads

$$\sum_{i=1}^{I} p_{it}^{\omega} + \sum_{k=1}^{K} p_{kt}^{\omega} + \sum_{j=1}^{J} (s_{jt}^{\omega} - w_{jt}^{\omega}) = d_t^{\omega}, \quad t = 1,\ldots,T,\ \omega \in \Omega. \tag{20}$$

Fuel costs can be divided into start-up costs and operation costs for the power units. There are no direct fuel costs for pumped storage plants, although the latter have indirect impact on fuel costs by the pumping energy needed to establish the necessary levels in the upper dams. Start-up costs for thermal units depend on the preceding down time of the block. Here, we will neglect this dependence and assume constant costs a_i, a_k for start-ups in coal fired blocks and gas turbines, respectively. Total start-up costs for the coal fired units then compute as

$$\sum_{t=2}^{T} \sum_{i=1}^{I} a_i \max\{u_{it} - u_{it-1}, 0\}. \tag{21}$$

Denoting by $\mathbb{E}_\omega$ expectation with respect to $\mathbb{P}$, the expected value of total start-up costs for the gas turbines reads

$$\mathbb{E}_\omega \Big[\sum_{t=2}^{T} \sum_{k=1}^{K} a_k \max\{u_{kt}^{\omega} - u_{kt-1}^{\omega}, 0\} \Big]. \tag{22}$$

For reasons of clarity we prefer the (nonlinear) maximum term in (21) and (22). For computations, of course, these are transformed in the usual way into linear terms by introducing further variables and including additional linear constraints. Fuel costs of coal and gas fired thermal units in operation are assumed to be affinely linear with coefficients c_i, c_i^0 and c_k, c_k^0, respectively. Total expected operation costs compute as

$$\mathbb{E}_\omega \Big[\sum_{t=1}^{T} \Big(\sum_{i=1}^{I} u_{it}(c_i p_{it}^{\omega} + c_i^0) + \sum_{k=1}^{K} u_{kt}^{\omega}(c_k p_{kt}^{\omega} + c_k^0) \Big) \Big].$$

Here, nonlinearities can be removed using (12), (13), and we obtain

$$\sum_{t=1}^{T} \sum_{i=1}^{I} c_i^0 u_{it} + \mathbb{E}_\omega \Big[\sum_{t=1}^{T} \sum_{i=1}^{I} c_i p_{it}^{\omega} + \sum_{t=1}^{T} \sum_{k=1}^{K} c_k p_{kt}^{\omega} + \sum_{t=1}^{T} \sum_{k=1}^{K} c_k^0 u_{kt}^{\omega} \Big]. \tag{23}$$

Altogether, we have the following optimization problem:

$$\begin{aligned}\text{Minimize}\quad & \sum_{t=1}^{T}\sum_{i=1}^{I} c_i^0 u_{it} + \sum_{t=2}^{T}\sum_{i=1}^{I} a_i \max\{u_{it} - u_{it-1}, 0\} + \\ & + \mathbb{E}_\omega\Big[\sum_{t=1}^{T}\sum_{i=1}^{I} c_i p_{it}^\omega + \sum_{t=1}^{T}\sum_{k=1}^{K} c_k p_{kt}^\omega + \sum_{t=1}^{T}\sum_{k=1}^{K} c_k^0 u_{kt}^\omega + \\ & + \sum_{t=2}^{T}\sum_{k=1}^{K} a_k \max\{u_{kt}^\omega - u_{kt-1}^\omega, 0\}\Big] \qquad (24)\\ \text{subject to}\quad & (12)-(20) \text{ and } u_{it}, u_{kt}^\omega \in \{0,1\}, i = 1,\ldots,I,\ k = 1,\ldots,K,\\ & t = 1,\ldots,T,\ \omega \in \Omega.\end{aligned}$$

The algorithm from Section 3 was coded in Fortran using NOA 3.0 by Kiwiel [16, 17] and the CPLEX Callable Library [9] and run on realistic problem instances. The power system of VEAG comprises 17 coal fired blocks, 8 gas fired units and 7 pumped storage plants. For a typical weekly unit commitment problem involving 168 time periods, each of the scenario subproblems has approximately 14000 constraints and 16000 variables, of which 4200 are binary. A characteristic feature of the power system met at VEAG is that several of the coal-fired thermal blocks are identical. This allows us to reduce the size of the scenario subproblems by aggregation of these units. The start-up/shut-down decisions for these units are then represented by one integer variable, namely the number of units which are turned on. In the following we present results for both formulations. We generated 16 scenarios for uncertain load and build 3 instances with 4, 10 and 16 scenarios for each of the examples. For simplicity we assume the scenarios in all 3 instances have probabilities according to a uniform distribution. The load profiles for a 4-scenario instance are depicted in Figure 1. Seen as a mixed-integer linear program of the type (6), the stochastic program then involves up to 180000 constraints and 172000 variables of which 21000 are integers.

The branch-and-bound scheme from Section 3 is the algorithmic guideline for our implementation. In addition there are some enhancements and specifications: To save computation time, the Lagrangian dual is solved at the root node only, and the obtained multipliers are then used for solving the Lagrangian relaxation of non-root nodes. There are no NOA iterations at non-root nodes. The "Without NOA" column in Table 1 reports on tests where even at the root node no iterations with NOA were carried out. Instead, the feasible point $\lambda = 0$ was used for the lower bound. The table shows that this leads to gaps which are about 10 times wider than the gaps obtained when using NOA in the described way. Hence, although quite costly, it pays to use NOA for heading towards the maximum of the function $D(\lambda)$ and improving the lower bound this way.

As already mentioned in Section 3, candidates for feasible solutions can easily be found in our case due to the simple structure of the non-anticipativity constraints. We used the average $\bar{x}$ and rounded all components to the nearest integer. For all our

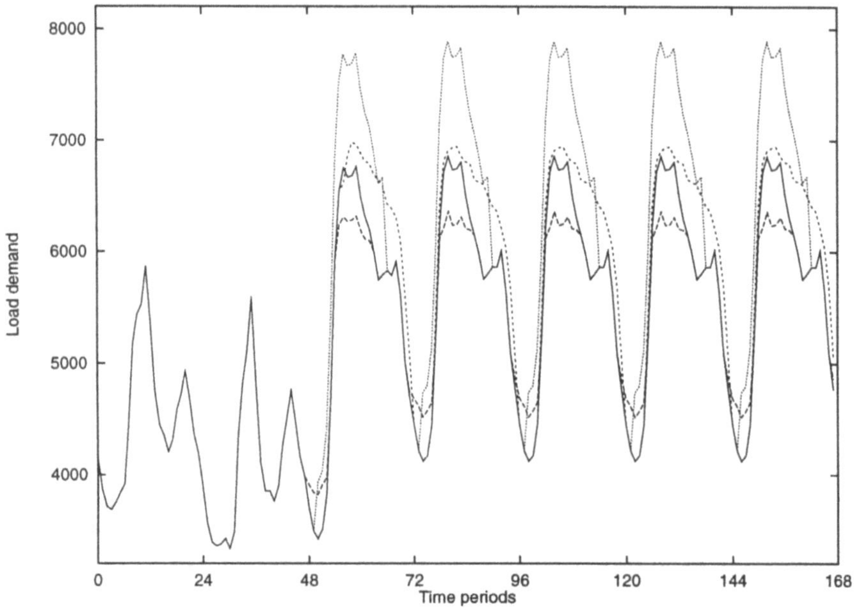

Figure 1. Profiles for uncertain load in 4-scenario instance

test runs good feasible solutions were obtained this way already in the root node of the branching algorithm.

The main workload of our algorithm consists of solving the scenario subproblems and it is therefore crucial to solve them as fast as possible. To this end, we store and update optimal bases for the LP-relaxations of all scenario subproblems every 5 iterations of the NOA code. The bases are used to start the solution process of the scenario subproblems. During NOA-iterations we employ primal Simplex for solving LP-relaxations since primal feasibility is maintained and multipliers affect cost coefficients, only. For non-root nodes dual feasibility is maintained since variable bounds are being fixed, thus dual Simplex is employed during the branching procedure for solving LP-relaxations.

Special attention has been paid to the branching and node selection criteria of our branch-and-bound algorithm. For problems with only binary integer variables we take the average $\bar{x}$ and branch on the component for which the fractional part is closest to 0.5. Intuitively, this corresponds to the component for which the nonanticipativity condition is violated most. For the formulation with general integer variables we calculate the sum of squared deviations, $\sum_{\nu=1}^{r} \pi^{\nu}(x^{\nu} - \bar{x})^2$, and choose the component for which this dispersion measure is largest.

In the bounding step of our algorithm we check whether nodes in the branching tree can be fathomed. However, since the subproblems are not solved exactly, φ_{LD} will not necessarily be a lower bound for the optimal value of the initial problem if it is computed on the basis of *approximate* solutions to the subproblems. Instead of approximate solutions we use the subproblem (lower) bounds returned by CPLEX when computing φ_{LD}. This guarantees that the φ_{LD} are indeed lower bounds, however, not necessarily the best possible, so a larger branching tree may result. As node selection strategy we use a best-bound search to quickly improve the lower bound obtained from the Lagrangian dual of the root node problem.

The computational results are given in Table 1.

Formulation	Scenarios	NOA Steps	Best solution	Lower Bound	Gap	Without NOA
Binary	4	30	3.6598	3.6411	0.5%	1.4%
	10	10	3.6955	3.5781	3.3%	8.3%
	16	5	3.6225	3.5276	2.7%	10.1%
Integer	4	100	3.6579	3.6527	0.1%	4.1%
	10	40	3.6195	3.6080	0.3%	3.1%
	16	25	3.5698	3.5556	0.4%	2.5%

Table 1. Computational results for uncertain-load instances

5 Test Sets in Integer Programming

In the remainder of the present paper we will pursue an alternative idea of decomposition. This idea involves the consideration of Graver test sets which are objects containing essential solution information, and that turn out to be decomposable if the underlying optimization problem has the block structure of (6). We will confine ourselves to the pure-integer case. In the present section we collect some necessary prerequisites on test sets for general integer programs. In particular, we will present an algebraic procedure for computing test sets (Algorithms 8 and 9 below). We already announce that the pattern behind that procedure will reappear in a different context in Algorithms 15 and 16 in Section 6. These algorithms enable a direct computation of building blocks that arise as results of the mentioned test set decomposition for optimization problems with the block structure of (6).

For given rational (l, d)-matrix A consider the family of optimization problems

$$(IP)_{p,b}: \qquad \min\{p^\intercal z : Az = b,\ z \in \mathbb{Z}^d_+\}$$

as $p \in \mathbb{R}^d$ and $b \in \mathbb{R}^l$ vary.

Definition 5 (Test set). A set $\mathcal{T}_p \subseteq \mathbb{Z}^d$ is called a test set for the family of problems $(IP)_{p,b}$ as $b \in \mathbb{R}^l$ varies if

1. $p^\intercal t > 0$ for all $t \in \mathcal{T}_p$, and

2. for every $b \in \mathbb{R}^l$ and for every non-optimal feasible solution $z_0 \in \mathbb{Z}_+^d$ to $Az = b$, there exists a vector $t \in \mathcal{T}_p$ such that $z_0 - t$ is feasible. Such a vector is called an improving vector or an improving direction.

A set $\mathcal{T}$ is called a universal test set for the family of problems $(\mathrm{IP})_{p,b}$ as $b \in \mathbb{R}^l$ and as $p \in \mathbb{R}^d$ vary if it contains a test set $\mathcal{T}_p$ for every $p \in \mathbb{R}^d$.

Once a finite test set $\mathcal{T}_p$ is computed or given, we may apply the following augmentation algorithm in order to solve the optimization problem $(\mathrm{IP})_{p,b}$.

Algorithm 6 (Augmentation algorithm).

Input: a feasible solution z_0 to $(\mathrm{IP})_{p,b}$, a test set $\mathcal{T}_p$ for $(\mathrm{IP})_{p,b}$
Output: an optimal point $z_{\min}$ of $(\mathrm{IP})_{p,b}$

```
while there is t ∈ T_p with pᵀt > 0 such that z0 − t is feasible do
    z0 := z0 − t
end while
return z0
```

Under the assumption that the optimization problem is solvable, this augmentation process always terminates with an optimal solution to $(\mathrm{IP})_{p,b}$ [11,31,33]. Thus, if we were given a finite universal test set for $(\mathrm{IP})_{p,b}$ then for any given right-hand side b and for any given cost function vector p an optimal solution to $(\mathrm{IP})_{p,b}$ could be easily found as long as an initial feasible solution is available.

Without elaborating this in more detail, we assume an initial feasible solution to be given (cf., e.g., [13] for a test set augmentation algorithm to find such a solution).

Naturally, all the vectors in the integer kernel $\ker(A) := \{v \in \mathbb{Z}^d : Av = 0\}$ of A form an infinite universal test set. However, for all rational matrices A there always exist finite universal test sets as well. In the following we present a particular universal test set, the Graver test set, and show how to compute it. For proofs and further reading we refer to the seminal paper of Graver [10] and to [11,29,30,33].

Let C be a polyhedral cone with rational generators. A finite set

$$H = \{h_1, \ldots, h_t\} \subseteq C \cap \mathbb{Z}^d$$

is called a Hilbert basis of C if every $z \in C \cap \mathbb{Z}^d$ has a representation of the form $z = \sum_{i=1}^t \lambda_i h_i$, with non-negative integral multipliers $\lambda_1, \ldots, \lambda_t$. Every pointed, rational cone has a unique Hilbert basis that is minimal with respect to inclusion [24]. Let $\mathbb{O}_j$ be the jth orthant of $\mathbb{Z}^d$ and $H_j(A)$ be the unique minimal Hilbert basis of the pointed rational cone $\{v \in \mathbb{R}^d : Av = 0\} \cap \mathbb{O}_j$.

Lemma 7. *$\mathcal{G}(A) := \bigcup H_j(A) \setminus \{0\}$ is a universal test set, called the Graver test set or Graver basis, for the family of problems $(\mathrm{IP})_{p,b}$ as $b \in \mathbb{R}^l$ and $p \in \mathbb{R}^d$ vary.*

Now let $u \sqsubseteq v$ iff $u^+ \leq v^+$ and $u^- \leq v^-$, where, for $t \in \mathbb{R}^d$, we denote $t^+ := \max\{t, 0\}$ and $t^- := \max\{-t, 0\}$. Then the elements of $\mathcal{G}(A)$ are minimal in $\ker(A) \setminus \{0\}$ with respect to the partial ordering $\sqsubseteq$ on $\mathbb{Z}^d$. Graver test sets exist, are finite, and can be computed using the algorithmic pattern of a completion procedure [4].

Algorithm 8 (Algorithm to Compute Graver Test Sets).

Input: $F = \bigcup_{f \in F(A)} \{f, -f\}$, where $F(A)$ is a set of vectors generating $\ker(A)$ over $\mathbb{Z}$
Output: a set G which contains the Graver test set $\mathcal{G}(A)$

```
{forming S-vectors:}
C := ⋃_{f,g∈G} {f + g}
while C ≠ ∅ do
  s := an element in C
  C := C \ {s}
  f := normalForm(s, G)
  if f ≠ 0 then
    {adding S-vectors:}
    C := C ∪ ⋃_{g∈G} {f + g}
    G := G ∪ {f}
  end if
end while
return G.
```

Behind the function normalForm(s, G) there is the following algorithm.

Algorithm 9 (Normal Form Algorithm).

Input: a vector s, a set G of vectors
Output: a normal form of s with respect to G

```
while there is some g ∈ G such that g ⊑ s do
  s := s − g
end while
return s
```

We aim at computing $\sqsubseteq$-minimal vectors of $\ker(A)$. This motivates to say that s can be reduced by g to $s - g$ if $g \sqsubseteq s$. In this case, s, g, and $s - g$ all belong to the same orthant.

The Graver test set is the set of all elements $z \in G$ which are irreducible with respect to $G \setminus \{z\}$.

The Graver test set algorithm works correctly for arbitrary choices of the element $s \in C$ in the Graver test set algorithm and of the element g in the normal form algorithm.

Lemma 10. *Algorithm 8 terminates and computes a set containing the Graver test set.*

6 Decomposition of Test Sets in Stochastic Integer Programming

Theoretically, the procedure from Section 5 can be used to solve stochastic integer programs (6) as well. However, due to the huge amount of stored information, Graver test sets are quite large already for small problems. Therefore, a direct test set approach to (6) should be avoided.

The block angular structure of the problem matrix in (6) now induces a symmetry structure on the elements of the Graver basis, telling us, that these test set vectors are formed by a comparably small number of building blocks. We will show that these building blocks can be computed without computing the Graver test set of (6) itself and that we can reconstruct an improving vector to a given non-optimal feasible solution to (6), scenario by scenario, using building blocks only.

To study Graver test sets of (6) we assume that $X = \mathbb{Z}^n_+, Y = \mathbb{Z}^m_+$. This differs from the setting in (1), where X, Y denote (integer) points in polyhedra rather than just in orthants. Here we assume that the matrices T, W capture the remaining polyhedral conditions. We consider the matrix

$$A_N := \begin{pmatrix} T & W & 0 & \cdots & 0 \\ T & 0 & W & \cdots & 0 \\ \vdots & \vdots & \vdots & \ddots & \vdots \\ T & 0 & 0 & \cdots & W \end{pmatrix}$$

together with the objective function vector

$$p = (p_0, p_1, \dots, p_N)^\intercal := (c, \pi_1 q, \dots, \pi_N q)^\intercal$$

and the right-hand-side $b = (h^1, \dots, h^N)^\intercal$, where N corresponds to the number of scenarios. Problem (6) then may be written as $\min\{p^\intercal z : A_N z = b, z \in \mathbb{Z}^d_+\}$ with $d = n + Nm$. We assume all entries in T and W to be rational.

When referring to components of z, the notation $z = (u, v_1, \dots, v_N)$ will be used throughout. Herein, u corresponds to the first-stage and is always of dimension n, whereas $v_1, \dots, v_N$ are the second-stage vectors whose dimension is m.

The following simple observation is the basis for the decomposition of test set vectors.

Lemma 11. $(u, v_1, \dots, v_N) \in \ker(A_N) \iff (u, v_1), \dots, (u, v_N) \in \ker(A_1)$.

Thus, by permuting the v_i we do not leave $\ker(A_N)$. Moreover, a $\sqsubseteq$-minimal element of $\ker(A_N)$ will always be transformed into a $\sqsubseteq$-minimal element of $\ker(A_N)$. Thus, a Graver test set vector is transformed into a Graver test set vector by such a permutation. This leads us to the following definition.

Definition 12 (Building blocks). Let $z = (u, v_1, \dots, v_N) \in \ker(A_N)$ and call the vectors $u, v_1, \dots, v_N$ the building blocks of z. Denote by $\mathcal{G}_N$ the Graver test set associated with A_N and collect into $\mathcal{H}_N$ all those vectors arising as building blocks of some $z \in \mathcal{G}_N$. By $\mathcal{H}_\infty$ denote the set $\bigcup_{N=1}^{\infty} \mathcal{H}_N$.

The set $\mathcal{H}_\infty$ contains both n-dimensional vectors u associated with the first-stage in (6) and m-dimensional vectors v related with the second-stage in (6). For convenience, we will arrange the vectors in $\mathcal{H}_\infty$ into pairs (u, V_u). For fixed $u \in \mathcal{H}_\infty$, all those vectors $v \in \mathcal{H}_\infty$ are collected into V_u for which $(u, v) \in \ker(A_1)$. In what follows, we will employ this arrangement into pairs to arbitrary sets of n- and m-dimensional building blocks, not necessarily belonging to $\mathcal{H}_\infty$.

The set $\mathcal{H}_\infty$ is of particular interest, since, by definition, it contains all building blocks of test set vectors of (6) for an arbitrary number N of scenarios. Next, we will address finiteness of $\mathcal{H}_\infty$, computation of $\mathcal{H}_\infty$, and reconstruction of improving vectors using $\mathcal{H}_\infty$.

Definition 13. We say that $(u', V_{u'})$ reduces (u, V_u), or $(u', V_{u'}) \sqsubseteq (u, V_u)$ for short, if the following conditions are satisfied:

- $u' \sqsubseteq u$,
- for every $v \in V_u$ there exists a $v' \in V_{u'}$ with $v' \sqsubseteq v$,
- $u' \neq 0$ or there exist vectors $v \in V_u$ and $v' \in V_{u'}$ with $0 \neq v' \sqsubseteq v$.

Theorem 14. *Given rational matrices* T *and* W *of appropriate dimensions, and let* $\mathcal{H}_\infty$ *be defined as above. Then* $\mathcal{H}_\infty$ *is a finite set.*

Proof. First we show that there are only finitely many pairs $(u, V_u) \in \mathcal{H}_\infty$. Then we show that each V_u is finite.

Consider the set $\mathcal{H}' := \mathcal{H}_\infty \setminus \{(0, V_0)\}$. We associate with each $(u, V_u) \in \mathcal{H}'$ the monomial ideal $I(u, V_u) \in \mathbb{Q}[x_1, \ldots, x_{2n+2m}]$ generated by all the monomials $x^{(u^+, u^-, v^+, v^-)}$ with $v \in V_u$. Call $\mathcal{I}(\mathcal{H}')$ the family of all these monomial ideals.

Clearly, $(u', V_{u'}) \not\sqsubseteq (u, V_u)$ for all different pairs $(u', V_{u'}), (u, V_u) \in \mathcal{H}_\infty$. That is, either $u' \not\sqsubseteq u$, or there is some $v \in V_u$ such that there is no $v' \in V_{u'}$ with $v' \sqsubseteq v$, or both. In all of these three cases we find a monomial $x^{(u^+, u^-, v^+, v^-)} \in I(u, V_u)$ with $v \in V_u$ such that there is no monomial generator $x^{(u'^+, u'^-, v'^+, v'^-)}$ of $I(u', V_{u'})$ which divides $x^{(u^+, u^-, v^+, v^-)}$. Therefore, $I(u, V_u) \not\subseteq I(u', V'_u)$.

Thus, no ideal $I \in \mathcal{I}(\mathcal{H}')$ is contained in another ideal $J \in \mathcal{I}(\mathcal{H}')$. Therefore, by Maclagan's Theorem [18, 19], the family $\mathcal{I}(\mathcal{H}')$ is finite, which proves that $\mathcal{H}_\infty$ contains only finitely many pairs (u, V_u).

It remains to show that each pair $(u, V_u) \in \mathcal{H}_\infty$ is finite, that is, that V_u is finite. Fix some arbitrary $N \in \mathbb{Z}_+$.

If $u = 0$ then any Graver basis element $z = (0, v_1, \ldots, v_N)$ contains exactly one non-zero building block v_i. If, on the contrary, v_i and v_j were both non-zero, then we could construct a non-zero vector $z' \in \ker(A_N)$ with $z' \sqsubseteq z$ and $z' \neq z$ by replacing v_i by 0 in z. This contradicts the $\sqsubseteq$-minimality of the Graver basis element z. Since there is only one non-zero building block v_i in z, this building block has to be $\sqsubseteq$-minimal in $\ker(W)$, that is, v_i belongs to the Graver basis of W. Thus, $V_0 = \mathcal{G}(W) \cup \{0\}$ is finite.

If $u \neq 0$ then any Graver basis element $z = (u, v_1, \ldots, v_N)$ can contain only $\sqsubseteq$-minimal solutions v of $Wv = -Tu$ as building blocks $v_i, i = 1, \ldots, N$. If v_j were not such a $\sqsubseteq$-minimal solution, then there exists some $v'_j \neq v_j$ with $Wv'_j = -Tu$

and $v'_j \sqsubseteq v_j$. Thus, replacing in z the building block v_j by v'_j we obtain a non-zero vector $z' \in \ker(A_N)$ with $z' \sqsubseteq z$ and $z' \neq z$ contradicting the $\sqsubseteq$-minimality of the Graver basis element z. The set $M(u)$ of $\sqsubseteq$-minimal solutions of $Wv = -Tu$, however, is finite. To see this, apply the Gordan-Dickson-Lemma (see, e.g., [8]) to the set $\{(v^+, v^-) : v \in M(u)\}$. □

Using the finiteness of $\mathcal{H}_\infty$, we could find this set by computing the Graver test set $\mathcal{G}_N$ for sufficiently large N and by decomposing its elements into their building blocks. Even when disregarding that we do not know in advance how big N then has to be taken, this approach is not very practical, due to the size of $\mathcal{G}_N$. The idea now is to retain the pattern of Graver test set computation from Algorithm 8, but to work with pairs (u, V_u) instead, and to define the two main ingredients, normalForm and S-vectors, appropriately. In what follows, the objects f, g, and s all are pairs of the form (u, V_u).

Algorithm 15 (Algorithm to compute $\mathcal{H}_\infty$).

Input: a generating set F of $\ker(A_1)$ in (u, V_u)-notation to be specified below
Output: a set G which contains $\mathcal{H}_\infty$

```
{forming S-vectors}
C := ⋃_{f,g∈G} {f ⊕ g}
while C ≠ ∅ do
  s := an element in C
  C := C \ {s}
  f := normalForm(s, G)
  if f ≠ (0, {0}) then
    {adding S-vectors}
    C := C ∪ ⋃_{g∈G∪{f}} {f ⊕ g}
    G := G ∪ {f}
  end if
end while
return G.
```

Behind the function normalForm(s, G) there is the following algorithm.

Algorithm 16 (Normal form algorithm).

Input: a pair s, a set G of pairs
Output: a normal form of s with respect to G

```
while there is some g ∈ G such that g ⊑ s do
  s := s ⊖ g
end while
return s
```

It remains to define an appropriate input set, the sum $\oplus$ and the difference $\ominus$ of two pairs (u, V_u) and $(u', V_{u'})$. To get good candidates we may think of a computation of $\mathcal{G}_N$ where every vector is decomposed into its building blocks.

Lemma 17. *Let* F *be a generating set for* $\ker(A_1)$ *over* $\mathbb{Z}$ *which contains a generating set for* $\{(0, v) : Wv = 0\} \subseteq \ker(A_1)$ *consisting only of vectors with zero first-stage component. Moreover, let* F_N *be the set of all those vectors in* $\ker(A_N)$ *whose building blocks are also building blocks of vectors in* $F \cup \{0\}$*. Then, for any* N*, the vectors* F_N *generate* $\ker(A_N)$ *over* $\mathbb{Z}$*.*

Proof. cf. [13]. □

This lemma suggests the following input set.

Definition 18. We define the building blocks of all vectors in $F\cup -F\cup\{0\}$ in (u, V_u)-notation to be the input set to the above algorithm. Herein, F is a generating set for $\ker(A_1)$ over $\mathbb{Z}$ which contains a generating set for $\{(0, v) : Wv = 0\} \subseteq \ker(A_1)$ consisting only of vectors with zero first-stage component.

Definition 19. Let

$$(u, V_u) \oplus (u', V_{u'}) := (u + u', V_u + V_{u'}),$$

where

$$V_u + V_{u'} := \{v + v' : v \in V_u, v' \in V_{u'}\}.$$

Moreover, let

$$(u, V_u) \ominus (u', V_{u'}) := (u - u', \{v - v' : v \in V_u, v' \in V_{u'}, v' \sqsubseteq v\}).$$

Remark 20. In $(u, V_u) \ominus (u', V_{u'}) := (u - u', \{v - v' : v \in V_u, v' \in V_{u'}, v' \sqsubseteq v\})$ it suffices to collect only one difference $v - v'$ for every $v \in V_u$. It will be elaborated in the proof of the subsequent proposition that Algorithm 15 still terminates and works correctly if we defined $(u, V_u) \ominus (u', V_{u'})$ this way.

Proposition 21. *If the input set, the procedure normalForm, and* $f \oplus g$, $s \ominus g$ *are defined as above, then Algorithm 15 terminates and satisfies its specifications.*

Proof. In the course of the algorithm, a sequence of pairs in $G \setminus F$ is generated, let us denote it by $((u_1, V_{u_1}), (u_2, V_{u_2}), \ldots)$, with the property that $(u_i, V_{u_i}) \not\sqsubseteq (u_j, V_{u_j})$ whenever $i < j$. We will show that the subsequences with $u = 0$ and with $u \neq 0$ are both finite. Therefore, Algorithm 15 terminates.

Let $((0, V_1), (0, V_2), \ldots)$ be a sequence of pairs such that $(0, V_i) \not\sqsubseteq (0, V_j)$ whenever $i < j$. We associate with each pair $(0, V_i)$ the monomial ideal $I(0, V_i) \in \mathbb{Q}[x_1, \ldots, x_{2m}]$ generated by all monomials $x^{(v^+, v^-)}$ with $v \neq 0$ and $v \in V_i$. Then $(0, V_i) \not\sqsubseteq (0, V_j)$ whenever $i < j$ implies that there is some nonzero vector $v' \in V_j$ such that there is no $v \in V_i$ with $v \neq 0$ and $v \sqsubseteq v'$. Thus, there exists a generator $x^{(v'^+, v'^-)}$ of $I(0, V_j)$ which is not divisible by any generator $x^{(v^+, v^-)}$ of $I(0, V_i)$.

But this means that $I(0, V_j) \not\subseteq I(0, V_i)$ whenever $i < j$. Maclagan's Theorem [18, 19] now implies that the sequence $(I(0, V_1), I(0, V_2), \dots)$ terminates. Therefore, the subsequence $((0, V_1), (0, V_2), \dots)$ generated by Algorithm 15 is finite.

Let $((u_1, V_{u_1}), (u_2, V_{u_2}), \dots)$ be a sequence of pairs such that $u_i \neq 0$ for all $i = 1, 2, \dots$, and such that $(u_i, V_{u_i}) \not\sqsubseteq (u_j, V_{u_j})$ whenever $i < j$. We associate with (u_i, V_{u_i}), $u_i \neq 0$, the monomial ideal $I(u_i, V_{u_i}) \in \mathbb{Q}[x_1, \dots, x_{2n+2m}]$ generated by all the monomials $x^{(u_i^+, u_i^-, v^+, v^-)}$ with $v \in V_{u_i}$. Consider the sequence $(I(u_1, V_{u_1}), I(u_2, V_{u_2}), \dots)$ of monomial ideals. $(u_i, V_{u_i}) \not\sqsubseteq (u_j, V_{u_j})$ whenever $i < j$ implies that $u_i \not\sqsubseteq u_j$ or that there is some $v' \in V_{u_j}$ such that there is no $v \in V_{u_i}$ with $v \sqsubseteq v'$. Thus, there exists a generator $x^{(u_j^+, u_j^-, v'^+, v'^-)}$ of $I(u_j, V_{u_j})$ which is not divisible by any generator $x^{(u_i^+, u_i^-, v^+, v^-)}$ of $I(u_i, V_{u_i})$. Hence, $I(u_j, V_{u_j}) \not\subseteq I(u_i, V_{u_i})$ whenever $i < j$. Maclagan's Theorem [18, 19] now implies that the sequence $(I(u_1, V_{u_1}), I(u_2, V_{u_2}), \dots)$ terminates. Therefore, the subsequence $((u_1, V_{u_1}), (u_2, V_{u_2}), \dots)$ generated by Algorithm 15 is finite. We conclude that Algorithm 15 terminates and it remains to prove its correctness.

To show that $\mathcal{H}_\infty \subseteq G$, we have to prove that $\mathcal{H}_N \subseteq G$ for all $N \in \mathbb{Z}_+$. Fix N and start a Graver test set computation (Algorithm 8) with $\bar{F} := \{(u, v_1, \dots, v_N) : (u, V_u) \in G, v_i \in V_u\}$ as input set. $\bar{F}$ generates $\ker(A_N)$ over $\mathbb{Z}$ for all $N \in \mathbb{Z}_+$, since $F_N \subseteq \bar{F}$ by the assumption on the input set to Algorithm 15.

We will now show that all sums $z + z'$ of two elements $z, z' \in \bar{F}$ reduce to 0 with respect to $\bar{F}$. In this case, Algorithm 8 returns the input set $\bar{F}$ which implies $\mathcal{G}_N \subseteq \bar{F}$. Therefore, $\mathcal{H}_N \subseteq G$ as desired.

Take two arbitrary elements $z = (u, v_1, \dots, v_N)$ and $z' = (u', v_1', \dots, v_N')$ from $\bar{F}$, and consider the vector $z + z' = (u + u', v_1 + v_1', \dots, v_N + v_N')$.

In the above algorithm, $(u, V_u) \oplus (u', V_{u'})$ was reduced to zero by elements $(u_1, V_{u_1}), \dots, (u_k, V_{u_k}) \in G$. From this sequence we can construct a sequence $z_1, \dots, z_k$ of vectors in $\bar{F}$ which reduce $z + z'$ to zero as follows.

$(u_1, V_{u_1}) \sqsubseteq (u, V_u) \oplus (u', V_{u'})$ implies that $u_1 \sqsubseteq u + u'$ and that there exist $v_{1,1}, \dots, v_{1,N} \in V_{u_1}$ such that $v_{1,i} \sqsubseteq v_i + v_i'$ for $i = 1, \dots, N$. Therefore, $z_1 := (u_1, v_{1,1}, \dots, v_{1,N}) \sqsubseteq z + z'$ and $z + z'$ can be reduced to $z + z' - z_1$. Moreover, $z_1 \in \bar{F}$ and all the building blocks of $z + z' - z_1$ lie in $((u, V_u) \oplus (u', V_{u'})) \ominus (u_1, V_{u_1})$.

But $((u, V_u) \oplus (u', V_{u'})) \ominus (u_1, V_{u_1})$ was further reduced by $(u_2, V_{u_2}), \dots, (u_k, V_{u_k}) \in G$. Therefore, we can construct from (u_2, V_{u_2}) a vector $z_2 \in \bar{F}$ with $z_2 \sqsubseteq z + z' - z_1$. Therefore, $z + z' - z_1$ can be further reduced to $z + z' - z_1 - z_2$. Repeating this construction, we obtain in the kth step $z + z' - z_1 - \dots - z_{k-1}$ whose building blocks lie in $((u, V_u) \oplus (u', V_{u'})) \ominus (u_1, V_{u_1}) \ominus \dots \ominus (u_{k-1}, V_{u_{k-1}})$. The latter can be reduced to $(0, \{0\})$ by the pair $(u_k, V_{u_k}) \in G$. Therefore, there exists a vector $z_k \in \bar{F}$ such that $z_k \sqsubseteq z + z' - z_1 - \dots - z_{k-1}$ and $0 = z + z' - z_1 - \dots - z_k$. □

To solve the optimization problem $\min\{p^\intercal z : A_N = b, z \in \mathbb{Z}_+^{n+Nm}\}$ by Algorithm 6 we have to reconstruct improving vectors from building blocks in $\mathcal{H}_\infty$. The following lemma shows how to accomplish this.

Lemma 22. *Suppose there exists no pair* $(u', V_{u'}) \in \mathcal{H}_\infty$ *with the properties*

1. $u' \leq u$,
2. *for all* $i = 1, \ldots, N$ *there exists* $\bar{v}_i \in V_{u'} : \bar{v}_i \leq v_i$,
3. $p^\top z' > 0$, *where* $z' = (u', v'_1, \ldots, v'_N)$ *and* $v'_i \in \operatorname{argmax}\{c_i^\top \bar{v}_i : \bar{v}_i \leq v_i, \bar{v}_i \in V_{u'}\}$ *for* $i = 1, \ldots, N$.

Then $z_0 = (u, v_1, \ldots, v_N)$ *is optimal for* $\min\{p^\top z : A_N z = b, z \in \mathbb{Z}_+^d\}$.

If there exists such a pair $(u', V_{u'}) \in \mathcal{H}_\infty$ *then* $z_0 - z'$ *is feasible and it holds* $p^\top(z_0 - z') < p^\top z_0$.

Proof. Suppose that z_0 is not optimal.

Then there is some vector $z'' = (u'', v''_1, \ldots, v''_N) \in \mathcal{G}_N$ so that $z_0 - z''$ is feasible and $p^\top(z_0 - z'') < p^\top z_0$. Feasibility of $z_0 - z''$ implies $z_0 - z'' \geq 0$, hence $z'' \leq z_0$. Therefore, $u'' \leq u$ and $v''_i \leq v_i$, $i = 1, \ldots, N$, the latter implying that for any $i = 1, \ldots, N$ there exists a $\bar{v}_i \in V_{u''}$ such that $\bar{v}_i \leq v_i$. Let $z' := (u'', v'_1, \ldots, v'_N)$ where $v'_i \in \operatorname{argmax}\{p_i^\top \bar{v}_i : \bar{v}_i \leq v_i, \bar{v}_i \in V_{u''}\}$.

Now $p^\top(z_0 - z'') < p^\top z_0$ implies that $p^\top z'' > 0$. Moreover, $p^\top z' \geq p^\top z'' > 0$. In conclusion, the pair $(u'', V_{u''})$ fulfills conditions 1.–3. proving the first claim of the lemma.

With $z' = (u', v'_1, \ldots, v'_N)$ according to 3. we obtain $p^\top(z_0 - z') < p^\top z_0$. Moreover $v'_i \leq v_i$, $i = 1, \ldots, N$, and $u' \leq u$ together imply $z' \leq z_0$, and $z_0 - z' \geq 0$. Finally, $(u', v'_1, \ldots, v'_N) \in \ker(A_N)$, and therefore $A_N(z_0 - z') = A_N z_0 + 0 = b$ which completes the proof. □

The reconstruction procedures in the above lemma yield an improving vector in linear time with respect to the number N of scenarios. Accordingly, we observed in test runs that the method is fairly insensitive with respect to growing of the number N of scenarios. Of course, this becomes effective only after $\mathcal{H}_\infty$ has been computed.

Algorithm 15 together with an initialization procedure and the augmentation procedure from Lemma 22 have been implemented. The current version of that implementation can be obtained from [12]. To indicate the principal behaviour of our method, we report on test runs with an academic example. The algorithmic bottleneck of the method is the completion procedure in Algorithm 15. Therefore the sizes of the matrices T and W are very moderate in this initial phase of testing. On the other hand, the method is fairly insensitive with respect to the number of scenarios.

Consider the two-stage program

$$\min\{35x_1 + 40x_2 + \frac{1}{N}\sum_{\nu=1}^{N} 16y_1^\nu + 19y_2^\nu + 47y_3^\nu + 54y_4^\nu :$$

$$\begin{aligned} x_1 + y_1^\nu + y_3^\nu &\geq h_1^\nu, \\ x_2 + y_2^\nu + y_4^\nu &\geq h_2^\nu, \\ 2y_1^\nu + y_2^\nu &\leq h_3^\nu, \\ y_1^\nu + 2y_2^\nu &\leq h_4^\nu, \\ x_1, x_2, y_1^\nu, y_2^\nu, y_3^\nu, y_4^\nu &\in \mathbb{Z}_+\}. \end{aligned}$$

Here, the random vector $h \in \mathbb{R}^s$ is given by the scenarios $h^1, \ldots, h^N$, all with equal probability $1/N$. The realizations of (h_1^ν, h_2^ν) and (h_3^ν, h_4^ν) are given by uniform grids (of differing granularity) in the squares $[300, 500] \times [300, 500]$ and $[0, 2000] \times [0, 2000]$, respectively. Timings are given in CPU seconds on a SUN Enterprise 450, 300 MHz Ultra-SPARC.

It took 3.3s to compute $\mathcal{H}_\infty$ altogether consisting of 1464 building blocks arranged into 25 pairs (u, V_u). $\text{Aug}(\mathcal{H}_\infty)$ then gives the times needed to augment the solution $x_1 = x_2 = y_1^\nu = y_2^\nu = 0$, $y_3^\nu = \xi_1^\nu$, and $y_4^\nu = \xi_2^\nu$, $\nu = 1, \ldots N$ to optimality.

(h_1, h_2)	(h_3, h_4)	scenarios	variables	optimum	$\text{Aug}(\mathcal{H}_\infty)$	CPLEX	scendec
5×5	3×3	225	902	$(100, 150)$	1.52	0.63	> 1800
5×5	21×21	11025	44102	$(100, 100)$	66.37	696.10	–
9×9	21×21	35721	142886	$(108, 96)$	180.63	> 1 day	–

Although further exploration is necessary, the above table seems to indicate linear dependence of the computing time on the number N of scenarios, once $\mathcal{H}_\infty$ has been computed. Existing methods in stochastic integer programming are much more sensitive on N. The scenario decomposition from Section 3, for instance, involves a non-smooth dual optimization in a space of dimension $(N-1)n$. This explains the failure reported in the column "scendec". Tackling the full-size integer linear program (6) by CPLEX directly is possible, but, as shown in the respective column, becomes inferior when N becomes large.

7 Conclusions

Two-stage stochastic integer programs are possible modeling alternatives for optimization problems where real-time features arise due to incompleteness of data information at the moment of decision. From structural viewpoint, stochastic integer programs bear nonconvexities which prevents straightforward extension of algorithms known from stochastic programming without integer requirements.

With discrete probability distributions, linear two-stage stochastic integer programs can be reformulated as large-scale, mixed-integer linear programs. The latter are amenable for decomposition. We have proposed two such methods. The first one combines Lagrangian relaxation of nonanticipativity and a branch-and-bound scheme. This method works for general mixed-integer linear two-stage stochastic programs, which we have illustrated at an example from unit commitment under uncertainty.

Our second decomposition method works with test sets instead with the problem itself. We have demonstrated that the Graver test set of a pure-integer linear two-stage stochastic program can be decomposed into building blocks. The set of these building blocks stabilizes with growing number of scenarios. It can be computed by a critical-pair/completion procedure, directly, without advance knowledge of the Graver test set. The building blocks finally lead to an augmentation algorithm for stochastic integer programs. According to our preliminary testing, this algorithm is far less sensitive to the number of scenarios than hitherto stochastic programming methods are.

REFERENCES

1. B. Bank, R. Mandel: Parametric Integer Optimization, Akademie-Verlag, Berlin 1988.
2. J. R. Birge, F. Louveaux: Introduction to Stochastic Programming, Springer, New York, 1997.
3. C. E. Blair, R. G. Jeroslow: The value function of a mixed integer program: I, Discrete Mathematics 19 (1977), 121-138.
4. B. Buchberger: History and basic features of the critical-pair/completion procedure, Journal of Symbolic Computation, 2 (1987), 3-38.
5. C. C. Carøe, R. Schultz: Dual decomposition in stochastic integer programming, Operations Research Letters 24 (1999), 37-45.
6. C. C. Carøe, R. Schultz: A two-stage stochastic program for unit commitment under uncertainty in a hydro-thermal system, Schwerpunktprogramm "Echtzeit-Optimierung großer Systeme" der Deutschen Forschungsgemeinschaft, Preprint 98-13, 1998, revised 1999.
7. C. C. Carøe, J. Tind: L-shaped decomposition of two-stage stochastic programs with integer recourse, Mathematical Programming 83 (1998), 451-464.
8. D. Cox, J. Little, D. O'Shea: Ideals, Varieties, Algorithms, Springer-Verlag, 1992.
9. Using the CPLEX Callable Library, CPLEX Optimization, Inc. 1998.
10. J. E. Graver: On the foundation of linear and integer programming I, Mathematical Programming, 9 (1975), 207-226.
11. R. Hemmecke: On the positive sum property of Graver test sets, Preprint SM-DU-468, University of Duisburg, 2000, available for download[1].
12. R. Hemmecke: Homepage on test sets, URL: `http://www.testsets.de`.
13. R. Hemmecke, R. Schultz: Decomposition of Test Sets in Stochastic Integer Programming, Preprint SM-DU-475, University of Duisburg, 2000, available for download[2].
14. J. B. Hiriart-Urruty, C. Lemaréchal: Convex Analysis and Minimzation Algorithms, Springer-Verlag, Berlin 1993.
15. P. Kall, S. W. Wallace: Stochastic Programming, Wiley, Chichester, 1994.
16. K. C. Kiwiel: Proximity control in bundle methods for convex nondifferentiable optimization, Mathematical Programming 46 (1990), 105-122.
17. K. C. Kiwiel: User's Guide for NOA 2.0/3.0: A Fortran Package for Convex Nondifferentiable Optimization, Systems Research Institute, Polish Academy of Sciences, Warsaw, 1994.
18. D. Maclagan: Antichains of monomial ideals are finite, Electronic preprint, University of California at Berkeley, 1999, available for download[3].
19. D. Maclagan: Structures on sets of monomial ideals, PhD thesis, University of California at Berkeley, 2000.
20. G. L. Nemhauser, L. A. Wolsey: Integer and Combinatorial Optimization, Wiley, New York, 1988.
21. A. Prékopa: Stochastic Programming, Kluwer, Dordrecht, 1995.

[1] `http://www.uni-duisburg.de/FB11/disma/ramon/articles/preprint2.ps`

[2] `http://www.uni-duisburg.de/FB11/disma/ramon/articles/preprint3.ps`

[3] `http://front.math.ucdavis.edu/math.CO/9909168`

22. A. Ruszczyński: Some advances in decomposition methods for stochastic linear programming, Annals of Operations Research 85 (1999), 153-172.
23. W. Römisch, R. Schultz: Multistage stochastic integer programs: an introduction, this volume.
24. A. Schrijver: Theory of Linear and Integer Programming, Wiley, Chichester, 1986.
25. R. Schultz: Continuity properties of expectation functions in stochastic integer programming, Mathematics of Operations Research 18 (1993), 578-589.
26. R. Schultz: On structure and stability in stochastic programs with random technology matrix and complete integer recourse, Mathematical Programming 70 (1995), 73-89.
27. R. Schultz: Some aspects of stability in stochastic programming, Annals of Operations Research 100 (2001), (to appear).
28. R. Schultz, L. Stougie, M. H. van der Vlerk: Solving stochastic programs with integer recourse by enumeration: A framework using Gröbner basis reductions, Mathematical Programming 83 (1998), 229-252.
29. B. Sturmfels: Gröbner Bases and Convex Polytopes, American Mathematics Society, Providence, Rhode Island, 1995.
30. B. Sturmfels, R. R. Thomas: Variation of cost functions in integer programming, Mathematical Programming, 77 (1997), 357-387.
31. R. R. Thomas: A geometric Buchberger algorithm for integer programming, Mathematics of Operations Research, 20 (1995), 864-884.
32. R. M. van Slyke, R. J.-B. Wets: L-shaped linear programs withapplication to optimal control and stochastic programming, SIAM Journal of Applied Mathematics 17 (1969), 638-663.
33. R. Weismantel: Test sets of integer programs, Mathematical Methods of Operations Research, 47 (1998), 1-37.

Modeling of Uncertainty for the Real-Time Management of Power Systems

Nicole Gröwe-Kuska[1], Matthias P. Nowak[2], and Isabel Wegner[1]

[1] Institut für Mathematik, Humboldt-Universität zu Berlin, Germany
[2] SINTEF, N-7465 Trondheim, Norway

Abstract A major issue in the application of multistage stochastic programming to model the cost-optimal generation and trading of electric power is the approximation of the underlying stochastic data processes by tree-structured schemes. We present a methodology for the generation of scenario trees for the stochastic load process from historical load profiles. The statistical modeling of the load process exploits the decomposition of the load process into a daily mean load process and a mean-corrected load series. The probability distribution of the load process over the optimization horizon is derived by using a time series model for the daily mean load process and regression models for the mean-corrected load series. We utilize the explicit representation of the distribution to compute approximate load scenarios and their probabilities. In a final step we reduce the number of load scenarios by a scenario deletion procedure. We report on the application of our approach to the cost-optimal generation of electric power in the hydro-thermal generation system of a German power utility.

1 Introduction

In industrial practice, mathematical models for the efficient generation, transmission, and distribution of electric power have been proved indispensable. The ongoing liberalization of electricity markets stimulates the interest of power utilities in developing modeling and optimization techniques for operating power systems and trading electricity under uncertainty. Utilities participating in deregulated markets observe increasing uncertainty in load (i.e., demand for electric power), inflows to reservoirs and prices for fuel and electricity on spot and contract markets. The mismatched power between actual and predicted demand may be supplied by the power system or by trading activities. The competitive environment forces the utilities to rate alternatives within a few minutes.

In the present paper we develop approximate tree-structured schemes for the stochastic load process entering a multistage mixed-integer stochastic program. It models the weekly cost-optimal generation and trading of an electric hydro-thermal based utility under data uncertainty. The relevant uncertain data may comprise electric load, stream flows to hydro units, and fuel and electricity prices. For solving the stochastic power management model a stochastic Lagrangian relaxation algorithm [15] has been designed. With a state-of-the-art bundle method for solving the dual, specialized subproblem solvers and Lagrangian heuristics, the stochastic version of classical Lagrangian relaxation becomes fairly efficient.

The stochastic power management model has emerged from a collaboration with the German utility Vereinigte Energiewerke AG (VEAG). The VEAG generation system consists of 25 (coal-fired or gas-burning) thermal units and 7 pumped-storage hydro units. Its total capacity is about 13,000 megawatts (MW), including a hydro capacity of 1,700 MW; the system peak loads are about 8,600 MW. In contrast to other hydro-thermal based utilities the amount of installed pumped-storage capacity enables the inclusion of pumped-storage plants into the optimization. An additional feature of that system is that for a planning period of one week, the changes of reservoir levels in the pumped-storage hydro units caused by stream inflows are negligible.

The paper is organized as follows. In §2 we describe the stochastic power management model and the stochastic Lagrangian relaxation approach for its solution. The model selection for the electric load is addressed in §3.2. Our procedures for generating and reducing load scenario trees are presented in §3.3 and §3.4, respectively. In §3.5 we report on numerical tests.

2 Power System Modeling

2.1 Modeling

We consider a power generation system comprising thermal units, pumped storage plants and contracts for delivery and purchase, and describe a model for its cost-optimal operation under uncertainty in electrical load (i.e., demand), stream flows in hydro units and prices for fuel or electricity.

The scheduling horizon for unit commitment is typically discretized into uniform (e.g., hourly) intervals. Accordingly, the load, stream flows and prices are assumed to be constant within each time period. The scheduling decisions for thermal units are: which units to commit in each period, and at what generating capacity. The decision variables for hydro plants are the generation and pumping levels for each period. Contracts for delivery and purchase are regarded as special thermal units. The schedule should minimize the total generation costs, subject to the operational requirements.

We use the following notation. There are T time periods. I and J are the numbers of thermal and hydro units, respectively. For a thermal unit i in period t, $u_{it} \in \{0, 1\}$ is its *commitment* (1 if on, 0 if off), and p_{it} its *production*, with $p_{it} = 0$ if $u_{it} = 0$, $p_{it} \in [p_{it}^{min}, p_{it}^{max}]$ if $u_{it} = 1$, where p_{it}^{min} and p_{it}^{max} are the minimum and maximum capacities. Additionally, there are *minimum up/down-time requirements*: when unit i is switched on (off), it must remain on (off) for at least $\bar{\tau}_i$ ($\underline{\tau}_i$, resp.) periods. For a hydro plant j, v_{jt} and w_{jt} are its *generation* and *pumping* levels in period t, with upper bounds v_{jt}^{max} and w_{jt}^{max} respectively, and l_{jt} is the *storage* volume in the upper dam at the end of period t, with upper bound l_{jt}^{max}. The water balance relates l_{jt} with $l_{j,t-1}$, v_{jt}, w_{jt} and the *water inflow* γ_{jt}, using the *pumping efficiency* η_j. The initial and final volumes are specified by l_j^{in} and l_j^{end}.

The basic system requirement is to meet the electric load. Another important requirement is the spinning reserve constraint. To maintain reliability (compensate sudden load peaks or unforeseen outages of units) the total committed capacity should exceed the load in every period by a certain amount (e.g., a fraction of the demand). The *load* and the *spinning reserve* during period t are denoted by d_t and r_t, respectively. Typical load curves exhibit daily and weekly cycles with lows during nights and weekends, and morning and early evening peaks. Efficient operation of pumped storage hydro plants exploits such cycles by generating during peak load periods and pumping during off-peak periods.

Since the operating costs of hydro plants are usually negligible, the total system cost is given by the sum of startup and operating costs of all thermal units over the whole scheduling horizon. The *fuel cost* C_{it} for operating thermal unit i during period t has the form

$$C_{it}(p_{it}, u_{it}) := \max_{l=1,\dots,\bar{l}} \{a_{ilt} p_{it} + b_{ilt} u_{it}\}, \tag{1}$$

with coefficients a_{ilt}, b_{ilt} such that $C_{it}(\cdot, 1)$ is convex and increasing on R_+; note that $C_{it}(0,0) = 0$. The startup cost of unit i depends on its downtime; it may vary from a maximum cold-start value to a much smaller value when the unit is still relatively close to its operating temperature. This is modeled by the *startup cost*

$$S_{it}(u_i) := \max_{\tau=0,\dots,\tau_i^c} c_{i\tau} \left(u_{it} - \sum_{\kappa=1}^{\tau} u_{i,t-\kappa} \right), \tag{2}$$

where $0 = c_{i0} < \dots < c_{i\tau_i^c}$ are fixed cost coefficients, τ_i^c is the cool-down time of unit i, $c_{i\tau_i^c}$ is its maximum cold-start cost, $u_i := (u_{it})_{t=1}^T$, and $u_{i\tau} \in \{0,1\}$ for $\tau = 1 - \tau_i^c, \dots, 0$ are given initial values.

In electric utilities, schedulers forecast the electric load for the required time span. Since the load is mainly driven by meteorological parameters (temperature, cloud cover, etc.), the actual load deviates from its prediction. Of course, the load uncertainty increases with the length of the planning horizon. Other sources of uncertainty are generator outages, stream flows in hydro units, and prices of fuel and electricity.

To formulate a power generation model that incorporates fluctuations in stream inflows in hydro plants, and fuel and electricity prices in addition to the load uncertainty, we use a probabilistic description of uncertainty. Thus

$$\{\xi_t := (d_t, r_t, \gamma_t, a_t, b_t, c_t)\}_{t=1}^T \tag{3}$$

is assumed to be a discrete-time stochastic process on a probability space $(\Omega, \mathcal{F}, \mathcal{P})$, where d_t, r_t and γ_t represent the load, the spinning reserve and the water inflows in period t, while a_t, b_t and c_t collect the cost coefficients of (1) and (2).

The scheduling decisions for period t are made *after* learning the realization of the stochastic variables for that period. Denote by $\mathcal{F}_t \subseteq \mathcal{F}$ the σ-field generated by $\{\xi_\tau\}_{\tau=1}^t$, i.e., the events observable till period t. Since the information on ξ_1 is complete, $\mathcal{F}_1 = \{\emptyset, \Omega\}$, i.e., ξ_1 is deterministic. By assuming $\mathcal{F}_T = \mathcal{F}$ we require that full information be available at the end of the planning horizon. The sequence of scheduling decisions $\{u_t, p_t, v_t, w_t\}$ also forms a stochastic process on $(\Omega, \mathcal{F}, \mathcal{P})$, which is assumed to be adapted to the filtration of σ-fields, i.e., *nonanticipative*.

Nonanticipativity means that the decisions (u_t, p_t, v_t, w_t) may depend only on the data observable till period t, or equivalently that (u_t, p_t, v_t, w_t) is $\mathcal{F}_t$-measurable.

We now assume that we have a *discrete* distribution of the data process $\{\xi_t\}_{t=1}^T$ (cf. (3)). Its support consists of *scenarios* (i.e., realizations of $\{\xi_t\}_{t=1}^T$) that form a *scenario tree* based on a finite set of nodes $\mathcal{N}$. The *root* node $n = 1$ stands for period $t = 1$. Every other node n has a unique *predecessor* node n_- and a *transition* probability $\pi_{n/n_-} > 0$, which is the probability of n being the successor of n_-. The successors to node n form the set $\mathcal{N}_+(n)$; their transition probabilities add to 1. The probability π_n of each node n is generated recursively by $\pi_1 = 1$, $\pi_n = \pi_{n/n_-}\pi_{n_-}$ for $n \neq 1$. Nodes n with $\mathcal{N}_+(n) = \emptyset$ are called *leaves*; they constitute the *terminal* set $\mathcal{N}_T$. A scenario corresponds to a path from the root node to a leaf. The probabilities $\{\pi_n\}_{n\in\mathcal{N}_T}$ provide a distribution for the set of all scenarios. Conversely, given such scenario probabilities, the remaining node and transition probabilities are generated recursively by $\pi_n = \sum_{n_+\in\mathcal{N}_+(n)} \pi_{n_+}$, $\pi_{n_+/n} = \pi_{n_+}/\pi_n$ for $n_+ \in \mathcal{N}_+(n)$.

Let $\mathrm{path}(n)$ denote the path from the root to node n. Then node n corresponds to a set of realizations of $\{\xi_t\}_{t=1}^T$ that coincide until the period $t(n) := |\mathrm{path}(n)|$ associated with node n; their common value $\xi_{t(n)}$ is denoted by $\xi^n := (d^n, r^n, \gamma^n, a^n, b^n, c^n)$. Let the decisions for period t be made after learning the realization of $\{\xi_t\}_{\tau=1}^t$. The scheduling decisions (u^n, p^n, v^n, w^n) assigned to nodes n in $\mathcal{N}_t := \{n : t(n) = t\}$ are realizations of the stochastic decisions (u_t, p_t, v_t, w_t); note that $\sum_{n\in\mathcal{N}_t} \pi_n = 1$. We denote by $\mathcal{N}_{\text{first}} := \cup_{t=1}^{t_1}\mathcal{N}_t$ the set of *first-stage* nodes, where t_1 is the maximal period such that the data process $\{\xi_t\}_{t=1}^{t_1}$ is deterministic.

Let $u_i^{\mathrm{path}(n)} := (u_i^\nu)_{\nu\in\mathrm{path}(n)}$. We use the following notation for the sequence of predecessors of any node $n \in \mathcal{N}\setminus\{1\}$: $n_{-1} := n_-$, $n_{-(\kappa+1)} := (n_{-\kappa})_-$ if $t(\kappa) > 1$; note that $t(n_{-\kappa}) = t(n) - \kappa$ for $\kappa = 1, \dots, t(n) - 1$. To handle initial values of the commitment decisions u_i we let $n_\kappa := \kappa - t(n)$ for $\kappa = t(n) + \tau_{\text{ini}}, \dots, t(n)$, where $\tau_{\text{ini}} := 1 - \max_{i=1,\dots,I}\{\tau_i^c, \bar{\tau}_i - 1, \underline{\tau}_i - 1\}$, and assume that fixed initial values $u_i^{1-\kappa}$ for $\kappa = \tau_{\text{ini}}, \dots, 0$, are given. Then (cf. (1) and (2))

$$C_i^n(p_i^n, u_i^n) := \max_{l=1:\bar{l}} \{ a_{il}^n p_i^n + b_{il}^n u_i^n \}$$

and

$$S_i^n\left(u_i^{\mathrm{path}(n)}\right) := \max_{\tau=0:\tau_i^c} c_{i\tau}^n \left(u_i^n - \sum_{\kappa=1}^{\tau} u_i^{n_{-\kappa}}\right) \tag{4}$$

are the fuel and startup costs of unit i at node n.

The *scenario-tree formulation* of the power generation problem reads:

$$\min \sum_{n\in\mathcal{N}} \pi_n \sum_{i=1}^{I} \left[C_i^n(p_i^n, u_i^n) + S_i^n\left(u_i^{\mathrm{path}(n)}\right)\right] \quad \text{s.t.} \tag{5}$$

$$p_{it(n)}^{\min} u_i^n \le p_i^n \le p_{it(n)}^{\max} u_i^n, \quad u_i^n \in \{0,1\}, \quad n \in \mathcal{N},\ i = 1, \dots, I, \tag{6a}$$

$$u_i^{n-\kappa} - u_i^{n-(\kappa+1)} \leq u_i^n, \quad \kappa = 1, \dots, \bar{\tau}_i - 1,\ n \in \mathcal{N},\ i = 1, \dots, I, \tag{6b}$$

$$u_i^{n-(\kappa+1)} - u_i^{n-\kappa} \leq 1 - u_i^n, \quad \kappa = 1, \dots, \underline{\tau}_i - 1,\ n \in \mathcal{N},\ i = 1, \dots, I, \tag{6c}$$

$$0 \leq v_j^n \leq v_{jt(n)}^{\max},\ 0 \leq w_j^n \leq w_{jt(n)}^{\max},\ 0 \leq l_j^n \leq l_{jt(n)}^{\max},\ n \in \mathcal{N},\ j = 1, \dots, J, \tag{7a}$$

$$l_j^n = l_j^{n-} - v_j^n + \eta_j w_j^n + \gamma_j^n, \quad n \in \mathcal{N},\ j = 1, \dots, J, \tag{7b}$$

$$l_j^0 = l_j^{\text{in}}, \quad l_j^n = l_j^{\text{end}}, \quad n \in \mathcal{N}_T,\ j = 1, \dots, J, \tag{7c}$$

$$\sum_{i=1}^{I} p_i^n + \sum_{j=1}^{J} (v_j^n - w_j^n) \geq d^n, \quad n \in \mathcal{N}, \tag{8a}$$

$$\sum_{i=1}^{I} (u_i^n p_{it(n)}^{\max} - p_i^n) \geq r^n, \quad n \in \mathcal{N}. \tag{8b}$$

Here, (5) is the expected cost, (6) describes the operating ranges and minimum up/down-time requirements of thermal units, (7) models the operating ranges and dynamics of hydro units (with l_j treated as state variables), and (8) imposes the load and reserve requirements. The nonanticipativity constraint is handled *implicitly* (i.e., it is ensured automatically) by the tree-based model. Note that the model

S	N	Variables		Constraints	Nonzeros
		binary	continuous		
1	168	4200	6652	13441	19657
20	1176	29400	45864	94100	137612
50	2478	61950	96642	198290	289976
100	4200	105000	163800	336100	491500

Table 1. Size of the scenario-tree model (5)–(8) depending on the numbers of scenarios and nodes for $T = 168$, $I = 25$ and $J = 7$

(5)–(8) forms a large-scale linear mixed-integer program. For $N := |\mathcal{N}|$ nodes and $S := |\mathcal{N}_T|$ scenarios, this model involves IN binary and $(I + 2J)N$ continuous decision variables, and $(2 + J)N + JS$ (in)equality constraints and $(I + 3J)N$ bounds for continuous variables (without taking into account the constraints of type (6b)–(6c) and the objective function). Table 1 shows how the size of the scenario-tree model (5)–(8) increases with the number of nodes and scenarios.

2.2 *Lagrangian Relaxation*

Recent algorithmic approaches to large-scale mixed-integer stochastic programs [1, 4, 5, 24, 31] are based on a successive decomposition into finitely many smaller subproblems. The stochastic programming model (5)–(8) is almost separable with respect to generation decisions of individual units, since only the constraints (8) couple different units. Hence, when dualizing these coupling constraints, the corresponding Lagrangian dual decomposes into much smaller subproblems, namely, into power generation models for single (thermal or hydro) units. Such a Lagrangian decomposition approach is discussed in [28] for general multistage stochastic integer programs under the label *component decomposition*. In the following, we give a brief description of this approach and of a Lagrangian relaxation algorithm for solving (5)–(8). For a more detailed presentation we refer to [7, 15, 25].

Let $z := (u, p, v, w)$, $N := |\mathcal{N}|$ and let $\lambda = (\lambda_1, \lambda_2) := (\lambda_1^n, \lambda_2^n)_{n \in \mathcal{N}} \in \mathbb{R}_+^N \times \mathbb{R}_+^N$ denote the Lagrange multiplier in scenario-tree form. Then (cf. [7, §4]) with the *Lagrangian*

$$L(z;\lambda) := \sum_{n\in\mathcal{N}} \pi_n \left\{ \sum_{i=1}^{I} \left[C_i^n(p_i^n, u_i^n) + S_i^n \left(u_i^{\text{path}(n)} \right) \right] \right. \tag{9}$$
$$+ \lambda_1^n \left[d^n - \sum_{i=1}^{I} p_i^n - \sum_{j=1}^{J} (v_j^n - w_j^n) \right]$$
$$\left. + \lambda_2^n \left[r^n - \sum_{i=1}^{I} (u_i^n p_{it(n)}^{\max} - p_i^n) \right] \right\},$$

and the *dual function*

$$D(\lambda) := \min_x \{ L(z;\lambda) \quad \text{s.t. constraints (6)–(7)} \}, \tag{10}$$

the *dual problem* reads

$$\max \left\{ D(\lambda) : \lambda \in \mathbb{R}_+^{2N} \right\}. \tag{11}$$

Under the assumptions made on the fuel costs, the dual function D is concave and polyhedral. The dual problem is separable and its solution may be found by solving stochastic single unit subproblems. Specifically, the dual function

$$D(\lambda) = \sum_{i=1}^{I} D_i(\lambda) + \sum_{j=1}^{J} \hat{D}_j(\lambda_1) + \sum_{n\in\mathcal{N}} \pi_n (\lambda_1^n d^n + \lambda_2^n r^n), \tag{12}$$

decomposes into the *thermal subproblems*

$$D_i(\lambda) = \min_{u_i} \left\{ \sum_{n\in\mathcal{N}} \pi_n \left[\min_{p_i^n} \{ C_i^n(p_i^n, u_i^n) - (\lambda_1^n - \lambda_2^n) p_i^n \} \right. \right. \tag{13}$$
$$\left. \left. - \lambda_2^n u_i^n p_{it(n)}^{\max} + S_i^n \left(u_i^{\text{path}(n)} \right) \right] \text{ s.t. (6)} \right\},$$

and the *hydro subproblems*

$$\hat{D}_j(\lambda_1) = \min_{(v_j, w_j)} \left\{ \sum_{n\in\mathcal{N}} \pi_n \lambda_1^n (w_j^n - v_j^n) \text{ s.t. (7)} \right\}. \tag{14}$$

Both subproblems represent multistage stochastic programming models for the operation of a single unit. While the thermal subproblem (13) is a combinatorial multistage program involving stochastic costs, the hydro subproblem (14) is a linear multistage model with stochastic costs and stochastic right-hand sides.

In short, our method for solving the stochastic programming model (5)–(8) consists of the following ingredients:

(a) Solving the dual problem (11) by a proximal bundle method using function and subgradient information;
(b) Efficient solvers for the single unit subproblems: dynamic programming for (13) and a special descent algorithm for (14);
(c) Lagrange heuristics for determining a nearly optimal first-stage decision.

Thus, the approach is based on the same, but *stochastic*, ingredients as in the classical case: a solver for the nondifferentiable dual, subproblem solvers, and a Lagrangian heuristics. The interaction of these components is illustrated in Figure 1. They have been extensively described in previous studies [15,24] and are now briefly

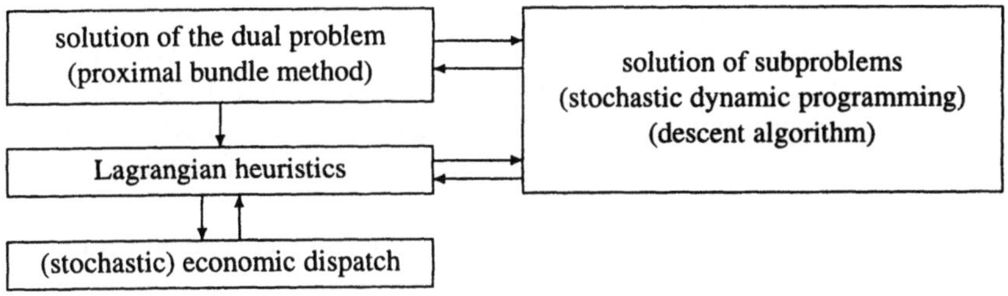

Figure 1. Structure of the (stochastic) Lagrangian relaxation algorithm

discussed. For a single unit, the hydro subproblem (14) can be solved by stochastic linear programming techniques that are presently available, see, e.g., [2, 29]. However, a specialized descent method [24] is found to be more efficient for these problems. The outer minimization of the thermal subproblem (13) with respect to the commitment state u_i is solved by dynamic programming, while the minimization with respect to p_i is carried out explicitly. Since the dual function D is non-differentiable, the dual problem (12) has to be attacked by a subgradient-type method for concave nondifferentable maximization. The reason for employing the proximal bundle method [20] in our algorithm are its very strong convergence properties. The optimal value $D(\lambda^*)$ for (12) delivered by the bundle method provides a lower bound for the optimal cost of the model (5)–(8). In general, however, the "dual optimal" scheduling decisions $z(\lambda^*) = (u(\lambda^*), p(\lambda^*), v(\lambda^*), w(\lambda^*))$ violate the load and reserve constraints (8) such that a low-cost primal feasible solution has to be determined by a *Lagrangian heuristic*. Two Lagrangian heuristics (cf. [15,24]) have been developed that determine nearly optimal first stage decisions

$\{(u^n, p^n, v^n, w^n)\}_{n \in \mathcal{N}_{\text{first}}}$ starting from the optimal multiplier λ^* and $z(\lambda^*)$. While the first heuristics provides a nearly optimal decision only at nodes $n \in \mathcal{N}_{\text{first}}$, the result of the second one is a nearly optimal solution at every node in $\mathcal{N}$.

3 Modeling of Uncertainty

3.1 Introduction

Since the stochastic power management model (5)–(8) uses a set of scenarios to model data uncertainty, new questions are raised on generating approximate scenario-based data processes. Recent scenario tree generation methods [8] may essentially be classified into two categories: (a) approaches that are embedded in the solution procedure of stochastic programs [6, 11, 14, 17, 19], and (b) approaches that control the goodness-of-fit of the approximation by certain distances [18,27,32]. For power management under uncertainty discrete time stochastic models are calibrated from historical time series for the load and stream flows [13, 16, 26, 31, 32]. The calibrated models can be used to simulate or select a large number of sample paths, which are combined into scenario trees. The algorithmic approaches in (a) are computationally demanding, but allow possible updates of the scenario tree structure as part of the solution procedure in the case of linear or convex stochastic programs without integrality constraints. The tree building procedures in (b) generate a tree-structured discrete distribution that minimizes the selected distance (Wasserstein distance of probability measures in [27], ℓ_1- and ℓ_2-distances of certain parameters of the distributions in [18] and [32], resp.).

Our approach to load scenario tree generation for the stochastic power management model proceeds according to the following steps:

1. Identify a statistical model of the load.
2. Generate an initial load tree.
3. Reduce the number of scenarios in the tree optimally.

These steps are explained in the following subsections.

3.2 A Statistical Model for the Electric Load

Description of the Data

The identification of a statistical model for the electric load of the VEAG generation system is based on an hourly load profile for a period of three years (1098 days). A plot of the hourly load data is displayed in Figure 2.

The historical load records show seasonal variations caused by meteorological factors like temperature, cloud cover, etc. The periodic patterns complete themselves within the calendar year and are then repeated on a yearly basis. In the weekly and monthly load data there are recurring patterns of length 24 (one day) and of length 168 (one week). Interruptions of this regularity are caused by customs like public holidays or the start/end of the daylight saving time. Thus, in principle the electric

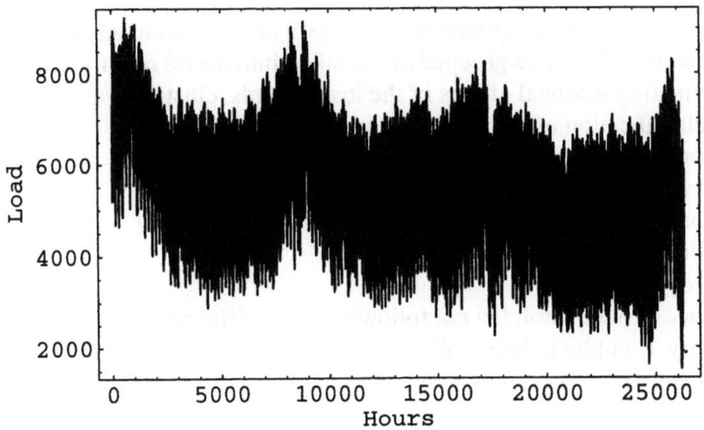

Figure 2. Hourly load data versus time

load depends on the category of the day (Monday, . . . , Sunday, public holiday, etc.) and on the season. Figure 3 highlights the periodic components of our historical data.

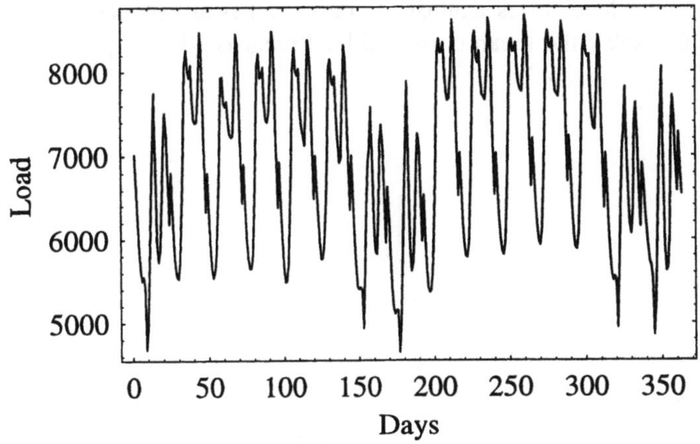

Figure 3. Hourly load data for a period of two weeks

In a first step days of a similar load pattern are identified using daily load records (24 load data of a day). To each record there we assign a day category (1 if it is a Monday, . . . , 7 if it is a Sunday, 8 for a public holiday following a working day, 9 for days between holidays and weekends, 10 for a public holiday following a weekend or a holiday). Clustering methods from [30] are applied to answer the question whether the records can be grouped or classified into useful or informative clusters. After eliminating seasonal effects of the load records, clustering and ANOVA-tests lead to a classification of the load records into 8 categories:

category 1: Monday or working day after a public holiday
category 2: working day (Tuesday, Wednesday, Thursday)
category 3: Friday or working day before a public holiday
category 4: Saturday
categroy 5: Sunday
category 6: public holiday not following days of the categories 2,3
category 7: public holiday following days of the categories 2,3
category 8: working day between days of the categories 4-7

Transformation of Data

The statistical modeling of the load process exploits the decomposition of the load process into a *daily mean load process* and a *mean-corrected load series*. The component series are treated separately. To do this, let $x_{j\tau}$ be the observed load at time step $\tau = 1, \ldots, 24$ of day j (record j of the data base). Due to the daily mean load $\overline{d}_j := (24)^{-1} \sum_{\tau=1}^{24} x_{j\tau}$ and the mean-corrected load record $(d_{j\tau} := x_{j\tau} - \overline{d}_j)_{\tau=1}^{24}$, the historical load records are decomposed according to

$$x_{j\tau} = d_{j\tau} + \overline{d}_j \quad (\tau = 1, \ldots, 24;\ j = 1, \ldots, 1098). \tag{15}$$

The daily mean load series against the day number are plotted in Figure 4. Figure 5 and 6 display the mean-corrected load records $(d_{j\tau})_{\mathrm{cat}(j)=k}$ for days of the categories $k = 1, \ldots, 5$.

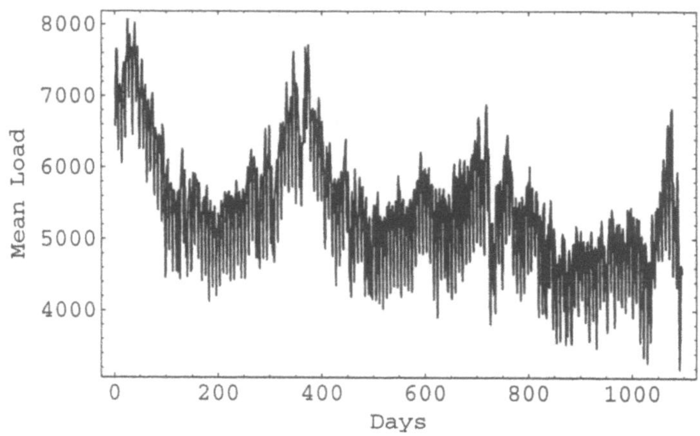

Figure 4. Daily mean load versus the day number

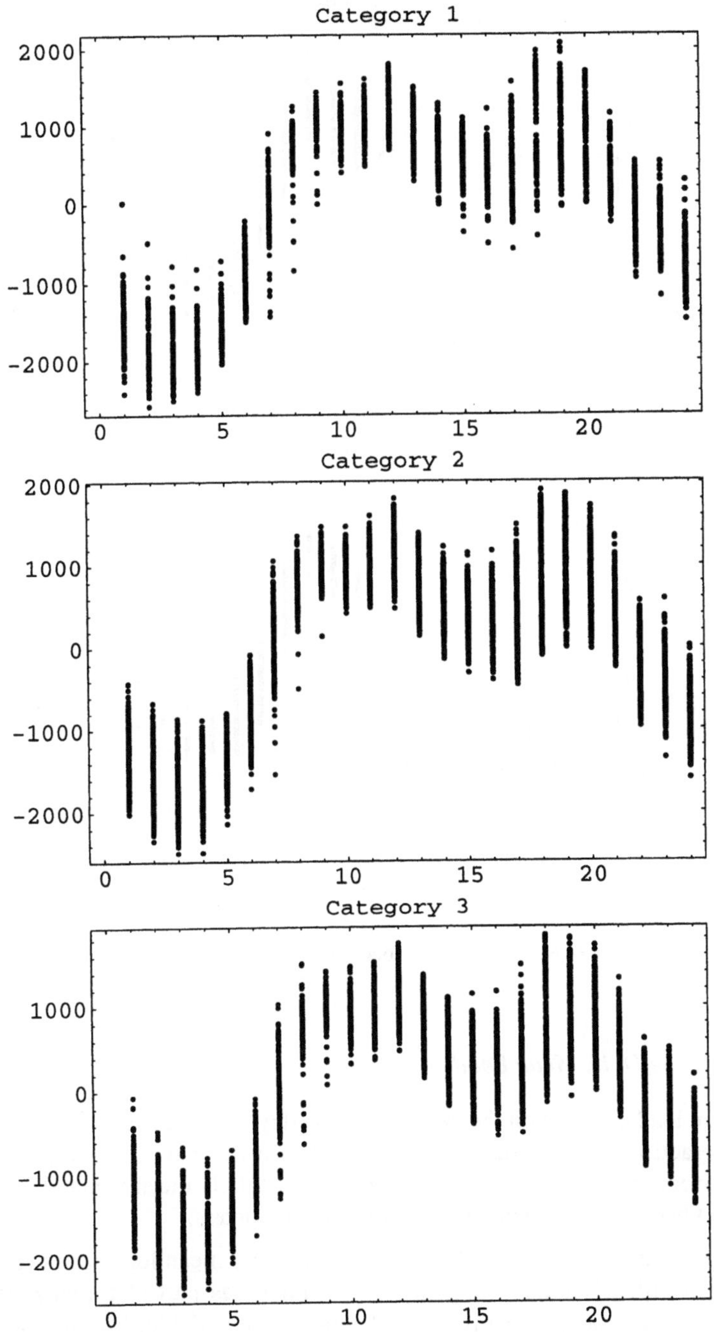

Figure 5. Mean-corrected load records for days of category 1, 2, 3

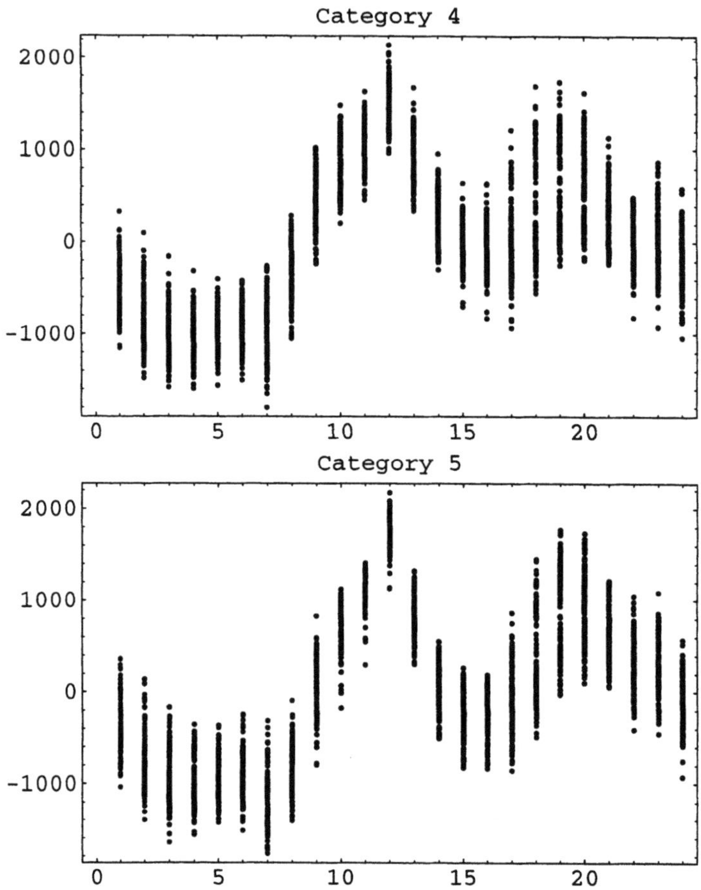

Figure 6. Mean-corrected load records for days of category 4, 5

Model for the Daily Mean Load

The mean load depends on the category of the day and on the season. Further, there is an interaction between the mean load and meteorological factors like temperature, cloud cover etc. The meteorological impact on the daily mean demand could not be modelled because of missing meteorological parameters.

To select a suitable class of models for the daily mean load series $\{\bar{d}_j\}_{j\in J}$ with $J \subset \mathbb{Z} := \{0, \pm 1, \pm 2, \ldots\}$, $\{\bar{d}_j\}_{j\in J}$ is considered as part of a realization of the stochastic mean load process $\{\bar{d}_j\}_{j\in \mathbb{Z}}$. A *time series model* for $\{\bar{d}_j\}_{j\in J}$ is a specification of the joint distributions of $\{\bar{d}_j\}_{j\in \mathbb{Z}}$. (The term "time series" is frequently used for both the observed data and the corresponding stochastic process.) We now recall some concepts of time series analysis.

A complete time series model for a stochastic process $\{X_t\}_{t\in\mathbf{Z}}$ should specify the distribution of any random vector $(X_{t_1},\ldots,X_{t_l})$. Often the analysis focuses on the second-order properties of $\{X_t\}$: the expected values EX_t and the covariances $\mathrm{cov}(X_t,X_s) := E[(X_t - EX_t)(X_s - EX_s)]$ for all t,s. In the particular case of *Gaussian time series* all random variables X_t are normally distributed. Therefore, all the joint distributions are multivariate normal and completely characterized by the second-order properties of $\{X_t\}$. Classical time series analysis relies on the concept of stationarity. Recall that $\{X_t\}$ is *stationary* if $EX_t^2 < \infty$, EX_t is constant and $\mathrm{cov}(X_r,X_s) = \mathrm{cov}(X_{r+t},X_{s+t})$, $\forall r,s,t \in \mathbb{Z}$.

To select an appropriate class of time series models for observed data, their properties are analyzed first. In particular, the data graph is searched for any *seasonal* (periodic) or *trend* (nonconstant mean) components, outlying observations or sharp changes in behavior. Then suitable transformations are applied to the data to get a new stationary series (*residuals*) with zero mean and unit variance. The trend and seasonal components may be removed by estimating these components and subtracting them from the data; this is the classical decomposition model incorporating trend, a seasonal component and random noise. Another transformation is called *differencing*; it replaces $\{X_t\}$ by $\{Y_t := X_t - X_{t-s}\}$ for some *lag* $s \in \mathbb{N}$, thus eliminating a seasonal component of period s.

In the daily mean load series $\{\overline{d}_j\}_{j\in J}$ there is clearly a recurring pattern with the seasonal period of 365 (one year). There are further periodic components of length 7 (one week) and change points due to the start/end of the daylight saving time. Irregularities of the weekly patterns have been removed from the time series by replacing outlying observations by the value of the nearest day of the same category.

Most approaches for fitting a time series to the deseasonalized data rely on linear models. *Autoregressive moving average* (ARMA) models are characterized by finite-order linear difference equations with constant coefficients. The process $\{X_t\}$ is called *ARMA*(p,q) if it is stationary and

$$X_t - \phi_1 X_{t-1} - \cdots - \phi_p X_{t-p} = Z_t + \theta_1 Z_{t-1} + \cdots + \theta_q Z_{t-q} \quad \forall t, \tag{16}$$

where $(\phi_k)_{k=1}^p$ and $(\theta_l)_{l=1}^q$ are real coefficients and $\{Z_t\}_{t\in\mathbf{Z}}$ is the *white noise* process $\mathrm{WN}(0,\sigma^2)$ with zero mean and variance σ^2, i.e., $EZ_t = 0$, $EZ_t^2 = \sigma^2$, $\forall t \in \mathbb{Z}$, and $EZ_rZ_t = 0$ if $r \neq t$. Using the *backward shift operator* B defined by $B^\ell X_t := X_{t-\ell}$ for $t,\ell \in \mathbb{Z}$, the ARMA equations (16) can be rewritten as

$$\phi(B)X_t = \theta(B)Z_t, \quad \forall t \in \mathbb{Z}, \quad \{Z_t\} \sim \mathrm{WN}(0,\sigma^2),$$

where ϕ and θ denote the polynomials $\phi(z) = 1 - \phi_1 z - \cdots - \phi_p z^p$, $\theta(z) = 1 + \theta_1 z + \cdots + \theta_q z^q$. An ARMA$(p,q)$ process $\{X_t\}_{t\in\mathbf{Z}}$ is said to be *causal* (or *future-independent*) if there exists a real sequence $\{\psi_\ell\}$ such that $\sum_{\ell=0}^\infty \psi_\ell < \infty$ and

$$X_t = \sum_{\ell=0}^{\infty} \psi_\ell Z_{t-\ell}, \quad \forall t \in \mathbb{Z}.$$

If the differenced series $\{Y_t = (1 - B^s)X_t\}_{t\in\mathbf{Z}}$ is an ARMA(p,q) process then the model for the original series $\{X_t\}$ reads $\phi(B)(1-B^s)X_t = \theta(B)Z_t$; further, $\{X_t\}$ belongs to the class of *seasonal autoregressive integrated moving average* (SARIMA) processes if $\{Y_t\}$ is causal. General SARIMA processes are defined as follows. The process $\{X_t\}_{t\in\mathbb{Z}}$ is said to be a *SARIMA*$(p, d, q) \times (P, D, Q)_S$ *process with period* s if the differenced process $Y_t := (1-B)^d(1-B^S)^D X_t$ is the causal ARMA process

$$\phi(B)\Phi(B^s)Y_t = \theta(B)\Theta(B^s)Z_t, \quad \{Z_t\} \sim WN(0, \sigma^2),$$

where $\phi(z) = 1 - \cdots - \phi_p z^p$, $\Phi(z) = 1 - \cdots - \Phi_P z^P$, $\theta(z) = 1 + \cdots + \theta_q z^q$ and $\Theta(z) = 1 + \cdots + \Theta_Q z^Q$. Then the model for $\{X_t\}_{t\in\mathbb{Z}}$ reads $\phi(B)\Phi(B^S)(1-B)^d(1-B^S)^D X_t = \theta(B)\Theta(B^S)Z_t$.

To identify a suitable SARIMA model for the given time series, the differencing orders d, D, the model orders p, P, q, Q and the length S of the seasonal component must be identified. They can be discovered by inspecting the empirical autocorrelation function, the empirical counterpart of the *autocorrelation function* $\mathrm{cov}(X_\ell, X_0)/\mathrm{var}(X_0)$, $\ell \in \mathbb{Z}$; see, e.g., [3]. The model coefficients $(\phi_\ell)_{\ell=1}^p$, $(\Phi_\ell)_{\ell=1}^P$, $(\theta_\ell)_{\ell=1}^q$, $(\Theta_\ell)_{\ell=1}^Q$ and the white noise variance σ^2 can be estimated via parameter estimation procedures for ARMA processes. The maximum likelihood method produces the most efficient estimates in the particular case of Gaussian time series. Initial values for the model coefficients can be obtained by the Hannan-Rissanen algorithm (cf. [3, §5]) which solves the problem of order selection and parameter estimation for ARMA processes simultaneously.

In our case, we obtained stationary residuals after three differencing operations (two lag-364 differencing operations followed by one lag-1 differencing). The residuals were treated as part of a realization of the stochastic process

$$\{Y_j := \overline{d}_j - \overline{d}_{j-1} - 2\overline{d}_{j-364} + 2\overline{d}_{j-365} + \overline{d}_{j-728} - \overline{d}_{j-729}\}.$$

For $\{Y_j\}$ the Hannan-Rissanen algorithm from the *Mathematica Time Series Pack* [33] selected an ARMA(1,1) model that served as an initial model for the maximum likelihood method. The resulting maximum likelihood estimates for the model coefficients and random noise process are

$$\hat{\phi}_1 = 0.357756, \quad \hat{\theta}_1 = -0.639978, \quad \{Z_j\} \sim N(0, 15533.88), \quad j \in \mathbb{Z}.$$

The time series model for $\{Y_j\}$ reads $Y_j - \hat{\phi}_1 Y_{j-1} = Z_j + \hat{\theta}_1 Z_{j-1}$, $j \in \mathbb{Z}$. Accordingly, the time series model for the daily mean load process $\{\overline{\mathbf{d}}_j\}_{j\in\mathbb{Z}}$ is

$$(1-B)(1-B^{364})^2(1-\hat{\phi}_1 B)\overline{\mathbf{d}}_j = (1+\hat{\theta}_1 B)Z_j. \tag{17}$$

The above SARIMA$(1,1,1) \times (0,2,0)_{364}$ model can be converted to the following ARMA$(730, 1)$ model:

$$\begin{aligned}\overline{d}_j &- (1+\hat{\phi}_1)\overline{d}_{j-1} + \hat{\phi}_1\overline{d}_{j-2} - 2\overline{d}_{j-364} + 2(1+\hat{\phi}_1)\overline{d}_{j-365} - 2\hat{\phi}_1\overline{d}_{j-366} \\ &+ \overline{d}_{j-728} + (\hat{\phi}_1 - 1)\overline{d}_{j-729} + \hat{\phi}_1\overline{d}_{j-730} = Z_j + \hat{\theta}_1 Z_{j-1}, \quad j \in \mathbb{Z}.\end{aligned} \tag{18}$$

Model for the Mean-Corrected Load Records

We have selected polynomials to model the time dependence of the mean-corrected load records corresponding to days of the same category k, $k = 1, \ldots, 7$. More specifically, we fit models of the form

$$d_{k\tau} = \beta_{k0}+\beta_{k1}\tau+\cdots+\beta_{km_k}\tau^{m_k}+\epsilon_{km_k}, \quad (k=1,\ldots,5;\tau=1,\ldots,24), \tag{19}$$

where the *error term* e_{km_k} is normally distributed with zero mean and variance $\sigma^2_{m_k}$. These models are known as *linear* or *polynomial regression models* (cf. [10]). From (19) we obtain the *predicted load records*

$$\hat{d}_{k\tau} = \beta_{k0} + \beta_{k1}\tau + \cdots + \beta_{km_k}\tau^{m_k}, \quad (k=1,\ldots,5;\tau=1,\ldots,24). \tag{20}$$

For model fitting, regression diagnostics, and forecasting we used the statistical package *S-PLUS* [30]. It remains to answer the question how we selected the degree m_k of the polynomials. This will be done in the following subsection.

Model for the Load

The statistical model for the load is obtained by combining the models for the daily mean load and the mean-corrected load records according to (15). Regression models for the mean-corrected load records corresponding to different day categories may be included into (15) by using *dummy or artificial variables* D_k, $k=1,\ldots,7$. These variables are defined as follows.

$$D_k := \begin{cases} 1, & \text{if the record corresponds to a day of category } k, \\ 0, & \text{otherwise,} \end{cases} \qquad (k=1,\ldots,7).$$

With these definitions (15) may be rewritten as

$$x_{j\tau} = D_1 d_{1\tau} + \cdots + D_7 d_{7\tau} + \bar{d}_j, \quad (\tau=1,\ldots,24; j=1,\ldots,1098). \tag{21}$$

The different time scales for the historical load records and the load process can be synchronized by an index transformation:

$$d_t = D_1 d_{1,t\%24} + \cdots + D_7 d_{7,t\%24} + \bar{d}_{[t/24]}, \quad (t \in \mathbb{Z}). \tag{22}$$

(By $[t/24]$ and $t\%24$ we denote the (rounded down) integral part and the remainder of dividing t by 24.)

Inserting (18) and (19) into (22) we obtain the *statistical model for the load*:

$$\begin{aligned} d_t = {} & D_1 \sum_{l=0}^{m_1} \beta_{1l}(t\%24)^l + \cdots + D_7 \sum_{l=0}^{m_7} \beta_{7l}(t\%24)^l \\ & + (1+\hat{\phi}_1)\bar{d}_{[t/24]-1} - \hat{\phi}_1\bar{d}_{[t/24]-2} + 2\bar{d}_{[t/24]-364} - 2(1+\hat{\phi}_1)\bar{d}_{[t/24]-365} \\ & + 2\hat{\phi}_1\bar{d}_{[t/24]-366} - \bar{d}_{[t/24]-728} - (\hat{\phi}_1 - 1)\bar{d}_{[t/24]-729} - \hat{\phi}_1\bar{d}_{[t/24]-730} \\ & + D_1\epsilon_{1m_1} + \cdots + D_7\epsilon_{7m_7} + Z_j + \hat{\theta}_1 Z_{[t/24]-1}, \quad (t \in \mathbb{Z}). \end{aligned} \tag{23}$$

To select the degrees m_k of the regression polynomials we measured the squared distance between (23) and the historical load data for the third year. The best fit we obtained for $m_k = 10$, $k=1,\ldots,7$.

3.3 Generation of Load Scenario Trees

An important initial decision is the choice of the number of stages and of the branching scheme for the scenario tree, i.e., the number and positions of branching levels and the branching degree in every node. We choose the following initial structure of the load scenario tree:

- A balanced tree with $K = 6$ branching periods t_k, $k = 1, \dots, K$. The branching periods t_k, $k = 2, \dots, K$, are equidistant within the time span $t = t_1, \dots, T$, i.e., $t_k := t_1 + (T - t_1)(k-1)/K$, $k = 2, \dots, K$.
- $|\mathcal{N}_+(n)| = \begin{cases} 2, & n \in \mathcal{N}_{t_k} = \{n : t(n) = t_k\},\ k = 1, \dots, K, \\ 1, & \text{otherwise.} \end{cases}$

Thus, the tree consists of $S := 2^6 = 64$ scenarios. The first branching period t_1 is defined by the length of the planning period within the scheduling horizon. The branching periods t_k, $k = 2, \dots, K$, should correspond to the (normally fixed) times when already observable meteorological and load data provide the possibility to re-adjust the unit commitment. For the scheduling horizon of one week with an hourly discretization, $t_k = 24k$ for $k = 1, \dots, 6$ is a reasonable choice for the generation system of the utility VEAG. For longer scheduling periods, non-equidistant branching periods would be preferable in order to restrict the number of scenarios. By assigning two successors to any node n in $\mathcal{N}_{t_k}$, $k = 1, \dots, K$, we may distinguish the events "low load" and "high load" for periods $t = t_k + 1, \dots, t_{k+1}$, where $t_{K+1} := T$. An additional event such as "medium load" would increase the scenario number to $S = 3^6 = 729$.

It remains to specify the realizations (scenario values) and their probabilities. Suppose the power utility supplies starting values $(d_t)_{t=-729}^{1}$. The realizations for nodes of the first stage period are given by

$$d^n := D_1 \hat{d}_{1t(n)\%24} + \cdots + D_7 \hat{d}_{7t(n)\%24} + \overline{d}_1, \quad t(n) = 1, \dots, t_1,$$

their probabilities $\pi^n := 1$. To assign realizations and probabilities to the remaining nodes we first compute the probability distribution of $(d_t)_{t=t_1+1}^{T}$. By (23), $(d_t)_{t=t_1+1}^{T}$ has a $(T - t_1)$-variate normal distribution with mean $\mu := (Ed_t)_{t=t_1+1}^{T}$ and covariance matrix $\Sigma := (\text{cov}(d_t, d_s))_{s,t=t_1+1}^{T}$. The following properties reduce the computational effort. First,

$$E(d_t - \sum_{l=1}^{7} D_l \hat{d}_{lt\%24}) = E(d_{t_k} - \sum_{l=1}^{7} D_l \hat{d}_{lt_k\%24}) \tag{24}$$

for $t = t_{k-1} + 1, \dots, t_k$, $k = 2, \dots, K$. Second, the random vector $(\nu_t := d_t - D_1 \hat{d}_{1t\%24} - \cdots - D_7 \hat{d}_{7t\%24})_{t_1+1}^{T}$ has a $(T - t_1)$-variate normal distribution with mean $\hat{m} := \mu - \sum_{l=1}^{7} D_l \hat{d}_{lt_k\%24}$ and covariance matrix Σ. By (23) and (24), $\hat{m}$ and Σ are completely determined by Ed_{t_k} and $\text{cov}(d_{t_k}, d_{t_l})$ for $k, l = 1, \dots, K$. With $t_k = 24k$, $Ed_{t_k} = E\overline{d}_k$ and

$$\text{cov}(d_{t_k}, d_{t_l}) = \text{cov}(\sum_{\nu=1}^{7} D_\nu e_{\nu m_\nu}, \sum_{\xi=1}^{7} D_\xi e_{\xi m_\xi}) + \text{cov}(\overline{d}_k, \overline{d}_l)$$

for $k, l = 1, \dots, K$. To compute $E\bar{d}_k$ and $\text{cov}(\bar{d}_k, \bar{d}_l)$ we use the following lemma:

Lemma 1. *Let random variables* X_t, $t \geq 0$, *be defined by the ARMA equation (16) with starting values* $(X_{-p}, \dots, X_{-1}) \in \mathbb{R}^p$ *and* $Z_t \sim N(0, \sigma^2)$, $t \geq -q$. *For* $s, t \geq 0$ *the expected value* EX_t *and the covariance* $\text{cov}(X_s, X_t)$ *are given by*

$$EX_t = \begin{cases} \sum_{k=0}^{t} a_{t-k} \left(\sum_{m=1+k}^{p} \Phi_m X_{-m+k}\right), & t = 0, \dots, p-1 \\ \sum_{k=0}^{p-1} a_{t-k} \left(\sum_{m=1+k}^{p} \Phi_m X_{-m+k}\right), & t \geq p \end{cases},$$

$$\text{cov}(X_s, X_t) = \sigma^2 \sum_{k=0}^{t} \sum_{m=0}^{s} a_{t-k} a_{s-m} \sum_{l=0}^{q-|k-m|} \Theta_l \Theta_{l+|k-m|},$$

where $\{a_t\}_{t=0}^{\infty}$ *is defined by*

$$a_0 := 1, \quad a_k := \begin{cases} \Phi_1 a_{k-1} + \Phi_2 a_{k-2} + \dots + \Phi_k a_0, & 1 \leq k \leq p \\ \Phi_1 a_{k-1} + \Phi_2 a_{k-2} + \dots + \Phi_p a_{k-p}, & k > p \end{cases}.$$

Proof. By induction.

Using the computer algebra system *Mathematica* [33], the mean $\hat{m}$ and the covarince matrix Σ of the transformed random vector $(v_t)_{t=t_1+1}^{T}$ can be computed within a few seconds. The mean μ of $(d_t)_{t=t_1+1}^{T}$ can be obtained by the transformation $\mu = \hat{m} + \sum_{l=1}^{7} D_l \hat{d}_{l t_k \% 24}$.

After computing the distribution of $(d_t)_{t=t_1+1}^{T}$, a scenario tree for its approximation is constructed in a standard way. Suppose that we already assigned realizations and probabilities to all nodes of a path from the root to node $n \in \mathcal{N}$. We distinguish two cases.

(a) $t(n) \neq t_k, k = 1, \dots, K$.
Then there is a single successor n_+ to n with probability $\pi_{n+} := \pi_n$ and realization $d^{n+} := E(d_{t(n)+1} | d^n \dots d^1)$, the conditional mean of $d_{t(n)+1}$ given the past realizations $d_1 = d^1, \dots, d_{t(n)} = d^n$.
(b) $t(n) = t_k$, for some $k \in \{1, \dots, K\}$.
Let $\mathcal{N}_+(n) = \{n_1, n_2\}$, i.e., n_1, n_2 are the successors to node n. The conditional (normal) distribution of d_{t_k+1} given $d_1 = d^1, \dots, d_{t_k} = d^n$ is approximated by a discrete distribution with the two realizations

$$d^{n_1} = d^n + \delta_n + \sum_{l=1}^{7} D_l (\hat{d}_{l(k+1)} - \hat{d}_{lk}), \tag{25}$$

$$d^{n_2} = d^n - \delta_n + \sum_{l=1}^{7} D_l (\hat{d}_{l(k+1)} - \hat{d}_{lk}). \tag{26}$$

A criterion for determining the innovation δ_n and the transition probability $\pi_{n_1/n}$ ($\pi_{n_2/n} = 1 - \pi_{n_1/n}$) is that their choice preserves the first two moments of the conditional distribution for d_{t_k+1} given the past realizations $d_1 = d^1, \dots, d_{t(n)} = d^n$ [23].

With the conditional mean $\mu_n := E(d_{t_k+1}|d^n \dots d^1)$ and the conditional variance $\sigma_n^2 := \text{var}(d_{t_k+1}|d^n \dots d^1)$, the innovation and the transition probability satisfies

$$\delta_n = \sqrt{(\mu_n - d^n)^2 + \sigma_n^2}, \quad \pi_{n_1/n} = \frac{(\mu_n - d^n)^2 + \sigma_n^2 + \delta_n(\mu_n - d^n)}{2\delta_n^2}.$$

The node probabilites π^{n_1} and π^{n_2}, resp., are recursively computed from the relations $\pi^{n_1} = \pi_{n_1/n}\pi^n$ and $\pi^{n_2} = (1 - \pi_{n_1/n})\pi^n$.

Figure 7 and 8 show load scenario trees for a planning horizon of one week with an hourly discretization and branching periods $t_k = 24k, k = 1, \dots, 6$. For the summer week (Figure 7) the scenario probabilities vary between 0.37 and 10^{-10}, for the winter week (Figure 8) between 0.7 and 10^{-13}. The generation of one load scenario tree took less than two minutes on an HP 9000 (780/J280) Compute-Server with 180 MHz frequency and 768 MByte main memory under HP-UX 10.20.

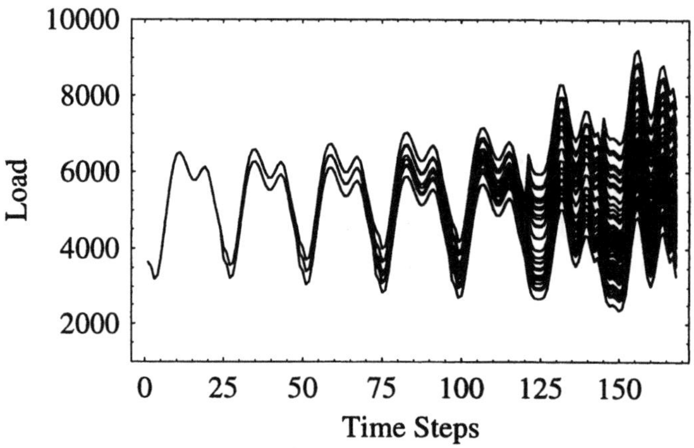

Figure 7. Load scenario tree for one week in summer

A few comments on the tree construction procedure are in order. First, the construction of the load scenario tree is consistent with the normality assumption imposed on $(d_t)_{t=t_1+1}^T$ by the statistical model (23). $\mathcal{N}$ preserves the mean and covariance matrix. In particular, the scenario tree resembles the correlation structure of the time series model. Second, in general the two transistion probabilities for a branching point are different. As a consequence, the probabilities of the scenarios in the tree differ. Third, the construction of the load scenario tree does not prevent unrealistic ("too large") load values. Load values may exceed the maximum capacity of the thermal system. Empirical results have shown that they are related to very small node probabilities ($< 10^{-10}$). In order to avoid computational difficulties these load values are replaced by the maximum capacity of the thermal system.

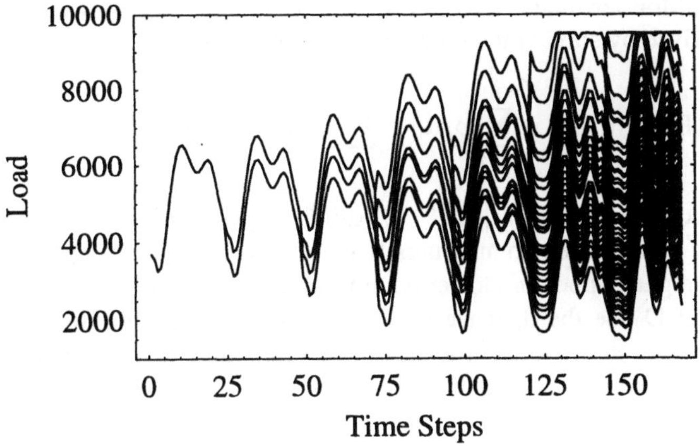

Figure 8. Load scenario tree for one week in winter

3.4 *Optimal Reduction of Load Scenario Trees*

The initial scenario tree generated in Section 3.3 may contain too many nodes to solve the power management model (5)–(8) within acceptable time. Therefore, one often incorporates a procedure to reduce the number of scenarios of the initial tree [9, 34].

The load scenario tree provides a first approximation of the distribution of the discrete-time stochastic load process. Our reduction concept determines a scenario subset of prescribed cardinality and a probability measure based on this set that is closest to the initial approximation in terms of a natural (or canonical) probability metric. Quantitative stability results for stochastic programs (cf. [9]) indicate that the Fortet-Mourier metrics ζ_h, $h \geq 1$, are canonically associated for a multistage stochastic program like (5)–(8).

Let δ_ω denote the probability measure on $\mathbb{R}^T$ having unit mass at $\omega \in \mathbb{R}^T$. The initial scenario tree represents a discrete probability distribution P carried by scenarios $\omega_i \in \Omega$ with weights $p_i > 0$, $\sum_{i=1}^{S} p_i = 1$. A reduced scenario tree is obtained by deleting all scenarios ω_j, $j \in J$ belonging to some index set $J \subset \{1, \ldots, S\}$ and by assigning new probabilistic weights q_j to each scenario ω_j, $j \notin J$. Let Q denote the corresponding probability distribution, i.e.,

$$P = \sum_{i=1}^{S} p_i \delta_{\omega_i}, \quad Q = \sum_{j \notin J} q_j \delta_{\omega_j}.$$

The Fortet-Mourier metric ζ_h, $h \geq 1$, of the discrete probability measures P, Q may be derived by solving the dual transportation problem

$$\zeta_h(P, Q) = \max \left\{ \sum_{i=1}^{S} p_i u_i + \sum_{j \notin J} q_j v_j \; : \; u_i + v_j \leq c_h(\omega_i, \omega_j) \right\},$$

where $c_h(\omega, \omega') := \max\{1, \|\omega\|^{h-1}, \|\omega'\|^{h-1}\}\|\omega - \omega'\|$ and $\|\cdot\|$ is the Euclidean norm on $\mathbb{R}^T$. Since the two measures P and Q have the same support $\{\omega_s\}_{s=1}^S$, but different weights, upper and lower bounds for $\zeta_h(P, Q)$ can be derived [9].

Now, let D_J be the distance of P to a closest probability distribution having support $\{\omega_s : s = 1, \ldots, S, \; s \notin J\}$, i.e., corresponding to deleting all scenarios of P belonging to some index set J:

$$D_J := \min\{\zeta_h(\sum_{i=1}^{S} p_i \delta_{\omega_i}, \sum_{j \notin J} q_j \delta_{\omega_j}) \; : \; q_j \geq 0, \sum_{j \notin J} q_j = 1\}.$$

Then we have (cf. [9])

$$D_J = \sum_{i \in J} p_i \min_{j \notin J} c_h(\omega_i, \omega_j).$$

The optimal weights $(q_j)_{j \notin J}$ for the scenarios remainimg in the reduced tree are

$$\bar{q}_j = p_j + \sum_{\substack{i \in J \\ j(i) = j}} p_i \;\; \forall j \notin J \tag{27}$$

where $j(i) \in \arg\min_{j \notin J} c_h(\omega_i, \omega_j)$ for $i \in J$.

An optimal rule for reducing P to a measure Q with a prescribed number $\tilde{S}$ of scenarios is given as the solution of the combinatorial optimization problem

$$\min\{D_J = \sum_{i \in J} p_i \min_{j \notin J} c_h(\omega_i, \omega_j) \; : \; J \subset \{1, \ldots, S\}, \#J = S - \tilde{S}\}. \tag{28}$$

Explicit solutions to (28) are available for the cases $\tilde{S} = S - 1$ (single scenario deletion) and $\tilde{S} = 1$ (keeping only one scenario) [9]. Upper and lower bounds for (28) yield heuristic reduction strategies for the general case. The *forward selection* algorithm recusively determines the indices $u_j, j = 1, \ldots, \tilde{S}$ for the scenarios in the reduced tree. It uses the lower bound

$$\min\{D_J : J \subset \{1, \ldots, S\}, \#J = S - \tilde{S}\} \leq \sum_{i \in J_u} p_i \min_{j \notin J_u} c_h(\omega_i, \omega_j),$$

where $J_u = \{1, \ldots, S\} \setminus \{u_1, \ldots, u_{\tilde{S}}\}$. The indices u_j, $j = 1, \ldots, \tilde{S}$ are chosen recursively such that

$$u_j \in \arg \min_{l \notin \{u_1, \ldots, u_{j-1}\}} \sum_{\substack{i=1 \\ i \notin \{u_1, \ldots, u_{j-1}, l\}}}^{S} p_i \min_{u \in \{u_1, \ldots, u_{j-1}, l\}} c_h(\omega_u, \omega_i).$$

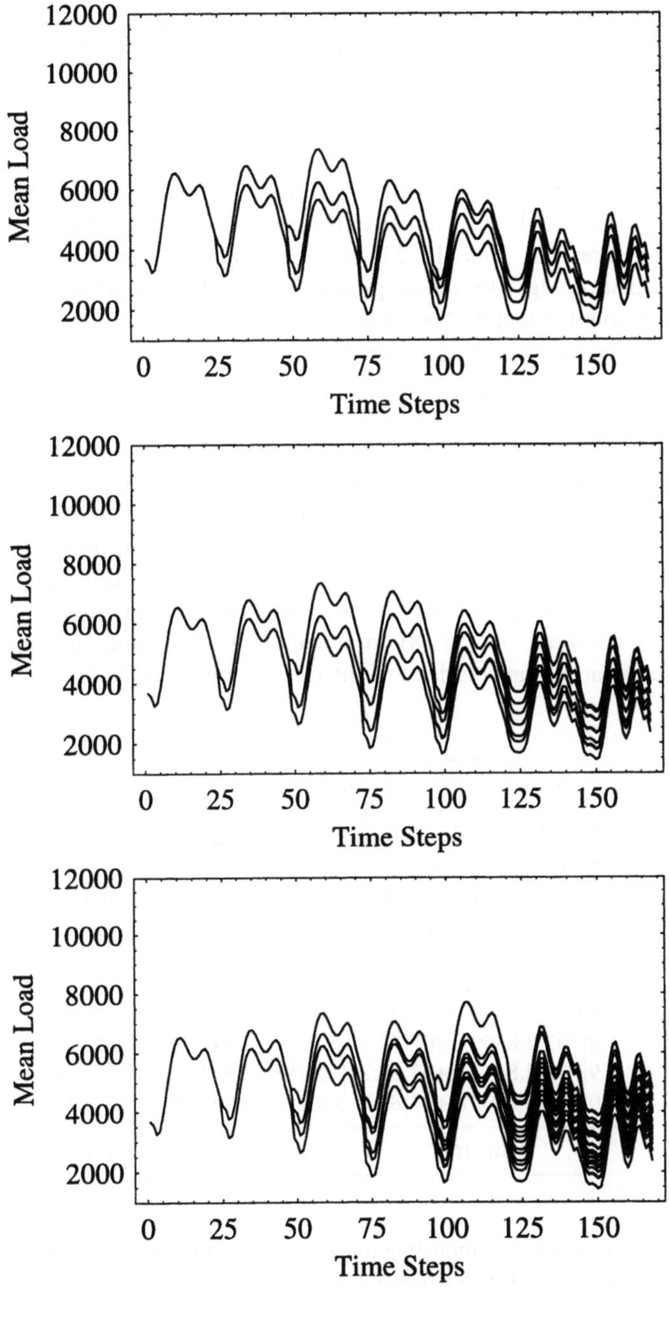

Figure 9. Reduced load scenario tree with 5, 10, and 20 scenarios

Optimal probabilistic weights $(q_j)_{j \notin J_u}$ are computed from (27) (with J replaced by J_u). Figure 9 shows the reduced load scenario trees with 5, 10 and 20 scenarios obtained by the forward selection algorithm with $c_h(\omega, \omega') := \|\omega - \omega'\|$.

3.5 Numerical Testing Results

For numerical tests we considered the hydro-thermal power system of VEAG (with $T = 168$, $I = 25$ and $J = 7$) under uncertain load (i.e., the remaining data were deterministic). Previous tests [15, 24] of our implementation of the stochastic Lagrangian relaxation algorithm have proved its potential for solving the stochastic power management model within a few minutes.

In this testing, we combined the stochastic Lagrangian relaxation algorithm with the tree generation and reduction technique. The methods for solving the subproblems and the heuristics are kept identical. Again, the test runs were performed on an HP 9000 (780/J280) Compute-Server with 180 MHz frequency and 768 MByte main memory under HP-UX 10.20. First we generated a load scenario tree (cf. §3.3) for an hourly discretized time horizon of one week ($T = 168$) with branching points $t_k = 24k$, $k = 1, \ldots, 6$ (cf. Figure 8). The initial number of scenarios $S = 64$ was reduced by applying the forward scenario selection rule of §3.4 with $c_h(\omega, \omega') := \|\omega - \omega'\|$. Due to the nonconvexity of the underlying stochastic programs, primal solutions for (5)–(8) are obtained by a heuristic method. This is the reason for comparing the solutions of the dual problem. Table 2 reports the objec-

$\tilde{S}$	N	objective	time [s]	$\tilde{S}$	N	objective	time [s]
1	168	2.83589e+07	10.83	15	1166	2.84102e+07	109.40
2	313	2.81617e+07	28.41	20	1387	2.84189e+07	141.59
3	386	2.82691e+07	28.71	25	1656	2.84164e+07	186.07
4	411	2.83139e+07	35.32	30	1829	2.84159e+07	186.32
5	532	2.83874e+07	40.33	35	2122	2.84146e+07	259.52
6	581	2.83942e+07	57.84	40	2295	2.84248e+07	301.57
7	678	2.83964e+07	55.95	45	2492	2.84219e+07	276.72
8	703	2.83999e+07	58.71	50	2689	2.84149e+07	321.04
9	728	2.84124e+07	63.24	55	2838	2.84207e+07	379.66
10	801	2.84071e+07	83.61	60	2987	2.84181e+07	374.44
		Initial tree:		64	3111	2.84210e+07	487.27

Table 2. Objective values and computing times of the dual stochastic problem (11) for different numbers of scenarios ($\tilde{S}$) in the reduced tree

tive and computing time of the dual stochastic problem (11) for different numbers of scenarios ($\tilde{S}$) in the reduced tree, each having a different number of nodes (N).

The reduced trees with $\tilde{S} = 5, 10, 20$ have been presented in Figure 9. It can be seen that the optimal values of (11) for reduced scenario trees slowly converge to the objective of the full scenario tree problem ($S = 64$). Rough approximations of the full scenario tree generally require only short computing times. For $\tilde{S} < 10$ the scenarios are highly concentrated and the true objective is therefore underestimated. The approximation of the initial scenario tree improves for $\tilde{S} > 20$, resulting in a better approximation of the objective.

ACKNOWLEDGEMENT

The authors wish to thank W. Römisch (Humboldt-University Berlin) for many useful suggestions. We are grateful to G. Schwarzbach, J. Thomas and J. Krause (VEAG Vereinigte Energiewerke AG Berlin) for their outstanding cooperation over many years. Further thanks are due to J. Dupačová (Charles University Prague) and her former students A. Henclová, M. Kotyzová and J. Muková for the collaboration in modeling load profiles and to K.C. Kiwiel (Polish Academy of Sciences, Warsaw) for the permission to use the NOA 3.0 package.

REFERENCES

1. L. Bacaud, C. Lemaréchal, A. Renaud, C. Sagastizábal: Bundle methods in stochastic optimal power management: A disaggregated approach using preconditioners. Comput. Optim. Appl. (to appear)
2. J. R. Birge: Stochastic programming computation and applications. INFORMS J. Comput. **9** (1997) 111–133
3. P. J. Brockwell, R. A. Davies: Introduction to Time Series and Forecasting. Springer, New York, 1996
4. C. C. Carøe, R. Schultz: Dual decomposition in stochastic integer programming. Oper. Res. Lett. **24** (1999) 37–45
5. P. Carpentier, G. Cohen, J.-C. Culioli, A. Renaud: Stochastic optimization of unit commitment: A new decomposition framework. IEEE Trans. Power Systems **11** (1996) 1067–1073
6. G. Consigli, M. A. H. Dempster: Dynamic stochastic programming for asset-liability management. Ann. Oper. Res. **81** (1998) 131–161
7. D. Dentcheva, W. Römisch: Optimal power generation under uncertainty via stochastic programming. Stochastic Programming Methods and Technical Applications, K. Marti and P. Kall, eds., Lecture Notes in Economics and Mathematical Systems 458, Springer-Verlag, Berlin, 1998, pp. 22–56
8. J. Dupačová, G. Consigli, S. W. Wallace: Scenarios for multistage stochastic programs. Ann. Oper. Res. **100** (2001) (to appear)
9. J. Dupačová, N. Gröwe-Kuska, W. Römisch: Scenario reduction in stochastic programming: An approach using probability metrics. Preprint 00-09, Institut für Mathematik, Humboldt-Univ. Berlin, Berlin, Germany, 2000 and submitted to Math. Programming
10. N. Draper, H. Smith: Applied Regression Analysis. Hafner, New York, second ed., 1981
11. N. C. P. Edirisinghe: Bound-based approximations in multistage stochastic programming: Nonanticipativity aggregation. Ann. Oper. Res. **85** (1999) 173–192

12. S. Feltenmark, K. C. Kiwiel: Dual applications of proximal bundle methods, including Lagrangian relaxation of nonconvex problems. SIAM J. Optim. **10** (2000) 697–721
13. S.-E. Fleten, S. W. Wallace, W. T. Ziemba: Hedging electricity portfolios via stochastic programming. IMA Volumes in Mathematics and its Applications, Springer-Verlag (to appear)
14. K. Frauendorfer: Barycentric scenario trees in convex multistage stochastic programming. Math. Programming **75** (1996) 277–293
15. N. Gröwe-Kuska, K. C. Kiwiel, M. P. Nowak, W. Römisch, I. Wegner: Power management in a hydro-thermal system under uncertainty by Lagrangian relaxation. Preprint 99-19, Institut für Mathematik, Humboldt-Universität Berlin, 1999 and IMA Volumes in Mathematics and its Applications, Springer-Verlag (to appear)
16. A. Henclová, M. Kotyzová, J. Muková: Analysis of the demand for electric energy. Semainar work (in Czech), Department of Probability and Mathematical Statistics, Charles University Prague, 2000
17. J. L. Higle, S. Sen: Stochastic Decomposition; A Statistical Method for Large Scale Stochastic Linear Programming. Kluwer, Dordrecht, 1996
18. K. Høyland, S. W. Wallace: Generating scenario trees for multi-stage decision problems. Management Sci. (to appear)
19. G. Infanger: Monte Carlo (importance) sampling within a Bender's decomposition for stochastic linear programs. Ann. Oper. Res. **39** (1992) 69–95
20. K. C. Kiwiel: Proximity control in bundle methods for convex nondifferentiable minimization. Math. Programming **46** (1990) 105–122
21. K. C. Kiwiel: User's guide for NOA 3.0: A Fortran package for convex nondifferentiable optimization. Tech. rep., Systems Research Institute, Warsaw, 1994
22. C. Lemaréchal: Lagrangian decomposition and nonsmooth optimization: Bundle algorithm, prox iteration, augmented Lagrangian. Nonsmooth Optimization, Methods and Applications, F. Giannessi, ed., Gordon and Breach, Philadelphia, 1992, pp. 201–216
23. A. C. Miller, T. R. Rice: Discrete approximations of probability distributions. Management Sci. **29** (1983) 352–362
24. M. P. Nowak: Stochastic Lagrangian Relaxation in Power Scheduling of a Hydro-Thermal System Under Uncertainty. PhD thesis, Institut für Mathematik, Humboldt-Univ. Berlin, Berlin, Germany, 2000
25. M. P. Nowak, W. Römisch: Stochastic Lagrangian relaxation applied to power scheduling in a hydro-thermal system under uncertainty. Ann. Oper. Res. **100** (2001) (to appear)
26. A. B. Philpott, M. Craddock, H. Waterer: Hydro-electric unit commitment subject to uncertain demand. European Journal of OR 125(2000), 410–424
27. G. C. Pflug: Scenario tree generation for multiperiod financial optimization by optimal discretization. Math. Programming **89** (2001) 251–271
28. W. Römisch, R. Schultz: Multistage stochastic integer programs: An introduction. submitted to the present volume
29. A. Ruszczyński: Decomposition methods in stochastic programming. Math. Programming **79** (1997) 333–353
30. S-PLUS User's Guide Version 4.5. Data Analysis Products Division, MathSoft, Seattle, 1998
31. S. Takriti, B. Krasenbrink, L. S.-Y. Wu: Incorporating fuel constraints and electricity spot prices into the stochastic unit commitment problem. Oper. Res. **48** (2000) 268–280
32. B. Vitoriano, S. Cerisol, A. Ramos: Generating scenario trees for hydro inflows. Proceedings of the 6th International Conference Probabilistic Methods Applied to Power Systems PMAPS 2000, Volume 2, INESC Porto, 2000

33. Wolfram Research: Mathematica Time Series Pack: Reference and User's Guide. Champaign, IL, 1995
34. S. A. Zenios, M. S. Shtilman: Constructing optimal samples from a binomial lattice. J. Inform. Optim. Sci. **14** (1993) 125–147

Online Scheduling of Multiproduct Batch Plants under Uncertainty

Sebastian Engell[1], Andreas Märkert[2], Guido Sand[1], Rüdiger Schultz[2], and Christian Schulz[3]

[1] Fachbereich Chemietechnik, Universität Dortmund, Germany
[2] Fachbereich Mathematik, Gerhardt-Mercator-Universität Duisburg, Germany
[3] Process Systems Enterprise Ltd., London, UK

Abstract In this contribution, we propose a telescopic decomposition approach for solving scheduling problems from the chemical processing industries online. The general concept is realized for a real-world benchmark process by a two-level algorithm, which comprises a planning step with explicit consideration of uncertainties and a scheduling step where non-linearities are include in the model. Both steps constitute optimization problems, which are modeled and solved by mathematical programming techniques. Besides conceptual considerations concerning online scheduling, we present the two mathematical models and their problem specific solution algorithms with some numerical results.

1 Introduction

In the chemical processing industries supply chain management received increasing attention during the last years: faster changing demands for an increasing variety of products have to be met in a marketplace which grows from a local into a global one. It is an undisputed challenge for future research to develop strategies to increase the efficiency of the material flows within the supply chains from the raw material suppliers to the customers. An important next step in this direction is the improvement of the currently used strategies for the operation and flexible adaptation of individual plants. Major issues are to improve the dynamics of the processing systems, to handle disturbances in an active manner and to master the computational complexity by a feedback control schemes and decomposition approaches [3].

Within the growing market for specialty chemicals, the multiproduct batch plant is a widespread concept that points to the future for producing small volumes of high-valued products in several complicated synthesis steps [13, 30]. Multiproduct batch plants consist of a number of units which are grouped in stages with several parallel units per stage. They are used for the manufacture of products with similar recipe structures, e.g. modifications of one type of polymer, such that each product undergoes a similar sequence of processing tasks. To operate flexible batch plants efficiently, the resources (e.g. reactors or storages) have to be assigned to the processing tasks in order to match certain production goals. To make these assignments properly, known as scheduling, has a large economic impact on process operations [30, 39]. Recent surveys of batch scheduling in the chemical processing industries can be found in e.g. [2, 15, 24, 38].

In this contribution, we present an online scheduling algorithm for a real-world multiproduct batch plant, which provides solutions for the three issues above. The remainder is organized as follows. First, we give an outline of the mentioned properties of the considered benchmark process under scheduling aspects (Section 2). Motivated by this example, in Section 3 we elaborate a new online scheduling approach, and we state the various aspects that have to be taken into account. The solution concept for the specific example, presented in Section 4, is based on a two-level telescopic decomposition framework, which utilizes mathematical programming techniques to solve the emerging optimization problems. The framework consists of a long-term planning problem under uncertainty and a deterministic short-term scheduling problem. The models and the solution algorithms for theses two problems are discussed in Sections 5 and 6, respectively. The contribution closes with suggested research directions for current and future work (Section 7).

2 Benchmark Process

The production of expandable polystyrene (EPS) is used here as a benchmark process. It exhibits, in addition to common properties of recipe-driven multiproduct batch processes (e.g. limited capacity of equipment items, shared and non-shared intermediates, different storage policies) some features which give rise to special difficulties. In the plant which is shown schematically in Figure 1, two EPS-types (A and B) with five grain size fractions each are produced from several raw materials (E). The plant consists of a preparation stage operated in batch mode, a polymerization stage with four batch reactors and two continuously operating finishing lines.

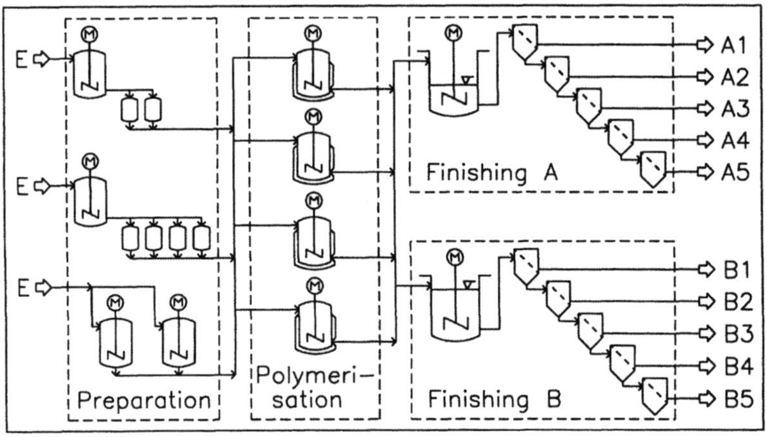

Figure 1. EPS-Process

2.1 Process Structure

Recipes. One EPS-batch is produced from one batch of each intermediate (two dispergators and one organic phase). It contains portions of all five grain size fractions of one EPS-type, which are separated in the finishing lines. There are five recipes for each of the EPS-types, which give rise to different grain-size distributions. The composition of the organic phase is for each recipe specific, whereas the same dispergators are used in all batches.

Preparation Stage. For the scheduling task, unlimited raw material supply can be assumed. Each of the dispergators is produced in one reactor which has a capacity sufficient for several batches. After the reaction is finished, the intermediate has to be transferred immediately into the tanks (no-wait storage policy), which can store one batch each for a limited period of time. One batch of organic phase is produced in one out of two reactors and may be stored in the reactors for an unlimited period of time.

Polymerization Stage. The polymerization stage comprises four reactors with equal capacities. One batch of polymer is produced from three batches of intermediates. Between two starts of polymerizations there has to be a safety interval of a fixed duration. When a polymerization is terminated the product is transferred to a mixing tank immediately.

Finishing Lines. Each finishing line is exclusively assigned to one EPS-type. Each line consists of a mixer tank and a separation stage. The mixers are driven semicontinuously with instantaneous inflow of batches from the polymerization stage (negligible transfer times) and continuous but variable outflow of feed to the separation stages. The mixing process is assumed to be, which leads to non-linear relations for the concentrations of the grain size fractions in the feed, see Section 6.3. The separation units are driven continuously, and the separation process takes a feed-rate independent amount of time. As long as a finishing line is running, the mixer content and the feed-rate are constrained by upper and lower bounds. If a mixer runs empty the corresponding separation stage, and consequently the whole finishing line, has to be shut down. Shut-down and start-up procedures are time consuming and expensive, so that on- and off-duty intervals must not be shorter than several days.

Degrees of Freedom. The following degrees of freedom have to be controlled by a scheduler:

- Starting times and batch sizes for the production of both dispergators,
- starting times and choices of recipes for the production of the organic phase,
- starting times and choices of recipes for the polymerizations, and
- feed-rates for the separation units for both finishing lines. This implicitly determines the start-up and shut-down times of the finishing lines.

2.2 *Production Goals*

A basic requirement for a scheduling algorithm for a technical process is to ensure the feasibilities of the decisions despite disturbances. For technical applications like this one, feasibility is essential and much more important than optimality in a mathematical sense. With respect to (mathematical) optimization, the primary production goal of a demand-driven process is to fulfill all demands without any or with minimal delay. Second to the primary production goal, further goals result from the need to drive the plant as cost efficiently as possible. Key targets are to avoid production of not demanded fractions and to shut down the finishing lines rarely, if at all.

2.3 *Uncertainties*

There are different sources of uncertainties which affect the scheduling activities.

1. *Demands.* The demands are not completely known in advance. They are announced between a few days and a few weeks in advance, and the demanded amounts and the due dates are subject to changes.
2. *Processing times.* The processing times of the dispergator production and of the polymerizations may vary.
3. *Product yields.* With a certain (small) probability, a polymerization runs astray, which leads to a bimodal grain size distribution (with a second maximum at large grains) and a shortened processing time. The grain size distributions are in general only reproducible with a certain variance.
4. *Breakdowns.* Each of the reactors or storage tanks of the preparation and polymerization stage may break down.

3 Online Scheduling: General Remarks

A heuristic analysis of the benchmark problem shows that the preparation stage does not constrain the main stages (polymerization stage, finishing lines), even under consideration of uncertainties. But it also shows that the various main decisions concerning the polymerizations and the finishing lines may interact over a finite horizon of some days with respect to feasibility, and they may interact for an infinite horizon with respect to the production goals. Because of the long term-term effects of the decisions, there is a large number of degrees of freedom, which are continuous or discrete in nature and interact via complex linear and non-linear constraints, possibly subject to stochastic changes.

The classical approach for solving large-scale scheduling problems is a multi-scale decomposition over the entire time horizon under the assumption of complete information (see Figure 2). The decisions are assigned to hierarchically structured layers and optimized on the basis of deterministic models with different degrees of temporal aggregation. The aim is to generate detailed decisions over the entire horizon by solving the models from coarser to finer scales with possible backtracking to avoid infeasibilities [4,41]. If the problem of uncertainty is addressed, this is usually done either by recomputation of nominal schedules, which are computed based on deterministic data depending on the actual situation, or by the offline generation of robust schedules, which are insensitive to a priori defined uncertainties [17,27].

Efficient scheduling of chemical batch processes necessitates the ability to react quickly to new information while the process is running. Since the mentioned approaches are based on offline views on the problem, they disregard the ability to react to new information in the future, which leads to a loss of optimization potential. The scheduling decisions should be generated as late as possible based on the information available before the time of the decision under explicit representation of the uncertainties. An online algorithm should generate only such decisions here-and-now, which are actually supposed to be implemented, whereas all other coupled decisions should be regarded as a recourse for the effects of realized uncertainties.

The above implies that the scheduling task must be treated as a decision problem on a moving horizon. A control scheme with feedforward and feedback should be applied, conceptually known from model predictive control (see e.g. [1]). New information about the demands or the process state is passed to the scheduler (feedback) which generates the next decisions (feedforward). In real-time applications generating detailed decisions over the full horizon does not make sense in general, since uncertainties make detailed decisions obsolete soon after they were obtained. By generating superfluous information, optimization effort is wasted, which reduces the overall process efficiency in real-time applications. Due to the strong coupling it is inevitable to include decisions which may never be implemented, but they should be regarded as a recourse. The time of the decision influences the solution of the decision problem itself, since there is a trade-off between the gain of information and the loss of performance. The response time of the scheduling algorithm enters indirectly into the performance because it determines the delay after which new information leads to new decisions. Online scheduling goes beyond solving optimization problems, since there is always a trade-off between the accuracy of the model, the accepted optimality gap and the response time. This means firstly that proper problem identification and adequate modeling are crucial solution steps and secondly that earlier sub-optimal answers may be more efficient than later optimal ones.

In the following, we propose a solution concept for the EPS scheduling problem which deals with the special structural aspects of real-time applications.

4 Solution Concept

4.1 *Problem Size*

It has been shown for several examples that mathematical programming techniques offer appropriate methods to solve complex constrained problems which arise in the scheduling of chemical plants [2, 3, 38]. An ideal scheduler would be based upon a monolithic model of the decision problem for a horizon of several weeks, parameterized by the data available online about the demands and the process state as well as by probability distributions of the uncertain parameters. To guarantee the

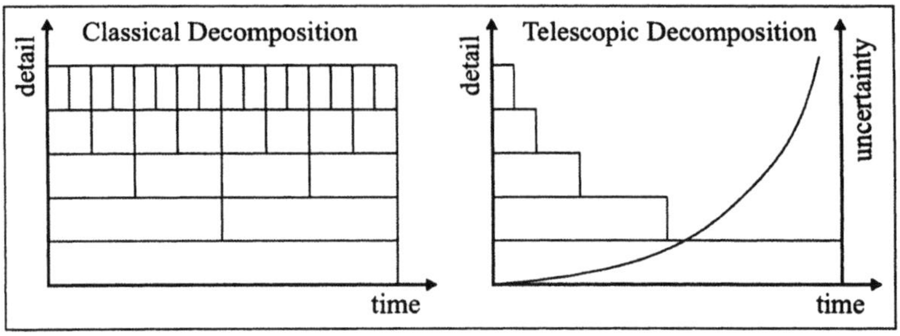

Figure 2. Decomposition Approaches

feasibility of the decisions, a temporal resolution of not less than one hour and immediate responses to disturbances within a few minutes are necessary. Considering these aspects, this constitutes a large-scale mixed-integer andnon-linear real-time optimization problem. To solve real-world scheduling problems efficiently, approximating strategies are necessary, as e.g. the formulation and solution of simplified substitute problems [2, 3].

4.2 Telescopic Decomposition

For the solution of online scheduling problems we propose a telescopic decomposition approach with a number of layered sub-models of different degrees of temporal aggregation (see Figure 2). The sub-models should reflect the alternating sequence of making decisions and receiving new information adequately. This approach avoids the generation of superfluous information, since the level of detail of the integral overall model decreases with increasing distance from the known situation depending on the sensitivity of the modeled effect. The sub-models communicate by implementing results on the coarser scales as targets or constraints on the finer scales. Due to usually strictly constrained response times on each of the levels, backtracking steps from finer to coarser scales are prohibitive. New information is processed adequately on the various levels by updating a scale after a few decisions on the next finer scale either time- or event-triggered. However, despite the lack of backtracking within the scheduler, this iterative strategy implements an information flow from the lower to the higher aggregated levels by the update of the respective models.

Multiple scales only make sense if there are additional degrees of freedom available on the finer scales to cope with the more detailed constraints and to use the additional information. This implies that either there is room for improvement relative to the goals set on the coarser scales or the target cannot be met due to additional constraints or uncertainties. Ideally, feasibility on the coarser scales should imply feasibility on the finer scales; if this cannot be ensured, optimization on the finer scales can be a means to achieve feasibility.

There is an urgent need for designing multi-scale real-time algorithms for specific types of problems with adequate formulations of uncertainty; in particular, there are no such approaches in the literature for the online scheduling of multiproduct batch plants.

4.3 EPS Scheduling Problem

With respect to the process dynamics and the size of the problem, a two-level algorithm is proposed to schedule the EPS-process. The overall problem is hierarchically structured into a long-term, stochastic linear planning problem and a short-term, deterministic non-linear scheduling problem, implemented in the feedback structure shown in Figure 3 (see also [33, 34]). This approach on one hand exploits that the scheduling decisions can be divided into those with long lasting and others with short lasting effects; on the other hand it takes into account that the degree of uncertainty increases with increasing distance in time. Not reflecting the uncertainties explicitly on the lower level leads to a smaller model with shorter computing times and makes online reactions possible.

The planning algorithm uses an aggregated process model for a horizon of two to four weeks and generates scheduling guidelines for a horizon of four to eight days. It is formulated as a two-stage stochastic program, which can – under certain assumptions – be transformed into a mixed-integer linear program (MILP) and be solved by a problem specific decomposition algorithm. The model highlights the discrete long-term decisions of the main stages of the process and uses a linear approximation for concentrations in the mixing tanks (see Section 5). Each optimization run is based on deterministic data of the actual process state and the demand profile as well as probability distributions of the uncertain parameters. A set of guidelines is valid for the next 24 hours.

The scheduler is based on a detailed deterministic process model, which comprises all scheduling decisions (including the preparation stage) and the non-linear model of the mixing process (26). This modeling approach results in a MINLP (Mixed-Integer Non-Linear Program) which is solved by a problem specific depth-first search (see Section 6). It is event-triggered and completely re-schedules the process with respect to the guidelines after new information about the demands or the process state is obtained. The process dynamics require response times of less than five minutes. In the moving horizon approach only those scheduling decisions are actually implemented on the plant which can be realized before the next stochastic event occurs. To enable the scheduler to react flexibly to new information and to ensure feasibility of the decisions, the guidelines from the long-term stochastic optimization do not have to be kept strictly, but their violation is punished by the objective function.

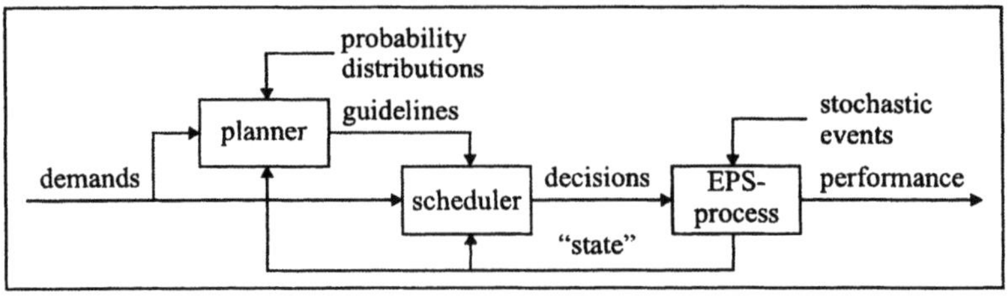

Figure 3. Feedback Structure

5 Long-Term Planning under Uncertainty

5.1 Modeling Framework

A study of modeling frameworks under uncertainty from the literature showed that a two-stage stochastic program (stochastic program with recourse) is a valid choice for formulating the planning problem [32]. This framework has the potential to reflect both the need to make some of the decisions here and now and the possibility to react to realizations by recourse decisions. The deterministic equivalent of a two-stage stochastic program is defined as follows, cf. [6, 16, 29]:

$$\min_x \{c^T x + \mathcal{Q}(x) \ \text{ s.t. } x \in X\} \tag{1}$$

where

$$\mathcal{Q}(x) = \mathbb{E}_\xi \min_y \{q^T(\omega)y \ \text{ s.t. } W(\omega)y + T(\omega)x \geq h(\omega), \ y \in Y\}, \tag{2}$$

$c \in \mathbb{R}^n$, $\xi : \Omega \to \mathbb{R}^m \times \mathbb{R}^{m,s} \times \mathbb{R}^s \times \mathbb{R}^{n,s}$, $\xi(\omega) = \big(q(\omega), W(\omega), h(\omega), T(\omega)\big)$, is a random variable on a probability space $(\Omega, \mathcal{A}, \mathbb{P})$, and $\mathbb{E}_\xi$ denotes the expectation w.r.t. the distribution of ξ. We distinguish first-stage (x) and second-stage (y) variables in terms of their dependence on the random experiment. Components of both types of variables may be restricted to integer values.

For computational reasons, the probability distribution of ξ is assumed to be discrete or is approximated discretely [9]. If one introduces additional second-stage variables y_i for each scenario $i = 1, \ldots, N$ problem (1) can be transformed into the large-scale deterministic optimization problem

$$\begin{aligned} \min_{x, y_1, \ldots, y_N} \{c^T x + \sum_{i=1}^{N} p_i q_i^T y_i \ \text{ s.t. } & T_i x + W_i y_i \geq h_i, \\ & x \in X, y_i \in Y, \ i = 1, \ldots, N\}, \end{aligned} \tag{3}$$

whereby N denotes the number of mass points (scenarios) of the probability distribution of ξ and p_i the corresponding probabilities[1].

The problem is stated as a mixed-integer *linear* program (MILP) for the sake of computational efficiency [14].

5.2 Single-Scenario Model

Modeling Approach

The two-stage stochastic model is derived from a MILP-formulation for one scenario. In order to be able to consider a planning horizon of some weeks and to meet the computing time limitations at the same time, the degree of accuracy of the model is reduced relative to the scheduling model following a problem specific approach. It can be characterized by three key features:

1. The model highlights decisions with long-term effects. These are, on the one hand, the discrete mixer states (on-duty/ off-duty), because the change-over intervals of several days are beyond the length of the scheduling horizon. On the other hand, the choices of the recipes for the polymerizations are optimized for the entire horizon. Due to the coupled production of all grain size fractions, one polymerization may cover demands with very different due dates.
2. The model is based on an aggregated time representation with time intervals of equal lengths. Consequently, similar decisions (mixer states, polymerizations) in one interval are grouped and modeled by a single variable. So the scheduling decisions are not assigned to points of time but only to intervals of one or two days length.
3. The non-linear mixing effects are approximated linearly by constant delays for each input batch. This approximation causes only small errors and ensures feasible planning decisions because the mass balances for the mixers are still satisfied. However, it should be noted that the results of the decisions, namely the amount of material which is produced, may not be feasible for the scheduler, because they are not only affected by aggregation but also by the linear approximation of the mixer. The non-linear mixing are smoothed out over a period of several days, so they can be neglected for planning horizons of several weeks and aggregation intervals of several days. In addition, the mixing effects are only relevant for the production goal "fulfill demands with minimal delay" but not for the other production goals.

Problem Formulation

A single-scenario instance of the planning problem can be characterized as follows. For each product $p \in P$ we are given a set of EPS-types $P = \{1, \ldots, p^{max}\}$, a set of fractions $F_p = \{1, \ldots, f_p^{max}\}$ and recipes $R_p = \{1, \ldots, r_p^{max}\}$ as well as a set of time

[1] Due to the large number of symbols and their colliding definitions in the stochastic programming and in the scheduling literature, the symbols may have different meanings in different sections.

intervals $I = \{1, \dots, i^{max}\}$ covering the planning horizon. We denote by $N_{i,p,r} \in \mathbb{N}$ a variable indicating the number of polymerizations in interval i of type p and recipe r and by $x_{i,p}$ a $\{0,1\}$-variable indicating the state of finishing line p in interval i (off-duty/ on-duty, see Section 2.1). Furthermore, we introduce variables $C_{i,p} \in \mathbb{R}^+$ for the contents of the mixer tanks and parameters $B_{i,p,f} \in \mathbb{R}^+$ which specify customers' demands at the interval endings.

Polymerization Stage. The number of polymerizations which can be started in one aggregation interval is constrained not only by the number of available reactors but also by the number of polymerizations that are still running at the beginning of the interval. Consequently, the capacity in one interval depends on the used capacity in the preceding intervals, which is not known a priori. This dynamic dependency is modeled by constraining the number of polymerization starts for all intervals i to k by

$$\sum_{j=i}^{k} \sum_{p \in P} \sum_{r \in R} N_{j,p,r} \leq N_{i,k}^{max} \qquad \forall i \in I,\ \forall k \in \{i, \dots, i^{max}\}. \tag{4}$$

The right hand side parameters $N_{i,k}^{max}$ can be determined before the optimization run.

Finishing Lines. The mixers are fed with batches at points of time which are shifted against the polymerization starts by the processing time. So, the horizon of the mixers is shifted against the polymerization horizon by this fixed delay. The mixer contents at the beginning of the planning horizon are regarded as known, and the contents at the interval boundaries then follow from the mass balances around the mixers. The mixer input is modeled by the variables $N_{i,p,r}$, which are now interpreted as the number of polymerizations which end in the interval i on the mixer horizon. The feeds are restricted by G_p^{min} and G_p^{max} and thus

$$C_{i,p} \leq C_{i-1,p} + \sum_{r \in R_p} N_{i,p,r} - x_{i,p} G_p^{min} \qquad \forall i \in I,\ \forall p \in P \tag{5}$$

and

$$C_{i,p} \geq C_{i-1,p} + \sum_{r \in R_p} N_{i,p,r} - x_{i,p} G_p^{max} \qquad \forall i \in I,\ \forall p \in P. \tag{6}$$

The binary variables x enforce that the feed equals 0 in off-duty intervals.

The mixer tanks may not contain any material in and at the boundaries of idle intervals. We indicate the state of mixer p at the ending of interval i with the variable $y_{i,p} \in \{0, 1\}$. It is logically constrained by

$$y_{i,p} = x_{i,p} \wedge x_{i+1,p} \qquad \forall i \in I,\ \forall p \in P, \tag{7}$$

which can easily be transformed into linear constraints [28]. The capacity constraints of the mixer tanks are stated as

$$C_{i,p} \geq y_{i,p} C_p^{min} \qquad \forall i \in I,\ \forall p \in P \tag{8}$$

and

$$C_{i,p} \leq y_{i,p} C_p^{max} \qquad \forall i \in I,\ \forall p \in P. \tag{9}$$

To guarantee a smooth operation of the finishing lines with at least $\delta \in \mathbb{N}$ subsequent off-duty and $\varepsilon \in \mathbb{N}$ on-duty intervals, the filter constraints

$$x_{i-j,p} - x_{i-j+1,p} + x_{i,p} \leq 1 \qquad \forall i \in I,\ \forall p \in P,\ \forall j \in \{1, \ldots, \delta\} \tag{10}$$

and

$$x_{i-j,p} - x_{i-j+1,p} + x_{i,p} \geq 0 \qquad \forall i \in I,\ \forall p \in P,\ \forall j \in \{1, \ldots, \varepsilon\} \tag{11}$$

are introduced.

Production Goals. The production goals "fulfill demands with minimal delay" and "avoid production of not demanded fractions" may be translated into the aim to maximize the coincidence of the demand and production profiles. A prerequisite for such a formulation is a relaxed mass balance around the product storages. We propose the following two types of relaxations:

$$\sum_{j=1}^{i} \sum_{r \in R_p} \rho_{p,f,r} N_{j,p,r} \geq \sum_{j=1}^{i} B_{j,p,f} - M_{i,p,f}^{-} \qquad \forall i \in I,\ \forall p \in P,\ \forall f \in F_p \tag{12}$$

and

$$\sum_{j=1}^{i} \sum_{r \in R_p} \rho_{p,f,r} N_{j,p,r} \geq \sum_{j=1}^{i} \sum_{k \in I} u_{k,j} B_{k,p,f} \qquad \forall i \in I,\ \forall p \in P,\ \forall f \in F_p \tag{13}$$

with

$$\sum_{j \in I} u_{k,j} = 1 \qquad \forall k \in I \tag{14}$$

On the left-hand-sides of the inequalities the production profile is calculated as the product of the number of performed recipes $N_{i,p,r}$ and the relative amount $\rho_{p,f,r}$ of type p and fraction f in a polymerization batch, produced according to recipe r. To fulfill all demands $B_{i,p,f}$ for interval i without delay, the production profiles should not be smaller than the demand profiles. In (12) infeasibilities are avoided by introducing amounts of under-production $M_{i,p,f}^{-} \in \mathbb{R}^{+}$, which leads to a profile shift in material direction. In contrast, the indicator variables $u_{k,j} \in \{0, 1\}$ in (13) allow for shifting the profile in time direction.

To minimize the number of start-ups and shut-downs of the finishing lines they have to be counted; indication variables $w_{i,p}^{+}$ and $w_{i,p}^{-} \in \mathbb{R}^{+}$ are introduced and constrained as follows:

$$x_{i-1,p} - x_{i,p} = w_{i,p}^{+} - w_{i,p}^{-} \qquad \forall i \in I,\ \forall p \in P \tag{15}$$

For any fixed values of $x_{i,p}$ the optimal values (e.g. according to (16)) of $w_{i,p}^{+}$ and $w_{i,p}^{-}$ are 1 for start-ups and shut-downs, respectively, and 0 otherwise.

The objective function models a trade-off between the three production goals and is stated as a weighted sum with appropriate weights α, β, γ^+ and γ^-:

$$\min_{\substack{M^-,N,\\ w^+,w^-}} \left\{ \sum_{\substack{i\in I\\ p\in P}} \left(\sum_{f\in F} \alpha_{i,p,f} M^-_{i,p,f} + \sum_{r\in R} \beta_{i,p,r} N_{i,p,r} + \gamma^+_{i,p} w^+_{i,p} + \gamma^-_{i,p} w^-_{i,p} \right) \right\} \tag{16}$$

This formulation is based on the mass balance (12) and models the goal "avoid production of not demanded fractions" by minimizing the sum of performed recipes. A formulation based on the mass balance (13) can be stated analogously.

Numerical Results

The model was implemented in and solved with GAMS/ CPLEX 6.5, which comprises an implementation of a branch-and-bound algorithm based on LP-relaxation [7]. The most important setting concerning the solution algorithm is the sequence of branching variables: assigning the highest branching priority to the mixer states leads to the shortest average CPU-times for given integrality gaps. It turned out that (16) is the objective function with the best numerical condition. In terms of numerical efficiency, the weights α, β, γ^+ and γ^- should not be too different to avoid large integrality gaps, and $\beta_{i,p,r}$ should be a non-linear function of i, p and r to avoid multiple solutions (see [42]). A reasonable choice is e.g.

$$\begin{aligned}
\alpha_{i,p,f} &= 10,\\
\beta_{i,p,r} &= 3 - \left(\frac{r^{max}\cdot p^{max}\cdot(i-1) + r^{max}\cdot(p-1) + (r-1)}{r^{max}\cdot p^{max}\cdot i^{max}}\right)^{1.2},\\
\gamma^+_{i,p} &= 3,\\
\gamma^-_{i,p} &= 3.
\end{aligned}$$

The definition of $\beta_{i,p,r}$ as a non-linear function of i, p and r ensures for two different vectors N that the corresponding objective function sums are different, even if the sum of polymerizations is equal, and that a larger total number of polymerizations leads to a larger objective function sum. As a consequence, the solution algorithm excludes branches from the search tree earlier and closes the optimality gap faster. The exponent 1.2 turned out to be numerically efficient.

The model size mainly depends on the number of aggregation intervals i^{max}. Table 1 gives an impression on the size and the numerical performance of some model instances with $p^{max} = 2$, $r^{max} = f^{max} = 5$ and at least three subsequent intervals with the same states of the finishing lines ($\delta = \epsilon = 3$). The CPU-time was limited to ten seconds.

The integrality gaps are given as intervals, since they significantly depend on the allocated plant capacity. Larger gaps are observed if the plant runs close to full capacity, whereas demand profiles which constantly under- or overload the plant lead to inherent complexity reductions. A further extension of the CPU-time does not lead to substantial improvements of the integrality gaps, and under practical aspects they can be regarded as satisfactory. The gaps indicate that the potential for

Intervals	6	8	10	12	14
No. of Variables	181	241	301	361	421
No. of Integers	72	96	120	144	168
No. of Constraints	174	241	312	387	466
Integrality Gaps [%]	0 – 4.3	0 – 5.8	0 – 6.6	0 – 12.1	0 – 17.8

Computations performed on a SUN Ultra 2 1300.

Table 1. Single-Scenario Planning Model

improvement is bounded by the equivalent of switching the recipes or timings of one or two out of up to more than 100 polymerizations.

5.3 Two-Stage Stochastic Model

First and Second Stage

In two-stage stochastic programming, the involved random variable separates the decision variables. First-stage decisions have to be taken before the realization of the random variable, second-stage decisions can be taken afterwards. A basic modeling aspect is therefore the assignment of the model variables to these groups.

The planning model is designed to transform long-term information into information that can be handled by the short-term scheduling algorithm. One should note in this context that a feasible first-stage solution of the two-stage stochastic programs (3) – if it exists – is feasible for all possible scenarios, i.e. if first-stage and second-stage variables are separated in terms of the time intervals they belong to, we are able to pass short-term information (first-stage solution) to the scheduling taking long-term information (second-stage solution) into account.

We propose two different choices for the grouping of the variables. Due to the time horizon of the scheduling model, information about the state of the finishing stages has to be generated by the planning model (see also Section 6.1). If the emphasis is placed on qualitative information about the production process only, the variables $x_{i,p}$ $\forall i \in \{1, \ldots, \hat{i}\}$, $\hat{i} \le i^{max}$, $\forall p \in P$ representing the operation mode of the finishing units should be chosen as first-stage variables. In this case the first-stage solution determines the operation and idle time intervals for the single products similar to the model used in [8].

If the information the planning model provides is desired to have a more quantitative character, in addition to the finishing states $x_{i,p}$ the variables $N_{i,p,r}$ $\forall i \in \{1, \ldots, \hat{i}\}$, $\forall p \in P$, $\forall r \in R_p$ should form the first-stage vector. By fixing the polymerization variables the operation modes of the finishing units are fixed as well.

The latter approach offers two ways of constraining the scheduler, to prescribe the generated (first-stage) scheme of polymerizations or to fix the corresponding production profile. The time interval $\hat{i} \in I$ can be chosen such that it corresponds to the length of scheduling horizon.

We will refer to the two proposed models as FIN and POLY. Note that the size of the planning model increases with the number of first-stage variables. The model POLY generates larger problems than model FIN in terms of the number of constraints, the number of first-stage variables, and consequently in terms of the number of multipliers of the resulting Lagrangian dual (see [9] and Section 5.4).

Scenarios

In Section 2.3, a number of sources of uncertainty influencing the EPS production process was listed. Due to the complexity of the model and the rare numerical experiments reported in the literature, we decided to perform a step by step inclusion of uncertainties. To date, our numerical experiments take into account stochastic demand and stochastic polymerization yields.

The most substantial source of uncertainty is customers' demand. The uncertainty about future demand scenarios applies to many other production processes in chemical engineering as well [3]. In our model, we made several assumptions about the structure of demand scenarios that were derived from the knowledge of the demand structure on the market for the products considered here.

- The total amount of demand which is accepted does not exceed the capacity of the plant.
- The demand discrepancy between different scenarios increases with time.
- There are only two sources of uncertainty in the demand data:
  - Shifts of single demands to another time interval.
  - Changes in the amounts of single demands.

Scenarios for the production yield (grain size distribution) were generated from the fixed distribution used in the single-scenario case. For each recipe three equally likely outcomes were assumed.

In the formulation of the two-stage stochastic program (1), stochastic demands imply stochastic right-hand sides $h(.)$ and stochastic production yields imply a stochastic matrix $T(.)$ as well as a stochastic recourse matrix $W(.)$.

We assume the independence of demand and production yield, i.e. the stochastic independence of the corresponding components of the random variable.

5.4 Solution Algorithm

Problem (3) is a large-scale mixed-integer linear program (MILP). The size of the problem depends on the size of the single-scenario problems but also on the dimension of the first-stage vector and the number of scenarios used to represent the uncertainties. In general, standard mixed-integer solvers fail to produce acceptable solutions when the size of the problem is large (Table 2). We therefore use the decomposition method proposed in [9] and explained in detail in [14]. By means of Lagrangian relaxation, the method splits the problem into a number of subproblems which correspond to the single-scenario case. The Lagrangian dual is a concave non-smooth program that is tackled by bundle methods [19, 20].

5.5 Numerical Results

We extended the easiest single-scenario model from Section 5.2 by multiple scenarios as mentioned above and set $\hat{\imath} = 3$ (no. of first-stage intervals). Table 2 reports the results obtained by CPLEX for a model-instance POLY with $\imath^{max} = 7$ (no. of intervals) after four hours CPU-time.

No. of Scenarios	10	100
No. of Variables (integer | binary | continuous)	430 | 566 | 2,240	4,030 | 5,606 | 22,400
No. of Constraints	3,379	33,709
Optimality Gaps [%]	10.71	64.49
First Feasible Solution [s]	30	3,200

Computations performed on a SUN Ultra Enterprise 450.

Table 2. Multi-Scenario Planning Model Solved by CPLEX

The decomposition algorithm was implemented in C and FORTRAN77. We use CPLEX 7.0 [10] to solve the subproblems and NOA 3.0 [21] to obtain lower bounds for the master problem. Upper bounds are generated by heuristics based on the solutions of the subproblems (frequency of occurrence, distance to average, rounding, best solution etc.). The current implementation as well as the decomposition algorithm can be used for general MILP's.

Instance	FIN		POLY	
No. of Scenarios	10	100	10	100
No. of Integer Variables	700	7,000	700	7,000
No. of Binary Variables	620	6,200	620	6,200
No. of Continuous Variables	2,240	22,400	2,240	22,400
No. of Constraints	3,514	35,194	3,730	37,570
No. of Multipliers	54	594	324	3,564
Solution Time [s]	710	9,806	14,400	14,400
Optimality Gap [%]	0	0	3.7	5.9

Computations performed on a SUN Ultra Enterprise 450.

Table 3. Multi-Scenario Planning Model with $\imath^{max} = 7$

Tables 3 and 4 report the size of several problem instances in decomposition structure, i.e. including N copies of the first-stage vector (see [14] for more detail), and the worst run out of five with different scenario sets. Since neither the demand

Instance	FIN		POLY	
No. of Scenarios	10	100	10	100
No. of Integer Variables	1,400	14,000	1,400	14,000
No. of Binary Variables	1,180	11,800	1,180	11,800
No. of Continuous Variables	4,480	44,800	4,480	44,800
No. of Constraints	7,364	73,694	7,600	76,270
No. of Multipliers	54	594	324	3,564
Solution Time [s]	1,234	14,400	14,400	14,400
Optimality Gap [%]	0	5.9	4.8	7.7

Computations performed on a SUN Ultra Enterprise 450.

Table 4. Multi-Scenario Planning Model with $\iota^{max} = 14$

nor the production-yield scenarios influence the feasibility of a solution, i.e., of a sequence of polymerizations and finishing-stage states, each solution of a subproblem is a solution of the master problem. A feasible solution for the master problem can therefore be found in the root node of the branch-and-bound tree. Problem FIN could partially be solved to optimality, since the small number of first-stage variables enabled us to completely enumerate the solution space. A major difficulty results from the relatively bad numerical properties of the single-scenario problems. Solving these problems to optimality or with a gap of, say, less than $10^{-3}\%$ is very time consuming. Therefore we impose time and node limits for the subproblems and have to be satisfied with approximate solutions.

6 Deterministic Scheduling

6.1 *Interfaces*

A detailed process model including the preparation stage is used here, since the scheduler has to generate feasible decisions for the entire process. To couple the planner with the scheduler the guidelines generated on the higher level have to be implemented in the lower one, either by constraints or in the objective function. As the durations of the discrete mixer states (up/ down) may be outside the scheduling horizon, the scheduler has to be given information about the start-up and shut-down strategy of the finishing lines as constraints. Furthermore, the guidelines should include information about what and when to produce to achieve a high long-term efficiency. According to the possible first-stage decisions of the planning problem mentioned in Section 5.3 there are two approaches:

Decision Oriented Approach. The discrete decisions of the planning step, namely the number and type of recipes in the aggregation intervals, are passed to the scheduler. According to Section 5, these guidelines can be kept strictly despite any modeled disturbances. If they are defined as hard constraints, the scheduler is only responsible for optimizing the exact timing of the polymerization starts and the choice of the flow rates. This restricted degree of freedom leads to reduced optimization potential on the scheduling level but enables fast reactions and ensures a high long-term efficiency. The lack of short-term efficiency can be avoided by interpreting the set of recipes as targets, which may be subject to changes. However, since the planner does not generate any information about this reaction to disturbances, the long term effects of such changes are not under control.

Target Oriented Approach. The aggregated production profiles an the known demands are given to the scheduler as predicted demand profiles. In contrast to the decision oriented approach, the aggregated amount of material may not be feasible for the detailed model (cf. Section 5). So these guidelines have to be modeled as targets to ensure feasibility of the decisions. The scheduler does not affect the long-term targets, since it has the freedom to react on realizations (e.g. runaway reactions) by choosing a recipe immediately.

6.2 Model Formulation

In our modeling effort, we tried to develop compact, generic models for the type of problems under consideration. The crucial decision to be made is the choice of the representation of time, where, roughly speaking, two concepts can be distinguished. In the first, the planning horizon is divided into intervals of fixed length and the scheduling of task and resource usage is relative to this fixed grid. The second approach is to use a continuous representation of time, where the duration of all intervals is determined by the optimization algorithm. This approach may lead, depending on the process characteristics, to a smaller number of intervals and thus a smaller number of variables in the scheduling problem.

In the models, we exploit specific properties of the process in order to reduce the problem size. In particular for the continuous representation of time, modifications in the synchronization mechanism and the use of a priori knowledge about the minimal period of time between events leads to a significant reduction of the problem size.

The resulting problems are large nonconvex MINLPs with non-linearities which can neither be linearized nor convexified exactly. Therefore, solution algorithms which exploit the problem structure had to be developed to produce solutions within reasonable times (approx. one minute). For further details see [23, 36, 37].

6.3 Continuous-Time Representation

The main reason for developing mathematical models based on a continuous representation of time was the observation that for models with fixed durations of the intervals the number of events in the model by far exceeds the number of possible events in the process, cf. e.g. [31]. Since each possible event is associated with several variables in the model, the reduction of the number of events can be expected to lead to a better performance of the solution algorithms.

Recently, several types of continuous-time models were suggested, e.g. [25,35, 43]. One main difference among the models is how events which occur in different parts or stages of the process are handled. One approach is to define a common reference grid with which all events are synchronized and where all resource balances are calculated. Another approach is to define one grid for each stage and to synchronize only the stages which are directly connected, e.g. via mass balances. An advantage of the latter approach is that one can easily compute the maximum number of events for each reference grid, whereas the first approach avoids the additional effort of synchronization. A common feature is, however, that external events as e.g. supplies, demands and changes in resource availability must be synchronized with the stages affected.

For the process under consideration, it turned out to be effective to use two groups of stages each with a common reference grid.

The first group consists of the preparation stage together with the raw materials, the second comprises the polymerization stage, the finishing lines and the final storage. Forming the second group, we exploit the fact that almost all events in this part are driven by the events in the polymerization stage. Besides, almost all events in the stages after the polymerization are related by a constant shift because all operations in the polymerization stage have the same duration. This allows us to use a single reference grid for this group where all internal events are inherently synchronized by a constant shift. The only events which have to be considered as external events are due dates for customer orders and changes in the resource availability. Changes in the mode of operation of the finishing lines are also referenced against this common grid, which does not restrict the degrees of freedom, since the throughput is defined as an integral quantity.

Another key issue in continuous-time models is how to determine or to bound the maximum number of events, since this number has a major impact on the problem size, esp. on the number of binary (or integer) variables. A bound for one type of events, the starts of the polymerizations, is given by the minimum offset between two starts; this, however, does not limit the overall number of events if it is still required that the start and the end of all operations must coincide with points of the reference grid. This condition is usually required in order to formulate resource (or capacity) balances in a uniform manner. For the specific process considered here, one can drop this condition, since the capacity constraints can be represented in an alternative manner (see below). The maximum number of - internal - events then equals the maximum number of polymerizations which can be started in the planning horizon.

Polymerization Stage. To state the mathematical formulation of the reactor group, we introduce the following variables:

t_n	$\in \mathbb{R}^+$	: time of event $n \in N$ ($N = \{1, \ldots, n^{max}\}$) in the reactor group,
$W_{n,i}$	$\in \{0, 1\}$	: equals 1 if polymerization $i \in I$ ($I = \{1, \ldots, i^{max}\}$) is started at event n, 0 otherwise,
d_n	$\in \mathbb{R}^+$	: duration of interval n.

The time of each event has an upper bound of $H^{max} \in \mathbb{R}^+$ which is larger than the scheduling horizon considered, since we allow operations which are started at the end of the scheduling horizon. We enforce that at each event, exactly one operation is started:

$$\sum_{i \in I} W_{n,i} = 1 \qquad \forall n \in N \tag{17}$$

For each event, it must be ensured that the offset between the operations d_n is larger than the minimal offset $q \in \mathbb{R}^+$:

$$t_{n+1} = t_n + d_n, \quad q \leq d_n \qquad \forall n \in N \tag{18}$$

Capacity constraints are usually imposed by summing up the starting and finishing of operations at each point of the grid. For our problem, we use two different approaches, a simple one which is applied for constant resource availability and a more complex one which has to be applied when the maximum number of available reactors, $N_R^{max} \in \mathbb{N}$, changes over the scheduling horizon. In the first case, the resource constraints can be fulfilled by the conditions

$$t_n - t_{n-N_R^{max}} \geq d_p \qquad \forall n \in \{N_R^{max}+1, \ldots, n^{max}\}, \tag{19}$$

where $d_p \in \mathbb{R}^+$ denotes the duration of the polymerizations. In the second case, we have to synchronize an external event, a change of the resource availability, with the internal events, since it is not known a priori how the external and the internal events are ordered relatively to each other. The resource availability is modeled similar to the approach in [43], where a sequence of intervals is used during which the availability remains constant. We define intervals $[t_r^R, t_{r+1}^R[$, $r \in \mathbb{N}$ during which the maximum capacity $C_r^R \in \mathbb{N}$ is given. Further, the number of operations running at each event has to be calculated. Therefore, the following binary variables are defined:

$X_{n,n'} \in \{0,1\}$: equals 1 if $t_n - t_{n'} \leq d_p$, 0 otherwise.
$Y_{r,n}^R \in \{0,1\}$: equals 1 if $t_r^R \leq t_n$, 0 otherwise.

The capacity restriction then can be represented as

$$\sum_{\substack{n'<n \\ n-n'<d_p/q+1}} X_{n,n'} \leq \sum_r C_r^R \left(Y_{r,n}^R - Y_{r+1,n}^R\right) \qquad \forall n \in N, \tag{20}$$

where the left term denotes the number of running operations and the right term acts as a filter which calculates the capacity at the time of event n. In the sum on the left, the number of terms is reduced to the necessary minimum; the number of the binary variables $Y_{r,n}^R$ can also be restricted by calculating a maximum duration of each event: if a certain duration is exceeded, one (or both) of the finishing lines will run empty, regardless of the amount stored initially in the mixing stage. The definition of these bounds is omitted due to the limited space, as well as the set of equations defining the binary variables $Y_{r,n}^R$ and $X_{n,n'}$ (for details of the latter cf. [43]).

Finishing Lines. In the finishing lines, each event occurs with a constant offset relative to the polymerization stage, we can thus use the same index n for each event. Before stating the equations for the mass balances, the necessary variables and parameters are introduced:

$M_{n,k} \in \mathbb{R}^+$: Total mass in mixer $k \in K$ ($K = \{1, \dots, k^{max}\}$)
$m_{n,s} \in \mathbb{R}^+$: Mass of each grain size fraction $s \in S_k$ ($S_k = \{1, \dots, s_k^{max}\}$)
$F_{n,k} \in \mathbb{R}^+$: Integral feed into separation stage k
$f_{n,s} \in \mathbb{R}^+$: Feed of fraction s
$p_{n,s} \in \mathbb{R}^+$: Mass of product s
$b \in \mathbb{R}^+$: Batch size of each polymerization (constant)
$\rho_{i,s} \in \mathbb{R}^+$: Relative amount of fraction s in polymerization batch, produced according to recipe i.

The mass balances then follow directly as

$$m_{n,s} = m_{n-1,s} + b \sum_i \rho_{i,s} W_{n,i} - f_{n-1,s} \quad \forall n \in N,\ \forall k \in K,\ \forall s \in S_k, \tag{21}$$

$$M_{n,k} = \sum_{s \in S_k} m_{n,s} \quad \forall n \in N,\ \forall k \in K, \tag{22}$$

$$F_{n,k} = \sum_{s \in S_k} f_{n,s} \quad \forall n \in N,\ \forall k \in K, \tag{23}$$

together with the bounds

$$M^{min} \le M_{n,k} \le M^{max} \quad \forall n \in N,\ \forall k \in K, \tag{24}$$

$$d_n F^{min} \le F_{n,k} \le d_n F^{max} \quad \forall n \in N,\ \forall k \in K. \tag{25}$$

In the mixer, the mass from several polymerizations is mixed; thus, we cannot assume that each batch is directly transferred to the finishing lines. Instead, we must calculate the concentration of each fraction in the mixers and ensure that a feed with this concentration is fed into the finishing lines:

$$\frac{m_{n,s}}{M_{n,k}} = \frac{f_{n,s}}{F_{n,k}} \quad \forall n \in N,\ \forall k \in K,\ \forall s \in S_k \tag{26}$$

Since the amount of each fraction s in the feed is added to the amount in the final storage after passing the finishing lines, it can be calculated by the expression

$$p_{n,s} = p_{n-1,s} + f_{n-1,s} \quad \forall n \in N,\ \forall k \in K,\ \forall s \in S_k. \tag{27}$$

Preparation Stage. Each line in the preparation stage operates independently of the other lines, the only coupling occurs at the connection to the polymerization reactor into which the intermediates are transferred. For the reasons mentioned above, each line $j \in J$ ($J = \{1, \dots, j^{max}\}$) has its own reference grid $t_l \in \mathbb{R}^+$ which is synchronized with the reference grid of the reactor group. Since all intermediates must be used up before the next batch of intermediates is produced, we filter out the polymerizations which take place between two such productions. This can be accomplished by introducing another set of binary variables:

$Y_{j,n,l} \in \{0,1\}$: equals 1 if $t_{l,j} \leq t_n$, 0 otherwise.

The mass balance of the intermediates then is

$$m^I_{l,j} = m^I_{l-1,j} + B_{l,j} - b\rho_j \sum_n (Y_{j,n,l} - Y_{j,n,l+1}) \qquad \forall l \in L,\ \forall j \in J \quad (28)$$

with

$m^I_{l,j} \in \mathbb{R}^+$: Mass of intermediate j at event $l \in L$ ($L = \{1, \ldots, l^{\max}\}$),
$B_{l,j} \in \mathbb{R}^+$: Batch size of intermediate j at event l,
$\rho_j \in \mathbb{R}^+$: Relative amount for polymerisation.

The bounds of all quantities as well as the capacity restrictions for the vessels are straightforward. Furthermore, the number of the necessary grid points l and the range of the sum in (28) can be restricted if we take the consumption of intermediates by the polymerizations into account.

Production Goals. To reflect the production goals, the fulfillment of the due dates together with minimum overproduction, we do not directly calculate the lateness of each order. Instead, for each order, the over- and underproduction is calculated at the due date and the objective then is to minimize a weighted sum of over- and underproduction.

Since each due date has to be regarded as an external event, it must be synchronized with the internal events. The synchronization follows the same principle as above; thus, for each of the $u \in U$ due dates $t^L_u \in \mathbb{R}^+$ a binary variable is defined:

$Y^D_{u,n} \in \{0,1\}$: equals 1 if $t^L_u \leq t_n$, 0 otherwise.

The amount of each product at the due date can then be calculated by a filter similar to the one in (20), but an additional complication has to be taken into account: the amount of each fraction in the final storage increases between two consecutive events because the finishing lines continuously feed into the final storage. Thus, the intermediate amount, denoted by p^D_m, has to be considered in the filter equation:

$$p^D_u = \sum_n (Y^D_{n+1,u} - Y^D_{n,u}) \left(p_{n,s} + f_{n,s} \frac{t^D_u - t_n}{d_n} \right) \qquad \forall u \in U \quad (29)$$

The objective function then can be stated as

$$\text{ggmin} \sum_u \alpha_u \max\left(0,\ D_u - p^D_u\right) + \beta_u \max\left(0,\ p^D_u - D_u\right), \quad (30)$$

which can easily be represented by linear terms.

6.4 Fixed-Grid Representation

Due to the limited space, we can only give an outline of this model and highlight the main differences to the continuous-time representation. It basically follows the ideas developed in [22], i.e. one common reference grid is used for the whole plant. Then, at each point of the grid, any of the operations of the intermediate and polymerization stage can be started but no start is required. The mass balances, e.g. (21) - (23), (26) - (28) and the objective function (30), then are formulated at all points of the grid.

The main difference to the previous model is that no synchronization is necessary, since all events, internal and external ones, can be mapped on the grid; the remaining binary decision variables only represent the decision for the choice and the timing of the operations.

6.5 Solution Strategy

Both models presented above are large, nonconvex MINLPs with non-linearities ((26) for both models and (29) for the continuous-time representation) that cannot be eliminated or transformed into convex representations.

The size of the problems, esp. the number of binary variables (cf. Table 5 below), creates problems for general purpose algorithms like DICOPT++ [40] or branch-and-bound algorithms. The application of branch-and-bound algorithms is problematic due to the nonconvexity of the problem, because good bounds cannot be derived or require a large computational effort for solving and tightening convex relaxations. In the algorithms presented here, however, the continuous relaxation is used as a basis for the scheduling decisions, since its solution provides good hints for the scheduling decisions.

The strategy used in our algorithms is to find an integer feasible solution of the problem by a depth-first search by repeated solutions of relaxed MINLP on each search level. The key question for a depth-first search is the choice and the setting of the binary variables which differ for the two types of problems due to the different types of binary variables in the model, although they both represent the same scheduling decisions.

In general branch-and-bound algorithms, the choice of the branching variable usually depends on numerical properties, e.g. the fractional value or the reduced costs etc. This is inappropriate for our problem, since we wish to reduce the likelihood of frequent and deep backtracking due to infeasibilities, because of the computational effort which is spent for the solution of each relaxed MINLP. Furthermore, we try to fix as many binary variables as possible in each step by rounding heuristics.

The core decisions of the scheduling algorithms are

- to choose the recipes of the polymerizations and their timing,
- to schedule the operations in the preparation stage, and
- to determine the feed rates for the two separation stages.

We first describe the algorithm for the fixed-grid representation and then the modifications which are necessary to set the binary synchronization variables of the continuous-time representation.

Instead of the criteria mentioned above, the choice of the binary variables starts at the beginning of the scheduling horizon, subsequently scheduling the polymerizations. The scheduling of the preparation stage is performed after each of the scheduling decisions by setting the affected binary variables up to the starting time for the currently scheduled polymerization.

The choice of the polymerization which is scheduled next is based on the largest nonzero value of the next relaxed binary variables; "next" in this context means the time interval starting at the current scheduling time.

Before solving the next relaxed MINLP, all binary variables which can be fixed by applying the capacity restrictions of the affected resources are set. This is done as far forward in time as possible. Since the search starts with the beginning of the scheduling horizon, we can thus ensure that no capacity restrictions of the reactors are violated, not even for the intermediates. The only source of infeasible subproblems are violations of the capacity constraints of the mixers and the separation stages which are, however, inevitable and cause backtracking.

In order to limit the amount of backtracking, the next choice after a backtracking step is determined by the reason for the infeasibility. The constraint violation reported by a NLP-solver can however not reveal this reason. Instead one gets a good guess from partially simulating the production plan which allows to detect e.g. overflows. This partial simulation is performed before solving the relaxed MINLP because the determination of infeasibilities by simulation is computationally much less expensive than by an NLP-solver.

When backtracking, one has the following choices:

- schedule a polymerization which produces into another line at the same point of time,
- schedule the last polymerization earlier or later or
- go back in time and perform the above possibilities for the previous polymerization.

At each stage, the scheduling of the production of intermediates has to be recalculated. The search is stopped when the end of the scheduling horizon is reached.

A similar algorithm is used for the solution of the continuous-time model with the exception that the second choice for backtracking is not available because the event times are determined by the NLP-solver. In order to find a branching strategy for the additional binary synchronization variables, several strategies were tested: to branch on these variables along with branching on the variables denoting the start of the operations, to fix them before and to fix them after the scheduling choices were made. It turns out, however, that these strategies fail because they dramatically increase the amount of backtracking because infeasibilities introduced by branching on one of these variables are mostly detected far down in the tree. This is due to choices concerning conflicting external events.

The strategy which turned out to be effective branches on one group of synchronization variables before the scheduling decisions are made: all synchronization variables which belong to the polymerization stage, reflecting both internal and external events. The remaining synchronization variables belong to the due dates of the customer orders and are fixed along with the scheduling decisions.

In order to fix the first group, a linear substitute problem is formed which abstracts from the single fractions and takes into account only the overall masses in the mixers and the finishing lines. The integer values of a solution of this problem are always feasible values for the complete problem and are used as fixed values during the solution of the remaining non-linear problem.

6.6 *Numerical Results*

To investigate the relative merits of the two models and algorithms, tests for several scheduling horizons were performed. For each of these tests, the capacity of the polymerization stage was assumed to change over time. Table 5 lists the size of the problems.

Due to the nonconvexity of the problem, a global optimum cannot be determined in general. To examine the quality of the solution, we used satisfiable demands in the objective function which were determined by simulation runs. Thus the optimal value of the objective function always equals 0. Furthermore, the overall sum of the demands is shown which serves as a relative measure of the delays in the schedule, since overproduction is weighted by a factor of 1/10 relative to underproduction and, as an analysis has shown, the main part of the value of the objective function results from underproduction.

The numerical results shown in Table 5 are only an excerpt but give a good indication of the overall picture. The results obtained by both algorithms are very similar but the solution times required by the fixed-grid model are much longer than for the continuous-time model which is due to the larger problem size.

	Continuous-Time			Fixed-Grid		
Horizon [days]	6	10	14	6	10	14
No. of Variables	1,747	3,554	5,678	5,965	10,977	15,999
No. of Binary Variables	316	635	1,021	1,060	2,020	2,980
No. of Constraints	1,437	2,997	4,886	4,347	8,015	11,693
Cost Function	0.49	1.07	2.06	0.56	1.13	2.33
Demand Sum	20.3	52.3	128.4	20.3	52.3	128.4
Solution Time [s]	37	346	1,301	239	1,311	4,294

Computations performed on a SUN Ultra 2 1300.

Table 5. Scheduling Model

As a horizon of six days is sufficient if the scheduler is combined with the stochastic planning algorithm, the proposed solution is sufficiently efficient for reactive rescheduling.

7 Summary and Perspectives

In this contribution we presented a multi-level decomposition approach for online scheduling of multiproduct batch plants under uncertainty. In contrast to classical approaches, the telescopic decomposition avoids the generation of superfluous information and reflects the ability to react to new information an the modeling level. The proposed approach is realized for a real-world benchmark process by means of a two-level model. The higher level sub-model is formulated as a two-stage stochastic program, whose deterministic equivalent can be stated as a large mixed-integer linear program (MILP). The problem is solved by a decomposition algorithm specific for multi-stage stochastic programs. The lower-level sub-model is a deterministic mixed-integer non-linear program (MINLP) with a continuous representation of time. It is solved by problem specific algorithms based on a depth-first search. Numerical experiments for both algorithms were performed and show the applicability of the proposed approach for real-world problems.

Building on these promising results, the current and future research will be on three main points. Firstly, we will design, implement and evaluate different interfaces between the planning and the scheduling level. The focus will be on the question how to utilize and transmit second order information, i.e. information about sensitivities of scheduling decisions. Secondly, additional sources of uncertainties will be integrated in the models on both levels. Therefore, the properties of the uncertainties will have to be analyzed with respect to recourse as well as short-term and long-term effects.

Thirdly, we will develop models that are able to reflect risk aversion of decision makers. In the classical approach of two-stage stochastic programming one minimizes the sum of here-and-now costs (first-stage costs) and the *expected value* of recourse costs (second stage), but the optimization of the expected-value is often not satisfactory. There may be realistic scenarios for which the performance is not acceptable, and a human scheduler would prefer a more cautious policy, e.g. keeping a larger stock. In multiproduct batch plants the huge input of capital suggests that solutions which lead to excessive losses for certain scenarios should be excluded (cf. [32]). In financial mathematics this approach has been studied for many years, starting with the models of Markowitz [26]. Other models are due to Duffie et al. [11], Eppen et al. [12] and King et al. [18]. We follow an approach introduced by Bereanu [5]. The idea is to minimize the probability of a "disastrous event", i.e. to find a solution that performs acceptably for (almost) all scenarios.

References

1. F. Allgöwer, T. A. Badgwell, J. S. Qin,J. B. Rawlings, S. J. Wright: Nonlinear predictive control and moving horizon estimation – an introductory overview. In: P. M. Frank (Ed.): Advances in Control (1999) 391–449
2. G. Applequist, O. Samikoglu, J. Pekny, G. Reklaitis: Issues in the use, design and evolution of process scheduling and planning systems. ISA Transactions (1997)
3. T. Backx, O. Bosgra, W. Marquardt: Towards intentional dynamics in supply chain concious process operations. In: J. F. Pekny, G. E. Blau, B. Caranham. (Eds.): Proc. Foundations of Computer Aided Process Operations (FOCAPO98), CACHE Publications, Michigan, (1998) Suppl.
4. M. H. Bassett, J. F. Pekny, G. V. Reklaitis: Decomposition techniques for the solution of large-scale scheduling problems. AIChE Journal **42** (1996) 3373–3387
5. B. Bereanu: Minimum risk criterion in stochastic optimization. Econ. Comput. Econ. Cybern. Stud. Res. **2** (1981) 31–39
6. J. R. Birge, F. V. Louveaux: Introduction to stochastic programming. Springer, New York (1997)
7. A. Brooke, D. Kendrick, A. Meeraus, R. Raman: GAMS – A user's guide. GAMS Development Corporation, Washington (1998)
8. C. C. Carøe, R. Schultz: A two-stage stochastic program for unit commitment under uncertainty in a hydro-thermal power system. Priority Programme of the Deutsche Forschungsgemeinschaft "Real-Time Optimization of Large Systems", Preprint 98-13[2], (1998, revised 1999)
9. C. C. Carøe, R. Schultz: Dual decomposition in stochastic integer programming. Oper. Res. Lett. **23** (1999)
10. CPLEX Optimization Inc.: Using the CPLEX callable library. CPLEX Optimization[3], (1989–2000)
11. D. Duffie, J. Pan: An overview of value at risk. Journal of Derivatives **4** (1997) 7–49
12. G. D. Eppen, R. K. Martin, L. Schrage: A scenario approach to capacity planning. Oper. Res. **37** (1989) 517–527
13. S. T. Harding, C. A. Floudas: Global optimization in multiproduct and multipurpose batch design under uncertainty. Ind. Eng. Chem. Res. **36** (1997) 1644–1664
14. R. Hemmecke, R. Schultz: Decomposition in two-stage stochastic integer programs for real-time optimization. This volume.
15. S. J. Honkomp, S. Lombardo, O. Rosen, J. F. Pekny: The curse of reality – why scheduling problems are so difficult in practice. Comput. Chem. Eng. **24** (2000) 323–328
16. P. Kall, S. Wallace: Stochastic Programming. Wiley, New York (1994)
17. K. B. Kanakamedala, G. V. Reklaitis, V. Venkatasubramaniam: Reactive schedule modification in multipurpose batch chemical plants. Ind. Eng. Chem. Res. **33** (1994) 77–90
18. A. J. King, S. Takriti: Issues in risk modeling for multi-stage systems. IBM Research Report, RC 20993, Illinois (1997)
19. K. C. Kiwiel: Proximity control in bundle methods for convex nondifferentiable minimization. Math. Programm. **46** (1990) 105–122
20. K. C. Kiwiel: Exact penalty functions in proximal bundle methods for constraint convex nondifferentiable minimization. Math. Programm. **52** (1991) 285–302
21. K. C. Kiwiel: User's guide for NOA 3.0: a fortran package for convex nondifferentiable optimization. System Research Institute, Polish Academy of Sciences, Warsaw (1994)

[2] `http://www.zib.de/dfg-echtzeit/Publikationen/Preprints/Preprint-98-13.html`

[3] `http://www.ilog.com`

22. E. Kondili, C. C. Pantelides, R. W. H. Sargent: A general algorithm for short-term scheduling of batch operations. Part I: MILP formulation. Comput. Chem. Eng. **17** (1993) 211–227
23. T. Löhl, C. Schulz, S. Engell: Sequencing of batch operations for a highly coupled production process: genetic algorithms vs. mathematical programming. Comput. Chem. Eng. **22** (1998) 579–585
24. J. F. Pekny, G. V. Reklaitis: Towqards the convergence of theory an practice: a technology guide for scheduling/ planning methodology. In: J. F. Pekny, G. E. Blau, B. Caranham. (Eds.): Proc. Foundations of Computer Aided Process Operations (FOCAPO98), CACHE Publications, Michigan (1998) 91–111
25. J. M. Pinto, I. E. Grossmann: A continuous time MILP model for short term scheduling of batch plants with pre-ordering constraints. Proc. Sixth European Symposium on Computer Aided Process Engineering (ESCAPE–6) (1996) 1197–1202
26. H. M. Markowitz: Portfolio selections: efficient diversification of investments. Wiley, New York (1959)
27. D. J. Mignon, S. J. Honkomp, G. V. Reklaitis: A framework for investigation schedule robustness under uncertainty. Comput. Chem. Eng. **S19** (1995) 615–620
28. G. L. Nemhauser, L. A. Wolsey: Integer and combinatorial optimization. Wiley, New York (1988)
29. A. Prekopa: Stochastic programming. Kluwer, Dordrecht (1995)
30. J. Rauch (Ed.): Mehrproduktanlagen. Wiley, Weinheim (1998)
31. G. V. Reklaitis: Scheduling approaches for the batch process industries. ISA Transactions **34** (1995) 349–358
32. G. Sand: Planning model for real-time optimization of multiproduct batch plants under uncertainty (in German). Priority Programme of the Deutsche Forschungsgemeinschaft "Real-Time Optimization of Large System", Preprint 98-28[4], (1998)
33. G. Sand, S. Engell, A. Märkert, R. Schultz, C. Schulz: A hierarchical approach to real-time scheduling of a multiproduct batch plant with uncertainties. In: S. Pierucci (Ed.): Computer-Aided Chemical Engineering, Vol. 8: European Symposium on Computer-Aided Process Engineering–10 (2000) 1075–1080
34. G. Sand, S. Engell, A. Märkert, R. Schultz, C. Schulz: Approximation of an ideal online scheduler for a multiproduct batch plant. Comput. Chem. Eng. **24** (2000) 361–367
35. G. Schilling, C. C. Pantelides: A simple continuous-time process scheduling formulation and a novel solution algorithm. Proc. Sixth European Symposium on Computer Aided Process Engineering (ESCAPE–6) (1996) 1221–1226
36. C. Schulz, S. Engell, R. Rudolf: Scheduling of a multiproduct polymer batch plant. In: J. F. Pekny, G. E. Blau, B. Caranham (Eds.): Proc. Foundations of Computer Aided Process Operations (FOCAPO98), CACHE Publications, Michigan (1998) 224–230
37. C. Schulz: PhD thesis: Modeling and optimization of a multiproduct batch plant (in German). University of Dortmund, Dortmund (2001) (in preparation)
38. N. Shah: Single- and multisite planning and scheduling: current status and future challenges. In: J. F. Pekny, G. E. Blau, B. Caranham. (Eds.): Proc. Foundations of Computer Aided Process Operations (FOCAPO98), CACHE Publications, Michigan (1998) 75–90
39. D. E. Shobrys, D. C. White: Planning, scheduling and control systems: why can they not work together. Comput. Chem. Eng. **24** (2000) 163–173
40. J. Viswanathan, I. E. Grossmann: A combined penalty function and outer approximation for MINLP optimization. Comput. Chem. Eng. **14** (1990) 769–782

[4] `http://www.zib.de/dfg-echtzeit/Publikationen/Preprints/Preprint-98-28.html`

41. S. J. Wilkinson, N. Shah, C. C. Pantelides: Aggregate modelling of multipurpose plant operation. Comput. Chem. Eng. **19** (1995) S583–S588
42. H. P. Williams: Model building in mathematical programming. John Wiley & Sons, New York (1994)
43. X. Zhang: PhD thesis: Algorithms for optimal process scheduling using nonlinear models. University of London, London (1995)

VIII

Combinatorial Online Planning in Transportation

*Combinatorial Online Optimization in Real Time

Martin Grötschel[1], Sven O. Krumke[1], Jörg Rambau[1], Thomas Winter[2], and Uwe T. Zimmermann[3]

[1] Konrad-Zuse-Zentrum für Informationstechnik Berlin (ZIB), Germany
[2] Information and Communication Mobile Networks, Siemens AG, Germany
[3] Abteilung Mathematische Optimierung, Technische Universität Braunschweig, Germany

Abstract Optimization is the task of finding a best solution to a given problem. When the decision variables are discrete we speak of a combinatorial optimization problem. Such a problem is online when decisions have to be made before all data of the problem are known. And we speak of a real-time online problem when online decisions have to be computed within very tight time bounds. This paper surveys the art of combinatorial online and real-time optimization, it discusses, in particular, the concepts with which online and real-time algorithms can be analyzed.

1 INTRODUCTION

Models and methods from Combinatorial Optimization provide powerful tools for solving highly complex problems from a broad spectrum of industrial and other applications. The traditional optimization techniques assume, in general, knowledge of all data of a problem instance. There are many cases in practice, however, where decisions have to be made before complete information about the data is available. In fact, it may be necessary to produce a part of the problem solution as soon as a new piece of information becomes known. We call this an *online situation*, and we say that an algorithm *runs online* if it makes a decision (computes a partial solution) whenever a new piece of data requests an action.

Practice may be even more demanding. The online algorithm may indeed be required to deliver the next piece of the solution within a very tight time bound. In this case, we speak of a *real-time problem* (or real-time system), i.e., a problem where an online algorithm is required to react in real-time.

How tight do time bounds have to be in order to turn an online problem into a real-time problem? There is no general rule. A standard answer is: The required reaction time of the algorithm must be short compared to the "time frame of the system", i.e., the definition depends on problem-specific settings. For example, we all expect telecommunication and computer systems to react within a few seconds or faster. Thus, real-time algorithms that, e.g., decide about routing, switching, capacity, or paging must answer within milliseconds. Real-time algorithms controlling chemical reactions or other production processes may be given a few seconds for the computation of a solution, while in transportation or traffic a few minutes lead time could be acceptable. In fact, what could be considered real-time or not may also depend on the complexity of the mathematical model applied, the importance of the decision, and other problem-specific items.

Online and real-time problems have been around in continuous optimization (e.g., control of airplanes, re-entry of a spacecraft) for quite a long time, while combinatorial optimizers have neglected this issue to a large extent. With a few exceptions, systematic investigation of combinatorial online problems started only about 15 years ago. Initially, research was mainly driven by applications in computing and communication machinery. The emergence of new paradigms for the analysis of online algorithms particularly fostered this "combinatorial online research". Interesting and important additional applications broadened its scope.

Why is new theory necessary? Isn't it possible to transfer online results from continuous optimization to combinatorial optimization? The (unfortunate) truth is that continuous and discrete optimization are very different in nature. Combinatorial decision making is, in general, non-convex and non-continuous. Continuous techniques rarely apply to discrete models.

In this paper we will discuss many of the models that have been proposed in the recent years for the analysis of online algorithms. These models usually differ in the way information becomes available to the online algorithm. We will describe the by far most common online paradigms, the *sequence model* and the *time-stamp model*, in greater detail.

Despite significant research efforts in recent years, combinatorial online optimization is not in a mature state yet. Compared to this, combinatorial real-time optimization is even still in its infancy. No commonly accepted tools and concepts for the analysis of combinatorial real-time algorithms that take both, solution quality and time requirements, into account have been established yet. We will address this topic in Section 3.

It is, however, important to note that practical applications have become a driving force in this area. And, thus, we may hope to see new success stories on both, the theoretical and the practical side, in the near future.

1.1 The Sequence Model

An online problem in the *sequence model* can be described as follows. An algorithm ALG, we call it the *online algorithm*, is confronted with a finite *request sequence* $\sigma = r_1, r_2, \ldots$. The requests must be served in the order of their occurrence. More precisely, when serving request r_i, the online algorithm ALG does not have any knowledge of requests r_j with $j > i$. When request r_i is presented to ALG it must be served by ALG according to the specific rules of the problem. The action taken by ALG to serve r_i incurs a cost and the overall goal is to minimize the total service cost.[1] The decision by ALG of how to serve r_i is irrevocable. Only after r_i has been served, the next request r_{i+1} becomes known to ALG. In some cases the appearance of the last request is announced, in some not.

We begin with sketching a very basic decision problem that occurs in various forms frequently in everyday life. We phrase it as a ski rental problem. Despite its

[1] It is also possible to define online profit-maximization problems. For those problems, the serving of each request yields a profit and the goal is to maximize the total profit obtained.

simplicity the ski rental problem will enable us to point out some of the subtleties in the modeling and analysis of online algorithms.

Example 1 (Ski Rental Problem). Suppose that a woman goes skiing for the first time in her life. She is faced with the question of whether to buy skis for $B \gg 1$ Euro or to rent skis at the cost of 1 Euro per day. Of course, if the woman knew how many times she would go skiing in the future, her decision would be easy. But unfortunately, she is in an online situation where the number of skiing days only becomes known at the very last day. □

The above situation can be modeled as an online problem in the sequence model. In the *Ski Rental Problem* each request r_i is a day the woman goes skiing. Each request can be "served" in three different ways: (i) rent skis at the cost of 1 Euro, (ii) buy skis at the cost of B Euro, (iii) use the skis that she already owns at the cost of 0 Euro (where of course this option is only available in case she already bought skis when serving some request r_j with $j < i$). Request r_{i+1} (that is, the next skiing day, if there is any) only becomes known to the woman after r_i has been served. The overall goal is to minimize the total rental/buying cost.

Some comments apply to the ski rental problem. We have formulated the problem in such a way that the skiing woman does neither have any lookahead (that is knowledge about a certain number of subsequent requests) nor any statistical information about the future. This is in accordance with the basic sequence model. If we want to incorporate any of these additional information into the problem then the sequence model must be augmented.

Example 2 (Paging Problem). Consider a two-level memory system (e.g., of a computer) that consists of a small fast memory (the cache) with k pages and a large slow memory consisting of a total of N pages. Each request specifies a page in the slow memory, that is, $r_i \in \{1, \dots, N\}$. In order to serve the request, the corresponding page must be brought into the cache. If a requested page is already in the cache, then the cost of serving the request is zero. Otherwise one page must be evicted from the cache and replaced by the requested page at a cost of 1. A paging algorithm specifies which page to evict. An online algorithm must base its decisions when serving r_i only on the requests $r_1, \dots, r_i$ without any knowledge of future requests. The objective is to minimize the total cost of processing the sequence of page requests. □

1.2 The Time Stamp Model

In the *time stamp model* requests become available over time at their *arrival* or *release dates*. The release date $t_i \geq 0$ is a nonnegative real number and specifies the time at which request r_i is released (becomes known). An online algorithm ALG must determine its behavior at a certain moment t in time as a function of all the requests released up to time t. Again, we are in the situation that an online algorithm ALG is confronted with an input sequence $\sigma = r_1, \dots, r_n$ of requests which is given in order of non-decreasing release times and the service of each request incurs a cost for ALG. The difference to the sequence model is that the online algorithm is allowed to wait and to revoke decisions. Waiting incurs additional costs, typically depending on the elapsed time. Previously made decisions may, of course, only be revoked as long as they have not been executed.

Example 3 (Online Machine Scheduling with Jobs Arriving Over Time).
In scheduling one is concerned with the distribution of jobs (activities) to a number of machines (the resources). In our example, one is given m identical machines and is faced with the task of scheduling independent jobs on these machines. The jobs become available at their release dates, specifying their processing times. An online algorithm learns the existence of a job only at its release date. Once a job has been started on a machine the job may not be preempted and has to run until completion. However, jobs that have been scheduled but not yet started may be rescheduled. The objective is to minimize the average flow time of a job, where the *flow time* of a job is defined to be the difference between the completion time of the job and its release date. □

The above problem can be modeled as an online problem in the time stamp model. Request r_i (corresponding to job i) is a pair $r_i = (t_i, p_i)$, where t_i is the release time of job i and p_i is the processing time. An online algorithm must make its decisions at point t in time only based on the jobs released up to time t. The online algorithm may leave some of its machines idle for some time even if unprocessed jobs that have already been released exist. (Using a small amount of idle time can actually be beneficial in order to gather information about potential new jobs).

Example 4 (Online Traveling Salesman Problem). An instance of the *Online Traveling Salesman Problem* consists of a metric space $M = (X, d)$ with a distinguished origin $o \in M$ and a sequence $\sigma = r_1, \ldots, r_n$ of requests. Each request is a pair $r_i = (t_i, x_i)$, where t_i is the time at which request r_i is released (becomes known), and $x_i \in X$ is the point in the metric space requested to be visited. A server is located at the origin o at time 0 and can move at unit speed. A feasible online/offline solution is a route for the server which serves all requested points, where each request is served not earlier than the time it is released, and which starts and ends in the origin o. The cost of such a route is the time when the server has served the last request and has returned to the origin (if the server does not return to the origin at all, then the cost of such a route is defined to be infinity). This objective function is also called the *makespan* in scheduling.

It is assumed here that an online algorithm does neither have information about the time when the last request is released nor about the total number of requests. An online algorithm must determine the behavior of the server at a certain moment t of time as a function of all the requests released until time t. □

Notice that the Online Traveling Salesman Problem differs from its famous relative, the Traveling Salesman Problem (see Example 17 in Section 3.2), in certain aspects: First, the cost of a feasible solution is not the length of the tour but the total travel-time needed by the server. The total travel time is obtained from the tour length plus the time during which the server remains idle. Second, due to the online nature of the problem it may be unavoidable that a server reaches a certain point in the metric space more than once.

A delicate issue arises when designing an online algorithm for the Online Traveling Salesman Problem: Suppose that at some moment in time all known requests have been served. If the algorithm wants to produce a solution with finite cost, then its server must return to the origin after a finite amount of waiting time. But how long should this waiting time be? If the server returns immediately, then a new request might become known and all the traveling to the origin has been in vain. However, a too large waiting time before returning to the origin increases the cost of the solution unnecessarily.

2 Competitive Analysis

Combinatorial online problems and algorithms had been studied in the sixties to eighties rather sporadically. Broad systematic investigation started when Sleator and Tarjan [46] suggested comparing an online algorithm to an *optimal offline algorithm*, thus laying the foundations of *competitive analysis*. The term "competitive analysis" was coined in the paper [33].

We call an algorithm *deterministic* if its actions are uniquely determined by the input. A *randomized* algorithm may, in contrast, execute random moves, i.e., one and the same input given to such an algorithm twice may result in two different outputs. For the analysis of deterministic and randomized algorithms, of course, different tools are needed.

2.1 Deterministic Algorithms

Let ALG be a deterministic online algorithm. Given a request sequence σ denote by $\mathsf{ALG}(\sigma)$ the cost incurred by ALG when serving σ and denote by $\mathsf{OPT}(\sigma)$ the optimal offline cost (the optimal offline algorithm OPT knows the entire request sequence in advance and hence can serve it with minimum cost).

Definition 5 (Competitive Algorithm, Deterministic Case). Let $c \geq 1$ be a real number. A deterministic online-algorithm ALG is called c-*competitive* if

$$\mathsf{ALG}(\sigma) \leq c\,\mathsf{OPT}(\sigma) \tag{1}$$

holds for any request sequence σ. The *competitive ratio* of ALG is the infimum over all c such that ALG is c-competitive. □

We want to remark here that the definition of c-competitiveness varies in the literature. Often, an online algorithm is called c-competitive if there exists a constant b such that

$$\mathsf{ALG}(\sigma) \leq c\,\mathsf{OPT}(\sigma) + b$$

holds for any request sequence. Some authors even allow b to depend on some problem or instance specific parameters. Thus, whenever c-competitiveness is addressed one should check which definition is applied. We will stick to the definition given above since, in the examples we consider, requiring $b = 0$ is the natural choice.

Observe that, in the above definition, there is no restriction on the computational resources of an online algorithm. The only scarce resource in competitive analysis is information. In many practical applications, severe restrictions on the computation time of an online algorithm apply. We address this issue in Section 3.

Competitive analysis of online-algorithms can be imagined as a game between an *online player* and a malicious *offline adversary*. The online player uses an online algorithm to process an input which is generated by the adversary. If the adversary knows the (deterministic) strategy of the online player, he can construct a request sequence which maximizes the ratio between the player's cost and the optimal offline cost.

We illustrate competitive analysis of deterministic online algorithms on two examples.

Example 6 (A Competitive Algorithm for the Ski Rental Problem). Due to the simplicity of the Ski Rental Problem *all possible* deterministic online algorithms can be specified. A generic online algorithm ALG_k rents skis until the woman has skied $k-1$ times for some $k \geq 1$ and then buys skis on day k. The value $k = \infty$ is allowed and means that the algorithm never buys. Clearly, each such algorithm is online. Notice that on a specific request sequence σ algorithm ALG_k might not get to the point that it actually buys skis, since σ might specify less than k skiing days. We claim that ALG_k for $k = B$ is c-competitive with $c = 2 - 1/B$.

Let σ be any request sequence specifying n skiing days. Then our algorithm has cost $\mathsf{ALG}_B(\sigma) = n$ if $n \leq B-1$ and cost $\mathsf{ALG}_B(\sigma) = B-1+B = 2B-1$ if $j \geq B$. Since the optimum offline cost is given by $\mathsf{OPT}(\sigma) = \min\{n, B\}$, it follows that our algorithm is $(2 - 1/B)$-competitive. □

Example 7 (A Bad Algorithm for the Paging Problem). The algorithm LFU – least frequently used – for the Paging Problem given in Example 2 works as follows: For each page p from the main memory, LFU maintains a counter on the number of times that p has been requested so far. Upon a request r_i which is currently not in the cache, LFU evicts the page from the fast memory which has been requested least frequently in the past.

The algorithm LFU is not competitive. Indeed: suppose that $X = \{p_1, \dots, p_k\}$ is the initial cache contents and p_{k+1} is one additional page from the slow memory. Let $\ell \geq 1$, and consider the sequence $\sigma = p_1^\ell, p_2^\ell, \dots, p_{k-1}^\ell, (p_{k+1}, p_k)^\ell$. Here p_i^ℓ means that page p_i is requested ℓ times in a row and $(p_{k+1}, p_k)^\ell$ states that p_{k+1} and p_k are requested alternatingly ℓ times. Starting with the $\ell(k-1)+1$st request, LFU has cost 1 for every subsequent request, which gives $\mathsf{LFU}(\sigma) = 2\ell$. On the other hand, OPT can process the sequence at cost 1 by evicting page p_1 upon the first request to p_{k+1}. Since ℓ can be chosen arbitrarily large, it follows that LFU is not competitive. □

Example 8 (Negative Result in Machine Scheduling). The online scheduling problem in Example 3 is notoriously difficult. It can be shown that even in the case of a single machine any deterministic online algorithm has a competitive ratio that grows with the number of jobs presented in the input sequence. More precisely, any deterministic online algorithm has a competitive ratio of at least $n - 1$, where n is the number of jobs. (see [24, Chapter 9]). □

Example 9 (Competitive Algorithms for the Online TSP). Probably the most obvious algorithm for the Online TSP (see Example 4 for the definition) is given by the following "REPLAN"-strategy: If a new request becomes known, plan a shortest route starting at the current position, serving all yet unserved requests and ending in the origin. It can be shown that this algorithm is 5/2-competitive (see [8, 9]). However, there are more complicated algorithms which achieve a competitive ratio of 2 (see [6, 8, 9]) in general metric spaces. For the special case that the metric space is the real line, a 7/4-competitive algorithm is presented in [8, 9]. □

2.2 *Randomized Algorithms*

So far we have only considered deterministic online algorithms. The definition of competitiveness for randomized algorithms is a bit more subtle. In the case of a deterministic online algorithm, the adversary has complete knowledge about his opponent and can exploit this knowledge. For randomized algorithms we have to be precise in defining what kind of information about the online player is available to the adversary. This leads to different adversary models which are explained below. For an in-depth treatment we refer to [15, 39].

An *oblivious adversary* (OBL) must choose the entire request sequence in advance. He does neither have knowledge about the outcome of the random experiments of the online algorithm ALG nor about the specific actions taken by ALG as a result of the random decisions. However, the oblivious adversary knows the online algorithm ALG itself including the probability distributions guiding ALG's decisions.

An *adaptive adversary* can choose each request in the input sequence based on knowledge of all actions taken by the randomized algorithm so far, and of the outcome of all random experiments. One distinguishes different adaptive adversaries depending on how the adversary himself must serve the input sequence.

The *adaptive offline adversary* (ADOFF) defers serving the request sequence until he has generated the last request. He then uses an optimal offline algorithm. The *adaptive online adversary* (ADON) must serve the input sequence (generated by himself) online. Notice that in case of an adaptive adversary ADV, the adversary's cost $\mathrm{ADV}(\sigma)$ for serving σ is a random variable.

Definition 10 (Competitive Algorithm, Randomized Case). A randomized algorithm ALG is c-*competitive against an adversary of type* ADV$\in$ {OBL, ADON, ADOFF} for some $c \geq 1$, if

$$\mathbb{E}[\mathrm{ALG}(\sigma) - c\,\mathrm{ADV}(\sigma)] \leq 0 \tag{2}$$

for all request sequences σ. Here, the expectation on the left hand side is taken over all random choices made by ALG. □

In case of an oblivious adversary, the adversary's cost $\mathsf{ADV}(\sigma) = \mathsf{OBL}(\sigma)$ does not depend on any random choices made by the online algorithm. Hence, a randomized online algorithm ALG is *c-competitive* against an oblivious adversary, if for any request sequence the inequality $\mathsf{E}[\mathsf{ALG}(\sigma)] \leq c\,\mathsf{OPT}(\sigma)$ holds.

The power of a randomized algorithm depends on the adversary it competes with. Relations between the adversaries have been studied in a general model called *request-answer games* (see [15]). It turns out that randomization does not help against an adaptive offline adversary. More precisely, it can be shown that the existence of a c-competitive algorithm against an adaptive offline adversary implies the existence of a deterministic algorithm which is c-competitive (see [15]). However, against an oblivious adversary, a randomized algorithm can "hide" its current configuration from the adversary which might enable him to achieve a better competitive ratio.

Example 11 (Ski Rental Problem Revisited). We look again at the Ski Rental Problem given in Example 1. It is easy to see that any deterministic algorithm has a competitive ratio at least $(2 - 1/B)$. Any competitive algorithm must buy skis at some point in time. The adversary simply presents skiing requests until the algorithm buys and then ends the sequence. A straightforward calculation shows that this forces a ratio of at least $2 - 1/B$ between the online and the offline cost.

We consider the following randomized algorithm RANDSKI against an oblivious adversary. Let $\rho := B/(B-1)$ and $\alpha := \frac{\rho-1}{\rho^B-1}$. At the start RANDSKI chooses a random number $k \in \{0, \dots, B-1\}$ according to the distribution $\Pr[k = x] := \alpha\rho^k$. After that, RANDSKI works completely deterministic, buying skis after having skied k times. We analyze the competitive ratio of RANDSKI against an oblivious adversary. Note that it suffices to consider sequences σ specifying at most B days of skiing. For a sequence σ with $n \leq B$ days of skiing, the optimal cost is clearly $\mathsf{OPT}(\sigma) = n$. The expected cost of RANDSKI can be computed as follows

$$\mathbb{E}[\mathsf{RANDSKI}(\sigma)] = \sum_{k=0}^{n-1} \alpha\rho^k(k+B) + \sum_{k=n}^{B-1} \alpha\rho^k n$$

A lengthy computation shows that

$$\mathsf{E}[\mathsf{RANDSKI}(\sigma)] = \frac{\rho^B}{\rho^B - 1}\mathsf{OPT}(\sigma).$$

Hence, RANDSKI is c_B-competitive with $c_B = \frac{\rho^B}{\rho^B-1}$. Since $\lim_{B\to\infty} c_B = e/(e-1) \approx 1.58$, this algorithm achieves a better competitive ratio than any deterministic algorithm whenever $2B - 1 > e/(e-1)$, that is, when $B > (2e-1)/2(e-1)$. □

Example 12 (Paging Revisited). It can be shown that no deterministic algorithm for the Paging Problem (see Example 2) can achieve a competitive ratio smaller than k, the size of the cache. However, there exists a randomized algorithm which is $2H_k$, competitive, where $H_k = 1 + 1/2 + \cdots + 1/k$ is the kth harmonic number. Proofs and the algorithm can be found in [15]. □

Example 13 (Machine Scheduling Revisited). The scheduling problem from Example 3 remains difficult even for randomized algorithms. Every randomized algorithm has a competitive ratio of $\Omega(\sqrt{n})$ against an oblivious adversary where again n denotes the number of jobs given in the request sequence (see [49]). □

2.3 Alternatives to Competitive Analysis

Competitive analysis is a type of worst-case analysis. It has (rightly) been criticized as being overly pessimistic. The competitive ratios observed in practice are usually much smaller than the pessimistic bounds provable from a theoretic point of view. Often the offline adversary is simply too powerful and allows only trivial competitiveness results. This phenomenon is called "hitting the triviality barrier" (see [24]). To overcome this unsatisfactory situation various extensions and alternatives to pure competitive analysis have been investigated in the literature.

In *comparative analysis* the class of algorithms where the offline algorithm is chosen from is restricted. This concept has been explored in the context of the Paging Problem [34] and the Online TSP [14]. Another approach to strengthen the position of an online algorithm is the concept of *resource augmentation* (see, e.g. [10, 41, 42, 46]). Here, the online algorithm is given more resources (more or faster machines in scheduling) to serve requests than the offline adversary. The *diffuse adversary* [34] model deals with the situation where the input is chosen by an adversary according to some probability distribution. Although the online algorithm does not know the distribution itself, it is given the information that this distribution belongs to a specific class of distributions. Other approaches to go beyond pure competitive analysis include the *access graph model* for paging [16, 17, 32] and the *statistical adversary* [18]. We refer to [24, Chapter 17] for a comprehensive survey.

All of the extensions and alternatives to competitive analysis have been proven to be useful for some *specific problem* and powerful enough to obtain meaningful results. However, none of these approaches has yet succeeded in replacing competitive analysis as *the* standard tool in the theoretical analysis of online algorithms. Hence, it is particularly irritating that competitive analysis can only give substantial decision support for a few "real-world problems".

3 Real-Time Issues

In real-time systems (cf. section 1), an algorithm has to deliver a solution within prescribed time constraints. The behavior of a real-time system depends of course on the quality of the solution but it depends as well as on the time needed for producing the solution. A solution provided too late may be useless or, in some cases, even dangerous because it does not fit to the current system parameters which may vary over time.

For instance, if a decision support system watching the stock market needs a long time to propose buying or selling a certain share, the price of the share (especially in a volatile market) may have changed so much that this action is no longer reasonable. If, however, the decision support system of a pilot takes long to suggest the right action in case of an emergency the result may be fatal.

In our context, the notion *time* emphasizes the fact that the system significantly depends on the time in which answers to requests are produced. The notion *real* indicates that the system's reaction to external events must occur instantaneously. In other contexts, *real-time* reaction is just a synonym for fast reaction to external events. We have to be more precise, the speed of *real-time* reaction must correspond to the specific time requirements of the systems environment and the problem setting. The time available for computation may vary, e.g., from milliseconds to minutes. The general objective of real-time optimization is to match the problem specific timing requirements of each task and to produce a best possible solution within the incurred time constraints. Since the solution is based on the information available at the beginning of the computation, it may be necessary to check its feasibility for the state of the system at the end of the computation.

3.1 Real-Time Decision Support Systems

Real-time algorithms are often integrated into computerized decision support systems, see [44] for such examples in local transport.

Decision support can be based on the knowledge of a previously forecasted development of the real-time system. In our context, such forecasts may be obtained via offline computations of optimal solutions of some combinatorial optimization problem for real-world data describing the standard situation of the real-time system. We will thus call the presently available forecasted development of the real-time system briefly the *current solution*. Real-time decision support systems provide proposals for "quick" reactions to external unforseen events which change the current solution. Real-time decisions usually have to be made subject to and despite of severe limitations of resources: hardware, time, and information. Some fundamental components of such a decision system (according to [44]) are:

1. Information management: current update of incoming information.
2. Situation assessment: evaluation of the situation, decision whether or not a reaction of the system is required (or should be proposed).
3. Evaluation of alternatives: checking possible actions for the real-time event.
4. Decision: determining an action (or choosing to do nothing).

Real-time decision support systems for complex real-time systems are (more or less) semi-autonomous systems that support and assist human operators. Due to efficiency, responsibility, and security issues, human operators are seldom replaced by such systems. On the other hand, these systems usually require highly qualified personnel.

Decision support systems may propose actions with different degree of influence on the development of the real-time system. Three different types of decisions with increasing impact [44] are *reactive planning*, *incremental planning*, and *deliberative planning*. In reactive planning, the current solution is only locally adapted to some real-time event. Incremental planning already results in a more global update of the current solution. Deliberate planning is a complete revision of the current solution. This is advisable when the observed situation significantly differs from the predicted state so that the current solution becomes ineffective or even infeasible. The choice of reactions on real-time events depends on the time available for computation and on the observed effects of the real-time event.

Example 14 (Dispatching Trams in Local Transport). In municipal tram dispatch, trams start from a certain depot for serving scheduled round trips. In the depot, trams are stored in several sidings one behind the other. The dispatcher has to assign the trams to a sequence of round trips each requiring a certain type of tram [13, 50, 51].

Due to unforeseen external events, e.g., delays, the pre-calculated current feasible assignment of trams to round trips has to be replaced by a new one. The dispatcher needs to find the new feasible assignment as fast as possible within a few minutes. His objective is to minimize (or prevent) shunting of trams. Otherwise, new delays may be generated or more tram drivers may be required for moving trams. □

For results on competitive analysis for online versions of tram dispatch problems, unfortunately mainly negative observations have been made, we refer to [50, 51].

In the following sections we survey some of the prominent methods for solving offline-optimization problems and comment on their usability in a real-time context.

3.2 *Exact Solution Methods for Combinatorial Optimization Problems*

Online and real-time algorithms try, of course, to make use of the existing machinery of combinatorial optimization. Core ingredients are, thus, fast solvers for linear, integer and mixed-integer programs.

Mixed-integer programming (MIP) provides effective tools for solving combinatorial optimization problems which arise from industrial applications. Constraints from combinatorial optimization can often easily be reformulated in terms of linear MIP-constraints though it may turn out to be difficult to find a computationally effective formulation. Modelling software like AMPL [25] or GAMS [19] is available which support modelling and problem solving. Powerful state-of-the-art solvers for linear and mixed integer programming problems such as CPLEX [12] have successfully been applied to such formulations of industrial applications.

Definition 15 (Mixed integer programming). In (linear) mixed integer programming the given (linear) objective function

$$c^T x + d^T y \tag{3}$$

has to be minimized subject to the given (linear) constraints

$$A_1 x + A_2 y = b_1 \tag{4}$$

$$A_3 x + A_4 y \leq b_2 \tag{5}$$

for integer valued vectors x and real valued vectors y. □

Solving MIPs is difficult in theory (NP-hard) and, in general, hard in practice. Nevertheless, MIP formulations and solution techniques may help under real-time constraints. Here are two examples of the MIP approach.

Example 16 (Load Balancing on Identical Machines). Consider the *Load Balancing Problem* (also called *makespan minimization*) arising in machine scheduling. One is given a sequence $I = (1, \dots, n)$ of jobs where job i has processing time p_i. The task is to distribute the jobs on m identical machines such that the maximum load of a machine is minimized. Here, the load of a machine is defined to be the sum of job processing times assigned to the machine.

The above problem can be formulated as the following Integer Linear Program:

$$\begin{aligned} &\text{minimize } M \\ &\text{subject to} \\ &\sum_{i=1}^{n} p_i x_{ij} \le M && \text{for } j = 1, \dots, m && (6) \\ &\sum_{j=1}^{n} x_{ij} = 1 && \text{for } j = 1, \dots, n && (7) \\ &x_{ij} \in \{0, 1\} && \text{for all } i, j && (8) \end{aligned}$$

The binary (decision) variable x_{ij} has the following meaning: $x_{ij} = 1$ if and only if job i is assigned to machine j. Constraints (7) ensure that each job is assigned to exactly one machine, constraints (6) ensure that M is greater or equal to the load of any of the m machines. Since M is minimized it follows that in an optimal solution M will be exactly the maximum load of a machine. □

Example 17 (Offline Traveling Salesman Problem). In the famous symmetric *Traveling Salesman Problem* one is given a complete undirected graph $G = (V, E)$ on n vertices $V = \{1, \dots, n\}$ with (symmetric) edge weights d_{ij} for each edge $ij \in E$. The problem consists of finding a shortest tour starting and ending at the same vertex and visiting each other vertex exactly once. The cost of a solution is the total length of all edges in the tour. □

We formulate the Traveling Salesman Problem as an Integer Linear Program. To this end define, for a subset $S \subseteq V$, the set $\delta(S) := \{ ij \in E : i \in S, j \notin S \}$ of edges incident with S. Then using the decision variables x_{ij}, with $x_{ij} = 1$ if and only if edge ij is contained in the tour, we can write the TSP as the following Integer Linear

Program

$$\text{minimize} \sum_{i,j=1}^{n} d_{ij} x_{ij}$$

subject to

$$\sum_{ij \in \delta(\{i\})} x_{ij} = 2 \qquad \text{for } i = 1, \ldots, n \qquad (9)$$

$$\sum_{ij \in \delta(S)} x_{ij} \geq 2 \qquad \text{for all } S \subseteq V, 2 \leq |S| \leq |V|/2 \qquad (10)$$

$$x_{ij} \in \{0, 1\} \qquad \text{for all } i, j \qquad (11)$$

A proof that the feasible solutions to the above Integer Linear Program are in fact exactly (incidence vectors of) tours, can be found, e.g., in [22, 36].

As already noted, there are important differences between the objectives in the Offline TSP and the Online TSP specified in Example 4. However, an algorithm for the Online TSP can make use of an (exact or approximate) algorithm for the Offline TSP to solve the following sub-problem: For a set of known but yet unserved requests R find a shortest route which serves all requests in R and returns to the origin.

Integer programming formulations are quite flexible and general. While adding or cancelling of constraints and/or variables in a MIP may severely change the complexity of the model, it still remains a MIP and thus basic methods for solving MIP's still apply. Combinatorial algorithms specially designed and tuned for some combinatorial optimization problem usually break down when such changes become necessary.

For example, integer programming methods have successfully been applied to real-time problems in transport and logistics. If solving a complete real-time model turns out to be too time-consuming, it may be decomposed into smaller parts which can be solved fast enough. Trading computing time versus solution quality helps to adapt the problem setting to the changing requirements in real-time applications.

Example 18 (Dispatching Trams in Local Transport Revisited). The task of finding shunting free assignments for the tram dispatching problem of Example 14 can be modelled as a 0-1-quadratic assignment problem [50, 51].[2] Shunting-free assignments correspond to assignments that obey certain additional side constraints [13]. After exact linearization and some model tuning, the resulting integer programming model can be solved within reasonable time [50, 51]. □

A similar approach has proved to be useful in the context of container logistics [47].

Exact solution methods, here based on mixed integer programming formulations of the combinatorial optimization model, are essential for pre-calculation of good or

[2] See [20] for a definition of the quadratic assigment problem and a comprehensive survey of solution approaches.

optimal solutions to real-time optimization problems. While computation times are reasonable, matching tight real-time requirements in real-time applications may enforce tradeoffs between solution quality and computation time. However, even then, exact methods provide indispensable information on the quality of other approaches.

3.3 Approximation Algorithms

If exact methods fail to produce answers in real-time the next step is to look for suboptimal solutions which have a guaranteed quality. *Approximation algorithms* for offline minimization problems are closely related to competitive online algorithms.

Definition 19 (Approximation algorithm). A deterministic algorithm ALG is α-*approximative* if

$$\mathsf{ALG}(I) \leq \alpha\, \mathsf{OPT}(I) \tag{12}$$

holds for any problem instance I. The quantity $\alpha - 1$ provides a worst case bound on the relative error of the approximation. The infimum of all values of α for which ALG is α-approximative is called *performance ratio* of ALG. (The remarks made on variants of the definition of c-competitiveness also apply here.) □

By the above definition, a c-competitive online algorithm is c-approximative. Conversely, if a c-approximate algorithm is also online, it is also c-competitive. In view of applications, in the design of approximation algorithms speed is of first priority since here computation time is the scarce resource. Thus, one usually restricts approximation algorithms to the class of polynomial time algorithms.[3] In contrast, time complexity is not an issue in competitive analysis: there is (at least in theory) no bound on the computation time for an answer generated by an online algorithm.

Many approximation algorithms have a simple structure and are in fact online. For NP-hard problems, polynomial time approximation algorithms offer a way to trade solution quality for computation time. Polynomial time approximation algorithms have intensively been considered within the last years. Comprehensive surveys on approximation algorithms can be found in [7, 31, 38, 48].

Example 20 (Load Balancing on Identical Machines revisited). Consider again the load balancing problem described in Exampe 16. Graham [27, 28] proposed the following greedy-type heuristic LIST: Consider the jobs in order of their occurence in the input sequence I. Always assign the next job to the machine currently with the least load (breaking ties arbitrarily). Clearly, LIST can be implemented to run in polynomial time. Moreover, LIST is also an online algorithm for the online version of the problem where jobs are revealed to an online algorithm according to the sequence model.

We are now going to analyze the performance of LIST. Obviously, the optimum load is at least as large as any job processing requirement resp. at least as large the

[3] In the literature often the notion of an approximation algorithm includes the property of the algorithm being polynomial time.

average processing time for each machine, i.e.:

$$\text{OPT}(I) \geq p_i \quad \text{for } i = 1, \ldots, n \qquad \text{and} \qquad \text{OPT}(I) \geq \frac{1}{m} \sum p_i. \tag{13}$$

Consider the machine j where LIST generates the maximum load when processing I. Let p_i be the load of the last job assigned to machine j and let L be the load of j before job i was assigned. With these notations we have $\text{LIST}(I) = L + p_i$.

By definition of LIST, at the moment job i was assigned to machine j all other machines had load at least L. Hence, the total sum of job sizes is at least $mL + p_i$. Hence from the second inequality in (13) we get $\text{OPT}(I) \geq 1/m(mL + p_i) = L + p_i/m$. This results in

$$\text{LIST}(I) = L + p_i \leq \text{OPT}(I) + \left(1 - \frac{1}{m}\right) p_i \leq \left(2 - \frac{1}{m}\right) \text{OPT}(I),$$

where for the last inequality we have used the first inequality in (13).

This proves that LIST is $(2 - 1/m)$-approximative. Since we have already remarked that LIST is in fact an online algorithm, LIST is also $(2-1/m)$-competitive for the online variant of the problem in the sequence model. □

For $m \geq 2$, Albers [1] describes an online scheduling algorithm which is 1.923-competitive. Her algorithm tries to prevent schedules which distribute the load uniformly on all machines by keeping some machines with a "low" load whereas the other machines have a "high" load. For $m \geq 80$, Albers [1] derives a lower bound of 1.852 on the competitive ratio of deterministic online algorithms for the machine scheduling problem.

In the offline case, LIST can easily be improved by taking advantage of the information about I. In worst-case examples for LIST, the last job has a very long processing time. By sorting the jobs in non-increasing order according to their processing times, i.e., processing jobs with the longest processing times first, a better approximation ratio of $\frac{4}{3} - \frac{1}{3m}$ can be achieved [28]. Since sorting is quite fast, this algorithm may still be applied to real-time versions of machine scheduling problems where several jobs arrive simultaneously.

Example 21 (Offline Traveling Salesman Problem Revisited). It is easy to see that for the Traveling Salesman Problem (see Example 17), polynomial-time approximation algorithms with constant performance ratio can only exist if the edge weights satisfy the triangle inequality [26]. In this case, a 2-approximative algorithm can be constructed using a minimum spanning tree in the graph [40]. Christofides' algorithm [21] also starts with a minimum spanning tree. For the nodes with odd degree in this tree a shortest perfect matching is computed. Then, a tour following a Eulerian walk in the multi-graph formed by the spanning tree and the perfect matching is constructed. The solution found this way is $\frac{3}{2}$-approximative.

For the special case that the vertices in the input graph corresponds to points in the Euclidean plane and the edge lengths are given by the Euclidean distances,

Arora [2] and independently Mitchell [37] have devised polynomial time approximation schemes.[4] However, no practical implementation of these fairly complicated algorithms has been reported yet. □

Real-time applications require that answers are computed online *and* within tight time windows. The length of this time window is closely connected to the arrival times of the requests. In view of the discussion of polynomial approximation algorithms, one may define a real-time algorithm as an online algorithm that generates answers in constant or, at least, in "suitably" low polynomial time. A concept for the evaluation of the performance of real-time algorithms, that combines approximation aspects and time requirements in a convincing manner, would be of great value for real-time applications. Up to now, no convincing concept has been proposed.

3.4 Offline Heuristics (without Provable Worst-Case Performance Guarantees)

In some applications, optimal or approximate solutions even for small problem instances cannot be computed within the tight required real-time bounds. The typical approach in this case is to look for algorithms that quickly produce a feasible solution and iteratively keep on improving the solution. There are general principles, such as *local search* (or more fashionable: *meta heuristics*), that can be adapted to special applications and have indeed successfully been applied to many real-world applications. For a comprehensive introduction to local search we refer to [23,43].

Local search for a combinatorial optimization problem proceeds in the following way. Let $\mathcal{F}$ be the set of all feasible solutions (also called *solution space*). For each feasible solution $x \in \mathcal{F}$, one defines a *neighborhood* $N_x \subseteq \mathcal{F}$ containing all feasible solutions which are "close" to x and which can be reached from x by applying certain modifications to x. The solution space $\mathcal{F}$ is covered by the collection of neighborhoods $\{N_x : x \in \mathcal{F}\}$.

Starting from an initial feasible solution, local search moves from one feasible solution to another, while storing the best solution found so far. Local search can thus be stopped at any time and will always provide a feasible solution. In our context, local search algorithms may thus be called *any-time algorithms*. The initial solution for a local search algorithm is usually generated using a starting heuristic. The local search algorithm terminates either after a certain number of steps (in the context of real-time computation this may also be after a user-defined time threshold) or according to some other stopping criterion with respect to the objective function value.

The basic local search paradigm leaves open how a successor $x' \in N_x$ to the current solution x is selected. Different rules to select a successor lead to different incarnations of local search. For instance, a *Greedy-type local search* would always choose the solution with the best objective function value among all solutions in N_x.

[4] An approximation scheme consists of a collection $\{\mathsf{ALG}_\varepsilon : \varepsilon > 0\}$ of algorithms where ALG_ε ϵ-approximate and has polynomial running time.

However, since such an algorithm can get stuck at local optima, various other approaches have been suggested in the literature. In *Simulated Annealing* one accepts also successors with worse objective function value but only with a certain probality which decreases over time. *Tabu Search* is another implementation of local search which attempts to avoid a breakdown at local optima. Other approaches include so called *improvement heuristics* like k-opt. We refer to [23, 43] for comprehensive survey.

Example 22 (Dispatching Trams in Local Transport by Local Search). An example for a powerful local search algorithm is the reactive tabu search (RTS) heuristic developed by Battiti and Tecchiolli [11]. In RTS, the next solution in the neighborhood is chosen at random while recording whether and how often this solution has been visited before. If a re-visiting counter exceeds a threshold, some random steps are executed in order to leave the previously visited neighborhood in which RTS threatens to stall. RTS is known to be very effective for instances for quadratic assignment problems.

The real-time tram dispatch problem introduced in Example 14, requires to compute tram assignments within two minutes. Reactive tabu search provided optimal solutions for more than 80 percent of the considered real-world instances as well as for randomly generated instances within these tight time bounds [50, 51]. □

4 General-Purpose Online-Heuristics

There are general principles which can be used to design an online algorithm.

4.1 First-In-First-Out (FIFO)

The FIFO-strategy does only make sense in the time stamp model. This approach to control the order in which requests are served completely works without regard of efficiency issues: FIFO strictly serves requests in the order of appearance.

Although it is clear form the definition that this strategy is almost never cost-efficient it is incredibly popular in production-planning and control. One reason for this might be that FIFO has a desired side-effect: items in a production environment are delivered according in the order of production, so that no newer items are sold (or used) before the old items are cleared. One other reason is that a FIFO-heuristic is sometimes hidden in a control system based on priority rules. These systems usually employ FIFO as a tie-breaker inside the priority classes. Whenever there are many requests in one priority class the efficiency problem will take effect. Thus, a major problem for such controls is catching up after system break-downs.

In principle it is possible to use any of the following strategies as a tie-breaker in a priority-based control system. Therefore, FIFO– if not explicitely required – is usually an inferior strategy whenever there is a substantial number of requests available for planning.

4.2 Greedy

The greedy algorithm is a well defined algorithm in the context of matroids or independence systems in combinatorial optimization. In terms of online optimization the notion of an algorithm being "*greedy*" is used for all kinds of algorithms which have in common the following strategy: Make a "locally most promising" decision how to process the next request.

In the sequence model the GREEDY principle amounts to serving the next request that is revealed to the online algorithm by such an action that has least service cost. In the time stamp model at any time GREEDY serves that request (among the yet unserved requests) next that can be served with the least cost with respect to the current system state. This is one extremal case of local optimization: GREEDY only decides upon the next request to be served, i.e., it does not plan into the future or does not consider the system state after the service. That means, even when no other request arrives, GREEDY is very likely to be sub-optimal. Moreover, GREEDY does not take into account possible future requests.

Although the above GREEDY-strategy is very shortsighted and the solutions produced maybe sub-optimal it is very popular because it is

- easy to implement,
- usually real-time compliant, and
- it produces a stable, predictable behavior since no decision is revised.

However, if cost-efficiency is the main-goal one usually needs a more sophisticated approach.

4.3 Replan

The REPLAN-strategy for an online problem in the time stamp model assumes that we have a method that computes an optimal (or almost optimal) solution to the static optimization problem (the corresponding *offline-problem*) at a specific point in time. Note that, in a realistic environment, this imposes the restriction of real-time compliance on the algorithms used to compute the optimum of the offline-problem (see Section 3).

While the GREEDY-approach 4.2 acted as locally as one could think, for REPLAN we find the other extreme case: at any time REPLAN tries to be "as globally optimal as possible", given the information it has at that point.

More specific: REPLAN maintains a "plan" containing the information on how to serve the already known requests. This plan is followed as long as no relevant event happens. Whenever a relevant event happens (a new request arrives, a change of the system state gives rise to a new cost of the current solution, etc.), REPLAN computes a cheapest solution of all known request in the current system state. Due to its nature, REPLAN is also called REOPT in the literature.

At any point in time we compute an optimal solution that is globally optimal at that particular moment. However, with respect to the complete instance the current solution is yet only locally optimal. Whenever a new request arrives the plan maybe revised, and the global efficiency of the old plan is never really exploited since only the first couple of requests have been served according to that plan.

A more serious problem, however, is the fact that REPLAN can completely revise all decisions for which this is still possible. This often leads to an unpredictable behavior over time. One can even produce "oscillating" solutions. This means the following: assume, e.g., at some point in time, we find an optimal solution serving some request r of type A before another request r' of type B. Before we can serve r, a new request of type B arrives. Now the optimal plan may suggest to serve r' prior to r. But then there might arrive a new request of type A changing the plan back, and so on.

4.4 Ignore

The IGNORE-strategy also assumes that we are working in the time stamp model and that we have a way of computing (sub-) optimal solutions to the static (offline) version of the problem. The main idea of this method is to make sure that the efficiency of an optimal offline-solution computed at a certain point in time be exploited completely. More important even: once we computed an optimal plan it is absolutely predictable how the system will work in the near future.

The way it works is the following: IGNORE again maintains a plan. In contrast to REPLAN, the strategy IGNORE will stubbornly serve requests according to this plan until the plan is finished. Upcoming requests are temporarily ignored and collected in a buffer. When the current plan is finished IGNORE computes a new plan optimizing the service of all not yet served requests.

Although IGNORE might give away optimization potential by temporarily ignoring requests it still exploits optimization. Moreover, the upcoming requests that fit "very well" into the old plan (i.e., with no cost) can be incorporated with no harm.

We illustrate the general purpose strategies FIFO, GREEDY, IGNORE and REPLAN for the Online Traveling Salesman Problem:

Example 23 (Online TSP Revisited). Applying FIFO to the Online Traveling Salesman Problem leads to a tour that visits all cities in the order of appearance. If not required by other constraints this is certainly not the best choice.

The GREEDY-heuristic for the Online Traveling Salesman Problem means the following: At any point in time visit the closest city next. It is known that this can lead to a very inefficient solution. In some practical applications of the Online Traveling Salesman Problem, however, the experience shows that even this simple heuristics is acceptable.

Whenever a new city becomes known the REPLAN-heuristic computes an optimal tour (according to the objective function used to model the cost) visiting all cities known so far. This tour is followed until the next city pops up.

The observed performance depends heavily on the application. In practical instances it maybe necessary to cope with the problem of *system break-downs*: the salesman has to interrupt his work at some point. During the break the number of unserved requests increases, and so does the gain of offline-optimization of all unserved requests. For instance, an automatic storage system where transportation tasks are served by a stacker crane can be modeled as an asymmetric Online Traveling Salesman Problem (see [4,5]). In this specific application unexpected system break-downs of the automatic storage system may occur. During the forced idle period of the server a lot of requests pile up. All these requests can be taken into account by REPLAN when the server resumes. Thus, it is plausible that REPLAN yields a good *recovery method.*

The effort to get the necessary offline-solutions is usually large and not always real-time compliant. This, however, could be achieved in cases where good approximation algorithms (see Section 3.3) exist, like in the metric case (see Example 21).

The method IGNORE waits for the first city to be "released". Then it moves its salesman to that city. Once arrived, it computes an optimal tour through all the cities that have been released during the time the salesman was underway. Then this tour is completely traveled. At the end of the tour, the cities that have become known in the meantime are planned.

Again, the success of this method in realistic systems modeled by variants of this problem is application dependent. Simulation experiments show that in single server systems there usually is a substantial gain in stability and predictability of the system behavior over REPLAN. □

4.5 Chasing the Offline Optimum and Balancing Costs

Suppose that there are n (system-) states $s_1, \dots, s_n$ in which an algorithm can be and that the service cost for a request depends (only) on the current state. Moreover, there is a cost for changing states. (This situation can be stated more formally as a *Metrical Task System*, see [15]).

Upon arrival of a new request r_i, the strategy of *chasing the offline optimum* changes to that state s_j in which the offline optimum for the sequence $r_1, \dots, r_i$ would process r_i. A *balancing costs* type algorithm would change from the current state s to that state s' which minimizes some function of the following two values: (i) the charge for changing from s to s', and (ii) the cost of serving r_i in s'. The most famous representative of the latter class is the *work function algorithm* which has been successfully applied to the theoretical analysis of the Paging Problem and the k-Server Problem [15,35].

5 SIMULATION

One can view simulation as a method of checking industrial system layouts and associated algorithms by an organized sequence of computer based experiments and evaluations. This takes place, of course, on the border line of mathematics and engineering. Therefore, we cannot hope for exact mathematical definitions of all relevant objects in the realm of simulation.

In this section we informally describe the method of *discrete event based* simulation and address issues that may come up during the process of modeling and computing. More elementary information can be found in [45].

5.1 *Why Simulation?*

The theoretical background surveyed in Section 2 leads to mathematical problems of substantial difficulty, even for seemingly easy online optimization problems. On the other hand, the performance guarantees achieved by these methods are often very poor. This renders competitive analysis problematic for most industrial purposes. In this case, evaluation and comparison of the practical performance of online algorithms are necessary.

There are new theoretical developments – one of them in this volume – that provide some improvements in this area; the final decision about which algorithm to choose in practice, however, is usually done on the basis of simulation experiments.

5.2 *Discrete Event Based Simulation*

Simulating an aspect of the real world on a computer requires a quantitative definition of the relevant part of the real world. This is referred to as the *system*. The system may consist of several *components*. In order to investigate waiting time distributions in a supermarket consider, e.g., the check-out area in that supermarket as the system. This system consists of several cashiers, waiting queues, etc.

The part of the real world outside the system is usually called *environment*. Sometimes there is a feed-back between system and environment, and it is at times a difficult modeling issue to find a suitable separation. The system interacts with the environment by producing an output of the system for an input of the environment.

In the area of online optimization we are usually concerned with *dynamic systems*, i.e., the system parameters change over time. For example, the lengths of the lines at the cashiers in the supermarket are not constant. Moreover, in the realm of combinatorial online optimization it is usually possible to find discrete points in time where the system changes its state. Such systems are called *time-discrete*. In the sequel we restrict ourselves to time-discrete systems.

A *simulation model* is a translation of the relevant parameters of the system into mathematical language so that the behavior of the system over time can be investigated by a computer calculation. In this step it is necessary to specify the components and their *attributes* that one would like to keep track of. Very important attributes are *strategies* or *algorithms* that hold information about how components react on system events. Some attributes are time-dependent, some are not. In the supermarket example we could specify a component "cashier" and a component "customer". The attributes for a cashier, e.g., could be *open/closed*, *operator speed*, *length of line*.

The changes of a time-discrete system over time is described by *Events*. First, an event specifies a system transition function that assigns to every possible system state a new system state. Second, it defines a successor function that assigns to every system state a set of succeeding events together with their time of occurrence.

In the supermarket example the event "customer arrives in line at cashier i" can be formalized as follows: for all current system states the new system state incorporates the following changes: the queue at cashier i contains a new customer, and the set of succeeding events is empty.

The event "customer is being served at cashier i" can be defined as follows: the customer is no longer in the corresponding queue, and there is one successor event, namely "customer leaves the system" in five seconds times number of items in shopping cart from now.

Simulation means computing the output of the system (over time) for a given input (over time) of the environment. In *Discrete Event Based Simulation* this is done by dynamically *processing events*, i.e., computing the system states and the successor events until no events are left or a specified time is over. To start the simulation one uses environment input events modeling the input of the environment to the system.

An example of a discrete event based simulation system is the library AMSEL [3]. It was used in the investigations in [29].

5.3 Issues for the Practitioner

The quality of an evaluation of algorithms by means of simulation experiments heavily depends on the input data used. The following ways of generating input data are common:

- Generate data according to a probability distribution (random data).
  Advantages: It is possible to generate an arbitrarily large set of test data.
  Draw-backs: A realistic probability distribution maybe hard to come by.
- Compile data in the system under consideration.
  Advantages: One can adjust parameters of the simulation model by comparing the outcome of the simulation experiments with the outcome in the real world operation.
  Draw-backs: Compiling the data is extremely time-consuming, often it is not clear whether the compilation contains typical or unusual data.

Although we are advertising here the use of simulation for the performance evaluation of online algorithms we are aware of the fact that simulation experiments may be misguiding. It is a nontrivial matter to come up with meaningful and representative simulation tests.

6 Conclusion

More and more industrial decision makers appear to understand the issues coming up in online and real-time systems. Solution techniques are requested in a range of applications which will certainly improve research and development in online and real-time algorithms.

We have introduced competitive analysis as a mathematical method for the evaluation of combinatorial online-algorithms resulting in provable performance guarantees. A shortcoming of this approach is that it does not take into account the real-time requirements that are present in many real-world systems. Moreover, for complex systems and complicated algorithms a rigorous competitive analysis is in most cases impossible.

Thus, using this method on elementary problems that are similar to the given complex problems seems to be the right utilization: it is possible to get an idea about what kind of strategies are promising for real-world systems and why.

There are new developments in the area of theoretical evaluation of online-algorithms [30]; this field is, however, still in its childhood.

Most online-strategies caring about cost efficiency employ offline-algorithms. Here the need for real-time compliant methods is apparent. Theoretical concepts to get a hand on the issue that a solution is computed under circumstances that might have changed when the computation finishes are not yet available. Some achievements are presented in this volume.

After all, up to now there is no way to replace the experience in simulation experiments completely by a purely theoretical concept for evaluation of combinatorial online-algorithms.

REFERENCES

1. S. Albers, *Better bounds for online scheduling*, Proceedings of the 24th Annual ACM Symposium on the Theory of Computing, 1997, pp. 130–139.
2. S. Arora, *Polynomial-time approximation schemes for euclidean TSP and other geometric problems*, Proceedings of the 38th Annual IEEE Symposium on the Foundations of Computer Science, 1997, pp. 2–11.
3. N. Ascheuer, *Amsel – a modelling and simulation environment library*, Online-Documentation available[5].
4. N. Ascheuer, *Hamiltonian path problems in the on-line optimization of flexible manufacturing systems*, Ph.D. thesis, Technische Universität Berlin, 1995.
5. N. Ascheuer, M. Grötschel, N. Kamin, and J. Rambau, *Combinatorial online optimization in practice*, Optima – Mathematical Programming Society Newsletter (1998), no. 57, 1–6.
6. N. Ascheuer, S. O. Krumke, and J. Rambau, *Online dial-a-ride problems: Minimizing the completion time*, Proceedings of the 17th International Symposium on Theoretical Aspects of Computer Science, Lecture Notes in Computer Science, vol. 1770, Springer, 2000, pp. 639–650.
7. G. Ausiello, P. Crescenzi, G. Gambosi, V. Kann, A. Marchetti-Spaccamela, and M. Protasi, *Complexity and approximation. combinatorial optimization problems and their approximability properties*, Springer, 1999.
8. G. Ausiello, E. Feuerstein, S. Leonardi, L. Stougie, and M. Talamo, *Competitive algorithms for the traveling salesman*, Proceedings of the 4th Workshop on Algorithms and Data Structures, Lecture Notes in Computer Science, vol. 955, August 1995, pp. 206–217.

[5] `http://www.zib.de/ascheuer/AMSEL.html`

9. G. Ausiello, E. Feuerstein, S. Leonardi, L. Stougie, and M. Talamo,, *Algorithms for the on-line traveling salesman*, Algorithmica (2001), To appear.
10. B. Awerbuch, Y. Bartal, and A. Fiat, *Distributed paging for general networks*, Proceedings of the 7th Annual ACM-SIAM Symposium on Discrete Algorithms, 1996, pp. 574–583.
11. R. Battiti and G. Tecchiolli, *The reactive tabu search*, ORSA journal on computing **6** (1994), no. 2, 126–140.
12. R. E. Bixby, M. Fenelon, Z. Gu, E. Rothberg, and R. Wunderling, *MIP: Theory and practice closing the gap*, ILOG Technical Report, Presented at 19th IFIP TC7 Conference on System Modelling and Optimization, Cambridge, England, July 1999.
13. U. Blasum, M. R. Bussieck, W. Hochstättler, H. H. Scheel, and T. Winter, *Scheduling trams in the morning*, Mathematical Methods of Operations Research **49** (1999), no. 1, 137–148.
14. M. Blom, S. O. Krumke, W. E. de Paepe, and L. Stougie, *The online-TSP against fair adversaries*, Proceedings of the 4th Italian Conference on Algorithms and Complexity, Lecture Notes in Computer Science, vol. 1767, Springer, 2000, pp. 137–149.
15. A. Borodin and R. El-Yaniv, *Online computation and competitive analysis*, Cambridge University Press, 1998.
16. A. Borodin, S. Irani, P. Raghavan, and B. Schieber, *Competitive paging with locality of reference*, Proceedings of the 23th Annual ACM Symposium on the Theory of Computing, 1991, pp. 249–259.
17. A. Borodin, S. Irani, P. Raghavan, and B. Schieber, *Competitive paging with locality of reference*, Journal of Computer and System Sciences **50** (1995), 244–258.
18. A. Borodin, J. Kleinberg, P. Raghavan, M. Sudan, and D. P. Williamson, *Adversarial queueing theory*, Proceedings of the 23rd Annual ACM Symposium on the Theory of Computing, 1996, pp. 376–385.
19. A. Brooke, D. Kendrick, A. Meeraus, and R. Raman, *GAMS - a user's guide*, GAMS Development Corporatio, 1998.
20. E. Çela, *The quadratic assignment problem. theory and algorithms*, Kluwer Academic Publishers, Dordrecht, 1998.
21. N. Christofides, *Worst-case analysis of a new heuristic for the traveling salesman problem*, Tech. report, Graduate School of Industrial Administration, Carnegie-Mellon University, Pittsburgh, PA, 1976.
22. W. J. Cook, W. H. Cunningham, W. R. Pulleyblank, and A. Schrijver, *Combinatorial optimization*, Wiley Interscience Series in Discrete Mathematics and Optimization, John Wiley & Sons, 1998.
23. J. K. Lenstra E. H. L. Aarts, *Local search in combinatorial optimizatio*, Wiley, 1997.
24. A. Fiat and G. J. Woeginger (eds.), *Online algorithms: The state of the art*, Lecture Notes in Computer Science, vol. 1442, Springer, 1998.
25. R. Fourer, D. M. Gay, and B. W. Kernighan, *AMPL: A modeling language for mathematical programming*, Duxbury Press, Brooks/Cole Publishing Company, 1993.
26. M. R. Garey and D. S. Johnson, *Computers and intractability (a guide to the theory of NP-completeness)*, W.H. Freeman and Company, New York, 1979.
27. R. L. Graham, *Bounds for certain multiprocessing anomalies*, Bell System Technical Journal **45** (1966), 1563–1581.
28. Ronald L. Graham, *Bounds on multiprocessing timing anomalies*, SIAM Journal on Applied Mathematics **17** (1969), 263–269.
29. M. Grötschel, S. O. Krumke, and J. Rambau, *Forschungsartikel*, ch. This book, Springer, 2001.

30. D. Hauptmeier, S. O. Krumke, and J. Rambau, *The online dial-a-ride problem under reasonable load*, Proceedings of the 4th Italian Conference on Algorithms and Complexity, Lecture Notes in Computer Science, vol. 1767, Springer, 2000, pp. 125–136.
31. D. S. Hochbaum (ed.), *Approximation algorithms for NP-hard problems*, PWS Publishing Company, 20 Park Plaza, Boston, MA 02116–4324, 1997.
32. S. Irani, A. Karlin, and S. Phillips, *Strongly competitive algorithms for paging with locality of reference*, Proceedings of the 3rd Annual ACM-SIAM Symposium on Discrete Algorithms, 1992, pp. 228–236.
33. A. Karlin, M. Manasse, L. Rudolph, and D. D. Sleator, *Competitive snoopy caching*, Algorithmica **3** (1988), 79–119.
34. E. Koutsoupias and C. Papadimitriou, *Beyond competitive analysis*, Proceedings of the 35th Annual IEEE Symposium on the Foundations of Computer Science, 1994, pp. 394–400.
35. E. Koutsoupias and C. Papadimitriou, *On the k-server conjecture*, Journal of the ACM **42** (1995), no. 5, 971–983.
36. E. L. Lawler, J. K. Lenstra, A. H. G. Rinnooy Kan, and D. B. Shmoys (eds.), *The traveling salesman problem*, Wiley-Interscience series in discrete mathematics, John Wiley & Sons, 1985.
37. J. S. B. Mitchell, *Guillotine subdivisions approximate polygonal subdivisions: A simple polynomial-time approximation scheme for geometric tsp, k-mst, and related problems*, SIAM Journal on Computing **28** (1999), no. 4, 1298–1309.
38. R. Motwani, *Lecture notes on approximation algorithms: Volume I*, Tech. Report CS-TR-92-1435, Department of Computer Science, Stanford University, Stanford, CA 94305-2140, 1992.
39. R. Motwani and P. Raghavan, *Randomized algorithms*, Cambridge University Press, 1995.
40. C. H. Papadimitriou and K. Steiglitz, *Combinatorial optimization*, Prentice-Hall, Inc., 1982.
41. C. Phillips, C. Stein, E. Torng, and J. Wein, *Optimal time-critical scheduling via resource augmentation*, Proceedings of the 29th Annual ACM Symposium on the Theory of Computing, 1997, pp. 140–149.
42. K. Pruhs and B. Kalyanasundaram, *Speed is as powerful as clairvoyance*, Proceedings of the 36th Annual IEEE Symposium on the Foundations of Computer Science, 1995, pp. 214–221.
43. C. R. Reeves, *Modern heuristic techniques for combinatorial problems*, McGraw-Hill, 1995.
44. S. Séguin, J.-Y. Potvin, M. Gendreau, T. G. Crainic, and P. Marcotte, *Real-time decision problems: An operational research perspective*, Journal of the Operational Research Society **48** (1997), 162–174.
45. H.-J. Siegert, *Simulation zeitdiskreter systeme*, Oldenbourg, München, Wien, 1991.
46. D. D. Sleator and R. E. Tarjan, *Amortized efficiency of list update and paging rules*, Communications of the ACM **28** (1985), no. 2, 202–208.
47. D. Steenken, T. Winter, and U. T. Zimmermann, *Stowage and transport optimization in ship planning*, (2001).
48. V. Vazirani, *Approximation algorithms*, Springer, 2001.
49. A. P. A. Vestjens, *On-line machine scheduling*, Ph.D. thesis, Eindhoven University of Technology, Eindhoven, The Netherlands, 1994.
50. T. Winter, *Online and real-time dispatching problems*, Ph.D. thesis, Technical University Braunschweig, 1999.

51. T. Winter and U. T. Zimmerman, *Real-time dispatch of trams in storage yards*, Annals of Operations Research **96** (2000), 287–315.

Online Optimization of Complex Transportation Systems

Martin Grötschel[1], Sven O. Krumke[1], and Jörg Rambau[1]

Konrad-Zuse-Zentrum für Informationstechnik Berlin (ZIB), Germany

Abstract This paper discusses online optimization of real-world transportation systems. We concentrate on transportation problems arising in production and manufacturing processes, in particular in company internal logistics. We describe basic techniques to design online optimization algorithms for such systems, but our main focus is decision support for the planner: which online algorithm is the most appropriate one in a particular setting? We show by means of several examples that traditional methods for the evaluation of online algorithms often do not suffice to judge the strengths and weaknesses of online algorithms. We present modifications of well-known evaluation techniques and some new methods, and we argue that the selection of an online algorithm to be employed in practice should be based on a sound combination of several theoretical and practical evaluation criteria, including simulation.

1 Introduction

The strategic planning of complex transportation systems such as public transportation networks, automatically guided vehicles in warehouses, etc. has received a considerable amount of attention in the last decade. *Strategic planning* is the stage of system design where an object (e.g., a telecommunication network) is designed that will remain static for a certain planning period (the network topology, and edge capacities will not change) such that a few control parameters (e.g., routing and switching) will allow an (almost) optimal handling of all input data (within a certain realistic or predicted range). The system itself is usually not yet operational when this "strategic optimization" takes place. Here, methods of *offline optimization* apply. The increasing computing power and significant advances in traditional optimization techniques have resulted in substantial savings of resources in this area.

Despite many successes of this approach, e.g., for transportation systems, it has turned out that achieving savings also requires an *optimized operational control*. Such a control involves actions to be executed while a system is running; i.e., input data arise over time, have to be processed, and (irrevocable) decisions have to be made before all input data are known. This means that methods of *online optimization* have to be employed. In many cases decision making has to satisfy certain *real-time requirements*: every decision has to be made within strict time limits.

In this paper, we survey some new methods (beyond standard competitive analysis) to obtain decision support for the choice of online algorithms in real-world transportation systems. In each case, we are looking for a "good" online control on the basis of online algorithms. The methods discussed are *competitive analysis against restricted adversaries* (a variant of competitive analysis where the offline adversary is given less power), *analysis under Δ-reasonable load* (we compare the cost of the online algorithm to a certain property of the input), *a-posteriori-analysis* (we perform an approximate, instance wise competitive analysis to compute a lower bound on the unavoidable cost), and *comparative simulation* (we compare algorithms that run simultaneously in simulation experiments).

These concepts are employed along with standard competitive analysis in real-world examples. We indicate which combination of methods could support decisions best.

The rest of the paper is structured as follows: We start out by sketching our real-world examples in Section 2. In Section 3 we present the above mentioned evaluation methods for online algorithms. In Section 4 the applications of traditional and new methods to real-world systems is discussed. Section 5 summarizes what we consider the key findings of our research.

2 Four Real-World Examples and why they Raise Questions

In this section we introduce four real-world online optimization problems. One common feature is the difficulty to evaluate online algorithms for them.

The first example is the automated stacker crane in a production plant of Siemens Nixdorf Informationssysteme AG (SNI). The question is in which order the stacker crane should perform storage and retrieval operations so as to minimize the unloaded travel time. We show that for the related objective "minimize the makespan" (the time the system needs to serve a set of requests) we find a $5/2$-competitive algorithm. This is the REPLAN-heuristics already discussed in [10]. This algorithm is, however, not competitive with respect to the minimization of the total unloaded travel distance. Shall we use this algorithm anyway?

The second example studies a system of automated guided vehicles for commissioning greeting cards in a large distribution center of Herlitz PBS AG, Falkensee, one of the main distributors of office supply in Europe. Orders specifying a combination of greeting card sets have to be assigned to vehicles. These must stop at the shelf positions where the corresponding cards have to be collected ("order picking"). The question is how orders should be assigned to vehicles so that the total number of stops over all vehicles is minimized. It turns out that competitive analysis tells us nothing about which algorithm to choose in practice. For a greatly simplified problem we show that competitive analysis is even in favor of an intuitively senseless algorithm. Is there an evaluation method that proves dumb algorithms to be dumb?

The next example is a pallet elevator in the same distribution center. In this case it was already difficult to isolate a single objective function to be optimized. We decided to consider several objectives: We want to guarantee a small average and/or a small maximal flow time over a set of pallets requesting transportation. It turns out that for these objectives there is no competitive algorithm, mainly because the cost of an offline solution cannot be bounded from below. Can we evaluate algorithms without using a lower bound on the offline cost?

Finally, we investigate the integrated elevator system plus conveyor belt that distributes the pallets among the elevators. We find out that a similar analysis as in the single elevator case is still valid; the improvements, however, are leveled off by the conveyor control. Therefore, we studied an integrated optimization model for the combined elevator-conveyor system. Does this help to improve the overall performance of the system?

We will give answers to the above questions in Section 4 after we have introduced our "evaluation toolbox" in Section 3. Some of the answers are quite satisfactory, others show the need for further research.

3 MODELING AND EVALUATION TECHNIQUES

In this section, we present a sequence of methods to analyze the performance of online algorithms. The methods are ordered by decreasing mathematical strength, that is to say, the first method – if successful – yields the most rigorous analysis of the ones in this section, the last one is merely experimental. (Classical competitive analysis as described in [10] would belong to the very beginning of the section.)

3.1 Competitive Analysis with Restricted Adversaries

In restricting the class of algorithms for the adversary, one attempts to deal with the (justified) objection – frequently encountered against competitive analysis – concerning the unrealistic power of the adversary against which performance is measured. In standard competitive analysis the adversary is an optimal offline algorithm which has complete knowledge about the whole input in advance. There have been a number of approaches in the literature to devise "more realistic" adversary models for specific problems than the omnipotent standard offline adversary.

For the exposition we consider the Online Traveling Salesman Problem ONLINETSP on the non-negative real numbers $\mathbb{R}_0^+$ endowed with the Euclidean metric (see [10, Example 4] for the ONLINETSP in general metric spaces). The origin of the salesman is the point 0. In the ONLINETSP requests for visits to cities (points in a metric space) arrive online while the salesman is traveling. The salesman moves at unit speed and starts and ends his work at the origin 0. The objective is to find a route for the salesman which finishes as early as possible.

Each *request* is a pair $\sigma_i = (t_i, x_i)$, where $t_i \in \mathbb{R}$ is the time at which request σ_i is released (becomes known to an online algorithm), and $x_i \in \mathbb{R}_0^+$ is the point requested to be visited. It is assumed that an online algorithm does neither have information about the time when the last request is released nor about the total number of requests. An online algorithm must base its decisions at time t solely on the requests released up to time t.

Notice that the offline version of the ONLINETSP in $\mathbb{R}_0^+$ can be solved very easily even in the presence of release times (the problem is almost trivial!). However, in the online case, there does not exist an algorithm that always finds an optimal solution. More specifically, it can be shown that there is no deterministic online algorithm that achieves a competitive ratio smaller than $3/2$. The competitive ratio of 3/2 is achieved by the following very natural and simple strategy MRIN (see [7] for the proofs):

Strategy MRIN("Move-Right-If-Necessary") If a new request is released and the request is to the right of the current position of the server operated by MRIN, then the MRIN-server starts to move right. The server continues to move right as long as there are yet unserved requests to the right of the server. If there are no more unserved requests to the right, then the server moves towards the origin 0. □

In the lower bound construction the offline adversary abuses his power in the sense that he can move to points where he knows a request will pop up without revealing the request to the online server before reaching the point. This has motivated the concept of a "fair adversary" in the ONLINETSP: A fair adversary always keeps its server within the convex hull of the requested points released so far. As shown in [7] this adversary model indeed allows for lower competitive ratios. For instance, the above mentioned 3/2-competitive algorithm MRIN against the conventional adversary is 4/3-competitive against the fair adversary. In addition, one can prove the following:

Theorem 1 ([7]). *There exists an online algorithm for the* ONLINETSP *in* $\mathbb{R}_0^+$ *with competitive ratio* $\frac{1+\sqrt{17}}{4} \approx 1.28$ *against a fair adversary. Moreover, no deterministic online algorithm can achieve a competitive ratio smaller than* $\frac{1+\sqrt{17}}{4}$ *against the fair adversary.* □

The use of a restricted adversary falls within the concept of *comparative analysis*, which was introduced by Koutsoupias and Papadimitriou [14]. The authors compare the performance of an online algorithm for the Paging Problem with that of the best paging algorithm having limited lookahead. Let Π be a minimization (online) problem. The *comparative ratio* of an algorithm ALG for Π relative to a class $\mathcal{B}$ of algorithms is defined as the worst case ratio between the solution cost produced by ALG and the best solution produced by an algorithm in $\mathcal{B}$. If $\mathcal{B}$ is the class of all offline algorithms for Π, then the comparative ratio reduces to the standard competitive ratio.

The comparative ratio has also been studied in the context of online financial problems. For most of these problems the standard adversary also appears to be too strong. To obtain meaningful (theoretical) results about the performance, e.g., of online portfolio selection algorithms, a comparison with a restricted class of offline algorithms is used. We refer to [8, Chapter 14] for details.

3.2 Reasonable Load

This concept was motivated by the problem of minimizing the maximal or average flow time of pallets transported by an elevator. Such a system can be modeled by the so-called *online dial-a-ride problem* ONLINEDARP. The concept of reasonable load also works in a more general setting. However, we do not want to go too much into abstraction in this paper, and we restrict our attention to ONLINEDARP, which we explain in the sequel.

We are given a metric space (X, d) with a special point $o \in X$ (the origin). Requests are triples $r = (t, a, b)$, where a is the start point of a transportation task, b its end point, and t its release time, which is – in this context – the time where r becomes known to an online algorithm. A *transportation move* is a quadruple $m = (t, x, y, R)$, where x is the starting point, y the end point, and t the starting time, while R is the set (possibly empty) of requests the server has loaded during the move. We say in this case, the move m *carries* R. The *arrival time* of a move is the sum of its starting time and $d(x, y)$. A *(closed) transportation schedule* is a sequence $(m_1, m_2, \dots)$ of transportation moves such that

- the first move starts in the origin o;
- the starting point of m_i is the end point of m_{i-1};
- the starting time of m_i carrying R is no earlier than the maximum of the arrival time of m_{i-1} and the release times of all requests in R (it may be later, though);
- the last move ends in the origin o.

An *online algorithm* for ONLINEDARP has to move a server in X so as to fulfill all released transportation tasks without preemption (i.e., once an object has been picked up it is not allowed to be dropped at any other place than its destination), while it does not know anything about requests that come up in the future. In order to plan the work of the server, the online algorithm may maintain a preliminary (closed) transportation schedule for all known requests, according to which it moves the server. A posteriori, the moves of the server induce a complete transportation schedule that may be compared to an offline transportation schedule that is optimal with respect to some objective function (competitive analysis). For a detailed set-up see [4].

Recall that the *flow time* of a request is the difference between its completion time and its release time, while the *waiting time* is the difference between its service starting time and its release time. In the sequel, we are concerned with the following objectives:

- Minimize the *makespan* (also called the *completion time*) for the given set of requests. This is the time the server needs to fulfill all the transportation tasks.
- Minimize the *maximal flow time* (or *waiting time*) of the requests.
- Minimize the *average flow time* (or *waiting time*).

We will consider the online heuristics REPLAN and IGNORE from [10]. Since we did not choose a particular objective function yet we need to specify according to which objective function REPLAN and IGNORE will solve the corresponding offline problems. We will evaluate REPLAN- and IGNORE-heuristics that use a different objective for the local optimization than the one that is to be minimized globally in the online problem.

Thus, for an arbitrary objective function *obj* we denote by REPLAN^{obj} resp. IGNORE^{obj} the following online heuristics:

REPLAN^{obj} Follow the current plan. Whenever a new request becomes available compute a new plan minimizing *obj* starting at the current position.

IGNORE^{obj} Follow the current plan; while executing it collect upcoming requests in a buffer. When done and there are non-served requests in the buffer compute a new plan for all these requests minimizing *obj*.

The motivation to consider the concept of reasonable load in this situation was two-fold.

First, competitive analysis of ONLINEDARP provides the following [4]:

- The two online heuristics IGNORE^{makespan} and REPLAN^{makespan} are both 5/2-competitive for the goal of minimizing the *makespan* of the schedule.
- For the tasks of minimizing the *maximal (or average) waiting time* or the *maximal (or average) flow time* there can be no algorithm with constant competitive ratio.
- In particular, the algorithms IGNORE^{makespan} and REPLAN^{makespan} that repeatedly minimize the *makespan* of all known requests have an unbounded competitive ratio for the overall task of minimizing the maximal or average *flow time*.

Second, in simulation studies a fundamental difference in the behavior of IGNORE and REPLAN was observed: the maximal flow times on similar inputs produced by REPLAN varied a lot while the ones produced by IGNORE were better predictable. The concept of reasonable load was developed to find a mathematical explanation of this phenomenon.

We start with some useful notation.

Definition 2. The *offline version* of a request $r = (t, a, b)$ is the request

$$r^{offline} := (0, a, b).$$

The *offline version* of a request set R is the request set

$$R^{offline} := \left\{ r^{offline} \ : \ r \in R \right\}. \square$$

An important characteristic of a request set with respect to system load considerations is the time period in which it is released.

Definition 3. Let R be a finite request set for ONLINEDARP. Let the release time of a request r be denoted by $t(r)$. The *release span* $\delta(R)$ of R is defined as

$$\delta(R) := \max_{r \in R} t(r) - \min_{r \in R} t(r). \square$$

Provably good algorithms exist for the makespan and the weighted sum of completion times. How can we make use of these algorithms in order to get performance guarantees for minimizing the maximum (average) waiting (flow) times? We suggest a way of characterizing request sets which we want to consider "reasonable".

In a continuously operating system we wish to guarantee that work can be accomplished at least as fast as it is presented. In the following we propose a mathematical set-up that models this idea in a worst-case fashion. Since we are always working on finite subsets of the whole request set the request set itself may be infinite, modeling a continuously operating system.

We start by relating the release spans of finite subsets of a request set to the time we need to fulfill the requests.

Definition 4. Let R be a request set for the ONLINEDARP. A weakly monotone function

$$\mathsf{f}\colon \begin{cases} \mathbb{R} \to \mathbb{R}, \\ \delta \mapsto \mathsf{f}(\delta); \end{cases}$$

is a *load bound* on R if, for any $\delta \in \mathbb{R}$ and any finite subset S of R with $\delta(\mathsf{S}) \leq \delta$, the makespan $\mathsf{OPT}^{makespan}(\mathsf{S}^{offline})$ of the optimum schedule for the offline version $\mathsf{S}^{offline}$ of S is at most $\mathsf{f}(\delta)$. In formula:

$$\mathsf{OPT}^{makespan}(\mathsf{S}^{offline}) \leq \mathsf{f}(\delta).\square$$

Remark 5. If the whole request set R is finite then there is always the trivial load bound given by the makespan of R. For every load bound f, we may set $\mathsf{f}(0)$ to be the maximum completion time we need for a single request, since nothing better can be achieved. □

A stable situation would be characterized by a load bound equal to the identity on $\mathbb{R}$. In that case we would never get more work to do than we can accomplish, even if we had an optimal offline algorithm at hand.

If R has a load bound equal to a function id/ρ, where id is the identity and where $\rho \geq 1$, then ρ measures the "tolerance" of the request set: An algorithm that is by a factor ρ worse than optimal will still accomplish all the work that it gets. However, we cannot expect that the identity (or any linear function) is a load bound for ONLINEDARP because of the following observation: a request set consisting of one single request has a release span of 0 whereas in general it takes non-zero time to serve this request. In the following definition we introduce a parameter describing how far a request set is from being load-bounded by the identity.

Definition 6. A load bound f is called *(Δ,ρ)-reasonable* for some $\Delta, \rho \in \mathbb{R}_+$ if

$$\mathsf{f}(\delta) \leq \frac{\delta}{\rho} \quad \text{for all } \delta \geq \Delta$$

A request set R is *(Δ,ρ)-reasonable* if it has a (Δ,ρ)-reasonable load bound. For $\rho = 1$, we say that the request set is *Δ-reasonable*, and we call a request set or a load bound *reasonable* if it is (Δ,ρ)-reasonable for some $\Delta, \rho \in \mathbb{R}_+$. □

In other words, a load bound is *(Δ,ρ)-reasonable*, if it is bounded from above by $1/\rho \cdot id(x)$ for all $x \geq \Delta$ and by the constant function with value $1/\rho\Delta$ otherwise.

Remark 7. If Δ is sufficiently small so that all request sets consisting of two or more requests have a release span larger than Δ then the first-come-first-serve strategy suffices to ensure that there are never more than two unserved requests in the system. Hence, the request set does not require "scheduling" the requests in order to provide for a stable system. (By "stable" we mean that the number of unserved requests in the system does not become arbitrarily large.) □

Resonable load is a plausible restriction:

Observation 8 (Justification of Reasonable Load). *Assume that a request set for* ONLINEDARP *is not reasonable. Then the following holds: For all* $\Delta \geq 0$ *there is a request set with release span at least* Δ *whose offline makespan is larger than its release span.*

In other words: no matter how long one collects requests there is provably no method to accomplish their service in a time equal to the collection time.

Finally, we state the theorem that mathematically shows the (somewhat surprising) fundamental difference of IGNORE^{makespan} and REPLAN^{makespan} on ONLINEDARP. (See [11] for a proof.)

Theorem 9. *For* ONLINEDARP *under* Δ*-reasonable load,* IGNORE^{makespan} *guarantees a maximal and an average flow time of at most* 2Δ*, whereas the maximal and the average flow time of* REPLAN^{makespan} *are unbounded.*

In Sections 4.3 and 4.4 we present practical applications where an analysis under reasonable load is possible.

3.3 A-Posteriori-Analysis

Competitive analysis – even in the case of existing competitiveness results – does often not provide performance guarantees that appear convincing in an efficiency oriented industrial environment. Consider a statement such as "The solution produced is in each and every situation at most 3 times worse than the optimum". Will a user be happy to hear that? Such a result is too weak in terms of the performance ratio and too strong in the sense that it covers too many (from a customer's point of view probably irrelevant) situations.

The same problem occurs in the framework of approximation algorithms: a performance guarantee for all instances of a problem is often not required. One approach that made combinatorial optimization methods have impact in real life was the delivery of *instance-wise* performance guarantees via the computation of so-called *lower bounds* for the very special instance of the minimization problem to be solved in a particular situation.

Lower bounds can usually be derived by relaxing side constraints of a problem. The most prominent relaxation technique in combinatorial optimization is to relax the integrality constraints, thereby transforming notoriously difficult (Mixed) Integer Programs into efficiently solvable Linear Programs [10]. Optimal solutions of these may be easier to come by, and an optimal solution of the original problem cannot be cheaper than the one of the relaxed problem. On the other hand, the value of any feasible solution to the original problem yields an *upper bound* for the optimal solution. The gap between lower and upper bound at any stage of the optimization process provides, thus, an instance specific quality guarantee: the difference between the objective function values of the current feasible solution and a presently unknown optimal solution is not bigger than this gap.

In this framework the role of fast approximation algorithms is to provide for good feasible starting solutions. Good initial solutions often help to close the gap between lower and upper bounds fast and, thus, help to speed up the optimization process.

One can similarly compute lower and upper bounds for a special instance of an online optimization problem. This leads to an instance-wise competitive analysis. Since, in an online situation, a special instance is not known in advance, this kind of analysis can only be applied after all decisions have been made. Therefore, this approach is called *a-posteriori analysis*.

We now state an observation that shows what a-posteriori-analysis can achieve. We concentrate on online optimization in the time stamp model (see [10] for details). We may assume w.l.o.g. that all time stamps are positive, and we assume also that the way how a request sequence is served by an online algorithm does not influence this sequence. (This assumption is not always satisfied in real systems, since after observing how an algorithm has handled the first elements of a request sequence, the remaining requests may be altered or their order may be changed.)

Suppose that I is an instance of an online optimization problem in the time stamp model and that A is an online algorithm for this problem. Denote by $\mathsf{A}(I)$ the value of the solution A produces on I. Denote by J the corresponding instance of the offline optimization problem induced by I where all requests are known in advance and where a feasible solution has to respect all release times. Denote by K the corresponding instance of the offline optimization problem induced by I where all time stamps are removed (set to zero). Denote the optimal solution values of J and K by $\mathsf{OPT}(J)$ and $\mathsf{OPT}(K)$, respectively.

Then the following simple observation can be made.

Observation 10 (Justification of A-posteriori Analysis). *Let* I, J, K *be as above. Then, under the above assumptions, there exist real numbers* $c(I,\mathsf{A})$ *and* $c'(I,\mathsf{A})$, *depending on* I *and on the online algorithm* A, *satisfying* $c'(I,\mathsf{A}) \geq c(I,\mathsf{A}) \geq 1$, *such that*

$$\mathsf{OPT}(K) \leq \mathsf{OPT}(J) \leq \mathsf{A}(I) = c(I,\mathsf{A})\mathsf{OPT}(J) = c'(I,\mathsf{A})\mathsf{OPT}(K).$$

The above chain of equations and inequalities yields two versions of instance-wise competitive analysis depending on the chosen relaxations J or K. Usually, the quality guarantee $c(I,\mathsf{A})$ is reported as the relative gap

$$J\text{-}\mathrm{GAP}_{\mathsf{A}} := c(I,\mathsf{A}) - 1 = \frac{\mathsf{A}(I) - \mathsf{OPT}(J)}{\mathsf{OPT}(J)}.$$

It is, however, not clear how the instance J can be solved. The corresponding combinatorial offline problems may, in fact, be hard. Even worse, it is often not apparent how to formulate these offline problems properly. The reason is that online problems in real life may come along with implicit restrictions that are difficult to model. In this sense, online problems coming from practice are sometimes "ill-posed". In such cases, one has to relax further side constraints in addition to assuming full knowledge of the input sequence in the beginning. The resulting offline problems may then turn out to be useless in practice because of rather poor instance specific gaps. (Even for the relaxation K this is often the case.)

Thus, a-posteriori analysis is often used as follows: find relaxations between J and K that model the online restrictions as faithfully as possible and replace $\mathsf{OPT}(J)$ by the optimal objective function value of this modified problem. An example of this technique can be found in 4.1.

3.4 Comparative Simulation

The draw-back of a-posteriori analysis is that all decisions have irreversibly been made when the analysis of these decisions is available. One way out is testing the system behavior in a simulation experiment. An a-posteriori analysis can be made for every possible online algorithm. If the data used for the runs of the simulation system is "typical enough" then one can hope that a strategy whose gap in the sense of Observation 10 is convincingly small will behave well in reality.

Sometimes even this is too much to ask for: even in an instance-wise analysis the optimal offline algorithm may be too strong in the sense that the computed gap is quite large for every conceivable, non-clairvoyant online strategy. Then we are left with a comparison of online algorithms in simulation experiments.

One feature that makes this (somewhat "soft") method valuable is that evaluation is not limited to the computation of a single scalar objective function. Visualization of the system behavior may, e.g., help to grasp the influence of various online strategies from different perspectives: efficiency, stability, predictability, maybe others. Some of these aspects are very difficult to hard-code in a mathematical model so that the evaluation of simulation experiments by experienced human operators is still one of the most commonly accepted ways of evaluating online algorithms. We describe simulation experiments in all of our applications in Section 4.

4 The Toolbox in Action

In this section we apply the methods outlined in Section 3 including standard competitive analysis to the real-world problems sketched in Section 2. We describe the systems and the corresponding mathematical models in more detail, show that classical methods of evaluation of online algorithms are not sufficient, and apply combinations of the methods from Section 3. Where a greater level of detail is beyond the scope of this paper we provide references to the original research articles.

4.1 Automated Stacker Cranes

Siemens Nixdorf Informationssysteme AG (SNI) maintains a production plant in which all their personal computers (PCs) and related products are assembled. Parts are brought into one of six automatic storage systems (AUSS). The AUSS serve as material buffers between the receiving area and the assembly lines located at each side of the AUSS. For each of the AUSS, there is one stacker crane fulfilling transportation tasks between the receiving buffer, the storage locations, and the buffers for the assembly line. (For a more detailed description of the layout, see [1].) The goal is to minimize the unloaded travel time of the stacker crane.

Mathematical Models

If we were to minimize the total travel time (makespan) of the stacker crane then our problem would be known as the *online stacker crane problem* ONLINESCP, a special case of the ONLINEDARP, explained in Section 3.2. Here we are concerned with a slightly different objective function.

The offline problem without release times can be modeled as an Asymmetric Traveling Salesman Problem (ATSP). An instance of ATSP consists of a complete directed graph $D = (V, A_n)$. Each node in V corresponds to a transportation task, and the the weight of the arc from v to w corresponds to the travel time from the end point of task v to the starting point of task w.

If release times have to be taken into account we are concerned with an *Asymmetric Traveling Salesman Problem* with *release times*, a special case of the Asymmetric Traveling Salesman Problem with *time windows*, the ATSPTW: here, for each request r, there is a time window $[e_r, \ell_r]$ given with a release time (*e*arliest possible start of service) e_r and a deadline (*l*atest possible completion of service) ℓ_r.

We investigated the ATSPTW because there were given deadlines for the service of requests anyway.

Time windows impose precedence constraints on the order in which the requests are served. Relaxing the time windows of an instance of the ATSPTW to the corresponding precedence constraints yields an instance of the so-called *Sequential Ordering Problem* SOP: here we are given a partial order on the set of requests and we try to find a shortest tour through all requests respecting the given partial order.

The problems ATSP, SOP, and ATSPTW are NP-hard. While much attention had already been paid to the investigation of the ATSP a thorough polyhedral study of SOP and ATSPTW was carried out for the first time in [6]. Those results were later strengthened in [1].

The goal was the design of a branch&cut algorithm able to solve on the one hand typical instances of the ATSP used for the REPLAN online heuristic and on the other hand the larger SOP resp. ATSPTW instances used for the a-posteriori analysis of several online heuristics (see Section 3.3). In the following we summarize the achievements for the SOP as an example for the polyhedral investigations contained in this article.

There are related results for the ATSPTW. We do not include them here since they are of similar nature and their statement would not shed more light on the principle situation. The interested reader may want to check again [1].

Let us start with the graph theoretic formulation of SOP. Recall that we are given a complete directed graph $D = (V = \{1, \dots, n\}, A_n)$ on n nodes with non-negative arc costs $c_{ij} \geq 0$. Moreover, in the SOP we are given a partial order "$\prec$" on V with $1 \prec i \prec n$ for all $i \in V$, w.l.o.g. A feasible solution to the SOP is a set of arcs forming a path that visits all nodes in V exactly once and that visits node i before node j whenever $i \prec j$. The goal is to find a feasible solution with minimal total arc costs.

There are several possibilities to formulate the SOP as an integer program. The polyhedral model chosen here is the following. We define the *feasible arc set* A as follows:

$$A := A_n \setminus (\{(j,i) \in A_n \ : \ i \prec j\} \cup \{(i,k) \in A_n \ : \ \exists j \in V : i \prec j \vee j \prec k\}).$$

For all feasible arcs $(i,j) \in A$ we introduce binary arc variables x_{ij} meaning that $x_{ij} = 1$ if and only if arc (i,j) is chosen to be in the solution.

With the notation

$$\begin{aligned} x(B) &:= \sum_{(i,j)\in B} x_{ij} && \text{for } B \subseteq A,\\ A(W) &:= \{(w,w') \in A \ : \ w,w' \in W\} && \text{for } W \subseteq V,\\ \delta^+(i) &:= \{(i,j) \ : \ j \in V \setminus \{i\}\},\\ \delta^-(i) &:= \{(j,i) \ : \ j \in V \setminus \{i\}\},\\ x(i : W) &:= \{(i,w) \in A \ : \ w \in W\},\\ x(W : i) &:= \{(w,i) \in A \ : \ w \in W\} \end{aligned}$$

an integer programming formulation of the SOP can be stated as follows:

$$\begin{aligned} &\min c^T x\\ \text{s.t.}\quad & x(\delta^-(i)) = 1 && \forall i \in V \setminus \{1\} && (1)\\ & x(\delta^+(i)) = 1 && \forall i \in V \setminus \{n\} && (2)\\ & x(A(W)) \leq |W| - 1 && \forall W \subset V, 2 \leq |W| && (3)\\ x(j : W) + x(A(W)) + x(W : i) &\leq |W| && \forall i \prec j, W \subseteq V \setminus \{i,j\}, W \neq \emptyset && (4)\\ & x_{ij} \in \{0,1\} && \forall (i,j) \in A. && (5) \end{aligned}$$

The object of study is the *sequential ordering polytope* SOP defined as

$$\text{SOP}(n, \prec) := \text{conv}\left\{x \in \mathbb{R}^A \ : \ x \text{ satisfies (1)–(5)}\right\}$$

This polytope had already been studied in [6], where new inequalities such as the predecessor/successor inequalities were derived. The following theorem summarizes the new results achieved in our project group. For details see [1].

Theorem 11 (Offline Problems – Polyhedral Study). *For* $\mathrm{SOP}(n, \prec)$ *the following hold:*

(i) If "$\prec$" is obeys a certain regularity condition then the dimension of $\mathrm{SOP}(n, \prec)$ *equals* $|A| - 2n + 3 + |F|$, *where* F *is the set of nodes whose position in the path is fixed by "$\prec$".*
(ii) There are three types of new valid inequalities for $\mathrm{SOP}(n, \prec)$*: the strengthened* D_3*-inequalities, the strengthened* T_k*-inequalities, and the strengthened two-matching constraints.*

We refrain from explicitely listing the inequalities here because the overhead in notation would not pay off given the purpose of this article. The corresponding results on the ATSPTW can be found in [2].

Evaluation of Algorithms

The online version ONLINEATSP of the ATSP is defined in the same way as the ONLINETSP, except that the distances are not symmetric. A competitive analysis of the ONLINEATSP with the objective to minimize unloaded travel time cannot provide additional insight. The reason for this is the following: one can find request sequences that can be served by an offline algorithm without unloaded travel time and that incur a positive cost for any online algorithm. Thus, the competitiveness ratio would be infinite: not particularly helpful.

If one, however, replaces the objective "minimize total unloaded travel time" by the objective "minimize total travel time (makespan)" then – as we mentioned already – we are concerned with a special case of the ONLINEDARP. Note that these objective functions are equivalent in the sense that their function values only differ by an additive constant and that, therefore, the sets of optimal solutions are equal. From the point of view of competitive analysis, however, this change in the objective makes a huge difference.

As an application of a result in [4] we mention the following (see Theorem 20):

Theorem 12 (Competitive Analysis). REPLAN *is* $5/2$*-competitive for the problem of minimizing the makespan of the stacker crane.*

Such a performance guarantee does not really help a decision maker. Therefore, it does make sense to evaluate the REPLAN-strategy by other means. An a-posteriori analysis was also made: we investigated the ATSP, the SOP, and the ATSPTW as relaxations in the spirit of Section 3.3.

Observation 13 (A-Posteriori Analysis). *Real data sets from SNI provided the following a-posteriori analysis for the* REPLAN*-strategy repeatedly solving the* ATSP *of all known requests:*

(i) The online solution is 46%–120% worse than an optimal a-posteriori solution for the ATSP.
(ii) The online solution is 24%–98% worse than an optimal a-posteriori solution for the SOP.

(iii) The online solution is 3%–72% worse than an optimal a-posteriori solution for the ATSPTW.

Since these gaps are not small enough to convince decision makers to use the REPLAN-heuristics, simulation experiments were made.

Observation 14 (Comparative Simulation). *On real data sets,* REPLAN *slightly outperforms other online heuristics such as best insertion heuristics. The at SNI previously used priority strategy with* FIFO *as a tie breaker performs roughly 50% worse than* REPLAN*; the* FIFO *priority algorithm is no better than a random sequencing of request.*

Thus, the conclusion was to implement the REPLAN-heuristic.

Implemented Solution and Practical Impact

Although the subproblems to be solved within the REPLAN-heuristics are NP-hard, there are codes available so that the REPLAN-heuristics can be used in real-time situations in practice. In order to obtain an any-time algorithm [10], we implemented an optimization process working in three phases:

Phase 1: Perform cheapest insertion (BESTFIT).
Phase 2: Run a random insertion. Then pick the winner of Phase 1 and 2.
Phase 3: Solve the ATSP to optimality (branch&bound from [2]) and replace the old sequence completely by the optimal one (REPLAN).

Phase 1 runs in time linear in the number of requests and is always completed. For the typical problem sizes that occur in our application (the number of requests is less than 60) this is done in fractions of a second. Even Phase 3 could always be completed within a few seconds. If the stacker crane becomes idle before Phase 3 is finished, the optimization process is interrupted, and the best sequence found so far is passed to the stacker crane.

Our simulation experience showed that REPLAN empirically gives the best results on average. SNI provided data for one week of production. During this period on one AUSS each generated task and each move of the stacker crane were recorded. It turned out that in heavy load periods the times needed for unloaded moves could be reduced by approx. 30%.

As a result the optimization package was put in use on five AUSS, and the results were confirmed in everyday production.

4.2 Commissioning of Greeting Cards

One of the commissioning departments in the distribution center of Herlitz is devoted to greeting cards. The cards are stored in four parallel shelving systems. In accordance with the customers' orders, the different greeting cards are collected in boxes that are eventually shipped to the recipient. Order pickers on eight highly automated guided vehicles collect the orders from the storage system, following a circular course. The vehicles are unable to pass each other. Moreover, due to security reasons, only two vehicles are allowed to be in the middle aisles at the same time, whereas three are allowed in the first and last aisle.

At the loading zone, each vehicle is logically "loaded" with up to 19 orders from a pool that changes over time. A dispatcher decides when to send a vehicle onto the course. After leaving this area the vehicles automatically stop at a position where cards have to be picked from the shelf according to the logical load. The goal is the minimization of the makespan of all requests generated on one day subject to some side constraints explained in [3, 13]. Congestion among the vehicles should be avoided. This is important because congestions lead to undesirable side-effects (that are very difficult to evaluate mathematically). These include human order pickers leaving for an extra-break when their vehicles run into congestions. (For more details consult [13].)

Mathematical Models

For the theoretical analysis it is necessary to provide a proper mathematical formulation of the problem under consideration. We remark again that the modeling phase may already result in a heuristic approach because the practical problem comes in day-to-day terms that have no straight-forward mathematical translations. The Commissioning Vehicle Routing Problem (CVRP) to be considered in the competitive analysis in Section 4.2 is the following.

An instance of CVRP consists of a set $L = \{1, 2, \ldots, m\}$, the pick positions, and a set of empty vehicles $v_1, \ldots, v_q$, each with capacity C. A request sequence $\sigma = r_1, r_2, \ldots$ consists of a chronologically ordered collection of sets of pick positions.

A vehicle to which C requests have been assigned is replaced by a new empty vehicle. In the online situation we require that request r_i is permanently assigned to vehicle $v(r_i)$ before r_{i+1} becomes known and that the length of the sequence is unknown until the last request comes in. That means, CVRP is an online problem in the sequence model (see [10] for basic facts on online problems).

For a sequence of requests, a solution to the CVRP is an assignment of every request r_i to a vehicle $v(r_i)$ so that the number of requests assigned to each vehicle does not exceed C. The objective is to minimize the total number of pick positions assigned to the vehicles. In [13] it was shown that the offline version of CVRP with no release times is already an NP-hard problem. In fact, solving the corresponding integer program in reasonable time turned out to be out of reach for commercial software packages like CPLEX.

One explanation for the intrinsic difficulty of this variation of a capacitated assignment problem is given by the following result that we state informally here (see [13] for details):

Theorem 15 (Offline Problem). *The optimal solution of a certain linear-programming relaxation of* CVRP *corresponds to the evenly distributed fractional assignment, i.e., every request is partially assigned to each available vehicle.*

This observation yields that the linear programming relaxation does not provide any exploitable information on how to assign requests to vehicles.

Evaluation of Algorithms

In [13] the following result was shown:

Theorem 16 (Competitive Analysis). *The following hold for the* CVRP*:*

(i) Any rule for the assignment of requests to vehicles yields a C*-competitive algorithm for the* CVRP*, where* C *is the capacity of a vehicle.*
(ii) No online algorithm for the CVRP *can be better than* 2*-competitive.*
(iii) The algorithm BESTFIT*– set* $v(r_i)$ *to the vehicle whose number of pick positions gets the least increase – is no better than* C*-competitive.*

In other words: competitive analysis does not provide much insight. In particular, the intuitively "reasonable" BESTFIT-heuristic is, from a competitive analysis point of view, not better than any stupid rule.

Even worse: recent investigations showed that even for a substantially simplified version of the CVRP we run into the odds of competitive analysis. In the following excursion into theoretical online optimization we sketch the result.

Consider the following *Online Bin Coloring Problem* ONLINEBC: We are given a natural number $q > 0$, infinitely many numbered bins with volume capacity C, and a sequence of requests $r_1, r_2, \ldots$ consisting of colored items of unit volume. We have to place the items into the bins so that, at any time, no more than q bins contain more than zero and less than C items. We have to stuff r_i into a bin before we get to know r_{i+1} (sequence model). The goal is to minimize the number of colors in the most colorful bin, i.e., the maximum number of items of distinct colors in a bin over all bins.

This translates to the language of commissioning as follows: every request has only one stop position, and we try to minimize over all vehicles the maximal number of stops of a vehicle, rather than the total number of stops. (This is a useful objective that helps to balance the vehicle load and, thus, to reduce congestion).

Consider the following online algorithms for ONLINEBC:

- Algorithm ONEBIN puts all items into one single bin until it is full. Then it picks another bin etc. (This is a truly dumb algorithm.)
- Algorithm BESTFIT puts every item into the bin that already contains that color, if such a bin exists. Otherwise, it puts the item into the bin with the least number of colors so far, with ties broken arbitrarily.

The following theorem shows that standard competitive analysis is problematic for this class of problems:

Theorem 17 (Competitive Analysis). *The following hold for the* ONLINEBC*:*

(i) BESTFIT *is* $\min\{C, 2q + \lfloor (qC - 3q + 1)/C \rfloor\}$*-competitive.*
(ii) ONEBIN *is* $\min\{C, (2q - 1)\}$*-competitive.*
(iii) BESTFIT *is no better than* $2q$*-competitive whenever* $C \geq 2q^3 - q^2 - q + 1$.
(iv) No deterministic online algorithm can be better than $O(q)$*-competitive.*

This proves that competitive analysis does not provide any hint as to which algorithm should be chosen in practice, even in the restricted models of this section.

Implemented Solution and Practical Impact

Several heuristics that reduce the total number of stops and distribute them evenly among the vehicles were implemented. These are versions of the BESTFIT algorithm, together with local exchange heuristics. The computation times of these algorithms are short so that they can be run in a real-time situation.

We implemented a detailed simulation model for the whole commissioning area in which we compared our approach to the one used so far. Herlitz provided production data from a period of about six weeks, which were the basis for the comparison. The main results are the following:

- A significant improvement with respect to the completion times of the orders can be achieved.
- The number of vehicles, used at Herlitz, can be reduced from eight to six without any negative impact on the system performance.
- Congestions over a few seconds can be avoided completely.

We conclude that BESTFIT– although not distinguished in the competitive analysis – was the basis for significant improvements in practice. The simulation results convinced Herlitz to test a prototype of the simulation program as a decision support tool for the dispatcher.

4.3 Elevators

The automated pallet transportation system in the Europe-wide distribution center of Herlitz PBS AG has been designed to handle all pallet transportation taks from/to the receiving docks, the production and commissioning departments, the automated shelf system, and the loading dock from where the products are shipped to the customers by trucks. This pallet transportation network runs on nine floors and is quite complex. The overall goal is to run the operations "smoothly", a mathematically not well-defined term that means something like: each individual transportation task should be executed quickly, time windows (existing for some of the tasks) should be observed, and the whole system should be congestion free. The last objective may be in conflict with the others, and a difficulty is to find an appropriate balance.

We address here the elevators, one of the building blocks of the pallet transportation system. There are two systems of five elevators. Each elevator can carry at most one pallet. Transportation requests occur (unpredictably) throughout the day and are somehow distributed to the elevators. Congestion does frequently occur at the entry points and should be avoided by running the elevators "well". Of course, congestion depends on both the assignment of requests to the elevators and on the control of the elevators. We discuss here the second issue.

At Herlitz, each elevator is controlled independently from the others; there is no "master control" watching over the whole elevator system simultaneously. It is therefore clear that optimizing the individual elevators may not result in the desired congestion-free system, but it will at least help running the system faster. We decided to investigate the following problems for individual elevators and systems of elevators (compare to Section 3.2):

- Minimize the *makespan* for the given set of requests.
- Minimize the *maximal flow time* of the requests.
- Minimize the *average flow time* of the requests.

While the makespan is a measure for how fast the system is as a whole the other two objectives are rather a measure for the speed of the system as "experienced" by the individual pallets. Note that in contrast to the makespan the maximal and average flow times also make sense in a continuously operating system, i.e., with infinite request sets.

Mathematical Models

The basic model chosen for investigating algorithms for the control of elevators is the ONLINEDARP, which we introduced in Section 3.2. In the sequel we first investigate the control of a single elevator. Briefly, this is the problem of how to serve online transportation requests in a metric space which is a path, where the server is assumed to have capacity one.

In the context of pallet transportation there is a subtle additional side constraint involved: we do not have random access to the pallets waiting on a particular floor. That means that requests from the same floor need to be scheduled in their order of appearance, while requests on different floors can still be shuffled. This leads to the problem ONLINEFIFODARP. Here the subset of requests occuring at a particular point in the metric space must be served in the order of appearance.

As an extension of ONLINEDARP we also investigated the corresponding problem with capacity larger than one, the ONLINECDARP.

In order to be able to use REPLAN- or IGNORE-heuristics for any of the online problems in real-time we need to find efficient algorithms for the corresponding offline problems. In the following theorem we summarize the results:

Theorem 18 (Offline Problems). *The following complexity results hold:*

(i) There is a polynomial time algorithm for DARP *on paths.*
(ii) DARP *on trees (even on so-called caterpillars) is NP-hard.*
(iii) There is a polynomial time algorithm for FIFODARP *on paths.*
(iv) CDARP *is NP-hard on paths.* □

Theorem 19 (Offline Problems). *The following approximation results hold:*

1. *There is a* 5/3*-approximation algorithm for* FIFODARP *on trees.*
2. *There is a* 9/4*-approximation algorithm for* FIFODARP *on general graphs.*
3. *There is a* 3*-approximation algorithm for* CDARP *on paths.* □

The observed performances of the approximation algorithms for FIFODARP on instances occuring in the online situation (e.g., while applying the REPLAN-heuristics) are much better. Therefore, these approximation algorithms can be used to produce a starting solution for a branch&bound procedure to find reasonably good offline solutions to feed the REPLAN-heuristics in real-time.

Evaluation of Algorithms

Motivated by results on the ONLINETSP in [5] we carried out a competitive analysis for ONLINEDARP for the minimization of the makespan. The results are the following:

Theorem 20 (Competitive Analysis). *For the problem of makespan minimization in* ONLINEDARP *the following hold:*

(i) No deterministic online algorithm can be better than 2-competitive. (This follows easily from [5].)
(ii) The REPLAN- *and the* IGNORE-*heuristic are 5/2-competitive.*
(iii) There is a 2-competitive algorithm (called SMARTSTART *in [4]).*

In other words, we found one optimally competitive online algorithm for our problem.

For the other objective functions, the approach via competitive analysis yields the strongest conceivable negative result, i.e., no decision support at all:

Observation 21 (Competitive Analysis). *There are no competitive algorithms for the tasks of minimizing the maximal or average flow times in* ONLINEDARP.

The concept of reasonable load (see 3.2) was developed to get at least a weaker performance evaluation. We have already seen two canonical online heuristics in that section: $\text{REPLAN}^{makespan}$ and $\text{IGNORE}^{makespan}$. Recall that both work by repeatedly minimizing the makespan: while $\text{REPLAN}^{makespan}$ computes a new plan whenever a new request becomes available, $\text{IGNORE}^{makespan}$ does not compute a new plan before the old plan is completely served. What about $\text{REPLAN}^{maxflow}$, $\text{REPLAN}^{avgflow}$, $\text{IGNORE}^{maxflow}$, $\text{IGNORE}^{avgflow}$? What about the problems ONLINECDARP and ONLINEMDARP (more than one server)? Some answers are collected in the following theorem:

Theorem 22 (Analysis Under Reasonable Load). *For all of* ONLINEDARP, ONLINEFIFODARP, ONLINECDARP, ONLINEFIFOCDARP, ONLINEMDARP, ONLINEFIFOMDARP *the following hold under* Δ*-reasonable load:*

(i) The maximal and average flow times of $\text{IGNORE}^{makespan}$ *are at most* 2Δ.
(ii) The maximal and average flow times of $\text{REPLAN}^{makespan}$ *maybe arbitrarily large.*
(iii) The maximal and average flow times of $\text{REPLAN}^{avgflow}$ *maybe arbitrarily large.*

We do not know the performance of $\text{REPLAN}^{maxflow}$ at present. We have, however, found another provably good algorithm that imposes additional restrictions on the repeatedly computed plans. We assume that this algorithm, called DELTAREPLAN, knows Δ. The algorithm DELTAREPLAN follows the current plan. Whenever a new request comes up DELTAREPLAN computes a new plan minimizing the makespan subject to the condition that all requests in the plan have a flow time of no more than 2Δ. If the optimal plan is shorter than Δ then it is accepted as the new plan. Otherwise it is rejected, and the algorithm proceeds with the old plan. When the old plan is done, a new plan is accepted in any case.

We could prove the following in [9]:

Theorem 23 (Analysis Under Reasonable Load). *For all of* ONLINEDARP, ONLINEFIFODARP, ONLINECDARP, ONLINEFIFOCDARP, ONLINEMDARP, ONLINEFIFOMDARP *the following holds under* Δ*-reasonable load:*

The maximal and average flow times of DELTAREPLAN *are at most* 2Δ.

This theorem motivates the problem of finding out the Δ while working on a Δ-reasonable request set. Observe that for, e.g., IGNORE^{makespan} it is not necessary to have information on the correct Δ.

Assume that DELTAREPLAN dynamically computes and uses an approximation $\tilde{\Delta}$ of Δ while working on a Δ-reasonable request set. If always $\tilde{\Delta} = 0$ then we observe that all plans are rejected and the algorithm behaves like IGNORE^{makespan}, thus the performance guarantee in Theorem 22 takes effect. More general: whenever we underestimate Δ then DELTAREPLAN achieves the same performance guarantee as in Theorem 23.

In the following we define a modification DYNDELTAREPLAN of DELTAREPLAN that needs not know the real Δ. Algorithm DYNDELTAREPLAN works similar to DELTAREPLAN except that it computes a dynamically changing $\tilde{\Delta}$. This $\tilde{\Delta}$ is defined to be the makespan of the latest accepted plan. The first value for $\tilde{\Delta}$ is the length of the first plan computed. Whenever a new request occurs DYNDELTAREPLAN computes a potential new plan with all flow times at most $2\tilde{\Delta}$. If the makespan of the potential plan is at most $\tilde{\Delta}$ then DYNDELTAREPLAN accepts it as the new plan.

The following result could be achieved.

Theorem 24 (Analysis Under Reasonable Load). *For all of* ONLINEDARP, ONLINEFIFODARP, ONLINECDARP, ONLINEFIFOCDARP, ONLINEMDARP, ONLINEFIFOMDARP *the following holds under* Δ*-reasonable load:*

The maximal and average flow times of DYNDELTAREPLAN *are at most* 2Δ.

A "heuristic reason" for the correctness of this result is the following: whenever we underestimate Δ we may get fewer accepted new plans. But whenever no new plan is accepted and the old plan is accomplished we are working like IGNORE^{makespan}, which is fine because of Theorem 22.

In order to get some idea how the investigated algorithms behave on the average with respect to speed, stability, and predictability we carried out simulation studies for the basic elevator control problem. In addition to our algorithms we tested the heuristics FIFO and NN. The latter one always serves the nearest request next. Moreover, we included the heuristic NN-MAXAGE. This heuristic works like NN except that whenever a request is older than a maximal age parameter this request has to be served next. These three heuristics are implemented as possible elevator controls in the Herlitz system.

Observation 25 (Comparative Simulation). *A simulation experiment on several random data sets for the* ONLINEDARP *yielded the following results:*

- *The* FIFO*-heuristic is suitable only for very low load situations. Otherwise, the maximal and the average flow times explode; heavy system congestion is apparent.*
- *The* NN*-heuristic produces very low average flow times on the average. The maximal flow times are – especially in medium load situations – unpredictable, i.e., sometimes very high.*
- *The* NN-MAXAGE*-heuristic cures the problem of unreliability of* NN *only in low load situations. In high load situations it suddenly behaves like the* FIFO*-heuristic and leads to heavy system-congestion.*
- *The* REPLAN^{makespan}*-heuristic shows mostly good average flow times. Its maximal flow times are comparable to* NN*, i.e., at times very bad.*
- *The* IGNORE^{makespan}*-heuristic produces slightly worse average flow times than* NN *or* REPLAN^{makespan}*. The maximal flow times, however, are among the best for all load situations. This heuristic is in a sense self-calibrating.*
- *The* DYNDELTAREPLAN*-heuristic behaves like* IGNORE^{makespan} *but shows on the average a little bit worse maximal flow times and slightly better average flow times.*

The additional benefit of the simulation studies over a mere evaluation of an objective function is the possibility of watching the system behavior as a whole. The algorithm that is chosen eventually depends on the preferences of the administrator of the system under consideration. At Herlitz, there is a strong focus on stability over mere speed so that IGNORE^{makespan} and related heuristics seem suitable.

4.4 *Integrated Elevator Systems*

We mentioned in the previous section that the software at the Herlitz plant does not support a so-called synchronized pallet transportation. This means the controls for the individual elevators make their decisions without taking into account each other's and the conveyor system's states. Thus, we investigated the control of single elevators as discussed in the previous section. The interplay between these modules of the transportation system is not negligible, though.

In simulation studies where the conveyor belts from and to the elevators were included in the simulation system we found out that many effects observed for single elevators are leveled out. This motivated the investigation of the integrated system of conveyor belts and multiple elevators. Since the software base of the transportation system at Herlitz cannot be changed easily; research results in this area do not have direct bearing in practice.

Having this in mind we simplified the layout of the combined conveyor/elevator system in order to approach an integrated system control in reasonable steps. At Herlitz, on each floor, the conveyor system lets the pallets move on a circular belt with one entry and one exit to the production and commissioning area. There are five elevators in the interior of the circle. The pallets can reach and leave the corresponding waiting slots via switches. The waiting/leaving slots have capacity one.

A pallet may move to the waiting slot only if the corresponding leaving slot on its destination floor is empty.

The coupling in this system is very difficult to model. Moreover, one question that arises in this context is whether layouts of this type are suitable for efficient control. Thus, we started our investigation on the basis of the following hypothetical layout: Pallets line-up in a waiting queue of infinite capacity. Behind that queue they enter separate waiting queues in front of the elevators. We call this problem *online multi server sequential ordering problem*, ONLINE m-COST-SOP for short. The task is to distribute pallets online to the elevator queues and to control the elevators so that the maximal or average flow times are minimal.

The idea is to use a variant of the IGNORE^{makespan}-heuristic. This requires minimizing the makespan in the corresponding offline problem. Here, the makespan is the time when the last elevator has finished. In contrast to the case of single elevators, not all types of REPLAN-heuristics can be employed (at least not in a straightforward form) because of the following problem: Once a set of pallets is distributed among the elevator queues the pallets will immediately move into their queues. Because the pallets cannot change the elevator the decision which elevator a particular pallet should take can not be revised.

Mathematical Models

The main idea is to model the problem as an ATSP on the request digraph (cf. Section 4.1) with two modifications: first, there is more than one server. Second, the pallets in the waiting queue at a particular elevator on some floor need to be served in a FIFO order. Each of these generalizations of the ATSP has been studied already in the literature: the first one in the case of a single server type was reduced to the single server case in [15]; in a more general form for servers with distinct properties (m-COST-ATSP) it was studied in [12]. The second one was already discussed in Section 4.1. We decided to investigate the combined problem m-COST-SOP: the *multi server sequential ordering problem.*

There is one further subtlety involved: since in the m-COST-SOP-model the maximal length of a tour in the request graph over all servers is minimized we need to take into account the loaded travel time in the arc costs. Otherwise we might get a solution where all the servers have similar unloaded travel times but their total travel times (makespans) may vary a lot and the makespan of the whole system is not optimal at all. That means: in the case of more than one server minimizing the makespan and minimizing the unloaded travel times are no longer equivalent.

Having this in mind, our model is almost the same as the m-COST-ATSP in [12] except that it also contains the corresponding precedence forcing constraints. These look like the constraints (4) in 4.1. We do not want to reproduce the complete model here. We just state that several properties of the SOP and the m-COST-ATSP survive in their common generalization m-COST-SOP.

Theorem 26 (Offline Problems – Polyhedral Study). *The following hold for the* m-COST-SOP*-polytope:*

- *The dimension of the* m-COST-SOP-*polytope for regular precedences equals* $m(n^2 - |R|) - n$, *where* R *is the set of comparable pairs of nodes.*
- *Modified versions of the so-called* σ, π, ν_0-σ *and* ν_0-π-*inequalities are valid for the* m-COST-SOP-*polytope.*
- *Facets of the one-server subproblem of* m-COST-SOP *can be lifted to facets of the* m-COST-SOP.

Computational experiments have shown that the integrated optimization of all servers yields an improvement in the unloaded travel times of 50% on the average.

Evaluation of Algorithms

It turns out that, also for the ONLINE m-COST-SOP, the analysis under reasonable load is analogous to the previously discussed cases.

Theorem 27 (Analysis Under Reasonable Load). *The maximal and average flow times of* IGNORE^{makespan} *for the* ONLINE m-COST-SOP *under* Δ-*reasonable load are at most* 2Δ.

This theoretical result is hard to implement in a real-time compliant way: the m-COST-SOP turned out to be very difficult. It rarely happens that one can find optimal solutions for instances with 20 requests in less than a minute. The real-time restrictions on an elevator control rather require answers within seconds. Thus, only heuristic solutions can be used in the online situation. Evaluation of such heuristics is research in progress.

There is another strong argument against using the unmodified IGNORE^{makespan}: all servers but one would very frequently wait idle for the last server to finish its part of the plan. This can be by-passed by, e.g., letting the servers work on some requests inbetween. Still, the theoretical analysis matches reality much less than in the single server case.

Preliminary simulation studies on the basis of simple heuristics for the m-COST-SOP and on modified IGNORE- and NN-heuristics are no longer in favor for the IGNORE-approach for certain parameter settings.

This shows among other things that it is quite hard to find a well-performing online control of an integrated transportation system.

5 CONCLUSION

We have discussed various evaluation methods for online optimization problems on the basis of four real-world examples. I turns out that, usually, only a combination of such methods is able to deliver convincing advice to decision makers.

To meet real-time requirements fast offline optimization algorithms are needed, in general, as building blocks for the online heuristics such as IGNORE and REPLAN. We have, e.g., introduced fast approximation algorithms for DARP that enable us to run these heuristics in real-time in the elevator control problem.

We have shown that, for the evaluation of online algorithms, classical competitive analysis may lead to either void conclusions (all algorithms are equally bad for the minimization of flow times for ONLINEDARP) or may even be in favor of a senseless algorithm (ONEBIN is best possible for ONLINEBC). New methods such as analysis under reasonable load provide new insight in some of these cases. For example, we could tell which of the two online heuristics IGNORE and REPLAN is more suitable with respect to the minimization of flow times for the ONLINEDARP.

The observation of the system behavior as a whole in simulation experiments is still unavoidable because, this way, it is possible to monitor more complex effects than the projection to a one-dimensional objective function can possibly detect.

REFERENCES

1. N. Ascheuer, *Hamiltonian path problems in the on-line optimization of flexible manufacturing systems*, Ph.D. thesis, Technische Universität Berlin, 1995.
2. N. Ascheuer, M. Fischetti, and M. Grötschel, *Solving the asymmetric travelling salesman problem with time windows by branch-and-cut*, Preprint SC 99-31, Konrad-Zuse-Zentrum für Informationstechnik Berlin, 1999.
3. N. Ascheuer, M. Grötschel, S. O. Krumke, and J. Rambau, *Combinatorial online optimization*, Proceedings of the International Conference of Operations Research (OR'98), Springer, 1998, pp. 21–37.
4. N. Ascheuer, S. O. Krumke, and J. Rambau, *Online dial-a-ride problems: Minimizing the completion time*, Proceedings of the 17th International Symposium on Theoretical Aspects of Computer Science, Lecture Notes in Computer Science, vol. 1770, Springer, 2000, pp. 639–650.
5. G. Ausiello, E. Feuerstein, S. Leonardi, L. Stougie, and M. Talamo, *Algorithms for the on-line traveling salesman*, Algorithmica (2001), To appear.
6. E. Balas, M. Fischetti, and W. Pulleyblank, *The precendence constrained asymmetric traveling salesman polytope*, Technical Report 15213, Carnegie Mellon University, Pittsburgh, 1992.
7. M. Blom, S. O. Krumke, W. E. de Paepe, and L. Stougie, *The online-TSP against fair adversaries*, Proceedings of the 4th Italian Conference on Algorithms and Complexity, Lecture Notes in Computer Science, vol. 1767, Springer, 2000, pp. 137–149.
8. A. Borodin and R. El-Yaniv, *Online computation and competitive analysis*, Cambridge University Press, 1998.
9. B. Glück, *Online-Steuerungen automatischer Transportsysteme bei vertretbarer Belastung*, Diplomarbeit, Technische Universität Berlin, 2000.
10. M. Grötschel, S. O. Krumke, J. Rambau, T. Winter, and U. T. Zimmermann *Combinatorial Online Optimization in Real Time*, this volume.
11. D. Hauptmeier, S. O. Krumke, and J. Rambau, *The online dial-a-ride problem under reasonable load*, Proceedings of the 4th Italian Conference on Algorithms and Complexity, Lecture Notes in Computer Science, vol. 1767, Springer, 2000, pp. 125–136.
12. C. Helmberg, *The m-cost ATSP*, Proceedings of the 7th Mathematical Programming Society Conference on Integer Programming and Combinatorial Optimization, Lecture Notes in Computer Science, vol. 1610, Springer, 1999, pp. 242–258.
13. N. Kamin, *On-line optimization of order picking in an automated warehouse*, Ph.D. thesis, Technische Universität Berlin, 1998.
14. E. Koutsoupias and C. Papadimitriou, *Beyond competitive analysis*, Proceedings of the 35th Annual IEEE Symposium on the Foundations of Computer Science, 1994, pp. 394–400.

15. G. Reinelt, *The traveling salesman – computational solutions for tsp applications*, Lecture Notes in Computer Science, vol. 840, Springer, 1994.

Stowage and Transport Optimization in Ship Planning

Dirk Steenken[1], Thomas Winter[2], and Uwe T. Zimmermann[3]

[1] Abteilung DC/DV EDI, Hamburger Hafen- und Lagerhaus AG, Germany
[2] Information and Communication Mobile Networks, Siemens AG, Germany
[3] Abteilung Mathematische Optimierung, Technische Universität Braunschweig, Germany

Abstract We consider the ship planning problem at maritime container terminals where containers are loaded onto and discharged from ships using quay cranes. The container transport between the ships and the yard positions in the terminal is carried out by a fleet of straddle carriers. Based on a stowage plan provided by the shipping company, the dispatcher assigns containers to specified bay positions. Then, subject to operational and stability constraints, he schedules containers in order to avoid waiting times at the quay cranes. We propose an approach combining stowage planning and the selection of "good" loading and transport sequences. For a just-in-time scheduling model, we present computational results based on real-world data of a German container terminal. Moreover, we discuss some real-time and online influences on the daily dispatch situation.

1 Introduction

Within the last years, the rate of containerization increased by approximately 8 percent per year. Shipping a larger number of containers around the world requires matching efficiency improvements in maritime container terminals. Besides the introduction of computer-aided decision systems and infra-structural improvement, the complete logistic chain has to be examined in order to increase the container handling rates. In this article, we focus on a particular problem arising at the quay side.

In maritime container terminals, a large number of containers is handled day by day. The containers arrive at the terminal by truck, ship, or train. Before leaving the terminal, containers are usually stored in the terminal's yard area. In the yard's storage blocks, the containers are arranged in stacks, one beside the other in several rows. Transport between the storage positions in the yard and the terminal's exit points is usually handled by straddle carriers, by automated guided vehicles, or by transtainers. In this article, we only consider straddle carriers as, e.g., used at the terminal "Burchardkai" terminal in Hamburg, Germany. The turnover at the terminal "Burchardkai", operated by the Hamburg Port and Warehouse Company (HHLA), increased from 1.1 million container units (TEU) in 1992 to 1.6 million TEU in 1998 and 2.1 million TEU in 2000. It is expected that the number of container units will soon reach the actual maximum number of 2.6 million TEU which can be handled at Burchardkai. This increase requires improved, intelligent logistics. At "Burchardkai", more than 3200 vessel calls are operated per year. Loading and discharging is carried out by quay cranes whereas the transport is performed by a fleet of straddle carriers. The complete dispatch process consists in about 10000 container movements per day.

Combinatorial optimization models apply for instance when assigning vessels to berths, when planning the tours for each transport vehicle, or when computing good storage positions for the containers. The berth planning problem is modelled by Lim [10] as a rectangular packing problem with side constraints. Lim presents a heuristic based on (heuristically) computing longest paths in a graph model. An alternative network flow approach is due to Chen [6]. Different versions of tour planning models for straddle carriers have been considered by Steenken et al. [11–13]. A linear sum assignment model of the dispatch of straddle carriers for discharging and loading trucks is iteratively solved in real-time [12, 13]. A travelling salesman model combining various hinterland operations is heuristically solved in [11].

In this article, we discuss the following combination of stowage and transport of containers to be loaded to certain container vessels, named *export* containers. At first, an export container is moved to a respective quay crane. Then, the quay crane loads the container into a suitable position in the bay currently served.

For each export container, the corresponding loading position is specified in accordance with the stowage plan. This stowage plan is derived from information provided by the shipping company. For each bay position, the shipping company defines properties for a container which may be stored at this position. In particular, the shipping company specifies the discharge port, the container type, and its weight. Even restrictions on stored goods may apply.

Today, the transport of export containers to the quay cranes is not taken into account when deciding on the final bay position for a container aboard the vessel. The ship planning process starts two days before the vessel arrives at the terminal. At that time, the responsible dispatcher prepares a stowage plan based on the following information: the onboard storage situation at the previous port and a preliminary list of export containers. In particular, potential information on transportation times is not used.

In section 2, we propose a just-in-time scheduling formulation for combined stowage and transport planning and we introduce a corresponding mixed integer model as well as exact and heuristical methods to solve it. Moreover, we consider different objectives.

In section 3 we discuss how the proposed approach extends to real-time requirements. Particularly flexible update techniques allow adaptation of previously computed schedules with regard to real-time requirements.

2 Ship Planning in Container Terminals

Maritime container terminals form important links in the transport chain of containers. Import and export containers are temporarily stored in the terminal area. Ship planning is very important for the productivity of a container terminal. Ship planning is based on preliminary information provided by the shipping company. The first information, submitted two days before the arrival of a container vessel, consists of a map of the current storage situation at the previously visited port. At the same time, information on *import* containers, i.e., containers which have to be discharged, is made available. For each bay position, the shipping company defines properties for a container which may be stored at this position. In particular, the shipping company specifies the discharge port, the container type, and its weight. Before the vessel's arrival, information about the export containers is updated from time to time.

From the information provided by the shipping company, the dispatcher derives a stowage plan which assigns a particular export container for each loading position. After discharging the vessel, these export containers are moved to the quay cranes by straddle carriers. For each bay, the chosen loading strategy implies a loading sequence of containers for each quay crane. In order to avoid waiting times during the loading process, the transport sequences of the straddle carriers have to match the loading sequence of the respective crane.

For a large number of containers arriving by truck, the exact delivery times are unknown. Up to 30 percent of the export containers arrive at the terminal after the beginning of the loading process. Due to lack of complete information and due to tight timing constraints, the dispatcher has to handle online and real-time versions of the above problems [15].

2.1 Stowage Planning on Container Vessels

In [1,2], a stowage plan model for a container vessel visiting several ports is presented. For each bay position, this stowage plan specifies the destination port for the container to be loaded in any given port. Hence, it could be used as the preliminary stowage plan which is provided by the shipping company. In this model, the weights of containers are not taken into consideration. However, for stability reasons, container weights must be considered (see for instance [3,4]): heavy containers should be stored below containers having less weight.

A potential stowage plan for one bay is presented in Figure 1. For each bay position, a container type is specified. Hence, at this position, only a container with prescribed weight and destination can be stored. The required size of the container is defined by the type of the bay. Usually, a bay is restricted to 20' containers or to 40' containers. Some bays may contain both types of containers, whereas all the 20' containers should stand on top of the 40' containers.

In the combined stowage and transport problem, an *abstract* container type is described by: the container's discharge port, the container weight including the weight of the stored goods, the type of the container, i.e., its size (20' or 40') as well as special equipment attributes, the kind of goods stored in the container, and the delivery time of the container.

We distinguish between the above abstract container types described above and the *real world* container types named in the following list which does not claim to be comprehensive: general purpose container, hardtop container, high cube general purpose container, high cube hardtop container, flat container, open top container,

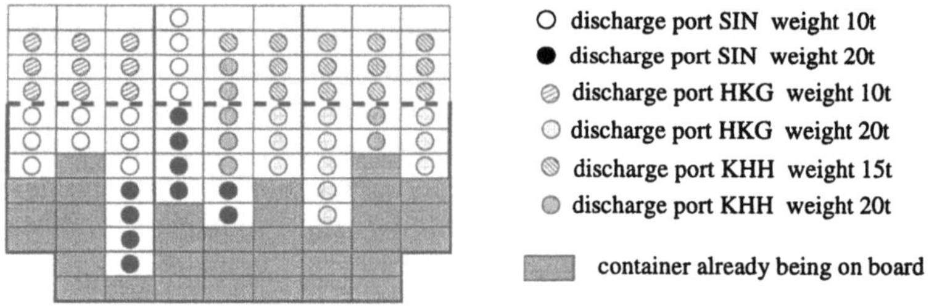

Figure 1. An example for a stowage plan provided by the shipping company

high cube flat container, platform, insulated container, ventilated container, reefer container, bulk container, high cube reefer container, or tank container. Some types require that the container must be stored at a specially equipped position. For instance, reefer containers should be kept cool and must be supplied with electricity. High cube containers differ in height from the standard general purpose containers and will probably occupy two stowage positions.

In the following, we use the notion of the abstract container types which may be defined as refinement of the real container types. Furthermore, we suppose that we know a preliminary stowage plan specifying a container type for each bay position. We assume, that the number of export containers of a specific type exactly matches the number of bay positions of the same container type.

2.2 Stowage Planning in Container Terminals

Ship planning (or stowage planning) in container terminals differs from stowage planning for container vessels. As discussed in the previous section, for container vessels it suffices to specify a certain container type for each bay position. This preliminary type-based stowage plan provided by the shipping company and the list of export containers form the basis for the dispatcher's work at the container terminal. The dispatcher prepares a final stowage plan which assigns to each bay position a particular export container with matching type.

As mentioned above, a large number of export containers arrives after the beginning of the loading process. The dispatcher has to take such difficulties into consideration when assigning containers to bay positions. Additionally, the dispatcher must take into account that containers are stored in stacks (cf. Figure 2). Containers on top of a stack should be moved before a container at a bottom position is required. In order to minimize unnecessary container shifts, stacks of containers of identical type are preferable. Obviously, this may be impossible. In fact, up to 30 percent of the stacks contain containers of differing types.

Nowadays, a dispatcher subsequently assigns export containers in inverse order of ports to be visited. First, he chooses a bay. Then, he marks all free positions for containers of the currently considered type. For these positions, the decision support

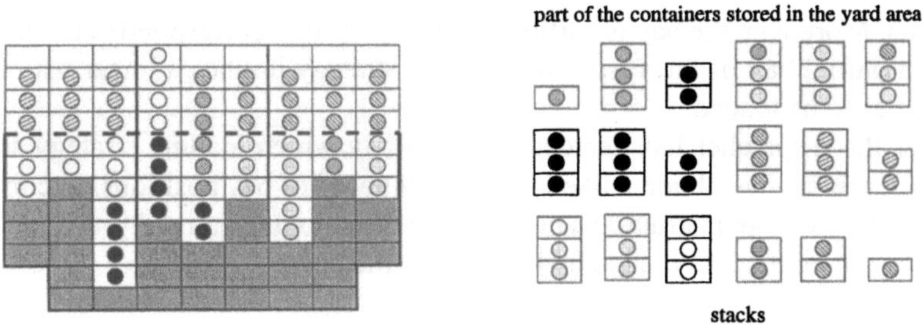

Figure 2. An example for the storage situation in the yard

system offers a list of not yet assigned export containers of that type. The dispatcher selects some containers from this list. The final assignment is determined by a simple heuristic in accordance with a specified loading strategy and with regard to container weights. When all containers are assigned, the stowage plan is transmitted to the shipping company which may accept the plan or may ask for some changes.

2.3 *Combining Ship and Transport Planning*

By now, the stowage plan is generated ignoring loading and transport sequences. In particular, containers are assigned to bay positions without consideration of the necessary transportation times between storage positions in the yard and the quay cranes. As mentioned before, for each bay position, the preliminary stowage plan only assigns a container type. For each bay, the dispatcher chooses a loading strategy which specifies a linear order of export container types. Since bays consists of stacks, there are two straight-forward strategies used in real-world ship planning: loading column-wise or loading layer by layer. For reasons of visibility, the quay cranes always start with the bay positions at the water-side of the vessel. This fixes a loading sequence for both strategies. Two examples of these common loading strategies and the resulting loading sequences are presented in Figure 3.

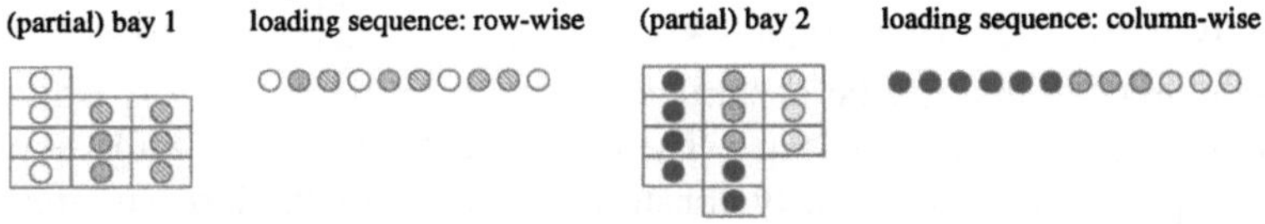

Figure 3. An example of two loading strategies and the resulting loading sequences

Each bay may be partitioned into some partial bays which are considered separately. These partial bays correspond to the bay positions on deck or in the hold of the vessel. Moreover, the bay is partitioned into areas that correspond to the hatches. For each partial bay of the vessel, the loading strategy implies a linear list of container types to be loaded.

After the dispatcher has decided for each bay which loading strategy will be used, for each bay we obtain a fixed loading sequence of bay positions. In combination with the stowage instructions provided by the shipping company, this results in a sequence of container types to be loaded into the bays.

2.4 *The Crane Split*

Next, the bays of a vessel are partitioned into bay areas. Each bay area will be served by one quay crane. This step is called crane split. Based on availability information of cranes, a crane split can be computed by solving a partitioning problem with some operational side constraints. Since the number of cranes available for the loading process is small, an optimal solution of this partitioning problem can be computed within acceptable time.

More formally, we are given a set of bays (or partial bays) $\{1, \ldots, B\}$. b_i denotes the number of containers to be loaded into bay i, $b_i > 0$. This number is defined by the stowage plan provided by the shipping company. The vessel will be loaded using C quay cranes each of which has capacity q_i, $1 \le i \le C$. The capacity corresponds to the time the crane will be available.

We search for a partition $\mathcal{Q}_1, \mathcal{Q}_2, \ldots, \mathcal{Q}_C$ of $\{1, \ldots, B\}$ where each bay area $\mathcal{Q}_i$ contains only consecutive bays, i.e., $\mathcal{Q}_i = \{j_i, j_i + 1, \ldots, j_i + k_i\}$ for all $1 \le i \le C$. Obviously, for a given partition, the resulting absolute load is $Q_i = \sum_{j \in \mathcal{Q}_i} b_j$, i.e. the total number of export containers for bay area $\mathcal{Q}_i$. A good choice of a partition may balance the resulting relative loads Q_i/q_i for all quay cranes $1 \le i \le C$ as much as possible. Minimizing the maximum relative imbalance, we find

$$\min \max_{i,j \in \{1,\ldots,C\}} \left| \frac{Q_i}{q_i} - \frac{Q_j}{q_j} \right| .$$

For C, a value between 2 and 6 is reasonable for real world container terminals. The number of bays may vary between 20 and 50. Consequently, we may solve this partitioning problem by straight-forward enumeration. An initial upper bound can be derived from the weighted average loads $\mu_i = \frac{q_i}{\sum_{i=1}^{C} q_i} \sum_{j=1}^{B} b_j$. A corresponding "partition" may recursively be constructed. Let $\mathcal{Q}_1 := \{1, 2, \ldots, k_1\}$ where k_1 is chosen minimal such that $Q_1 \ge \mu_1$. Then, $\mathcal{Q}_2$ is defined by $Q_2 \ge Q_1 + \mu_2$. The remaining partition is analogously constructed except for the last bay area which contains all remaining bays. Better upper bounds may easily be obtained by slightly varying the values of k_i.

Combining the Loading Strategies

For each quay crane and for each bay, we obtain a loading sequence of container types (cf. Figure 4). A loading sequence is served by the straddle carriers available for the crane. We assume that a certain fixed number of straddle carriers is available for each crane. These straddle carriers move containers from their current stowage position in the yard to the crane. Here, pooling of straddle carriers is not considered but may help to stay within real-time bounds at a particular crane where more straddle carriers are required.

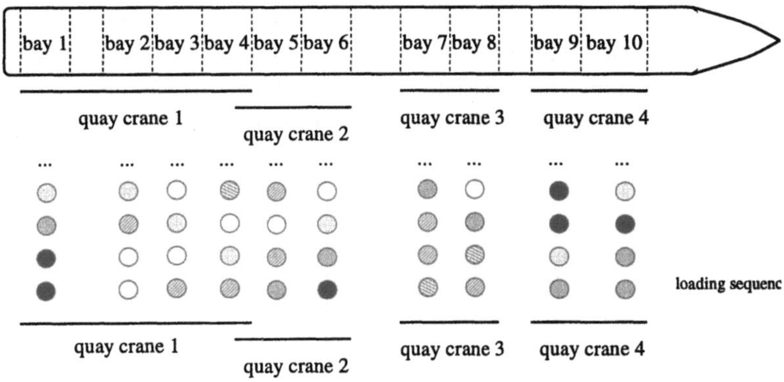

Figure 4. A crane split and the corresponding loading sequences

For each loading event of a loading sequence of a quay crane, an export container of the required type is moved to the crane. At the crane, the containers should arrive in the order defined by the loading sequence. If a straddle carrier with a container for a subsequent loading event arrives too early, it may have to wait until all the predecessors of that container have been handled since there is only limited buffer space close to cranes. Usually, at most one or two containers can be placed in this buffer area.

Consequently, only a careful assignment of transportation duties to straddle carriers will optimize the overall loading process. There are several objectives which may be considered. Minimizing the loading time of the last export container corresponds, in the notation of scheduling problems, to the latest completion time. For further improvement it will also be interesting to find the bottleneck of a loading process. Empirical studies and discussions with HHLA showed that the time requirements for the loading process strongly depends on the effectiveness of the quay cranes. Therefore, another promising approach is to minimize the waiting times of the quay cranes, or in other words, to avoid the late deliveries of export containers.

In all experience of HHLA, a quay crane requires between 80 and 120 seconds to load a single container to its bay position. Thus, 30 to 45 containers may be loaded per hour and crane. A more or less regular sequence of loading events ensures a smooth loading process which helps to avoid waiting times. This observation suggests loading events every 80 to 120 seconds and defines reasonable due dates for each loading event. Thus, container transports with m straddle carriers may be modelled as a parallel m-machine scheduling problem with due dates, minimizing late deliveries. Obviously this problem is NP-hard, since it contains parallel machine scheduling problem with due dates (cf. for instance [7, 9]).

We assign export containers to each loading event of the loading sequence. An export container will be moved by some specified straddle carrier. The respective transportation time corresponds to the distance between the stack position of the container and the position of the quay crane. An example for a schedule of straddle carriers is presented in Figure 5. The respective transportation times are represented by the thick strokes behind the containers. In particular, the first and second straddle carriers will at first move the first and second containers in the loading sequence. Due to the different transportation times, the second straddle carrier should start with some delay in order to arrive at the quay crane later than the first straddle carrier.

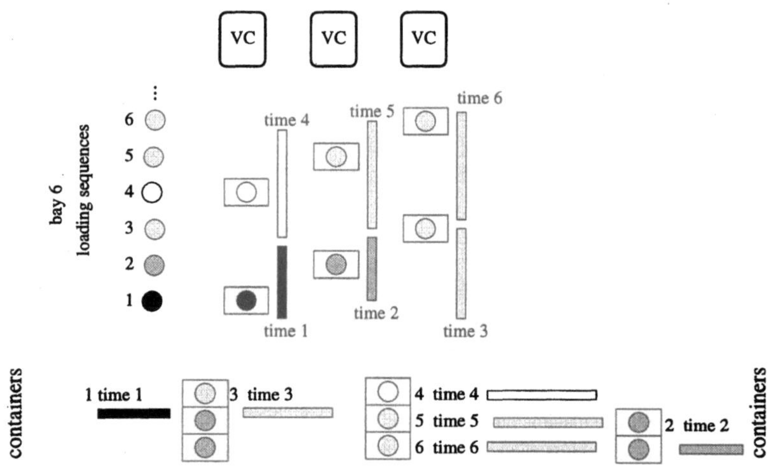

Figure 5. The scheduling model for the straddle carriers. The corresponding transport times for each container stored in the stacks are illustrated by the rectangles of different lengths. Six containers are chosen and assigned to the three straddle carriers (VC). Since the arrival of container two at the quay crane should be after the arrival of container one the transport of container two is delayed. The containers are assigned to the transport jobs in accordance with the container types given by the loading sequence

2.5 *Just-in-Time Transport of Containers*

Just-in-time scheduling problems have been applied in production by Steiner and Yeomans [14], for a single machine by Liaw [9] and for parallel machines by Chen and Powell [7] who assume a large common due date. For a related introduction to scheduling we refer to [8].

We present a mixed integer model for just-in-time container scheduling with one quay crane. Here, N denotes the number of export containers, i.e., the length of the loading sequence. $\mathcal{L}$ denotes the set of loading events in the loading sequence. For $1 \leq i \leq N$, the i-th container of requested type $t(i)$ is delivered at the crane at time T_i. $\mathcal{C}$ denotes the set of eligible export containers. For $c \in \mathcal{C}$, the transportation time of container $c \in \mathcal{C}$ from its yard position to the crane is denoted by p_c and its type is denoted by $t(c)$. The set of available straddle carriers is denoted by $\mathcal{V} = \{1, 2, \dots, V\}$.

A solution of this problem is a schedule assigning the straddle carriers to container transports. By the one-to-one correspondence of container transport and loading jobs, the assignment implies a stowage plan for all bays considered. We call this just-in-time container scheduling problem the *combined container stowage and transport planning problem.* It is NP-hard, since it contains the scheduling problem introduced in [7]. We consider the following mixed integer programming formulation (CSTP) of the combined container stowage and transport planning problem:

$$\min \sum_{i \in \mathcal{L}} L_i \tag{1}$$

$$\text{s.t.} \sum_{i \in \mathcal{L}} \sum_{v \in \mathcal{V}} x_{civ} \leq 1 \quad \text{for all } c \in \mathcal{C} : t(c) = t(i) \tag{2}$$

$$\sum_{c \in \mathcal{C}} \sum_{v \in \mathcal{V}} x_{civ} = 1 \quad \text{for all } i \in \mathcal{L} : t(i) = t(c) \tag{3}$$

$$\sum_{j \in \mathcal{L}: j \leq i} \sum_{c \in \mathcal{C}} p_c \, x_{cjv} \leq T_i + L_i \quad \text{for all } i \in \mathcal{L}, v \in \mathcal{V} \tag{4}$$

$$x_{civ} \in \{0, 1\} \quad \text{for all } c \in \mathcal{C}, i \in \mathcal{L}, t(c) = t(i), v \in \mathcal{V} \tag{5}$$

$$L_i \geq 0 \quad \text{for all } i \in \mathcal{L} \tag{6}$$

The schedule is defined by assignment variables x_{civ}, where $x_{civ} = 1$ if and only if container $c \in \mathcal{C}$ is assigned to loading event $i \in \mathcal{L}$ and moved by straddle carrier $v \in \mathcal{V}$. An assignment x may imply that the i-th container arrives later at the quay crane than required, i.e., later than at time $T_i = T + (i-1)\frac{3600}{r}$ where r denotes the loading rate per hour in a regular loading sequence. The value of the variable L_i carries the resulting lateness. Since the CSTP contains no precedence constraints, the j-th container may arrive earlier than the i-th container despite of $i < j$. Here, we presume sufficient buffer space at the quay crane. In our computational results for real-world data buffer space for two containers was sufficient.

We discuss computational results for real-world data about four vessels provided by HHLA (cf. Table 1). For each quay crane, we solve the CSTP for different lengths of the loading sequence $\mathcal{L}$. Here, the startup offset value is $T = 90$, and the loading rate $r = 40$ implies a regular loading time of 90 seconds for all loading events of $\mathcal{L}$. We display results for three real-world instances. The lengths of $\mathcal{L}$ vary from 20 to 60 loading events which are the typical lengths of loading sequences dispatched in real-time at container terminals. We apply the standard MIP solver CPLEX 6.6 to the resulting instances of CSTP. We compare the best feasible solution determined within a real-time computation limit of 60 seconds to the final optimum solution. Within this one minute limit, we obtain quite good solutions which are in fact optimum solutions in most cases. We will take advantage of this good computational performance in section 3 where we describe an integrated approach for solving CSTP in a real-time setting. Smaller CSTP will iteratively be solved for each quay crane and for each part of the loading sequence. The number of iterations depends on the length of the partial loading sequence considered in one step. The length of a partial loading sequence strongly depends on the computation time available as well as on the real-time effects influencing the incumbent solution.

Instance	$\|\mathcal{L}\|$	Constraints	Variables	Nonzeros	1 min. UB	Optimum	CPU sec.
1	20	191	6680	83310	20	20	5.99
	30	231	10020	174915	20	20	21.98
	40	271	13360	299820	20	20	16.48
	50	311	16700	458025	60	20	916.27
	60	351	20040	649530	156	20	6037.04
2	20	130	335	3666	43	43	0.07
	30	199	627	8844	43	43	0.34
	40	248	961	17451	43	43	0.84
	50	298	1607	31797	43	43	2.72
	60	346	1881	49524	43	43	26.50
3	20	186	1844	17571	2894	2894	201.39
	30	396	3219	45306	3074	2894	463.17
	40	443	6274	104138	3002	2894	1289.70
	50	516	6893	170967	3201	2894	1821.24
	60	586	7236	241494	3178	2894	1737.07

Table 1. Computational results for CSTP within a one minute time limit applying CPLEX 6.6 MIP solver on a Pentium-III PC with 700 MHz and 1 GByte core memory. 1 min. UB is the lateness $\sum L_i$ of the best solution obtained within the one minute computation time limit. Opt. is the lateness of an optimum solution, as was proved after CPU sec

Precedence Constraints due to the Container Stacks

As mentioned in section 1, the containers are stored in stacks on the yard. In the considered terminal, these stacks consists of one, two, or three containers. Since straddle carriers can lift containers only up to layer three, a loaded straddle carrier cannot pass a stack of height three. Therefore, some care is necessary when using layer three. In particular, third layer containers are stored in such a way that no deadlocks occur and the third layer containers in a row have the same type. Third layers are only used for short time periods.

The question whether or not a feasible assignment of containers to loading events without rearranging stacks exists, is equivalent to a certain tram scheduling problem in depots in local transport which is known to be NP-complete [5]. The tram scheduling problem is described in example of [8] in this volume. In [5] a dynamic programming approach is used in order to decide whether or not a linear sequence of type-constrained events (tram departures) can be served by items (trams) stored in stacks (sidings). The resulting algorithm is polynomial in the number of departures (here: containers) and exponential in the number of stacks. For stacks of height two, the problem of minimizing the number of rearrangements can be reduced to a minimum weight perfect matching problem in a related graph and, therefore, it is polynomially solvable [16]. Due to the small height of container stacks, the related rearrangement problem in container terminals is solvable in reasonable time. For more details on the above mentioned problems, we refer to [5] and [16].

We may model necessary rearrangements of stacks in CSTP by increasing the transportation times of the affected containers so that additional time needed for the rearrangement is covered. Of course, this simple modification is correct only if the final rearrangement is already known. Alternatively, we may add penalty constraints for rearranging stacks and stack related precedence constraints (c.f. in [16]) to the CSTP. However, the raised complexity of modified enlarged CSTP reduces its applicability in real-time decision support systems.

Heuristic Approaches for more than one Quay Crane

The following best-fit heuristic offers an alternative to the exact algorithms solving the above mixed integer program. The best-fit heuristic can be applied in parallel for all quay cranes available.

Container-Best-Fit (CBF)

For each quay crane and each i-th loading event, we select an available straddle carrier and a previously not assigned container c of matching type minimizing the time delay $\max(\Theta - T_i, 0)$ to the actual delivering time Θ of c. We always prefer containers with $\Theta \geq T_i$.

Computational results for CBF are displayed in Table 2. We apply CBF to two real-world instances for different values for the loading rate and the (average) speed of the straddle carriers. We observe that a loading rate of about 40 containers per hour results (i.e., in intervals of 90 seconds) in a reasonable value of cumulative lateness and makespan. Simulation studies covering more side constraints promise a reduction of the time needed to load a vessel.

3 Real-Time Ship Planning

Ship planning in the real world has to handle uncertain, changing and missing data as well as general real-time influences. For example, decision support systems must provide proposals within sometimes quite tight time bounds. A short introduction to the general difficulties of combinatorial online optimization in real time can be found in this volume [8].

Here, based on the incomplete information available before the container ship enters the port, a stowage plan is prepared. This stowage plan is sent to the shipping

		570 containers		758 containers	
Loading rate	VC speed [$\frac{m}{s}$]	Lateness	Last event	Lateness	Last event
45	1.6	3 h 35'	4 h 46'	6 h 18'	6 h 57'
40	1.6	2 h 28'	4 h 53'	4 h 54'	7 h 09'
36	1.6	1 h 46'	5 h 05'	3 h 35'	7 h 21'
33	1.6	1 h 14'	5 h 23'	2 h 24'	7 h 37'
45	1.8	2 h 11'	4 h 20'	4 h 21'	6 h 21'
40	1.8	1 h 54'	4 h 33'	3 h 02'	6 h 33'
36	1.8	57'	4 h 51'	1 h 52'	6 h 50'
33	1.8	33'	5 h 11'	1 h 15'	7 h 05'
45	2.0	1 h 25'	4 h 04'	2 h 52'	5 h 52'
40	2.0	51'	4 h 22'	1 h 42'	6 h 09'
36	2.0	28'	4 h 42'	1 h 07'	6 h 25'
33	2.0	19'	5 h 03'	47'	6 h 49'
45	2.2	54'	3 h 55'	1 h 44'	5 h 32'
40	2.2	27'	4 h 14'	1 h 02'	5 h 48'
36	2.2	17'	4 h 35'	42'	6 h 11'
33	2.2	13'	4 h 57'	18'	6 h 35'

Table 2. Computational results for CBF for different values of loading rate and straddle carrier velocity. Lateness compared with the time of the last loading event ("makespan")

company querying for acceptance. When the vessel has arrived at its berth position, the quay cranes start discharging import containers and those containers that must be reloaded later on, possibly to a different bay position. When a quay crane finishes the discharge process, the loading process starts as described in the above accepted stowage plan. According to the corresponding loading sequences for the bay currently served, export containers are moved from the yard to the quay crane.

In particular, containers should be moved by straddle carriers as defined in the previously computed optimal or approximative assignment solution of the combined stowage and transport planning problem. Unfortunately, the stowage plan was generated using only incomplete information which is now out of date. Real-time effects influence the performance of the loading process and require a partial or complete update of previous assignments of containers and straddle carriers. Some examples of such real-time influences are:

- delay of a container's delivery to the terminal
- unavailability of a container due to customs regulations
- delay of a container's delivery to the quay crane due to high yard traffic
- delays in the loading process due to unavailable quay cranes

Due to real-time influences transportation times used in the model may differ substantially from the current transportation times. Due to delays, assigned containers may be not available on time. Then, if possible, different containers should be assigned to the loading sequence. Consequently, assignments of containers and straddle carriers must be adapted in real-time to the different online situations. Since the accepted stowage plan should not be changed, the resulting update problem is a just-in-time scheduling problem with due dates for each container.

Update algorithms working in real-time for changing time limits require a high flexibility. We shortly describe a possible algorithmic scheme. Whenever changed transportation times require an update of assignments of containers and straddle carriers, we re-optimize the next Δ, say twenty to thirty, assignments of containers and straddle carriers. The size of the update problem is chosen subject to the real-time requirements, i.e., we may use as much new information as possible in the available computation time.

Then, the complete remaining assignment of containers and straddle carriers is updated accordingly. In an update, we may apply a MIP solver for CSTP, dynamic programming, or heuristics like CBF. Furthermore, we may generate exact solutions to smaller update problems (less new information) or we may generate approximate solutions of larger update problems (more new information). In this way such algorithmic schemes allow to choose the amount of new information with regard to the real-time requirement. Here, as a result, the length of the adapted part of the assignment varies according to the available computation time. Similar "Δ-REPLAN" techniques have previously been proposed in [16, 17] for dispatch problems in local transport.

Failing availability of a quay crane is a severe online event requiring a more global update. Besides technical failure, a quay crane may be withdrawn in order to serve another vessel. In any case, the crane split has to be recomputed and bay areas will be redistributed among the remaining quay cranes. Crane split computation is very fast and can be performed within usual real-time requirements. Of course, updates for the assignment of containers and straddle carriers are required, too. In this way, the proposed combined stowage and transport planning approach allows to handle such failures, too.

4 Conclusion

In this article, we propose an integrated approach for combined stowage and transport planning in container terminals. The basic underlying concept of the resulting model is similar to a certain model for tram dispatch. In ship planning, containers are partitioned into classes of types. The shipping company defines type requirements for stack positions in the bays of a vessel. The dispatcher has to assign the export containers to matching stack positions in the bays. Contrary to tram dispatch in [5, 16], containers do not arrive in a completely predefined sequence. However, the set of export containers is stored in stacks, implying a partial order on the set of containers which may be modelled by precedence constraints. We propose a just-in-time scheduling model (CSTP) combining the stowage plan for the quay cranes and the transportation schedule for the straddle carriers. The resulting model as well as the proposed algorithms for solving the model are particularly suitable for real-time planning in maritime container terminals where various online and real-time influences require flexible response in order to guarantee and improve the overall performance of the terminal.

Acknowledgements

We would like to thank the Hamburger Hafen- und Lagerhaus AG (HHLA) for many helpful and informative discussions and, in particular, for providing real-world data which form the necessary basis for a useful evaluation of any practicable real-time approach.

REFERENCES

1. M. Avriel and M. Penn. Exact and approximate solutions of the container ship stowage problem. *Computers and Industrial Engineering*, 25(1-4):271–274, 1993.
2. Mordecai Avriel, Michal Penn, Naomi Shpirer, and Smadar Witteboon. Stowage planning for container ships to reduce the number of shifts. *Annals of Operations Research*, 76:55–71, 1998.
3. A. H. Aslidis. *Combinatorial Algorithms for Stacking Problems*. PhD thesis, Massachusetts Institute of Technology, January 1989.
4. A. H. Aslidis. Minimization of overstowage in containership operations. *Operational Research*, 90:451–471, 1990.
5. U. Blasum, M. R. Bussieck, W. Hochstättler, H. -H. Scheel and T. Winter. Scheduling Trams in the Morning. *Mathematical Methods of Operations Research*, 49 (1):137–148,1999.
6. Chuen-Yih Chen and Tung-Wei Hsieh. A time-space network model for the berth allocation problem. Presented at the 19 th IFIP TC7 Conference on System Modelling and Optimization, 1999.
7. Zhi-Long Chen and W. B. Powell. A column generation based decomposition algorithm for a parallel machine just-in-time scheduling problem. *European J. Oper. Res.*, 116:220–232, 1999.
8. M. Grötschel, Sven O. Krumke, Jörg Rambau, Thomas Winter and Uwe T. Zimmermann. Combinatorial Online Optimization in Real Time. appears in the same volume.
9. Ching-Fang Liaw. A branch-and-bound algorithm for the single machine earliness and tardiness scheduling problem. *Comput. Oper. Res.*, 26:679–693, 1999.
10. Andrew Lim. The berth planning problem. *European J. Oper. Res.*, 22:105–110, 1998.
11. Dirk Steenken, Andreas Henning, Stefan Freigang, and Stefan Voß. Routing of straddle carriers at a container terminal with the special aspect of internal moves. *OR Spektrum*, 15(3):167–172, October 1993.
12. D. Steenken. Integrierte DV-Systeme im Container-Umschlag. *Deutsche Verkehrs Zeitung (DVZ)*, 12, Dezember 1992.
13. Dirk Steenken. Fahrwegoptimierung am Containerterminal unter Echtzeitbedingungen. *OR Spektrum*, 14:161–168, 1992.
14. G. Steiner and S. Yeomans. Level schedules for mixed-model, just-in-time processes. *Management Sci.*, 39(6):728–735, 1993.
15. T. Winter and U. T. Zimmermann. Discrete Online and Real-Time Optimization. *Proceedings of the 15th IFIP World Computer Congress, Budapest/Vienna*, 1998.
16. T. Winter. Online and Real-Time Dispatching Problems. *PhD thesis. TU Braunschweig*, 1999.

17. T. Winter and U. T. Zimmermann. Real-time dispatch of trams in storage yards. *Annals of Operations Research* 96:287–315, 2000.

IX

Real-Time Annealing in Image Segmentation

*Basic Principles of Annealing for Large Scale Non-Linear Optimization

Joachim M. Buhmann[1] and Jan Puzicha[2]

[1] Institut für Informatik, Rheinische Friedrich-Wilhelm Universität Bonn, Germany
[2] Department of Computer Science, University of California, Berkeley, USA

Abstract *Computational Annealing*, a class of optimization heuristics that are inspired by statistical physics of phase transitions has been demonstrated to be highly effective for large, non-linear combinatorial optimization problems. In many applications in computer vision and pattern recognition one encounters non-linear objective functions with a very large number of discrete and possibly additional continuous variables. Typical cases of such problems are clustering, grouping and image segmentation or assignment problems in motion or stereo analysis or in object recognition. For this type of problems, standard integer programming techniques are not applicable and one has to resort to optimization heuristics that are fast, yet avoid a possibly exponential number of unfavorable local minima. A particularly powerful, generic class of algorithms is provided by simulated or deterministic annealing techniques. Simulated annealing and the Gibbs sampler are discussed first to present the basic concepts; then, the theory of deterministic annealing is presented in great detail and the relation to continuation methods are discussed.

1 Introduction

Heuristic optimization techniques are promising candidates to find at least approximately optimal solutions to very large combinatorial and mixed combinatorial optimization problems. Stochastic optimization methods which are inspired by statistical physics like *Simulated Annealing* (SA) or their deterministic variants *Deterministic Annealing* (DA) mimic the ordering process in solids during carefully controlled cooling. Kirkpatrick et al. [35] and, independently, Černy [10] have proposed in two seminal papers to apply SA as stochastic search strategy to large scale combinatorial and discrete optimization problems. The essential idea is to treat the variables of an optimization problem as random variables and to define a stochastic process which is controlled by the quality of solutions, i.e., it has to converge to solutions with low costs. The basic strength of these techniques is their flexibility and the ease of adaptation to new problems including non-linear and constrained cost functions. While convergence in probability to the global minimum has been established [18], SA techniques are often inherently slow because of their randomized local search strategy.

Two generic, closely related methods are reviewed in this chapter: a Monte Carlo algorithm known as the Gibbs sampler [18] and a deterministic variant known as *mean-field annealing* [3]. The two approaches estimate the random optimization variables either by sampling (SA) or by analytical calculation/approximation of their expectation value (DA). The degree of approximation is parameterized by a computational temperature in both methods, which offer a number of advantages:

1. They are general enough to cover a large class of linear and non-linear integer programming objective functions,
2. they yield *scalable* algorithms (in terms of the complexity-quality trade-off), and
3. they provide *robust* algorithms which are insensitive to the specific data instances and which, therefore, exhibit increased generalization performance.

While SA provides a universally applicable optimization principle, stochastic techniques are often inherently slow compared to deterministic algorithms. DA combines the advantages of a temperature controlled continuation method with a fast, purely deterministic computational scheme. To stress the possibility to canonically derive efficient optimization algorithms for general grouping and quantization problems, results are presented which apply to a large class of grouping objective functions.

The underlying rationale of DA is to replace the search through a discrete search space of a combinatorial optimization problem by an optimization problem over the space of probability distributions over the discrete solution space. That means that we calculate exactly or approximate the stationary probability distribution of the stochastic search process by analytical techniques rather than by sampling. This program is carried out by analytically calculating the characteristic function of these distributions which is known as the free energy in statistical physics. The characteristic function yields a complete characterization of the statistical equilibria. While DA has originally been motivated as an analytic approximation of the simulated annealing algorithm, it is analyzed from a purely mathematical viewpoint providing additional insight in its algorithmic structure. The DA scheme relies on two major ideas.

(i) A *relaxation* is introduced, i.e. an embedding of the combinatorial search space in a continuous optimization space. The main reason for embedding the original discrete search space $\mathcal{C}$ into a continuous search space is to avoid the integer constraints inherent in $\mathcal{C}$. In DA, the space of probability distributions $\mathcal{P}$ or a suitably restricted subspace provide the relaxation spaces. In the probabilistic relaxation, the original cost function F is replaced by the expected costs under a probability distribution.

(ii) A *homotopy* is defined to enable *continuation methods* [1] for optimization. The underlying rationale of a homotopy is the smooth deformation of a non-convex functional realized by a one-parametric family of cost functions. These cost functions become convex for one end of the homotopy parameter range and converge to the original cost function at the other end of the parameter range. Continuation methods proceed by tracking the solution computed for the convex functional while varying the homotopy parameter. In DA, the inverse temperature parameterizes the homotopy induced by the generalized free energy. The idea of annealing in the context of DA refers to the tracking of solutions from

high temperatures where the free energy is convex, to zero temperature where the original cost function is recovered.
The main components of any continuation method are the *predictor* and the *corrector step*. In the predictor step, a new value of the variable configuration is extrapolated for a small variation of the homotopy parameter. In the corrector step, a new variable configuration is computed by minimizing the cost function corresponding to the new homotopy parameter. In DA, the step size is controlled by the *annealing schedule*, i.e., how fast the temperature is decreased as a function of iteration steps. The predictor is chosen as a constant extrapolation, i.e. the variable state of the last iteration is simply kept as initial condition. Usually, iterative update equations derived by differentiation of the generalized free energy provide a suitable, efficient corrector method.

DA methods are applicable to non-linear cost functions and yields efficient algorithms which possess a favorable scaling behavior in terms of computational complexity. DA has empirically shown to compute optimal or near-optimal solutions [26, 30, 46], which makes it a promising optimization procedure for large problem instances. Global optimality, however, has not yet been established even for careful annealing. Convergence of DA to a local optimum of the effective cost function is established for a large class of assignment cost functions [50].

The presentation proceeds as follows: First, the class of clustering optimization problems is introduced and motivated with k-means and pairwise clustering. Then several fundamental definitions are formally introduced, followed by a discussion of the maximum entropy principle, the Metropolis algorithm and the simulated annealing optimization technique. In the subsequent sections, the deterministic annealing optimization strategy is introduced and the mathematical derivation is presented.

2 The Class of Clustering Problems

For simplicity, we focus the following discussion of annealing techniques on the class of *grouping* or *clustering* problems. Let $\mathcal{U} = \{u_1, \ldots, u_n\}$ denote a set of n (abstract) objects. The number of groups k is assumed to be fixed. The grouping task is formalized as a *disjunctive partition* of the set of objects $\mathcal{U}$ into groups $\mathcal{G}_c \subset \mathcal{U}$, $c = 1, \ldots, k$ with $\dot{\bigcup}_{c=1}^{k} \mathcal{G}_c = \mathcal{U}$. As a suitable coding of such a data partition, an assignment function $c : \mathcal{U} \to \{1, \ldots, k\}$ is introduced. $c(u) = c$ denotes an assignment of object u to cluster c, hence $c(u) = c$ if and only if object $u \in \mathcal{G}_c$. The space of all possible assignments is denoted by

$$\mathcal{C}_{n,k} = \bigotimes_{u \in \mathcal{U}} \{1, \ldots, k\} \subset \mathbb{N}^n . \tag{1}$$

For notational convenience an explicit dependency on n and k is often dropped if it is obvious from the context. To motivate the general definitions and assumptions on clustering cost functions we first discuss the most widely used clustering criterion of grouping vectorial data into k clusters.

```
INPUT vectorial data X, number of clusters k
INITIALIZE c randomly
repeat
  RECALCULATE Y according to (4)
  RECALCULATE c according to (5)
until CONVERGENCE()
```

Algorithm 1: k-means Clustering

Central Clustering. Vectorial data $\mathbf{X} = (\mathbf{x}_u)_{u \in \mathcal{U}} \in \mathcal{X}^{vec}$, $\mathbf{x}_u \in \mathbb{R}^d$ is the most common data type. Many algorithms have been developed for grouping of vectorial data, which are known as *Central Clustering* in data analysis or as *Vector Quantization* in coding. The most important optimization model is the k-means cost function defined by

$$H^{km}(c, \mathbf{Y}; \mathbf{X}) = \sum_{u \in \mathcal{U}} \|\mathbf{x}_u - \mathbf{y}_{c(u)}\|^2. \tag{2}$$

Here, the $\mathbf{y}_c \in \mathbb{R}^d$ denote d-dimensional cluster prototypes and they constitute a continuous parameter space $\mathcal{Y}_k \subseteq \mathbb{R}^{d \cdot k}$. The k-means criterion[1]

$H^{km}(c, \mathbf{Y}; \mathbf{X})$ defines a monotone clustering criterion, i.e., the global minimum of a k partition is always an upper bound on the global minimum of a $k+1$ partition. Intuitively, this property is important in our view, since a clustering criterion should always make use of additional structure, e.g., an additional cluster parameter. Clustering cost functions which lack this property are biased towards preferred cluster numbers, but these preferences should be inferred from the data by appropriate inference principles rather than being encoded in the bias of a search criterion.

A slightly more general form of k-means clustering is introduced by considering an additional object-specific scalar weighting $w(u) \geq 0$, $(\sum_u w(u) = 1)$ for each object u,

$$H^{gkm}(c, \mathbf{Y}; \mathbf{X}) = \sum_{u \in \mathcal{U}} w(u) \|\mathbf{x}_u - \mathbf{y}_{c(u)}\|^2. \tag{3}$$

The minimization of the cost function (3) w.r.t. $\mathbf{y}_c \in \mathcal{Y}_k$, $c \in \{1, \dots, k\}$ can be solved for arbitrary partitions leading to the generalized centroid equations

$$\mathbf{y}_c = \frac{\sum_{u \in \mathcal{G}_c} w(u)\, \mathbf{x}_u}{\sum_{u \in \mathcal{G}_c} w(u)}. \tag{4}$$

For constant weights, the prototypes are the centroids of the associated data objects $\mathbf{y}_c = \sum_{u \in \mathcal{G}_c} \mathbf{x}_u / |\mathcal{G}_c|$. The optimal grouping solution for fixed centroids $\mathbf{y}_c$ is given by

$$c(u) = \arg\min_c w(u) \|\mathbf{x}_u - \mathbf{y}_c\|^2. \tag{5}$$

[1] We use the notation for conditioning widely used in statistics that data are separated by a semicolon in the argument of (2).

This rule leads to a simple alternating minimization algorithm known as k-means algorithm, which converges to a local minimum of (2). The k-means algorithm is summarized in Alg. 1.

Graph Partitioning. Frequently, an explicit representation of data as vectors in a d-dimensional Euclidian space is not available for data analysis but objects are implicitly characterized by their mutual dissimilarities $\mathbf{D} \in \mathcal{X}^{\text{prox}}$. Here, $\mathbf{D}$ denotes a matrix of pairwise dissimilarity scores between objects $u, v \in \mathcal{U}$. More generally, sparse representations are often necessary due to the scaling behavior of $\mathbf{D}$ which is quadratic in the number of objects. Therefore, we introduce a graph notation $(\mathcal{U}, \mathcal{E})$ with an edge set $\mathcal{E} \subset \mathcal{U}^2$ for convenience. $(u, v) \in \mathcal{E}$ denotes that there exists a dissimilarity measurement $D_{uv} = D_{vu}$ for the objects u and v. By $\mathbf{D} = (D_{uv})_{(u,v)\in\mathcal{E}}$ the sparse matrix of existing dissimilarity measurements is summarized. Note, that reflexive graphs are permitted and, thus, self-similarities D_{uu} are not explicitly excluded. For a given object u the *graph neighborhood* $\mathcal{N}_u \subset \mathcal{U}$ is defined as the set

$$\mathcal{N}_u = \{v \in \mathcal{U} : (u, v) \in \mathcal{E} \vee (v, u) \in \mathcal{E}\}. \tag{6}$$

The most popular cost function for proximity data [31] is the *graph partitioning* cost function

$$H^{gp}(c; \mathbf{D}) = \frac{1}{n} \sum_{\substack{(u,v)\in\mathcal{E}: \\ c(u)=c(v)}} D_{uv}. \tag{7}$$

For notational convenience, the following sets of objects are introduced for a given, fixed partition c. Let

$$\mathcal{G}_c(u) = \{v \in \mathcal{G}_c : v \in \mathcal{N}_u\}, \tag{8}$$
$$\mathcal{E}_{cd} = \{(u, v) \in \mathcal{E} : (u \in \mathcal{G}_c \wedge v \in \mathcal{G}_d) \vee (u \in \mathcal{G}_d \wedge v \in \mathcal{G}_c)\}. \tag{9}$$

Thus, $\mathcal{G}_c(u)$ denotes the set of objects that are in the graph neighborhood $\mathcal{N}_u$ of u and belong to the cluster c. $\mathcal{E}_{cd}$ denotes the set of all edges with one vertex belonging to cluster c and one vertex belonging to d. $\mathcal{E}_c := \mathcal{E}_{cc}$ denotes the set of all edges with both vertices in cluster c. The graph partitioning cost function can then be rewritten as

$$H^{gp}(c; \mathbf{D}) = \frac{1}{n} \sum_{c=1}^{k} \sum_{(u,v)\in\mathcal{E}_c} D_{uv}. \tag{10}$$

Thus, a cluster specific score is computed simply by adding up all known dissimilarities between objects in that cluster. The final value is obtained by summing over all cluster scores. One should note that the graph partitioning cost function is applicable solely to ratio scale data where positive and negative dissimilarities must be carefully balanced. A negative dissimilarity is interpreted as a vote to join both objects in the same cluster while positive dissimilarities provide an indicator to separate the objects. Therefore, graph partitioning does not define a monotone clustering criterion as can be seen from the case of purely negative dissimilarities. For all $D_{uv} < 0$ the global minimum of $H^{gp}(c; \mathbf{D})$ is achieved by assigning all objects to one cluster. The opposite limit of $D_{uv} \gg 0$ favors partitions with exactly the same number of objects despite the fact that data sources might generate objects with different frequencies.

Pairwise Clustering. This observation of the sensitivity on the mean dissimilarity motivates a normalized cost function which is given by

$$H^{pc}(c; \mathbf{D}) = \sum_{c=1}^{k} |\mathcal{G}_c| \, \bar{D}_c \quad \text{with} \quad \bar{D}_c = \frac{1}{|\mathcal{E}_c|} \sum_{(u,v) \in \mathcal{E}_c} D_{uv}. \tag{11}$$

The variable $\bar{D}_c$ denotes the average dissimilarity of the cluster c. This cost function is invariant to the shifts of the mean dissimilarity score, at the expense of a normalization factor which makes it impossible to linearize the costs $H^{pc}(c; \mathbf{D})$. This case is a typical instance of a non-linear integer program encountered in many pattern recognition applications.

General Partitioning Problem. A theory of optimization is developed in the following for a generic cost (objective) function $F_{n,k}(c, \mathbf{Y}; \mathbf{X})$ with n discrete k-state assignment variables summarized by $c \in \mathcal{C}_{n,k}$ and a k-tuple of d-dimensional, real-valued optimization parameter $\mathbf{Y} \in \mathcal{Y}_k \subset \mathbb{R}^{k \cdot d}$. Here, the cost function is parametrized by a fixed observation $\mathbf{X} \in \mathcal{X}(n)$ from some (abstract) measurement space $\mathcal{X}(n)$.

Definition 1. Objective Function Let $\mathcal{Y}_k$ be a compact subset of $\mathbb{R}^{k \cdot d}$. The set

$$\mathcal{F}_{n,k} = \{F : \mathcal{C}_{n,k} \times \mathcal{Y}_k \to \mathbb{R} : \; \exists b \in \mathbb{R} \, \forall c, \mathbf{Y} \; F(c, \mathbf{Y}) \geq b > -\infty\} \tag{12}$$

is called the space of objective functions defined over $\mathcal{C}_{n,k}$ and $\mathcal{Y}_k$.

An objective function has to be bounded from below by some constant b to ensure the existence of a global minimum and thus a *globally optimal variable state*

$$(c^*, \mathbf{Y}^*) = \arg \min_{(c,\mathbf{Y})} F(c, \mathbf{Y}). \tag{13}$$

The continuous parameters $\mathbf{Y}$ are typically group-specific parameters and they characterize complete cluster properties. In the k-means case the parameters $\mathbf{Y}$ are the centroids. They solely simplify the representation and the algebraic structure of the cost function since they can be canonically removed from the formulation by defining

$$F(c) = \min_{\mathbf{Y} \in \mathcal{Y}_k} F(c, \mathbf{Y}), \tag{14}$$

which introduces a cost function depending solely on c. This problem formulation is equivalent to (13) in the sense that both cost functions possess identical minima in $\mathcal{C}_{n,k}$. For most of the typical grouping objective functions (14) can be solved analytically. Next, the notion of a clustering criterion is introduced, which basically formalizes a class of algebraically related cost functions.

Definition 2. (Clustering Criterion) A family $\mathcal{H}$ of objective functions

$$\mathcal{H} = \{\mathsf{H}_{n,k}(\mathbf{c}, \mathbf{Y}; \mathbf{X}) \in \mathcal{F}_{n,k} \; : \; \mathbf{X} \in \mathcal{X}(n), \; 2 \leq n < \infty, \; 1 \leq k \leq n\} \tag{15}$$

is called a clustering criterion. A clustering criterion is called monotone, if for all $\mathbf{X}$ and for all $k \geq 1$

$$\min_{\mathbf{c},\mathbf{Y}} \mathsf{H}_{n,k+1}(\mathbf{c}, \mathbf{Y}; \mathbf{X}) \leq \min_{\mathbf{c},\mathbf{Y}} \mathsf{H}_{n,k}(\mathbf{c}, \mathbf{Y}; \mathbf{X}) \tag{16}$$

Thus, a clustering criterion provides a unique objective function for each $n \geq 2$, each $1 \leq k \leq n$ and each measurement instance $\mathbf{X} \in \mathcal{X}(n)$. For a monotone clustering criterion, solutions with decreasing costs are obtained as optimal solutions, if the number of groups is increased. As discussed in the k-means case monotonicity avoids a preference (bias) of a particular number of clusters without reference to the data.

3 Algorithms for Clustering

A cost function F defines an unconstrained, mixed combinatorial optimization problem, which is typically a non-linear function in its optimization variables $(\mathbf{c}, \mathbf{Y}) \in \mathcal{C} \times \mathcal{Y}$. Therefore, the development of efficient, yet *global* optimization algorithms is essential, i.e. algorithms which avoid getting trapped in local minima at least to some degree. The following definitions can be generalized to cases where the assignment configuration space $\mathcal{C}$ is replaced by an abstract configuration space $\mathcal{Z} = \bigotimes_{u \in \mathcal{U}} \mathcal{Z}_u$, with a local configuration space $\mathcal{Z}_u$ being associated with each site u as a minimal requirement. The more general notation becomes necessary for the development of deterministic annealing algorithms, which typically replace $\mathcal{C}$ by a probabilistic search space. Elements of $\mathcal{Z}$ are denoted by $\mathbf{Z} = (\mathbf{z}(u))_{u \in \mathcal{U}}$.

The notion of a locally optimal state is well-defined for continuous variables as $\mathcal{Y}$ inherits the natural topology of $\mathbb{R}^{k \cdot d}$. For discrete spaces, the definition of a local minimum depends on an additional *topology in optimization space*. Let

$$\mathsf{M}(\mathbf{Z}, \mathbf{Z}') = |\{u \in \mathcal{U} : \mathbf{z}(u) \neq \mathbf{z}'(u)\}| \tag{17}$$

define a metric on the space $\mathcal{Z}$ which is known as the *Hamming distance*. This metric induces a topology in the search space $\mathcal{Z}$. The set $\mathcal{N}^l_{\mathbf{Z}} = \{\mathbf{Z}' : \mathsf{M}(\mathbf{Z}, \mathbf{Z}') \leq l\}$ is called the l-neighborhood of $\mathbf{Z}$. In the following, the attention is restricted to 1-neighborhood optimality. This motivates the following definition.

Definition 3. (Minimum) A variable state $(\mathbf{Z}^*, \mathbf{Y}^*)$ is called a local minimum of F, if it is 1-neighborhood optimal. It is called a global minimum, if the condition holds:

$$(\mathbf{Z}^*, \mathbf{Y}^*) = \arg \min_{(\mathbf{Z},\mathbf{Y})} \mathsf{F}(\mathbf{Z}, \mathbf{Y}; \mathbf{X}) \tag{18}$$

The main focus of this chapter is on the development of numerical algorithms to compute a solution of (18). Formally, the notion of a (possibly stochastic) optimization algorithm is introduced as a sequence of states $(\mathbf{Z}^{(t)}, \mathbf{Y}^{(t)})$.

Definition 4. (Optimization Algorithm) A map

$$A_F : \mathcal{Z} \times \mathcal{Y} \times \mathcal{X} \times \mathcal{R}^N \longrightarrow (\mathcal{Z} \times \mathcal{Y})^N \quad (19)$$

$$(\mathbf{Z}^{(0)}, \mathbf{Y}^{(0)}, \mathbf{X}, (r^{(t)})) \longmapsto \left((\mathbf{Z}^{(t)}, \mathbf{Y}^{(t)})\right)_{t \in N} \quad (20)$$

is called an optimization algorithm for F, where $r^{(t)}$ is a real-valued, uniformly distributed random number generated by a source $\mathcal{R}$. The state $(\mathbf{Z}^{(0)}, \mathbf{Y}^{(0)})$ is called the initial configuration which might itself be a random variable. A variable state $(\mathbf{Z}^{(t)}, \mathbf{Y}^{(t)})$ may depend on $r^{(t)}$ and on $(\mathbf{Z}^{(t')}, \mathbf{Y}^{(t')})$ with $t' < t$, but not on $r^{(t')}$.

- An algorithm A_F is called deterministic, if $(\mathbf{Z}^{(t)}, \mathbf{Y}^{(t)})$ does not depend on $r^{(t)}$ for any $t > 0$. Otherwise it is called stochastic.
- A_F is called convergent, if $(\mathbf{Z}^*, \mathbf{Y}^*) = \lim_{t \to \infty} (\mathbf{Z}^{(t)}, \mathbf{Y}^{(t)})$ exists and $(\mathbf{Z}^*, \mathbf{Y}^*)$ is a local minimum of F.
- A_F is called an alternating minimization algorithm, if $\mathbf{Y}^{(t+1)}$ is computed by

$$\mathbf{Y}^{(t+1)} = \arg \mathrm{Min}_{\mathbf{Y} \in \mathcal{Y}} F(\mathbf{Z}^{(t+1)}, \mathbf{Y}). \quad (21)$$

  The deterministic minimization operator[2] Min computes the next local minimum of F with respect to $\mathbf{Y}$ for the initial state $\mathbf{Y}^{(t)}$.

For a deterministic algorithm, the initial state provides the only source of randomness, while for a stochastic algorithm, each *state transition* $(\mathbf{Z}^{(t)}, \mathbf{Y}^{(t)}) \to (\mathbf{Z}^{(t+1)}, \mathbf{Y}^{(t+1)})$ can be stochastic in nature. Convergence is a natural requirement for any algorithm. In the numerical implementation, convergence is implemented by two thresholds ϵ_Z and ϵ_Y. The numerical convergence of an algorithm is then tested by the simultaneous validity of the conditions $M\left(\mathbf{Z}^{(t)}, \mathbf{Z}^{(t-1)}\right) < \epsilon_Z$ and $||\mathbf{Y}^{(t)} - \mathbf{Y}^{(t-1)}|| < \epsilon_Y$. Alternating minimization provides the key concept to address mixed combinatorial optimization problems, as it effectively decouples the joint problem in a purely combinatorial and a purely continuous part. Alternating minimization is employed in all algorithms developed in the sequel. As seen from the definition, for the continuous minimization only local optimization strategies are employed.

The configuration space for grouping problems $\mathcal{Z} = \mathcal{C} = \bigotimes_{u \in \mathcal{U}} \mathcal{C}_u$ naturally decomposes into single site configurations $\mathcal{C}_u$. The cardinality of $\mathcal{C}$ grows with $\mathcal{O}(k^n)$ which is exponential in the number of sites. Thus, exhaustive search becomes prohibitive even for medium size problems and alternative search strategies have to be employed. This motivates the definition of a local algorithm.

Definition 5. (Local Algorithm) An algorithm A_F is called local (with respect to the 1-neighborhood), if at most one site configuration is modified in each step, i.e. if

$$\forall t \in \mathbb{N} : \; M\left(\mathbf{Z}^{(t)}, \mathbf{Z}^{(t+1)}\right) \leq 1. \quad (22)$$

[2] The capitalized notation Min is used to distinguish this local operator from the global minimization operator min.

– A map $V : \mathbb{N} \longrightarrow \mathcal{U}$ is called a site visitation schedule for the local algorithm A_F, if for all $t \in \mathbb{N}, u \in \mathcal{U}, (r^{(t)}) \in \mathcal{R}^{\mathbb{N}}$

$$u \neq V(t) \Longrightarrow \mathbf{Z}^{(t+1)}(u) = \mathbf{Z}^{(t)}(u) \tag{23}$$

and if all sites are visited infinitely often.

Markovian algorithms provide an especially important class of local, stochastic algorithms. In this section we will define several concepts from information theory which allow us to characterize the properties of annealing algorithms. $c_{u\to d}$ denotes a locally modified assignment vector, which is obtained by setting $c(u) = d$ and by keeping all other function values of c unchanged.

Definition 6. (Markovian Algorithm) A local algorithm A_F is called Markovian, if $c^{(t)}$ is a Markov chain, i.e. if there exists a site visitation schedule V and a state transition probability $P_{u,t}(d \,|\, c, \mathbf{Y})$ such that $c^{(t+1)} = c^{(t)}_{V(t)\to d}$ and d is distributed according to $P_{V(t),t}(\cdot \,|\, c^{(t)}, \mathbf{Y}^{(t)})$.

Thus, the outcome of a local Markovian algorithm depends only on the previous state and a single random number distributed according to the state transition probability. Local Markovian algorithms are stochastic algorithms, which only locally change the state of the assignment variable. The site visitation schedule simply selects which assignment variable does change for a given algorithmic time step t.

To further analyze stochastic algorithms over $\mathcal{C}$, we study the space $\mathcal{P}$ of all probability distributions over $\mathcal{C}$,

$$\mathcal{P} = \mathcal{P}(\mathcal{C}) = \left\{ P : \mathcal{C} \longrightarrow [0,1] : \sum_{c\in\mathcal{C}} P(c) = 1 \right\}. \tag{24}$$

The expectation of a function F w.r.t. P is denoted by $\mathbf{E}_P[F]$. Moreover, $P_c \in \mathcal{P}$ denotes the Dirac distribution at c, i.e. $P_c(c) = 1$.

An entity, which is of special interest for the design of stochastic algorithms, is the entropy of a probability distribution. The entropy measures the information content of a probability distribution. This topic has been extensively studied in Information and Communication Theory (see [12] for an axiomatic justification).

Definition 7. (Entropy) The functional

$$\mathcal{S}(P) = -\mathbf{E}_P[\log P] = -\sum_{c\in\mathcal{C}} P(c) \log P(c) \tag{25}$$

is called the entropy of a distribution P, where $0 \log 0 = 0$ by continuation.

The entropy is a measure of the uncertainty of a random variable distributed according to P. Moreover, the entropy is bounded by $\log_2 \mathcal{C} \geq \mathcal{S}(P) \geq 0$, where $\mathcal{S}(P) = 0 \Longleftrightarrow P = P_c$ for some c. The maximal value $\log_2$ is obtained if all configurations $c \in \mathcal{C}$ are assumed with the same probability $P(c) = |\mathcal{C}|^{-1}$.

A fundamental family of probability distributions associated with a cost function F is the Gibbs distribution P_F.

Definition 8. (Gibbs Distribution) The probability distribution

$$P_{F,\beta,\mathbf{Y}}(c) = \frac{1}{\mathcal{Z}_{F,\beta,\mathbf{Y}}} \exp\left(-\beta F(c,\mathbf{Y})\right) \tag{26}$$

is called a Gibbs distribution associated with F. The normalization constant $\mathcal{Z}_{F,\beta,\mathbf{Y}}$ is called the partition function, the free parameter $T = 1/\beta$ is called computational temperature.

The notation is chosen by analogy to statistical physics where these distributions have been studied since Boltzmann's ground-breaking work on gas dynamics. The Gibbs distribution is parameterized by the inverse temperature β and the state of the continuous variable $\mathbf{Y}$. For notational convenience, the explicit dependency is dropped in the sequel, whenever unmistakable. The Gibbs distribution can be motivated by the *maximum entropy* inference principle [32] as the probability distribution P with maximal entropy for a fixed expectation $\mathbf{E}_P[F]$. Alternatively, the Gibbs distribution minimizes the expected costs $\mathbf{E}_P[F]$ constrained to all distributions with a fixed entropy. The computational temperature then plays the role of a Lagrange parameter. More formally, the Gibbs distribution with inverse temperature β minimizes the Lagrangian known as *generalized free energy* over $\mathcal{P}$.

Definition 9. (Generalized Free Energy) The functional

$$\mathcal{F}_\beta(P,\mathbf{Y}) = \mathbf{E}_P[F] - \frac{1}{\beta}\mathcal{S}(P) = \sum_{c\in\mathcal{C}} P(c)F(c,\mathbf{Y}) + \frac{1}{\beta}\sum_{c\in\mathcal{C}} P(c)\log P(c) \tag{27}$$

defined on $\mathcal{P}$ is called the generalized free energy w.r.t. P. The value of $\mathcal{F}_\beta(P_{F,\beta,\mathbf{Y}})$ which is the minimum of the generalized free energy is simply called the free energy and is given by $\mathcal{F}_\beta(P_{F,\beta,\mathbf{Y}}) = -\frac{1}{\beta}\log \mathcal{Z}_{F,\beta,\mathbf{Y}}$.

The functional $\mathcal{F}$ is called generalized free energy again in reminiscence to the nomenclature used in statistical physics. It can be seen from its functional form that the minimizing Gibbs distribution optimizes a mixture of the expected costs $\mathbf{E}[F]$ and of the entropy, while the inverse computational temperature balances these competing effects. The generalized free energy plays a fundamental role in the derivation of deterministic annealing algorithms.

4 Annealing Algorithms

Alternating minimization as defined in Def. 4 provides an efficient way to decouple mixed combinatorial problems into a continuous and a purely combinatorial task. While standard local optimization routines are applicable for continuous optimization, it is a central question how to design optimization procedures for combinatorial problems $F(c)$ over $\mathcal{C}$, which are capable to avoid local minima intrinsic in almost all combinatorial optimization problems.

Simulated Annealing. The underlying mathematical idea can be stated as follows. The Gibbs distribution P_F converges to the uniform distribution over $\mathcal{C}$ for $\beta \to 0$ and it converges to the uniform distribution on the set of global minima of $F(c)$ for $\beta \to \infty$. Thus, finding a global optimum of F can be understood as sampling from the Gibbs distribution for $\beta \to \infty$. While this is difficult without solving the optimization problem directly, sampling the approximately uniform distribution for small β is feasible. The key idea of annealing is to sample $P_{F,\beta}$, but to gradually increase the inverse temperature β during the sampling process. Monte Carlo sampling provides a generic possibility to sample from the Gibbs distribution for arbitrary, but fixed β.

Definition 10. (Monte Carlo Sampler) A local algorithm is called a Monte Carlo sampling scheme for a probability distribution P, if

$$\lim_{t \to \infty} P(c^{(t)}) = P(c). \tag{28}$$

A Monte Carlo sampler is called a Monte Carlo Markov chain (MCMC), if it is implemented as a local Markovian algorithm.

A valid MCMC scheme can be mathematically characterized by two properties, the irreducibility of the underlying Markov chain [38] and the *detailed balance* condition

$$P_t(c|c')P_F(c') = P_t(c'|c)P_F(c), \tag{29}$$

where the state transition probabilities $P_t(c|c')$ for the algorithmic time step t are induced by the transition probabilities of the local Markovian algorithm as defined in Def. 5.

The *Gibbs sampler* provides an example of an MCMC method, which is particularly efficient, if the conditional distribution of one site given fixed assignments for all other sites can be calculated efficiently. This property holds for the cost functions mentioned so far. The Gibbs sampler is implemented as a local Markovian algorithm, which performs only state transitions between configurations, which differ in the assignment of at most one site. Following Def. 5, it is completely specified by a site visitation schedule V and a state transition probability $P_{u,t}(d|c)$.

Definition 11. (Gibbs Sampler) For a fixed site visitation schedule V the Gibbs sampler is defined as the local Markovian algorithm with the state transition probabilities

$$P_{u,t}(d|c) = \frac{\exp[-\beta(t) g_u(d)]}{\sum_{e=1}^{k} \exp[-\beta(t) g_u(e)]}, \tag{30}$$

where $g_u(d) = F(c_{u \to d})$ and $u = V(t)$.

The Gibbs sampler draws a new state for site $V(t)$ from the conditional distribution $P_{u,t}(d|c)$ given the assignments at all other sites $\{v : v \neq V(t)\}$. The entities $g_u(d)$ are called Gibbs fields by analogy to statistical physics. The Gibbs sampler can be

```
INPUT c^(0), β^start, β^final, it^max, site visitation schedule V
SET β = β^start
repeat
  for i = 1,...,it^max, do
    CALCULATE g_V(t)(d) = F(c_V(t)→d) for all d
    SAMPLE d according to (30)
    SET c(V(t)) = d
    SET t = t + 1
  end for
  SET β = SCHEDULE()
until β > β^final
```

Algorithm 2: Simulated annealing

efficiently implemented, if and only if the Gibbs fields can be calculated efficiently. For constant $\beta(t)$, the Markov chain defined by (30) fulfills the detailed balance condition (29) and, therefore, converges towards its equilibrium distribution.

The basic idea of annealing is to use Monte Carlo sampling, but to gradually raise the inverse temperature $\beta(t)$, on which the transition probabilities depend. This forces the system into solution states with low costs. The effect of the temperature can be seen as a random force with an amplitude inversely proportional to β. Cost differences smaller than $1/\beta$ smear out and vanish in the stochastic search. For a logarithmic annealing schedule $\beta(t) = s(1 + \log t)$ the Gibbs sampler converges in probability to the uniform distribution on the global minima of F [18,24]. Of course, in practice β is increased too fast to guarantee convergence to a global minimum. Large deviation estimates applied to exponential schedules are discussed in [7–9]. SA offers a generally applicable, heuristic random search strategy. Despite its success it has the reputation of being slow compared to deterministic optimization techniques. The main reason is the random walk behavior of the sampling process. As one of the major motivations deterministic annealing tries to overcome this deficiency. The SA algorithm is summarized in Alg. 2.

In the infinite inverse temperature limit $\beta \to \infty$, only state transitions are accepted which do not increase the configuration costs. In this limit, a deterministic greedy optimization algorithm known as *Iterated Conditional Mode* (ICM) is obtained. The probabilistic sampling step $P_{u,t}(d|c)$ of (30) degenerates to a simple minimum search.

Definition 12. (ICM) For a fixed site visitation schedule V the local, deterministic algorithm with $c^{(t+1)} = c^{(t)}_{V(t)\to d}$ where

$$d = \arg\min_{e} g_{V(t)}(e), \tag{31}$$

is called the Iterated Conditional Mode (ICM) algorithm.

The ICM algorithm thus simply selects the state of maximal conditional probability instead of sampling. Again, the efficiency of the ICM algorithms critically depends on an efficient evaluation of the Gibbs fields. The ICM algorithm may serve as an algorithmic step in the alternating minimization scheme as described in Def. 4, i.e. a complete ICM algorithm is run on $H(c, \mathbf{Y}^{(t)})$ with $c^{(t)}$ as initial condition to compute $c^{(t+1)}$. The ICM algorithm for a purely combinatorial problem is summarized in Alg. 3.

```
INPUT c^(0), site visitation schedule V
repeat
  CALCULATE g_V(t)(d) = F(c_V(t)→d) for all d
  UPDATE c(V(t)) according to (31)
  SET t = t + 1
until CONVERGENCE()
```

Algorithm 3: Iterated conditional mode (ICM)

```
INPUT Y^init, β^start, β^final,
SET β = β^start, Y = Y^init
repeat
  repeat
    CALCULATE average costs E [F(c, Y; X)] w.r.t. P_F,β,Y
    MINIMIZE E [F(c, Y; X)] w.r.t. Y
  until CONVERGENCE()
  SET β = SCHEDULE()
until β > β^final
```

Algorithm 4: Deterministic annealing

Deterministic Annealing. While SA performs a stochastic search over a discrete space, DA is a deterministic optimization procedure with probabilistic search space. DA determines a probability distribution which minimizes the generalized free energy $\mathcal{F}_\beta(P, \mathbf{Y})$ (27) over the space of all distributions $\mathcal{P}$ defined on $\mathcal{C}$ or a suitably restricted subspace of distributions $\mathcal{Q}$. For the non-restricted case $\mathcal{Q} = \mathcal{P}$, the minimum of the free energy with respect to $\mathcal{Q}$ is known to be the (temperature dependent) Gibbs distribution $P_{F,\beta,\mathbf{Y}}$. Then, alternating minimization by averaging the cost function F with respect to $P_{F,\beta,\mathbf{Y}}$ and minimizing the average costs with respect to the continuous parameters $\mathbf{Y}$ provides the corrector step. Deterministic annealing in this abstract form is depicted in Alg. 4. However, depending on the algebraic form of F the necessary averages $\mathbf{E}_{P_{F,\beta,\mathbf{Y}}}[F]$ are computationally intractable since there is no efficient way known to calculate the exponentially large sum $\mathcal{Z}_{F,\beta,\mathbf{Y}}$. For the general case of $\mathbf{Q} \subset \mathcal{P}$ several important algorithmic issues have to be addressed in order to define a tractable procedure.

- The definition of a *proper relaxation space*. The deterministic annealing approach relies on the possibility of minimizing the generalized free energy. As a consequence, deterministic annealing only results in a tractable procedure for $\mathcal{P}$, if an explicit summation over $\mathcal{C}_{n,k}$ can be avoided in calculating the averages. Otherwise the calculation of assignment probabilities would require an exhaustive overall evaluation of F. In such cases, the space $\mathcal{P}$ must be restricted to a suitable subspace $\mathcal{Q}$. As the premium choice, in Section 5.1 the space of factorial distributions is introduced, where the minimization of $\mathcal{F}_\beta$ can be carried out efficiently for a large class of objective functions. The main properties of factorial distributions are outlined in this section.
- The design of the *corrector step*. In Section 5.2, a generic algorithm known as *mean-field approximation* is described to carry out the minimization of $\mathcal{F}_\beta$ over the space of factorial distributions for fixed β. Its convergence to a local minimum of the free energy is established and the inherent connection to the Gibbs sampler is clarified.
- The *validity of the homotopy* includes the correspondence between local and global minima of F and the minima found for $\mathbf{E_Q}[F]$ in the relaxation space $\mathcal{Q}$ as well as the convergence of minimal points of $\mathcal{F}_\beta(\mathbf{Q})$ to minimal points of $\mathcal{F}_\infty(\mathbf{Q}) = \mathbf{E_Q}[F]$. These theoretical issues are addressed in Section 5.4. Therefore, the mean-field approximation can be combined with the concept of annealing and alternating minimization to compute a minimum of $\mathbf{E_Q}[F] = \mathcal{F}_\infty(\mathbf{Q}, \mathbf{Y}) = \lim_{\beta\to\infty} \mathcal{F}_\beta(\mathbf{Q}, \mathbf{Y})$ and thus a minimum of F.

Thus, all mathematical details are provided which are necessary to understand and implement a generic mean-field annealing procedure. Section 6 illustrates additional properties of deterministic annealing, e.g., the phase transition behavior.

5 Approximation Techniques in Annealing

5.1 *Factorial Distributions*

It is well known that the minimum of the generalized free energy is assumed by the associated Gibbs distribution. The deterministic algorithm as presented in Alg. 4 only results in a tractable procedure if averages

$$\mathbf{E}[F(c, \mathbf{Y}; \mathbf{X})] = \sum_{c \in \mathcal{C}_{n,k}} F(c, \mathbf{Y}; \mathbf{X}) P_{F,\beta,\mathbf{Y}}(c) \tag{32}$$

can be evaluated efficiently. An explicit summation over $\mathcal{C}_{n,k}$ should be avoided since the cardinality $|\mathcal{C}_{n,k}|$ grows exponentially in the number of objects n. If the algebraic structure of $P_{F,\beta,\mathbf{Y}}$ is too complex, e.g., it contains an exponentially large sum, then one possibility is to restrict the space of probability distributions $\mathcal{P}$ to a suitable subset $\mathcal{Q}$. The discrete search space $\mathcal{C}_{n,k}$ has a canonical embedding in $\mathcal{P}$ by mapping each $c \in \mathcal{C}_{n,k}$ to the distribution P_c, where P_c defines the Dirac distribution on c, i.e. $P_c(c) = 1$. This provides a canonical requirement for any subspace $\mathcal{Q} \subset \mathcal{P}$.

Definition 13. (Probabilistic Relaxation) A set $\mathcal{Q} \subset \mathcal{P}$ defines a permissible relaxation space, if $P_c \in \mathcal{Q}$ for all c.

Factorial Gibbs distributions are a particularly simple case for which the averages (32) can be carried out efficiently, and deterministic annealing becomes tractable. This motivates the following definition.

Definition 14. (Factorial Distribution) The space $\mathcal{Q} \subset \mathcal{P}$ with

$$\mathcal{Q} = \prod_{u \in \mathcal{U}} \mathcal{Q}_u = \left\{ \mathbf{Q} \in \mathcal{P} : \mathbf{Q}(c) = \prod_{u \in \mathcal{U}} q_u(c(u)),\ q_u(d) \in [0,1] \right\}. \tag{33}$$

is called the space of factorial distributions. A deterministic annealing procedure over the space of factorial distributions is called mean-field annealing.

Thus a factorial distribution over $\mathcal{C}_{n,k}$ is specified by $n \cdot k$ continuous parameters $q_u(d)$. It is easily verified that the normalization conditions $\sum_d q_u(d) = 1$ must be valid for all $u \in \mathcal{U}$ in order to define a valid probability distribution. For notational convenience, the distribution $\mathbf{Q}$ is identified with its parameters, $\mathbf{Q} = (q_u(d))_{u \in \mathcal{U}}$, $1 \leq d \leq k$. The parameter vector $\mathbf{q}_u = (q_u(d)) \in \mathcal{Q}_u$ defines a generic probability distribution over $\mathcal{C}_k = \{1, \ldots, k\}$. The space of factorial distributions is a permissible probabilistic relaxation space, since $P_c = (q_u(d))$ with $q_u(d) = \delta_{d,c(u)}$. The relationship between Gibbs distributions and factorial distributions is clarified by the following proposition.

Proposition 15. *Let*

$$F^0(c) = \sum_{u \in \mathcal{U}} h_u(c(u)) \tag{34}$$

be a linear cost function and denote by $P_{F^0} = \exp\left[-\beta F^0\right] / \mathcal{Z}^0$ *the associated Gibbs distribution with respective partition function* $\mathcal{Z}^0$. *Then*

- P_{F^0} *is factorial, and*
- *every factorial distribution* $\mathbf{Q}$ *can be expressed as a Gibbs distribution with a linear cost function.*

Thus, the set of Gibbs distributions with a linear Hamiltonian and the space of factorial distributions are in fact isomorphic.[3] The parameters $h_u(c)$ are often called *mean-fields* by analogy to statistical physics. Some important properties of factorial distributions are summarized in the following proposition.

Proposition 16. *Let* $\mathbf{Q}$ *be a factorial distribution. Then*

1. *the parameter value* $q_u(d)$ *equals the probability of assigning object* u *to cluster* d,

$$\mathbf{Q}\,(c(u) = d) = q_u(d), \tag{35}$$

[3] Note, that the space of factorial distributions is not isomorphic to the set of all linear cost functions, since a constant offset on the cost function yields an identical Gibbs distribution.

2. *all correlations w.r.t.* $\mathbf{Q}$ *vanish for assignment variables of different objects. More precisely, for all* $c \in \mathcal{C}_{n,k}$, $\mathcal{V} \subset \mathcal{U}$ *and for all functions* $f_u(d)$ *depending on the assignment of a single object* u

$$\mathbf{E_Q}\left[\prod_{u\in\mathcal{V}} f_u(c(u))\right] = \prod_{u\in\mathcal{V}} \mathbf{E}_{\mathbf{q}_u}\left[f_u(d)\right]. \quad (36)$$

This property of factorial distributions is extremely valuable as it ensures that the expectation can be carried out separately for all objects u for a large class of cost functions F.

For discrete search spaces, a local minimum has been introduced as a 1-neighborhood optimal variable state. The space of factorial distributions can be considered as the canonical topologically equivalent probabilistic embedding space, as it introduces a single coordinate axis for each object. For a given variable state $\mathbf{Q}^* = (\mathbf{q}_u^*) \in \mathcal{Q}$ each coordinate axis defines an one-dimensional, localized search space $\mathcal{Q}_u(\mathbf{Q}^*) = \left(\bigotimes_{v\neq u} \mathbf{q}_v^*\right) \otimes \mathcal{Q}_u$ which is obtained from $\mathbf{Q}^*$ by varying only the variable associated with a single object u. Thus, the choice of factorial distributions arises naturally, when considering solely local algorithms. This motivates the following definition.

Definition 17. (One-change Optimality) The distribution $\mathbf{Q}^* = (\mathbf{q}_u^*) \in \mathcal{Q}$ is called (strictly) one-change optimal w.r.t. F and $\mathbf{Y}$, if $\mathbf{Q}^*$ is a (strict) local minimum of $\mathbf{E_Q}[F(c, \mathbf{Y})]$ w.r.t. all subspaces $\mathcal{Q}_u(\mathbf{Q}^*)$.

5.2 *Mean-Field Approximation*

The original cost function $\mathcal{F}_\infty(\mathbf{Q}) = \mathbf{E_Q}[F]$ is embedded into the one-parametric family of cost functions $\mathcal{F}_\beta(\mathbf{Q})$. It is thus a key algorithmic issue to compute optimal states of $\mathcal{F}_\beta(\mathbf{Q})$. The solution for the space of factorial distributions $\mathcal{Q}$ is known as *mean-field approximation*. Differentiation w.r.t. $\mathbf{Q}$ yields the following characteristic equations of the critical points of $\mathcal{F}_\beta(\mathbf{Q})$.

Theorem 18. *Let* F *be an arbitrary partitioning cost function. The factorial distributions* $\mathbf{Q}^* \in \mathcal{Q}$*, which minimize the generalized free energy* $\mathcal{F}_\beta$ *over* $\mathcal{Q}$*, are characterized by the stationarity conditions*

$$q_u^*(c) = \frac{\exp[-\beta h_u(c)]}{\sum_{d=1}^{k} \exp[-\beta h_u(d)]}, \qquad h_u(c) = \mathbf{E}_{\mathbf{Q}^*}[g_u(c)] = \mathbf{E}_{\mathbf{Q}^*_{u\to c}}[F]. \quad (37)$$

Here, $\mathbf{Q}^*_{u\to c}$ denotes the probability distribution obtained by replacing the u-th row with the unit vector e_c, i.e. $\mathbf{Q}^*_{u\to c}$ is defined by setting $q_u(c) = 1$, $q_u(d) = 0$ for all $d \neq c$ and keeping all other parameters $q_v(c)$, $v \neq u$. For a proof of the theorems in this section see App. 7. The theorem establishes an intrinsic relationship between the Gibbs sampler and the mean-field approximation scheme. The *mean-fields* $h_u(c)$ in (37) are the $\mathbf{Q}^*$-averaged versions of the local costs (Gibbs weights)

$g_u(c)$ defined by (30). This fact enables a constructive derivation of efficient mean-field equations if an efficient implementation of the Gibbs sampler is available. The equations for stationary points motivate a deterministic, local algorithm where one object is updated while the assignment probabilities of all other objects are kept fixed. Its convergence properties are clarified by the following theorem.

Theorem 19. *For any schedule $V(t)$ and arbitrary initial conditions, the following local, deterministic algorithm converges to a one-change optimal minimum of the generalized free energy $\mathcal{F}_\beta$:*

$$q_u^{t+1}(c) = \frac{\exp\left[-\beta h_u^{t+1}(c)\right]}{\sum_{d=1}^{k} \exp\left[-\beta h_u^{t+1}(d)\right]}, \quad \textit{where} \tag{38}$$

$$h_u^{t+1}(c) = \mathbf{E}_{\mathbf{Q}^t}\left[g_u(c)\right] = \mathbf{E}_{\mathbf{Q}^t_{u\to c}}\left[\mathcal{F}\right] \quad \textit{and} \quad u = V(t). \tag{39}$$

Note that Prop. 15 allows us to efficiently evaluate the averages $\mathbf{E}_{\mathbf{Q}^t_{u\to c}}[\mathcal{F}]$ for polynomial cost functions, while for non-polynomial $\mathcal{F}$ some approximations have to be introduced. The theorem establishes the convergence to a (one-change optimal) local minimum of the free energy for the complete class of grouping objective functions. The mean-field approximation (MA) algorithm is summarized in Alg. 5.

INPUT $\mathbf{Q}^{\text{init}}$, β, object visitation schedule V, objective function $\mathcal{F}$
repeat
  CALCULATE all partial costs $h^t_{V(t)}(d)$ according to (39)
  CALCULATE all $q^t_{V(t)}(d)$ according to (38)
  SET $t = t + 1$
until CONVERGENCE()

Algorithm 5: Mean-Field Approximation (MA)

5.3 Exemplary Mean-Field Equations

k-means: For the k-means cost function (2) the associated Gibbs-distribution is already factorial and the mean-field approximation is exact. The Gibbs weights equal the mean-fields and are simply given by

$$h_u^{km}(c) \quad = \quad g_u^{km}(c) = w(u)\|\mathbf{x}_u - \mathbf{y}_c\|^2. \tag{40}$$

The equations for stationary points in (38) can then be solved without iteration in one step.

Proximity-Based Clustering. To obtain an efficient implementation of the Gibbs sampler, one has to find a way to efficiently calculate $g_u(c)$ under single object changes. The mean-field algorithm in addition requires to perform the $\mathbf{Q}$-averages $\mathbf{E}_\mathbf{Q}[g_u(c)]$. A straightforward implementation of the Gibbs sampler would partition the cost function into a sum of clique potentials and recalculate at each step all potentials of cliques to which object u belongs [18]. For the normalized clustering objective functions H^{pc} as given by (11) this procedure is highly inefficient, since the assignments enter in the denominator. It is a key observation that all additive terms of $H(c_{u\to c})$ and $(\mathbf{E}_{Q_{u\to c}}[H])$, which do not depend on $c(u)$ ($q_u(d)$ for any d), simply cancel in (30) and (37). They can be neglected for efficiency reasons. Consequently, we propose a different implementation, defining $g_u(c) = H(c_{u\to c}) - H^{-u}(c)$, where the costs $H^{-u}(c)$ of the reduced system without object u have been subtracted. Let

$$f_c = \sum_{u\in\mathcal{G}_c} f_c(u) \quad \text{with} \quad f_c(u) = \sum_{v\in\mathcal{G}_c(u)} D_{uv} \tag{41}$$

define the sum over known dissimilarities in the same cluster and $f_c(u)$ the restriction of this sum to the neighborhood of u. The Gibbs fields for the graph partitioning and the pairwise clustering criteria are then given by

$$g_u^{gp}(c) = f_c^{-u}(u), \tag{42}$$

$$g_u^{pc}(c) = \left(|\mathcal{G}_c^{-u}| + 1\right) \frac{f_c^{-u} + f_c^{-u}(u)}{|\mathcal{E}_c^{-u}| + |\mathcal{G}_c^{-u}(u)|} - |\mathcal{G}_c^{-u}| \frac{f_c^{-u}}{|\mathcal{E}_c^{-u}|}. \tag{43}$$

Here, for a set $\mathcal{A}_c$ the notation $|\mathcal{A}_c^{-u}|$ denotes the set size after removing object u from the statistics,

$$|\mathcal{G}_c^{-u}| = |\mathcal{G}_c| - \delta_{c(u),c}, \tag{44}$$

$$|\mathcal{E}_c^{-u}| = |\mathcal{E}_c| - |\mathcal{G}_c(u)|, \tag{45}$$

$$|\mathcal{G}_c^{-u}(u)| = |\mathcal{G}_c(u)| + \delta_{u\in\mathcal{N}_u}\left(1 - \delta_{c(u),c}\right). \tag{46}$$

Similarly, for the cluster dissimilarity score of u

$$f_c^{-u}(u) = f_c(u) + \delta_{u\in\mathcal{N}_u}\left(1 - \delta_{c(u),c}\right) D_{uu}, \tag{47}$$

$$f_c^{-u} = f_c - f_c(u). \tag{48}$$

The quantities f_c, $f_c(u)$, $|\mathcal{G}_c|$, $|\mathcal{E}_c|$, and $|\mathcal{G}_c(u)|$ are used as *bookkeeping quantities* to achieve a fast evaluation of $g_u(c)$. These bookkeeping quantities must only be updated after changing the assignment of an object.

The remaining technical difficulty in calculating the mean-field equations are the averages of the normalization constants, especially their inverse proportional dependency on functions of sets of assignment variables. Although a polynomial normal form exists, which would in principle eliminate the involved denominator, some approximations have to be made to avoid an exponential number of conjunctions. These approximations are implemented by independently averaging the numerator and the normalization in the denominator in (43),

$$h_u(c)(\mathbf{Q}) = \mathbf{E}_{\mathbf{Q}}[g_u(c)(c)] \approx g_u(c)(\mathbf{E}_{\mathbf{Q}}[c]). \tag{49}$$

The compact notation $\mathbf{E}_{\mathbf{Q}}[c]$ used here implies rewriting the cost function in terms of indicator functions $\delta_{c(v),d}$ by replacing cluster assignment conditions in the argument of a sum. In the mean-field approximation, all indicator functions $\delta_{c(u),d}$

are replaced by the probabilities $q_u(d) = \mathbf{E_Q}\left[\delta_{c(u),d}\right]$ of assigning object u to cluster d. This approximation is exact in the limit of $\beta \to \infty$ for any n and in the thermodynamic limit of $n \to \infty$ for arbitrary β. General bounds as well as higher order corrections of the approximation error can be obtained by a Taylor expansion around $\mathbf{E_Q}[c]$ [27]. The technical details are omitted here for the sake of readability. To illustrate the general formula (49) the mean-field equations for the normalized pairwise clustering cost function are depicted here:

$$h_u^{pc}(c)(\mathbf{Q}) = \left(\sum_{\substack{v\in\mathcal{U}:\\ v\neq u}} q_v(c)+1\right) \frac{\sum\limits_{\substack{(v,w)\in\mathcal{E}:\\ v\neq u\wedge w\neq u}} q_v(c)q_w(c)D_{vw} + \sum\limits_{\substack{v\in\mathcal{N}_u:\\ v\neq u}} q_v(c)D_{uv} + \delta_{u\in\mathcal{N}_u}D_{uu}}{\sum\limits_{\substack{(v,w)\in\mathcal{E}:\\ v\neq u\wedge w\neq u}} q_v(c)q_w(c) + \sum\limits_{\substack{v\in\mathcal{N}_u:\\ v\neq u}} q_v(c) + \delta_{u\in\mathcal{N}_u}}$$

$$- \left(\sum_{\substack{v\in\mathcal{U}:\\ v\neq u}} q_v(c)\right) \frac{\sum\limits_{\substack{(v,w)\in\mathcal{E}:\\ v\neq u\wedge w\neq u}} q_v(c)q_w(c)D_{vw}}{\sum\limits_{\substack{(v,w)\in\mathcal{E}:\\ v\neq u\wedge w\neq u}} q_v(c)q_w(c)}. \tag{50}$$

Similar bookkeeping entities as for the Gibbs sampler and ICM are used for efficient implementation.

5.4 *Mean-Field Annealing*

Theorem 19 establishes the convergence of the mean-field approximation to a (one-change optimal) local minimum of the free energy $\mathcal{F}_\beta(\mathbf{Q},\mathbf{Y})$ over $\mathcal{Q}$ for fixed inverse temperature. Alternating the minimization of $\mathcal{F}_\beta(\mathbf{Q},\mathbf{Y})$ with respect to $\mathbf{Y}$ and the mean-field approximation, the algorithm converges to a (local) minimum of $\mathcal{F}_\beta(\mathbf{Q},\mathbf{Y})$. These results motivate a predictor-corrector method, which uses a constant predictor and tracks the trivial solution at low inverse temperature for $\beta \to \infty$. However, several theoretical issues arise when varying the parameter β. The most fundamental one is concerned with the convergence of minimal points of $\mathcal{F}_\beta(\mathbf{Q},\mathbf{Y})$ to minimal points of $\mathcal{F}_\infty(\mathbf{Q},\mathbf{Y})$. The sufficient condition for the convergence of minima is the *uniform convergence* of $\lim_{\beta\to\infty}\mathcal{F}_\beta(\mathbf{Q},\mathbf{Y})$.

Proposition 20 (Uniform Convergence). *The free energy $\mathcal{F}_\beta(\mathbf{Q},\mathbf{Y})$ over $\mathcal{Q}$ converges uniformly to $\mathcal{F}_\infty(\mathbf{Q},\mathbf{Y}) = \mathbf{E}[F]$.*

The proofs of all propositions in this section are collected in App. 7. A second crucial question concerns the relation between global and local minima of the combinatorial optimization problem and the respective minima of the probabilistic relaxation. For global minima the following relation is valid.

Proposition 21. *Let $\mathcal{Q}$ be a permissible probabilistic relaxation space and let $c^* \in \mathcal{C}_{n,k}$ be a global minimum of F. Then P_{c^*} is a global minimum of $\mathcal{F}_\infty$ over $\mathcal{Q}$.*

Convergence to the global minimum, however, should not be expected in the general case for two reasons. First, there might not exist a connected component in the set of local minima for all values of the parameter β defining the homotopy. Second, if such a path exists, it is not a priori guaranteed that the minimum reached for $\beta \to \infty$ corresponds to a global minimum of the original cost function.

The following proposition clarifies the relation between local minima of F and strictly one-change optimal states of $\mathbf{E_Q}[F]$ over the space of factorial distributions.

Proposition 22. *Let a factorial distribution $\mathbf{Q}^* \in \mathcal{Q}$ and a cost function F be given.*

1. *If $\mathbf{Q}^*$ is strictly one-change optimal w.r.t. F, then $\mathbf{Q}^* = P_c$ for some c.*
2. *P_{c^*} is strictly one-change optimal w.r.t. F iff c^* is strictly one-change optimal w.r.t. F.*

The proposition establishes a one-to-one correspondence between the strictly local minima of the combinatorial problem and its probabilistic relaxation, which justifies the choice of the space of factorial distributions as a proper relaxation space. For critical points, which are not strict local minima, the situation is less satisfactory. Non-strict minima c^* of F induce a corresponding continuum of minima over $\mathbf{E_Q}[F]$. They can be easily recovered for a given one-change optimal $\mathbf{Q}^*$, however, $\mathbf{Q}^*$ is not unique. A second type of critical points are saddle points of $\mathcal{F}_\infty$, which frequently occur as minima of $\mathcal{F}_\beta$ traced for $\beta \to \infty$ and which become unstable while raising β. They correspond to symmetric superpositions of different local minima of F and must be circumvented by adding noise during the annealing process.

The convexity of $\mathcal{F}_\beta$ for $\beta \to 0$ is the second important condition for a valid homotopy. The convexity of $\mathcal{F}_\beta$ for small β is ensured by the following proposition.

Proposition 23. *$\mathcal{F}_\beta$ is strictly convex over $\mathcal{Q}$ for β sufficiently small.*

This ensures that deterministic annealing over the space of factorial distributions, which is referred to as *mean-field annealing*, defines a valid continuation algorithm. The complete mean-field annealing algorithm is summarized in Alg. 6.

6 Properties of Deterministic Annealing

Phase Transitions and Cluster Splits. At an inverse temperature $\beta = 0$, the effective number of clusters of the optimal solution degenerates to one, i.e. the positions of the prototypes, e.g., for the k-means cost function are identical for all clusters and objects are assigned to all prototypes with uniform probability. The reason for this phenomenon is the dominance of the entropy as opposed to the distortion costs measured by the cost function. The collapse of clusters occurs for all grouping and quantization cost functions. To determine the effective number of groups, clusters c and d are identified if they have identical assignment probabilities $q_u(c) = q_u(d)$

INPUT β^{start}, β^{final},
SET $\beta = \beta^{\text{start}}$
INITIALIZE $q_u(c) = \frac{1}{k}$
repeat
  add a small random perturbation to $\mathbf{Q} = (q_u(c))$
  repeat
    MINIMIZE $\mathbf{E}\,[F(c, \mathbf{Y}; \mathbf{X})]$ w.r.t. $\mathbf{Y}$
    CALL MA $(\mathbf{Q}, \beta, F)$ as given by Alg.5
  until CONVERGENCE()
  SET β = SCHEDULE ()
until $\beta > \beta^{\text{final}}$

Algorithm 6: Mean-Field Annealing

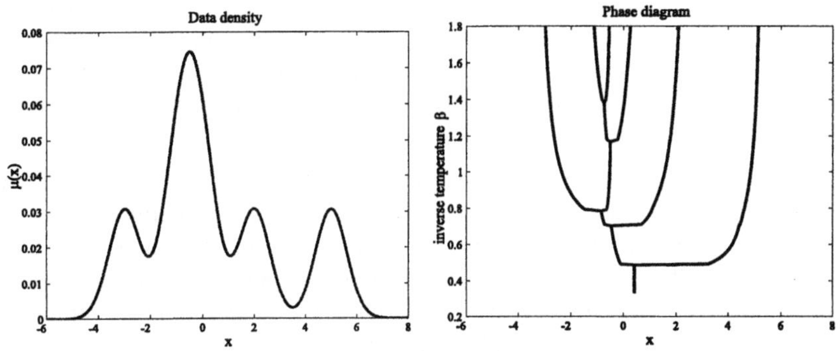

Figure 1. Cluster splits and phase transition behavior in DA. Depicted is (a) the one-dimensional input density of the data samples and (b) the computed optimal prototype positions depending on β for the k-means clustering criterion with $k = 6$ different clusters. The input density is a linear superposition of $k = 6$ Gaussian distributions with identical weights and identical variance. Note that the centers of the Gaussian distributions are approximately recovered by the cluster prototypes

for all $u \in \mathcal{U}$. While all annealing algorithms exhibit a similar behavior it is particularly easy to measure cluster degeneracies in deterministic annealing.

The annealing process exhibits a series of bifurcations, also called phase transitions for the large data limit $n \to \infty$ in Statistical Physics, where cluster assignments successively split. In a bifurcation, symmetric superpositions of optimal cluster configurations become unstable. This phenomenon is illustrated in Figure 1 for a simple one-dimensional k-means example, but similarly occurs for all objective functions. As demonstrated in [52] for k-means, the first bifurcation occurs along the first principal axis at a critical temperature $1/2\lambda_{\max}$ which depends on the largest absolute eigenvalue $\lambda_{\max}$ of the data co-variance matrix.

7 Bibliographic Notes

Statistical Physics. The concept of entropy first emerged in the context of thermodynamics and plays a fundamental role in information and communication theory [55]. Shannon axiomatically derived the entropy as a unique measure of uncertainty and identified the entropy as a lower bound of the expected coding length for a stochastic source. The maximum entropy principle has been first proposed by Jaynes [32, 33] and can be seen as an application of the Laplace principle of *insufficient reasoning* [39]. The maximum entropy principle has been justified in terms of repeated random trials, culminating in the entropy concentration theorem [34]. An axiomatization of maximum entropy inference has been presented in [13].

The free energy plays an important role in statistical physics, as it summarizes all relevant thermodynamic properties of a system defined by a Hamiltonian F. Depending on the specific structure of $\mathcal{C}$, Gibbs expectation values can be calculated simply by differentiation of the free energy. It is referred to [43] for a deeper treatment. Annealing as discussed in this paper emphasizes the discrete search space $\mathcal{C}$. It is possible to define annealing algorithms by entropic smoothing over $\mathcal{Y}$ as developed in [36] for *M*ulti*D*imensional *S*caling.

There is a vast literature on MCMC methods, e.g., [25, 38]. The idea of Monte Carlo sampling dates back to [41], who proposed a sampling scheme now known as the Metropolis sampler. Other popular choices include the heat bath acceptance rule [25] and the described Gibbs sampler [18]. Important improvements have been made recently in the theory of Markov chain methods to speed up stochastic sampling for special systems, but are not generally applicable [21, 56].

Simulated Annealing. The idea of simulated annealing has first been introduced in [10, 35]. A first proof for the convergence of the annealing process to the uniform distribution on the global minima of F has been established in [18] for logarithmic annealing schedules. These results have been sharpened by B. Hajek [24]. Structured introductions to simulated annealing are found in [37, 38]. An annotated bibliography is found in [11]. Simulated annealing has been extended to continuous search spaces [17, 19], but the theoretical results are less well established.

The iterated conditional mode (ICM) algorithm, though already applied e.g., in [18], is usually attributed to [2]. Both the simulated annealing and the ICM have been applied in numerous vision applications, especially in the context of Markov random field models. The reader is referred to [40, 58] for an overview and further bibliographic cues. While the ICM is fast, it suffers from its inherent locality, i.e. the ICM is frequently getting trapped in local minima. One class of techniques, which combines stochastic with gradient-based methods are known as hybrid Monte Carlo methods [14,42]. Hybrid methods define a stochastic dynamics to replace the memoryless random walk behavior. Cutting plane algorithms as developed in [22, 23] provide global optimization algorithms in the strict sense. However, they are restricted to linear cost functions. Moreover, it is not clear how to extend them to mixed combinatorial objective functions.

Deterministic Annealing. Deterministic annealing has originally been proposed as an efficient approximation to simulated annealing [15, 45, 46]. In the sequel, deterministic annealing has been applied to the traveling salesman problem [46], graph partitioning [57], quadratic assignment and graph matching [20, 59], vector quantization [6, 51–53], surface reconstruction [16], image restoration [4, 61], and edge detection [60]. More specifically, *mean-field theory* as an approximation principle [3, 15, 45, 62] has been used to obtain computationally tractable algorithms. In the context of distributional clustering, deterministic annealing has been applied by [44]. A deterministic annealing approach for clustering and visualization of complete proximity data has been presented in [26]. For sparse proximity matrices it has been introduced in [28–30]. Its application for color quantization has been proposed in [47]. Astonishingly, deterministic annealing algorithms have only been derived independently for highly specific optimization instances despite these widespread research activities. The generic scheme presented here maximally encapsulates the dependency on the cost function. It has first been published in [29, 30, 48]

Convergence of mean-field annealing has been established for a very specific problem instance by [62]. The general results presented in this section including the general convergence results have been published in [30, 49]. The analysis of the relaxation and the correspondence between global and local minima has been presented in [48].

The minimization of $\mathcal{F}_\beta$ over $\mathcal{Q}$ can be seen as an approximation procedure for a problem, which is intractable over the full search space $\mathcal{P}$. It can be shown that minimizing the generalized free energy is equivalent to minimizing the cross entropy to the Gibbs distribution. A proof can be found, e.g., in [3].

Continuation Methods. A theory of continuation methods has been developed in the context of nonlinear systems of equations [1]. A popular example in computer vision is the *graduated non-convexity algorithm* (GNC) developed in [5]. Deterministic annealing can be interpreted as a continuation method in that it defines a linear homotopy by adding entropic contributions to a given objective function and in that it globally solves the optimization problem for low β and then tracks the solution for $\beta \to \infty$. Optimization with respect to the reduced set $\mathcal{Q}$ has been introduced as a variational approximation to a continuation method, which is intractable in $\mathcal{P}$. However, it should be noted that minimization with respect to $\mathcal{Q}$ yields a valid homotopy in its own right.

Conventional continuation methods have only been designed for low-dimensional, continuous optimization problems [1] and it is by no means obvious how to extend them to high-dimensional, discrete search spaces. As the temperature defines a natural resolution parameter the maximum entropy rationale seems to be a natural framework to derive efficient and robust continuation algorithms. This has been confirmed by the excellent empirical results in numerous independent studies as cited above. However, it has to be admitted though that there is still a lack of theoretical understanding. Especially, in general no convergence to the global optimum of a cost function should be expected [54]. Nothing is theoretically known about the average quality of local minima found by deterministic annealing.

REFERENCES

1. E. Allgower and K. Georg, *Numerical Continuation Methods*, vol. 13 of Springer Series in Computational Mathematics, Springer Verlag, 1990.
2. J. Besag, *On the statistical analysis of dirty pictures*, Journal of the Royal Statistical Society, Series B, 48 (1986), pp. 25–37.
3. G. Bilbro and W. Snyder, *Mean field approximation minimizes relative entropy*, Journal of the Optical Society of America, 8 (1989).
4. G. Bilbro, W. Snyder, S. Garnier, and J. Gault, *Mean field annealing: A formalism for constructing GNC-like algorithms*, IEEE Transactions on Neural Networks, 3 (1992).
5. A. Blake and A. Zisserman, *Visual Reconstruction*, MIT Press, 1987.
6. J. Buhmann and H. Kühnel, *Vector quantization with complexity costs*, IEEE Transactions on Information Theory, 39 (1993), pp. 1133–1145.
7. O. Catoni, *Rough large deviation estimates for simulated annealing: Applications to exponential schedules*, Annals of Probability, 20 (1992), pp. 1109–1146.
8. O. Catoni, *Rough large deviation estimates for simulated annealing: Applications to exponential schedules*, Journal of Complexity, 12 (1996), pp. 595–623.
9. O. Catoni, *Erratum*, Journal of Complexity, 13 (1997), p. 384.
10. V. Cerny, *Thermodynamical approach to the travelling salesman problem*, Journal of Optimization Theory and Applications, 45 (1985), pp. 41–51.
11. N. Collins, R. Eglese, and B. Golden, *Simulated annealing - an annotated bibliography*, American Journal of Mathematical and Management Science, 8 (1988), pp. 209–308.
12. T. Cover and J. Thomas, *Elements of Information Theory*, John Wiley & Sons, 1991.
13. I. Csiszar, *Why least squares and maximum entropy - an axiomatic approach to inference for linear inverse problems*, Annals of Statistics, 19 (1991), pp. 2032–2066.
14. S. Duane, A. Kennedy, B. Pendleton, and D. Roweth, *Hybrid Monte Carlo*, Physics Letters B, 195 (1987), pp. 216–222.
15. D. Geiger and F. Girosi, *Coupled markov random fields and mean field theory*, in Advances in Neural Information Processing Systems 2, 1990, pp. 660–667.
16. D. Geiger and F. Girosi, *Parallel and deterministic algorithms from MRF's: Surface reconstruction*, IEEE Transactions on Pattern Analysis and Machine Intelligence, (1991), pp. 401–412.
17. S. Gelfand and S. Mitter, *Simulated annealing type algorithms for multivariate optimization*, Algorithmica, 6 (1991), pp. 419–436.
18. S. Geman and D. Geman, *Stochastic relaxation, Gibbs distributions, and the Bayesian restoration of images*, IEEE Transactions on Pattern Analysis and Machine Intelligence, 6 (1984), pp. 721–741.
19. S. Geman and C. Hwang, *Diffusion for global optimization*, SIAM Journal of Control and Optimization, 24 (1986), pp. 1031–1043.
20. S. Gold and A. Rangarajan, *A graduated assignment algorithm for graph matching*, IEEE Transactions on Pattern Analysis and Machine Intelligence, 18 (1996), pp. 377–388.
21. J. Goodman and A. Sokal, *Multigrid Monte-Carlo method. Conceptual foundations*, Physical Review D, 40 (1989), pp. 2025–2071.
22. M. Grötschel and Y. Wakabayashi, *A cutting plane algorithm for a clustering problem*, Mathematical Programming, 45 (1989), pp. 59–96.
23. ——, *Facets of the clique partitioning polytope*, Mathematical Programming, 47 (1990), pp. 367–387.
24. B. Hajek, *Cooling schedules for optimal annealing*, Mathematics of Operation Research, 13 (1988), pp. 311–324.

25. W. Hastings, *Monte Carlo sampling methods using Markov chains and their applications*, Biometrika, 57 (1970), pp. 97–109.
26. T. Hofmann and J. Buhmann, *Pairwise data clustering by deterministic annealing*, IEEE Transactions on Pattern Analysis and Machine Intelligence, 19 (1997), pp. 1–14.
27. T. Hofmann, J. Puzicha, and J. Buhmann, *A deterministic annealing framework for textured image segmentation*, IAI-TR 96-2, Institut für Informatik III, 1996.
28. T. Hofmann, J. Puzicha, and J. Buhmann, *Unsupervised segmentation of textured images by pairwise data clustering*, in Proceedings of the IEEE International Conference on Image Processing (ICIP'96), 1996, pp. III: 137–140.
29. T. Hofmann, J. Puzicha, and J. Buhmann, *Deterministic annealing for unsupervised texture segmentation*, in Proceedings of the International Workshop on Energy Minimization Methods in Computer Vision (EMMCVPR'97), Lectures Notes in Computer Science, Springer Verlag, 1997, pp. 213–228.
30. T. Hofmann, J. Puzicha, and J. Buhmann, *Unsupervised texture segmentation in a deterministic annealing framework*, IEEE Transactions on Pattern Analysis and Machine Intelligence, 20 (1998), pp. 803–818.
31. A. Jain and R. Dubes, *Algorithms for Clustering Data*, Prentice Hall, 1988.
32. E. Jaynes, *Information theory and statistical mechanics*, Physical Review, 106 (1957), pp. 620–630.
33. E. Jaynes, *Information theory and statistical mechanics II*, Physical Review, 108 (1957), pp. 171–190.
34. E. Jaynes, *On the rationale of maximum-entropy methods*, Proceedings of the IEEE, 70 (1982), pp. 939–952.
35. S. Kirkpatrick, C. Gelatt, and M. Vecchi, *Optimization by simulated annealing*, Science, 220 (1983), pp. 671–680.
36. H. Klock and J. M. Buhmann, *Data visualization by multidimensional scaling: A deterministic annealing approach*, Pattern Recognition, 33 (2000), pp. 651–669.
37. P. v. Laarhoven, *Theoretical and Computational Aspects of Simulated Annealing*, CWI Tracts, 1988.
38. P. v. Laarhoven and E. Aarts, *Simulated Annealing: Theory and applications*, Reidel Publishing Company, 1987.
39. P. Laplace, *Theorie Analytique des probabilites*, Courcier, Paris, 1812.
40. S. Li, *Markov Random Field Modeling in Computer Vision*, Springer, 1995.
41. N. Metropolis, A. Rosenbluth, M. Rosenbluth, A. Teller, and M. Teller, *Equation for state calculations by fast computing machines*, Journal of Chemical Physics, 21 (1953), pp. 1087–1092.
42. R. Neal, *Probabilistic inference unsing Markov chain Monte Carlo methods*, Tech. Rep. CRG-TR-93-1, Department of Computer Science, University of Toronto, Canada, 1993.
43. G. Parisi, *Statistical Field Theory*, Addison Wesley, Redwood City, Ca., 1988.
44. F. Pereira, N. Tishby, and L. Lee, *Distributional clustering of English words*, in 30th Annual Meeting of the Association for Computational Linguistics, Columbus, Ohio, 1993, pp. 183–190.
45. C. Peterson and J. Anderson, *A mean field theory learning algorithm for neural networks*, Complex Systems, 1 (1987), pp. 995–1019.
46. C. Peterson and B. Söderberg, *A new method for mapping optimization problems onto neural networks*, International Journal of Neural Systems, 1 (1989), pp. 3–22.
47. J. Puzicha, M. Held, J. Ketterer, J. Buhmann, and D. Fellner, *On spatial quantization of color images*, IEEE Transactions on Image Processing, 9 (2000), pp. 666–682.

48. J. Puzicha, T. Hofmann, and J. Buhmann, *Deterministic annealing: Fast physical heuristics for real time optimization of large systems.*, in Proceedings of the 15th IMACS World Congress on Scientific Computation, Modelling and Applied Mathematics, 1997.
49. J. Puzicha, T. Hofmann, and J. Buhmann, *A theory of proximity based clustering: Structure detection by optimization*, Pattern Recognition, 33 (2000), pp. 617–634.
50. J. Puzicha, T. Hofmann, and J. M. Buhmann, *A theory of proximity based clustering: Structure detection by optimization*, Pattern Recognition, 33 (2000), pp. 617–634.
51. K. Rose, E. Gurewitz, and G. Fox, *A deterministic annealing approach to clustering*, Pattern Recognition Letters, 11 (1990), pp. 589–594.
52. K. Rose, E. Gurewitz, and G. Fox, *Statistical mechanics and phase transition in clustering*, Physical Review Letters, 65 (1990), pp. 945–948.
53. K. Rose, E. Gurewitz, and G. Fox, *Vector quantization by deterministic annealing*, IEEE Transactions on Information Theory, 38 (1992), pp. 1249–1257.
54. M.-A. Sato and S. Ishii, *Bifurcations in mean-field-theory annealing*, Physical Review E, 53 (1996), pp. 5153–5168.
55. C. Shannon, *A mathematical theory of communication*, Bell System Tech. Journal, 27 (1948), pp. 379–423, 623–659.
56. R. Swendsen and J. Wang, *Nonuniversal critical dynamics in Monte Carlo simulations*, Physical Review Letters, 58 (1987), pp. 86–88.
57. D. van den Bout and T. Miller, *Graph partitioning using annealed neural networks*, IEEE Transactions on Neural Networks, 1 (1990), pp. 192–203.
58. G. Winkler, *Image Analysis, Random Fields and Dynamic Monte Carlo Methods: A Mathematical Introduction*, Springer, 1995.
59. A. Yuille, *Generalized deformable models, statistical physics and matching problems*, Neural Computation, 2 (1990), pp. 1–24.
60. J. Zerubia and R. Chellappa, *Mean field annealing using compound Gauss-Markov random fields for edge detection and image estimation*, IEEE Transactions on Neural Networks, 4 (1993), pp. 703–709.
61. J. Zhang, *The mean field theory in EM procedures for blind Markov random fields*, IEEE Transactions on Image Processing, 2 (1993), pp. 27–40.
62. J. Zhang, *The convergence of mean field procedures for MRF's*, IEEE Transactions on Image Processing, 5 (1996), pp. 1662–1665.

Appendix: Proofs

Theorem 18: Let F be an arbitrary partitioning cost function. The factorial distributions $\mathbf{Q}^* \in \mathcal{Q}$, which minimize the generalized free energy $\mathcal{F}_\beta$ over $\mathcal{Q}$, are characterized by the stationary conditions

$$q^*_u(c) = \frac{\exp[-\beta h_u(c)]}{\sum_{d=1}^k \exp[-\beta h_u(d)]}, \qquad h_u(c) = \mathbf{E}_{\mathbf{Q}^*}[g_u(c)] = \mathbf{E}_{\mathbf{Q}^*_{u\to c}}[\mathsf{F}] . \quad (51)$$

Proof. Introducing Lagrange parameters λ_u to enforce the normalization

$$\sum_{c=1}^k q_u(c) = 1$$

and taking derivatives of the Lagrangian of the generalized free energy (27) results in

$$\frac{\partial}{\partial q_u(c)}\left[\mathbf{E}_{\mathbf{Q}}[H] - \frac{1}{\beta}\mathcal{S}(\mathbf{Q}) + \sum_{v\in\mathcal{U}}\lambda_v\sum_{d=1}^{k} q_v(d)\right] \tag{52}$$

$$= \frac{\partial \mathbf{E}_{\mathbf{Q}}[H]}{\partial q_u(c)} + \frac{1}{\beta}\frac{\partial}{\partial q_u(c)}\sum_{v\in\mathcal{U}}\sum_{d=1}^{k} q_v(d)\log q_v(d) + \lambda_u \tag{53}$$

$$= \frac{\partial \mathbf{E}_{\mathbf{Q}}[H]}{\partial q_u(c)} + \frac{1}{\beta}\log q_u(c) + \lambda_u + \frac{1}{\beta}. \tag{54}$$

Setting the derivatives equal to zero establishes stationary conditions.

$$q_u^*(c) = \frac{\exp\left(-\beta h_u(c)\right)}{\sum_{d=1}^{k}\exp\left(-\beta h_u(d)\right)}, \qquad \text{where}$$

$$h_u(c) = \frac{\partial \mathbf{E}_{\mathbf{Q}^*}[H]}{\partial q_u(c)}. \tag{55}$$

Performing the derivatives gives

$$h_u(d) = \sum_{c\in\mathcal{C}} H(c)\frac{\partial \mathbf{Q}^*(c)}{\partial q_u(d)} = \sum_{c\in\mathcal{C}} H(c)\delta_{c(u),d}\prod_{v\neq u} q_v^*(c(v))\ \mathbf{Q}^*(c) \tag{56}$$

$$= \sum_{c\in\mathcal{C}}\frac{\delta_{c(u),d}}{q_u^*(d)}H(c) = \mathbf{E}_{\mathbf{Q}^*}[g_u(d)] = \mathbf{E}_{\mathbf{Q}^*_{u\to d}}[H]. \tag{57}$$

□

Theorem 19: For any schedule $V(t)$ and arbitrary initial conditions, the following local, deterministic algorithm converges to a one-change optimal minimum of the generalized free energy $\mathcal{F}_\beta$:

$$q_u^{t+1}(c) = \frac{\exp\left[-\beta h_u^{t+1}(c)\right]}{\sum_{d=1}^{k}\exp\left[-\beta h_u^{t+1}(d)\right]}, \qquad \text{where} \tag{58}$$

$$h_u^{t+1}(c) = \mathbf{E}_{\mathbf{Q}^t}[g_u(c)] = \mathbf{E}_{\mathbf{Q}^t_{u\to c}}[F] \tag{59}$$

and $u = V(t)$.

Proof. By differentiating (54) the Jacobian of the generalized free energy can be expressed as

$$\frac{\partial^2 \left(\mathcal{F}_\beta(\mathbf{Q}) + \sum_{\nu\in\mathcal{U}} \lambda_\nu \sum_{d=1}^{k} q_\nu(d)\right)}{\partial q_u(c)\ \partial q_\nu(d)} \tag{60}$$

$$= \frac{\mathbf{E}_\mathbf{Q}\left[\delta_{c(u),c}\delta_{c(\nu),d}H\right]}{q_u(c)q_\nu(d)} + \left[\frac{1}{\beta q_u(c)} - \frac{\mathbf{E}_\mathbf{Q}\left[\delta_{c(u),c}H\right]}{q_u(c)^2}\right]\delta_{u,\nu}\delta_{c,d} \tag{61}$$

$$= \begin{cases} 1/(\beta q_u(c)) & \text{for } u=\nu \text{ and } c=d \\ 0 & \text{for } u=\nu \text{ and } c\neq d \\ \mathbf{E}_\mathbf{Q}\left[\delta_{c(u),c}\delta_{c(\nu),d}H\right]/(q_u(c)q_\nu(d)) & \text{otherwise} \end{cases} \tag{62}$$

which is positive definite, when considering only the subspace $\mathcal{Q}_u \otimes \bigotimes_{\nu\neq u,c} q_\nu(c)$ spanned by one site u. This reduced configuration space is obtained by varying $q_u(c)$ keeping the configurations $q_\nu(c), \nu \neq u$ fixed. An asynchronous step (37) minimizes $\mathcal{F}_\beta$ with respect to $\mathcal{Q}_u \otimes \bigotimes_{\nu\neq u,c} q_\nu(c)$. As H and thus $\mathcal{F}_\beta$ are bounded from below by definition, this ensures convergence of the asynchronous update scheme to a local minimum. □

Proposition 20 (Uniform Convergence): The free energy $\mathcal{F}_\beta(\mathbf{Q},\mathbf{Y})$ over $\mathcal{Q}$ converges uniformly to $\mathcal{F}_\infty(\mathbf{Q},\mathbf{Y}) = \mathbf{E}_\mathbf{Q}[F]$.

Proof. It must be shown that for all ϵ there exists β_0 such that for all $\beta \geq \beta_0$, $\mathbf{Q} \in \mathcal{Q}$

$$|\mathcal{F}_\beta(\mathbf{Q},\mathbf{Y}) - \mathcal{F}_\infty(\mathbf{Q},\mathbf{Y})| \leq \epsilon. \tag{63}$$

The difference can be uniformly bounded by

$$|\mathcal{F}_\beta(\mathbf{Q},\mathbf{Y}) - \mathcal{F}_\infty(\mathbf{Q},\mathbf{Y})| \leq \frac{1}{\beta} n \log k \tag{64}$$

which can be made arbitrarily small for large β. □

Proposition 21: Let $\mathcal{Q}$ be a permissible probabilistic relaxation space and let $c^* \in \mathcal{C}_{n,k}$ be a global minimum of F. Then P_{c^*} is a global minimum of $\mathcal{F}_\infty$ over $\mathcal{Q}$.

Proof. Let $c^* \in \mathcal{C}$ be a global minimum of F. Then for all $c \in \mathcal{C} : F(c^*) \leq F(c)$ and, therefore,

$$F(c^*) = \mathbf{E}_{P_{c^*}}[F] = \sum_{c\in\mathcal{C}} F(c)P_{c^*}(c) \tag{65}$$

$$\leq \sum_{c\in\mathcal{C}} F(c)P(c) = \mathbf{E}_P[F] \tag{66}$$

for any P. □

Proposition 22: Let a factorial distribution $\mathbf{Q}^* \in \mathcal{Q}$ and a cost function F be given.

1. If $\mathbf{Q}^*$ is strictly one-change optimal w.r.t. F, then $\mathbf{Q}^* = \mathsf{P}_{\mathsf{c}}$ for some c.
2. $\mathsf{P}_{\mathsf{c}^*}$ is strictly one-change optimal w.r.t. F, iff c^* is strictly one-change optimal w.r.t. F.

Proof. The proof relies on Theorem 19. Let $\mathbf{Q}^*$ be one-change optimal and let $h_u(c) = \mathbf{E}_{\mathbf{Q}^*_{c \to u}}[\mathsf{H}]$. Then, according to (37), $q^*_u(c) > 0$ implies that $\forall d : h_u(c) \leq h_u(d)$ and, therefore, according to (37) and (56):

$$\begin{aligned}\mathbf{E}_{\mathbf{Q}^*_{u \to c}}[\mathsf{H}] = h_u(c) &\leq \sum_d q^*_u(d) h_u(d) = \sum_d q^*_u(d) \sum_{c \in \mathcal{C}} \frac{\delta_{c(u),d}}{q^*_u(d)} \mathsf{H}(c)\mathbf{Q}^*(c) \\ &= \sum_{c \in \mathcal{C}} \left(\sum_d \delta_{c(u),d} \right) \mathsf{H}(c)\mathbf{Q}^*(c) = \mathbf{E}_{\mathbf{Q}^*}[\mathsf{H}] \,. \end{aligned} \tag{67}$$

Strict one-change optimality in conjunction with the convexity of $\mathbf{E}_{\mathbf{Q}}[\mathsf{H}]$ over the subspace $\mathcal{Q}_u(\mathbf{Q}^*)$ as established by Theorem 19 implies $\mathbf{Q}^* = \mathbf{Q}^*_{u \to c}$. □

Proposition 23: $\mathcal{F}_\beta$ is strictly convex over $\mathcal{Q}$ for β sufficiently small.

Proof. The proof is a direct consequence of (62) since the Jacobian of the generalized free energy becomes diagonal dominant for $\beta \to 0$. □

Multiscale Annealing and Robustness: Fast Heuristics for Large Scale Non-linear Optimization

Joachim M. Buhmann[1] and Jan Puzicha[2]

[1] Institut für Informatik, Rheinische Friedrich-Wilhelm Universität Bonn, Germany
[2] Department of Computer Science, University of California, Berkeley, USA

Abstract *Multiscale Annealing* is an extension of the idea of deterministic annealing which link the approximation quality of the algorithms to their effective spatial resolution. The optimization variables of a particular scale are linked together to reduce the spatial resolution, and, thereby, to simplify the computational complexity of the optimization task. Robustness and efficiency issues are discussed in this article on research questions of annealing techniques.

1 Introduction

Annealing techniques for optimization have steadily gained acceptance in the pattern recognition, computer vision and speech processing community since their invention by Kirkpatrick et al. [19] and, independently, by Černy [6]. These stochastic optimization techniques are well adapted to probabilistic modeling concepts like Markov Random Fields [10], Hidden Markov Models [35], probabilistic neural networks [36] and, more general, the class of graphical models [18]. The popularity of the methods is partially based on the observed ease of implementation and the experimentally observed robustness against noise distortions in the data. Active research areas are concerned with questions of provably efficient implementations and with a theoretical understanding of robustness.

Deterministic Annealing (DA) [5] shows an increased the computational efficiency compared to the Gibbs sampler. It is, however, still too slow by an order of magnitude for the typical demands of low-level computer vision tasks. Large scale optimization problems as they occur, e.g., in image segmentation need to be solved within at most a few seconds. Many of these problems can be phrased in a grouping, clustering or quantization framework, which result in non-linear, large optimization problems. To gain additional efficiency in optimization, a novel real-time approach[1] to global optimization of several grouping and quantization criteria (cost functions) is developed. It relies on the concept of *Multiscale Optimization* [14]. Available (image) neighborhood information is used to significantly accelerate computation by exploiting the fact, that nearby objects belong with high probability to the same cluster. The optimization problem is redefined on different scales such that the original cost function is minimized over a suitable nested sequence of subspaces in a

[1] Note that a weak notion of real-time processing is adopted, i.e., *any time algorithms* are developed with processing times that are simply short enough, i.e. at most a few seconds.

coarse-to-fine manner. As the major advantage this procedure reduces the number of variables and thus significantly saves computational effort.

Multiscale techniques are combined with the deterministic annealing algorithm. The annealing process is tightly coupled with the coarse-to-fine optimization by a uniform convergence criterion derived from statistical learning theory. The proposed algorithm, which is referred to as *Multiscale Annealing*, couples the resolution hierarchy in image and in cluster space with the resolution in optimization space introduced by annealing with a computational temperature. The essential structure of the grouping solution is detected on the coarse scales using global optimization techniques and is then propagated into the details using high-speed low or zero temperature algorithms. Large deviation bounds adapted from statistical learning theory yield a criterion for a finite stopping temperature.

This chapter is organized as follows: In Section 2 the canonical multiscale operator is introduced, and equations for multiscale optimization for different grouping criteria are formally derived. We continue by linking optimization resolution and spatial resolution using a statistical learning theory approach in Section 3, followed by real-world texture segmentation experiments in Section 4. Here, we demonstrate the efficiency of the proposed optimization techniques yielding acceleration factors from 1.5–50 in typical applications. This chapter makes heavy use of the notation and the material covered in the introductory chapter on deterministic annealing in this volume [5].

2 Multiscale Annealing: Efficiency Issues

A theory of optimization is developed in the following for a generic cost function $F_{n,k}(c, \mathbf{Y}; \mathbf{X})$ with n discrete k-state assignment variables summarized by $c \in \mathcal{C}_{n,k}$ and a k-tuple of d-dimensional, real-valued optimization parameter $\mathbf{Y} \in \mathcal{Y}_k \subset \mathbb{R}^{k \cdot d}$. Here, the cost function is parametrized by a fixed observation $\mathbf{X} \in \mathcal{X}(n)$ from some (abstract) measurement space $\mathcal{X}(n)$.

Definition 1. Objective Function Let $\mathcal{Y}_k$ be a compact subset of $\mathbb{R}^{k \cdot d}$. The set

$$\mathcal{F}_{n,k} = \{F : \mathcal{C}_{n,k} \times \mathcal{Y}_k \to \mathbb{R} : \ \exists b \in \mathbb{R}\ \forall c, \mathbf{Y}\ \ F(c, \mathbf{Y}) \geq b > -\infty\} \tag{1}$$

is called the space of objective functions defined over $\mathcal{C}_{n,k}$ and $\mathcal{Y}_k$.

Typical objective functions include *central clustering* (vector quantization) given by

$$R^{km}(c, \mathbf{Y}; \mathbf{X}) = \sum_{u \in \mathcal{U}} \|\mathbf{x}_u - \mathbf{y}_{c(u)}\|^2 \tag{2}$$

with vector-valued observations $\mathbf{x}_u$, where the $\mathbf{y}_c \in \mathbb{R}^d$ denote additional real-valued, d-dimensional cluster prototypes to be optimized. For indirect observations, i.e. similarity measurements $D_{uv} = D_{vu}$ between the objects u and v, the graph partitioning cost function

$$R^{gp}(c; \mathbf{D}) = \frac{1}{n} \sum_{\substack{(u,v) \in \mathcal{E}: \\ c(u) = c(v)}} D_{uv}. \tag{3}$$

and the normalized graph partitioning cost function

$$\mathcal{R}^{pc}(\mathbf{c};\mathbf{D}) = \sum_{c=1}^{k} |\mathcal{G}_c| \bar{\mathbf{D}}_c \quad \text{with} \quad \bar{\mathbf{D}}_c = \frac{1}{|\mathcal{D}_c|} \sum_{(u,v)\in\mathcal{D}_c} D_{uv}. \tag{4}$$

are frequently used cost functions. See the introductory chapter [5] in this volume for nomenclature and a detailed motivation of these cost functions.

2.1 The Canonical Multiscale Operator

In the following presentation, a generic objective function $F \in \mathcal{F}_{n,k}$ is assumed to be given. Let the number of objects n and the number of clusters k be fixed for now. The objects which should be clustered are assumed to be arranged as a two- or three-dimensional grid which corresponds to the pixel or voxel structure in computer vision. There are multiple possibilities to define the coarse grids and the respective prolongation and restriction operators, i.e. coarse-to-fine and fine-to-coarse maps of optimization variables. In the usual grouping application, the number n of (discrete assignment) variables $c(u)$ is extremely large, while only a few continuous variables $\mathbf{Y} \in \mathcal{Y}$ enter the optimization problem. Therefore, only coarsening operations on $\mathcal{U}$ are considered. Denote by $\mathcal{U}^0 = \mathcal{U}$ the original set of sites and assume a set of grid sites $\mathcal{U}^l = \{1, \ldots, n^l\}$ be given for each coarse grid level l. To each coarse grid, a reduced set of optimization variables $c^l \in \mathcal{C}^l$ is assigned,

$$\mathcal{C}^l = \left\{ \left(c^l(u)\right)_{u\in\mathcal{U}^l} : c^l(u) \in \{1, \ldots, k\} \right\}. \tag{5}$$

These grids as well as the corresponding variable sets are linked by a *coarsening map* $I_l : \mathcal{U}^l \longrightarrow \mathcal{U}^{l+1}$ defined on the sets of sites. Thus, any fine grid point is linked to a single coarse grid point, which gathers the information. In image analysis, typically $s^l = 4$ sites are combined in a single coarse site by the two-dimensional index operation $(\lceil i/2 \rceil, \lceil j/2 \rceil)$. A typical multiscale hierarchy is visualized in Figure 1 which contains an exemplary illustration of I_l and I_l^{-1}. As I_l is a many to one map the inverse I_l^{-1} is a subset of the fine grid sites, $I_l^{-1}(u) \subset \mathcal{U}_l$. Recursive application of I_l^{-1} yields the set or block of sites $\mathbf{B}(u)$ represented by u. For notational convenience the following summarizing notation is introduced.

Definition 2. (Coarsening Structure) The family $(\mathcal{U}^l, I_l)$ is called the coarsening structure relative to which all coarsening operations are defined. $s^l(u) = |I_l^{-1}(u)|$ denotes the set sizes for the inverse coarsening operation, i.e. the number of fine grid sites connected to u. If $s^l(u)$ is constant, i.e. independent of u, then the coarsening structure is called regular, otherwise it is called irregular.

The coarsening map I connects the variable sets between grids and thereby defines the *configuration subspace* on the fine grid that is spanned by the coarse grid variables. The corresponding *prolongation map* is formally given by

$$\begin{aligned} P : \mathcal{U}^{l+1} &\longrightarrow \mathcal{U}^l \\ c^{l+1} &\mapsto P\left[c^{l+1}\right] = c^l \quad \text{with} \quad c^l(u) = c^{l+1}(I_l(u)). \end{aligned} \tag{6}$$

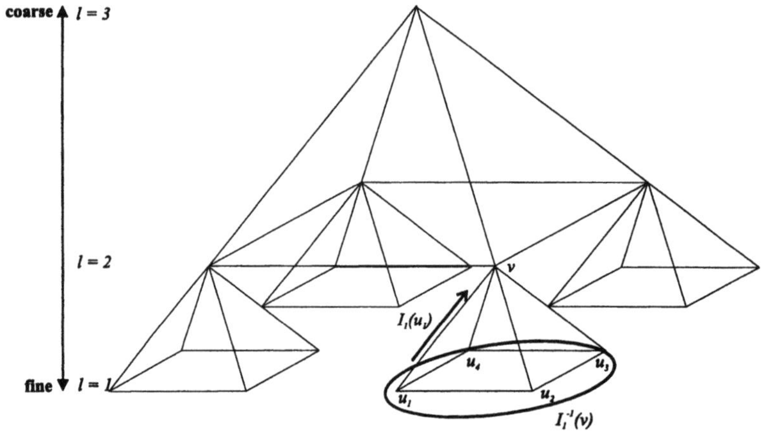

Figure 1. Exemplary multiscale hierarchy with three level: Represented in detail is the relation between the coarse site v and the corresponding fine grid sites u_1, u_2, u_3, u_4: $v = I_l(u_i)$, $\{u_1, u_2, u_3, u_4\} = I_l^{-1}(v)$

Each fine grid variable is thus set to the value of the corresponding coarse grid entity. Now, a family (F^l) of coarse grid cost functions can be canonically derived for a given cost function F^0.

Definition 3. (Multiscale Operator) The recursively defined operator Γ

$$\begin{aligned} F^{l+1}\left(c^{l+1}\right) &= \Gamma\left[F^l\right] \\ &= F^l\left(P[c^{l+1}]\right). \end{aligned} \quad (7)$$

is called the coarsening operator relative to the coarsening structure $(\mathcal{U}^l, I_l)$.

Thus the value of the coarsened cost function is given by the value of the original cost function at the prolongated point. It has to be emphasized that the multiscale coarsening operator is well-defined for arbitrary cost functions F.

But even for cost functions of simple algebraic structure the corresponding coarse cost functions can become arbitrarily complex [13]. Thus it is a valuable property of a cost function, if the algebraic structure does not change under the coarsening structure. This motivates the following definition.

Definition 4. (Closure under Coarsening) A family of objective functions $\mathcal{R}$ is closed under multiscale coarsening, if $\forall\, R \in \mathcal{R}$ there exist constants s_1, s_2 such that $s_1 \cdot \Gamma[R] + s_2 \in \mathcal{R}$.

Thus, a family of cost functions is closed under coarsening, if either the coarse version of a cost function itself or some affine transformation is again a clustering objective function of the same algebraic type. Note that an affine transformation does not change the position of the minimum of a cost function. Closure under coarsening is an extremely valuable property as it guarantees highly efficient optimization of the coarse cost functions and thus improved performance.

2.2 Multiscale Equations for Grouping

In this subsection, multiscale equations are derived for the objective functions discussed earlier.

Central Clustering. Multiscale equations for central clustering as defined by the generalized k-means objective function with centroids $\mathbf{y}_{c(u)}$ and cluster weights $\mathbf{w}(u)$ (see [5])

$$R^{gkm}(c, \mathbf{Y}; \mathbf{X}) = \sum_{u \in \mathcal{U}} \mathbf{w}(u) \|\mathbf{x}_u - \mathbf{y}_{c(u)}\|^2. \tag{8}$$

are derived by recursively applying (7), thus constraining the original cost function to a constant label for each $\mathbf{B}(u)$.

Theorem 5. *For arbitrary coarsening structures the generalized* k*-means clustering criterion* R^{gkm} *is closed under multiscale coarsening. More specifically, using the recursive definition*

$$\mathbf{w}^0 = \mathbf{w}, \quad \mathbf{w}^{l+1}(u) = \sum_{v \in I_l^{-1}(u)} \mathbf{w}^l(v), \tag{9}$$

$$\mathbf{x}^0 = \mathbf{x}, \quad \mathbf{x}_u^{l+1} = \frac{1}{\mathbf{w}^{l+1}(u)} \sum_{v \in I_l^{-1}(u)} \mathbf{w}^l(v)\, \mathbf{x}_v^l$$

the coarse grid cost functions

$$R^l(c^l, \mathbf{Y}) = \sum_{u \in \mathcal{U}^l} \mathbf{w}^l(u) \left\|\mathbf{x}_u^l - \mathbf{y}_{c^l(u)}\right\|^2 + G^l(\mathbf{x}), \tag{10}$$

are obtained where G^l *is independent of* c^l *and* $\mathbf{Y}$.

Therefore, optimization problems of identical algebraic structure are obtained for all coarse levels. The coarse grid data entities are computed as the weighted centroid of the associated data vectors of the fine grid.

Proof. Inserting the generalized k-means cost function (8) into recursive definition (7) of the canonical multiscale operator yields (10) by setting $G^l(\mathbf{X})$ to

$$G^{l+1}(\mathbf{X}) = \sum_{u \in \mathcal{U}^{l+1}} \left[-\mathbf{w}^{l+1}(u) \left(\mathbf{x}_u^{l+1}\right)^t \mathbf{x}_u^{l+1} + \sum_{v \in I^{-1}(u)} \mathbf{w}^l(v) \|\mathbf{x}_v^l\|^2 \right] + G^l(\mathbf{X}). \tag{11}$$

Note that G^l does not depend on c^l or $\mathbf{Y}$. □

The usual k-means clustering criterion (8) is not closed under irregular coarsening. However, for regular coarsening structures, the coarse grid weight $\mathbf{w}^l(u)$ equals the number of associated fine grid sites and is, therefore, independent of u.

Corollary 6. *For regular coarsening structures the k-means clustering criterion R^{km} is closed under coarsening. The coarse grid cost functions are obtained by*

$$R^l(c^l, \mathbf{Y}) = \left(\prod_{m=1}^{l-1} s^m \right) \sum_{u \in \mathcal{U}^l} \left\| \mathbf{x}_u^l - \mathbf{y}_{c^l(u)} \right\|^2 + G^l(\mathbf{x}), \tag{12}$$

where G^l is independent of c^l and $\mathbf{Y}$.

As the minimum of a cost function is invariant under affine transformation, the parameters c^l and s^l can be dropped during optimization. Again optimization problems of identical algebraic structure are obtained for all coarse levels.

Sparse Pairwise Clustering. Considering any of the proximity-based clustering optimization problems over sparse random graphs immediately raises the question how to define sparse graphs on coarse grids and how to obtain consistent graph structures across resolution levels. One possibility is to start with a graph on the finest level and to recursively define the coarser levels by

$$\mathcal{E}^{l+1} = \left\{ (u, v) \in \mathcal{U}^{l+1} \times \mathcal{U}^{l+1} : \exists (u', v') \in \mathcal{E}^l,\ (u, v) = (I_l(u'), I_l(v')) \right\}. \tag{13}$$

Thus $\mathcal{E}^{l+1}$ simply contains all edges for which there exists at least one corresponding edge on the fine graph.

As a major drawback the sparseness property of a random graph is lost under coarsening. Starting with a sparse graph defined on a coarse grid provides a solution but is more complicated to realize. There is no canonical way to propagate edges to finer resolution, where consistency provides only a necessary requirement. There still remain numerous possibilities to define a suitable fine grid graph. This degree of freedom can be easily exploited for deriving random graph families, which are *uniformly sparse* over all scales and sufficiently randomized to avoid the introduction of artificial structure.

Definition 7. (Uniform Sparseness) A family of graphs is called uniformly sparse, if the degree of a node at level 0 does not change under the coarsening operation.

The normalized pairwise clustering criterion [8,15] $R^{pc}(c; \mathbf{D})$ is not closed under coarsening. By recursively applying (7) to the pairwise clustering criterion R^{pc} a family of coarse grid cost functions is defined, which have a slightly more general

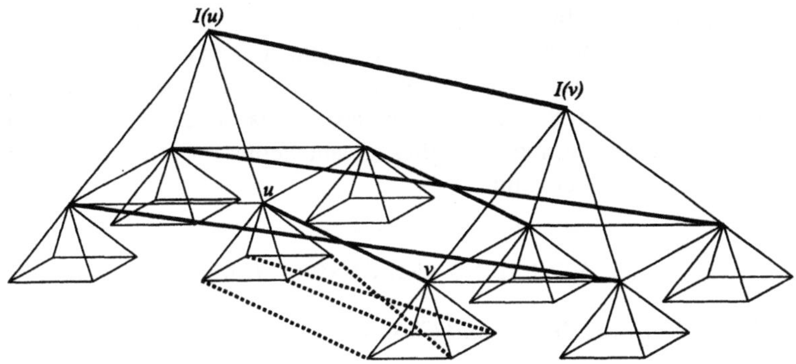

Figure 2. Uniformly sparse random graph: Illustrated are the descendents of the edge $(I_1(u), I_1(v))$. At level 0 only the edges below u and v are depicted

functional form. They are obtained using the following recursive definitions:

$$r_u^0 = 1, \qquad r_u^{l+1} = \sum_{v \in I_l^{-1}(u)} r_v^l, \tag{14}$$

$$s_{uv}^0 = 1, \qquad s_{uv}^{l+1} = \sum_{(u',v') \in I_l^{-1}(u,v)} s_{u'v'}^l, \tag{15}$$

$$\mathbf{D}_{uv}^0 = \mathbf{D}_{uv} \cdot \delta_{(u,v)\in\mathcal{E}}, \qquad \mathbf{D}_{uv}^{l+1} = \sum_{(u',v') \in I_l^{-1}(u,v)} \mathbf{D}_{u'v'}^l. \tag{16}$$

Here, $I_l^{-1}(u,v)$ abbreviates the set $I_l^{-1}(u) \times I_l^{-1}(v)$. Then

$$R^l = \sum_{c=1}^{k} \left[\sum_{u \in \mathcal{G}_c} r_u^l \right] \frac{\sum_{(u,v)\in\mathcal{D}_c^l} \mathbf{D}_{uv}^l}{\sum_{(u,v)\in\mathcal{D}_c^l} s_{uv}^l}. \tag{17}$$

Therefore, a slightly more general type of optimization problem is obtained, that can be efficiently optimized, if $\mathcal{E}^l$ is sufficiently sparse. Assuming a regular coarsening structure and a uniformly sparse family of random graphs, (17) simplifies to the functional form in (4), as $r_u^l = s_{uv}^l = s^l$. This yields the following theorem.

Theorem 8.

1. *For arbitrary coarsening structures the clustering criterion R^{gp} and the generalized pairwise clustering criterion (17) are closed under coarsening.*
2. *For a regular coarsening structure and a uniformly sparse family of random graphs the normalized clustering criterion R^{pc} is closed under coarsening.*

The multiscale equations for R^{gp} are obtained simply by applying the recursive definition (16) to the (dis)similarity matrix. For all proximity-based clustering schemes equations of identical algebraic structure are obtained at all grids. From (16), it follows that (dis)similarity scores between coarse grid entities are simply obtained by adding up all known (dis)similarity values between corresponding sites in the fine grid. A proof of Theorem 8 is straight forward and is omitted here. A uniformly sparse family of random graphs is highly advantageous for efficient optimization for all pairwise clustering cost functions.

3 Annealing and Generalization

3.1 Statistical Learning Theory and Clustering

Multi-scale optimization coarsens the data space to facilitate the search for good clustering solutions. Annealing algorithms coarsen the quality of solutions by treating the optimization variables as random variables with fluctuations controlled by a computational temperature [5]. Evidently, both coarsening concepts should be interleaved to achieve the optimal speedup (see Section 3.3) since high temperature solutions at a fine resolution scale as well as zero temperature solutions at a coarse data scale are inefficient and might lead to suboptimal solutions. To understand how both coarsening concepts can be linked we investigate DA from the point of statistical learning theory to bound the maximally achievable approximation accuracy in the presence of noise in the data.

By choosing different finite stopping temperatures $1/\beta^{\text{final}} > 0$ in annealing, a complete family of different estimation procedures is defined. The natural question arises, which of the methods provides the "best" estimate and, more fundamentally, according to which criterion should the quality be compared, since all algorithms optimize different effective cost functions — the generalized free energies (see [5]). As a key observation in this context, grouping algorithms should be *robust* with respect to measurement noise in the data recording process and should not be affected by modeling uncertainty and the natural within-class variability. More specifically, algorithms should abstract from the peculiarities of the specific problem instance and should only extract *significant* grouping structure. Thus, they should *generalize* to similar problem instances. We present the related theoretical issues in two steps: (i) the generalization issues related to k-means clustering are discussed for the case of nearest neighbor assignments, i.e., the vector quantization case; (ii) we sketch the essential ideas for the general case allowing arbitrary assignments.

Deterministic annealing with finite stopping temperature $\beta^{\text{final}} < \infty$ can be utilized to efficiently prevent the algorithm to over-fit to a given data (image) instance. As a key observation, factorial cost functions like the k-means criterion (8) can be understood as an empirical estimate of an underlying *risk functional* R defined over the unknown true data distribution.

Definition 9. (Expected and Empirical Risk) The measurements $\mathbf{X} = (\mathbf{x}_u) \in \mathcal{X}(n)$ are assumed to be drawn independently and identically distributed (i.i.d.) according to the distribution $\mathbf{P}^{\text{true}}\{\mathbf{x}\}$. $p(\mathbf{x})$ denotes the density of $\mathbf{P}^{\text{true}}\{\mathbf{x}\}$. Given is a set of prototypes $\mathbf{Y} = (\mathbf{y}_c) \in \mathcal{Y}$ and the nearest neighbor rule $\hat{c}(\mathbf{x}) =$

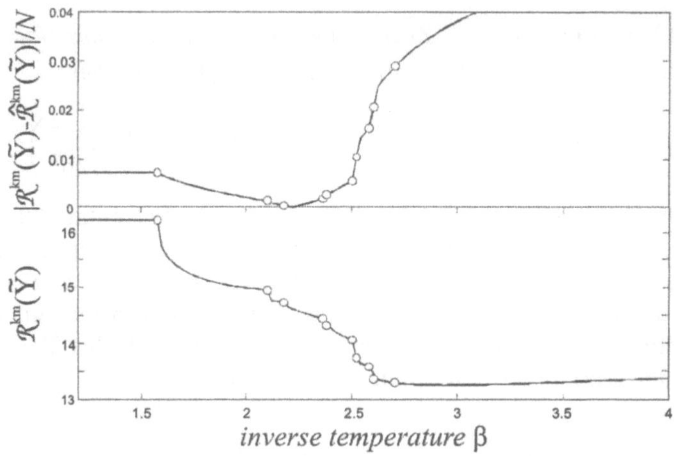

Figure 3. Overfitting results of a 15-means clustering results for different inverse temperatures β of $|\mathcal{U}| = 150$ data vectors in $d = 15$ dimensions. The circles indicate cluster splits. $\tilde{\mathbf{Y}}$ are the cluster centers which have been estimated by deterministic annealing

$\arg\min_c \mathbf{w}(c)\, \|\mathbf{x} - \mathbf{y}_c\|^2$ for the discrete variables. Then

$$\mathsf{R}^{\mathrm{km}}(\mathbf{Y}) = \int_{\mathbf{x}\in\mathbb{R}^d} \left\|\mathbf{x} - \mathbf{y}_{\hat{c}(\mathbf{x})}\right\|^2 p(\mathbf{x})d\mathbf{x} \tag{18}$$

is called the expected risk of k-means clustering and

$$\hat{\mathsf{R}}^{\mathrm{km}}(\mathbf{Y}) = \sum_{u\in\mathcal{U}} \left\|\mathbf{x}_u - \mathbf{y}_{\hat{c}(\mathbf{x}_u)}\right\|^2 . \tag{19}$$

is known as the empirical risk of k-means clustering. $\hat{\mathbf{Y}} = \arg\min_{\mathbf{Y}} \hat{\mathsf{R}}(\mathbf{Y})$ denotes the minimizer of the empirical risk and $\mathbf{Y}^* = \arg\min_{\mathbf{Y}} \mathsf{R}(\mathbf{Y})$ is the minimizer of the expected risk.

From the statistical point of view, $\mathbf{Y}^*$ defines the optimal solution. But as the density $p(\mathbf{x})$ is typically unknown, $\mathbf{Y}^*$ is not algorithmically computable. The inductive principle of *Empirical Risk Minimization* [40] proposes to use $\hat{\mathbf{Y}}$ instead of $\mathbf{Y}^*$. Naturally, one is interested in the approximation error $\mathsf{R}^{\mathrm{km}}(\hat{\mathbf{Y}}) - \mathsf{R}^{\mathrm{km}}(\mathbf{Y}^*)$. As $\hat{\mathsf{R}}^{\mathrm{km}}(\mathbf{Y})$ and thus $\hat{\mathbf{Y}}$ are random variables, the deviation should be bounded in probability, i.e.

$$\mathbf{P}\left\{\mathsf{R}^{\mathrm{km}}(\hat{\mathbf{Y}}) - \mathsf{R}^{\mathrm{km}}(\mathbf{Y}^*) > \epsilon\right\} \tag{20}$$

should be small for a given, fixed precision ϵ. Then the deviation (20) is uniformly bounded independently of $p(\mathbf{x})$ by the following theorem due to Linder et al. [22].

Theorem 10 (Linder, Lugosi & Zeger). *Let $x_1, \ldots, x_n$ be values of an i.i.d. random variable X such that* $\mathbf{P}\{\|X\| \leq b\} = 1$ *and* $n \left(\frac{\epsilon}{8b}\right)^2 \geq 2$. *Then*

$$\mathbf{P}\left\{R^{km}(\hat{\mathbf{Y}}) - R^{km}(\mathbf{Y}^*) > \epsilon\right\} \leq 4(2n)^{k(d+1)} \exp\left(\frac{-n\epsilon^2}{512b^2}\right). \tag{21}$$

Thus, the theorem establishes *consistency*, i.e. the error probability (20) converges to zero $\forall \epsilon > 0$ in the limit of an infinite data set ($n \rightarrow \infty$). For a proof of the theorem the reader is referred to [22]. The theorem presents a zero-temperature result typical for statistical learning theory, i.e. the discrete variables are interpreted as optimization variables. It, however, provides no answer whether $\hat{\mathbf{Y}}$ is an optimal estimate in the sense of (20) for a *finite* number of measurements.

3.2 Statistical Learning Theory and Annealing Methods

There exists both empirical and theoretical evidence that DA with an optimally selected finite stopping temperature provides uniformly better estimates [3] than solutions which minimize the empirical risk, i.e., solutions at zero temperature. Figure 3, e.g., shows a typical k-means clustering result as derived by deterministic annealing at different temperatures. The centroids $\tilde{\mathbf{Y}}$ have been estimated by deterministic annealing for k-means clustering at a finite inverse temperature β using algorithm 5.2 of [5]. The figure clearly demonstrates that the empirical risk is decreasing faster than the expected risk for $\beta \geq 2.2$ – a phenomenon called overfitting. The monotonicity of the k-means cost function yields a decreasing expected risk of the empirically estimated centroids $\hat{\mathbf{Y}}$ when more centroids are used after a cluster split. However, only the first three splits decrease the deviation between expected and empirical risk which is normalized with the variance of the loss function $\|\mathbf{x} - \mathbf{y}_{c(\mathbf{x})}\|^2$.

How can we understand such a behavior in statistical estimation? Intuitively, it is clear that estimates with too many clusters and with too few data samples per cluster reflect more the noise in the data than the signal. Therefore, averaging over clustering solutions which are statistically indistinguishable should ameliorate the noise influence and leads to robust estimates. To describe this averaging effect in precise mathematical terms we extend the inference principle of Empirical Risk Minimization to a new inference principle called *Empirical Risk Approximation* (ERA). Rather than finding the set of solutions with minimal empirical risk we determine as the solution set all functions which do not exceed the minimal empirical risk by more than γ in costs. We then can either (i) select an arbitrary solution from this solution set or (ii) we average over all such solutions. The first choice samples from solutions which are statistically equivalent to the global empirical minimum, whereas the second procedure averages models in the Bayesian spirit. The following definitions allow us to characterize the bias-variance tradeoff which causes overfitting in clustering.

Definition 11.

- A training data set $\mathbf{Z}_1$ and a test data set $\mathbf{Z}_2$ are given.

- The index $\alpha \in \Lambda$ enumerates the different clustering solutions, e.g., $\alpha = (c(.), \mathbf{Y})$ enumerates the different k-means clustering solutions.
- $r_u(\alpha; \mathbf{Z})$ defines the partial risk of object u, e.g., for k-means clustering $\mathbf{Z} = \mathbf{X} \in \mathbb{R}^{d \cdot n}$ and $r_u(\alpha; \mathbf{X}) := \|\mathbf{x}_u - \mathbf{y}_{c(u)}\|^2$.
- The risk functional $R(\alpha; \mathbf{Z}) = \frac{1}{n} \sum_{u \in \mathcal{U}} r_u(\alpha; \mathbf{Z})$ measures the total risk or the quality of clustering solutions.
  The abbreviations $R_1(\alpha) := R(\alpha; \mathbf{Z}_1)$ and $R_2(\alpha) := R(\alpha; \mathbf{Z}_2)$ are used for the training costs and test costs.
- The distance between two clustering solutions $\alpha, \tilde{\alpha}$ is defined by the l_1 metric $d(\alpha, \tilde{\alpha}) = \frac{1}{n} \sum_{u \in \mathcal{U}} |r_u(\alpha; \mathbf{Z}) - r_u(\tilde{\alpha}; \mathbf{Z})|$

Definition 12 (Approximation Set). Let $\alpha_1 := \arg\min_{\alpha \in \Lambda} R(\alpha; \mathbf{Z}_1)$ denote the best clustering solution on the training instance defined by $\mathbf{Z}_1$.
$\mathcal{L}_\gamma := \{\alpha : R_1(\alpha) - R_1(\alpha_1) \le \gamma\}$ defines the set of approximative solutions on the training instance.

Definition 13 (γ-cover). A subset $\Lambda_\gamma \subset \Lambda$ such that for all $\alpha \in \Lambda$ there exists a solution $\tilde{\alpha} \in \Lambda_\gamma$ with $d(\alpha, \tilde{\alpha}) \le \gamma$ denotes a γ-cover. The minimal cardinality of a γ-cover is called the γ-covering number. The function in the γ-cover with the smallest test costs is denoted by $\alpha_{2,\gamma} := \arg\min_{\alpha \in \Lambda_\gamma} R_2(\alpha)$.

The γ-cover replaces the original solution space of a clustering problem with a coarsened version of it. The approximation quality of solutions is controlled by the coarsening parameter γ. The following analysis which generalizes an argument of Vapnik & Chervonenkis to γ-covers will yield an upper bound on the optimal γ value.

Lemma 14. *Let $\alpha_\gamma \in \mathcal{L}$ be a γ-bounded approximation of the lowest training costs and let $\Delta R_2(\alpha_\gamma) := R_2(\alpha_\gamma) - R_2(\alpha_2)$ be the difference between α_γ's test costs and the minimal test costs. Then*

$$\Delta R_2(\alpha_\gamma) \le 2 \sup_{\alpha \in \Lambda_\gamma \cup \mathcal{L}} |R(\alpha; \mathbf{Z}_1) - R(\alpha; \mathbf{Z}_2)| + 2\gamma \tag{22}$$

Proof.

$$\begin{aligned}
\Delta R_2(\alpha_\gamma) &= R_2(\alpha_\gamma) - R_1(\alpha_1) + R_1(\alpha_1) - R_2(\alpha_2) \\
&\le R_2(\alpha_\gamma) - R_1(\alpha_1) + R_1(\alpha_{2,\gamma}) - R_2(\alpha_{2,\gamma}) + \gamma \\
&\le R_2(\alpha_\gamma) - R_1(\alpha_\gamma) + \sup_{\alpha \in \Lambda_\gamma} |R_1(\alpha) - R_2(\alpha)| + 2\gamma \\
&\le \sup_{\alpha \in \mathcal{L}} |R_1(\alpha) - R_2(\alpha)| + \sup_{\alpha \in \Lambda_\gamma} |R_1(\alpha) - R_2(\alpha)| + 2\gamma \\
&\le 2 \sup_{\alpha \in \Lambda_\gamma \cup \mathcal{L}} |R_1(\alpha) - R_2(\alpha)| + 2\gamma \quad \square
\end{aligned}$$

The inequality (22) reflects the wellknown bias-variance tradeoff in statistics. If γ is very small or vanishes then the bias 2γ decays to zero. On the other hand, the supremum has to be taken over a large set of functions which increases the variance. In the opposite limit of large γ we decrease the variance at the expense of an increasing bias. $\Delta R_2(\alpha_\gamma)$ depends on the training data and the test data. To achieve an estimate on the generalization performance $\Delta R_2(\alpha_\gamma)$ we have to derive an upper bound for the probability $\mathbf{P}\{\Delta R_2(\alpha_\gamma) > 2\epsilon\}$ which results from the concentration-of-measure inequality by Bernstein.

Theorem 15 (Bernstein (1946)). *Let $X_1, \ldots, X_n$ be independent real valued random variables with zero mean and assume that $|X_i| \le c$ with probability one. Let $\sigma^2 = \frac{1}{n}\sum_{i=1}^{n} \mathbf{V}\{X_i\}$. Then for any $\epsilon > 0$*

$$\mathbf{P}\left\{\left|\frac{1}{n}\sum_{i=1}^{n} X_i\right| > \epsilon\right\} \le 2\exp\left(-\frac{n\epsilon^2}{2\sigma^2 + c\epsilon}\right). \tag{23}$$

Large deviations of the generalization performance $\mathbf{P}\{\Delta R_2(\alpha_\gamma) > 2\epsilon\}$ can be bounded by using the Vapnik Chervonenkis inequality (22) and the union bound.

Theorem 16. *Given is a clustering risk $R(\alpha; \mathbf{Z}) = \frac{1}{n}\sum_{u \in \mathcal{U}} r_u(\alpha; \mathbf{Z})$ where the loss $r_u(\alpha; \mathbf{Z})$ of object u for fixed α is an independent distributed random variable bounded by $|r_u(\alpha; \mathbf{Z})| \le c$. Let $\sigma_\alpha^2 = \frac{1}{n}\sum_{u \in \mathcal{U}} \mathbf{V}\{r_u(\alpha; \mathbf{Z})\}$. Then*

$$\mathbf{P}\{\Delta R_2(\alpha_\gamma) > 2\epsilon\} \le 2|\Lambda_\gamma \cup \mathcal{L}| \sup_{\alpha \in \Lambda_\gamma \cup \mathcal{L}} \exp\left(-\frac{n(\epsilon-\gamma)^2}{2\sigma_\alpha^2 + c(\epsilon-\gamma)}\right) \tag{24}$$

We set the right side of (24) equal to the selected confidence level δ. In the proof we first replace the generalization performance $\Delta R_2(\alpha_\gamma)$ by the uniform convergence criterion using the VC-inequality (14). Then the union bound allows us to replace the large deviation of the supremum term by a sum of probabilities which then is bounded by the largest element times the cardinality of the respective function set:

Proof.

$$\begin{aligned}
\mathbf{P}\{\Delta R_2(\alpha_\gamma) > 2\epsilon\} &\le \mathbf{P}\left\{\sup_{\alpha \in \tilde{\Lambda} \cup \Lambda_\gamma} |R_1(\alpha) - R_2(\alpha)| \ge \epsilon - \gamma\right\} \\
&\le \sum_{\alpha \in \tilde{\Lambda} \cup \Lambda_\gamma} \mathbf{P}\{|R_1(\alpha) - R_2(\alpha)| \ge \epsilon - \gamma\} \\
&\le 2|\Lambda_\gamma \cup \mathcal{L}| \sup_{\alpha \in \Lambda_\gamma \cup \mathcal{L}} \exp\left(-\frac{n(\epsilon-\gamma)^2}{2\sigma_\alpha^2 + \tau_\alpha \sigma_\alpha(\epsilon-\gamma)}\right) \quad \square
\end{aligned}$$

This large deviation inequality weighs two competing effects in the estimation problem, i. e. the probability of a large deviation exponentially decreases with growing sample size n, whereas a large deviation becomes increasingly likely with growing cardinality of the γ-cover of the hypothesis class. According to the bound (24) the sample complexity $n_0(\gamma, \epsilon, \delta)$ with confidence value δ is defined by

$$\log|\Lambda_\gamma \cup \mathcal{L}| - \sup_{\alpha \in \Lambda_\gamma \cup \mathcal{L}} \frac{n_0(\epsilon-\gamma)^2}{2\sigma_\alpha^2 + c(\epsilon-\gamma)} + \log\frac{2}{\delta} = 0. \tag{25}$$

The optimal coarsening of the hypothesis class according to the bound (24) is achieved if we fulfil the condition $d\epsilon/d\gamma = 0$. To establish the link between the large deviation bound (25) and annealing algorithms we interpret the log-cardinality $\log|\mathcal{L}|$ of the approximation set $\mathcal{L}$ as microcanonical entropy and the approximation parameter γ as energy which completely corresponds to the microcanonical approach to statistical physics. Furthermore, we bound $|\Lambda_\gamma \cup \mathcal{L}| \leq |\Lambda_\gamma| + |\mathcal{L}| \approx |\Lambda|/|\mathcal{L}| + |\mathcal{L}|$ and we approximate

$$\frac{d}{d\gamma} \log\left(\frac{|\Lambda|}{|\mathcal{L}|} + |\mathcal{L}|\right) = -\frac{1 - |\mathcal{L}|^2/|\Lambda|}{1 + |\mathcal{L}|^2/|\Lambda|} \frac{d}{d\gamma} \log|\mathcal{L}| \approx -T^{-1} \tag{26}$$

We have used the well-known relation from statistical physics $\frac{d\,\text{entropy}}{d\,\text{energy}} = T^{-1}$ to derive the stop temperature in (26). Neglecting the small correction term $|\mathcal{L}|^2/|\Lambda|$, the stop temperature is given by

$$\frac{1}{T^{\text{stop}}} \leq -\frac{d}{d\gamma} \sup_{\alpha \in \Lambda_\gamma \cup \mathcal{L}} \left.\frac{n_0\,(\epsilon - \gamma)^2}{2\sigma_\alpha^2 + c\,(\epsilon - \gamma)}\right|_{\frac{d\epsilon}{d\gamma}=0} . \tag{27}$$

With probability $1 - \delta$ the deviation of the empirical risk from the expected risk is bounded by $(\epsilon^{\text{opt}} - \gamma)$. Averaging over all γ-close functions to the empirical minimum or sampling from this set yields a robust clustering solution. The remaining key task is to evaluate the right side of (27) and to use problem dependent information to calculate or bound σ_α^2.

3.3 Multiscale Optimization and Annealing

Algorithms like k-means efficiently minimize R^{km} by *splitting techniques* to obtain successive solutions for a growing number of clusters. Deterministic Annealing with its sequence of cluster bifurcations (see Figure 1 in [5]) supports this optimization strategy. On the other hand, enough data points have to be available for a reliable estimate of a given number k of clusters as outlined in Section 3. Since the number of effective data points available drastically reduces at coarser resolution levels, splitting strategy and coarse-to-fine optimization should be interleaved. The question of how many effective data points are needed to distinguish k clusters has been addressed for the k-means cost function in the context of *uniform convergence of empirical means to their expectations* [22, 30].

As a main result, the deviation between empirical and expected risk should be bounded to obtain robust and reliable results. Bounds for the deviation of the empirical costs $R^{km}(c, \mathbf{Y}; \mathbf{X})$ from the expected costs of a grouping solution can be derived independently of the underlying distribution and are given by Theorem 10. Given n data points, an accuracy ϵ and a bound δ for the large deviation probability $\mathbf{P}\{R(\hat{\mathbf{Y}}) - R(\mathbf{Y}^*) > \epsilon\}$ an upper bound for the maximal number of clusters $k_{\max}$ is obtained from (21) by

$$k_{\max} \leq \frac{\epsilon^2}{512 b^2 (d+1)} \frac{n}{\log 2n} + \frac{\log\delta - \log 4}{d+1} \frac{1}{\log 2n}. \tag{28}$$

This bound has been derived by a worst case analysis independent of specific assumptions on the probability distribution of the data. Although the bound is impractically low as a consequence of this distribution independence it is assumed that it exhibits the correct asymptotic behavior. This assumption yields the selection criterion

$$k_{\max} < s \frac{n}{\log 2n} \tag{29}$$

for some empirically selected constant s. Thus, an approximately linear dependency of the maximal number of clusters on the number of data points is conjectured, which roughly corresponds to the statistical rule-of-thumb to use at least 10 data points for each parameter to be estimated.

```
Input: objective function F, β^start, β^final, site visitation schedule V, l^max
Output: clustering solution

for l = 1, ..., l^max do
   compute R^l = Γ[R^(l-1)] recursively by (9),(14)-(16)
end for
β = β^start, l = l^max
while l ≥ 0 AND β > β^final do
   {//— Annealing loop}
   call MA (Q^l, Y, β, F^l, V) as given by Alg. 2.5
   determine effective number of clusters k
   if k > k_max then
      propagate Q^l to Q^(l-1) by q^(l-1)_(u^(l-1))(c) = q^l_(I(u^(l-1)))(c)
      l = l - 1
   else
      β = SCHEDULE()
   end if
end while
c^l(u) = arg max_d q^l_(u^l)(d)
while l > 0 do
   {//— Zero temperature refinement}
   propagate c^l to c^(l-1)
   l = l - 1
   call ICM (c^l, Y, F^l, V)
end while
```

Algorithm 7: Multiscale Annealing (M-DA)

One of the key advantages of the DA approach is the *inherent splitting behavior*, as clusters degenerate at high temperature and successively split in bifurcations or phase transitions when β is raised. As demonstrated in [5] Section 2.9, there is a monotone relationship between the computational temperature and the effective number of clusters. Therefore, at a specific temperature scale β an easily measurable[2] effective number k_β of clusters can be observed. Thus, for a given resolution level l it is proposed to anneal until k_{β^*} exceeds the bound $s \frac{n^l}{\log 2n^l}$ for some critical temperature β^*. After prolongation to level $l-1$ the DA optimization is continued at the same temperature β^*. The essential structure of the grouping solution is detected on the coarser scales using global optimization techniques and is then refined at the finer scales using fast low or zero temperature algorithms. This scheme, which is referred to as *multiscale annealing* (M-DA), works very well for all proximity-based clustering criteria as well as spatial quantization, even though in these cases the concept of uniform convergence is theoretically not yet well understood. The complete multiscale annealing algorithm is summarized in Alg. 7.

Multiscale annealing couples the resolution hierarchies in object and label space in a systematic fashion with the resolution in optimization space introduced by the computational temperature. The object resolution determines the maximally allowed number of clusters by (29), which in turn determines the maximal value for β allowed at a specific resolution. Alternatively, given the number of data points n one could estimate the optimal stopping temperature as in [4] which in turn determines the effective number of clusters. Thus, the link between clustering resolution and computational temperature is an intrinsic property of deterministic annealing, while the object scale at one hand and the clustering and optimization scale at the other hand are coupled by the results of statistical learning theory, i.e. the demands of robustness in algorithm design.

4 Empirical Results

The unsupervised segmentation of textured images is now used as a prototypical application to evaluate the performance of the different clustering models and the optimization algorithms. The unsupervised segmentation of images is a typical computer vision problem and serves nicely to evaluate and benchmark the designed algorithms. It can be phrased as a grouping or clustering problem in many ways, including central clustering [23], graph partitioning [9] and normalized graph partitioning [16] as treated in this chapter. For completeness we provide results for two additional approaches, the normalized cut [37] and histogram clustering [33], which also result in grouping criteria that are closed under coasening. Thus the experimental data covers a large number of modeling approaches, multiscale annealing improving all of them from a computational perspective. A typical example is presented in Figure 4 with $k = 16$ different segments.

4.1 Pre-Processing

As outlined in Figure 5, the different grouping approaches need different feature extraction and pre-processing stages, which are now specified.

---

[2] Two clusters c and d are identified, if $\|q_u(c)-q_u(d)\|^2 = \frac{1}{n}\sum_{u\in\mathcal{U}}(q_u(c)-q_u(d))^2 < \tau$ for some small threshold parameter τ. In the experiments, $\tau = 0.01$ has been used.

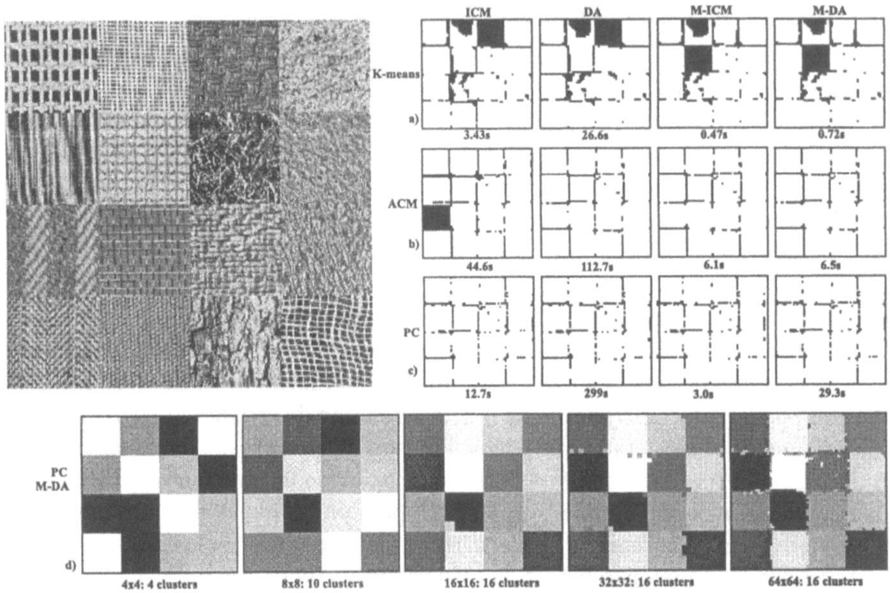

Figure 4. Typical segmentation result with $k = 16$ for three different models and four different optimization algorithms. For the proximity-based algorithms a random graph with an average node degree 150 has been used. Misclassified blocks are depicted in black. The image captions show the optimization time needed. In (d) a typical course-to-fine optimization is illustrated for R^{pc}

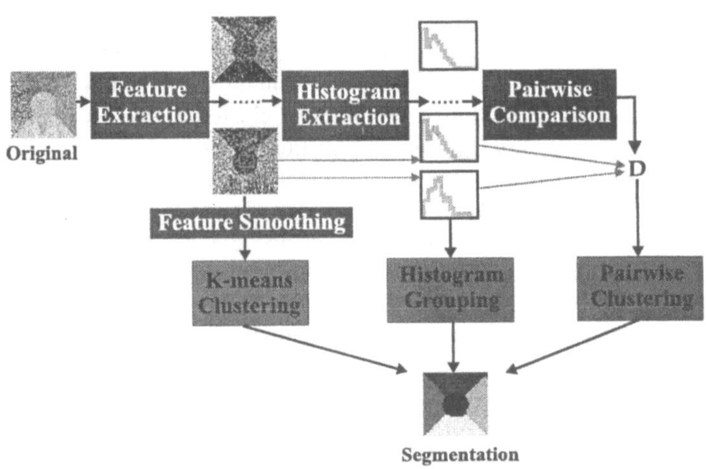

Figure 5. Overview of the unsupervised segmentation pipeline. Illustrated are the pre-processing and feature extraction stages necessary for the different grouping models

	Multiscale DA	Multiscale ICM	DA	ICM (K-means)
R^{km}	6.8 % / 5.7 % 0.44s / 0.94s / 5 %	6.8 % / 5.6 % 0.11s / 0.33s / 5 %	6.8 % / 6.0 % 4.0s / 27.1s / 7%	6.8 % / 6.0 % 0.41s / 2.1s / 9 %
R^{acm}	5.1 % / 3.7 % 1.6s / 3.5s / 4 %	5.1 % /3.7 % 1.0s / 3.1s / 5 %	5.0 % / 3.9 % 21.1s / 89.8s / 6 %	5.3 % / 3.8 % 4.6s / 20.7s / 6 %
R^{gp}	11.0 % / 5.3 % 4.0s / 4.8s / 18 %	12.0 % / 5.6 % 0.35s / 1.6s / 16 %	13.4% / 7.1 % 42.2s / 219.0s / 18 %	19.2 % / 9.9 % 0.8s / 4.0s / 28 %
R^{pc}	5.9 % / 3.7 % 2.4s / 3.6s / 8 %	5.9 % / 3.6 % 0.54s / 1.6s / 11 %	5.8 % / 3.6 % 16.1s / 308.6s / 9 %	5.9 % / 3.7 % 1.1s / 5.9s / 8 %
R^{nc}	5.9 %/ 4.7 % 2.4s / 2.5s / 8 %	6.6 % / 5.3 % 0.44s / 1.3s / 17 %	5.8 % / 4.2 % 18.8s / 155.5s / 10 %	7.2 % / 5.1 % 1.0s / 5.0s / 18 %

Table 1. Quality and run-time of different grouping methods for unsupervised texture segmentation on `Brodatz-5-86`. First row: Median error [%] / Median error [%] after noise removal. Second row: Mean run-time [sec] for 64 × 64 / Mean run-time for 128x128 [sec] / Segmentations with more than 20% error rate after noise removal [%]. Errors are obtained by comparison with ground truth for segmenting 100 randomly generated images for a segmentation resolution of 128x128

- All segmentations are based on a filter bank of 12 Gabor filters using 4 orientations (0°, 45°, 90°, 135°) and 3 scales, separated by octaves. A wavelength of 2 pixels has been chosen for the smallest scale. If not specified differently, segmentations have been obtained on a regular grid of 128 × 128 image sites.
- For K-means, each Gabor channel is spatially smoothed using a Gaussian kernel with standard deviation proportional to the filter scale.
- For histograms, the marginal feature distribution at a site is estimated in a 16 × 16 window for the finest resolution. The distribution is estimated using 16 bins adapted to the dynamic range of the filter output.
- For the proximity-based clustering methods, a sparse reflexive graph was chosen at a coarse grid of 16 × 16 sites consisting of the 4 nearest neighbors and on average 80 randomly selected sites. Then, a uniformly sparse graph structure for all grids was constructed. This graph structure has been employed in all experiments. The sparse dissimilarity matrix has been computed applying the χ^2 test statistic.
- For R^{gp} the matrix has been normalized to a maximal dissimilarity of 1 and shifted to a mean dissimilarity of 0.25. This optimal shift was obtained after extensive bench-marking. For the normalized cut R^{nc} the dissimilarity data has first been normalized to unit maximum and then converted to similarity scores applying the transformation $\mathbf{D}_{uv}^{new} = \exp(-\mathbf{D}_{uv}/s)$ as suggested by Shi et al. [37] with a parameter $s = 0.09$ obtained after extensive bench-marking.

The processing times needed for the feature extraction are summarized in Tab. 2. The processing times for the Gabor transformation can be further accelerated by a pyramidal implementation of the Gabor transformation or by implementation on special purpose hardware. All times mentioned in the next subsections solely denote the processor time needed during optimization.

Feature extraction	Processor time 64 × 64	Processor time 128 × 128
Gabor transformation	6.5s	6.5s
Feature smoothing (k-means only)	4.6s	10.5s
Histogram extraction	3.8s	13.7s
Pairwise comparison	9.1s	34.3s

Table 2. Processor time spent on a PC Pentium II, 300 MHz for different feature extraction schemes and for a resolution of 512×512 pixels and 64×64 / 128×128 sites. In brackets run-time on the special-purpose hardware DataCube, if applicable. Execution times for pairwise comparisons are measured for an average node degree of 80. The implementation of the histogram extraction for the multivariate case has not been optimized

4.2 Acceleration

To empirically evaluate the acceleration, that is obtained when using multiscale optimization as opposed to its single-scale counter parts, the test set `Brodatz-5-86` is utilized which consists of 100 patchwork images composed of 86 different micro-textures taken from the Brodatz photographic album, each image is composed of 5 different, randomly chosen textures. The main question examined in detail in this section addresses the benefits of multiscale techniques w.r.t. run-time. A typical coarse-to-fine optimization is illustrated in Figure 4 (d). The essential structure is detected during early optimization on the coarse-grids and is then propagated to the finer levels. Figure 4 also highlights the interleave between annealing schedule and coarse-grid optimization, where jumps between grids occur whenever a maximum number of effective clusters is reached. This illustrates the any-time characteristic of the algorithms, where approximative intermediate coarse results are obtained already after a small fraction of the overall run-time.

As seen from Tab. 1 the multiscale ICM optimization is 2–4.5 times faster than ordinary ICM (k-Means) for $k = 5$ and a segmentation resolution of 64 × 64 pixels, while the speed gain for the global multiscale annealing (M-DA) optimization procedure is 6.5–13 compared to its single resolution counterpart DA, where mean optimization times of 0.44s for K-means, 1.6s for ACM, 4.0s for graph partitioning (GP), 2.4s for normalized pairwise clustering (PC) and 2.4s for the normalized cut (NC) have been obtained using multiscale annealing. Thus it is possible for all clustering schemes to compute globally optimal solutions within a few seconds.

The performance of multiscale techniques is even better for a larger number of segments resulting in a speed gain of 4–7.5 for ICM and 10–40 for DA for $k = 16$. The same result holds, when the segmentation resolution is increased, as seen from Figure 6. The gap between multiscale and ordinary optimization increases for larger number of optimization variables, resulting in a gain of 3–9 for ICM and 20–90 for DA for a problem size of 327680 Boolean variables at resolution 256 × 256. In addition, the speed gain of multiscale ICM (M-ICM) compared to M-DA significantly shrinks for large problem sizes.

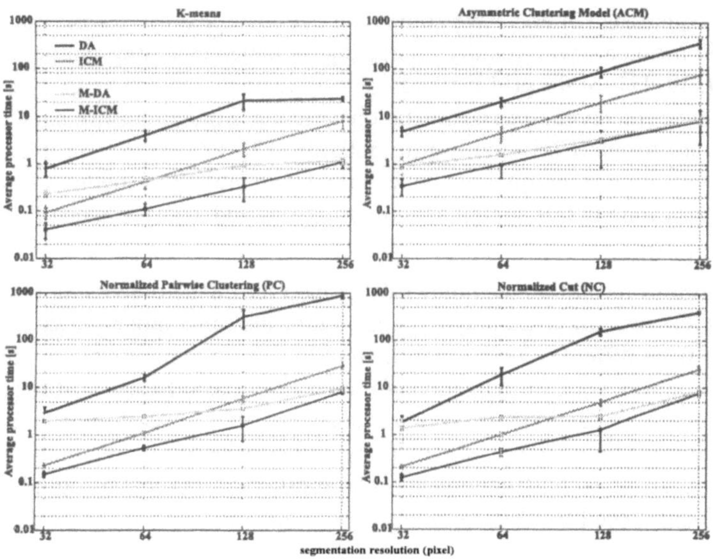

Figure 6. Average optimization time in seconds on a logarithmic scale depending on the segmentation resolution: (a) k-means, (b) ACM, (c) PC and (d) NC. The error-bars show the empirical standard deviation. The results are based on the database `Brodatz-86-5`

5 BIBLIOGRAPHIC NOTES

Multiscale Optimization. Multiscale optimization in the strict sense, i.e. the canonical multiscale operator, has been first proposed in [14] for a visual reconstruction application. Since then, it has been developed and refined for clustering [31,32] and color image quantization [34]. In addition, multiscale optimization has been applied to visual reconstruction [26] and motion estimation [29, 38]. It should be emphasized that in contrast to all other cost functions discussed in literature which are not closed under coarsening, no approximations have to be made for all grouping and quantization cost functions discussed in this paper since the multiscale coarsening yields cost functions of identical algebraic form for all grids. This is a very valuable property as it enables highly efficient optimization with acceleration rates far beyond the results reported, e.g., for visual reconstruction [14].

The multiscale annealing scheme has first been proposed in [31] in the context of texture segmentation where previous approaches solely relied on local optimization techniques like iterative conditional mode (ICM) [7,17,20,23,27,41] or single scale annealing techniques [9, 16]. Multi-resolution optimization techniques have been used only occasionally [1].

For other vision applications, several optimization approaches relying on coarse versions of a cost function have been proposed in the past. Technically similar are *multi-grid* methods, which have first been developed for the solution of partial differential equations [2] and have since then been adopted to a broad range of optimization problems with locally interacting variables including image processing tasks [24, 39, 42]. Multi-grid methods rely on incremental coarse grid corrections and, therefore, on *continuous* optimization variables. Similar in spirit but technically different are Monte Carlo multi-grid methods [12]. For *discrete* problems multiscale optimization techniques are better suited than multi-grid methods [14]. Multiscale optimization in this context is sometimes referred to as the Block Label (BL–) heuristics [25].

Most multi-resolution techniques developed for image processing tasks are *semantic* multi-resolution techniques. They adopt the same model class for features extracted from different *image scales*. Note that there is no guarantee that coarse grid solutions are good initializations for finer levels, as different cost functions are optimized and interactions between variables are often severely resolution dependent. Multiscale optimization significantly outperformed semantic multi-resolution methods in visual reconstruction and motion estimation problems both in terms of accuracy and speed [14, 38].

The most principled multiscale methods are renormalization group approaches [11,28]. Renormalization group algorithms rely on coarsening the Gibbs distribution which is associated with a given objective function. The transformation is not only order preserving (for small temperature parameters) but also preserves the expectation of arbitrary functions and thus the statistics inherent in the cost function. Thus the renormalized cost function summarizes the coarse scale statistics of a model while multi-grid and multiscale cost functions are merely computational artifacts. However, computing exact renormalized cost functions is notorious difficult and even in very simple examples considerable approximations like the cummulant expansion proposed in [11] have to be made. It has been conjectured that one should not expect to derive renormalized cost functions for any reasonably large class of functions [25]. Moreover, from the optimization perspective one is not interested in maximally preserving the statistics of a system. Solely the minima of the coarse system should be close to the minima of original cost function. It is the computational complexity, that provides the main characteristic which should guide the selection of a coarsening method. For renormalization group approaches acceleration factors of approximately 1.5 have been obtained [11]. As shown, multiscale annealing is closed under coarsening for most grouping problems and thus enables highly efficient optimization with typical acceleration factors of $2 - 90$. It is thus preferable from the optimization perspective. It has been established in [25] that the canonical multiscale operator is an excellent approximation to the renormalization group approach in the zero temperature limit and can thus be expected even theoretically to yield very good results.

Closely related to renormalization group approaches are multi-resolution Markov Random Field (MRF) models [21] which effectively rely on subsampling of random fields [13]. It has been established that most multi-resolution MRF as well as most renormalized cost functions loose their local structure at coarse levels and substantial approximations have to be made in order to obtain tractable models [13].

6 Conclusion

Multi-scale Annealing methods for grouping have been demonstrated in a number of studies as efficient and easily applicable methods to approximately optimize large combinatorial optimization problems. The method combines the acceleration idea of optimization on different scales with the stochastic approximation of annealing strategies. Deterministic Annealing has been characterized as a continuation method which establishes a homotopy between combinatorial optimization problems and their corresponding space of probability distributions. The homotopy parameter is known as the inverse temperature β and it controls the degree of relaxation of the discrete problem.

Another important advantage of annealing methods is their empirically reported robustness to distortions in the problem instance, e.g., by noise. In noisy optimization problems as they arise in computer vision or statistical pattern recognition the main goal is to minimize the expected costs of a solution and not the empirical costs defined by the available sample set. A controlled approximation can avoid overfitting and yields solutions which perform better on future problem instances drawn from the same probability distribution than the empirical risk minimizer. The robustness requirement yields a maximal value of the homotopy parameter β which can be determined by numerical techniques or by analytical bounds from Statistical Learning Theory.

References

1. C. Bouman and B. Liu, *Multiple resolution segmentation of textured images*, IEEE Transactions on Pattern Analysis and Machine Intelligence, 13 (1991), pp. 99–113.
2. W. Briggs and S. McCormick, *Introduction*, in Multigrid Methods, S. McCormick, ed., Frontiers in Applied Mathematics, Society for Industrial and Applied Mathematics, 1987, ch. 1, pp. 1–31.
3. J. Buhmann, *Empirical risk approximation: An induction principle for unsupervised learning*, IAI-TR 98-3, Institut für Informatik III, 1998.
4. J. M. Buhmann and J. Puzicha, *Unsupervised learning for robust texture segmentation*, in Performance Characterization and Evaluation of Computer Vision Algorithms, R. Klette, S. Stiehl, and M. Viergever, eds., Kluwer Academic Publishers, 2000, pp. 211–225.
5. J. M. Buhmann and J. Puzicha, *Basic Principles of Annealing for Large Scale Non-linear Optimization*, Springer, 2001, this volume.
6. V. Cerny, *Thermodynamical approach to the travelling salesman problem*, Journal of Optimization Theory and Applications, 45 (1985), pp. 41–51.
7. B. Chaudhuri and N. Sarkar, *Texture segmentation using fractal dimension*, IEEE Transactions on Pattern Analysis and Machine Intelligence, 17 (1995), pp. 72–77.
8. R. Duda and P. Hart, *Pattern Classification and Scene Analysis*, Wiley, 1973.
9. D. Geman, S. Geman, C. Graffigne, and P. Dong, *Boundary detection by constrained optimization*, IEEE Transactions on Pattern Analysis and Machine Intelligence, 12 (1990), pp. 609–628.
10. S. Geman and D. Geman, *Stochastic relaxation, Gibbs distributions, and the Bayesian restoration of images*, IEEE Transactions on Pattern Analysis and Machine Intelligence, 6 (1984), pp. 721–741.

11. B. Gidas, *A renormalisation approach to image processing problems*, IEEE Transactions on Pattern Analysis and Machine Intelligence, 11 (1989), pp. 164–180.
12. J. Goodman and A. Sokal, *Multigrid Monte-Carlo method. Conceptual foundations*, Physical Review D, 40 (1989), pp. 2025–2071.
13. F. Heitz and P. Perez, *Restriction of a Markov random field on a graph and multiresolution statistical image modeling*, IEEE Transactions on Information Theory, 42 (1996), pp. 180–190.
14. F. Heitz, P. Perez, and P. Bouthemy, *Multiscale minimization of global energy functions in some visual recovery problems*, CVGIP: Image Understanding, 59 (1994), pp. 125–134.
15. T. Hofmann and J. Buhmann, *Pairwise data clustering by deterministic annealing*, IEEE Transactions on Pattern Analysis and Machine Intelligence, 19 (1997), pp. 1–14.
16. T. Hofmann, J. Puzicha, and J. Buhmann, *Unsupervised texture segmentation in a deterministic annealing framework*, IEEE Transactions on Pattern Analysis and Machine Intelligence, 20 (1998), pp. 803–818.
17. A. Jain and F. Farrokhnia, *Unsupervised texture segmentation using Gabor filters*, Pattern Recognition, 24 (1991), pp. 1167–1186.
18. M. Jordan, ed., *Learning in Graphical Models*, Kluwer Academic Publisher, 1998.
19. S. Kirkpatrick, C. Gelatt, and M. Vecchi, *Optimization by simulated annealing*, Science, 220 (1983), pp. 671–680.
20. A. Laine and J. Fan, *Frame representations for texture segmentation*, IEEE Transactions on Image Processing, 5 (1996), pp. 771–779.
21. S. Lakshmanan and H. Derin, *Gaussian Markov random fields at multiple resolutions*, in Markov Random Fields: Theory and Application, R. Chellappa and A. Jain, eds., Academic Press, 1993, pp. 131–157.
22. T. Linder, G. Lugosi, and K. Zeger, *Rates of convergence in the source coding theorem, in empirical quantizer design, and in universal lossy source coding*, IEEE Transactions on Information Theory, 40 (1994), pp. 1728–1740.
23. J. Mao and A. Jain, *Texture classification and segmentation using multiresolution simultaneous autoregressive models*, Pattern Recognition, 25 (1992), pp. 173–188.
24. E. Mjolsness, C. Garrett, and W. Miranker, *Multiscale optimization in neural nets*, IEEE Transactions on Neural Networks, 2 (1991), pp. 263–273.
25. G. Nicholls and M. Petrou, *Multiresolution representation of Markov random fields*, VSSP-TR-3/93 ACT-ST-272-89, Department of Electronic and Electrical Engineering, University of Surrey, 1993.
26. G. Nicholls and M. Petrou, *On multiresolution image restoration*, in Proceedings of the International Conference on Pattern Recognition (ICPR'94), vol. III, 1994, pp. 63–67.
27. D. Panjwani and G. Healey, *Markov random field models for unsupervised segmentation of textured color images*, IEEE Transactions on Pattern Analysis and Machine Intelligence, 17 (1995), pp. 939–954.
28. M. Petrou, *Accelerated optimization in image processing via the renormalization group transfrom*, in Proceedings of the Conference on Complex Stochastic Systems and Engineering, D. Titterington, ed., 1993, pp. 105–120.
29. M. Petrou, M. Bober, and J. Kittler, *Multiresolution motion segmentation*, in Proceedings of the International Conference on Pattern Recognition (ICPR'94), 1994, pp. I: 379–383.
30. D. Pollard, *Strong consistency of* k*-means clustering*, The Annals of Statistics, 9 (1981), pp. 135–140.
31. J. Puzicha and J. Buhmann, *Multiscale annealing for real-time unsupervised texture segmentation*, IAI-TR 97–4, Institut für Informatik III, 1997.

32. J. Puzicha and J. Buhmann, *Multiscale annealing for real-time unsupervised texture segmentation*, in Proceedings of the International Conference on Computer Vision (ICCV'98), 1998, pp. 267–273.
33. J. Puzicha and J. Buhmann, *Multiscale annealing for grouping and unsupervised texture segmentation*, Computer Vision and Image Understanding, 76 (1999), pp. 213–230.
34. J. Puzicha, M. Held, J. Ketterer, J. Buhmann, and D. Fellner, *On spatial quantization of color images*, IEEE Transactions on Image Processing, 9 (2000), pp. 666–682.
35. L. Rabiner, *A tutorial on hidden markov models and selected applications in speech recognition*, IEEE, (1989).
36. B. Ripley, *Pattern Recognition and Neural Networks*, Cambridge University Press, 1996.
37. J. Shi and J. Malik, *Normalized cuts and image segmentation*, in Proceedings of the IEEE Conference on Computer Vision and Pattern Recognition (CVPR'97), 1997, pp. 731–737.
38. C. Stiller, *Object-based estimation of dense motion fields*, IEEE Transactions on Image Processing, 6 (1997), pp. 234–250.
39. D. Terzopoulos, *Image analysis using multigrid relaxation methods*, IEEE Transactions on Pattern Analysis and Machine Intelligence, 8 (1986), pp. 129–139.
40. V. Vapnik, *The Nature of Statistical Learning Theory*, Springer Verlag, 1995.
41. C. S. Won and H. Derin, *Unsupervised segmentation of noisy and textured images using Markov random fields*, CVGIP: Graphical Models and Image Processing, 54 (1992), pp. 308–328.
42. K. Zhou and C. Rushforth, *Image restoration using multigrid methods*, Applied Optics, 30 (1991), pp. 2906–2912.

AUTHOR INDEX

MIX
Papier aus verantwortungsvollen Quellen
Paper from responsible sources
FSC® C105338

If you have any concerns about our products,
you can contact us on
ProductSafety@springernature.com

In case Publisher is established outside the EU,
the EU authorized representative is:
Springer Nature Customer Service Center GmbH
Europaplatz 3, 69115 Heidelberg, Germany

Printed by Libri Plureos GmbH
in Hamburg, Germany